M000084860

Biology
Fundamentals

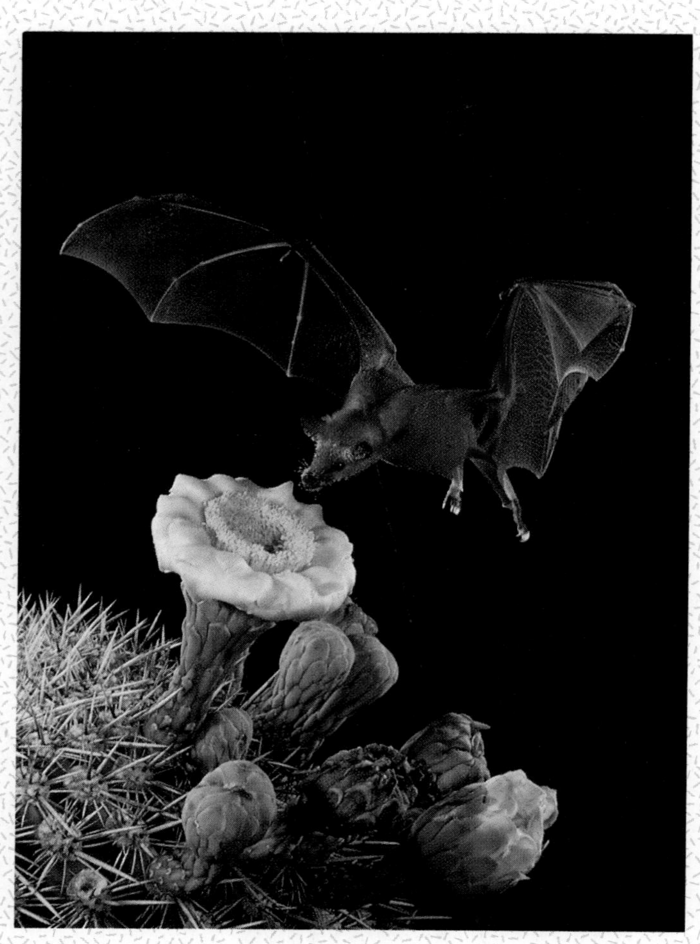

Biology
Fundamentals

▼ ▼ ▼

Gil Brum

California State Polytechnic University, Pomona

Larry McKane

California State Polytechnic University, Pomona

Gerry Karp

Formerly of the University of Florida, Gainesville

John Wiley & Sons, Inc.

New York / Chichester / Brisbane / Toronto / Singapore

Acquisitions Editor Sally Cheney
Developmental Editor Rachel Nelson
Associate Marketing Manager Rebecca Hershler
Marketing Manager Catherine Faduska
Senior Production Editor Katharine Rubin
Cover/Text Designer Karin Gerdes Kincheloe
"Steps to Discovery" Art Illustrator Carlyn Iverson
Assistant Manufacturing Manager Mark Cirillo
Cover Photography Editor Hilary Newman
Photo Editor Charles Hamilton
Photo Researchers Hilary Newman, Pat Cadley, Lana Berkovitz
Director of Photo Department Stella Kupferberg
Senior Freelance Illustration Coordinator Edward Starr
Text Illustrations Network Graphics/Blaize Zito Associates, Inc.
Cover Photo Courtesy Merlin D. Tuttle, Bat Conservation International

This book was set in New Caledonia by Ruttle, Shaw & Wetherill, Inc. and printed and bound by Von Hoffmann Press, Inc. The cover was printed by The Lehigh Press, Inc. The color separations were prepared by Color Associates, Inc.

Library of Congress Cataloging in Publication Data

Brum, Gilbert D.
Biology fundamentals / Gil Brum, Larry McKane, Gerry Karp.
p. cm.
Includes bibliographical references.
ISBN 0-471-59401-6 (pbk.)
1. Biology. I. McKane, Larry. II. Karp, Gerald. III. Title.
QH308.2.B785 1995
574--dc20 94-46228
CIP

Printed in the United States of America

10 9 8 7 6 5 4 3 2

Dedication

To our mentors [Rob Schlising (CSU Chico), Frank Vasek (University of California), Joseph Ferretti (University of Oklahoma), Arthur Whiteley (University of Washington)] for your inspiration, guidance, and for passing on an appreciation of the power of the scientific process.

Preface to the Instructor

ABOUT THE COVER

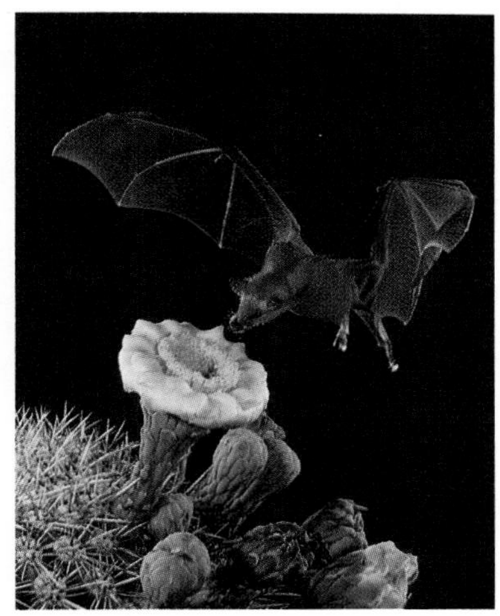

The life of the lesser long-nosed bat (*Leptonycteris curasoae*) and that of the majestic saguaro cactus (*Carnegiea gigantea*) are intimately linked in a relationship that benefits both organisms. The bat harvests needed energy and nutrients for its life activities from the cactus flower and, in the process, transfers the saguaro's gametes for sexual reproduction.

During spring and early summer, large saguaro flowers open precisely at dusk, a time when bats are most active. Each saguaro flower produces energy-rich nectar at the flower base. But to reach this nectar reservoir, the long-nosed bat must plunge its head into the flower. The bat's face emerges dusted with thousands of saguaro pollen grains, each of which contains the cactus's sperm gametes. When finished at one saguaro flower, the bat flies off to another saguaro and repeats its feeding ritual. Some of the pollen adhering to its face is transferred to the female part of the cactus flower, making it possible for the cactus's egg and sperm to fuse in fertilization to begin a new generation of saguaro offspring.

▼ ▼ ▼

Why did we choose to put a bat and saguaro on the cover of our new text? Our hope is the photo will arouse student curiosity, prompting such questions as Why is the bat hovering over that flower? What kind of bat is it? What kind of flower is it? How do bats fly? We believe that by piquing curiosity, students become motivated to inquire more about the living world, and through inquiry, they will discover the fundamental concepts and many fascinating wonders of modern biology.

This philosophy—curiosity triggers inquiry, and inquiry leads to discovery —forms not only the pedagogical foundation for our brief textbook, but, as every biologist knows, is also the way science advances. We design this text to take advantage of our students' natural curiosity about themselves and the surrounding world of life. It is the same curiosity that motivates scientists to investigate the world of life, making discoveries that lead to understanding and often producing revolutionary technological advances.

The cover photograph of the bat and cactus also underscores two central biological themes that form the foundation of this text. First, the physical match between the size and shape of the saguaro flower and the size and shape of the bat's face and tongue is the result of evolution—the fundamental process of change that leads to adaptations that ultimately increase an organism's chances of survival and reproduction. Second, the bat–saguaro relationship illustrates the interdependency of organisms, a fundamental theme that extends beyond bats and saguaros. All organisms ultimately depend on other organisms for the necessities of life. Humans depend on plants for food and oxygen; on other animals for food, work, and compansionship; and on bacteria and fungi for recycling essential nutrients. As John Muir put it in his book *My First Summer in the Sierra,* "When we try to pick out anything by itself, we find it hitched to everything else in the universe."

Our primary goals in developing this book are to help college students regardless of their background in biology, understand the excitement and importance of scientific investigation and discovery; to provide them with the essential biological facts and concepts they will need to be informed citizens in an increasingly technological world; and to encourage that each student continue to build a better understanding and appreciation of the natural world.

THE APPROACH

The elements of the approach are described in the upcoming section "To the Student: A User's Guide." The pedagogical features are embedded in a book that is written in an informal, accessible style that invites the reader to explore the process of biology. In addition, we have tried to help students make connections between the scientific facts and their everyday lives. One way to do this is to relate the fundamentals of biology to humans, revealing the human perspective in each biological principle, from biochemicals to ecosystems. With each such insight, students take a substantial step forward becoming the informed citizens that make up responsible voting public.

We hope that, through this textbook, we can become partners with the instructor and the student. The biology teacher's greatest asset is the basic desire of students to understand themselves and the world around them. Unfortunately, many students have grown detached from this natural curiosity. Our overriding objective in creating this book was to arouse the students' fascination with exploring life, building knowledge and insight that will enable them to make real-life judgments as modern biology takes on greater significance in everyday life.

THE ART PROGRAM

Each photo was picked specifically for its relevance to the topic at hand and for its aesthetic and instructive value in illustrating the narrative concepts. The illustrations were carefully crafted under the guidance of the authors for accuracy and utility as well as aesthetics. The value of illustrations cannot be overlooked in a discipline as filled with images and processes as biology. Through the use of cell icons, labeled illustrations of pathways and processes, and detailed legends, the student is taken through the world of biology, from its microscopic chemical components to the macroscopic organisms and the environments that they inhabit.

SUPPLEMENTARY MATERIALS

In our continuing effort to meet all of your individual needs, we have developed an integrated supplements package that helps the instructor bring the study of biology to life in the classroom and that will maximize the students' use and understanding of the text.

The *Instructor's Manual,* developed by Michael Leboffe of San Diego City College, contains lecture outlines, transparency references, suggested lecture activities, sample concept maps, and answers to study guide questions.

Michael Leboffe developed the test bank, which consists of four types of questions: fill-in questions, matching questions, multiple-choice questions, and critical thinking questions. A computerized bank is also avaiable.

A comprehensive visual ancillary package includes four-color transparencies (183 figures from the text), *Process of Science* transparency overlays that break down various biological processes into progressive steps, a video library consisting of tapes from Coronet MTI, and the *Bio Sci* videodisk series from Videodiscovery, covering topics in biochemistry, botany, vertebrate biology, reproduction, ecology, animal behavior, and genetics. Suggestions for integrating the videodisk material in your classroom discussions are available in the instructor's manual.

A comprehensive study guide and lab manual are also available and are described in more detail in the User's Guide section of the preface.

Acknowledgments

When an author team joins forces with a publishing company, it is not without an ample dose of anxiety, for the textbook that authors envision may not be the one that the company ultimately produces. Fortunately, we found John Wiley & Sons a company devoted to developing and producing the most outstanding textbooks available, and a company committed to improving higher education. The staff at Wiley shared our vision for a new type of biology textbook and devoted all their resources to assure that our vision became a reality. We wish we could acknowledge each of the many people at Wiley who contributed to *Biology: Fundamentals.* The book you now hold in your hands is a tribute to their talent, dedication, artistry, and willingness to pour their hearts and souls into this project.

One of these exceptionally talented people is Stella Kupferberg, director of the photo department at Wiley. We treasure your friendship and your exceptional talent for unearthing the highest quality and most effective photo images. Stella also provided us with two other important assets, Charles Hamilton and Hilary Newman. Stella and Charles tirelessly applied their skill, resources, and sharp eyes to help us find images of incomparable beauty and teaching effectiveness. Hilary's diligence and organizational skills helped insure that there were no oversights.

We would also like to acknowledge the contributions of Rachel Nelson for her meticulous editing, Katharine Rubin for expertly (and patiently) guiding this project through the complicated stages of production, Karin Kincheloe for an effective text design, Ishaya Monokoff and Ed Starr for orchestrating a brilliant art program, and Network Graphics, whose artists translated our original drawings into illustrations that combine beauty and style with conceptual strength and remarkable clarity. Thanks also to Carlin Iverson for visually distilling our "Steps to Discovery" episodes into images that help bring the process of science to life.

Steve Kraham, Cathy Faduska, and Rebecca Hershler deserve thanks for their creative flare in helping instructors see the value of letting their students experience this new way of exploring biology.

We appreciate the invaluable contributions of Diana Lipscomb of George Washington University for her insightful additions to the evolution chapters, of Judy Goodenough of the University of Massachusetts–Amherst, who created the initial text for the Animal Behavior chapter, and of Dorothy Rosenthal for contributing the critical thinking questions at the end of each chapter.

Most of all, we would like to thank Sally Cheney, our biology editor at Wiley, whose influence on the development of this book has been immeasurable. Sally's commitment to us and her belief in our new approaches to teaching biology helped assure that this text received the support needed for our vision to become reality. With Sally's uncommon ability to think both like a biologist and an editor, she quickly recognized that the differences in this textbook represent a very real departure from the traditional encyclopedic approach to teaching biology and fully appreciated the improved teaching value of our innovations. Without her dedication and unflagging support, this project could never have developed into the powerful teaching instrument we struggled so hard to create. Thank you, Sally.

Thanks also to all the reviewers on this project, who provided insightful feedback that helped assure that this is of the highest quality textbook possible, as well as the best teaching tool it can possibly be. We appreciate your thoughtful guidance and your dedication to your students.

Dennis Anderson, Oklahoma City Community College
Sarah Barlow, Middle Tennessee State University
Robert Beckman, North Carolina State University
Timothy Bell, Chicago State University
Arthur Bender, College of the Desert
David F. Blaydes, West Virginia University
Richard Bliss, Yuba College
Richard Boohar, University of Nebraska, Lincoln
Clyde Bottrell, Tarrant County Junior College
J. D. Brammer, North Dakota State University
Peggy Branstrator, Indiana University, East
Allyn Bregman, SUNY, New Paltz
Daniel Brooks, University of Toronto
Gary Brusca, Humboldt State University
Jack Bruk, California State University, Fullerton
Marvin Cantor, California State University, Northridge
Jerry Carpenter, Northern Kentucky University
Richard Cheney, Christopher Newport College
Larry Cohen, California State University, San Marcos
David Cotter, Georgia College
Robert Creek, Eastern Kentucky University
Ken Curry, University of Southern Mississippi
Judy Davis, Eastern Michigan University
Loren Denny, Southwest Missouri State University
Captain Donald Diesel, U.S. Air Force Academy
Tom Dickinson, University College of the Cariboo
Mike Donovan, Southern Utah State College
Patrick Doyle, Middle Tennessee State University
Lee Drickamer, Southern Illinois University at Carbondale
Robert Ebert, Palomar College

Robert Elgart, SUNY at Farmingdale
Thomas Emmel, University of Florida
Joseph Faryniarz, Mattatuck Community College
Alan Feduccia, University of North Carolina, Chapel Hill
Eugene Ferri, Bucks County Community College
Victor Fet, Loyola University, New Orleans
Bob Ford, Ball State University
David Fox, Loyola University, New Orleans
Mary Forrest, Okanagan University College
Michael Gains, University of Kansas
S. K. Gangwere, Wayne State University
Dennis George, Johnson County Community College
Bill Glider, University of Nebraska
Paul Goldstein, University of North Carolina, Charlotte
Judy Goodenough, University of Massachusetts, Amherst
Roger Goos, University of Rhode Island
Nels Granholm, South Dakota State University
Nathaniel Grant, Southern Carolina State College
Mel Green, University of California, San Diego
Dana Griffin, Florida State University
Barbara L. Haas, Loyola University of Chicago
Richard Haas, California State University, Fresno
Fredrick Hagerman, Ohio State University
Tom Haresign, Long Island University, Southampton
Wallace Harmon, California State University
W. R. Hawkins, Mt. San Antonio College
Vernon Hendricks, Brevard Community College
Paul Hertz, Barnard College
Howard Hetzel, Illinois State University
Walter Hewitson, Bridgewater State College
Ronald K. Hodgson, Central Michigan University
W. G. Hopkins, University of Western Ontario
Thomas Hutto, West Virginia State College
Alice Jacklet, University of Albany
Duane Jeffrey, Brigham Young University
John Jenkins, Swarthmore College
Claudia Jones, University of Pittsburgh
R. David Jones, Adelphi University
J. Michael Jones, Culver Stockton College
Florence Juillerat, Indiana University, Purdue University at Indianapolis
Gene Kalland, California State University, Dominiquez Hills
Arnold Karpoff, University of Louisville
Judith Kelly, Henry Ford Community College
Richard Kelly, SUNY, Albany
Richard Kelly, University of Western Florida
Dale Kennedy, Kansas State University
Miriam Kittrell, Kingsborough Community College
John Kmeltz, Kean College New Jersey
Robert Krasner, Providence College
Eliot Krause, Seton Hall University
Susan Landesman, Evergreen State College
Anton Lawson, Arizona State University
Lawrence Levine, Wayne State University
Jerri Lindsey, Tarrant County Junior College
Diana Lipscomb, George Washington University
James Luken, Northern Kentucky University
Ted Maguder, Univeristy of Hartford
Jon Maki, Eastern Kentucky University
Charles Mallery, University of Miami
Carol Mapes, Kutztown University
William McEowen, Mesa Community College
Ricard McGuire, San Juan College
Craig Milgrim, Grossmont College
Roger Milkman, University of Iowa
Helen Miller, Oklahoma State University
Elizabeth Moore, Glassboro State College
Janice Moore, Colorado State University
Eston Morrison, Tarleton State University
Dave Mullet, Green River Community College
John Mutchmor, Iowa State University
Jane Noble-Harvey, University of Delaware
Steve Novak, Boise State University
Douglas W. Ogle, Virginia Highlands Community College
Joel Ostroff, Brevard Community College
Charles Paulson, Lake Forest College
James Lewis Payne, Virginia Commonwealth University
Gary Peterson, South Dakota State University
MaryAnn Phillippi, Southern Illinois University, Carbondale
R. Douglas Powers, Boston College
Rudolph Prins, Western Kentucky University
Robert Raikow, University of Pittsburgh
Charles Ralph, Colorado State University
Aryan Roest, California State Polytechnic Univ., San Luis Obispo
Robert Romans, Bowling Green State University
Raymond Rose, Beaver College
Richard G. Rose, West Valley College
Donald G. Ruch, Transylvania University
Lynette Rushton, South Puget Sound University
A. G. Scarbrough, Towson State University
Gail Schiffer, Kennesaw State University
John Schmidt, Ohio State University
John R. Schrock, Emporia State University
Marvin Scott, Longwood College
Prem Sehgal, East Carolina University
Marilyn Shopper, Johnson County Community College
John Smarrelli, Loyola University of Chicago
Deborah Smith, Meredith College
Gordon Snyder, Schoolcraft College
Donald Stearns, Rutgers University—Camden
Guy Steucek, Millersville University
Gail Stratton, Albion College
Ralph Sulerud, Augsburg College
Jill Targett, Villanova University
Tom Terry, University of Connecticut
James Thorp, Cornell University
W. M. Thwaites, San Diego State University
Michael Torelli, University of California, Davis
Michael Treshow, University of Utah
Terry Trobec, Oakton Community College
Len Troncale, California State Polytechnic University, Pomona
Ella Turner Gray, Canada College
Richard Van Norman, University of Utah
David Vanicek, California State University, Sacramento
Terry F. Werner, Harris-Stowe State College
David Whitenberg, Southwest Texas State University
P. Kelly Williams, University of Dayton
Robert Winget, Brigham Young University
Steven Wolf, University of Missouri, Kansas City
Harry Womack, Salisbury State University
William Yurkiewicz, Millersville University

Gil Brum Larry McKane Gerry Karp

Brief Table of Contents

Contents

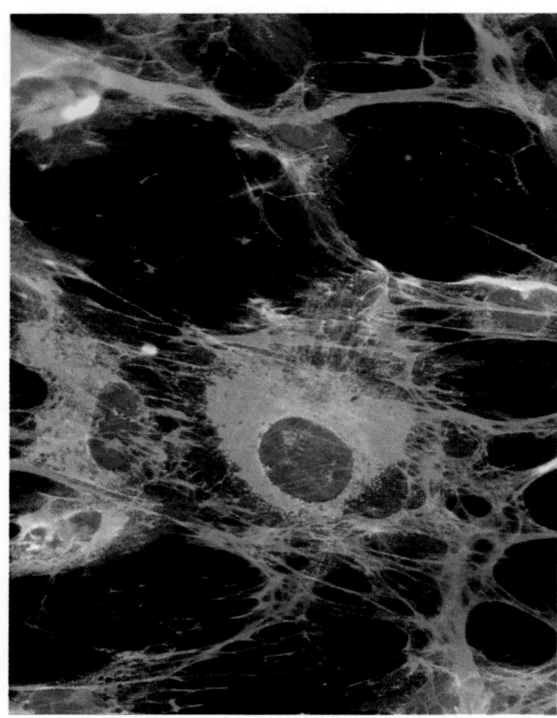

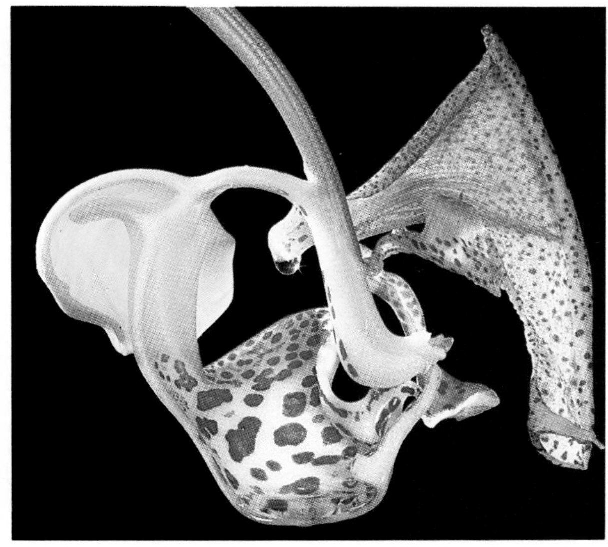

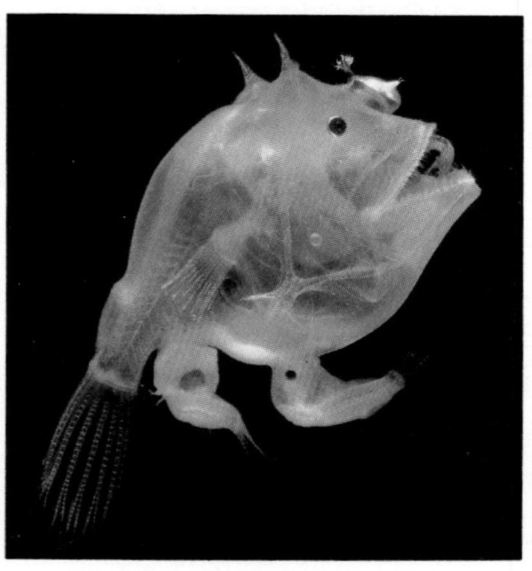

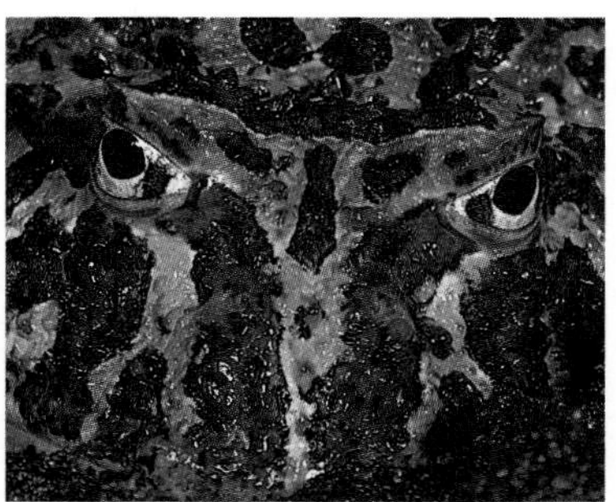

To The Student: A User's Guide

As citizens of the twenty-first century, your decisions will shape the future of life on earth. We hope this textbook will build a foundation of knowledge and insight to help you make some of these critical decisions.

Steps to Discovery

*T*he process of science enriches all parts of this book. We believe that students, like biologists, themselves, are intrigued by scientific puzzles. Every chapter is introduced by a "Steps to Discovery" narrative, the story of an investigation that led to a scientific breakthrough in an area of biology which relates to that chapter's topic. The "Steps to Discovery" narratives portray biologists as they really are: human beings, with motivations, misfortunes, and mishaps, much like everyone experiences. We hope these narratives help you better appreciate biological investigation, realizing that it is understandable and within your grasp.

Throughout the narrative of these pieces, the writing is enlivened with scientific work that has provided knowledge and understanding of life. This approach is meant not just to pay tribute to scientific giants and Nobel prize winners, but once again to help you realize that science does not grow by itself. Facts do not magically materialize. They are the products of rational ideas, insight, determination, and, sometimes, a little luck. Each of the "Steps to Discovery" narratives includes a painting that is meant primarily as an aesthetic accompaniment to the adventure described in the essay and to help you form a mental picture of the subject.

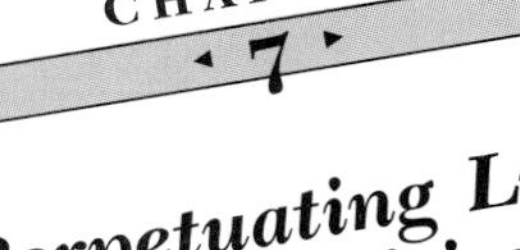
CHAPTER
7
Perpetuating Life:
Cell Division

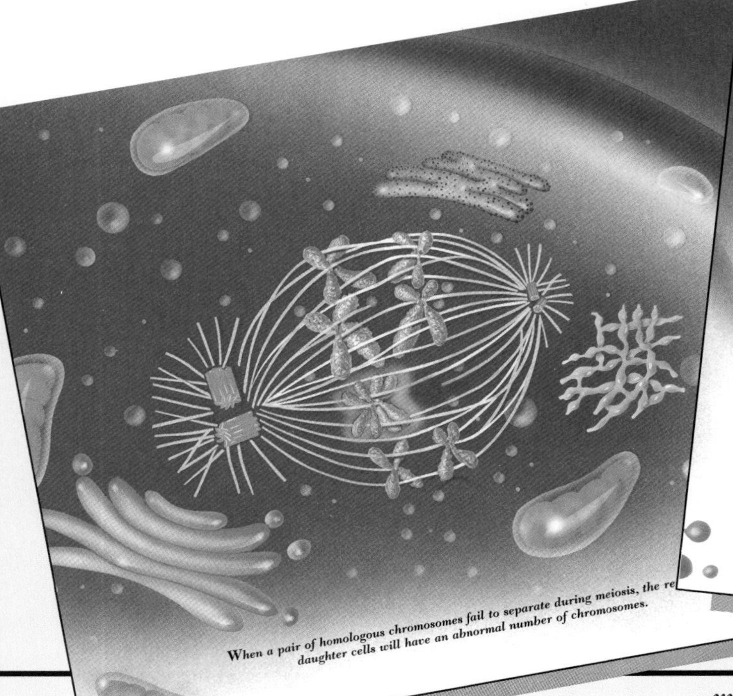
When a pair of homologous chromosomes fail to separate during meiosis, the re
daughter cells will have an abnormal number of chromosomes.

STEPS TO DISCOVERY
Counting Human Chromosomes

"...*Thus man has 48 chromosomes.*"
This quote was taken from a biology book published
in 1952. As you will learn in this chapter, the cells of
humans have *46* chromosomes, not 48. Before you condemn the authors of this book for their inaccuracy, consider that, up until 1955, biologists accepted the "fact" that
human cells contained 48 chromosomes.
Why did it take so long to discover the right number?
Prior to the 1950s, chromosome numbers were determined primarily by examining sections of tissue in which
cells were occasionally "caught" in the process of cell division. Trying to count several dozen chromosomes
crammed together within a microscopic nucleus is like
trying to count the number of rubber bands present in a
cellophane package without opening the package. It is actually quite remarkable that biologists got as close as 48!
Before 1922, when Theophilus Painter, a geneticist at the
University of Texas, arrived at this number, previous
guesses had ranged from 8 to more than 50; 48 chromosomes then became the accepted value for more than 30
years until a remarkably simple discovery changed everything.
In 1951, times were tough in academia, and many new
PhDs couldn't find teaching positions. Having just completed his PhD on insect chromosomes at the University
of Texas, T. C. Hsu fell into this distressing category. Reluctantly, Hsu accepted a postdoctoral research position in
Galveston working on mammalian chromosomes, which
were notoriously difficult to study. After several months of
frustration, he was examining a new batch of cells with a
microscope when, in Hsu's words "I could not believe my
eyes when I saw some beautifully scattered chromosomes
in these cells. I did not tell anyone, took a walk around the
building, went to the coffee shop, and then returned to
the lab. The beautiful chromosomes were still there. I
knew they were real."
He tried to repeat the work, but his preparations regained their "normal miserable appearance." For many
months after that, every attempt to discover what he had
done "wrong" to get that beautiful preparation failed. Finally, Hsu tried pretreating the cells with a hypotonic
(more dilute) salt solution. The cells expanded like balloons
and exploded onto the slide, releasing the chromosomes
and spreading them out so each was separated from its
neighbors (see Figure 7-7*a*). The earlier preparation of
cells must have been accidently washed with a dilute saline
solution to which someone had failed to add enough salt.
Since then, Hsu's hypotonic technique for treating cells
has become a standard part of preparing chromosomes for
microscopic examination. Ironically, Hsu did not use his
technique to reexamine the question of the human chromosome number; Painter had been one of his mentors at
the University of Texas, and Hsu never questione
chromosome count.
Within 3 years, however, Albert Levan and
Tijo, working in the United States and Sweden,
new hypotonic pretreatment technique on hum
treated with a drug called colchicine. Colchicine di
bles the machinery needed for cell division causi
to remain "frozen" in the process, providing man
dividing cells to observe. Levan and Tijo carefully c
the chromosomes of these cells and found only 46
repeated their observations and cautiously conclude
at least in lung cells, the human chromosome numb
46. The number was soon confirmed by other investi
on other human cell types and has been the accepted
ever since.
Meanwhile, Jerome Lejeune, a French clinician
been studying children with Down syndrome for
years. Such children are characterized by a short st
stature, distinctive folds of the eyelids (which gave ri
the earlier name "mongolism"), and mental retarda
After hearing a lecture by Jo Hin Tijo, who described
chromosomes could be counted using newly develo
techniques, Lejeune decided to examine the chromosor
from a few of his Down syndrome patients. These child
showed such a wide range of abnormalities that some
teration in the chromosomes was very likely responsib
Lejeune had never been involved in this type of researc
however, and, in fact, did not even possess a microscop
He finally located one that had been discarded by th
bacteriology lab. The microscope had been used so muc
that it would not remain in focus. In order to use it, Le
jeune inserted a piece of tinfoil from a candy wrapper to
hold the focusing knob in place. In 1959, Lejeune and two
colleagues who had helped in cell preparation published a
2 page paper indicating that the cells of nine different
patients with Down syndrome all had 47 chromosomes,
rather than 46.
Lejeune's paper opened the door to a new field of
medical genetics. It was soon followed by a number of
other reports in which patients with various types of disorders were shown to have an abnormal number of chromosomes. Examination of cells from fetuses that had spontaneously aborted revealed that many had extra sets of
chromosomes in their cells. Thus, an abnormal fetal chromosome number was discovered to be a common cause of
miscarriage. Taken together, these insights into the effects
of chromosome abnormalities revealed how important it
was that the cells of developing embryos contained precisely the correct number of chromosomes. It soon became
obvious that the formation of gametes with an abnormal
number of chromosomes could be traced to a defect occurring during meiosis—a subject of this chapter.

Evolution and Adaptation: Tying It Together

*T*he last section of each chapter, entitled "Evolution and Adaptation: Tying It Together," revisits the theme of evolution and adaptation and shows how it emerges and is exemplified within the context of the chapter's concepts and principles. This section is intended to show how evolution and adaptation play a role at all levels of biological organization, from the molecular and cellular aspects to the global characteristics of biology. It provides an opportunity to summarize the concepts introduced in the chapter within the context of the text's overarching theme.

310 • PART 4 / *Form and Function of Plant and Animal Life*

The body uses bones as levers, and joints as fulcrums, to accomplish similar functions. When you stand on your tiptoes, for example, an upward force is generated by the muscle in your calf using the ball joint of the toes as the fulcrum (Figure 17-10*b*) and the bones of the foot as a lever. The load, in this case, is the entire weight of your body.

EVOLUTION AND ADAPTATION: TYING IT TOGETHER

An animal's integument is the protective barrier that shields the organism from its external environment. Accordingly, many of the properties of the integument can be understood by considering the environmental challenges the animal must face. This feature can be illustrated by briefly surveying the types of integuments found in vertebrates, all of which are constructed on a similar epidermal–dermal plan.

The earliest vertebrates were jawless, bottom-dwelling fishes that were clothed in heavy, bony armor that protected them from predators. During subsequent evolution, fishes moved away from the ocean bottom, becoming more buoyant and mobile. The thick plates of bone along the sides of the body were no longer adaptive and became reduced to the thin, familiar bony *scales* that are scraped away when a fish is "cleaned." As vertebrates moved out of the water and onto the land, the integument became adapted to terrestrial habitats. The bony scales of the ancestral fishes were lost, and the dermis became a more fibrous, flexible layer. In amphibians—animals that live both in water and on land—the skin is usually moist and permeable, facilitating oxygen absorption across the body surface (Figure 17-11*a*). Among reptiles and other land vertebrates, the epidermis has become a tough, impervious layer that prevents water loss in harsh, dry, terrestrial environments (Figure 17-11*b*). The role of oxygen uptake in land vertebrates with impermeable integuments was taken over by the lungs.

(a)

(b)

FIGURE 17-11
Contrasting integuments. The yellow spotted salamander (an amphibian) has thin, moist, permeable skin ***(a)***, while the iguana (a reptile) has thick, dry, impermeable skin ***(b)***.

The Human Perspective

Students will naturally find many ways in which the material represented in any biology course relates to them. But it is not always obvious how you can use biological information for better living or how it might influence your life. Your ability to see yourself in the course boosts interest and heightens the usefulness of the information. This translates into greater retention and understanding.

To accomplish this desirable outcome, the entire book has been constructed with you—the student—in mind. Perhaps the most notable feature of this approach is a series of boxed essays called "The Human Perspective" that directly reveals the human relevance of the biological topic being discussed at that point in the text. You will soon realize that human life, including your own, is an integral part of biology.

532 • PART 6 / *Ecology and Animal Behavior*

◁ THE HUMAN PERSPECTIVE ▷

Acid Rain and Acid Snow: Global Consequences of Industrial Pollution

Two alarming trends were recently documented in the biomes of North America and Europe. The first was a change in the color of several lakes, from murky green to crystal clear. Sounds good, right? Not really, for a green lake is a biologically active lake, teaming with microscopic algae, which are eaten by small aquatic animals, which, in turn, are eaten by fish. A clear lake is biologically sterile, devoid of aquatic life. How widespread is lake sterility? In eastern Canada, nearly 100 lakes have become sterile, as have more than 1,000 lakes in the northeastern United States and approximately 20,000 lakes in Sweden.

The second trend was the premature death of an excessive number of trees, especially those on high slopes that face prevailing winds. More than 17 million acres (7 million hectares) of trees in North America and Europe look as if they've been burned, but there have been no fires. Over 50 percent of the forests in Germany alone are affected in this way, impacting more than 1.2 million acres (500,000 hectares).

Just how are dead trees and dead lakes related? The destruction of both trees and lakes is caused by acid rain or runoff from acid snow. Acid deposition not only destroys forests and kills lakes, it also damages crops, alters soil fertility, and erodes statues and buildings.

The chemicals that create acid rain and snow (sulfur oxides and nitrogen oxides) come primarily from human activities. Although sulfur oxides are released during volcanic eruptions, forest fires, and from bacterial decay, quantities of sulfur oxides from human activities far exceed those that come from natural sources. Nearly 70 percent of sulfur oxides comes from electrical generating plants, most of which burn coal. Most nitrogen oxides come from motor vehicles and industries, including electrical generation. When sulfur and nitrogen oxides mix with the water in the air, they form acids:

$$SO_2 + H_2O \rightarrow H_2SO_4 \text{ (sulfuric acid)}$$
$$NO_2 + H_2O \rightarrow HNO^{-3} \text{ (nitric acid)}$$

Acid rain or acid snow has a pH below 5.7, the pH of unpolluted rain. Over the past 25 years, rains in the northeastern United States dropped to an average pH 4.0. The lowest recorded pH for rainfall was 2.0, reported in Wheeling, West Virginia. The rain in Wheeling was more acidic than lemon juice!

Acids that create acid rain and snow remain airborne for up to 5 days, during which time they can travel over great distances. For example, the acid rainfall that killed many of the lakes in Sweden was caused by pollutants that were released in England. The acid rain that is damaging trees and lakes in the Adirondack Mountains of New York originated in the upper Mississippi and Ohio River Valleys. Since these acids circulate in large air masses, acid deposition is widespread, spreading from Japan to Alaska, from New Jersey to Canada, to name just a few places.

The rate of destruction caused by acid rain and snow is increasing. A 1988 survey of U.S. lakes lists 1,700 lakes as having high acidity. Another 14,000 lakes were identified as becoming acidified. Scientists estimate that by the turn of the century, over half of the 48,000 lakes in Quebec, Canada, will have been destroyed.

In 1979, the U.S. Congress passed the "Acid Precipitation Act" to identify sources of acid deposition. Congress is also considering taking steps to cut sulfur oxide emissions by nearly 50 percent, and nitrogen oxides by 10 percent by the year 2000. Such steps might include: (1) installing scrubbers on power plant smoke stacks; (2) using coal that is low in sulfur; (3) using coal that has been pretreated to remove sulfur; or (4) reducing auto and truck use.

Additional Pedagogical Features

We have worked to assure that each chapter in this book is an effective teaching and learning instrument. In addition to the pedagogical features discussed above, we have included some additonal tried-and-proven-effective tools.

Synopsis

The synopsis section offers a convenient summary of the chapter material in a readable narrative form. The material is summarized in concise paragraphs that detail the main points of the material, offering a useful review tool to help reinforce recall and understanding of the chapter's information.

Review Questions

Along with the synopsis, the Review Questions provide a convenient study tool for testing your knowledge of the facts and processes presented in the chapter.

Stimulating Critical Thinking

Each chapter contains as part of its end material a diverse mix of Critical Thinking Questions. These questions ask you to apply your knowledge and understanding of the facts and concepts to hypothetical situations in order to solve problems, form hypotheses, and hammer out alternative points of view. Such exercises provide you with more effective thinking skills for competing and living in today's complex world.

Energy and Life: Respiration and Fermentation / CHAPTER 6 • 111

SYNOPSIS

The first stage in glucose disassembly is glycolysis. Glycolysis occurs in the cytoplasm. Here glucose is converted to two molecules of pyruvic acid, four ATPs, and two NADHs (which can be cashed in later for up to six ATPs). Glycolysis costs the cell two ATPs to activate the glucose molecule before it is split into two fragments.

Under anaerobic conditions, cells conduct fermentation as a means to regenerate NAD^+ from the NADH formed during glycolysis. More than 90 percent of the chemical energy contained in glucose remains in the discarded end products of fermentation.

Aerobic respiration continues the disassembly of glucose in the presence of oxygen. The two pyruvic acids generated by glycolysis are completely oxidized to six carbon dioxide molecules. The Krebs cycle occurs in the matrix of mitochondria in eukaryotes or in the cytoplasm of prokaryotes, and generates reduced coenzymes, NADHs and $FADH_2$. For the two pyruvic acids produced at the end of glycolysis, a total of two ATPs, eight NADHs, and two $FADH_2$s are produced by the Krebs cycle. The high-energy electrons carried by NADH and $FADH_2$ are used to generate ATP via the electron transport system. Three ATPs are generated for each NADH and two for each $FADH_2$. As electrons are passed down the electron transport chain, a proton gradient is formed across the inner mitochondrial membrane. As protons flow through the ATP synthase channel, ADP is phosphorylated to form ATP.

Glycolysis, the Krebs cycle, and electron transport are central metabolic pathways that provide virtually all aerobic cells with energy and raw materials. A diverse array of molecules (from proteins to lipids) are broken down and introduced into glycolysis or the Krebs cycle for further metabolism. Intermediate molecules in these pathways can be diverted and used to form various biochemicals, depending on the needs of the cell at the time.

Review Questions

1. Arrange the following molecules in the sequence they would be found during glycolysis and the Krebs cycle:

pyruvic acid	succinic acid
acetyl CoA	citric acid
oxaloacetic acid	PGAL
glucose	DPG

2. Under what conditions do muscle cells form lactic acid? Why do biologists consider this to be adaptive?
3. How is the proton gradient formed in mitochondria during aerobic respiration similar and how is it different from the gradient formed in chloroplasts during photosynthesis?
4. What is the role of NAD^+ and FAD in the Krebs cycle?
5. Rank the following compounds in terms of energy content: pyruvic acid, glucose, carbon dioxide, lactic acid, PGAL.

Critical Thinking Questions

1. Why would isolated vesicles in the supernatant of disrupted mitochondria in Racker's experiment still be able to oxidize glucose?
2. Burning fuel in gasoline-powered engines is about 25 percent efficient whereas releasing the energy in glucose by cells is about 40 percent efficient. What accounts for this difference in efficiency?
3. If you held a spoon of sugar over a flame, it would break down into carbon dioxide and water. Why is activation not necessary when sugar is burned in this way, yet is necessary when sugar is broken down by cells during respiration?
4. Why do organisms that have respiratory equipment ever resort to fermentation? Why can't you—as a human being—switch to fermentation to sustain yourself indefinitely?
5. How do you personally benefit from the fact that your metabolic pathways are interconnected? How does this fact benefit people who have had to subsist on a starvation diet for long periods of time?

Careers in Biology

The appendices of this edition include "Careers in Biology," a frequently overlooked aspect of our discipline. Although many of you may be taking biology as a requirement for another major (or many have yet to declare a major), some of you are already biology majors and may become interested enough to investigate the career opportunities in life sciences. This appendix helps students discover how an interest in biology can grow into a livelihood. It also helps the instructor advise students who are considering biology as a life endeavor.

Appendix • D-1

APPENDIX

D

Careers in Biology

Although many of you are enrolled in biology as a requirement for another major, some of you will become interested enough to investigate the career opportunities in life sciences. This interest in biology can grow into a satisfying livelihood. Here are some facts to consider:

- Biology is a field that offers a very wide range of possible science careers
- Biology offers high job security since many aspects of it deal with the most vital human needs: health and food
- Each year in the United States, nearly 40,000 people obtain bachelor's degrees in biology. But the number of newly created and vacated positions for biologists is increasing at a rate that exceeds the number of new graduates. Many of these jobs will be in the newer areas of biotechnology and bioservices.

Biologists not only enjoy job satisfaction, their work often changes the future for the better. Careers in medical biology help combat diseases and promote health. Biologists have been instrumental in preserving the earth's life-supporting capacity. Biotechnologists are engineering organisms that promise dramatic breakthroughs in medicine, food production, pest management, and environmental protection. Even the economic vitality of modern society will be increasingly linked to biology.

Biology also combines well with other fields of expertise. There is an increasing demand for people with backgrounds or majors in biology complexed with such areas as business, art, law, or engineering. Such a distinct blend of expertise gives a person a special advantage.

The average starting salary for all biologists with a Bachelor's degree is $22,000. A recent survey of California State University graduates in biology revealed that most were earning salaries between $20,000 and $50,000. But as important as salary is, most biologists stress job satisfaction, job security, work with sophisticated tools and scientific equipment, travel opportunities (either to the field or to scientific conferences), and opportunities to be creative in their job as the reasons they are happy in their career.

Here is a list of just a few of the careers for people with degrees in biology. For more resources, such as lists of current openings, career guides, and job banks, write to Biology Career Information, John Wiley and Sons, 605 Third Avenue, New York, NY 10158.

A SAMPLER OF JOBS THAT GRADUATES HAVE SECURED IN THE FIELD OF BIOLOGY*

Agricultural Biologist
Agricultural Economist
Agricultural Extension Officer
Agronomist
Amino-acid Analyst
Analytical Biochemist
Anatomist
Animal Behavior Specialist
Anticancer Drug Research Technician
Antiviral Therapist
Arid Soils Technician
Audio-neurobiologist
Author, Magazines & Books
Behavioral Biologist
Bioanalyst
Bioanalytical Chemist
Biochemical/Endocrine Toxicologist
Biochemical Engineer
Pharmacology Distributor
Pharmacology Technician
Biochemist
Biogeochemist
Biogeographer
Biological Engineer
Biologist
Biomedical Communication Biologist
Biometerologist
Biophysicist
Biotechnologist
Blood Analyst
Botanist
Brain Function Researcher
Cancer Biologist
Cardiovascular Biologist
Cardiovascular/Computer Specialist
Chemical Ecologist
Chromatographer
Clinical Pharmacologist
Coagulation Biochemist
Cognitive Neuroscientist
Computer Scientist
Dental Assistant
Ecological Biochemist
Electrophysiology/ Cardiovascular Technician
Energy Regulation Officer
Environmental Biochemist
Environmental Center Director
Environmental Engineer
Environmental Geographer
Environmental Law Specialist
Farmer
Fetal Physiologist
Flavorist
Food Processing Technologist
Food Production Manager
Food Quality Control Inspector
Flower Grower
Forest Ecologist
Forest Economist
Forest Engineer
Forest Geneticist
Forest Manager

Study Guide

Written by *Michael Leboffe* of San Diego City College, the *Study Guide* has been designed with innovative pedagogical features to maximize your understanding and retention of the facts and concepts presented in the text. Each chapter in the *Study Guide* contains the following elements.

Go Figure!

In each chapter, questions are posed regarding the figures in the text. Students can explore their understanding of the figures and are asked to think critically about the figures based on their understanding of the surrounding text and their own experiences.

Self-Tests

Each chapter includes a set of matching and multiple-choice questions. Answers to the Study Guide questions are provided.

Concept Map Construction

The student is asked to create concept maps for a group of terms, using appropriate connnector phrases and adding terms as necessary.

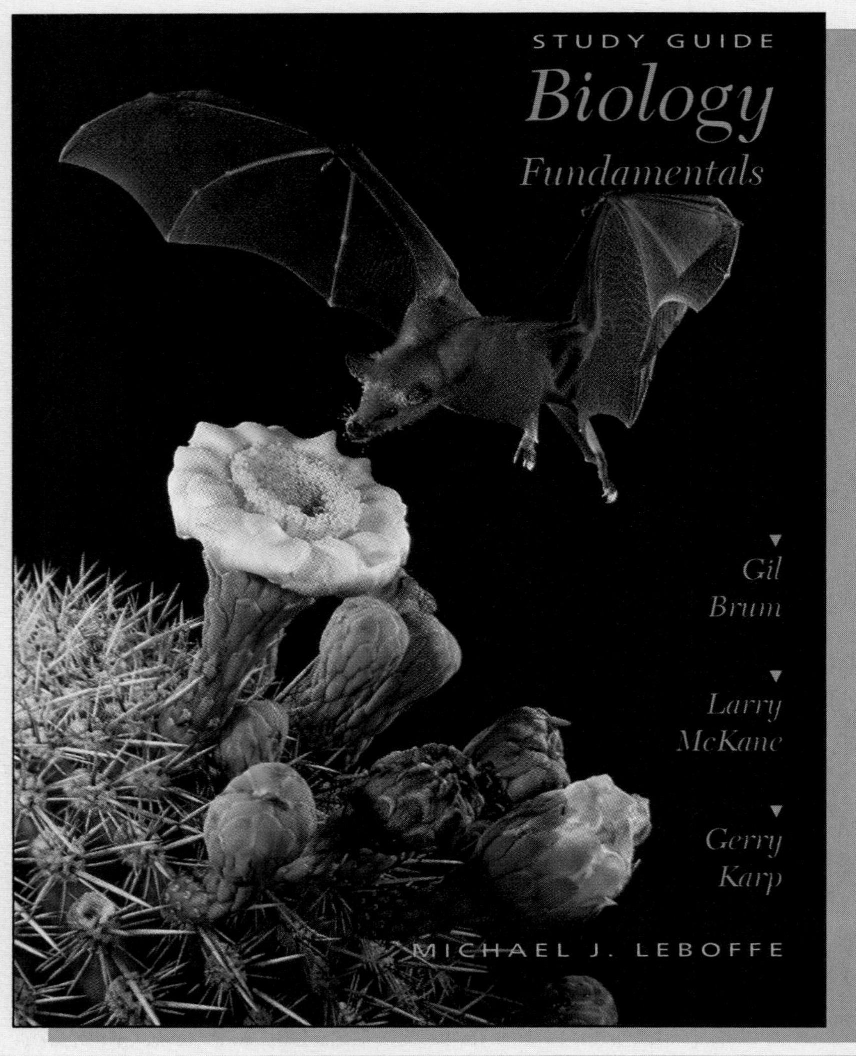

Laboratory Manual

*B*iology Fundamentals is supplemented by a comprehensive *Laboratory Manual* containing approximately 60 lab exercises chosen by the text authors from the National Association of Biology Teachers. These labs have been thoroughly class-tested and have been assembled from various scientific publications. They include such topics as

- Chaparral and Fire Ecology: Role of Fire in Seed Germination *(The American Biology Teacher)*
- A Model for Teaching Mitosis and Meiosis *(American Biology Teacher)*
- Laboratory Study of Climbing Behavior in the Salt Marsh Snail *(Oceanography for Landlocked Classrooms)*
- Down and Dirty DNA Extraction *(A Sourcebook of Biotechnology Activities)*
- Bioethics: The Ice-Minus Case *(A Sourcebook of Biotechnology Activities)*
- Using Dandelion Flower Stalks for Gravitropic Studies *(The American Biology Teacher)*
- pH and Rate of Enzymatic Reactions *(The American Biology Teacher)*

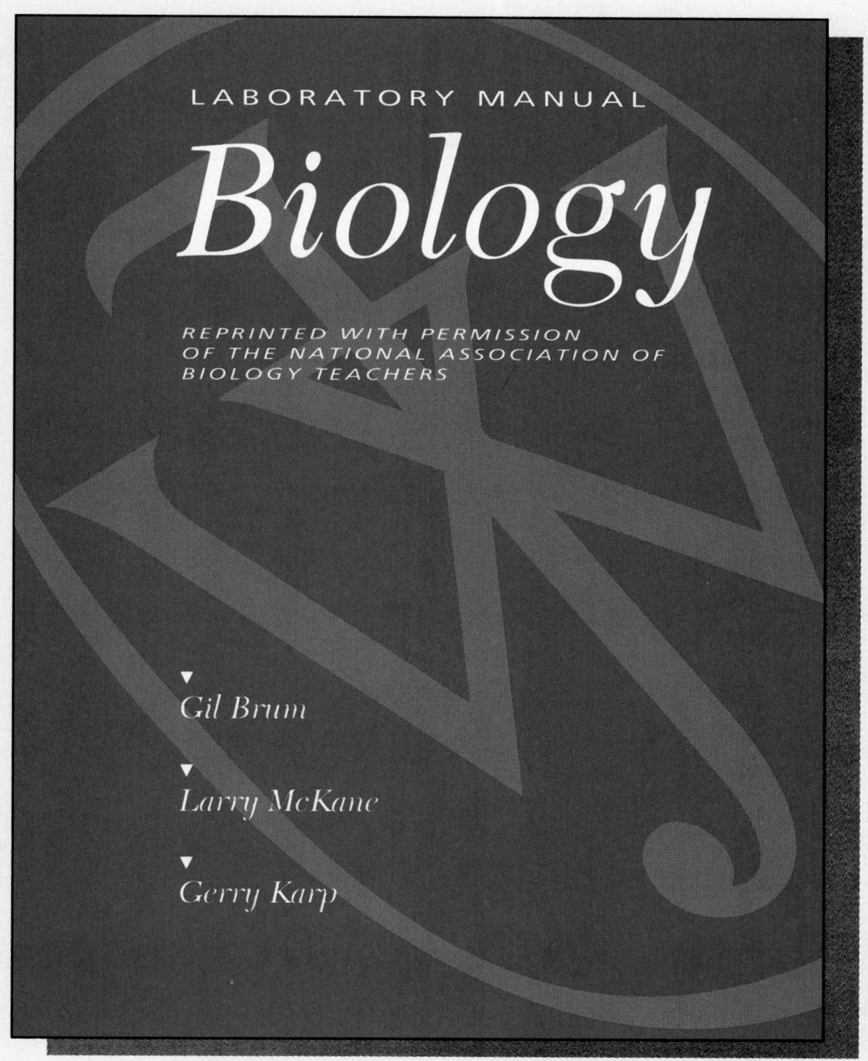

PART 1

Biology: The Study of Life

To understand life you must explore the obvious and the subtle, as well as all levels in between. This translucent jellyfish represents the organismal level of biological organization. But to understand how a jellyfish or any other organism survives, grows, and reproduces, biologists must study all levels of organization, including the organs, tissues, cells, and even molecules that make up an organism, as well as the population, community and ecosystem in which the organism lives.

CHAPTER

1

Biology: The Study of Life

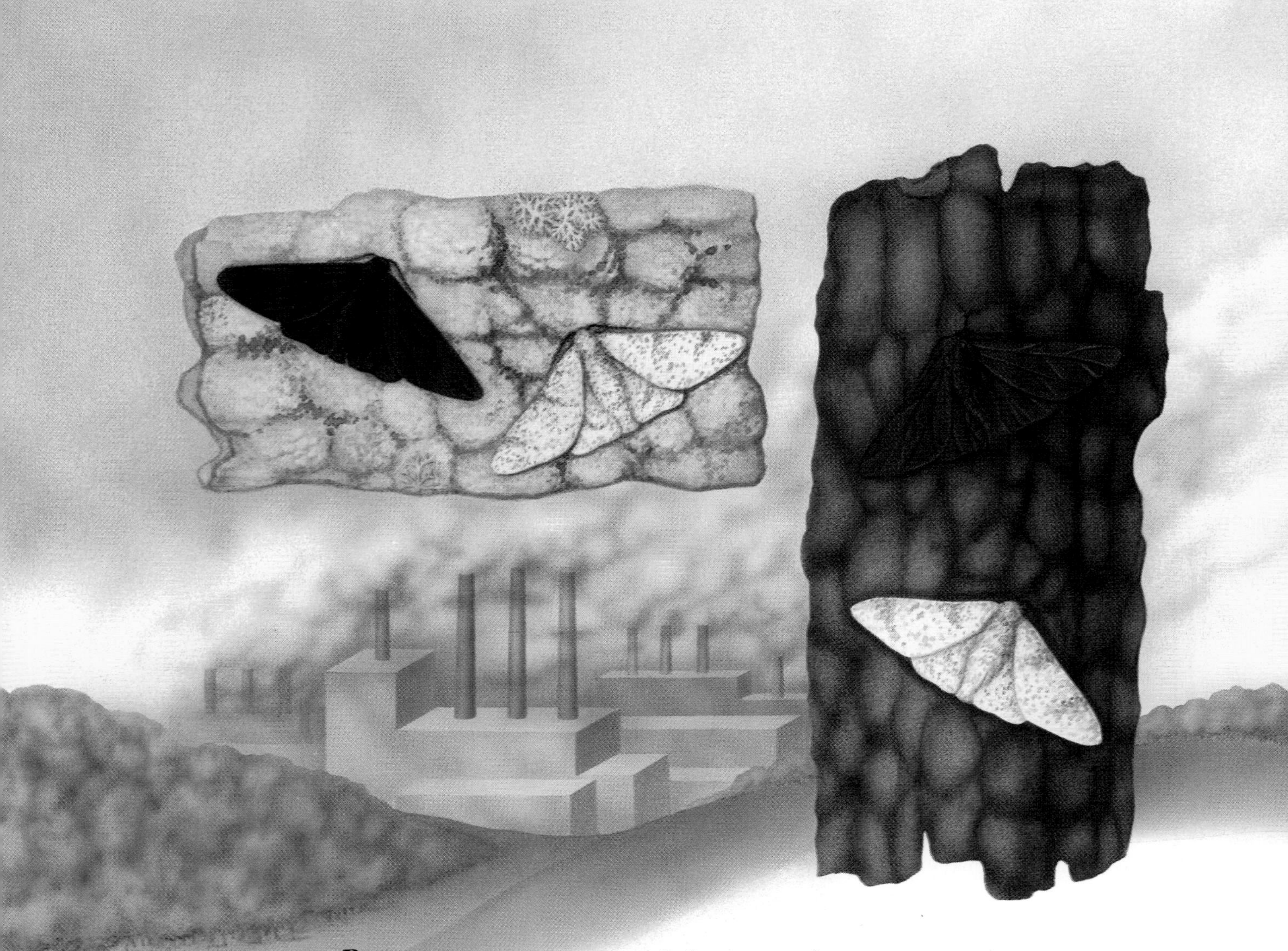

Before the Industrial Revolution, dark pepper moths were eaten more frequently by birds than white moths as they rested on the light bark of trees. As tree trunks were darkened by industrial smoke, the situation reversed.

STEPS TO DISCOVERY
Exploring Life—The First Step

Biology is not magic. It does create some very impressive illusions, however, such as its amazing "disappearing acts." A leaf-shaped butterfly lands on an oak branch and "disappears," becoming indistinguishable from the real leaves of the tree. A succulent plant growing close to the ground looks more like a rock than a living organism. Unseen by hungry animals, these organisms escape being eaten by hiding in plain sight. In England before the 1800s, where most trees were covered with light-colored lichens, the common white peppered moth was very adept at such disappearing acts. All it had to do was land on one of these mottled white tree surfaces, and the moth became virtually invisible. But then, disaster struck for these light-colored moths. The Industrial Revolution gained its full stride, and the white peppered moth performed a different kind of disappearing act, one that almost lasted forever.

Blackened by industrial smoke, their bright landing places changed to dark, sooty surfaces. The white peppered moths were now easily spotted by birds as they "hid" on their former sanctuaries. In these new conditions, they no longer had a competitive edge, and their numbers plunged toward the vanishing point. Yet, as the lighter-colored moths were eliminated, something unusual began to happen. Rarely seen before the Industrial Revolution, dark-colored peppered moths grew in numbers. Soon the population of peppered moths was back to its former prevalence, but the new moths were dark, perfectly camouflaged on the newly blackened trees of industrialized England. Somehow the peppered moth species had "switched colors," and the species continued to survive.

To understand how the moths changed in response to their environment, we first must discuss what they *did not* do. These moths were not like chameleons—an individual moth could no more change its color to match its background than you can. It was the *species,* not the individual moths, that changed color so that, by the mid-1950s, virtually all peppered moths in industrialized Britain were dark. If they could not recognize their black background and change colors, where did the black moths come from? The answer is *genetic variability.*

Most of the traits among the individuals in any species are similar, but there are also many genetic differences. People, for example, have two eyes, a nose, fingernails, and hundreds of other features that illustrate our similarities. Yet we all look different from one another, even from our own parents, because of genetic variability—that is, differences in the genes possessed by different organisms. Genes are coded bits of information in cells that determine an organism's traits. Copies of these genes are passed on from parents to their offspring, who therefore inherit traits characteristic of their parents. Occasionally a spontaneous change, a *mutation,* will occur in a gene, which causes the offspring to inherit a new trait. In a population of billions of light-colored moths, for example, a few black offspring will inevitably be produced by mutation of a pigment gene. Before the Industrial Revolution, however, few of these black varieties were ever found because they were easily seen against the light surfaces and snatched up by hungry birds. But around the mid-1800s, environmental conditions changed. The light moths had become the easy dinners, and the black moths became the "invisible" and predominant variety. Had there not been genetic variation, there would have been no black peppered moths, and the species would now be extinct in these industrialized areas.

There is another surprising chapter in this story. Today if you go to industrialized England, you will once again see the white peppered moth. Modern pollution controls have cleaned up the air, and the surfaces of the trees are once again brightly colored. The lighter moths once again have the competitive edge, and the black moth is rarely found. The process of change continues in the peppered moth, and in all other types of organisms, according to the dictates of environmental conditions.

The case of the peppered moth illustrates how a species can change over time. The process that has changed the population of peppered moths is the same one responsible for generating the millions of species produced by three and a half billion years of biological evolution. Thus, the study of the peppered moth provides a vivid portrayal of how changes in the environment can change the genetic composition of a species. The mechanism by which new species evolve is discussed later in this chapter, and some of the powerful evidence for its occurrence is described in Chapters 22 and 23.

Our modern knowledge of evolution helps us understand life. It enables us to answer such childlike questions as "Why do we have houseflies?" as well as better address more global questions, such as the possibility of human extinction. Evolution enriches the study of life by making rational sense of it. It helps us understand where organisms and their properties come from and why they exist. As you progress through this book, you will find that evolution illuminates all areas of biology. It will help you piece together this information into a satisfying body of understanding that will enable you make better sense of the world and what happens among living things.

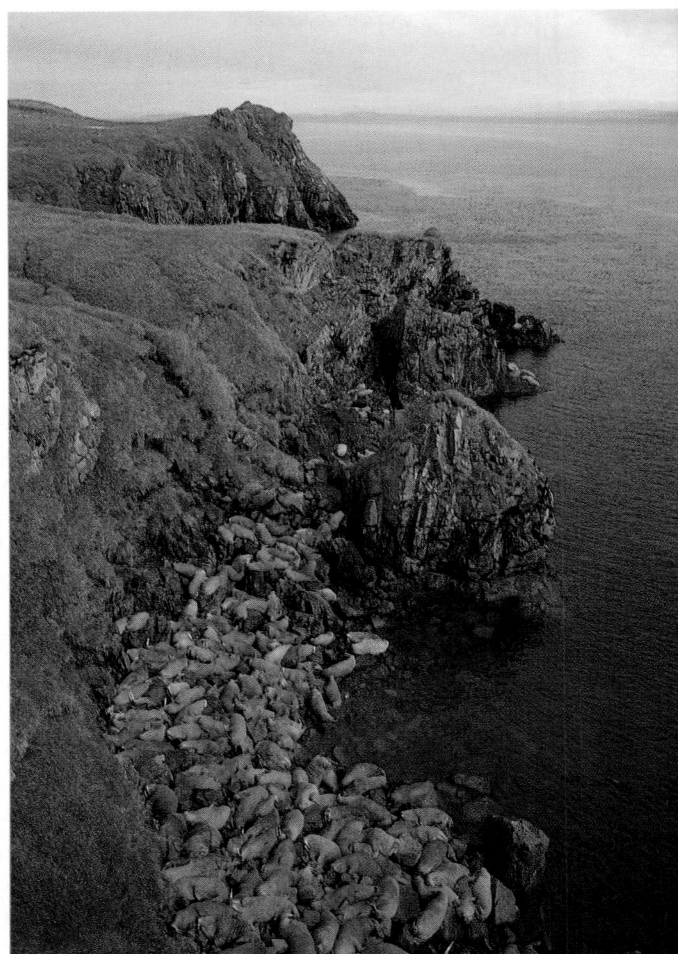

FIGURE 1-1
Defining life. As simple as it is to identify each object in the photograph as either alive or inanimate, *justifying* your choices can be much more difficult. Most characteristics popularly associated with living things (growth, movement, reproduction, consumption of food) may also be properties of nonliving entities. Clouds and mineral crystals grow; rivers, air, and clouds move; fire consumes "food" (fuel) as it "grows" and "reproduces" giving rise to "progeny" fires. Yet none of these is alive. Defining "life" requires a combination of many properties, no one of which can, by itself, be considered *the* criterion of life.

***B*iology,** the study of life, is a multidimensional, dynamic, creative activity that replaces mystery with understanding. No list of terms or facts can produce understanding, any more than a pile of unassembled gears and springs can explain how a clock works. Individual biological facts reveal little about how life works; they are mere threads that must be woven together by concepts or principles to arrive at an understanding of living phenomena. Biologists are detectives who use bits of information as clues to solve the complex mysteries of life. Their work also yields practical bonuses, such as controlling diseases, increasing crop yields, and proposing measures to preserve our environment.

▼ ▼ ▼

DISTINGUISHING THE LIVING FROM THE INANIMATE

Most of us have little trouble identifying something as being alive or inanimate. The diverse organisms shown in Figure 1-1 are easily distinguished from nonliving environmental components. But try to define what distinguishes the two, and you may find yourself at a loss. You'll discover that many of the properties that flag an object as being alive cannot be found in all organisms, or may be exhibited by some nonliving things. For example, you may have selected the ability to move as a basic "characteristic of life." But is a redwood tree an inanimate object simply because you cannot observe any movement? Or, conversely, is a river alive because its movement is evident? Because of this difficulty, rather than trying to define life, biologists describe it—usually as a list of properties that characterize all living things.

As you examine these properties in more detail in the following list, bear in mind that together they not only "define" life, they are inseparably linked to its success—its nonstop presence on this planet for more than 3.5 billion years. Mountains crumble, continents collide, climates change, yet life on earth persists in spite of the changes (Figure 1-2).

1. **Organisms Are Highly Complex And Organized.** *Complexity* is a measure of the number of different types of parts that make up an object and the precision by which the parts are *organized.* An automobile is more complex than a bicycle and less complex than a space shuttle. These differences in complexity reflect the relative capabilities of these objects. Living organisms (Figure 1-3*a*) are vastly more complex than any space ship, and they are capable of a vastly greater variety of activities.
2. **Organisms Are Composed Of Cells.** As we will see in Chapter 3, cells are the functional units of life—all living organisms are composed of cells. For some **unicellular** organisms—those that consist of one cell—the cell *is* the organism (Figure 1-3*b*). Most organisms, however, are **multicellular,** consisting of hundreds to trillions of cells, depending on the organism's size and complexity.

3. **Organisms Acquire and Use Energy.** Manufacturing and maintaining complexity requires the constant input of *energy*. Living organisms require an input of energy to build and maintain their tissues and fuel their activities. Some organisms, such as plants, acquire the necessary energy by trapping sunlight and converting it to a form they can use. Eating these plants provides energy for other organisms, which in turn surrender their energy to other organisms when they are eaten. For example, the spider in Figure 1-3*c* is closing in on the energy stored in the body of a fly. The original source of energy for both animals was sunlight absorbed by plants.
4. **Organisms Produce Offspring Similar to Themselves.** New life is generated by the process of *reproduction*. Reproduction not only creates more organisms, it produces organisms that are very similar to their parent(s). Humans always produce humans, and octopuses always produce octopuses. Similar individuals are produced generation after generation, assuring that the characteristics and functions of a particular kind of organism persist long after the parents have died (Figure 1-3*d*).
5. **Organisms Are Built According to Internal Genetic Instructions.** Offspring resemble their parents because they *inherit* a set of *genetic instructions* as part of the process of reproduction. The genetic instructions that we inherited from our mothers and fathers consist of a vast collection of individual **genes,** the stable units of genetic information that determine the shapes of our faces, the intricate interconnections among our billions of nerve cells, our propensity to form social units, and thousands of other traits. The extent to which the genetic blueprint dictates biological activity can be appreciated by watching a spider as it plays out the intricate, genetically programmed behavior responsible for constructing a web (Figure 1-3*e*).

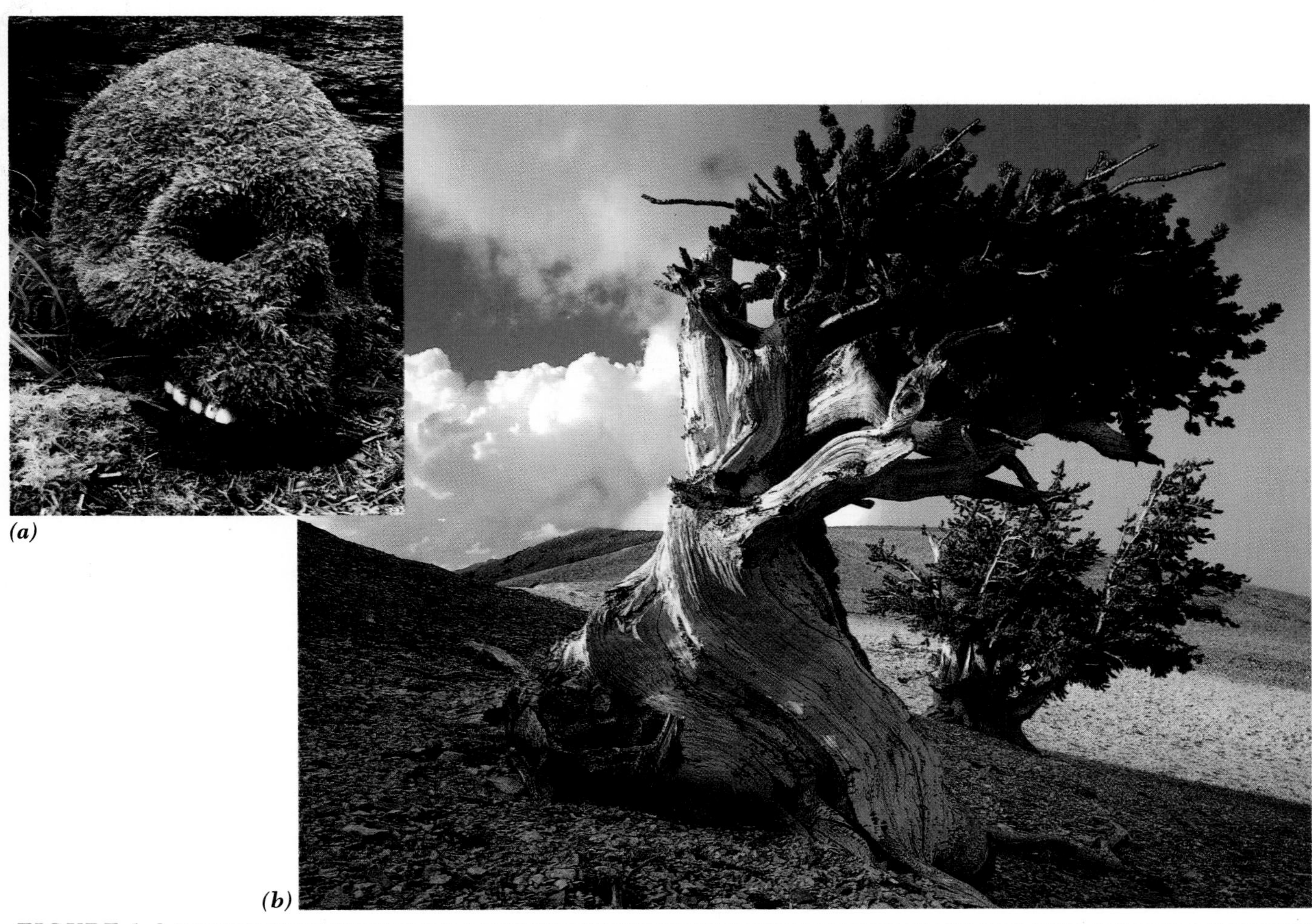

FIGURE 1-2
The enduring nature of life. ***(a)*** This moss-covered skull testifies that life goes on in spite of the death of an individual. ***(b)*** This bristlecone pine has endured cold, lack of soil, and slashing winds in this same spot for thousands of years.

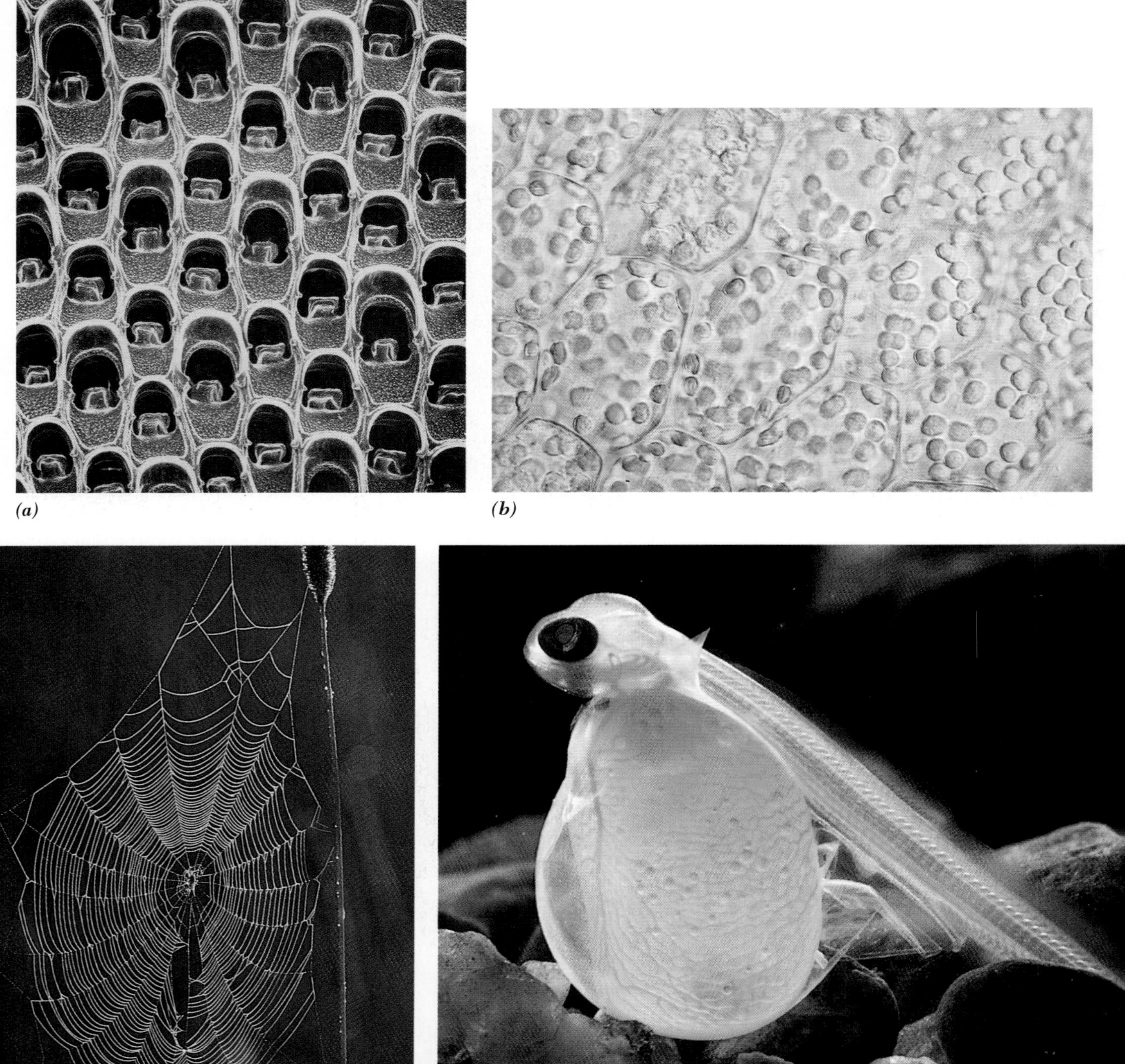

FIGURE 1-3

A number of life's properties illustrated. ***(a)*** *Organization.* The skeletal remains of these marine bryozoans reveal a high degree of order and organization. ***(b)*** *Cells.* A multicellular organism is a constellation of cells, each of which can perform all the activities associated with life. ***(c)*** *Acquisition of energy.* This spider is about to harvest energy from a housefly. ***(d)*** *Reproduction.* The resemblance within this family of swallow-tailed bee eaters is the dual product of reproduction and heredity. ***(e)*** *Genetic heredity.* The construction of a complex spiderweb is a task that requires no learning—each spider constructs its web on the very first attempt. Like the spider's color, shape, and other characteristics, web-spinning behavior is a genetically programmed trait inherited from the spi-

(c) ***(d)*** ***(g)*** ***(h)*** ***(i)***

der's parents. ***(f)*** *Growth and development.* All organisms develop from a single cell. This week-old salmon embryo is acquiring the form of an adult fish, but still obtains all its nutrients from the yolk sac protruding from its midsection. ***(g)*** *Responsiveness.* The final form of these sculptured tree roots reflects their ability to respond to the presence of a boulder. ***(h)*** *Metabolism.* The glow produced by this marine animal (a ctenophore) is due to the metabolic release of energy from its food molecules, one of thousands of chemical activities that occur in an organism. ***(i)*** *Homeostasis.* Even when exposed to subzero air temperatures, thermal homeostasis is maintained in these snow monkeys, whose body temperatures remain steady in spite of the cold.

6. **Organisms Grow in Size and Change by Development.** Organisms cannot survive and perpetuate their kind without growing at some time during their life. *Growth,* an increase in size, is usually accompanied by *development,* a change in an organism's form and capabilities (Figure 1-3*f*). An acorn develops into an oak tree; a caterpillar changes into a butterfly; a fertilized human egg develops into a person capable of contemplating his or her own nature and origin.
7. **Organisms Respond to Stimuli.** The ability to respond to *stimuli* (changes within an organism or in its external environment) is literally a matter of life or death. Responses may help an organism escape predators, capture prey, optimize its exposure to sunlight, move away from detrimental environmental conditions, move toward a source of water or other resources, locate mates, change growth patterns according to season, and perform many other activities necessary for survival. (Figure 1-3*g*).
8. **Organisms Carry Out a Variety of Chemical Reactions.** Even the simplest bacterial cell is capable of hundreds of different chemical transformations. Virtually all chemical changes in organisms require **enzymes,** molecules that increase the rate at which a chemical reaction occurs. Some enzymes participate in the breakdown of food molecules; others mediate the controlled release of usable energy; still others contribute to the assembly of substances required to build more tissue. The total of all the chemical reactions occurring within an organism is called **metabolism.** (Figure 1-3*h*).
9. **Organisms Maintain a Relatively Constant Internal Environment.** An organism can remain alive only as long as the properties of its internal environment remain within a certain range. If, for example, an organism's cellular fluids become too salty, too warm, or too acidic, or if they retain too high a level of toxic waste products, the cells die, and consequently so does the entire organism. Organisms possess self-regulatory mechanisms that allow them to maintain a relatively constant internal environment despite being bombarded by changing external conditions. This characteristic of life is known as **homeostasis.** (Figure 1-3*i*).

LEVELS OF BIOLOGICAL ORGANIZATION

An object, living or inanimate, is composed of relatively simple building blocks that are assembled into increasingly complex subunits, which ultimately combine to form the final complex structure. This book, for example, consists of ink and paper, that combine to form letters printed on pages (a higher level of organization). Individual letters combine to form a more complex structure (a word), which groups with other words to form the next level of complexity, a sentence. These levels of organization increase in steps until the final book is formed. Living organisms can also be placed within such a *hierarchy of organization* in which simpler structures combine to form the more complex structures of the next level of organization, which in turn interact to form even more complex units (Figure 1-4) .

All matter is built of protons, electrons, and neutrons. These three kinds of subatomic (smaller than atoms) particles combine to form many different types of atoms. Atoms, in turn, can combine with one another to form a virtually limitless variety of molecules. In living organisms, molecules join together to form subcellular components (including structures called *organelles*) that are assembled into *cells.* Cells occupy a very special level of organization; a cell is the least complex unit in the biological hierarchy that is, in itself, alive.

The attainment of the cellular organization reveals something profound and fascinating about hierarchies of organization. *Increasing complexity often creates new properties that exceed the sum of the parts used to form the structure.* To return to our book analogy, when individual letters are combined to form a word, the structure at the next level of organization, an additional property emerges: the word's meaning. No new materials were added—the new property is strictly a function of higher organization, something that didn't exist when the same parts had no ordered arrangement.

Life itself is such an extra property, a phenomenon that emerges at the level of the cell and that exceeds the sum of the parts. If not organized to form a cell, the components of a cell are incapable of generating or sustaining life. Subcellular organelles and cytoplasm removed from one another simply deteriorate—they cannot maintain their ordered state, reproduce, or respond to stimuli. They are not alive. Yet together, as an intact cell, all the properties of life emerge.

As highly ordered as organisms are, they too are subunits of more complex levels of organization. Individuals of the same species inhabiting the same area constitute a **population.** Different populations in a particular area interact with one another to form **communities,** which form part of a particular **ecosystem.** Ecosystems consist of the community in an area plus the nonliving environment, for example the water, rocks, and mud, together with the bacteria, plants, and animals at the bottom of a lake. All the world's ecosystems combine to form the **biosphere.**

This hierarchy of life provides one of the foundations on which this book is organized. We begin at the atomic and molecular levels, work our way through a variety of subjects at the cellular level, turn to a discussion of organs and organ systems, then move on to an appreciation of the diverse forms of organisms on earth, and finally examine the ways organisms interact with one another and their environment.

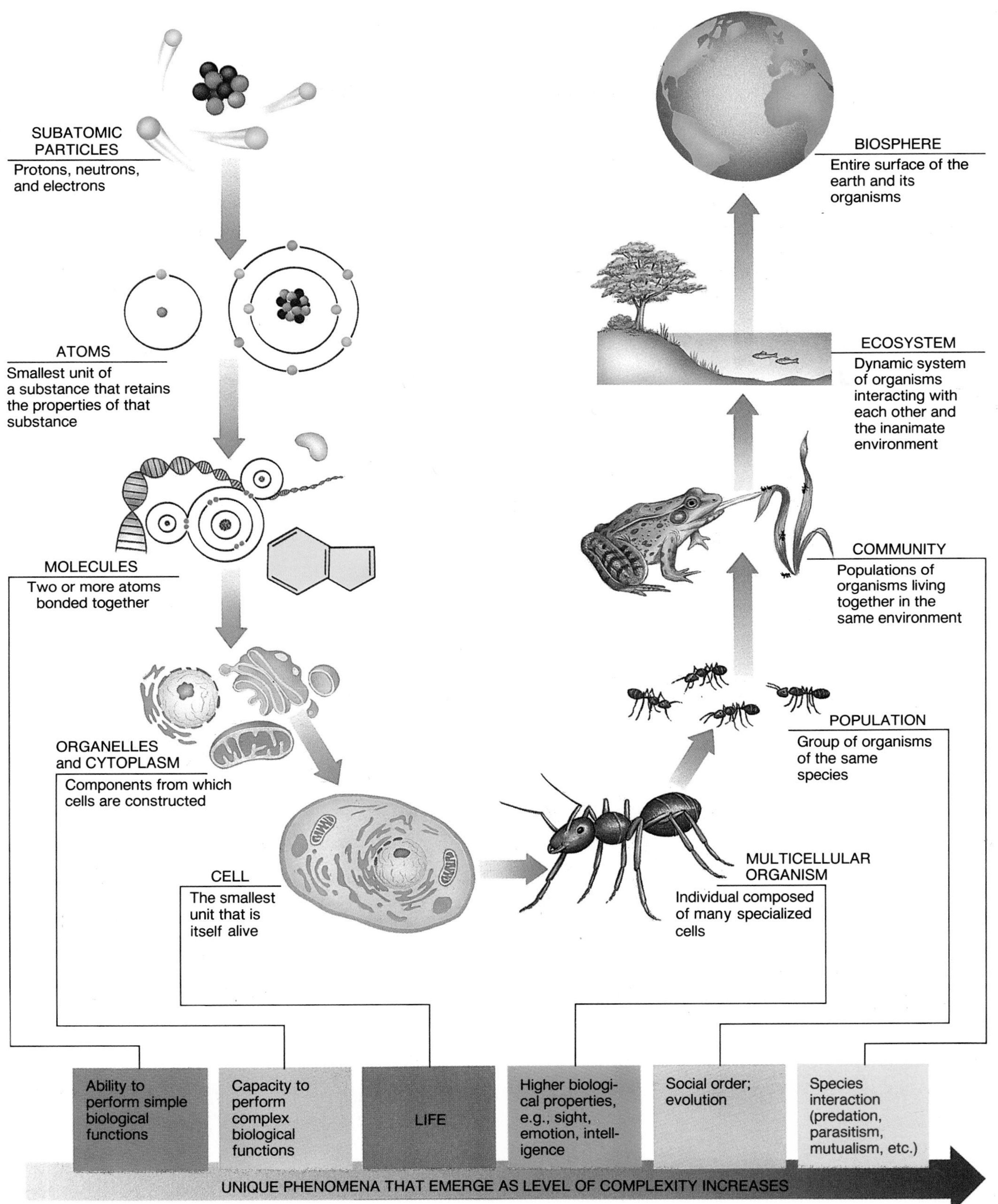

FIGURE 1-4

Levels of biological organization. Each single-step jump increases structural complexity and may also generate a unique property distinct from the structure.

WHAT'S IN A NAME

Another hierarchy exists in biology, this one needed to help us organize and make sense of an otherwise bewildering array of organisms and their characteristics. The formal system of naming, cataloguing, and describing organisms is the science of **taxonomy.**

ASSIGNING SCIENTIFIC NAMES

All known species are assigned a name, a label that doesn't vary from biologist to biologist. In addition, no two species have the same name. Each name consists of two latinized words according to our **binomial system of nomenclature** (binomial loosely means "two names"). Latinizing these names may strike you as overly formal and difficult, but it would be infinitely more confusing if each scientific name reflected the native language of the biologist who provided the label. Some scientific names would be in English, others would be written with Chinese characters, and still others would contain only Arabic letters. The use of Latin standardizes the language of nomenclature so that all species receive names expressed in the same language, using the same alphabet.

The first word in an organism's pair of names always identifies its *genus,* a group that contains closely related species. The second word, called the *specific epithet,* singles out from within that genus one kind of organism, the species. For example, *Streptococcus pyogenes,* the bacteria responsible for many diseases, from "strep throat" to the recently sensationalized "flesh-eating" infections, is just one species in a genus that mostly consists of harmless members.

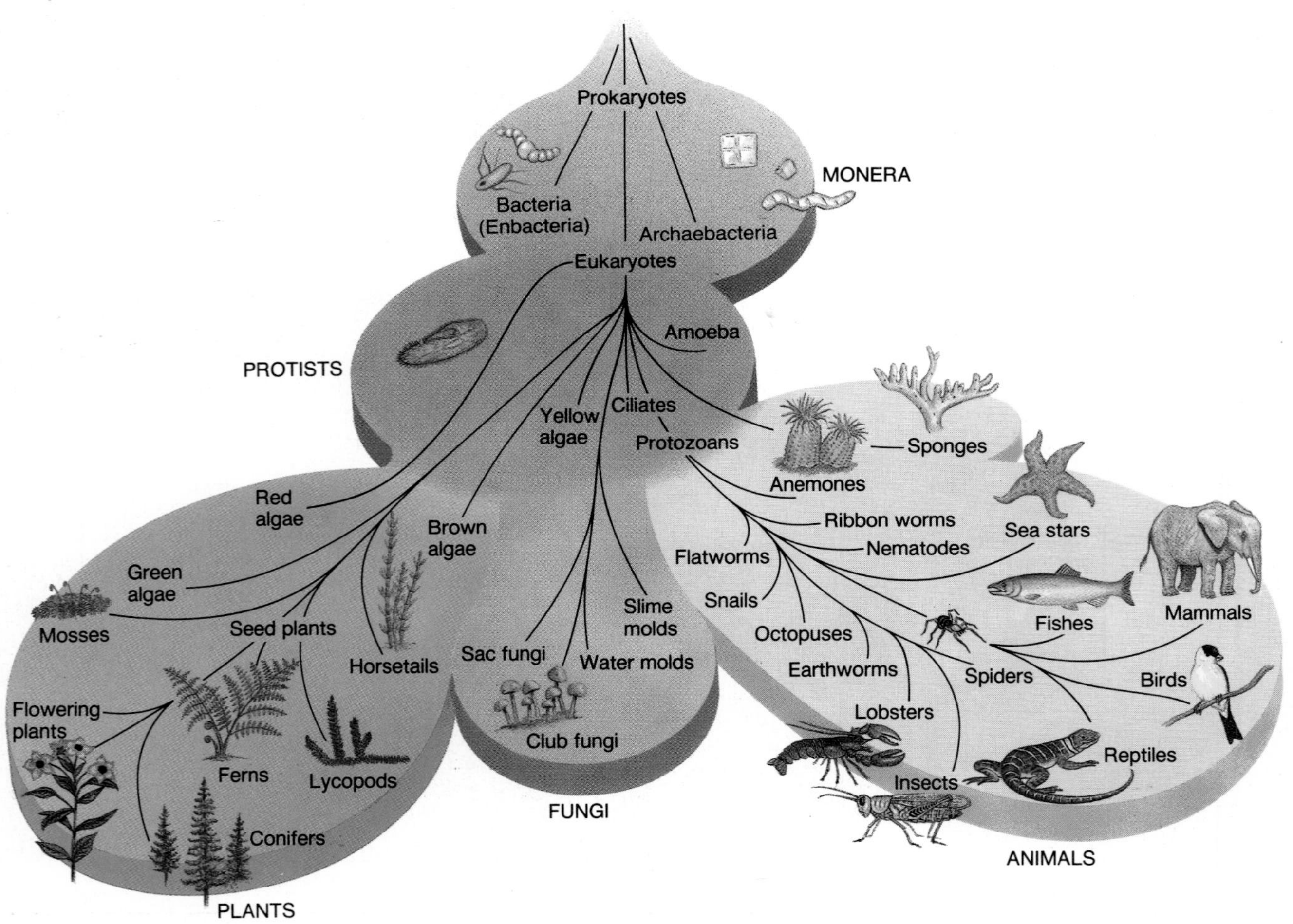

FIGURE 1-5

The five kingdoms of biological classification. This diagram shows the relationship between the five kingdoms of life. The animal kingdom, with its 1.3 million species, is the largest kingdom; in other words, it contains the greatest diversity.

CLASSIFICATION OF ORGANISMS

Taxonomists classify organisms by sorting them into groups according to traits that reveal **phylogenetic** relationships. This means that the organisms in the same group have a more similar ancestral history and are therefore more closely related than are organisms in different groups. ("Phylogenetic" refers to *evolutionary* relationships, revealing how recently two types of organisms shared a common ancestor). These relationships are found in all taxonomic categories, from the lowest level of category (the species) to the highest level of biological classification, a kingdom.[1] The currently accepted scheme of classification maintains that all organisms on earth belong to one of five kingdoms—monera, protists, fungi, plants, or animals (Figures 1-5 and 1-6). The simplest organisms, bacteria and their relatives, belong to the monera kingdom; the most complex belong to the animal kingdom (the characteristics of each kingdom and the evolutionary relationships within each are discussed in Chapters 25 through 27).

[1] A newly proposed taxonomic group, called a *domain*, contains kingdoms that possess similar cell types. Three domains have taxonomically solid foundations but, at the time of writing, have not yet been officially adopted.

FIGURE 1-6

Representative organisms from each kingdom. ***(a)*** rod-shaped bacteria—Monera Kingdom; ***(b)*** *Paramecium,* a unicellular animal-like organism—Protista (protist) Kingdom; ***(c)*** poisonous *Amanita* mushroom—Fungi (fungus) Kingdom; ***(d)*** pink moccasin flower—Plantae (plant) Kingdom; ***(e)*** a pair of young caribou—Animalia (animal) Kingdom.

EVOLUTION AND ADAPTATION: BIOLOGY'S CENTRAL "UNIFIER"

As you progress through your introductory biology course, you will learn that, in spite of its diverse array of subjects, there is one principle that permeates all areas of the biological sciences, constituting the most important unifying principle in biology. This principle, *evolution and adaptation,* provides the rational basis for understanding life's presence and life's forms. It explains why the biosphere is populated by millions of species rather than just one type of organism. At the same time, this principle accounts for the unmistakable unity that exists among even the most diverse organisms. The principle of **evolution** states simply that species change over time. As a result of evolutionary changes, new species of organisms emerge while older forms of life may disappear. In other words, life evolves.

Evolution explains how all life on earth, both living and dead, are members of one extended, genetically related "family" of organisms. You and a mushroom are similar in so many fundamental ways because you shared the same ancestor far back in time, and both of you retain many of the same molecular, genetic, and cellular endowments provided by that ancestor. All of the diverse species on earth are descendants of primitive cells that were the earliest life forms. It is this common ancestry that accounts for the unity of life.

Biological success for a particular type of organism depends on how well suited the organism is to its environment. An organism can survive only in environments that supply all its essential needs, and then only if the organism possesses the "equipment" needed for acquiring those environmental resources. Even then, it can succeed only if it can survive the adverse conditions to which it is exposed.

FIGURE 1-7

Evolution and adaptation. The concentric bands of color in this aerial photograph of a hot sulfur pool at Yellowstone National Park are due to the growth of microbes adapted to different temperatures. The pond, which is hottest in the center and coolest at its rim, is colored by enormous populations of microorganisms concentrated in each ring-shaped thermal zone. Each distinctly pigmented type of microorganism is specifically adapted to (and limited to) one narrow temperature range.

In other words, an organism must be *adapted* to its particular set of environmental conditions (Figure 1-7). **Adaptations** are traits that improve the suitability of an organism to its environment.

Every topic we discuss in this book has evolutionary foundations. As we describe certain characteristics of various organisms, keep in mind how these features better suit the organism to its environment, increasing the chance for survival of both the individual and the species. We will see how some of these adaptations have become modified over evolutionary time as environmental conditions have changed and new species evolved. For example, the tiny bones in your middle ear, which transmit sound through a recess in your skull, evolved from bones that were originally parts of the jaws of ancestral fishes that lived long before the first four-legged animal ever set forth on dry land.

Evolution is the cornerstone of our understanding of biological phenomena, biology's central theme that you will revisit in each of this book's chapters.

A VOYAGE THAT REVISED OUR CONCEPT OF LIFE'S ORIGINS

When a 22-year-old graduate named Charles Darwin was appointed "naturalist" for a scientific expedition around the world, the world would be plunged into a heated debate that continues today. It was 1831, and Charles Darwin (Figure 1-8) accepted the commission on the *HMS Beagle*, a small surveying ship that was to map the coastlines and harbors of South America. As the ship's naturalist, it was Darwin's job to collect and organize thousands of specimens from around the world.

Darwin, who had recently decided to become a clergyman in the Church of England, started his voyage with the firm belief that all of the world's plants and animals
l ... reated directly by the hand of God in the form
... As he traveled and explored new
... estion this view of a static world
... had been fixed since the time of
... few days of the voyage, as Darwin
... ring from seasickness, he read the
... *les of Geology*, a textbook written
... s book meticulously presented ev-
... was much older than previously be-
... gradually changed over long periods
... s as mountain building, erosion, vol-
... earthquakes.
... gan collecting evidence that paralleled
... Argentina, he found a fossil bed con-
... als—a giant sloth, a hippopotamus-like
... armadillo. Darwin saw the clear rela-
... the appearance of these extinct animals
... day and wondered how it was that these
... ad disappeared. He suggested that they

FIGURE 1-8
Charles Darwin and the voyage of the HMS Beagle. Portrait of the young Charles Darwin.

had fallen to competition and predation from animals that had invaded following the formation of a land connection between North and South America.

But Darwin's most important observations came from his explorations of the Galapagos islands, 900 kilometers off the coast of Ecuador. On these islands, he made one of the most important scientific observations in history, observations concerning the now famous Darwin's finches. Of the 13 species of Galapagos finches Darwin identified, all were similar to one another in overall body form, but they differed in their "lifestyle" and the shape of their beaks. Some of the species lived on the ground, others in the trees. Some had strong, thick beaks adapted for crushing seeds, while others had beaks that were especially suited for feeding on flowers or insects. Among the insect eaters was the so-called "woodpecker finch," which digs insects out of the tree bark using a cactus spine held in its beak. Why do birds with such different feeding habits look so much alike? Darwin concluded (although years later) that all of the island finches were descendants of one of the mainland species of ground finch that had drifted several hundred kilometers to one of the islands. Because the islands were essentially devoid of competitors, the immigrant birds established a thriving population. Eventually the individuals of the original finch species gave rise to the various species now present on the islands. In other words, 13 species of finches gradually evolved from one species. To explain how evolution works, Darwin proposed his theory of natural selection.

DARWIN'S THEORY OF EVOLUTION BY NATURAL SELECTION

Darwin realized that any species of animal or plant has the reproductive potential to generate enough individuals in a relatively short period of time to cover the earth many times over. Yet, the numbers of individuals of most species remain relatively constant from one year to the next. Darwin concluded that the number of individuals of a species was limited by competition with others, and a struggle to survive the potentially adverse factors in their environment, such as lack of food or water, predation, parasitism, disease, cold, heat, flooding, and salinity.

What then determines which members of the population survive the "struggle" and which are eliminated? Darwin was aware that there is **variability** within the members of a plant or animal population, much like that found among members of a human population (Figure 1-9). Variation among animals may be reflected in the color of the coat, body size, an ability to withstand high temperature, the choreography of a mating dance, or any other type of characteristic. It became clear to Darwin that some members of the population possessed characteristics that gave them an increased chance of survival relative to other members of the population. The survivors might have a more efficient style of gathering food, or a particularly high resistance to a common parasite, or a little faster gait. These animals would be better adapted to their environment and thus more likely to survive.

But survival alone was not the most important consideration; rather, surviving *to successfully reproduce* was the critical factor. Those organisms that are best suited to survive long enough to reproduce tend to produce a greater number of offspring. According to Darwin, the environment "selects" those organisms that will survive and reproduce, while it eliminates those organisms that do not (they simply fail to survive and reproduce). Darwin called this process **natural selection.** Natural selection results in the production of more offspring by individuals that are better adapted to their environment.

Darwin also knew that offspring tend to inherit the characteristics of their parents (although he didn't know how). Consequently, the offspring of survivors will inherit the advantageous traits that promoted the success of their parents. After several generations a population will tend to acquire more organisms with genetic traits that make its members better adapted to its environment. Individuals with traits poorly suited would be less likely to successfully compete and would therefore leave fewer (if any) offspring. This change in genetic composition of a population from generation to generation is the very essence of the evolutionary process. Giraffes, for example, have long necks because individuals that happened to have longer necks were more successful in obtaining food than their short-necked competitors. Longer-necked members of the population were more likely to survive and have offspring which, like their parents, would have longer necks. The result is adaptation of a species to the environment.

Environments do not remain constant over long periods of time, however. Consequently, an animal that is successful in a particular habitat at one time might be poorly adapted at some other time, after a severe climate change, for example. Darwin realized that the natural variation that exists in a population provides the basis for evolutionary change that accompanies environmental change. Consider, for example, a hypothetical population of fleas that fed on the blood of the bison that roamed the American plains during the early nineteenth century. As long as the bison were plentiful, natural selection favored those individuals who were most efficient at finding and piercing the skin of these large herbivores. But as the numbers of bison sharply diminished as they were overhunted, bison-seeking characteristics would no longer be adaptive. Instead, those fleas in the population that could locate and survive on the blood of horses or dogs might be

FIGURE 1-9
Genetic variation in the human population.

favored by natural selection and the population would gradually shift toward this new variety of insect. The gradual accumulation of small genetic differences among individuals could radically transform a species' characteristics. If the populations of a species become so isolated from one another that they cannot interbreed, then a strong likelihood exists that the populations will evolve into separate species. After several billion years, this divergence of species has produced the diversity of life that we see today, including the human species.

Darwin published *The Origin of the Species* in 1859. The response was thunderous. All 1250 copies of the book were sold on the first day it was available. Discussions, protests, and personal attacks followed—continuing to this very day.

BIOLOGY AND MODERN ETHICS

Our modern understanding of biological concepts and how to apply these concepts is already changing the planet and all its inhabitants, and the influence grows dramatically as we approach the next century. While these changes hold tremendous promise of improved living conditions and hopes for a healthier environment for all organisms, new biological technologies come with their share of ethical controversy. Some applications may seem clearly unethical, such as using technology to develop biological weapons (strains of viruses or bacteria that incapacitate or kill people). Yet many of these questions are not so easy to identify as ethical or unacceptable. Consider the following questions:

- Should scientists be allowed to patent life?
- Do the benefits of genetic engineering outweigh the potential risks that this new technology poses?
- Do economic freedoms and needs justify habitat destruction and subsequent extinction of "obscure" species?
- Although treating disease by replacing faulty genes with good ones is not so controversial, is it ethical to use the same technology to introduce genes into healthy people, genes that enhance a particular desirable characteristic such as intelligence or athletic prowess?
- Should biological research be subjected to external regulation by political institutions?
- Should persons with AIDS be allowed to work in health-care professions?

These questions represent just the tip of the bioethical iceberg. An informed public is the best protection against clearly unethical applications or abuses of biological knowledge. Information and open discussion are also essential before the more debatable issues can be resolved. Regardless of how they are resolved, these bioethical decisions will permanently influence the world and all its inhabitants.

BIOLOGY AND THE PROCESS OF SCIENCE

Consider the time-honored saying that if you get wet on a cold day, before long you will be suffering from a cold. For many of us, our experience seems to verify this statement—we get caught in a cold rain or step in a puddle, and within a couple of days we have a cold. If this has happened to you, can you conclude that getting chilled causes colds? In other words, if two events occur close together, is that proof that the first event caused the second?

The answer is "no." In fact there is no relationship between getting cold and catching colds. The reason these two events often seem to occur together is because both tend to happen with the onset of wintry weather. Colds are more common in the winter because people spend more time in buildings when the weather is cold or rainy, and being indoors increases close exposure to other people, many of whom have colds and are shedding the virus. In addition, cold wintry air dries the mucous layer that lines the nasal passages, exposing the underlying cells to viral attack. But getting chilled does *not* make you more vulnerable to the common cold. It's a popular myth.

This example, plus countless other similar types of conclusions that find their way into our lives, suffers a serious shortcoming in reasoning. They are all products of **anecdotal evidence**—information based on personal experiences and testimonials about those experiences. Such stories lead many people to jump to conclusions of cause and effect between events that may be related by simple coincidence or clustered by some underlying and less obvious set of events.

For example, is the longevity of George Burns reliable evidence that tobacco smoke doesn't shorten life? Scientific studies (as opposed to anecdotal evidence) have proven that smoking is the leading cause of preventable death in the United States and other industrialized countries. We can explain George Burns' longevity in the same way we explain how some rare people have survived falls out of airborne planes without parachutes—they were very lucky. We certainly do not conclude that falling out of airplanes without a parachute is harmless simply because someone lives through it. Again, anecdotal evidence fails to reveal the critical information, that is, how long George Burns would live had he not smoked cigars.

Anecdotal evidence is the heart of misinformation. It often leads to absurdly irrational beliefs, such as black cats causing bad luck or ostriches sticking their heads in the ground to escape danger (such behavior would quickly lead to a severe ostrich shortage). It has also misled people in their attempts to understand themselves and the nature of life. One faulty argument asserts that the very fact that we don't understand all aspects of life proves that life is a bewildering mystery beyond human understanding.

Scientists, however, believe that all phenomena in the universe have rational, verifiable explanations, even though we haven't discovered all such explanations. In fact, we will likely never know everything there is to know about life and the universe, but we continue to expand our understanding by making careful observations, asking questions, and seeking answers. Yet none of these activities alone has provided us with our current understanding of how organisms work. Observations alone, like anecdotal evidence, provide descriptive information, but they rarely reveal the mechanisms responsible for biological activities, or the ways in which the intricate processes of biology affect one another. For example, the changing of seasons from warm summer-like conditions to cold, wintry days and nights is accompanied by a profound change in the biological activity of countless organisms. Tree leaves turn brightly colored and fall; many animals hibernate, while others migrate to warmer climate; still other animals develop a thick winter coat of fur. But is it the onset of cold that triggers these biological changes or the shortening of the days as winter approaches? In other words, what is the *cause* of this observed effect? Furthermore, *how* does that factor bring about the observed change?

The answers to such questions are not obtained by simple observation. They require the combined input of several different methods for acquiring and examining information, assembled into what amounts to a scientific approach. Using scientific methods provides us with a way of not only understanding how life and the universe are put together and how they work, but a way of verifying the accuracy of the information we discover and the explanations we propose. While this approach helps scientists evaluate and explain nature, it can also be used to solve problems in our personal lives and help us understand for ourselves.

THE SCIENTIFIC APPROACH

"The skies were overcast, and the city was expecting heavy rain and frogs." That might have been a typical forecast in the mid-1600s, when it was commonly believed that frogs developed from falling drops of rain. The notion that living organisms arose directly from inanimate materials was known as **spontaneous generation** and was popular among scientists and nonscientists alike for hundreds of years. It seemed particularly evident that flies arose directly from decayed meat, since everyone had observed rotting meat covered with maggots—the larval stage of flies.

One of the first public doubters of spontaneous generation was Francesco Redi, an Italian physician and naturalist. By recognizing that spontaneous generation might be an erroneous concept, Redi had taken the first step on the path to scientific discovery. This is one way scientists begin an investigation—they learn about the information currently available on a subject, and they make relevant observations on their own. Redi had observed that maggots tended to be located in places where adult flies could also be found. He proposed that the origin of maggots was not the flesh of dead animals, but eggs from living parent insects. A tentative proposal of this type is called a **hypothesis.** Two hallmarks of a good hypothesis are that

- it is consistent with all observations collected up to that point (a hypothesis that contradicts a known observation is already shown to be inaccurate); and
- its accuracy can be tested by experimentation (or further observation).

Redi came up with an experimental plan by which he could test his hypothesis that maggots arose only from eggs deposited by flies and not from meat.

In his experiment, Redi put a dead snake, some fish, some eels, and a slice of veal separately into four large, wide-mouthed vessels and covered the openings (Figure 1-10). He then placed the same materials into another set of four vessels which he left open to the air (and flies). He initially sealed the jars with wax, but soon recognized a serious flaw in this approach. If maggots failed to develop in the sealed vessels, was it because no flies could get to the meat or because no fresh air could? It is essential in a scientific experiment to determine *cause and effect.* In Redi's initial experiment, the two sets of vessels differed by more than one **variable**—a condition that changes in a specified way. In this case, the two variables were the presence or absence of flies and the presence or absence of air. To determine which of these two variables was the *cause* of the results, Redi refined his experiment. He once again set up two sets of vessels containing meat, but this time both vessels were left open to the air. One was left totally uncovered, however, while the other was covered with a fine layer of gauze with holes too small to allow the passage of flies.

In the design of a proper experiment it is essential to allow only one condition to vary—this is the one variable whose role the investigator is attempting to evaluate (in this case, access by flies). Otherwise, it is impossible to determine which variable is causing the result. In modern scientific terminology, the gauze-covered set of vessels is called the **control group** because the variable being tested—access by flies—is absent. The uncovered vessels constitute what is called the **experimental group** because they were subjected to the variable. If Redi's conclusions about his first experiment were correct, maggots would not develop in the gauze-covered vessels, even though the meat had been fully exposed to air, because that one crucial variable—access by flies—was being restricted.

Within a matter of days, Redi observed that the decaying meat within each of the open vessels was teeming with maggots, and flies were observed coming and going

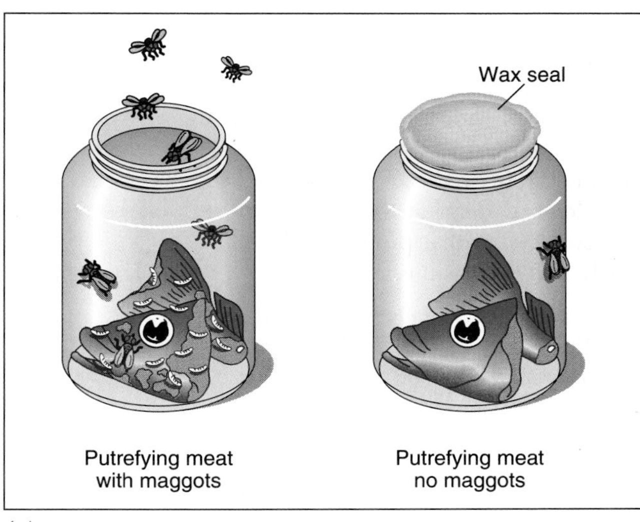

(a)

(b)

FIGURE 1-10

Redi tests the validity of spontaneous generation. *(a)* In his initial experiment, Redi placed meats in two groups of vessels. The vessels in one group was left open to the air, while the vessels in the other group were sealed with wax. Maggots appeared only in the vessels exposed to air. ***(b)*** To eliminate the possibility that maggots failed to appear in the sealed vessels due to the absence of fresh air, Redi conducted a second series of experiments. Again he prepared two groups of vessels containing meat, but instead of sealing one group of vessels, he covered the openings with a layer of fine gauze through which flies could not pass. Once again, maggots appeared in the vessels in which flies could reach the decaying meat but not in the meat isolated from flies (conclusion—flies do not spontaneously generate; it takes flies to make flies).

at will. In contrast, the meat in the sealed vessels was in the same state of decay, but no maggots were evident. Redi concluded that the closed vessels failed to produce maggots because flies could not reach the meat. In other words, the results of the experiment supported his hypothesis that maggots appeared only from the deposited eggs of flies.

Formulating and Testing New Hypotheses

Scientists typically go beyond the specific results obtained from an experiment or a series of observations and use the data to explain a more general phenomenon. In other words, they formulate a more comprehensive hypothesis, which can be subjected to additional testing. Based on his experiments with flies, for example, Redi extended his observations, hypothesizing that *all living beings* come from a parent organism.

Hypotheses form the foundation on which science grows, the fuel that drives new investigations. For Redi's expanded hypothesis to be correct, it would have to apply to all types of organisms, not just flies living in Florence. If someone were to demonstrate even one clear exception—one clear case of spontaneous generation—then Redi's hypothesis would have to be rejected or significantly modified.

Additional Tests and Confirmation

Another important characteristic of scientific findings is their *repeatability.* When papers are published, scientists present their methods so that other investigators will know exactly how the procedures were carried out and can then subject the reported findings to independent tests. They in turn publish whether their investigations confirmed the initial report or if the results could not be repeated. Because of the repeatability of Redi's experiments, the scientific world was convinced that macroscopic organisms, such as frogs and flies, arose from the eggs of parents.

The discovery of microorganisms about 300 years ago, however, rekindled the notion of spontaneous generation. These simpler microscopic life forms were believed to arise spontaneously from the remains of dead organisms or from other nonliving materials, substances in pond water, rain puddles, or a bowl of chicken soup. Did Redi's hypothesis hold true for microorganisms as well? During the mid-1700s, Lazzaro Spallanzani, an Italian naturalist, performed experiments to answer this question. He boiled beef broth to kill all microorganisms, then sealed the vessel to keep out "parental" microorganisms. After several days there was no evidence of bacterial growth in the sealed vessel but ample growth in a second unsealed vessel with boiled broth. Although Spallanzani's results seemed to disprove spontaneous generation, his critics argued that he had sealed the boiled flask, preventing air from entering the flask. Unlike Redi, he devised no way to eliminate this extra variable.

In 1860, Louis Pasteur devised an ingenious approach to eliminate this additional variable. The experiment was similar in principle to that performed by Redi 200 years earlier. Pasteur added broth to a flask, then melted the glass neck of the flask and molded it into a long S-shape (Figure 1-11). The contents of the flask and the glass neck were sterilized by heat, cooled, and allowed to remain open to the outside environment. Even though fresh air could pass through the neck into the flask, all particles of matter suspended in the air settled in the "trap" of the S-curved tube, so no airborne organisms could reach the broth. Under these conditions, microorganisms failed to grow in the flask. After several days of no growth, he tilted the flask so that some of the sterilized broth ran into the S-shaped neck and contacted the bacteria that had collected in the trap. Within 18 hours, the broth was swarming with bacteria. Thus, absence of bacterial growth was not due to the alteration of the broth by heat or to the absence of air. The lack of growth was due to the absence of contamination by airborne bacteria. Pasteur had shown that living organisms do not spontaneously appear; it takes an organism to make an organism.

FROM HYPOTHESIS TO SUPPORTED THEORY

When a hypothesis has been repeatedly verified by observation and experimentation and combines with other confirmed hypotheses to help explain an important aspect of a field of investigation, the collection of related hypotheses are considered a supported **theory.** Most scientists define the term theory much differently than does the general

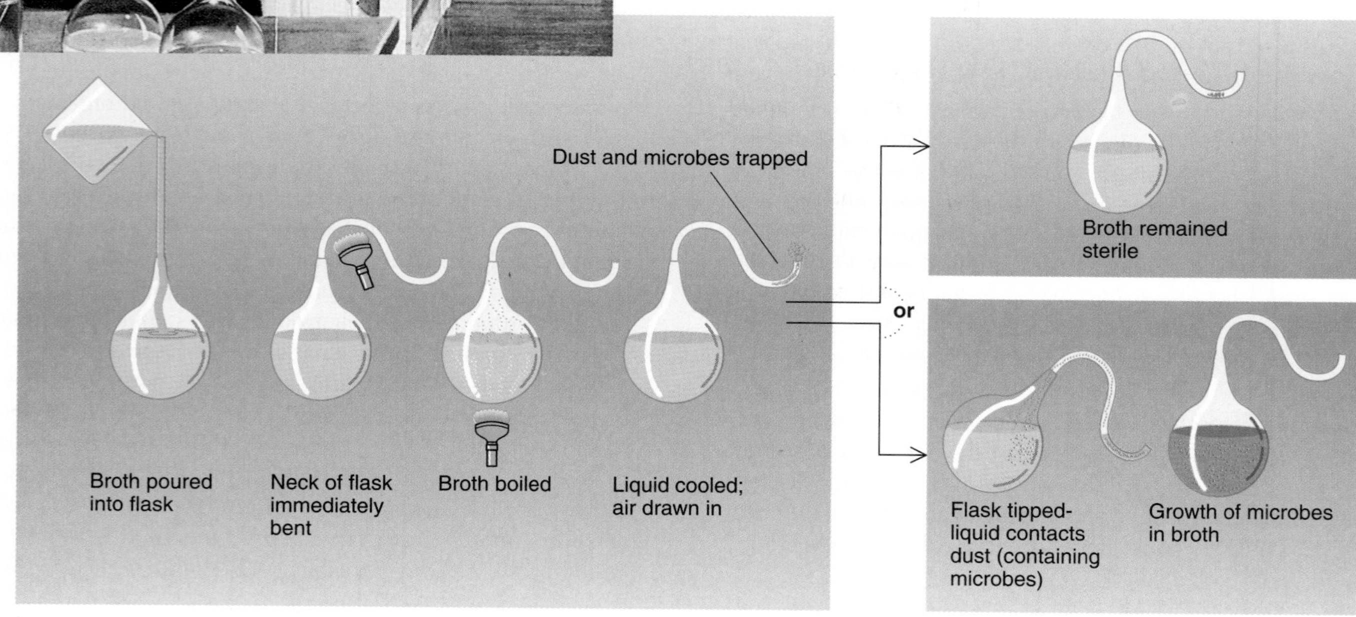

FIGURE 1-11
Pasteur's experiment disproves spontaneous generation. Pasteur poured the nutrient broth into the flask, boiled it extensively, and allowed it to cool. The flask remained open to the outside air but remained free of microorganisms. When Pasteur carried out the same procedure, then tilted the flask so that the sterilized broth would run into the neck and contact airborne particles that collected there, the medium became densely populated with microorganisms after several hours.

public, which often speaks of theories as "speculations" or "guesses." Yet even among scientists, there is debate over the nature of theories. For example, should the spontaneous generation notion be called a theory? Many scientists would say "yes—it is a *disproved* theory." In this book, we will distinguish between supported theories and disproved or unsubstantiated theories. Because supported theories are backed by repeated confirmation, they are held with great confidence by the scientific community (they are not "mere hunches" or guesses).

For instance, one of biology's most important supported theories is the theory of evolution—the theory that species change over time. Although one argument against evolution states that "it is just a theory," this tact incorrectly uses the popular interpretation of the word theory as meaning "speculation." The theory of evolution is no less reliable than is the atomic "theory." Just as we cannot directly see atoms, we cannot *directly* observe the production of new species by evolution. Biologists have gathered a tremendous amount of evidence in support of biological evolution, however, and not one major piece of evidence has ever been obtained that suggests it has not occurred. Conversely, the so-called "theory" of creationism is not really a scientific theory at all. It proposes no testable hypotheses and thus cannot be proven right or wrong. Proponents simply accept it on faith. The theory of evolution, on the other hand, makes many predictions, and, although many types of evidence could prove it incorrect, no such contradictory evidence has ever been discovered. In the words of novelist and biologist H. G. Wells, "The order of (ancestral) descent is always observed."

APPLICATIONS OF THE SCIENTIFIC PROCESS

There is another side to scientific research. It often leads to practical applications, some of which have improved our lives and others that threaten the life-supporting capacity of the biosphere. Pasteur's work, which demonstrated that relatively simple measures of sterilization could prevent bacterial growth, had almost immediate practical ramifications. The British surgeon Sir Joseph Lister grasped the importance of Pasteur's work and instituted the revolutionary procedure of cleansing surgeons' hands, instruments, and surgical sites prior to their performing an operation. The procedure produced a striking drop in the incidence of fatal post-surgical infections and was quickly adopted by other medical professionals. As a result, the entire practice of medicine took a giant stride toward becoming safer and more effective.

SCIENTIFIC METHODS—AN OVERVIEW

An overview of the scientific method as presented here is summarized in Figure 1-12. It is important to realize, however, that this is a general description of *one way* scientists approach their studies. Even though it is systematic and

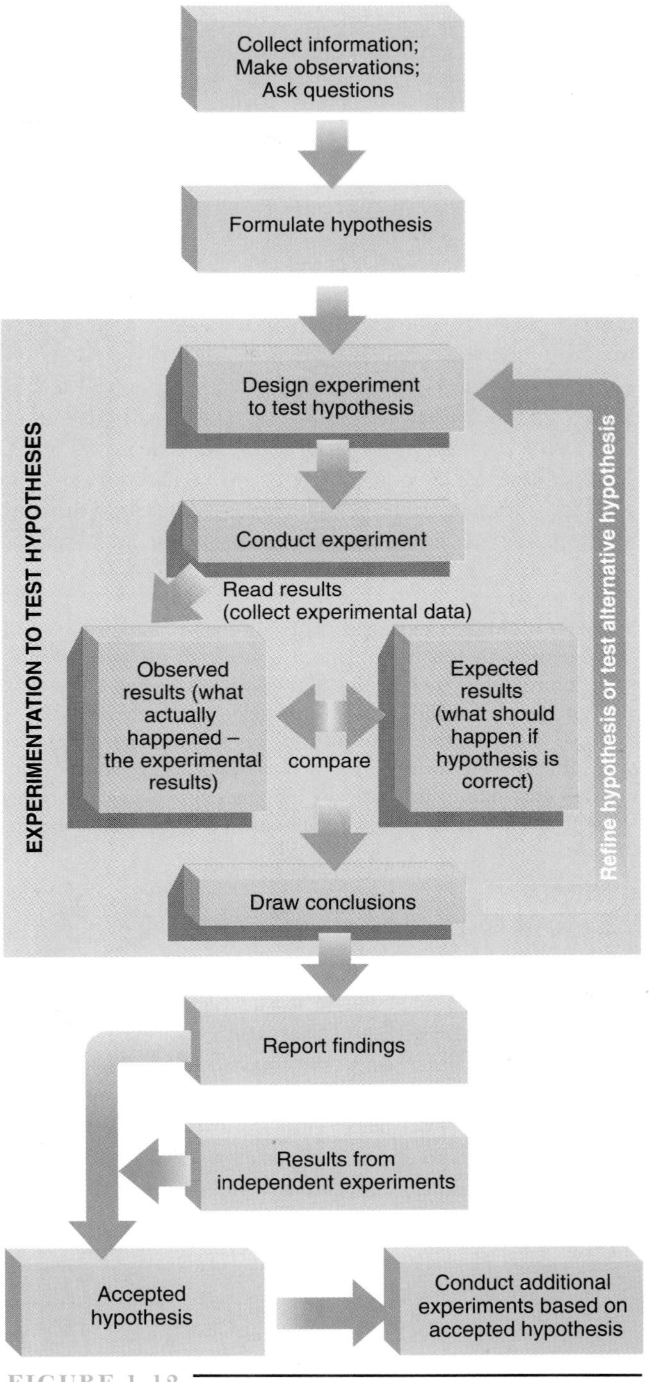

FIGURE 1-12

The scientific method. Many scientific investigations start with an observation of a phenomenon, a discussion with colleagues, or ideas triggered by listening to the presentation of a scientific paper or by reading a research article in a scientific journal. Such activities stimulate the scientist to formulate one or more hypotheses to explain a particular set of observations. These hypotheses lead to a plan for gathering experimental data that either support a hypothesis or refute it. Failure of one hypothesis often leads to another hypothesis, which is then tested through additional experimentation.

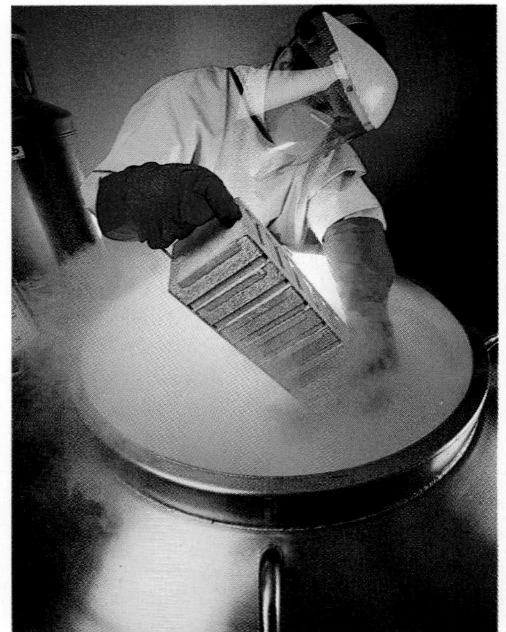

(a)

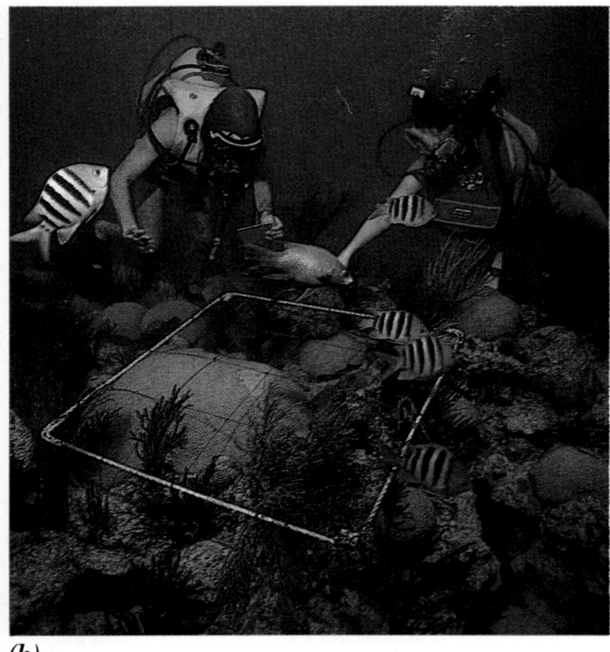

(b)

FIGURE 1-13

Modern biologists exploring life. Although biologists rarely become politicians or economic leaders, they nonetheless help shape the future of the world. Some scientists actively seek solutions to specific problems, such as this biologist investigating the virus that causes AIDS ***(a)***. Others seek answers simply for the sake of knowledge and increased understanding—these marine biologists, for example, studying the ecology of the ocean floor ***(b)***. Discoveries by such "basic researchers," however, often have hundreds of practical applications. As you will see throughout this book, scientific findings continue to dramatically influence our lives, from providing products of huge economic value to informing us how we may alter our global lifestyles in order to preserve the planet's life-supporting capacity.

allows for creativity and flexibility, there is no absolute path for discovering the answer to a question. Some scientific studies might start with an observation; others might start with a question based on previous work. Sometimes a scientific study might begin with a hypothesis to be tested, while other times an accidental discovery may trigger a finding that leads to a hypothesis. As you will see throughout this book, scientific facts and concepts have their own unique and often fascinating stories of exploration and discovery.

Regardless of which path is taken, scientific knowledge grows over time, with new discoveries blossoming from the seeds planted by previous researchers. Thousands of questions, both practical and theoretical, propel scientific inquiry as modern biologists continue to investigate the world of life (Figure 1-13).

SYNOPSIS

Organisms possess a combination of properties that is not found anywhere in the inanimate world. Organisms are highly organized. They are composed of one or more cells; they acquire and use energy; they produce offspring similar to themselves; they contain a genetic blueprint that dictates their characteristics; they grow in size and change in appearance and abilities; they respond to changes in their environment; they carry out a variety of controlled chemical reactions; and they maintain a relatively constant internal environment.

Organisms can be described at various levels of organization. At the lowest, simplest levels, life consists of subatomic particles organized into specific atoms, and atoms organized into specific molecules. At the other end of the spectrum, communities made up of various species are organized into ecosystems which, in turn, make up the biosphere. With each step-increase in complexity, new properties emerge. Life, for example, emerges at the cellular level.

Evolution forms a foundation for understanding life. Species change over time, and new species are generated. This process of biological evolution produces organisms that are well suited (adapted) to the conditions of their environment at that time. Evolution not only explains how organisms change, it also helps us understand the underlying principles of life that will be discussed throughout this book.

Biologists and the scientific method. Scientists believe that the world is understandable and that rational explanations exist for all phenomena. The basic objective of science is to understand life and the universe in a way that can be verified.

Scientific investigations commonly follow an investigative strategy we call the scientific method. The process includes making careful observations, learning what is known about the subject, and asking questions; formulating hypotheses that might explain a phenomenon or answer a question; designing and conducting experiments to test the accuracy of a particular hypothesis; analyzing results; and drawing conclusions as to whether to accept, reject, or modify the hypothesis being tested. Experiments performed with proper controls determine the effect of a single experimental variable. Controlled experiments help distinguish cause and effect relationships from coincidence, and reduce the influence of human bias. A hypothesis is not accepted as verified until it is successfully repeated by other scientists in independent studies. When a hypothesis has been repeatedly reinforced by experimentation, it may become a supported theory, a term that indicates its widespread acceptance. Unlike a single hypothesis, a theory combines many ideas into a unifying principle. For a theory's continued acceptance, it must remain compatible with new evidence. Practical applications of scientific findings are one form of support for a theory's validity.

Review Questions

1. Describe the various properties that characterize life. Identify some inanimate objects that exhibit one or more of these properties. Can you identify any inanimate objects that possess *all* of these properties combined?
2. Rearrange the following list from simpler to more complex: ecosystems, subatomic particles, cells, organs, the biosphere, populations, molecules.
3. Describe Darwin's explanation for the presence of similar-looking finches with different types of beaks and feeding habits on the Galapagos islands. Use this example to discuss the principle of natural selection.
4. Why is it important to have a control group when carrying out experiments? Why was Redi's initial experiment using sealed vessels an inappropriate control?

Critical Thinking Questions

1. In the opening Steps to Discovery story, the question was posed, "Why do we have houseflies?" Use your understanding of the mechanisms of evolutionary adaptation to provide a rational explanation for the existence of houseflies.
2. The disappearance of any one of the characteristics of life is fatal for an organism or its species. Describe what would happen to you if one of these properties of life malfunctioned. Repeat the exercise for each of the nine characteristics.
3. Using a familiar process, create an analogy that reveals how novel properties emerge as levels of order increase. Begin by drawing an organizational hierarchy for one of the following: composing a song; building a house; writing a novel. Identify the intangible "extra" properties that emerge as the level of complexity increases. Why would the structures be of little value without these intangible "extras"?
4. Considering the process of natural selection, how does greater variation within a species strengthen the chance that the species will avoid extinction following the appearance of some new deadly environmental condition (such as a new virus)?
5. Classify the following statements as products of simple observation, speculation, anecdotal evidence, or the scientific approach. How might each statement that falls into the first three categories be scientifically tested?
   —The sun rises in the east.
   —Since the sun always rises in the east, it will rise in the east tomorrow.
   —The hormone insulin is essential to the uptake of the sugar glucose by the cells of a person's body.
   —I took vitamin E everyday this year and improved my grades. Therefore vitamin E increases mental effectiveness.
   —Persons with protein deficiencies are more vulnerable to infectious diseases.

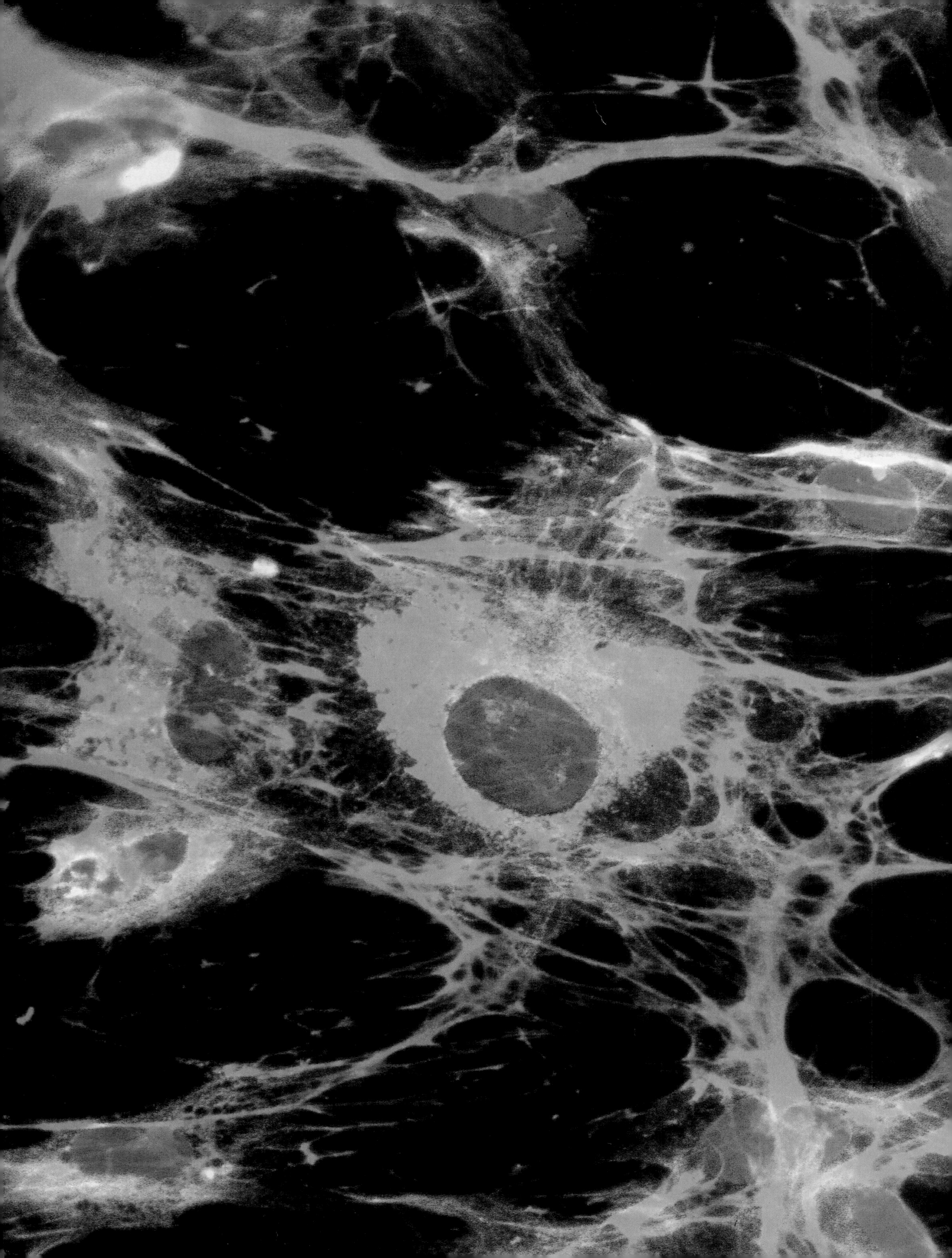

PART

◄ 2 ►

Chemical and Cellular Foundations of Life

The multitude of functions necessary for life require complexity, organization, and specialization—all of which are embodied in cells. Part of the cellular organization required for life is revealed in this photograph of a human skin cell. The green threads are microtubules, which help give the cell its shape and organization and participate in cell division. The blue threads are the cell's DNA. The remaining threads are proteins on the cell surface that enable it to interact with its environment.

CHAPTER

◂ 2 ▸

The Chemical Basis of Life

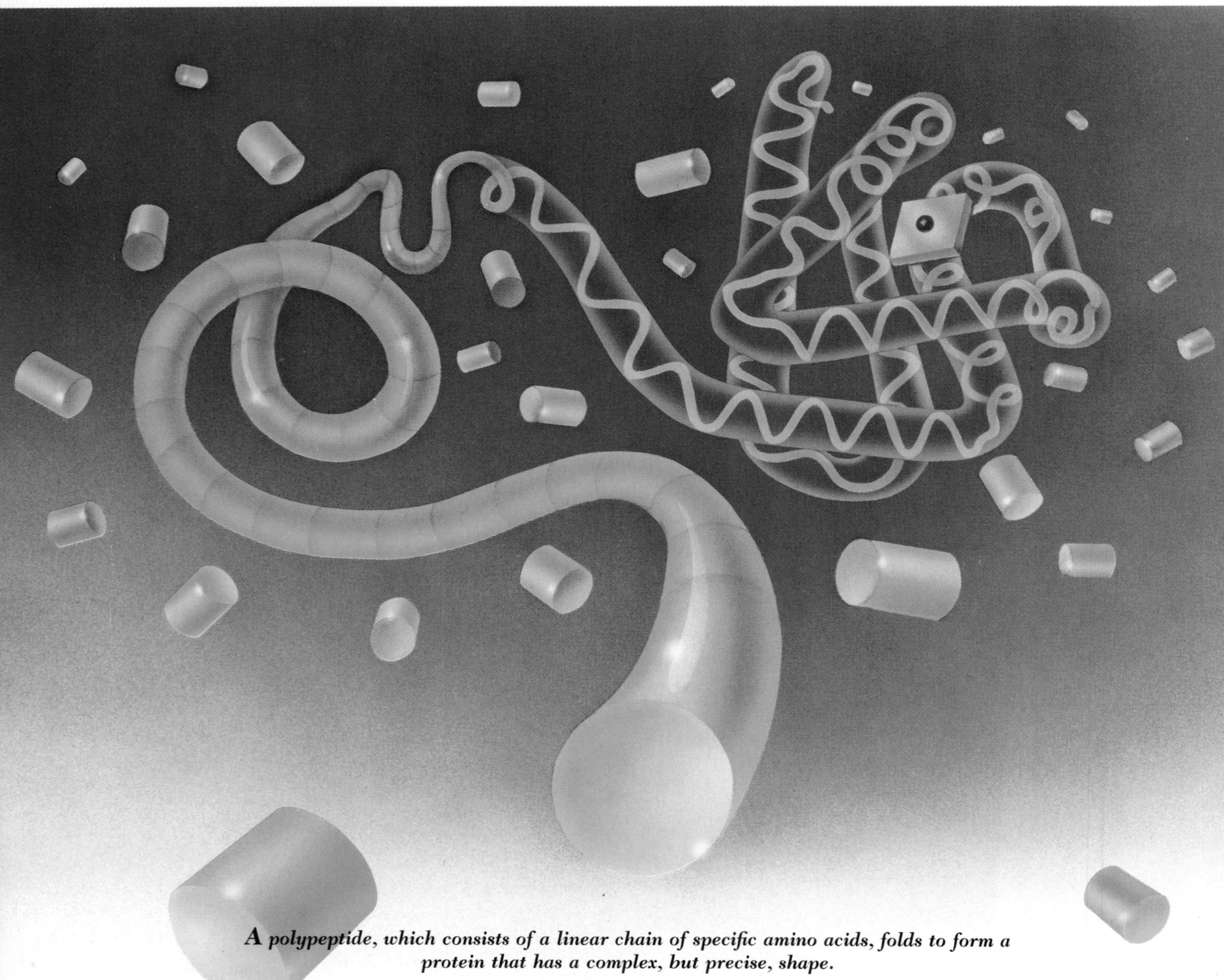

A polypeptide, which consists of a linear chain of specific amino acids, folds to form a protein that has a complex, but precise, shape.

STEPS TO DISCOVERY
Determining the Structure of Proteins

Considering all the molecules that make up the fabric of living organisms, proteins have the most complex structure and carry out the most demanding functions. At the end of World War II, in 1945, we still knew very little about the structure of these key molecules. Just 15 years later, however, the scientific world had an accurate, detailed picture of the way these molecular giants were constructed and how many of them worked. During this remarkable decade and a half, a handful of investigators using very different experimental techniques, not only changed our view of proteins, but dramatically improved our understanding of life at the molecular level.

Scientists in the mid-1940s knew that proteins were made up of small molecules, called *amino acids*, linked together to form chains called *polypeptides.* What they didn't know was whether or not the sequence of amino acids in a polypeptide chain was of critical importance, and if so, *how* was it important? Frederick Sanger, a young protein chemist at Cambridge University in England, began a study to determine the order of the amino acids that make up the protein insulin. Insulin was a natural choice. Not only was it one of the few proteins available commercially (it was being extracted from the pancreas of domestic pigs and used in the treatment of diabetes), it was also a much easier protein to sequence since it contained fewer amino acids than most proteins (Sanger eventually found it was composed of only 51 amino acids).

When Sanger began his studies, determining the sequence of amino acids linked together in a chain required a tedious, chemical process. Most importantly, only short chains of about five amino acids could be sequenced at a time. Consider an analogy in which a chain of letters is linked together forming a sequence such as

TCEJBUSETIROVAFYMSYGOLOIB

How would you determine this sequence if you were able to identify pieces containing only five letters in a row? If you were able to cut the chain of letters at random positions, forming overlapping fragments (such as tir*ov*, *ov*afy), you should be able to figure out the entire sequence.

TCE**JB**
  JBU*SE*
    *SE***TIR**
      TIR*OV*
        *OV*AFY

TCEJBUSETIROVAFY ← the deciphered sequence (up to this point)

This is precisely the approach taken by Sanger. He chemically cut each purified polypeptide into random fragments, then *sequenced* the fragments and pieced together the linear order of amino acids in the entire chain. From start to finish, the work took Sanger and his colleagues nearly 10 years.

Sanger's sequences showed that the insulin molecule had a precise order of amino acids. Examination of other proteins showed them to have their own stable amino acid sequences, each of which differed from that of insulin. There were no predetermined rules of chemistry that dictated the sequence of amino acids in polypeptides; each sequence was unique and could only be revealed experimentally. Sanger had provided the foundation for the future understanding of protein function based on the chemical properties of their amino acids.

While Sanger was working on amino acid sequences in England, Linus Pauling of the California Institute of Technology was working on the three-dimensional organization of the amino-acid chains in a protein molecule. Pauling used molecular models to determine how they could stably fit together, much like assembling pieces of a jigsaw puzzle. One day in 1948, while he was visiting in England, Pauling was suffering from a bad cold and decided to stay in his room. Using a pencil, paper, and straightedge, he drew the atoms and chemical bonds of a polypeptide chain. As he was folding the paper in various positions, he discovered that when the polypeptide chain was arranged in a structure resembling a "spiral staircase," the hydrogen bonds from the amino acids of the chain fit perfectly into place, providing the chain with maximum stability. Pauling called the spiral structure an *alpha helix* (refer to Figure 2-23*b*, page 44).

Meanwhile, John Kendrew and Max Perutz of Cambridge University were attempting to determine the shape of an entire protein molecule. They relied on a technique called X-ray crystallography, in which crystals of a purified substance are bombarded with X-rays. The X-rays are deflected by the molecules in the crystal and strike a photographic plate, producing a pattern of dots which, when interpreted mathematically, reveals the structure of the molecule that produced the pattern.

The first protein chosen by Kendrew and Perutz for study by X-ray crystallography was *myoglobin,* a protein that stores oxygen in muscle tissue. The picture revealed a compact molecule whose polypeptide chain was folded back on itself in a complex, irregular arrangement. It was discovered that not only do proteins have a unique amino-acid sequence (as Sanger had demonstrated in the case of insulin), they also have a unique three-dimensional shape that fits their specific function—in myoglobin's case, oxygen binding. In addition, eight segments of the polypeptide chain were composed of Pauling's alpha helices, which together accounted for approximately 70 percent of the amino acids in the protein (refer to Figure 2-23*c*, page 44). Pauling was delighted. He was also delighted to receive the Nobel Prize for Chemistry in 1954. Four years later, Sanger received a Nobel Prize of his own, followed by Perutz and Kendrew in 1962.

To explore life adequately, one must venture into an invisible realm—the realm of chemicals and chemical reactions, where minuscule particles combine with one another to forge everything in the universe, including that which is alive. Chemistry provides an explanation for biological properties at their most fundamental and essential level, that which ultimately accounts for what we are and what we do as living entities. Probing the chemical basis of life is a first step in understanding biology. It reveals how different particles can be arranged and rearranged to form an unlimited variety of organisms.

▼ ▼ ▼

THE NATURE OF MATTER

When we look around at objects in our environment we find a seemingly endless diversity of materials. All materials, however, are composed of a limited number of fundamental substances, or **elements.** The diverse nature of materials stems from the variety of ways these elements combine with one another to form more complex substances with new and different properties. Scientists have identified 109 elements; 92 of these occur in nature, the remaining 17 have been synthesized by scientists. Four of these elements (carbon, hydrogen, oxygen, and nitrogen) comprise over 95 percent of your body's weight.

Elements are designated by either one or two letters derived from their English or Latin names. The 12 elements shown in Table 2-1 make up over 99 percent of the living matter of all the diverse organisms on earth (Figure 2-1).

FIGURE 2-1
Different but similar. Each of these strikingly different organisms is composed almost entirely of the same group of chemical elements.

TABLE 2-1
MOST COMMON ELEMENTS IN THE HUMAN BODY

	Percentage by weight	*Atomic Number*	*Atomic Mass*[a]
Oxygen (O)	65	8	16
Carbon (C)	18	6	12
Hydrogen (H)	10	1	1
Nitrogen (N)	3	7	14
Calcium (Ca)	2	20	40
Phosphorus (P)	1.1	15	31
Potassium (K)	0.35	19	39
Sulfur (S)	0.25	16	32
Sodium (Na)	0.15	11	23
Chlorine (Cl)	0.15	17	35
Magnesium (Mg)	0.05	12	24
Iron (Fe)	0.004	26	56

[a] Atomic number and atomic mass are discussed in the following section of the text.

THE STRUCTURE OF THE ATOM

In 1810, John Dalton, an English schoolteacher, formulated the atomic theory of the structure of matter. Dalton conceived of matter as being composed of **atoms,** tiny spheres much too small to be directly observed. We now know that atoms are not the smallest particles in matter, but are the smallest units unique to a particular element (an atom of carbon is different than an atom of oxygen, for example). Today we can actually verify the existence of atoms microscopically (Figure 2-2), but our modern understanding of atoms is very different than that of Dalton. For example, we also know that atoms are not solid spheres, as Dalton imagined them to be, but are actually 99 percent space. Each atom contains 3 *subatomic* particles—positively charged (+) **protons,** negatively charged (−) **electrons,** and neutral (no electrical charge) **neutrons.** The atomic nucleus contains the protons and neutrons and therefore carries a positive charge—one unit of charge for each proton. Although electrons are essential to the atom's properties and behavior, each electron is about 1/1,800th the mass of a proton and contributes little to the mass of the atom. An atom is held together as a unit by the electrical attraction exerted between the oppositely charged protons in the nucleus and electrons in orbit around the nucleus.

It's a striking fact that every material in the world—your hair, a brick wall, water—are built from the same three building materials: protons, electrons, and neutrons.

HOW ATOMS DIFFER FROM ONE ANOTHER

Atoms of different elements differ from each other in the number of protons and electrons they possess, and usually in their neutron number as well. These differences account for the unique properties of the corresponding element.

The Number of Protons

An atom of one element differs from those of all other elements by its **atomic number,** that is, the number of protons in its nucleus. For example, the atomic number of hydrogen is 1—all hydrogen atoms have a single proton, and, conversely, all atoms with one proton are hydrogen atoms. All carbon atoms have six protons; all oxygen atoms have eight protons, and so forth.

The Arrangement of Electrons in an Atom

One of the properties of an atom determined by its proton number is its number of electrons. The classical view of electron movement around protons in the nucleus resembles the way the planets of our solar system move around the sun, forming what is called the Bohr's atom (Figure 2-3). Niels Bohr, a Danish physicist, proposed that each

FIGURE 2-2
How do we know atoms exist? After centuries of scientific evidence to support their existence, now we can see them (or at least pictures of them). These rows of individual atoms of the element antimony, for example, are clearly visible using a recently developed microscope called a *scanning tunneling electron microscope.*

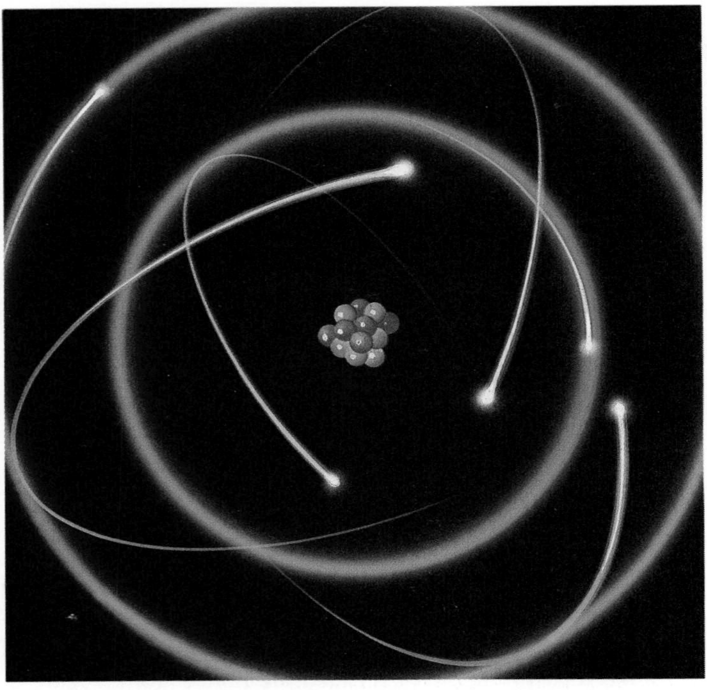

FIGURE 2-3
Simplified version of a typical atom. The nucleus of this carbon atom contains six protons and six neutrons (2 electrons in the inner shell; 4 electrons in the outer shell). Six electrons orbit the nucleus in two shells. Although useful in understanding atomic structure and the chemical reactions of an atom, this simplified model (called a *Bohr atom* after the Danish physicist who proposed it) fails to show the enormous relative distance between nucleus and electrons and inaccurately represents the shape of the shells. According to modern physics, the shells consist of orbitals of spherical or dumbbell shape, in which an electron has a high probability of being located at any given instant. These orbitals are more like "clouds" encompassing the area occupied by the electrons traveling around the nucleus at about the speed of light.

electron orbit—or **shell**—has a maximum number of electrons it can hold. When a shell becomes filled, any additional electrons must go into the next shell further away from the nucleus. The innermost shell can contain only two electrons. The next two shells can contain eight each. Thus, the six electrons of a carbon atom are distributed in two shells in a 2:4 arrangement. That is, the innermost shell holds a pair of electrons, and the second shell holds the other four electrons. Sulfur's 16 electrons form a 2:8:6 arrangement; this formation differs from that of chlorine (17 electrons) by having one less electron in its third shell. The atomic number (and resultant electronic distribution) of a number of common atoms is illustrated in Figure 2-4. The chemical reactivity of an atom is largely determined by the number of electrons residing in its outer shell.

The oversimplistic Bohr's model of the atom has been updated to show that electrons sweep around the nucleus in loosely defined "clouds," called **orbitals.** Orbitals, which may have a spherical or dumbbell shape, are roughly defined by their boundaries. Each orbital is capable of containing a maximum of two electrons. The innermost shell contains a single orbital; the second shell contains four orbitals (thus eight electrons); the third shell also contains four orbitals (thus eight electrons); and so forth.

The very existance of life depends on the fact that *electrons contain energy.* Those of inner shells contain less energy than do those of outer shells. Furthermore, electrons can temporarily move from an inner to an outer shell by absorbing additional energy (Figure 2-5). This is what occurs, for example, when a molecule of the plant pigment chlorophyll absorbs light energy during photosynthesis (Chapter 5), arguably the most important chemical reaction on earth. Virtually every organism depends on the energy captured by these electrons during photosynthesis.

Isotopes—Same Element, Different Mass

The combined number of neutrons and protons in the nucleus make up the weight of an atom, or its **atomic mass.** Not all atoms with the same atomic number have the same atomic mass. These differing versions of an element are called **isotopes.** Hydrogen, for example, exists as three different isotopes containing zero, one, or two neutrons. (The atomic mass of these isotopes is denoted as ^{1}H, ^{2}H, and ^{3}H, respectively.) The predominant natural isotope of hydrogen is ^{1}H (no neutrons), and only one in every 6,000 hydrogen atoms contains a neutron (this isotope is deuterium, ^{2}H, used to form "heavy water"). Hydrogen atoms with two neutrons are unstable isotopes called *tritium.* Such unstable isotopes are **radioactive**; that is, they tend to disintegrate spontaneously.

Radioisotopes have opened the door to worlds that humans could not otherwise effectively explore. Radioisotopes have the chemical properties of ordinary isotopes, but reveal their presence by the radiation they emit. Thus, a chemical containing radioisotopes can be easily tracked using a radiation-sensitive measuring device. Chemicals, such as sugars or amino acids, can be made with radioactive atoms and then introduced into living cells to see how they are metabolized by animals. Growing plants in an atmos-

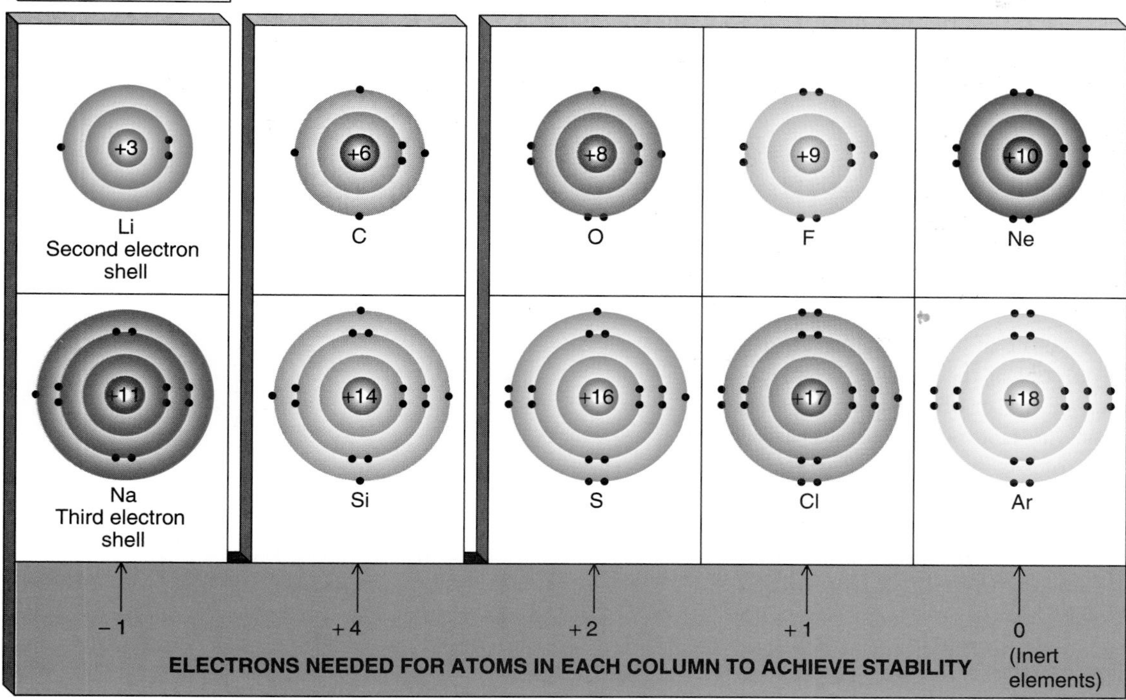

FIGURE 2-4

A representation of the arrangement of electrons in a number of common atoms. Each shell has a limited number of electrons it can hold. Electrons in each shell are grouped in pairs to illustrate that each orbital of a shell can hold two electrons. The number of outer-shell electrons is a primary determinant of the properties of elements. Atoms with similar number of outer-shell electrons have similar properties. Lithium (Li) and sodium (Na), for example, have one outer-shell electron, and both are highly reactive metals. Carbon (C) and silicon (Si) atoms can each bond with four different atoms simultaneously as described in an upcoming section.

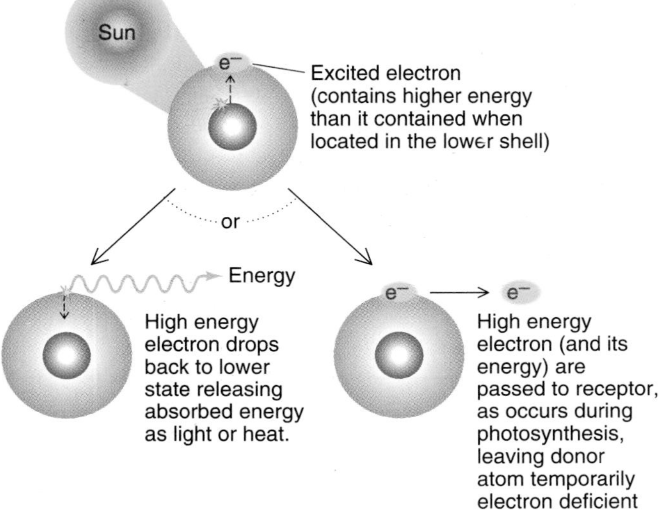

FIGURE 2-5

Electron shells and energy content. When an atom absorbs energy, an electron may jump to a shell located farther from the nucleus. This "excited" electron has two possible fates: It can drop back to a lower shell, releasing the absorbed energy back into the environment, or it can be passed on to another atom which acts as an "electron acceptor." This latter option occurs when light energy is captured during photosynthesis.

phere enriched with radioactive carbon dioxide ($^{14}CO_2$) revealed the details of photosynthesis that enable plants to change simple chemicals into the complex molecules that make up the plant's leaves, stems, and roots.

Radioisotopes are also recruited to help diagnose and treat disease. Radioactive tracers are used to detect abnormal thyroid function or to "see" a person's blood flow (and pinpoint dangerous circulatory obstructions). Many sensitive laboratory tests for diagnosing allergies, cancers, and other disorders also use radioisotopes. Other radioactive isotopes, those that emit lethal amounts of energy, are used to kill abnormal cells, such as rampantly growing cancers. Many cancer patients are alive today because their lethal tumors were showered with radiation.

Radioactive isotopes also provide accurate geological "clocks" that reveal the approximate age of very old objects, such as a fossil. Each type of radioisotope has a characteristic **half-life**—the time required for half of the radioisotope to decay into its stable, nonradioactive form. For example, ^{14}C has a half-life of 5,730 years, after which only half of the original amount of ^{14}C remains. During the next 5,730 years, 50 percent of the remaining amount "decays," and so on. The age of an object is determined by measuring the ratio of one isotope to another. In all *living* organisms, the ratio of ^{14}C (radioactive carbon) to ^{12}C (the most abundant—and nonradioactive—isotope of carbon) is the same as that of atmospheric carbon dioxide. When an organism dies, however, it stops accumulating carbon. As the dead organism's ^{14}C decays, the relative proportion of ^{14}C to ^{12}C declines steadily according to the isotope's half-life. Because the rate of decay is constant, the isotopic ratio identifies the object's age. With a half-life of several thousand years, ^{14}C is useful in dating rocks and fossils that are less than 50,000 years old. Older objects can be dated using other radioisotopes, such as ^{40}K, which decays much more slowly (half-life of 1.3 billion years). Radioactive dating techniques help biologists reconstruct evolutionary events that occurred even billions of years ago.

CHEMICAL BONDS

The atoms of most elements can interact with other atoms to form larger, more complex structures. The force of attraction that holds one atom to another is a **chemical bond.** Chemical bonds fall into two broad categories: covalent and noncovalent.

COVALENT BONDS

Most atoms tend to establish stable partnerships with other atoms, forming larger complexes called **molecules.** A molecule containing more than one type of element (as most do) is called a **compound.** The two or more atoms that make up a molecule are joined together by **covalent**

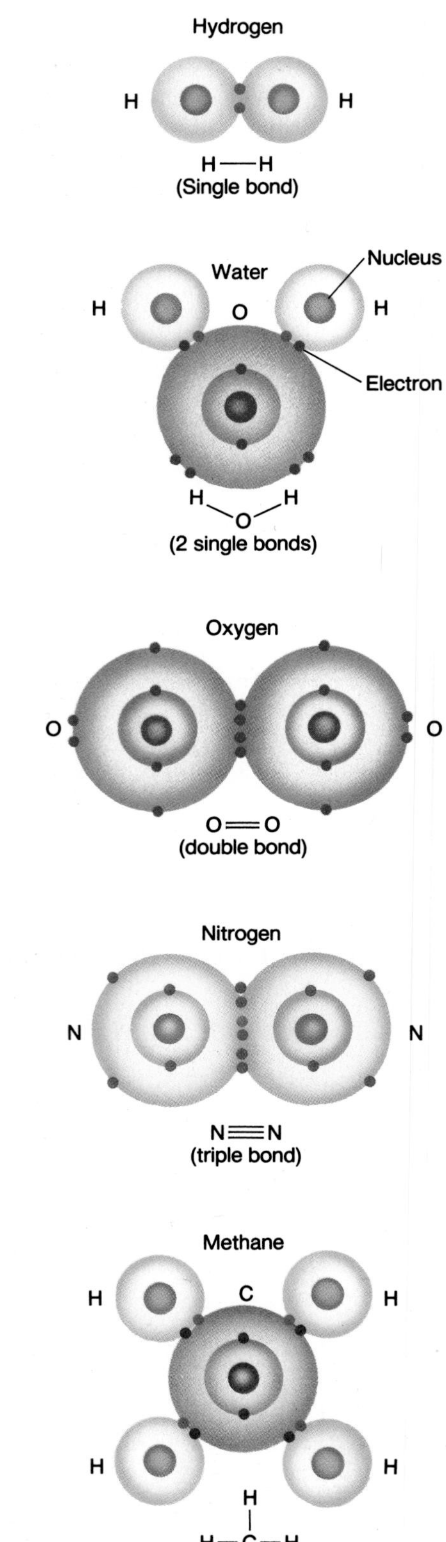

FIGURE 2-6
Examples of covalent bonding. The shared electron pairs orbit around the nuclei of both atoms forming the bond. (The color of the electrons in the shared pair reflects the atom that donated them.)

bonds, a type of bond in which pairs of electrons are *shared* between atoms. Examples of covalent bonding are presented in Figure 2-6. The formation of a covalent bond between two atoms is governed by a fundamental principle—namely, that an atom is most stable when its outermost electron shell is filled.

The number of bonds an atom can form depends on the number of electrons needed to fill its outer shell (and therefore achieve increased stability). Two hydrogen atoms can form a tandem, each atom sharing its single electron with the other, thereby filling its outer shell by forming a hydrogen molecule (H_2). The two hydrogen atoms are linked together by a *single* covalent bond. In a single covalent bond, both electrons of the shared pair orbit around both nuclei of the bonded atoms, thus satisfying each atom's requirement for a filled outer shell. An oxygen atom can satisfy its two-electron deficiency by combining with two hydrogen atoms, forming water (H_2O) and releasing energy. Oxygen can also bond with an identical oxygen atom, forming an oxygen molecule (O_2). The two atoms of molecular oxygen are bound together by a *double bond,* in which two pairs of electrons are shared. Some elements can even form *triple bonds,* sharing three pairs of electrons, as illustrated by molecular nitrogen (N_2). Although quadruple bonds do not exist, some atoms can form four covalent bonds, although it requires more than one partner to do so. Carbon, for example, shares its electrons with four atoms of hydrogen to form methane (CH_4), a flammable gas produced by a variety of swamp-dwelling microbes as well as by the bacteria inhabiting your large intestine. Methane is also known as natural gas.

Polar versus Nonpolar Molecules. In some molecules, the nucleus of one atom is more electron attracting than its partner and exerts a greater pull on the shared electrons. Among the atoms found in biological molecules, nitrogen, oxygen, phosphorus, and sulfur exert the greatest electron "appetites."

The effect of atoms with unequal electron attractiveness combining with one another is apparent when we examine a molecule of water (H_20):

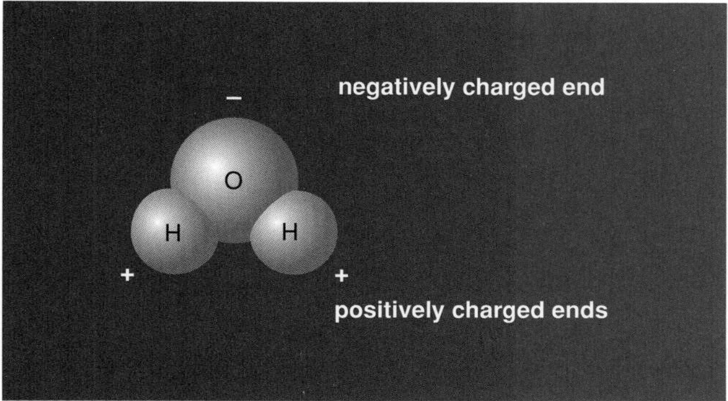

Water's single oxygen atom attracts electrons much more forcefully than do either of its hydrogen atoms. Molecules that contain an unequal charge distribution, such as water, are referred to as **polar** molecules; they possess distinct positive and negative regions, or *poles.* In contrast, **nonpolar** molecules lack regions of electric charge. The difference between polar and nonpolar molecules is fundamental to the properties of biologically important molecules. As we will see shortly, for example, the polarity within the water molecule is the reason water can dissolve so many different substances, a property that is essential for life.

NONCOVALENT BONDS

Noncovalent bonds do not depend on shared electrons but on attractive forces between positive and negative charges. Because they are weak, noncovalent bonds require very little energy to be broken. Noncovalent bonds play a key role in maintaining the intricate three-dimensional shape and function of such large molecules as DNA—life's genetic material—and of proteins. Noncovalent bonds include *ionic bonds, hydrogen bonds,* and *hydrophobic interactions.*

Ionic Bonds: Complexes of Oppositely Charged Atoms

Some atoms are so electron-attracting that they can capture electrons from other atoms. For example, when the elements sodium (a silver-colored metal) and chlorine (a toxic gas) are mixed, the single electron in the outer shell of each sodium atom migrates to an electron-attracting chlorine atom. As a result, these two materials are transformed into sodium chloride—neither a gas nor a metal, but ordinary table salt.

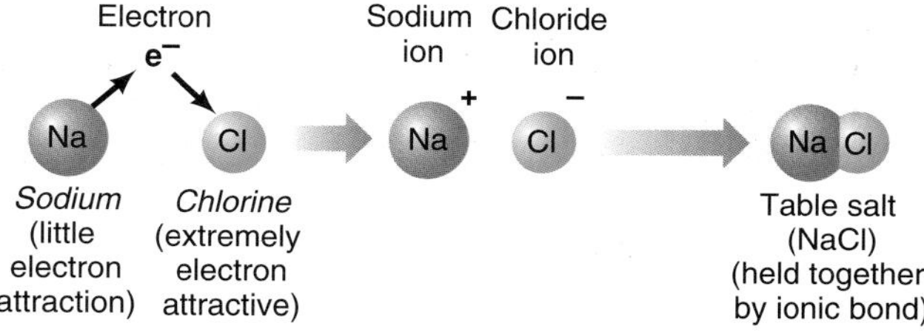

Since the chloride atom has an extra electron (relative to the number of protons in its nucleus), it has a single negative charge (Cl^-). The sodium atom, which has lost an electron (leaving it with an extra proton relative to the number of electrons) has a single positive charge (Na^+). Such electrically charged atoms are called **ions.** Oppositely charged ions attract one another, forming a noncovalent linkage called an **ionic bond.** Ionic bonds are essential to life, for example, by holding together large, complex biological molecules.

Hydrogen Bonds: Sharing Hydrogen Atoms

Another kind of chemical bond holds polar molecules close to one another. This linkage—the **hydrogen bond**—is formed when two molecules share an atom of hydrogen. Water clearly illustrates how this attractive force works

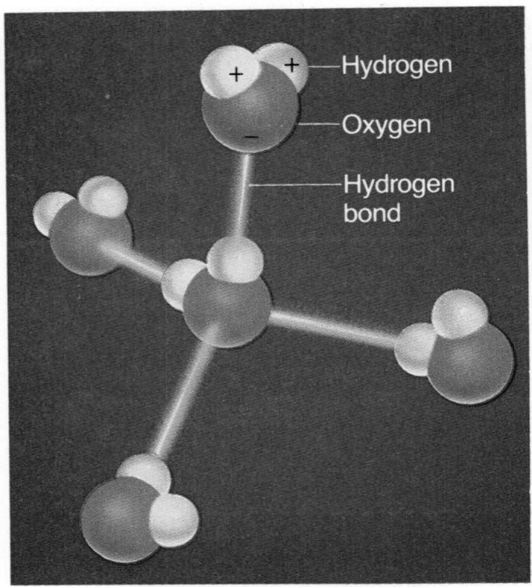

(a) *(b)*

FIGURE 2-7

Hydrogen bond formation between adjacent water molecules. ***(a)*** Positively charged hydrogens of one polar water molecule align next to negative portions of another water molecule. These weak bonds are really attractions between oppositely charged portions of adjacent molecules. In each of the hydrogen bonds depicted, a hydrogen atom is being shared between two oxygen atoms. The length and strength of a bond are related properties. The hydrogen bond between an H and O atom is longer than the covalent bond between an H and O atom because it is a weaker bond. ***(b)*** The cohesion between water molecules creates "film" at the surface of water, a film of water molecules that resists being separated. This property of water, called *surface tension,* allows a water strider to seemingly defy gravity and walk across the surface of a pond.

(Figure 2-7). The hydrogen bond is only about one-twentieth the strength of the covalent bonds that hold together the water molecule, but its intermolecular pull is essential to all life on earth.

Without ionic and hydrogen bonding, important biological molecules would fall apart (Figure 2-8). Water itself would be just another liquid on an inanimate earth.

THE LIFE-SUPPORTING PROPERTIES OF WATER

Water is more than just another liquid. It constitutes about 70 percent of our body weight and is essential to the existence of all forms of life. The life-supporting properties of water can be traced to the polarity of its O—H bonds; its positively and negatively charged regions allow water molecules to form hydrogen bonds.

WATER AS A SOLVENT

The pull of water's polar regions make it life's most important solvent. A **solvent** is a substance in which another material—the **solute**—dissolves by dispersing as individual molecules or ions. The resulting product is a **solution.** Most biologically important ions and molecules are highly soluble in water.

Water is such an efficient solvent that by the time a raindrop completes its fall, it has become a solution containing a number of dissolved gases, mainly nitrogen, oxygen, and carbon dioxide, extracted from air. The molecular oxygen (O_2) that dissolves in lakes, streams, and oceans is sufficient to supply huge communities of fish and other oxygen-dependent underwater dwellers with oxygen (Figure 2-9).

The polarity of water-soluble molecules make them **hydrophilic,** literally "water-loving" (*hydro* = water, *philic* = loving). The electrical charges on polar molecules are attracted to the opposite charges on the water molecule.

WATER AS AN "ANTISOLVENT" OF NONPOLAR MOLECULES

The familiar tendency of oil to separate from water stems from the interactions between the molecules. Oils and fats are nonpolar molecules that are insoluble in water because they lack charged regions and are therefore not attracted to the poles of water molecules. In other words, nonpolar substances are **hydrophobic,** or "water-fearing" (*phobia* = fearing). When nonpolar compounds are mixed with water, the polar solvent forces the substances into aggre-

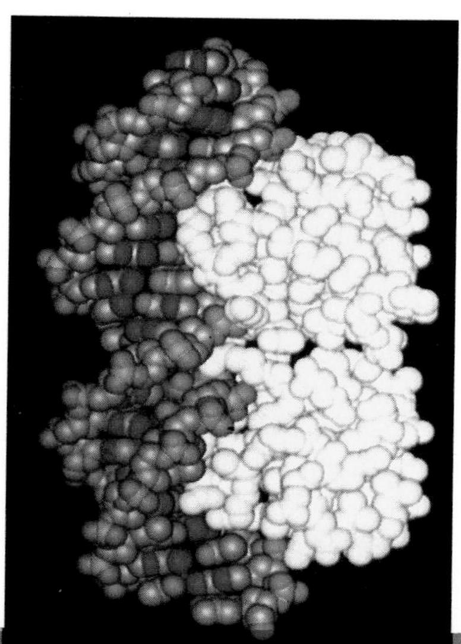

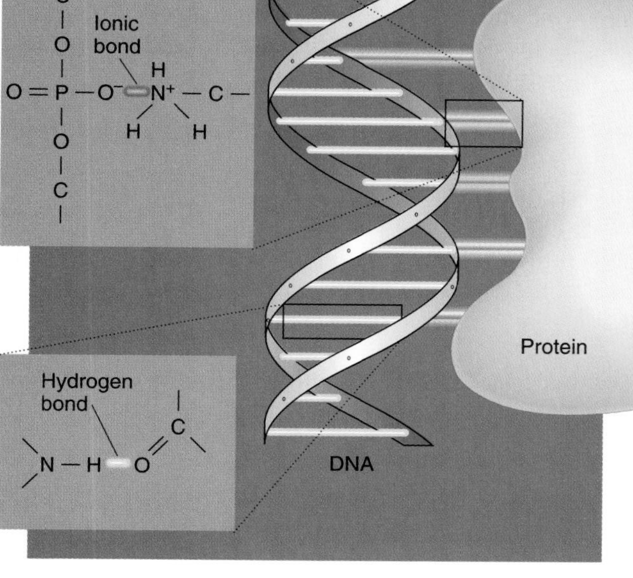

FIGURE 2-8

Noncovalent bonds play an important role in holding two or more molecules together into a complex. In the computer-simulated model depicted here, a molecule of protein is bound to a molecule of DNA by noncovalent ionic bonds. The ionic bond forms between a positively charged nitrogen atom and a negatively charged oxygen atom. The DNA molecule itself consists of two separate strands held together by noncovalent hydrogen bonds. While a single noncovalent bond is relatively weak and easily broken, large numbers of such bonds between two molecules, as between two strands of DNA, makes the overall complex quite stable. The life of every organism depends on such noncovalent interactions.

FIGURE 2-9

External gills atop these striking nudibranchs (sea slugs) extract enough dissolved oxygen from water to allow them to live permanently in the ocean without having to surface for air.

gates, such as fat droplets, which minimizes their exposure to their polar surroundings (Figure 2-10). Hydrophobic interactions are also critical to life, holding certain molecules together in the sheetlike membranes needed by even the simplest cells. (These membranes are discussed later in this chapter.)

THE THERMAL PROPERTIES OF WATER

Water's extensive hydrogen bonding makes it an excellent heat absorber. Much of the energy that is absorbed by water as it is heated is absorbed in breaking hydrogen bonds, and is not used to increase the liquid's temperature. Because of this property, organisms living in large bodies of water are protected from rapid, potentially lethal changes in body temperature despite sudden changes in the temperature of the air.

Water is also an efficient cooling agent. In order to evaporate, molecules of water must absorb a large amount of thermal energy. This removes energy from the remaining liquid, thereby cooling it (and the underlying surfaces). This "evaporative cooling" property of water explains the advantage of sweating when you are hot. As the sweat evaporates, it lowers the body temperature. Sweating is especially effective in dry environments that encourage rapid evaporation. In humid conditions, sweat accumulates rather than evaporating, preventing it from cooling. This accounts for the common complaint, "It's not the heat, it's the humidity."

ACIDS, BASES, AND BUFFERS

Any molecule that releases *hydrogen ions (H^+, a proton)* is defined as an **acid.** Acetic acid (vinegar), for example, has a weak grip on one of its hydrogens, which escapes the molecule as a proton. Conversely, any molecule that can *accept* a hydrogen ion is defined as a **base.** Most bases do this by generating a hydroxyl ion (OH^-), which is the molecular component that accepts the hydrogen ion. For example, in water, the strong base sodium hydroxide ($NaOH$) dissociates into Na^+ and OH^-. Since the solution contains more OH^- than H^+, it is an *alkaline* (basic) solution ("alkaline" is the opposite of "acidic"). Water is neither alkaline nor acidic since it dissociates into equal number of hydrogen and hydroxide ions:

$$H_2O \xrightarrow{\text{dissociates}} H^+ + OH^-$$

Water is therefore a *neutral* compound.

Different acids vary considerably in how readily they give up their protons. The more easily the proton is lost,

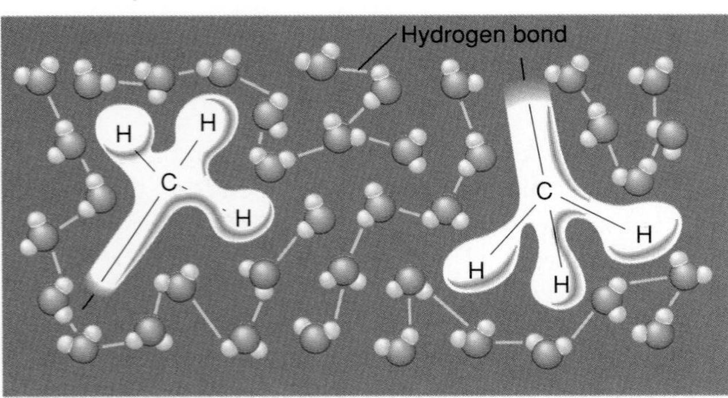

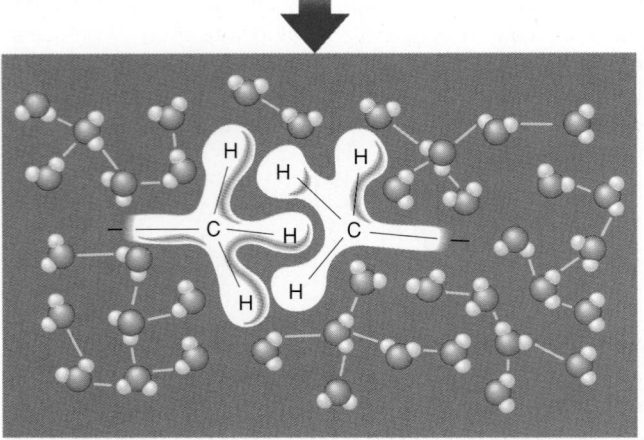

FIGURE 2-10
In a hydrophobic interaction, the nonpolar (hydrophobic) molecules are pushed together into globules, literally squeezed out of the water by the force of attraction between polar water molecules.

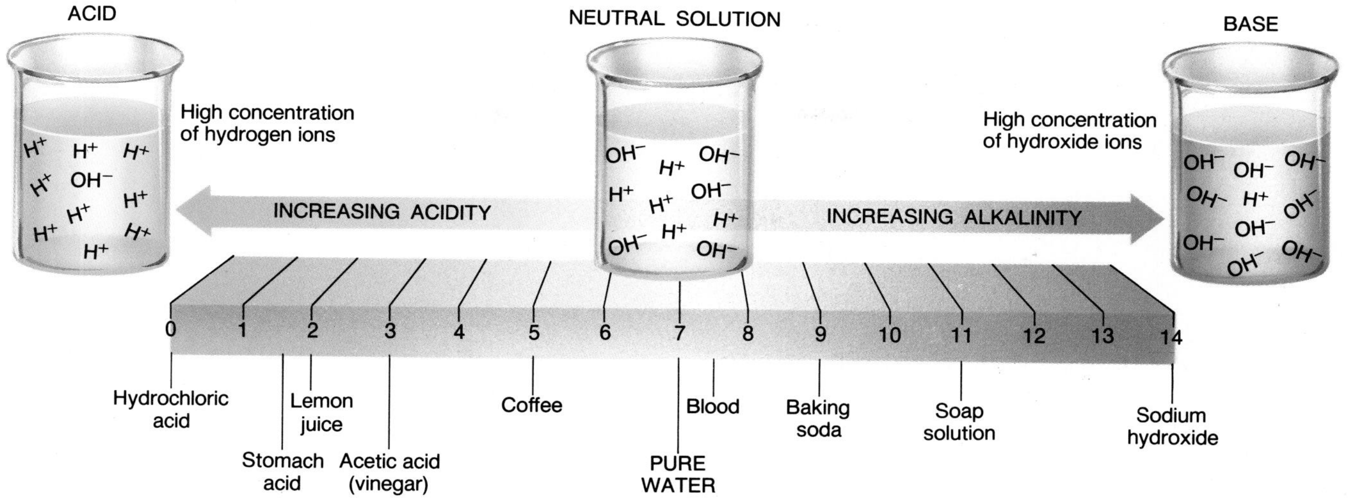

FIGURE 2-11

The pH scale. The acidity or alkalinity of a solution is indicated by a value between 0 and 14. The number is a measure of the concentration of hydrogen ions in a solution; the lower the number, the higher the H^+ concentration. The concentration of hydrogen ions in solution is always inversely related to the concentration of hydroxyl ions; as one goes up, the other goes down. The reason is that the two ions can combine to form water. If the OH^- concentration were to go up, for example, these ions would combine with the small number of H^+ ions, thus lowering the hydrogen ion concentration. The concentration of these two ions are equal at one point on the scale, namely pH 7, which is termed a *neutral* solution. Acidic solutions have pH values below 7; alkaline (basic) solutions are greater than 7. Each unit on the scale represents a tenfold change in the hydrogen ion concentration. For example, lemon juice (pH = 2) is ten times more acidic than is vinegar (pH = 3).

the stronger the acid. For example, hydrogen chloride, the acid produced in your stomach, is a very strong acid and almost completely dissociates into H^+ and Cl^-. Most acids produced by organisms, however, are relatively weak, such as acetic acid (vinegar) or citric acid (the acid in lemons, oranges and other citrus fruits).

The acidity of a solution is measured by the concentration of hydrogen ions and is expressed in terms of **pH** (Figure 2-11). The pH scale ranges from 0 to 14. Pure water has a pH of 7.0, which is the pH value of a *neutral* solution. If an acid is added to pure water, the overall hydrogen ion concentration increases, causing the solution to become acidic, which is measured as a *lower* pH. Conversely, if a base, such as sodium hydroxide, is added to pure water, the net hydrogen ion concentration decreases, causing the solution to become basic (alkaline), which is measured as a *higher* pH.

THE SIGNIFICANCE OF ACIDS AND BASES TO ORGANISMS

Most biological processes are acutely sensitive to pH, as evidenced by the devastating effects that acid rain is having on the forests in many parts of the world. One reason for this is that the shape of most biological molecules (and therefore their biological function) depends on their being in a neutral environment. Even slight changes in pH can change the shapes of these molecules and destroy their life-sustaining activities. A slight drop in the pH of your blood, for example, could kill you. Organisms protect themselves from pH fluctuations with **buffers**—chemicals that resist changes in pH by binding with free hydrogen and hydroxide ions. When the hydrogen ion concentration of our blood begins to rise (as occurs during exercise), blood buffers immediately combine with the H^+ before it can acidify the blood. Without buffers to protect our internal fluids from becoming too acidic or alkaline, we would not survive.

THE IMPORTANCE OF CARBON IN BIOLOGICAL MOLECULES

Atoms and molecules comprise the structure of organisms and dictate their life-sustaining activities. Some of these chemicals are gigantic structures that are also the most complex compounds on earth, perhaps in the universe.

FIGURE 2-12

Synthetic organic compounds. The sails of these boats represent just one of thousands of organic compounds that could never be synthesized in organisms. The invention of such synthetic organic compounds required scientists to redefine the term "organic" (which originally meant "manufactured in a organism") to mean simply "carbon containing."

Whether small or large, simple, or complex, all of these biological molecules share a common characteristic—a chemical "skeleton" composed of carbon.

At one time, scientists referred to molecules that could be manufactured only in living organisms as *organic chemicals*, in contrast to *inorganic chemicals*, which can exist independent of living organisms. Chemists eventually discovered that all of these so-called "organic" compounds contained the element carbon and that many of them could be synthesized in the laboratory (Figure 2-12). These developments led to the redefining of **organic** molecules to encompass virtually any molecule that contains carbon.[1] Naturally occurring organic molecules, those that are produced by living organisms, are referred to as **biochemicals** to distinguish them from the vast array of organic molecules now synthesized by organic chemists.

Carbon's chemical properties make it the ideal central element on which life is based. Carbons have the ability to bond with one another to form long chains. These chains of carbon atoms provide the chemical backbone of biological molecules, a backbone that may be linear, cyclic, or branched:

```
          C—C—C—C—C—C
               Linear

    C                        C
  /   \                      |
 C     C                     C
 |     |                     |
 C     C           C—C—C—C—C
  \   /              |
    C                C               Branched
  Cyclic             |
                     C—C—C
                     |
                     C
```

Because each carbon atom has four unpaired electrons in its outermost shell it can form four covalent bonds. Consequently, even after two of these bonds are tied up forming a chain, each carbon in the skeletons depicted above is still capable of bonding to other atoms. These may be hydrogen atoms or a **functional group** (a particular grouping of atoms which gives an organic molecule its chemical properties and reactivity). Most functional groups are charged or highly polar, making organic molecules much more soluble in aqueous solutions and more chemically reactive than are those composed of only carbon and hydrogen. In addition to the complex shapes provided by carbon chains, functional groups help account for the great

[1] The most important exception to this definition of organic is carbon dioxide (CO_2), one of the few inorganic forms of carbon.

diversity of organic molecules; for example, whether a molecule is an acid, a protein, or an alcohol is determined by its functional groups.

MOLECULAR GIANTS

The chemicals that form the structure and perform the functions of life are highly organized molecules. Compared to small molecules like those just mentioned, many of these organic compounds are enormous, containing hundreds to millions of carbon atoms. These enormous chemicals are called **macromolecules** (*macro* = large). Giant biological molecules possess special properties absent in their smaller relatives. Because of their size and the intricate shapes they can assume, these molecular giants can perform complex tasks with great precision and efficiency. Without these large, complex biochemicals, complicated biological activities would not be possible.

Macromolecules are constructed by assembling together small molecular subunits, often in a linear process that resembles coupling railroad cars onto a train (described in Figure 2-13*a*). This coupling process is referred to as **condensation.** Each subunit is called a **monomer** (*mono* = one; *mer* = part), and the macromolecule is referred to as a **polymer** (*poly* = many). The reverse process, in which the polymer is disassembled into its individual monomers, is called **hydrolysis** (*hydro* = water, *lysis* = split) because the bond that joins two monomers in a chain is split by the insertion of a water molecule between the two units (Figure 2-13*b*).

FOUR BIOCHEMICAL FAMILIES

Macromolecules fall into four fundamental families of organic compounds: *carbohydrates, lipids, proteins,* and *nu-*

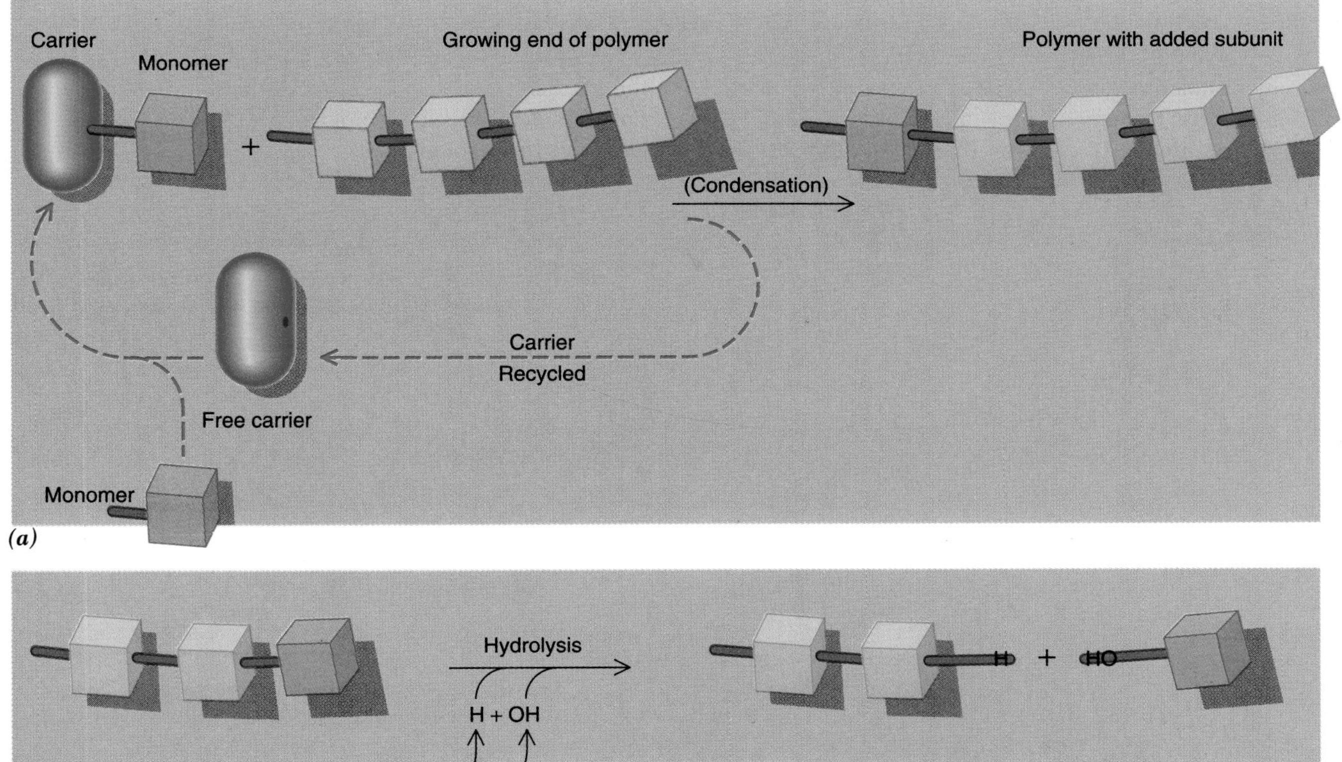

FIGURE 2-13

Monomers and polymers. ***(a)*** Biological macromolecules consist of monomers (subunits) linked together by covalent bonds. The assembly of macromolecules does not occur simply as a result of reactions between free monomers. Rather, each monomer is first activated by attachment to a "carrier" molecule that subsequently transfers the monomer to the end of the growing macromolecule. ***(b)*** Disassembly of a macromolecule occurs by hydrolysis of the bonds that join the monomers together. Hydrolysis is the splitting of a bond by water.

TABLE 2-2
MAJOR MACROMOLECULES FOUND IN LIVING SYSTEMS

Macromolecule	*Constituents*	*Some Major Functions*	*Examples*
Polysaccharide (large carbohydrates)	Sugars	Energy storage: physical structure	Starch; glycogen; cellulose
Lipid			
Triglycerides	Fatty acids and glycerol	Energy storage; thermal insulation; shock absorption	Fat; oil
Phospholipids	Fatty acids, glycerol, phosphate, and an R group[a]	Foundation for membranes	Plasma membrane
Waxes	Fatty acids and long-chain alcohols	Waterproofing; protection against infection	Cutin; suberin; ear wax; beeswax
Nucleic acid	Ribonucleotides; deoxyribonucleotides	Inheritance; ultimate director of metabolism	DNA; RNA
Protein	Amino acids	Catalysts for metabolic reactions; hormones; oxygen transport; physical structure	Hormones (oxytocin, insulin); hemoglobin; keratin; collagen; and a class of proteins called enzymes

[a] R group = a variable portion of a molecule.

cleic acids (Table 2-2). The basic structure and function of each family of macromolecule is very similar in all organisms, from bacteria to humans, providing another example of the unity of life. It is the way these monomers are assembled that provide the differences between organisms.

CARBOHYDRATES

Carbohydrates are a group of substances which includes *simple* sugars and all larger molecules constructed of sugar subunits. Because sugars are rich in chemical energy, carbohydrates are high-calorie molecules, most of which are used as storehouses of chemical energy (such chemical energy is measured in **calories,** each of which is the amount of heat needed to raise the temperature of 1 gram of water by 1 degree Celsius). Other carbohydrates, although rich in energy, are used as durable building materials to construct sturdy biological structures, such as the cell walls that protect plant cells.

Glucose is life's most common simple sugar (Figure 2-14). Individual sugars like glucose—called *monosaccharides* (*mono* = one, *saccharide* = sugar)—can be covalently linked together to form larger molecules. A molecule composed of only two sugar units is a *disaccharide* (*di* = two). Lactose, a disaccharide found in milk, provides a nursing baby with most of its energy. Lactose is composed of the two sugars, glucose and galactose. The best known (and sweetest) disaccharide is sucrose, common table sugar, which is composed of glucose and fructose. Sucrose is a major component of the sap of plants, and it carries energy from one part of the plant to another.

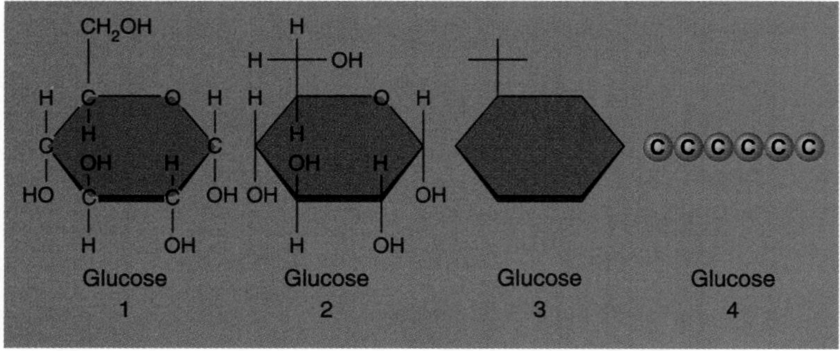

FIGURE 2-14
Four ways of representing simple sugars. *Simple sugars* have the general formula $(CH_2O)_n$, in which the value of n typically ranges from three to seven. Four structural conventions for representing glucose ($n = 6$). Sugars with five or more carbon atoms undergo a type of "self-reaction" where one end of the sugar molecule bonds with the other end, forming a closed, or ring-containing, molecule. In (1), all atoms of the molecule are shown; in (2), the carbons are omitted since their position is understood; in (3), only the skeleton is indicated, with no atomic detail shown; and (4) depicts the convention used in many illustrations throughout this textbook; each ball represents a carbon.

Polysaccharides: Macromolecular Carbohydrates

One problem with sugars is their size—they can escape a cell just as easily as they entered. After eating a meal, you have a surplus of glucose, and the unburned surplus sugar could return to the bloodstream and be removed by the kidneys. You would literally be flushing your energy reserves down the toilet. To prevent such a potentially fatal waste, your body joins together glucoses to form a **polysaccharide**—a long chain of sugar units joined together as a polymer. This polysaccharide is too large to escape the cells in which it is stored, where it remains until energy is needed. This polysaccharide, called glycogen, is disassembled when the body's glucose supplies grow scarce, the released sugars providing instaneous energy. Many different animals bank their surplus chemical energy in glycogen, a highly branched polysaccharide composed entirely of glucose monomers (Figure 2-15*a*). In humans, glycogen is stored and used as fuel in a wide variety of tissues,

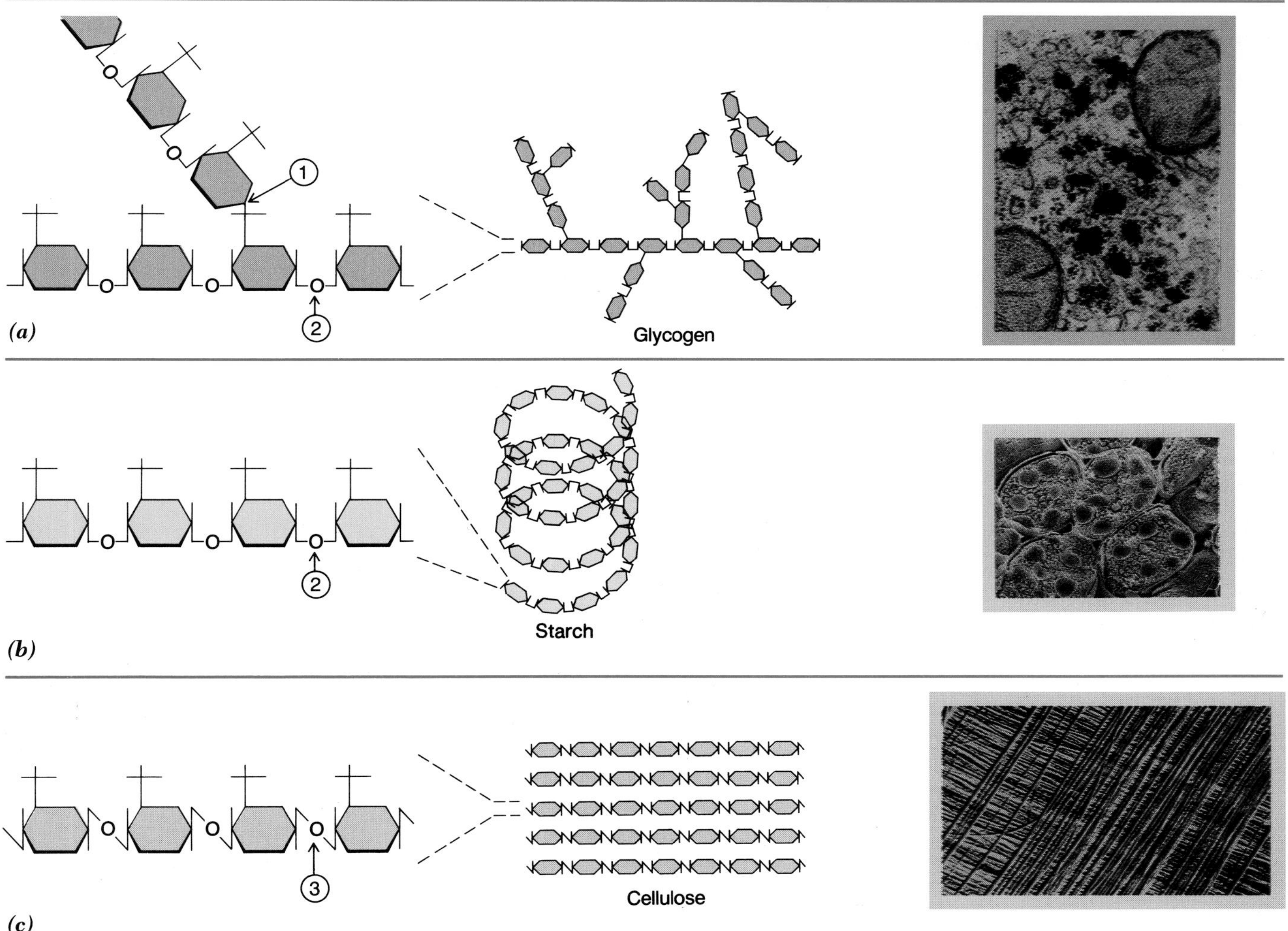

FIGURE 2-15

Three polysaccharides–identical sugar monomers, dramatically different properties. Glycogen ***(a)***, starch ***(b)***, and cellulose ***(c)***, are each composed entirely of glucose subunits, yet their chemical and physical properties (and thus their functions) are very different due to the distinct ways that the monomers are linked together (three different types of linkages are indicated by the circled numbers). Glycogen molecules are the most highly branched; starch molecules assume a helical (spiral) arrangement; and cellulose molecules are unbranched and highly extended and are bundled together to form very tough fibers suited for their structural role.

FIGURE 2-16
Strenuous exertion requires a constant supply of energy. Glycogen in muscles and the liver provides instant access to stored supplies of glucose.

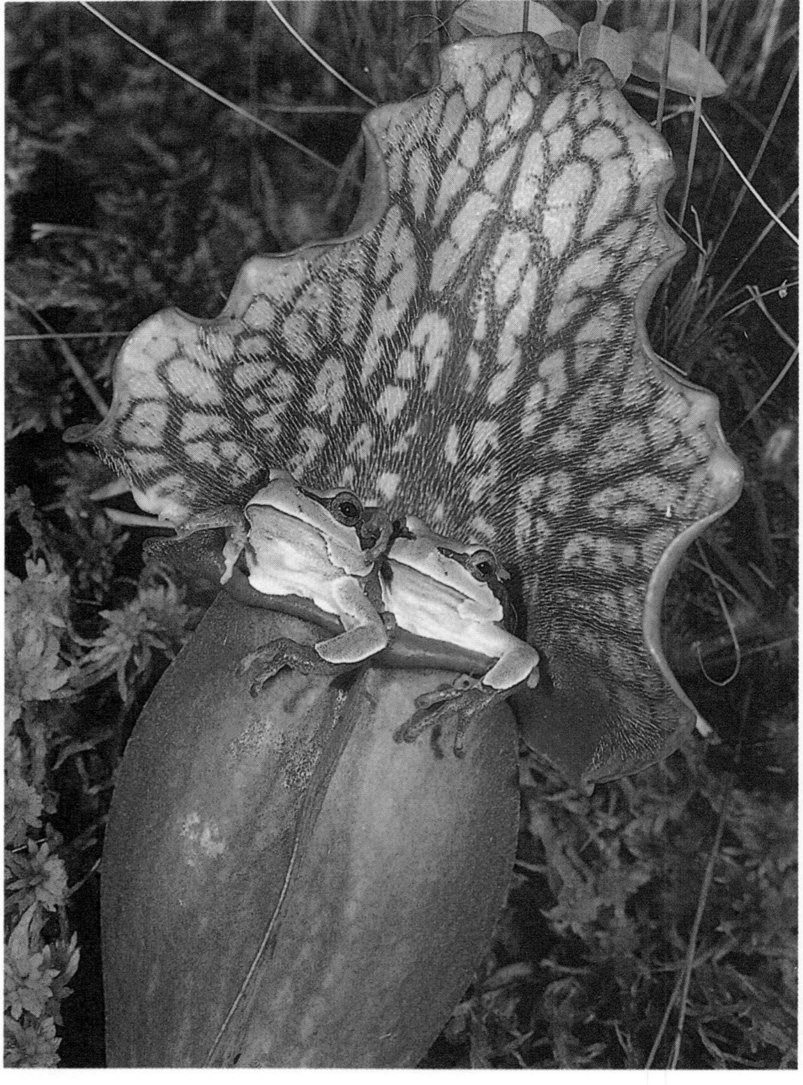

FIGURE 2-17
A showcase of structural polysaccharides. The organisms in this picture reveal a few of the natural roles of polysaccharides. The fibrous polysaccharide that supports and shapes the pitcher plant is cellulose. The internal organs and skin of these frogs are secured in place by structural polysaccharides that strengthen connective tissue. Although not shown, insects that fall prey to the carnivorous pitcher plant possess external skeletons consisting of the rigid polysaccharide chitin.

including muscles, but only the liver serves as a glucose supplier for the rest of the body. Muscles typically contain enough glycogen to fuel about 30 minutes of moderate activity (Figure 2-16).

Another polymer of glucose is *starch,* the polysaccharide most commonly used by plants for energy storage. Potatoes and cereals, for example, consist primarily of starch. Like glycogen, starch is a polymer composed entirely of glucose monomers, but it is much less branched (Figure 2-15*b*). Even though animals don't produce starch, they can readily digest it by hydrolysis. In fact, starch is the primary source of energy in the human diet in most parts of the world.

Some polysaccharides provide structure for an organism. Two of the most important of such structural polysaccharides are *cellulose* and *chitin* (Figure 2-17). Cellulose is the earth's most abundant polysaccharide, forming the tough fibers present in wood and plant cell walls (Chapter 3). Fabrics made of plant material (cotton, for example) owe their durability to the long, unbranched cellulose molecules (Figure 2-15*c*). Because cellulose is composed entirely of glucose, the same monomer found in starch, one of the most easily digested polymers, it represents a tremendous potential energy source. Ironically, animals lack any enzyme capable of digesting cellulose, so this vast reserve remains unavailable to them as a direct source of energy. Starch and cellulose differ in the way the glucose monomers are linked together, a seemingly minor difference that has spelled starvation for countless millions of people. Whereas the bonds of starch are readily hydrolyzed by an enzyme in our digestive tracts, cellulose molecules pass through intact, providing fiber that aids in the formation and elimination of feces but no energy. Unlike animals, a variety of microorganisms possess the necessary enzyme for digesting cellulose. If not for these microscopic cellulose decomposers, the world would be permanently littered with dead bodies of plants.

Another important structural polysaccharide—chitin—is a polymer of a nitrogen-containing sugar and is a major component of the outer covering (*exoskeleton*) of insects, spiders, and crustaceans, such as lobsters and crabs. Chitin gives the exoskeleton a tough, resilient quality, not unlike that of certain plastics. Insects and crustaceans owe much of their biological success to this highly adaptive polysaccharide covering.

LIPIDS

Lipids are a diverse group of organic molecules whose only common property is their inability to dissolve in water—a property which explains many of their varied biological functions. The hydrophobic character of lipids is revealed by their molecular structure—the carbon atoms of lipids are bonded almost exclusively to hydrogen atoms rather than to functional groups. Lipids include fats, oils, phospholipids, steroids, and waxes.

Fats and Oils

Fats consist of three fatty acids coupled to a single molecule of glycerol. As illustrated in Figure 2-18, **fatty acids** are long, water-insoluble chains composed primarily of —CH_2— units. Each fatty acid in a fat molecule is linked to one of the three carbons of the glycerol "backbone." Fats are very rich in chemical energy; a gram of fat contains more than twice the energy content of a gram of carbohydrate. Unlike carbohydrate, which functions primarily as a short-term, rapidly available energy source, fat reserves are utilized to store energy on a long-term basis. It is estimated that an average person contains about 0.5 kilo-

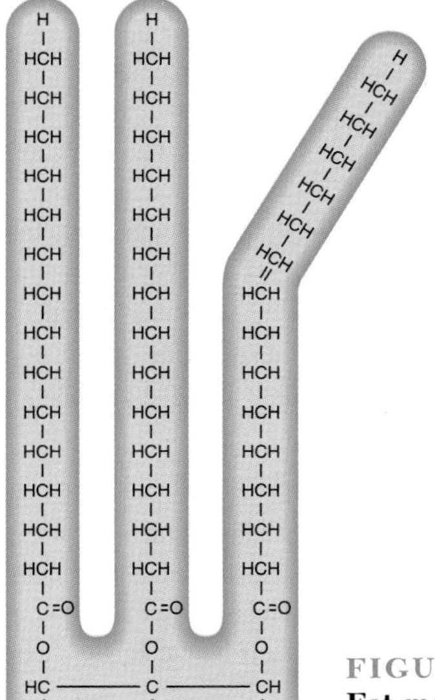

FIGURE 2-18
Fat molecule. The molecule consists of three long-chain fatty acids joined to a glycerol (lower portion of the fat molecule).

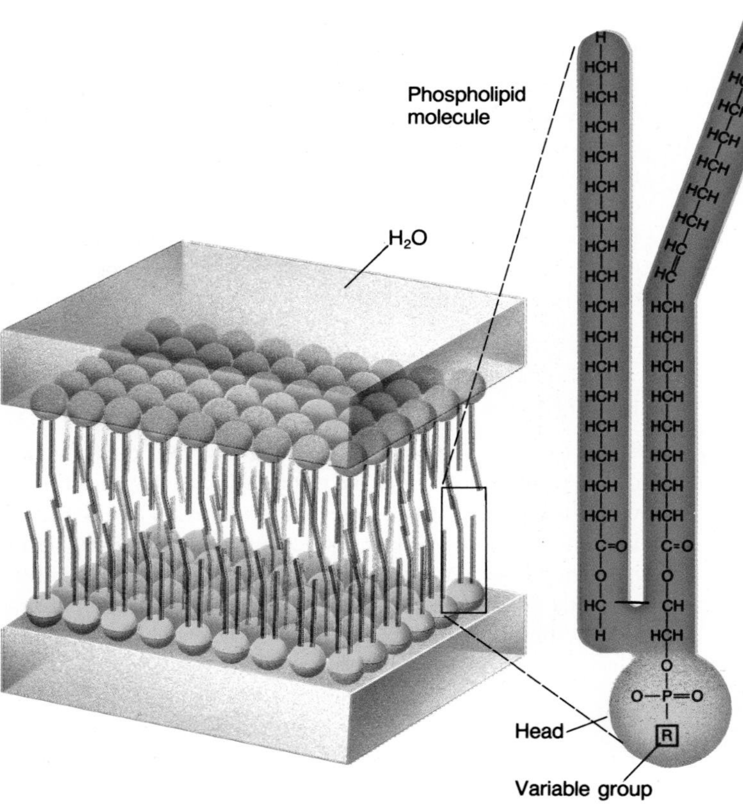

FIGURE 2-19
Phospholipids—molecules that are both soluble and insoluble in water. The polar phosphate "head" is attracted to water, while the nonpolar fatty acid tails extend in the opposite direction, away from water. This "molecular schizophrenia" produces molecules that, in cells, align to form double-layered sheets. All life depends on the ability of these membrane sheets to form cell membranes.

grams (kg) of glycogen. During the course of a strenuous day's exercise, a person can virtually deplete his or her body's entire store of glycogen unless it is replenished. In contrast, the average person contains approximately 16 kg of fat which the body taps much less readily than its reserves of carbohydrate, one reason why it takes much longer to reduce our supplies of stored fat (see The Human Perspective: Obesity and the Hungry Fat Cell).

Lipids occur in either a solid or a liquid state (if they are liquids at room temperature they are termed **oils;** fats are solid). In general, the greater the number of *unsaturated* (double) bonds that exist between the carbons of the fatty-acid chains, the less well these long chains can be packed together. This lowers the temperature at which the lipid melts. The profusion of double bonds in vegetable oils accounts for their liquid state—both in the plant cell and on the grocery shelf—and for their being labeled as "polyunsaturated" (many double bonds). In contrast, almost all the linkages between the carbons of animal fats are single bonds, that is, they are *saturated* with hydrogens, causing the material to remain a solid at room temperature. Diets that contain polyunsaturated lipids are less likely to promote cardiovascular disease than are diets high in saturated animal fats, such as those found in lard and butter.

Phospholipids

Most **phospholipids** are similar in structure to fats except that the glycerol is attached to only two fatty acid chains instead of three. The third glycerol carbon is joined to a phosphate group ($-PO_3^-$), which, in turn, is linked to one of a variety of small polar groups. The end of the phospholipid containing the phosphate and polar group is soluble in water, while the opposite end containing the fatty acid tails is hydrophobic and "shuns" the aqueous medium. This bipolar nature of phospholipids plays a key role in their function as a major component of cell membranes (see Figure 2-19).

Steroids

Steroids are molecules that are built around a characteristic four-ringed skeleton (Figure 2-20). One of the most common steroids is *cholesterol,* a component of animal cell membranes (the absence of cholesterol from plant cell

◁ THE HUMAN PERSPECTIVE ▷

Obesity and The Hungry Fat Cell

FIGURE 1
Actor Robert DeNiro in *(left)* a scene from the movie *Raging Bull* and *(right)* a recent photograph.

Being overweight—obese—clearly poses increased risk of serious health problems, including heart disease and cancer. By most definitions, a person is obese if he or she is about 20 percent above "normal," or desirable, body weight. This places about 35 percent of U.S. adults in the obese category, twice as many as at the turn of the century. Among young adults, high blood pressure is five times more prevalent and diabetes three times more prevalent in a group of obese people than in a group of people who are at normal weight. Why, then, are so many of us so overweight? Our more sedentary lifestyles and fondness for high-calorie foods provide some answers, but why do people so readily return to their previous weights after they have lost unwanted fat?

Excess body fat is stored in fat cells (*adipocytes*) located largely beneath the skin. These cells can change their volume more than a hundred-fold, depending on the amount of fat they contain, accounting for the bulging, sagging body shape associated with obesity. If the person's fat cells approach their maximum fat-carrying capacity, they release chemical messages that trigger the formation of new fat cells. Once a fat cell is formed, it may expand or contract in volume, but it appears to remain with us for the rest of our lives.

Each person apparently has a particular weight that his or her body's regulatory machinery acts to maintain. This particular value—whether 40 kgs (80 pounds) or 200 kgs (400 pounds)—is referred to as the person's **set-point.** Obese individuals may have a higher set-point than do persons of normal weight. In many cases, the set-point value appears to have a strong genetic component. For instance, there is no correlation between the body mass of adopted children and their adoptive parents, but there is a clear weight relationship between adoptees and their biological parents, with whom they have not lived.

The existence of a body weight set-point is particularly evident when the body weight of a person is "forced" to deviate from the regulated value. Persons of normal body weight who eat large amounts of high-calorie foods under experimental conditions tend to gain increasing amounts of weight. If these people cease their energy-rich diets, however, they return quite rapidly to their previous levels, at which point further weight loss stops. This is illustrated by the actor Robert DeNiro, who gained about 50 pounds for the filming of the movie *Raging Bull* (Figure 1) and then lost the weight prior to his next acting role. Conversely, a person who is put on a strict, low-calorie diet will lose weight. The drop in body weight soon triggers a decrease in the person's resting metabolic rate; that is, the amount of calories "burned" when the person is not engaged in physical activity. The decreased metabolic rate is apparently the body's compensatory measure for the decreased food intake, which tends to halt further weight loss. This effect is particularly pronounced among obese people who diet and lose large amounts of weight. Their pulse rate and blood pressure drop markedly; their fat cells shrink; they tend to feel continually hungry. If these individuals return to a *normal* diet, they tend to regain the lost weight rapidly. These formerly obese persons tend to keep increasing their food intake in response to chemical signals emanating from "hungry" fat cells that have shrunk below their previous size.

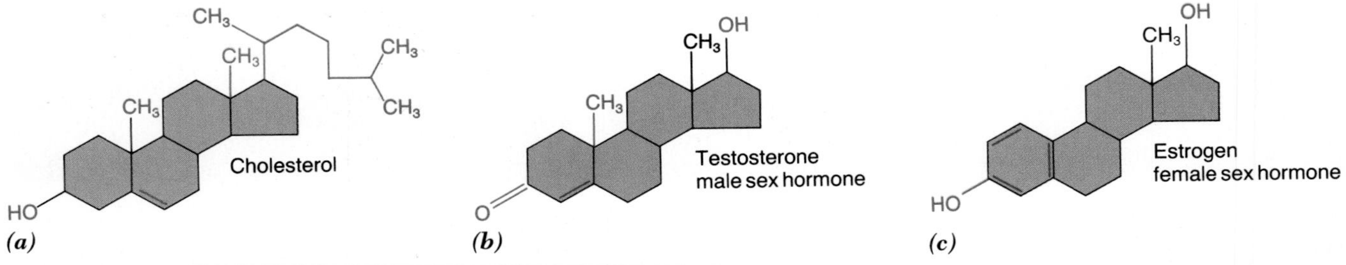

FIGURE 2-20

The structure of steroids. ***(a)*** All steroids share the basic four-ring skeleton *(gray brown)*. (Although not shown, a carbon atom occupies the point of each angle.) The orange parts of the molecule are unique to the individual steroid, in this case, cholesterol. In humans, the seemingly minor differences in chemical structure between testosterone ***(b)*** and estrogen ***(c)*** generate profound biological differences. Testosterone induces male characteristics, such as a deep voice and facial hair; estrogen stimulates the development of female characteristics, such as breast enlargement.

membranes explains why vegetable oils are "cholesterol-free," an important consideration since high cholesterol levels in the blood is associated with an increase likelihood of developing heart disease). Cholesterol is also a precursor for the synthesis of a number of *hormones*—chemical messengers sent from one part of the body to other parts, orchestrating many of the body's processes. For example, sexual maturation is coordinated by steroid hormones; *testosterone* (in males) and *estrogen* (in females).

Waxes. Waxes are similar in structure to fats, but they contain many more fatty acids linked to a longer-chain backbone. Waxes provide a waterproof covering over the leaves and stems of many plants, preventing the organism from losing precious water. Animals also make waxes— to build the honey-containing compartments of a bee hive; as a protective material in the human ear canal; and as a waterproofing material spread over the feathers of birds.

PROTEINS

If you were to evaporate all the water from an organism, more than half the dried remains would consist of protein. It is estimated that the typical mammalian cell has at least 10,000 different proteins. Many of them provide physical structures. For example, proteins are the major structural component of hair, feathers, fingernails, skin, ligaments, tendons, and thousands of other biological structures (Figure 2-21). Proteins are also the macromolecules of the cell that make things happen. They determine much of what gets into and out of each cell, they regulate the expression of genes, they enable organisms to move, and they perform thousands of other functions essential to life. One group of proteins—enzymes—are the mediators of metabolism. Thousands of different enzymes collaborate to direct the development and maintenance of every organism (see page 77).

Amino Acids: The Building Blocks of Proteins

Proteins are polymers made of amino-acid monomers. Each protein has a unique sequence of amino acids that gives that molecule its unique properties. Many of these capabilities can be understood by examining the properties of a protein's amino acids. The same 20 amino acids are found in all proteins, whether in a virus or a human. There are two aspects of amino-acid structure to consider: that which is common to all of them, and that which is unique to each.

Every unbound amino acid has a carboxyl group and an amino group, separated by a single carbon atom (Figure 2-22). The remainder of each amino acid—the **R group**—is the variable part of the monomer, different for each of the 20 building blocks. Depending on the R group, an amino acid may be polar (hydrophilic) or nonpolar (hydrophobic), electrically charged (+ or −) or neutral, have large bulky configurations or very small sizes. The properties of its amino acids are all important determinants as to the shape and function of a protein.

Let's take another look at Figure 2-22. Proteins are synthesized by joining each amino acid to two other amino acids, forming a long, continuous, unbranched polymer called a **polypeptide.** Adjacent amino acids are joined to each other by **peptide bonds**—linkages formed by joining the carboxyl group of one amino acid to the amino group of its neighbor, with the elimination of a molecule of water. The backbone of a polypeptide chain (illustrated by the cubes) is composed of the common parts of the string of amino acids, while the R groups (illustrated by the colored

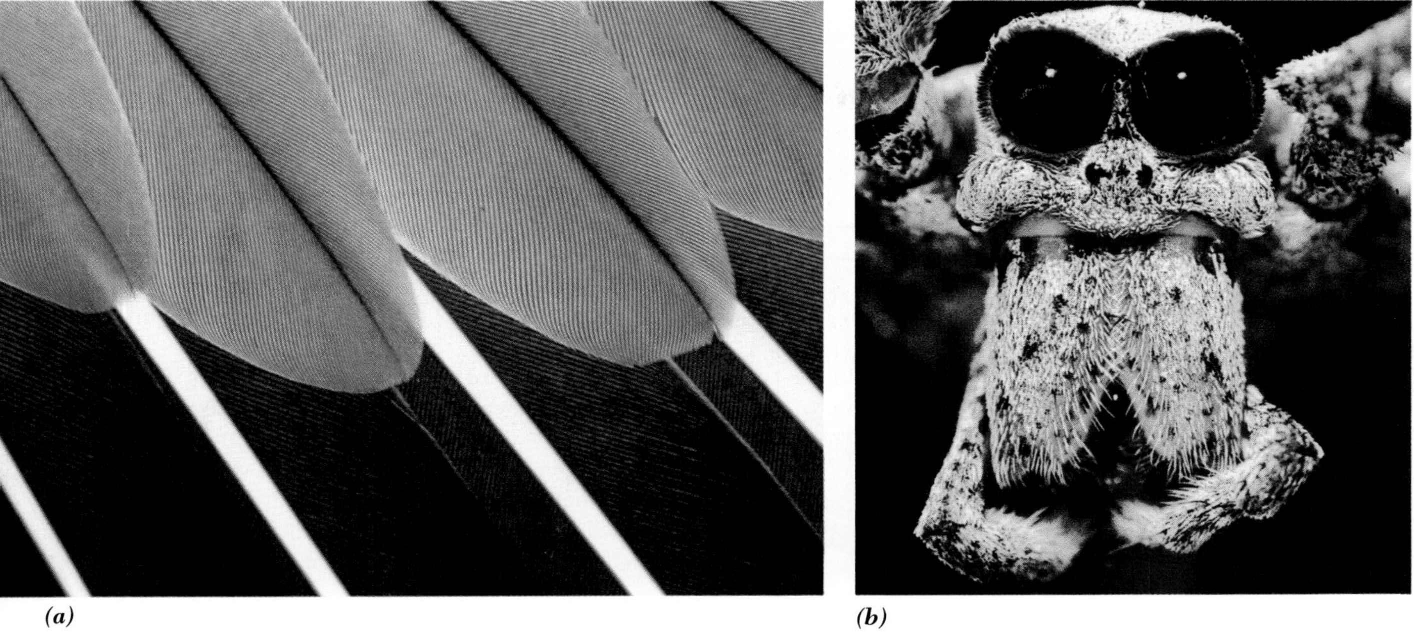

FIGURE 2-21

This protein sampler shows two of the thousands of biological structures composed predominantly of protein. These include: ***(a)*** the fabric of feathers used for thermal insulation, flight, and sex recognition among birds, and ***(b)*** the lenses of eyes, as in this wolf spider.

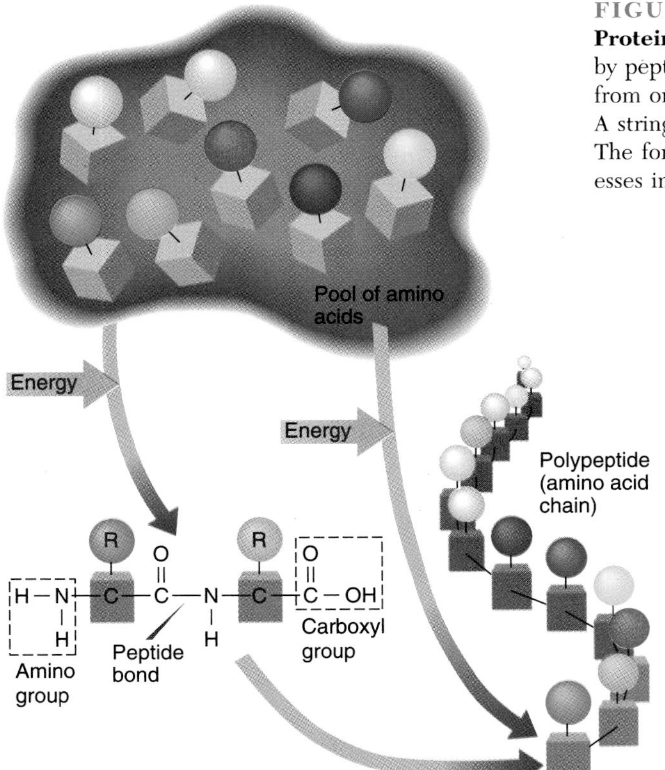

FIGURE 2-22

Protein assembly. Proteins are assembled from amino acids that are joined by peptide bonds. Each peptide bond forms by the linkage of an amino group from one amino acid and a carboxyl group from the neighboring amino acid. A string of amino acids joined by peptide bonds is called a polypeptide chain. The formation of a polypeptide is one of the most complex molecular processes in biology and involves the participation of many different components.

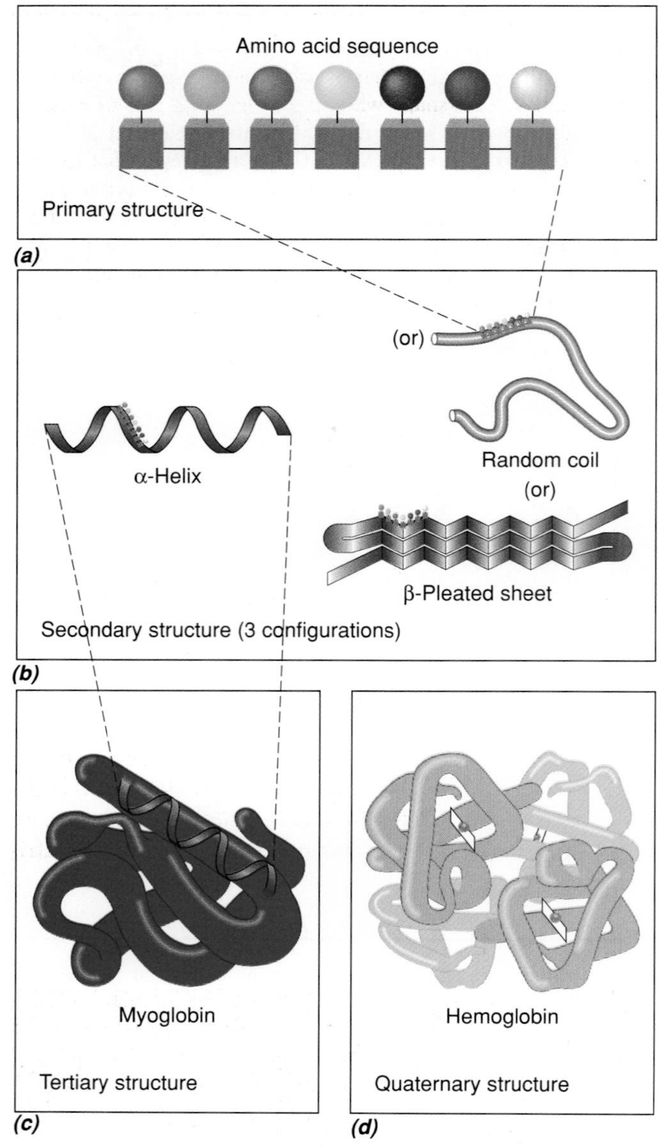

FIGURE 2-23

Four levels of protein structure. Primary structure is the specific sequence of amino acids of a polypeptide chain ***(a)***. Secondary structure is the conformation of a portion of the polypeptide chain ***(b)***. Tertiary structure is the manner in which an entire polypeptide chain is folded ***(c)***. Proteins consisting of more than one polypeptide chain have quaternary structure, which describes the way the chains are arranged within the protein ***(d)***.

spheres) project out from the chain. A protein acquires its shape and function by the way these exposed R groups interact.

The Structure of Proteins

Proteins are large molecules, most containing at least 100 amino acids, and some containing as many as 20,000. The protein's function is determined by its shape, in much the same way the function of a wrench is dictated by its shape. Protein structure is described at four levels of organization: primary, secondary, tertiary, and quaternary (Figure 2-23).

Primary Structure. The *primary structure* of a protein is the specific linear sequence of amino acids that make up its polypeptide chain(s) (Figure 2-23*a*). Different proteins have different primary structures and, consequently, different functions. The primary structure of proteins is determined by information passed on from parents to offspring, encoded in the genetic material.

Secondary Structure. *Secondary structure* determines the way the polypeptide chain initially folds or bends. Three major types of secondary structure are recognized: alpha helix, beta-pleated sheet, and random coil (Figure 2-23*b*).

1. In an *alpha* (α)*helix,* the backbone of the polypeptide assumes the form of a spiral held together by hydrogen bonds. *Alpha-keratin,* the major protein of wool, consists largely of alpha-helix, which gives the material extendibility. When a fiber of wool is stretched, the hydrogen bonds are broken, allowing the helix to be stretched, like pulling a spring. When the tension is relieved, the hydrogen bonds can reform, and the fiber snaps back to its original length.
2. In a *beta* (β)*-pleated sheet,* two or more sections of the polypeptide chain lie side by side, forming an accordion-like sheet that is held together by hydrogen bonds. The major protein of silk is composed largely of pleated sheets stacked on top of one another. Unlike the fibers of wool, those of silk cannot be stretched because the polypeptide chains are already extended.
3. Any portion of a polypeptide chain not organized into a helix or a sheet is said to be in a *random coil.* For example, the sites in the myoglobin where the polypeptide makes a sharp turn are regions of random coil. The random coils tend to be the most flexible portions of a polypeptide and often represent sites of greatest activity.

Tertiary Structure. *Tertiary structure* describes the final shape of an entire polypeptide chain. Like myoglobin, the polypeptide chains of most proteins are folded and twisted to form a globular-shaped molecule (Figure 2-23*c*). Each protein has a precise shape that enables it to carry

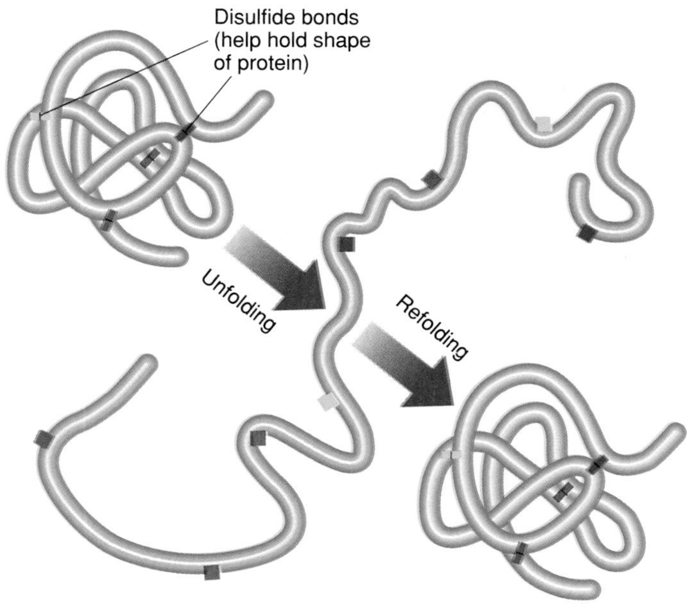

FIGURE 2-24
Self-assembly. When the polypeptide chain of a particular enzyme is experimentally unfolded, it spontaneously reforms its natural three-dimensional shape when the agents that caused denaturation are removed, a shape determined by amino-acid sequence. This experiment provided the first evidence that the sequence of amino acids in a polypeptide dictates the way the polypeptide becomes folded to form the active molecule. It should be noted that not all protein denaturation is reversible. A hard-boiled egg (heat denatured egg protein) does not return to its liquid state when heat is removed.

out a precise function, a shape maintained largely by noncovalent bonds (ionic bonds, hydrogen bonds, and hydrophobic interactions).

Quaternary Structure. Many proteins are composed of more than one polypeptide chain. The spatial arrangement of the combined chains describes these proteins' *quaternary structure* (Figure 2-23*d*). Hemoglobin, for example, is an iron-containing blood protein consisting of four polypeptide chains, each of which can bind and transport a single molecule of oxygen.

Determination of Protein Shape

When protein is treated with agents that cause the protein molecules to unfold, they lose their secondary and tertiary structure, a disruption known as **denaturation** (Figure 2-24). With some denaturizing agents, however, the disorganized protein molecules regain their activity by *spontaneously* refolding into their pre-denatured shape as soon as the agent is removed. The results of these experiments indicate that the three-dimensional shape of a protein is determined directly by the protein's primary structure. The linear sequence of amino acids directs the protein to fold into its proper shape.

The importance of amino acid sequence in determining the form (and function) of a protein is dramatically illustrated by the consequences of changing just one amino acid in hemoglobin. Substituting a hydrophobic, uncharged amino acid (valine) for a hydrophilic, negatively charged amino acid (glutamic acid) distorts the protein and seriously impairs its ability to carry oxygen. Under certain conditions, the distorted hemoglobin elongates red blood cells into a sickle shape (Figure 2-25) that clogs vessels. This potentially fatal disease is called *sickle cell anemia,* an inherited disorder that occurs primarily in people of African descent.

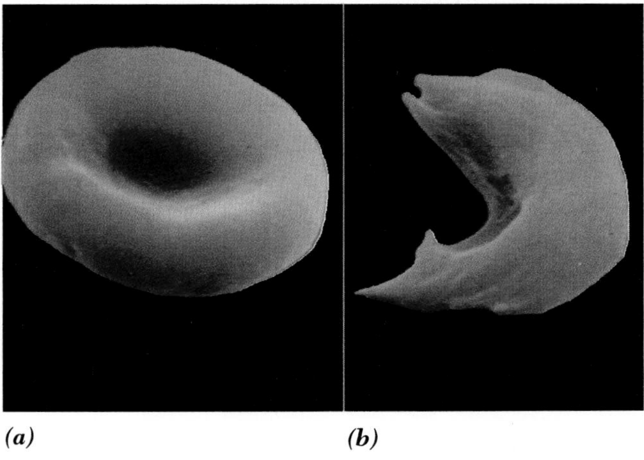

FIGURE 2-25
The effects of a single amino-acid change in a polypeptide chain of hemoglobin. ***(a)*** A normal human red blood cell. ***(b)*** A red blood cell from a person with sickle cell anemia. The cell becomes sickle-shaped when the level of oxygen in the blood drops. These abnormally shaped cells can clog small blood vessels, causing pain and life-threatening crises. The condition is due to the erroneous substitution of a single amino acid in hemoglobin.

NUCLEIC ACIDS (DNA AND RNA)

One of the properties distinguishing the living from the inanimate is the ability of the living to reproduce offspring that have characteristics similar or identical to those of their parent(s). The instructions required to "build" a new individual are transmitted to the offspring encoded in the polymer **deoxyribonucleic acid (DNA)**—the "genetic material" of which genes are made (Figure 2-26). DNA is primarily a storehouse of genetic information. A similar polymer, **ribonucleic acid (RNA),** is needed by cells to translate the encoded information into expressed genetic traits. Both DNA and RNA are **nucleic acids**—macromolecules constructed as a long chain (strand) of **nucleotide** monomers. (Their chemical structure and dynamics are discussed in Chapter 9.) Nucleotides function as chemical "letters" that form the genetic alphabet used to encode genetic instructions. The information stored in nucleic acid is a function of the linear sequence of the nucleotides in the linear molecule, just as the information in this sentence is encoded by its sequence of letters. The sequence of nucleotides in a gene's DNA determines the linear sequence of amino acids in a protein. Change the nucleotide sequence, and the corresponding protein's function will change accordingly.

Nucleotides also participate in life-sustaining activities other than formation of nucleic acids. Most of the energy being put to use at this very instant within your body is derived from the nucleotide **adenosine triphosphate (ATP).** The structure of ATP and its role in cellular metabolism are discussed in the next chapter.

EVOLUTION AND ADAPTATION: TYING IT TOGETHER

The basic structure and function of each family of macromolecules is very similar in all organisms, from bacteria to humans, one of the many similarities that reveal the evolutionary relatedness of all organisms. Chemicals, like organisms, evolve. In fact, macromolecules are biochemical adaptations that are subject to natural selection and evolutionary change in the same way as are other types of

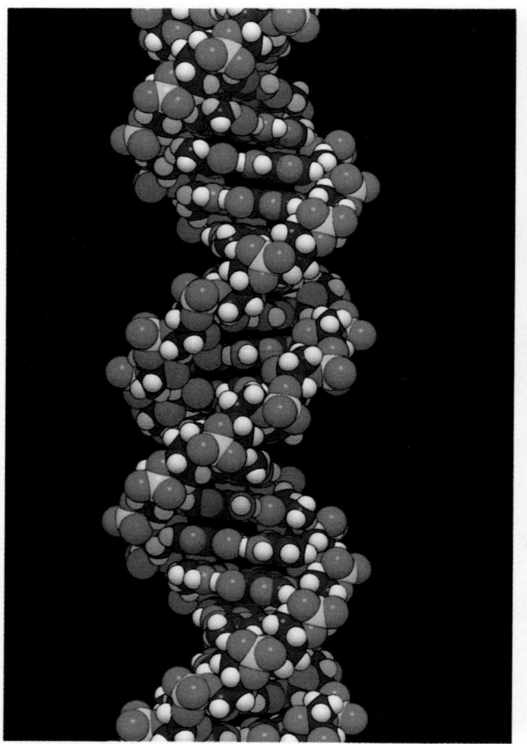

(a)

(b)

FIGURE 2-26

DNA: the material of genes. ***(a)*** Computer-generated model of the DNA molecule. ***(b)*** Each of the physical characteristics of these Japanese cranes developed under the direction of information stored in DNA. Even complex behaviors, such as defending territory (as this crane is doing) and mate-seeking behavior, is encoded within the DNA that makes up the animal's genes. All organisms encode their genetic traits in DNA.

characteristics. Many of these biochemicals have been around for billions of years. DNA for example, is used for encoding genetic information by every organism on earth, a legacy inherited from an ancestor shared by all living organisms far back in evolutionary history (ironically, that legacy of DNA is "written" in DNA). Similarly, virtually all animals store surplus glucose as glycogen. In contrast, some molecules show striking differences from organism to organism, revealing a fairly recent history of divergence in these molecules. For example, although hemoglobin is the oxygen-carrying protein of all vertebrates (animals with backbones), other animal groups may use a totally different protein for acquiring oxygen. Octopus blood, for example, contains a bright blue, copper-containing protein called *hemocyanin*, only distantly related to hemoglobin.

The evolutionary relatedness of organisms that have the same protein can be determined by modern techniques of amino-acid sequence analysis. Hemoglobin, for example, is much more similar between humans and monkeys—organisms that are closely related—than between humans and turtles, who are much less related. In fact, the evolutionary tree that emerges when comparing the structure of specific proteins from various animals very closely matches that previously constructed from fossil evidence.

Perhaps the most striking evolutionary unifier is the similarities in the biochemical basics of all organisms. Why is it that all organisms use DNA for information storage and heredity, use enzymes to drive specific chemical reactions, use the same 20 amino acids to manufacture proteins, have cell membranes composed of phospholipid bilayers, and are biochemically similar in hundreds of ways? The probability of these similarities occurring "by coincidence" is astronomically improbable. Such biochemical similarities highlight those chemical adaptations that have proven so successfully adaptive that natural selection has favored their conservation throughout the history of life, while newer biochemicals show that, while some aspects of life remain the same, life is forever changing, constantly evolving.

SYNOPSIS

Organisms are composed of a small number of elements whose atoms are built from protons, neutrons, and electrons. An atom contains a positively charged central nucleus of protons and neutrons encircled by negatively charged electrons. The number of protons in an atom determines the identity of the atom and matches the number of electrons. Pairs of electrons are organized into orbitals, which exist as part of electron shells of varying energy levels. If an atom gains or loses an electron, it becomes a charged ion. Distinct forms of the same element that differ from each other only in their number of neutrons are called *isotopes.*

Covalent bonds join atoms together to form molecules, whereas many noncovalent bonds join molecules (or different parts of the same molecule). Covalent bonds are stable partnerships formed when atoms share their outer-shell electrons, each participant gaining a filled shell. If electrons shared in a covalent bond show more of an affinity for one of the component atoms than the other(s), the molecule has a polar character. In nonpolar molecules, each bond's electrons are shared equally.

Noncovalent bonds are formed by attractions between positive and negative regions of nearby molecules. Noncovalent bonds include ionic bonds (formed between oppositely charged ionic groups), hydrogen bonds (formed when two atoms share a hydrogen atom), and hydrophobic interactions (formed when nonpolar, hydrophobic molecules are forced together by attractions between surrounding water molecules).

Water has unique properties on which life depends. The covalent bonds that make up a water molecule are highly polarized. As a result, water is an excellent solvent that can form hydrogen bonds with virtually all polar molecules. In addition, the hydrogen bonding of water molecules to one another provides water with important thermal properties. For example, water absorbs unusually large amounts of heat per degree of temperature rise; the evaporation of water requires unusually large amounts of heat; and water becomes less dense as it approaches the freezing point.

When dissolved in water, acids increase the relative proportion of hydrogen ions (which lowers pH), and bases decrease the relative proportion of hydrogen ions (which elevates pH). Neutral solutions have a pH of 7.0, the same as pure water. Some substances resist changes in pH; these buffers help protect cytoplasm and tissue fluids from pH fluctuations.

Carbon is the central element of life. Carbon, whose presence defines organic molecules, has such life-supporting properties as its ability to form four covalent bonds, making it a hub for joining with other carbons to form straight, branching, or ring-shaped chains. In these complex configurations, carbon still has bonds available to attach to various functional chemical groups, that help determine the activity of the organic molecule.

Some biochemicals are molecular giants. Macromolecules constitute the fabric of an organism and conduct its life-sustaining functions. Most macromolecules are polymers constructed by linking together the same class of subunits—monomers—into long chains. Macromolecules are disassembled by hydrolysis.

Four families of biochemicals are essential to life. Carbohydrates include simple sugars and larger molecules (polysaccharides) formed by linking together many sugars. Some polysaccharides store energy, others contribute to an organism's structure. Lipids are a diverse array of hydrophobic molecules—fats (triglycerides) and oils for long-term energy storage; phospholipids that contain a hydrophilic phosphate group making it the basis for membrane formation; four-ringed steroids such as cholesterol and many types of hormones; and waxes that form waterproof coverings. Proteins consist of chains of amino acids. The sequence of amino acids in a protein determines the shape of the protein, which in turn determines the protein's biological role, whether to form an essential structure or perform a particular function. Nucleic acids are informational molecules that consist of linear strands of nucleotide monomers. In both types of nucleic acids, DNA and RNA, the precise sequence of nucleotides in a chain "spells out" the information content of the molecule. Changes in the sequence of nucleotides in DNA lead to changes in traits, one of the forces that drives evolution.

Review Questions

1. Explain what is *wrong* (if anything) with the description of each of the following key terms:

element	all atoms with the same combined number of protons and neutrons
radioisotope	the most stable form of an element
shell	always filled to its electron capacity
ionic bond	chemical partnership in which atoms share electrons
covalent bonds	may be single, double, triple, or quadruple
hydrophobic compounds	polar molecules that form hydrogen bonds with each other and dissolve in water
pH	measurement of acidity or alkalinity; the neutral pH is 0.0

2. Oxygen atoms have eight protons in their nucleus. How many electrons do they have? How many electrons are in the inner electron shell? How many electrons are in the outer shell? How many more electrons can the outer shell hold before it is filled? Do all oxygen atoms have the same number of neutrons? How many neutrons does a radioactive ^{18}O atom have?

3. You have three glasses of liquid. One holds pure water, one holds water to which table salt (NaCl) has been added, and one holds water to which salad oil has been added. In which beaker(s) would you expect hydrogen bonds? Ionic bonds? Hydrophobic interactions?

4. What is the relationship between macromolecules and monomers?

5. Discuss how carbon's ability to form four chemical bonds is critical to life.

6. Complete the following table:

Compound	*Class of Compound*	*Monomer(s)*	*Function(s)*
____	polysaccharide	____	____
ribonuclease	____	____	____
____	____	unsaturated fatty acids & glycerol	____
DNA	____	____	____

Critical Thinking Questions

1. Fredrick Sanger worked out the amino-acid sequence of beef insulin. Would you predict this sequence would be more comparable to insulin found in buffaloes or that found in fish? Why?
2. If you were comparing atoms having 9, 10, and 11 protons in their nuclei, which atom would you expect to be the least reactive? Which would you expect would form a positively charged ion? A negatively charged ion?
3. Explain why the two hydrogen atoms in an H_2 molecule are bound together by a single bond, whereas the two oxygen atoms in an O_2 molecule are bound together by a double bond.
4. What would happen to life on earth if you could kill all microorganisms without directly harming plant or animals?
5. Bacteria can change the kinds of fatty acids they produce as the temperature in which they are living changes. What types of fatty acids changes would you expect in membranes of bacteria living in cold water? Explain the adaptive value of these fatty acid changes.

CHAPTER 3

Cell Structure and Function

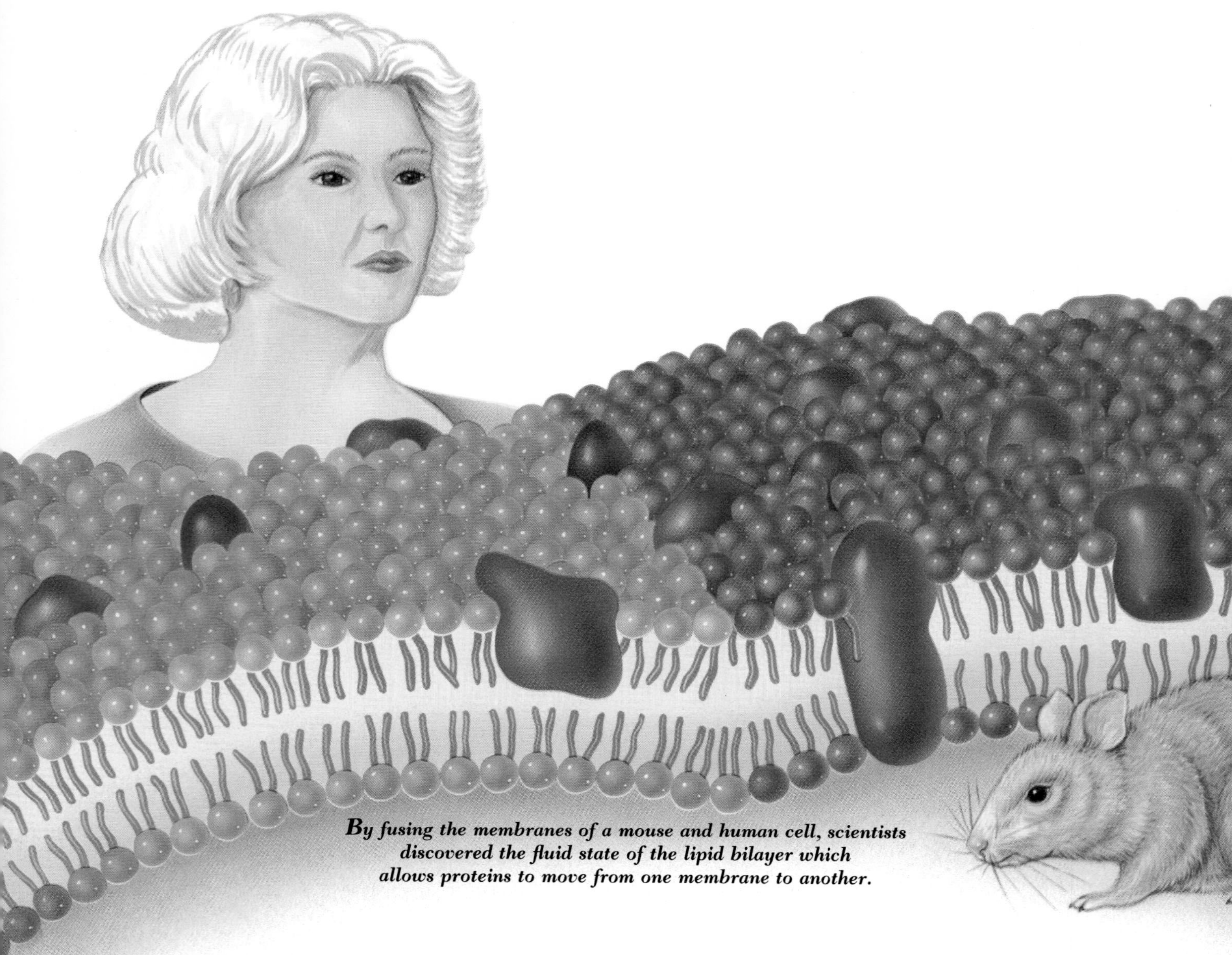

By fusing the membranes of a mouse and human cell, scientists discovered the fluid state of the lipid bilayer which allows proteins to move from one membrane to another.

STEPS TO DISCOVERY
The Nature of the Plasma Membrane

Observing the parts of a cell through a light microscope is like looking for cars and people through the window of an airplane at 35,000 feet—in both situations, the objects are simply too small to be seen. Before the advent of the electron microscope, information on cell structure depended largely on indirect techniques and subsequent interpretations. This was particularly true of the plasma membrane—the delicate structure at the outer edge of all cells which is so thin that not even a hint of its presence is revealed by the light microscope.

In the 1890s, a German physiologist, Ernst Overton, thought he could obtain information about the structure of a cell's outer boundary layer by analyzing the types of substances that passed from the outside environment through the "invisible" barrier into the cell. Overton placed living plant cells into solutions containing various types of solutes. He found that the more nonpolar the solute, the better it was able to penetrate the cell boundary and enter the cell. Overton concluded that the dissolving power of a cell's outer layer matched that of a fatty oil. He hypothesized that the cell possessed a "lipid-containing membrane" that separated its living contents from the outer world.

In 1925, two Dutch scientists, E. Gorter and F. Grendel, designed an experiment to answer two questions: (1) Does the plasma membrane contain lipid? (2) If so, how much? Gorter and Grendel extracted the lipid present in human red blood cells and concluded that the amount of lipid in each cell was just enough to form a layer two molecules thick—a **lipid *bi*layer**. This conclusion assumed that all of the lipid of the cell was part of the plasma membrane. This is not an unreasonable assumption to make when it comes to human red blood cells since they are essentially a "bag of dissolved hemoglobin" with virtually no internal structure, not even a nucleus.

The lipid bilayer was soon shown to be composed of phospholipids (Fig. 2-19). Phospholipids have a "split personality"—a hydrophilic end (the phosphate and polar R group) that can form bonds with water, and a hydrophobic end (the two nonpolar fatty acid tails) that shuns the watery medium. As a result of their structure, phospholipids become aligned into a bilayered sheet, with the hydrophilic ends of the lipid molecules facing out toward the water and the fatty acid tails facing inward toward each other and away from the water (see Figure 2-20). Gorter and Grendel concluded that cells were surrounded by a lipid bilayer that acts as a barrier to protect the cell's internal contents. A plasma membrane is more than just lipid, however; it also contains protein. In 1935, James Danielli of Princeton University and Hugh Davson of University College in London proposed a model for membrane structure that became the focal point of experimentation for 30 years. In the Davson–Danielli model, as it became known, the protein was present as a layer of globular molecules on both sides of the lipid bilayer, not unlike two slices of bread surrounding a double layer of cheese in a sandwich.

Nearly 40 years later, in 1972, S. Jonathan Singer and Garth Nicolson of the University of California proposed a new model of plasma membrane structure, which they named the *fluid-mosaic model* (see Figure 3-3). According to this model, the phospholipid bilayer of a membrane exists in a liquid (*fluid*) phase, having a viscosity similar to that of light machine oil. In the fluid-mosaic model, proteins are embedded in the membrane, penetrating into or passing completely through the lipid bilayer. The scattered distribution of the proteins within the lipid bilayer gave rise to the "mosaic" component of the fluid-mosaic model. What led Singer and Nicolson to view the membrane so differently from the earilier models?

The "fluid" component of the model was based on an earlier experiment employing the technique of *cell fusion*. In these experiments, cells taken from humans and mice were fused to form single cells containing both human and mouse cell nuclei surrounded by a common cytoplasm and a continuous human-mouse plasma membrane (depicted in the illustration that accompanies this essay). Locations of human and mouse membrane proteins were mapped at various times after fusion. At the instant of fusion, the two types of proteins were located in their respective separate portions of the membrane. Within 40 minutes, however, the membrane proteins from the two cells were completely intermixed. This experiment was the first to suggest that membrane proteins were not necessarily fixed in place but were capable of diffusing laterally within the membrane. For this to happen, the lipid bilayer of the membrane must be in a fluid state. Since its initial proposal, the fluid mosaic model has been confirmed for virtually all cell membranes, regardless of their location in the cell or the type of organism in which they are found, once again emphasizing the unity of life.

The Singer–Nicolson model of membrane structure has led to an appreciation of the dynamic quality of membranes which, as we will see, participate in nearly all cellular activities.

Figure 3-1
The discovery of the cell. Microscope used by Robert Hooke, with lamp and condensor for illumination of the object. Inset shows Hooke's drawing of a thin slice of cork, showing the honeycomblike network of "cells."

*I*n these times of burgeoning computer technology, we have grown accustomed to hearing how smaller and smaller microchips are able to store and process greater and greater amounts of information. A 75-volume encyclopedia, for instance, can now be stored on a single chip and "read" by a computer in less than 1 second. As remarkable as this may be, there is another microscopic package that can accomplish this feat and even more—the cell. In addition to storing an enormous amount of information—from the contours of a thumb print to the rate of hair growth—the cell is the center of life itself and is responsible for screening, sheltering, organizing, and coordinating a multitude of life-sustaining chemical reactions.

▼ ▼ ▼

THE CELL THEORY

The discovery of the cell is generally credited to Robert Hooke, an English microscopist who, at age 27, was awarded the position of curator of the Royal Society, England's foremost scientific academy. Three years later, in 1665, Hooke published *Micrographia,* a book in which he described his microscopic observations on such far-ranging subjects as fleas, feathers, and cork. Hooke wondered why stoppers made of cork were so well suited to holding air in a bottle. While viewing thin slices of cork that he cut with his pen knife, Hooke saw rows of tiny compartments resembling a honeycomb (Figure 3-1). He called these compartments "cells" because they reminded him of the cells that housed monks living in a monastery.

Another early microscopist to achieve notoriety was Anton van Leeuwenhoek, a Dutchman who earned a living selling clothes and buttons and spent his spare time grinding lenses and constructing microscopes. For 50 years, Leeuwenhoek sent letters to the Royal Society of London describing his microscopic observations—along with a rambling discourse on his daily habits and the state of his health. Leeuwenhoek was the first to examine a drop of pond water and, to his amazement, to observe the teeming microscopic "beasties" that darted back and forth before his eyes. He was also the first to describe various bacteria which he obtained from scrapings of his own teeth. His initial letters to the Royal Society describing this unseen world were met with such skepticism that the Society dispatched its curator, Robert Hooke, to confirm the observations. Hooke did just that, and Leeuwenhoek, an untrained amateur scientist, was soon a worldwide celebrity, receiving visits in Holland from Peter the Great of Russia and the Queen of England.

As microscopes gradually improved, biologists began to see a consistent pattern: Cells were present in all types of plants and animals. In 1838, Matthias Schleiden, a German lawyer-turned-botanist, proposed that all plants were constructed of cells. The following year, Theodor Schwann, a German zoologist, extended this generalization to include animals. Summarizing the new observations gained from studying the microscopic structure of life, Schwann proposed the first two tenets of the **cell theory**.

1. *All organisms are composed of one or more cells.*
2. *The cell is the basic organizational unit of life.*

More than 10 years later, after studying cell reproduction, Rudolf Virchow, a German physician, proposed the third tenet of the cell theory.

3. *All cells arise from preexisting cells.*

The cell theory is one of the greatest unifying concepts in biology; no matter how diverse organisms appear, all are made up of one or more cells. The cell is the most fundamental structure to harbor and sustain life as well as to give rise to new life. Unlike the parts of a cell which simply deteriorate if isolated, cells can be removed from a plant or animal and kept alive and healthy in a culture dish for long periods of time. These cultured cells have all the necessary regulatory mechanisms to maintain a homeostatic condition. They take up nutrients, digest them, and excrete waste products; they take up oxygen and release carbon dioxide; they maintain a particular water and salt content; they are capable of growth, reproduction, and movement; they respond to external stimulation; they expend energy to carry out their activities; they inherit a genetic program from their parents and pass it on to their offspring; and, finally, they die. These are the characteristics of life, and they are all exhibited by individual cells.

Cells come in a variety of shapes and sizes, ranging from enormously extended nerve cells that connect your spinal cord to your big toe (while very long, these cells have a very small diameter), to minute bacteria so small that more than a thousand would be needed to fill the dot in the letter *i*. The evolution of the first primitive cells was one of the most crucial—and poorly understood—steps in the entire course of biological evolution on this planet. For all of these reasons, understanding the cell is one of the cornerstones of biology.

TWO FUNDAMENTALLY DIFFERENT CLASSES OF CELLS

All cells are either prokaryotic or eukaryotic (Figure 3-2), distinguishable by the types of internal structures, or **organelles**, they contain. The existence of two distinct classes of cells, without any known intermediates, represents one of the most fundamental evolutionary discontinuities that exists in the biological world. Both types of cells—prokaryotic and eukaryotic—are bounded by a similar type of membrane, and both may be surrounded by a nonliving cell wall. In other respects, the two classes of cells are very different. **Eukaryotic cells** are the larger, more complex cell type. The genetic material (DNA) is housed in a special compartment, the **nucleus**, which is separated from the rest of the cellular space, the **cytoplasm**, by a complex, double-layered membrane, the **nuclear envelope**. The cytoplasm of a eukaryotic cell is filled with a variety of membranous and nonmembranous organelles that are specialized for various activities. Eukaryotic cells make up eukaryotic organisms, including fungi, protists, plants, and animals. In contrast, prokaryotic cells are found only among bacteria. **Prokaryotic cells** are smaller, with a much simpler internal structure. These cells lack a membrane-bound nucleus so that the genetic material is not clearly separated from the surrounding cytoplasm. In addition, the cytoplasm of prokaryotic cells is essentially devoid of organelles (other than ribosomes) and is capable of much less complex activities than that of eukaryotic cells. Prokaryotic cells are thought to represent the vestiges of an early stage in the evolution of life which was present before the more complex eukaryotes evolved. Nearly all cells—prokaryotic and eukaryotic—are microscopic; that is, they are too small to be seen without a microscope.

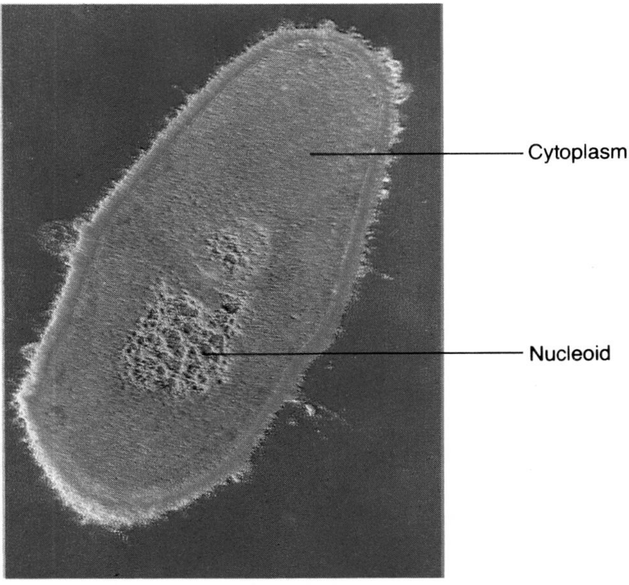

(a)

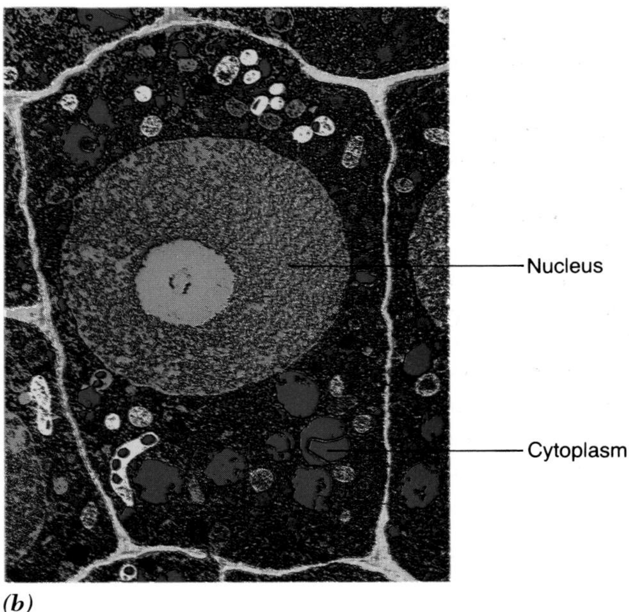

(b)

Figure 3-2

A comparison of prokaryotic *(a)* and eukaryotic *(b)* cells, the two fundamental cell types. ***(a)*** A bacterial cell showing the cytoplasm devoid of visible organelles and the DNA housed within an ill-defined area called the *nucleoid*. ***(b)*** A plant root tip cell showing the varied cytoplasmic organelles and the membrane-bound nucleus which houses the cell's DNA.

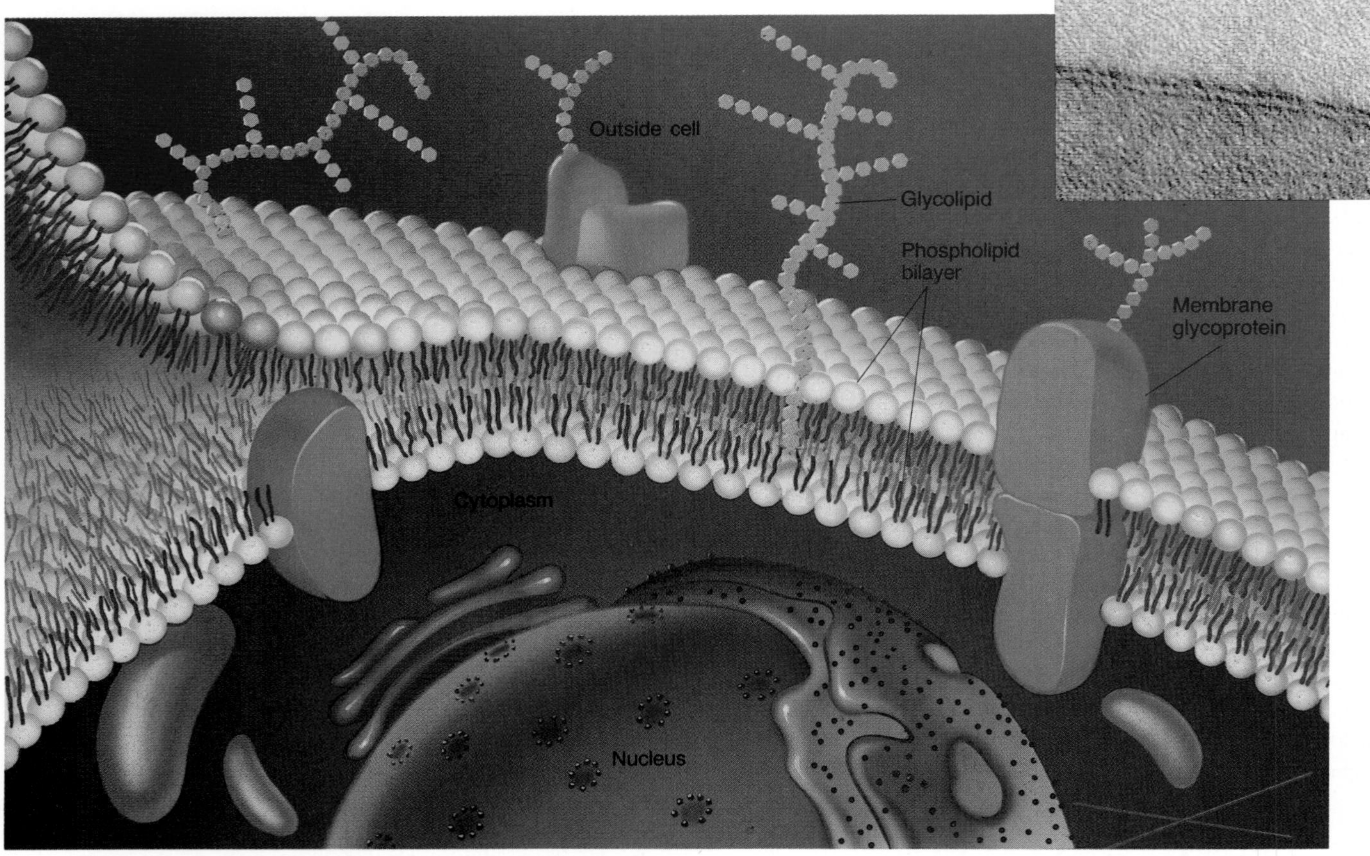

Figure 3-3
The structure of the plasma membrane. All cells have a plasma membrane of similar structure. Proteins penetrate into or through the lipid bilayer. Chains of sugars are covalently bonded to the outer surface of most, if not all, of the proteins, and to a small percentage of the phospholipids. Membranes appear in the electron micrograph (inset) as three-layered structures having two dense outer layers and a less dense middle layer.

THE PLASMA MEMBRANE: SPECIALIZED FOR INTERACTION WITH THE ENVIRONMENT

All cells are surrounded by an exceedingly thin **plasma membrane** of remarkably similar construction in all organisms—from bacteria to humans. Among its numerous functions, the plasma membrane (1) forms a protective outer barrier for the living cell and (2) helps maintain the internal environment of the cell by regulating the exchange of substances between the cell and the outside world. As discussed in the Steps to Discovery that opened the chapter, the plasma membrane is a thin sheet consisting of a phospholipid bilayer pierced by proteins (Figure 3-3). Unlike most thin sheets, such as plastic wrap or waxed paper, which exist in the solid state, the lipid bilayer is a viscous fluid whose molecules are able to move from place to place.

Most of the proteins of the plasma membrane are inserted into the fluid lipid bilayer, much like pegs can be inserted into the holes of a pegboard. Those portions of the membrane proteins that protrude beyond the lipid bilayer generally contain attached carbohydrates, making them *glycoproteins*. A person's blood type (A, B, AB, or O), for example, is determined by the particular sugars that make up the carbohydrate chains that project from the surface of a person's red blood cells.

The plasma membrane is able to recognize and interact with certain substances in the cell's environment. These specific interactions are mediated by those parts of the glycoproteins that protrude from the outer surface of the membrane. Membrane proteins having this role are termed **receptors**. In eukaryotes, each type of specialized cell has its own set of receptors, which allows the cell to respond to particular ions, hormones, antibodies, and other circulating molecules.

THE EUKARYOTIC CELL: ORGANELLE STRUCTURE AND FUNCTION

You have several hundred different types of cells in your body, each recognizably different from the others. Yet, virtually all of these different cells are composed of the same types of organelles (Table 3-1). Figure 3-4 shows a "generalized" plant and animal cell, each containing a com-

TABLE 3-1
THE PRIMARY FUNCTIONS OF CELL COMPONENTS OF EUKARYOTIC CELLS

Component	*Primary Functions*
Plasma membrane	Boundary of cell, exchange with environment
Nucleus	Storage of hereditary information, control of cell activities
Nuclear envelope	Exchange between nucleus and cytoplasm
Nucleolus	Ribosome synthesis
Chromosomes	Storehouse of genetic information
Endoplasmic reticulum (ER)	Synthesis of protein, steroids, lipids; storage of Ca^{2+}; detoxification
Ribosomes	Sites of protein synthesis
Mitochondria	Chemical energy conversions for cell metabolism
Chloroplasts (plant cells only)	Conversion of light energy into chemical energy, storage of food and pigments
Golgi complex	Synthesis, packaging, and distribution of materials
Lysosomes	Digestion, waste removal, discharge
Central vacuoles (plant cells only)	Storage, excretion
Microfilaments, microtubules, intermediate filaments	Cellular structure, movement of internal cell parts, cell movement
Cilia and flagella	Locomotion, production of currents
Vesicles	Storage and transport of materials
Cell wall (plant cells only)	Protection, fluid pressure, support

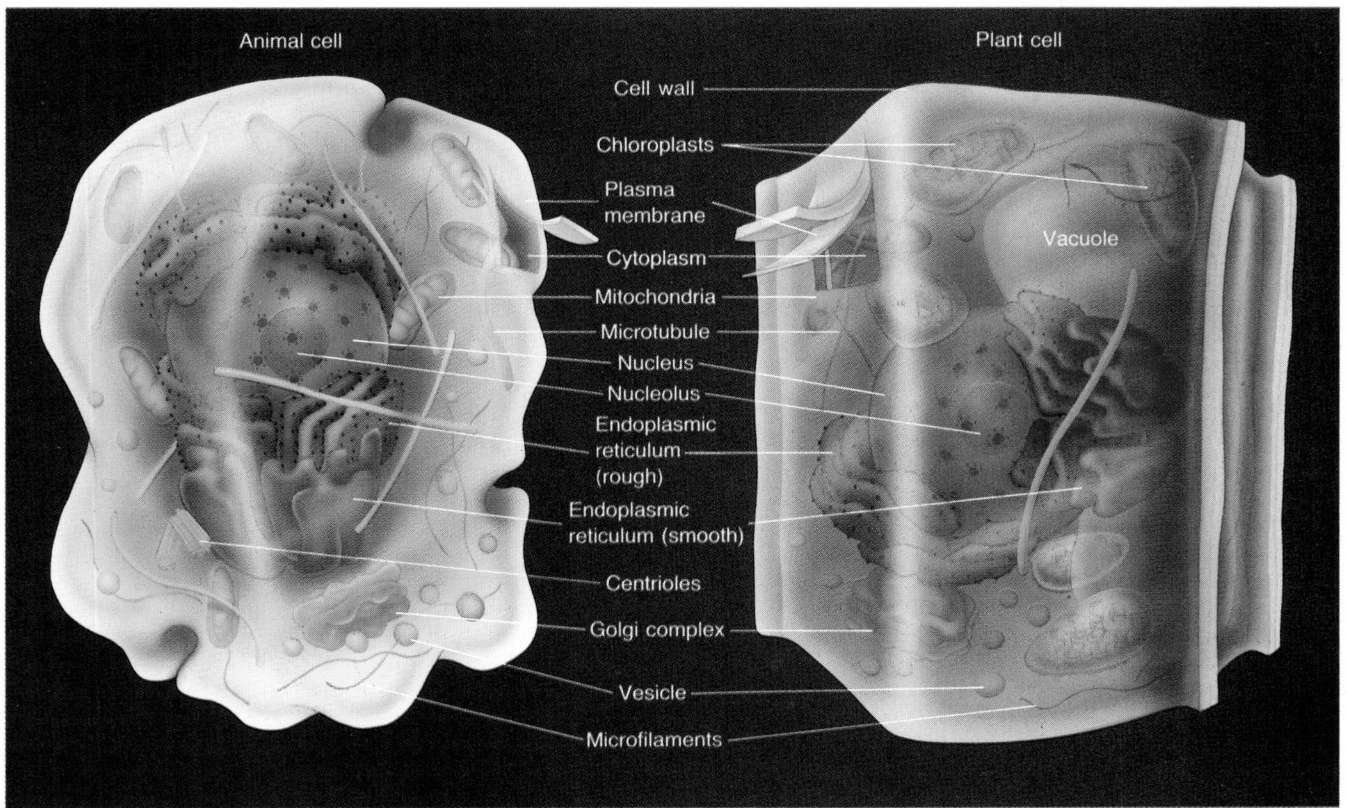

Figure 3-4
Generalized structure of eukaryotic cells. Both plant and animal cells have a plasma membrane, nucleus, and cytoplasmic organelles. They differ in some types of organelles and by the presence of cell walls just outside of plant cells, which animal cells do not have.

bination of organelles typically found in eukaryotic cells. Keep in mind that there is really no such thing as a "generalized" or "typical" cell. The appearance and distribution of cellular organelles vary greatly from a nerve cell, to a bone cell, to a gland cell. The analogy might be made to a variety of orchestral pieces: All are composed of the same notes, but varying arrangement gives each its unique character and beauty.

Organelles are specialized structures or compartments inside the cell in which specific activities take place without interference from other events. Viewed in this way, organelles maintain order within a cell, preventing biochemical chaos. A eukaryotic cell could no more operate without its organelles than a restaurant could operate without a separate kitchen, dining room, garbage bin, and restrooms.

THE NUCLEUS: GENETIC CONTROL CENTER OF THE CELL

Typically, the most prominent structure in a eukaryotic cell is the nucleus (Figure 3-5). The nucleus is the residence of the cell's genetic material—its DNA. Genetic instructions leave the nucleus in the form of RNA and enter the cytoplasm, where they direct the synthesis of specific proteins. Because of its role as genetic headquarters, the nucleus is often thought of as the "control center" of the cell.

The genes of a cell reside in the **chromosomes**, structures composed of DNA molecules and bound proteins. During most of the life of a cell, the material that makes up the chromosomes is unraveled to form highly elongated threads. It is only when the cell gets ready to divide that the threads become packaged into visible chromosomes (Figure 3-6). Also situated within the nucleus is one or more irregular-shaped structures called **nucleoli** (singular *nucleolus*), which consist of densely packed, minuscule granules. Nucleoli are the sites for manufacturing **ribosomes**, structures that consist of RNA and protein and function in the cytoplasm as workbenches for the assembly of proteins.

The nucleus is surrounded by a complex **nuclear envelope** that separates the nuclear contents from the cy-

Chromatin (DNA-protein threads)

Nuclear envelope

Nucleoplasm

Nucleolus

(a)

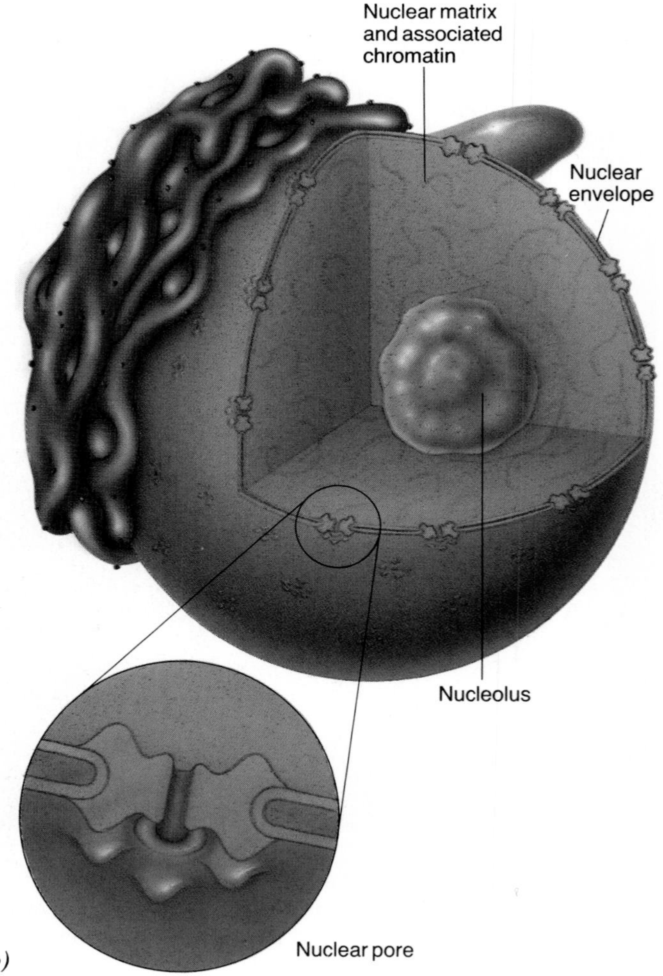

(b)

Figure 3-5

The nucleus. ***(a)*** Electron micrograph of a section through the nucleus of a eukaryotic cell. The nucleolus is visible as a dense internal structure. The scattered clumps of stained material consist of DNA-protein fibers (*chromatin*) that make up the chromosomes. ***(b)*** Cut-away diagram showing the nuclear envelope that bounds the nuclear compartment and the nucleolus. The nuclear envelope is studded with pores through which materials are exchanged between the fluid of the nucleus (*nucleoplasm*) and cytoplasm. The protein fibers of the *nuclear matrix* provide mechanical support.

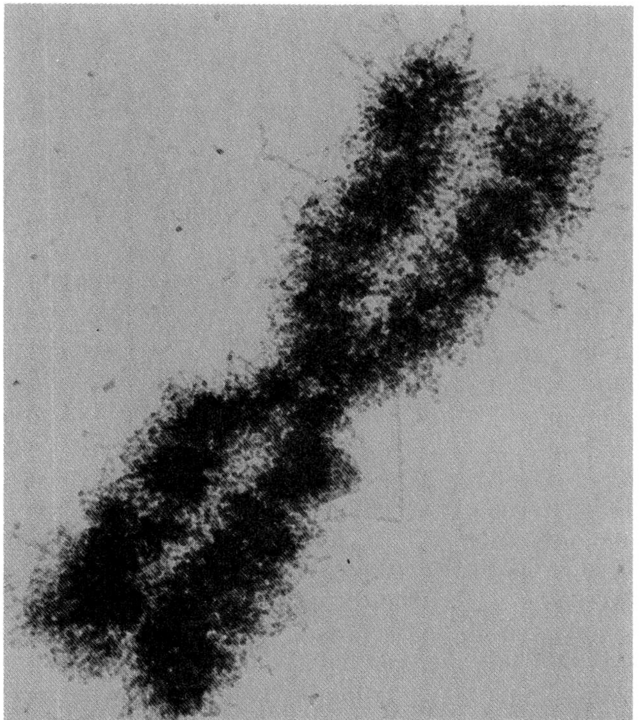

Figure 3-6
A chromosome. During cell division in a eukaryotic cell, each chromosome becomes greatly compacted to form the type of structure seen in this electron micrograph.

toplasm (Figure 3-5*b*). The nuclear envelope is a double membrane pierced by complex pores that act as passageways for materials entering or exiting the nucleus. The nuclear envelope is an important gateway for regulating traffic and maintaining essential differences between the two major regions of the cell.

MEMBRANOUS ORGANELLES OF THE CYTOPLASM: A DYNAMIC INTERACTING NETWORK

The cytoplasm of most eukaryotic cells is filled with membranous structures that extend to every nook and cranny of the cell's interior. Included among the cytoplasmic membranes are small, spherical containers (*vesicles*) of varying diameter, long interconnected channels, and flattened membranous sacs. Although the membranes of the cytoplasm may appear disconnected in an electron micrograph, in reality, they form a highly interdependent membranous network that includes the endoplasmic reticulum, the Golgi complex, and lysosomes.

Endoplasmic Reticulum

Coursing through most cells is an elaborate system of folded, stacked, and tubular membranes known as the **endoplasmic reticulum**, or simply the **ER**. The membranes of the ER divide the cytoplasm into two compartments: one within the confines of the ER membranes, and one outside the ER membranes. As a result, materials confined within the ER space are segregated from the remainder of the cytoplasm and can be shipped to various parts of the cell.

The ER occurs in two different forms: rough ER and smooth ER (Figure 3-7). **Rough ER** (**RER**) appears bumpy (rough) in electron micrographs because of its many attached ribosomes. These ribosomes are sites where proteins are assembled, one amino acid at a time. The newly synthesized protein passes through the ER membrane and is then segregated within the ER compartment. These proteins have specific destinations: Some are exported out of the cell; some end up as membrane

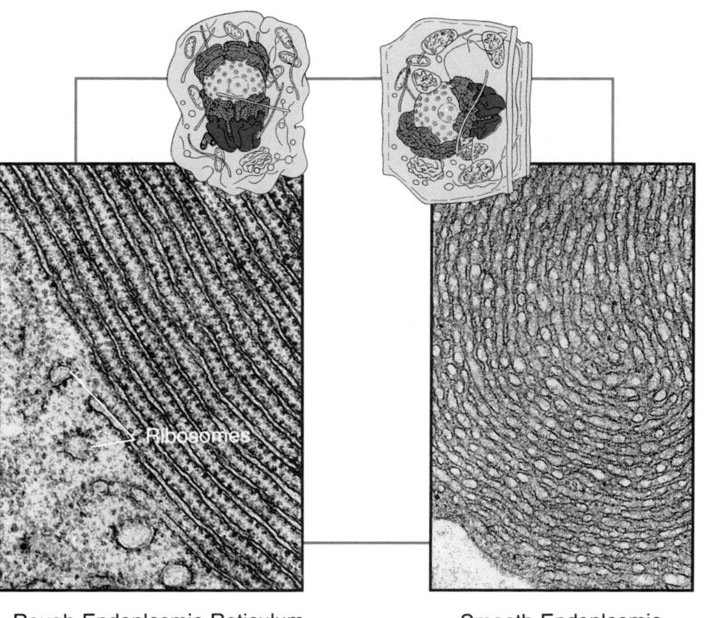

Figure 3-7
Endoplasmic reticulum. ***(a)*** Portion of a pancreatic cell showing the rough endoplasmic reticulum (RER) where digestive enzymes are assembled. ***(b)*** Portion of a cell from the testis showing the smooth endoplasmic reticulum (SER) where steroid hormones are synthesized.

protein; and some become part of other cytoplasmic organelles. The RER is most highly developed in cells that export (*secrete*) large quantities of protein, such as the cells of the pancreas, which secrete digestive enzymes, and the salivary glands, which secrete salivary proteins. The **smooth ER (SER)**, which lacks ribosomes, has several diverse functions, including the synthesis of steroids (such as the sex hormones estrogen and testosterone), destruction of toxic materials (such as barbiturates), and storage of calcium ions.

Golgi Complex

In 1898, an Italian biologist, Camillo Golgi, was working with a new type of metallic stain when he discovered a dark yellow network near the nuclei of nerve cells. This network was later named the **Golgi complex** and helped earn its discoverer the 1906 Nobel Prize. The Golgi complex is usually present as a stack of flattened membranous sacs and associated vesicles (Figure 3-8). The sacs are arranged in an orderly pile, resembling a stack of "hollow" pancakes. The Golgi complex serves as a way station in the movement of materials from the ER through the cell (Figure 3-9). For example, proteins move from their site of synthesis in the rough ER to the Golgi complex within small vesicles. These vesicles fuse with the membranes of the Golgi complex, and the proteins are modified in various ways. For example, the insulin that controls the level of sugar in your blood is originally synthesized in the RER as a much larger protein that is cut down to its final size in the Golgi complex. Similarly, the sugars that determine your blood type are added as the glycoproteins pass through the Golgi complex.

Materials processed in the Golgi complex are packaged into vesicles that bud from the lateral edges of the Golgi sacs. Some vesicles remain in the cell, storing important chemicals until they are needed; others move to the cell surface and discharge their contents to the outside via a process called **exocytosis** (Figure 3-9, inset). Exocytosis occurs when the membrane of a vesicle fuses with the overlying plasma membrane, thus opening the vesicle and allowing its contents to be discharged.

In addition to carrying enclosed materials, the movement of materials from the ER to the Golgi complex to the plasma membrane provides a mechanism for building plasma membrane since the membrane of the cytoplasmic vesicles become *incorporated* into the plasma membrane during exocytosis (Figure 3-9). This, in fact, is how the plasma membrane is formed—from vesicles whose membranes were originally produced in the ER.

Lysosomes

Some of the vesicles that bud from the Golgi complex remain in the cytoplasm as **lysosomes**—storage vesicles that contain dozens of powerful hydrolytic enzymes capa-

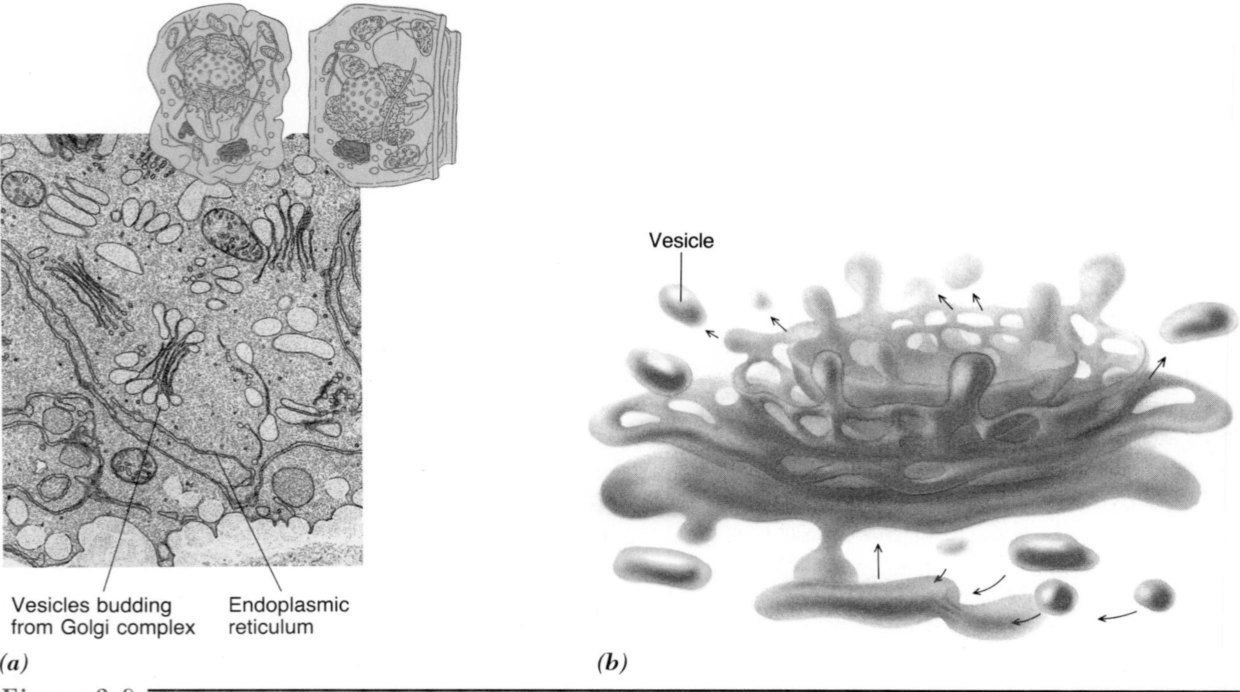

Figure 3-8

The Golgi complex. ***(a)*** Electron micrograph of the Golgi complex of a plant cell. Vesicles are seen budding from the lateral edges of the Golgi sacs. ***(b)*** Diagrammatic view of a Golgi complex. A single cell of the human pancreas, churning out enzymes needed to digest three meals a day, may contain thousands of such Golgi complexes.

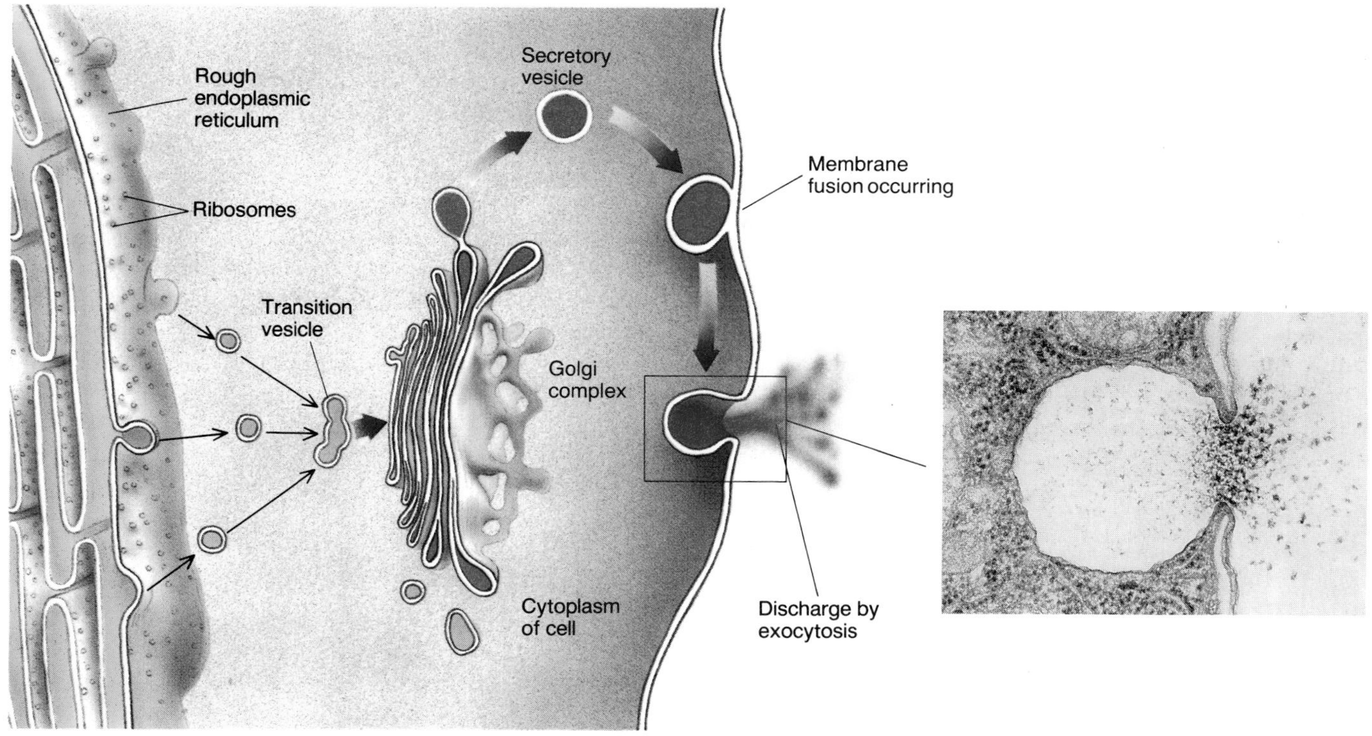

Figure 3-9
The path through the cell. Proteins destined for secretion (such as the digestive enzymes of the pancreas or the proteins of salivary mucus) are synthesized on ribosomes situated on the outer surface of the ER membranes. The proteins are then passed into the ER compartment, packaged in a transition vesicle, and transferred to the Golgi complex, where the proteins are concentrated and often modified before being sent on their way in secretory vesicles. The vesicles may be stored in the cytoplasm before fusing with the plasma membrane (exocytosis), allowing the vesicle contents to be discharged into the external medium (shown in inset).

ble of digesting virtually every type of macromolecule in the cell. The membrane at the outer edge of a lysosome keeps these lethal enzymes safely sequestered. Lysosomes play a key role in maintaining constancy and order within the cell; they function as digestion chambers for destroying dangerous bacteria and foreign debris, as well as sites for disposing of impaired or worn-out organelles. In a few cases, an organism's own cells are "deliberately" digested by lysosomal enzymes.

A number of diseases can be traced to defective lysosomes, which is not surprising considering the destructive nature of the enzymes they contain. For example, a miner's disease known as *silicosis* results from the uptake of silica fibers by wandering cells in the lungs. The fibers become enclosed within lysosomes but cannot be digested; instead, they cause the lysosomal membrane to leak, spilling the contents of digestive enzymes into the cell and damaging the tissue of the lungs.

Just as there are diseases resulting from excessive lysosomal activity, there are also serious consequences associated with a lack of lysosomal enzymes. Tay-Sachs disease, for example, is due to a missing lysosomal enzyme that normally destroys certain plasma membrane lipids found in nerve cells. In the absence of this enzyme, the lipids accumulate in the nerve cells, causing severe mental retardation and death by about age 5.

Vacuoles

Vacuoles are essentially large vesicles; that is, large, membrane-bound, fluid-containing sacs. Vacuoles are particularly prominent in mature plant cells, where they may occupy more than 90 percent of the cell's volume (see Figure 3-4). In addition to containing water, the fluid in the plant vacuole may contain gases (oxygen, nitrogen, and/or carbon dioxide), acids, salts, sugars, pigments (which account for some of the colors of flowers and leaves), and even toxic wastes. Unlike animals, who have elaborate excretory systems for expelling toxic wastes, plants apparently "excrete" such substances into their own vacuoles, safely partitioning the toxins from the plant's cytoplasm. Plant vacuoles also maintain high internal water pressure, which aids in the support of the plant. When plants lose water from their vacuole, as occurs, for example, by evaporation on a hot day, some of this support is lost, and the plant wilts.

MITOCHONDRIA AND CHLOROPLASTS: ACQUIRING AND USING ENERGY IN THE CELL

Mitochondria can be likened to miniature "power plants" located within the cytoplasm of plant and animal cells. Whereas power plants convert the energy stored in energy-rich fuels (such as oil and coal) into electricity—a form usable to the consumer—**mitochondria** (singular *mitochondrion*) convert the energy stored in energy-rich macromolecules (such as fats and polysaccharides) into adenosine triphosphate (ATP)—a form usable by the cell in running virtually all of its immediate activities (Chapter 6).

Mitochondria are typically sausage-shaped organelles (Figure 3-10) constructed of two membranes—an *outer membrane* surrounding an elaborately folded *inner membrane*. The inner mitochondrial membrane contains dozens of different components arranged in the precise spatial order required for the formation of a cell's ATP. As with other organelles, the number and location of mitochondria depend on the activities of the particular cell. A particularly striking arangement of mitochondria is seen in many sperm cells, where the mitochondria are distributed in a spiral in the middle of the cell (Figure 3-10*c*). The movements of the sperm are powered by the ATP produced in these mitochondria.

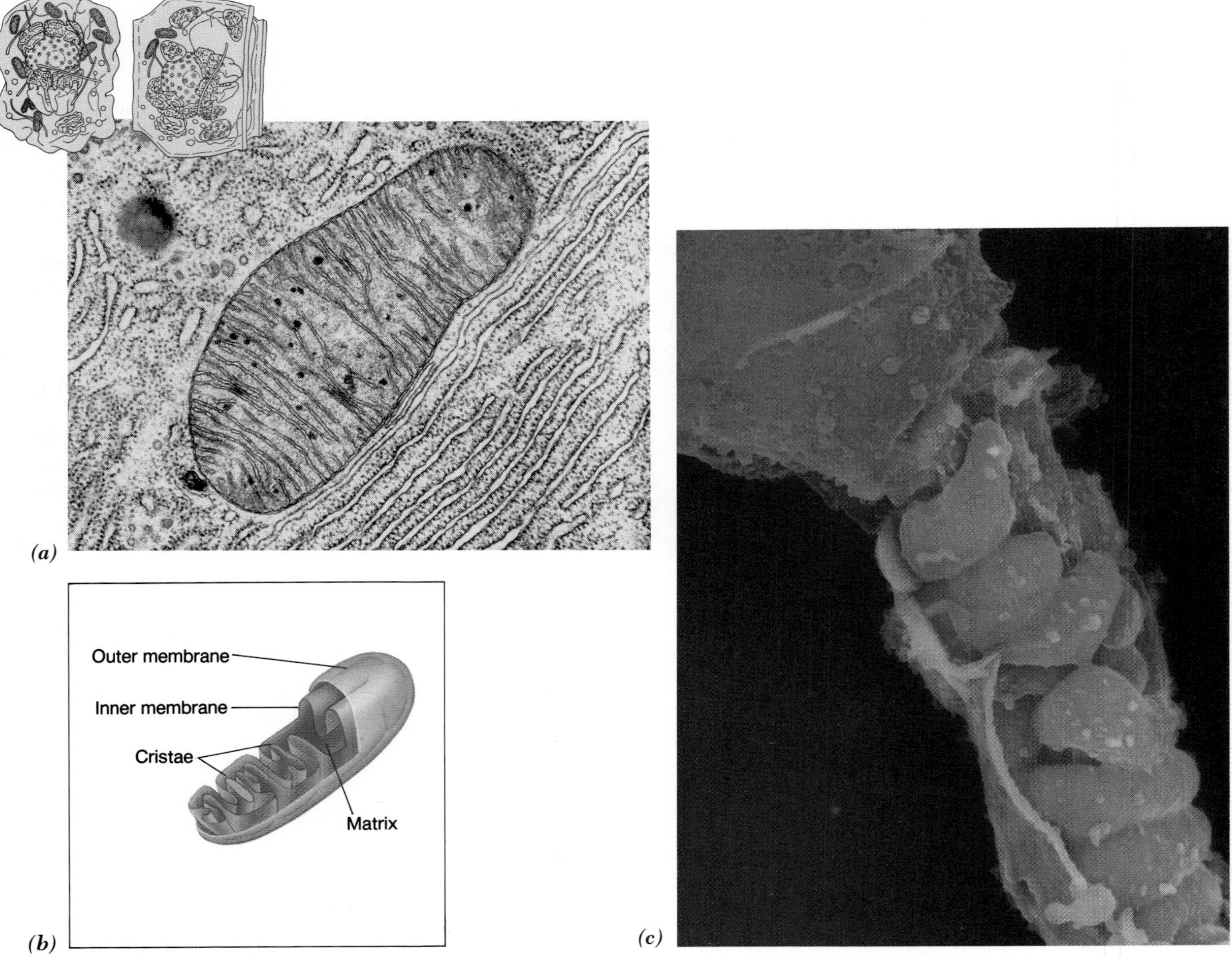

Figure 3-10

Mitochondria. Electron micrograhs ***(a, c)*** and drawing ***(b)*** of mitochondria showing the outer and inner membranes. The labyrinth of convolutions created by the inner membrane forms the *cristae*, which project into an internal, semifluid compartment, the *matrix*. The matrix also contains DNA and ribosomes. The electron micrograph in part ***(c)*** shows the way in which mitochondria (orange structures) are packaged in the middle portion of a mammalian sperm.

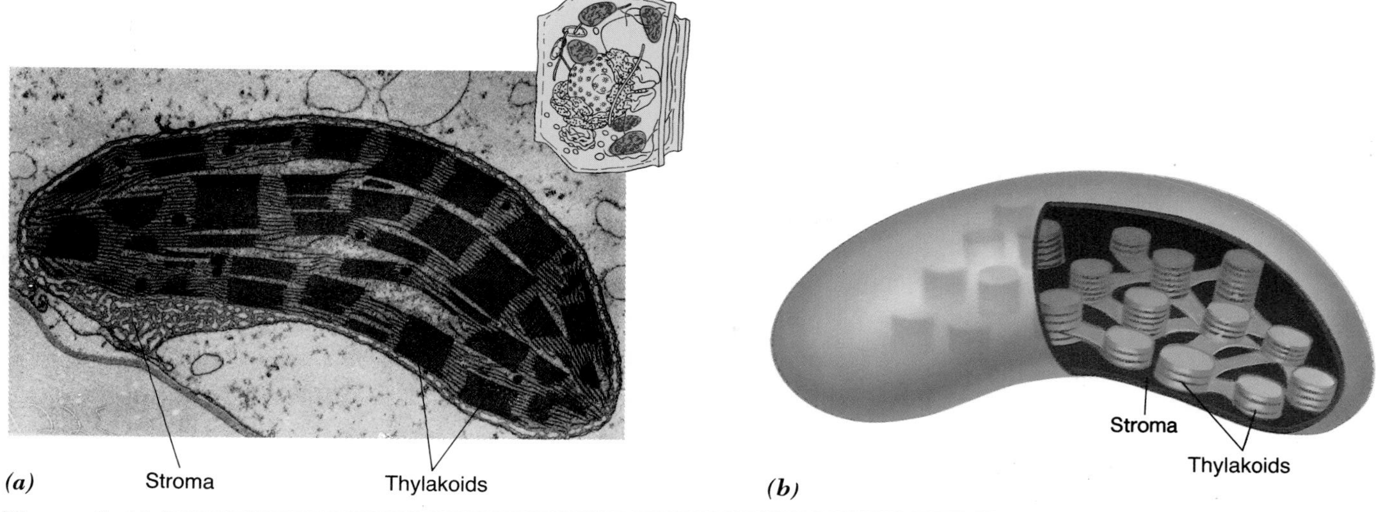

Figure 3-11

Chloroplasts. *(a)* An electron micrograph, and *(b)* a three-dimensional diagram of a chloroplast. A chloroplast contains a double-membrane envelope as well as stacks of membranous thylakoids that are contained in the central *stroma*. The stroma also contains DNA and ribosomes.

Chloroplasts (Figure 3-11) are found only in plants and certain photosynthetic protists and are sites of photosynthesis, a complex process during which light energy is captured and used to construct complex biochemicals (Chapter 5). Virtually all life on earth depends on the energy captured by chloroplasts. Each chloroplast is bounded by a double-membrane envelope and contains an elaborate internal system of flattened membranous discs, called **thylakoids**, in which light-capturing pigments, such as chlorophyll, are precisely arranged.

THE CYTOSKELETON: PROVIDING SUPPORT AND MOTILITY

The human skeleton consists of hardened parts of the body which support the soft tissues and play a key role in mediating body movements. The cell also has a "skeletal system"—a **cytoskeleton**, with analogous functions. The cytoskeleton serves two interrelated activities:

1. as a scaffold, providing structural support, maintaining the shape of the cell, and organizing the internal contents of the cytoplasm; and
2. as part of the machinery required for the movement of materials within the cell or for the movement of the cell itself.

The cytoskeleton (Figure 3-12) includes three components: hollow, cylindrical microtubules; solid, thinner, and more flexible microfilaments; and tough, ropelike intermediate filaments. These three elements are typically interconnected to form an elaborate interactive network of fibers. The elements of the cytoskeleton are a dynamic

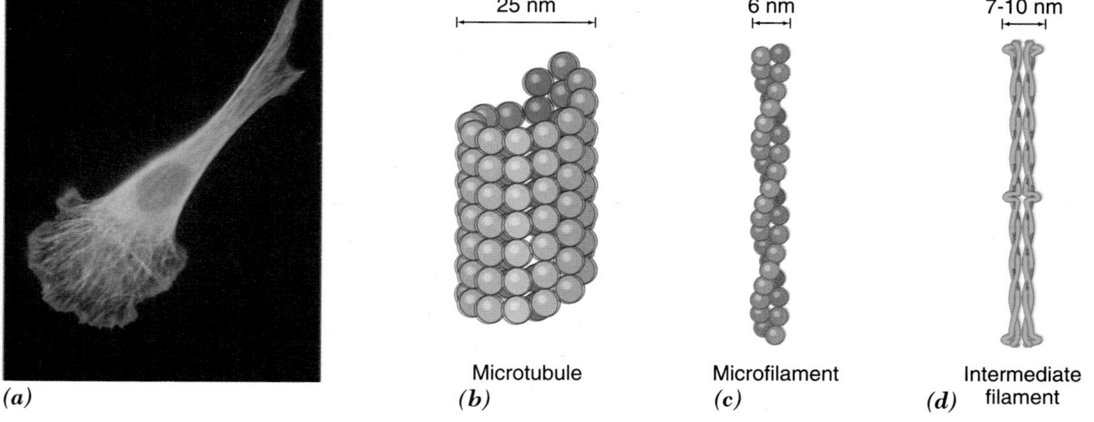

Figure 3-12

The cytoskeleton. *(a)* This cell was caught in the act of moving over the surface of a culture dish. The cell has been stained to reveal the distribution of microfilaments and microtubules. The rounded edge of the cell is leading the way; the clusters of microfilaments at the leading surface are sites where the forces required for movement are generated. *(b–d)* Microtubules, microfilaments, and intermediate filaments are composed of different types of protein subunits that become arranged in characteristic patterns.

group of structures capable of rapid and dramatic reorganization. For example, at one moment, the protein subunits that make up the microtubules may be organized into fibers that separate the chromosomes during cell division, while a few minutes later, these fibers may have completely disassembled and the protein subunits used to assemble a different group of microtubules involved in cell movement.

Cilia and Flagella

Microscopic organisms typically dart through their watery environment, powered by tiny locomotor organelles called cilia and flagella, which are really two versions of the same structure. **Cilia** are short, hairlike organelles that project from the surface of many small eukaryotic organisms; they act like oars to propel the organism through the water (Figure 3-13*a*). In larger organisms, such as mammals, cilia often line the surfaces of various tracts (Figure 3-13*b*), where they help propel materials through the channel.

Flagella are longer than cilia and are present in fewer numbers. Flagella beat with an undulating motion that pushes or pulls the cell through the medium. In addition to powering microscopic organisms, flagella provide the propulsive force for the movement of sperm in most animals. Both cilia and flagella contain an internal skeleton of microtubules. The microtubules are arranged in a specific pattern called a 9 + 2 *array*—nine pairs of microtubules encircling two central microtubules (Figure 3-13*c*). This same 9 + 2 array is found in the cilia and flagella of virtually every eukaryotic organism, from fungi to humans, another of many reminders of the unity of life.

CELL WALLS

In the beginning of this chapter, we described how Robert Hooke was the first person to have observed a cell. In actual fact, the compartments Hooke described were the empty cell walls of dead cork tissue which had originally been produced by the living cells they once surrounded. **Cell walls** of one type or another are present in bacteria, fungi, and plants, but are absent in animals. These walls form a rigid outer casing that provides support, slows dehydration, and prevents a cell from bursting when internal pressure builds due to an influx of water. Plant cell walls (Figure 3-14) consist of a nonstructured polysaccharide matrix in which tough cables (*microfibrils*) of cellulose are embedded. The matrix resists forces that might crush the enclosed cell, while the cellulose cables act to resist forces that might pull the cell apart.

(a)

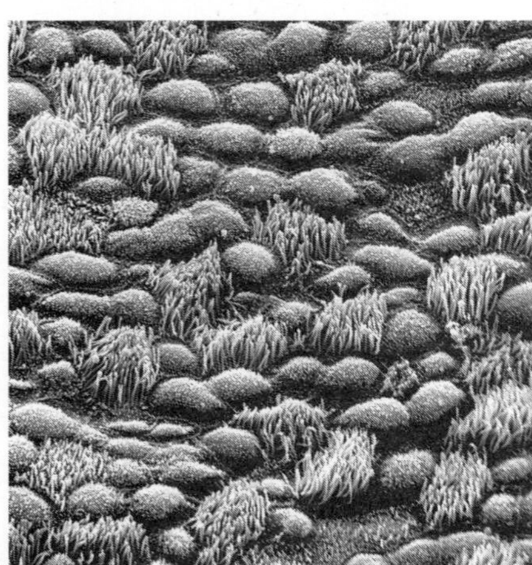

(b)

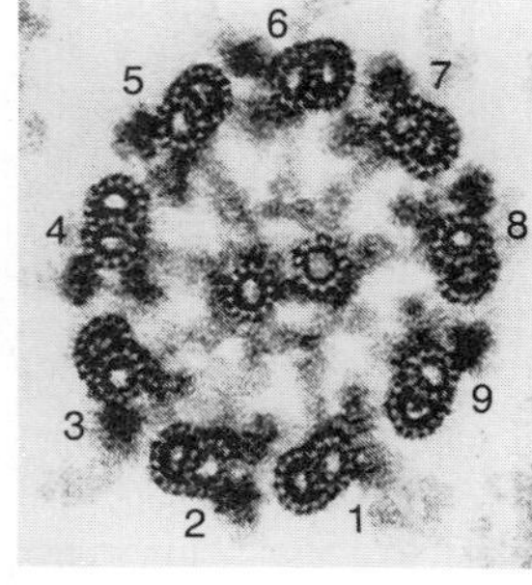

(c)

Figure 3-13

Cilia. ***(a)*** The numerous cilia that cover the external surface of this protist, *Paramecium*, propel the cell through the water while creating currents that channel food into its "mouth." ***(b)*** The coordinated beating of the cilia lining the mammalian trachea moves mucus and its trapped dust particles out of the respiratory tract into the throat. ***(c)*** Cross section through a single cilium showing the 9 + 2 array of microtubules.

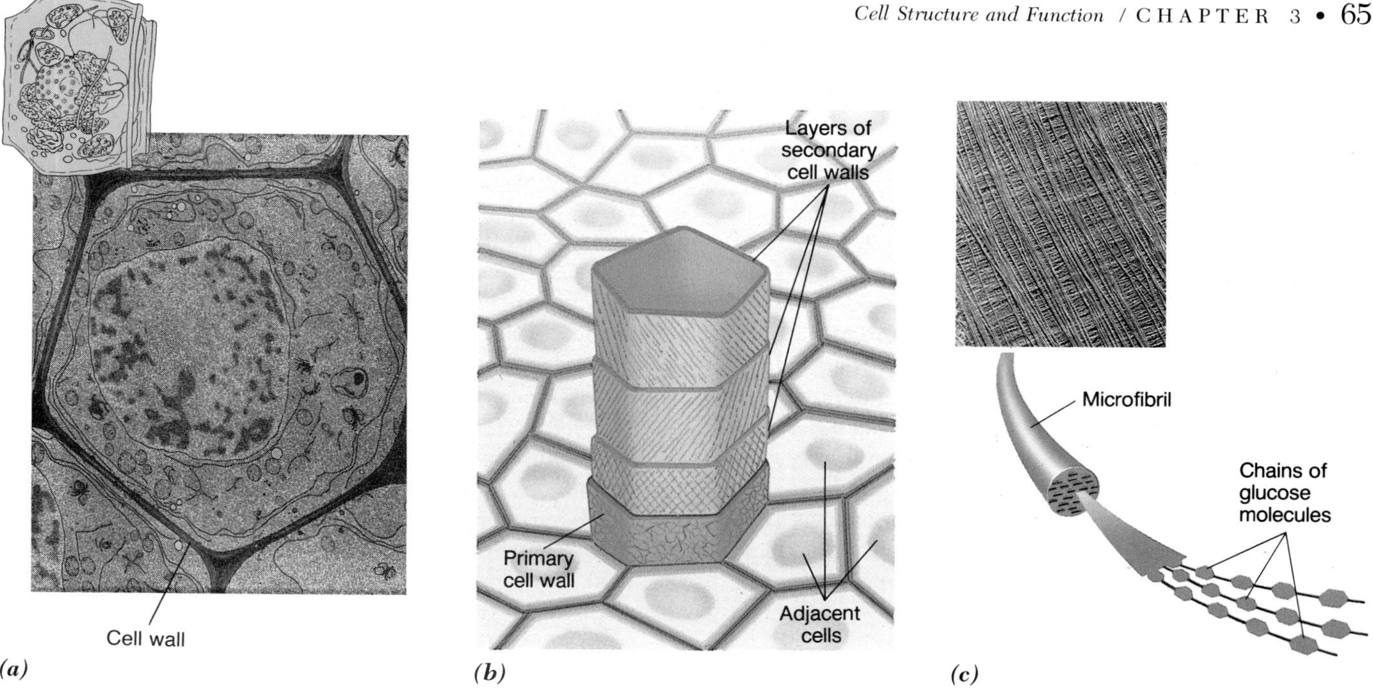

Figure 3-14
Plant cell walls. *(a)* Electron micrograph of a plant cell surrounded by its cell wall. ***(b)*** A diagrammatic plant cell wall, telescoped to reveal the primary cell wall, layers of the secondary cell wall, and the arrangement of cellulose microfibrils. ***(c)*** A surface view of the layers of parallel microfibrils in a cell wall. Each microfibril is a complex of cellulose chains, held together by hydrogen bonds between adjacent glucose molecules to form flat, sheetlike strips. Cellulose molecules are assembled from glucose monomers by proteins embedded in the plasma membrane. Plant cell walls provide support and resist external forces.

MOVEMENT OF MATERIALS ACROSS MEMBRANES

A cell's internal order is protected by the plasma membrane—a structure so thin it would take approximately 40,000 of them stacked one on top of the other to equal the thickness of one page of this book. The plasma membrane keeps essential substances inside the cell and regulates the passage of materials between a cell and its environment. Of the two major components of the plasma membrane—the lipid bilayer and the embedded proteins—it is the lipid that acts primarily as a barrier, and the proteins that act primarily as selective gates. Let us look more closely at this dual function of the membrane's major components.

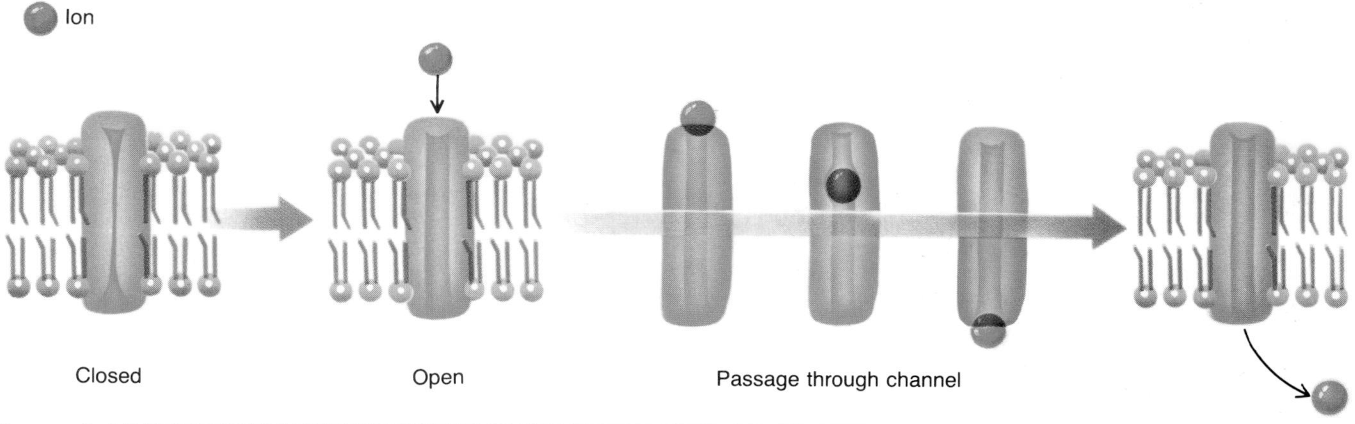

Figure 3-15
Membrane channels consist of pores situated within the center of membrane proteins or within clusters of such proteins. Some channels possess a "gate" that can exist in either an open or a closed conformation, depending upon conditions in the cell.

For the most part, those substances that are able to pass unimpeded, directly through the lipid bilayer have either a small size, such as water, or a hydrophobic (lipid soluble) structure, such as steroids. Most of a cell's solutes, such as its ions, sugars, amino acids, and phosphates, are not able to pass through the lipid bilayer, but, instead, must make their way through special protein-lined channels in the membrane (Figure 3-15). These protein channels are highly selective. Just as enzymes bind specific substrates, these membrane proteins bind specific solutes and promote their movement across the membrane. Together, the phospholipd bilayer and embedded proteins make the plasma membrane **selectively permeable**; that is, they allow certain substances to cross, while restricting the passage of others. Selective permeability enables a cell to import and accumulate essential molecules to concentrations high enough for normal metabolism. It also enables the cell to export wastes and other substances that might interfere with metabolism. There are four primary routes by which substances are able to pass across the plasma membrane and enter a cell: simple diffusion, facilitated diffusion, active transport, and endocytosis.

SIMPLE DIFFUSION

All dissolved ions and molecules exist in a state of continuous, random (that is, unpredictable) movement. As a result of this movement, ions or molecules tend to move from a region where they are present at higher concentration to a region where they are present at lower concentration; in other words, *down a concentration gradient*. This process is called **simple diffusion**.

Simple diffusion is readily demonstrated by dropping colored dye into a glass of water. The dye molecules are localized at first in one region, which appears dark in color. After time, however, the color spreads until the dye molecules are distributed rather evenly throughout the glass. A similar process occurs if a substance is present at different concentrations on two sides of a membrane. *As long as the membrane is permeable to the substance*, the molecules will move back and forth across the membrane until, eventually, the concentration of the substance on the two sides of the membrane will be equal. Simple diffusion plays a key role in the movement of many substances within the body. When you breathe, for example, oxygen moves by diffusion from the lungs into the bloodstream. Then, when the oxygen-rich blood moves through the tissues, oxygen again moves by diffusion out of the bloodstream and into the cells, where it is utilized.

Osmosis

The diffusion of water through a semipermeable membrane is of particular importance in biology and has a special name—**osmosis** (Figure 3-16). When cells are placed into solutions having a lower solute concentration than that of their own cytoplasm (a *hypotonic* solution), water tends to enter the cell, causing the cell to swell. In contrast, when cells are placed into solutions having a higher solute concentration (a *hypertonic* solution), water tends to leave the cell, causing the cell to shrink. It is only when the solute concentrations are the same (*isotonic*) inside and outside the cell that the volume of the cell remains the same.

Osmosis is a fact of life with which cells—and entire organisms—have to contend. For example, as organisms moved from the sea to inhabit fresh waters, they moved from an environment that was either isotonic or hypertonic to their own body fluids into one that was very hypotonic. These organisms evolved adaptations that allowed them to expel the excess water that flooded into their bodies. In contrast, plants are relatively unaffected by the osmotic uptake of water because they are surrounded by a rigid cell wall that resists expansion. Instead of causing the plant cell to swell, the osmotic influx of water causes the pressure inside the cell to build. Plant cells utilize the water pressure that develops due to osmosis to maintain their rigidity. The loss of water due to osmosis (or evaporation in a dry, terrestrial habitat) causes plants to lose their support and wilt.

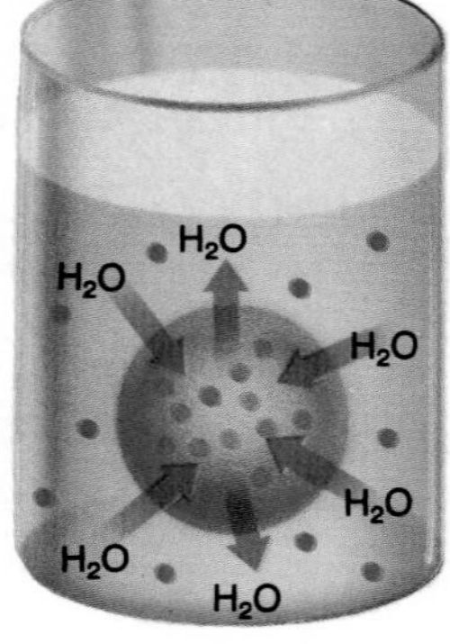

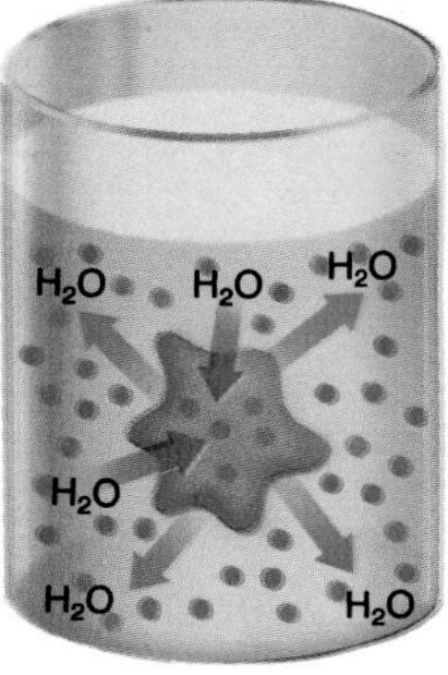

Figure 3-16 **The effects of hypotonic, hypertonic, and isotonic solutions on osmosis.** ***(a)*** A cell placed in a hypotonic solution (one having a lower solute concentration than the cell) swells because of a net gain of water by osmosis. ***(b)*** A cell in a hypertonic solution (one containing a higher solute concentration than the cell) soon shrinks because of a net loss of water by osmosis. ***(c)*** A cell placed in an isotonic solution (one containing the same solute concentration as the cell) neither swells nor shrinks because it gains and loses equal amounts of water.

◁ THE HUMAN PERSPECTIVE ▷

LDL Cholesterol, Endocytosis, and Heart Disease

Cholesterol is a relatively small, hydrophobic molecule (see Figure 2-21*a*) needed by virtually all human cells as a component of their plasma membranes. While many cells synthesize their own cholesterol, they are able to remove additional cholesterol from the blood. Because cholesterol, and other lipids, are so insoluble, they must be carried through the blood in special envelopes, called **lipoproteins**. Cells take up lipoproteins by endocytosis. To do this, the plasma membrane of the cell contains special receptors on its surface that bind the lipoproteins and initiate their uptake (see Figure 3-18*b*).

Cholesterol is carried primarily by two different types of lipoproteins, called *low-density lipoproteins (LDLs)* and *high-density lipoproteins (HDLs)*. Both types of lipoproteins are spheres that contain a core of cholesterol molecules packaged in a sac of phospholipids and protein that keeps the enclosed cholesterol suspended in the blood. HDLs and LDLs have very different roles in the body and very different effects on human health. LDL serves primarily to carry cholesterol molecules from the liver, where they are packaged, through the blood to the body's cells, where they are taken up via *LDL receptors* on the cells' surfaces. HDL, in contrast, carries cholesterol in the opposite direction—from the body's cells to the liver, where it is taken up by endocytosis and excreted as part of the bile.

LDL has received a great deal of attention in the past decade because high blood levels of this lipoprotein are associated with the development of *atherosclerosis*, a serious condition characterized by the narrowing of arteries due to accumulation of cholesterol-containing plaque along the inner arterial wall (Figure 1). Most seriously affected by atherosclerosis are the coronary arteries which supply blood to the heart. In addition to decreasing the flow of blood through the arteries, the presence of plaque in the coronary arteries acts as a site for the formation of blood clots which totally block blood flow, triggering a heart attack and heart-tissue damage.

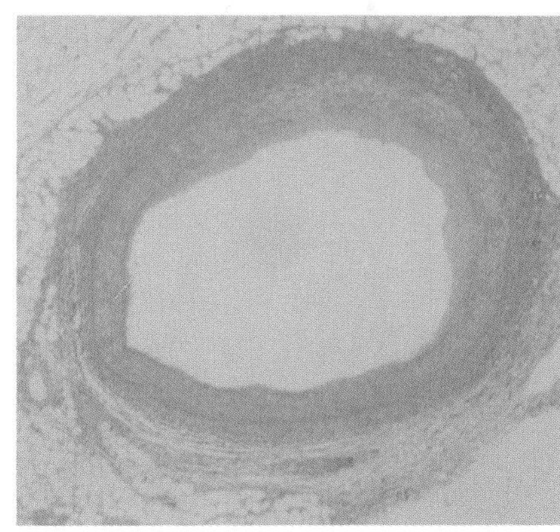

(a)

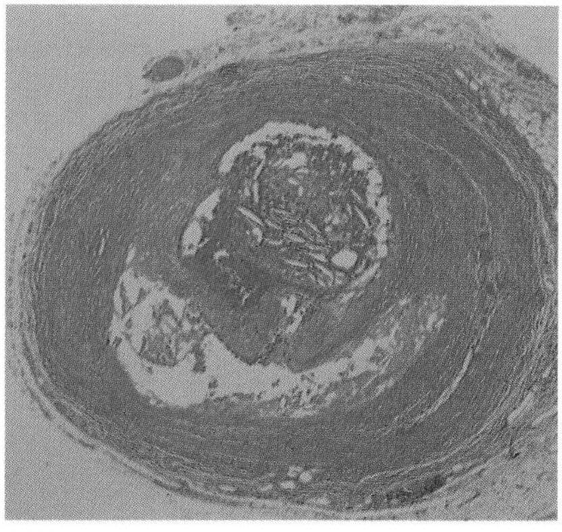

(b)

FIGURE 1

Comparison of a normal coronary artery ***(a)*** and an artery whose channel is almost completely closed as the result of atherosclerosis ***(b)***.

Persons with high levels of LDL have more cholesterol circulating in their bloodstreams than their cells can use and, consequently, some of the blood-borne cholesterol becomes deposited on arterial walls. High LDL levels, in turn, may depend on the numbers of LDL receptors present on the surfaces of cells; the greater the number of receptors, the more LDL is removed from the blood, and the less cholesterol is left to stick to the walls of arteries. This relationship is supported by the existence of a rare disease *familial hypercholesterolemia* whose victims have defective LDL receptors. Individuals with this disorder have extraordinarily high levels of blood cholesterol and typically suffer heart attacks before the age of 20.

Just as high blood levels of LDL are associated with increased risk of heart disease, high blood levels of HDL are associated with decreased risk. HDL has this effect because it facilitates removal of cholesterol from the blood by the liver. Because of their seemingly opposite effects on health, LDL has become known as the "bad" cholesterol and HDL as the "good" cholesterol. One approach that many clinicians use to assess a person's blood-cholesterol profile is the ratio of total serum cholesterol to HDL cholesterol. Current guidelines suggest that it is desirable to maintain total cholesterol below 200 milligrams/deciliter (generally given simply as a value of 200) and HDL cholesterol *above* 50 milligrams/deciliters. Thus, a ratio of 200/50, or 4.0 would predict that an individual has a healthy blood-lipid composition. It is estimated that a drop of 1 unit in this ratio may translate into a 50 percent decrease in the risk of cardiovascular disease. Studies suggest that the best way to raise one's HDL level is to exercise regularly, a habit that also tends to lower total blood cholesterol levels.

FACILITATED DIFFUSION

While diffusion of a substance across a membrane always occurs from a region of higher concentration to a region of lower concentration, the penetrating molecules do not always move through the membrane unaided. Many plasma membranes contain *carrier proteins* that bind specific substances and facilitate their diffusion across the membrane. This process is called *facilitated diffusion* and is particularly common in the inward movement of sugars and amino acids, substances that would be unable to penetrate the hydrophobic lipid bilayer directly.

ACTIVE TRANSPORT

A frog sitting in a shallow freshwater pond has internal fluids with a solute concentration 40 to 50 times greater than its aqueous environment, which is not much different from pure water. How is this frog able to maintain a salt concentration that is so much higher than its surrounding medium? It can do so because its skin contains proteins capable of "pumping" salts from the environment into the body *against* a very large concentration gradient (Figure 3-17). Movement of substances against a concentration gradient—that is, from a region of lower concentration to one of higher concentration—is called **active transport**. This process requires the input of considerable energy, which is either directly or indirectly supplied by the hydrolysis of ATP.

One of the most widely occurring active transport proteins is the *sodium-potassium pump*. This protein transports sodium ions out of cells and potassium ions into cells, both against steep concentration gradients. Human nerve cells, for example, have internal concentrations of potassium that are 30 times greater than that of the extracellular fluid, due to the pumping activities of this protein. As we will see in Chapter 15, these ionic gradients are utilized in the transmission of nerve impulses that travel along the membrane of nerve cells.

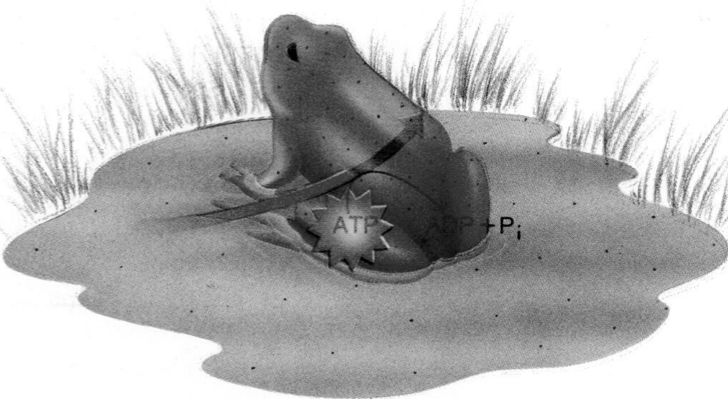

Figure 3-17
An example of active transport. Animals that live in freshwater habitats tend to lose salts from their body to their environment by diffusion. Some of this salt is regained in their diet, but most freshwater animals, including this frog, possess mechanisms to take back salts from their environment against a concentration gradient. This process requires the input of energy and is called *active transport*. Active-transport proteins in this frog are located in the plasma membranes of its skin cells.

ENDOCYTOSIS

Both diffusion and active transport can move smaller-sized solutes directly through the plasma membrane, but what of the uptake of materials that are too large to penetrate a membrane, regardless of its nature? We saw on page 60 how materials stored in cytoplasmic vesicles can be released from cells by exocytosis. Cells can also carry out a reverse process, called **endocytosis**, in which large-molecular-weight materials are enclosed within invaginations of the plasma membrane, which subsequently pinch off to form cytoplasmic vesicles. There are two forms of endocytosis:

1. **phagocytosis**, in which a cell ingests large particles, such as bacteria or pieces of debris, and
2. **pinocytosis**, in which a cell ingests liquid and/or dissolved solutes and small suspended particles.

Both types of endocytosis play an important role in human biology. For example, certain white blood cells are capable of leaving the bloodstream and wandering through the tissues, ingesting "foreign" materials by phagocytosis (Figure 3-18*a*). Similarly, many cells take up dissolved solutes or suspended particles by pinocytosis. Many of the materials carried through the blood are taken into cells by endocytosis. One of the most important of these is cholesterol (Figure 3-18*b*), which is the subject of The Human Perspective: LDL Cholesterol, Endocytosis, and Heart Disease.

AN OVERVIEW OF CELL STRUCTURE AND FUNCTION

In this chapter, we have dissected the cell into its most obvious components and have described the most salient features of each. In a real sense, however, a cell is greater than the sum total of its parts; more than what appears visible in an electron micrograph (see Figure 3-5*a*). An electron micrograph of a cell is like a still picture of a ballet; it fails to convey the dynamic nature of the subject being examined. A cell is a bustling center of activity, with materials moving from one part to another, organelles changing shape and distribution, and structures being disassembled and reassembled. While it is easy to discuss the function of each organelle, one after another, it is difficult to describe their interdependent activities. Each organelle depends on the continuing function of other cellular components; they do not function in isolation. For example, a

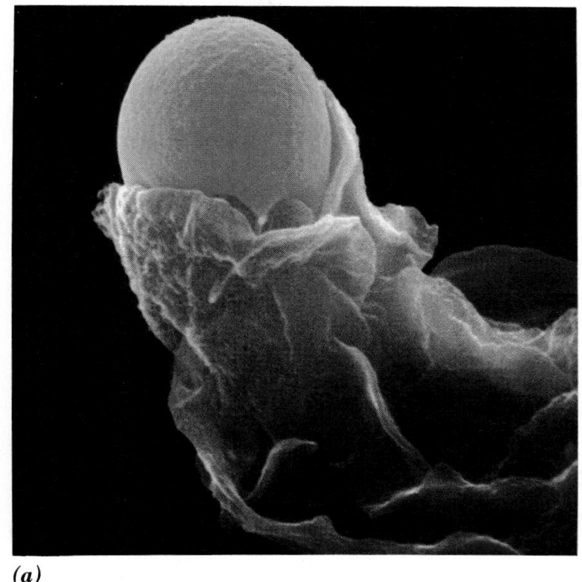

(a)

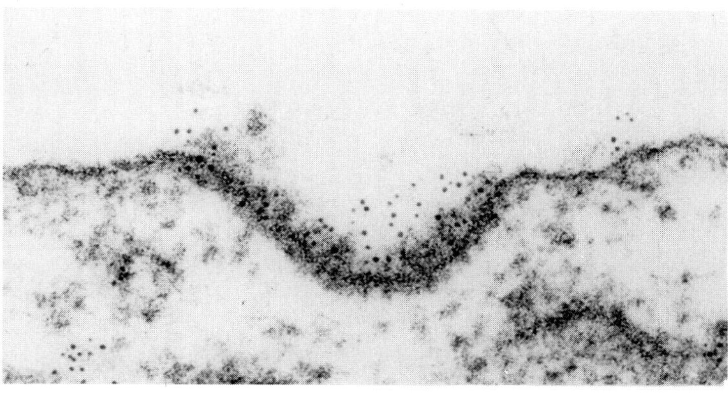

(b)

Figure 3-18
Endocytosis. ***(a)*** Cell ingesting a synthetic particle. ***(b)*** LDL particles (indicated by black dots) bind to receptors situated in pits in the plasma membrane. Following binding to the LDL receptor, the LDL particles are taken into the cell as part of a vesicle.

mitochondrion depends on energy-rich substrates (fats and polysaccharides) provided by its surrounding and on proteins that are synthesized on cytoplasmic ribosomes using instructions from genes that reside in the nucleus. The cell's membranous organelles form a continuum, both in time and space. Rough ER may be formed from membranes that are budded from the nuclear envelope; lysosomes form from the Golgi complex; plasma membrane is formed by the fusion of cytoplasmic vesicles during exocytosis and, conversely, is an important contributor via endocytosis to the formation of internal cytoplasmic structure. As is illustrated by a number of genetic disorders, disruption of one activity in a cell often affects the smooth functioning of the entire cell and, in many cases, can lead to the death of the entire organism.

EVOLUTION AND ADAPTATION: TYING IT TOGETHER

All living cells—from the simplest bacterium to the most highly complex nerve or muscle cell—share a wealth of common features, including a plasma membrane of similar construction, a common mechanism for the storage and expression of genetic information, and a comparable set of metabolic pathways. All cells possess these shared properties because all have evolved from a common ancestral cell. In other words, all of the cells present on earth today can, theoretically, be traced back through a series of countless cell divisions to a single primitive prokaryotic cell that lived on earth nearly 5 billion years ago. As difficult as this may be to comprehend, there is no other explanation to account for the basic similarities of life at the cellular level; they could not have evolved independently.

Eukaryotic cells evolved about 2 billion years ago, presumably from simpler, prokaryotic ancestors. With specialized cytoplasmic organelles and a nucleus that sequestered genetic material, eukaryotic cells were more highly organized than were prokaryotes and were capable of conducting more complex functions. One of the questions that has intrigued biologists for decades is how the transition from the prokaryotic to eukaryotic state took place. Since there are no organisms living today that are intermediate in complexity between prokaryotes and eukaryotes, the question has been the subject of considerable speculation. Two modern hypotheses offer possible explanations.

The **membrane invagination hypothesis** (Figure 3-19*a*) proposes that the plasma membrane of the ancestral prokaryotic cell gradually folded in on itself, forming pockets, which pinched off from the surface membrane to form organelles in which particular enzymes or cellular materials could be concentrated. Over long periods of time, these simple membrane-bound compartments evolved into the highly specialized organelles found in modern eukaryotic cells.

It is easy to see how membrane invagination could have formed the endoplasmic reticulum, Golgi complex, and even the nuclear envelope. However, the evolution of complex DNA-containing organelles, such as mitochondria and chloroplasts, is usually explained by the **endosymbiosis hypothesis** (Figure 3-19*b*), which proposes that these organelles were originally derived from small prokaryotic cells that took up residence *inside* a larger cell. (The name of this hypothesis uses the word *symbiosis*, a

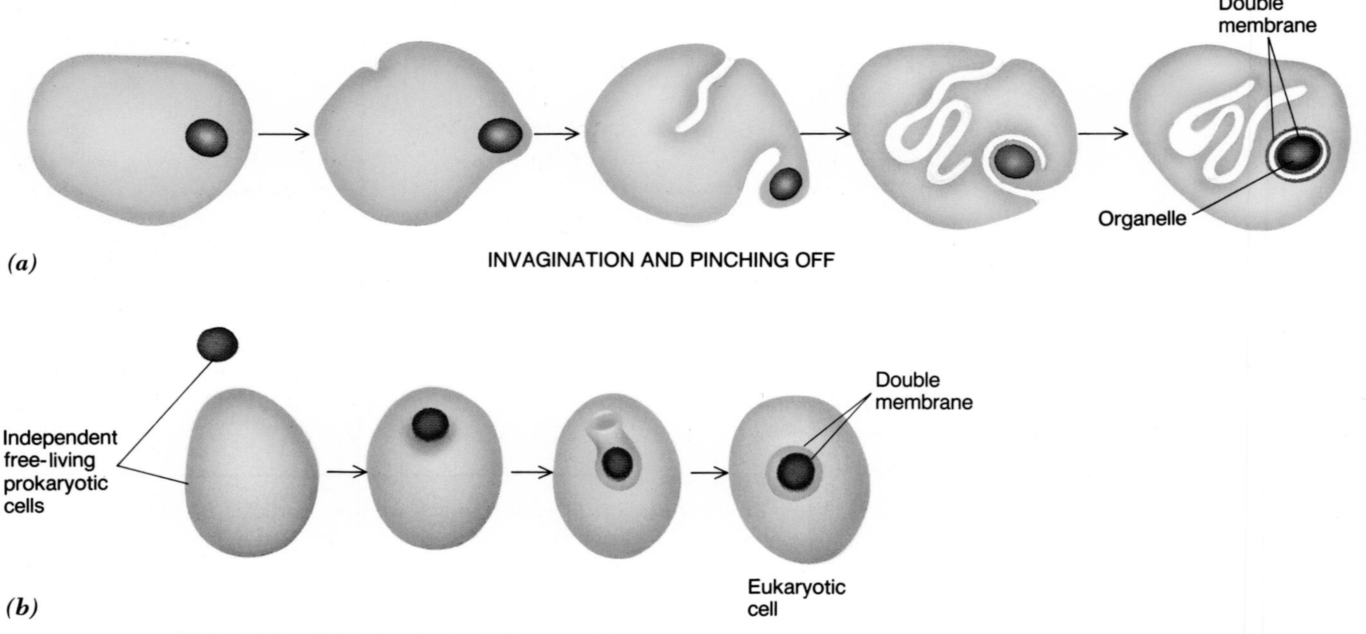

Figure 3-19

Proposals for the origin of the organelles of a eukaryotic cell. ***(a)*** Membrane invagination may have been responsible for generating many of the cytoplasmic membranous organelles and the nuclear envelope. ***(b)*** Endosymbiosis may have been responsible for the evolution of mitochondria and chloroplasts.

term that refers to a close association between different kinds of organisms.) For example, chloroplasts would have evolved from a small photosynthetic bacterium, and mitochondria from a nonphotosynthetic bacterium containing the machinery to use oxygen in the formation of ATP (Chapter 6). Included among the evidence used to support the endosymbiosis hypothesis are the following:

- Some bacteria form symbiotic partnerships with eukaryotic cells today. These bacteria resemble mitochondria in structure and carry out the same steps in oxygen utilization as do mitochondria.
- Mitochondria and chloroplasts contain their own nucleic acids and ribosomes, both of which resemble the types found in bacterial cells.

SYNOPSIS

The cell theory has three parts. (1) All organisms are composed of one or more cells; (2) the cell is the basic organizational unit of life; and (3) all cells arise from preexisting cells.

Cells are either prokaryotic or eukaryotic. Prokaryotic cells are found only among bacteria; all other organisms are composed of eukaryotic cells. Both types of cells are surrounded by a plasma membrane consisting of a viscous, fluid lipid bilayer and embedded proteins. Unlike prokaryotic cells, eukaryotic cells contain a true membrane-bound nucleus and a variety of distinct cytoplasmic organelles.

Eukaryotic cells contain a common collection of organelles. The plasma membrane serves as a barrier between the cell and the outside world and regulates exchanges between the two. The plasma membrane contains receptors that recognize and interact with specific substances in the external medium. The nucleus is bounded by a complex double-layered membrane and houses the chromosomes, which contain the genes. Genetic messages move into the cytoplasm through pores in the nuclear envelope. The cytoplasm contains an interconnected, interfunctional network of membranous organelles. Membrane proteins, secreted proteins, and the proteins of certain organelles are synthesized on ribosomes bound to the outer surface of the ER membranes, passed through the ER membrane to the internal space, and then sent off in vesicles to the Golgi complex. Once in the Golgi complex, materials are modified and then packaged for transport either to specific vesicles, such as lysosomes, or for discharge outside the cell during exocytosis. Lysosomes contain a variety of hydrolytic enzymes that digest macromolecules. Plant cells often contain large vacuoles that store various substances. Mitochondria are specialized for the transfer of energy from stored macromolecular fuels to ATP for use in the cell. Chloroplasts, which are not found in animal cells, are organelles in which light energy is captured and used to manufacture biochemicals during photosynthesis. The cytoskeleton consists of several distinct elements whose combined function maintains the shape of a cell, organizes its internal components, and provides the machinery necessary for cell movement. Plant cells are surrounded by complex cell walls that contain cellulose fibers embedded in a nonstructured matrix.

The plasma membrane directs the exchange of materials between the cell and the extracellular medium. There are several pathways by which substances enter a cell. These are (1) simple diffusion, which depends on the random movements of molecules across the membrane; (2) facilitated diffusion, which employs a membrane carrier protein to aid the diffusion process; (3) active transport, which moves substances against a concentration gradient at the expense of cellular energy; and (4) endocytosis, in which materials are enclosed within folds of the plasma membrane and enter the cytoplasm within vesicles.

Review Questions

1. List the similarities and differences between prokaryotic and eukaryotic cells; between plant and animal cells.
2. Trace the path of a secretory protein from synthesis to discharge from the cell.
3. Compare the structure and functions of chloroplasts and mitochondria.
4. Describe the role of the cytoskeleton in the life of a cell.
5. Why does water stop moving into a plant cell even if the cell remains hypertonic to its environment?

Critical Thinking Questions

1. Gorter and Grendel used human red blood cells for their determination of a cell's lipid content. Would their results have differed if they had used a more complex cell, such as those of a salivary gland or the pancreas? What do you think they would have been able to conclude about plasma membrane structure?
2. In 1674, Anton van Leeuwenhoek used a single glass lens (a "simple" microscope) to look at a drop of lake water. He discovered a new world, one of tiny organisms he called animalcules ("little animals"). Imagine you are van Leeuwenhoek. Write a letter to the Royal Society of London in which you try to convey your feelings about your momentous discovery.
3. Fossil evidence indicates that prokaryotic cells evolved before eukaryotic cells. If you did not know this, what evidence from the structure of these two main types of cells would lead you to the same conclusion?
4. If you were designing instruments to look for signs of life on Mars, would you include a search for a selectively permeable membrane, such as that found in living cells on Earth? Adopt a position on this question and defend it.
5. For each case given below, indicate whether it illustrates simple diffusion, facilitated diffusion, or active transport.
   (1) Some marine algae contain much higher concentrations of iodine than does the surrounding sea water.
   (2) A single-celled freshwater organism is hypertonic to the surrounding water, and water moves into the organism.
   (3) Sugars generated by digestion in the small intestine are able to enter the cells that line the epithelium even though they are highly insoluble in lipids. The movement of these sugars is from higher concentration in the lumen to lower concentration in the cell cytoplasm.

CHAPTER 4

The Metabolism of Life

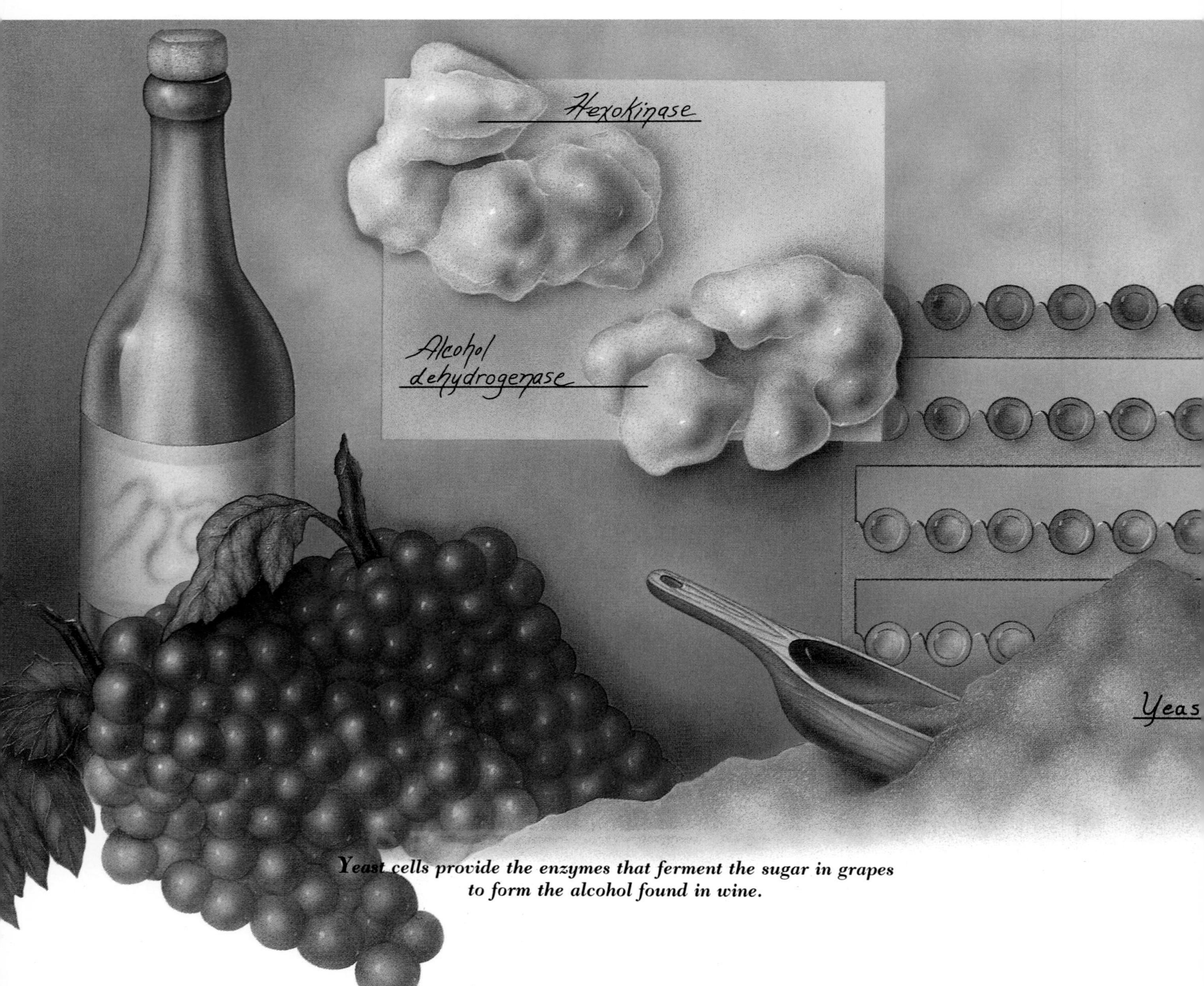

Yeast cells provide the enzymes that ferment the sugar in grapes to form the alcohol found in wine.

STEPS TO DISCOVERY
The Chemical Nature of Enzymes

In 1779, the French Academy of Science offered 1 kilogram (2.2 pounds) of gold to anyone who could unravel the mystery by which sugars present in grape juice are converted to alcohol during the formation of wine. The prize was never collected.

A hundred years later, two conflicting explanations for the nature of alcohol formation (a process known as *fermentation*) prevailed. On one side of the argument were the biologists, led by Louis Pasteur, a biologist working on behalf of the French wine industry. Pasteur correctly believed that the conversion of sugars to alcohol was a process carried out by living yeast cells. In fact, he applied this scientific fact to the faltering French wine industry, thereby restoring it to its previous station as the best winemaker in the world, simply by manipulating the organisms used for fermentation. In adopting this position, however, Pasteur hypothesized that fermentation required a "vital force" that could only be supplied by an intact, highly organized, living organism. Pasteur rejected the notion that life processes, such as fermentation, could be reduced to simple chemical reactions.

On the other side of the argument were the organic chemists of the period, including the aggressive debater Justus von Liebig, who ridiculed the suggestion that yeast cells were responsible for fermentation. Liebig and his fellow chemists proposed that fermentation was simply an organic reaction that occurred on its own, no different from those reactions they had been studying in the test tube. The battle lines were drawn, biologists versus chemists.

Then came Eduard Buchner. Buchner was both a chemist—teaching chemistry at a German university—and a biologist—working in his brother's bacteriology laboratory. In 1897, 2 years after Pasteur's death, Buchner was preparing "yeast juice"—an extract made by grinding yeast cells with sand grains and then filtering the mixture through filter paper. Buchner was planning to use this "yeast juice" in a series of medical studies, but he wanted to preserve it for later work. After trying to preserve the extract with antiseptics and failing, Buchner attempted to protect his cell juice from spoilage by adding sugar, the same procedure used to preserve jams and jellies. Instead of preserving the solution, the sugar produced gas, which bubbled continuously for days. Instead of discarding the solution and searching for another way to maintain the yeast juice, Buchner's scientific curiosity was aroused by this unexpected occurrence. Analysis revealed that fermentation was occurring, producing alcohol and bubbles of carbon dioxide. All this was taking place without a single living yeast cell in the "soup," a severe blow to Pasteur's hypothesis that fermentation required living organisms.

Buchner had accidentally discovered that chemical agents (later identified as *enzymes*) promote biological reactions, and can do so *outside* of the cells in which they were originally produced. Investigators could now purify enzymes and study their activities without interference from other cellular ingredients. The door to modern biochemistry had swung wide open, raising the question: What was the nature of these remarkable molecular mediators?

One prominent scientist to turn his attention to the study of enzymes was the German chemist Richard Willstater, who had earlier won a Nobel Prize for his work on the structure of chlorophyll, the central plant pigment in photosynthesis. Willstater set out to purify plant enzymes but found that his most active preparations contained so little material they could not be chemically identified. Today, we understand that enzymes are so efficient that they are active at extremely low concentrations—too low to be characterized using 1920s methods. To Willstater, it was bewildering. Because he was unable to detect any protein in the purified enzyme preparations, Willstater erroneously concluded that enzymes could not be made of protein.

Then, in 1926, James Sumner, an American biochemist, prepared the first crystals of an enzyme (urease) which he had purified from the seeds of a tropical plant; he determined that the crystals were indeed composed of protein. He also demonstrated that protein-denaturing treatments destroyed the preparation's enzymatic activity and concluded that the enzyme was a protein. Sumner's finding was not greeted with much acclaim. Willstater argued that the crystals may have been partly composed of protein but that the enzyme was only a minor contaminant of the preparation and it was not being detected in the chemical analysis. A few years later, however, another American, John Northrop, crystallized a number of different enzymes (including pepsin, the digestive enzyme of the stomach) and showed conclusively that all were made of protein. As a result of their work, Buchner won the Nobel Prize in chemistry in 1907, while Sumner and Northrop shared the same prize in 1946.

■

Every organism is a precisely coordinated collection of chemicals, each of which is created through biochemical reactions. Every detail of your body, from the hair on your head to the nails on your toes, is the product of such reactions. Your eyes, for example, owe their color to pigments manufactured by metabolic activity. The chemical changes that construct these pigments don't happen automatically or haphazardly; they require the tight supervision provided by enzymes, which may be thought of as "metabolic traffic directors." Most series of reactions—whether they lead to the production of pigment or of some other material—also require a source of chemical energy. The topic of energy is a good starting point in a discussion of **metabolism,** a term that describes all of the chemical reactions that occur in a living organism.

▼ ▼ ▼

(a)

(b)

FIGURE 4-1

Biological transfers of energy. Animals that feed on plants, such as a hare ***(a)***, utilize the chemical energy in starch to fuel their diverse energy-requiring activities. Plant-eating organisms (herbivores), such as the hare, serve as energy sources for other animals, such as this lynx ***(b)***. In each case, however, most of the stored energy in the food matter escapes back into the environment as unusable energy.

ACQUIRING AND USING ENERGY

A living cell bustles with activity. Materials constantly enter and exit the cell; genetic instructions flow from the nucleus to the cytoplasm, where proteins and other substances are synthesized or degraded. To maintain such a high level of activity a cell needs energy. For most of us, the word "energy" conjures up a rather vague concept of being energetic, such as playing baseball, feeling enthusiastic, or making an effort. For scientists, **energy** has a more precise definition: It is the capacity to do work; that is, the capacity to change or move something. There are several different forms of energy. Chemical energy can change the structure of molecules; mechanical energy can move objects; light energy can boost electrons to an outer shell; thermal energy (heat) can increase the motion of molecules; and electrical energy can move electrically charged particles.

THE LAWS OF THERMODYNAMICS

Much of what we know about energy is summarized in two basic laws of nature—the laws of thermodynamics.

- *The first law of thermodynamics* states that energy can neither be created nor destroyed. Energy can be converted from one form to another, however. For example, electrical energy is converted to mechanical energy when we plug in a clock. Living organisms are also capable of energy conversion. Green plants perform *photosynthesis,* a process described in Chapter 5 by which solar energy is converted into chemical energy that is stored in organic chemicals, such as starch or cellulose.
- *The second law of thermodynamics* expresses the concept that nature is "wasteful." Any time energy is exchanged, some *usable* energy is inevitably lost. In other words, the energy *available to perform additional work* decreases. For example, when a hare browses on the leaves of a tree (Figure 4-1*a*), or a lynx preys on the hare (Figure 4-1*b*), most of the chemical energy in the food is inevitably lost to the animal having the meal. This lost or unusable energy has a special name, *entropy.*

THE CONCEPT OF ENTROPY

Entropy can be measured in terms of an increased state of randomness or disorder in the universe, which is often a result of the release of heat. Entropy can be illustrated by innumerable familiar activities. Suppose you were to walk into a library and search for books by three of your favorite authors. You find the books, take them to a desk, browse through them, and leave them on a table. You have just increased the entropy of the universe. Prior to your visit, the books were in an ordered location; there is only one proper place for each book in the library. In contrast, after you left the library, the books were left on a table that was selected essentially at random; you could just as well have left them in any one of a hundred or more places. As a result of your activities, the library has become more disordered; that is, its entropy has increased. Energy must be expended to reverse the effects of entropy. A librarian, for example, must expend energy to replace the books in their proper locations. Another example of entropy is shown in Figure 4-2.

If every event increases the disorder in the universe, how is it possible for organisms to maintain their high degree of order? The fact that entropy in the universe always increases does not mean that *every part* of the universe has to become more and more disordered. On the contrary, each organism represents a temporary departure from the relentless march toward disorganization. Maintaining a state of low entropy requires the input of energy, which lowers the energy content of the remainder of the universe. Consider just one molecule of DNA located in one cell in your liver. That cell has dozens of different enzymes whose sole job it is to patrol the DNA, looking for damage and repairing it. Without this expenditure of energy, the ordered arrangement of nucleotides in the cells of your body would literally disappear overnight. Energy is expended to maintain order at every level of biological organization—from molecules to ecosystems.

THE USE OF ATP TO DRIVE ENERGY-REQUIRING REACTIONS

The key compound of energy metabolism in all cells is *adenosine triphosphate,* or *ATP* (Figure 4-3). ATP is a nucleotide (Chapter 2); it is made up of a sugar (ribose), a nitrogenous base (adenine), and three phosphate groups linked to one another. ATP is the molecule in which all cells temporarily store the chemical energy used to run virtually all their activities. The amount of ATP present in a cell is surprisingly small; the entire human body has only enough ATP "on hand" to last about 20 seconds. Conse-

FIGURE 4-2

A victim of entropy. A house left on its own for a number of years will gradually fall into a state of disorder, increasing the entropy of the universe. All organized structures require a constant input of energy to maintain their complexity and order; otherwise, they slowly deteriorate. The same is true for organisms; without the input of energy, they die and decompose.

quently, ATP supplies must be continually replenished from the energy stored in large, energy-rich organic molecules, particularly polysaccharides and fats. A cell's polysaccharides and fats can be likened to money banked in a savings account, while its ATP supply can be thought of as money in the cell's "pocket," available to be "spent" on its needs at that very instant.

The breakdown (hydrolysis) of ATP

$$\text{ATP} + H_2O \rightarrow \text{ADP} + P_i \text{ (inorganic phosphate)}$$

is an energy-releasing reaction (Figure 4-3). Because it releases energy, the hydrolysis of ATP is a thermodynamically favored chemical reaction; that is, one that can occur spontaneously. Reactions of this type are described as *exergonic,* meaning "energy out." Unlike ATP hydrolysis, many chemical reactions that occur in a cell are not favored at all; they are thermodynamically unfavorable, or *endergonic,* meaning "energy in." An example of a thermodynamically unfavorable reaction is the synthesis of the amino acid glutamine

$$\text{glutamic acid} + \text{ammonia} \rightarrow \text{glutamine}$$

If the conversion of glutamic acid into glutamine cannot occur spontaneously, how does a cell produce this essential amino acid? The answer was revealed in 1947, when it was shown that the formation of glutamine is *coupled* to the hydrolysis of ATP. **Coupling** is accomplished, in this case, by combining the energy-requiring and energy-releasing processes into two new, thermodynamically favorable (energy-releasing) reactions.

First reaction:
glutamic acid + ATP → glutamyl phosphate + ADP
Second reaction:
glutamyl phosphate + ammonia → glutamine + P_i
Net reaction:
glutamic acid + **ATP** + ammonia → glutamine + **ADP** + $\mathbf{P_i}$

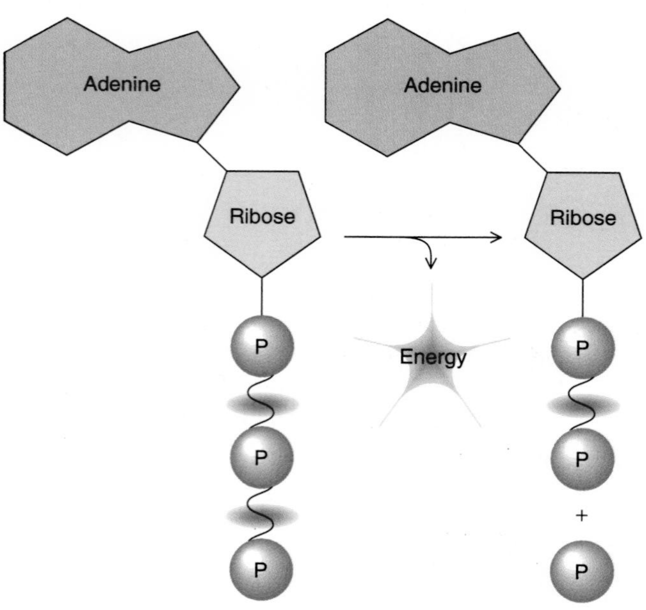

FIGURE 4-3

ATP. ATP is a nucleotide consisting of a sugar (ribose), a nitrogenous base (adenine), and three phosphate groups joined to one another. When hydrolyzed, the molecule is split into two products: ADP and inorganic phosphate (P_i). The squiggly lines connecting the second and third phosphate groups represent the bonds that can be broken when ATP is hydrolyzed.

ATP hydrolysis is used to drive virtually every energy-requiring process within the cell, including chemical reactions such as the formation of glutamine, the assembly of macromolecules (Figure 4-4), the concentration of ions across a membrane, and the contraction of a muscle cell.

ENZYMES

Left to themselves, biochemical reactions usually take place at extremely slow rates, even if the reactions are energy-releasing. Herein lies the value of enzymes. Enzymes are biological **catalysts**—substances that greatly increase the rate of *particular* chemical reactions. An enzyme may cause a reaction to proceed billions of times faster than it would otherwise occur. Consider what would happen, for example, if you were to add a teaspoon of glucose to a test tube, seal the tube, and sterilize the solution. The glucose would remain essentially intact for years. If you were to add bacteria to the solution, however, the sugar molecules would be taken into the cells and degraded in seconds by the enzymes of the bacteria.

Because of their powerful catalytic activity, enzymes are effective in small numbers. Even a tiny bacterial cell has space for hundreds of different enzymes, all catalyzing different reactions. Just as importantly, enzymes convert reactants into specific products needed by the cell. In contrast, when these same reactions are attempted in the laboratory by organic chemists without the benefit of enzymes, many different byproducts are usually produced. In a cell, formation of such unwanted byproducts would disrupt the progress of other reactions and kill the cell.

THE CATALYTIC ACTIVITY OF ENZYMES

In order for substances to undergo a chemical reaction, the reactant molecules must possess a certain amount of

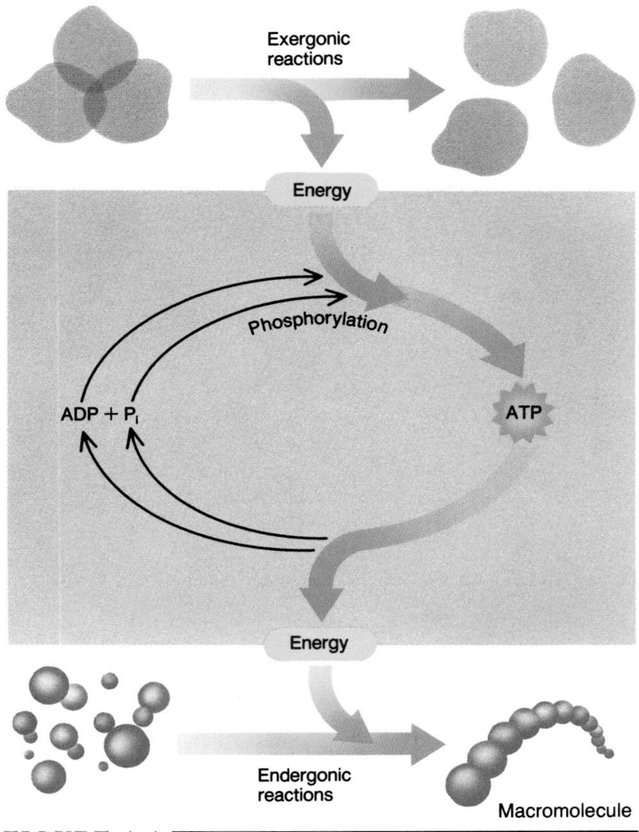

FIGURE 4-4

Producing and "spending" the cell's energy currency. Energy-releasing (exergonic) and energy-requiring (endergonic) processes are coupled by ATP, the common denominator in energy production and utilization. Catabolic pathways, such as glucose disassembly, generate ATP, whereas anabolic pathways, such as the assembly of macromolecules, utilize ATP.

energy, called **activation energy.** In Figure 4-5, the activation energy is represented by the height of the barrier. Enzymes act by lowering a reaction's activation energy; consequently, molecules are able to undergo a reaction in the presence of an enzyme when they possess less energy than would otherwise be required.

Enzymes lower activation energy by physically binding to the reacting molecules, called **substrates,** and facilitating the reaction. A simple analogy may help explain how enzymes are able to increase the rate of a reaction. Suppose you were to place a handful of nuts and bolts in a bag and shake the bag for 15 minutes. It is very unlikely that any of the bolts would end up with a nut firmly attached to its end when you stopped shaking the bag. In contrast, if you were to pick up a bolt in one hand and a nut in the other, you could guide the bolt into the nut very rapidly. Like the placement of a nut on a bolt, enzymes hold their substrates in a particular orientation that forces them together in the proper spatial relationship.

The area on the enzyme that binds the substrate(s) is called the **active site,** and it is often situated within a groove or cleft of the enzyme (Figure 4-6*a*). The active site and the substrate(s) have complementary shapes, enabling them to bind together with a high degree of precision, like pieces of a jigsaw puzzle (Figure 4-6*b*). The complementarity in shape between the active site of the enzyme and the substrate accounts for the high degree of *specificity* of enzyme activity. Specificity is a term used to describe the high degree of selectivity and precision with which biological molecules interact; it is a key ingredient

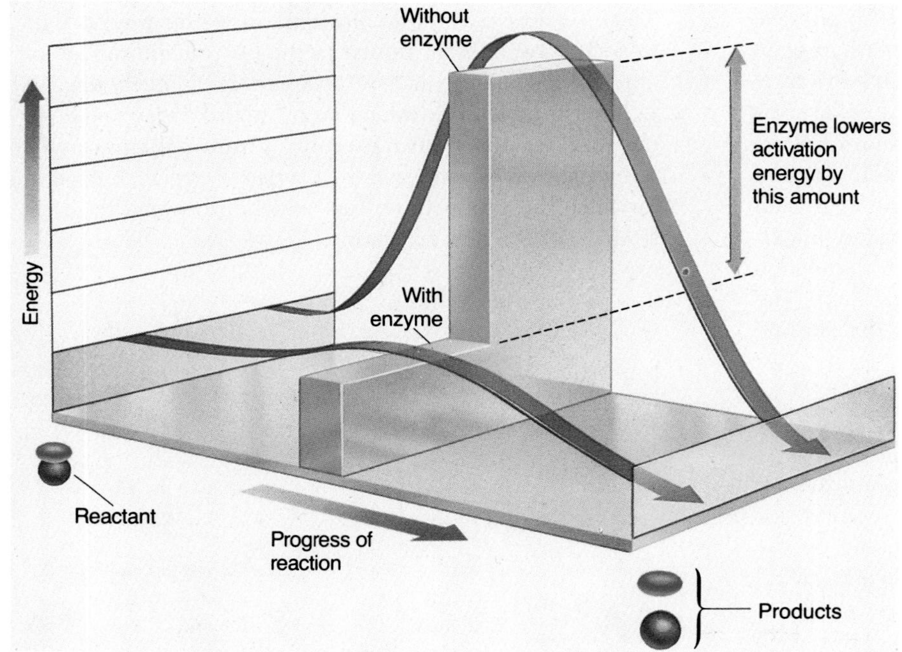

FIGURE 4-5

Activation energy and enzymes. Activation energy is the energy required to initiate a chemical reaction—to overcome the reactants' tendency to resist change. This energy barrier must be hurdled before a chemical reaction can occur. Enzymes reduce activation energy, so a reaction proceeds at a much faster rate than would occur in their absence.

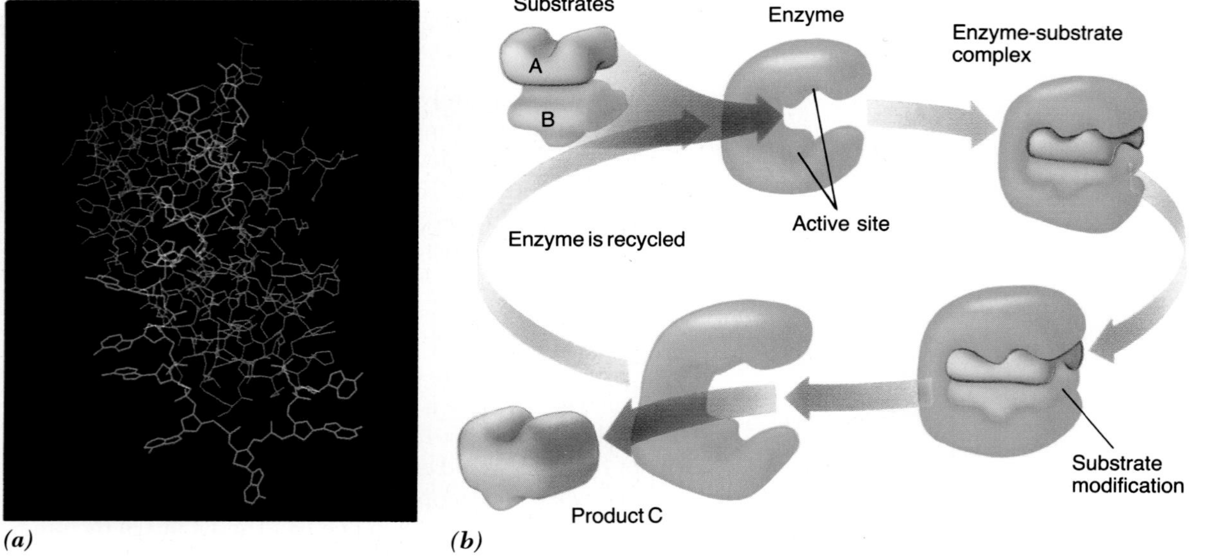

FIGURE 4-6

An enzyme in action. ***(a)*** The closeness of the enzyme–substrate fit is realistically revealed by this computer-generated version showing the RNA substrate (in green) being attacked and disassembled by the enzyme ribonuclease (in purple). ***(b)*** Substrates A and B are enzymatically altered to form product C. The reaction occurs following the binding of the substrates within the enzyme's active site. Enzymes are recycled—following completion of a reaction and the dispersal of the product, the enzyme binds a new pair of substrates and catalyzes another reaction.

in maintaining biological order. Regarding enzyme activity, for example, only the proper substrate can fit into the active site of an enzyme. As a result, the course of an enzymatic reaction is not affected by the hundreds of other types of molecules present in the cell at the time. However, the activity of an enzyme can be affected by adding a substance whose structure is very similar to that of a substrate (see The Human Perspective: Saving Lives by Inhibiting Enzymes).

The activity of enzymes is also affected by physical changes. For example, excessive heat disrupts hydrogen bonding and other forces that stabilize protein structure, causing the enzyme to change its shape and lose activity. Proper cooking destroys (*denatures*) the proteins of many disease-causing microbes residing in food, which is one of the reasons it is safer to eat cooked meats and eggs than raw ones. Enzyme activity is also altered by changes in pH. A rise or fall in H^+ concentration can change the charge of many of the amino acids of a protein, thus affecting the structure of the active site and the ability of the enzyme to bind a substrate and facilitate the reaction.

COFACTORS: NONPROTEIN HELPERS

Although enzymes are proteins, many enzymes utilize nonprotein helpers, called **cofactors,** which are required for the enzyme to carry out its function. Depending on the enzyme, the cofactor may be an organic molecule, called a *coenzyme,* or a metal atom. Cofactors typically carry out chemical activities for which amino acids are not well suited. The importance of coenzymes is illustrated by your daily need for vitamins. For the most part, these nutritional supplements act as essential coenzymes that you cannot synthesize for yourself. Similarly, most of the trace elements required in the diet, such as copper and manganese, serve as required components of enzymes.

METABOLIC PATHWAYS

The formation of complex biochemicals within a cell occurs in a stepwise fashion, similar to the way an automobile or appliance is built on an assembly line. Each biochemical is either assembled or dismantled by an orderly sequence of chemical reactions which comprise a **metabolic pathway.** Each pathway is catalyzed by a series of specific enzymes in which the product(s) of one reaction become the substrate(s) of the next reaction:

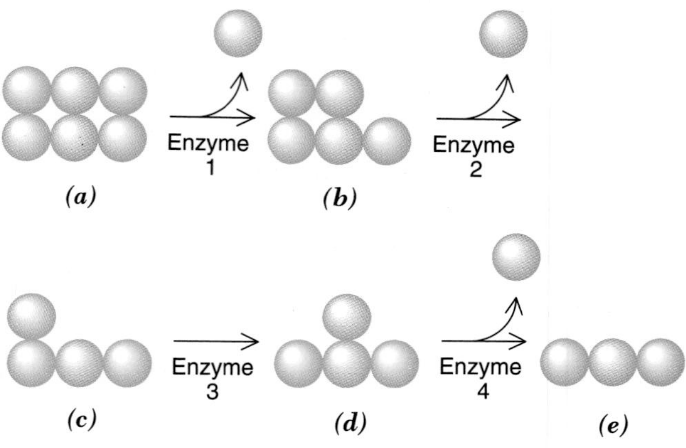

◁ THE HUMAN PERSPECTIVE ▷

Saving Lives by Inhibiting Enzymes

Many of the diseases that ravaged our ancestors are treatable today with chemicals that selectively block the action of enzymes present in disease-causing bacterial cells but are absent from human cells. Many of these substances, called *inhibitors,* are able to block the activity of the bacterial enzyme because their structure is similar enough to the enzyme's normal substrate that they can compete with the substrate in binding to the enzyme's active site (Figure 1*a*). In this state, the enzyme ceases to function.

Sulfa drugs—compounds that have saved countless human lives—provide an example of this type of *competitive inhibitor.* These agents are given to people to fight bacteria that cause diseases ranging from urinary bladder infections to pneumonia. The drugs block the ability of bacteria to transform a particular chemical in their cells—*para-aminobenzoic acid* (PABA)—to the essential coenzyme *folic acid.* The structural similarity between sulfa drugs and PABA (Figure 1*b*) creates metabolic confusion in the bacteria. Although many critical reactions in the human body also depend on folic acid, we cannot naturally manufacture this coenzyme. Instead, all of our folic acid is obtained as a vitamin in our diet. For this reason, we are not affected by the sulfa drugs in the same way as the bacteria, which must make their own folic acid. As a result, we survive the treatment unimpaired, but the bacteria do not.

Enzymes may also be inactivated by heavy metals, such as lead, silver, mercury, and arsenic. Unlike sulfa compounds and other antibiotics that compete with an enzyme's natural substrate, these poisonous substances bind nonspecifically to many proteins. Some of these inhibitors inactivate enzymes by altering their shape; others bind in place of required cofactors. Since these metals act as inhibitors of essential human enzymes, as well as those of bacterial cells, they are not useful for treating human diseases. Before the widespread use of penicillin during the 1940s, however, arsenic-containing solutions were commonly employed as "home remedies" and may have been responsible for the gradual poisoning of large numbers of people who used these drugs. The bacteria responsible for syphilis are particularly sensitive to an arsenic compound developed by the microbiologist Paul Ehrlich in 1910. The drug—which was the first compound discovered that would kill bacteria without also killing the human taking the drug—was the 606th arsenic compound that Ehrlich had methodically synthesized in the laboratory and tested.

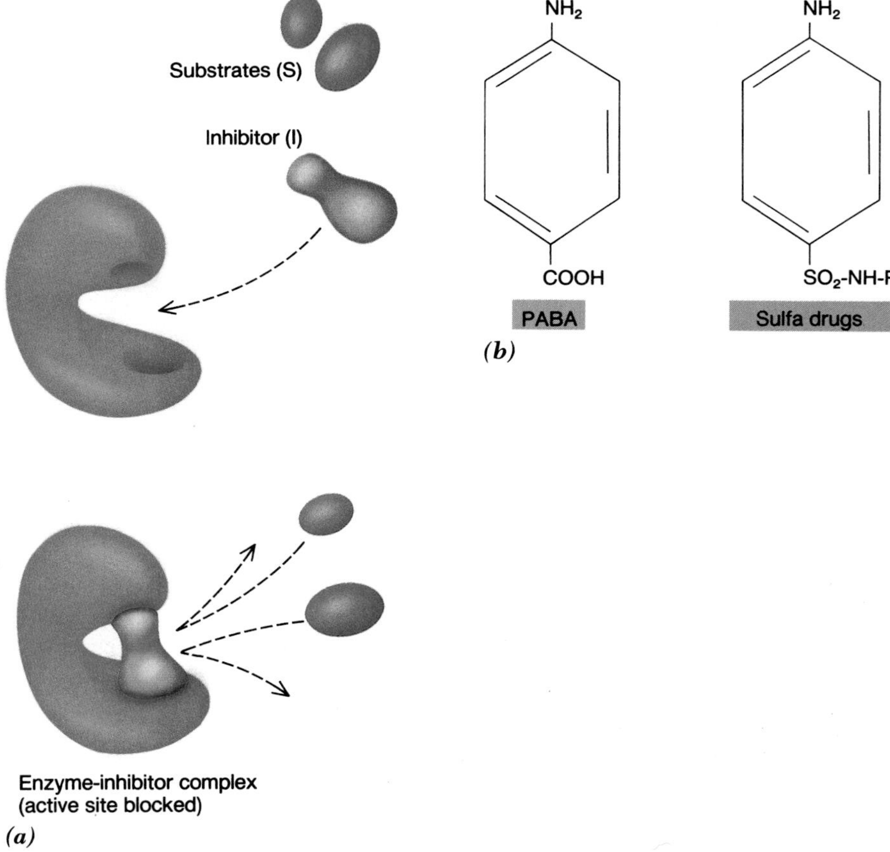

FIGURE 1

Competitive inhibition. ***(a)*** Competitive inhibition of enzyme activity is due to the structural similarity between the inhibitor and the substrate(s). The enzyme's active site recognizes both the inhibitor (I) and the substrate (S), but the inhibitior cannot be converted to products. ***(b)*** Due to its similar structure, the sulfa drug acts as a competitive inhibitor of PABA, the enzyme's normal substrate.

Metabolic pathways fall into two categories, depending on whether the products are more or less complex than the original substrates. *Catabolic pathways* degrade complex compounds into simpler molecules, releasing building materials and chemical energy. Therefore, catabolic pathways are energy-releasing. In contrast, *anabolic pathways* lead to the biosynthesis of complex molecules from simpler components. Since molecular construction requires energy, anabolic pathways are energy-requiring. Another key feature of both catabolic and anabolic reactions is the transfer of electrons, which brings us to the matter of oxidation and reduction.

OXIDATION AND REDUCTION: A MATTER OF ELECTRONS

Oxidation and reduction describe the relationship between atoms and electrons. When a carbon atom is covalently bonded to a hydrogen atom, the carbon atom has the strongest pull on the shared electrons. Because of its greater "possession" of electrons, carbon atoms bonded to hydrogen atoms are said to be in a *reduced state.* In contrast, if a carbon atom is bonded to a more electron-attracting atom, such as an oxygen or a nitrogen atom, the electrons are pulled away from the carbon atom; the carbon is said to be in an *oxidized state.* But the state of reduction or oxidation of a carbon atom is not an all-or-nothing condition. Since carbon has four outer-shell electrons it can share with other atoms, it can exist in various oxidation states. This is illustrated by a series of one-carbon molecules in Figure 4-7.

The oxidation state of an organic molecule is a measure of its energy content. The compounds that we use as chemical fuels to run our furnaces and automobiles are highly reduced organic molecules, such as natural gas (CH_4) and petroleum derivatives. Energy is released when these molecules are burned in the presence of oxygen, converting the carbons to more oxidized forms, primarily carbon dioxide (CO_2). The energy-rich molecules in a living organism are its carbohydrates and fats. Carbohydrates are rich in chemical energy because they contain strings of (H—C—OH) units. Fats contain even greater energy per unit weight because they contain strings of more reduced (H—C—H) units.

Catabolic pathways release the energy contained in carbohydrates and fats by removing hydrogen atoms from the substrate. These hydrogen atoms contain energy-rich electrons, which are used in the formation of ATP (Chapter 6). In contrast, anabolic pathways lead to the formation of more complex, reduced molecules by including reactions in which energy-rich electrons are added to the substrate. This feature can be illustrated by examining one of the key reactions of the anabolic pathway of photosynthesis. Don't be concerned with the chemical structures of the molecules in these reactions. Your attention is best directed to the red carbons and the transfer of the electrons indicated by the circles.

DPG ($O{=}C(OPO_3^{2-})$—H—C—OH—H—C—OPO_3^{2-}—H) + NADP(:H) → PGAL ($O{=}C$(:H)—H—C—OH—H—C—OPO_3^{2-}—H) + $NADP^+$ + PO_4^{3-} (P_i)

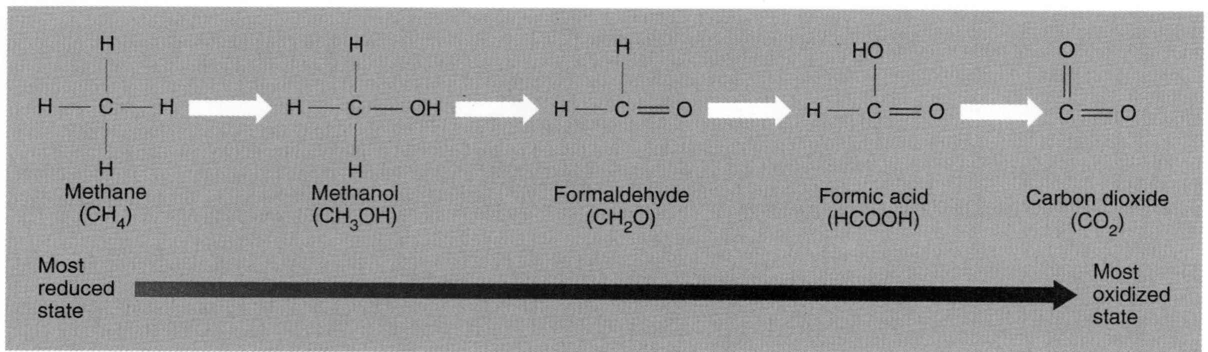

FIGURE 4-7

The oxidation state of a carbon atom. The oxidation state of a carbon atom depends on the other atoms to which it is bonded. Each carbon atom can form a maximum of four bonds with other atoms. This series of simple, one-carbon molecules illustrates the various oxidation states in which a carbon atom can exist, from its most reduced state in CH_4 to its most oxidized state in CO_2.

In this reaction, a pair of electrons (together with a proton) are transferred from NADPH (nicotinamide adenine dinucleotide phosphate) to the substrate DPG. NADPH is an important molecule in many anabolic reactions that occur in all types of organisms, from bacteria to humans. NADPH functions primarily as an electron donor, raising the state of reduction—and thus the energy level—of the substrate. A cell's reserve of NADPH is referred to as its *reducing power,* which is one measure of a cell's energy supply. In the reaction shown above, energy has been transferred in the form of electrons from the coenzyme NADPH to the substrate, leaving the substrate with a higher energy content.

DIRECTING METABOLIC TRAFFIC

Just as organisms must adapt to changes in external conditions, cells must respond to changes in internal conditions. One of the mechanisms by which cells maintain an ordered internal environment is by directing materials into the metabolic pathways that satisfy the cell's needs at that particular moment. Glucose, for example, can be directed along several different metabolic routes. It can be dismantled into carbon dioxide and water to generate energy; it can be linked to other glucose molecules to form a polysaccharide; it can be partially disassembled into fragments used to build lipids; or it can be modified to form amino acids or nucleotides. Even the simplest cell can accurately "evaluate" these needs and direct the fate of glucose (and the thousands of other molecules for which multiple options exist). How can cells "decide" which of these options best satisfies its needs? The answer is they can't, but there is no need for "decisions" because a cell has built-in mechanisms that regulate enzyme activities according to the cell's needs. Changing an enzyme's activity is usually accomplished by modifying the enzyme in a way that alters the shape of its active site. Two of the most common mechanisms for doing so are

1. **covalent modification,** whereby a small chemical group, such as a phosphate, is covalently added to the enzyme by another enzyme, and
2. **feedback inhibition,** whereby the end product of a metabolic pathway binds to a key enzyme in that pathway, decreasing its activity.

   A hypothetical example of feedback inhibition is shown in Figure 4-8.

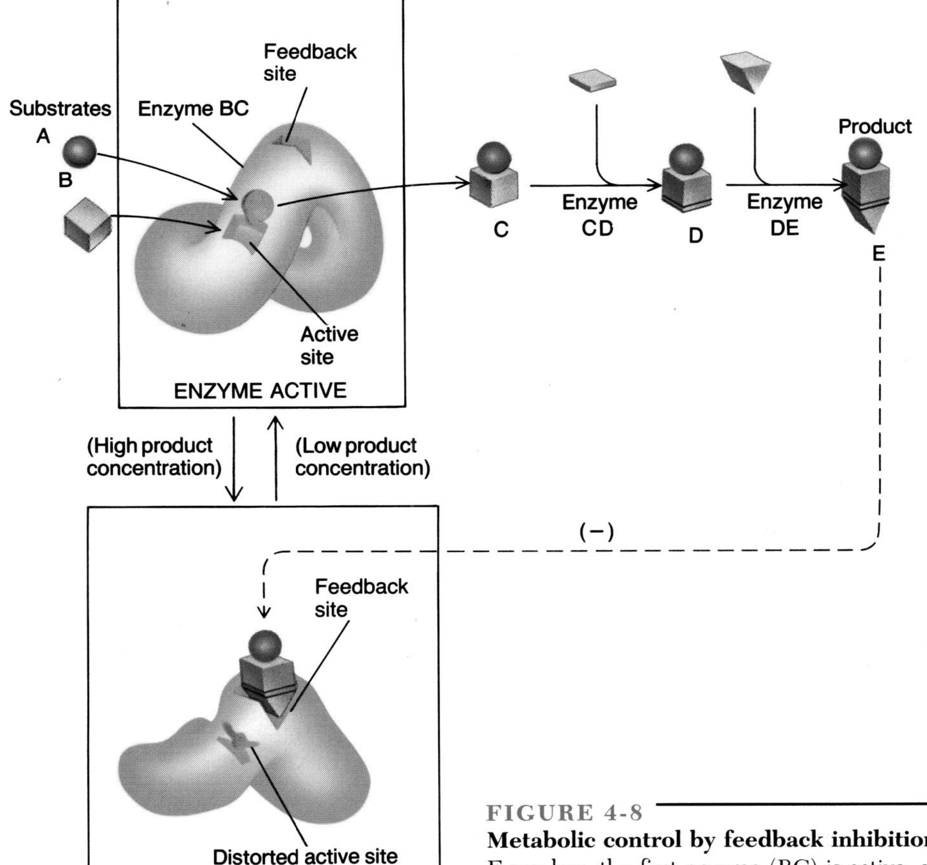

FIGURE 4-8
Metabolic control by feedback inhibition. When concentrations of the end product E are low, the first enzyme (BC) is active, and the metabolic pathway proceeds to completion. As the end product (E) accumulates, it binds to the enzyme's feedback site, changing the shape of the enzyme's active site, decreasing further production of the end product.

EVOLUTION AND ADAPTATION: TYING IT TOGETHER

While diverse organisms may be very different anatomically, they share a remarkable number of enzymes and metabolic pathways. For example, the same enzymes and metabolic pathways are utilized to disassemble glucose in bacteria, mushrooms, trees, and humans. It is evident that these enzymes arose at an early stage in biological evolution and have been retained for billions of years. Even though many of the same enzymes are found throughout the biological world, the molecular structures of these enzymes have undergone dramatic change over the course of evolution. This is not unexpected since, just as organisms are adapted to specific habitats, so too are enzymes.

"Fingerprints" left by evolution can be revealed by comparing the same enzyme in similar organisms that live under very different environmental conditions. Consider two species of bacteria, one living in a hot spring in Yellowstone Park at temperatures above 90°C, and another living in a nearby pond fed by cool spring water. The enzymes present in these two types of bacteria have their optimal activity at drastically different temperatures. For example, the purified, DNA-synthesizing enzyme from the bacterium that lives in the hot springs is active in the test tube at 90°C, while the corresponding enzyme from the other bacterium is totally inactive at this temperature. The increased temperature resistance of the enzymes of the hot-springs bacterium can be traced to structural features that increase the protein's stability. It would appear that as a species becomes adapted to life at a higher and higher temperature, those individuals with more temperature-resistant enzymes are selected for, while individuals whose enzymes are more sensitive to elevated temperature are selected against.

SYNOPSIS

Energy is the capacity to do work. Energy can occur in various forms, including chemical, mechanical, light, electrical, and thermal, which are interconvertible. Whenever an exchange of energy occurs, the total amount in the universe remains constant, but there is a loss of usable energy—energy available to do additional work. This loss of usable energy, measured as entropy, results from an increase in the randomness and disorder of the universe. Living organisms are systems of low entropy, maintained by the constant input of external energy—energy ultimately derived from the sun. The hydrolysis of ATP is an example of an energy-releasing (exergonic) reaction. Many energy-requiring (endergonic) reactions that would normally fail to occur in a cell are driven by coupling them to ATP hydrolysis.

Enzymes are proteins that vastly accelerate the rate of specific chemical reactions by binding to the reactant(s) and increasing the likelihood that they will be converted to products. Enzymes act by lowering the activation energy—the energy required by reactants to undergo reaction. The specificity and catalytic activity of enzymes is due to the complementary structure of the active site that binds the substrate(s) and exerts its influence. Many enzymes also contain nonprotein "helpers," which may be organic coenzymes or metals. Enzyme structure—and thus function—is affected by temperature, pH, and the presence of inhibitors.

Enzymatic reactions are organized into metabolic pathways in which the product of one reaction becomes the substrate of a subsequent reaction. Catabolic pathways lead to less complex products, with the accompanying release of energy. In contrast, anabolic pathways build more complex products at the expense of cellular energy. Many reactions include the transfer of electrons from one molecule to another. When a molecule receives one or more electrons it becomes reduced. Electrons carry with them chemical energy; thus, the more reduced an organic molecule, the higher its energy content. The activity of many key enzymes is under cellular control and can be altered by interaction with certain metabolic end products or by covalent modification, such as the addition of a phosphate group.

Review Questions

1. Contrast the first and second laws of thermodynamics, using both a biological and a nonbiological example.
2. Describe how enzymes decrease activation energies; how they are able to bind only specific substrates; and why their activity is sensitive to temperature and pH.
3. Describe how energy-releasing and energy-requiring reactions can be coupled.
4. Compare the mechanisms by which a competitive inhibitor, such as a sulfa drug, and a noncompetitive inhibitor, such as arsenic, act to decrease the activity of an enzyme.
5. Why does NADPH provide a cell with reducing power?

Critical Thinking Questions

1. Suppose that Eduard Buchner had added sugar to his yeast-juice preparation and failed to observe the occurrence of fermentation. Would this have proven that fermentation could only occur in a living yeast cell? Why or why not?
2. Using the energy principles discussed in this chapter, explain why it is a more efficient use of agricultural land for people to eat grains rather than meat from grain-fed cattle.
3. Lead was formerly used in the manufacture of house paints. Young children who live in homes where lead paint still exists unintentionally ingest flakes of it, which may result in serious and permanent brain damage. Using your knowledge of enzymes, explain how even minute quantitites of lead can have such disastrous consequences.
4. How is it that feedback inhibition can block the formation of a particular product, such as an amino acid, even though it binds to only one enzyme in the entire pathway?
5. The enzyme lactase digests lactose, a disaccharide found in milk, to glucose and galactose, two monosaccharides. When lactase is not present, lactose is not digested and can cause intestinal disturbances in humans. Lactase is normally present in young children and in adults who continue to eat a diet that includes milk products, but it is not present in adults who do not consume milk. What do you think was the adaptive value of this enzyme for primitive humans?

CHAPTER

5

Energy and Life: Photosynthesis

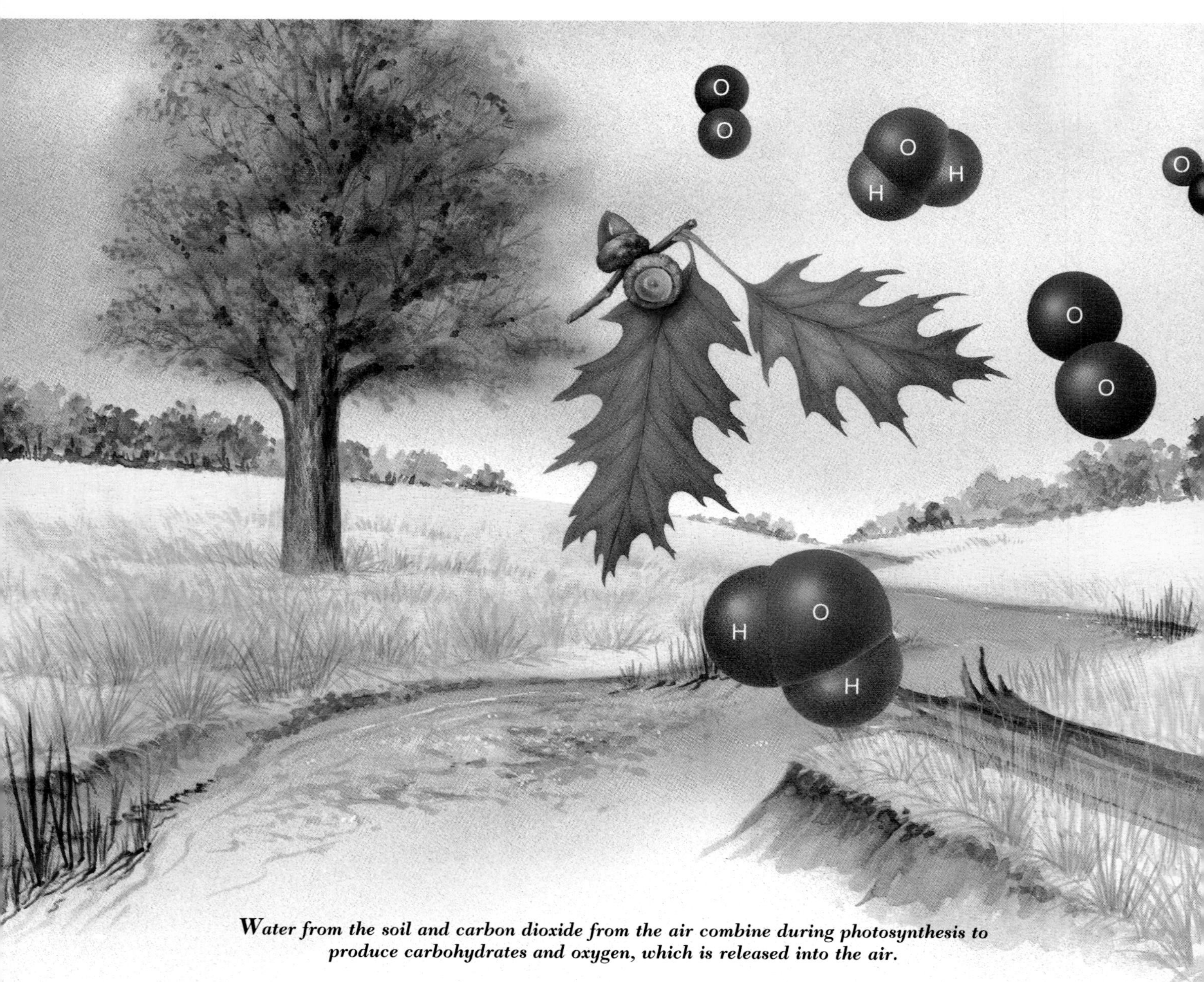

Water from the soil and carbon dioxide from the air combine during photosynthesis to produce carbohydrates and oxygen, which is released into the air.

STEPS TO DISCOVERY
Turning Inorganic Molecules into Complex Sugars

It was the late 1930s, and Martin Kamen had just received his Ph.D. in nuclear physics. Kamen decided to accept a position at the University of California at Berkeley, partly because Berkeley had an important new instrument not found anywhere else—a cyclotron. A cyclotron accelerates subatomic particles to high speeds and directs them at the nuclei of atoms, where the subsequent crash changes the atom's structure. Kamen was interested in using the cyclotron to produce new isotopes, such as $^{11}CO_2$, a radioactive isotope of carbon.

Soon after his arrival at Berkeley, Kamen met a chemist by the name of Sam Ruben. Ruben pointed out that radioactive carbon could be used to help decipher the steps of metabolic pathways in organisms. Since very little was known about metabolic pathways at that time, Kamen and Ruben teamed up to work out the details of one of the least understood, yet most important, metabolic pathways in all biology—photosynthesis. Scientists already knew that carbon dioxide and water were converted to carbohydrates, such as glucose, but just how this remarkable conversion took place was a complete mystery.

Kamen and Ruben planned to grow barley plants or algae cultures in the presence of radioactive carbon dioxide ($^{11}CO_2$). After increasing periods of exposure to $^{11}CO_2$, they would kill the cells with acidified, boiling water and then extract radioactive intermediate compounds in the photosynthetic pathway. By determining the chemical identity of the radioactive intermediates, Kamen and Ruben felt they could assemble a diagram of the metabolic pathway. The approach was sound, but the technical problems proved to be overwhelming.

The first problem had to do with the isotope itself. ^{11}C has a half-life of only 21 minutes, meaning that within 21 minutes of the onset of the experiment, only half of the starting amount remained; the other half had already disintegrated. Before the experiment was an hour old, only 15 percent of the original radioactivity remained. Because of the isotope's short lifespan, the entire procedure—from producing the isotope at the cyclotron, to carrying out the experiment, to measuring the radiation with a counter—had to be performed impossibly fast. Kamen and Ruben were seen running across campus in the middle of the night between the cyclotron and their laboratory. A second problem facing Kamen and Ruben was the absence of a method for separating radioactive organic molecules. After 3 years of intense research, they had made very little progress on the project.

Then, in 1940, Kamen and Ruben produced a longer-lived isotope of carbon. By bombarding nitrogen atoms with neutrons, the researchers produced a new isotope of carbon—^{14}C—which has a half-life of 5,700 years. Unfortunately, just as they were preparing to use ^{14}C to identify the steps of photosynthesis, World War II broke out. Kamen was quickly assigned to work on war-related isotope research, while Ruben was put to work on the development of chemical weapons. Ruben was later killed in a laboratory accident.

By the end of the war, the problem of how to separate radioactive organic molecules had been solved with a surprisingly simple technique. Archer Martin and Richard Synge, both British biochemists, discovered that closely related organic compounds could be separated by a technique they called *paper chromatography*. In this procedure, a sample of a mixture containing organic compounds is applied to a piece of filter paper, and one end of the paper is submerged in a solvent. The solvent migrates through the paper in much the same way that a paper towel "sucks up" water. As the solvent moves through the sample, it dissolves the organic molecules and carries them along at different speeds—the smaller, more soluble compounds moving fastest. At the end of the "run," all the compounds are found at different locations on the paper strip. The isolated spots of compounds can then be cut out of the strip and dissolved in a solvent to obtain a purified sample of each compound.

Following World War II, another team of scientists at Berkeley, headed by the biochemist Melvin Calvin, continued where Ruben and Kamen had been forced to stop. Using ^{14}C and paper chromatography, Calvin and his colleagues, James Bassham and Andrew Benson, grew algae in the presence of $^{14}CO_2$, killed the cells after a time, and extracted the labeled compounds with boiling alcohol. The Calvin team reasoned that, following a very brief exposure to the radioactive $^{14}CO_2$, the radioactive carbon would become incorporated into the initial compounds formed during photosynthesis. Exposing the cells to ^{14}C a little longer allowed the next compounds in the pathway to become radioactive, and so on, until all the intermediates in the photosynthetic pathway would be labeled. The Calvin team was successful in tracing the route by which carbon travels from CO_2 to carbohydrate. For his work in unraveling the pathway of the photosynthetic conversion of CO_2 to carbohydrate, Calvin received the Nobel Prize in Chemistry in 1961.

*H*ow does a tiny willow seed eventually increase its size several thousand times to become a giant tree with heavy branches and deep, massive roots? Where does the growing seed get the tremendous quantities of nutrients necessary for such remarkable growth? In the early 1600s, Jan Baptista van Helmont, a Belgian physician, conducted a simple experiment that disproved a long-accepted belief that plants absorbed all their food and nutrients from the soil. After planting a 5-pound willow tree in 200 pounds of dry soil, van Helmont added only rainwater to the soil for a period of 5 years. When he reweighed the soil and the willow tree, the tree had gained 164 pounds, yet the soil had lost only 2 ounces—not the 164 pounds that would be expected of a soil-eating plant. Van Helmont concluded that the willow's weight gain must have come from the added water, a conclusion we now know is only partly correct.

Willows, like the majority of other plants, gather the carbon needed to build their tissues from the air, not from the soil. Through the process of **photosynthesis,** plants use the energy of sunlight to combine the carbon in atmospheric carbon dioxide to form the backbone of new organic compounds that are fashioned into the cellular material required for growth. The soil provides only a few minerals needed for growth, which accounts for the 2-ounce weight loss measured by van Helmont.

Like the willow, all organisms depend on their environment to provide them with raw materials and energy necessary for life. Depending on how they acquire food and energy, organisms are classified into two general groups: autotrophs and heterotrophs. **Autotrophs** (*auto* = self, *troph* = feeding) convert simple inorganic molecules, such as carbon dioxide and water, into all of the complex, energy-rich materials they need to sustain their

(a)

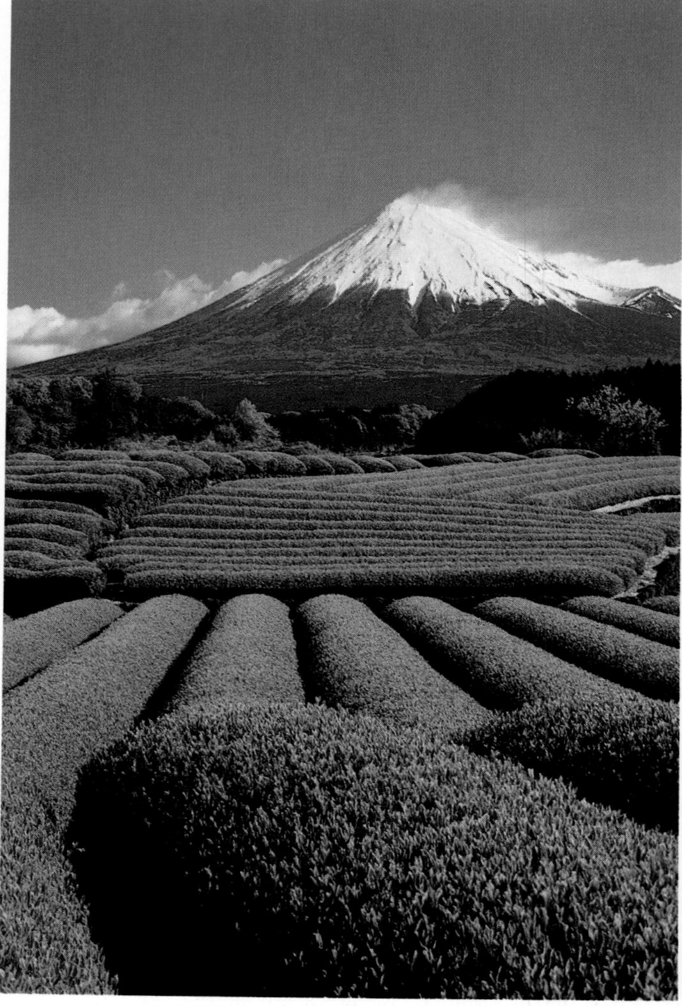

(b)

FIGURE 5-1

Life is centered around photosynthesis. ***(a)*** Plants supply energy for virtually all organisms in the biosphere by transforming the energy in sunlight into chemical energy. Every year plants manufacture about 150 billion tons of energy-rich sugars, a weight equivalent to 54 times the combined weight of all people living on earth today. ***(b)*** Photosynthetic plants are essential to human existence and to the existence of virtually all other animals. Although about 3,000 species of plants have been cultivated throughout human history, today only 15 plant species supply most of the world's human population with food.

life. This feat requires the input of energy. The vast majority of autotrophs harness energy from the sun through photosynthesis (Figure 5-1). There are, however, a number of autotrophic bacteria that acquire their energy from inorganic substances through chemosynthesis, a process we discuss later in the chapter.

All organisms depend on the metabolic activity of autotrophs, yet only plants, some protists, and some bacteria are autotrophs. The remainder of organisms—more than 80 percent of living species—are unable to carry out either photosynthesis or chemosynthesis. These organisms are called **heterotrophs** (*hetero* = other) because they must rely on *other* organisms (ultimately autotrophs) for food and energy. Thus, autotrophs not only provide themselves with food, but they provide food to a vast array of heterotrophs as well—including you.

AN OVERVIEW OF PHOTOSYNTHESIS

Photosynthesis in eukaryotes takes place inside chloroplasts (page 63). Chloroplasts contain an elaborate collection of membranes in which clusters of light-absorbing pigments are embedded. These pigments absorb solar energy, which is used to transform carbon dioxide and water into energy-rich sugars. These sugars not only provide a store of chemical energy but serve as the raw materials for producing thousands of other molecules needed by the plant. Oxygen, which is vital to energy utilization by both plants and animals, is released as a byproduct of photosynthesis.

The overall chemical equation for photosynthesis makes the process appear rather simple:

$$CO_2 + H_2O \xrightarrow[\text{chlorophyll}]{\text{light energy}} \underset{\substack{\text{unit of}\\ \text{carbohydrate}}}{(CH_2O)} + O_2$$

You might conclude from this equation that the energy in light is used to split carbon dioxide, releasing molecular oxygen (O_2) and transferring the carbon atom to a molecule of H_2O to form a unit of carbohydrate (CH_2O). In fact, this was the prevailing line of thought as late as 1941, the year Ruben and Kamen reported their results of an experiment using a labeled isotope of oxygen (^{18}O) as a replacement for the common isotope (^{16}O). The results showed that labeled water produced labeled oxygen. Therefore, it wasn't the carbon dioxide that was being split into its two atomic components, it was the water.

Botanists—biologists who specialize in the study of plant life—group the reactions of photosynthesis into two general stages: light-dependent reactions and light-independent reactions (Figure 5-2). During the **light-dependent reactions,** energy from sunlight is absorbed and converted to chemical energy, which is stored in two energy-rich molecules: ATP and NADPH. This process involves the splitting of water. The second stage of photosynthesis, the **light-independent reactions** (or "dark reactions"), does not require light. During the light-independent reactions, carbohydrates are synthesized from the carbon dioxide gathered from the atmosphere by using the energy stored in the ATPs and NADPHs formed in the light-dependent reactions.

LIGHT-DEPENDENT REACTIONS

The first evidence that photosynthesis occurred in chloroplasts was revealed in an ingenious demonstration conducted by Theodor Engelmann in 1881. Engelmann illuminated only certain cells of an aquatic plant. He noticed that actively swimming bacteria only concentrated immediately outside the lighted cell, near a chloroplast. Engelmann correctly concluded that the bacteria were using the oxygen released from photosynthesis in the chloroplast to stimulate their own energy metabolism.

Recall from Chapter 3 that chloroplasts contain stacks of membrane disks, or thylakoids, surrounded by a semifluid stroma. The light-dependent reactions occur within these thylakoid membranes. These reactions begin when the energy in sunlight is captured by light-absorbing pigments.

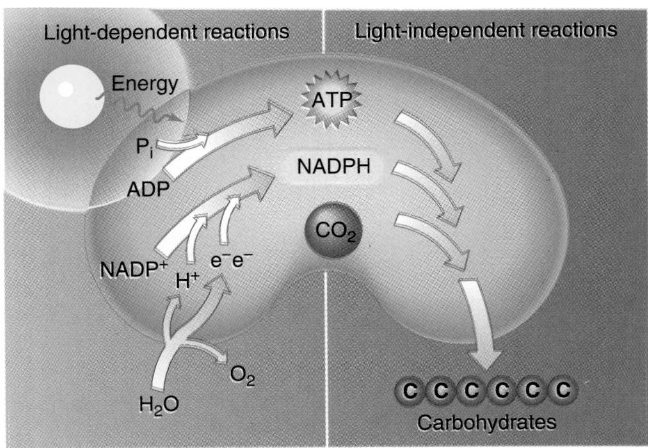

FIGURE 5-2

Photosynthesis: a two-stage process. Photosynthesis is divided into light-dependent reactions and light-independent (dark) reactions. During the light-dependent reactions, the energy of sunlight provides the power to generate ATP and NADPH from ADP, P_i, $NADP^+$, and H_2O. Oxygen gas, derived from water, is given off as a byproduct. In the light-independent reactions, ATP and NADPH from the light-dependent reactions provide the energy and electrons to convert low-energy CO_2 molecules into energy-rich carbohydrates.

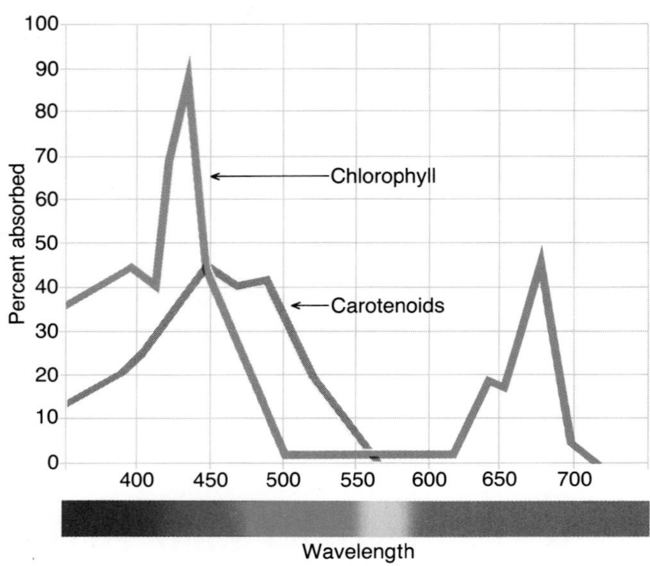

FIGURE 5-3
Absorption spectrum of photosynthetic pigments. The greatest absorption of light energy by photosynthetic pigments takes place at the peaks of each graph; the valleys indicate wavelengths that are reflected by the pigments. All plants use chlorophylls and carotenoids as photosynthetic pigments. Chlorophylls absorb primarily violet-blue and orange-red wavelengths, while carotenoids absorb in the violet-blue range of the spectrum. Differences in absorption from one chlorophyll molecule to the next depend largely on differences in the proteins to which the pigments are bound.

PHOTOSYNTHETIC PIGMENTS: CAPTURING LIGHT ENERGY

Plants are able to absorb a wide range of wavelengths because they contain a wide variety of pigment molecules, each with a slightly different structure and absorption property. All together, photosynthetic pigments absorb light energy with wavelengths between 400 nanometers (nm), violet light, and 700 nm, red light (Figure 5-3).

Plant chloroplasts contain two major groups of light-capturing pigment: chlorophylls and carotenoids. **Chlorophylls** are complex, magnesium-containing compounds that absorb light of red and blue wavelengths (Figure 5-3). Plants are typically green because chlorophyll, which is by far the predominant plant pigment, does not absorb light of green wavelengths. Instead, green light is reflected to our eyes, which is why we see the green color. **Carotenoids** absorb light of blue and green wavelengths (Figure 5-3). Yellow, orange, and red wavelengths are reflected by carotenoids, producing the characteristic colors of carrots, oranges, tomatoes, and the leaves of some plants during the fall (Figure 5-4).

ORGANIZATION OF PHOTOSYNTHETIC PIGMENTS

Capturing light energy and converting it into usable chemical energy requires a highly ordered biological structure (Figure 5-5). Photosynthetic pigments are embedded in the thylakoid membranes of chloroplasts in precise clusters

FIGURE 5-4
Fall Colors. As leaves stop producing chlorophyll in cooler weather, the gold and red carotenoids become more visible, which accounts for the brightly colored landscapes of autumn.

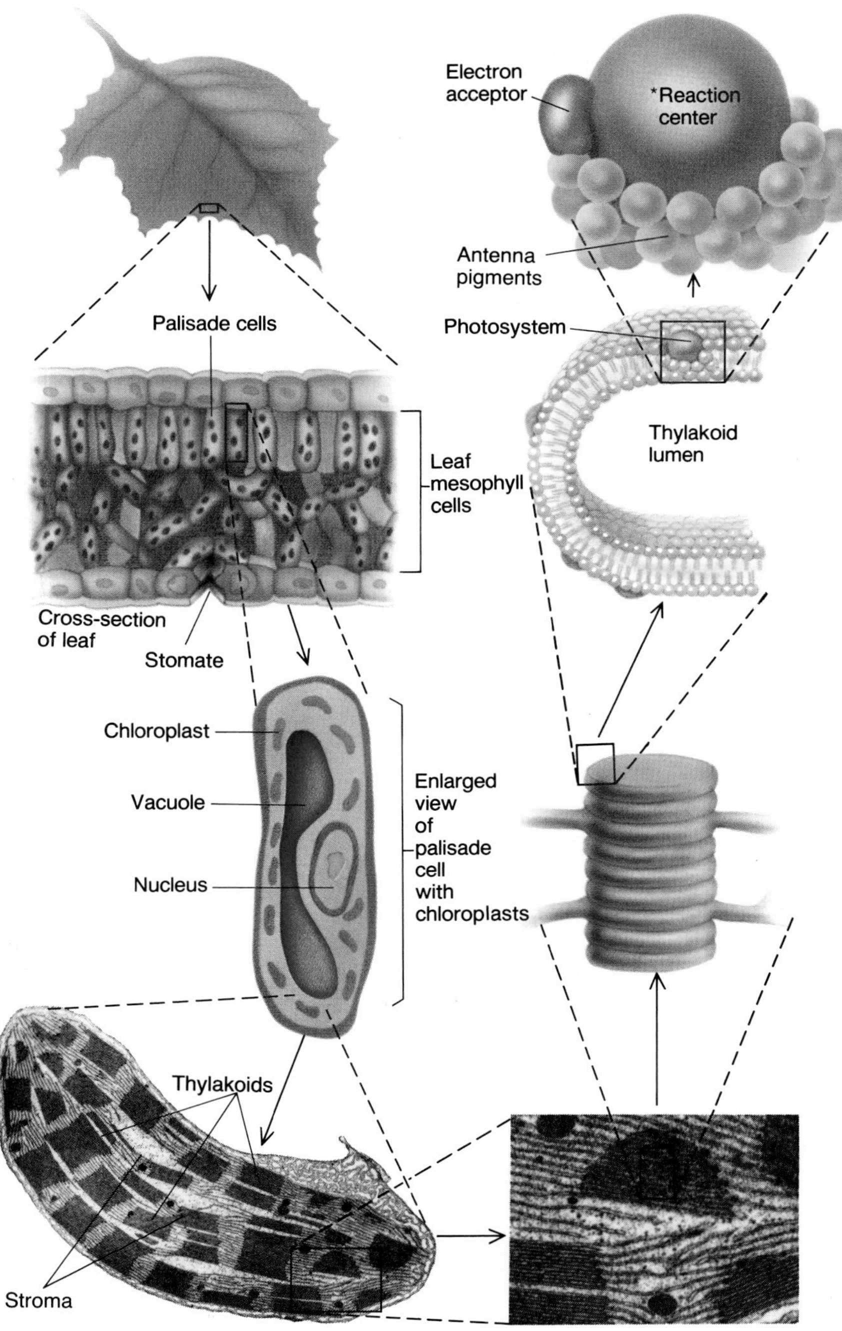

FIGURE 5-5
Organization for photosynthesis is evident at many different levels. Photosynthesis takes place in the chloroplasts of plant cells. A leaf cell may contain as many as 60 chloroplasts. The thylakoid system of the chloroplast is the site of the light-dependent reactions, whereas the stroma is the site of the light-independent reactions. Photosynthetic pigments are precisely arranged in the thylakoid membranes to form light-harvesting photosystems. Within a photosystem, light is absorbed by antenna pigments and the energy is transferred to the reaction center pigment boosting an electron to a higher energy level. The photoexcited electron is passed to an electron-acceptor molecule. This starts a chain of chemical reactions that leads to the formation of ATP and NADPH, and, ultimately, energy-rich carbohydrates.

called **photosystems.** Each photosystem is organized in the same way (Figure 5-6). A special chlorophyll molecule, called the **reaction center,** is surrounded by 250 to 350 **antenna pigments,** including both chlorophyll and carotenoid molecules. Antenna pigments are so named because they gather light energy between 400 and 700 nm and then channel this energy into the reaction center, much as an antenna receives and relays many radio and television waves. As a result, the reaction center always ends up with the absorbed energy. Thus, it is the reaction center molecule of a photosystem alone that is capable of transferring this accumulated energy; that is, of converting light energy into chemical energy.

Each chloroplast contains several thousand photosystems. Yet, despite this large number of photosystems, biologists have found only two different types of photosystems—**Photosystem I** and **Photosystem II.** Photosystem I and II differ in the kind of reaction center chlorophyll molecule they contain (Figure 5-6). The reaction center of Photosystem I is referred to as *P700*—"P" standing for "pigment," and "700" for the wavelength of light that this particular chlorophyll molecule absorbs most strongly. The reaction center of Photosystem II is referred to as *P680*, for the same reasons.

When sunlight strikes the thylakoid membrane, the energy is absorbed simultaneously by the antenna pigments of both Photosystems I and II. The absorbed energy is funneled into both the P700 and P680 reaction centers. The concentrated energy boosts electrons of the reaction-center pigments to an outer orbital, and each photoexcited electron is then passed to an electron-acceptor molecule. The transfer of electrons out of the photosystems leaves the two reaction-center pigments missing an electron; at that point, they are positively charged ($P700^+$ and $P680^+$).

SPLITTING WATER

The positively charged $P680^+$ of Photosystem II has a very strong attraction for electrons—strong enough to pull tightly held (low-energy) electrons from water (Figure 5-

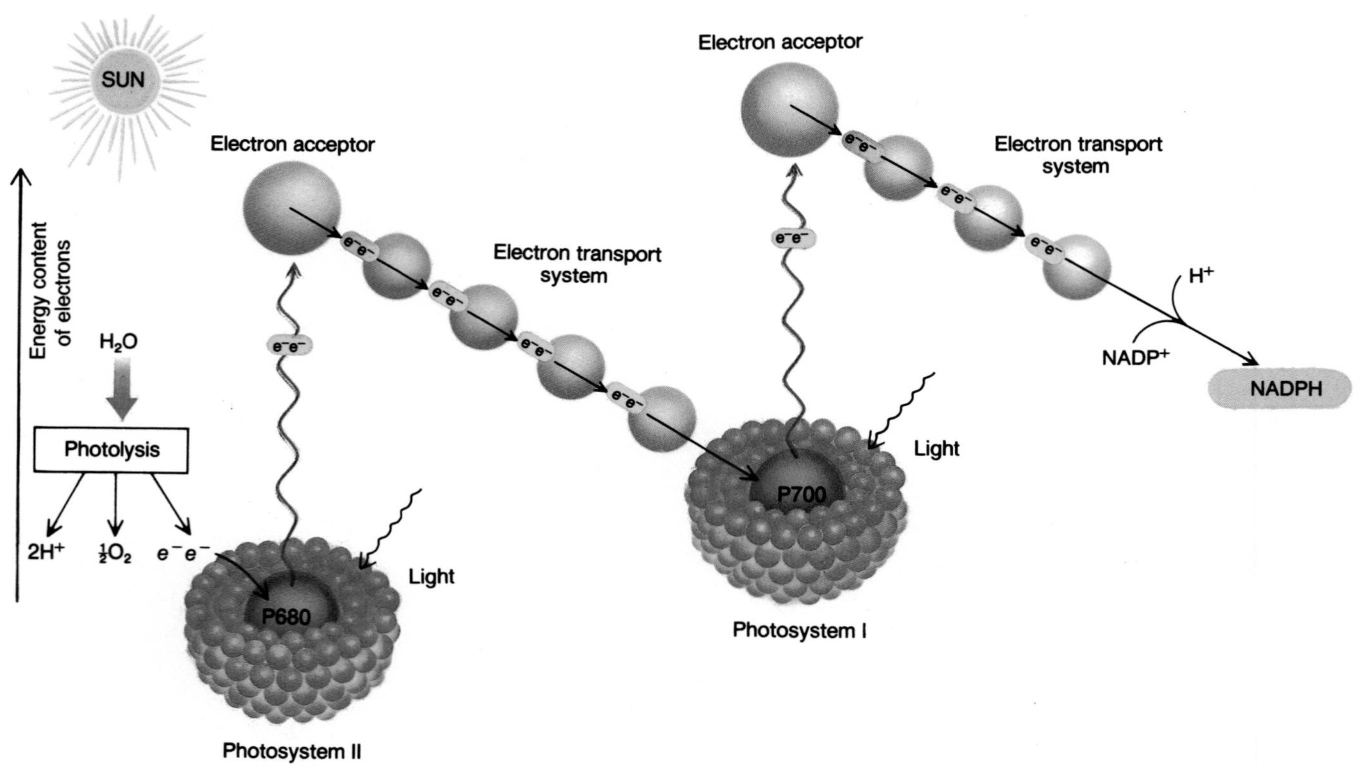

FIGURE 5-6

Electron flow from H_2O to $NADP^+$. Photoexcited electrons from the P700 and P680 reaction centers are passed to electron acceptors, which in turn transfer electrons to an electron transport system. The electron vacancy created by photoexcitement in P680 is filled with electrons released during the photolysis of water. The electron vacancy in P700 is filled with electrons that originated in P680. The photoexcited electrons from P700 pass down a second electron transport system until they react with $NADP^+$ and H^+ to form NADPH. The overall consequence of the process is to remove low-energy electrons from water, boost them to a higher energy level, and transfer the energized electrons to $NADP^+$ which provides the cell with reducing power.

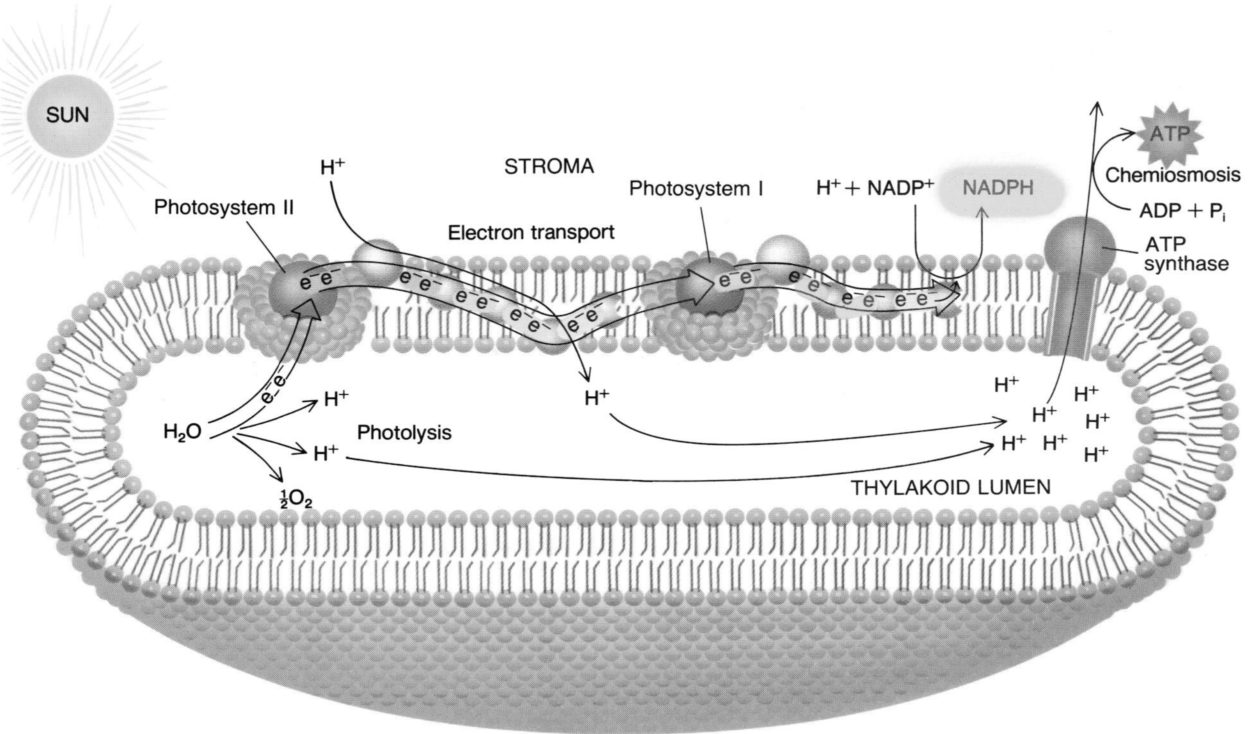

FIGURE 5-7

Photosynthesis and the thylakoid membrane. The carriers that transport electrons during the light-dependent reactions of photosynthesis are embedded within the thylakoid membrane. The combination of transporting energized electrons along an electron transport system and the splitting of water during photolysis concentrates protons inside the thylakoid lumen. This proton gradient drives the phosphorylation of ADP to ATP as protons diffuse through special membrane channels called the ATP synthase complex.

7). Pulling away electrons causes the water molecule to split, a process called *photolysis* (*photo* = light, *lysis* = splitting):

$$\text{energy} + H_2O \longrightarrow 2H^+ + 1/2\ O_2 + 2e^-$$

Photolysis generates three products: (1) oxygen, which is released into the atmosphere; (2) electrons that travel to the P680 reaction center, returning it to the uncharged state; and (3) protons (H^+), which are released into the thylakoid lumen. We will discuss the important fate of these protons shortly. But first, let's consider the fate of the photoexcited electrons that had been transferred earlier from P680 to an electron-acceptor molecule.

TRANSPORTING ELECTRONS AND FORMING NADPH

Each electron that is passed by the P680 reaction center to an electron acceptor is transferred again and again through a number of electron carrier molecules. In other words, electrons are transferred along an *electron transport system* (Figure 5-7). This process of shuttling electrons along an electron transport chain of molecules is analogous to the passage of water along a bucket brigade. The molecules in the electron transport system are precisely arranged within the thylakoid membrane to facilitate electron passage. At the end of this electron transport chain lies the $P700^+$ reaction center of Photosystem I. When $P700^+$ accepts an electron, it returns to its uncharged state (P700). Thus, the electron deficiency that resulted from P700 passing its photoexcited electron to an electron-acceptor molecule is filled by the electrons routed from P680 via an electron transport system. But what happens to the photoexcited electrons that had originated from P700? These electrons are passed through another set of electron carriers and on to $NADP^+$ to form energy-rich NADPH (Figure 5-7).

We have now followed the process in which electrons flow from water to Photosystem II, from Photosystem II to Photosystem I via an electron transport system, and from Photosystem I to $NADP^+$ via another electron transport system. The NADPH formed by this process provides

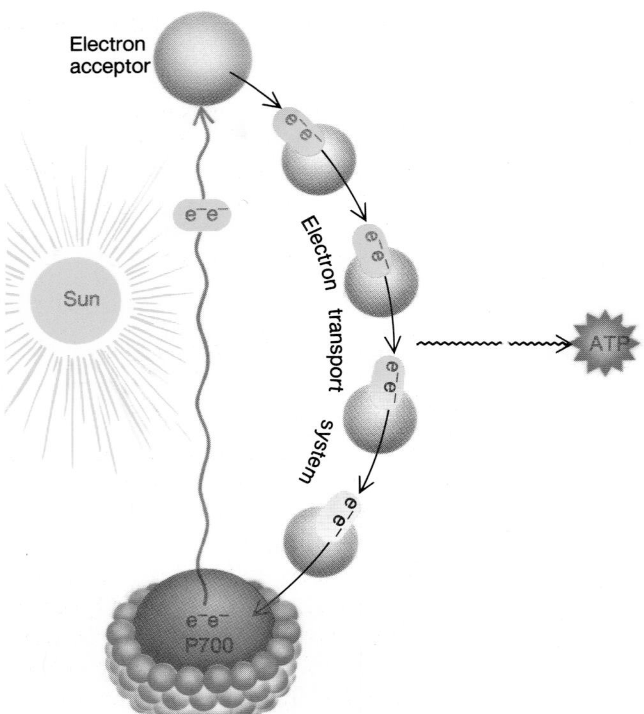

FIGURE 5-8
Electron flow during cyclic photophosphorylation. During cyclic photophosphorylation, light energy is absorbed by Photosystem I. Photoexcited electrons from the P700 reaction center are transferred to an electron acceptor and passed down an electron transport system which returns the electron to the P700 pigment. During this cyclic flow of electrons, protons are shuttled across the membrane generating a proton gradient used to form ATP. Cyclic photophosphorylation does not involve Photosystem II and does not generate NADPH.

the cell with reducing power, which is a form of stored energy. Because electrons only move in one direction, botanists call this process **noncyclic photophosphorylation.**

Plants have a second type of electron flow, one that returns electrons to the same reaction center from which they originated. Unlike noncyclic photophosphorylation, which requires the participation of both Photosystem I and II, only Photosystem I participates in the cyclic routing of electrons. Not surprisingly, the cyclic flow of electrons is called **cyclic photophosphorylation** (Figure 5-8).

During cyclic photophosphorylation, light energy is absorbed by antenna pigments and funneled into the P700 reaction center. Electrons pass from P700 to an electron acceptor and are then relayed through an electron transport system that returns the electrons back to the same $P700^+$ reaction center.

Both the cyclic and noncyclic flow of electrons along electron transport systems build a proton gradient necessary to generate ATP, the second energy-rich molecule formed by the light-dependent reactions. Indeed, the term "photophosphorylation" describes the fact that light energy is used to drive a reaction in which ATP is formed, the subject of the next section.

MAKING ATP

Electron transport is a spontaneous process in which electrons drop to lower and lower energy levels as they are passed from one molecule to another in an electron transport system. The energy that is released by electron flow is used to manufacture ATP. For decades, scientists investigated how the "downhill" process of electron transport was coupled with the "uphill" process of making ATP. In 1961, Peter Mitchell, a British physiologist, formulated what at first seemed to be a radical hypothesis to explain this phenomenon. Mitchell proposed that the energy released as the electrons fell to lower and lower energy levels during electron transport produced a proton gradient across the thylakoid membrane. A proton gradient is established when hydrogen ions (H^+) are present at higher concentration on one side of the membrane than the other side. In chloroplasts, then, a proton gradient would be established across the thylakoid membranes. It is this proton gradient that is used to power the energy-requiring reaction in which ADP combines with P_i to form ATP. Mitchell called his mechanism for ATP formation **chemiosmosis.** Although the proposal initially met with skepticism because of a lack of direct evidence, experimental support accumulated over the next decade to an overwhelming degree, culminating in Mitchell's receiving the Nobel Prize in 1978.

A proton gradient becomes established across a thylakoid membrane as a result of two events: (1) the splitting of water (photolysis) releases protons directly into the thylakoid lumen; and (2) protons are released into the thylakoid lumen as a result of electron transport (Figure 5-8). But how does the energy stored in a proton gradient lead to the formation of ATP?

ATP is synthesized by a large enzyme complex called *ATP synthase*, whose shape resembles a ball sitting on a stalk (Figure 5-7). The stalked portion of the enzyme crosses through the membrane and contains an internal channel that allows the protons massed in the lumen to flow through the thylakoid membrane toward the region of lower concentration in the stroma. The flow of protons through these channels is analogous to water flowing through a bathtub drain, flowing to a lower, more stable energy level. The active site of the enzyme—where ADP and P_i come together to form ATP—is located in the ball-shaped portion at the end of the proton channel. The "downhill" movement of protons through the channel drives the "uphill" synthesis of ATP at the enzyme's active site.

LIGHT-INDEPENDENT REACTIONS

You might be wondering, since both ATP and NADPH are forms of chemical energy, why do photosynthesizers go on to manufacture another form of chemical energy—energy-rich sugars—through the light-independent reactions? The answer is simple: Neither ATP nor NADPH can be stored or translocated from one part of a plant to another, but sugars can. Transport and storage of energy-rich compounds are critical to the survival of a plant. For example, glucose produced in leaves is converted to sucrose and then moved to several growing regions of roots and stems, areas where photosynthesis may not take place. Thus, the function of the light-independent reactions is to use the energy of ATP and electrons of NADPH to manufacture glucose. To supply the carbon they need to manufacture glucose, plants gather carbon dioxide from the air.

To date, scientists have identified three variations of the light-independent reactions—C_3, C_4, and *CAM synthesis*. All three types of light-independent reactions begin with **carbon dioxide fixation,** the combining of carbon dioxide with a carbon-acceptor molecule, and all share the same cyclic pathway for producing glucose; they differ in the way carbon dioxide is first combined with a carbon-acceptor molecule. In certain environments, this seemingly small difference gives some plants a metabolic edge over others.

C_3 PLANTS AND THE CALVIN-BENSON CYCLE

The light-independent reactions progress from carbon dioxide fixation to a common cyclic pathway, the details of which were worked out by a team of researchers at the University of California, Berkeley, headed by Melvin Calvin. When Calvin and his colleagues, J. Bassham and A. Benson, began their studies using $^{14}CO_2$, one of their first objectives was to determine the nature of the CO_2 acceptor. The Calvin team observed that when the cells were exposed to $^{14}CO_2$ for very short periods—5 seconds or less—the first product to be formed was a three-carbon compound, phosphoglyceric acid (PGA), hence the designation *C_3 pathway,* or the *Calvin-Benson cycle.* Calvin concluded that the single carbon from carbon dioxide reacted with a two-carbon acceptor molecule to form the three-carbon PGA. A search for the mysterious two-carbon compound began. Subsequent studies revealed that carbon dioxide is not fixed to a two-carbon molecule but to a five-carbon molecule, **ribulose biphosphate (RuBP),** by the enzyme *RuBP carboxylase,* nicknamed "Rubisco" (Figure 5-9). However, the six-carbon molecule formed by this union is so unstable that it "immediately" splits into two molecules of PGA, accounting for the early detection of PGA.

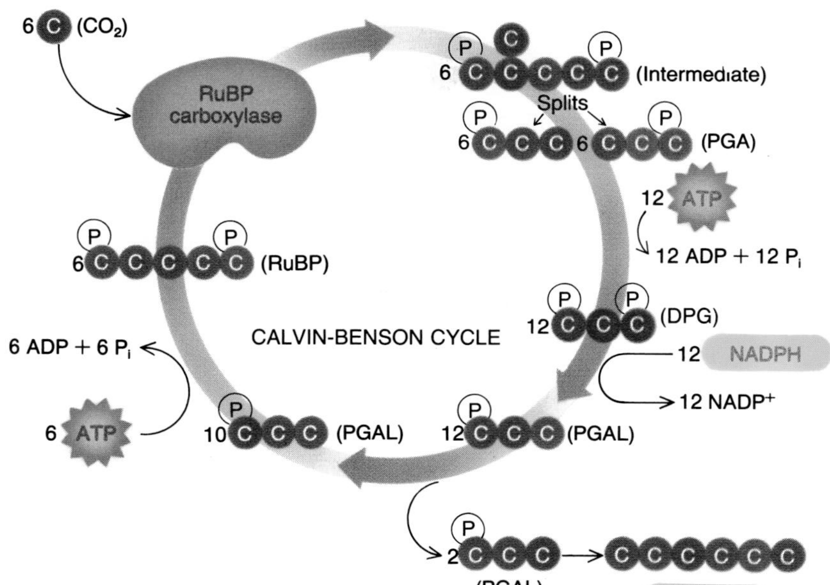

FIGURE 5-9

The Calvin-Benson cycle. During the formation of glucose from carbon dioxide, six CO_2 molecules are combined with six RuBPs to form six unstable, six-carbon carbohydrates. Each carbohydrate immediately splits into two, producing a total of 12 molecules of PGA. The PGAs are converted to DPGs by the transfer of phosphate groups from ATP. This is one of the places where the ATPs produced by the light-dependent reactions are utilized. NADPHs formed by the light-dependent reactions are used in the following step in which the 12 DPGs are reduced to 12 PGALs by the transfer of electrons. Two of the PGALs are "drained" away to form a molecule of glucose, whereas the remaining 10 PGALs (representing a total of 30 carbon atoms) recombine to regenerate six molecules of RuBP to begin another round of the Calvin-Benson Cycle.

PGA is not an energy-rich molecule. In fact, two molecules of PGA contain less energy than does a single molecule of RuBP, the actual carbon dioxide acceptor. This explains why carbon dioxide fixation is spontaneous and does not require energy. However, producing energy-rich molecules from PGA requires the energy from ATP and electrons from NADPH formed during the light-dependent reactions. ATP and NADPH are used to reduce each PGA to a more energy-rich molecule, PGAL. Two PGAL molecules can ultimately give rise to one glucose (Figure 5-9).

C_4 PLANTS

Although the vast majority of plants use the C_3 pathway exclusively for constructing glucose, the existence of alternate means of fixing carbon dioxide was uncovered by Hugo Kortschak in the early 1960s while working with sugar cane in Hawaii. Kortschak was surprised to find that when sugar cane was provided ^{14}C-labeled carbon dioxide, radioactivity first appeared in organic compounds containing four carbons rather than three, thus the name C_4 synthesis. It soon became apparent that these plants utilize a different mechanism for CO_2 fixation which sometimes gives them an adaptive advantage over C_3 plants in hot, dry habitats. For example, at high temperatures [45° C (103° F), or higher] plants with C_4 synthesis produce as much as six times more carbohydrate than do C_3 plants. C_4 plants, which, in addition to sugar cane, include corn, Bermuda grass, and sorghum, frequently do better than C_3 plants in environments in which water is limited because C_4 plants are able to fix carbon dioxide at very low concentrations and thus conserve water by closing leaf pores, or *stomates*. Another survival advantage of C_4 plants stems from their resistance to *photorespiration*, a phenomenon whereby high levels of oxygen actually interfere with photosynthesis by binding to RuBP carboxylase, thereby inhibiting carbon-dioxide-fixing activity. C_4 plants are less susceptible to photorespiration because they are able to photosynthesize under lower carbon dioxide and oxygen levels.

CAM PLANTS

The third synthesis reaction, CAM synthesis, is found in about 5 percent of plants, which includes plants such as cacti, pineapple, jade plant, and some lilies and orchids. (CAM is an acronym for *C*rassulacean *A*cid *M*etabolism, named for the plant family Crassulaceae in which it was first discovered.) While C_3 and C_4 plants open their leaf pores (stomates) and fix carbon dioxide only during the daytime, CAM plants open their stomates to take in carbon dioxide at night, when the rate of water vapor loss is greatly reduced. At sunrise, stomates close (conserving water), and sunlight powers ATP and NADPH formation. Attempts to manipulate C_3, C_4, and CAM plants using biotechnological techniques in the laboratory to improve crop production are discussed in The Human Perspective: "Biochemical Flexibility: Adjusting to Environmental Conditions."

CHEMOSYNTHESIS

Although the predominant means of generating energy-rich organic molecules is through photosynthesis, a number of bacteria depend on chemosynthesis for a supply of chemical energy. During photosynthesis, electrons and protons come from the splitting of water, while during chemosynthesis, they are stripped from reduced inorganic substances, such as ammonia (NH_3) and hydrogen sulfide (H_2S). The equation for chemosynthesis using hydrogen sulfide is:

$$CO_2 + 2\ H_2S \longrightarrow (CH_2O) + H_2O + 2\ S$$

The removal of electrons (oxidation) provides the energy needed for the formation of ATP and NADPH.

Until recently, chemosynthetic bacteria did not receive much attention from the scientific community be-

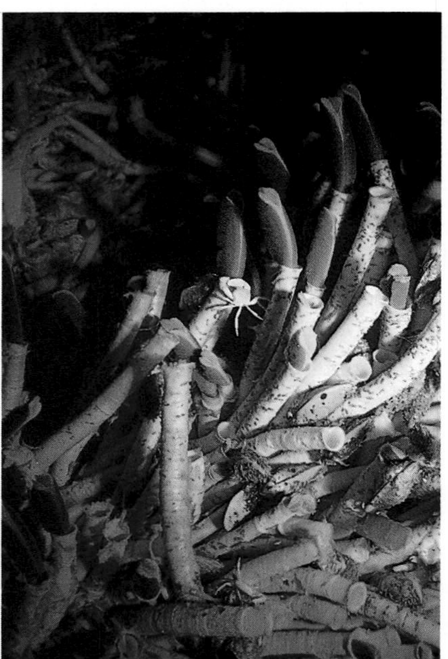

FIGURE 5-10
Giant red tube worms, several times larger than their closest relatives, and several other unusual marine animals have been found living on the deep ocean bottom near hydrothermal vents. The entire community depends on the autotrophic activity of chemosynthetic bacteria.

◁ HUMAN PERSPECTIVE ▷

Biochemical Flexibility: Adjusting to Environmental Conditions

Biologists are currently investigating whether some plants are naturally able to shift among C_3, C_4, and/or CAM pathways, depending on environmental conditions. Biologists are also investigating whether it is possible to genetically engineer "flexible" plants that shift between two, or possibly all three, synthesis pathways. Shifting between different pathways would give these "designer" plants a definite advantage over those with only one pathway. For example, a flexible "designer" plant could be in the CAM pathway when conditions are very hot and dry and then shift into C_3 or C_4 following irrigation or a rainfall, enabling it to grow many times faster than it could in CAM.

So far, scientists have not found a plant capable of switching among all three synthesis pathways. But a few species have been described as natural C_3–C_4 intermediates, and preliminary results indicate that some intermediates show the higher C_4 photosynthesis rates. Other investigations have revealed natural C_3–CAM intermediates.

As for genetically engineered flexible plants, scientists have engineered C_3–C_4 hybrid plants. Unfortunately, none of the hybrids had increased rates of photosynthesis over the C_3 level yet. But investigations will continue, for if biologists can eventually engineer biochemically flexible crop plants, much of the world's hunger problems might be solved—assuming there is no corresponding increase in the size of human population.

cause they were not believed to be an important component of any ecosystem. But that all changed in 1977, when an entirely new type of community was discovered deep in the ocean off the coast of South America, a community that depended entirely on chemosynthesis by microscopic prokaryotes. Scientists discovered several unusual marine animals, including sea anemones, clams, mussels, crabs, and giant-sized tube worms, living near deep, hydrothermal vents on the ocean floor (Figure 5-10). Since light could not penetrate to these depths (2,500 meters, or 8,000 feet), the entire community of organisms depended on the chemosynthetic bacteria that used hydrogen sulfide as an energy source. Remarkably, the organisms of the hydrothermal vents form a self-sufficient community completely independent of life elsewhere on earth.

EVOLUTION AND ADAPTATION: TYING IT TOGETHER

The first organisms on earth—bacteria—were prokaryotes that absorbed free organic compounds directly from the surrounding primordial sea. These "chemical heterotrophs" depended entirely on organic molecules that were synthesized abiotically (not by organisms). With an abundant supply of these nutrients, prokaryotic bacteria proliferated in the primordial seas. But eventually the prokaryotes consumed the organic molecules more quickly than they could be replaced (the rate of abiotic production of organic molecules not only depends on a supply of available energy (which was plentiful at the time) but on the density of precursors; that is, the density must be high enough to allow precursors to interact spontaneously). As demand exceeded supply, natural selection favored cells that could manufacture their own organic compounds. Such organisms could be freed from the limitations of abiotically synthesized organic molecules. These were the earth's first autotrophs.

The first autotrophs appeared about 3 billion years ago. They not only supplied themselves with organic molecules but also became food for heterotrophic organisms. The first autotrophs stripped electrons and hydrogens from inorganic substances, and, using the released energy or energy from the sun, produced their own organic molecules. Because water was such an abundant resource, the autotrophs that stripped electrons and hydrogen from water quickly flourished. But splitting water not only releases electrons and hydrogen, but oxygen as well. Thus, as these autotrophs prospered, they released vast amounts of oxygen into the water and air. The pattern and pace of evolution was forever altered.

Evolution continues today, as photosynthetic plants metabolically adapt to the unique rigors of their environment. Much of natural selection has operated on the light-independent reactions, whereby carbon dioxide fixation varies in ways that helps preserve water in hot, dry habitats, reduce metabolic waste (photorespiration), or give the plant an edge over its competitors.

SYNOPSIS

Organisms are categorized as autotrophs if they manufacture organic nutrients from carbon dioxide and other inorganic precursors, or as heterotrophs if they depend on other organisms for their supply of organic materials. Most autotrophs are photosynthesizers, converting light energy from the sun into chemical energy, which is then used to manufacture carbohydrates from carbon dioxide and water.

During the light-dependent reactions, light energy is absorbed by photosynthetic pigments embedded in the thylakoid membranes of chloroplasts. These pigments are clustered into photosystems, each with antenna pigments and a single reaction-center chlorophyll that ultimately receives the energy. There are two types of photosystems, I and II, which operate in conjunction during noncyclic photophosphorylation. Photoexcited electrons are transferred from each reaction center (P700 of Photosystem I and P680 of Photosystem II) to an electron acceptor molecule. The P680 reaction center of Photosystem II passes its photoexcited electrons to an electron transport system which relays electrons to the $P700^+$ reaction center of Photosystem I. The electron-deficient reaction center of Photosystem II ($P680^+$) pulls electrons from water, splitting it into protons, electrons, and oxygen atoms. The photoexcited electrons from P700 pass along another electron transport system to $NADP^+$, forming NADPH. During electron transport, protons move into the thylakoid lumen.

Protons accumulate in the thylakoid lumen as the result of the splitting of water and electron transport, building a proton gradient. ATP formation results from the movement of protons back across the thylakoid membrane through a channel created by an ATP-synthesizing complex.

During the light-independent reactions, the chemical energy stored in ATP and NADPH is used to synthesize carbohydrates from carbon dioxide. In C_3 plants, the majority of plants on earth, carbon dioxide is fixed to a five-carbon molecule, RuBP, to form an unstable six-carbon compound, which immediately splits into two molecules of PGA. In C_4 and CAM plants, carbon dioxide is fixed to a three-carbon molecule first, and then to RuBP, giving these plants a metabolic advantage in some habitats. In all photosynthesizers, the NADPH and ATP synthesized during the light-dependent reactions are then used to convert PGA into glucose and to regenerate RuBP.

Chemosynthesis occurs in certain bacteria. These autotrophs obtain energy by oxidizing inorganic substances, such as ammonia and hydrogen sulfide, to form NADPH and ATP needed to synthesize carbohydrates.

Review Questions

1. Rearrange the order of the following terms to match the correct sequence of reactions during photosynthesis:

glucose synthesis	PGA production
photolysis	absorption of light
electron transport	carbon dioxide fixation
noncyclic flow	PGAL production

2. Why do the light-independent reactions depend on the light-dependent reactions having already occurred?

3. In what way does light absorption in Photosystem II differ from light absorption in Photosystem I? How does the splitting of water contribute to ATP formation?

4. Describe the process of NADPH formation. Why are both Photosystems I and II involved?

5. Why would C_4 and CAM plants be able to flourish in hot, dry environments?

Critical Thinking Questions

1. If Melvin Calvin had been working with a C_4 or CAM plant, what compound would he have found most heavily labeled after exposure to radioactive carbon dioxide?
2. Today, only about 15 plant species supply most of the world's human population with food. Can you name five of these critical crop plants? What would be the impact on human population growth and health of expanding the number of plant species used for crops?
3. Supposed you discovered a mutant plant having thylakoids that lacked Photosystem II. How would this affect carbohydrate production? Do you suppose this plant would be able to produce ATP from converted light energy?
4. Organization is one of the characteristics of all living things. Explain how plant-cell organization contributes to the process of photosynthesis.
5. Suppose somebody were to tell you that all life is dependent on energy from the sun and that the "sudden death" of the sun would quickly extinguish all life forms. Would you agree with their conclusion? Why or why not?

CHAPTER

6

Energy and Life: Respiration and Fermentation

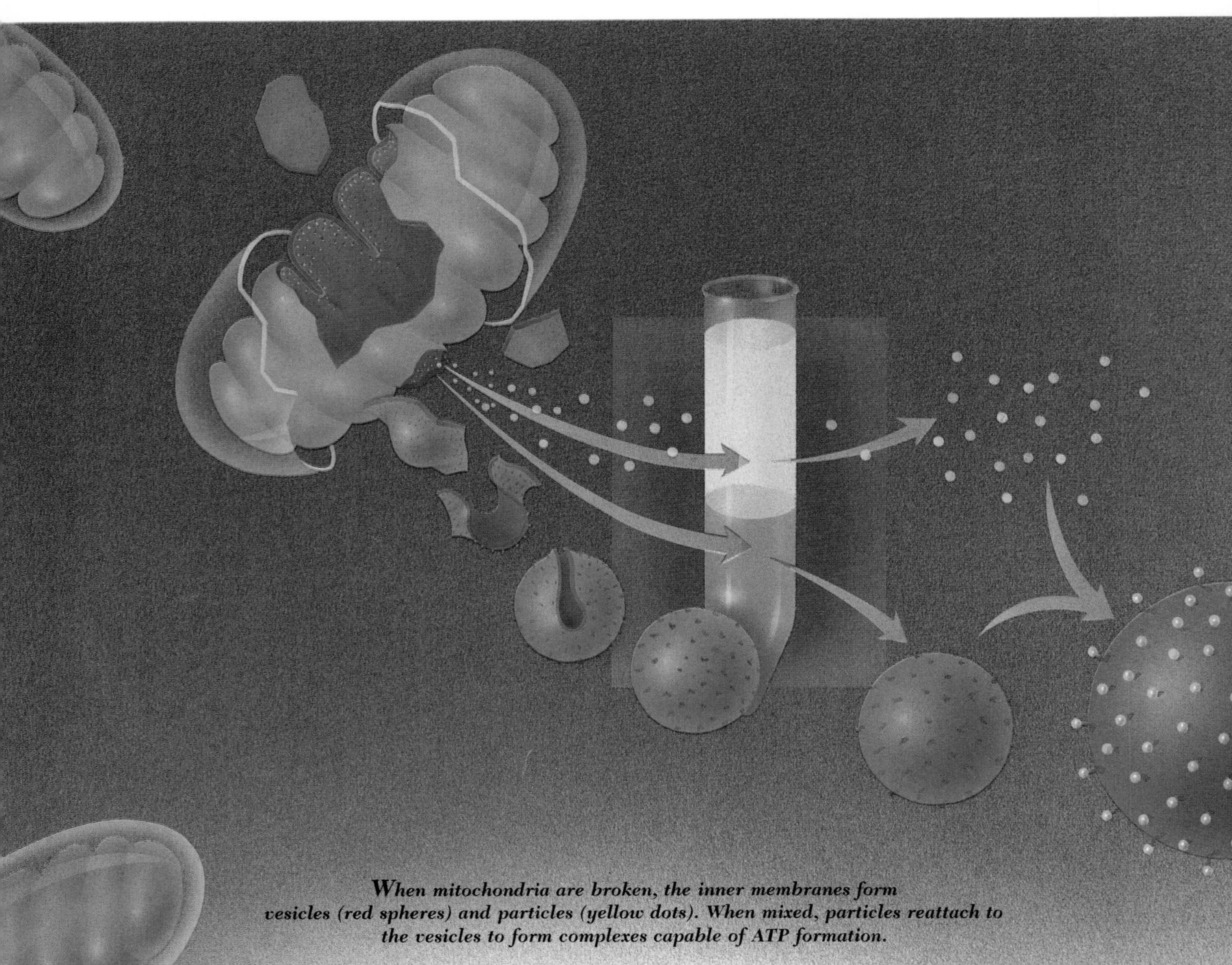

When mitochondria are broken, the inner membranes form vesicles (red spheres) and particles (yellow dots). When mixed, particles reattach to the vesicles to form complexes capable of ATP formation.

STEPS TO DISCOVERY
The Material Responsible for ATP Synthesis

With improved light microscopes, nineteenth century biologists began to see thread-like particles within the cytoplasm of cells. They named the organelle "mitochondrion," from the Greek meaning "thread granule." Despite the prevalence of mitochondria in cells, their function remained a complete mystery until scientists perfected techniques to remove and isolate organelles from cells. The isolation of mitochondria depended on the development of a new instrument—the "ultracentrifuge"—a device that spins special tubes in a circular path at speeds high enough to generate centrifugal forces (forces that pull an object outward when it is rotating around a center) over 100,000 times the force of gravity.

To separate the organelles inside cells, cells are first broken open to release their contents and then the suspension is spun in an ultracentrifuge tube at different speeds. At slower speeds, larger organelles, such as cell nuclei, settle at the bottom of the tube. The supernatant is then removed and recentrifuged at even higher speeds, causing mitochondria to settle at the bottom.

In 1948, Eugene Kennedy and Albert Lehninger, then at the University of Chicago, purified mitochondria using this technique and demonstrated that these organelles were the cell's "chemical power plants." The purified mitochondria were able to oxidize organic compounds (such as fatty acids) and to use the released energy to make ATP.

To determine which part of the mitochondrion was responsible for synthesizing ATP, Efraim Racker and his colleagues in New York spun the suspension of broken mitochondria using high-speed centrifugation. This procedure divided the material into two portions; the lower portion consisted of membrane vesicles that went to the bottom of the centrifuge tube, the other material remained in the liquid supernatant. The research team determined that the isolated vesicles at the bottom could still oxidize organic substrates but were unable to make ATP. But, when the researchers mixed these same vesicles with the liquid supernatant, they acquired full ATP-generating capacity. Clearly, the supernatant contained some factor needed by the vesicles for ATP production. Racker named this factor F_1.

At the same time, another researcher, Humberto Fernandez-Moran of Massachusetts General Hospital in Boston, examined mitochondrial membranes using a new electron microscope technique called negative-staining. In negative-staining, objects are brightly outlined against a dark background, making small objects much easier to see. Using this technique, Fernandez-Moran revealed never-before-seen rows of particles protruding from the inner mitochondrial membrane. Each particle was attached to the membrane by a thin stalk.

Racker used the electron microscope to examine the supernatant formed from disrupted mitochondria to see if his F_1 could be visualized. Racker observed particles in the supernatant that appeared identical to the stalked particles Fernandez-Moran had discovered attached to the mitochondrion's inner membrane. Furthermore, when Racker added the F_1 particles to the membrane vesicles formed from disrupted mitochondria, the particles became attached to the membranes by "little stalks." Racker concluded that the released F_1 particles were the same ones seen on the inner wall of mitochondria. After more studies, Racker discovered that these particles were, in fact, the sites in mitochondria where ATP was synthesized. He called the entire complex (sphere, stalk, and membrane base) *ATP synthase*.

Similar types of stalked particles have been found attached to the thylakoid membranes of chloroplasts and to the plasma membrane of bacteria (which have no mitochondria). In all of these organisms, the stalked particles are the sites where energy from substrate oxidation drives the synthesis of ATP. Once again, the unity of life reveals the common evolutionary origins of all organisms. Bacteria, plants, and animals are all descended from the same ancestor, one from which they inherited their energy-generating mechanism.

*I*t may be the most catastrophic occurrence in the history of life on earth. Billions of individual organisms died, and countless species were driven to extinction, banished forever by one of the most poisonous substances the world had ever known. This unwelcome toxic intruder is still around today; in fact, it is the second major component of today's atmosphere. We are referring to molecular oxygen (O_2).

Molecular oxygen gradually appeared on earth beginning about 3 billion years ago. It was produced by a then new type of microorganism, cyanobacteria (formerly known as "blue-green algae"), the prokaryotes that "invented" oxygen-evolving photosynthesis. Up to that time, none of the earth's organisms required molecular oxygen, and none of the gas was present in the atmosphere. The world was populated by *anaerobes*—organisms that utilize oxygen-free (anaerobic) metabolism to disassemble nutrients and obtain energy. Following the rise of the cyanobacteria, however, the oceans and atmosphere became infused with molecular oxygen, forever altering the history of life on earth.

When it first appeared, molecular oxygen fatally oxidized the cytoplasm of all but a few types of microorganisms. Eventually, species evolved that not only withstood the poisonous effects of molecular oxygen but actually became dependent on it, using it to achieve enormous increases in their metabolic efficiency. These oxygen-dependent organisms are called *aerobes.* These evolutionary innovators—the first users of oxygen—were the pioneers of an evolutionary line that ultimately produced all oxygen-dependent organisms, including humans.

In many animals, breathing draws in life-supporting oxygen and expels the waste product carbon dioxide. You need molecular oxygen in order to release the energy stored in the chemical fuel you consume as food. Without oxygen, the metabolic furnace that powers your life is extinguished (Figure 6-1).

▼ ▼ ▼

FERMENTATION AND AEROBIC RESPIRATION: A PREVIEW

Even such distantly related organisms as a bacterium, a garden weed, and a human being share many common biological characteristics. For example, all organisms harvest chemical energy by oxidizing organic molecules, generating raw materials, and releasing energy that is trapped momentarily in the form of ATP. Moreover, they accomplish this shuffling of chemical energy from one compound to another by virtually the same set of metabolic reactions that arose at a very early stage of evolution. The starting point for most of these metabolic reactions is glucose, a sugar molecule with a six-carbon backbone ($C_6H_{12}0_6$).

Glucose is an energy-rich molecule, but it cannot directly power biological activities. A glucose molecule in a cell is like a person carrying around a $500 bill and trying to use it to purchase a gallon of gasoline; both the cell and the gas station require smaller denominations. Cells therefore must convert energy-rich glucose into other compounds containing more usable quantities of energy, mainly ATP.

The *complete* oxidation of a gram of glucose releases 3,811 calories. Not all organisms can take full advantage of glucose's energy content, however. The extent to which glucose is dismantled to release its energy is the basis for distinguishing between two fundamental processes—fermentation and aerobic respiration. As we will see shortly, **fermentation** is an incomplete breakdown of glucose,

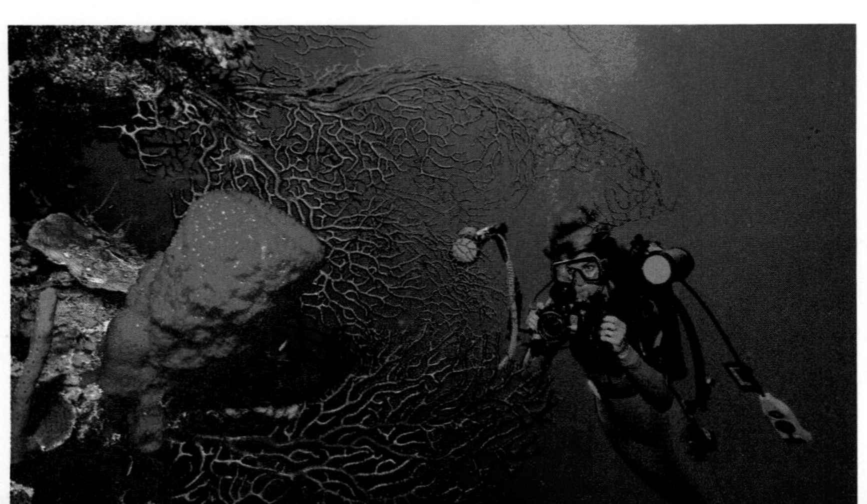

FIGURE 6-1

Oxygen—an essential commodity. Without a supply of oxygen, this scuba diver would only be able to remain submerged for a minute or two. Like all aerobes, this diver needs oxygen to extract energy for life from food.

which occurs in the absence of oxygen and extracts only a small portion of the sugar's energy content. In contrast, **aerobic respiration** completely disassembles glucose, step by step; pairs of high-energy electrons are stripped from the substrate, transferred to an electron transport system, and eventually passed to molecular oxygen.

Regardless of the strategy (fermentation or aerobic respiration), tapping the energy stored in glucose begins with **glycolysis** (*glyco* = sugar, *lysis* = split), a universal pathway whereby glucose is split in two, both portions of which are converted to pyruvic acid.

GLYCOLYSIS

In 1905, two English chemists, Arthur Harden and W. J. Young, were studying glucose breakdown by yeast cells, a process that generates bubbles of carbon dioxide. Harden and Young observed that the bubbling eventually slowed and stopped, even though there was plenty of glucose left to metabolize. Apparently, some component of the broth was being exhausted. After experimenting with a number of substances, the chemists found that adding inorganic phosphates started the reaction going again. They concluded that the reaction was exhausting phosphate, the first clue that phosphate played a role in metabolic pathways. It would be several decades before biochemists would understand that the inorganic phosphate was being used to form ATP, which is used in glucose disassembly, the first reaction of glycolysis.

Glycolysis begins with the linkage of two phosphate groups to glucose at the expense of two molecules of ATP (Figure 6-2). This process activates the sugar, making it reactive enough to be split into two, three-carbon fragments, each with a phosphate. The loss of a pair of ATPs can be considered the cost of getting into the glucose-oxidation business.

The next two reactions are particularly important because they generate ATP. The first of these reactions is the conversion of phosphoglyceraldehyde (PGAL) to 1,3-diphosphoglycerate (DPG). This reaction releases "high-energy" electrons from PGAL, which are transferred to NAD^+, forming NADH. (NAD^+—nicotinamide adenine dinucleotide—is a coenzyme derived from the vitamin niacin, which we readily obtain from various types of meats and leafy vegetables.) In respiring organisms, the "high-energy" electrons carried by NADH are cashed in for ATP at a later stage in the process involving an electron transport system.

In the next reaction of glycolysis, a phosphate group is transferred directly from DPG to ADP to form a molecule of ATP. This "direct" production of ATP formation is referred to as *substrate-level phosphorylation* because a phosphate group is transferred directly from a substrate molecule and does not require an electron transport chain. A second substrate-level phosphorylation occurs later in glycolysis during the formation of pyruvic acid (Figure 6-2).

We can now look back over glycolysis and total up the energy profits. Four ATPs are produced directly for each glucose entering the pathway. But because two ATPs must be spent to get glycolysis "rolling," the *net* yield is two ATPs. In addition to a net gain of two ATPs, glycolysis also generates two NADHs, each carrying a pair of "high-energy" electrons.

The molecular remains from glycolysis are two molecules of pyruvic acid. This is where anaerobic and aerobic strategies diverge. They differ in the way they solve a common problem: how to transfer electrons from NADH formed during glycolysis to another molecule, regenerating NAD^+. If NADH donates its electrons directly to an organic substrate, which is simply excreted as a waste product, the process is called *fermentation*. If NADH passes its electrons to an electron transport chain in which oxygen is the final electron acceptor, the process is called *aerobic respiration*.

FERMENTATION

During fermentation, the electrons from NADH are transferred to pyruvic acid or to a compound formed from pyruvic acid. The end products of fermentation (Figure 6-3) vary according to the organism but always lead to the regeneration of NAD^+. In the most familiar type of fermentation—*alcoholic fermentation*—yeast cells convert pyruvic acid to ethyl alcohol, the alcohol consumed in beer, wine, and spirits.

Alcoholic drinks are the products of converting pyruvic acid to ethyl alcohol by brewer's yeast. Alcoholic fermentation also releases carbon dioxide which, if not allowed to escape, becomes trapped in the liquid, producing the natural carbonation of beer and champagne. During the leavening of bread and pastries, the carbon dioxide forms the expanding bubbles that cause the dough to rise; the dough literally inflates with the gas. The alcohol is driven off by baking, which explains why you can eat bread without becoming inebriated.

During fermentation, bacteria generate a wide variety of products, of which lactic acid is perhaps the most common. This slightly sour acid imparts flavor to yogurt, rye bread, and some cheeses. It also lowers the pH of food below that which is tolerable by many spoilage microbes. Dairy products that contain lactic acid are therefore more resistant to spoilage than is the milk from which they were made. Swiss cheese is produced by other types of bacteria that convert pyruvic acid to propionic acid and carbon dioxide. The acid imparts a nutty flavor to the cheese, while the gas creates the large bubbles that form its characteristic holes. Vinegar is a product of acetic acid-producing bacteria that grow in apple cider or grape mash. The tangy

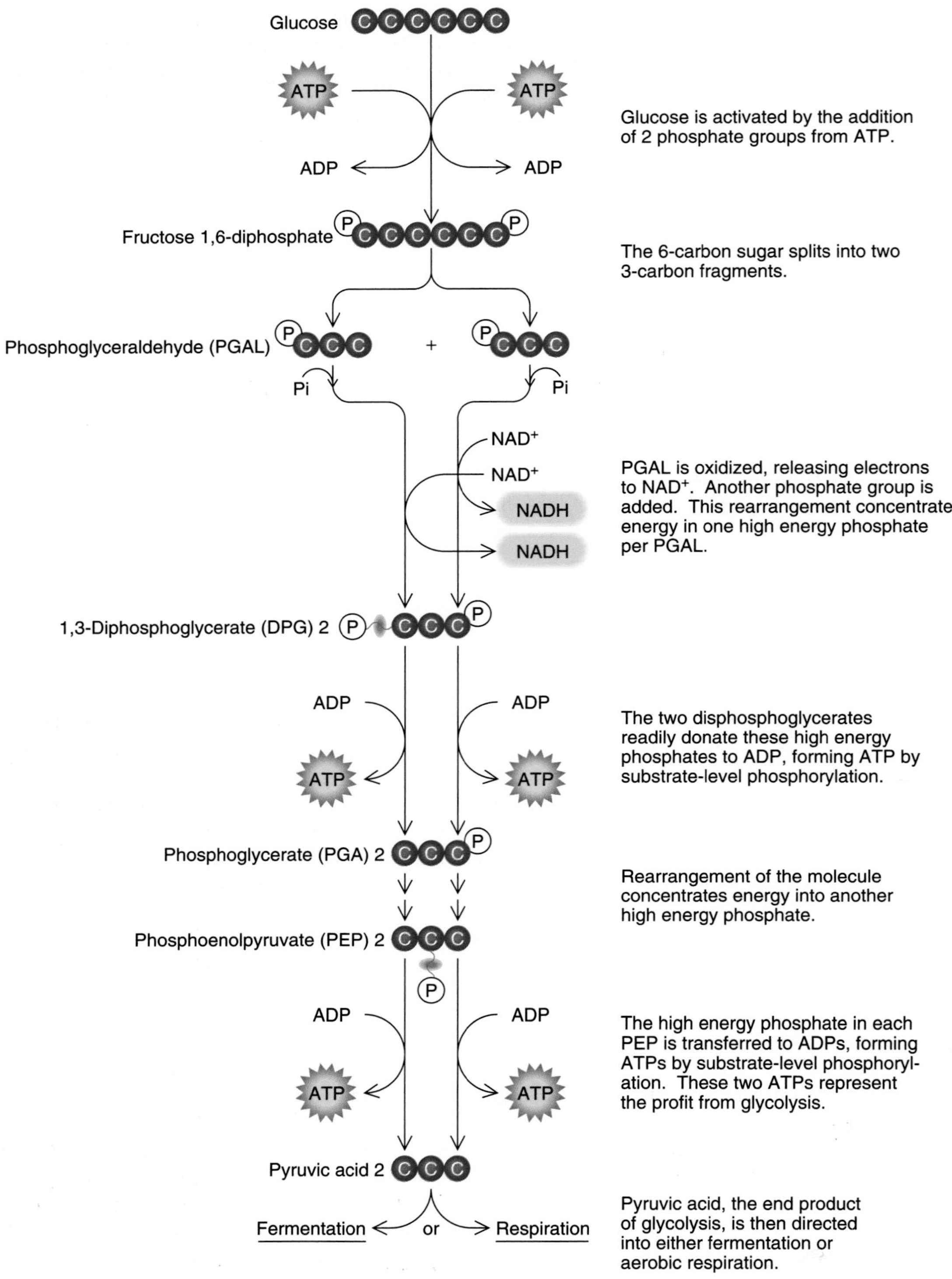
Glucose
ATP
ATP
ADP
ADP
Glucose is activated by the addition of 2 phosphate groups from ATP.
Fructose 1,6-diphosphate
The 6-carbon sugar splits into two 3-carbon fragments.
Phosphoglyceraldehyde (PGAL)
Pi
Pi
NAD+
NAD+
NADH
NADH
PGAL is oxidized, releasing electrons to NAD+. Another phosphate group is added. This rearrangement concentrates energy in one high energy phosphate per PGAL.
1,3-Diphosphoglycerate (DPG) 2
ADP
ADP
ATP
ATP
The two disphosphoglycerates readily donate these high energy phosphates to ADP, forming ATP by substrate-level phosphorylation.
Phosphoglycerate (PGA) 2
Rearrangement of the molecule concentrates energy into another high energy phosphate.
Phosphoenolpyruvate (PEP) 2
ADP
ADP
ATP
ATP
The high energy phosphate in each PEP is transferred to ADPs, forming ATPs by substrate-level phosphorylation. These two ATPs represent the profit from glycolysis.
Pyruvic acid 2
Fermentation
or
Respiration
Pyruvic acid, the end product of glycolysis, is then directed into either fermentation or aerobic respiration.

FIGURE 6-2

A condensed version of glycolysis. Although some of the ten reactions have been omitted for clarity, the overall activities of the pathway are evident in the oxidation of glucose to two molecules of pyruvic acid with the release of two NADHs and a profit of two ATPs. The overall reaction can be written:

Glucose + 2 ATPs + 2 $NAD^+ \longrightarrow$ 2 pyruvic acid + 4 ATPs + 2 NADH.

Note: Each gray ball represents a carbon atom.

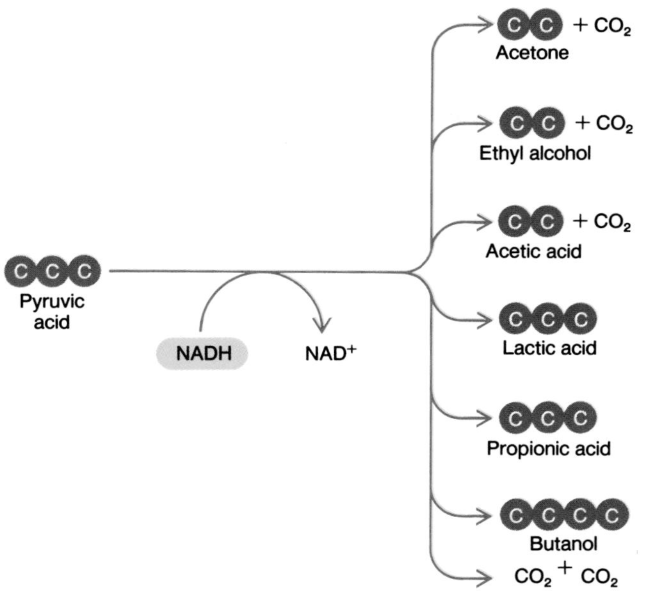

FIGURE 6-3
Some metabolic products of fermentation. Pyruvic acid may be converted to several end products, depending on the organism and, for some organisms, the absence of oxygen. In all cases, the final compound contains electrons donated by NADH. This recycles the coenzyme NAD^+, allowing it to continue its essential role as electron acceptor in glycolysis.

taste of pickles, sauerkraut, and olives reflects the presence of lactic and acetic acids, microbial byproducts that provide a flavorful twist to these foods. Soy sauce requires an 8- to 12-month fermentation period before the mash of soy beans acquires its characteristic flavor.

LACTIC ACID FERMENTATION

Fermentation does not only take place in unicellular yeasts and bacteria; it also occurs in the cells of multicellular organisms. In animal muscles, for example, fermentation of pyruvic acid produces lactic acid. In 1907, the British biochemist Frederick Hopkins demonstrated that when isolated frog muscles were stimulated to contract in an anaerobic (oxygen-devoid) environment, the muscles produced large amounts of lactic acid.

In humans, lactic acid builds when muscles are subjected to strenuous activity and cannot obtain enough oxygen to oxidize the pyruvic acid fully (see The Human Perspective: "The Role of Anaerobic and Aerobic Metabolism in Exercise").

To reiterate, all types of fermentation, whether it leads to ethyl alcohol or lactic acid production, are anaerobic processes. During the early stages of life on earth, before oxygen was available, glycolysis and fermentation were probably the primary metabolic pathways by which energy was extracted from sugars by primitive prokaryotic cells. Today, many organisms live in environments lacking oxygen, such as in deep soil or inside the body of an animal; for them, glycolysis and fermentation remain the only method of obtaining energy.

AEROBIC RESPIRATION

As the earth's atmosphere became infused with molecular oxygen from the activity of early photosynthetic prokaryotes, organisms evolved a new metabolic strategy of *aerobic respiration*, which could completely oxidize the two pyruvic acids generated by glycolysis (instead of fermenting them) and obtain more than 30 additional ATPs. In order to accomplish this, the complete oxidation of glucose requires three sequential groups of reactions—glycolysis, the Krebs cycle, and electron transport (Figure 6-4). Glycolysis always takes place in the cell's cytoplasm. In eukaryotic cells, the Krebs cycle and electron transport occur in mitochondria.

THE KREBS CYCLE

As discussed earlier, very little oxidation occurs during glycolysis. The oxidation of the substrate increases rapidly during the Krebs cycle, however, as (1) hydrogen atoms are stripped from several of the carbons and their "high-energy" electrons transferred to electron-carrying coenzymes (either NAD^+ or FAD), and (2) the carbons are split from the rest of the molecule in the form of carbon dioxide, the most oxidized state of a carbon atom.

The very first reaction (step 1, Figure 6-5) connects glycolysis and the Krebs cycle. During this reaction, the 3-carbon pyruvic acid is split into a 2-carbon and a 1-carbon fragment. The 1-carbon fragment loses a pair of electrons to NAD^+ and is released as carbon dioxide. The 2-carbon

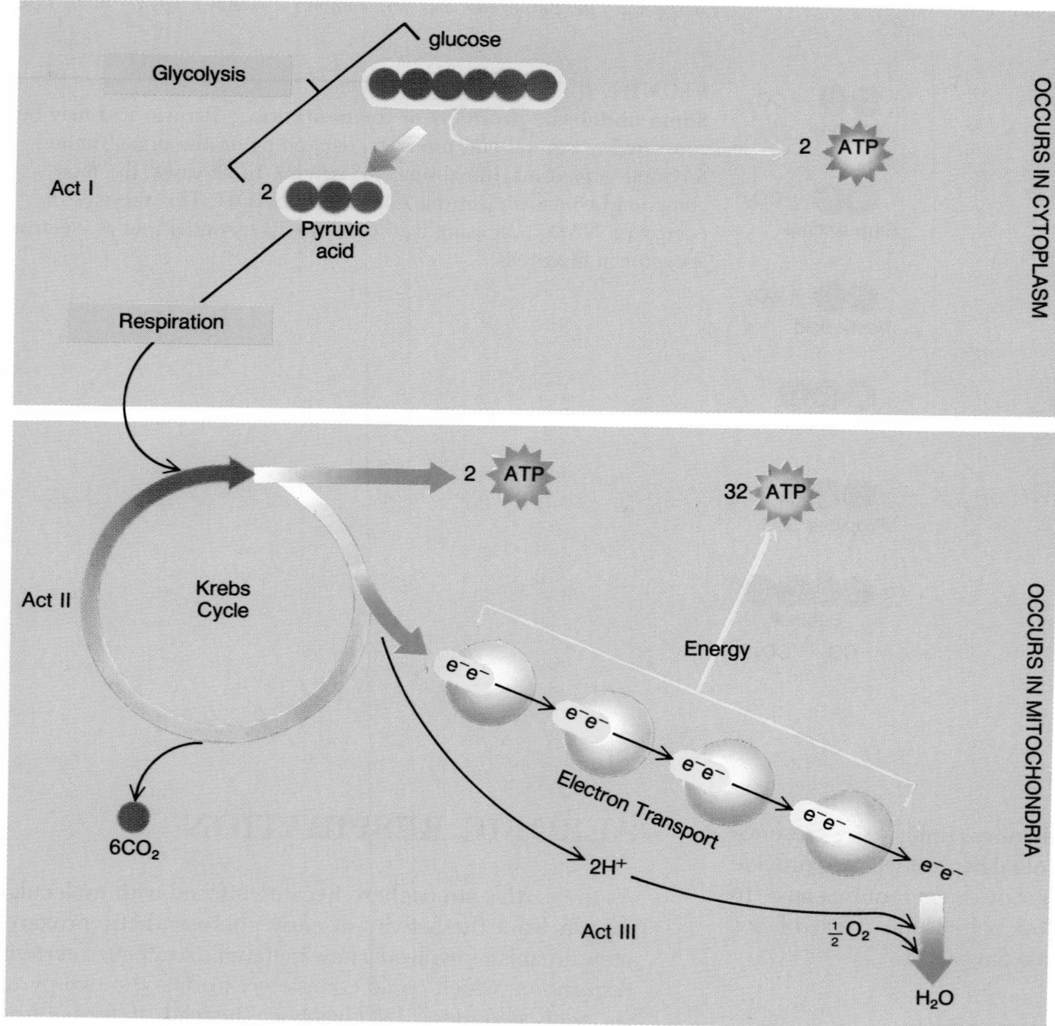

FIGURE 6-4

An overview of respiration. Respiration is a three-part process: 1—Glycolysis; II—The Krebs Cycle; III—Electron Transport. During glycolysis, glucose is converted into two molecules of pyruvic acid with a net formation of two ATPs. As substrates pass through the Krebs cycle, carbons are stripped away and released as carbon dioxide, and high-energy electrons are stripped away and transferred to the coenzymes NAD^+ and FAD. The electron transport system allows the energy of the electrons to be used in the formation of large numbers of ATP. Electron transport depends on the availability of oxygen. Oxygen combines with electrons and protons to form water. Without oxygen, only glycolysis continues and pyruvate is converted to fermentation byproducts such as lactic acid or ethyl alcohol.

acetyl fragment is temporarily coupled with a carrier molecule, coenzyme A (CoA), forming the complex *acetyl coenzyme A* (acetyl CoA). The discovery of acetyl CoA by Fritz Lipmann in 1951 was the last piece in the puzzle of glucose oxidation.

The **Krebs cycle,** named after the British biochemist Hans Krebs who worked out the cyclic metabolic pathway in the 1930s, begins when the two carbons of acetyl CoA are joined with a four-carbon molecule, *oxaloacetic acid* (step 2 in Figure 6-5), releasing CoA for reuse. The product of step 2 is a six-carbon compound *citric acid*, the substance that gives the Krebs Cycle its alternate name, the "citric-acid cycle" and also gives citric fruits (lemons, grapefruits, oranges) their tart flavor.

During the next reactions (steps 3 and 4 of Figure 6-5), two carbons are fully oxidized to carbon dioxide, with the electrons being transferred to two NAD^+ molecules, forming 2 NADHs. One ATP (formed from GTP) is generated directly. This happens as the final carbon dioxide is removed in step 4, leaving a four-carbon compound (*succinic acid*).

The last of the "high-energy" electrons are removed in the final steps of the Krebs cycle, at which point succinic acid is converted to oxaloacetic acid. First, electrons and

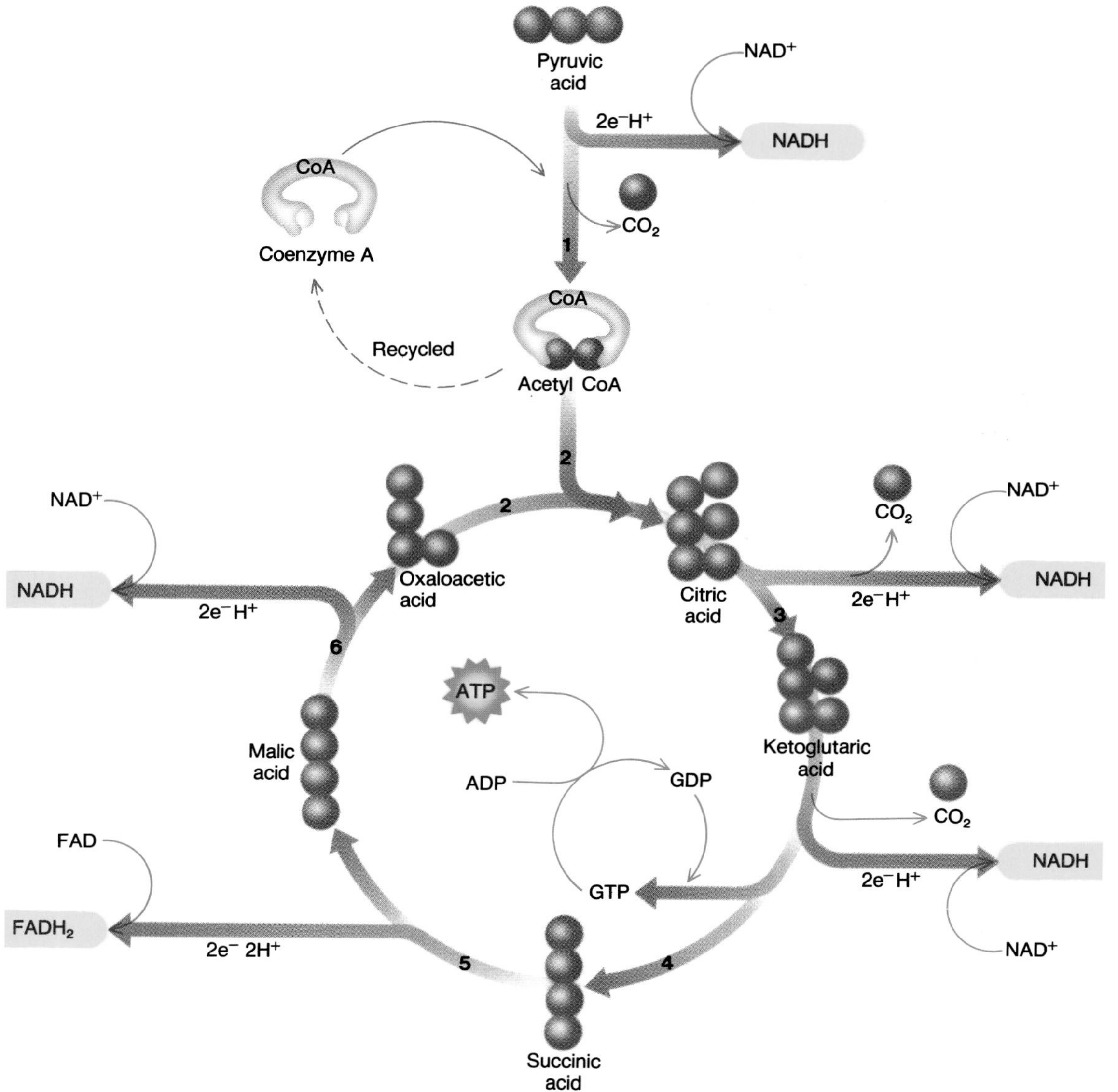

FIGURE 6-5

An abbreviated version of the Krebs cycle. As a result of the reactions shown in this figure, one molecule of pyruvic acid is completely oxidized to three molecules of carbon dioxide. In the process, four NADHs, one $FADH_2$, and one GTP are formed. GTP (guanosine triphosphate) is a high-energy compound that is readily converted to ATP. The energy in the NADH and $FADH_2$ coenzymes can be cashed in for ATPs via the electron transport system and chemiosmosis. Because glycolysis generates *two* pyruvic acids, oxidation of a single glucose molecule produces *twice* the amounts shown here.

two protons (H^+) are transferred to a different coenzyme FAD (*flavin adenine dinucleotide,* which is derived from the vitamin riboflavin, found in meats, fruits, and leafy vegetables), forming the reduced $FADH_2$ (step 5, Figure 6-5). Like NADH, $FADH_2$ is able to feed electrons into the electron transport system to generate ATP. In the subsequent step (step 6), electrons (and a proton) are transferred to NAD^+, forming NADH. Oxaloacetic acid, the last product of the Krebs cycle, is then free to react with another acetyl CoA to form another citric acid, starting the cycle over again. For their work in unraveling the details of the pathway by which pyruvic acid is oxidized, Krebs and Lipmann shared the 1953 Nobel Prize in physiology and medicine.

In summary, each "turn" of the Krebs Cycle oxidizes pyruvic acid to its simplest, most oxidized form, carbon dioxide, gradually releasing energy stored in the original glucose molecule. Two turns of the cycle are required for each glucose molecule. Most of the energy is transferred to NADH and $FADH_2$. In most organisms, the key that unlocks this treasury of energy is a small, electron-attracting molecule called oxygen.

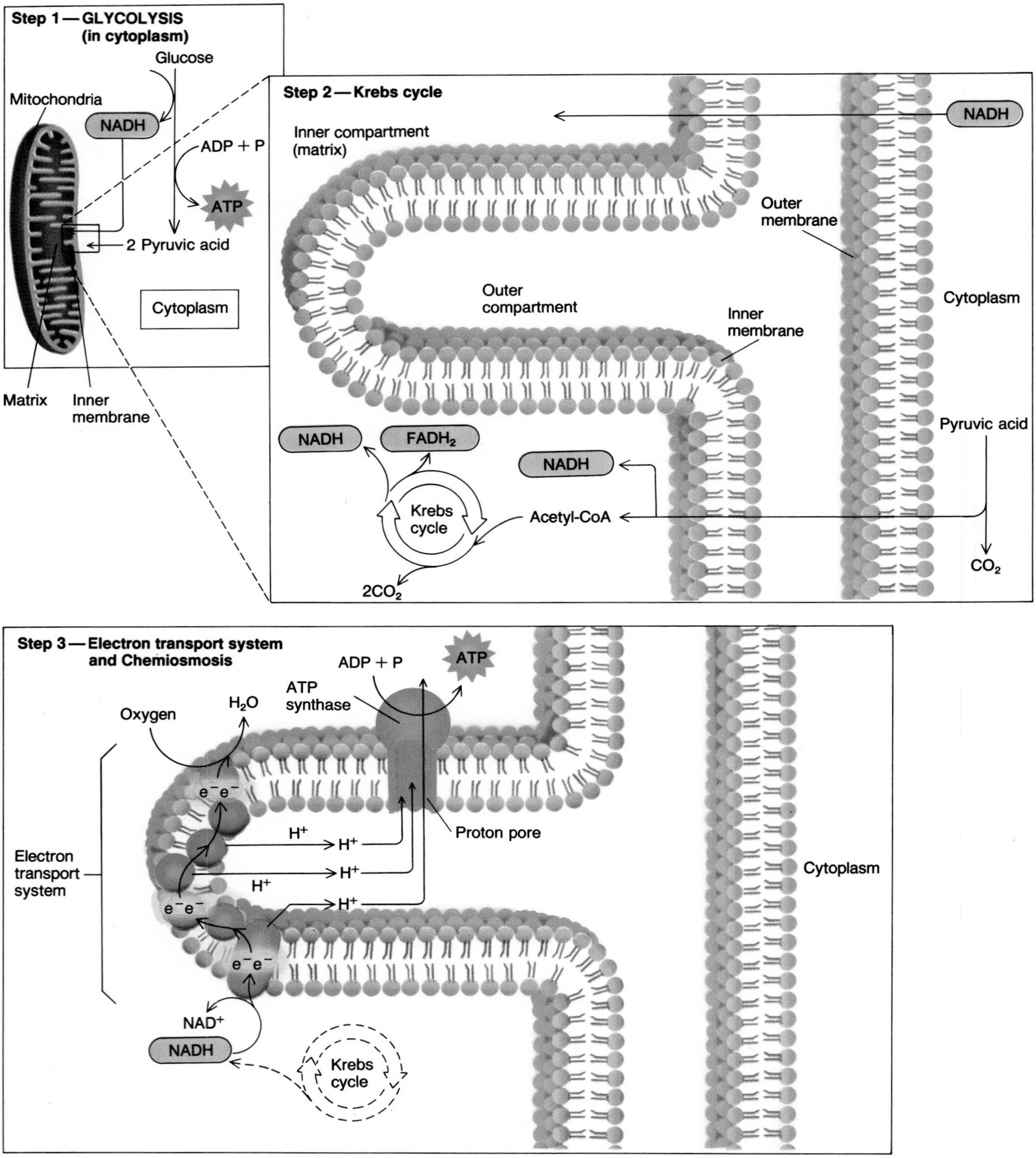

Step 1—GLYCOLYSIS (in cytoplasm)
Glucose
Mitochondria
NADH
ADP + P
ATP
2 Pyruvic acid
Cytoplasm
Matrix
Inner membrane

Step 2—Krebs cycle
Inner compartment (matrix)
NADH
Outer membrane
Outer compartment
Inner membrane
Cytoplasm
Pyruvic acid
NADH
$FADH_2$
NADH
Krebs cycle
Acetyl-CoA
CO_2
$2CO_2$

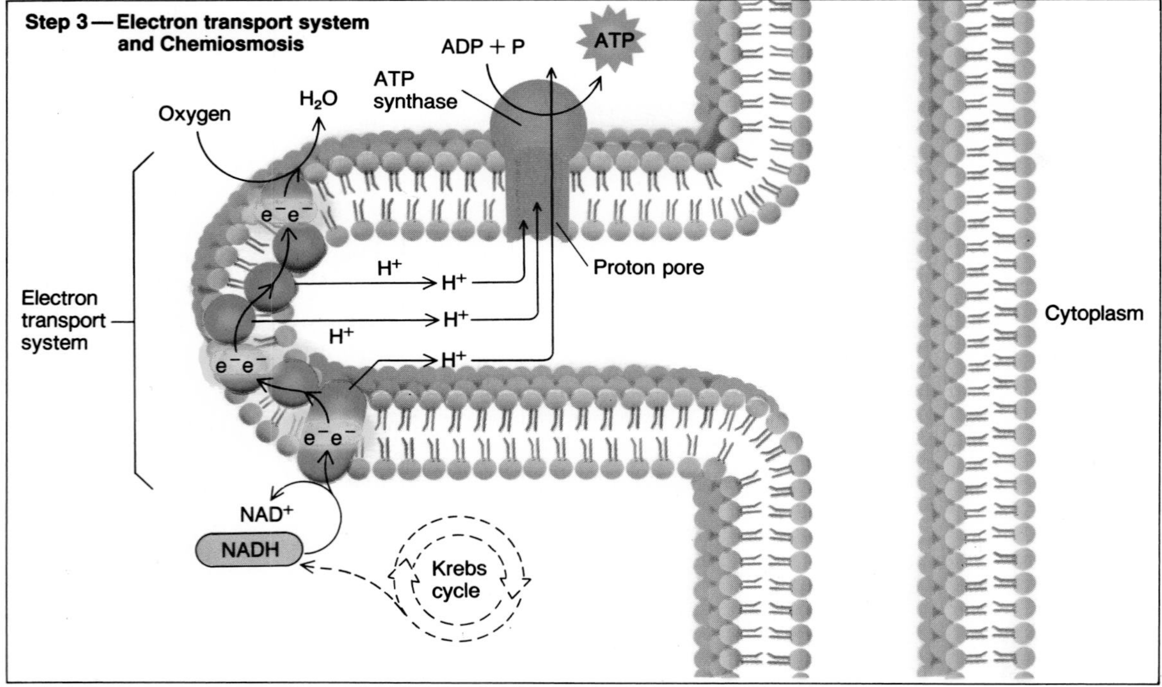

FIGURE 6-6

A closer look at glucose oxidation in eukaryotes. Step 1: Glycolysis converts glucose to pyruvic acid in the cytoplasm, generating NADH and a small amount of ATP as well. Step 2: Pyruvic acid is moved into the inner compartment of the mitochondrion, where it is converted to acetyl CoA and then completely oxidized via the Krebs cycle to carbon dioxide, generating molecules of NADH and $FADH_2$. Step 3: The energized electrons carried by NADH and $FADH_2$ are transported from carrier to carrier along the electron transport system which generates a proton gradient. As protons pass through the ATP synthase channel, they drive the phosphorylation of ADP to form ATP.

THE ELECTRON TRANSPORT SYSTEM

Fermentation taps less than 3 percent of the energy stored in glucose, whereas aerobic respiration captures about 40 percent—enough to form nearly 40 molecules of ATP per molecule of glucose oxidized. Only a small number of ATPs are produced directly (a net of two ATPs during glycolysis and two during the Krebs cycle), the remainder is formed when the energized electrons carried by NADH and $FADH_2$ are processed by the electron transport system (Figure 6-6).

The electrons that had been removed from the organic molecules of the Krebs cycle are high-energy electrons. As the high-energy electrons pass down an electron transport chain, protons move from the inner compartment of the mitochondrion (the matrix) to the outer compartment (Figure 6-6), building a proton gradient across the membrane. This proton gradient drives the formation of ATP by the same mechanism of chemiosmosis that accompanies photosynthesis (see page 94). With their energy released, the electrons are eventually passed to oxygen.

Now you know why you must breathe—to provide a powerful oxidizing agent, oxygen, capable of removing the electrons from the last carrier of the electron transport chain. A common poison, cyanide, blocks the transfer of electrons to oxygen, stopping ATP production. The lethal effects of this poison reemphasizes the essential role of cellular respiration to all aerobic organisms, from bacteria to humans. Without cellular respiration, ATP production drops to practically zero, leading to energy starvation and fatal deterioration of biological order.

BALANCING THE METABOLIC BOOKS

The complete oxidation of one pyruvic acid via the Krebs cycle yields four NADHs, one $FADH_2$, and one GTP (=ATP) (Figure 6-7). Because each glucose yields two pyruvic acids, we need to double these numbers. The reduced coenzymes generated by the Krebs cycle can be "cashed in" at the rate of three ATPs for each NADH and two for each $FADH_2$. Thus, the total profit from oxidizing both pyruvic acids is 30 ATPs (24, or 3 × 8, from NADH, four, or 2 × 2, from $FADH_2$, and the two that are formed directly).

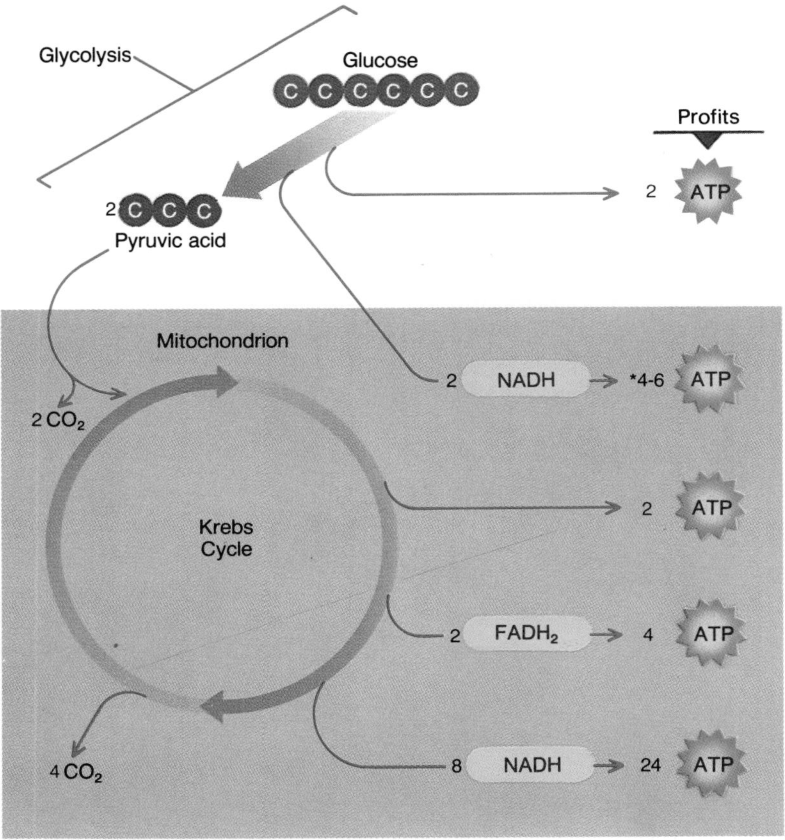

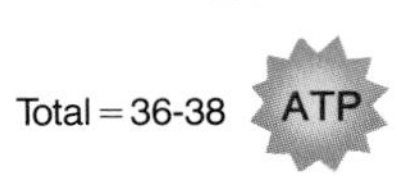

FIGURE 6-7 **A metabolic ledger** reveals the energy profits from respiratory oxidation of one glucose molecule. More than 90 percent of the ATPs are produced as a result of electron transport.

◃ THE HUMAN PERSPECTIVE ▹

The Role of Anaerobic and Aerobic Metabolism in Exercise

Most of you have tried lifting a barbell or doing "push-ups" at some time. You may have noticed that the more times you repeat the exercise, the more difficult it becomes, until you are no longer able to perform the activity. The failure of your muscles to continue to work can be explained by oxidative metabolism. Skeletal muscles (the muscles that move the bones of the skeleton) consist of at least two types of fibers (photo insets): "fast-twitch" fibers, which can contract very rapidly, and "slow-twitch" fibers, which contract more slowly. Fast-twitch fibers are nearly devoid of mitochondria, which indicates that these cells are unable to produce much ATP by aerobic respiration. In contrast, slow-twitch fibers contain large numbers of mitochondria—sites of aerobic ATP production.

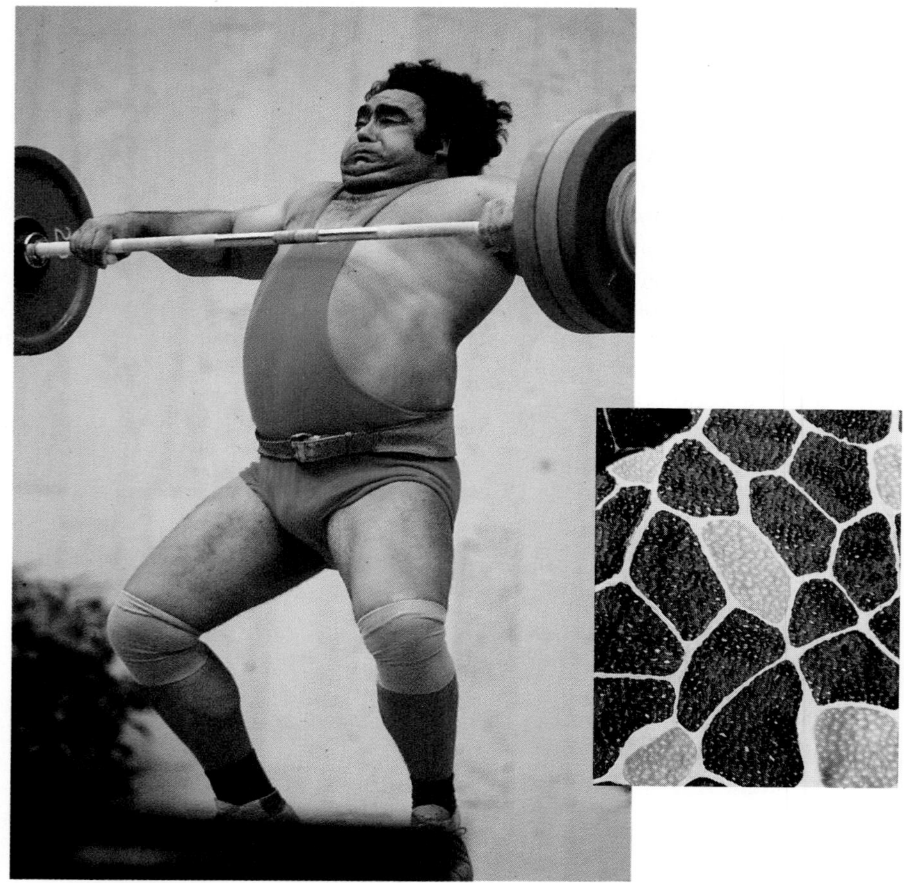

Figure 1
Skeletal muscles (photo insets) contain a mix of fast-twitch fibers (darkly stained) and slow-twitch fibers (lightly stained). The muscles of world-class weight lifters typically have a higher-than-average percentage of anaerobic, fast-twitch fibers, whereas marathon runners tend to have a higher-than-average percentage of aerobic, slow-twitch fibers.

These two types of skeletal muscle fibers are suited for different activities (Figure 1). Lifting weights or doing push-ups depends primarily on fast-twitch fibers, which are able to generate more force than are their slow-twitch counterparts. Fast-twitch fibers produce nearly all of their ATP anaerobically as a result of glycolysis. The problem with producing ATP by glycolysis is the rapid use of the fiber's available glucose and the production of an undesirable end product, lactic acid, producing pain, cramps, and the sensation of muscle fatigue.

If, instead of lifting weights or doing push-ups, you engaged in "aerobic" exercise, such as jumping jacks or fast walking, you would be able to continue to perform the activity for much longer periods of time without feeling muscle pain or fatigue. Aerobic exercises, as their name suggests, are exercises designed to allow your muscles to continue to perform aerobically; that is, to continue to produce the necessary ATP by electron transport. Aerobic exercises depend largely on the contraction of the slow-twitch fibers of skeletal muscles; these muscles are able to generate less force but can continue to function for long periods as a result of continued aerobic ATP production and without the buildup of lactic acid.

The ratio of fast-twitch to slow-twitch fibers in a particular muscle is genetically determined, a factor that may play a role in enabling particular individuals to excel in certain sports. For example, athletes who require short bursts of exertion, such as weight lifters or sprinters, have a higher proportion of fast-twitch fibers in their muscles than do long-distance runners who excel in events that require endurance (Table 1).

TABLE 1

TYPICAL MUSCLE FIBER COMPOSITION IN ELITE ATHLETES REPRESENTING DIFFERENT SPORTS AND IN NONATHLETES

Sport	*% Slow-twitch Fibers*	*% Fast-twitch Fibers*
Distance Runner	60–90	10–40
Track sprinters	25–45	55–75
Shot-putters	25–40	60–75
Nonathletes	47–53	47–53

From S. K. Powers and E. T. Howley, Exercise Physiology: Theory and Application to Fitness and Performance. 1990. W. C. Brown, Dubuque, Iowa. All Rights Reserved. Reprinted by permission.

In addition to these 30 ATPs, there are still the profits from glycolysis. In prokaryotes, electron transport yields an additional six ATPs from the two NADH molecules. In eukaryotic cells, however, these two NADHs yield only four ATPs because two ATPs are needed to transfer electrons from the fluid of the cytoplasm where glycolysis occurs into the mitochondrion where the electron transport machinery is located. If the two ATPs that were formed directly during glycolysis are included, the aerobic respiration of one glucose molecule releases enough energy to produce 36 (in eukaryotes) or 38 (in prokaryotes) ATPs.

Although some of the energy present in the original glucose molecule inevitably is lost, the formation of 36 to 38 ATPs represents an energy-capture efficiency of about 40 percent (the actual percentage varies with the conditions in the cell at the time). Compare this to a typical automobile engine that runs at less than 25 percent efficiency.

COUPLING GLUCOSE OXIDATION TO OTHER PATHWAYS

It is unlikely that you have ever sat down to a meal of pure glucose. What about the hamburger and French fries that may be fueling your biological processes right now? How is that energy utilized?

The respiratory pathway we have just described (glycolysis, Krebs cycle, electron transport) not only extracts energy from glucose but is the central pathway for the breakdown of a diverse variety of organic molecules (Figure 6-8). For example, the protein in hamburger is converted to amino acids in your intestine, absorbed into your bloodstream, carried to the liver, and converted to molecules that are part of the respiratory pathway. The three-carbon amino acid alanine, for example, is converted to pyruvic acid simply by removing its amino group (NH_2). The pyruvic acid then enters the Krebs cycle for complete oxidation. Similarly, the long fatty-acid chains that make up fat molecules in your French fries are broken down into two-carbon units that enter the Krebs cycle as acetyl CoA molecules.

A pyruvic acid or acetyl CoA is treated exactly the same, regardless of whether it came from glucose degradation or from some other carbohydrate, fat, or protein. Thus, any biochemical that can be degraded or converted to an intermediate of the respiratory pathway can be completely oxidized for its energy.

We have just seen how the breakdown products of all types of materials are fed into glycolysis and the Krebs cycle. This principle also works in reverse. That is, virtually

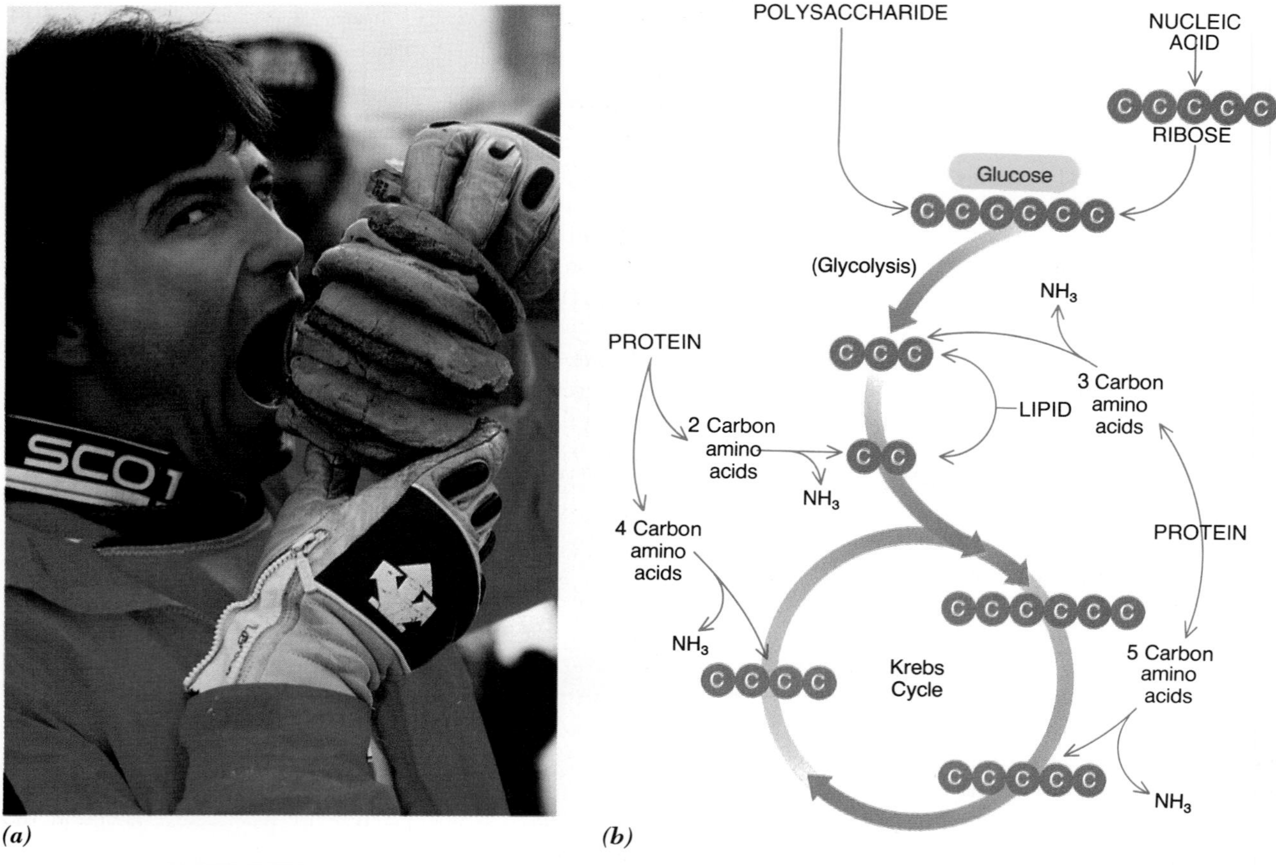

(a) *(b)*

FIGURE 6-8

Some common denominators of metabolism. *(a)* Most meals contain a variety of macromolecules. The meat and cheese in this hamburger are rich in protein and fat, while the bun is rich in polysaccharide. ***(b)*** Each of the catabolic pathways by which proteins, lipids, carbohydrates, and nucleic acids are broken down produce metabolic intermediates that are channeled into glycolysis or the Krebs cycle, where they are completely oxidized. Alternatively, the intermediates of glycolysis and the Krebs cycle provide raw materials that can be diverted into anabolic pathways leading to the synthesis of macromolecules.

any compound that an organism can synthesize can be manufactured by diverting metabolic intermediates of glycolysis or the Krebs cycle into the appropriate biosynthetic pathway. For example, much of the surplus acetyl CoA produced in your body when glucose is abundant will be diverted to fat synthesis for energy storage.

EVOLUTION AND ADAPTATION: TYING IT TOGETHER

Today, oxygen accounts for almost 21 percent of the earth's atmosphere. However during the early stages of life on earth (between 2.5 and 3.0 billion years ago), the earth's atmosphere was completely devoid of oxygen, so only anaerobic organisms were present. These ancient organisms must have gathered ATP directly from their surroundings and/or produced the ATP they needed for life's activities from glycolysis alone.

Two facts support the hypothesis that glycolysis was one of the first energy-harvesting pathways to evolve. First, glycolysis occurs entirely in the cytoplasm of cells and does not require membrane organelles, structures that did not evolve until 2.5 billion years after life first appeared on earth. Second, glycolysis is common to all present organisms, anaerobes and aerobes alike. Recall that glycolysis is a prelude to fermentation as well as to the Krebs cycle and electron transport of aerobic respiration. Because it is a common denominator to current energy-harvesting pathways, it was likely one of the original pathways to evolve on earth.

The invention of photosynthesis by early cyanobacteria at least 2.5 billion years ago changed conditions on earth by gradually infusing the earth's atmosphere and waters with oxygen, a byproduct of photosynthesis. The presence of oxygen paved the way for a new breed of organisms to evolve, organisms that not only could withstand the toxic effects of oxygen but possessed new metabolic pathways that took advantage of the ability of oxygen to attract electrons, and, in the process, to extract energy from organic substances. With greatly increased energy-harvesting efficiency, compared to glycolysis and fermentation, aerobes underwent rapid evolutionary proliferation.

SYNOPSIS

The first stage in glucose disassembly is glycolysis. Glycolysis occurs in the cytoplasm. Here glucose is converted to two molecules of pyruvic acid, four ATPs, and two NADHs (which can be cashed in later for up to six ATPs). Glycolysis costs the cell two ATPs to activate the glucose molecule before it is split into two fragments.

Under anaerobic conditions, cells conduct fermentation as a means to regenerate NAD^+ from the NADH formed during glycolysis. More than 90 percent of the chemical energy contained in glucose remains in the discarded end products of fermentation.

Aerobic respiration continues the disassembly of glucose in the presence of oxygen. The two pyruvic acids generated by glycolysis are completely oxidized to six carbon dioxide molecules. The Krebs cycle occurs in the matrix of mitochondria in eukaryotes or in the cytoplasm of prokaryotes, and generates reduced coenzymes, NADHs and $FADH_2$. For the two pyruvic acids produced at the end of glycolysis, a total of two ATPs, eight NADHs, and two $FADH_2$s are produced by the Krebs cycle. The high-energy electrons carried by NADH and $FADH_2$ are used to generate ATP via the electron transport system. Three ATPs are generated for each NADH and two for each $FADH_2$. As electrons are passed down the electron transport chain, a proton gradient is formed across the inner mitochondrial membrane. As protons flow through the ATP synthase channel, ADP is phosphorylated to form ATP.

Glycolysis, the Krebs cycle, and electron transport are central metabolic pathways that provide virtually all aerobic cells with energy and raw materials. A diverse array of molecules (from proteins to lipids) are broken down and introduced into glycolysis or the Krebs cycle for further metabolism. Intermediate molecules in these pathways can be diverted and used to form various biochemicals, depending on the needs of the cell at the time.

Review Questions

1. Arrange the following molecules in the sequence they would be found during glycolysis and the Krebs cycle:

pyruvic acid	succinic acid
acetyl CoA	citric acid
oxaloacetic acid	PGAL
glucose	DPG

2. Under what conditions do muscle cells form lactic acid? Why do biologists consider this to be adaptive?
3. How is the proton gradient formed in mitochondria during aerobic respiration similar and how is it different from the gradient formed in chloroplasts during photosynthesis?
4. What is the role of NAD^+ and FAD in the Krebs cycle?
5. Rank the following compounds in terms of energy content: pyruvic acid, glucose, carbon dioxide, lactic acid, PGAL.

Critical Thinking Questions

1. Why would isolated vesicles in the supernatant of disrupted mitochondria in Racker's experiment still be able to oxidize glucose?
2. Burning fuel in gasoline-powered engines is about 25 percent efficient whereas releasing the energy in glucose by cells is about 40 percent efficient. What accounts for this difference in efficiency?
3. If you held a spoon of sugar over a flame, it would break down into carbon dioxide and water. Why is activation not necessary when sugar is burned in this way, yet is necessary when sugar is broken down by cells during respiration?
4. Why do organisms that have respiratory equipment ever resort to fermentation? Why can't you—as a human being—switch to fermentation to sustain yourself indefinitely?
5. How do you personally benefit from the fact that your metabolic pathways are interconnected? How does this fact benefit people who have had to subsist on a starvation diet for long periods of time?

CHAPTER

7

Perpetuating Life: Cell Division

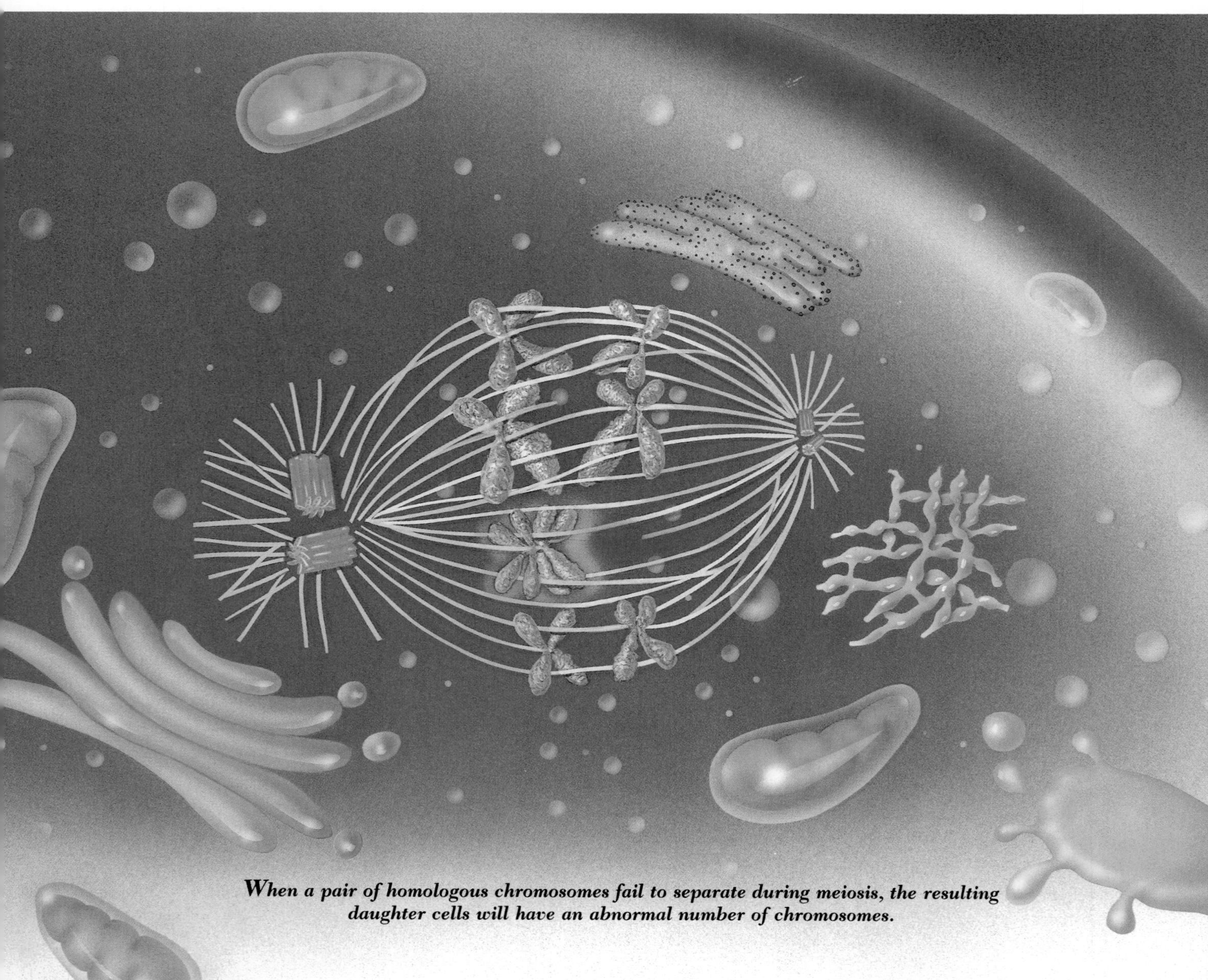

When a pair of homologous chromosomes fail to separate during meiosis, the resulting daughter cells will have an abnormal number of chromosomes.

STEPS TO DISCOVERY
Counting Human Chromosomes

"...*Thus man has 48 chromosomes.*"

This quote was taken from a biology book published in 1952. As you will learn in this chapter, the cells of humans have *46* chromosomes, not 48. Before you condemn the authors of this book for their inaccuracy, consider that, up until 1955, biologists accepted the "fact" that human cells contained 48 chromosomes.

Why did it take so long to discover the right number? Prior to the 1950s, chromosome numbers were determined primarily by examining sections of tissue in which cells were occasionally "caught" in the process of cell division. Trying to count several dozen chromosomes crammed together within a microscopic nucleus is like trying to count the number of rubber bands present in a cellophane package without opening the package. It is actually quite remarkable that biologists got as close as 48! Before 1922, when Theophilus Painter, a geneticist at the University of Texas, arrived at this number, previous guesses had ranged from 8 to more than 50; 48 chromosomes then became the accepted value for more than 30 years until a remarkably simple discovery changed everything.

In 1951, times were tough in academia, and many new PhDs couldn't find teaching positions. Having just completed his PhD on insect chromosomes at the University of Texas, T. C. Hsu fell into this distressing category. Reluctantly, Hsu accepted a postdoctoral research position in Galveston working on mammalian chromosomes, which were notoriously difficult to study. After several months of frustration, he was examining a new batch of cells with a microscope when, in Hsu's words "I could not believe my eyes when I saw some beautifully scattered chromosomes in these cells. I did not tell anyone, took a walk around the building, went to the coffee shop, and then returned to the lab. The beautiful chromosomes were still there. I knew they were real."

He tried to repeat the work, but his preparations regained their "normal miserable appearance." For many months after that, every attempt to discover what he had done "wrong" to get that beautiful preparation failed. Finally, Hsu tried pretreating the cells with a hypotonic (more dilute) salt solution. The cells expanded like balloons and exploded onto the slide, releasing the chromosomes and spreading them out so each was separated from its neighbors (see Figure 7-7*a*). The earlier preparation of cells must have been accidently washed with a dilute saline solution to which someone had failed to add enough salt. Since then, Hsu's hypotonic technique for treating cells has become a standard part of preparing chromosomes for microscopic examination. Ironically, Hsu did not use his technique to reexamine the question of the human chromosome number; Painter had been one of his mentors at the University of Texas, and Hsu never questioned the 48-chromosome count.

Within 3 years, however, Albert Levan and Jo Hin Tijo, working in the United States and Sweden, used the new hypotonic pretreatment technique on human cells treated with a drug called colchicine. Colchicine disassembles the machinery needed for cell division causing cells to remain "frozen" in the process, providing many more dividing cells to observe. Levan and Tijo carefully counted the chromosomes of these cells and found only 46. They repeated their observations and cautiously concluded that, at least in lung cells, the human chromosome number was 46. The number was soon confirmed by other investigators on other human cell types and has been the accepted value ever since.

Meanwhile, Jerome Lejeune, a French clinician, had been studying children with Down syndrome for many years. Such children are characterized by a short stocky stature, distinctive folds of the eyelids (which gave rise to the earlier name "mongolism"), and mental retardation. After hearing a lecture by Jo Hin Tijo, who described how chromosomes could be counted using newly developed techniques, Lejeune decided to examine the chromosomes from a few of his Down syndrome patients. These children showed such a wide range of abnormalities that some alteration in the chromosomes was very likely responsible. Lejeune had never been involved in this type of research, however, and, in fact, did not even possess a microscope. He finally located one that had been discarded by the bacteriology lab. The microscope had been used so much that it would not remain in focus. In order to use it, Lejeune inserted a piece of tinfoil from a candy wrapper to hold the focusing knob in place. In 1959, Lejeune and two colleagues who had helped in cell preparation published a 2 page paper indicating that the cells of nine different patients with Down syndrome all had 47 chromosomes, rather than 46.

Lejeune's paper opened the door to a new field of medical genetics. It was soon followed by a number of other reports in which patients with various types of disorders were shown to have an abnormal number of chromosomes. Examination of cells from fetuses that had spontaneously aborted revealed that many had extra sets of chromosomes in their cells. Thus, an abnormal fetal chromosome number was discovered to be a common cause of miscarriage. Taken together, these insights into the effects of chromosome abnormalities revealed how important it was that the cells of developing embryos contained precisely the correct number of chromosomes. It soon became obvious that the formation of gametes with an abnormal number of chromosomes could be traced to a defect occurring during meiosis—a subject of this chapter.

You were once just a fertilized egg—the product of the union of gametes: a sperm from your father and an egg from your mother. From this inauspicous beginning, you have grown into an organism consisting of trillions of cells. How did this remarkable transformation take place? As discussed in Chapter 3, new cells originate only from other living cells. The process by which this occurs is called **cell division**. For a multicellular organism like yourself, countless divisions of the fertilized egg result in an organism of astonishing cellular complexity and organization. Furthermore, cell division does not stop when you reach adult size but continues throughout life. Biologists estimate that more than 25 million cells are undergoing division each second of your life. This enormous ouput of cells is needed to replace those cells that have aged or died. Old, worn blood cells, for example, are removed and replaced by newcomers at the rate of about 100 million per minute. Not surprisingly, then, anything that blocks cell division, such as exposure to heavy doses of radiation, can have tragic effects. For example, many people who valiantly worked to seal the damaged nuclear reactor at Chernobyl, in the former Soviet Union, died because their bodies were unable to produce new healthy blood cells.

Each dividing cell is called a **mother cell**, and its descendants are appropriately called **daughter cells**. There is a reason for using these "familial" terms. The mother cell transmits copies of its hereditary information (DNA) to its daughter cells, which represent the next cell generation. Cell division is more than just a means of reproducing more cells; it is the basis for reproducing more *organisms*. Cell division, therefore, forms the link between a parent and its offspring; between living species and their extinct ancestors; and between humans and the earliest, most primitive cellular organisms.

▼ ▼ ▼

CELL DIVISION IN EUKARYOTES

The division of a eukaryotic cell into daughter cells occurs in two stages. First, the cell's nuclear contents are divided by either *mitosis* or *meiosis*, and then the cell is actually split into two by *cytokinesis*. As we will see below, mitosis and meiosis are similar in some ways and quite different in others. Most importantly, the two processes serve different roles in the lives of eukaryotic organisms. **Mitosis** produces nuclei that contain the same number of chromosomes as the nucleus of the mother cell. Mitotic cell divisions are responsible for generating all the cells of the body *except* those that will develop into reproductive cells (gametes). **Meiosis**, in contrast, produces nuclei that have half the number of chromosomes as the nucleus of the mother cell. Meiotic cell divisions are a key step in the formation of gametes, which are ultimately required for sexual reproduction.

THE CELL CYCLE

The life of a cell begins with its formation from a mother cell by division and ends when the cell divides to form daughter cells of its own or when it dies. The stages through which a cell passes from one cell division to the next constitute the **cell cycle** (Figure 7-1). The cell cycle is divided into two main phases: **M phase**, during which mitosis occurs; and **interphase**, which occupies the remainder of the cell's life. During interphase, a cell grows in volume, and normal metabolic functions, such as glucose oxidation or the production of proteins for export, are carried out. The divisions of interphase are indicated in Figure 7-1. It is during the S phase of interphase that the cell duplicates its DNA in anticipation of an upcoming mitosis.

We described in Chapter 1 how organisms utilize homeostatic mechanisms to maintain stable internal conditions. Controlling the rates at which cells divide is an important element in maintaining homeostasis; malfunctions in these controls can lead to one of the most dreaded types of diseases—cancer (see The Human Perspective: Cancer: The Cell Cycle Out of Control, page 122).

MITOSIS

The name "mitosis" comes from the Greek word *mitos*, meaning "thread." The name was first used in the 1870s to describe the threadlike chromosomes that appeared to "dance around" the cell just before it divided in two. Mitosis is a continuous process. For the sake of discussion, however, we will divide mitosis into four sequential phases: prophase, metaphase, anaphase, and telophase (Figure 7-2). Each phase is defined by the behavior of the chromosomes.

PROPHASE: PREPARING FOR CHROMOSOME SEPARATION

The first phase of mitosis, **prophase**, is the longest phase and includes a number of complex activities, including chromosome condensation and formation of the spindle apparatus (Figure 7-2).

Chromosome Condensation

Each chromosome contains a single molecule of DNA which, together, with its associated protein, is spread

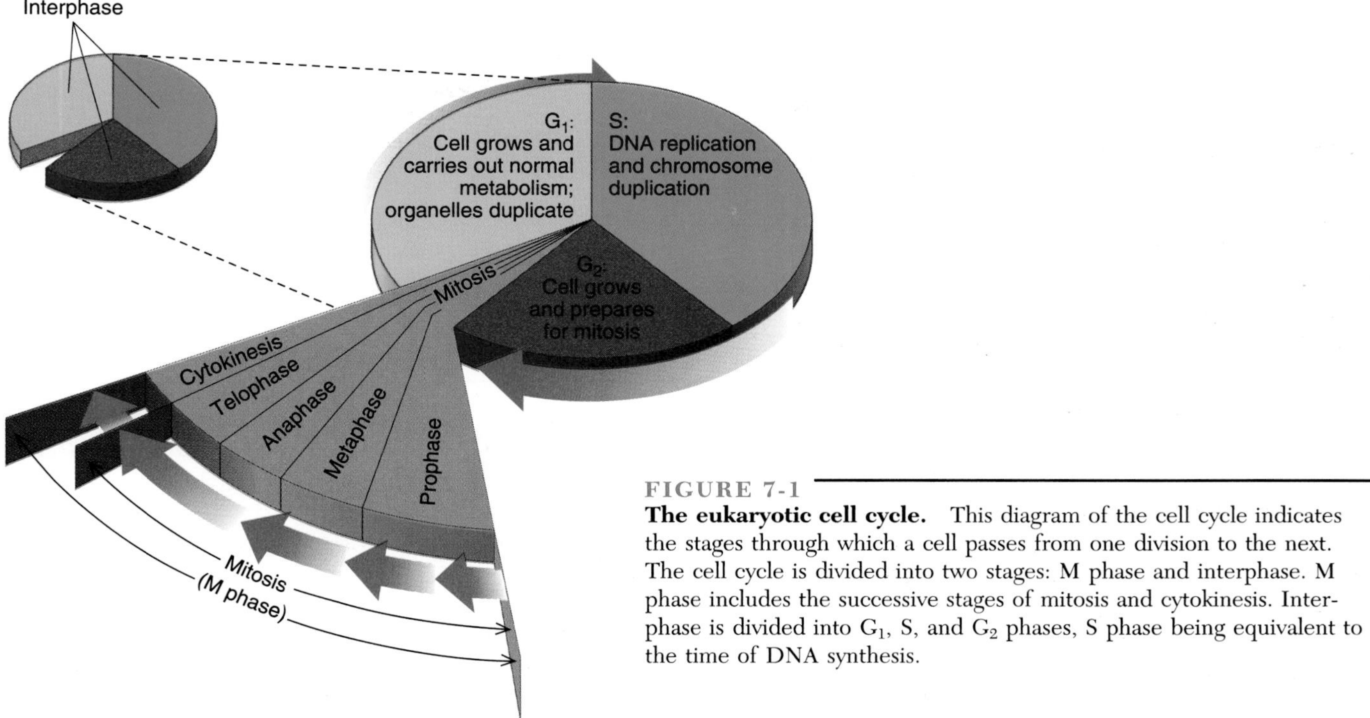

FIGURE 7-1
The eukaryotic cell cycle. This diagram of the cell cycle indicates the stages through which a cell passes from one division to the next. The cell cycle is divided into two stages: M phase and interphase. M phase includes the successive stages of mitosis and cytokinesis. Interphase is divided into G_1, S, and G_2 phases, S phase being equivalent to the time of DNA synthesis.

throughout the nucleus of a nondividing cell. As cell division approaches, these remarkably long DNA-protein fibers undergo a coiling process, in which they are packaged into compact **mitotic chromosomes**—structures that are ideally suited for the upcoming separation process.

The importance of chromosome coiling becomes clear when you consider that the DNA of a single human chromosome is 30,000 times longer in the extended state than when it is coiled. Imagine the molecular chaos that would reign inside a human cell during the separation of the duplicates of 46 *uncoiled* chromosomes. Inevitable tangling would pull the uncoiled DNA to pieces, forming a jumble of DNA fragments that would be unevenly sorted into the daughter cells.

As the condensation of the chromatin nears completion, it becomes evident that each mitotic chromosome consists of two identical partners, called **chromatids** (Figure 7-3). The chromatids are joined together at the **centromere**, a site where each chromosome is constricted. The two chromatids that make up a mitotic chromosome are genetically identical to each other; they are the visible result of the process of chromosome duplication that took place at a stage prior to the onset of mitosis.

Formation of the Spindle Apparatus

The precise separation of duplicated chromatids requires the activity of a cellular "machine" called the **spindle apparatus**. The spindle apparatus is constructed of fibers composed of bundles of microtubules. The "basket" of spindle fibers forms both a supportive scaffolding and a chromosome-pulling machine (Figure 7-4). Remarkably, the microtubules that make up the spindle apparatus are formed from the same subunits that a few minutes earlier might have been part of different microtubules having an entirely different function. The disassembly and reassembly of microtubules is analogous to demolishing an old brick building and then using the same bricks to construct a new building—all within a matter of minutes.

Assembly of the spindle apparatus takes place outside the nucleus at the same time the chromosomes are becoming compacted inside the nucleus. Toward the end of prophase, the nuclear envelope surrounding the compacted chromosomes becomes fragmented and disappears from view. At about the same time, the nucleoli disappear, and the chromosomes become attached to the ends of certain spindle fibers. By the end of prophase, the chromosomes have moved toward the center of the cell.

METAPHASE: LINING UP THE CHROMOSOMES

During **metaphase**, the chromosomes of the dividing cell are aligned in a plane that typically lies at the cell's "equator"; that is, midway between the spindle poles (Figure 7-2). The chromosomes are held in place by the spindle fibers that extend from the ends (poles) of the spindle apparatus to their attachment site at the centromeres of the chromosomes.

ANAPHASE: SEPARATING THE CHROMATIDS

Anaphase begins with the sudden, synchronous separation of the attached chromatids (which, after separation, are now called chromosomes). The duplicated chromosomes move away from each other, toward opposite spindle poles. Since each daughter cell receives one copy of every chromosome, the two daughters are genetically identical to each other as well as to the mother cell from which they arose.

TELOPHASE: PRODUCING TWO DAUGHTER NUCLEI

Telophase begins when the chromosomes reach their respective spindle poles (Figure 7-2). The events of telo-

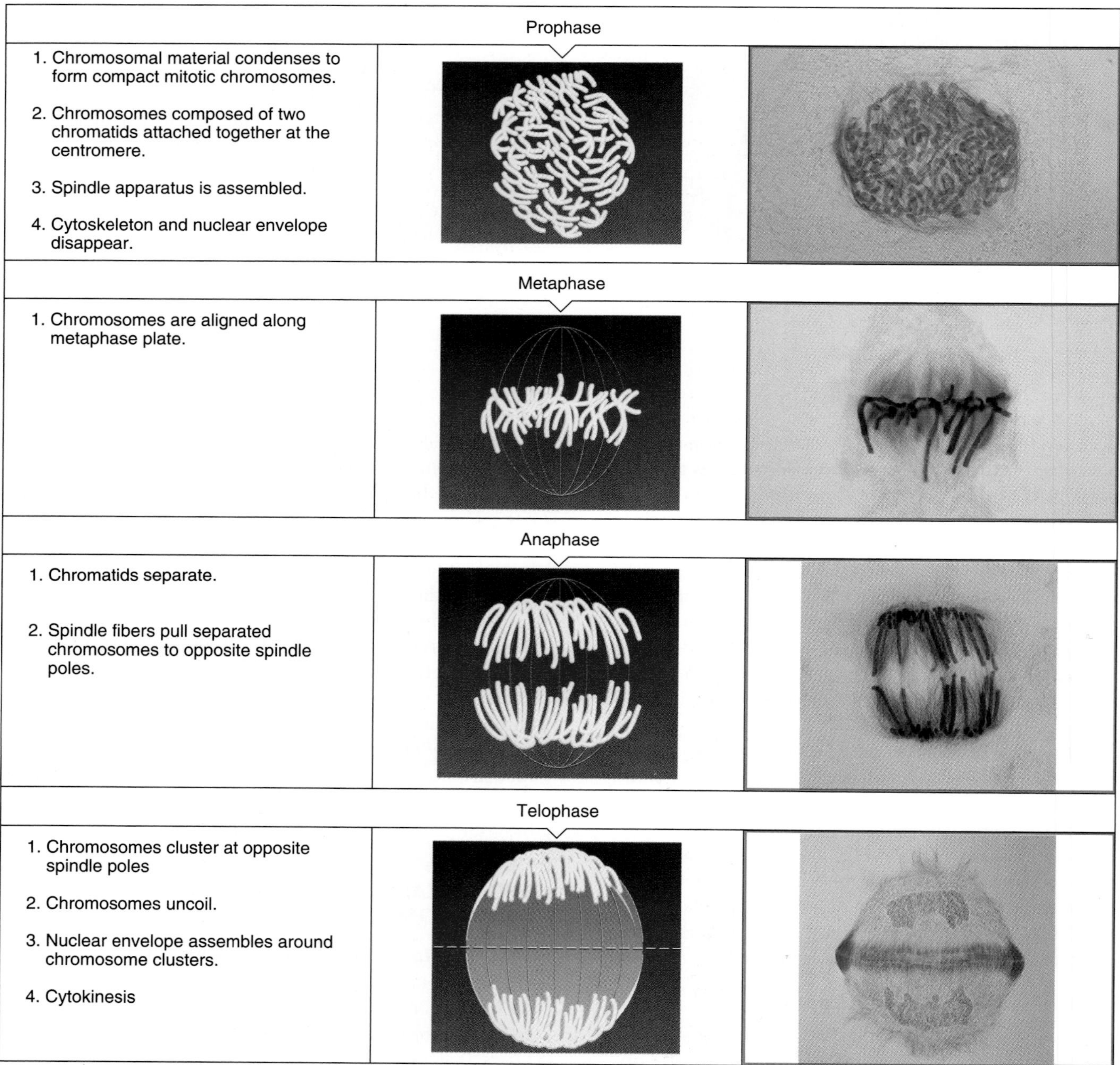

Prophase
1. Chromosomal material condenses to form compact mitotic chromosomes.
2. Chromosomes composed of two chromatids attached together at the centromere.
3. Spindle apparatus is assembled.
4. Cytoskeleton and nuclear envelope disappear.
Metaphase
1. Chromosomes are aligned along metaphase plate.
Anaphase
1. Chromatids separate.
2. Spindle fibers pull separated chromosomes to opposite spindle poles.
Telophase
1. Chromosomes cluster at opposite spindle poles
2. Chromosomes uncoil.
3. Nuclear envelope assembles around chromosome clusters.
4. Cytokinesis

FIGURE 7-2
The phases of mitosis.

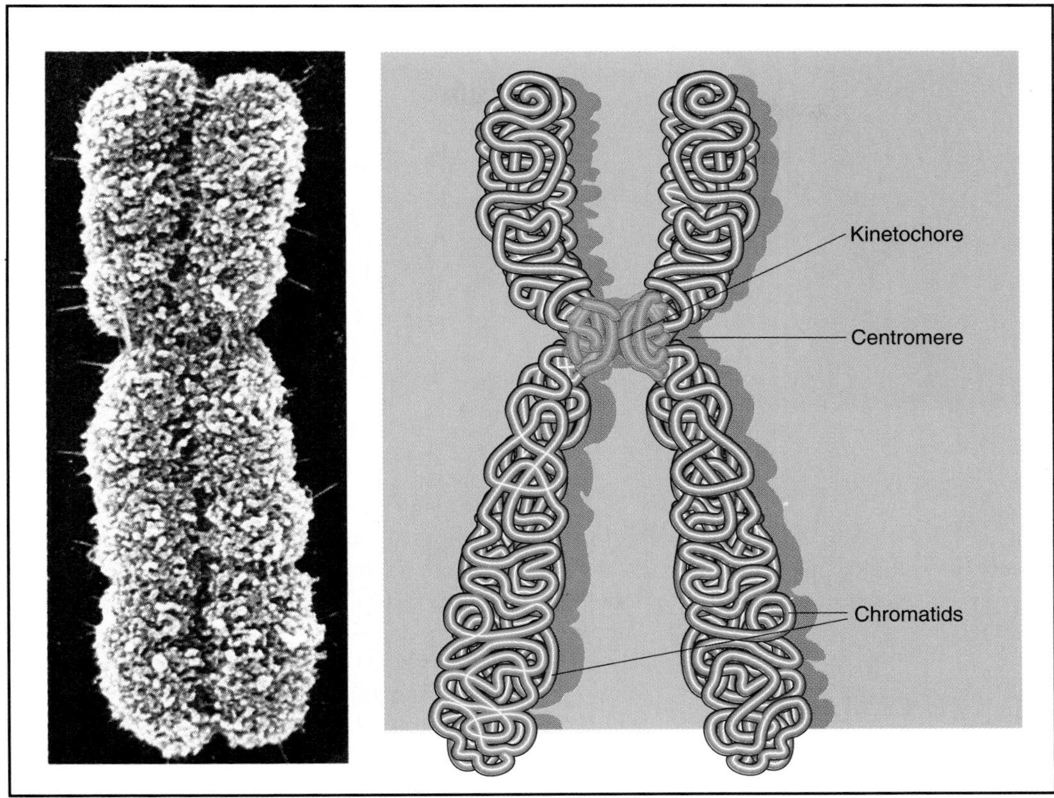

FIGURE 7-3
The structure of a mitotic chromosome. During mitosis, the DNA-protein fibers coil, as is indicated in both the drawing and the electron micrograph. Each chromatid remains distinct but is joined to the other in the indented region of the chromosome (the *centromere*). When the centromere is examined under the electron microscope, it is seen to contain a dense, protein-containing structure called the *kinetochore*.

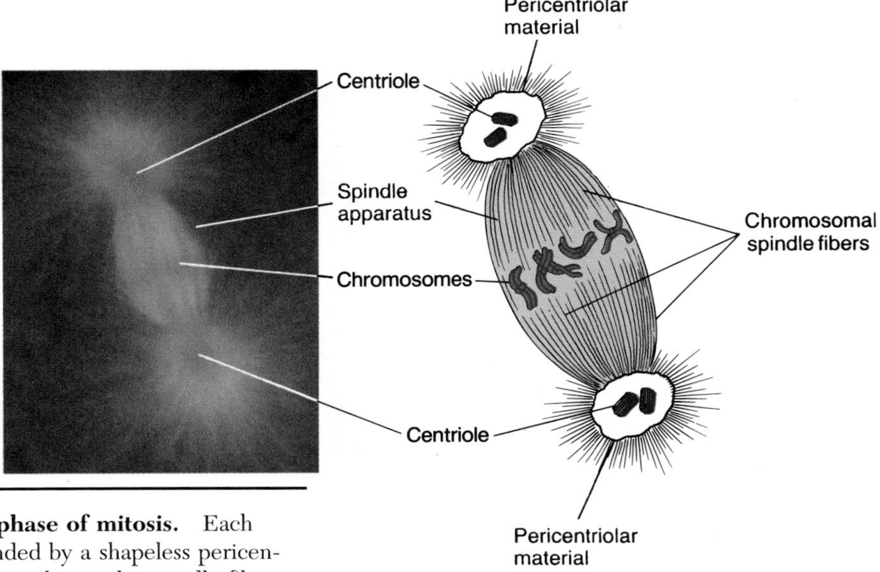

FIGURE 7-4
Spindle apparatus of an animal cell in metaphase of mitosis. Each spindle pole contains a pair of centrioles surrounded by a shapeless pericentriolar material from which the microtubules that make up the spindle fibers radiate. Some spindle fibers stretch from one pole to the other, while other spindle fibers attach to the kinetochores of the chromosomes.

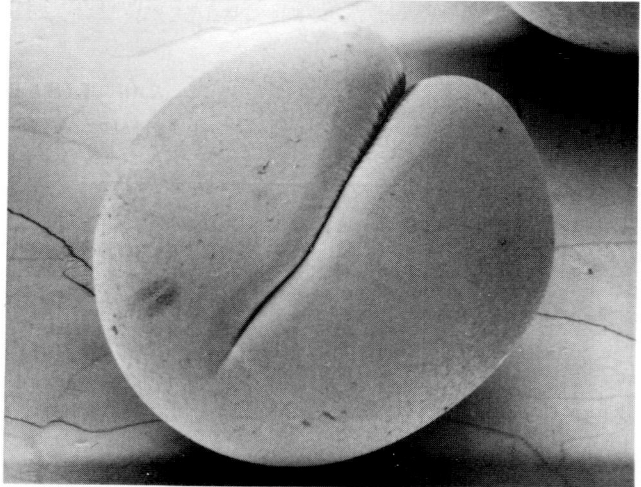

FIGURE 7-5
One cell splits into two. During cytokinesis, a frog egg divides into two cells as microfilaments contract and draw in the plasma membrane. Eventually, the microfilaments pull the edges of the plasma membrane together, pinching the cell's cytoplasm in two.

phase are virtually the reverse of those that occur during prophase. The chromosomes uncoil, the spindle apparatus is dismantled, nucleoli reappear, and nuclear envelopes form around each chromosome cluster.

CYTOKINESIS: DIVIDING THE CELL'S CYTOPLASM AND ORGANELLES

Although mitosis is very similar in plant and animal cells, the way in which the cytoplasm is divided during cytokinesis is very different. If, for example, you watch a sea urchin egg or frog egg undergoing its first cell division, you will see the first hint of cytokinesis as an indentation of the cell surface during late anaphase. As time passes, the indentation deepens and becomes a *cleavage furrow* that moves through the cytoplasm, pinching the cell in two (Figure 7-5). The plane of the furrow always lies in the same plane previously occupied by the metaphase chromosomes, which ensures that the separated chromosomes will be partitioned into different cells.

Cytokinesis in animal cells is the result of contractions of a ring of microfilaments (page 63), which assembles just beneath the plasma membrane. As the ring contracts, it pulls the overlying membrane inward, constricting the cell, much like a purse string closes the diameter of a purse's opening. The ring of microfilaments disassembles

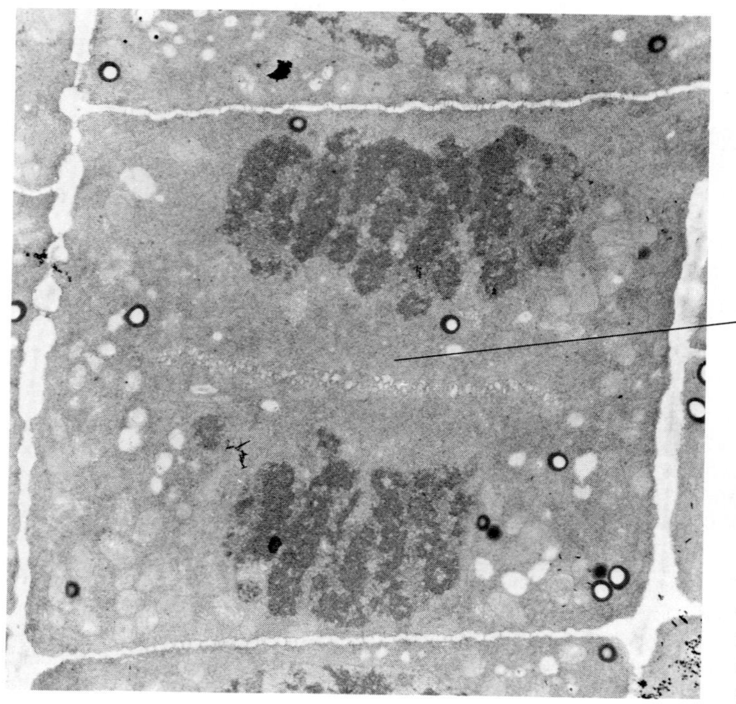

FIGURE 7-6
Cytokinesis in an onion root cell. Secretion vesicles from nearby Golgi complexes become aligned midway between newly formed daughter nuclei. The membranes of the vesicles will fuse to form the plasma membranes of the two daughter cells, while the contents of the vesicles will provide the material that forms the cell plate that will separate these cells. Additional materials for constructing a cell wall for each daughter cell will be delivered later by other vesicles.

after constriction is completed, and the subunits are put to use in the formation of microfilaments of the cytoskeleton.

Formation of a contractile ring would not be possible in plant cells because the rigid plant cell walls cannot be pulled inward. In a plant cell, cytokinesis begins with the formation of a **cell plate** between the daughter nuclei (Figure 7-6). The cell plate is formed by vesicles containing polysaccharides produced by the nearby Golgi complex (page 60). As these vesicles accumulate in the center of a dividing cell, their membranes fuse. The released polysaccharide forms the cell wall between the new cells, and the membranes of the fusing vesicles become incorporated into the plasma membranes of the adjoining daughter cells.

CHROMOSOME NUMBERS: HAPLOIDY AND DIPLOIDY

If you were to scrutinize carefully the chromosomes in a cell that was undergoing mitosis, you might notice that each chromosome had a "partner" that was identical in size and shape (Figure 7-7). The two similar-shaped chromosomes are described as **homologous chromosomes**,

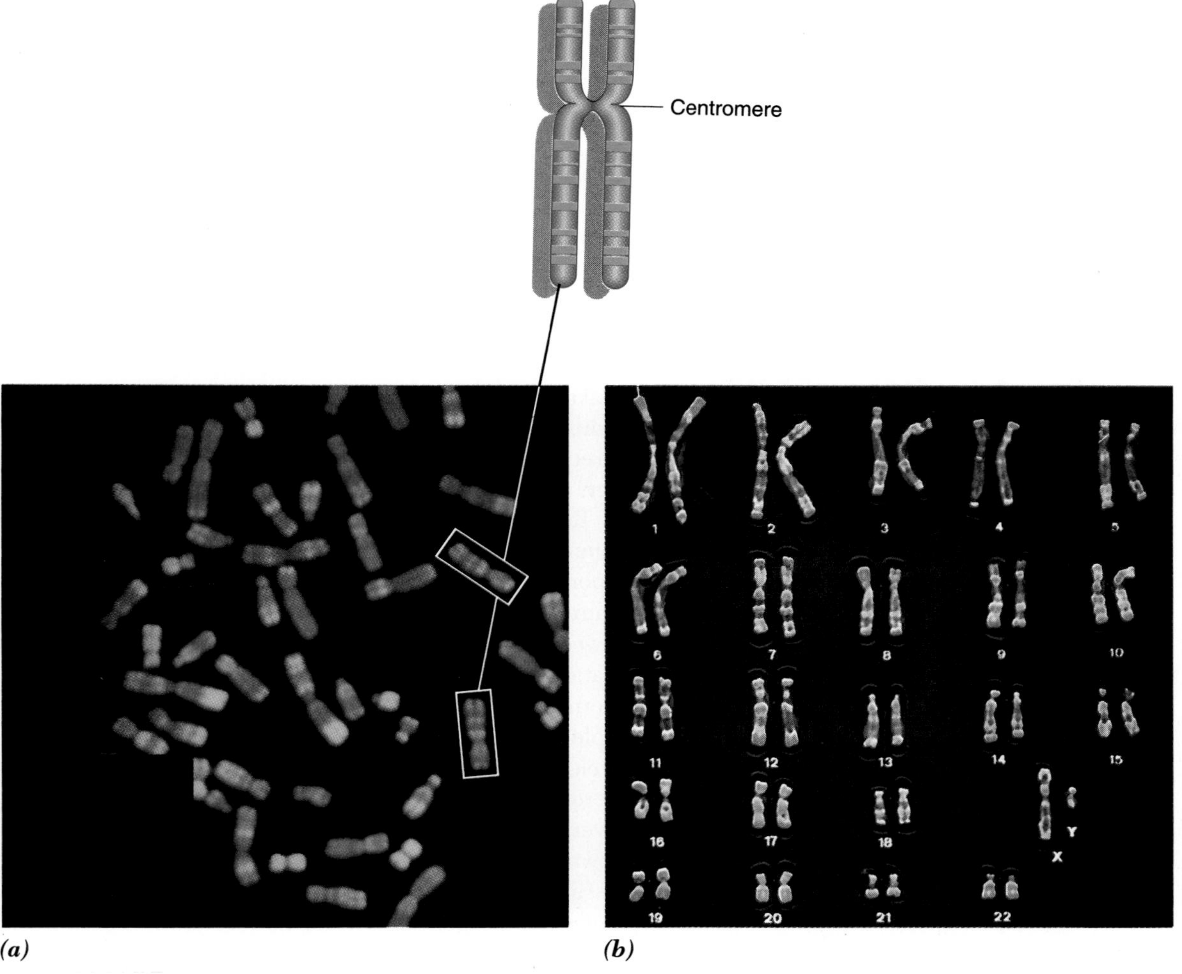

(a) ***(b)***

FIGURE 7-7

Human mitotic chromosomes. Top inset shows a stylized drawing of a stained chromosome. ***(a)*** Photograph of a cluster of mitotic chromosomes that spilled out of the nucleus of a single dividing human cell. The diploid set of 46 chromosomes has been stained with a dye that gives the chromosomes a banded pattern (see top inset). The two chromatids that make up each chromosome can be distinguished from each other. As discussed in the text, diploid cells contain pairs of homologous ("lookalike") chromosomes. One pair of homologous chromosomes is indicated by the boxes. ***(b)*** The chromosomes of a human male. In this figure, called a *karyotype,* homologous chromosomes are paired and arranged according to number (size). If the chromosome preparation had been made from the cells of a female, two X chromosomes would be seen, instead of an X and Y. Karyotypes are prepared from a photograph of chromosomes that have spilled out of a single nucleus. Each chromosome is cut out of the photograph and homologues are paired as shown.

or simply **homologues** (*homo* = same, *log* = writing) because each has the same sequence of genes along its length as the other. Even though homologues have the same genes—in the sense that they determine the same trait—they may have different *versions* of those genes. For example, consider a gene for height in plants. One homologue may code for a tall plant, while the other may code for a short one. Both chromosomes have a gene for height at the same location, but each may produce a different version of the trait.

Most eukaryotic cells carry two complete sets of homologues—a **2N** number of chromosomes; such cells are said to be **diploid**. In diploid cells, one set of homologous chromosomes was originally contributed by each parent during sexual reproduction. Every cell in your brain, for example, has 46 chromosomes—a set of 23 donated by your father and a homologous set of 23 from your mother. Put differently, the cells in your body contain 23 pairs of homologous chromosomes which can be displayed as a **karyotype** (Figure 7-7*b*). Each species has a characteristic diploid number, ranging from 2 in the horse roundworm and *Penicillium* fungus to a whopping 1,262 in Adder's tongue fern. Our closest relative, the chimpanzee, has 48 chromosomes, most of which are indistinguishable in appearance from our own.

In contrast, **haploid** cells contain only one set of homologues—a **1N** number of chromosomes. The haploid number for humans is 23. Except for haploid sperm or eggs produced by your testes or ovaries, the cells in your body are diploid.

MEIOSIS

There are many similarities between mitosis and meiosis, even the names of the stages are the same (prophase, metaphase, anaphase, and telophase). There are also some very important differences, however. Meiosis occurs in diploid cells, but not in haploid cells. Meiosis produces *four* daughter nuclei, each containing a haploid number of chromosomes. A mother cell undergoing meiosis is able to produce four daughters by duplicating its chromosomes prior to meiosis and then proceeding through two consecutive nuclear divisions—*meiosis I* and *meiosis II*. Each nuclear division is accompanied by a corresponding division of the cytoplasm, resulting in four haploid cells that will eventually develop into gametes (sperm or eggs). When two haploid gametes come together at fertilization, the fertilized egg (zygote) regains the diploid number.

The sequence of meiotic stages is shown in Figure 7-8.

MEIOSIS I

Meiosis I is called the **reduction division** of meiosis because the nuclei of the two daughter cells each contain half the number of chromosomes as does the nucleus of the mother cell. This reduction in the number of chromosomes is achieved by separating the members of each pair of homologous chromosomes into different nuclei. As a result, the daughter cells from meiosis I are haploid. In order to ensure that each of the daughter cells has only one member of each pair of homologues, an elaborate process of chromosome pairing occurs that has no counterpart in mitosis. During the period in which the homologous chromosomes are paired, some very important chromosome choreography takes place that increases genetic variability.

Prophase I

Prophase I begins when the diffuse threads of the interphase chromosomes begin coiling. When the chromosomes have become partially compacted, the members of each homologous pair recognize each other and become aligned together along their entire length. This process of chromosomal alignment is called **synapsis**. Since each chromosome is made up of two chromatids, a synapsed pair of homologous chromosomes is called a **tetrad**, which is a unit of four chromatids. Early observers noticed that synapsed homologous chromosomes actually become wrapped around each other (Figure 7-9). In 1909, F. A. Janssens proposed that this interaction might result in the breakage and exchange of pieces between different chromatids of the tetrad. Janssens' hypothesis proved to be correct; he had predicted the process now called **crossing over**.

Recall that homologous chromosomes have identical sequences of genes along their length, although the corresponding genes on the two chromosomes may code for different characteristics of a particular trait. Crossing over between homologous chromatids can produce gene combinations different from those originally found in either chromatid. This is illustrated in Figure 7-10. In this example, one chromosome (on top) of a plant directs the development of a tall individual with yellow flowers, whereas its homologue produces a short plant with white flowers. Crossing over can *reshuffle* these genes—a process known as **genetic recombination**. In our example, crossing over and recombination team the gene for a tall plant with the gene for white flowers on one chromatid and the gene for a short plant with the gene for yellow flowers on the homologous chromatid. Genetic recombination greatly increases the genetic variability of the offspring and, because of this, is one of the biological processes responsible for generating the diversity of life on earth.

Metaphase I, Anaphase I, and Telophase I

Near the end of prophase I, the nucleolus and nuclear envelope become dispersed; spindle fibers become connected to the chromosomes; and the tetrads become aligned, marking the start of metaphase I (Figure 7-8).

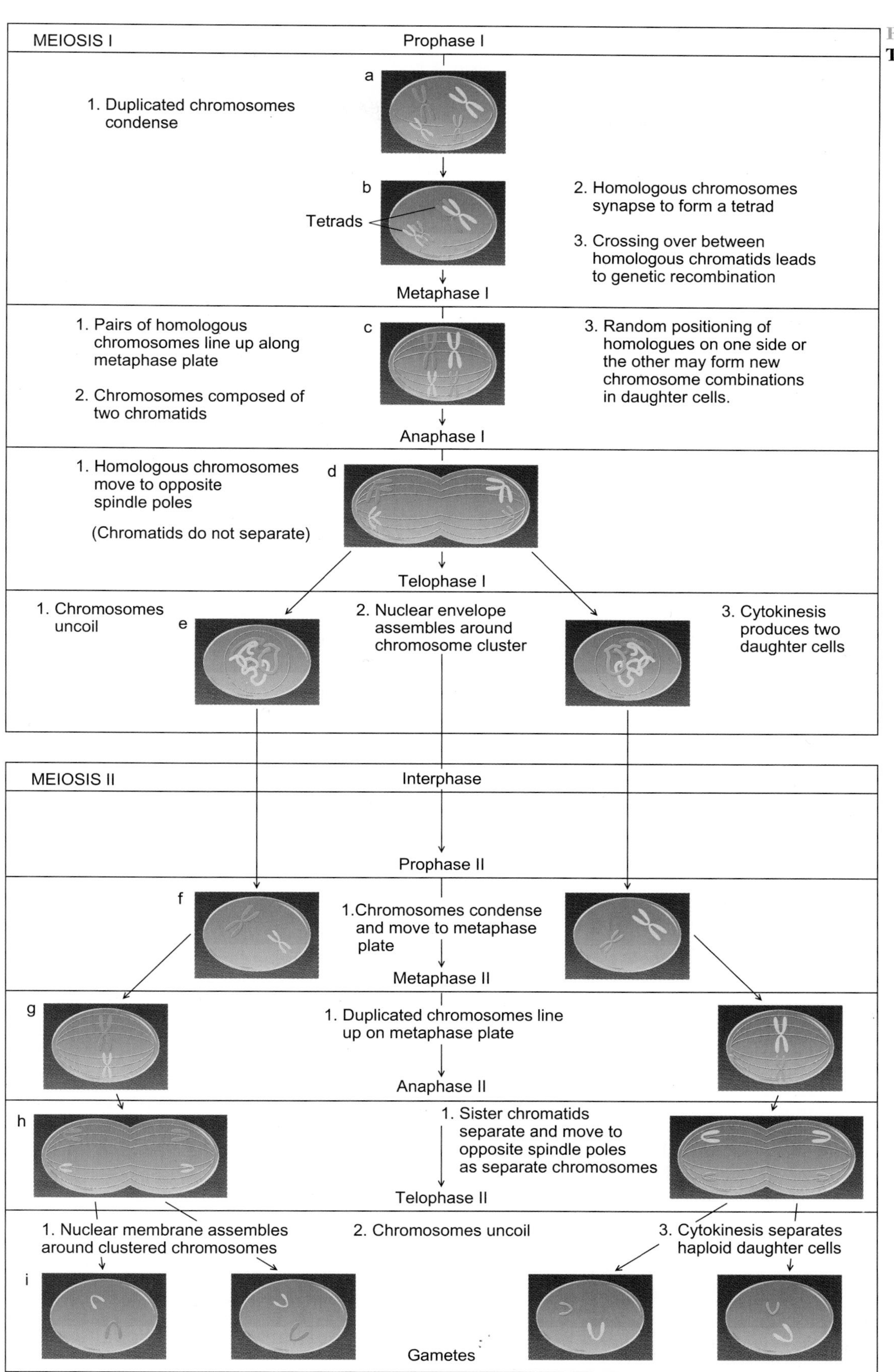

FIGURE 7-8
The stages of meiosis.

◁ THE HUMAN PERSPECTIVE ▷

Cancer: The Cell Cycle Out of Control

CANCER! The name alone strikes fear in all of us—and for good reason. Despite enormous research efforts, cancer remains a leading cause of death in the Western world. The development of better treatments for cancer depends on achieving a better understanding of the reasons why a normal cell becomes transformed into a malignant cell—one that is capable of destroying the organism of which it is a part.

Cancer is a disease that results from uncontrolled cell division. Normal cells may divide very rapidly, as occurs, for example, among the cells of the skin at the edge of a wound. These normal cells are closely regulated, however, and they stop dividing when the wound has grown over and healed. In contrast, cancer cells continue to grow and divide indefinitely. They are somehow freed from the normal metabolic checks and balances that would otherwise limit and coordinate their growth with other cells. Since cancerous cells invade and spread to other tissues—a process known as *metastasis*—many organs of the body can become adversely affected.

Why would a perfectly normal cell begin dividing wildly? We don't yet know the answer to this question, but it is becoming increasingly apparent that the genes involved with the cell cycle play an important role. Normal cells become transformed into cancer cells when something happens to certain genes, converting them to **oncogenes**, which causes them either to change their level of activity or to produce proteins with altered amino acid sequences (that is, mutant proteins). This is the basis of the action of *carcinogens*, such as cigarette smoke, ultraviolet radiation, X-rays, and more than 1,000 known chemicals, including numerous pesticides, household products, and food additives. Carcinogens act by causing alterations in the DNA.

The study of oncogenes has led to a better understanding of the genes that control normal cellular growth and division. The first oncogene discovered was subsequently shown to code for a protein kinase—an enzyme that adds phosphate groups to other proteins. (For their discovery of this oncogene in the mid 1970s, J. Michael Bishop and Harold Varmus of the University of California were awarded the 1989 Nobel Prize.) There are now several dozen known oncogenes, and the list continues to grow. Included on the list are genes that code for (1) growth factors—that is, substances that bind to cell receptors and stimulate the cell to divide; (2) receptors for growth factors—that is, the cell-surface protein that binds the growth factor and mediates its response; (3) regulatory proteins that bind to genes involved in cell growth; and (4) a number of protein kinases. Since all of these genes are normally involved in cell growth and division, it is easy to understand how changes in these proteins could make cells less responsive to the body's growth-control mechanisms. The more we know about the molecular mechanisms by which cells lose growth control, the greater the likelihood we will be able to develop effective means to enable cells to regain this control.

Anaphase I begins when the spindle fibers attached to each chromosome shorten, pulling homologous chromosomes of each tetrad to opposite spindle poles (Figure 7-8). As a result of anaphase I, each daughter cell receives one set of homologous chromosomes, each chromosome consisting of a pair of joined chromatids. The separation of homologous chromosomes from one another is known as *segregation*.

The separation of each pair of homologous chromosomes during anaphase I occurs independently of the separation of other pairs. Consequently, the chromosomes derived from each parent (the *maternal chromosomes* and *paternal chromosomes*) are sorted independently during meiosis I (Figure 7-11). This **independent assortment** of homologues produces gametes with new chromosome combinations. Depending on the number of chromosomes, independent assortment can produce an enormous potpourri of possible mixtures. For a human, this chromosome potluck works out to more than 8 million ways the 23 chromosomes originally donated by your father can be mixed with the 23 chromosomes from your mother during anaphase I. Add to this the gene exchanges that occur during crossing over, and it is easy to see why offspring may resemble one or both parents but, in most cases, look very different from either one of them.[1]

Telophase I of meiosis produces less dramatic changes than does telophase of mitosis. Although in many cases the chromosomes undergo some uncoiling, they do not reach the extremely extended state of the interphase nucleus. Telophase I is followed by cytokinesis, which splits the cell into two daughter cells.

[1]Keep in mind that we are considering chromosomes of three different generations: grandparents, parents, and children. A grandmother and grandfather provide maternal and paternal chromosomes, respectively, to a parent. These maternal and paternal chromosomes assort independently during meiosis and gamete formation in the parent. The gamete provided by the parent fuses at fertilization with a gamete from another individual to produce the chromosomes of the child.

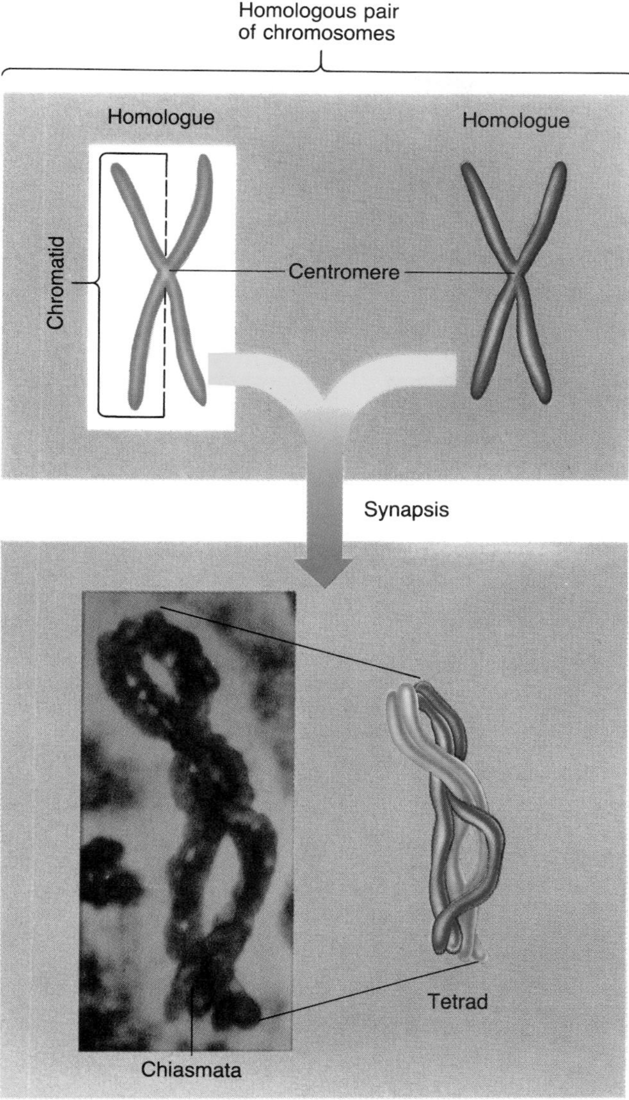

FIGURE 7-9
Homologous chromosomes forming a tetrad after synapsis. The points at which the two homologous chromosomes are in contact are called *chiasmata*. These are sites where crossing over is believed to have occurred at an earlier stage.

MEIOSIS II

Since sister chromatids become separated from one another into different nuclei during **meiosis II**, the events in meiosis II are similar to those that occur during mitosis. During prophase II, the chromosomes shorten and thicken and then become aligned during metaphase II. Anaphase II begins when the joined chromatids split apart and the newly independent chromosomes are pulled toward opposite spindle poles. During telophase II, a nuclear envelope is assembled around each of the four clusters of chromosomes, generating four nuclei. The two cells then divide by cytokinesis to produce a total of four haploid daughter cells. Not only do these daughter cells have half the number of chromosomes as the mother cell, the chromosomes that are present may be heavily altered due to crossing over.

THE RISKS OF MEIOSIS

On occasion, a mistake occurs during meiosis. Homologous chromosomes may fail to separate from each other during meiosis I (see drawing, page 112), or sister chromatids may fail to come apart during meiosis II. In either case, gametes are formed that contain an abnormal number of chromosomes—either an extra chromosome or a missing chromosome. If one of these gametes containing an abnormal number of chromosomes happens to fuse with a normal gamete during reproduction, a zygote with an abnormal chromosome number is formed. In humans, most such abnormal zygotes develop into embyros that die in the womb, causing a miscarriage. In a few cases, however, the zygote develops into an infant whose cells have an abnormal chromosome number, which causes the infant to develop characteristic deformities. The most common congenital disorder resulting from a meiotic mistake is Down syndrome, a condition in which individuals have three copies of chromosome number 21 in their cells, rather than two.

Persons with Down syndrome typically suffer from mental retardation, frequent circulatory problems, increased susceptibility to infectious diseases, a greatly increased risk of developing leukemia, and the early onset of Alzheimer's disease. Seventy years ago, children with Down syndrome had an average life expectancy of about

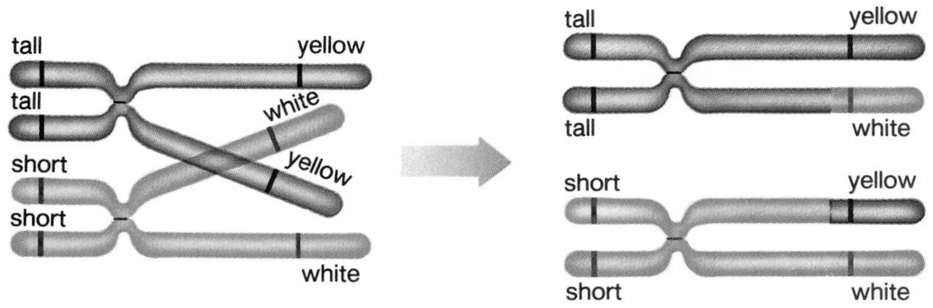

FIGURE 7-10
Genetic recombination. A pair of meiotic chromosomes each with two duplicates (sister chromatids) joined to one another. The sites on the chromosomes for the genes that govern two plant traits—height and flower color—are indicated. After crossing over, each chromosome contains one chromatid with the original combination of genetic characteristics and one chromatid with a mixture of the maternal and paternal characteristics.

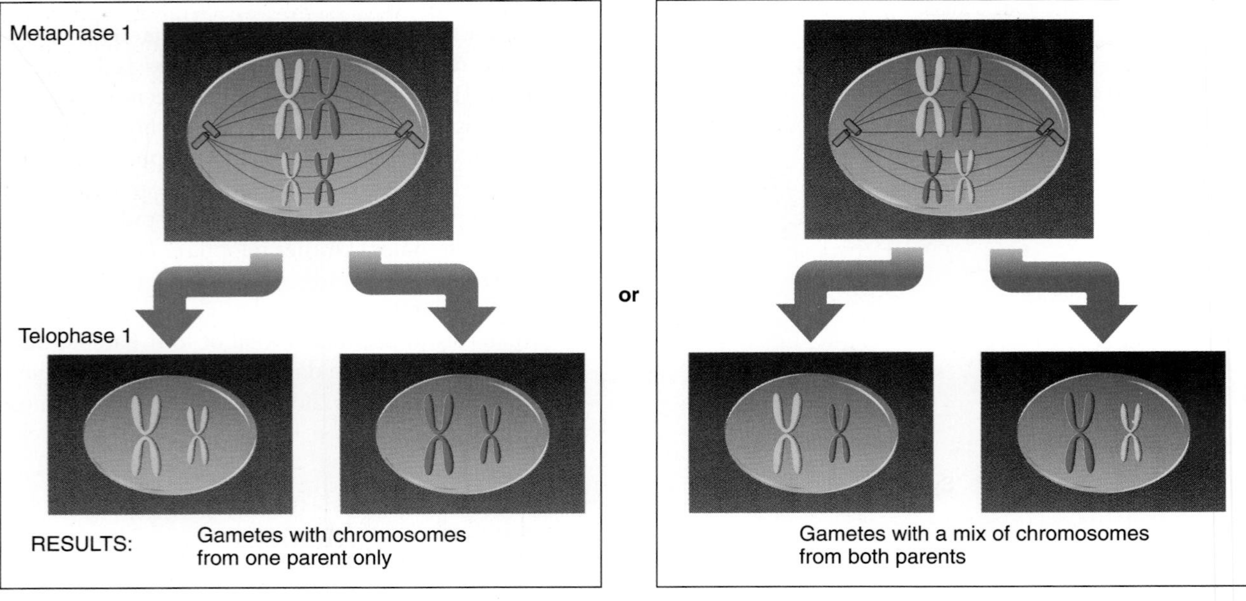

(a)

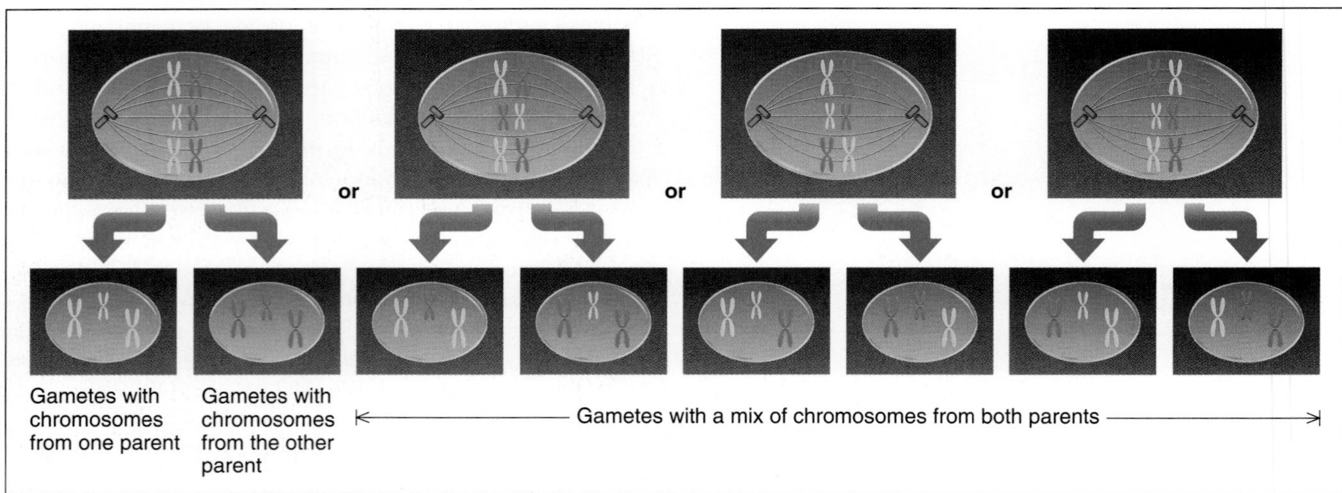

(b)

FIGURE 7-11

Independent assortment of homologous chromosomes during metaphase I of meiosis. Paired homologous chromosomes line up randomly along the metaphase plate. Because each pair of homologous chromosomes lines up independently of other pairs, new chromosome combinations are often produced in the daughter nuclei, leading to increased genetic variability in a species. (To follow independent assortment, the homologues originally derived from one parent are colored green, and those from the other parent are yellow.) ***(a)*** In a cell with two pairs of homologous chromosomes, there are four possible combinations of chromosomes and, thus, four genetically different daughter nuclei that can form. Either both chromosomes from the same parent go into the same daughter cell, or the chromosomes are mixed and each daughter cell receives one chromosome from each parent. ***(b)*** In a cell with three pairs of chromosomes, there are eight (2^3) genetically different daughter nuclei that can form. Because of independent assortment during meiosis I, the number of possible chromosome combinations increases with each additional chromosome. In human cells with 23 pairs of chromosomes, there are about 8.5 million (2^{23}) possible chromosome combinations, not including the added genetic variability introduced by crossing over.

10 years; most of them were shut away in mental institutions, where they withered and died. Today, most of these children are raised at home and are encouraged to develop to their full potential; many attend "regular" schools and grow up to become working adults.

EVOLUTION AND ADAPTATION: TYING IT TOGETHER

Mitosis and meiosis are two processes that occur by similar cellular mechanisms but play very different roles in the life of an organism. Both processes are characterized by chromosome coiling, the assembly of a spindle apparatus, and the separation of chromosomes into different cells. However, whereas mitosis preserves genetic composition and maintains chromosome number, meiosis shuffles genetic characteristics and reduces chromosome number (Figure 7-12). Mitosis produces the body's cells, while meiosis functions in the formation of gametes, the cellular vehicles of sexual reproduction.

Consider for a moment what would happen if meiosis did not occur and the gametes contained the same number of chromosomes as other cells. Every time two gametes

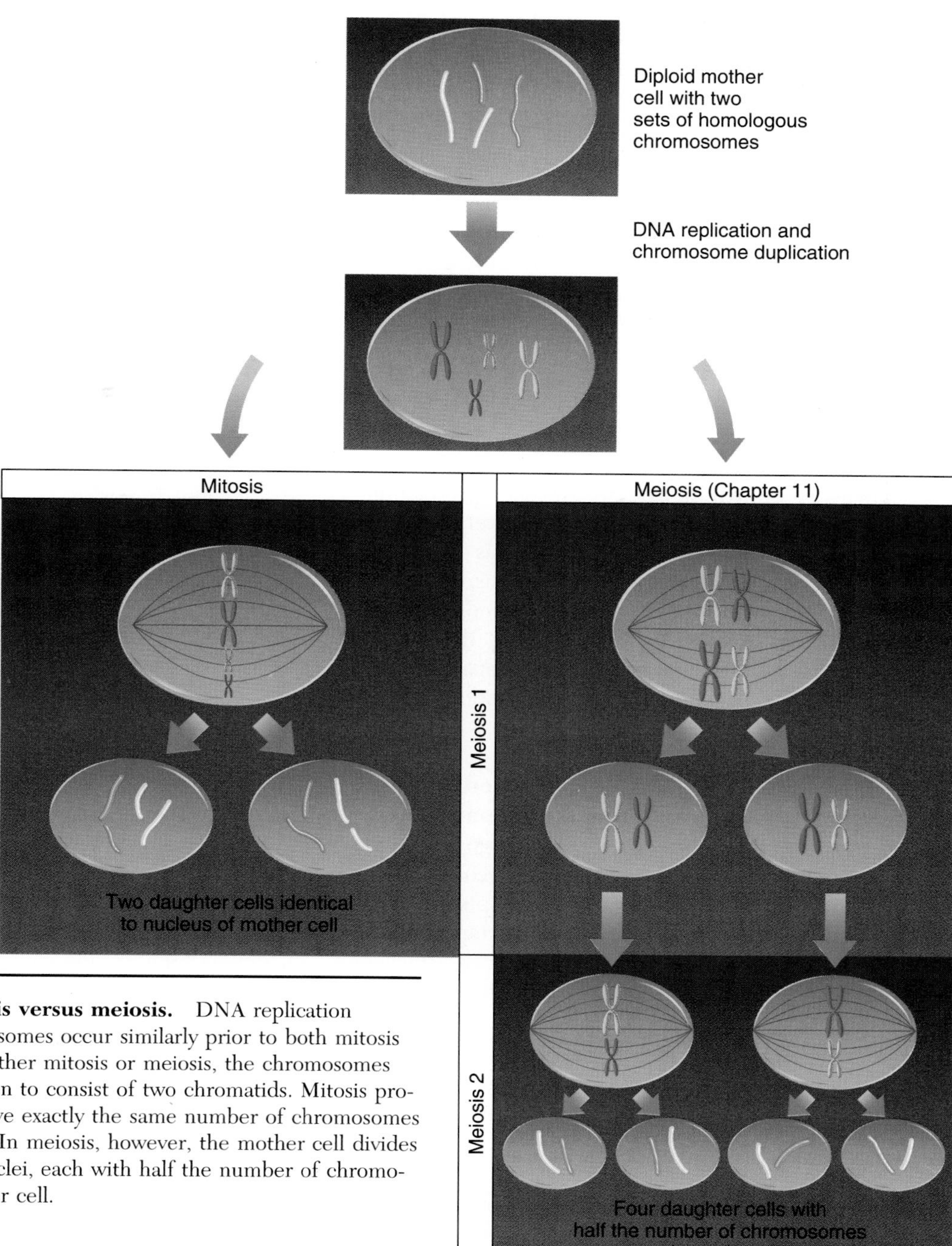

FIGURE 7-12

Schematic comparison of mitosis versus meiosis. DNA replication and the duplication of the chromosomes occur similarly prior to both mitosis and meiosis. When a cell enters either mitosis or meiosis, the chromosomes become condensed and can be seen to consist of two chromatids. Mitosis produces two daughter nuclei that have exactly the same number of chromosomes as the nucleus of the mother cell. In meiosis, however, the mother cell divides twice, producing four daughter nuclei, each with half the number of chromosomes of the nucleus of the mother cell.

came together at fertilization, the chromosome number of the offspring would be twice that of their parents. For humans, the first generation of offspring would have 92 chromosomes (46 from the sperm plus 46 from the egg); the second generation would have 184; and so forth. Cells would soon become overloaded with surplus chromosomes and die. Thus, for organisms that reproduce sexually, the formation of reproductive cells containing only one set of chromosomes is crucial for ensuring a stable chromosome number for a species from generation to generation.

In addition, meiosis builds genetic variability in a species. That is, meiosis increases the variety of combinations in characteristics among individuals that make up a species' population. A glance at a crowd of people instantly reveals the enormous variability that can exist within the human species. Different individuals may be better suited for different habitats. For example, shorter members of a population might be best suited for living in a forest whose trees have very low branches, while darker-skinned individuals may be better suited for living in bright, sunny climates, and so forth. Genetic variability forms the basis for natural selection. A greater range of characteristics in a population improves the chances that some individuals will survive environmental changes, thereby increasing the likelihood that the species will be perpetuated.

SYNOPSIS

Cells arise from other living cells by cell division. Cell division enables organisms to grow (by increasing the number of cells), reproduce, and repair or replace damaged or worn tissues.

Eukaryotic cells divide their nuclear contents by the complex processes of either mitosis or meiosis. Mitotic cell divisions produce genetically identical cells as part of the process of growth, repair, and asexual reproduction. Meiotic cell divisions produce haploid daughter cells containing one-half the number of chromosomes as the diploid mother cell. The daughter cells from meiosis eventually develop into sex cells (gametes) for sexual reproduction. The fusion of two haploid gametes restores the diploid number of chromosomes in the zygote, thus maintaining a stable number of chromosomes from generation to generation.

Mitosis is divided into several stages, based on activities relating to the separation of the chromosomes. During prophase, the chromosomes undergo a coiling process in which the DNA-protein fibers become highly compacted and the cell assembles a mitotic spindle apparatus consisting of microtubules. During metaphase, the chromosomes are aligned at the center of the cell, attached to spindle fibers. During anaphase, the chromatids split apart and move to opposite poles as separate chromosomes. During telophase, the cell returns to the nondividing state, which is called interphase. Interphase and M phase make up the cell cycle.

Cytokinesis in animal cells occurs by the contraction of a ring of microfilaments, which splits the mother cell in two. In plant cells, the mother cell is divided by formation of an intervening cell plate resulting from the fusion of Golgi-derived vesicles.

Meiosis includes two sequential nuclear divisions—meiosis I and meiosis II. The first division is called the reduction division because it separates (segregates) homologous chromosomes into different daughter nuclei. The second division separates attached chromatids, producing a total of four haploid nuclei.

Genetic variability among daughter cells of meiosis is ensured by two events: independent assortment and genetic recombination. Genetic recombination results from crossing over, which occurs during prophase I. In crossing over, homologous portions of maternal and paternal chromatids are exchanged with each other, producing chromosomes that contain both maternal and paternal genes. Since homologous chromosomes assort independently of one another during anaphase I, daughter cells receive mixtures of both maternal and paternal chromosomes.

Review Questions

1. Using pieces of colored yarn or crayons, draw a cell with a diploid number of 6. Draw the various stages of mitosis and meiosis, indicating how the chromosomes would be separated in each case.
2. How does the process of chromosome coiling facilitate cell division? During what stage does it occur? During what stage does DNA replication occur?
3. How would you describe the genetic relatedness of sister chromatids? Of two chromosomes that were split apart during anaphase and moved to different daughter cells? Of the two members of a pair of homologous chromosomes?
4. How do crossing over and independent assortment increase the genetic variability of a species?
5. Why is meiosis I (and not meiosis II) referred to as the reduction division?

Critical Thinking Questions

1. Why are disorders, such as Down syndrome, that arise from abnormal chromosome numbers, characterized by a number of seemingly unrelated abnormalities?
2. Taxol is a drug used in the treatment of certain cancers. The drug acts by stabilizing microtubules; that is, preventing their disassembly. What effect would you expect taxol to have on cell division? Why? Why would taxol be useful in cancer treatment?
3. The length of the cell cycle varies greatly from one cell type to another and for the same cell under different conditions. What do such differences tell you about the life span of the cell? In which parts of the human body would you expect to find cells with short life cycles?
4. Would you expect two genes on the same chromosome, such as yellow flowers and short stems, always to be exchanged during crossing over? How might they remain together *despite* crossing over?
5. Suppose paternal chromosomes always lined up on the same side of the metaphase plate of cells in meiosis I. How would this affect genetic variability of offspring? Would they all be identical? Why or why not?

PART

◂ 3 ▸

The Genetic Basis of Life

DNA stores the information that directs the development and maintenance of every organism. Such biological blueprints are fundamental to life itself. Changes in the DNA blueprint create new traits that propel evolution. Most of these random changes have no effect, are lethal, or create bizarre characteristics. Fruit flies, such as the one shown here, played an important role in exploring the genetic basis of life.

CHAPTER 8

Fundamentals of Inheritance

Mounting scientific evidence points to a genetic (chromosomal) basis for schizophrenia. The maze symbolizes being trapped in a world of distorted reality, one of the symptoms of schizophrenia.

STEPS TO DISCOVERY
The Genetic Basis of Schizophrenia

The human mind remains one of the most intriguing frontiers in the biological sciences. Let us consider a question pondered by social and natural scientists alike: "To what degree does a person's genetic 'makeup' determine whether he or she develops a serious psychological disorder, such as schizophrenia or manic depression?" In a sense, this question is part of the larger issue of how much of our personality is a result of our genes—nature—as opposed to influences from our environment—nurture.

Researchers unraveling the mysteries of schizophrenia—a disorder characterized by depression, delusions, hallucinations, and confusion—typify the scientists confronting the "nature versus nurture" debate. Early in this century, investigators tried to determine if schizophrenia ran in families, as would be expected if it were an inherited condition. The results unequivocally revealed that a person is ten times more likely to be schizophrenic if one parent has been diagnosed with the disease. If both parents are schizophrenic, nearly half of the offspring are likely to share the same fate.

Had these results been obtained in pea plants or mice, rather than in humans, only one conclusion would have been drawn: Schizophrenia is genetically determined. But many psychiatrists of the period argued vehemently that the data could just as well be interpreted to mean that schizophrenic parents created an *environment* that fostered the development of the disorder in their children. The debate has important implications. On one side, the argument that schizophrenia has its roots in environmental influences suggests that changing parental behavior could lessen the risk. On the other side, if the disease is genetically determined, perhaps a simple biochemical imbalance is the cause, and the condition could be remedied with drugs that correct the chemical irregularity. How could one of these two interpretations be confirmed?

In the 1960s, Leonard Heston of the University of Oregon and Seymour Kety and his colleagues at Harvard University simultaneously hit on a new approach to the question—the study of adoptees. Adoption provides a "natural experiment" in which the two variables, heredity and environment, can be separated. The natural parents provide the genes, while the adoptive parents (who had no knowledge of the adopted child's familial history of schizophrenia) provide the environment.

The results of the researcher's studies were clear-cut: Children of a schizophrenic mother who were raised by nonschizophrenic adoptive parents had as high an incidence of schizophrenia as if they had been reared in the home of their natural parents. Conversely, those adoptees who were born to a nonschizophrenic mother but were raised in a home with a schizophrenic parent had no increased risk of the disease, although they often complained of the "crazy home they were raised in."

These results supported the hypothesis that defective genes cause schizophrenia. But there is more data to consider. If one member of a pair of identical twins is a diagnosed schizophrenic, there is only a 50 percent likelihood that the other twin will also suffer from the "full-blown" disorder (although as many as 85 percent show some less severe schizophrenic behavior). Identical twins have identical genes; if schizophrenia is *strictly* determined by a person's genetic inheritance, both twins should have the identical disorder. There must be more to the story than which genes are acquired from the biological parents. Most mental-health experts believe that our genes provide a strong *predisposition* or vulnerability to the development of psychotic disorders, but that environmental factors influence whether or not the disease develops and its severity. In other words, nature and nurture work together.

Current investigations among biologists are focused on the specific genes that predispose a person to schizophrenia. For example, a recent paper in the journal *Nature* reported that newly available gene-locating techniques revealed a single gene (on chromosome number 5) that is associated with schizophrenia in a number of Icelandic and British families. Another investigative team reported that this particular gene was *not* associated with schizophrenia in members of a large Swedish family suffering from the disease. One interpretation of such conflicting results is that defects in several distinct genes can bring on the same symptoms. In other words, schizophrenia may result from different biochemical defects that generate a common disorder. The identification of these genes and the determination of the biochemical processes they control are two of the key investigative areas in clinical molecular biology.

*T*he story of modern genetics begins very painfully. The agony was not unlike the more celebrated torment suffered by Vincent Van Gough, who profoundly changed the art world but never sold a painting during his lifetime. In 1865 Gregor Mendel experienced a similar torment when he presented the results of his 8-year study of **heredity** (the passage of traits from one generation to the next) to a distinguished group of scientists, presenting them the first real tools for solving one of life's greatest puzzles—how parental traits are acquired by offspring. Yet this esteemed audience responded not with questions and interest but with polite applause. They simply did not understand that they had just witnessed the birth of modern genetics. Like Van Gogh, Mendel would not receive any recognition for his ground-breaking achievements during his lifetime.

Gregor Mendel's work emerged before its time, preceding the discovery of chromosomes, genes, diploidy, and meiosis, all of which would have provided a physical basis for understanding his principles. Mendel's experiments were published in 1865, but they generated no interest whatsoever until 1900. In that year, three different European botanists *independently* reached the same conclusions, and all three rediscovered Mendel's 35-year-old papers. It was only then that the scientific world began to recognize Gregor Mendel as the father of modern genetics.

▼ ▼ ▼

GREGOR MENDEL: THE FATHER OF MODERN GENETICS

Mendel's contemporaries supported the "blending" view of inheritance, according to which parents produce "hereditary fluids" that mix together during the formation of progeny, creating offspring with a mixture of the characteristics of their parents. (The term "blood relative" is a holdover from this misconception.) However, since blending would always produce offspring that were intermediate forms, halfway between each parent, it failed to explain why children are sometimes "chips off the old block," closely resembling one parent but not the other, or why certain characteristics disappear for a generation or two, only to reappear in subsequent generations. By the time Mendel had finished his work, he had answers for these enigmas. The answers did not include "blending."

A UNIQUE BEGINNING

Mendel grew up on a small farm where he learned methods for breeding plants. He had also studied natural science and mathematics for 2 years at the University of Vienna. This unique combination of skills in both plant breeding and mathematics set the stage for his eventual breakthroughs in genetics. Yet Mendel was a monk, lived in a monastery, taught mathematics, and worked in his laboratory (a small garden plot) in his spare time. It was here that Mendel developed a clear experimental goal—to mate pea plants that had different inheritable characteristics and to determine the pattern by which these characteristics were transmitted to the offspring. Mendel chose pea plants because breeding could be readily controlled (Figure 8-1). We now recognize that the genetic principles revealed by pea plants are in many ways identical to those of other organisms, including humans.

Mendel understood the need for experimental controls (see Chapter 1) to allow accurate and meaningful conclusions to be drawn from his experimental procedures. He also understood the value of quantifying (counting) his results so that his results could be objectively evaluated based on the numbers he obtained. Other investigators had approached the problem of inheritance by examining the multitude of traits in whole organisms. Such an unmanageable number of variables prevented any possibility of sorting out the principles governing inheritance and often led to subjective (and often biased) interpretations.

Mendel chose seven clearly definable traits, each of which occurred in one of two characteristic forms (Table 8-1). These two forms (dominant or recessive) are discussed below. He also developed stocks that *breed true* for a particular characteristic; that is, they always produce the same characteristic (round seeds, for example) following self-fertilization. In this way, Mendel could perform experiments that would answer the question, "What would you get if you crossed two true breeders that have the two opposite characteristics, for example, a true breeder for round seeds with a true breeder for wrinkled seeds?" In other words, what would these *hybrids* look

SEVEN TRAITS OF MENDEL'S PEA PLANTS

Trait	*Dominant Allele*	*Recessive Allele*
Height	Tall	Dwarf
Seed color	Yellow	Green
Seed shape	Round	Angular (wrinkled)
Flower color	Purple	White
Flower position	Along stem	At stem tips
Pod color	Green	Yellow
Pod shape	Inflated	Constricted

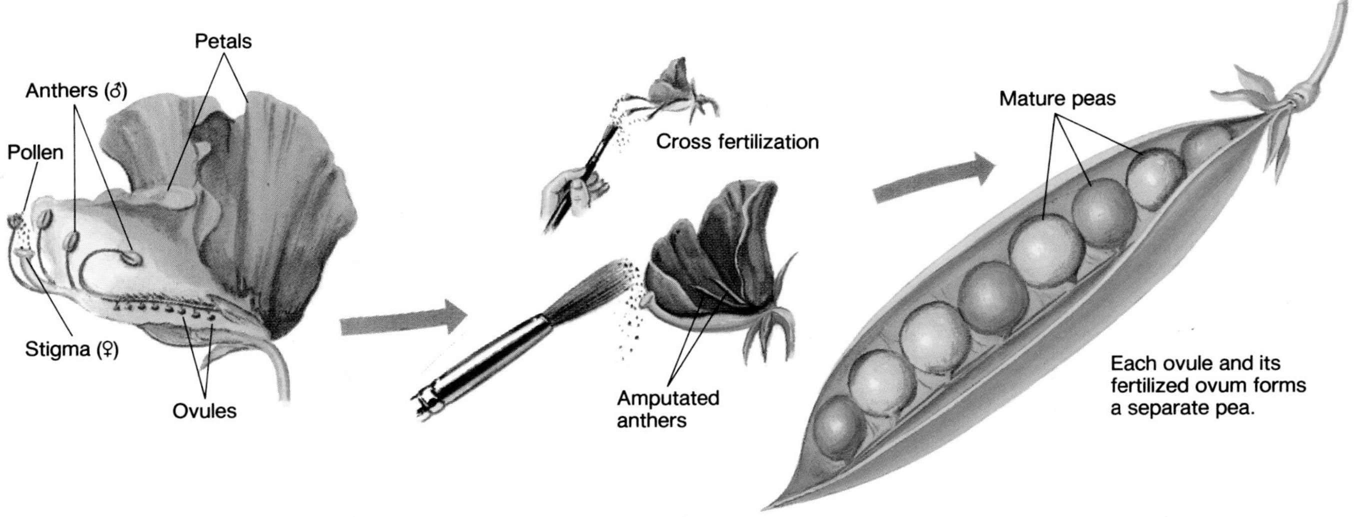

FIGURE 8-1

Reproduction in Mendel's pea plants. Mendel's choice of pea plants was strategic. He had two opportunities: (1) He could self-fertilize his plants by allowing pollen to fall from the *anthers* (the male reproductive organs) onto the *stigma* (the sticky area to which pollen adheres), and (2) he could prevent self-fertilization by amputating the pollen-producing anthers. (The eggs, or ova, are produced by the *ovules*, the female reproductive organs.) These pollenless flowers were then cross-fertilized by brushing on pollen from another plant (pollination and fertilization are discussed in Chapter 13). In this way, he was able to control which plants mated with each other. With accidental matings eliminated, Mendel was able to correlate parental traits with those of the progeny. Each pea seen in the pod is the product of a separate fertilization event (which is why both green and yellow peas can share the same pod). Seeds develop only following fertilization, so results are not complicated by "fatherless" offspring. Seeds are readily collected, grown into mature plants, and the inherited characteristics determined. All of these properties contributed to Mendel's scientific success.

like? The simplest of such single-trait experiments are called **monohybrid crosses,** mating experiments in which only one trait is followed. Mendel performed hundreds of such crosses for each of the seven traits. To simplify our discussion, we will focus on the results of just one set of crosses, that of seed shape.

When plants that produced wrinkled seeds were cross-fertilized with plants that produced round seeds, Mendel found that all the plants of the next generation had round seeds (Figure 8-2). Mendel called the original parental generation the **P generation** and the first generation of offspring the **F_1 generation** (F_1 = first filial). Mendel referred to the characteristic that appeared in the hybrids (in this case, round seeds) as the **dominant** characteristic, and the characteristic that had seemingly been lost (in this case, wrinkled seeds) as the **recessive** characteristic.

The following year, Mendel allowed the F_1 hybrids to self-fertilize. He then observed the nature of the subsequent offspring, which he called the **F_2 generation.** Surprisingly, the wrinkled characteristic that had previously disappeared in the F_1 generation had returned in a number of F_2 offspring. In fact, there were approximately three times as many F_2 plants with round seeds as with wrinkled ones (Figure 8-3). This same approximate 3:1 ratio was found for each of the seven traits he examined. Mendel concluded that hereditary patterns were not determined randomly but followed an orderly, predictable pattern.

INTERPRETING THE RESULTS

Mendel concluded that the potential to produce the wrinkled-seed characteristic was somehow carried by the F1 offspring, even though they could only produce round seeds. He proposed that hereditary traits were governed by particulate factors (or *units*) of inheritance (later termed *genes*) that remained intact from generation to generation. Each trait in a plant was determined by the presence of two such factors, one derived from each parent. The two factors could be of an identical or a nonidentical nature. Later, the term **allele** was used to refer to alternate forms of the same gene.

For each of the seven traits, Mendel found one of the two alleles to be dominant over the other, recessive allele. In the case of seed shape, the allele for round seeds is dominant over the allele for wrinkled seeds. Dominant alleles are usually represented by capital letters, recessive alleles by the same letter in lower case. For example, the two alleles for seed shape are denoted by the symbols *R* (round) and *r* (wrinkled). Thus, the F_1 plants had round

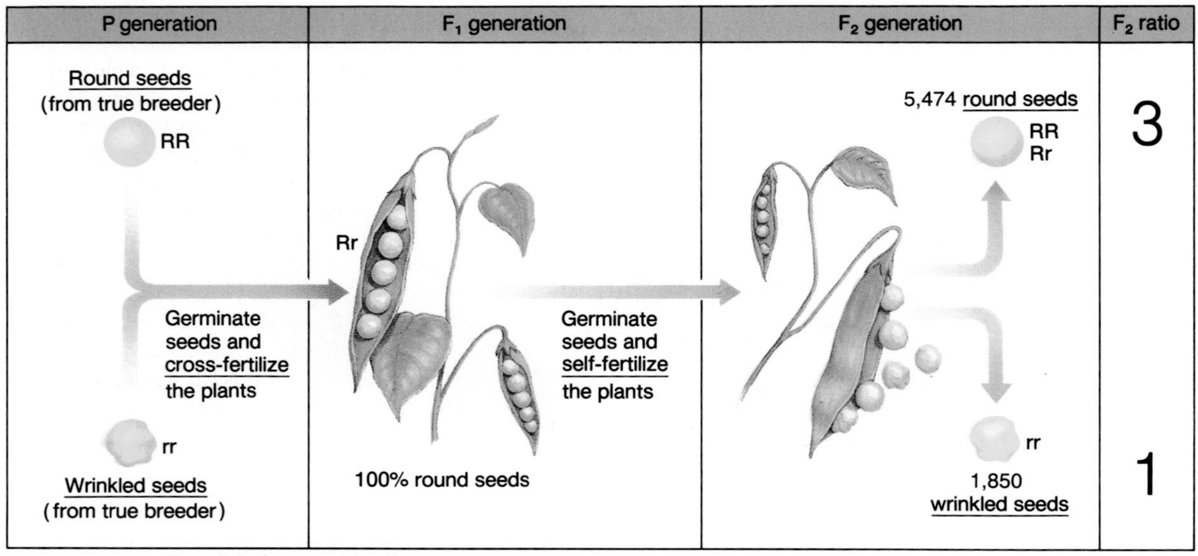

FIGURE 8-2

Mendel conducted monohybrid crosses to track the course of a single trait (in this case seed shape) through two generations. In this experiment, all plants in the F_1 generation had round seeds, while the F_2 generation included plants with both round and wrinkled seeds in a 3:1 ratio. He found a similar pattern of inheritance for each of the seven traits. From this pattern, Mendel concluded that there are paired genetic factors, one for each alternate characteristic of the trait (shown here as *R* and *r*), which segregate during gamete formation. The factor in one gamete pairs up again with another factor from another gamete during fertilization. (As discussed in the following section of the text, genetic characteristics, such as round and wrinkled seeds, are determined by alternate versions of a gene, called *alleles.* In this case, the alleles are *R* and *r.*)

seeds (rather than some intermediate, blended shape) because the *R* allele was dominant over the *r* allele.

The original, true-breeding parents had two identical forms of the gene for each trait; the parental plants with round seeds were designated *RR* and those with wrinkled seeds were labeled *rr.* This state is referred to as **homozygous** (*homo* = same); that is, both alleles for a particular trait are the same. In contrast to its parents, each F_1 plant had only one *R* and one *r* allele; consequently, they were **heterozygous** (possessing different alleles for a trait)

In modern terminology, we can say that the parental plants with round seeds and the F_1 hybrids had the same **phenotype**—physical appearance or measurable quality—but different **genotypes**—the underlying genetic makeup. Simply put, the genotype determines the phenotype. The genotype of the parent with round seeds was *RR,* and that of the F_1 was *Rr.* Mendel concluded that any plant exhibiting a wrinkled-seed phenotype must be homozygous for the recessive allele; that is, possessing a genotype of *rr.*

Mendel's Law of Segregation

Mendel concluded that each reproductive cell (gamete) could carry only one allele for each trait. Somehow, during formation of the gametes, each allele separated (*segregated*), going to different reproductive cells. This phenomenon is now referred to as Mendel's **Law of Segregation.** Years after his death, the discovery of homologous chromosomes and their separation during meiosis explained how this segregation of alleles occurs during gamete formation, generating a predictable pattern of inheritance (Figure 8-3).

PREDICTING PATTERNS OF INHERITANCE

Referring back to Mendel's F_1 cross (*Rr* x *Rr*), three genotypes in the F_2 generation are possible: homozygous dominant (*RR*), heterozygous (*Rr*), and homozygous recessive (*rr*). One way to visualize the results of such crosses is to use the **Punnett square method,** a tool for predicting the possible genotypes and phenotypes (and their expected ratios) when there is an equal chance of acquiring either of the two alleles. A Punnett square of each of Mendel's monohybrid crosses predicts a 3:1 ratio of dominant to recessive phenotypes, just as Mendel had observed (Figure 8-4).

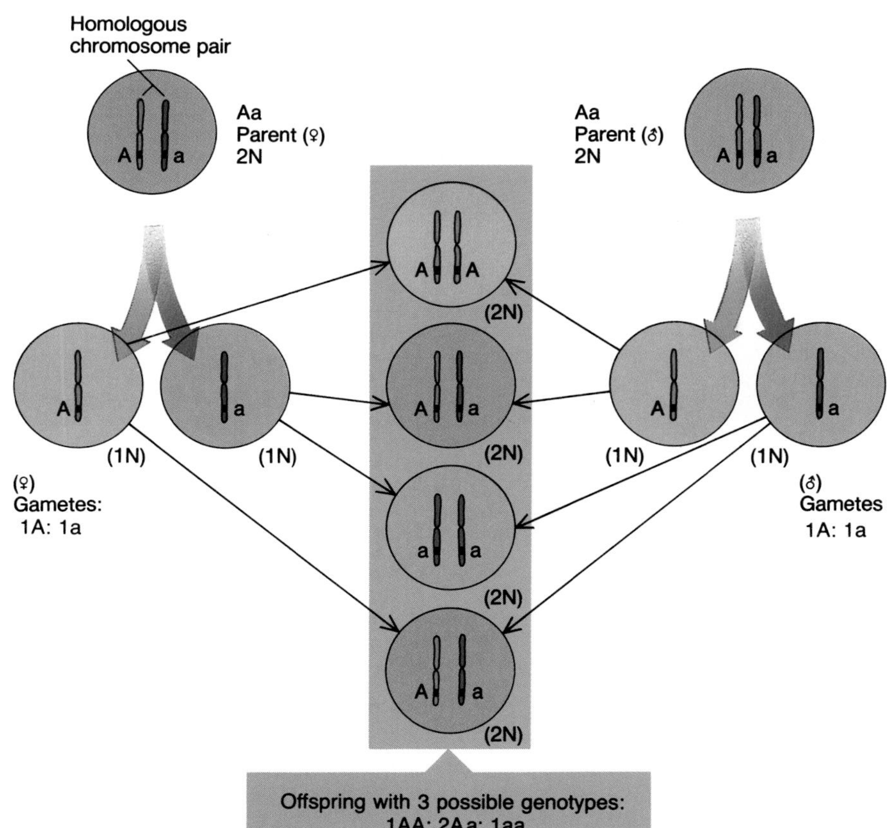

FIGURE 8-3

Meiosis explains Mendel's findings. Because they occur on separate homologous chromosomes, alleles (*A* and *a*) segregate from each other during meiosis to form gametes. Each gamete has an equal probability of receiving an "*A*" or an "*a*". Fertilization randomly reestablishes the diploid state, so each allele has an equal chance of joining with either of the others. Each offspring will therefore acquire one of the indicated genotypes. It will be twice as likely to be an *Aa* genotype than either an *AA* or *aa* since there are two chances of acquiring this heterozygous combination.

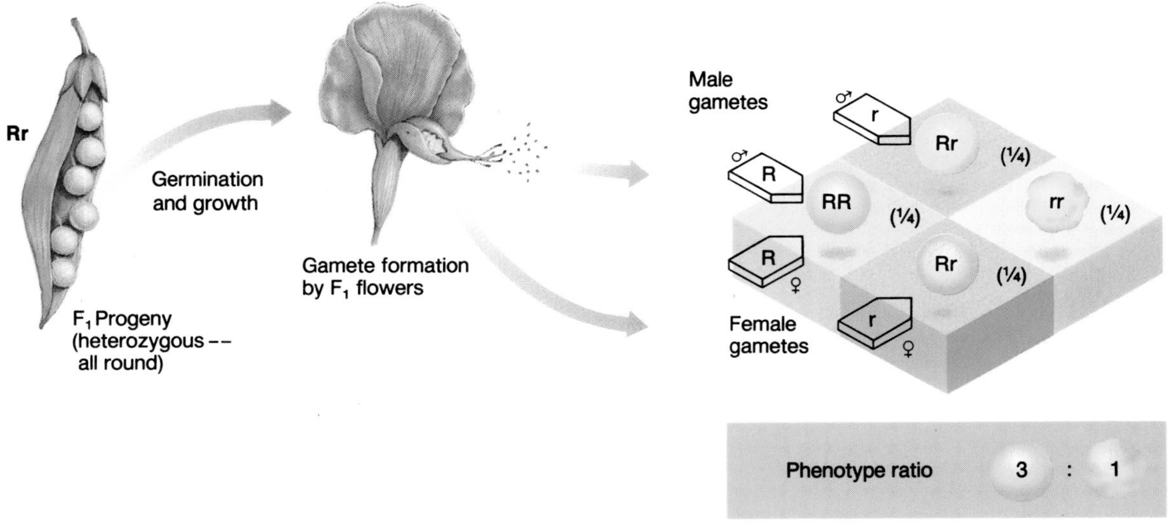

FIGURE 8-4

The Punnett square method of determining probability ratios. Following self-fertilization of a heterozygous pea plant, each F_1 gamete will contain either an *R* (round) or *r* (wrinkled) allele. In the Punnett-square representation, all possible male gametes are listed along the top and all female gametes along the side. The possible combinations of alleles following fertilization are shown in the boxes, each box representing the genotype created by the union of these two alleles. The genotype ratio is 1:2:1 (*RR*:*Rr*:*rr*), meaning that there are twice as many *Rr* boxes as either *RR* or *rr* boxes. The 3:1 phenotype ratio explains why recessive (wrinkled) traits reappeared in one-fourth of Mendel's F_2 progeny after disappearing from the F_1 generation.

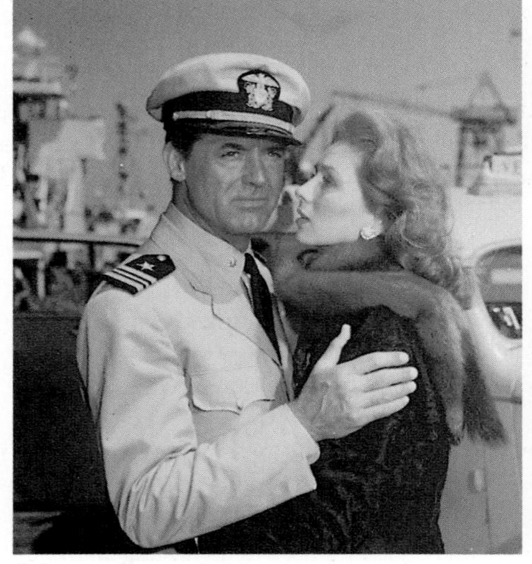
(a)

(b)

FIGURE 8-5
Dominant alleles. The dimple in Cary Grant's chin ***(a)*** and Marilyn Monroe's widow's peak hairline ***(b)*** are both expressions of dominant alleles.

Verifying such predictions, however, requires large numbers of offspring. The same is true for any random combination of events. For example, if you flipped a coin three times and happened to get "heads" all three times, the results could be attributed to chance variation. In other words, three flips are not enough to conclude that the probability of heads is 100 percent. But if you flipped the coin 100 times, the likelihood of it landing on heads every time would be very, very small. Instead, the frequency of obtaining heads would likely be close to 50 percent. In genetics, the larger the number of offspring examined, the closer the results should approach the predicted ratio.

Hereditary Patterns in Humans

Since the principles of dominance and segregation of alleles apply to all sexually reproducing diploid organisms, not just to garden peas, Mendel's findings can help you explain hereditary patterns in humans. A dimple in the chin, for example, is a trait determined primarily by dominant alleles. A person with at least *one* of these dominant alleles will develop the corresponding characteristic, regardless of the second allele (Figure 8-5). A child of two dimpled-chin parents that are both heterozygous for the trait has a 3 in 4 probability (a 75 percent chance) of acquiring the same dominant phenotype as that found in

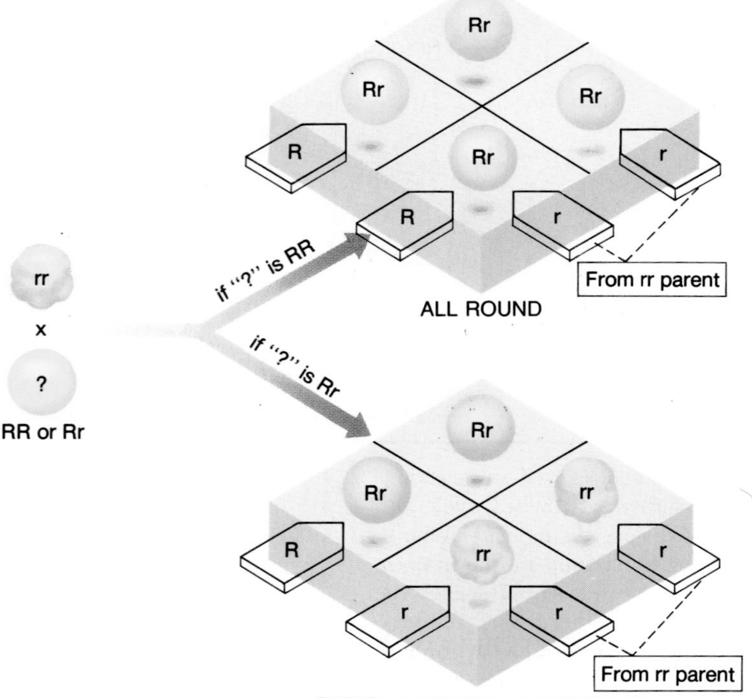

FIGURE 8-6
The testcross allows us to determine the genotype (*RR* or *Rr*) of an organism that shows the dominant trait. When crossed with an organism that shows the recessive trait (genotype of *rr*), a homozygous dominant (*RR*) individual will only yield progeny with the dominant phenotype since all the offspring will have a dominant allele. In contrast, if the test organism is a heterozygous individual (*Rr*), about half the progeny will possess the recessive phenotype.

the parents. This does *not* mean that individuals with a chin dimple are three times more numerous. For some traits, the recessive phenotype is far more common than the dominant. The absence of fingerprints is a dominant trait, but most of us *do* have fingerprints. The dominant allele is so rare that nearly everyone is homozygous recessive and has a complete set of fingerprints.

The Testcross

As we saw earlier, outward appearance of a dominant phenotype cannot reveal whether an individual is homozygous or heterozygous for that trait. A pea plant with purple flowers, for example, may be homozygous or heterozygous for the trait of flower color. Mendel concluded that if his concept of dominant and recessive factors were correct, then it should be possible to determine genotype by crossing an individual in question with a mate that shows the recessive trait. This strategy is known as a **testcross,** and it reveals the presence of a "hidden" recessive allele if the individual is heterozygous for that trait, as explained in Figure 8-6.

CROSSING PLANTS THAT DIFFER BY MORE THAN ONE TRAIT: MENDEL'S LAW OF INDEPENDENT ASSORTMENT

Mendel eventually expanded his experiments to **dihybrid crosses,** matings in which *two* traits are tracked. Mendel began by crossing true-breeding plants having round, yellow seeds (genotype of *RRYY*) with plants having wrinkled, green seeds (*rryy*). As expected, the seeds of the F_1 offspring showed only the dominant characteristics (Figure 8-7); that is, seeds of a round, yellow phenotype (*RrYy*

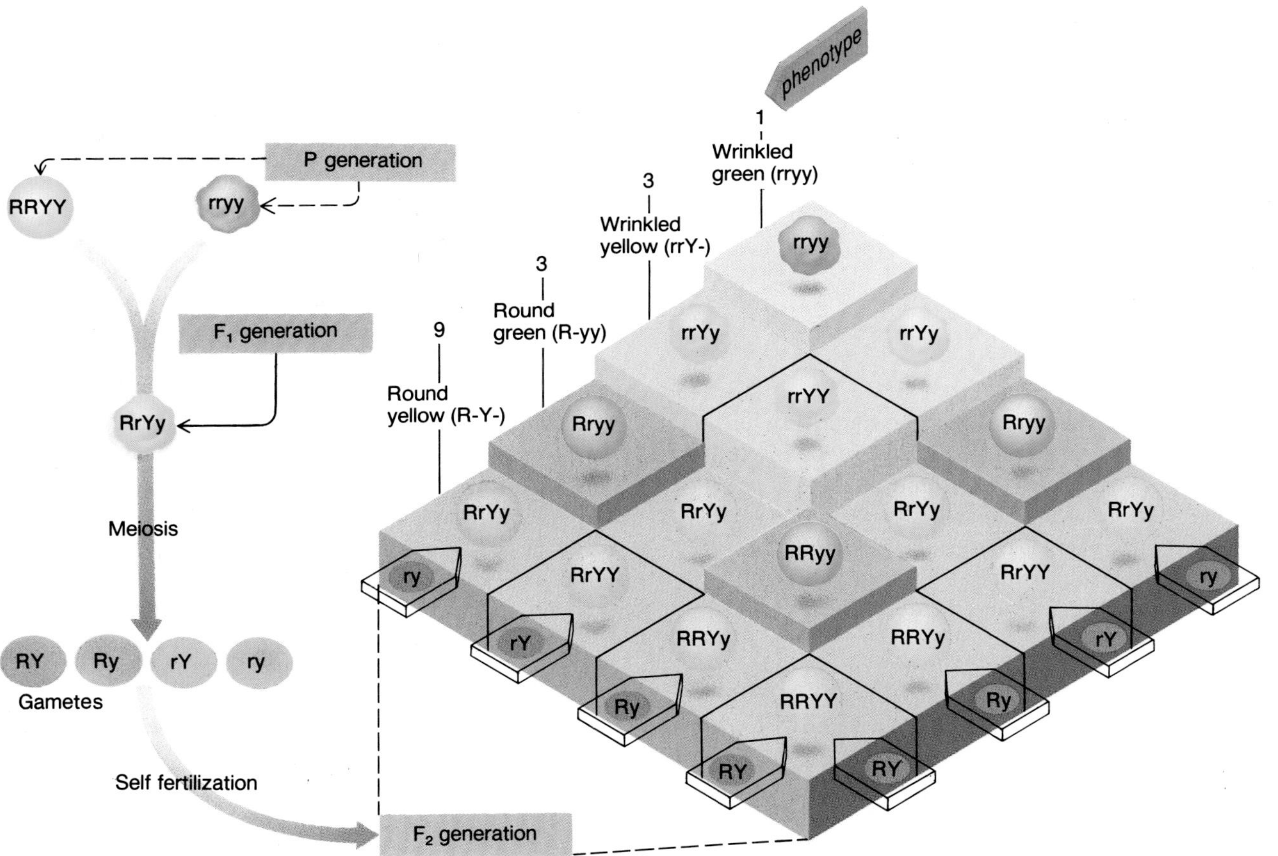

FIGURE 8-7

Mendel's dihybrid crosses. When plants that breed true for round, yellow (*RRYY*) seeds are crossed with those producing wrinkled, green (*rryy*) seeds, the F_1 progeny are heterozygous (*RrYy*) for both seed shape and color. Independent assortment allows either *R* or *r* to combine with either *Y* or *y*, generating gametes with four different possible combinations of alleles (*RY, Ry rY, ry*). Self-fertilization of an F_1 plant produces an F_2 generation with the nine genotypes and four phenotypes shown in the Punnett square. When Mendel performed this experiment with hundreds of peas, he obtained ratios very close to the 9:3:3:1 phenotype pattern predicted here. Mendel also determined the genotypes of the F_2 progeny by allowing them to self-fertilize, then examining their offspring. The F_2 generation consisted of all nine genotypes in a ratio that approximated the predicted 1:2:1:2:4:2:1:2:1 ratio.

genotype). When he allowed these F_1 hybrids to self-fertilize, however, four different types of seeds were found among the next generation. Most of the seeds (approximately 9 out of 16) were round and yellow, while approximately 1 out of 16 were wrinkled and green. In addition, this generation contained two types of seeds that had a combination of traits not seen in either parent—round, green seeds and wrinkled, yellow seeds (both represented by approximately 3/16 of the population). Thus, the ratio of these four phenotypes was approximately 9:3:3:1.

In order to account for these results, Mendel made a final conclusion: The segregation of the pair of alleles for one trait had no effect on the segregation of the pair of alleles for another trait; that is, alleles for different traits segregate *independently* of one another. Just because an F_1 individual inherited one R and one Y allele from one gamete did not mean that those two alleles must remain together when that F_1 plant formed its own gametes. Rather, either of the alleles for seed shape could team up with either of the alleles for seed color (Figure 8-7). This would produce gametes with four possible combinations (RY, Ry, rY, ry), each in equal proportion. This conclusion is referred to as Mendel's **Law of Independent Assortment** and, as we saw in the previous chapter, derives directly from the fact that homologous chromosomes become aligned independently in the metaphase plate during meiosis I.

MENDEL AMENDED

Even though Mendel's "laws" provided the foundation on which modern genetics was built, several exceptions to Mendel's principles were soon discovered as the science of genetics matured. For example, not all traits have two alternative forms of appearance or display the simple dominant–recessive relationship exhibited by the seven traits examined by Mendel. We will now take a closer look at some of the major variations on the "Mendelian pattern" of inheritance.

INCOMPLETE DOMINANCE AND CODOMINANCE

At first glance, some traits seem to follow the notion of "blending" rather than Mendelian inheritance. For example, when red-flowered snapdragons are crossed with white-flowered snapdragons, all of the progeny are pink, as though the traits had blended. But the next generation of plants clearly demonstrates that blending has *not* occurred. Self-pollination of pink snapdragons produces an interesting mix of offspring—50 percent pink, 25 percent red, and 25 percent white. If blending had occurred, red and white varieties could never have been recovered from pink parents. Although unknown for years after Mendel's experiments, this phenomenon is understood by the way some genes determine the corresponding trait.

In pea plants, the allele for purple flowers directs the plant to produce a purple pigment in the petals, whereas the allele for white flowers is a variant of this gene that has lost the ability to direct the production of *any* pigment. Enough pigment is produced by one good allele to create the final color, explaining why one purple flower allele is all that is needed to produce purple flowers. The white color appears only in the homozygous recessive configuration, and there is no intermediate color in heterozygous plants. In snapdragons, however, a single red pigment allele directs the production of a red color only half as intense as the two alleles of a homozygous dominant. The paler red color is actually pink. This phenomenon is known as **incomplete** (or **partial**) **dominance.**

Heterozygous individuals are also distinguishable in cases of **codominance,** in which both alleles are expressed simultaneously and are unmodified (undiluted) in the heterozygous individual. A classic example of codominance is seen in the ABO blood group antigens—proteins that are found on the surface of red blood cells and are used to

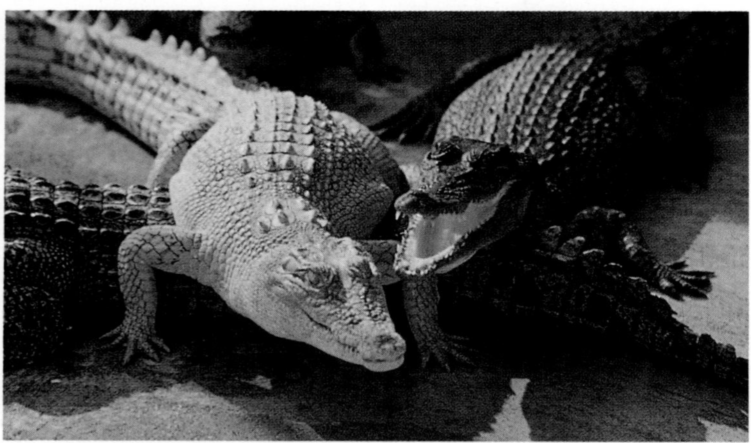

FIGURE 8-8
Albinism occurs throughout the vertebrates, including these crocodilians.

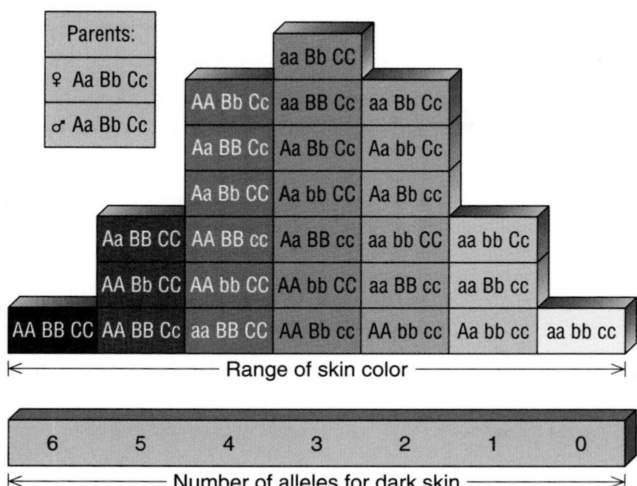

FIGURE 8-9

Polygenic inheritance and skin color. People don't come in two contrasting colors, dark and light; most of us lie somewhere between these extremes. There are at least three genes—*A, B, C*—located at three different loci, that determine the amount of pigment in human skin. At each locus, there are two possible alleles of the gene, one for maximum pigment (the capital letter) and one for no pigment (the lowercase letter). This figure shows the 27 genotypes and 7 phenotypes that could be produced by two people who are heterozygous for all three genes (*AaBbCc*). The 27 genotypes are indicated within each of the boxes, and the seven different phenotypes are indicated by the 7 columns indicating different shades of darkness. The darkness of skin is determined by the total number of alleles for dark pigment (capital letters) in the genotype. The darkest phenotype would have all six genes represented by the dominant allele, while the lightest would have all six genes represented by the recessive allele.

determine human blood type (page 56). The codominant alleles for A and B, written as I^A and I^B, direct the blood cell to manufacture the A antigen and the B antigen, respectively, giving the individual type AB blood. How, then, do people acquire type O blood? The answer lies in multiple alleles.

MULTIPLE ALLELES

The human ABO blood group system can be used to illustrate another departure from classical Mendelian genetics—the existence of more than two possible alleles for a single trait. Although a single diploid individual can possess a maximum of two alleles, the existence of additional alleles increases the number of possible genotypes in the population. There are three different alleles in the ABO system, for example: I^A, I^B, and I^O. Like the allele for white flowers in pea plants, the I^O allele is a variant that directs the formation of a defective protein, so a homozygous person with two I^O alleles would lack both A and B antigens. These individuals are blood type O. An allele such as I^O is recessive to both of the alternative alleles, which is why it is sometimes written as "*i*." Therefore, an I^AI^O heterozygous pair would produce the same phenotype as would the homozygous I^AI^A—that is, a person with type A blood.

COMBINED EFFECTS OF GENES AT DIFFERENT LOCI

The expression of some alleles is influenced by the genotype at another location (or **locus**) on the chromosomes. The distant genes may, for example, silence the expression of distant genes that would otherwise be fully expressed. This is dramatically expressed in albino individuals. Hair color, eye color, and skin color—traits determined by separate genes—have one feature in common: the more *melanin* (a dark pigment) that is produced, the darker the color. Melanin production is directed by alleles at the *A* locus, and as long as one dominant *A* gene is present, the individual will have the enzyme needed to manufacture melanin. Homozygous individuals, however, carry two recessive alleles (*aa*), and cannot produce melanin; consequently their hair, eyes, and skin will be devoid of color, regardless of the genotypes at the hair, eye, and skin color loci (Figure 8-8).

PLEIOTROPY

The product of one gene can have far-reaching secondary effects on many other characteristics, a condition called **pleiotropy.** The many serious symptoms of *cystic fibrosis,* for example, are due to a single defective gene that leads to the production of thickened mucus. The digestive, sweat, and respiratory mucous glands all express the thickened mucus, producing widespread disorder within the body. Digestive enzymes are poorly secreted, causing improper digestion and absorption of food. Mucus that lines the airways is too thick to be easily moved out of the lower airways, so the mucus, laden with trapped microbes, settles into the lungs, causing pneumonia and the other lung infections that claim the lives of cystic fibrosis victims.

POLYGENIC INHERITANCE

If people were like peas, we would come in two sizes—tall and short. In Mendel's garden peas, height is a **discontinuous** condition—phenotypes fall into distinct categories. But many traits, such as human height and skin coloration, show **continuous variation** within the population. Such traits are "polygenic," determined by a number of different genes present at different locations, rather than a single gene (Figure 8-9). The overall expression of a polygenic

FIGURE 8-10
A Siamese cat provides a visual portrayal of the influence of environment on phenotype.

trait is the sum of the contributing genes. For example, two people who differ in just one pair of alleles show only slight color differences, compared to two people who differ in *all* of the polygenic determinants.

ENVIRONMENTAL INFLUENCES

Nearly all phenotypes are subject to modification by the environment. A genetically tall person, for example, will fail to reach maximum height if deprived of adequate nutrition. In addition, a single phenotype may vary according to the environmental conditions, such as a light-skinned person acquiring a tan after a summer in the sun. Siamese cats provide another intriguing example of environmental effects on gene expression (Figure 8-10). In spite of its two-color coat pattern, all the cells in a Siamese cat are genetically identical at the color locus, including the cells that produce light fur and those that make dark fur. The *Siamese* allele (c^s) directs the synthesis of a "temperature-sensitive" variant of the enzyme that manufactures the dark pigment only on the cooler areas of the body (feet, snout, tip of the tail, and ears). The rest of the cat's body is warm enough to inactivate the enzyme, causing the fur to remain light.

MUTATION: A CHANGE IN AN INDIVIDUAL'S GENOTYPE

Mutations are rare but permanent changes in a gene, changes that can alter a gene's product. The original form of the gene is termed the *wild type,* to distinguish it from the newer *mutant* form. In a sense, we have been dealing with this topic since Chapter 1. Alleles arise by mutations in an existing gene. If there were no mutations, there would be no new alleles, *and there would be no biological evolution.*

The segregation of alleles and their independent assortment during meiosis increase genotype diversity by promoting new combinations of genes. But the shuffling of existing genes alone does not explain the presence of such a vast diversity of life. If all organisms descended from a common ancestor, with its relatively small complement of genes, where did all the genes present in today's millions of species come from? The answer is mutation.

Most mutant alleles are detrimental. They are much more likely to lead to loss of a critical biological function than to produce a gene that increases an organism's fitness. Occasionally, however, one of these stable genetic changes creates an advantageous characteristic. In this way, mutation provides the raw material for evolution and the diversification of life on earth.

GENES AND CHROMOSOMES

Mutations occur in the cellular structures that the nineteenth-century European microscopists first saw "dancing" around within the nucleus just before a cell divided. These structures are chromosomes, the bearers of an organism's genes.

Although Mendel provided convincing evidence that inherited traits were governed by inheritable units of information, he never observed anything under a microscope. The principles he discovered were finally explained by the presence of dual copies of a gene for a particular trait. These copies were located at the same position on each homologous chromosome (Figure 8-11). Genes were linearly arranged on a chromosome, much like pearls on a string.

But this finding posed a glaring problem. Each of the seven traits Mendel studied was inherited independently of the others—for example, inheriting purple flowers in no way influenced which allele for seed shape was inherited.

But what if the two traits were close together on the same chromosome? These alleles should be inherited as if they are *linked* to one another; that is, they should be part of the same **linkage group.**

Had Mendel selected traits that were linked together, he would never have discovered the principle of independent assortment. As it turned out, the garden pea has seven different pairs of homologous chromosomes, and each of the traits on which Mendel reported either occurred on a different chromosome or was so far apart on the same chromosome as to break apart from each other and act independently (discussed in the upcoming section, "Crossing Over—Gene Exchange"). Within a couple of years after confirming Mendel's findings, however, two traits (flower color and pollen shape) in sweet peas were shown to be tightly linked on the same chromosome and to assort together.

FRUIT FLIES: TEACHERS OF NEW GENETIC LESSONS

Geneticists studying domestic plants or animals suffer a serious handicap—one imposed by time. Experimental matings yield results only when offspring appear, often a year or more after the parents mate. Analysis of subsequent generations requires additional years of study. Geneticists needed an organism that was easy to maintain, produced multiple generations every year, and was governed by the same "rules" of genetic inheritance exhibited by more complex organisms. They found just such an organism swarming around rotting fruit. Fruit flies (of the species *Drosophila melanogaster*) could easily be raised by the thousands in small laboratory bottles, completing their entire reproductive cycle in just 10 days. In 1909, the fruit fly seemed like the perfect organism to Thomas Hunt Morgan of Columbia University, as he began the research that initiated the "golden age of genetics," an age that may never end.

Genetic Markers along the Chromosome

When Morgan began his work with fruit flies, there was only one type of fly available, the *wild type*. Inheritance can be charted only if there is detectable evidence of genetic *differences*. Breeding huge numbers of flies, Morgan isolated 85 different mutants. All of the mutations eventually were shown to represent four distinct linkage groups, a discovery that correlated perfectly with the eventual finding of four different chromosomes in the cells of *Drosophila*. Among such a profusion of evidence, one conclusion was unavoidable: Genes reside on chromosomes.

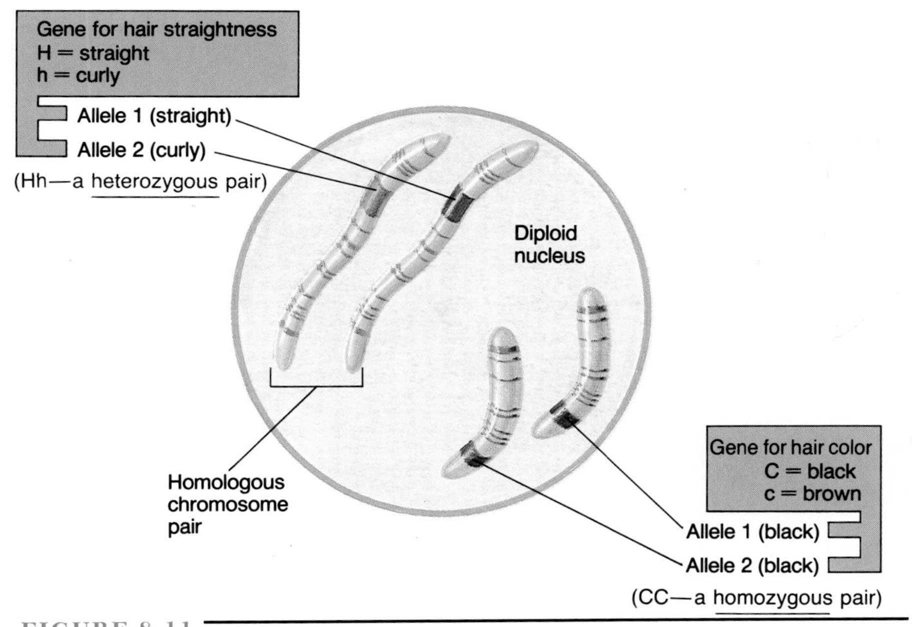

FIGURE 8-11
Stylized representation of alleles on homologous chromosomes in a human diploid nucleus (only 2 pairs of the 23 different human chromosomal pairs are shown here). Although each chromosome would actually contain thousands of genes at different loci, for simplicity, we have identified only one gene (two alleles) per pair of chromosomes in this illustration.

Crossing Over—Gene Exchange

Then Morgan discovered a contradiction to a contradiction. Linkage of genes on the same chromosome already contradicted Mendel's "independent assortment" principle for certain traits (Figure 8-12). Soon however, Morgan discovered a contradiction to this—alleles in a linkage group occasionally segregated independently, as though on different chromosomes, ending up in separate gametes instead of staying linked. To explain the "contradiction" Morgan proposed the phenomenon of **crossing over,** the exchange of genetic segments between homologous chromosomes during meiosis (Figure 8-13).

SEX AND INHERITANCE

Humans are a lot like fruit flies. Most of the genetic breakthroughs revealed by *Drosophila* apply to us as well as to most other diploid organisms. We even resemble fruit flies in the fundamentals of sex—our gender, for example, is determined by specific chromosomes.

GENDER DETERMINATION

Most of the chromosomes of males and females look identical, constituting a group of chromosomes called **autosomes.** However, one chromosomal pair, the **sex chromosomes,** come in distinct-looking pairs, whether in fruit flies or humans. The female's cells harbor a pair of identical sex chromosomes, called **X chromosomes,** whereas the male's cells possess one X chromosome and a smaller **Y chromosome** (illustrated in humans in Figure 7-7). These sex chromosomes determine an individual's gender.

During gamete formation in humans and other mammals, the X and Y chromosomes separate, half the sperm carrying the Y chromosome, the other half bearing the X chromosome. All gametes produced by an XX female will contain an X chromosome, so the sex of each offspring depends on whether the egg is fertilized by an X-bearing sperm (forming an XX *female* offspring) or a Y-bearing sperm (forming an XY *male* offspring). Gender in humans is thus determined by the male (Figure 8-14). King Henry VIII of England, who disposed of many wives because they failed to bear him a son, should have pointed his lethal finger squarely at himself. It was *his* royal gametes that determined the gender of his progeny!

Fruit flies and mammals are not the model for sex determination in all eukaryotic organisms, however. In some birds, the female's cells have an X and a Y chromosome, while the male's cells have two Xs. Most plants have only autosomes; consequently, each individual produces both male and female parts. In certain fish and alligators, gender is determined by environmental factors. Alligator eggs for example are heat sensitive—warm eggs become females, colder eggs male.

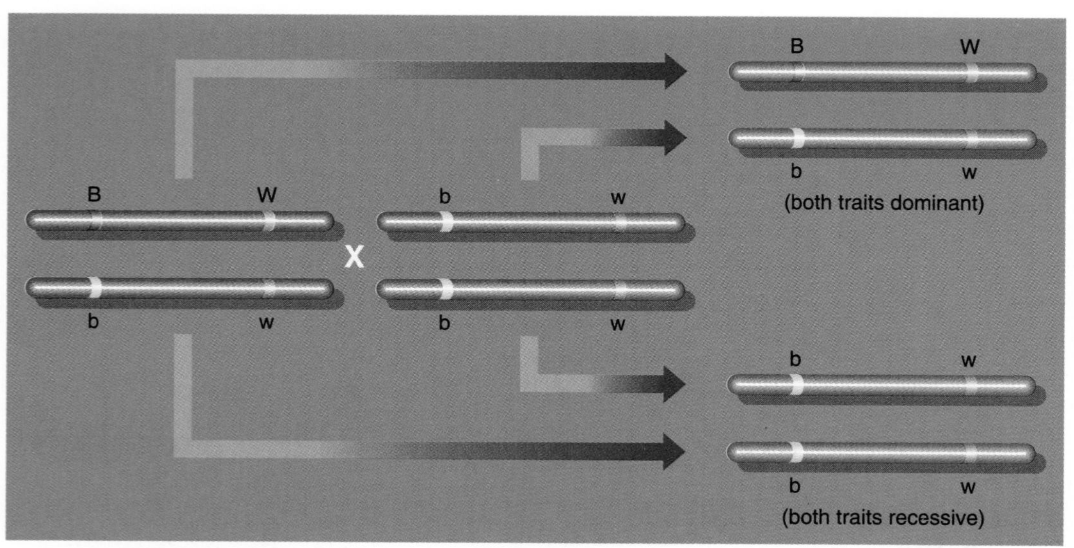

FIGURE 8-12

The pattern of inheritance if linkage were 100 percent complete. If two alleles on the same chromosome always remained together, then all of the offspring in the case depicted here would resemble one or the other parent. They would be either gray-bodied and long-winged (*BbWw*) *or* black-bodied and short-winged (*bbww*).

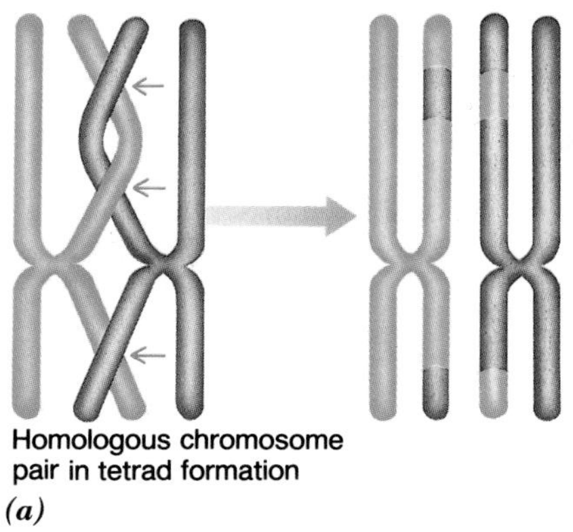

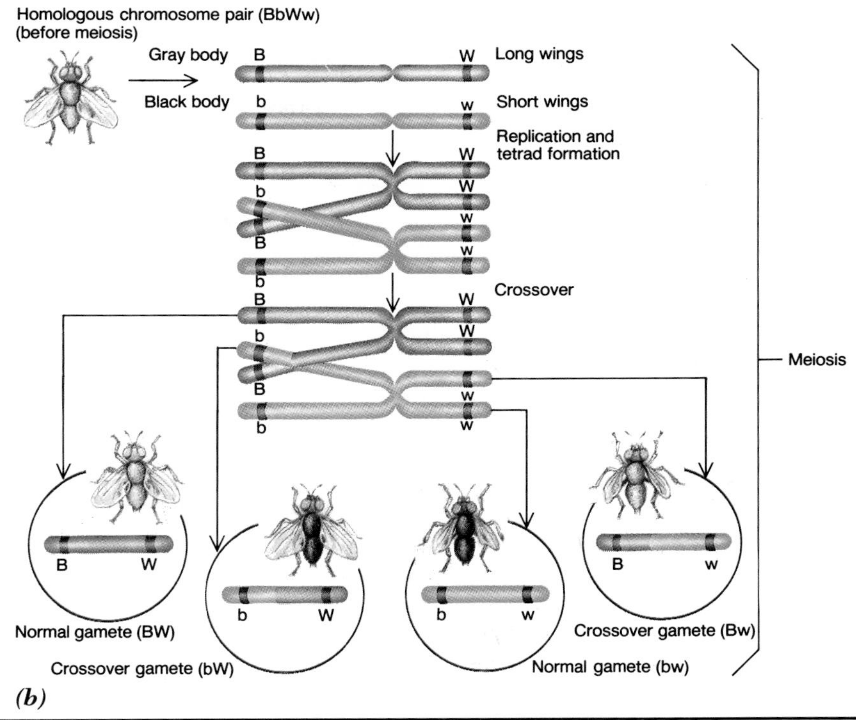

FIGURE 8-13

Crossing over provides the mechanism for reshuffling alleles between maternal and paternal chromosomes. ***(a)*** Tetrad formation during synapsis, showing three possible crossover intersections (*chiasmata*, indicated by red arrows). ***(b)*** Simplified representation of a single crossover in a *Drosophila* heterozygote (*BbWw*) at chromosome number 2 and the resulting gametes. If one of the crossover gametes participates in fertilization, the offspring will have a combination of alleles that was not present in either parent, an event that could not occur if the alleles remained linked.

SEX LINKAGE

For fruit flies and humans alike, there are hundreds of genes on the X chromosome that have no counterpart on the smaller Y chromosome. The phenotype of these traits *depends on* gender, since the recessive characteristic is more likely to be expressed in males. In females, a recessive allele on one X chromosome will be masked (and not expressed) if a dominant counterpart resides on the other X chromosome. In males, however, it only takes *one* recessive allele on the single X chromosome to determine the individual's phenotype since there is no corresponding allele on the Y chromosome. Inherited characteristics determined by genes that reside on the X chromosome are called **X-linked characteristics.**

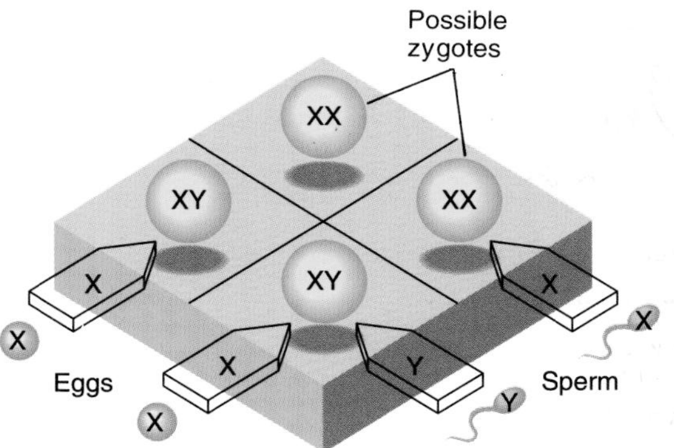

FIGURE 8-14

Sex determination. Whether in *Drosophila* or humans, males have an X chromosome and a Y chromosome needed to determine the gender of offspring (females have two X chromosomes). Males produce two types of gametes in equal numbers, while females produce only one. As depicted in this Punnett square, the sex of each offspring is determined by the chromosome composition of the fertilizing sperm.

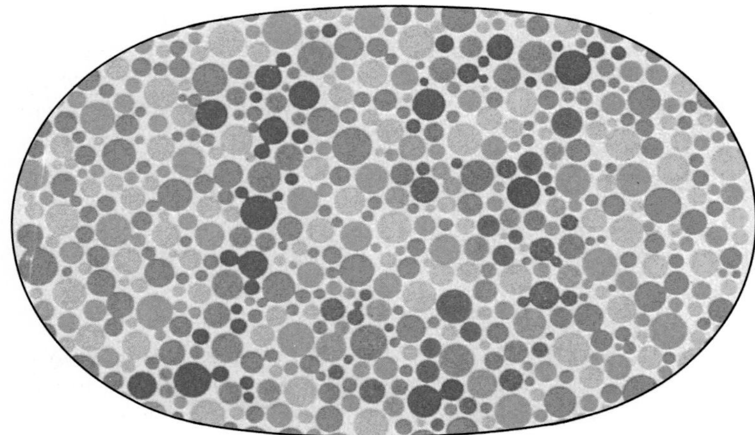

FIGURE 8-15 **Distinctively male.** Red-green colorblindness is one of hundreds of X-linked recessive characteristics that appear in males, but rarely in females, who would have to have the recessive allele on both X chromosomes. Males with this defect in vision cannot see the number "15" in the above pattern.

So far, some 200 human X-linked characteristics have been described, many of which produce disorders that are found almost exclusively in men. These include muscular dystrophy and red-green colorblindness (Figure 8-15). One X-linked disorder is *hemophilia,* or "bleeder's disease," a genetic disorder characterized by the inability to produce a clotting factor needed to quickly halt blood flow following an injury. Nearly all hemophiliacs are males (female hemophiliacs posses two recessive alleles for the disease, an extremely rare occurrence). Usually women who have acquired the rare defective allele are heterozygous *carriers* that show no signs of the disease because the normal allele on the other X chromosome directs formation of enough blood-clotting factor to assure a normal phenotype. Since the Y chromosome has no allele for producing the clotting factor, a boy who inherits the defective allele from his heterozygous mother develops hemophilia. X-linked characteristics typically skip a generation in their appearance. This tendency is depicted in Figure 8-16, which shows a typical pedigree for X-linked traits. (A **pedigree** is a diagram that shows the path by which genetic traits are inherited in a family).

ABERRANT CHROMOSOMES

In addition to mutations that alter the information content of a single gene, chromosomes may be subjected to more extensive alterations that typically occur during cell division. Pieces of a chromosome may be lost or exchanged between *nonhomologous* chromosomes, or extra segments may appear. Since these **chromosomal aberrations** follow chromosomal breakage, their incidence is increased by

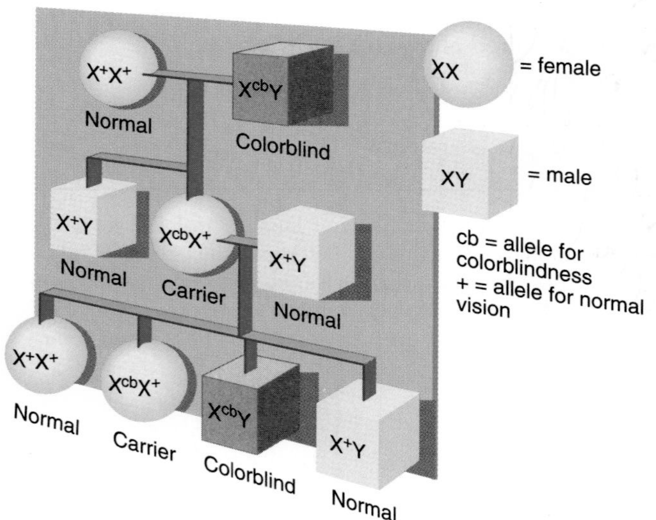

FIGURE 8-16 **The "generation gap."** Pedigrees (a record of a family's genetic history) show the tendency of X-linked traits—in this case, colorblindness—to skip a generation, producing a phenotypic gap—a generation of normal-visioned men and normal-visioned, carrier women, but no individuals showing the abnormal characteristic. Since the Y chromosome has no gene for color vision, the presence of a defective allele on the one X chromosome in the male produces the recessive phenotype. Hemophilia and other X-linked traits follow this basic inheritance.

exposure to agents that damage DNA, such as a viral infection, X-rays, or chemicals that can break the DNA backbone.

Many chromosome aberrations occur during meiosis, particularly as a result of abnormal crossing over, and can be transmitted to the next generation, where they occasionally confer advantageous traits. These structural modifications in chromosomes produce large-scale genetic changes that may propel evolution forward in giant steps. Table 8-2 summarizes each of the following chromosomal aberrations:

- *Deletions.* A **deletion** occurs when a portion of a chromosome is lost. Forfeiting a portion of a chromosome usually results in a loss of critical genes, producing severe, usually fatal, consequences.
- *Duplications.* A **duplication** occurs when a portion of a chromosome is repeated. Duplications have played a very important role in biological evolution. Many proteins are present as families consisting of a variety of closely related molecules. Consider the adult hemoglobin molecule, which consists of four polypeptides—two alpha-globin chains and two beta-globin chains. Examination of the amino-acid sequence of alpha- and beta-globin chains reveals marked similarities, indicating that they both evolved from a single globin polypeptide present in an ancient ancestor.

  The first step in the evolution of a protein family is thought to be the duplication of the gene. The two copies of the gene eventually evolve in different directions, generating polypeptides with different, but related, amino-acid sequences (as in hemoglobin).

TABLE 8-2
MODIFICATIONS IN CHROMOSOMAL STRUCTURE

Type of Alteration	Example of How Change May Occur	Some Possible Effects (Favorable / Harmful)
Deletion	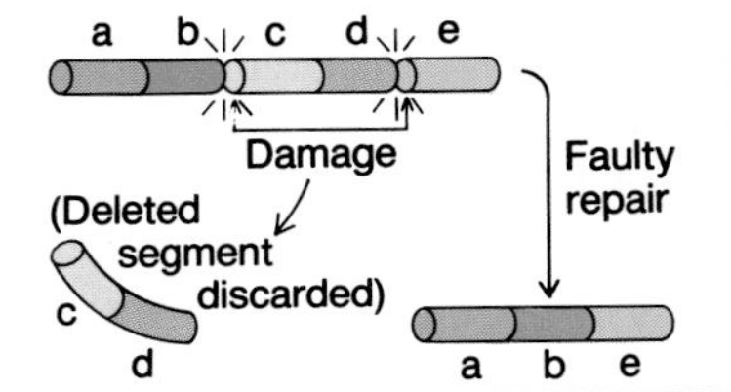 a b c d e; Damage; Faulty repair; (Deleted segment discarded) c d; a b e	Favorable: Rarely favorable; perhaps elimination of detrimental genes Harmful: Loss of critical genes is lethal; disrupts chromosome separation during meiosis
Duplication	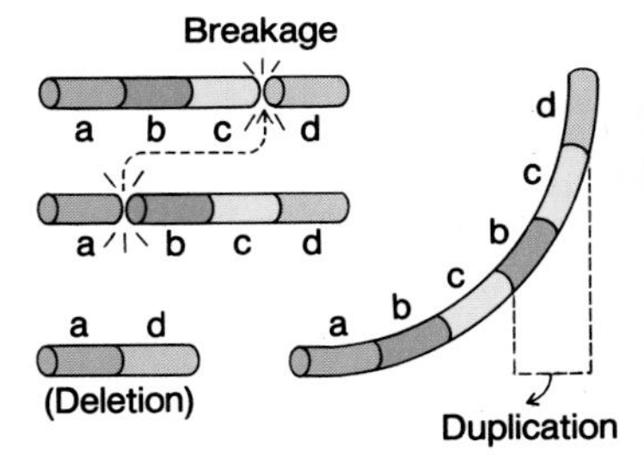 Breakage; a b c d; a b c d; a d (Deletion); a b c b c d Duplication	Favorable: Provides raw material for evolution of new proteins as part of a family of related proteins Harmful: May interfere with chromosome separation; may disrupt gene function if duplication occurs within a gene
Inversion	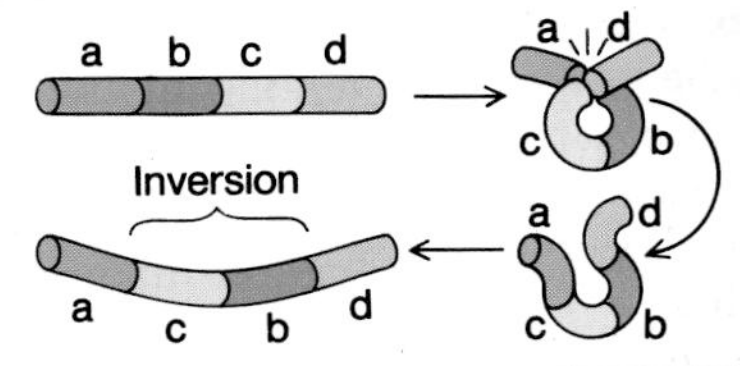 a b c d; Inversion; a c b d	Favorable: Increases genetic diversity by changing gene positions Harmful: Reduced fertility; loss of control of gene expression
Translocation	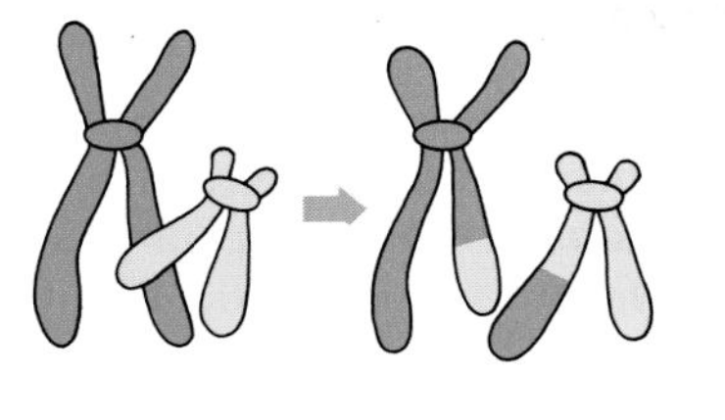	Favorable: Enormous genetic changes may generate rapid evolutionary advances Harmful: May activate genes that cause cancer; reduce fertility; may result in gain or loss of part or whole chromosome

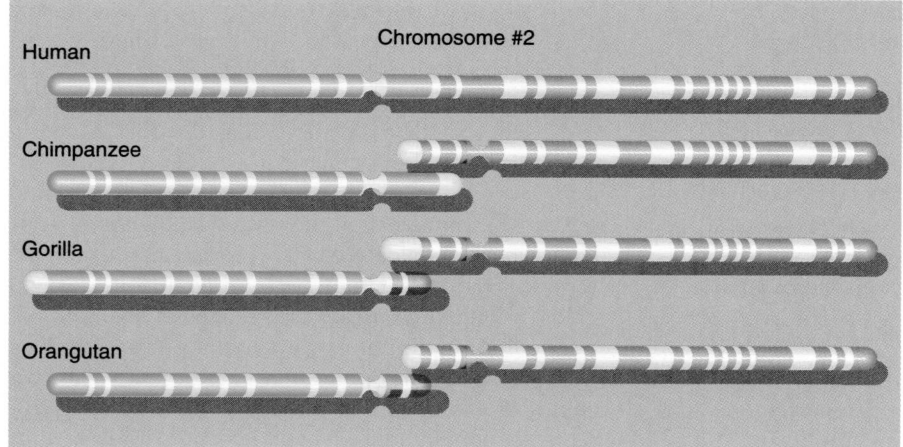

FIGURE 8-17 **Translocation and evolution.** Translocation may have played an important role in the evolution of humans from their apelike ancestors. If the only two ape chromosomes that have no counterpart in humans are "fused," they match human chromosome number 2, band for band.

- *Inversions.* Sometimes a chromosome is broken in two places, and the segment between the breaks becomes resealed into the chromosome in reverse order, generating an **inversion.** Sometimes inversions change gene position, separating a gene from its regulatory sequence (page 122). Such an inversion could disrupt the orderly expression of a gene, turning it into an uncontrolled oncogene (cancer-causing gene).
- *Translocations.* When all or a piece of one chromosome becomes attached to a nonhomologous chromosome, the aberration is called a **translocation.** Some translocations increase the incidence of cancer in humans. The best-studied example is the *Philadelphia chromosome,* which is found in individuals with certain forms of leukemia, a malignancy of the white blood cells. The Philadelphia chromosome, which is named for the city in which it was discovered in 1960, is a shortened version of human chromosome number 22. For years, it was believed that the missing segment represented a simple deletion, but with improved techniques for observing chromosomes, the missing genetic piece was found translocated to another chromosome (number 9). Although all the "normal genes" are present in the leukemia cell, the positions are different, apparently causing the cell to become malignant. Several other types of cancer are also associated with chromosomal aberrations (see The Human Perspective: Chromosome Aberrations and Cancer).

Translocations also play an important role in evolution, generating large-scale changes that may be pivotal in the branching of separate evolutionary lines from a common ancestor. Such a genetic incident probably happened during our own recent evolutionary history. A comparison of the 23 pairs of chromosomes in human cells with the 24 pairs of chromosomes in the cells of chimpanzees, gorillas, and orangutans reveals a striking similarity (Figure 8-17). But how did humans "lose" the twenty-fourth pair of chromosomes? A close examination of human chromosome number 2 reveals that we didn't really lose it at all. If the two ape chromosomes that have no counterpart in humans were joined together to form a single chromosome, they would form a perfect match, band for band, with human chromosome number 2. At some point during the evolution of humans, an entire chromosome was translocated to another, creating a single fused chromosome, reducing the haploid number from 24 to 23, perhaps generating the large-scale differences that distinguish humans from apes.

EVOLUTION AND ADAPTATION: TYING IT TOGETHER

The patterns of inheritance explain the apparent contradiction between trait *preservation* from generation to generation and trait *diversity*—the appearance of new types of organisms in spite of genetic stability. Similar patterns of inheritance emerge among all eukaryotes, whether humans or garden peas. All eukaryotic species have many traits that come in two or more forms, traits that show dominance and recessiveness, that segregate into separate gametes, and that assort independently of one another. These patterns of inheritance are themselves inherited, acquired from a common ancestor. This pattern has persisted for billions of years because it has selective advantages: It increases diversity and prevents the automatic doom of an organism that acquires a recessive lethal mutation (if heterozygous, the dominant allele will mask the lethal effect). In addition, all organisms also show linkage groups—that is, the hundreds of traits encoded by genes residing on the same chromosomes, fastened together like pearls on the same string. All organisms possess the ulti-

◁ THE HUMAN PERSPECTIVE ▷

Chromosome Aberrations and Cancer

Many malignant cells can be identified by the abnormal structure of their chromosomes, even before clinical evidence of cancer develops. The banding patterns of human karyotype chromosomes (review Figure 7-7) reveals malignant cells that possess observable cancer-causing aberrations that occur in about 100 specific bands. These may be the positions of *oncogenes,* which, when *activated,* cause uncontrolled proliferation of these cells, leading to tumors. A more recent discovery, the **tumor suppressor gene** (or anti-oncogene), *blocks* the formation of cancerous cells. The tumor suppressor gene on each homologous chromosome in a cell must be knocked out, their protective function eliminated, before a cell can be transformed to the malignant state.

A rare childhood cancer of the eye, called *retinoblastoma,* illustrates this tragic state. Children suffering from retinoblastoma apparently lack a small piece from the interior portion of chromosome pair #13. The cancer itself develops only in those individuals whose retinal cells lack *both* copies of the gene; that is, a retinal cell that has no tumor-suppressive capacity. The missing retinoblastoma genes qualify as a tumor-suppressor gene because it directs the formation of a protein that inhibits cell division.

The discovery of tumor-suppressor genes creates a whole new strategy for treating cancer. Rather than using toxic chemotherapeutic agents to kill malignant cells, malignant cells may respond to treatment with the missing protein or even gene replacement therapy (putting back a copy of the missing gene). Cancer biologists spend considerable time and effort identifying tumor supressor genes and their products.

mate similarity—genes that encode the information that determines what traits an organism will possess.

These are the successful genetic strategies we all use, strategies acquired from a common, shared ancestor. But what about the differences that distinguish organisms as they evolved from just a few types to the millions of species that have inhabited the earth? One of these diversifying forces is mutation, a change in genes that, when selected by the environment, generates new genes for species. Chromosomal mutations such as translocation accelerate the changes created by mutation, perhaps generating evolutionary "giant steps." Sex itself is more a diversifying strategy than it is a means of reproduction, mixing genes from different parents. Sex also creates the opportunity for chromosomal crossing over, creating new combinations that go beyond those possible by parental mixing of traits alone.

Together, the stabilizing principles of inheritance enable organisms to pass on their successful traits to their offspring while the diversifying principles of inheritance provide the changes needed for species to evolve. These contrasting principles of inheritance explain both the similarities and the differences between silkworms and sequoias.

SYNOPSIS

The primary pattern of inheritance was explained by paired alleles that separate and recombine during sexual reproduction. Mendel's principles of genetics evolved from his discovery that inherited traits were controlled by pairs of factors (genes). The two factors for a particular trait could be identical (homozygous) or different (heterozygous). In heterozygotes, one of the gene variants (alleles) is dominant over the other, recessive allele. Because of dominance, the heterozygote (*Aa*) has a phenotype for that trait that is identical to that of the dominant homozygote (*AA*). Only if an individual inherits two recessive alleles (*aa*) does it exhibit the recessive phenotype. Monohybrid crosses (those that examine only one trait) between heterozygous parents produce three times more offspring with the dominant characteristic than with the recessive characteristic.

Mendel's "Law of Segregation" explains the results of monohybrid crosses. Paired alleles segregate from each other during formation of the gametes, each gamete having an equal chance of receiving either of the two alleles.

Mendel's "Law of Independent Assortment" explains how dihybrid crosses (between heterozygous parents) can generate four phenotypes from two phenotypes. The ratios of the four phenotypes are approximately 9:3:3:1, which is exactly what would be expected if the four alleles of the two genes are pairing up in a random manner, as predicted by simple probability. These findings led Mendel to conclude that the segregation of the pair of alleles for one trait had no effect on the segregation of the pair of alleles for another trait. They were independent.

Not all inherited traits are transmitted according to Mendel's predictions.

- Some alleles show incomplete dominance—the pair of alleles "dilute" each other's effect in heterozygous individuals.
- Codominant alleles are both fully expressed in heterozygous individuals.
- Multiple alleles (three or more) for a particular trait at a single genetic locus can combine to form four or more genotypes and three or more phenotypes.
- Alleles present at one locus can effect the expression of genes at a different locus.
- A single gene can have many phenotypic effects.
- One trait may be due to multiple genes at several loci, leading to polygenic inheritance.
- Tracking the fate of alleles from generation to generation can be complicated by the effect of environment on phenotype.
- New alleles can emerge by mutations, changes in the genetic message of a gene. Such changes have produced a rich diversity of living organisms.

Genes reside on chromosomes in fixed linear orders. The physically connected set of alleles on the same chromosome constitutes a linkage group—a group of alleles that do not assort independently.

Linked genes don't always stay linked. Crossing over during meiosis shuffles alleles between homologous chromosomes, creating new combinations that are not found in the parents. Crossing over increases genetic diversity beyond that produced by independent assortment.

Gender is determined by sex chromosomes in most animals. In humans (and fruit flies), females possess two X chromosomes (XX), whereas males possess an X and a Y chromosome (XY). X-linked recessive characteristics, such as hemophilia and muscular dystrophy, are normally expressed when males inherit a recessive allele on the X chromosome from their mother.

Chromosomes are subject to genetic alterations. In addition to mutation, chromosomes can lose a segment (a deletion), acquire extra copies of its own genes (a duplication), reverse the order of genes in part of a chromosome (an inversion), and transfer a chromosome (or a part of a chromosome) to a nonhomologous chromosome (a translocation). Each of these aberrations can have serious effects on the cell and even the whole organism.

Review Questions

1. Why does understanding the Law of Independent Assortment require a grasp of the Law of Segregation?
2. Use Mendel's "rules of inheritance" to explain how a recessive trait may skip several generations then reappear.
3. How does each of the following increase genetic diversity? (a) meiosis, (b) fertilization, (c) mutation.
4. If a person has very light hair and skin but dark eyes, what might you conclude about the genotype of his or her melanism locus (*A*)? [Note: Practice the genetics problems provided in the study guide.]
5. Why do X-linked traits tend to skip a generation?

Critical Thinking Questions

1. As in determining the genetic basis of schizophrenia, the nature versus nurture argument provides perplexing questions to resolve. Explain some of the difficulties in distinguishing whether a human behavioral disorder is genetically based or the product of environmental influence.
2. Do any of the non-Mendelian patterns of inheritance discussed in the chapter provide a useful hypothesis for understanding the fact that one identical twin may not display the same magnitude of disease as his or her twin? Explain your answer.
3. Why might Mendel's conclusions about heredity have been different had he only been able to obtain 50 offspring from his dihybrid crosses?
4. Design a mating experiment that would demonstrate that incomplete dominance is not evidence of genetic blending. Explain how your proposed experiment proves that genes remain intact, even in organisms with the intermediate phenotype.
5. How might you determine if two proteins (of similar size but different function) were descended from the same ancestral protein? Describe a hypothetical situation in which such a situation occurs.

CHAPTER

9

The Molecular Basis of Genetics

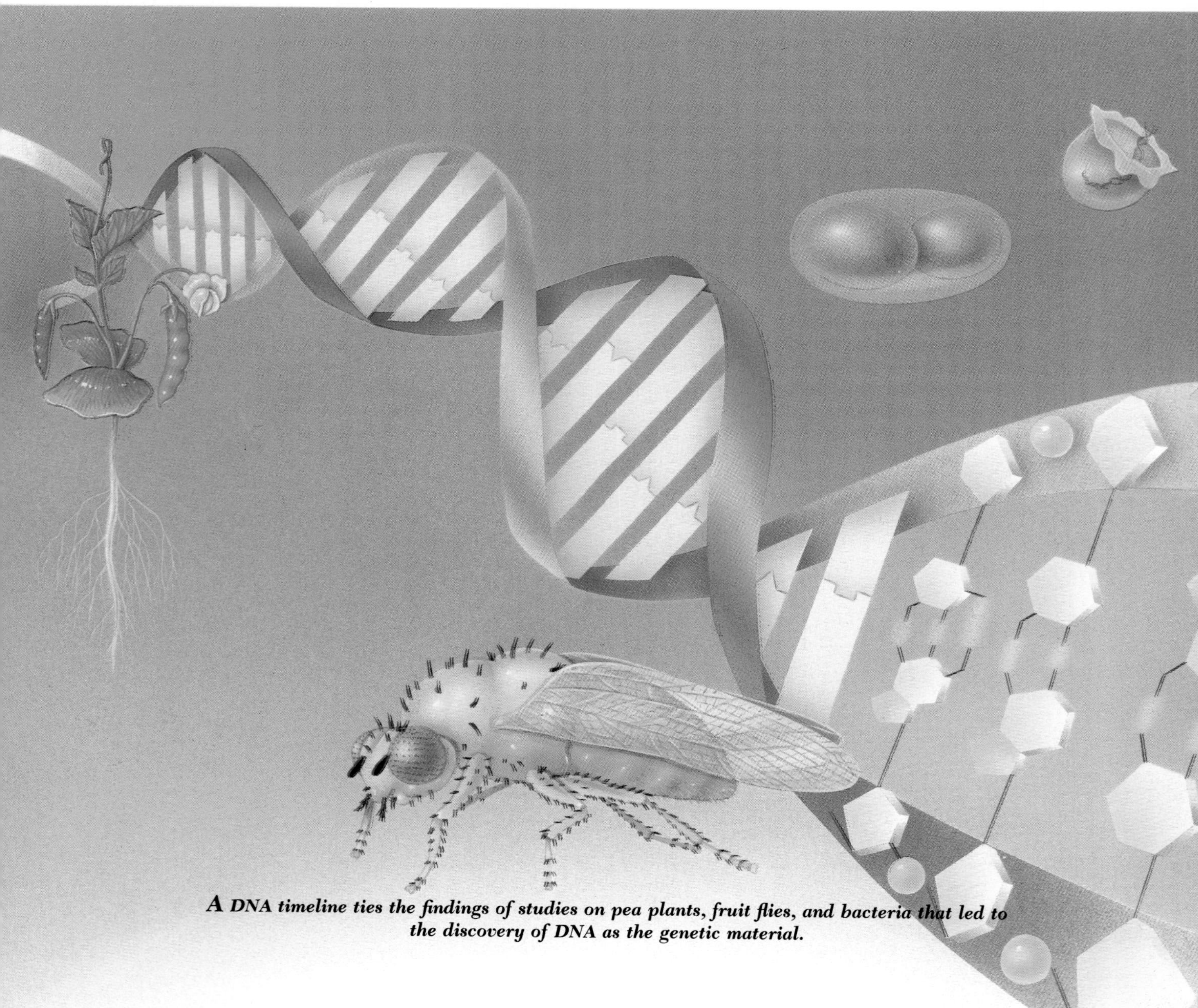

A DNA timeline ties the findings of studies on pea plants, fruit flies, and bacteria that led to the discovery of DNA as the genetic material.

STEPS TO DISCOVERY
The Chemical Nature of the Gene

By the end of the 1930s, biologists were well versed on the alignment of genes along chromosomes and the ways these genes were reshuffled during meiosis and transmitted from one generation to the next. By 1940, the major question confronting geneticists was very simple: What is the identity of the genetic material? Ironically, the answer had been hiding in "plain sight" for over 70 years.

When pus cells obtained from discarded surgical bandages were explored by the Swiss physician Friedrich Miescher in 1869, a "new" substance briefly captured the scientific limelight. Miescher had discovered and purified DNA. At that time, however, neither the nucleus nor its DNA contents were considered to be pieces of the genetic puzzle. Not even Mendel's findings hinted at the genetic significance of material from a cell's nucleus. This revelation awaited the turn-of-the-century discoveries of genes and mitosis, establishing the first associations between inheritance and the cell's nucleus.

Twenty-three years later, the German biochemist Robert Feulgen developed a procedure that specifically stained DNA in tissues. Observing stained cells under the microscope, Feulgen demonstrated the presence of Miescher's DNA in mitotic chromosomes. But chromosomes also contained abundant proteins, and most scientists believed that proteins, not DNA, were complex and versatile enough for this job of carrying genetic instructions. Unlike protein, DNA was a relatively simple molecule, a monotonous repeat of only four nucleotide subunits. A molecule of such simple construction could hardly be considered a candidate for the master molecule that directs life itself!

A decade later, Fred Griffith, a bacteriologist with the British Ministry of Health, was absorbed with exploring pneumonia-causing bacteria (called *pneumococcus*) when he made a discovery that would eventually be used to change our concept of the world. Griffith compared two genetically stable strains of pneumococcus. One was a dangerous version that enshrouded itself in capsules, coverings that enabled the bacteria to evade the internal defenses of mouse and human alike. Encapsulated pneumococci caused pneumonia in both these animals, whereas the second type of pneumococcus, a harmless, capsule-free variant, differed from its deadly counterpart only in the lack of this protective covering. Unlike the encapsulated bacteria, nonencapsulated versions of the same cell type were quickly destroyed by the host's defenses before they could cause disease. But even encapsulated bacteria could cause no damage if they were dead: Griffith found that heat-killed encapsulated bacteria were as harmless as the live, nonencapsulated strain when injected into mice. This expected result, however, was soon followed by a very surprising development.

When Griffith injected both these "harmless" preparations (nonencapsulated living bacteria and heat-killed encapsulated bacteria) into the same mouse, the animal subsequently contracted pneumonia and died. Furthermore, the dead mouse contained live, encapsulated bacteria, even though scrupulous experimental controls assured that the mice received none of these dangerous bacteria. Where did these living encapsulated strains come from? Griffith concluded that the presence of the dead encapsulated cells had *transformed* the nonencapsulated cells into an encapsulated strain. The transformed bacteria and their progeny continued to produce capsules; in other words, the change was permanent and *inheritable.* In 1928, however, the scientific community had yet to recognize that bacteria possessed genes, and the genetic significance of Griffith's results went unappreciated.

About the same time, a physician named Oswald Avery was shifting his attentions from his medical practice to conducting bacteriological research. Although initially skeptical of Griffith's results, Avery soon not only confirmed the findings, but discovered that the transformation of nonencapsulated cells to the encapsulated state could be accomplished in a culture dish (rather than in a living mouse) simply by adding a soluble extract of the encapsulated cells to the medium in which the nonencapsulated cells were growing. Because this cell-free transforming substance transferred a stable genetic trait, Avery reasoned that it must be the genetic material. For the next decade, Avery became preoccupied with purifying the substance responsible for transformation and determining its chemical nature.

Avery, along with colleagues Colin MacLeod and Maclyn McCarty, eventually identified the transforming substance as DNA. In addition, an enzyme (*DNase*) that selectively destroys DNA abolished the substance's transforming activity, whereas protein-degrading enzymes had no effect (countering the popular belief that genes were made of protein). Avery's now famous 1944 article on this research, however, was written with scrupulous caution and avoided even mentioning the conclusion that genes were made of DNA rather than protein. Although not given to speculation, Avery was clearly aware of the significance of his discovery. He had discovered "something that has long been the dream of geneticists."[1]

Avery's findings are no longer overlooked, but it took 8 long years before DNA was finally accepted as the genetic material. Acceptance arrived when additional investigators performed their own experiments that confirmed DNA's role as life's molecular director. Since then, Avery and his colleagues took their proper places in history, the initiators of perhaps the most monumental period in the scientific saga.

[1] Quote extracted from Avery's letters to his brother, a bacteriologist.

*I*f ever there was a fleeting period of discovery that changed the face of biology—indeed the world itself—it was the period between the early 1940s and the mid-1960s. Virtually all that you will read about in this chapter was discovered in that fertile period of about 25 years—a period of biological revolution. Before this time, we knew that genes were carried on chromosomes, and we knew the rules by which these genes were transmitted from generation to generation. But we knew little else about heredity. By the time the dust had settled, we understood the fundamentals of how it all worked.

▼ ▼ ▼

THE STRUCTURE OF DNA

DNA consists of two long molecular strands connected to each other, much like the rails of a ladder held together by the ladder's rungs. DNA's building blocks are mono-

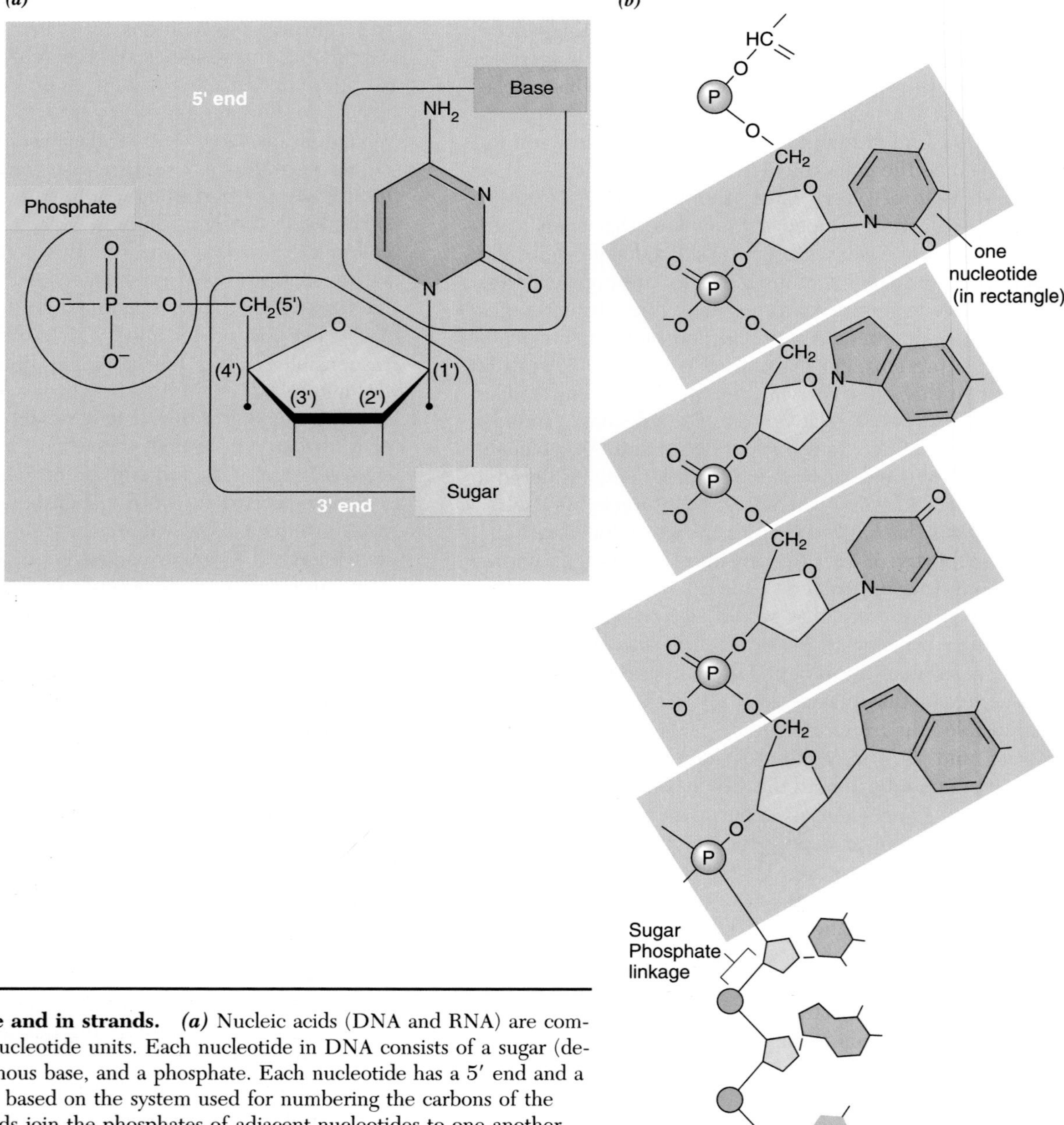

FIGURE 9-1

Nucleotides—alone and in strands. ***(a)*** Nucleic acids (DNA and RNA) are composed of repeating nucleotide units. Each nucleotide in DNA consists of a sugar (deoxyribose), a nitrogenous base, and a phosphate. Each nucleotide has a 5′ end and a 3′ end (numbers are based on the system used for numbering the carbons of the sugar). Covalent bonds join the phosphates of adjacent nucleotides to one another, forming the chemical backbone that holds the nucleotide "beads" together in a linear strand ***(b).***

mers called nucleotides, each of which contains three parts—a five-carbon sugar (in DNA this sugar is **deoxyribose),** a phosphate group, and a *nitrogenous (nitrogen-containing) base* (Figure 9-1*a*). Covalent bonds fasten the monomers together to form the linear sequences of nucleotides (Figure 9-1*b*).

Every nucleotide in DNA contains the same sugar and phosphate, but they differ from one another in their nitrogenous bases. Two of the bases, adenine (A) and guanine (G), are *purines,* while the other two bases, cytosine (C) and thymine (T), are *pyrimidines* (Figure 9-2).[2] Using molecular-shaped cutouts of these nucleotides, plus the knowledge that an adenine in one strand pairs with a thymine at the same point in the opposite strand, Watson and Crick constructed a helical model of the DNA molecule that, 40 years later, remains the centerpiece of molecular genetics (Figure 9-3).

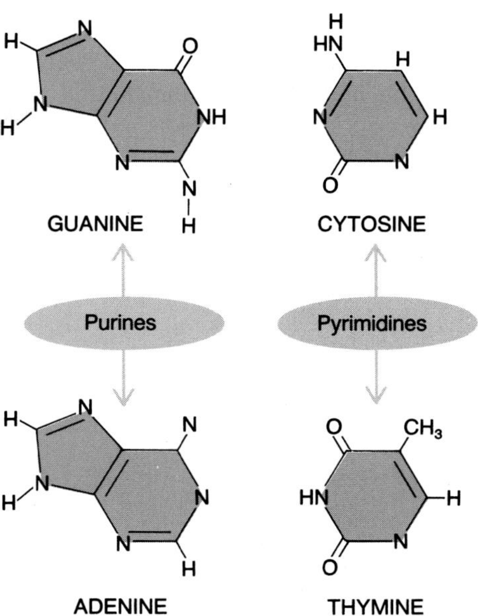

FIGURE 9-2
The four nitrogenous bases in DNA determine the identity of a nucleotide. Adenine and guanine are purines; thymine and cytosine are pyrimidines.

THE WATSON–CRICK PROPOSAL

A few of the main elements of the Watson-Crick model of DNA are listed below (refer to Figure 9-3 for a visual interpretation of these properties):

- A DNA molecule is composed of two chains of nucleotides that coil around each other to form a double helix.
- The two chains of a helix run in opposite directions, like two lines of people standing side by side, but facing opposite directions.
- The backbone (-sugar-phosphate-sugar-phosphate-) of each chain in DNA is located on the outside of the molecule, with the paired bases projecting toward the center, resembling the steps of a spiral staircase.
- The two chains are held together by hydrogen bonds that form between the nitrogenous bases. Unlike the covalent bonds that tie together the nucleotides of the linear backbone, the hydrogen bonds between nucleotides of opposite strands are relatively weak, allowing the double helix to separate. Such separation is necessary for inheritance of genetic information and the conversion of genetic information into expressed traits.
- An adenine on one chain always pairs with a thymine on the other chain, a phenomenon called **base pairing** (guanine and cytosine also form base pairs on the opposite chains of a DNA molecule). This complementary sequence of nucleotides in opposite relationship is known as **complementarity**—one entire strand in a DNA molecule is complementary to the other stand. Without complementarity, genetic information could neither be inherited nor expressed as traits.
- Each nucleotide has a *5′ end* (pronounced "5-prime" end) and a *3′ end.* All the nucleotides in a single strand face the same direction, much like people waiting in a single line: One end of the line—the front—is identifiable by a face, while the rear of the line is identifiable by the back of a head. For a strand of nucleic acid, one end is the 3′ end (the "face"), the other is the 5′ end (the "back of the head"). The significance of strand direction is discussed in the upcoming section on DNA replication.

[2] It might help to remember that the bases with the shorter name (purine) are larger than pyrimidines and have the more complex molecular structure (two rings, rather than one).

DNA: LIFE'S MOLECULAR SUPERVISOR

As the genetic material, DNA must fulfill three primary functions:

1. *Storage of genetic information.* DNA is the molecular "blueprint," a stored record of precise instructions that determine all the inheritable characteristics an organism can exhibit.
2. *Self-duplication and inheritance.* Because DNA contains all of an organism's genetic information, inheritance of these traits depends on an organism's ability to duplicate the molecule so each of the progeny gets

a complete set of genetic instructions. Duplicating the DNA provides the means by which genetic instructions can be doubled, then transmitted to the next generation.

3. *Expression of the genetic message.* Genes carry information for the formation of specific proteins, such as enzymes that direct the formation of an inherited trait (such as skin color). The stored information determines what polypeptides (and therefore what traits) an organism can possess.

Watson and Crick's model of DNA structure was a key to understanding the workings of each of these three essential functions.

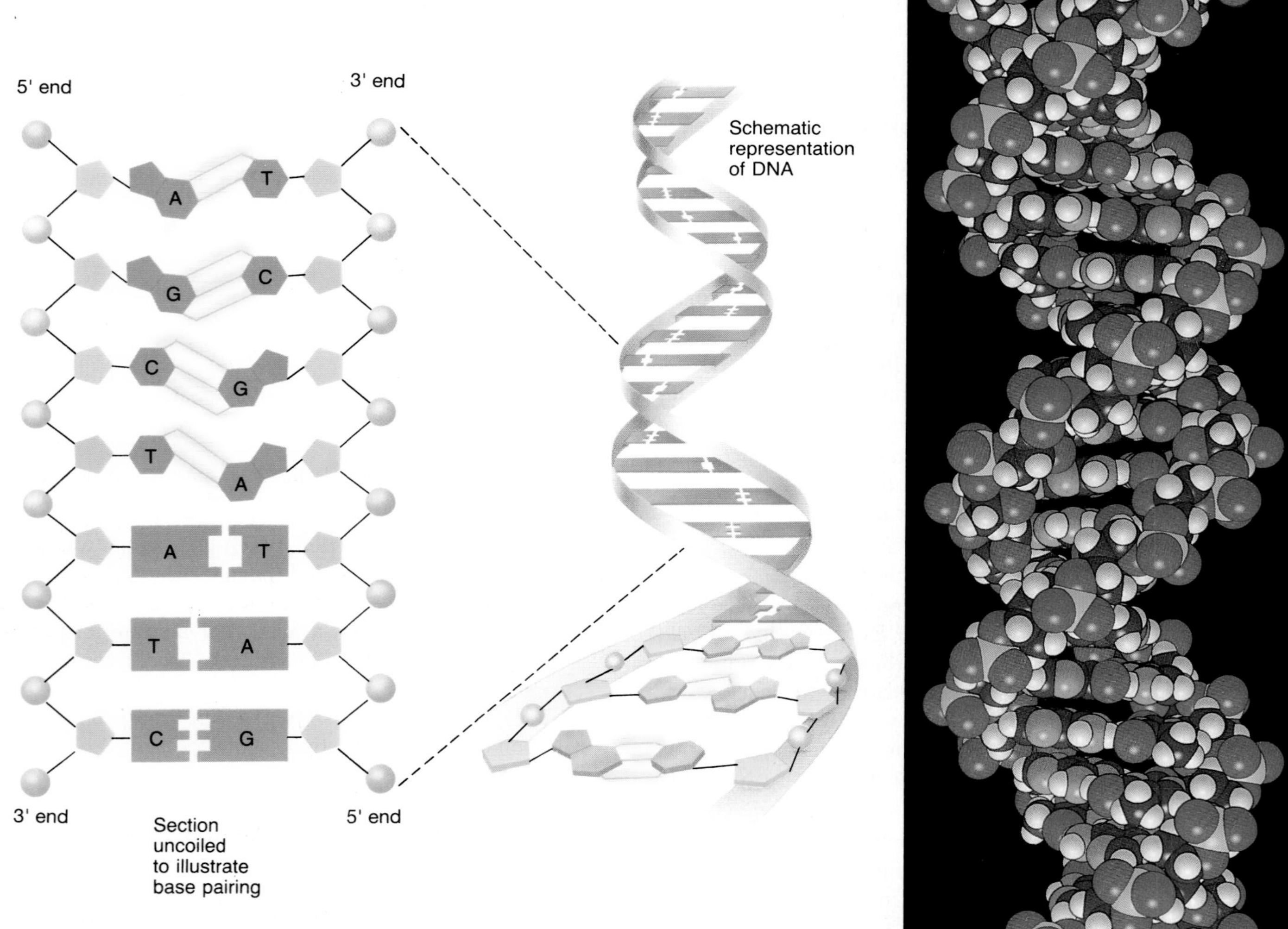

FIGURE 9-3

DNA—the double helix. Base-pairing creates a double-stranded molecule that twists into a helix, resembling a spiral staircase. The nitrogenous bases of the nucleotides form hydrogen bonds with each other; guanine (G) binds to cytosine (C) by three hydrogen bonds, whereas adenine (A) binds to thymine (T) by two hydrogen bonds. Because of such specific base-pairing, the nucleotide sequence in the two strands of the same molecule are complementary. Unlike the "stairway" image of DNA, however, the molecule's actual appearance is more like the computer-generated image in the photo.

FUNCTION 1: STORAGE OF GENETIC INFORMATION

Genetic information represents a kind of "instruction manual" situated within the DNA of an organism's genes, information that directs an organism's self-assembly project. *This information is encoded in the precise order (linear sequence) of the nucleotides in DNA,* just as the sequence of printed letters in this sentence encodes its meaning. Nucleotides are the chemical "letters" used to spell out an organism's genetic traits. Unlike the 26 letters in the English language, however, the genetic "alphabet" consists of only four letters—G, C, A, and T—the four types of nucleotides in DNA. But four "letters" are all that are needed to "write" an unlimited variety of genetic messages. For example, a portion of the double helix only ten nucleotides long can be arranged in more than a million different sequences. With 3 billion base pairs (the size of human DNA), the potential for encoded messages is virtually limitless.

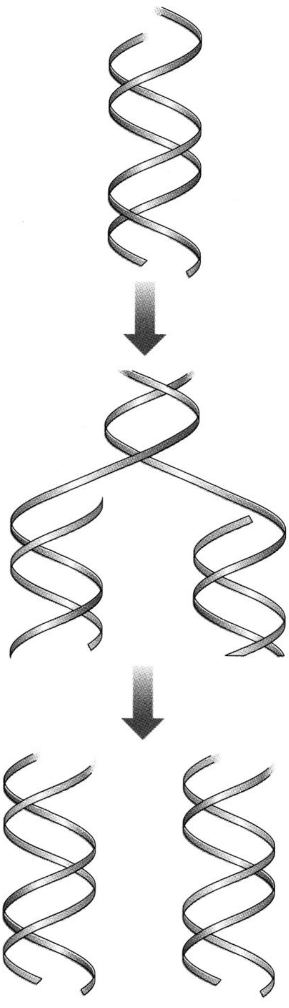

FIGURE 9-4
Semiconservative DNA replication. During replication, the double helix unwinds, and each of the strands serves as a template for the assembly of a new complementary strand. Following replication, each new DNA molecule consists of one strand from the original duplex and one newly constructed strand. This arrangement, which was first predicted by Watson and Crick in their first publication of the structure of DNA, is called *semiconservative replication.*

FUNCTION 2: PASSAGE OF GENETIC INFORMATION TO DESCENDANTS—DNA REPLICATION

Reproduction and heredity, two of the most fundamental properties of all living systems, depend on DNA **replication,** that is, duplication of the genetic material.

The Mechanism of Replication

During replication the hydrogen bonds holding the two strands of the DNA helix are sequentially broken, and the strands separate, much like the two halves of a zipper. Each of the separated strands, with its exposed nitrogenous bases, serves as a *template* (a mold or physical pattern), directing the order in which nucleotides are assembled to form the complementary strand. In this way, two identical molecules of double-stranded DNA are produced, each containing one strand from the original DNA molecule and one newly synthesized strand (Figure 9-4). This form of DNA synthesis is called **semiconservative replication,** because half the original DNA molecule is conserved in each "daughter" molecule. As a result of replication, two new DNA molecules are formed, each containing precisely the same genetic message as that stored in the original molecule. This is the mechanism by which genetic instructions can be passed on from generation to generation.

DNA replication is directed by the enzyme **DNA polymerase,** a protein that moves along each template strand of the open helix and lengthens the new strand by adding the appropriate nucleotides. But DNA polymerase can move in only one direction—from 3′ to 5′. This allows the enzyme to travel straight down the template strand *toward* the *replication fork* (the site where the DNA separates during replication). The strand is therefore constructed *continuously,* forming a complete strand. But the other strand runs in the opposite direction, so the enzyme must move *away* from the replication fork (Figure 9-5). The strand that grows away from the replication fork, therefore, must be constructed in pieces that are subsequently linked together, a process called *discontinuous* assembly.

The Accuracy of Replication

Maintaining biological order and stability from one generation to the next requires accurate transfer of genetic information to offspring. DNA polymerase is a remarkably accurate enzyme, making a mistake only once in every billion nucleotides it incorporates (in bacteria, this is less than one mistake for every 100 cycles of replication). One of the reasons for this extraordinary accuracy is that DNA polymerase is actually two enzymes in one; it has one active site for polymerization, and another active site for "proofreading." If the first active site happens to incorporate a noncomplementary monomer, the mistake is immediately recognized by the second active site, which removes the incorrect nucleotide. If the mistake happens to slip by the "proofreading" mechanism, it may result in a permanent change, or genetic *mutation,* in the information content of the DNA. The consequences of a mutation are discussed later in the chapter.

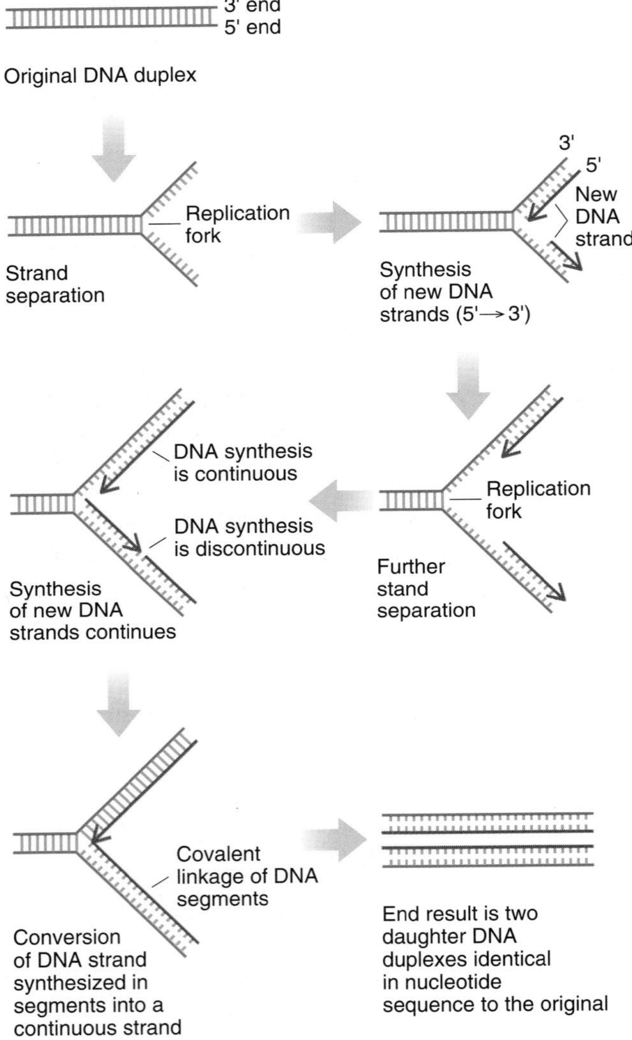

FIGURE 9-5

The mechanism of replication. Recall that a DNA duplex consists of two strands that run in opposite directions. One of the strands runs in the 3′ to 5′ direction, and the other runs in the 5′ to 3′ direction. DNA polymerase molecules are only capable of moving along a template in one direction, toward the 5′ end of the template. Consequently, polymerase molecules (and their associated proteins) move in opposite directions along the two strands. As a result, the two newly assembled strands also grow in opposite directions, one growing toward the replication fork, and the other growing away from it. One strand is assembled in continuous fashion, the other in segments that must be joined together by an enzyme.

FUNCTION 3: TRANSFORMING GENETICALLY ENCODED INFORMATION INTO REAL TRAITS

An organism is the manifestation of its particular constellation of proteins, each of which is made up of one or more polypeptide chains. *A single gene encodes the information for a single polypeptide chain,* the linear order of nucleotides in DNA determining the linear order of amino acids in a polypeptide chain. Such a transformation of information, however, requires an intermediate, a molecule of RNA. In other words, genetic information in a cell flows from DNA to RNA to protein.[3] This flow of encoded information is depicted in Figure 9-6. The cell first *transcribes* (copies) a gene's encoded instructions into a molecule of RNA, which is then sent to the "construction site" (ribosomes). These instructions direct the activities at the site, telling the "workers" (various proteins and RNA molecules) which polypeptide to build. The workers must be able to *translate* the instructions in the RNA into the exact gene product ordered for construction.

To understand this flow of genetic information, we must consider the structure of RNA. RNA differs structurally from DNA in three ways.

- The nucleotides in RNA contain *ribose,* a sugar that has one more oxygen atom than does DNA's sugar (*deoxy*ribose). This is an important difference because it allows enzymes to distinguish between the two types of nucleotides.
- As in DNA, RNA has four distinct nitrogenous bases, but one of them—*uracil*—is unique to RNA. Uracil replaces thymine as the base that is complementary to adenine.
- RNA is typically a single-stranded molecule, so its bases are generally exposed and available for hydrogen bonding with other molecules.

Transcribing the Message

The information stored in DNA is carried to the ribosomes by an RNA molecule aptly called **messenger RNA (mRNA).** RNA molecules are assembled by **transcrip-**

[3] This flow is reversed under certain circumstances—from RNA to DNA—as occurs in a cell infected by HIV, the virus that causes AIDS (Chapter 20).

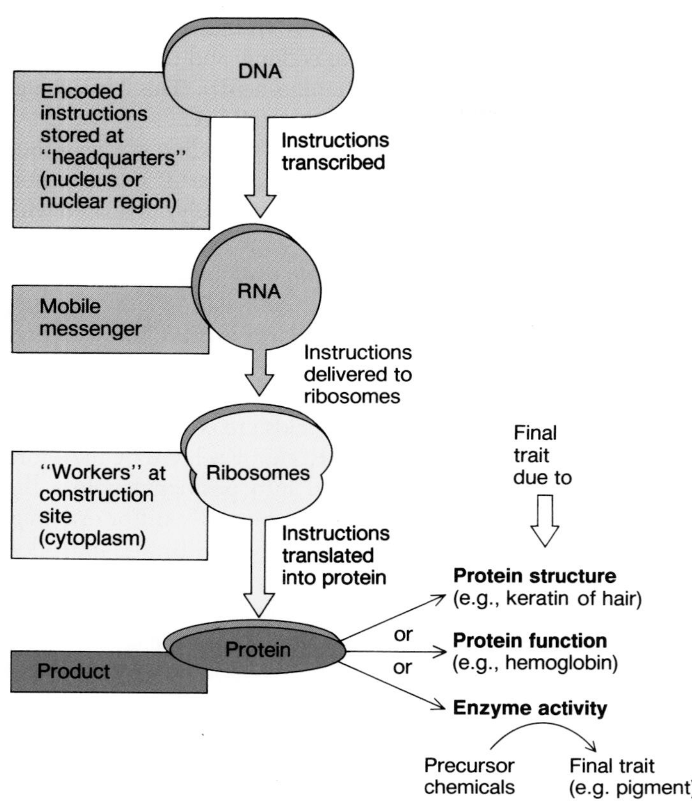

FIGURE 9-6

A summary of the flow of genetic information from DNA to RNA to protein.

tion, a process that in some ways resembles the synthesis of a new DNA strand during replication. During transcription (Figure 9-7), the double helix temporarily separates, and a complementary strand of mRNA assembles along one of the single DNA strands that acts as a template. **RNA polymerase,** the enzyme that directs the process, polymerizes only ribonucleotides into the growing chain. The enzyme also distinguishes between the two strands of the DNA molecule, selecting only the "sense" strand that encodes the appropriate sequence (the other strand would generate RNA containing biological gibberish). In this way, genetic information stored in a gene is transcribed into a molecule of mRNA that carries DNA's message by virtue of its complementary sequence of nucleotides. This mobile "messenger" leaves the DNA template and carries the encoded information to the ribosomes, where it directs the synthesis of a specific polypeptide. As we will see shortly, this process depends on two other types of RNA—transfer RNA and ribosomal RNA.

Translating the Message

In a sense, the genetic basis of life is a matter of "reading, copying, and following instructions" on a molecular scale. Just as a reader of a sentence deciphers an encoded message by translating a linear string of letters into a meaningful thought, cells use a language—a *genetic code*—which they translate into genetic characteristics. As you translate the line of characters in this sentence, you do so by recognizing groups of letters (words) that have specific mean-

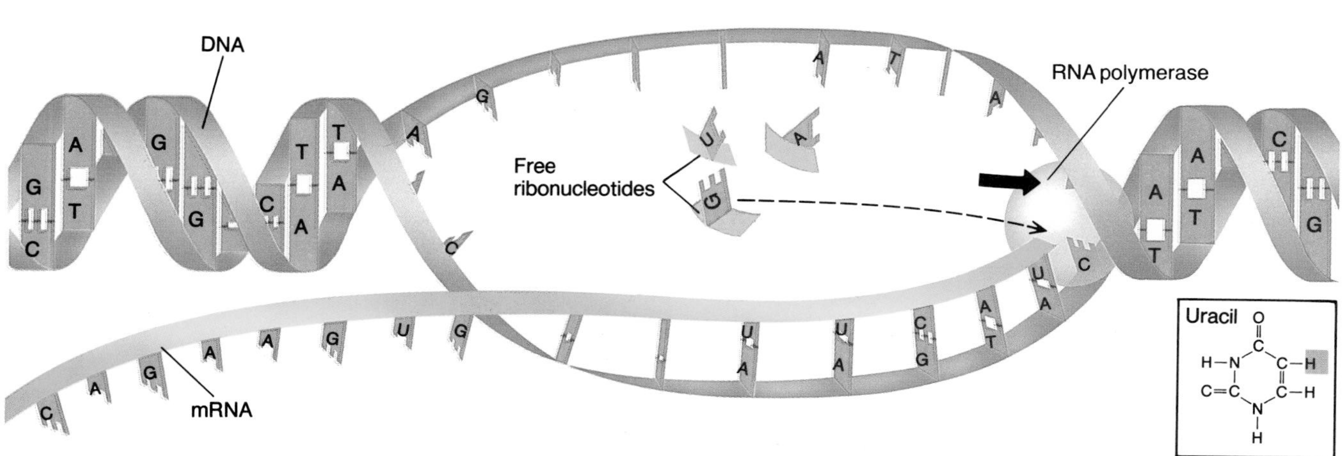

FIGURE 9-7

Transcription—dispatching the molecular messenger. Only one strand of DNA (called the "sense strand") encodes information in the message sent to the ribosomes. During transcription, the RNA polymerase binds to the DNA, the double helix is temporarily separated, the RNA polymerase recognizes the sense strand as the proper template, and the enzyme then assembles a complementary strand of RNA. (The uracil in RNA—see inset—that replaces thymine in DNA differs by the single chemical group shaded for emphasis.)

ings. In constructing proteins, the linear array of nucleotides in mRNA is also translated in groups of short nucleotide sequences, each of which represents a "molecular word," called a **codon.** Every codon is three nucleotides long, and each one (each sequence of three nucleotides) specifies the insertion of one—and only one—of the 20 different amino acids at a specific point in the polypeptide being built. The codon *CAC*, for example, always specifies the insertion of the amino acid histidine at that point in the protein being synthesized. The codon *CUA* always specifies leucine. In this way, the entire message is read codon by codon until the protein is complete. With a four-letter genetic alphabet (A, G, C, and U), 64 triplet combinations (4 × 4 × 4) are possible; 64 combinations are more than enough to assign at least one unique codon to each of the 20 different amino acids.

The mRNA-directed assembly of a polypeptide is called **translation** and is the cell's most complex synthesis process. Translation is the process of protein synthesis; it requires mRNA, amino acids, numerous enzymes, ribosomes, and chemical energy in the form of ATP (and GTP). It also requires another type of RNA that decodes mRNA's encoded message (written in codons) and translates it into the language of proteins (amino acids). This "molecular decoder" is called **transfer RNA (tRNA)**.

Transfer RNAs work like any decoding device—they recognize one set of symbols and convert it into another. This decoding function of tRNAs is closely correlated with their three-dimensional shape (Figure 9-8); the two opposite ends are exactly the same distance from each other regardless of the tRNA. At one end, a unique sequence of three nucleotides, called the tRNA's **anticodon,** can form base pairs with one of the mRNA codons. On the side opposite the anticodon on the tRNA is the site that carries a specific amino acid. Amino acids are covalently linked to tRNAs to form a "charged tRNA." Each tRNA therefore brings its unique amino acid into position anytime the corresponding codon appears in the mRNA. For example, the codon UCU in mRNA binds only with the anticodon AGA of the tRNA carrying the amino acid serine. Therefore, UCU in mRNA instructs the cell to insert serine at that point in the newly forming polypeptide. In this way,

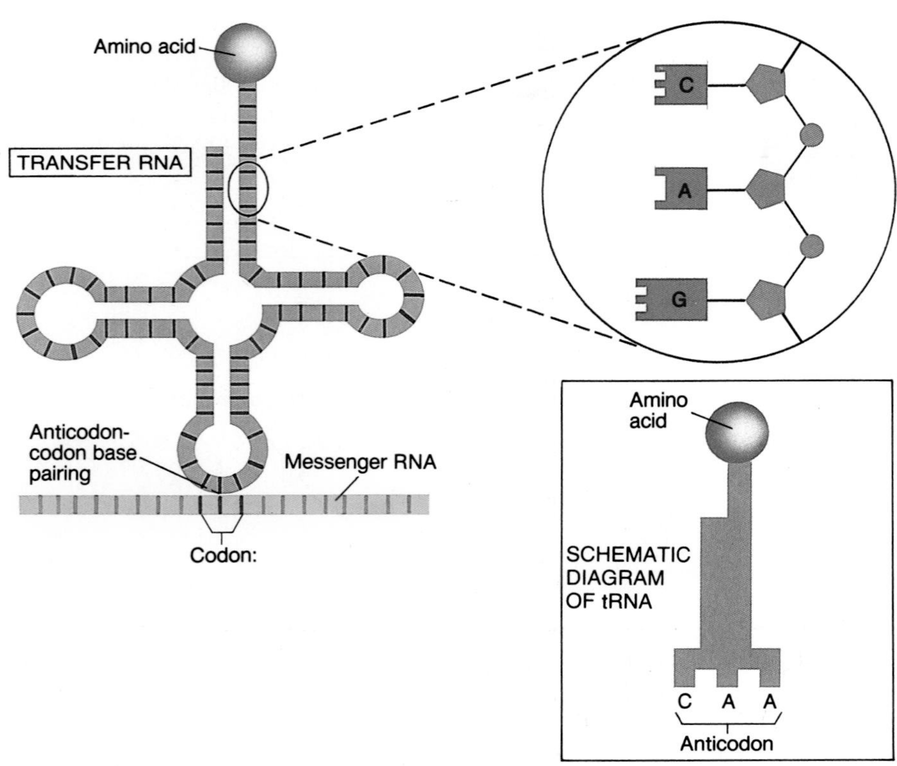

FIGURE 9-8
Molecular decoder. Transfer RNA molecules translate one "language" (a three-nucleotide codon "word") into another (a specific amino acid) by recognizing specific anticodons. Each anticodon in a tRNA "base pairs" with a single codon in mRNA. Similarly, each tRNA carries a specific amino acid. In this way, a particular tRNA associates the correct amino acid with its corresponding codon in mRNA. The "cloverleaf" structure illustrates how the anticodon base pairs with its complementary codon. The attachment of a specific amino acid to its corresponding tRNA is catalyzed by a set of amino acid "charging" enzymes.

tRNA bridges the language gap between the nucleotide codon and the amino acid.

FROM THE "LANGUAGE" OF GENES TO THE "LANGUAGE" OF PROTEINS

Although it took the first half of the 1960s, the codebreakers eventually deciphered the "genetic code," creating a chart that associates every molecular word (codon) with its corresponding amino acid (Figure 9-9). One of the most remarkable aspects of the genetic code is that it is a "universal" genetic language. The same codons specify the same amino acids, regardless of whether the organism is a bacterium, a yeast, a mushroom, a redwood tree, or a human. For example, in all of these organisms, the codon CAU will always specify the insertion of the same amino acid (tyrosine) at the corresponding point in the polypeptide being assembled. All organisms inherited the universal code from a common ancestor that we shared long before the first eukaryotic organisms evolved (the likelihood that all organisms independently developed the same genetic code is astronomically small).

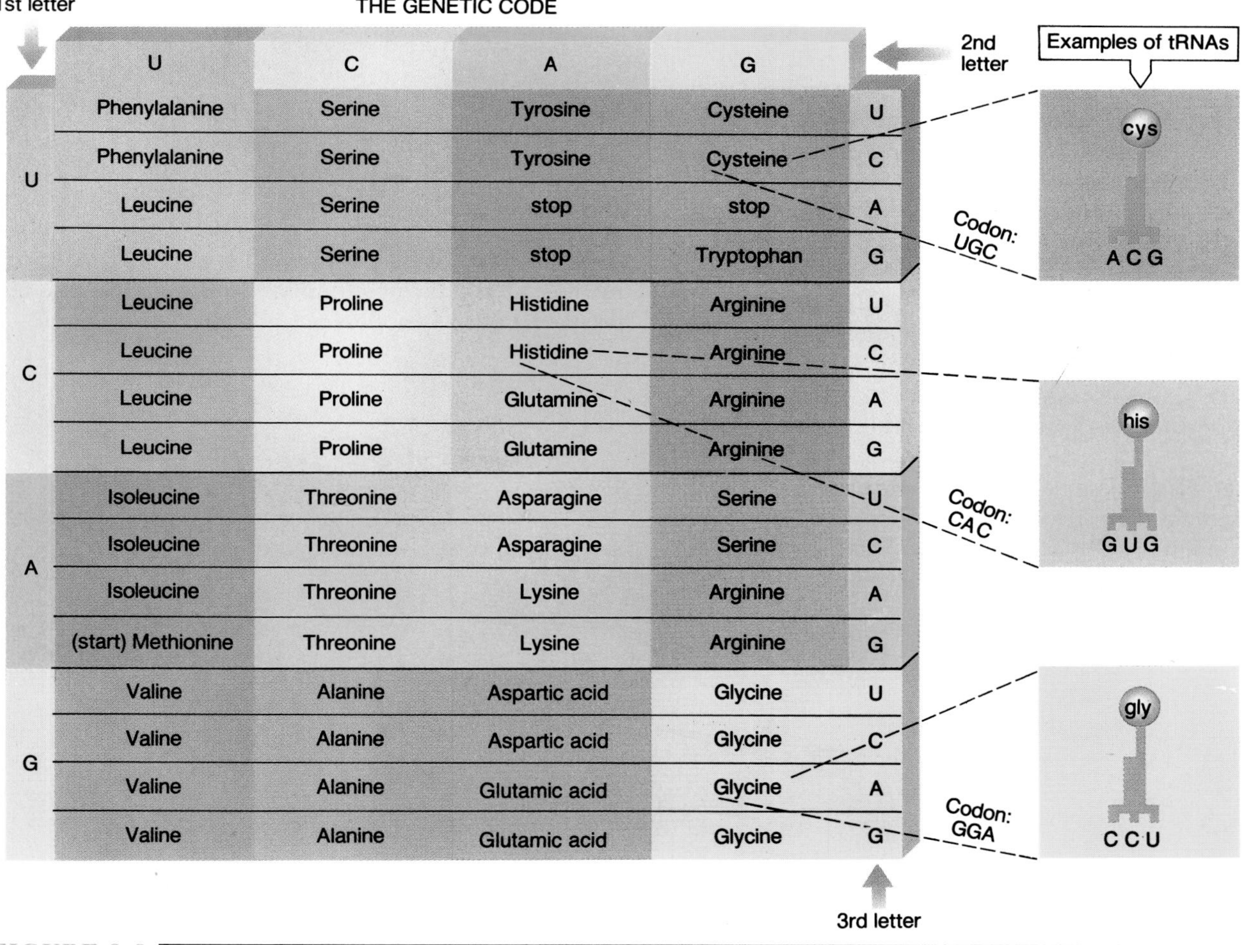

THE GENETIC CODE

1st letter	U	C	A	G	3rd letter
U	Phenylalanine	Serine	Tyrosine	Cysteine	U
U	Phenylalanine	Serine	Tyrosine	Cysteine	C
U	Leucine	Serine	stop	stop	A
U	Leucine	Serine	stop	Tryptophan	G
C	Leucine	Proline	Histidine	Arginine	U
C	Leucine	Proline	Histidine	Arginine	C
C	Leucine	Proline	Glutamine	Arginine	A
C	Leucine	Proline	Glutamine	Arginine	G
A	Isoleucine	Threonine	Asparagine	Serine	U
A	Isoleucine	Threonine	Asparagine	Serine	C
A	Isoleucine	Threonine	Lysine	Arginine	A
A	(start) Methionine	Threonine	Lysine	Arginine	G
G	Valine	Alanine	Aspartic acid	Glycine	U
G	Valine	Alanine	Aspartic acid	Glycine	C
G	Valine	Alanine	Glutamic acid	Glycine	A
G	Valine	Alanine	Glutamic acid	Glycine	G

FIGURE 9-9

The genetic code. The genetic code is a universal biological language. The correlation between codon and amino acid indicated in this decoder chart is the same in virtually all organisms. To use the chart to translate the codon UGC, for example, find the first letter (U) in the indicated row on the left. Follow that row to the right until you reach the second letter (G) indicated at the top; then find the amino acid that matches the third letter (C) in the row on the right. UGC specifies the insertion of cysteine. Each amino acid (except two) has several codons that order its insertion. These are genetic "synonyms," backup systems that reduce the danger of lethal mutations disrupting the cell. A change in a single nucleotide in the codon's third position, for example, would change which amino acid would be incorporated into the polypeptide, unless the new codon were synonymous with the original (which happens frequently). In that case, the same amino acid would be incorporated, and the mutation would have no phenotypic effect. As discussed in the following section, decoding in the cell is carried out by tRNAs, a few of which are illustrated in the right side of the figure.

Four of these 64 codons warrant special mention. Three (UGA, UAG, and UAA) have no corresponding amino acid. These triplets, called **stop codons** (or *nonsense codons*), spell "stop!" when mRNA is being translated into protein. They are used to terminate synthesis at the completion of a polypeptide. Another codon, AUG (specifying the amino acid methionine), is the "start" codon. AUG always appears at the beginning of the coding portion of an mRNA and initiates the assembly of amino acids into the polypeptide.

Translation occurs at ribosomes—complex particles composed of several dozen different proteins and a number of different RNA molecules. The RNAs of the ribosome, called **ribosomal RNAs (rRNAs)**, constitute the third major type of RNA made by cells. Whereas mRNAs carry encoded information, and tRNAs serve to decode this information, rRNAs are primarily structural molecules functioning as a scaffold to which the various proteins of the ribosome can attach. The proteins that make up the ribosome have varied functions: Some play a structural role in holding the particle together, while others bind to the mRNA or tRNAs.

The ribosome is a nonspecific component of the translation machinery—sort of a "workbench"—that can be used as a site to translate any mRNA. A functioning ribosome consists of a large and a small subunit (Figure 9-10). The ribosome's subunits assemble at the start of polypeptide synthesis, then separate when synthesis has been completed.

An assembled ribosome has a groove through which the mRNA molecule can travel. The ribosome also has sites that position two tRNA molecules so that their amino acids lie adjacent to each other on the large subunit. The large subunit contains an enzyme that covalently links the adjacent amino acids. As the ribosome moves along the mRNA strand, amino acids are incorporated into the growing polypeptide chain in the order specified by the mRNA. (Energy for incorporating each amino acid comes from the hydrolysis of ATP-like molecules.) A simplified version of the events of translation is depicted in Figure 9-11, although many of the enzymes and accessory molecules have been omitted for clarity.

The three stages of translation are initiation, chain elongation, and chain termination.

Step 1: Initiation. The process of translation begins when the mRNA binds to the small subunit of a ribosome. This binding always occurs at the initiation codon, AUG, followed by the attachment of the first tRNA (whose anticodon is UAC). This tRNA always inserts a methionine as the first amino acid of the new polypeptide. The binding of the AUG codon to the ribosome fixes the "reading frame," assuring that translation begins with the correct nucleotide. If the message were initiated one or two nucleotides over, all the remaining triplets would be incorrectly read, and the wrong amino acids would be inserted, producing a totally useless polypeptide.

```
correct   →        met leu his pro
mRNA      → — AUG CUG CAU CCA —
incorrect →          ala ala ser
```

The large subunit soon joins the complex, and the assembly of the polypeptide begins.

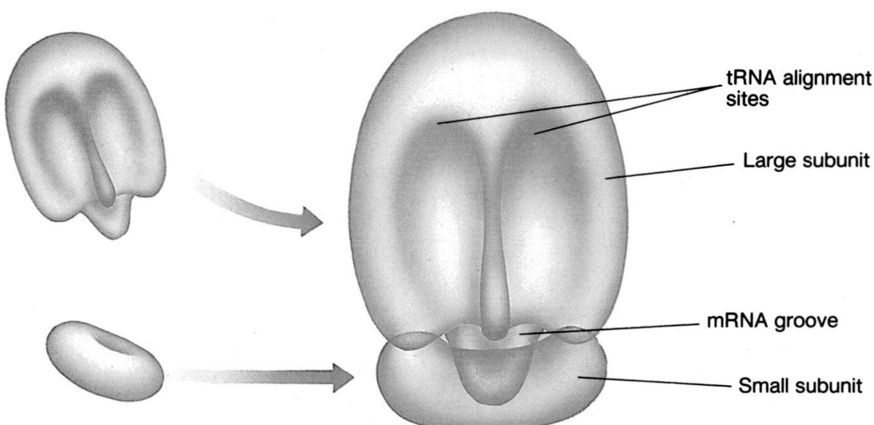

FIGURE 9-10

Assembly of a functional ribosome. Two subunits, one large and one small, fit together to form a groove for mRNA and create sites for accepting two tRNAs at a time. An enzyme (not shown) on the large subunit catalyzes the formation of a peptide bond that connects the two aligned amino acids. Thus, ribosomes are more than mere "workbenches" on which proteins are synthesized; they are more like "workers" that help assemble the protein.

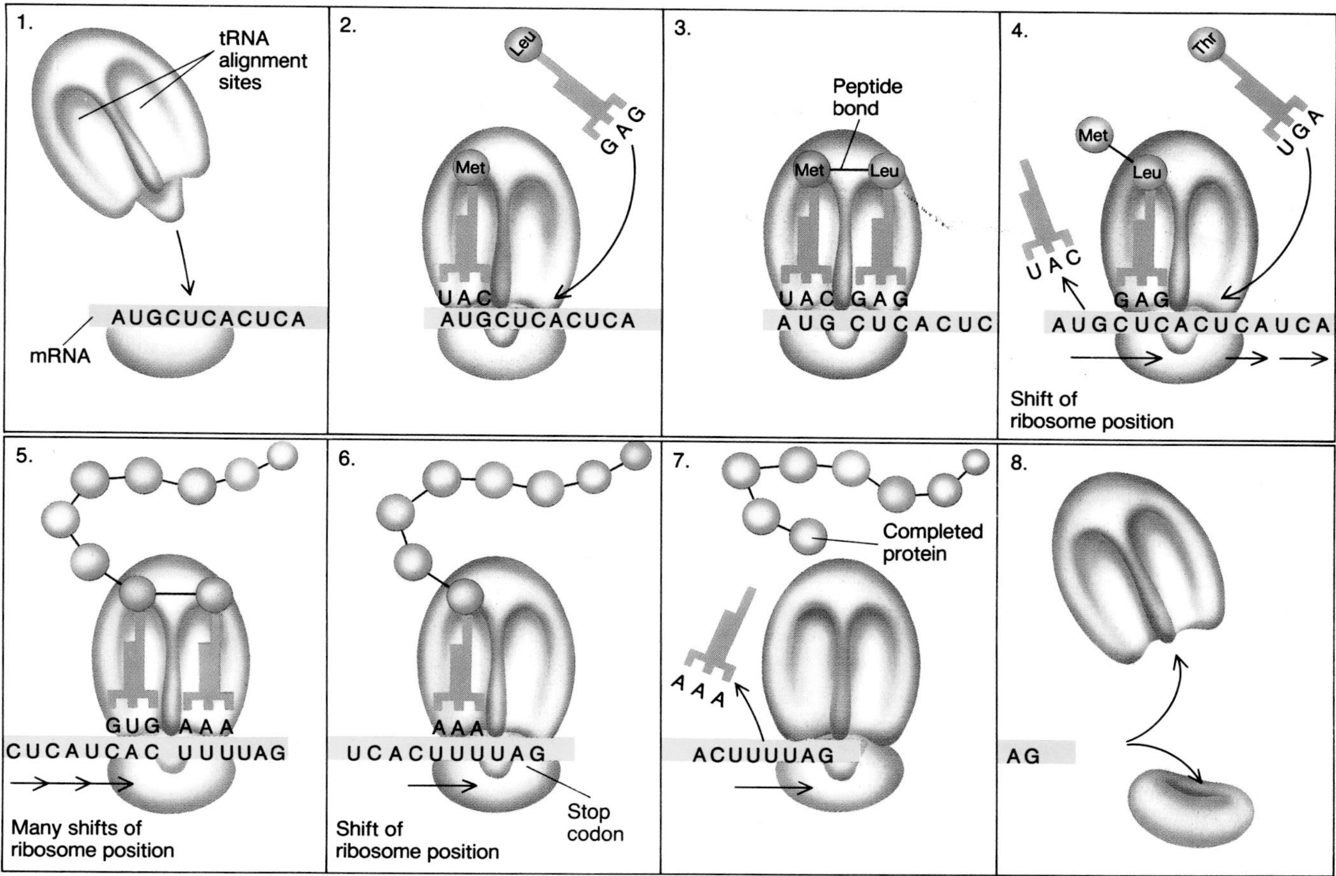

FIGURE 9-11

General steps in translation. Each of the three major steps in translation—initiation, elongation, and termination—is discussed in the text. (The mRNA in the figure is only a short portion of the entire molecule, which, for most proteins, would exceed 600 nucleotides in length.) The methionine inserted by the initiator tRNA as the first amino acid is usually clipped from the polypeptide by an enzyme.

Steps 2–5. Chain Elongation. With the initiator tRNA in place, a second free site remains on the ribosome, where another tRNA can align with the next codon in the mRNA (which is CUC in this illustration). CUC pairs with the anticodon GAG of the tRNA carrying the amino acid leucine. The two amino acids align next to each other, and the amino acid attached to the first tRNA is enzymatically transferred to the second amino acid, forming a covalent linkage, a peptide bond. The first tRNA, which has lost its amino-acid cargo, then departs, and the ribosome moves down the mRNA by three nucleotides, bringing the third codon into position. A tRNA molecule with a complementary anticodon binds to the third codon, orienting its amino acid next to the previous one, and a peptide bond now forms between them. The growing polypeptide chain is now three amino acids long. The second tRNA is then released, and the ribosome moves down the mRNA, bringing the fourth codon into position. This process continues, adding amino acids in the proper sequence, until the entire polypeptide chain is synthesized in an amino acid order dictated by the original genetic instructions.

Steps 6–8. Chain Termination. The completion of a polypeptide chain is signaled by the presence of a stop codon in the mRNA strand. Since these triplets do not specify amino acids, their presence produces a region on the mRNA to which no tRNA can bind. Therefore, the amino acid inserted just before the stop codon becomes the terminal member of the chain.

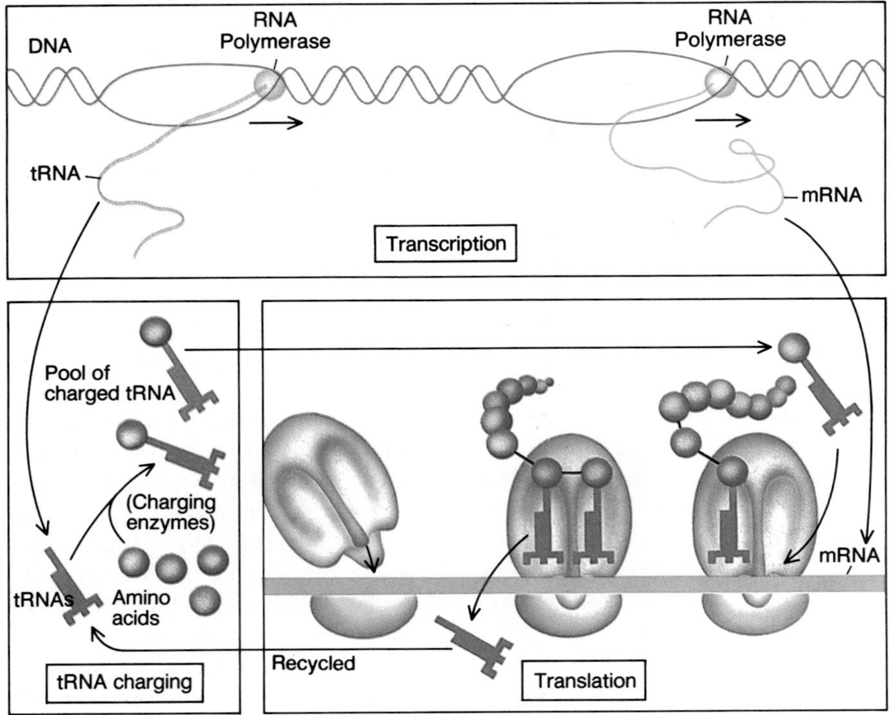

FIGURE 9-12
Overview of protein synthesis. All the processes depicted here occur in a single cell. The steps are compartmentalized for clarity. The genetic information in DNA ultimately dictates the amino-acid sequence in protein, explaining why cells are biologically "obedient" to their genes. A charged tRNA is a tRNA with an attached amino acid, and a charging enzyme is an enzyme that attaches a specific amino acid to the appropriate tRNA (one with the appropriate anticodon).

The ordered flow of genetic information from its stored form in DNA to its transcription into RNA and its expression as a specific protein is summarized in Figure 9-12.

THE MOLECULAR BASIS OF GENE MUTATIONS

Changes in the linear sequence of nucleotides alter the DNA's information content. This is the basis of a **gene mutation.** For example, sickle cell anemia is a genetic disease that results from the substitution of one amino acid (a valine) for another (a glutamic acid) in the oxygen-carrying protein hemoglobin (see page 47). A change from CTC to CAC in the DNA produces a change from GAG to GUG in the mRNA codon, which, in turn, produces a change from a glutamic acid to a valine in the corresponding polypeptide. This simple substitution of one nucleotide for another, thus one amino acid for another, generates a protein that no longer functions as it should.

The substitution of one base for another in the DNA is termed a **point mutation** because it only affects a single "point" in the gene. This type of mutation can occur during replication, when an incorrect nucleotide is incorporated into the growing DNA chain, or it may occur as the result of damage to the DNA of a nonreplicating cell. The likelihood of such mutations is greatly increased if the DNA happens to come into contact with a **mutagen**—a chemical or physical agent that induces genetic changes.

Any agent that can cause gene mutations is also a potential **carcinogen** (cancer-causing agent), since the alteration of certain types of genes can lead to the transformation of a normal cell into a malignant cell (see oncogenes, page 122, and tumor-suppressor genes, page 147). This relationship between mutation and cancer provides a mechanism to screen large numbers of substances for their carcinogenicity simply by determining their mutagenic effects on an especially convenient test organism—bacteria. In this test, developed by Bruce Ames of the University of California, bacteria are exposed to the chemical in question, incubated, and the number of mutant cells detected

◁ THE HUMAN PERSPECTIVE ▷

The Dark Side of the Sun

The sun "swings a double-edged sword." None of us could live without it; the energy of its rays fuels photosynthesis, providing food on which all heterotrophic organisms—you included—depend. But the sun also emits a constant stream of ultraviolet rays that ages and mutates the cells of our skin. Ultraviolet radiation damages DNA by causing adjacent thymine bases to become covalently bonded to one another, forming a "thymine dimer" (an abnormally linked pair of adjacent thymines that interfere with normal DNA function):

UV damaged DNA
(contains T-T dimer)

— C-C-T-A-T-T-A-G-C-A —

— G-G-A-T-A-A-T-C-G-T —

The region containing the dimer must be cut out of the damaged strand, and the original nucleotides replaced. Mistakes made during the repair process create mutations.

The hazardous effects of the sun are most dramatically illustrated by the rare recessive genetic disorder *xeroderma pigmentosum (XP)*, whose sufferers possess a deficient repair system that is unable to remove segments of DNA damaged by ultraviolet light. Because of their genes, children with this disease are forced to live in a continually dark environment. Even brief exposure to the ultraviolet rays in sunlight increases the danger of severe skin damage and promotes the formation of skin tumors and other fatal cancers. These children can die from enjoying what most of us take for granted, a day in the sun. They sleep during the day behind blackened windows, play indoors and go outside only at night when there is no possibility of exposure to ultraviolet light.

But even if you don't have xeroderma pigmentosum, exposure to sunlight still poses considerable risks. More than 600,000 people develop one of three forms of skin cancer every year in the United States and most of these cases are attributed to overexposure to the sun's ultraviolet rays. Fortunately, the two most common forms of skin cancer—basal cell carcinoma and squamous cell carcinoma—rarely spread to other parts of the body and can usually be surgically removed in a doctor's office. But malignant melanoma, the third type of skin cancer, is a killer. Melanomas develop from pigment cells in the skin. They may arise in an existing, noncancerous mole, or they may appear without such visual warnings. Each year, nearly 30,000 Americans are diagnosed with melanoma, and the number is climbing at an alarming rate due to the increasing amount of time people have been spending in the sun over the past few decades. Many scientists predict that the rate of melanoma will climb even more rapidly in the future if the UV-absorbing ozone layer of the atmosphere continues to deteriorate.

If a melanoma is detected and removed at an early stage, when it is small and has not yet penetrated into the deeper layers of the skin, the prognosis is very good. Unlike most tumors, melanomas appear on the surface of the skin, where they can be detected early if the person knows the warning signs (depicted in Figure 1).

Light-skinned individuals are particularly susceptible to developing this disease, as are those with close relatives who have had a melanoma or people who received a severe, blistering sunburn as a child. People who live in sunnier regions are also at greater risk. The highest incidence of melanoma is found in Australia, which has a sunny, tropical climate and is populated largely by light-skinned descendants of northern Europeans. Similarly, in the United States, the incidence of melanoma in Arizona is more than twice that found in Michigan. The best way to avoid developing melanoma is to avoid sunbathing and to wear sunblock creams when you plan to spend time in the sun. Avoid serious sunburns, and be sure your children do the same.

(a)

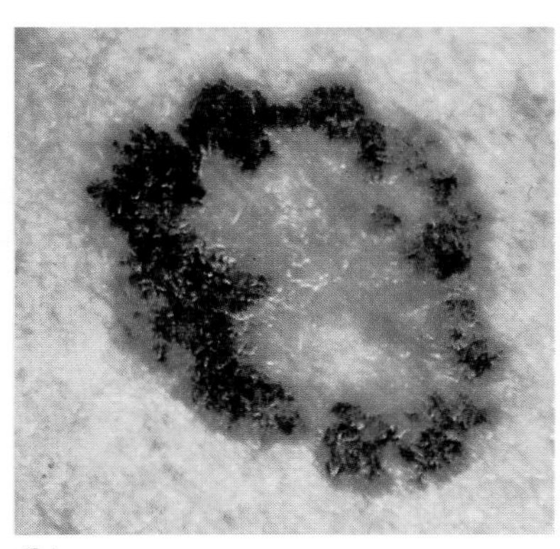
(b)

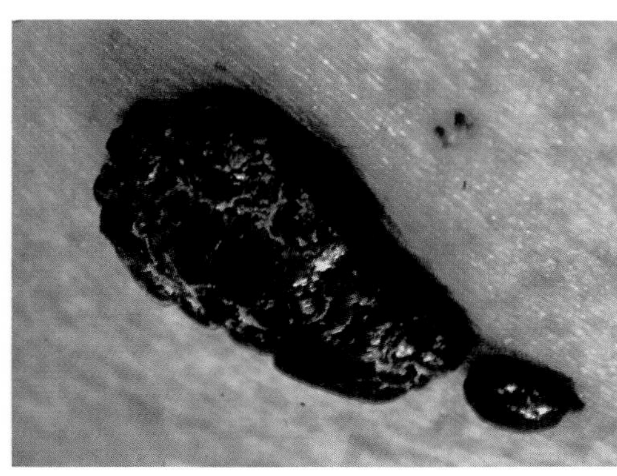
(c)

FIGURE 1 **Stages in the growth of a melanoma.** These malignant skin lesions are often characterized by rapid growth, an irregular boundary, variegation in color, and a tendency to become crusty and bleed.

compared to the number of mutants that appear in a control culture treated the same way but without exposure to the chemical being tested. An increased mutation rate in the chemically treated bacteria flags the substance as a potential human carcinogen. The Ames test has detected cancer-causing potential in many substances to which people are frequently exposed, including materials in hair dyes, cured meats, artificial food colors, and cigarette smoke.

DNA is also very susceptible to mutation following radiation. In fact, one of the most common mutagenic agents is ultraviolet radiation—the subject of The Human Perspective: The Dark Side of the Sun. Even in the absence of UV light or other external mutagens, mutation is inevitable. An average human cell loses several thousand nucleotide bases *every day*! Yet our cells maintain their nucleotide sequence in spite of the molecular punishment, thanks to a diverse array of DNA **repair enzymes** that patrol the DNA, fixing alterations soon after they occur.

Although most mutations are harmful, beneficial mutations occasionally arise, increasing an individual's chances of surviving and producing offspring. These mutations are the raw material of evolution that introduce new genetic information into a population, creating the new traits responsible for variation among organisms and biological evolution.

DNA ORGANIZATION IN PROKARYOTES AND EUKARYOTES

In the eukaryotic cells of your body about 2 meters (6 feet) of DNA are packed into every nucleus, a sphere 100,000 times smaller than the dot on the letter "i." The giant molecule must be packaged in an orderly way that allows the molecule to direct protein synthesis and to duplicate (and separate into daughter cells) without tangling. Eukaryotic chromosomes contain a rich supply of proteins, including a group of small basic (alkaline) proteins called **histones** (the alkaline amino acids facilitate DNA packaging). Two loops of DNA are always wrapped around a central cluster of eight histone molecules, forming a unit called a **nucleosome** (Figure 9-13). Nucleosomes are strung together like beads on a necklace.

Although winding the DNA around nucleosomes shortens its length by approximately one-sixth, the wound strand must be even further shortened to allow the chromosomal material to fit into the tiny nucleus of a nondividing cell. This is accomplished by coiling the wound fiber into thicker fibers, which, in turn, are bent into "looped domains." An additional series of compacting steps are required to prepare duplicated chromosomes for separation during mitosis (the final steps in Figure 9-13).

EVOLUTION AND ADAPTATION: TYING IT TOGETHER

The fact that DNA is the genetic substance in all organisms reveals at least three points of significance. First, it testifies to DNA's effectiveness in storing life's genetic instructions in an inheritable (and "translatable") form. Except for RNA in some viruses, no other genetic substances exist—if they did appear, they were eliminated by natural selection. Secondly, it reinforces the relatedness of all organisms. It is virtually impossible for millions of species to have "hit on" the DNA idea independently and succeeded with it. Rather, every organism inherited not only its traits, but also the very mechanism of trait-making and inheritance. Finally it explains how the traits that evolution molds into new species appeared in the first place, and how species maintain their traits through countless generations. Changes in DNA sequence—mutations—generate an advantageous new trait, and DNA replication assures that subsequent generations inherit the gene for that trait.

In addition to helping us understand evolution, molecular genetics provides powerful evidence for evolution. Perhaps the most striking piece of evidence is the universality of the genetic code—a particular codon directs the insertion of the same amino acid in all organisms, whether a human or a lilac bush. All organisms on earth inherited the same genetic language from the same source, from the common ancestor we all shared at some early point in the history of life.

Understanding DNA, its replication and how it directs the formation of a particular trait reinforces a deeply satisfying message: *life does not defy rational explanation.* On its most subtle level, life is governed by principles fully understandable by the rational mind, principles that not only explain how individual organisms work, but also how they evolve and how new species are formed.

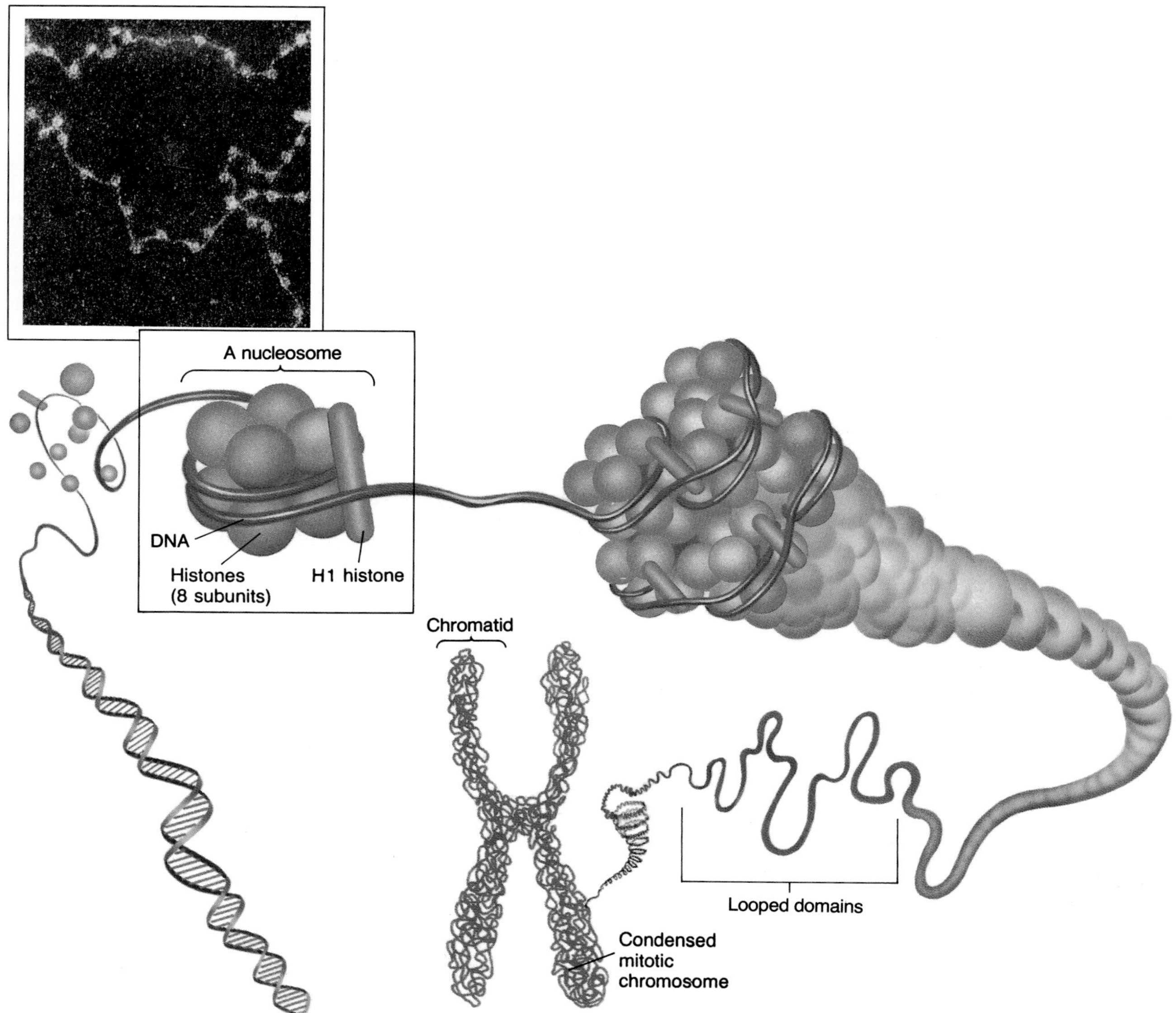

FIGURE 9-13

Packaging DNA into a eukaryotic nucleus. The nucleosome (boxed area) is the fundamental packing unit. Each nucleosome consists of eight histone molecules encircled by approximately two turns of DNA. A different type of histone (called H1) locks the nucleosome complex together so that the DNA cannot unwind from the histone core. The nucleosomes are then coiled into thicker fibers that bend into "looped domains." These thickened strands can coil even further, forming the arms of a condensed mitotic chromosome. Nucleosomes are visible in this electron micrograph of uncondensed chromatin (inset photo). The nucleosomes are the "beads" on the DNA "string."

SYNOPSIS

Experiments in the 1940s and 1950s established conclusively that DNA is the genetic material. DNA alone can transform bacteria from one genetic strain to another or carry the information that controls bacteriophage production in an infected cell. DNA is the substance of genes.

DNA is a double helix that encodes genetic information. Genetic information is encoded in the specific linear sequence of nucleotides that make up the two strands. Specific base-pairing aligns adenine on one strand with thymine on the other strand, and guanine with cytosine, creating complementary sequences of nucleotides in the two strands of a DNA molecule.

DNA replication creates identical duplicates of the original molecule. During semiconservative replication, the double helix separates into two single strands, each of which serves as a template for assembling a new, complementary strand by assembling free nucleotides along the single-stranded regions. The process is directed by the enzyme DNA polymerase, which moves along the single-stranded lengths and inserts the complementary nucleotide at each point on the growing half of the molecule. The result is two DNA molecules, each identical to the original molecule that was replicated.

Genetic information flows from DNA to RNA to protein. Each gene consists of a linear sequence of nucleotides that determines the linear sequence of amino acids in a polypeptide. This is accomplished in two major steps, transcription and translation.

During transcription, the information spelled out by the gene's nucleotide sequence is encoded into a molecule of messenger RNA (mRNA). The mRNA's nucleotide sequence is "read" as a series of codons, each codon consisting of three nucleotides. Of the 64 possible codons, 61 specify an amino acid, and the other three stop the process of protein synthesis.

During translation, the sequence of codons in the mRNA molecule determines the order in which amino acids are assembled to form a polypeptide. The information in mRNA is translated at ribosomes (the sites of protein synthesis), with the help of tRNAs, which serve as decoder molecules. Each tRNA has a particular anticodon at one end which binds to a complementary codon in the mRNA, and, at the other end, a specific amino acid that becomes incorporated into the growing polypeptide chain. In this way, amino acids are assembled in the precise order dictated by the DNA. This is how DNA directs the formation of an organism's every genetic trait.

Mutation is a change in the genetic message. Gene mutations may occur as a single nucleotide substitution, which changes the corresponding codon in mRNA, often resulting in the insertion of a different amino acid than originally encoded. Other types of mutations include the addition of one or two nucleotides, which throws off the reading frame of the ribosome as it moves along the mRNA, leading to the incorporation of incorrect amino acids "downstream" from the point of mutation. Exposure to mutagens increases the rate of mutation.

Review Questions

1. How is complementary base pairing important both to inheritance and to the expression of genetic messages?
2. Describe at least two similarities and two differences between the processes of DNA replication and transcription.
3. Using the genetic code chart in Figure 9-9, construct the short polypeptide chain coded for in this sense strand of DNA: TACGGATCGCCTACG—(remember to include the mRNA sequence).
4. If a portion of DNA had the nucleotide sequence AGCAGGCAGC, could you predict the next nucleotide in the sequence? Why or why not?

Critical Thinking Questions

1. The use of DNase was an important part of Avery's studies, pointing to DNA as the transforming principle. DNase is purified from the pancreas, an organ that produces a variety of digestive enzymes. With this in mind, can you think of any other possible explanations for Avery's results? Is there any experimental control that Avery might have run to eliminate this possibility?

2. Suppose there were only two types of nucleotides—G and C. Would this be sufficient in principle to encode genetic messages. If so, what effect would this have on the size of genes? The size of codons? The movement of the ribosome along a messenger RNA?

3. What would be the effect on the offspring if a DNA polymerase were absolutely foolproof in its proofreading activity? What would be the long-term effect on biological evolution?

4. You are a code-cracker. It is 1961 and your job is to decipher life's genetic code. You know how to manufacture a synthetic mRNA molecule made of nothing but adenines (-AAAAAAAAAA-). Design an experiment to determine which amino acid is inserted by the codon AAA.

CHAPTER

10

Gene Expression: Orchestrating Life

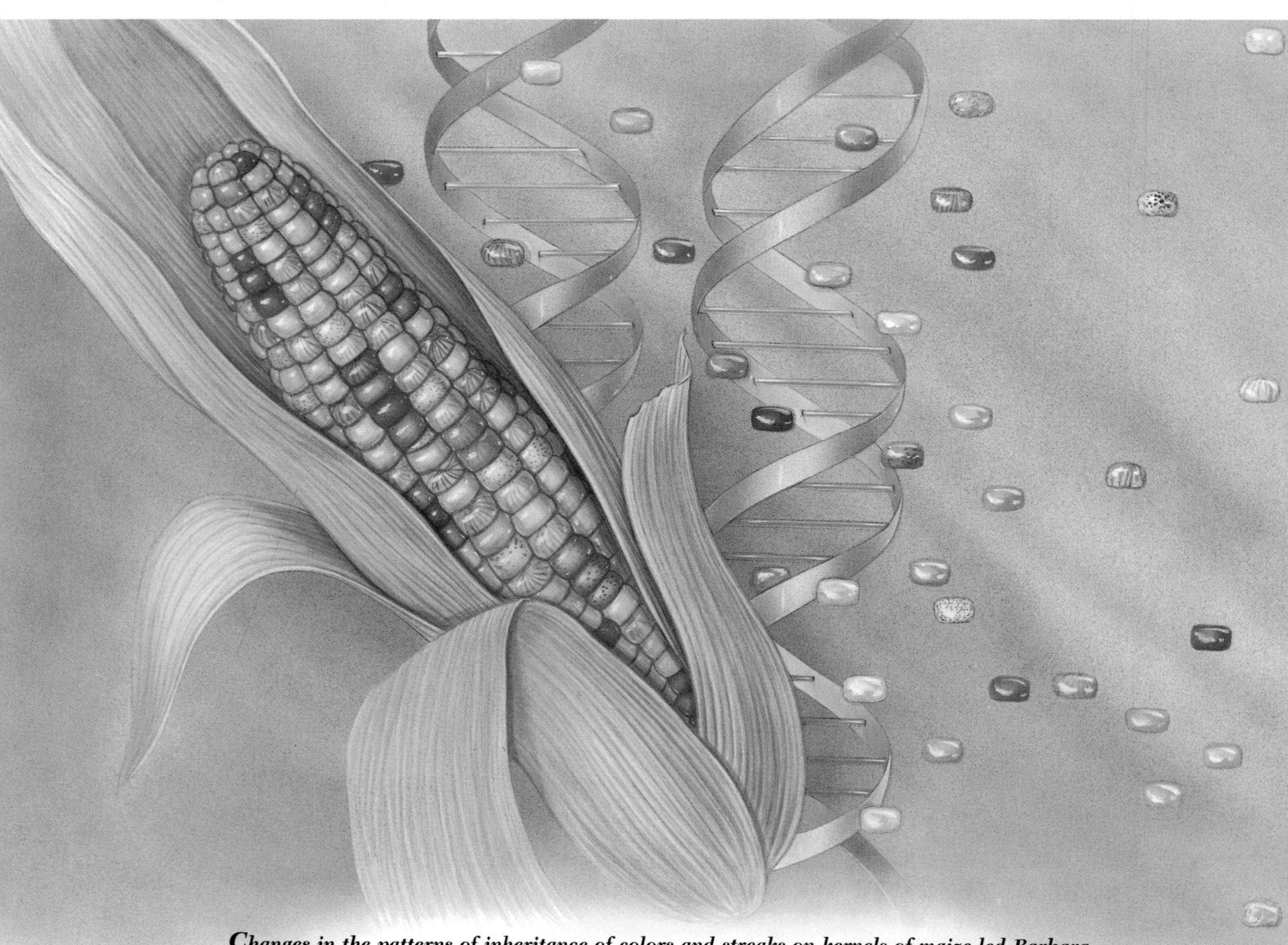

Changes in the patterns of inheritance of colors and streaks on kernels of maize led Barbara McClintock to the conclusion that genes were capable of changing their location on a chromosome.

STEPS TO DISCOVERY
Jumping Genes: Leaping into the Spotlight

During the first quarter of this century, it was very difficult for women to gain admission to graduate schools in science at major U.S. Universities. But Cornell University was an exception, having earned a reputation for its fair treatment of women. It was there that Barbara McClintock received her Ph.D. in botany in 1927. McClintock's research thesis was on the genetics of maize (*Zea mays*), the indigenous species from which agricultural corn was developed. McClintock was the first to identify and characterize the ten different chromosomes of the species and to localize a variety of genetic markers at specific sites on these chromosomes.

In 1931, McClintock and a student named Harriet Creighton published one of the classic papers in genetics that finally established direct visual "proof" that crossing over during meiosis actually involved the physical exchange of segments of homologous chromosomes. Until that time, it had been impossible to distinguish between maternal and paternal homologues under the microscope, even though the reshuffling of genetic characteristics in offspring pointed *indirectly* to chromosomal exchange. McClintock and Creighton developed strains in which maternal and paternal homologues carried extra pieces that enabled them to be microscopically distinguished. In this way, they were able to show the actual exchange of chromosomal segments.

By 1945, McClintock had become the President of the Genetics Society of America and only the third female member of the prestigious National Academy of Sciences. This was also the year she began a new series of experiments aimed at determining the mechanism by which gene expression was regulated. It was one year after Avery's report that DNA was the substance responsible for transformation, and within a decade most of the bright young geneticists had turned from fruit flies and maize to bacteria and bacterial viruses as their sources of new genetic insights. McClintock, however, continued her work with maize.

Genetic traits in maize are expressed primarily as changes in the patterns and markings in leaf and kernel coloration. McClintock discovered that certain mutations did not remain stable from generation to generation. She eventually concluded that these genes were moving from one place in a chromosome to an entirely different site. She called this movement of genetic elements *transposition* and suggested that these "jumping genes" were involved in regulation of gene expression.

Meanwhile, molecular biologists working with bacteria were finding no evidence of gene transposition. Bacterial genes appeared as stable elements situated in a linear array on the chromosome, an array that remained constant from one individual to another and from one generation to the next. McClintock's results were ignored. Despite the fact that her hypothesis concerning transposition was an elegant model backed by years of rigorous experimental evidence, her ideas were all but forgotten until the late 1960s and early 1970s. At that time, movable genetic elements were discovered in bacteria. Within a few years, movement of genetic elements within and between chromosomes was also found in fruit flies. Transposition also proved to be the mechanism by which antibody genes are assembled during the differentiation of cells of the immune system (Chapter 20) and was postulated to be one of the key mechanisms driving the formation of new genes during evolution. McClintock's work on transposition was finally recognized, and she received the Nobel Prize in 1983 at age 81.

Although mobile genetic elements do not appear to play a wide role in gene regulation, new evidence of their importance in other genetic functions continues to be uncovered. For example, Margaret Kidwell and her colleagues at the University of Arizona recently found that mites that parasitize insects may carry "jumping genes" from one species of host to another. The movement of genes across species' barriers provides the opportunity for the input of new genes into a species population and the possibility of rapid evolutionary changes. Haig Kazazian and his colleagues at Johns Hopkins University have subsequently discovered two patients with hemophilia whose disease is a result of a mobile genetic element that has "jumped" into the middle of one of the key genes involved in blood clotting. This is the first example of a human disease being caused by a mobile genetic element.

*I*f it happened to you, it would be a disaster. But to a salamander, losing a portion of a leg is a relatively small sacrifice to escape becoming a hungry predator's dinner. The salamander soon regenerates the missing limb by performing a feat of genetic "alchemy," turning one type of cell into another. For example, some of the muscle cells remaining in the stump transform into unspecialized cells, and then into cartilage cells in the regenerating limb. A muscle cell could only become a cartilage cell if it retained the genes needed for cartilage formation—genes that were "silenced" while the cell was part of a muscle.

The ability of a muscle cell to transform into a cartilage cell, however, is a small feat when compared to a plant cell that reverses its specialized commitment and gives rise to an entire plant (see page 174, The Human Perspective: Clones: Is There Cause for Fear?). The conversion of one type of cell into another type requires a major shift in gene expression, but these are exceptional cases. Let's begin with a discussion of the need for regulating gene expression under more typical conditions.

▼ ▼ ▼

WHY REGULATE GENE EXPRESSION?

No cell needs to use all of its genetic information at any given time. For example, the cells in a multicellular organism, such as yourself, have a complete set of genes in spite of their differences in form and function. Your liver cells are genetically identical to your nerve cells. The genetic information present in each of these cells can be compared to a book of blueprints prepared for the construction of a building. During the construction process, all of the blueprints will probably be needed, but only a small subset of this information needs to be consulted during the work on a particular part of the building. Similarly, electricians need not follow instructions written for plumbers, and vice versa. Similarly, cells carry many more instructions (genes) than they will use at any given time. Consequently, every cell contains mechanisms that allows it to express its genetic information *selectively*, following only those instructions needed for the tasks at hand. The distinct characteristics of each cell depend on which genes are expressed and which remain "silent" (Figure 10-1).

THE BASIS OF GENE REGULATION

Although fundamental differences exist in the ways prokaryotes and eukaryotes control gene expression, there is one unifying similarity: In both types of cells, specific genes are turned on or off as a result of direct physical interaction with **gene regulatory proteins.** In most cases, the regulatory sites in the DNA are located close to, but outside of, the gene itself.

Gene regulatory proteins have a precisely determined shape that enables them to recognize one specific sequence of nucleotides in the DNA while ignoring millions of other sequences constructed out of the same four nucleotides. Many of these regulatory proteins have "fingers" that fit into the grooves of the DNA at the site of recognition (Figure 10-2). These "fingers" allow the protein to hold on to the DNA in a manner analogous to the way your hand would use its fingers to grip a bowling ball. The evolution of sequence-specific, DNA-binding proteins was one of the important steps in the pathway leading to complex cellular life forms.

GENE REGULATION IN PROKARYOTES

Bacterial cells live in environments whose chemical composition may undergo drastic change. Consider the *Escherichia coli* bacteria thriving right now in your intestine, an environment that fluctuates according to your meals. If you drink only one glass of milk a day, for example, and limit your intake of dairy products, the bacteria would spend most of its time with little or no access to lactose (milk sugar). During these times, *E. coli* produces no lactose-digesting enzymes, thereby avoiding wasting its resources constructing unneeded gene products. Drink a milkshake, however, and these cells quickly respond to the environmental and nutritional change. Each cell would begin manufacturing thousands of lactose-digesting enzymes. The presence of lactose has *induced* their synthesis. How can a bacterial cell do this?

THE BACTERIAL OPERON

In bacteria, the genes that contain the information for the enzymes of a metabolic pathway are usually clustered together on the chromosome in a functional complex called an **operon.** All the genes of an operon are regulated together as a functional unit. A typical bacterial operon consists of structural genes, a promoter region, an operator region, and a regulatory gene (Figure 10-3).

- *Structural genes.* An operon's **structural genes** direct the synthesis of the enzymes of the pathway. They usually lie adjacent to one another, so RNA polymerase moves directly from one structural gene to the

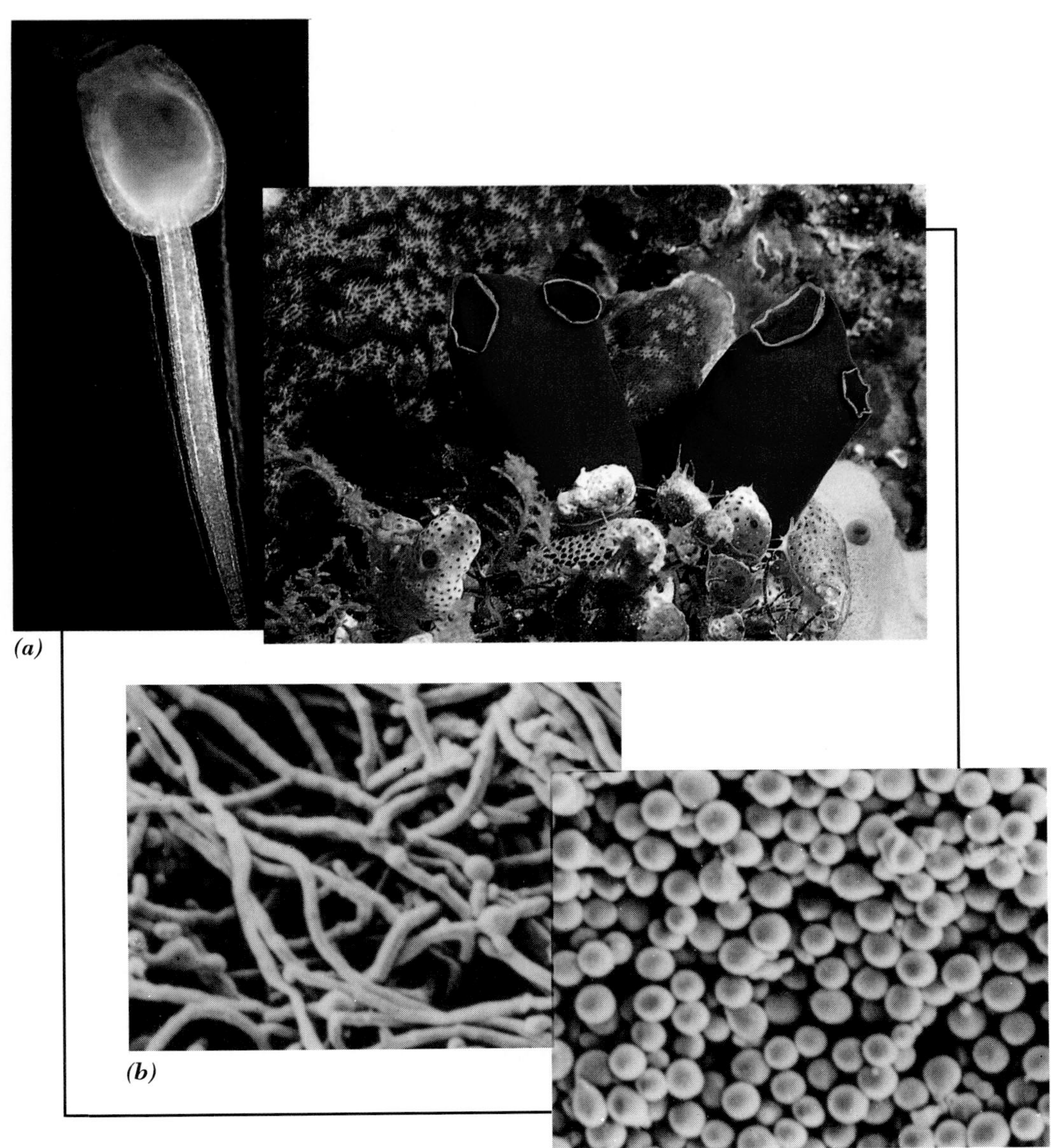

FIGURE 10-1

Two examples of gene orchestration. ***(a)*** *During normal development.* The life cycle of a sea squirt (a member of the same phylum containing humans) is characterized by two strikingly different body forms. The fertilized egg develops into a free-swimming larva that resembles the familiar tadpole. After swimming in the ocean for some time, the larva settles and metamorphoses into the adult body form shown on the right. The striking distinctions in form and function are due to the selection of different genes for expression as growth and development proceed. ***(b)*** *In response to environmental change.* Many fungi can "turn off" genes for growth in the mold form (fluffy and filamentous) and "switch on" a different block of genes, those for unicellular growth (the yeast phase). Moisture and temperature determine which block of genes is expressed.

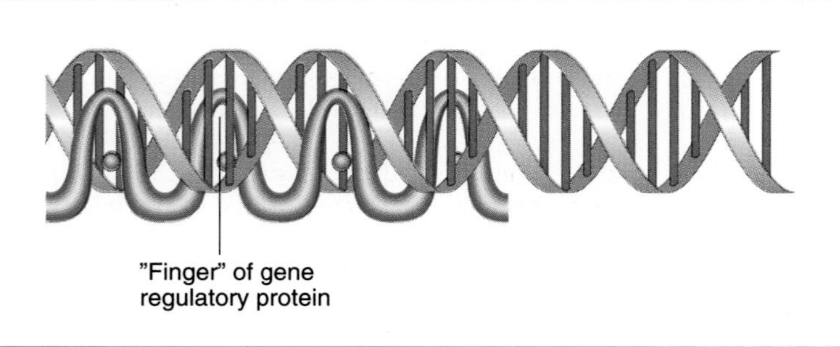

FIGURE 10-2

Interaction of a gene regulatory protein with the DNA double helix. The shape of this gene regulatory protein enables it to recognize a specific nucleotide sequence in the DNA. Many of these proteins, such as the one depicted here contain "fingers" that fit precisely and specifically into successive grooves in the DNA molecule.

next, transcribing them into a single continuous molecule of mRNA. This giant mRNA is then translated into the various enzymes. Consequently, turning on one gene turns on *all* the enzyme-producing genes of an operon.

- *A promoter and an operator region on the DNA.* The **promoter** is the site to which RNA polymerase binds prior to beginning transcription. The **operator** is the binding site for a regulatory protein, called the **repressor.** The operator sits between the promoter and the first structural gene, essentially forming a "bridge" that must be crossed before transcription can occur.
- *A regulatory gene.* The regulatory gene directs the formation of the repressor.

Attachment of the repressor to the operator (Figure 10-3) blocks the ability of mRNA polymerase to reach the structural genes, and gene expression is switched off. Release of the repressor reactivates the operon and the genes are expressed. Which of these outcomes occurs depends on the presence or absence of a key compound in the metabolic pathway being regulated. This key compound attaches to the repressor and, once attached, determines whether the repressor can attach to the operator. It is the concentration of this metabolic substance that determines if the operon is temporarily active or halted.

The interplay among these various elements is illustrated by the *lac operon*—the cluster of genes that regulates production of the enzymes needed to metabolize lactose in bacterial cells. The lac operon is an example of how the presence of the key metabolic substance (in this case, the milk sugar lactose) switches *on* transcription of the structural genes (Figure 10-4). In the absence of lactose, the repressor binds to the operator site, blocking the transcription of the structural genes and subsequent production of lactose-digesting enzymes. If lactose becomes available, the molecules enter the cell and bind to the repressor, which can no longer attach to the operator site. In this state, the mRNA is transcribed, the enzymes are synthesized, and the lactose molecules are consumed.

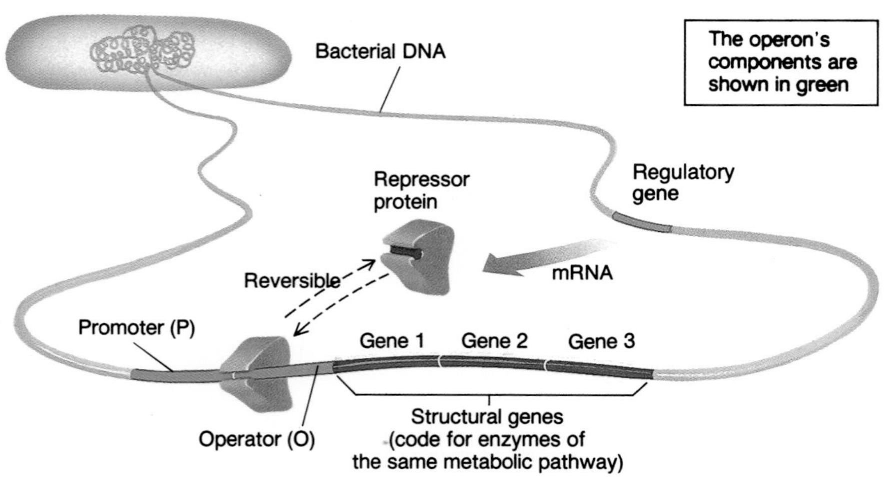

FIGURE 10-3

The bacterial operon—a model of the regulation of gene expression. Clustered structural genes (1, 2, and 3) lie "downstream" from a promoter (the mRNA polymerase attachment site) and an operator. A regulatory gene produces a repressor protein that can bind to the operator and block gene transcription. In this way, the repressor protein acts as the molecular "switch" that enables cells to turn genes on and off.

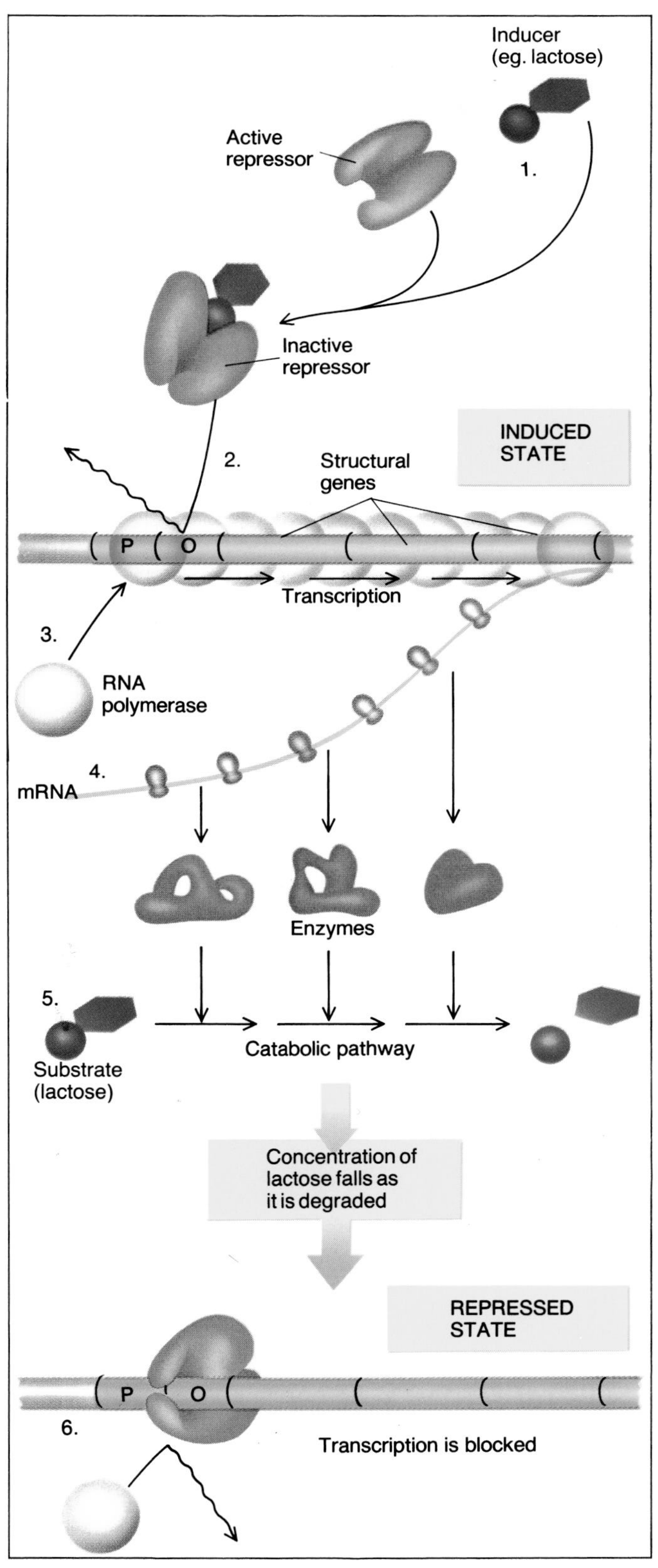

FIGURE 10-4

Gene regulation by operons. Operons work on the same principle: If the repressor can bind to the operator, genes are turned off; if the repressor cannot bind to the operator, genes are expressed. Here's one example of how it works: (1) In high concentration, the inducer (in this case, the disaccharide lactose) binds with the repressor protein and (2) prevents its attachment to the operator (O). (3) Without the repressor in the way, RNA polymerase attaches to the promoter (P) and (4) transcribes the structural genes. Thus, *when the lactose concentration is high, the operon is induced,* and the needed sugar-digesting enzymes are manufactured as needed. (5) Sugar is digested by the enzymes synthesized and, if not replenished, its concentration dwindles until the enzymes are no longer needed. (6) Now there is not enough lactose present to combine with the repressor, which then regains its ability to attach to the operator and prevent transcription. *When the inducer concentration is low, the operon is repressed (turned off),* preventing synthesis of unneeded enzymes.

◁ THE HUMAN PERSPECTIVE ▷

Clones: Is there Cause for Fear?

"The right Hitler for the right future. A Hitler for the 1980s, 90s, 2000s." These words weren't spoken by a Nazi official during World War II, but by Gregory Peck in his role as Josef Mengele in the movie *The Boys From Brazil.* Nazi hunter Ezra Lieberman has stumbled across two 14-year-old boys who look *exactly* alike, one living in Germany and the other in the United States. The boys are not simply identical twins, but *clones,* two of 94 boys genetically cloned from Hitler's cells by Mengele while hiding in the jungles of Brazil. This frightening account of science fiction, and more recently the cloning of dinosaurs in *Jurassic Park,* have rekindled discussions of possible misuse of cloning. But is there cause for fear in the real world?

Clones are asexually produced offspring that are genetically identical to their one-and-only parent. People have been cloning certain types of organisms for thousands of years; nature has been doing it for billions of years! Every time a "cutting" is taken from a plant for vegetative propagation, for example, an organism is being cloned. The resulting plant is genetically identical to the original from which it was cut, reproduced by mitotic cell divisions alone. Mitosis diligently preserves the organism's genetic makeup rather than scrambling it, as occurs during meiosis and fertilization. Cutting techniques long ago replaced the practice of growing oranges, apples, avocados, and dozens of other types of plants from seeds.

Scientists have also developed laboratory techniques for cloning organisms. Their primary goal has not been to produce identical organisms with particular genetic traits, but to answer a basic biological question: Does a cell that has acquired specialized properties, such as a leaf cell of a tree or a skin cell of a mammal, still contain all of the genetic information necessary to generate an entire individual? In 1958, Frederick Steward and his colleagues at Cornell University isolated root cells from a mature plant and placed the cells in an appropriate growth medium, where they proliferated into a tumor-like mass called a *callus.* The calluses could then be grown into a fully developed plant containing all the various cell types normally present. The original root cell is **totipotent**—capable of giving rise to any of the organism's cell types or even of generating a whole individual. It is evident that plants can be cloned by this technique, but is the same true for animals?

No one has been able to induce a differentiated animal cell to develop into a whole animal. Rather than trying to clone animals from isolated cells, scientists have focused on isolated nuclei, trying to demonstrate that the nucleus of a fully differentiated cell retains all of the genes originally present in the zygote. Individual nuclei are isolated and then transplanted into the cytoplasm of an egg cell that has no nucleus. (The egg's own nucleus is previously destroyed by irradiation with ultraviolet light.) The development of this egg then proceeds under the directions of the genes from the transplanted nucleus.

The result of one such experiment involving the transplantation of a nucleus from an adult skin cell is shown in Figure 1. The egg that receives this transplanted nucleus forms a well-developed tadpole, which contains a variety of fully differentiated cells, in addition to skin cells. This could happen only if the nucleus of the

GENE REGULATION IN EUKARYOTES

Gene regulation in eukaryotic cells cannot be explained by operon-type mechanisms for at least one fundamental reason: Operons are not found in eukaryotes. Although prokaryotes and eukaryotes employ the same fundamental mechanisms for storing, transmitting, and expressing genetic information, they have evolved different strategies for controlling gene expression.

CELL DIFFERENTIATION AND SPECIALIZATION

Multicellular eukaryotes contain a diverse array of cells possessing very different structures and functions. The human body contains several hundred recognizably different types of cells, each far more complex than a single bacterial cell, and each possessing a distinct set of proteins for carrying out its specialized activities. Yet all these cells contain the same genes—they all descended from the fertilized egg cell that divided by mitosis to produce the trillions of

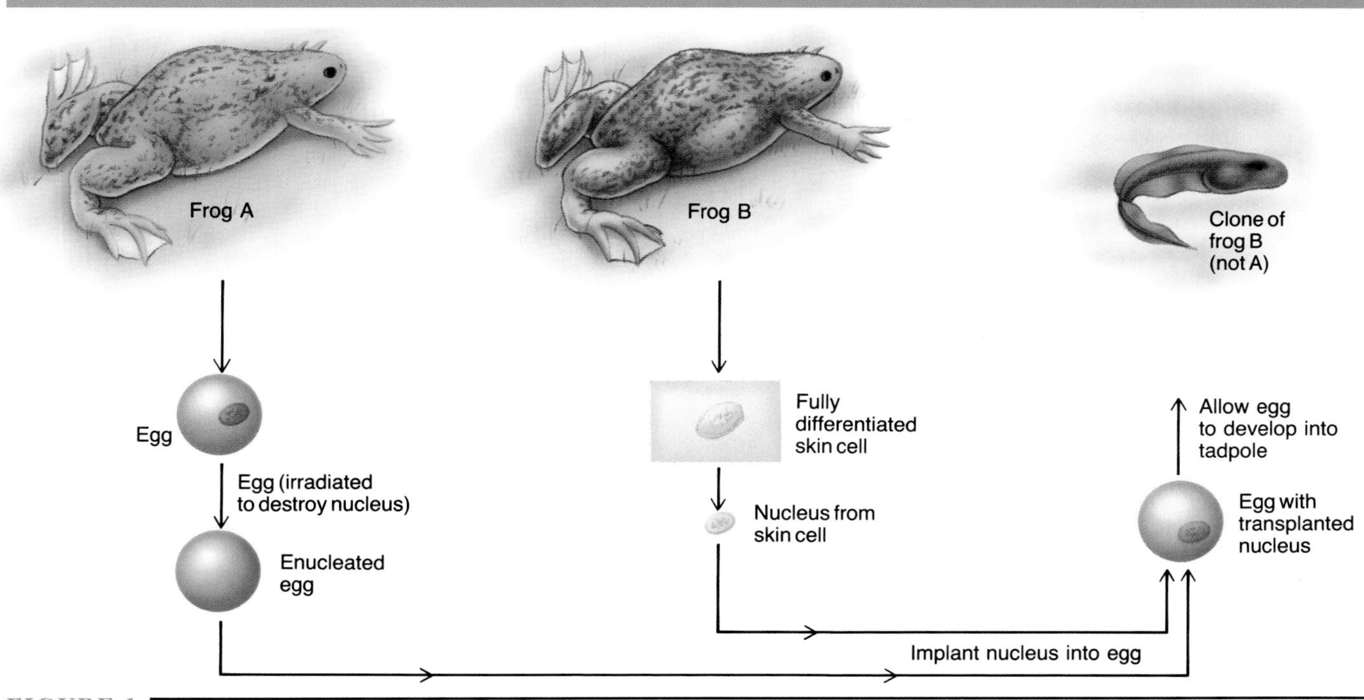

FIGURE 1

skin cell retained the genes needed to produce all other types of cells in the tadpole.

Scientists are currently banned from carrying out any type of nuclear transplantation studies using human embryos, but the ban does not apply to work on other mammals. Many researchers believe that nuclear-transplantation techniques can be developed to improve livestock, cloning thousands of copies of an especially hardy individual or a prolific milk producer. In spite of public concerns, most informed citizens oppose blocking avenues of research that can save lives or provide answers to basic biological questions. Constant vigilance among scientists and open discussion to ensure that genetic technologies are not put to an improper use would be a more productive safeguard. The average citizen's ability to influence the way in which governments use or abuse this technology depends on how well he or she understands both its benefits and its dangers.

cells that combine to form the human body. During early embryonic development, these rapidly dividing cells acquire their specialized forms and functions, a process called **cell differentiation.** Those cells that become liver cells express a specific set of "liver genes," while those that develop into nerve cells express a specific set of "nerve genes." Furthermore, even though a nerve cell may live for 80 years in a human brain, it will *never* express the genes it carries for hemoglobin, the oxygen-carrying protein found only in red blood cells. The transcriptional "silence" of most eukaryotic DNA in any particular cell is one of the multicellular eukaryotic organism's most important regulatory strategies.

Most of your DNA may never be expressed in *any* of your cells. While a human cell may have enough DNA to encode several million different average-sized polypeptides, less than 10 percent of this genetic information directs the formation of any human property. Although some of the remaining DNA clearly has a regulatory function, most of it is "forever silent." It might simply be "genetic garbage" that poses no selective disadvantage to an organism.

SPLIT GENES—EXONS SEPARATED BY INTRONS

In 1977, a startling discovery shattered the concept of the gene as an uninterrupted strip of genetic information transcribed into a molecule of mRNA of the same length as the gene. Some unexpressed DNA was found to reside smack in the middle of an individual gene, separating it

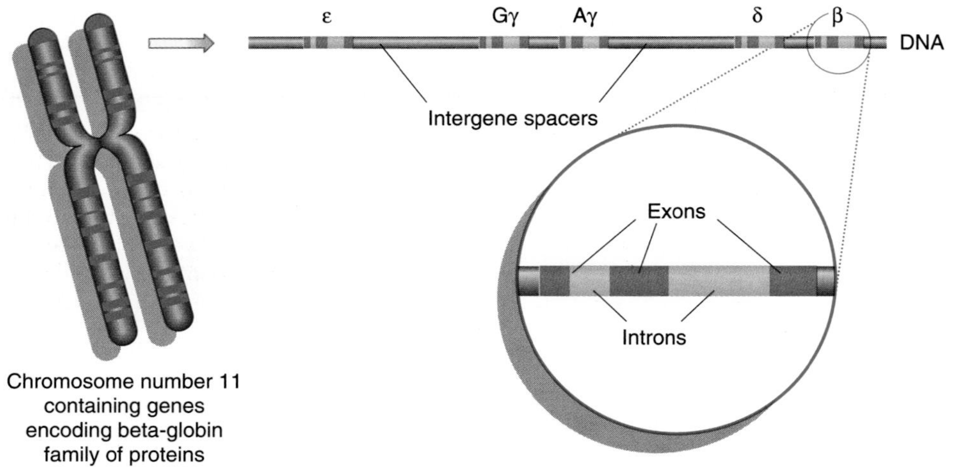

FIGURE 10-5

A map of human globin genes along a portion of the DNA of chromosome number 11, showing how much "space" exists both between and within the genes. Most of this DNA has no known function. The genes illustrated here code for a family of related proteins—the beta-globin family. Only the gene at the far right codes for a product found in adult hemoglobin molecules; the other four are synthesized during embryonic development and infancy. Each of the genes consists of three exons (the expressed sequences) and two introns (the intervening sequences). This arrangement is illustrated in the expanded view of the beta-globin gene. All five genes shown here have the same organization because they are presumed to have originated from duplication of a single ancestral globin gene that had this arrangement of exons and introns. Of the entire stretch of DNA shown in this drawing, only about 15 percent (corresponding to the red segments) actually codes for part of a polypeptide chain.

into discontinuous parts (Figure 10-5). It soon became apparent that most genes contain these *intervening sequences* that split the coding portion of the gene into separate sections (some genes are split into 50 to 100 different pieces). Investigators named the intervening sequences **introns.** The segments of DNA that are transcribed and translated into portions of the amino acid chain are called **exons** (expressed sequences).

At first, the existence of "split" genes seemed to make very little sense. What possible selective advantage could be gained by separating a gene into discontinuous pieces? The best explanation to date has been provided by Walter Gilbert, a Nobel laureate from Harvard University, who had earlier isolated the repressor of the lac operon. Gilbert proposed that introns facilitate evolution. The modular construction of eukaryotic genes allows the subunits (exons) to be randomly shuffled between different parts of the DNA during the course of evolution. This creates new combinations of exons that occasionally generate new advantageous genes. Evolution need not be limited by the slow accumulation of mutations, but can move ahead by "quantum leaps" as new proteins appear (literally) overnight from new combinations of exons.

LEVELS OF CONTROL OF EUKARYOTIC GENE EXPRESSION

Eukaryotic cells regulate their biochemical activities at three fundamental levels of gene supervision: the transcriptional level, the processing level, and the translational level.

1. **Transcriptional-level control.** These mechanisms determine whether a particular gene will be transcribed and how often.
2. **Processing-level control.** These mechanisms determine whether the transcribed RNA is converted (*processed*) into a messenger RNA that can be translated.
3. **Translational-level control.** These mechanisms determine whether a particular mRNA is used to direct the synthesis of a polypeptide.

Here's a closer look at each of these regulatory levels.

REGULATING GENE EXPRESSION AT THE TRANSCRIPTIONAL LEVEL

As in bacterial cells, control of gene transcription in eukaryotes depends on specific gene regulatory proteins. Unlike bacteria, however, eukaryotic cells contain DNA with huge numbers of sites that bind regulatory proteins (Figure 10-6). Transcription of a globin gene, for example, is controlled by at least five distinct regions of the DNA, located on both sides of the gene itself. In addition, some of these regulatory sites on the DNA may be thousands of base pairs away from the gene they regulate. One of the best-understood groups of gene regulatory proteins mediate the response of cells to steroid hormones, including testosterone, the male sex hormone.

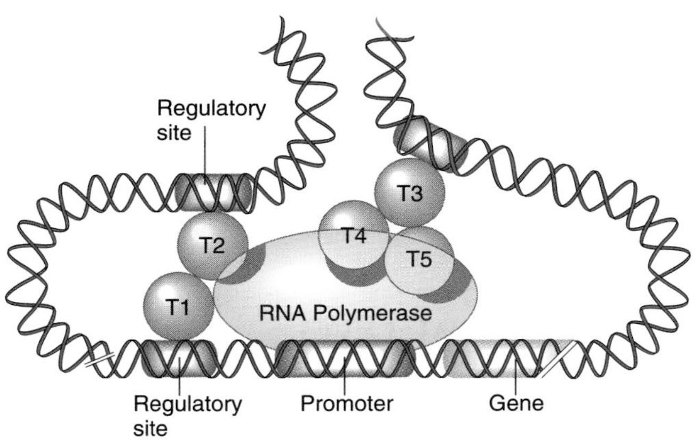

FIGURE 10-6
The interaction between gene regulatory proteins and DNA in eukaryotic cells. A hypothetical example whereby an interaction among several gene regulatory proteins (called transcription factors T1–T5) controls transcription. Several different proteins attach to different sites in the DNA, including the promoter (which binds the RNA polymerase prior to transcription). Proteins situated at widely spaced DNA sites can interact with one another as a result of the bending of the double helix into a loop.

Steroid Hormones and Gene Activation

Cells that respond to testosterone must contain a protein called the **testosterone receptor.** Normally, the receptor resides in the cytoplasm of specific target cells, including the cells of the male reproductive tract. When the testosterone level in the blood rises, hormone molecules diffuse into all cells, but only testosterone-sensitive cells contain the appropriate receptor, so only these cells can respond. The response of a cell to a steroid hormone, such as testosterone, is shown in Figure 15-17*b*.

A testosterone molecule binds to and changes the shape of the testosterone receptor. The receptor–testosterone complex binds to specific sites on the DNA, activating the expression of nearby genes. For example, hair cells on the face of both men and women possess testosterone receptors, but only men produce enough testosterone to bind with the receptor on the target cell in high enough quantities to activate the genes required for beard and mustache formation. Even though women possess competent hair follicles on their chins, these cells remain relatively inactive in the absence of testosterone, explaining why bearded women are so uncommon. It also explains why female athletes who use steroids to enhance their athletic performance often have to shave their faces (the steroid used is testosterone or a testosterone-like molecule).

Visual Evidence of Gene Activation: Chromosomal Puffs

The old adage "seeing is believing" is just as true for molecular biologists as it is for anyone else. When examined under a microscope, some chromosomes display a characteristic "puffing" at various points (Figure 10-7). These

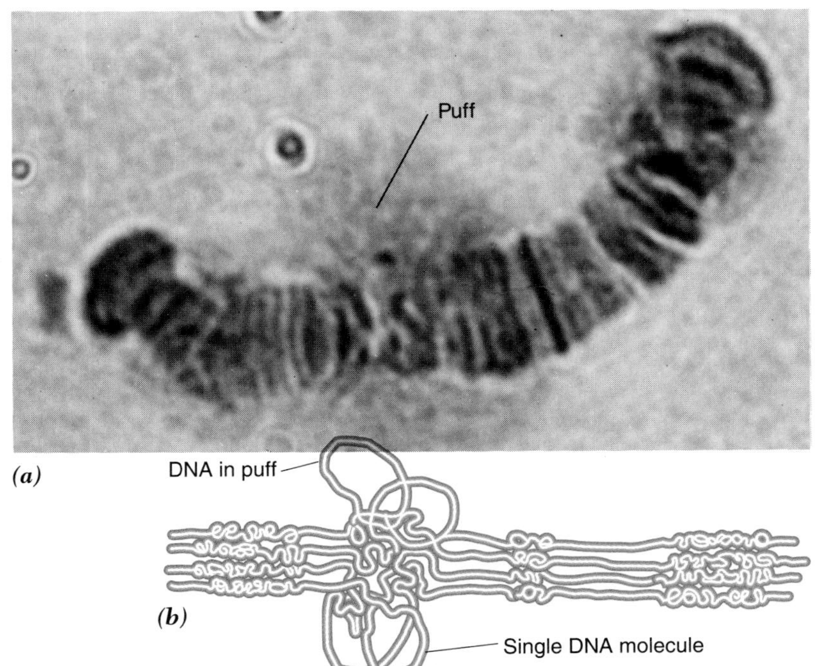

FIGURE 10-7
Puffing in giant chromosomes. ***(a)*** Photograph of a giant chromosome of a fly larva, taken with a light microscope. The image reveals the huge size of these structures. One large puff (and two smaller puffs) are evident, revealing areas of active transcription. ***(b)*** The DNA in the region of the puff is extended, providing sites accessible to RNA polymerase.

chromosomal puffs are sites where the DNA has unraveled and transcription is occurring. The presence of a puff provides direct visual evidence that the gene is being expressed, as occurs, for example, during metamorphosis. Like mammals, insects secrete steroid hormones that alter the patterns of gene expression. One of these hormones, *ecdysone,* triggers insect metamorphosis, from a pupa to an adult (Figure 10-8). When larval tissues are experimentally incubated in ecdysone, the puffing pattern along the chromosomes dramatically changes. Soon after, new proteins appear in the cytoplasm—products of the newly transcribed genes. These new proteins would normally participate in the activities that carry the insect beyond the larval stage.

REGULATING GENE EXPRESSION AT THE RNA PROCESSING LEVEL

If eukaryotic genes, with their introns and exons, were transcribed directly into mRNA, the intervening sequences in mRNA would be translated into a nonfunctional polypeptide containing extra stretches of amino acids. Unlike prokaryotic gene expression, the entire split gene of eukaryotes is transcribed into a giant RNA molecule, called a *primary transcript,* which is subsequently *processed* into the mature mRNA by removal of those segments that correspond to the introns (Figure 10-9).

Cells may exert control over gene expression by processing RNA transcripts differently. For example, in one type of cell, a primary transcript may be processed into a cytoplasmic mRNA molecule, whereas in another type of cell, the transcript is simply degraded in the nucleus without ever being translated. Cells can also process the same transcript in different ways; consequently, the same gene can direct the formation of distinct products in different cells or as conditions change.

REGULATING GENE EXPRESSION AT THE TRANSLATIONAL LEVEL

Even after the mRNA is processed and in the cytoplasm, cells can still control its expression by regulating translation. In some situations, mRNAs are temporarily "masked" by proteins, so they cannot be translated. Consequently, huge amounts of mRNA for certain proteins can accumulate in dormant cells, such as an unfertilized egg awaiting activation by a sperm. Once the dormant cell is activated, the mRNAs are "unmasked" by removal of the blocking proteins, initiating synthesis of the corresponding polypeptides.

Cells also regulate gene expression by controlling the lifespan of mRNAs. For example, protecting specific mRNAs from enzymatic degradation increases the number of times the mRNA can be translated into a polypeptide. Prolactin—the hormone that triggers milk production in mammary glands—operates by this mechanism. The cells of the mammary glands of a nursing mother must produce tremendous quantities of milk proteins in a sustained manner over a period of months or even years. It would be wasteful for these cells to synthesize continuously, and then destroy, the mRNAs that carry the message for milk proteins. Prolactin stabilizes these mRNAs, thus augmenting their translation. Prolactin concentrations drop dramatically when nursing stops, decreasing the stability of

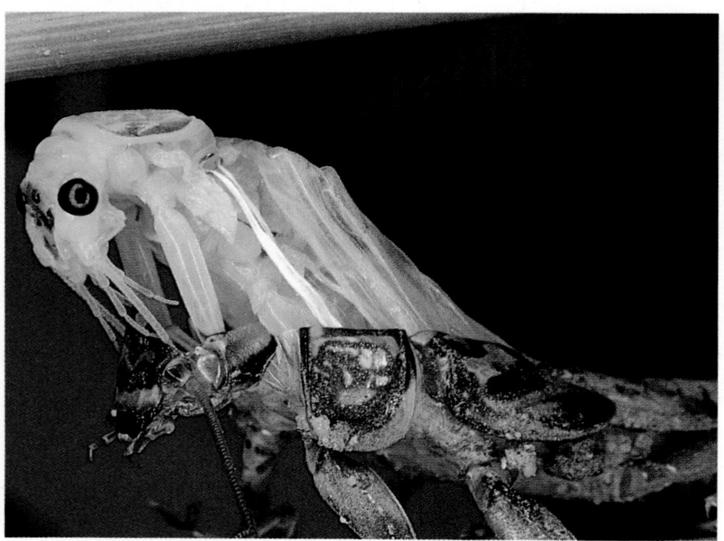

FIGURE 10-8
Metamorphosis of this stone fly was induced by the steroid hormone ecdysone.

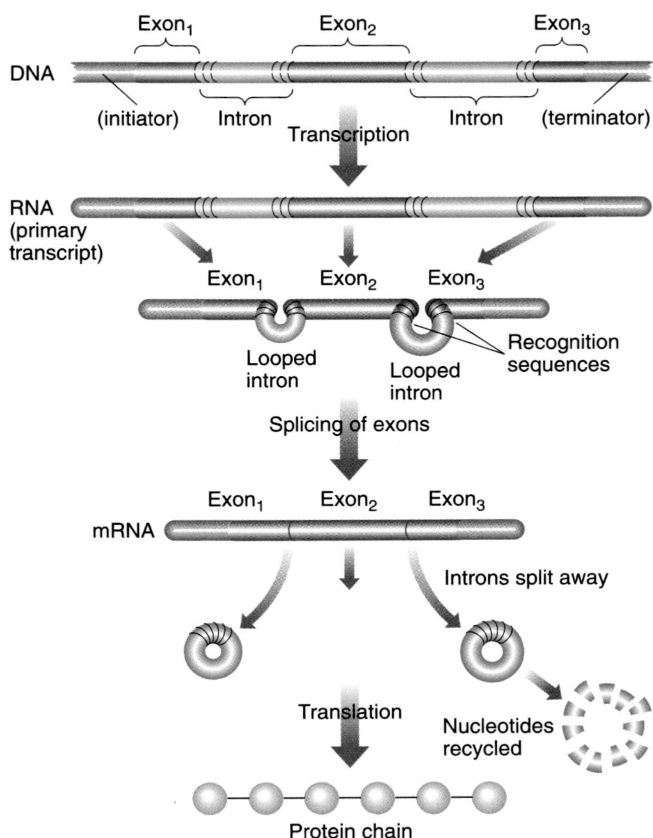

FIGURE 10-9

Editing a gene's message. As an RNA polymerase moves along a split gene it generates a primary transcript whose length is equivalent to that of the gene itself. The final mRNA product is produced by RNA processing, in which the sections corresponding to the introns are removed from the RNA transcript and the remaining sections corresponding to the exons are spliced together. RNA processing must occur with absolute precision so that the nucleotides that code for necessary amino acids are not accidentally removed from the exons. Specific nucleotide sequences identify the beginning and the end of each intron to be removed. These signals pinpoint the location where the RNA is to be cut. The resulting mRNA contains a continuous sequence of codons that specifies the sequence of amino acids for a polypeptide chain.

these mRNAs, which quickly deteriorate, causing milk-protein production to stop almost immediately.

EVOLUTION AND ADAPTATION: TYING IT TOGETHER

The genetic changes that drive evolution are more complex than simply the development of new advantageous traits by mutation. Most genes must also come equipped with a mechanism for regulating their expression if the adaptations are to succeed. If cells lacked such regulatory mechanisms, and all genes were expressed continually to the same degree, all semblance of biological order in a cell would be lost. For example, even the most advantageous traits would quickly disappear if the protein were equally expressed in all the cells of a multicellular organism. Regulatory mechanisms are therefore essential traits in themselves, subject to the forces of natural selection just as are the traits they regulate. Evolutionary breakthroughs such as the use of hemoglobin to transport oxygen would have quickly disappeared had the protein not been limited to red blood cells. Nerve cells packed with hemoglobin would be a distinct disadvantage.

Regulatory mechanisms give even the simplest organisms a competitive edge. Inducible operons, for example, prevent a bacterial cell from wasting resources producing enzymes when they are not needed, better enabling the cell to use its resources for growth and reproduction. Natural selection has favored these bacteria over those that have no means of regulating gene expression.

Split genes of eukaryotic cells, along with another potential regulatory mechanism, transposition of genes from one site to another, may also have played a profound role in evolution. Exons of existing genes can be shuffled to form new gene combinations for new traits that would likely never have developed without such an "evolutionary accelerator." Like shuffling a deck of cards and dealing them, countless "hands" (combinations of exons) are dealt, each subsequently eliminated except for those composite genes that beat the other combinations with which they are in competition.

SYNOPSIS

All cells regulate which part of their genetic endowment is expressed. Every cell has far more genes than it uses at any particular time. Cells possess mechanisms that determine which genes are turned on and which are turned off at a given time, providing a means of responding to the sudden appearance of a nutrient, or directing an embryonic cell to differentiate into a liver cell. The appropriate genes are turned on and expression of others are repressed.

In prokaryotes, the fundamental regulatory mechanism is the operon. Operons are clusters of related structural genes and the corresponding control elements (the operator, promoter, and repressor regions). All the structural genes are transcribed into a single mRNA, so their expression can be regulated in a coordinated manner. The level of gene expression is controlled by a metabolic compound key to that pathway, such as the inducer lactose, which attaches to a protein repressor and prevents it from binding to the operator site on the DNA. Without the repressor attached, the operon is expressed (induced).

The presence of introns in eukaryotic cells complicates gene expression. Although introns may have increased the rate of evolution by facilitating the formation of new genes by exon shuffling, the extra sequences must be removed from RNA during gene processing. This provides an additional target for controlling gene expression.

Eukaryotic gene expression is regulated primarily at three levels. (1) Genes are regulated by controlling the rate at which RNA is transcribed. Hormones such as testosterone regulate transcription of genes close to testosterone-binding sites. (2) Genes are transcribed into large primary transcripts that must be cut and spliced to form the mature mRNA. Control of these activities constitutes regulation at the processing level. (3) Gene expression is also regulated by determining whether mRNA is translated and how long the mRNA will survive.

Review Questions

1. Describe the cascade of events responsible for the sudden changes in gene expression in a bacterial cell following the addition of the milk sugar lactose.
2. Which of the following statements are true? (a) The regulatory gene produces the promoter. (b) Lactose binds to the operator site. (c) The repressor binds to DNA in the presence of lactose. (d) The structural genes are transcribed sequentially into one long mRNA.
3. How are the functions of a bacterial lac repressor and a human testosterone receptor similar? How are they different?
4. What is the primary difference between a primary transcript and the mRNA to which it gives rise?

Critical Thinking Questions

1. Taxol is a promising treatment for saving the lives of women with ovarian or breast cancer. Unfortunately the source of this life-saver is the Pacific yew, a scarce, slow-growing tree that produces too little taxol to supply the medical demand. An easy-to-grow fungus, however, has come to the rescue. This fungus, which lives in the yew tree, also synthesizes taxol, chemically identical to the cancer-fighter produced by the tree. Consider Barbara McClintock's discoveries and explain how this seeming remarkable "coincidence" may not be a coincidence at all.
2. What is the advantage of clustering structural genes so that all the enzymes for a metabolic pathway are regulated together rather than independently?
3. Hormones dramatically alter gene expression (as exemplified by the physical distinctions between men and women, which are fundamentally the result of sex hormones selecting different genes for activation). Create a model by which one hormone might temporarily activate a gene, and another hormone might turn off that gene later in development. Your model can use any of the mechanisms discussed in this chapter.
4. You have discovered a person with a genetic disorder characterized by the production of beta-globin molecules that are much shorter than are those of the normal protein. Describe at least two types of mutations that could account for this disorder?
5. Explain why it would be more likely over the course of evolution for recombination to occur between different genes having introns than between genes that lacked such intervening sequences.

CHAPTER

11

Genetics and Human Life

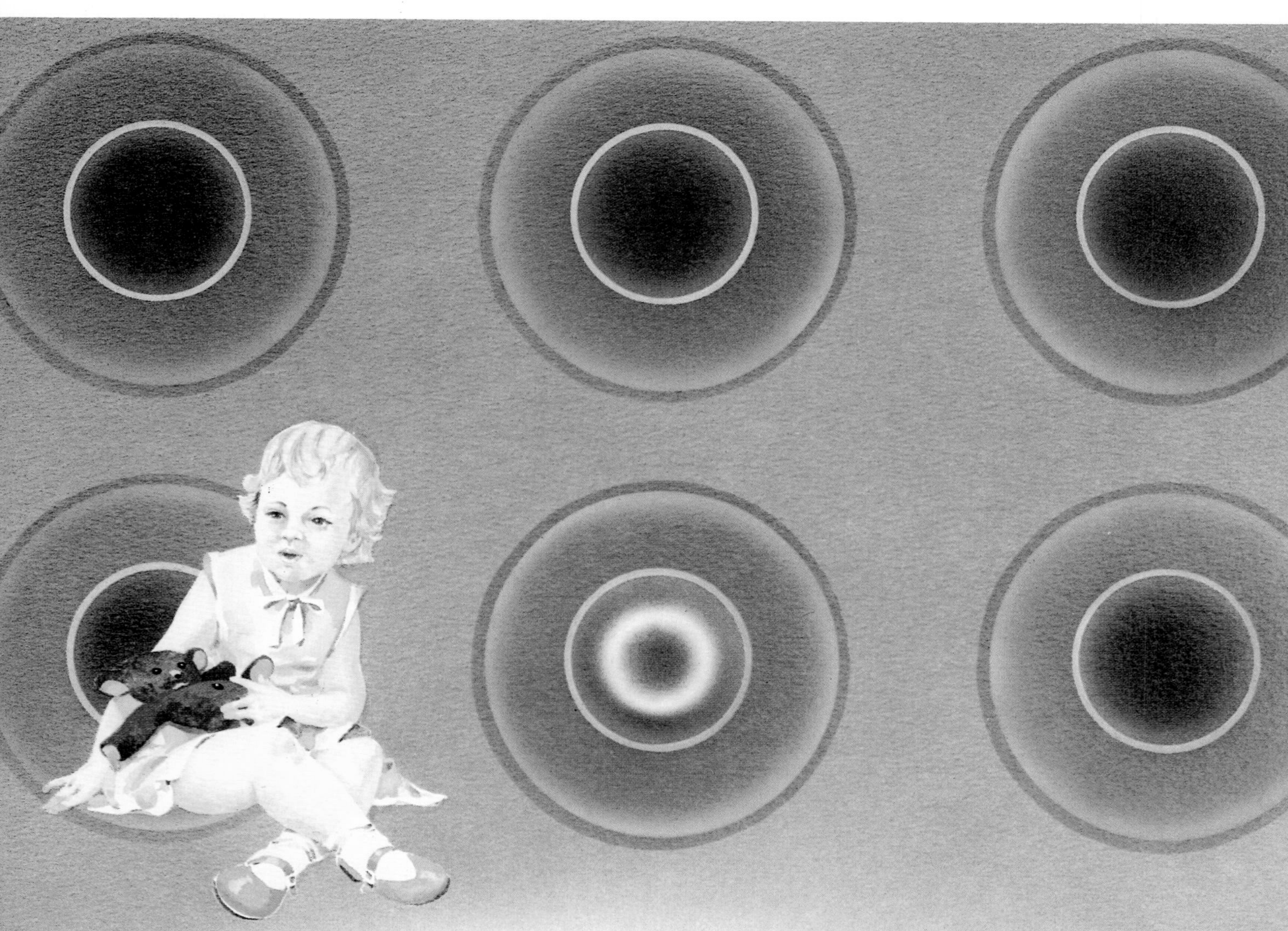

Phenylketonuria is detected in newborn infants when phenylalanine levels are high enough to promote the growth of bacteria around blood samples, producing a halo.

STEPS TO DISCOVERY
Developing a Treatment for an Inherited Disorder

Babies smell. There is little argument about it. But in 1934, the unusual musty odor continually emitted by two mentally retarded babies worried their mother enough for her to seek medical attention for them. Asbjorn Folling, a Norwegian physician, reported that chemical tests of the infants' diapers revealed the presence of high levels of phenyl ketones, toxic compounds formed from the metabolic breakdown of an essential amino acid, phenylalanine. Because of the presence of phenylalanine-derived ketones in the urine, the childrens' disease was called *phenylketonuria* (*PKU*). It was soon discovered that victims of PKU lack an enzyme that normally converts phenylalanine to another amino acid, tyrosine. The resulting accumulation of phenylalanine in the blood causes mental retardation.

In 1953, working at a British children's hospital, Evelyn Hickmans and co-workers hypothesized that it might be possible to treat PKU children by restricting their dietary intake of phenylalanine. The researchers tested their hypothesis on a 2-year-old PKU victim who was "unable to stand, walk, or talk ... and spent her time groaning, crying, and banging her head." The child was put on a diet containing only enough phenylalanine to support the synthesis of vital proteins. Over the next few months, the little girl improved dramatically; she learned to stand and climb on chairs, and she stopped crying and banging her head. Other studies soon followed, confirming the benefits of a low-phenylalanine diet for infants born with PKU. For treatment to be effective, however, early diagnosis is essential. Only then can the infant be placed on the diet before permanent damage to the nervous system occurs.

In 1961, Robert Guthrie of the University of Buffalo published a one-page "letter" outlining a procedure by which newborn infants could easily be screened for PKU. Guthrie's procedure took advantage of the fact that the blood of newborn infants with PKU contains about 30 times the level of phenylalanine as does that of normal infants. Using the Guthrie test, a drop of blood from a newborn infant is dried on a small piece of filter paper, which is then added to a well in a culture dish containing a low concentration of bacteria that require phenylalanine to grow. When the filter paper contains blood from an infant with PKU, the high phenylalanine content promotes the growth of the bacteria, producing a visible "halo" around the well. In contrast, blood from a normal infant does not promote the growth of the bacteria, so no halo will be observed.

Since the development of the Guthrie test, the vast majority of infants born in the United States and other western countries are automatically screened for PKU. Those infants who test positively for the inherited condition (about 1 in 18,000 newborns) can be placed on the prescribed diet. Once these children reach a certain age, their nervous systems are no longer susceptible to damage by high levels of phenylalanine in their blood, and they can begin eating a normal diet.

Although early diagnosis and treatment had virtually eliminated this rare form of mental retardation, PKU has returned to the spotlight in recent years. Successful screening and dietary treatment for PKU have allowed children with the disorder to develop into normal adults, who are having children of their own. Even though a high level of phenylalanine in the blood has little effect on the mother, it can produce terrible damage to the developing fetus. Damage to the fetus is best prevented by the mother's return to her strict, low-phenylalanine diet *before* she becomes pregnant to ensure that the developing baby will have a safe environment throughout the entire gestation period. This potentially serious problem illustrates the importance of genetic counseling for people at risk of having children with genetic diseases.

*H*uman genetics overflows with tales of tragedy and triumph. One strange tragedy comes in the form a bizarre disease called Lesch-Nyhan syndrome. Victims of this syndrome, who are almost all boys, begin to mutilate themselves during their second year of life by biting off their lips and fingertips. Although the behavior was once attributed to lunacy, or "demonic possession," scientific examination of pedigrees of victims and controlled analysis revealed that the underlying defect was not the effects of a dysfunctional family life or some childhood trauma but something completely biological—a deficiency in a single enzyme, one needed for normal purine metabolism. Without this enzyme, children accumulate a buildup of uric acid in their blood and urine. In fact, the urine of these children often contained "orange sand," the precipitated uric-acid crystals. The disorder was inherited from the children's mothers, carriers who showed no signs of the disorder.

Sadly, the triumph in this tragedy has yet to be realized, although through genetic counseling potential parents can discover their probability of having a baby with the syndrome. If a woman is pregnant, some of the methods described in this chapter can determine whether the incubating fetus is a victim of Lesch-Nyhan syndrome. Someday we may score a real triumph over this genetic-deficiency disease by developing the ability to replace the faulty gene with a fully functional copy. Such *gene therapy* represents what many consider to be the most promising threshold in the quest to understand human genetics.

▼ ▼ ▼

GENETIC DISORDERS IN HUMANS

Altogether, over 3,500 distinct genetic disorders have been described, 15 of which are described in Table 11-1. These disorders result from defects in a single gene, indeed even a single nucleotide. Others can be traced to more extensive genetic changes, such as the addition of an entire chromosome or a portion of a chromosome. Let's begin by examining some abnormalities in chromosome number.

DEFECTS IN NUMBER OF CHROMOSOMES

Homologous chromosomes occasionally fail to come apart during meiosis I, or sister chromatids may fail to detach from each other during meiosis II (see Chapter 7), resulting in gametes with either an extra chromosome or a missing chromosome (Figure 11-1). This type of abnormal separation is called **nondisjunction.** If one of these abnormal gametes fuses with a normal gamete, the results, although usually fatal, may produce a zygote that develops into a viable offspring whose cells have an abnormal chromosome number. The most common disorder resulting from nondisjunction is Down syndrome (also called *trisomy 21*), which is caused by the presence of three homologues of chromosome number 21 (See The Human Perspective, chapter 7).

Occasionally, a baby is born with an abnormal number of sex chromosomes due to meiotic nondisjunction. A zygote with three X chromosomes (XXX) develops into a

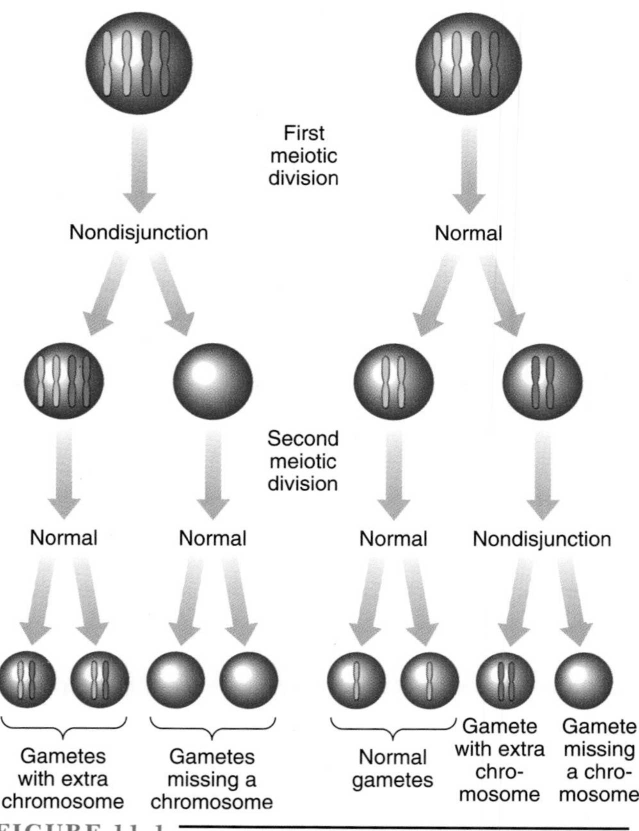

FIGURE 11-1
Meiotic nondisjunction occurs when chromosomes fail to separate from one another during meiosis. If the failure to separate occurs during the first meiotic division, all of the gametes will have an abnormal number of chromosomes. If nondisjunction occurs during the second meiotic division, only two of the four gametes will be affected.

relatively "normal" female, although she will likely have subaverage intelligence and experience menstrual irregularities. A zygote with only one X chromosome and no second sex chromosome (X0) develops into a female with *Turner syndrome,* in which genital development is arrested in the juvenile state, ovaries fail to develop, and body structure is slightly abnormal.

Persons with at least one Y chromosome develop as males. A male with an extra X chromosome (XXY) suffers from *Klinefelter syndrome,* characterized by mental deficiency, underdevelopment of genitalia, and the presence of feminine physical characteristics (such as breast enlargement). Alternatively, men with an extra Y chromosome (an XYY male) appear normal but will likely be taller than average. Considerable controversy has developed surrounding claims that XYY males tend to exhibit more aggressive, antisocial, and criminal behavior than do XY males, but this correlation has never been proven.

TABLE 11-1
GENETIC DISORDERS

Genetic Disorder	*Cause*	*Nature of Illness*	*Incidence*	*Inheritance*
Down syndrome	Extra chromosome number 21	mental retardation, body alterations	1 in 800	sporadic
Klinefelter's syndrome	Male with extra X chromosome	abnormal sexual differentiation	1 in 2,000	sporadic
Cystic fibrosis	Abnormal chloride transport	complications of thickened mucus	1 in 2,500 Caucasians	autosomal recessive
Huntington's disease	Unknown	progressive neurological degeneration	1 in 2,500	autosomal dominant
Duchenne muscular dystrophy	Deficient muscle protein dystrophin	progressive muscle degeneration	1 in 7,000 males	X-linked
Sickle-cell anemia	Abnormal beta-globin	weakness, pain, impaired circulation	1 in 625; mostly black	autosomal recessive
Hemophilia	Deficiency in one of a number of clotting factors	uncontrolled bleeding	1 in 10,000 males	X-linked
Phenylketonuria	Deficiency in enzyme phenylalanine hydroxylase	mental retardation	1 in 18,000	autosomal recessive
Tay-Sachs	Deficiency in enzyme acetylhexosaminidase	deposition of fatty materials in brain; infant death	1 in 3,000 Ashkenazic Jews	autosomal recessive
Lesch-Nyhan syndrome	Deficiency in enzyme HGPRT	mental retardation, self-mutilation	1 in 100,000 males	X-linked
Galactosemia	Deficiency in enzyme galactose transferase	mental retardation, digestive problems	1 in 60,000	autosomal recessive
Xeroderma pigmentosum	Deficiency in a UV repair enzyme	sensitivity to sunlight, cancer	1 in 250,000	autosomal recessive
Severe combined immunodeficiency (SCID)	Deficiency in enzyme adenosine deaminase, or others	absence of immune defenses	extremely rare	autosomal recessive
Ehlers-Danlos syndrome	Deficiency in collagen hydroxylating enzyme	abnormal connective tissues, joint problems	1 in 100,000	autosomal recessive
Familial hypercholesterolemia	Deficiency in LDL receptors	cardiovascular disease	1 in 100,000	autosomal recessive

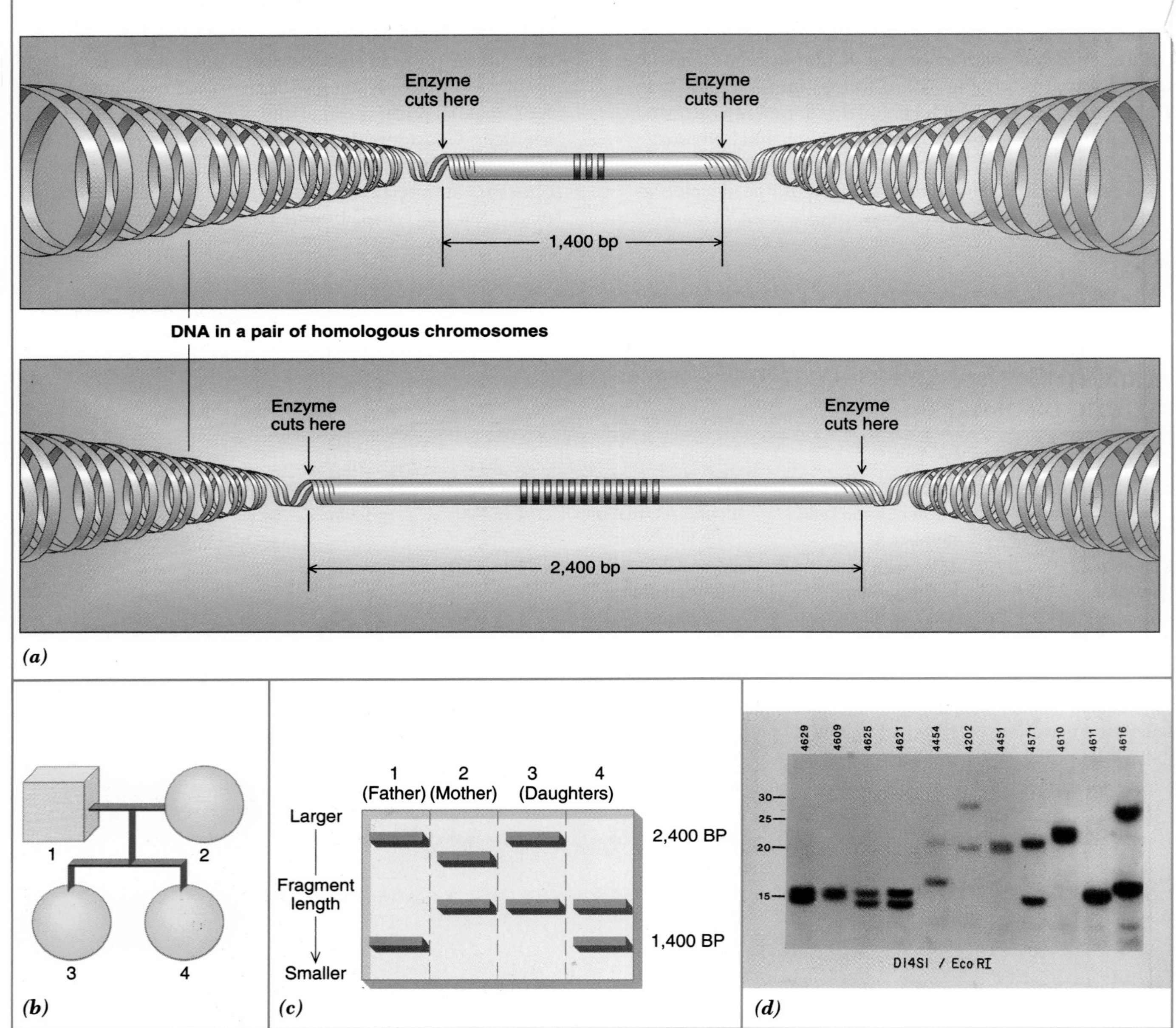

FIGURE 11-2

Restriction Fragment Length Polymorphisms (RFLPs). ***(a)*** Homologous segments of DNA from a pair of homologous chromosomes from one person. The arrows represent sites in the DNA which are attacked by a particular restriction enzyme. After enzyme treatment, one DNA molecule yields a 2,400-base-pair fragment, while the other DNA molecule yields a fragment of only 1,400 base pairs. This is an example of a restriction fragment length polymorphism (RFLP). ***(b)*** A short pedigree of four family members: a father, mother, and their two daughters. (The father's DNA is shown in part ***a***.) ***(c)*** The pattern of DNA fragments from the four family members. Each vertical lane has two bands—one from each homologous chromosome. The two bands in lane 1 identify the DNA fragments of 1,400 and 2,400 base pairs generated by enzyme treatment of the father's DNA from part ***a***. The DNA fragments of the mother are different from those of the father, reflecting differences in the nucleotide sequence of her DNA. Each of the daughters has inherited one chromosome from her mother and one from her father, generating the patterns in lanes 3 and 4. Such differences in RFLPs are being used to predict genetic disorders by two parents, since a chromosome carrying a mutant allele usually has different restriction sites than does the corresponding chromosome carrying the good allele.

INGLE GENE DEFECTS

en genes become altered so that they produce inactive ucts, the consequences are often serious. One of the in studying the properties of a defective gene is to ne the chromosomal location of the gene. This f gene mapping has undergone a revolution in rs, made possible by recent breakthroughs in logy.

NEW T...HNIQUES FOR MAPPING DEFECTIVE GENES

Over the past decade or so, molecular biologists have developed new techniques to locate specific genes in the DNA of particular chromosomes. DNA is first treated with **restriction enzymes** (enzymes that cut DNA only at a particular nucleotide sequence; for example, one restriction enzyme cuts DNA at every "TTAA" sequence). This creates a collection of different-sized DNA fragments (called *restriction fragments* because they are cut with restriction enzymes). The individual fragments can then be separated from each other by **electrophoresis** in which molecules move through a porous gel in response to an electric current. Differences in nucleotide sequences from one person to the next can be detected in this way; each person's DNA generates a unique and identifiable set of fragments as long as the same restriction enzyme is used. Although these differences play no apparent role in biological function, they have proven invaluable as markers for locating human genes. Here's how it works.

Differences in the number of nucleotides between two sites cut by the same restriction enzyme generate different sizes of restriction fragments. Such differences are called **restriction fragment length polymorphisms (RFLPs),**[1] or simply "riflips" (Figure 11-2*a*). RFLPs produce distinctive differences in banding patterns following electrophoresis (Figure 11-2*b* and *c*). These "genetic fingerprints" can be used to identify individuals or to track the inheritance of specific genes. Because each RFLP occurs at a fixed site on a chromosome, different forms of the RFLP at a particular site represent a visible marker that reveals that locus. Although the RFLP itself is not particularly meaningful, it can be used as a "signpost" that reveals the presence of nearby genes.

To illustrate, let's examine the inherited disease cystic fibrosis (CF), the most common debilitating inherited disease among Caucasians (about 1 per 2,500 newborns). Victims of CF suffer a variety of symptoms, many of which are related to the production of a thickened, sticky mucus that is very hard to propel out of the airways leading from the lungs. As a result, chronic lung diseases, including potentially fatal infections, are frequent among CF victims. Not long ago, children with cystic fibrosis faced near-certain death before the age of 5. Today new therapies to help clear congested airways (Figure 11-3) and antibiotics to fight infections have allowed these patients to live into their adult years. But RFLPs may be paving the way for a permanent solution.

Researchers began to seek the actual gene responsible for CF in hopes that its isolation could shed more light on the disease's ultimate cause and direct efforts to find a cure. The key to finding the CF gene was to locate a RFLP that was very close to the CF gene itself. Once a RFLP that resided within a million or so nucleotides of the CF gene had been identified, investigators used other techniques to move from the RFLP into the unknown adjoining regions of the DNA until they arrived at the CF gene itself. In the summer of 1989, a press conference was held to announce the isolation of the gene responsible for cystic fibrosis.

Once isolated, the gene could be sequenced; that is, its nucleotide order was determined. Almost three-fourths of the CF alleles contained the same genetic alteration—they were all missing three base pairs of DNA, so the amino acid phenylalanine was not inserted at the 508th

FIGURE 11-3
Coping with cystic fibrosis requires dislodging as much of the thickened mucus from the airways as possible, until recently by pounding the person's back. In 1992, a new experimental treatment was introduced in which the patient inhales a mist containing the DNA-digesting enzyme, DNase, which is manufactured using genetic engineering techniques (discussed later in this chapter). The enzyme degrades the DNA that contributes to the viscosity of the mucus. The DNA present in the mucus is derived from disintegrating inflammatory cells (white blood cells) that move into the respiratory tract.

[1] **RFLP**—"restriction fragments" (RF) because the DNA fragments are generated by restriction enzymes; "length polymorphisms" (LP) because the enzymes cut DNA fragments into different forms (lengths) from person to person.

position of the polypeptide chain, resulting in an abnormal protein. To determine how the abnormal protein of the CF gene differed from the corresponding protein in normal persons, researchers used a form of "reverse genetics." First they determined the nucleotide sequence of the CF gene; then they used this information to determine the amino-acid sequence of the corresponding protein using the universal genetic code. By comparing this sequence to the amino-acid sequence of other proteins whose function is known, the corresponding nonmutant protein can often be identified. The amino-acid sequence deduced for the CF gene suggested that the gene product was a membrane protein that, in normal people, moved chloride ions across the plasma membrane but could no longer do so in persons with CF. This inability to move chloride ions affects the movement of water, reducing the water content in bodily secretions (review osmosis, Chapter 3). The lower water content, combined with the presence of DNA (which is itself extremely viscous), contributes to the glue-like thickness of the mucus found in CF patients (see Figure 11-3).

The isolation of a disease-causing gene is a major step in understanding the underlying basis of the disease and may provide a foundation for new and innovative treatments, such as gene therapy (see The Human Perspective: Correcting Genetic Disorders).

GENETIC DISORDERS: PATTERNS OF INHERITANCE

The transmission of genetic characteristics in humans is similar to that described for pea plants and fruit flies. Disorders that result from mutations in specific genes are usually divided into three categories—autosomal recessive, autosomal dominant, and X-linked.

Autosomal Recessive Disorders

Autosomes include all the chromosomes except the sex chromosomes (X and Y). Because each diploid cell in the human body has two copies of each autosome, a recessive mutation on one autosome will be masked by the presence of a normal allele on the homologous autosome. These heterozygotes are referred to as **carriers** because they harbor the mutant gene without being adversely affected in most cases. In fact, each of us is probably a carrier for five to ten lethal recessive alleles. Fortunately, recessive alleles only produce a disease phenotype when they are present on both homologues; that is, in the homozygous recessive condition (Figure 11-4). Consequently, these disorders are usually rare. If two heterozygotes (carriers) have children, however, on the average, one in four of their offspring will suffer from the abnormal phenotype.

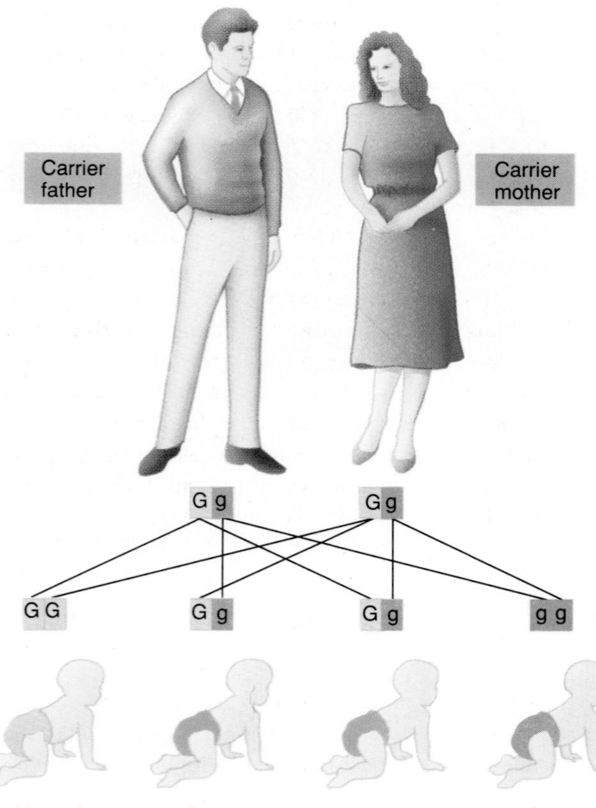

FIGURE 11-4

Inheritance of an autosomal recessive disorder. For these traits (including sickle cell anemia, phenylketonuria, and cystic fibrosis), both parents are carriers of the defective allele (indicated as g), but they do not show its ill effects due to the presence of the dominant normal allele (indicated as G). Each offspring has a one-in-four chance (25 percent) of inheriting two copies of the defective gene, resulting in the disease. Each offspring has a two-in-four chance of being a carrier, and a one-in-four chance of inheriting two normal alleles.

◁ THE HUMAN PERSPECTIVE ▷

Correcting Genetic Disorders by Gene Therapy

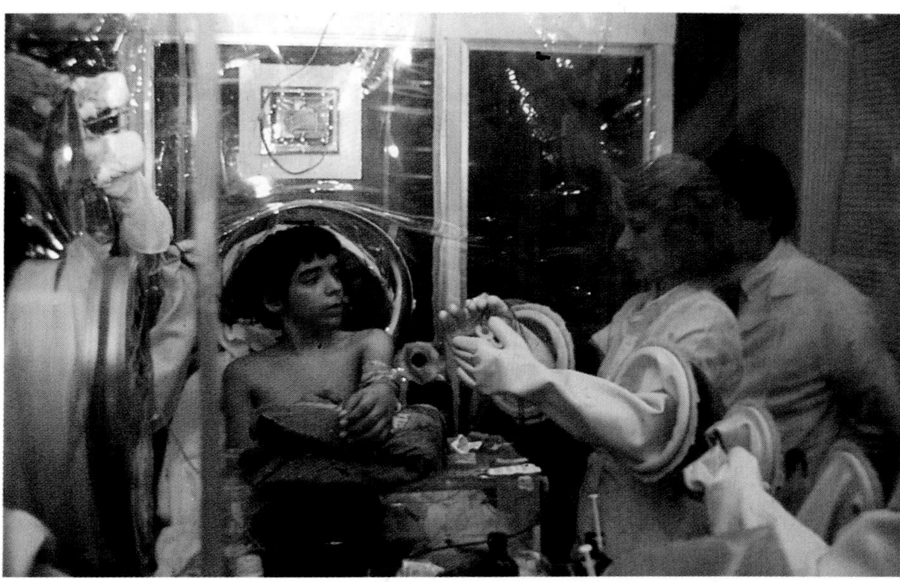

FIGURE 1

During the 1970s, a young boy named David captured the attention of the American public as "the boy in the plastic bubble" (Figure 1). The "bubble" was a sterile, enclosed environment in which David lived nearly his entire life. David required this extraordinary level of protection because he was born with a rare inherited disease called *severe combined immunodeficiency disease* (SCID) which left him virtually lacking an immune system—the system that protects us from invading microorganisms. The bubble protected David from viruses or bacteria that might infect and kill him, but it also kept him from any direct physical contact with the outside world, including his parents.

In approximately 25 percent of cases, SCID results from the hereditary absence of a single enzyme, adenosine deaminase (ADA). For a number of reasons, SCID is an excellent candidate for the development of **gene therapy,** the replacement of the faulty gene with a functional version. First, there is no cure for the disease, which inevitably proves fatal at an early age. Second, SCID results from the absence of a single gene product (ADA), the gene for which has been isolated and cloned and is available for treatment. Finally, the cells that normally express the ADA gene are white blood cells that are easily removed from a patient, genetically modified, and then reintroduced into the patient by transfusion.

In 1990, a 4-year-old girl suffering from SCID became the first person authorized by the National Institutes of Health and the Food and Drug Administration (FDA) to receive gene therapy. In September 1990, the girl received a transfusion of her own white blood cells which had been genetically modified to carry normal copies of the ADA gene. It was hoped that the modified white blood cells would provide the girl with the necessary armaments to ward off future infections. The patient's immune system was restored and she continues to do well. As the white blood cells die, however, the procedure must be repeated to maintain the patient's immune capacity.

Before long, however, this may not be the case. Researchers are trying to isolate the *stem cells* that give rise to both red and white blood cells. If these stem cells can be isolated, and their genotype modified, patients with genetic blood-cell diseases, such as SCID, will need only a single treatment—the genetically engineered stem cells will continue to provide healthy blood cells throughout the person's lifetime.

Another approach to gene therapy is possible for those diseases caused by the inability of the person to produce a particular gene product. Hemophilia, for example, is due to the lack of a particular clotting factor in the blood. Cells from the deep layer of the skin have been isolated, cultured, and genetically modified so that they carry extra genes for the clotting factor. If these genetically modified cells are reintroduced into the skin of a hemophiliac, they should secrete the clotting factor into the blood, where it can provide normal clotting. Trial studies for this procedure have succeeded in experimental animals.

All of the procedures discussed above, as well as all of those being contemplated, involve the modification of *somatic cells*—those cells of the body that are not destined to participate in gamete formation. Modification of somatic cells will affect only the person being treated, and the modified chromosomes cannot be passed on to future generations. If the germ cells of the gonads were altered, however, the modification would be inherited. Scientists researching new gene therapy approaches have agreed to avoid any modifications of human germ lines. Such modifications would risk the genetic constitution of future generations and raise serious ethical questions about scientists tampering with human evolution.

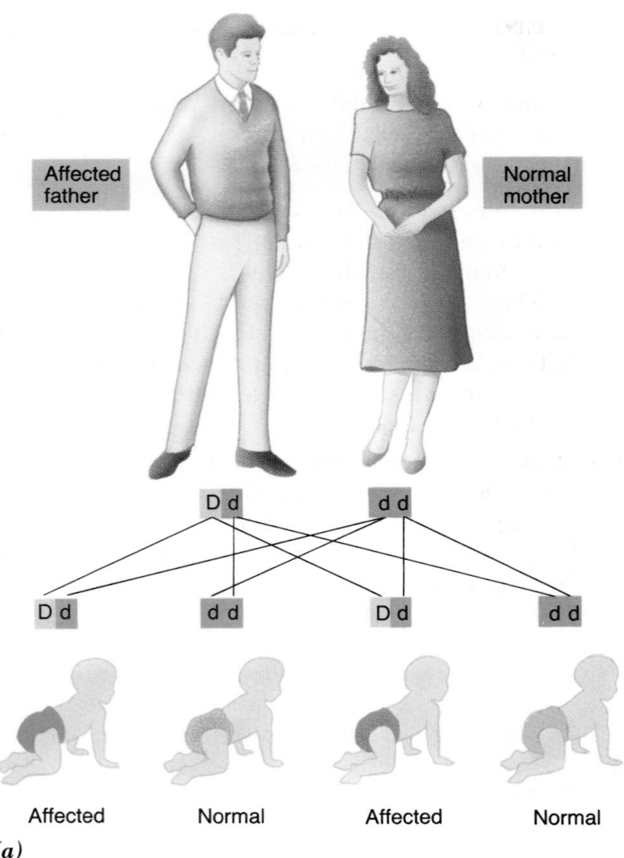

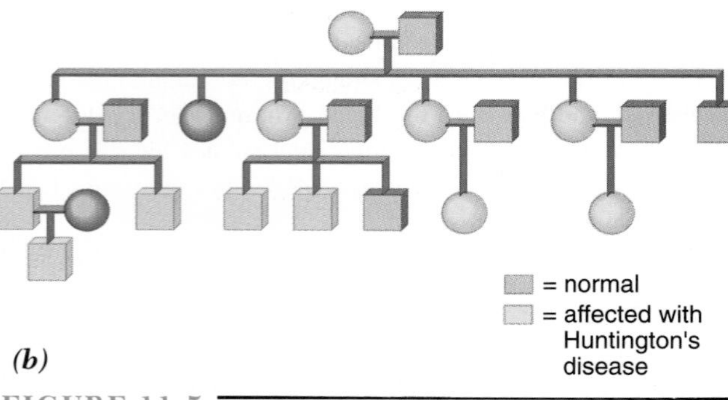

FIGURE 11-5

Inheritance of an autosomal dominant disorder. ***(a)*** For these traits, only one parent carries the defective allele (indicated as D), and that person exhibits the disease. On average, half the offspring will inherit the defective allele and develop the disorder. ***(b)*** Pedigree of a family with Huntington's disease. Individuals with the disease are indicated by solid shading.

Autosomal Dominant Disorders

Some genetic diseases result from *dominant* mutations in which the defective gene directs the production of a product that directly causes the diseased condition. Disease-causing dominant mutations can reside on either an autosome or a sex chromosome. Huntington's disease, a neurological deterioration disorder that causes no symptoms until after age 35, is an example of a condition that results from such an *autosomal dominant mutation* (Figure 11-5). Victims of Huntington's disease eventually show unusual involuntary movements, loss of memory, depression, and irrational behavior. The disease progresses steadily and unerringly toward dementia, loss of motor control, and, finally, death. The underlying biochemical defect causing the neurological deterioration still remains a mystery. Huntington's disease received media attention when legendary songwriter Woody Guthrie died from the disease in 1967. Until very recently, children of Huntington's victims (such as Guthrie's son, Arlo) found themselves in a horrible position. Knowing that they faced a 50-percent risk of developing the disease, they had to wait until they were well along in their adult life before learning their destiny and possibly that of their children. A team of researchers from several laboratories, assembled by Nancy Wexler of Columbia University, launched an unprecedented collaborative hunt for the Huntington gene. In 1993, the 10-year effort climaxed when one of the team members (James Gusella of Harvard) isolated the gene. The exact defect has now been characterized (the defective gene has an extra 100 or so nucleotides). The search for the normal gene product and new ways to treat the disorder continue today.

X-Linked Disorders

Some genetic disorders for which the defective allele is recessive are expressed in men even when only one copy of the gene is inherited. As we discussed on page 143, these **X-linked disorders** are due to faulty genes that lie on the X chromosome. They generally do not occur in women because the faulty allele is almost always masked by a dominant gene on the female's other X chromosome. X-linked disorders include hemophilia, red–green colorblindness, congenital night blindness, ichthyosis (hardening of the skin to scalelike consistency), some forms of anemia (low hemoglobin content in the blood), and muscular dystrophy (characterized by progressive muscle deterioration).

COMPLEX DISEASES

In addition to the somewhat rare congenital disorders discussed so far, many common diseases, including cancer, atherosclerosis, diabetes, Alzheimer's, manic depression, and even alcoholism, have strong genetic components. For example, we have discussed in earlier chapters how mutations in oncogenes (page 122) or tumor suppressor genes (page 147) can lead to the development of cancer. Because these mutant genes are inherited, women whose mothers have had breast cancer, for example, exhibit a much higher risk of developing the disease than members of the general population. It is becoming increasingly evident that the more we learn about human genetics, the better we will understand the complex diseases that threaten our health.

SCREENING HUMANS FOR GENETIC DEFECTS

The discovery of RFLPs (and their value in locating defective genes) has opened new vistas in the field of medical diagnostics. Couples can be tested so that they no longer have to wonder if they might be carriers for cystic fibrosis or sickle-cell anemia or if their unborn fetus will be afflicted with the disease.

SCREENING FETAL CELLS FOR GENETIC DISORDERS

Two commonly performed procedures allow physicians to obtain a sample of cells of an unborn fetus (Figure 11-6) to be screened for genetic deficiencies.

- **Amniocentesis** is the withdrawal of some amniotic fluid by inserting a hypodermic needle into the fluid-filled amniotic sac that surrounds and cushions the growing fetus (this procedure collects fetal cells that have been shed into the fluid). The cells are then cultured and analyzed for genetic defects. Amniocentesis is not usually performed before the fifteenth week of pregnancy.
- **Chorionic villus sampling (CVS)** is a newer procedure, whereby a small sample of tissue is removed from the developing placenta and the fetal cells analyzed. CVS can be performed earlier in the pregnancy than can amniocentesis.

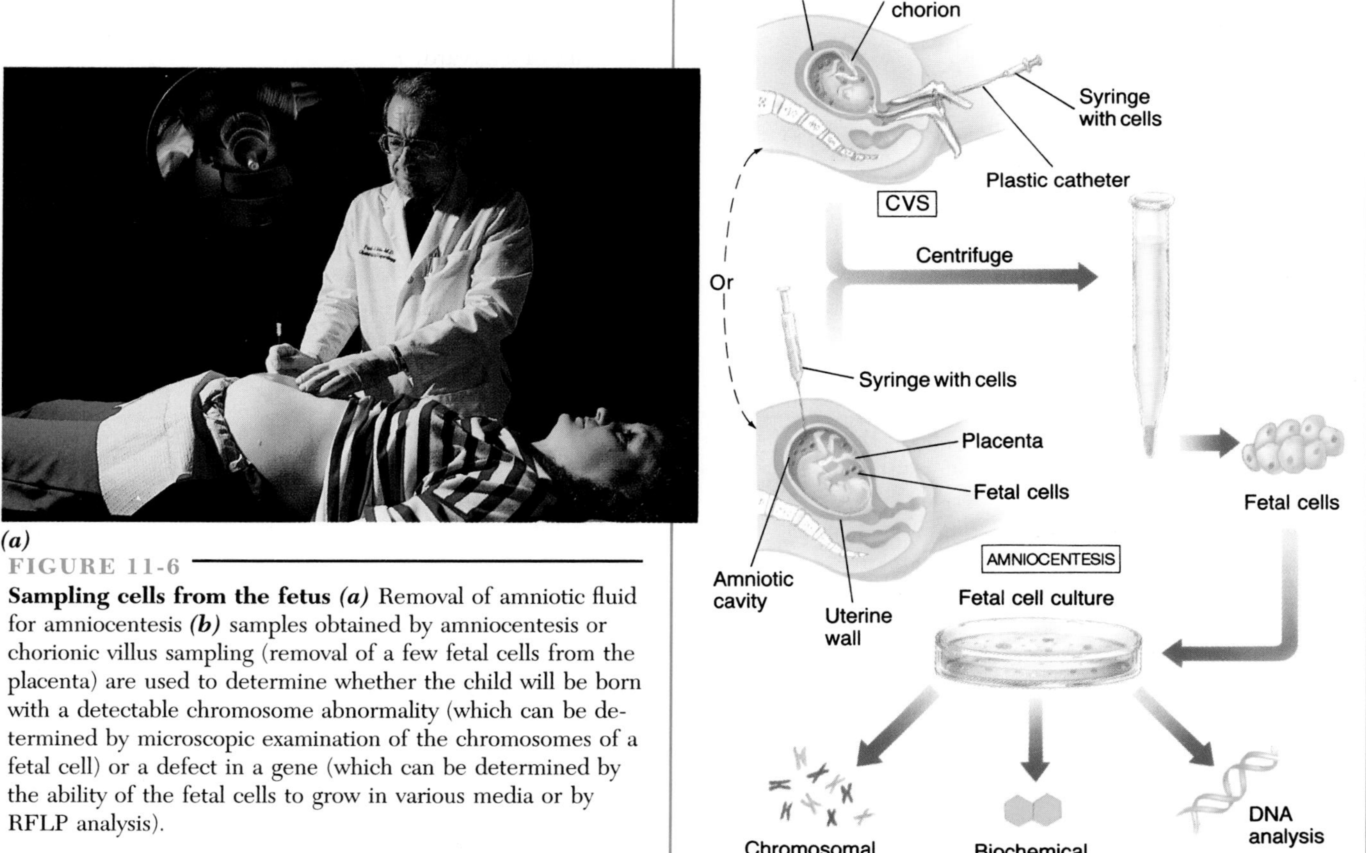

(a) *(b)*

FIGURE 11-6

Sampling cells from the fetus ***(a)*** Removal of amniotic fluid for amniocentesis ***(b)*** samples obtained by amniocentesis or chorionic villus sampling (removal of a few fetal cells from the placenta) are used to determine whether the child will be born with a detectable chromosome abnormality (which can be determined by microscopic examination of the chromosomes of a fetal cell) or a defect in a gene (which can be determined by the ability of the fetal cells to grow in various media or by RFLP analysis).

Newer, safer techniques are being developed that may soon enable physicians to obtain fetal cells directly from the mother's blood. Even though fetal cells are present in very small amounts in the mother's blood, techniques for separating them from the mother's cells have been developed. The fetal DNA can then be amplified to a usable quantity (described later in this chapter).

The ability to obtain fetal cells allows medical geneticists to make several types of determinations (Figure 11-6*b*):

1. **Detection of chromosomal abnormalities.** Amniocentesis and CVS are most commonly performed on older, pregnant women who, because of their age, have an increased likelihood of giving birth to a child with Down syndrome. Geneticists then search for an extra chromosome number 21, which is readily detected by karyotype analysis (review Figure 7-7). Other chromosomal aberrations can also be detected this way.
2. **Detection of metabolic deficiencies.** Tests for over 200 genetic disorders that affect metabolic processes can be detected in cultured fetal cells. If fetal cells are missing a particular enzyme, such as the enzyme whose deficiency causes Lesch-Nyhan syndrome, the condition may be revealed by the cells' inability to grow in certain culture media.
3. **Detection of defective alleles.** RFLP analysis is creating new tests that screen the DNA of fetal cells for genetic disorders not readily detected by a microscope or by biochemical tests. Sickle cell anemia, cystic fibrosis, and Huntington's disease are just a few disorders revealed by RFLP analysis of fetal cells.

ETHICAL CONSIDERATIONS

The merit of some of the new techniques for detecting genetic disorders is beyond dispute (saving PKU sufferers from severe mental retardation, for example). Others have created ethical dilemmas. Many people question the value of information provided by tests that reveal diseases for which there are no cures, for example. If you were a young adult with a 50 percent likelihood of developing Huntington's disease, would you want to know your genetic fate, or would you rather live with the uncertainty? Should you have children, knowing that you might be passing on the defective gene? What would you do if you discovered that both you and your fiancee carried the gene for cystic fibrosis, knowing that each of your children has a 25 percent chance of acquiring a disease that will be emotionally draining and expensive to treat and, in all likelihood, end in a premature death? Furthermore, who should have access to this information? Should you be forced to reveal results of genetic tests on applications for health or life insurance? As we develop diagnostic tests to detect an individual's predisposition to develop more genetically complex diseases, such as cancer, Alzheimer s, and heart disease, the bioethics argument will undoubtedly intensify.

GENETIC ENGINEERING

One of the most significant influences of genetics on human life emerges from our ability to strategically recombine specific genes in a single organism. This forms the core of modern *genetic engineering*, the modification of an

TABLE 11-2
SOME PRODUCTS OF RECOMBINANT DNA TECHNOLOGY

Interferons	Fight viral infection; boost the immune system; possibly effective against melanoma (a form of skin cancer) and some forms of leukemia; may help relieve rheumatoid arthritis.
Interleukin 2	Activates the immune system and may help in treating immune system disorders. Although it produces serious side effects, the drug is proving valuable in treating kidney cancer.
Tumor Necrosis Factor (TNF)	Attacks and kills cancer cells. Presently being used in the first experimental attempt to treat human cancer by introducing cells carrying foreign genes.
Erythropoietin	Stimulates red blood cell production; may be used to combat anemia.
Beta Endorphin	The body's "natural morphine": used to treat pain.
Metabolic enzymes	Perform a multitude of services, from catalyzing chemical reactions in the pharmaceutical industry to replacing defective human enzymes.
Vaccines (e.g., hepatitis B)	Stimulate the body's immunity to protect against disease causing viruses and bacteria.

organism's genotype by introducing new genes—often genes that have never before resided in the chromosomes of that particular species. This feat is accomplished using **recombinant DNA** molecules—molecules that contain DNA sequences derived from different biological sources which have been joined together in the laboratory. (Consequently, genetic engineering is often referred to as "*recombinant DNA technology.*") Before we discuss how specific pieces of DNA are isolated (or synthesized), recombined, and introduced into host cells, let's discuss some of the medical, agricultural, and industrial applications to which this technology is being put to use.

THE BENEFITS OF GENETIC ENGINEERING

Although curing genetic-deficiency diseases by replacing faulty genes with fully functional ones has yet to realize its full potential (see page 189 earlier in this chapter), someday gene therapy may provide effective weapons against diabetes and scores of other diseases, even AIDS. But the horizons of genetic engineering stretch far beyond gene therapy.

Human Proteins from Genetically Engineered Cells

The earliest successes in genetic engineering were achieved by the creation of strains of bacteria and yeasts that would act as microscopic "factories," churning out molecular products specified by a newly acquired gene. Today, genetically engineered cells are used to manufacture biological products ranging from drain cleaners to medicines. A deficiency in insulin production is the cause of one of the most common human ailments, diabetes mellitus. Insulin was one of the first usable human proteins to be produced in bacterial cells. The gene was first synthesized chemically, nucleotide by nucleotide, in 1978 and was subsequently introduced into bacterial cells, which produced the human insulin molecule and released it into the culture medium. Human insulin produced in bacterial cells is superior to the insulin previously obtained from the pancreas of pigs, since the human and pig hormones have a slightly different amino acid sequence. Even though it is produced in bacterial "factories," the product of the human gene is identical to human insulin and is used to treat tens of thousands of diabetics who suffer from a deficiency of the hormone. Unlike traditional insulin sources (pancreas of animals), the bacterial-produced drug is free of contaminating animal protein that caused allergic reactions in many recipients.

Dozens of proteins await approval by regulatory agencies, while others have become available in doctor's offices and hospitals throughout the world (see Table 11-2). Another product, a genetically engineered blood clotting factor, can now be administered to hemophiliacs, eliminating the risk of transmitting AIDS from viral contaminated blood-derived factor. Virtually any protein manufactured by any organism is a candidate for production by genetically engineered microbes or cultured cells.

FIGURE 11-7
Frost-bitten leaves. Ice formation is promoted by bacteria living on the leaf's surface. The "ice-minus bacterium" would have prevented formation of the ice crystals seen on these leaves.

Genetically Engineered Cells and Industrial Products

Fuel shortages may also be relieved by genetically engineered microbes that inexpensively store solar energy in the chemical bonds of combustible organic compounds, providing a virtually inexhaustible fuel source. In addition, genetic engineers have created "oil-producing" microorganisms that generate highly combustible organic compounds equivalent to petroleum in potential energy. The 900 million tons of paper based waste generated each year in the United States can be converted by genetically engineered microbes to one of two valuable fuels: ethanol (the combustible alcohol in "gasohol") or methane (natural gas).

Recombinant Organisms at Work in the Field

Some genetically engineered organisms must first get out of a biotechnology building and "into the world" before they can perform the task for which they were designed. The first genetically engineered microbe to be legally field tested is the "ice minus" bacterium, designed to prevent frost damage to crops and subsequent food loss. Bacteria that normally reside on plants produce an *ice-nucleating protein* that acts as "seed crystals" around which water freezes (Figure 11-7). Through recombinant DNA technology, a strain of the same bacteria has been developed that lacks the gene for producing the ice-nucleating protein and fails to promote ice formation. When sprayed on plants, the engineered "ice-minus bacteria" compete with normal ice-forming bacteria, displacing them and reducing crop loss.

Genetic engineers are also releasing their microbial work force against pollution and toxic wastes, which pose a serious threat to the biosphere. Many of today's toxic pollutants are new synthetic compounds. Organisms capable of decomposing these chemicals have not yet had a chance to evolve. Consequently, the compounds accumulate to dangerous levels in the environment. Genetic engineers are trying to accelerate evolution in the laboratory by developing bacteria that can quickly degrade hazardous substances; an oil spill, for example, would be one big meal for these bacteria.

Genetic Engineering of Domestic Plants

Bacteria are not the only targets of biotechnologists. Multicellular organisms have been genetically engineered, for example, to produce plants with improved resistance to drought, disease, poor soil conditions, chemical pesticides, and herbicides. Recombinant DNA researchers are also developing plants that produce insect-killing toxins, such as the protein toxins produced by the bacterium *Bacillus thuringiensis* (*BT*). In 1986, a plant that had been genetically engineered with a BT toxin gene successfully passed a field test, left undamaged by pests that ravaged the "unprotected" plants close by. In another effort, genetic engineers have recently developed a new strategy to improve the quality of tomato plants. "Mushy" tomatoes result from the presence of an enzyme that breaks down the plants' cell walls. To keep them firm, tomatoes are normally picked green, artificially reddened, and then transported to market. A tastier product could be marketed if the tomatoes were allowed to ripen on the vine and then kept from becoming mushy during transport. Genetic engineers have turned to "antisense technology" to accomplish this. Recall that only one strand of the DNA (the *sense strand*) is transcribed by an RNA polymerase (Chapter 9). This strand contains the information for the encoded polypeptide. Recombinant DNA technologists have created genes in which the "wrong" strand is transcribed, producing an "antisense" RNA that binds with the normal mRNA or the complementary gene and prevents its translation into protein. The mush-resistant tomato has just such an antisense gene that blocks the production of the troublesome enzyme.

These and several other recombinant DNA strategies are aimed at increasing the production of food for human consumption. As successful as many of these techniques may be, increases in food production are being obliterated by even greater increases in the number of people to feed. No technological advance in agriculture will be successful until worldwide population control has been achieved.

FIGURE 11-8
Transgenic mice. A pair of littermates at 10 weeks of age. The larger mouse carries in all of its cells copies of the growth-hormone gene from a rat, resulting in much greater growth than that of its littermate, who lacks the rat gene.

Genetic Engineering of Laboratory and Domestic Animals

In 1982, a litter of mice was born unlike any before; the chromosomes of some of these mice contained rat genes. Within a few weeks, these mice were huge compared to their littermates (Figure 11-8). The foreign genes directed the mice to produce excess quantities of growth hormone (GH), which led to their increased size and weight. To get the rat genes into the fertilized mouse egg, about 600 copies of the GH gene were injected and some of the genes stably integrated. These eggs were then implanted in the reproductive tracts of "surrogate mothers," where they developed into mice. Every cell of these mice contained the GH genes, including their gamete-producing cells. The GH trait was therefore transmitted to the next generation, which again exhibited exceptional growth. Animals that possess genes of a different species are called **transgenic** animals.

Transgenic mice that contain human genes are being used to study people in a way that would be otherwise impossible. Sickle cell anemia, for example, is a disease for which no animal models are available because humans are the only animal that can be afflicted with the disease. All that changed in 1990 when transgenic mice that carried the sickle cell version of the human globin gene were developed. A more common use for transgenic animals is as livestock, animals that have been developed to improve their value. For example, pigs born with foreign growth hormone genes incorporated into their chromosomes grow much leaner than do control animals lacking the genes. The meat of the former animals is leaner because the excess growth hormone stimulates the conversion of nutrients into protein rather than fat.

Transgenic animals (and transgenic plants) that are "living factories" for the production of human proteins are also being developed. The product of one scientific endeavor is sheep in which the human gene for factor IX (a blood-clotting factor absent in some hemophiliacs) has been joined to a sheep gene that codes for a milk protein. The sheep's milk contains sufficient quantities of the human protein to be purified for clinical use in treating hemophilia.

Now that we have explored a few of the advantages of DNA technology, let's turn to some of the specific techniques that have made these practical applications possible.

GENETIC ENGINEERING—SOME FUNDAMENTAL METHODS

The fundamental technique that underlies genetic engineering is gene **splicing,** insertion of the desired gene into the chromosome of another organism. To do this, the gene is often first spliced into a larger molecule of DNA that acts a vector, a transport vehicle that carries the desired gene into the host cell. Viruses are frequently used as vectors to introduce a gene into a host cell during infection, but the most common vectors are plasmids. **Plasmids** (Figure 11-9) are small, circular DNA molecules of bacteria that are distinct from the main bacterial chromosome. Plasmids often contain genes that endow bacterial cells with resistance to drugs such as ampicillin and tetracycline, antibiotics frequently used to treat bacterial infections.

Tools for Assembling Recombinant DNA Molecules

One of the first steps in constructing recombinant DNA molecules is to treat both the donor and recipient DNA with the same sequence-specific restriction enzyme (Fig-

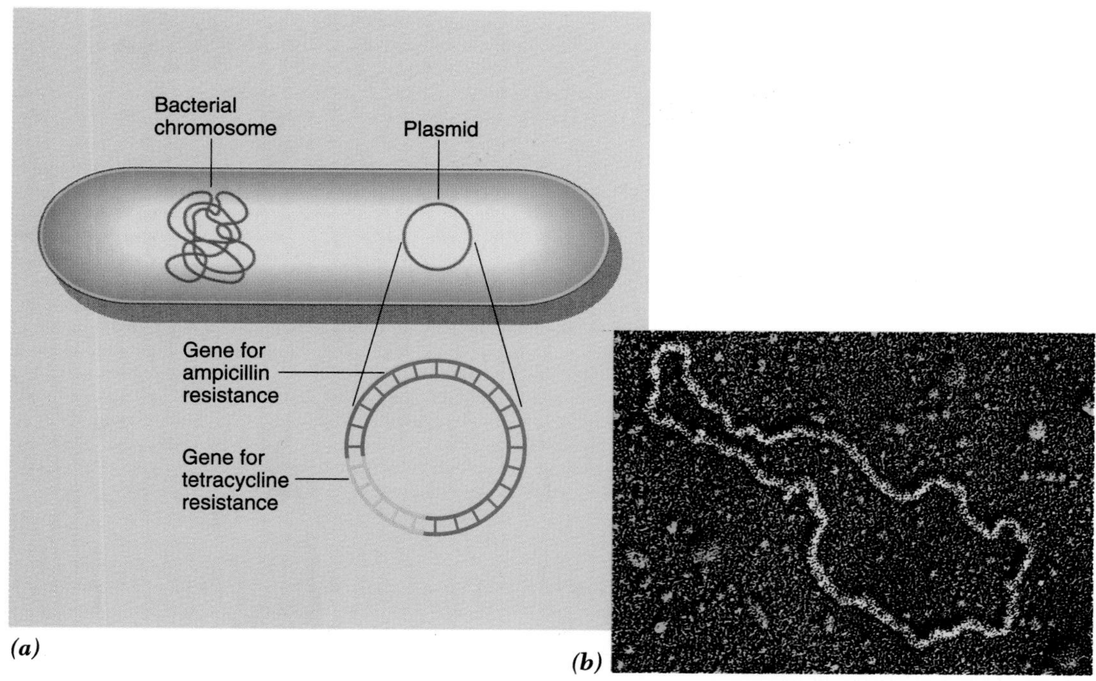

FIGURE 11-9

A bacterial plasmid. In addition to its chromosome, many bacteria contain small circular strands of DNA, called plasmids. The plasmid shown in part ***a*** contains genes that make the bacterial cells resistant to two antibiotics, tetracycline and ampicillin (a relative of penicillin). Consequently, a bacterium with this plasmid can live in the presence of these antibiotics, a property that helps isolate those bacteria that have received a foreign-gene-carrying plasmid from those that don't (only the plasmid-carrying cells will grow on media supplemented with tetracycline).

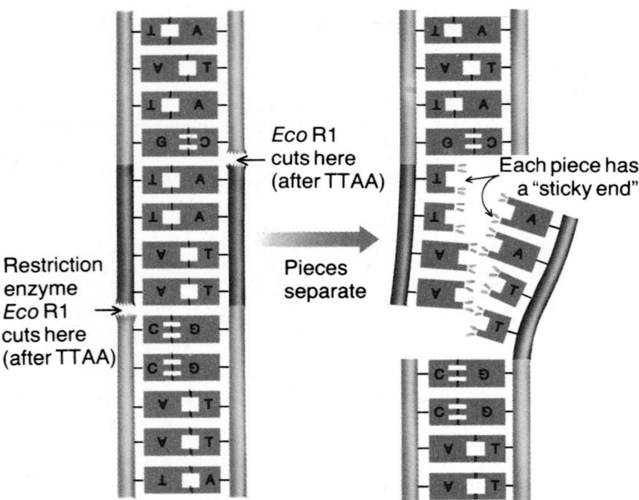

FIGURE 11-10

Restriction enzymes are the genetic engineer's DNA-cutting scissors. In the case depicted here, a restriction enzyme called *Eco* R1 cuts both strands of a DNA double helix at a particular DNA sequence (in this example, after every "TTAA" sequence). This restriction enzyme makes staggered cuts that generate DNA fragments that have "sticky ends." These sticky ends will join with complementary sticky ends on any DNA cut by the same restriction enzyme.

Bacterial plasmid

Human chromosome

Insulin gene

Same restriction enzyme splits both types of DNA

TTAA

AATT

Insulin gene joins with plasmid by sticky ends; nicks are sealed by enzyme

Plasmid

Insulin gene

EFFECT IS ANALOGOUS TO:

Plasmid's instructions

Cell's instructions

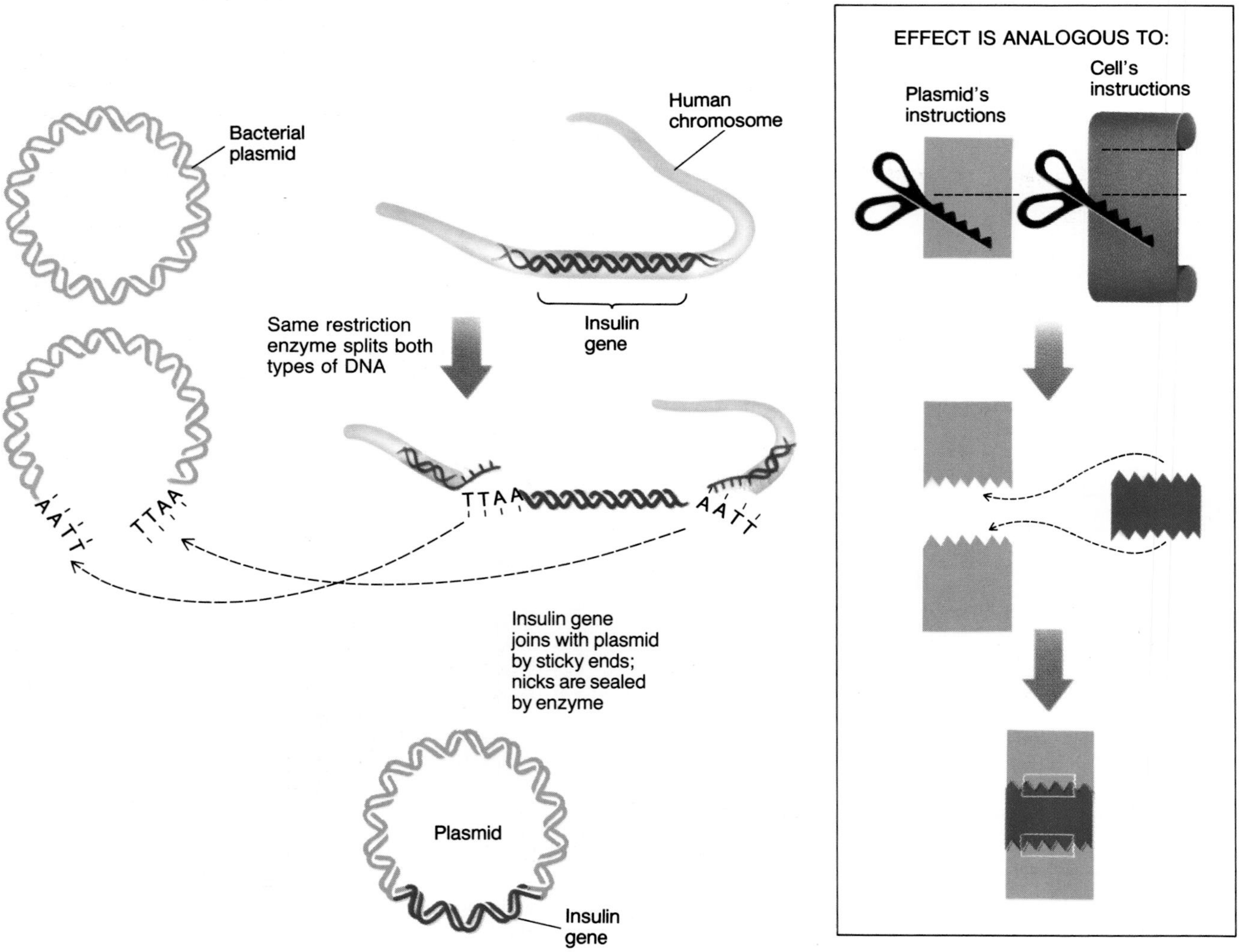

FIGURE 11-11

Formation of recombinant DNA molecules. In this example, both plasmid DNA and human DNA containing the insulin gene are treated with the same restriction enzyme so that the DNA from the two sources will have complementary "sticky ends." As a result, the two DNAs become joined to one another and are then covalently sealed by DNA ligase, forming a recombinant DNA molecule.

ure 11-10). The natural role of these enzymes is protective—bacteria use their restriction enzymes to destroy ("restrict") the DNA of invading viruses. Genetic engineers harvest these enzymes to use as molecular "scissors" to "cut and paste" DNA molecules obtained from different sources. These enzymes generate a "staggered cut" in the DNA that leaves short, single-stranded tails that act as "sticky ends" that can bind with a complementary single-stranded tail on another DNA molecule to restore a double-stranded structure.

Since a particular restriction enzyme always cuts a DNA molecule at the same site in a sequence, it creates the same sticky ends regardless of the organism that donated the DNA. For example, the sticky ends of a genetic segment cut from a human chromosome readily adhere to the complementary single-stranded tails of a bacterial plasmid that has been cut with the same enzyme (Figure 11-11). When these two DNA preparations are mixed together, the complementary sticky ends join. In many cases, the joining splices the isolated human genetic segment to the plasmid DNA, much as an extra paragraph may be taped into a sheet of instructions. As long as both papers are cut with scissors that make the same pattern, the cuts fit, and the pieces can be joined together. The "tape" that permanently bonds the joined DNA fragments together is **DNA ligase,** an enzyme that seals the gaps in the DNA backbone. The result is a recombinant DNA molecule—in this case, a bacterial plasmid containing a human protein gene.

Amplification of Recombinant DNAs by DNA Cloning

Before this recombinant DNA molecule can be put to use, it is necessary to generate a relatively large quantity of the incorporated gene(s). This DNA amplification process can be accomplished by a procedure called **gene cloning** (Figure 11-12). The recombinant molecule is taken up by bacteria, which then do all the work. The bacteria replicate the plasmid DNA molecule as they do their own chromosome, before the cell divides in two. Consequently, the number of recombinant DNA molecules increases as the bacterial culture grows in cell number. After a few hours, millions of copies of the recombinant plasmid are formed.

Expression of a Eukaryotic Gene in a Host Cell

In addition to serving as cellular "copying machines" for generating large quantities of recombinant DNA, host bacterial cells can transcribe and translate a human gene residing within a bacterial plasmid, producing large amounts of high-quality human protein. This process requires that the DNA encoding a human protein be placed in the proper position next to a bacterial promoter (otherwise, the bacterial RNA polymerase cannot attach to the DNA and transcribe it). The human DNA used in these processes must also be free of introns (page 176) since bacterial cells do not have the machinery required to remove the

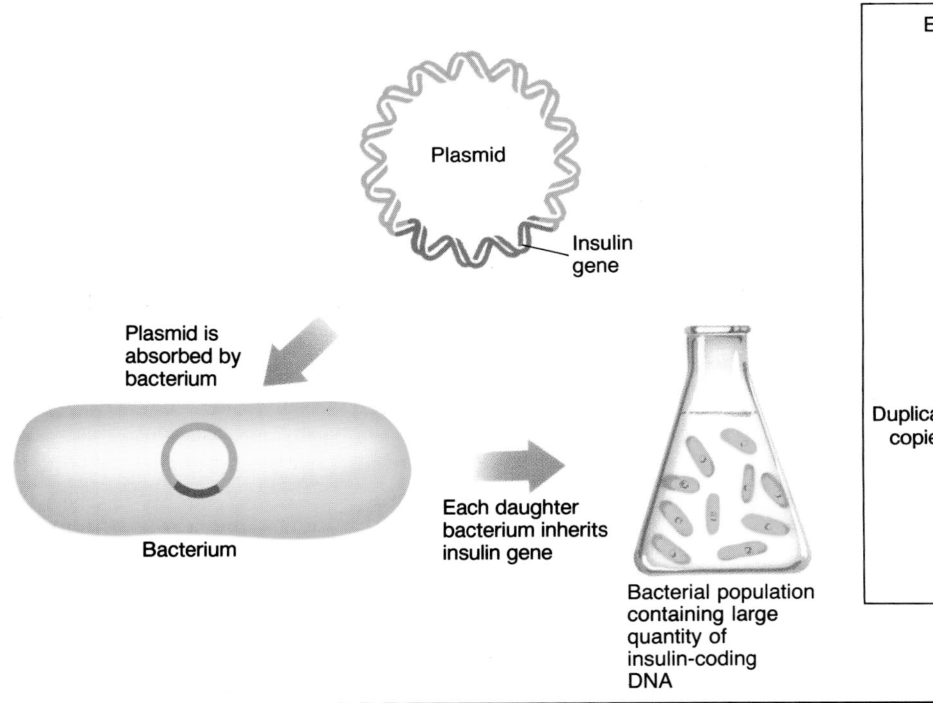

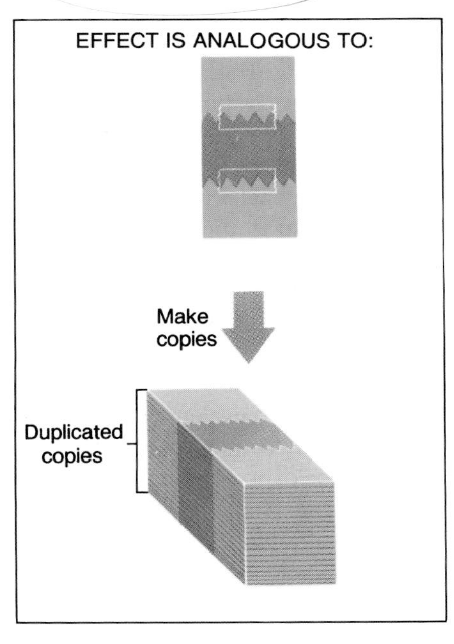

FIGURE 11-12

DNA cloning produces large quantities of a particular recombinant DNA molecule inside bacterial cells. The recombinant molecules are taken up by the bacterial cells and become amplified (cloned) as the bacterial cells proliferate.

intervening sequences from the transcribed message. When the introduced gene is artificially synthesized (as was the insulin gene), the interrupting introns are omitted.

The presence of introns is not a problem when eukaryotic cells are used as hosts for human genes. In fact, mammalian cells generally prove to be better suited than bacteria for manufacturing human gene products and, like bacteria, can be grown to high density in large vats, where they secrete human proteins into the medium.

BIOTECHNOLOGY THAT DOES NOT REQUIRE RECOMBINANT DNA MOLECULES

Although biotechnology is often equated with genetic engineering, several techniques that have helped launch the "biotechnology revolution" do not require the use of recombinant DNAs.

Determining DNA Nucleotide Sequences

The first nucleic acid to be sequenced required 7 years of painstaking (and Nobel Prize-winning) work and was only 77 nucleotides long. Today, a fragment of DNA several times this length can be sequenced within an hour using a totally automated apparatus. This technology has made it possible for scientists to launch the most ambitious scientific project in history, a full-scale offensive to determine the entire nucleotide sequence of a human being.

MAPPING THE HUMAN GENOME

In 1986, a group of scientists from around the world gathered at the Cold Spring Harbor Laboratories in Long Island, New York, to discuss the formation of a giant, international scientific collaboration to decode all of the genetic information stored in human chromosomes. Within a few years, governmental agencies in the United States, Europe, and Japan had decided to proceed with the effort, which has become known as HGP, the Human Genome Project. The term "genome" refers to the genetic information stored in all the DNA of a single haploid set of chromosomes, about 3 billion base pairs in the human genome. The magnitude of this project is staggering—if each base pair in the DNA were equivalent to a single letter on this page, the information contained in the human genome would produce a book approximately 1 million pages long.

The goal for the Human Genome Project is to sequence the DNA of all 24 human chromosomes (22 autosomes, plus the X and Y sex chromosomes) by the year 2005. Although the sequence doesn't automatically locate genes for inherited diseases or lead to conclusions about the functions of gene products, the information will shed light on virtually every aspect of human biology, from illuminating the biochemical basis of memory to determining what gene products are responsible for transforming a normal cell to a cancer cell.

The Human Genome Project is not without its critics. Some scientists believe the project is too costly, at least at our current level of DNA-sequencing expertise, and it will siphon money from other research endeavors. There are also questions about who will be "the human" who will genetically represent us all. Debate spans from fanciful paranoia (fear of using the information to clone humans) to realistic concerns about who will own the sequence.

Accelerated DNA Amplification

A new technique for amplifying a specific DNA fragment (such as the insulin gene) does not require its introduction into a bacterial cell. This technique, which is carried out in a machine instead of a living organism, uses the DNA-synthesizing enzyme (DNA polymerase) from hot-springs bacteria, an enzyme that is stable at temperatures of 90°C. This technique, called **polymerase chain reaction** (**PCR**), replicates any segment of DNA, inexpensively producing a limitless supply of identical segments even if only a single molecule is available (Figure 11-13).

PCR is also valuable in criminal investigations, generating large quantities of DNA from a spot of dried blood left on a crime suspect's clothing or even from the DNA present in part of a single hair follicle left at the scene of a crime (see The Human Perspective: DNA Fingerprinting and Criminal Law). In 1992, PCR was used to amplify DNA extracted from the exhumed bones of a man who had drowned several years earlier while swimming off the coast of Brazil. Some people claimed that the man was Dr. Josef Mengele, the infamous Nazi doctor who had carried out sadistic experiments on inmates at the Auschwitz concentration camp. Many skeptics doubted the identification, however. To confirm the man's identity, DNA from the bones was isolated, amplified by PCR, and compared to a DNA sample donated by Mengele's son, using DNA fingerprinting. The results were conclusive: The drowning victim was indeed Josef Mengele, closing the chapter on one of the most intensive international manhunts in history.

DNA Technology for Determining Evolutionary Relationships

DNA-sequencing also helps evolutionary biologists determine how closely related two species are. The longer the amount of time that passes since two organisms diverged from a common ancestor, the greater the differences in their DNA sequences (the more time for mutations to create differences). The technique has been used to help us answer a particularly interesting question: Who is our closest living relative? According to sequencing studies carried out by Michael Miyamoto and his colleagues at Wayne State University, the answer is chimpanzees. While this answer may not be surprising, Miyamoto's group made another, unexpected finding. It has generally been assumed that chimpanzees and gorillas are more closely re-

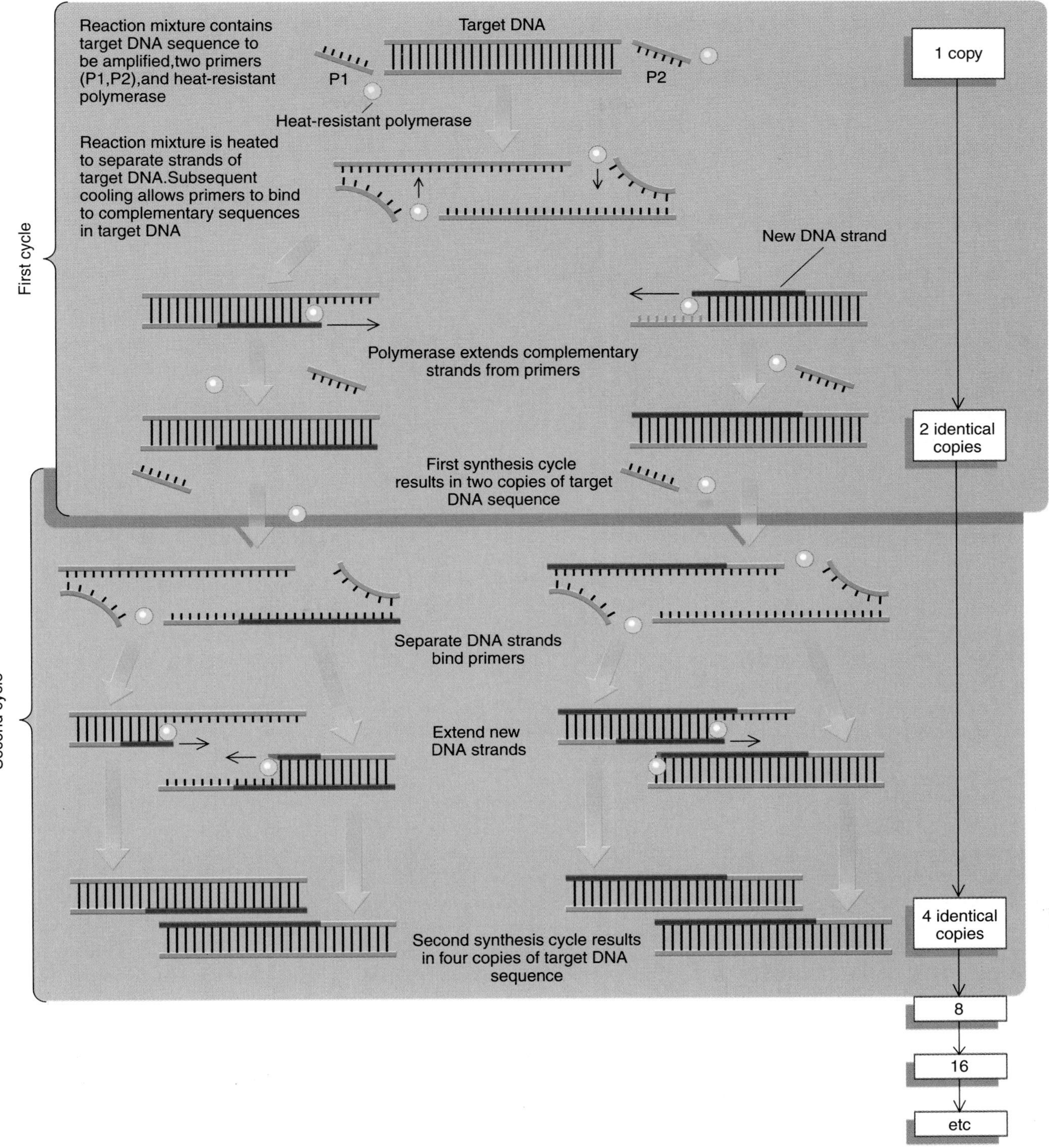

FIGURE 11-13

Polymerase chain reaction (PCR). Using this procedure, a single region of DNA present in a sample of minute proportions can be selectively replicated, producing large quantities of the gene. The DNA is mixed with short, synthetic DNA fragments (called *primers*) that are complementary to nucleotide sequences at either end of the region of the DNA to be amplified. Using heat to separate the DNA into single strands, the mixure is then cooled to allow the primers to bind to the flanking segments at the ends of the selected region. Another molecule in the mixure, a DNA replicating enzyme (DNA polymerase), then binds to the primers and selectively copies the intervening DNA segment. The cycle is repeated again and again each time doubling the amount of the specific DNA region. Billions of copies of a desired genetic segment can be produced in a few short hours.

◁ THE HUMAN PERSPECTIVE ▷

DNA Fingerprints and Criminal Law

In early February 1987, police were investigating a grisly murder. A woman and her 2-year-old daughter were stabbed to death in their apartment in the New York City borough of the Bronx. Following a tip, the police questioned a resident of a neighboring building. A small bloodstain was found on the suspect's watch, which was sent to a laboratory for DNA fingerprint analysis. The DNA from the white blood cells in the stain was amplified using the PCR technique and was digested with a restriction enzyme. The restriction fragments were then separated by electrophoresis, and a pattern of labeled fragments were identified with a radioactive probe. The banding pattern produced by the DNA from the suspect's watch was found to be a perfect match to the pattern produced by DNA taken from one the victims. The results were provided to the opposing attorneys. A pre-trial hearing was called in 1989 to discuss the validity of the DNA evidence.

During the hearing, a number of expert witnesses for the prosecution explained the basis of the DNA analysis. According to these experts, no two individuals, with the exception of identical twins, have the same nucleotide sequence in their DNA. Moreover, differences in DNA sequence can be detected by comparing the lengths of the fragments produced by restriction-enzyme digestion of different DNA samples. The patterns produce a "DNA fingerprint" that is as unique to an individual as is a set of conventional fingerprints lifted from a glass. In fact, DNA fingerprints had already been used in more than 200 criminal cases in the United States and had been hailed as a monumentally important development in forensic medicine (the application of medical facts to legal investigations). The widespread use of DNA fingerprinting evidence in court had been based on its general acceptability in the scientific community. According to a report from the company performing the DNA analysis, the likelihood that the same banding patterns could be obtained by chance from two *different* individuals in the Hispanic community was only one in 100 million.

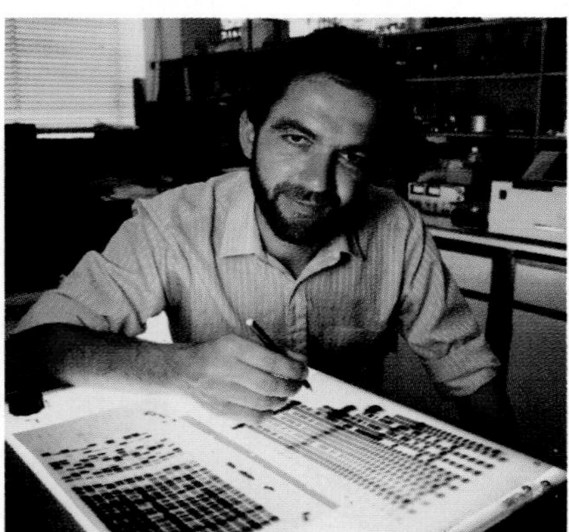

Alec Jeffreys of the University of Leicester, England, examining a DNA fingerprint. Jeffreys was primarily responsible for developing the DNA fingerprint technique and is the scientist who confirmed the death of Josef Mengele.

What made this case (known as the Castro case, after the defendant) memorable, and distinct from its predecessors, was that the defense also called on expert witnesses to scrutinize the data and to present their opinions. While these experts confirmed the capability of DNA fingerprinting to identify an individual out of a huge population, they found serious technical flaws in the analysis of the DNA samples used by the prosecution. In an unprecedented occurrence, the experts who had earlier testified *for the prosecution* agreed that the DNA analysis in this case was unreliable and should not be used as evidence! The problem was not with the technique itself, but in the way it had been carried out in this particular case. Consequently, the judge threw out the evidence.

In the wake of the Castro case, the use of DNA fingerprinting to decide guilt or innocence has been seriously questioned. Several panels and agencies are working to formulate guidelines for the licensing of forensic DNA laboratories and the certification of their employees. In 1992, a panel of the National Academy of Sciences released a report endorsing the general reliability of the technique but called for the institution of strict standards to be set by scientists.

The events surrounding the most famous use of DNA fingerprinting, the O. J. Simpson murder case in Los Angeles, transformed the issue from one regarded primarily by the scientific and legal professionals into one of great public concern. The world watched as the legal community put both O. J. Simpson and the use of DNA evidence on trial.

lated to each other than either is to the human species. However, Miyamoto found that there were fewer differences in the DNA nucleotide sequences between humans and chimpanzees (a 1.6 percent nucleotide-sequence difference) than between chimpanzees and gorillas (a 2.1 percent difference). According to this data, chimpanzees are actually more closely related to humans than they are to gorillas.

EVOLUTION AND ADAPTATION: TYING IT TOGETHER

The study of human genetics and biotechnology raises some evolutionary questions and reaffirms much of what we know about evolution. One question concerns the persistence of detrimental alleles in the human population. For example, why hasn't the gene for cystic fibrosis been eliminated by natural selection? One reason is the recessive nature of the allele, which is not detrimental in heterozygous carriers, who transmit the allele without any direct disadvantage to themselves or to their reproductive capacity. In fact, some of these detrimental alleles may be advantageous in heterozygous carriers. People with one recessive allele for sickle cell anemia, for example, are more resistant to malaria, so the allele would be an advantage in areas where malaria is endemic. Cystic fibrosis heterozygotes may be more resistant to cholera (a disease that causes severe diarrhea and death by dehydration) than are homozygotes that lack a CF allele. A person having one CF allele has a reduced capacity for chloride transport, which might result in a reduced loss of body water during a cholera infection.

Biotechnology provides us with modern ways of determining evolutionary relatedness by comparing similarities in DNA fingerprints. It also reaffirms one of the central principles of evolution—the common ancestry of all the earth's organisms. Genetic engineering is based on this evolutionary principle. Because we all inherited the same genetic code from an ancestor shared far back in evolutionary time, we can produce pharmaceuticals by implanting human genes into bacteria. The fact that, in bacteria, the human insulin gene directs the host bacteria to produce the identical human protein reveals a remarkable relatedness between people and bacteria. All living things "speak" the same genetic language.

SYNOPSIS

Human life is very sensitive to the proper number of chromosomes. Gametes containing an abnormal chromosome number are produced following nondisjunction during meiosis. In humans, zygotes having an abnormal chromosome number usually die in the womb or are born with characteristic abnormalities, such as Down syndrome.

Recent advances in gene mapping have allowed geneticists to locate the chromosomal position of genes, even when the gene product has not been characterized. Gene mapping takes advantage of differences in DNA nucleotide sequences from one person to another—differences that result in variations (polymorphisms) in the length of restriction fragments (RFLPs) produced when human DNA is treated with restriction enzymes. The location of a particular gene is discovered by identifying a specific RFLP that is always present in individuals that show the corresponding trait; that is, it is closely linked to the gene. Once a gene for a particular disease is isolated, its gene product can be deduced and the exact defect determined.

Genetic disorders are transmitted from generation to generation in several well-defined patterns. Disorders resulting from autosomal recessive alleles, such as cystic fibrosis and sickle cell anemia, occur only when both members of a pair of homologous autosomes carry the recessive alleles. Heterozygous carriers have a normal allele that compensates for the recessive allele. Disorders that result from autosomal dominant alleles require only one member of the pair to be defective. The abnormal gene product encoded by the allele causes the disorder, despite the presence of the normal product from the other homologue, so there are no carriers. Disorders that result from X-linked alleles, such as hemophilia and muscular dystrophy, occur almost exclusively in males. The allele producing the defective gene product is located on the X chromosome, and there is no compensating locus for the allele on the Y.

Genetic engineering depends on the formation of recombinant DNAs containing portions derived from different sources and joined together in the laboratory. Recombinant DNAs have been used to manufacture human proteins with clinical applications, such as insulin, growth hormone, and plasminogen activator. In such cases, a gene encoding the desired product is spliced into a bacterial plasmid, and the recombinant molecule is introduced into a bacterial or eukaryotic host cell, which follows the foreign gene's instructions and manufactures the gene product.

Genetic engineers have developed organisms that carry recombinant DNAs. Examples include plants containing a bacterial gene that directs the formation of an insect-killing protein, microbes that can produce enzymes to digest toxic compounds, drought-resistant crop plants, livestock animals with improved nutrient value, and animals that manufacture human proteins of clinical value.

The biotechnology revolution has been built on the development of new techniques in DNA technology. The isolation of particular genes of interest in medicine and agriculture is achieved by using restriction enzymes—DNA-cutting enzymes that recognize a particular stretch of nucleotides. Once the DNA is enzymatically fragmented, and the gene of interest is isolated, the gene is spliced into a suitable vector, such as a bacterial plasmid, the recombinant molecule is introduced into bacteria, and the gene product harvested.

DNA fragments can be separated from one another, and their nucleotide sequences determined. DNA fragments produced by treating DNA with a restriction enzyme are separated by electrophoresis. The pattern of fragments that migrate at different speeds forms the basis for comparing DNAs from different individuals of a species. For example, distinguishing the pattern of restriction fragments from different humans forms the basis for the forensic technique of DNA fingerprinting.

Recent advances in molecular genetics may revolutionize the treatment of genetic-based disorders. Not only has RFLP analysis provided ways to screen for a particular disorder, gene therapy has moved from the drawing boards to clinical trials, as researchers are attempting to replace the faulty genes with good ones. Information gained from sequencing the entire human genome will likely help in developing treatments for inherited diseases, to provide an opportunity to manufacture many new protein products, and to help us better understand the course of human evolution.

Review Questions

1. Review the pattern by which autosomal recessive, autosomal dominant, and X-linked disorders are inherited.
2. Why does a person with an XXY-chromosome complement develop into a male rather than a female? Which sex would you predict a person with XXXXY to exhibit should survival be possible?
3. Why do RFLPs make such useful genetic markers?
4. Distinguish between the following: bacterial plasmid and chromosome; conventionally produced insulin and insulin produced by genetically engineered *E. coli*; transgenic animal and a "normal" animal; a sense strand and an antisense strand of DNA; amplification by cloning and amplification by PCR.
5. Why is the production of "sticky ends" important in recombinant DNA technology?

Critical Thinking Questions

1. Diet soft drinks and other foods containing the artificial sweetener aspartame (trademark Nutrasweet) carry a warning for people with PKU. Aspartame is a dipeptide (two linked amino acids). Why do you suppose this warning might be necessary?
2. If you were a congressperson interested in drafting a law governing the use of human genetic data, what elements might you want to include in your law? How would you feel about having your own DNA screened for particular alleles?
3. If you were to isolate a gene for interferon from human DNA and add it directly to bacterial cells rather than first incorporating it into a plasmid, would this eukaryotic gene be transcribed in the host cell? Explain your answer.

PART

4

Form and Function of Plant and Animal Life

One of the most strikingly beautiful and complex plant structures is the flower. Flowers capture the attention of passing animals (mostly insects). As animals gather nectar, they transfer sperm-containing pollen from flower to flower for the plant's sexual reproduction.

CHAPTER 12

Plant Life: Form and Function

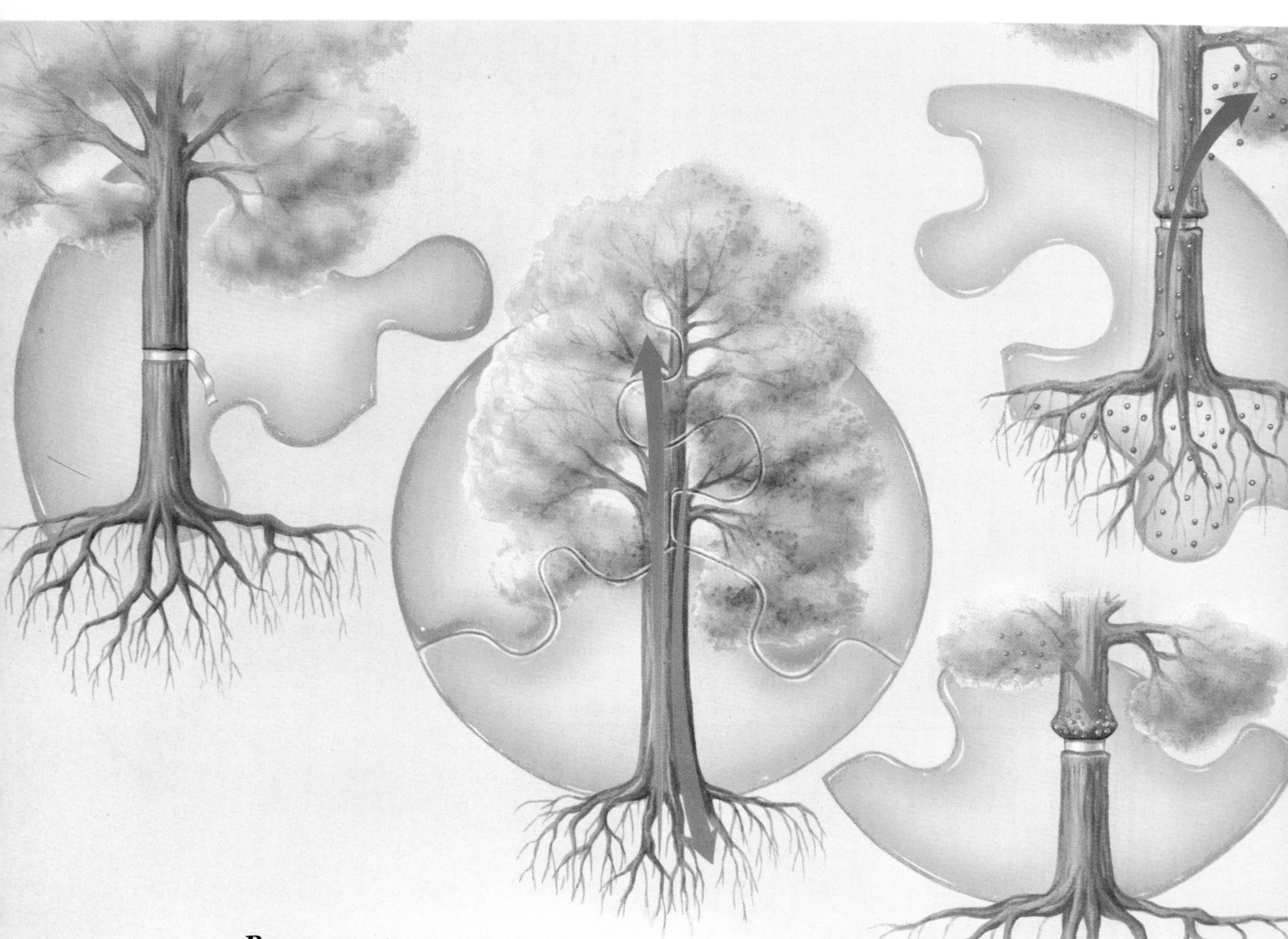

By stripping off a ring of bark, scientists solved the puzzle of how materials flow through plant tissues—essential nutrients flow down through the bark, while water streams up through internal tissues.

STEPS TO DISCOVERY
Discovering The Plant's Circulatory System

For 3 centuries, biologists have known that blood surges through the human body through vessels. In 1661, Marcello Malpighi, a professor of medicine at the University of Bologna, described the network of the smallest of these vessels, the capillaries. But Malpighi's interest extended beyond the circulatory system of humans; he was curious about how materials moved through the bodies of *all* organisms—animals and plants.

Malpighi reasoned that, like humans, larger plants must have some type of a circulatory system for carrying materials from one part to another; otherwise, how could materials move across long distances, such as between deep roots and the leaves on the tips of branches? Unlike animal research, little microscopic work had been done on plant tissues, so very little was known about how materials could be transported through a plant. Malpighi decided to test whether the bark of a tree contained the plant's circulatory system. He removed a complete strip of bark from around a tree's trunk, a technique called *girdling*. After a period of time, he observed that the bark *above* the girdle swelled; the tree eventually died. Because the bark swelled only above the girdle, Malpighi concluded that girdling blocked the downward movement of materials. The blockage caused materials to accumulate above the girdle, which then caused the bark to swell; had materials been moving upward, the bark would have swelled below the girdle. The fact that the tree eventually died led Malpighi to conclude that the materials that move downward through the bark are essential to the plant's life. But just what were these materials?

In 1928, two British plant physiologists, T. Mason and E. Maskell, repeated Malpighi's girdling experiment. After girdling the tree's trunk, the researchers measured the rate of water-vapor loss from its leaves. Mason and Maskell knew from earlier studies that most of the water that moves through a plant is lost by evaporation from the leaves. By measuring water-vapor loss from the leaves of a girdled plant, the researchers could determine whether the *main* pathway of water movement was through the bark or through tissues interior to the bark which were not destroyed by girdling. As in Malpighi's study, Mason and Maskell observed that the bark swelled only above the girdle. But their research determined that girdling did not change the rate of water-vapor loss from the plant's leaves. Mason and Maskell concluded, therefore, that the bulk of water moves up through tissues interior to the bark. To explain the swelling above the girdle, the researchers concluded that other materials move *down* through the bark itself. By the 1920s, botanists had named the distinctive tissue within the bark the phloem and the tissue interior to the bark the xylem.

During the 1940s, a new technique was developed that enabled researchers to label organic compounds with radioactive carbon and, consequently, to trace the movement of labeled molecules through an organism. To follow the movement of sugars produced during photosynthesis, for example, plants were exposed to radioactive carbon dioxide, which was absorbed and incorporated into photosynthesized sugars. Scientists then traced the movement of the labeled sugars and discovered that they were concentrated completely in the phloem tissues (in the bark) rather than in the xylem. The predominant direction of movement of labeled molecules was downward through the phloem from the leaves—the centers of photosynthesis—to all remaining plant tissues. Later studies using labeled compounds also revealed that virtually all organic molecules produced by a plant circulate through the phloem.

Nearly 300 years after Malpighi's girdling experiments, scientists had finally proven that plants—like most animals—have a circulatory system. But unlike animals, water and minerals are transported in plants through tissues that are separate from those through which organic molecules are transported.

At 8:30 A.M. on May 18, 1980, Mount St. Helens in southwestern Washington erupted with a force equal to that of 750 atomic bombs detonated simultaneously. The enormous blast shot 4 km^3 (1 $mi.^3$) of earth 20,000 kilometers (about 65,000 feet) into the air, lowering the elevation of Mount St. Helens by more than 300 km (1,000 feet). (The colossal volume of earth fired up into the air is approximately equal to the size of Mt. Rushmore, including the busts of all four presidents.) Where a dense forest once stood, a barren moonscape remained. Extending 4 kilometers (14 miles) out from this lifeless center was a progression of "damage" zones—areas populated only by the stumps of incinerated trees; heaps of toppled trees piled one on top of the other like toothpicks; and zones of scorched, standing trees. But within just 1 year, life began returning to the slopes of the volcano (Figure 12-1). Among the first organisms to return were more than 150 species of plants. Plants precede animals in all "new" environments, eventually creating the habitats and providing the food for animal life. Eventually, plants once again will become the most abundant form of life on the slopes of Mount St. Helens.

▼ ▼ ▼

THE BASIC PLANT DESIGN

The earth's 400,000 known plant species display an enormous range of external body forms and internal cell structures. In this chapter, we focus on the body construction of flowering plants, the most familiar, most evolutionarily advanced, and most structurally complex of any group in the plant kingdom. Botanists divide flowering plants into two main groups—**dicotyledons,** or dicots (*di* = two,

(a)

(b)

FIGURE 12-1

Mount St. Helens. (*a*) Within minutes, thousands of acres of forest were incinerated following the May 18, 1980 eruption. (*b*) Within only 1 year, first plants and then animals begin to return to the slopes of Mount St. Helens.

cotyledon = embryonic seed leaf), and **monocotyledons,** or monocots (*mono* = one). Table 12-1 illustrates the many differences that distinguish dicots from monocots.

The typical flowering plant grows through the soil and the air simultaneously, two very different habitats with very different conditions. As a result, the two main parts of the plant differ dramatically in form (anatomy) and function (physiology): The underground **root system** anchors the plant in the soil and absorbs water and nutrients, while the aerial **shoot system** absorbs sunlight and gathers carbon dioxide for photosynthesis (Figure 12-2). The shoot system also produces stems, leaves, flowers, and fruits. Interconnected **vascular tissues** transport materials between the aerial shoot system and underground root system. Vascular connections allow water and minerals absorbed by the root to be conducted to shoot tissues, and for food produced by the shoot to be transported to root tissues.

PLANT TISSUES

As in all multicellular organisms, groups of plant cells work together to form **tissues. Simple tissues,** such as those that store starch in a potato, contain all the same type of cell, whereas **complex tissues,** such as a plant's vascular tissues, are composed of different kinds of cells working together to carry out a particular function. Plants are composed of four simple tissues—*meristems, parenchyma, collenchyma,* and *sclerenchyma.*

PLANTS GROW FROM MERISTEMS

Meristems are clusters of cells dedicated to dividing by mitosis to produce new cells for increased growth. All shoots and roots have **apical meristems**—a group of meristematic cells located at their apexes, or tips (Figure 12-2). Divisions of apical meristem cells increase the *length* of these structures. All growth that originates from apical meristems is termed **primary growth.** *Annuals* (plants that live 1 year or less; e.g., corn, marigolds) experience only primary growth and are called *herbaceous* plants.

In addition to having apical meristems, *biennials* (plants that live for 2 years; e.g., carrots, Queen Anne's lace) and *perennials* (plants that live longer than 2 years; e.g., shrubs, trees) also develop lateral meristems called **cambia** (singular = *cambium*). Cambia are rings or clusters of meristematic cells that increase the *width* of a plant's stems and roots. Cambia produce **secondary growth,** and plants that experience such growth are called *woody* plants. Thus, a pine, oak, or any other perennial undergoes both primary and secondary growth at the same time (Figure 12-3).

Perennials develop two cambia. The **vascular cambium** produces new vascular tissues that increase transport capacity. The **cork cambium** produces cork cells, which make up the outer covering of the bark and repairs damaged tissues.

TABLE 12-1

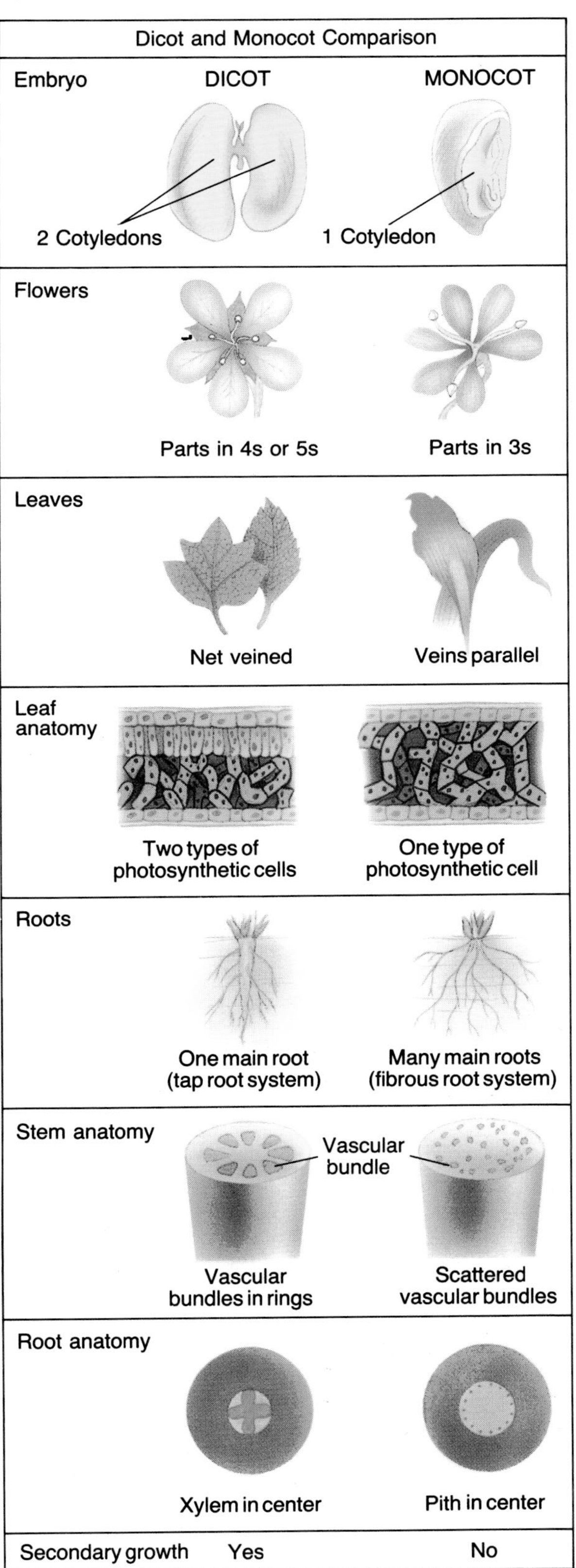

Dicot and Monocot Comparison		
	DICOT	MONOCOT
Embryo	2 Cotyledons	1 Cotyledon
Flowers	Parts in 4s or 5s	Parts in 3s
Leaves	Net veined	Veins parallel
Leaf anatomy	Two types of photosynthetic cells	One type of photosynthetic cell
Roots	One main root (tap root system)	Many main roots (fibrous root system)
Stem anatomy	Vascular bundles in rings	Scattered vascular bundles
Root anatomy	Xylem in center	Pith in center
Secondary growth	Yes	No

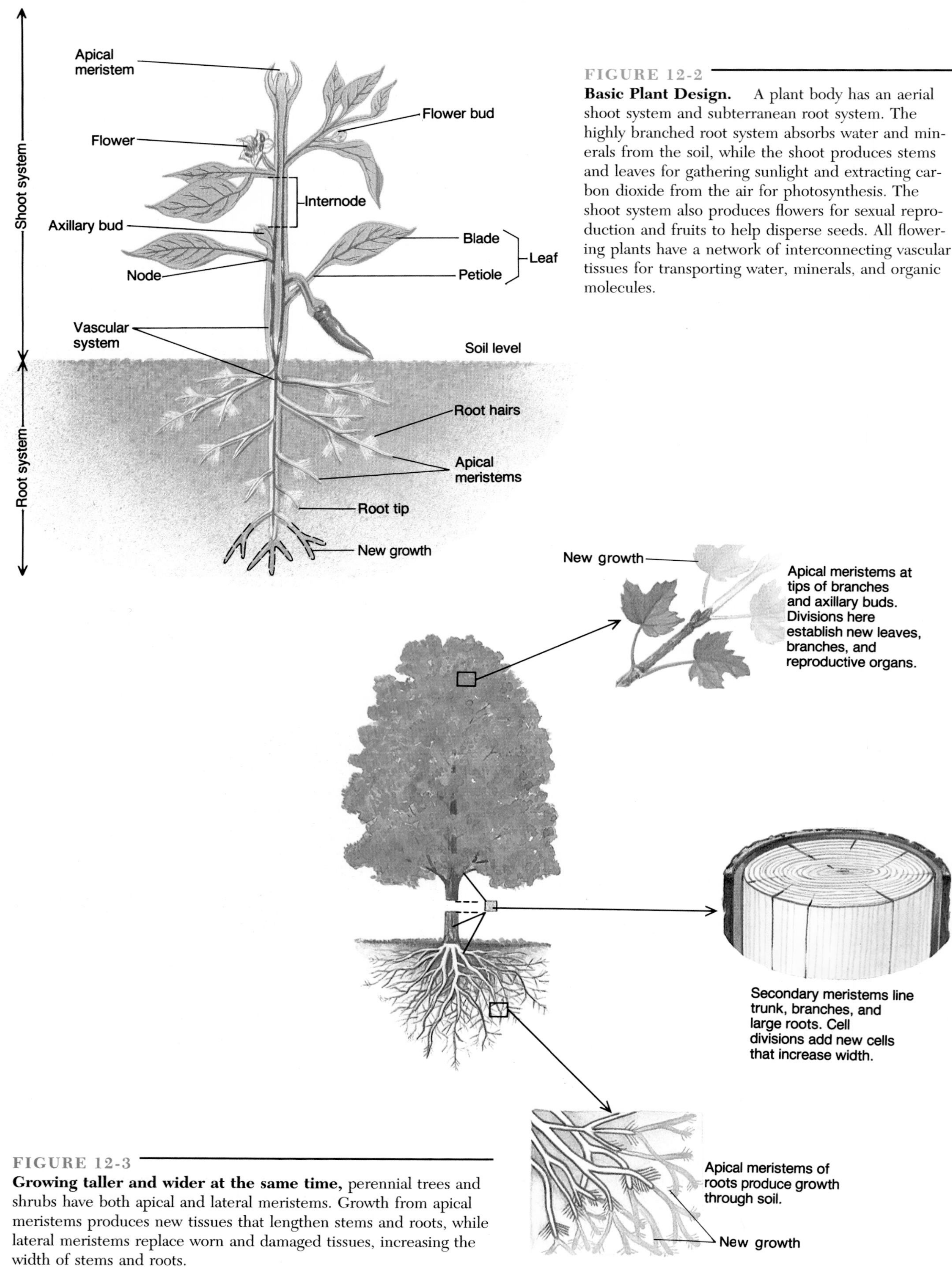

FIGURE 12-2

Basic Plant Design. A plant body has an aerial shoot system and subterranean root system. The highly branched root system absorbs water and minerals from the soil, while the shoot produces stems and leaves for gathering sunlight and extracting carbon dioxide from the air for photosynthesis. The shoot system also produces flowers for sexual reproduction and fruits to help disperse seeds. All flowering plants have a network of interconnecting vascular tissues for transporting water, minerals, and organic molecules.

FIGURE 12-3

Growing taller and wider at the same time, perennial trees and shrubs have both apical and lateral meristems. Growth from apical meristems produces new tissues that lengthen stems and roots, while lateral meristems replace worn and damaged tissues, increasing the width of stems and roots.

PARENCHYMA CELLS PROVIDE VERSATILITY

The most prevalent type of cell in an herbaceous plant is **parenchyma** (Figure 12-4). Parenchyma cells provide plants with both structural and functional flexibility, enabling them to adjust and respond to their immediate surroundings, as well as to changing conditions.

Parenchyma cells have many roles, including:

- photosynthesis
- storage of food, water, and pigments
- transport
- healing of wounds
- production of new roots, stems, and other plant parts.

With such a variety of functions, it's not surprising that parenchyma cells vary a great deal in size, shape, and cell-wall characteristics, although most are generally round and have thin walls (Figure 12-4). Parenchyma cells are unique in their ability to differentiate into any other kind of plant cell, making them one of the most versatile of all plant cells.

COLLENCHYMA CELLS SUPPORT STEMS AND LEAVES

The shape of a plant cell contributes to its function, just as the shape of a fork and a knife contribute to their function. In addition, the nature of the plant cell wall affects the cell's function; thick, hard walls offer much greater support than do thin, soft walls. **Collenchyma** are cigar-shaped cells with unevenly thickened cell walls, two characteristics that help make collenchyma effective support cells in a plant (Figure 12-4). The crunchy strands of a celery stalk are a prime example of collenchyma cells. In stems and leaves, collenchyma tissues lie just beneath the epidermis (the outer cell covering), helping the plant and its leaves remain upright. Collenchyma also contain chloroplasts and carry out photosynthesis.

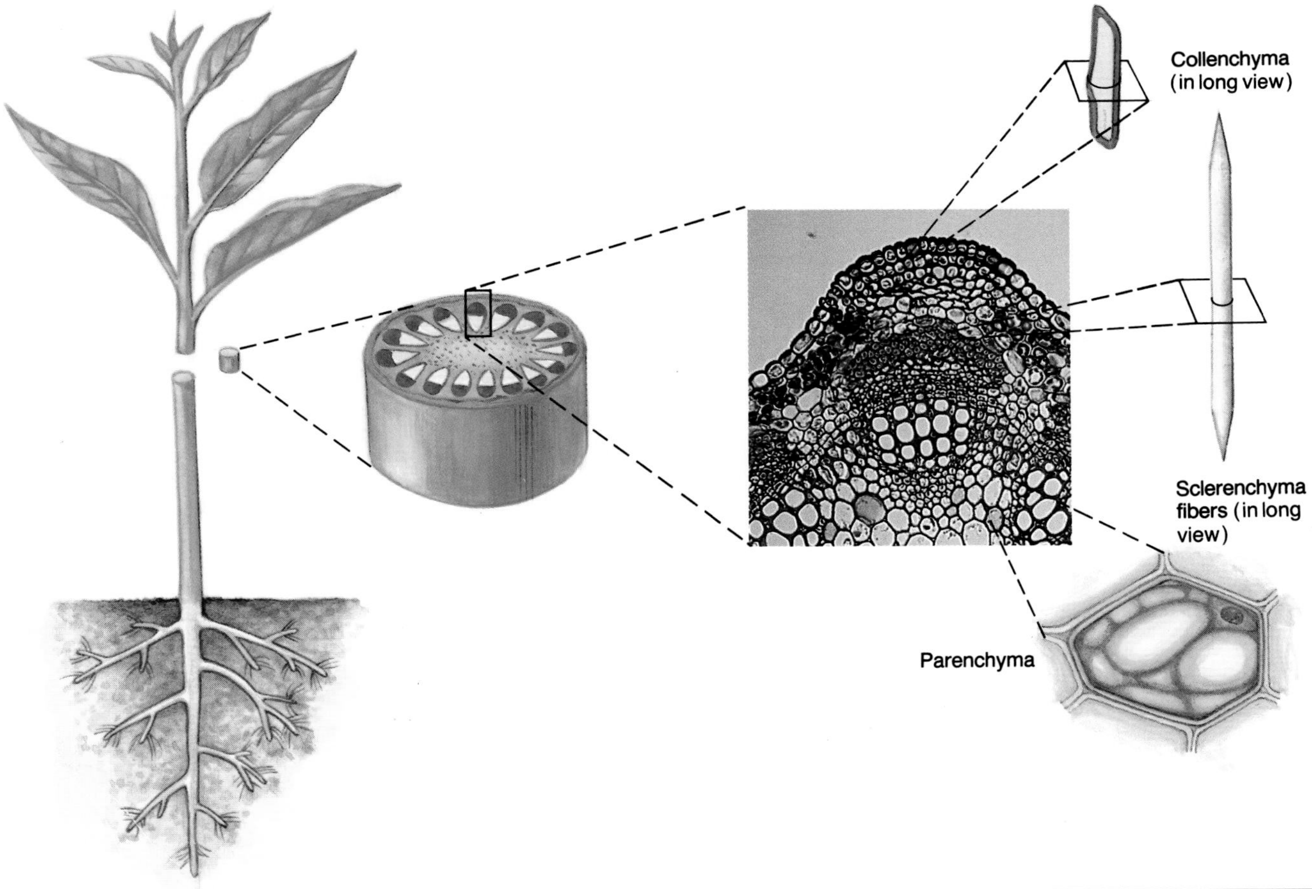

FIGURE 12-4
Simple Tissues. Roundish parenchyma cells fill most of the stem of an annual plant. Elongate collenchyma cells with thickened walls help support the stem and are found immediately under the epidermis. Dense collections of sclerenchyma fibers form a cap of supportive cells over each vascular bundle.

SCLERENCHYMA CELLS SUPPORT AND PROTECT

The thick secondary walls of **sclerenchyma** cells make the tissues they form even stronger and more supportive than collenchyma tissues. The critical component of a sclerenchyma cell is its wall, not the living cytoplasm inside. The secondary wall contains *lignin*, a hardening substance that gives the cell its toughness and strength. Because the cell wall is left behind after the cell dies, sclerenchyma tissues continue functioning even after all of the cells die.

Sclerenchyma fibers are extremely long, narrow cells with uniformly thickened walls (Figure 12-4). Their elastic cell walls allow the cell to bend and then spring back to its original position, similar to a rubber band springing back to its original size and shape after being stretched. Elasticity is particularly important in leaves, stems, and flowers for proper orientation. For example, after strong gusts of wind die down, sclerenchyma fibers enable a plant's leaves to reorient quickly, reexposing the maximum surface to sunlight for photosynthesis. We harvest hemp and flax to make rope and linen, a testimony to the strength of sclerenchyma fibers.

Not all sclerenchyma are long, thin fibers. Some sclerenchyma cells—called *sclereids*—come in various shapes and have hard, thick walls. For example, sharp, star-shaped sclereids not only help support loosely arranged leaf cells in plants that float on the surface of water, such as waterlilies, but they are also very effective in discouraging even the hungriest browsing animal by puncturing its gums and tongue as it nibbles (Figure 12-5). Densely packed, rectangular sclereids form the "shell" of a walnut, which protects the plant's embryo from injury as it falls from a tree or tumbles along the ground. The gritty texture of a pear is the result of clumps of stone-shaped sclereids that help support huge parenchyma cells filled with sweet pear juices.

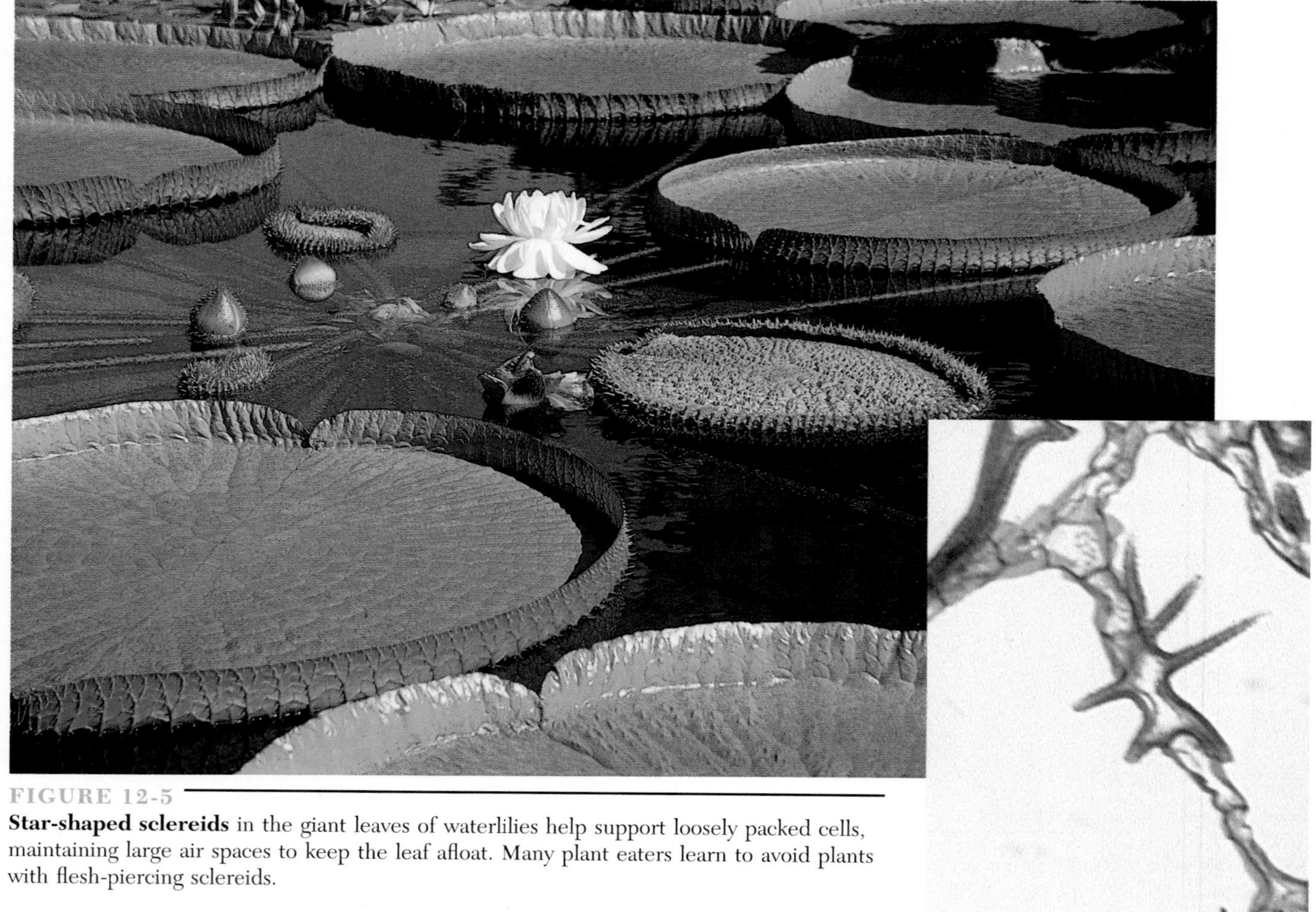

FIGURE 12-5

Star-shaped sclereids in the giant leaves of waterlilies help support loosely packed cells, maintaining large air spaces to keep the leaf afloat. Many plant eaters learn to avoid plants with flesh-piercing sclereids.

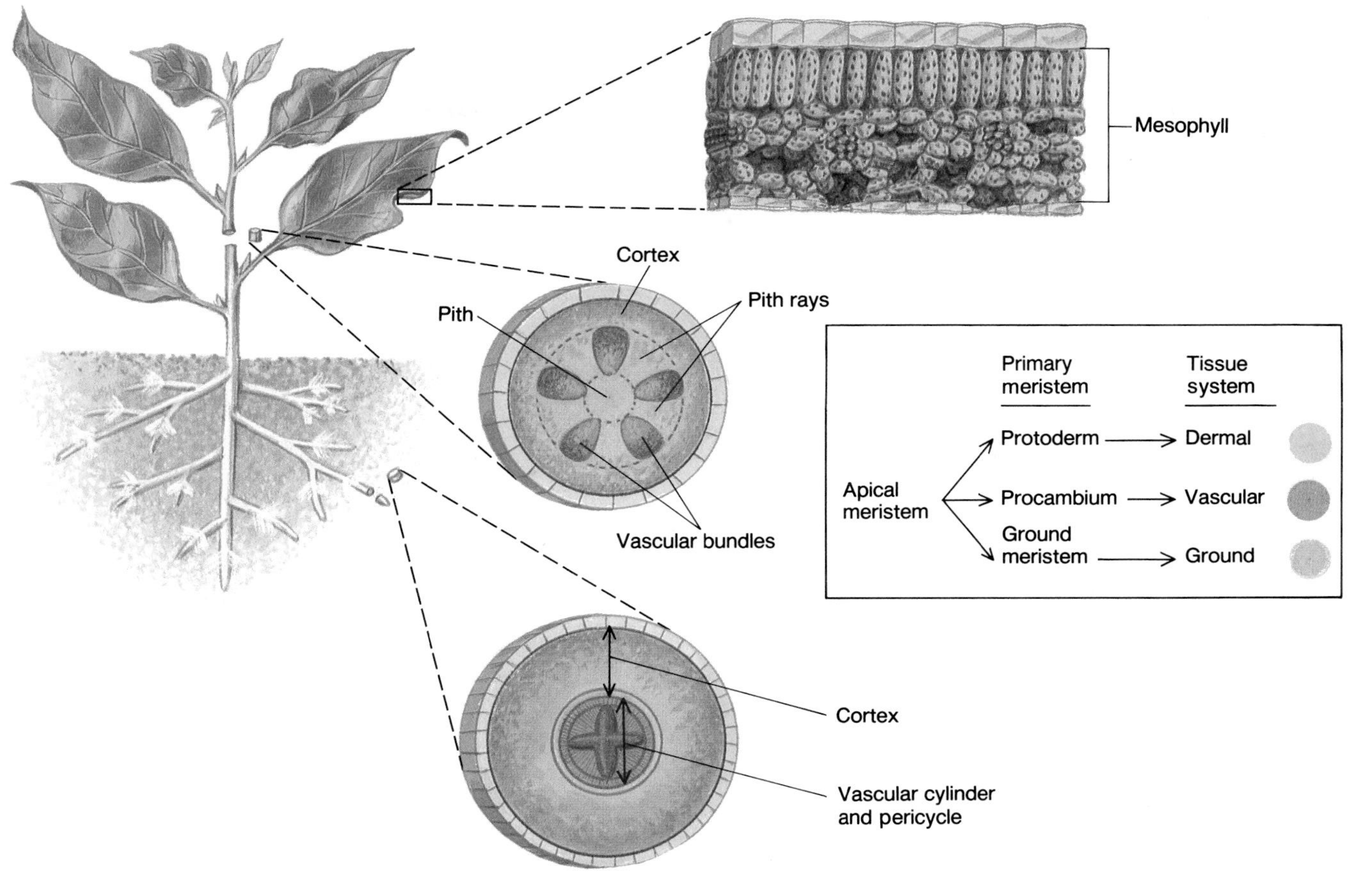

FIGURE 12-6

Plant tissue systems. Three tissue systems run continuously throughout the body of a plant—the Dermal Tissue System, Vascular Tissue System, and Ground Tissue System. The Ground Tissue System includes the cortex (the area between the epidermis and vascular tissues), pith (the area from inside the vascular tissues to the stem or root center), and pith rays between the vascular tissues in stems, the mesophyll in leaves, and the cortex in roots.

PLANT TISSUE SYSTEMS

Continuous sheets or columns of cells interconnect the plant's root, stem, leaf, and flowers to form one of three plant **tissue systems** (Figure 12-6). Plant tissue systems are complex tissues, composed of more than one type of cell.

- The protective **dermal tissue system,** or **epidermis,** forms the outer cell covering of the plant.
- The **vascular tissue system** contains conducting cells that transport food, water, and minerals throughout the plant.
- The **ground tissue system** comprises all remaining cells and tissues.

Each tissue system has a primary function: The dermal tissue system forms a seamless protective outer covering; the vascular tissue system transports materials between plant organs; and the ground tissue system provides support and storage continuity.

THE DERMAL TISSUE SYSTEM

The dermal tissue system (*derma* = skin) has two critical, yet opposing, functions: The epidermis must form a barrier to protect the plant's delicate internal tissues, yet it must also allow for exchanges of essential materials between the plant and its surrounding environment. The outer cell walls of the shoot epidermis are covered by a **cuticle**—a waxy layer that retards water loss and helps prevent dehydration. The cuticle is such an effective water barrier that the cuticle of the Brazilian wax palm is harvested to make carnuba wax for safeguarding furniture and cars.

The epidermis of the shoot is riddled with microscopic pores called **stomates** (Figure 12-7). Each stomate is

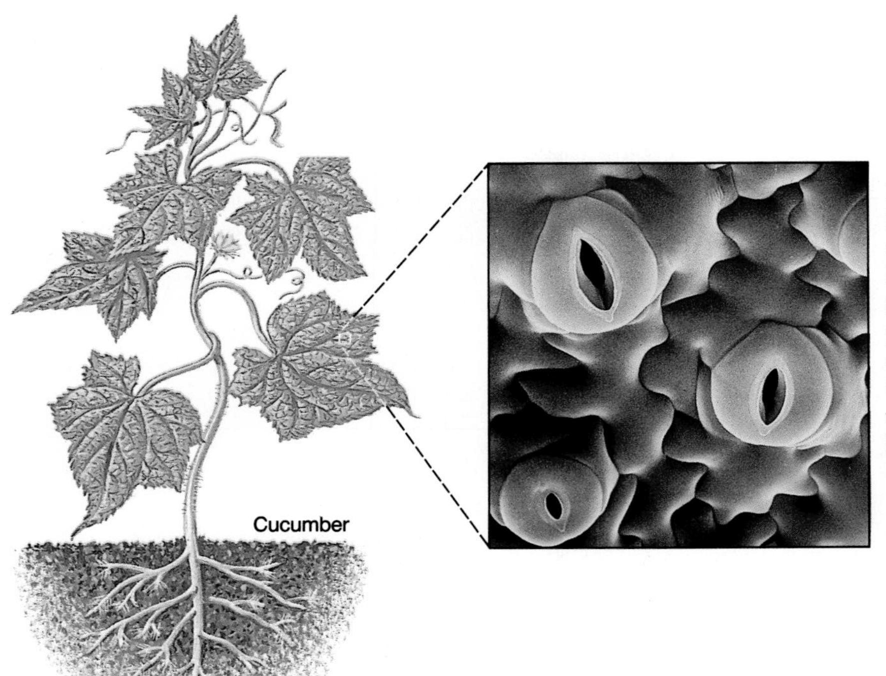

FIGURE 12-7

Stomatal pores between adjacent guard cells enable gases to diffuse into and out of air spaces between internal cells. Carbon dioxide mainly diffuses inward for photosynthesis, and water vapor usually diffuses out. As internal pressure increases, guard cells bend and enlarge the size of the stomatal pore; decreasing internal pressure within guard cells narrows or closes the pore, blocking further gas exchange.

flanked by two specialized **guard cells;** just like the doorman of a building regulates who enters and exits the building, guard cells regulate the rate of gas diffusion into and out of the plant: carbon dioxide for photosynthesis; oxygen for aerobic respiration; and water vapor for cooling. By regulating water vapor, guard cells not only moderate the temperature of a plant by preventing it from overheating, they also help prevent dehydration in plants that grow in hot, dry environments.

In woody plants, the **periderm** takes over the protective and regulating functions of the epidermis when it becomes disrupted as the stem and root increase in width. The periderm is commonly referred to as "bark," but, technically, bark includes all those plants tissues outside the wood (see Figure 12-11). The outer portion of the periderm contains cork cells, which continue to function even when dead (Figure 12-8). Each cork cell is composed of layers of secondary walls that are impregnated with a waterproof wax called suberin. These waxy cell layers effectively seal off internal tissues, protecting them against excess water-vapor loss, disease, extreme weather, and foraging insects. To enable oxygen to reach living cells beneath the periderm, small eruptions in the periderm, called **lenticels,** form air channels through which carbon dioxide and oxygen flow (Figure 12-8).

THE VASCULAR TISSUE SYSTEM

The vascular tissue system forms an internal circulatory network—the "veins and arteries" of a vascular plant (Fig-

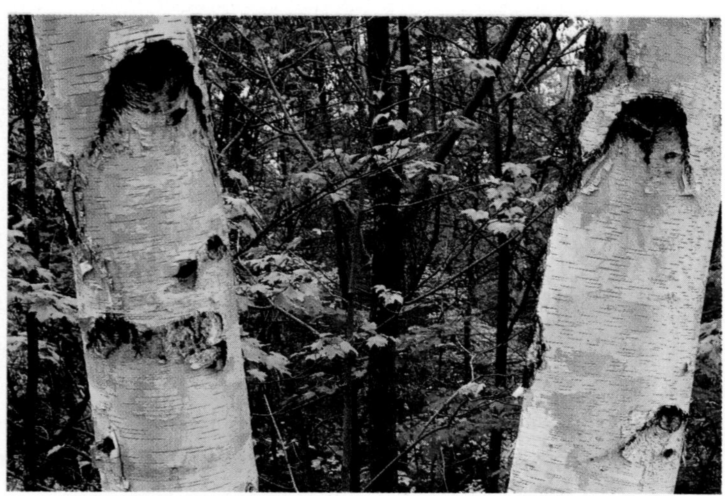

FIGURE 12-8

The periderm not only protects internal tissues during secondary growth but allows for gas exchange through lenticels.

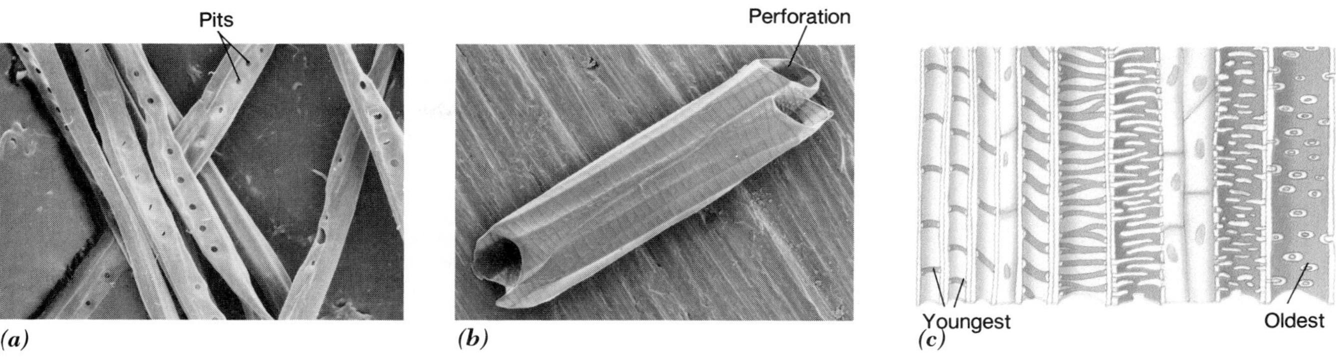

FIGURE 12-9
Conducting cells in the xylem transport water and dissolved minerals. Conducting xylem cells include long, thin tracheids ***(a)*** which are aligned precisely so that the openings (pits) of one match up with those of other tracheids, and vessel members ***(b, c)*** with pits and perforated ends.

ure 12-6). This tissue system contains two types of complex tissues—**xylem** and **phloem.** Water and dissolved minerals are carried primarily through the xylem; food (carbohydrates produced during photosynthesis) and other organic chemicals synthesized by the plant (such as hormones, amino acids, and proteins) move through the phloem.

Xylem: Water and Mineral Transport and Support

The main conducting cells of the xylem are **tracheids** and **vessel members** (Figure 12-9), both of which function as dead cells; water and minerals flow through their hollow interiors. The walls of each cell are riddled with pits—small holes in the secondary wall. Xylem cells line up so that the pits of one cell match up exactly with those of a neighbor, forming a pit pair. Water flows from one cell to the next through these pit pairs. Opposite ends of vessel members also have a *perforation plate,* an area where all or large portions of the end wall are absent. Vessel members stack one on top of another to form open **vessels** through which water readily travels (Figure 12-10).

Secondary xylem is called **wood** (Figure 12-11). New secondary xylem cells are produced by the vascular cambium. Toward the end of one growth period (usually at the end of summer), the width of xylem tracheids and vessel members gets progressively smaller. When water becomes plentiful again (usually not until the following spring), new tracheids and vessel members with large diameters are formed. The abrupt transition from xylem cells with narrow diameters to those with large diameters forms a distinct line, the border of a **growth ring.** Because one growth ring is usually produced each year, the number of growth rings is often used to estimate the age of a tree.

In most plants, vast amounts of water flow through the xylem. In fact, plants take in and transport far greater amounts of water than do most animals. A single sunflower plant, for example, will take in more than 17 times as much water as you will in 1 day. Plants demand such large amounts of water because many lose up to 98 percent of the water they absorb to the air as water vapor. The loss of water vapor from plant surfaces is called **transpiration.** Most water vapor is transpired from leaf surfaces through open stomates; the cuticle prevents evaporation from the surfaces of epidermal cells. Many plants that live in environments where water is limited have evolved adaptations

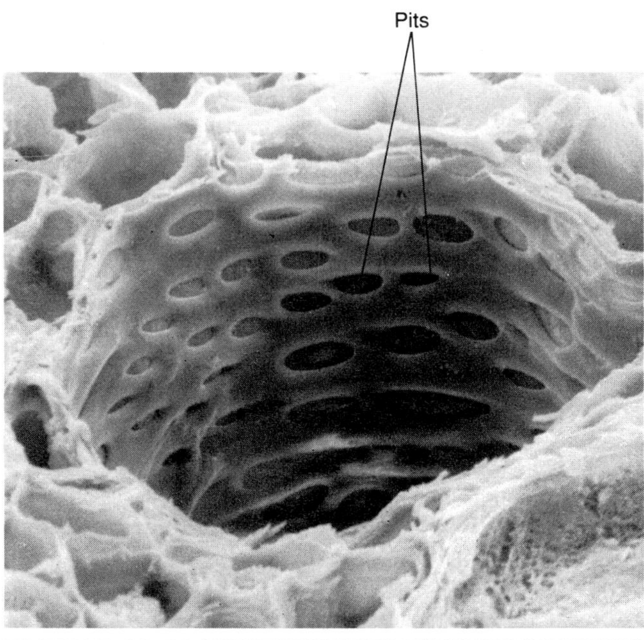

FIGURE 12-10
A vessel pipeline is formed by stacks of vessel members. Pits along side walls allow for lateral transport.

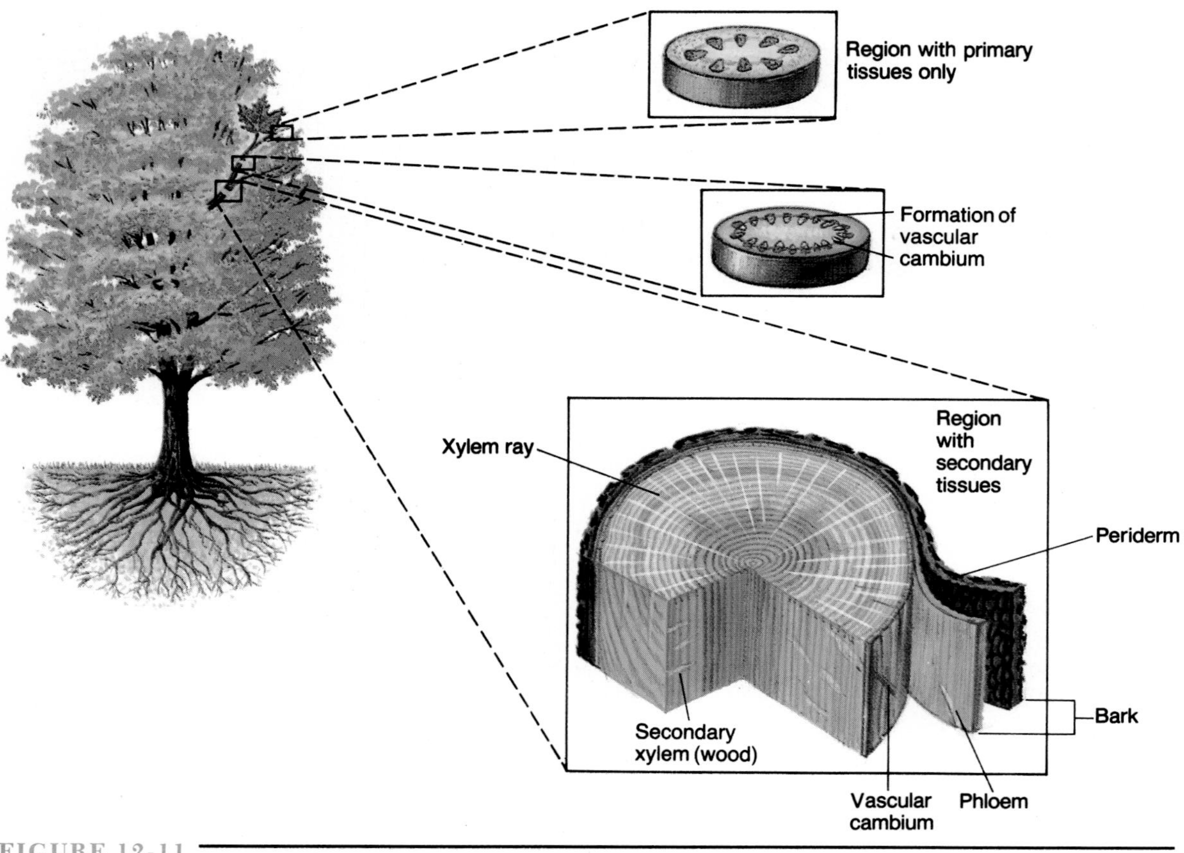

FIGURE 12-11

Sections of a stem and trunk of a tree. Although primary growth continues from apical meristems at the tips of stems, most of the tissue in a tree is secondary xylem, or wood. The vascular cambium divides to produce secondary xylem toward the inside and secondary phloem toward the outside.

that help reduce transpiration (Figure 12-12). Unless the plant's roots absorb enough water to replace the amount lost during transpiration, the plant will wilt and eventually die from dehydration.

Plant physiologists have now shown that the primary mechanism for water and mineral translocation through the xylem is **transpiration pull**—a process triggered by water-vapor loss. Transpiration pull results from a chain of events that starts when leaves begin absorbing solar radiation in the morning. Sunlight heats the plant's leaves, causing more water to evaporate from moist cell walls. The evaporated water is immediately replaced with water from inside the cell, which is replaced with water from neighboring cells deeper in the leaf, which, in turn, is replaced with water in the xylem. Within the xylem, liquid water molecules cling to one another by strong cohesive bonds, forming an unbroken water chain that extends all the way from the leaf vein, through the stem, and down into the root. Because of these strong bonds, as water molecules exit the xylem, they pull adjacent water molecules up, which, in turn, pull more adjacent water molecules up, and so on, down the xylem column to the roots below. In other words, cohesive bonds pull water up the xylem. This is the process of transpiration pull (Figure 12-13).

Because shoot surfaces are covered with a waxy cuticle, less than 10 percent of transpired water is lost directly from the surface of epidermal cells, unless something damages the cuticle layer. One reason biologists are so concerned about acid rain is that it dissolves the cuticle, destroying one of the plant's natural adaptations for controlling water-vapor loss. Acid rain also kills soil bacteria, which supply plants with needed nutrients, and increases the rate of leaching of nutrients from the soil.

Phloem: Food Transport

Phloem runs parallel with the xylem, transporting food and other organic chemicals throughout the plant (refer back to Table 12-1 and Figure 12-6). Unlike xylem tracheids and vessel members, however, phloem-conducting cells function as living cells. In flowering plants, phloem-con-

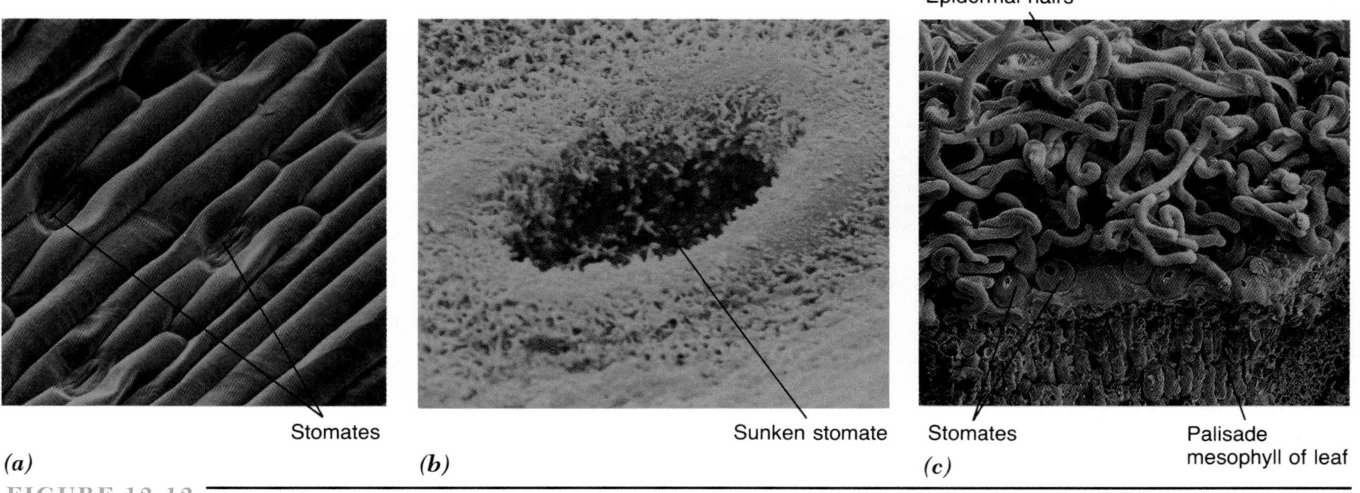

FIGURE 12-12

Plant adaptations that control transpiration rates. (*a*) **Number of stomates.** Some plants have many thousands of stomates per each square centimeter of leaf surface. Plants adapted to growing in moist habitats have stomates on both the top and bottom surfaces; plants adapted to arid areas have fewer stomates, which are restricted to the cooler, lower surfaces where the vapor gradient between the inside and outside of the leaf is less. (*b*) **Position of stomates**. Recessed stomates help some plants adapted to arid habitats conserve water. This placement doesn't affect carbon dioxide diffusion but retards water-vapor loss by forming pockets of high humidity. Sunken stomates also protect the plant from the drying effects of wind. (*c*) **Epidermal hairs**. Dense epidermal hairs not only reflect light, reducing the leaf's heat load, but they slow down transpiration by shielding stomate openings from drying winds.

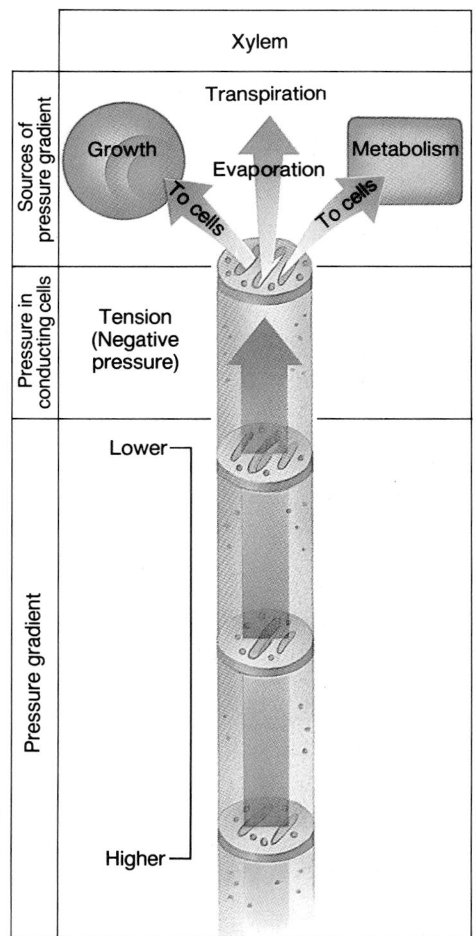

FIGURE 12-13

Xylem transport. Tension (negative pressure) builds in the xylem as water evaporates during transpiration or is used by cells for metabolism and growth. As water exits the xylem, cohesively bonded water molecules are pulled up.

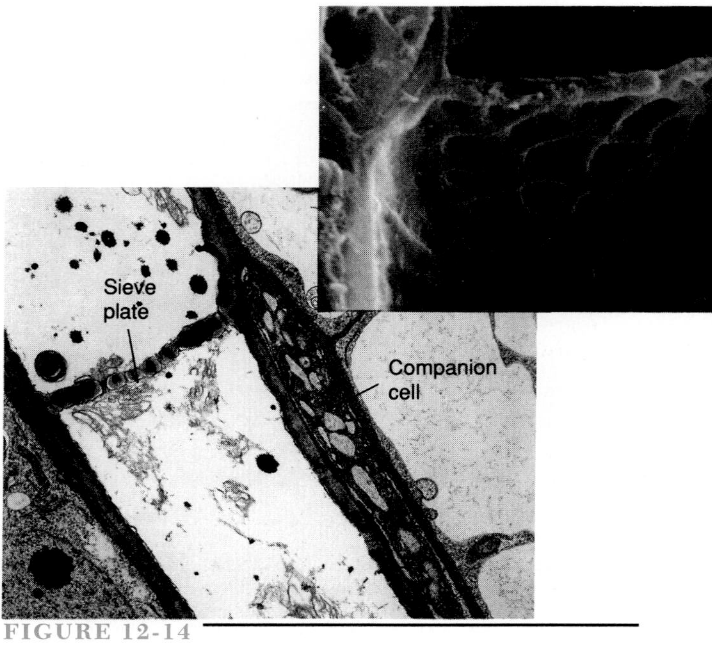

FIGURE 12-14
Sieve-tube members. Stacked end to end sieve-tube members form a sieve tube that conducts organic molecules throughout a plant. Each sieve-tube member has sieve plates on its end walls and at least one companion cell. A Scanning Electron Micrograph (SEM) of a sieve plate (inset) showing numerous pores through which organic materials pass from one sieve-tube member to the next. (Courtesy C. Y. Shih and R. G. Kessel, reproduced from *Living Images,* Science Books International, 1982.)

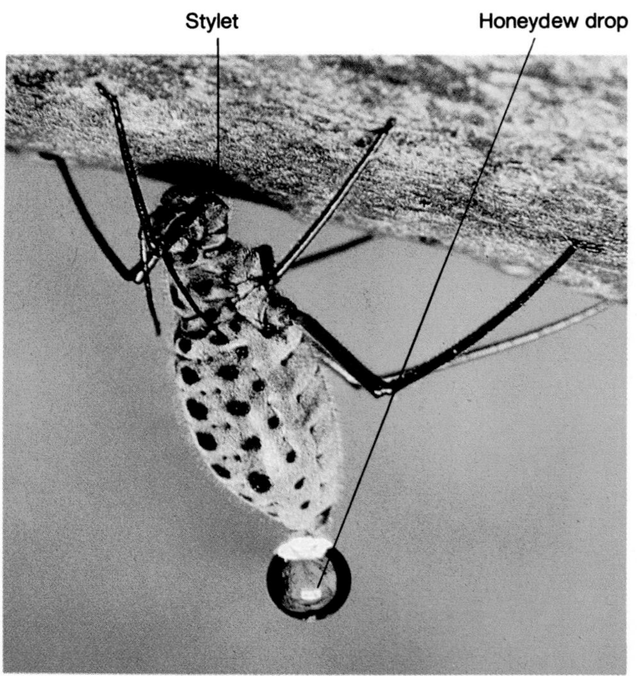

FIGURE 12-15
Aphids and phloem research. The stylet mouthparts of aphids enable these insects to penetrate phloem cells without damaging delicate conducting cells. Once the phloem is tapped, pressure in the phloem forces phloem sap through the digestive tract, forming a honeydew drop. When researchers remove the aphid's body, phloem sap flows out the insect's stylet for analysis.

ducting cells are called **sieve-tube members** because the clusters of pores in their cell walls resemble a strainer or sieve (Figure 12-14). These porous areas are located at opposite ends of the sieve-tube member and are called **sieve plates.** A row of sieve-tube members, stacked sieve plate to sieve plate, forms a sieve tube.

Although sieve-tube members are technically alive, they lack a nucleus and most of the other cellular organelles of a typical living cell. As these conducting cells mature, their organelles degenerate, leaving only a veneer of cytoplasm against the inside walls and a delicate mesh of protein that extends from sieve-tube member to sieve-tube member. To remain alive and active, sieve-tube members are continually nourished by an adjacent parenchyma cell called a **companion cell.**

It wasn't until 40 years ago that investigators actually identified which substances were being transported through the phloem. Before then, no matter how carefully researchers tried to extract the sap from phloem, the sieve tubes collapsed and plugged up, preventing researchers from getting a pure sample of phloem sap. Oddly enough, a common plant pest—aphids—provided scientists with the "tool" they needed to collect pure phloem sap. Researchers noticed that aphids gathered on growing stem tips and inserted their hollow, needlelike mouthparts (stylets) directly into the phloem, tapping into the plant's juices. The moment the aphid's stylet penetrates a sieve tube, its body balloons out, as sap gushes from the phloem-transport cells; sometimes the pressure in the sieve tube is so high it forces sap completely through the insect's digestive tract and out its anus, forming small droplets of honeydew (Figure 12-15). But researchers did not want to analyze the contents of honeydew drops because the drops would be "contaminated" with materials from the aphids digestive system. To collect "uncontaminated" phloem sap, investigators anesthetized the insects and then cut off their bodies, leaving the stylet in the phloem. The pure sap that exuded from the remaining stylets contained only materials flowing through the phloem—10 to 25 percent sugar (mainly sucrose), small amounts of amino acids, other nitrogen-containing substances, and plant hormones—all molecules synthesized by the plant.

Further studies revealed that phloem sap always flows from *sources* to *sinks* (Figure 12-16). Sources are locations where sugars and other organic substances are synthesized (such as in a photosynthesizing leaf cell) or where organic substances are stored (such as in parenchyma tissues). Sinks are plant tissues that use organic molecules, such as actively growing meristems, developing fruits and seeds,

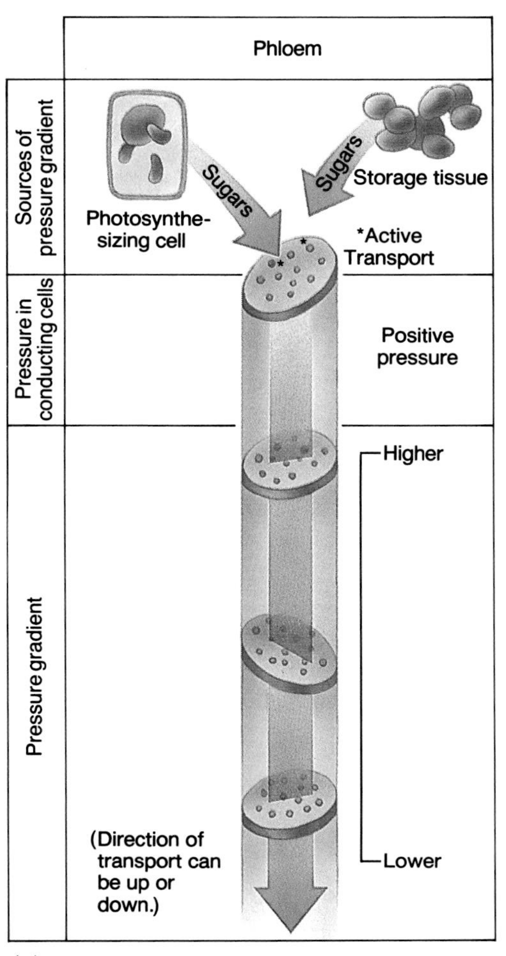

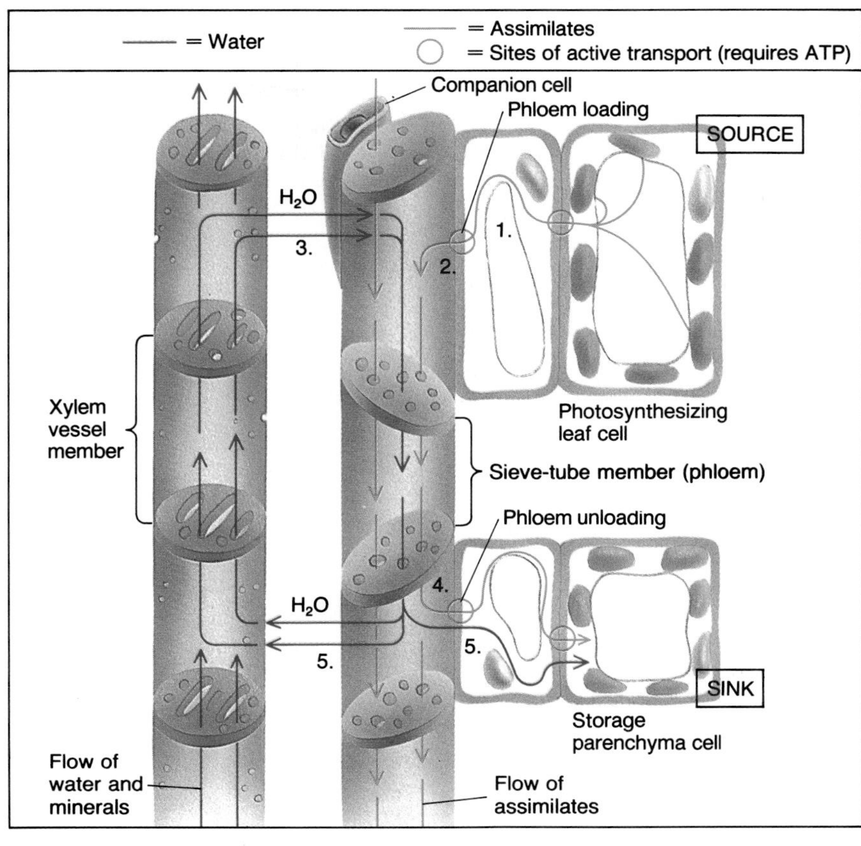

(a) *(b)*

FIGURE 12-16

Phloem transport by pressure flow. *(a)* Sugars and organic molecules are actively loaded into sieve elements, creating an osmotic gradient that draws water into phloem transport cells, thereby increasing internal pressure at the point of active loading. At the other end of a phloem tube, pressure falls as molecules are actively unloaded and water flows out of the phloem. This pressure gradient pushes materials through the phloem from sources to sinks. ***(b)*** Phloem sap flows from sources to sinks. (1) A photosynthesizing leaf cell is a source because it produces sugars. (2) Sugars are actively transported into a sieve tube during phloem loading. (3) As sugar concentration increases in the sieve tube, water is drawn in from nearby xylem vessels. This influx of water raises the pressure in the sieve tube at the source. (4) At sinks (in this example, a storage parenchyma cell is a sink because it stores sugars), sugars are actively unloaded, lowering sugar concentration in the sieve tube and (5) causing water to exit the phloem tube and be recycled back to the xylem or drawn into surrounding cells. The pressure differences between sieve-tube members at sources (high pressure) and sinks (low pressure) initiate the flow of an assimilate stream.

or storage tissues. (Notice that storage tissues can be both sources and sinks; they are sources when they export their stored molecules, and they are sinks when they import molecules to build reserves.)

Research indicates that materials flow through the phloem by **pressure-flow.** To initiate pressure flow, higher pressures are created at sources as organic molecules are actively transported into sieve-tube members; this process is called **phloem loading.** Phloem loading increases the concentration of solutes inside the sieve tube and creates an osmotic gradient that causes water to flow in from nearby xylem cells, building water pressure in the sieve tube at the source. At the sink end of the sieve tube, organic molecules are actively transported out of the phloem. This process, called **phloem unloading,** reduces solute concentration in the phloem, causing water to flow out of the phloem. As a result, water pressure drops at a sink. Since all sources and sinks are connected by columns of sieve-tube members, higher water pressure at sources pushes the phloem sap to lower water pressures at sinks.

THE GROUND TISSUE SYSTEM

With protection and exchange functions performed by the dermal tissue system, and long-distance transport accomplished by the vascular tissue system, all remaining plant functions are carried out in the ground tissue system, which is composed of a mixture of cells with different functions. The ground tissue system is made up of cells in the **cortex** (the region between the epidermis and the vascular tissues), the **pith** (the area from inside the vascular tissues to the center of the stem or root), and **pith rays** (the areas between bundles of vascular tissues) (refer to Figure 12-6). The photosynthetic cells between the upper and lower epidermis of leaves are also part of the ground tissue system.

The ground tissues are the principal sites of photosynthesis and food and water storage in a plant. Ground tissues also contain collenchyma and sclerenchyma, which provide support for shoot structures.

PLANT ORGANS

All the plant cells, tissues, and tissue systems we have discussed thus far are organized into four plant organs: the **stem, root, leaf,** and **flower.** By necessity, photosynthesis is restricted to aerial organs that are exposed to light, whereas water and minerals are absorbed by root tissues that grow in the water and mineral reservoirs found in soil.

THE STEM: GROWTH, SUPPORT, AND CONDUCTION

The **stem** is an exquisite example of nature's bioengineering. Not only do stems physically support all of a plant's leaves, flowers, fruits, and even its other stems, they even produce the structures they support.

At the tip of every stem is a dome-shaped apical meristem. The end of each stem is divided into three regions:

- The **meristematic region** contains apical meristem cells that divide by mitosis, producing young leaves and juvenile axillary buds (a bud that is directly above each leaf on a stem).
- The **region of elongation** lies just below the meristematic region. This is where cells enlarge and lengthen as they differentiate.
- The **region of maturation**, which lies immediately below the region of elongation, is where nearly all cells have completed enlarging and differentiating.

Monocot and dicot flowering plants differ in the way their stems' vascular tissues are arranged (refer to Table 12-1). Dicot vascular bundles are organized in distinct rings (Figure 12-17), whereas the more numerous monocot vascular bundles are scattered throughout the ground tissues. Monocots and dicots differ in another important way: Perennial and biennial dicots form cambia for secondary growth, whereas monocots do not.

THE ROOT: GROWTH, ABSORPTION, AND CONDUCTION

Much of a plant lies hidden beneath the ground. The extensive nature of a plant's root system is illustrated by a single, tiny rye plant. A rye plant only 25 cm tall (10 in.) has more than 14 million roots which, if placed end to end, would stretch over 609 km (380 miles). Knowing this, you might assume that a massive plant like a giant sequoia would have even more roots than would the rye; surprisingly, this is not the case. The sequoia has fewer roots than

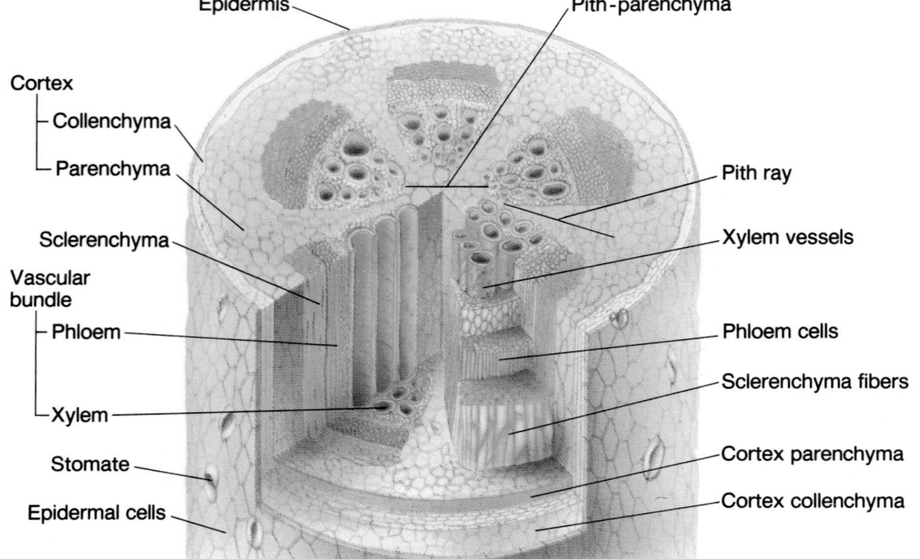

FIGURE 12-17
A three-dimensional, tiered view of the primary tissues in a dicot stem.

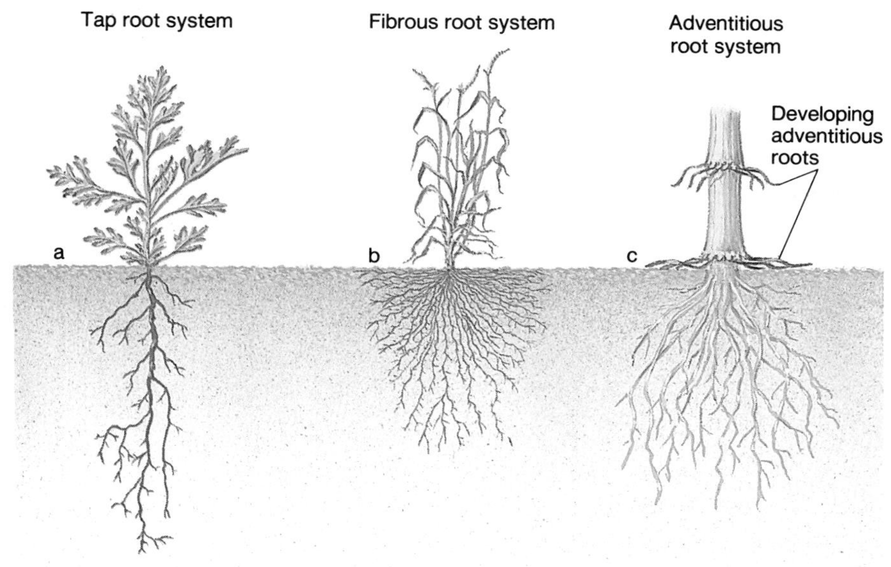

FIGURE 12-18

Three types of root systems in plants. ***(a)*** A tap root system consists of one main root with many branching lateral roots. ***(b)*** A fibrous root system has more than one main root, each with many branching lateral roots. Fibrous root systems are generally shallow and spread out from the stem base. ***(c)*** In an adventitious root system, roots arise from organs other than the root itself, mainly from stems or leaves. Adventitious roots develop when plant cuttings are made.

does the rye plant, but the entire root system of a giant sequoia is much larger than is that of the rye. This unexpected disparity exists because the rye and giant sequoia have different types of root systems. The giant sequoia has a **tap root system** (Table 12-1), which is similar to that of a carrot, with its one main root and many smaller branches, called **lateral roots** (Figure 12-18a). Like the giant sequoia, most dicots have a tap root system. In contrast, the rye is a monocot and has a **fibrous root system,** which is composed of many main roots—more than 140 for rye (Figure 12-18*b*). In both tap and fibrous root systems, main roots produce lateral roots; lateral roots form more lateral roots; and so on. This repeated branching of roots generates many root tips—locations of **root hairs,** extended surface cells through which water and minerals are absorbed (Figure 12-19).

Sometimes roots arise in unexpected places—on stems or even leaves. Roots that develop directly from shoot-

FIGURE 12-19

Root hairs give this radish seedling its fuzzy appearance. The enlargement (inset) shows a root cell beginning to expand to form a root hair.

tissues form an **adventitious root system.** For example, corn plants develop prop roots from the base of their stems (Figure 12-18*c*), which support the shoot when it is burdened with heavy clusters of fruits (what we call a "corn cob").

Unlike stem apical meristems, root apical meristems are not located at the very tip of each root (Figure 12-20). Instead, the tip of a root has a **root cap**—a protective cellular "helmet" that surrounds delicate meristematic cells and shields them from abrasion as the root grows through the soil. The root's apical meristem continuously regenerates its root cap as cells are worn away or pierced by sharp soil particles. Root-cap cells also help the root penetrate the soil by manufacturing and secreting a gelatinlike substance that helps lubricate the root. This secretion also creates a favorable habitat for certain soil bacteria that supply the plant with essential elements, especially nitrogen.

In addition to forming its own protective cap, the root apical meristem also produces all the cells that will differentiate into the root's primary tissue systems. As in stems, roots have a *meristematic region,* a *region of elongation,* and a *region of maturation* (Figure 12-20).

Roots contain two important internal tissues that are not found in stems—an endodermis and a pericycle—both of which are cylinders of cells that encircle the vascular tissues (Figures 12-20 and 12-21). The **endodermis** is the innermost layer of the cortex. Early in its development, each endodermal cell is encircled by a band of waxy suberin, forming the **Casparian strip.** Water-repelling Casparian strips of adjacent endodermal cells are aligned, creating a "gasket" that seals the space between adjacent cells so that water and minerals can only pass through living endodermal cells. By changing their solute concentration, endodermal cells control the uptake of water and minerals and prevent water from leaking out of the root and back into the sometimes dry or salty soil.

Like the endodermis, the **pericycle** is one cell layer thick. The pericycle lies immediately inside the endodermis (Figure 12-21) and divides to form lateral roots. In a perennial or a biennial plant, the pericycle also contributes to the formation of a vascular cambium in the root for secondary growth.

In general, plants absorb water and minerals through their root hairs. Since water and minerals are usually widely dispersed in the soil, plants often develop extensive root systems with many root tips to mine the soil for these scant resources. In addition, more than 90 percent of vascular

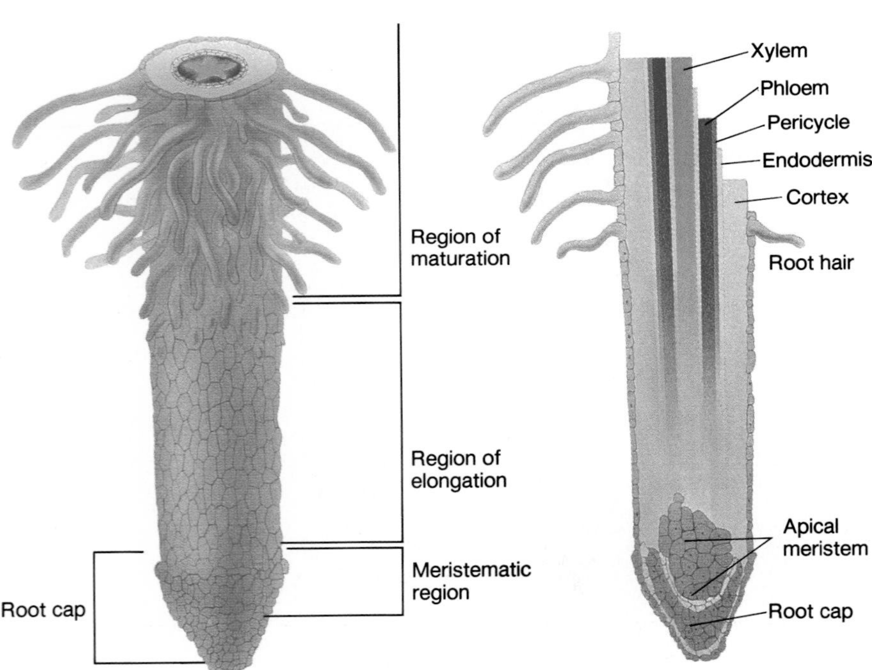

FIGURE 12-20
The root apex. As in stems, the root apex includes three regions—the meristematic region of active cell division, the region of elongation where cells enlarge in length as well as width and begin differentiating, and the region of maturation defining the area where most cells, including root hairs, have completed differentiation. Water and minerals are absorbed in the region of maturation.

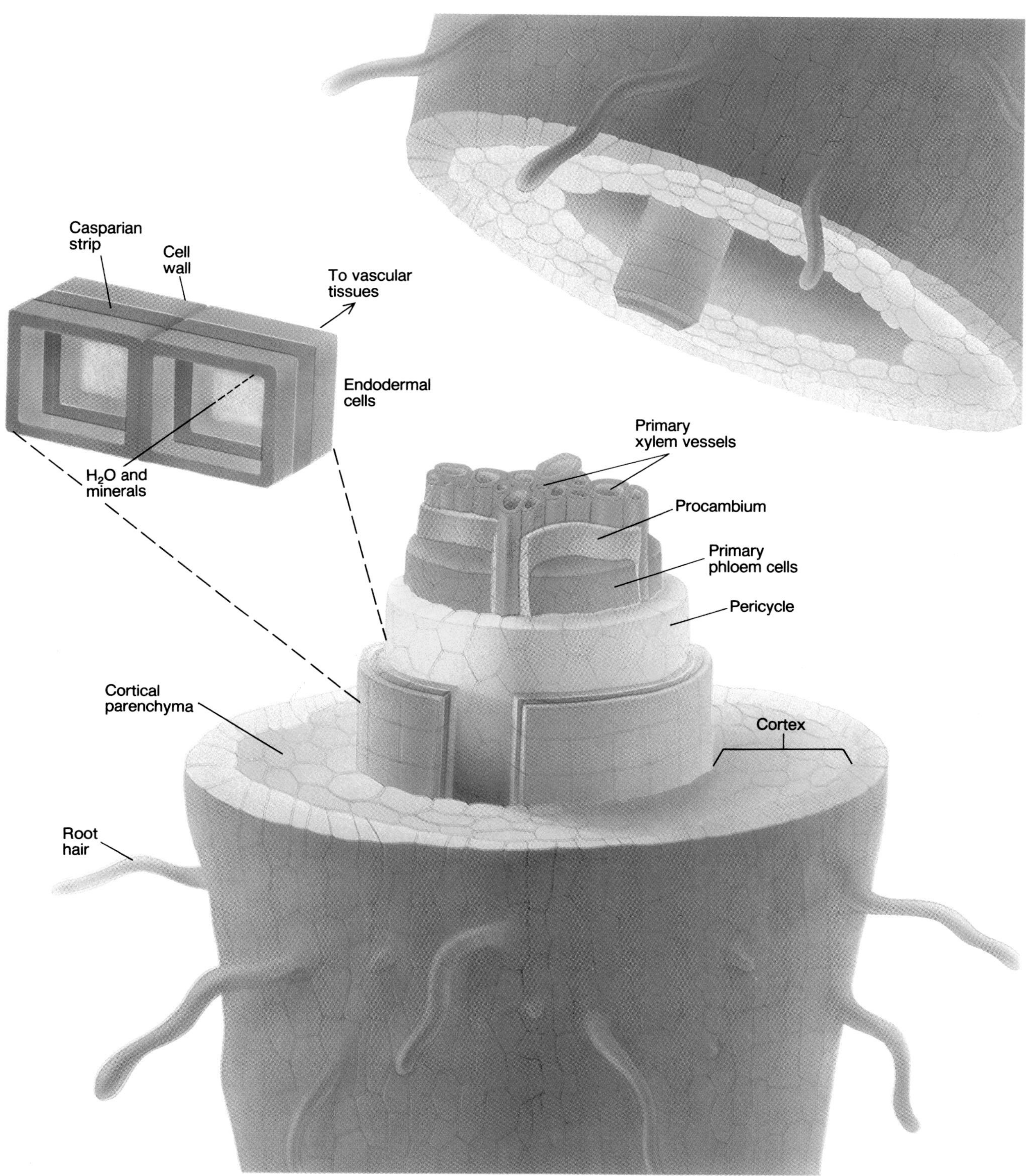

FIGURE 12-21

A three-dimensional, tiered view of the primary tissues in a dicot root. The innermost layer of cortex cells is the endodermis. A Casparian strip encircles each endodermal cell, enabling endodermal cells to control water movement. The pericycle is a cylinder of cells immediately inside the endodermis. The pericycle produces lateral roots or forms part of the vascular cambium during secondary growth.

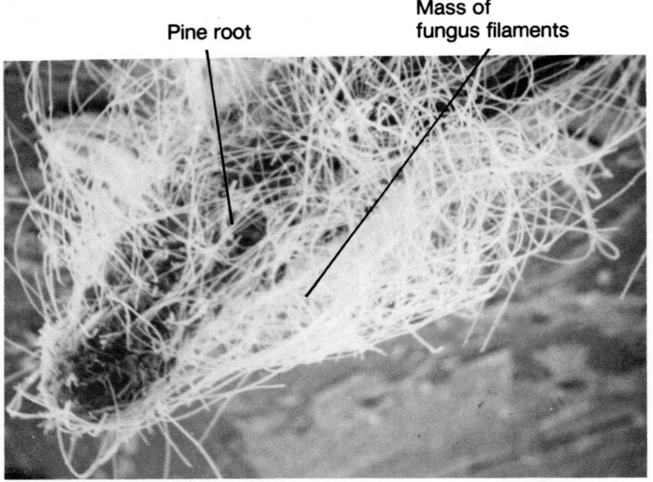

FIGURE 12-22

Mycorrhizae: An intimate association with mutual benefits. The filaments of a soil fungus join with the roots of plants to form mycorrhizae. The fungus absorbs some of the food (carbohydrates, amino acids) stored in the root, and the plant receives water and minerals absorbed by widespread fungal filaments. In most mycorrhizae, a mass of fungal filaments forms a tangled sheath around the root tip, with some penetrating root cells.

plants form root "partnerships" with soil fungi. In these fungus–root associations, called **mycorrhizae** (*myco* = fungus, *rhiza* = root), fungal filaments often form a tangled sheath around the root, with some hyphae penetrating into the root's cortex cells (Figure 12-22). Mycorrhizal associations increase a plant's ability to extract water and minerals from soil because fungal hyphae extend further into the surrounding soil than does the plant's root system.

Water enters a plant's roots by osmosis, whereas dissolved minerals are actively transported into root cells through the cell membrane. Once inside the root, water travels through the root cortex in one of two ways: along cell walls, or from cell to cell via connecting passages called *plasmodesmata* (Figure 12-23). Once water reaches the endodermis, however, aligned, waxy Casparian strips prevent further movement of water between or along cell walls. Water then moves through the endodermal cells as they actively transport minerals into interior cells; that is, water flows along an osmotic gradient. Water continues to follow the inbound path of actively transported minerals into a conducting cell in the xylem. Once inside the xylem,

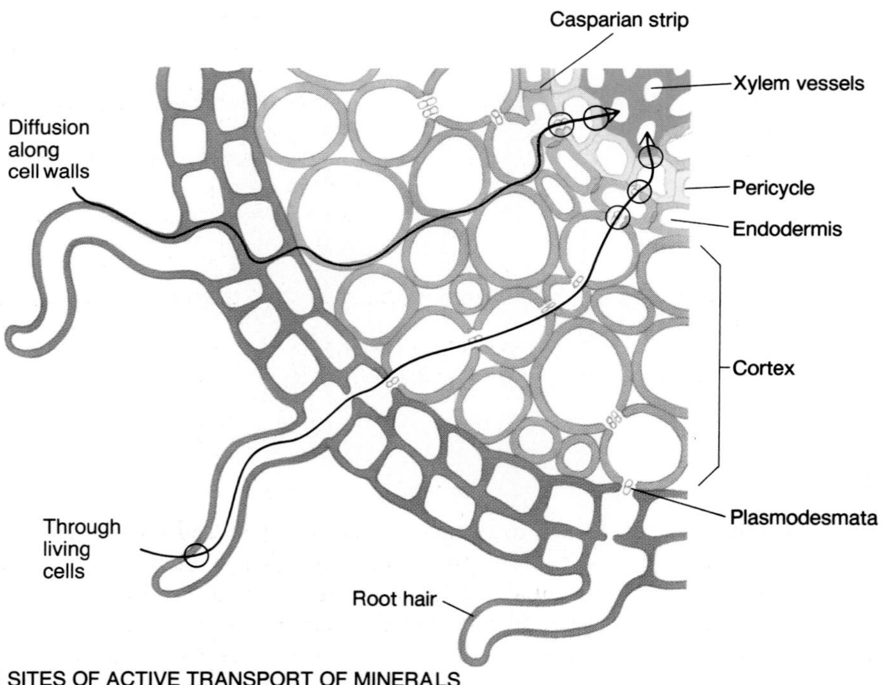

FIGURE 12-23

Roadmap to the xylem. Water enters roots by osmosis, whereas most dissolved mineral ions are actively transported into root cells. Water moves through living cortex cells or along cortex cell walls but is prevented from moving between endodermal cells by adjoining, waxy Casparian strips. Minerals are actively transported through the endodermis, the pericycle, the procambium, and into xylem-conducting cells, creating an osmotic gradient. Water flows along this gradient into the xylem.

TABLE 12-2

PRIMARY USES OF THE 16 ESSENTIAL PLANT NUTRIENTS

Macronutrients	*Function Symptoms*	°%	*Deficiency Dry Weight*
Carbon (C)	Major component of organic molecules	45.0	Severely impaired growth
Oxygen (O)	Major component of organic molecules	45.0	Severely impaired growth
Hydrogen (H)	Major component of organic molecules	6.0	Severely impaired growth
Nitrogen (N)	Component of amino acids, nucleic acids, chlorophyll, and coenzymes	1.0 - 4.0	Yellowing of older leaves; plant spindly
Potassium (K)	Component of proteins and enzymes; helps regulate opening and closing of stomata	1.0	Mottled, yellow older leaves which die back at tips and margins
Calcium (Ca)	Component of middle lamella; enzyme cofactor; influences cell permeability	0.5	Youngest leaves die back at tip and margins; deformed leaves
Phosphorus (P)	Component of ATP, ADP, proteins, nuceic acids, and phospholipids	0.2	Youngest leaves marked with purple
Magnesium (Mg)	Component of chlorophyll; enzyme activator, amino acid and vitamin formation	0.2	Yellowing between leaf veins of older leaves; leaves may turn orange
Sulfur (S)	Component of proteins and coenzyme A	0.1	Yellowing of young leaf veins bright red; stunted
Micronutrients			
Iron (Fe)	Catalyst for chlorophyll synthesis; component of cytochromes	0.01	Yellowing between veins of youngest leaves.
Chlorine (Cl)	Osmosis and photosynthesis	0.01	Wilting of leaf tips; yellow leaves with bronze color
Copper (Cu)	Enzyme activator and component; chlorophyll synthesis	0.0006	Dark green leaves with black spots at tips and margins
Manganese (Mn)	Enzyme activator; component of chlorophyll	0.005	Black spots on veins in youngest leaves
Zinc (Zn)	Enzyme activator; influences synthesis of hormone, chloroplasts, and starch	0.002	Reduced internodes; leaf margins puckered
Molybdenum (Mo)	Essential for nitrogen fixation	0.0001	Yellowing between veins of older leaves with black spots
Boron (B)	Important to flowering, fruiting, cell division, water relations, hormone movement, and nitrogen metabolism.	0.002	Black spots on young leaves and buds

° % Dry Weight: Proportion of weight after all water has been removed.
Memory Joggers:To help remember the macronutrients, use the phrase "C. HOPK(*i*)NS Ca*r is an* MG."
To help remember the micronutrients, use the phrase "*A* Fe*stive* MoB *comes in* (= CuMnZn) Cl*apping*."

water and minerals become part of a stream that conducts them through the plant.

In the nineteenth century, two independent researchers, J. Sachs and W. Knop, grew plants by immersing their roots in a solution of inorganic salts rather than soil (a technique now referred to as *hydroponics*). The studies proved for the first time that plants could receive all their needs from sunlight and inorganic elements. Since that time, researchers have used hydroponics to grow plants in solutions in which only one element is missing at a time. From these studies, botanists have identified 16 **essential nutrients**—elements that are absolutely necessary for plants to complete their life cycle and for normal growth and development to occur (Table 12-2). Nine of these essential nutrients are needed in very large amounts; as such, they are called **macronutrients.** The seven remaining **micronutrients** are needed only in small amounts.

Minerals are actively transported into roots, even when mineral concentration is lower in the root than it is in the soil or surrounding solution. Once inside the root, dissolved minerals are actively transported through cortical parenchyma cells, the endodermis, the pericycle, and, finally, into the xylem (Figure 12-23). Because of this one-way active transport path, minerals become concentrated in the xylem. This increase causes water to be drawn into the xylem by osmosis. The continuous influx of water can produce a positive **root pressure** that pushes water and dissolved minerals up a xylem column. Under certain conditions, and only in small plants, root pressure may build high enough to force water and minerals completely out the tips of leaves, a process called *guttation* (Figure 12-24).

Some plants, particularly legumes (peas, beans, clover, alfalfa) form *root nodules*—swellings that house nitrogen-fixing (capable of converting nitrogen gas to ammonia or nitrate) *Rhizobium* bacteria (Figure 12-25). As in the case of mycorrhizal associations, both organisms benefit from this association: The plant receives a supply of nitrogen, and the bacteria receive food, water, and living space.

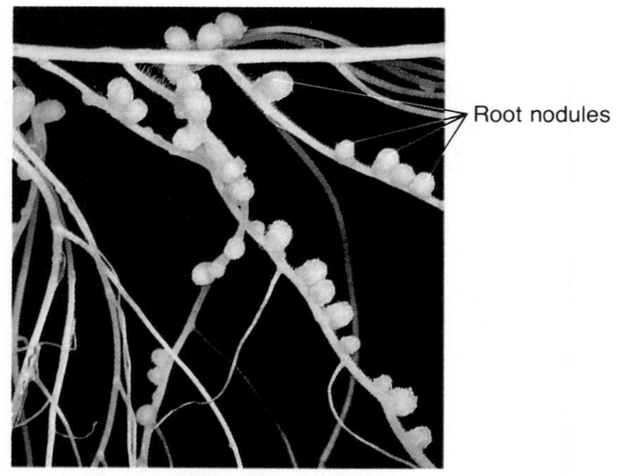

FIGURE 12-25
Root nodules. Some plants house nitrogen-fixing bacteria in lumplike root nodules. The bacteria benefit from this alliance by harvesting some of the plant's carbohydrate stores, and the plant benefits by receiving nitrogen from the bacteria in a form it can use.

THE LEAF: THE PLANT'S PRIMARY PHOTOSYNTHETIC ORGAN

Leaves are the plant's "solar-collectors," "energy generators," and "energy transmission lines," all rolled into one. Leaves capture solar energy; house chloroplasts for converting the energy in sunlight into chemical energy during photosynthesis; and transport materials via an extensive network of interconnecting vascular tissues.

FIGURE 12-24
Forcing the issue. Glistening droplets of water and dissolved minerals are forced out of leaf tips by positive root pressure during guttation.

The leaves of flowering plants typically are made up of a flattened **blade** that collects sunlight for photosynthesis, and a **petiole**—a stalk that connects the blade to the stem. Directly above the spot where a leaf joins the stem is an **axillary bud**—an underdeveloped cluster of cells that will either grow into a new stem with additional leaves and more axillary buds or develop into flowers.

Leaves have an upper and lower epidermis that sandwich the leaf's **mesophyll**—the layers of cells found between the two epidermises (Figure 12-26). In dicot leaves, the mesophyll contain two types of cells: column-shaped **palisade parenchyma** toward the top, and loosely packed **spongy parenchyma** toward the bottom. Air spaces between these mesophyll cells form corridors for diffusion of carbon dioxide for photosynthesis.

The extensive vascular leaf tissues form veins, which create a characteristic venation pattern for different plants. Monocots have leaves with larger veins that run parallel to each other, whereas dicots have highly branched venation (Table 12-1). Like the vascular tissues in stems and roots, the vascular tissues that form the veins of leaves transport water, minerals, and organic molecules throughout a leaf. In addition, the arrangement of vascular tissues in some cases increases the strength of leaves, helping them withstand the stresses of strong winds (see The Human Perspective: Agriculture, Genetic Engineering, and Plant Fracture Properties).

◁ THE HUMAN PERSPECTIVE ▷

Agriculture, Genetic Engineering, and Plant Fracture Properties

Attempts to breed or genetically engineer plants with larger fruits, stems, or leaves for improved productivity could fail if the fracture properties of plant structures is not considered. As a plant's structures increase in size, they are more likely to fracture. But in order to produce large fruits, stems, or leaves that will not fracture, they would have to be so "tough" (filled with supportive cells and tissues) that they may not be palatable.

Some botanists study the strength of plant structures as they are subjected to varying stresses in an attempt to reach some general conclusions about the durability of plant parts; such conclusions may have agricultural and horticultural applications. For example, the shoot system of a plant must sometimes withstand slashing winds and pelting rains. The way a plant resists fracturing under such conditions is the result of its mechanical design, from the cellulose in its cell walls, to specialized support and strengthening cells, to complexes of support tissues. Under normal conditions, the parallel arrangement of sclerenchyma fibers in a leaf petiole (such as a celery stalk) resists damage. If winds are strong enough to inflict damage, however, the cellular structure and arrangement control the fracture in a way that reduces injury, an advantage to an agricultural field of celery on a windy day.

Fracture studies point out a basic principle: The smaller the plant or plant part, the greater its inherent toughness and structural integrity. This principle can be of great importance in agriculture as we try to engineer crops that can grow in regions with severe winds and rains. The hope is that enlarging agricultural areas by growing crops in severe habitats that previously never supported agriculture will increase food supply and begin to solve the world hunger problem.

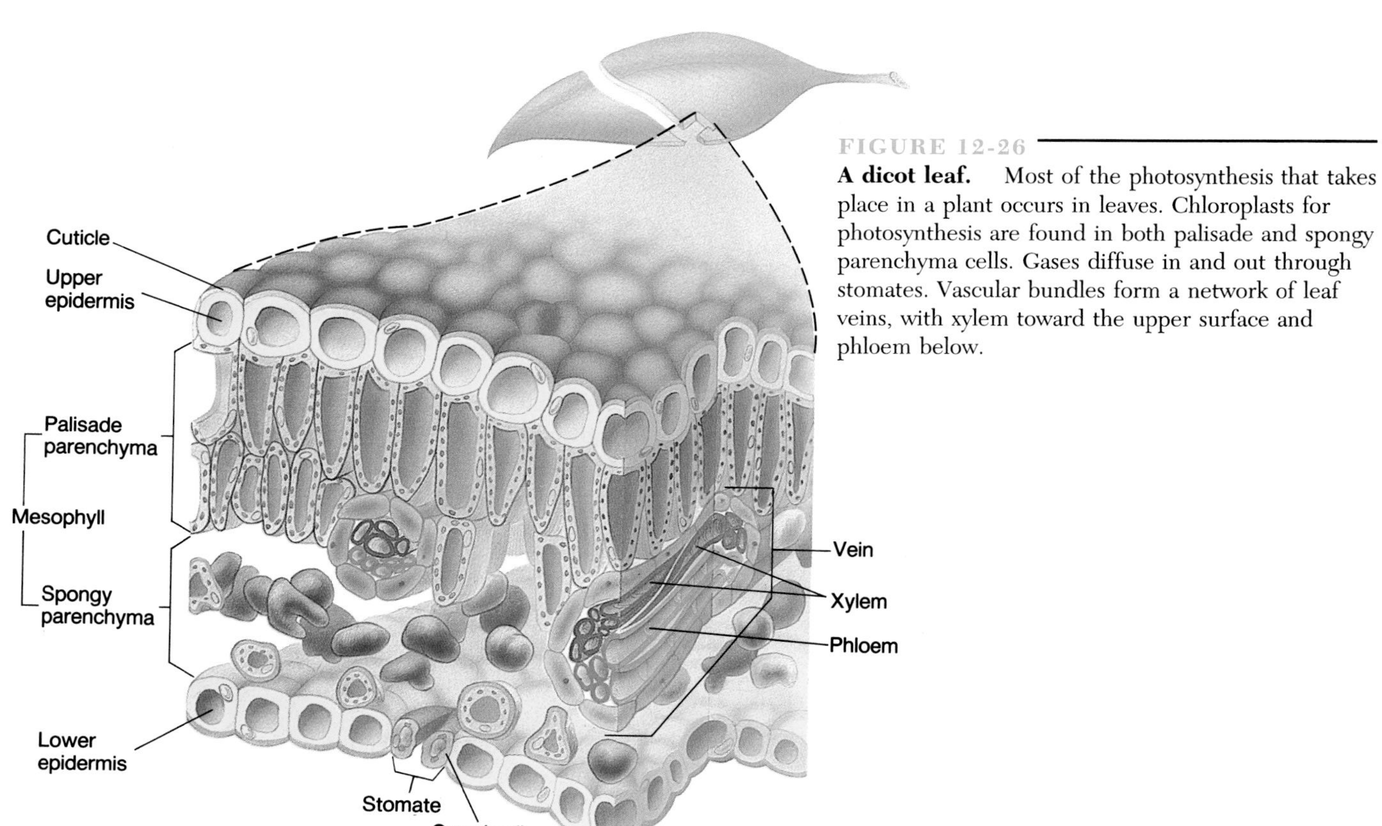

FIGURE 12-26

A dicot leaf. Most of the photosynthesis that takes place in a plant occurs in leaves. Chloroplasts for photosynthesis are found in both palisade and spongy parenchyma cells. Gases diffuse in and out through stomates. Vascular bundles form a network of leaf veins, with xylem toward the upper surface and phloem below.

THE FLOWER: THE SITE OF SEXUAL REPRODUCTION

The balance between form and function is most elegantly displayed in flowers, which are the site of sexual reproduction. Each part of a flower participates in the process of sexual reproduction by either producing gametes or influencing the transfer of gametes so that fertilization takes place. Following fertilization, all or part of the flower develops (ripens) into a fruit that contains seeds with newly formed plant embryos. The fruit protects these embryos and helps disperse them to new habitats. We devote much of the next chapter to discussing the form and functions of flowers and fruits.

EVOLUTION AND ADAPTATION: TYING IT TOGETHER

Flowering plants are considered to be the most evolutionarily advanced plant group. Judgments about whether organisms are "primitive" or "advanced" are based on many criteria, including, but not limited to, fossil record data, similarities and differences in structures, and the means by which efficient structures carry out specific functions. Flowering plants have a number of structures that are more efficient at completing their tasks than are those found in other plant groups. Examples include the flower, for sexual reproduction; fruits, for protecting and disseminating offspring; sieve-tube members, for translocating organic molecules; and vessel members, for transporting water and dissolved minerals and aiding in support of shoot structures.

Water is probably the most important environmental factor that affects the growth, distribution, and health of any land organism. Plants grow in many kinds of habitats, from those that are soaking wet all of the time (a tropical rain forest), to those that receive intense radiation and only scant amounts of rainfall (a desert). Plants in dry, hot environments must control water-vapor losses; otherwise, the plant can easily lose more water than is available from its environment. Guard-cell behavior, stomate number and placement, the presence of epidermal hairs, reflective surfaces, a hydrophobic cuticle layer, and photosynthetic pathways are all adaptations plants have evolved for regulating water-vapor loss.

SYNOPSIS

Flowering plants are the most evolutionarily advanced and structurally complex group in the plant kingdom. Whether they are dicots or monocots, flowering plants form a branched root system that absorbs and conducts water and dissolved minerals and helps support the aerial shoot system. The plant's shoot system produces stems with leaves and axillary buds that develop into more stems or into flowers for sexual reproduction. Leaves are adapted for photosynthesis.

Plants can be classified by their growth pattern. Short-lived annual plants experience only primary growth, whereas longer-lived biennials and perennials experience both primary and secondary growth simultaneously. Primary growth from meristems increases the length of stems and roots, whereas secondary growth from cambia increases stem and root width as older or dead cells are replaced with new, active cells.

Plant tissues form groups of cells that perform the same function. Meristems (apical and cambial) divide by mitosis to produce complex vascular tissues (xylem and phloem) and simple tissues (parenchyma, collenchyma, sclerenchyma). The cells that make up a plant's tissues vary in shape, size, and cell-wall characteristics, enabling the tissues to carry out specific functions.

Some plant cells, mainly xylem and sclerenchyma, function as dead cells. Because the nonliving cell wall is sometimes the most critical structure to a cell's function, cells that support or conduct water continue functioning even after its living contents die.

Parenchyma cells are the most prevalent type of cell in many plants. Versatile parenchyma cells carry out many functions, including photosynthesis and storage, and they have the ability to change into any other plant cell.

The organs of a flowering plant are interconnected by three tissue systems that extend throughout the plant's body. The outer epidermis forms the protective Dermal Tissue System. The internal network of conducting cells forms the Vascular Tissue System of xylem and phloem, which is surrounded by cells that make up the Ground Tissue System.

A vascular plant's circulatory system includes two sets of conduits. Water and dissolved minerals are conducted through the xylem primarily by transpiration pull; organic substances synthesized by the plant are translocated through the phloem by pressure flow.

Flowering plants contain four organs—root, stem, leaf, flower—each adapted for different functions. Although their overall functions may differ (roots absorb water and minerals, stems produce and support leaves for photosynthesis, and flowers are the sites of sexual reproduction), all four organs contain vascular tissues for the transport of water, minerals, and organic molecules throughout the plant's tissues.

Review Questions

1. Once a cell dies, it cannot carry out any metabolic functions. But plants contain cells, such as xylem vessels and sclerenchyma fibers, that continue to function even after the living part of the cell is dead. List the types of functions these dead cells carry out. What characteristics do dead, functioning cells have in common that enables them to continue to function even after the cell dies?
2. Why are parenchyma cells sometimes called "the most important type of cell in a plant"?
3. Using the following list of terms, identify with an "S" those that are found only in the shoot system, with an "R" those that are found only in the root system, and with a "B" those found in both the shoot and root systems.

apical meristem	collenchyma	sclerenchyma
parenchyma	node	vessel member
bundle sheath	Casparian strip	axillary bud
companion cell	lenticel	vascular cambium

4. Why do biennials and perennials experience simultaneous primary and secondary growth, while annual plants experience only primary growth?
5. List five differences and five similarities between the tissues in the stem and the tissues in a root.

Critical Thinking Questions

1. As described in Steps to Discovery, it was a common technique for researchers to strip off the bark of a tree—a technique referred to as girdling—to investigate how materials move through a plant. Why do girdled trees eventually die? Why don't they die immediately?
2. On a nature walk, you find an herbaceous plant that has leaves with parallel veins and flowers with 12 petals. There are no fruits or seeds, so it's impossible to determine whether the embryo has one or two cotyledons. You suspect that the plant is a monocot, but why? What other characteristics would you look for to verify your hypothesis?
3. Plants are either annuals, biennials, or perennials. What is the adaptive value for each of these lifespan strategies? Compare desert, seashore, and mountain environments. Would one strategy have increased survivability and reproduction over the others in each of these different environments?
4. Most plants will die if the water content of their cells drops below 60 percent. Discuss some of the possible causes of death from dehydration.
5. Transpiration rate is affected by many environmental conditions, including temperature, wind, light, relative humidity, and water availability in the soil. For the following environments, rank each of these factors from most important to least important in *increasing* the rate of transpiration: tropical rain forest, desert, and arctic tundra (refer to Chapter 28 for descriptions). Justify your rankings.

CHAPTER 13

Plant Life: Reproduction and Development

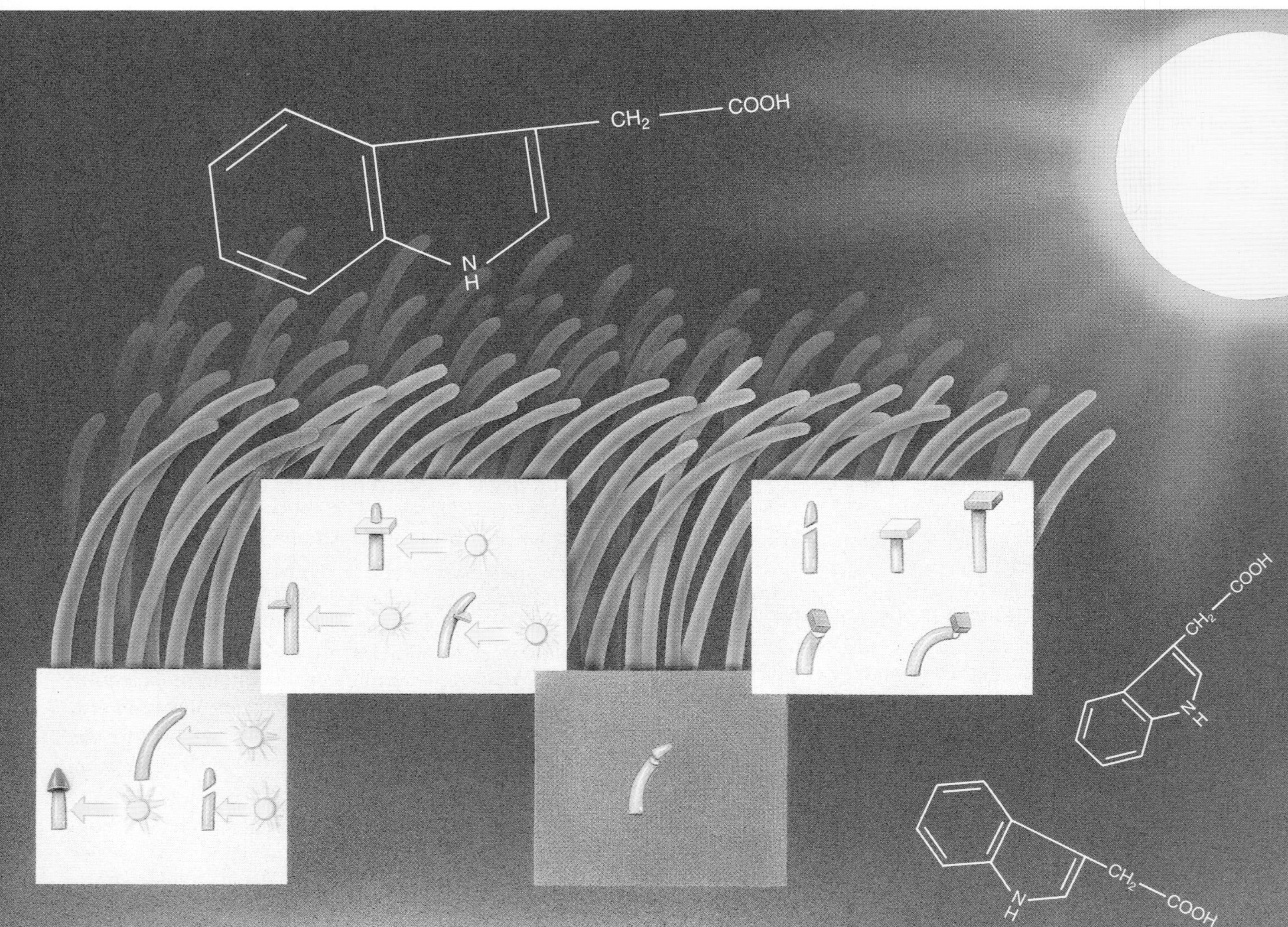

By varying experimental conditions depicted in the boxes, investigators discovered that a growth hormone produced at the tips of stems causes plants to bend toward light.

STEPS TO DISCOVERY
Plants Have Hormones

Although we are likely to associate Charles Darwin only with the theory of evolution and natural selection, Darwin actually completed a number of experiments that also added to our knowledge of plant growth and development. Some of his findings were published in 1881 in a book entitled *The Power of Movement in Plants.* Many of Darwin's experiments focused on the growth and movements of plants in response to gravity, touch, chemicals, and light, helping to confirm that plants—like animals—manufacture hormones.

In the 1880s, Darwin collaborated with his son Francis in a study on the bending of plants toward light. The Darwins tested the growth responses of young canary-grass seedlings to light shining from one direction.

The Darwins demonstrated that light coming from only one direction caused the young shoot to bend toward the light. Like many other grasses, canary-grass seedlings form a protective sheath (coleoptile) that covers the plant's young leaves and apical meristem. When the tip of the coleoptile was removed or covered with foil, the shoot did not bend. The Darwins concluded that the coleoptile tip somehow affected cell growth, causing the shoot to bend toward the light. The Darwins knew that most plant growth was the result of cell elongation, which takes place several millimeters below the coleoptile tip. In order to explain how the coleoptile tip influences elongation several millimeters away, the Darwins suggested that a growth signal produced in the coleoptile tip traveled down the shoot, where it triggered cells to enlarge differentially. It is the differential elongation of cells, they reasoned, that caused the young shoot to bend toward light. The identity of the signal and the way it affected cell elongation was beyond the scope of the Darwins' experiments, however.

In 1910, Peter Boysen-Jensen, a professor of plant physiology at the University of Copenhagen, designed studies that added new information on the nature of the growth signal. In two sets of experiments, Boysen-Jensen demonstrated that the growth signal was most likely a chemical that caused cells on the dark side of the illuminated shoot to elongate more than those on the lighted side. In his experiment, Boysen-Jensen removed the plant's coleoptile tips, positioned a block of water-permeable gelatin on the cut shoot, and then placed the coleoptile tips on top of the gelatin. When the seedlings were exposed to light, the shoots curved toward the direction of the light, demonstrating that the growth signal had traveled through the gelatin (otherwise, there would not have been curved growth). Because dissolved chemicals travel through gelatin, Boysen-Jensen concluded that the growth stimulus was a chemical.

In another experiment, Boysen-Jensen inserted pieces of water-insoluble mica (a thin piece of quartz or other silicate mineral) part way into either the lighted side or the dark side of the shoot. Mica blocks the movement of all dissolved substances. Only those shoots in which the mica was inserted on the lighted side bent toward the direction of light. Because none of the shoots bent when mica was inserted on the dark side, Boysen-Jensen concluded that the mica blocked the downward movement of the growth stimulus. Under normal circumstances, the growth-stimulating substance traveled down the dark side of the shoot. Therefore, stem bending was the result of the greater concentration of growth stimulus on the dark side than on the illuminated side. That is, a higher concentration of growth stimulus on the dark side triggered greater cell elongation so the shoot bent toward the light.

In 1918, A. Paal's work in Hungary provided further evidence that the growth-stimulating substance in the coleoptile tip was responsible for shoot bending. Paal removed coleoptile tips and then displaced them to one side of the shoot or the other. All plants were kept in the dark. It didn't matter on which side of the shoot Paal placed the severed tip; the shoot always bent in the opposite direction. Paal's study demonstrated that the growth stimulus originated from the coleoptile tip and promoted elongation of the cells immediately below the displaced tip, causing the shoot to bend.

The proof that the growth stimulus was a chemical was finally demonstrated in experiments conducted by Fritz Went in 1926. In his experiments, Went placed severed coleoptile tips onto agar (a gelatinous substance extracted from seaweed), allowing the growth substance ("juice") to diffuse into the agar and then discarded the tips. He then placed a small block of the agar with juice on a shoot that had had its coleoptile tip removed. As a control, he also placed plain agar blocks (no coleoptile juice) on decapitated shoots. None of the stems with plain agar elongated, while all of the stems with coleoptile juice in the agar grew straight up. When Went offset the agar blocks with coleoptile juice on the decapitated shoots, the shoots bent in the opposite direction. Went's explanation was that the growth substance had diffused into the agar, stimulating cell growth. If all the shoot's cells were stimulated, the shoot grew straight up; if cells on only one side of the shoot were stimulated, the shoot bended. Went named the growth-promoting chemical in the juice *auxin*.

*I*t could be a scene from a horror movie. A victim is lured and then knocked into a pool of water. The sides of the pool are so high and slippery, it is impossible to escape. Eventually, a narrow passageway is found at the back of the pool, and the victim squeezes through it and escapes. But the ordeal is far from over.

As the victim emerges from the passageway, two bulging sacs are attached to his back. As he begins to flee, the victim is lured to yet another pool, unable to resist the same powerful force. The events that follow are the same: The victim is dunked into the pool; he makes several unsuccessful attempts to climb out the sides; and he eventually escapes through a narrow exit. This time, however, the two sacs are gone.

This is not a scene from a movie but a real-life drama between a male euglossine bee (the "victim") and a bucket orchid flower, *Coryanthus leucocorys* (the "captor"). During its ordeal, the bee inadvertently transfers

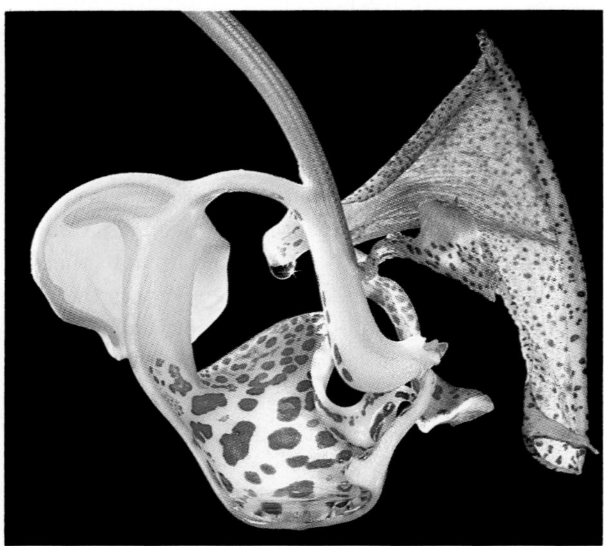

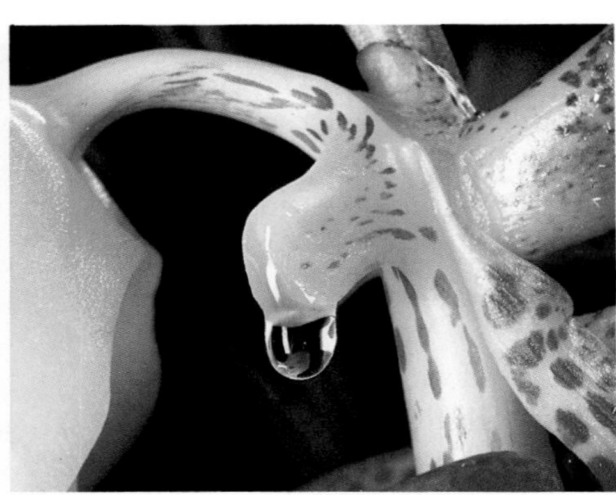

(a)

(b)

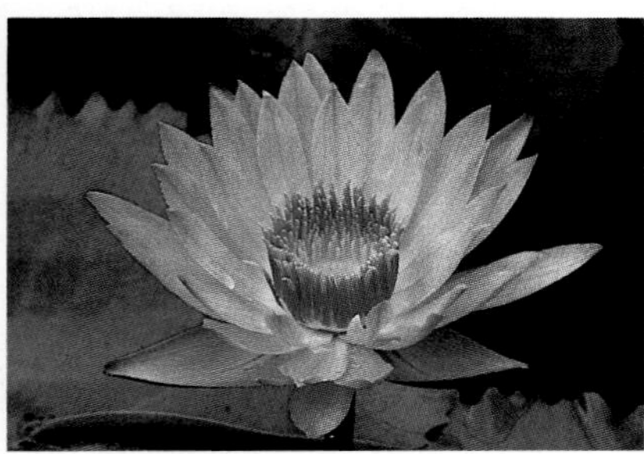

(c)

FIGURE 13-1

Flower power. The sizes, shapes, colors, and markings of flowers are adaptations that help improve the chances for successful pollen transfer for sexual reproduction. ***(a)*** The bucket orchid (*Coryanthus leucocorys*) attracts and dunks male bees into a pool of water. As the bee escapes through the back of the flower, it deposits or retrieves bags of pollen grains. ***(b)*** An African daisy is not a single flower but a miniature bouquet of tiny, densely packed flowers that forms a *composite.* The outer ray flowers attract bees and butterflies to the inner, inconspicuous flowers for pollination. ***(c)*** The South African waterlily attracts hoverflies, bees, and beetles which carry pollen from flower to flower. Pollen-laden animals are trapped and eventually drown, and the pollen washes off and settles to the bottom of the flower.

the orchid's male gametes from one flower to another (Figure 13-1*a*). Like other flowering plants, *Coryanthus* houses male gametes in **pollen grains**—the contents of the bulging sacs attached to the bees back. Male bees are attracted to the bucket orchid not only because of its alluring fragrance but also because the flower produces a waxy substance the male bee uses to makes its own sexy "perfume," which he uses to attract female euglossine bees. Thus, sexual reproduction of the orchid and the bee are inseparably intertwined; each depends on the other to form a critical line in a chain that leads to the sexual reproduction of its species.

Unlike most animals, most flowering plants have the ability to reproduce new individuals both sexually and asexually (without the fusion of egg and sperm gametes). Asexual, or vegetative, reproduction is the result of mitosis and, as a result, produces offspring that are always genetically identical to the parent plant. Asexual offspring form from *runners* or *stolons* (stems that grow over the soil surface), such as in strawberries and spider plants, from *rhizomes* (stems that grow at or below the soil surface), such as in crab grass and bamboo, or from vertical stems (African violets), leaves (piggyback plants), bulbs (garlic, daffodils, tulips), or tubers (potato, Jerusalem artichoke, tuberous begonias).

FLOWER STRUCTURE AND POLLINATION

The flower is the center of sexual activity for all flowering plants (Figure 13-1). Flowers come in an array of sizes, shapes, colors, color patterns, and arrangements, each representing an adaptation that promotes pollen transfer for sexual reproduction. Pollen grains contain the plant's male gametes, or sperm. The transfer of pollen grains between flowers is called **pollination.** Many plants rely on insects for pollination; others are adapted to utilizing wind, water, or other types of animals, such as birds and even a few mammals. Most animal-pollinated flowers produce food (usually nectar and pollen) that attracts the animal to the flower.

To understand how flowers accomplish sexual reproduction, we need to investigate the structure of the flower itself. A flower, such as the one illustrated in Figure 13-2, is a cluster of highly modified leaves attached to a shortened stem, called the **pedicel.** The flower parts emerge from a **receptacle—**the widened end of the pedicel which forms the flower base. A typical flower is made up of four groups of modified leaves—*sepals, petals, stamens,* and a *pistil*—each functioning in some way to achieve sexual reproduction.

Generally, the **sepals** surround and protect the flower bud as it develops. The collection of individual sepals make

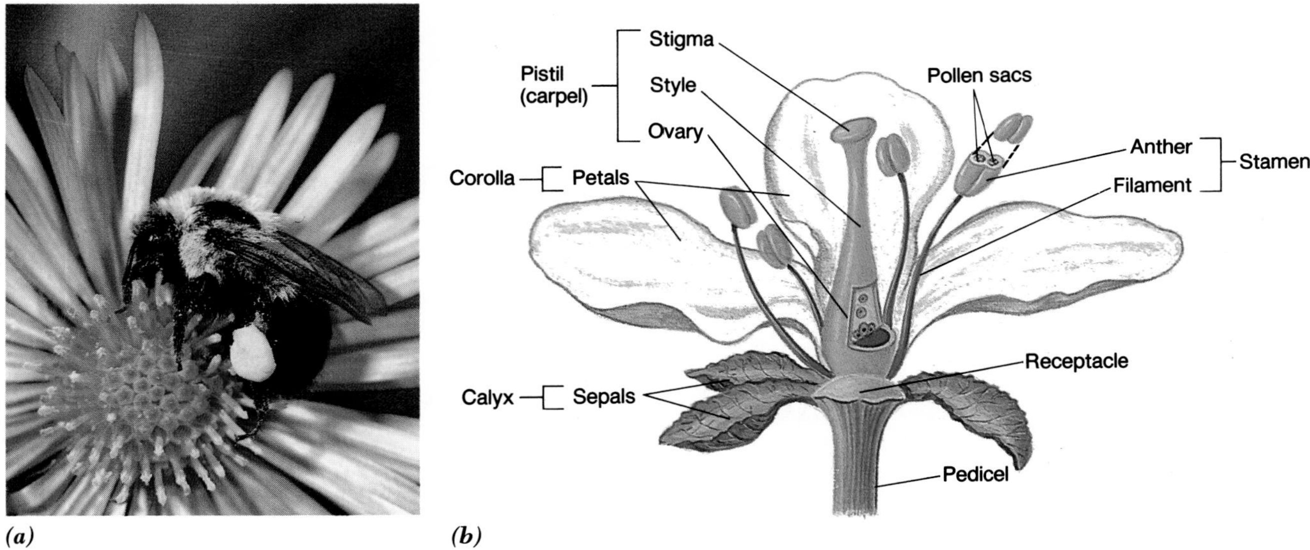

FIGURE 13-2

Structures of a typical flower. ***(a)*** Flowers produce the plant's gametes for sexual reproduction as well as attract pollinators that transport male gametes from flower to flower. ***(b)*** The anthers of stamens form pollen, which eventually produce the sperm gametes. The pistil is the female part of a flower; the stigma collects pollen, the style provides a channel that directs the growth of a pollen tube to the egg, which is produced in the ovary. Following fertilization, the ovary develops into a fruit that protects, nourishes, and helps disperse developing embryos contained in its seeds.

up the plant's *calyx*. To discourage hungry animals, the sepals of some flowers contain a distasteful latex, a milky white fluid that quickly deters animals from biting into the nutrient-rich flower bud. Bright or patterned **petals** distinguish the flower from its surroundings, catching the attention of pollen-transferring animals. The petals collectively make up the flower's *corolla.* Although some petal colors and markings are invisible to humans, they are readily spotted by the animals they attract (Figure 13-3).

The **stamens** are the flower's male reproductive structures; they produce pollen grains, which contain sperm. Most stamens are made up of a slender stalk—the **filament**—with a swollen end, called the **anther.** Pollen grains are produced inside the anther. When pollen matures, the anthers split open, releasing the pollen.

The **pistil** produces the plant's egg gametes. Each pistil, composed of one or more *carpels,* has three parts:

- the **stigma,** or sticky area to which pollen adheres;
- the **style,** or tube that connects the stigma to the ovary; and
- the **ovary,** or enlarged bottom part of the pistil, where eggs are produced and embryos and seeds develop.

In plants adapted to utilizing wind for pollination, the stigma is often enlarged and feathery, creating a large surface that gleans pollen from the air (Figure 13-4*a*). In flowers that utilize animals for pollination, the stigma is shaped in such a way that it contacts the pollen-laden areas of the animal's body (Figure 13-4*b*).

A flower that contains all four groups—sepals, petals, stamens, and pistil—is a **complete flower,** as opposed to an **incomplete flower,** which lacks one or more of these four basic parts. Roses and daisies are examples of complete flowers, whereas dogwood and palms have incomplete flowers. Some plants produce flowers with only stamens (male flowers) or only pistils (female flowers). A flower that lacks either stamens or pistils, or both, is called an **imperfect flower,** as opposed to a **perfect flower,** which has both stamens and pistils. In some plants, separate male and female flowers are produced on one plant, whereas other plants produce male and female flowers on separate individuals. For example, corn has tassels of male flowers and separate cobs of female flowers on the same plant, whereas eggplant and other squashes produce male or female flowers on separate individuals. Producing different-sexed flowers on different plants guarantees that eggs cannot be fertilized by sperm from the same individual, thus promoting genetic variability.

Many flowers have **nectaries**—glands that secrete sugary nectar that attract animals for pollination. Some animals collect both nectar and protein-rich pollen. As an animal laps up nectar and collects pollen, some of the

(a)

(b)

FIGURE 13-3

A Bee's-eye view. Humans and bees see different wavelengths of light, as these two photos of a marsh marigold (*Caltha palustris*) illustrate. ***(a)*** To us, this marsh marigold is almost uniformly yellow. ***(b)*** To bees, the petals have distinct lines that converge in the dark center where the nectar (and stamens and pistils) are found. These nectar guides absorb UV wavelengths, while the remaining parts reflect UV (we do not see UV wavelengths, but bees do).

(a)

(b)

FIGURE 13-4

Improving the odds. ***(a)*** Wind-pollinated plants, like this grass, produce tremendous quantities of pollen and form large feathery stigmas. Together, these characteristics improve the chances of pollination. ***(b)*** This two-lipped monkey flower stigma (*Mimulus*) "licks" pollen off the heads of entering bees and then quickly closes to exclude pollen from its own stamens as the bee exits.

pollen adheres to its body. Then, as the animal moves from flower to flower, pollen is carried from the animal's body to the flower's stigma, completing pollination and setting the stage for fertilization.

FORMATION OF GAMETES

Whether it takes place in an animal or a plant, gamete formation always begins with meiosis (Chapter 7). In plants, special diploid mother cells inside the anthers and ovaries of flowers divide by meiosis to form haploid *spores*. The spores then divide by mitosis to form a haploid, multicellular gametophyte, which produces the sperm or egg gametes for sexual reproduction. In the ovary, spores develop into **embryo sacs,** the female gametophytes. In the anther, spores develop into pollen grains, the male gametophytes.

Female gametophytes form within **ovules**—round masses of cells that are attached by a thin stalk to the inner surfaces of the flower's ovary. The outer cells of the ovule form one or two protective layers that leave a small opening, the *micropyle,* through which the pollen tube grows, to deliver sperm.

The production of the egg gamete begins when a diploid mother cell inside the ovule divides by meiosis (Figure 13-5). Three of the four resulting spores degenerate. The nucleus of the surviving spore divides three times by mitosis to produce the female gametophyte, which contains eight haploid nuclei. (The first mitotic division produces two nuclei; the second, four; and the third, eight.) The eight nuclei become partitioned into seven cells, one of which contains two nuclei; only one of the cells—that nearest the micropyle—becomes the egg. The cell that contains two nuclei is called an **endosperm mother cell,** which later proliferates into the **endosperm,** an important tissue that nourishes the embryo and the growing seedling.

Pollen grains form in the anthers of stamens (Figure 13-6). Diploid cells in the anther divide by meiosis to produce four spores, each of which divides once by mitosis, forming a two-celled pollen grain. One of the cells—the *generative cell*—is enclosed within the other—the *tube cell;* the generative cell divides by mitosis to produce two sperm.

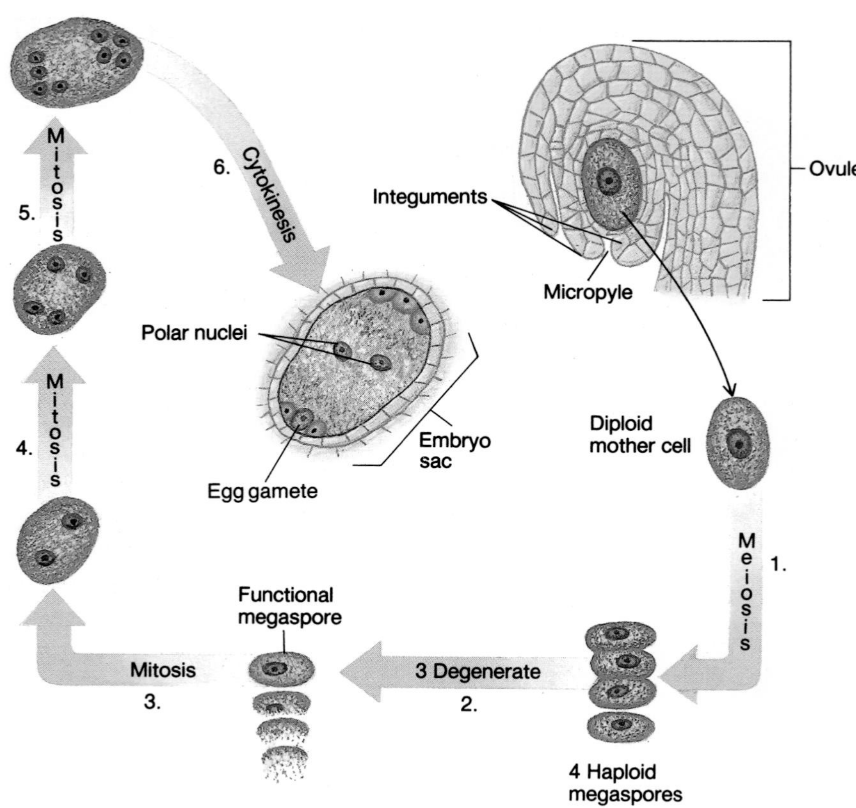

FIGURE 13-5
Formation of the female gametophyte is a six-step process that takes place inside an ovule. (1) A diploid mother cell divides by meiosis to produce four haploid spores. (2) Three spores degenerate, leaving a single functional spore. The nucleus of the surviving spore divides by mitosis three times (3, 4, and 5) to form an eight-nucleate gametophyte. Cytokinesis (6) partitions the nuclei into seven cells, producing the mature embryo sac with a single egg gamete.

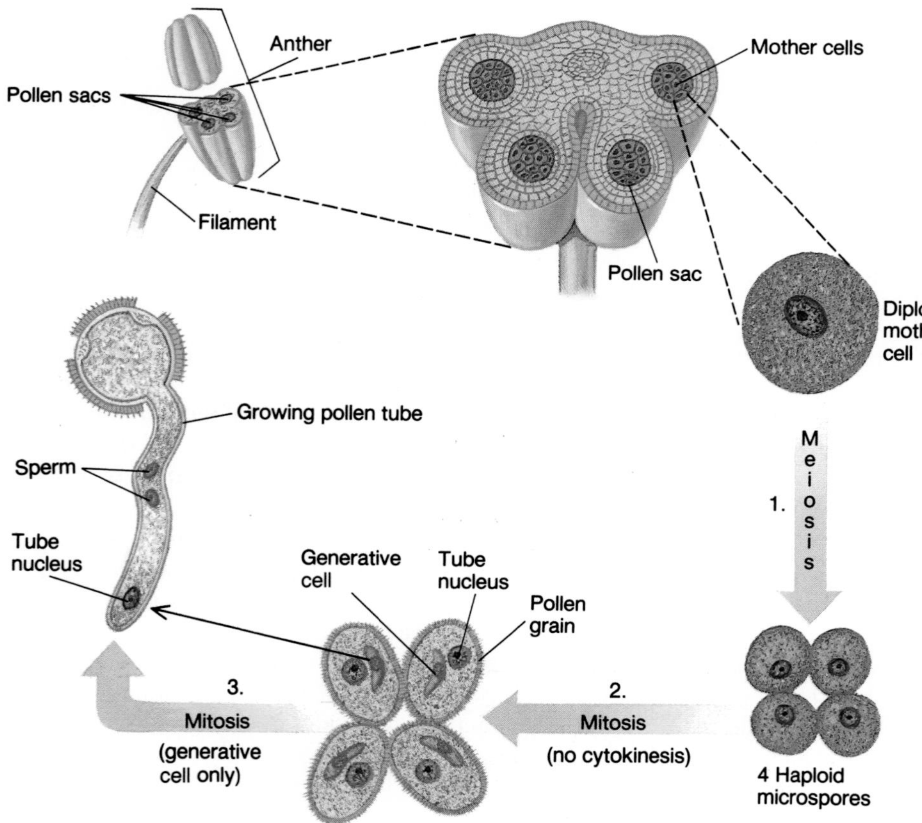

FIGURE 13-6
Formation of the male gametophyte is a three-step process that takes place inside the anthers of stamens. (1) A mother cell divides by meiosis to produce four haploid spores. (2) Each spore divides by mitosis to a pollen grain containing a tube cell and a generative cell. (3) When deposited on a stigma, the tube cell forms a pollen tube while the generative cell divides to form two sperm.

FERTILIZATION AND DEVELOPMENT

After pollination, the pollen grain lies on top of the stigma, while the egg is buried deep within the ovary. So how do the gametes ever get together? When pollen lands on a stigma, the tube cell of the pollen grain forms a pollen tube that grows down through the stigma, style, and ovary tissues (Figure 13-7), eventually worming its way through the micropyle of an ovule. Two sperm are discharged; one fuses with the egg to produce a diploid zygote, while the other fuses with the diploid endosperm mother cell to form a triploid (3N) *primary endosperm cell.*

Fertilization sparks dramatic transformations in the flower, including the following:

- the zygote develops into the embryo;
- the primary endosperm cell forms nourishing endosperm tissue;
- the ovule is transformed into a seed; and
- the ovary develops into a fruit.

Together, the endosperm, seed, and fruit nourish and protect the embryo as it develops and during its journey to a new habitat.

EMBRYO AND SEED DEVELOPMENT

A newly formed zygote divides by mitosis to produce the plant embryo (Figure 13-8). A plant embryo is a relatively simple structure, consisting of only three parts: one or two cotyledons, an epicotyl, and a hypocotyl. The **cotyledons,** or seed leaves, are the first structures to begin photosynthesis when the seedling emerges from the soil. In most

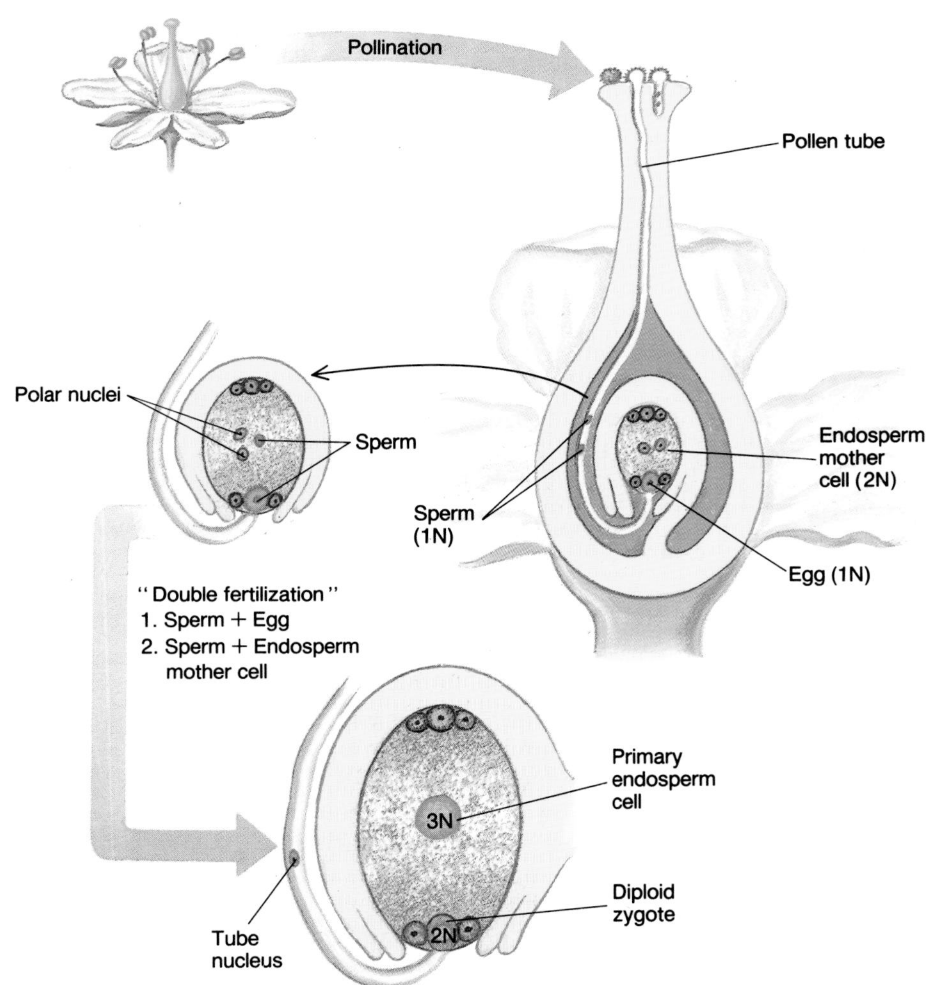

FIGURE 13-7
Pollination and fertilization: The beginnings of a new generation. Pollen is transferred from anthers to a stigma during pollination. The pollen grain grows a pollen tube down the style, into the ovary, and through the micropyle of an ovule. The generative cell divides to produce two sperm; one fertilizes the egg to form a zygote, the other fuses with the two nuclei of the endosperm mother cell to form a triploid primary endosperm cell.

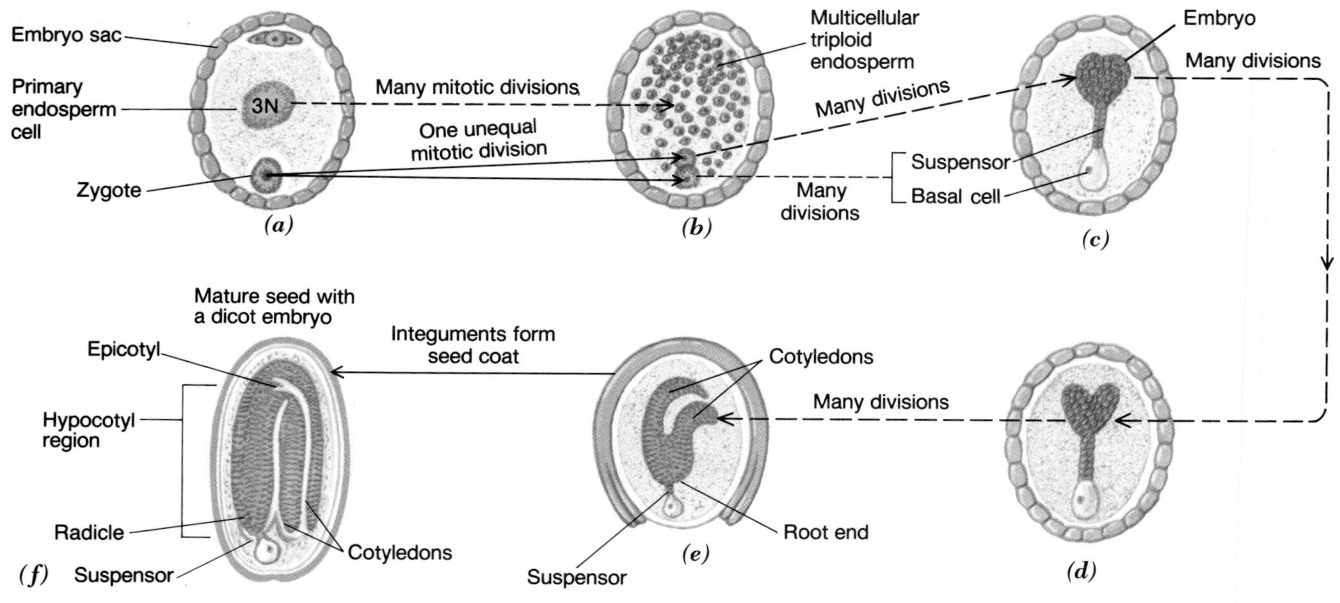

FIGURE 13-8

Development of a dicot embryo, *Capsella*. ***(a)*** A zygote and primary endosperm cell form following fertilization. ***(b)*** The triploid (3N) primary endosperm cell divides many times by mitosis to produce nutritive endosperm. The first mitotic division of the zygote produces different-sized daughter cells. ***(c)*** The smaller cell divides many times to produce the plant embryo. Divisions of the larger, basal cell produces the suspensor. ***(d)–(f)*** Continued mitotic divisions leads to the formation of a mature dicot embryo.

dicots, the endosperm (food) is located within the cotyledons and is part of the embryo; in most monocots, the endosperm surrounds the embryo and is part of the seed. As their names indicate, dicot embryos have two cotyledons, whereas monocot embryos have one (Figure 13-9). The **epicotyl** is the portion of the embryo above the cotyledons (*epi* = above, *cotyl* = cotyledon); it develops into the shoot system of a plant. The **hypocotyl** is the portion of the embryo below the cotyledons (*hypo* = below). At the tip of the hypocotyl is the **radicle,** which eventually develops into most or all of the plant's root system.

As the zygote develops into the embryo, the ovule develops into a seed, containing the embryo, endosperm,

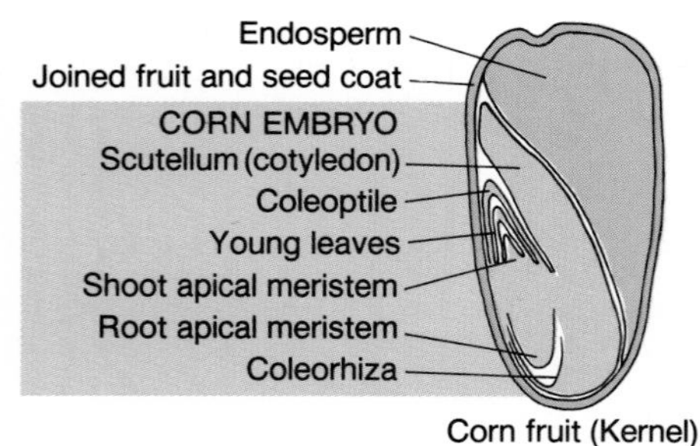

FIGURE 13-9

Corn (*Zea mays*) fruit (kernel) with mature embryo. A corn kernel is a combination of fruit and seed. The ovary of the pistil forms an outer "fruit + seed" coat. The single cotyledon digests and absorbs endosperm stored outside the embryo.

and a seed coat. The seed coat forms from the outer cell layers of the ovule, becoming tough and hard to protect the embryo from abrasion or other dangers as it is dispersed to new habitats. Some seed coats have projections that help absorb water; attach the seed to the fur of an animal for dispersal; or protect the embryo from hungry animals.

FRUIT DEVELOPMENT

Once sexual reproduction is complete, those parts of the flower that were involved in pollination (sepals, petals, stamens, stigma, and style) wither, their function completed. As embryos and seeds develop, the ovary of the pistil matures into the *fruit.* Many animal species, including humans, rely on plant fruits for nutrition. In fact, fruits have had a profound effect on human life, perhaps even sparking the development of human civilization (see The Human Perspective: The Fruits of Civilization).

Whether a fruit is green or brightly colored, juicy or dry, soft or hard, all (or part) of it formed from a ripened ovary. Some so-called "vegetables," such as cucumbers, tomatoes, squash, string beans, and grains, technically are fruits because they develop from an ovary. There are three categories of fruits (Figure 13-10)—simple, aggregate, and multiple—depending on whether a fruit develops from the ovary of one pistil of a single flower (**simple fruits**), such as a grape or plum; from many pistils in a single flower (**aggregate fruits**), such as a strawberry; or from the pistils of a number of flowers (**multiple fruits**), such as a pineapple.

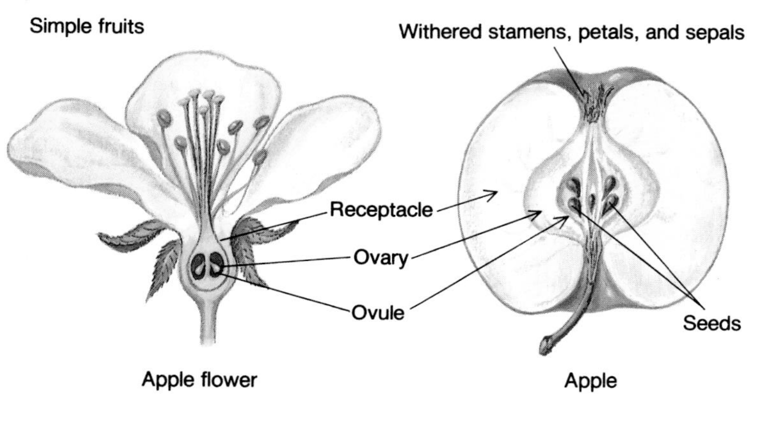

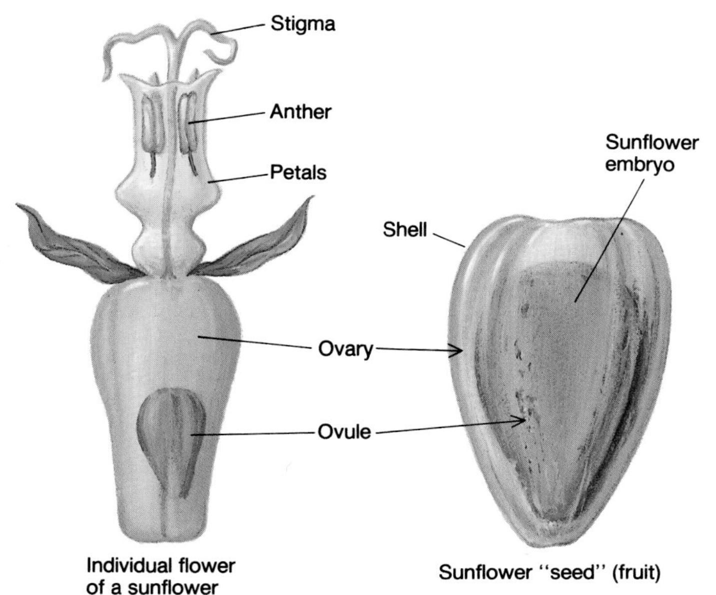

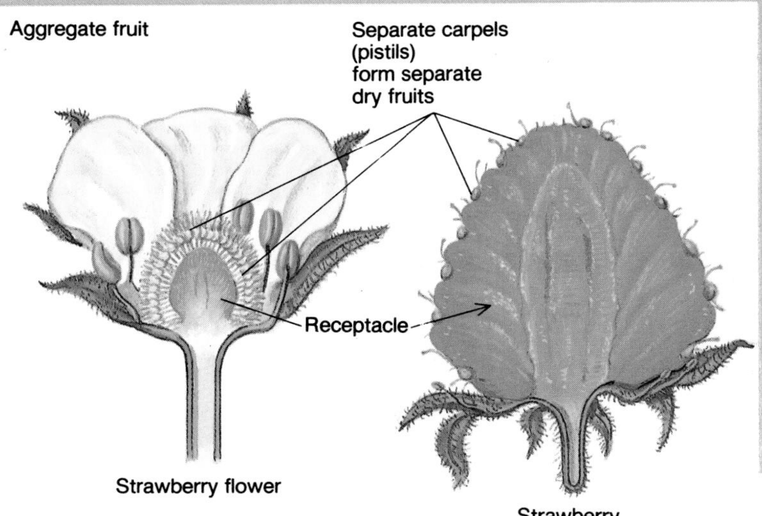

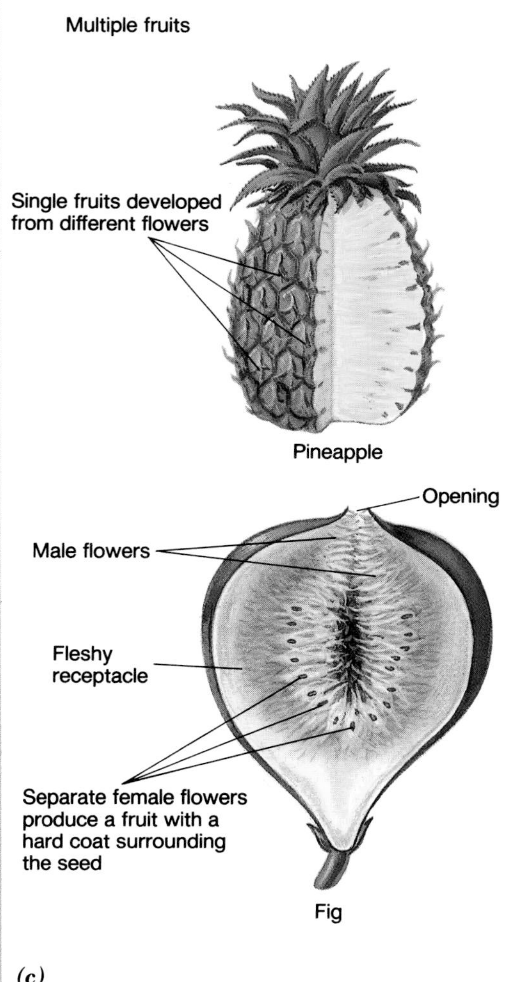

FIGURE 13-10

Three main fruit types. ***(a)*** Whether fleshy (apple) or dry (sunflower), simple fruits develop from one pistil. ***(b)*** Aggregate fruits, such as a strawberry, are formed from separate pistils in the same flower. ***(c)*** Multiple fruits develop as the ovaries of several, often densely packed flowers mature together. Multiple fruits form a single fruit (such as a pineapple) or develop from many flowers that are physically joined by other flower tissues. A fig, for example, is an enlarged receptacle filled with many tiny simple fruits, each of which developed from a separate flower.

◁ THE HUMAN PERSPECTIVE ▷

The Fruits of Civilization

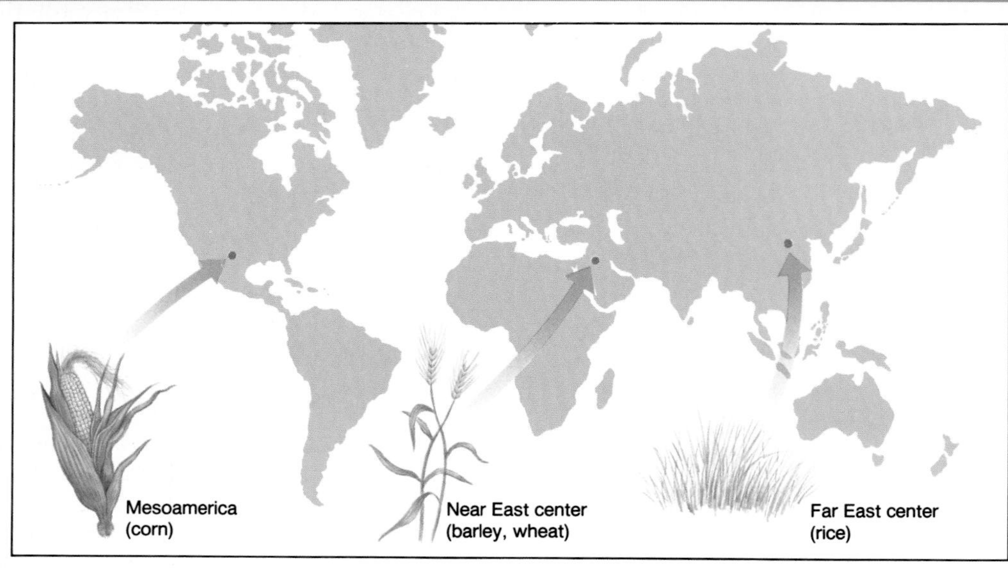

FIGURE 1

As humans, our existence depends on fruits and seeds, but our dependency goes beyond the obvious apple, orange, or peanut-butter sandwich. Flour for making bread is ground grain, the fruit of grasses; we feed domestic livestock fruits and grains for meat production. In fact, most of our food comes directly or indirectly from fruits and seeds.

The same was true for our early ancestors. Early humans traveled in nomadic tribes, forced to follow seasonal changes in plant growth in order to harvest enough food to feed their members. The perpetual search for food occupied most of the time and energy of every member of the tribe. There was no time for contemplation, no time or energy to develop art or science or even to begin the first scratchings of a written language. So how did these intellectual and civilized pursuits begin? Human civilization may owe its existence to one of the simplest kinds of fruits—grains.

Historians have identified three geographic regions where human civilization flourished and then radiated out to other areas of the world (Figure 1). Each of these centers of civilization had at least one naturally growing grain that likely provided the nutritional foundation for the development of civilization. The Near East Center had wheat and barley; the Far East Center had rice; and the Middle America Center had corn. Because of the abundance of these grasses and their high nutritional content, a few members of each tribe could gather, tend, and cultivate enough grain to feed the entire group. In addition, grass grains could be stored for long periods without spoiling, eliminating the need to travel from place to place to find a fresh supply of food. Some anthropologists speculate that as more and more members of the tribe were freed from the oppression of an endless, exhausting search for food, they had time to contemplate their surroundings and to develop the innovations that allowed civilization to progress. The development of plant and animal husbandry (production and care of domestic animals) practices freed even more people to specialize in other activities. Eventually, language, arts, and sciences began to flourish, leading to the social and technological advances we enjoy today.

FRUIT AND SEED DISPERSAL

In just one flowering season, a single plant can produce hundreds, thousands, and even millions of seeds, each containing an embryo capable of growing into a new plant. If a parent plant simply dropped all of its seeds onto the ground directly beneath its stems and branches, the resulting seedlings would compete with each other and with the parent plant for limited space, light, water, and nutrients as they grew. Such intense competition would lead to widespread malnourishment and massive numbers of deaths. *Seed dispersal* reduces competition by separating seeds. Dispersal also scatters seeds to new areas where conditions may be more favorable for growth and development.

Seeds can be dispersed over very long distances—across oceans to other continents or to distant islands—carried on a bird's feathers or in mud adhering to the feet of migrating birds. Some of the original plants that grew on the Hawaiian Islands were dispersed there in this way. Barbs, hooks, spines, or sticky surfaces that temporarily fasten fruits or seeds to an animal's fur aid dispersal of some plants to new locations.

GERMINATION AND SEEDLING DEVELOPMENT

When environmental conditions are favorable (proper temperature, sufficient water, and adequate oxygen, for example), a seed germinates, and a new generation of plants begin to grow and develop. During **germination,** the radicle of the hypocotyl is usually the first structure to thrust through the seed coat. The radicle then forms a root that anchors the new plant and begins absorbing water and minerals from the soil. As the root system develops, the epicotyl begins growing into the shoot system; photosynthesis begins when the shoot system emerges from the soil surface.

A young plant's (seedling's) battle for survival begins instantly. Producing a root system to obtain water is more important to survival than is breaking through the surface for photosynthesis. Nutrients stored in the seed can usually last for several days after germination, but without an immediate supply of moisture the seedling will quickly die of dehydration. Most plants establish mycorrhizal associations with soil fungi during the initial growth of the root system, improving their ability to absorb water and minerals.

Seedling development is slightly different for monocots than for dicots. During bean (a dicot) germination, for example, the radicle emerges and grows downward, producing many lateral roots from a single main root (Figure 13-11). Unequal cell growth causes the bean hypocotyl to bend, forming a hook that nudges its way up through the soil, carving a channel so that the delicate apical meristem, cotyledons, and young leaves can be drawn along without injury. Once the hypocotyl breaks the soil surface, it straightens, exposing the cotyledons and epicotyl to sunlight.

Up to this point, the seedling has depended on the nutrients stored in the endosperm of its cotyledons. Once the cotyledons and young leaves are exposed to light, however, the plant manufactures chlorophyll and other pigments, and photosynthesis begins. As food reserves are used up, the cotyledons wither and fall.

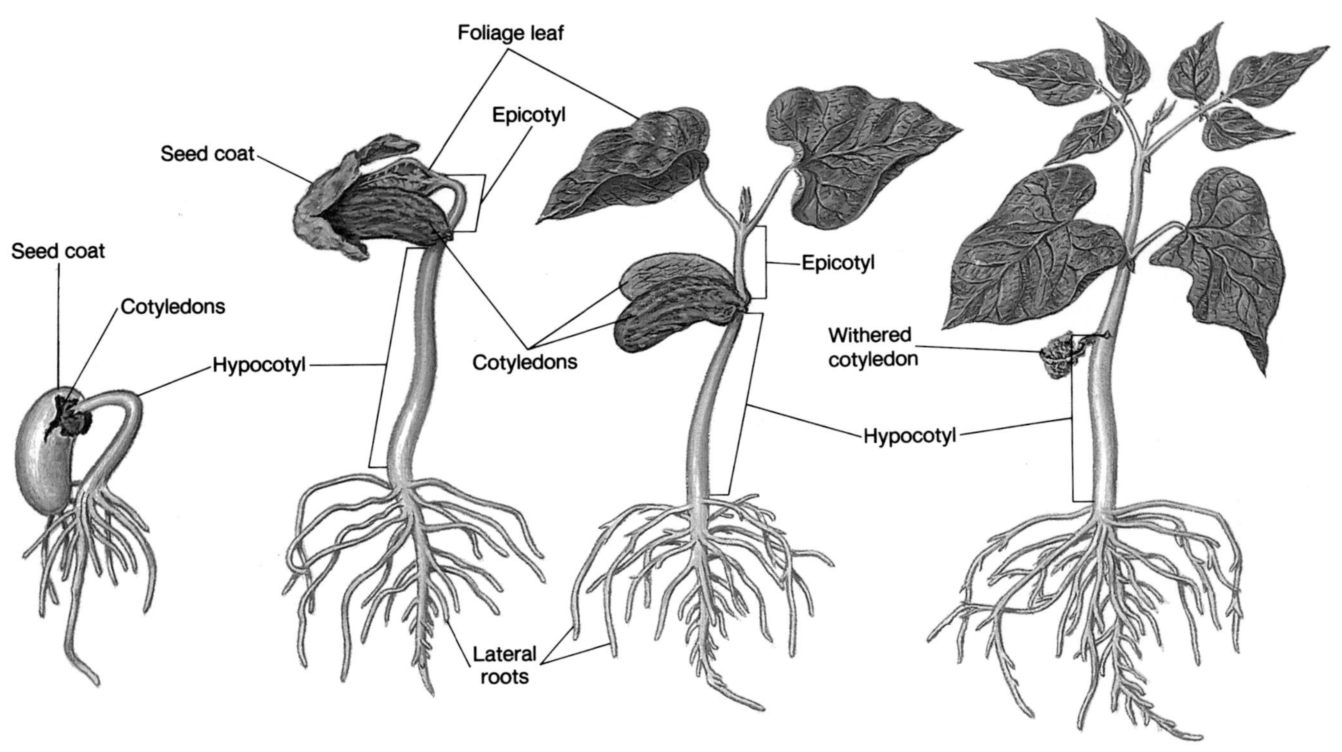

FIGURE 13-11
Dicot seed germination and seedling development: Bean (*Phaseolus vulgaris*).

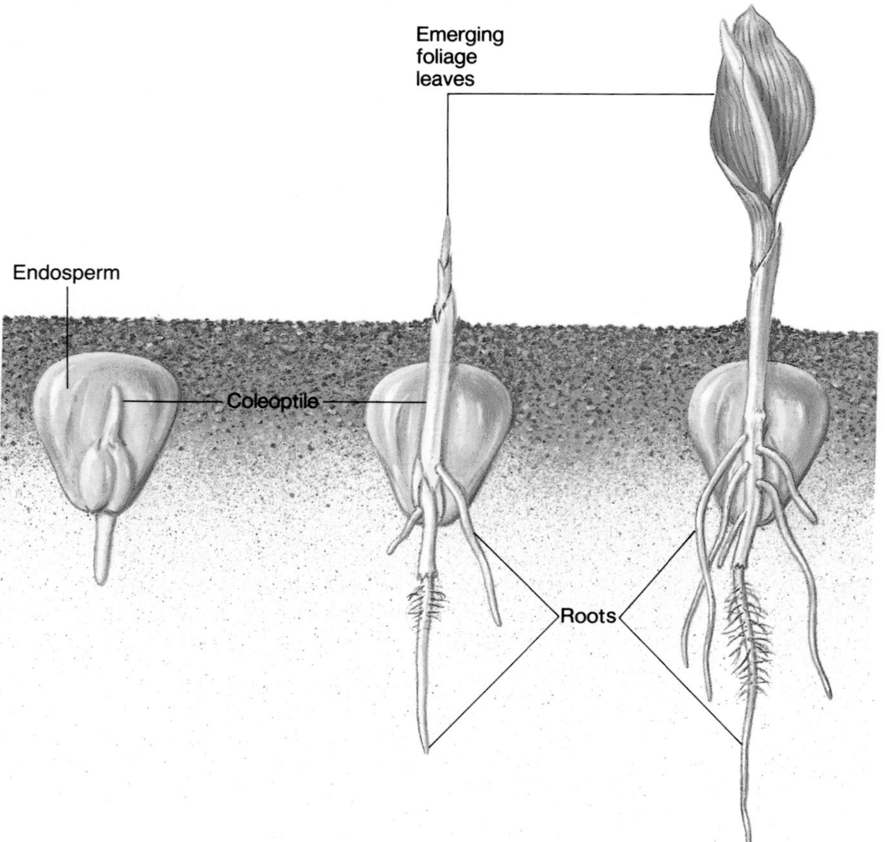

FIGURE 13-12
Monocot seed germination and seedling development: Corn (*Zea mays*).

Food reserves usually aren't stored in the single cotyledon of a monocot embryo. Instead, the cotyledon absorbs food from surrounding endosperm tissues, which fuels germination and initial seedling growth (Figure 13-12). When a corn kernel germinates, for example, the root tip breaks through the surrounding tissues, as the epicotyl forces its way up through the soil. As in dicots, once young monocot leaves are exposed to sunlight, photosynthesis begins, and the young plant begins its life of autotrophic independence.

FIGURE 13-13
Nature versus nurture. Persistent ocean winds (an environmental factor) have produced this picturesque Monterey cypress growing near the shoreline.

PLANT GROWTH AND DEVELOPMENT

Plant growth and development is the product of a complex interplay among three levels of control:

- *Intracellular level* (within a cell)—certain genes in the cells operate at precise times to code for specific characteristics and activities;
- *Intercellular level*—potent plant hormones coordinate growth and development in different parts of the plant;
- *Environmental level*—elements of the physical environment affect plant growth, such as the direction or intensity of light; wind (Figure 13-13); amount of moisture; rain acidity; extremes in temperatures; availability of minerals in the soil; and so on. Today, some human activities are dramatically affecting the earth's climate. For example, the rapid destruction of tropical rain forests leads to increased global temperatures and reduced rainfall (see The Human Perspective: Saving Tropical Rain Forests), environmental factors that directly impact plant growth and development.

◁ THE HUMAN PERSPECTIVE ▷

Saving Tropical Rain Forests

Satellite photographs reveal a perpetual cloud of smoke covering the forests of Brazil, a cloud the size of the state of Texas. Beneath the smoke, one of the earth's most critical life-sustaining habitats is being consumed. Every *minute,* 100 acres of tropical rain forest disappears, burned or hacked away by people seeking new land for cultivation. These are subsistence farmers whose lives depend on the crops grown on small plots of land. But their livelihood is doomed to a few years of marginal success; they then must clear more forest for their crops.

Nowhere is life more abundant or diverse than in the tropical rain forests, where more than half the world's species of plants and animals reside. But the abundance of life in the natural rain forest does not translate into lush crop growth once the forest has been destroyed. In fact, cleared rain forest land is paradoxically unproductive. Because the forest abounds with so much life, litter does not accumulate. Dead organisms are quickly consumed by other organisms, so little or no humus collects in the soil to condition it, and precious little nitrogen or phosphorous seeps into the ground to fertilize it. In other words, the topsoil is thin and poor. With the rain forest cleared, what few nutrients are present in the soil rapidly erode away with each rainfall. Within a few years of clearing, the land is abandoned, and more rain forest is burned to grow food for survival.

Each year, 27 million acres (11 million hectares) of tropical forest disappear, along with the habitats that support the most diverse range of species on earth. One of the most distressing aspects of tropical rain forest destruction is the extinction of thousands of species. Although extinction is part of the natural history of life, deforestation has accelerated the rate by 10,000 times. Extinction claims three to four more species *every day*!

The tropics provide humans with most of their edible plant species (a typical breakfast of bananas, orange juice, sugar and cornflakes, coffee or hot chocolate, and hash brown potatoes consists entirely of tropical plants). More than 40 percent of our medicines, from pain-relieving aspirin to treatments for malaria, have come from tropical plants. Cures for cancer and AIDS may be in plants on the brink of extinction, plants that could disappear before we ever discover them.

Destruction of the tropical rain forests also alters the earth's climate. Ecologists refer to tropical rain forests as "rain machines" because they recycle water back into the atmosphere (roots draw up the ground water, which evaporates from leaves, returning about 50 percent of rainwater to the atmosphere). With deforestation, almost 100 percent of the water runs off into streams, most of it flowing all the way to the the ocean, so deforested areas receive much lower amounts of rainfall. Some lush tropical areas have become desert-like following deforestation.

Deforestation plays a very important role in accelerating global warming. The forests are critical to the balance of carbon dioxide on earth. The plants of the forest remove carbon dioxide from the atmosphere through photosynthesis. In burning the rain forests, we are not only destroying our major carbon dioxide "sink," we are also turning the trees into carbon dioxide (about 1 billion tons of carbon dioxide are ejected into the atmosphere by burning). Because carbon dioxide in the atmosphere traps heat, this ecological "combination punch" will likely hasten global warming, which, in turn, will trigger unexpected and unwelcome changes.

With half the rain forests already destroyed, and at the current rate of destruction, virtually all the earth's rain forests could be eliminated (along with more than half the earth's species) within the next 50 years. What will the world be like then?

PLANT HORMONES

In multicellular plants, as in multicellular animals, coordinating the growth and activities of hundreds of thousands of cells requires hormones. But because plants are stationary organisms, their growth and development must also be adjusted to accommodate the peculiarities of their immediate surroundings. As a result, the production and distribution of plant hormones are very sensitive to environmental conditions.

Plant hormones, like those of animals, are chemicals that are synthesized in a particular part of the plant which stimulate or inhibit specific responses in target tissues. To date, botanists have identified three *groups* of plant substances (auxins, gibberellins, and cytokinins), and two *specific molecules* (ethylene gas and abscisic acid) that influence plant growth and development. As more research is completed, it becomes increasingly clear that hormonal control of plant growth and development is not simply the result of a single hormone acting alone but of a combination of hormones that triggers specific responses.

Auxins, gibberellins, and cytokinins stimulate plant growth by promoting cell division and/or cell elongation. **Auxins** trigger many growth responses, including causing stems to bend, orienting a plant more favorably to its environment (such as when a shaded plant grows out from under another plant, toward light); stimulating root development, thereby increasing water and mineral absorption; stimulating cell division in the vascular cambium, increasing xylem and phloem transport; and promoting flower and fruit development. Sometimes auxins inhibit development. For example, when auxins inhibit the development of axillary buds, a familiar, spindly growth pattern is produced (Figure 13-14). As horticulturists and gardeners will tell you, pinching off the tips of a plant's main stems removes the apical meristem, a center of auxin production. By removing this center, axillary buds begin to develop into new stems with more leaves and more axillary buds.

Between 1920 and 1990, researchers identified more than 70 compounds classified as **gibberellins.** When gibberellin is applied to dwarf plants, normal cell division and elongation is restored, suggesting that dwarf plants lack the gene needed to manufacture their own supply of gibberellins. In addition to affecting stem growth, gibberellins are critical to embryo and seedling development. In monocot seeds (including corn, orchids, and grasses), gibberellins stimulate mobilization of food reserves stored in the endosperm. Without such mobilization, stored energy and nutrients would not be available to the embryo. Gibberellins also inhibit seed formation; stimulate pollen tube growth; increase fruit size; initiate flowering; and end periods of dormancy in seeds and axillary buds.

In the 1950s and 1960s, researchers discovered that rapid growth could be stimulated in plant embryos and plant tissue cultures by adding coconut milk and yeast extract to the culture solution. Studies isolated the active ingredients in coconut milk and yeast extract, which were named **cytokinins** because they stimulated rapid growth by promoting cell division (cytokinesis). Injuries to plants (such as a cut or a torn branch) are quickly healed as a result of the production of cytokinins. Cytokinins also delay **senescence**—the aging and eventual death of an organism, organ, or tissue. In leaves, cytokinins slow senescence by delaying the breakdown of chlorophyll. In flowers and

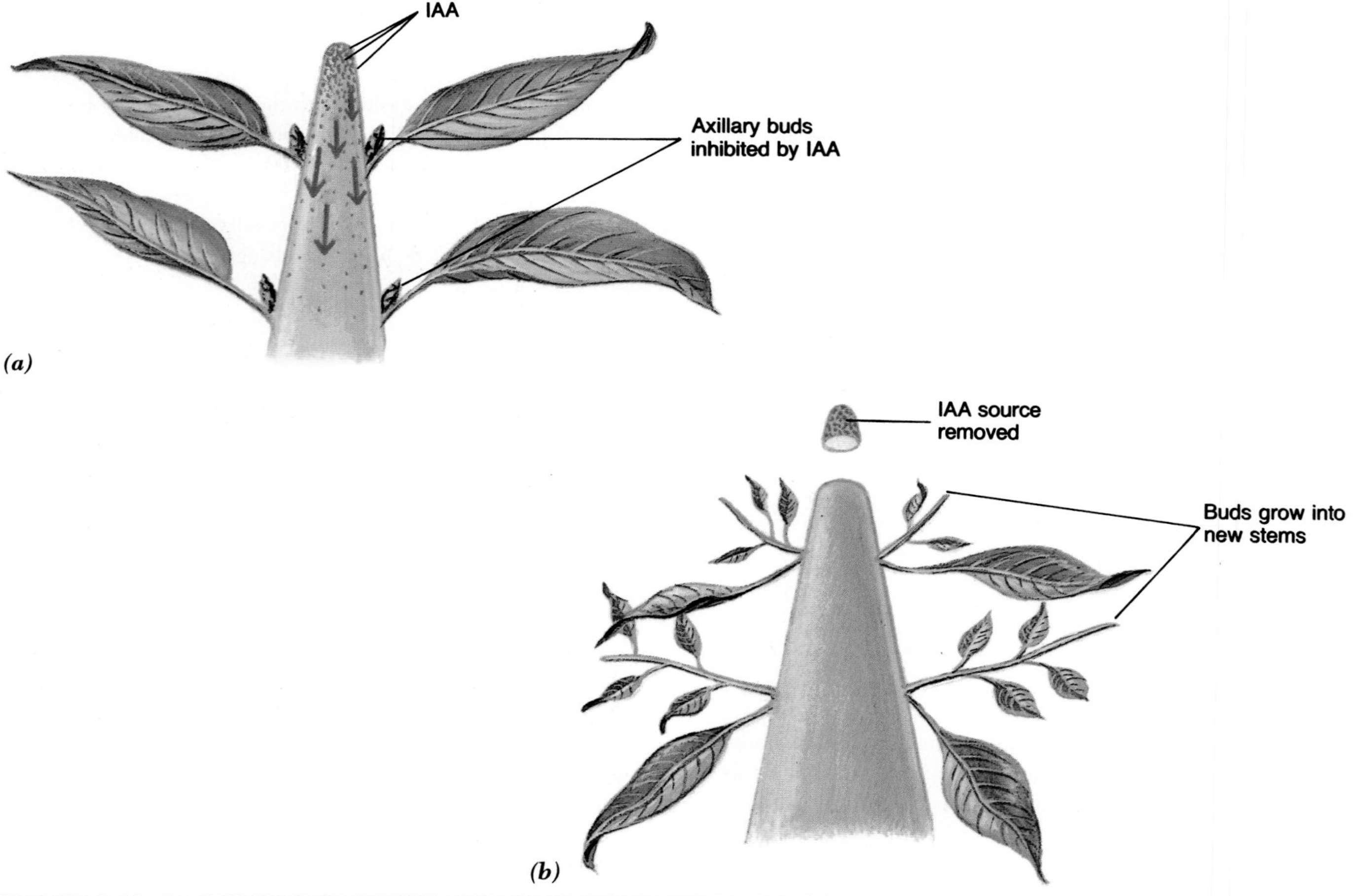

FIGURE 13-14

Controlling stem growth. Produced by apical meristems, auxin moves down the stem and inhibits axillary bud growth ***(a)***. When the shoot tip is clipped, the source of auxin is removed, and axillary buds grow into new stems ***(b)***.

fruits, cytokinins delay senescence by increasing nutrient transport.

One of only two specific molecules, **ethylene gas,** is a plant hormone that hastens fruit ripening. Enclosing fruits in a covered bowl or closed bag sets up a positive feedback system (ethylene breeds more ethylene), causing fruits to ripen faster and faster. Production of ethylene gas by overripe fruit explains why "one rotten apple can spoil the whole barrel." The discovery that ethylene gas promotes fruit ripening has dramatically lowered the cost of shipping fruits to markets. Tomatoes, bananas, apples, oranges, and many other fruits are usually picked when they are still green (reducing the chance of spoilage), transported in ventilated crates (to prevent ethylene-gas buildup), and then gassed with synthetic ethylene at distribution centers to promote last-minute ripening.

In addition to its role in fruit ripening, ethylene gas stimulates stems to grow thicker in response to physical stress. Trees that have been staked while growing (and thus remain protected from the bending stresses caused by winds) usually have weaker trunks and stems than do plants allowed to bend in the wind. High concentrations of ethylene gas can cause abnormal stem bending and leaf curling, discolored flowers, and stunted root growth. Ethylene gas interacts with other plant hormones to control leaf fall (ethylene gas + abscisic acid), to stimulate flower production (ethylene gas + auxins), and to control the ratio of male to female flowers on some monoecious plants (ethylene gas + gibberellins).

Abscisic acid (**ABA**), the other specific molecule, *inhibits* growth by either reducing the rate of cell division and elongation or halting the processes altogether. Abscisic acid was given its name because early investigators believed that ABA was involved in **abscission**—the process that caused flowers, fruits, and leaves to fall. While originally thought to be associated with both dormancy and abscission, abscisic acid has more recently been shown to have little or no role in either process. However, ABA-suspended growth can lead to plant dormancy, which can save a plant's life during periods of severe environmental conditions. During fall and winter, abscisic-acid levels remain high in the axillary buds of deciduous trees—plants that drop their leaves during unfavorable growth periods. High abscisic-acid levels counteract growth-promoting auxins and gibberellins, preventing cell division and elongation at a time when delicate, growing tissues would normally be exposed to lethal temperatures. ABA concentration drops at the onset of favorable growing conditions, releasing buds from their dormancy and causing plants to erupt with new growth (Figure 13-15).

FIGURE 13-15
Deciduous plants burst out with new growth after a period of dormancy, when conditions become favorable for growth.

Experimental evidence suggests there may be more plant hormones than the ones we just described. For example, studies suggest the presence of a flower-stimulating hormone named **florigen.** Other investigations suggest the existence of a root-growth hormone. And, recently, scientists have discovered that some plants produce chemicals that act like the growth hormones of insects. These plant chemicals mimic the insect's hormones so closely that they disrupt the insect's normal growth and development, preventing sexual reproduction.

TIMING GROWTH AND DEVELOPMENT

In many parts of the world, spring triggers a lavish surge of plant growth and development. But not all plants bloom or germinate in early spring. Flowering or germination is delayed in some plants until late spring, summer, or even fall. Clearly, plants are able to synchronize growth and development with the changing seasons; some, with great precision. In fact, Indian tribes in Arizona based their annual calender on the punctual flowering and fruiting of the saguaro cactus; a new year was marked by the appearance of the first saguaro fruits.

In addition to seasonal cycles, plants also have daily cycles. Plants are able to detect and respond to seasonal environmental changes by monitoring changes in temperature and **photoperiod**—the relative length of daylight to dark in a 24-hour period. Plants use an internal *biological clock* to measure the time of day. Together, temperature, photoperiod, and an internal biological clock regulate cell metabolism and hormone production so that plant growth and development remain synchronized with environmental changes, enabling plants to live in some of the earth's harshest environments.

Plants are sensitive to fluctuations and prolonged exposure to extreme temperatures, both of which can

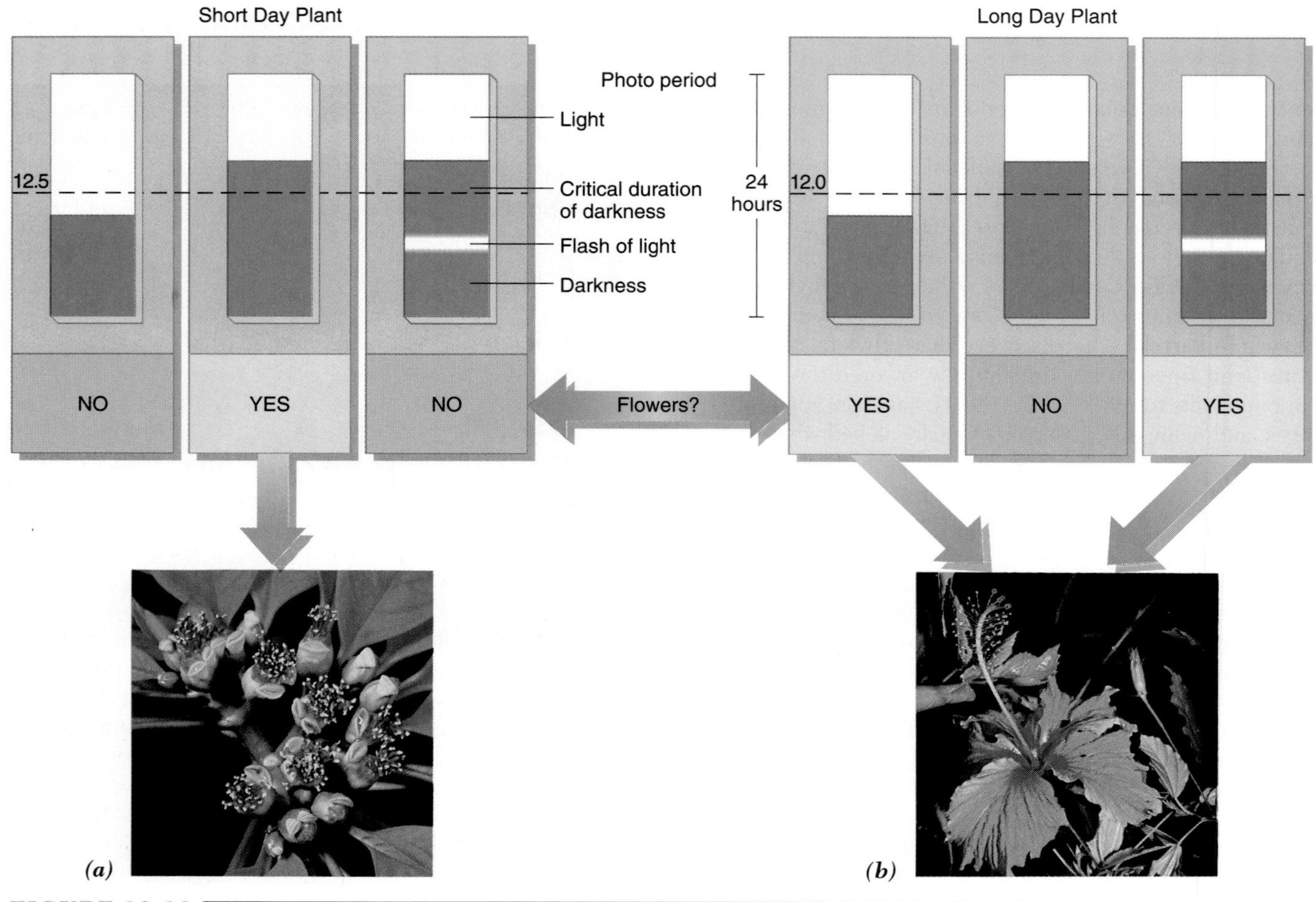

FIGURE 13-16

Photoperiod and flowering. Short-day and long-day plants flower in response to changes in photoperiod. ***(a)*** The poinsettia is a short-day plant that flowers when the critical daylength becomes *less* than 12.5 hours of daylight. ***(b)*** The hibiscus is a long-day plant that flowers when the critical daylength becomes *greater* than 12 hours.

threaten a plant's life. In dormant plants where metabolism is significantly reduced, plants are able to withstand greater temperature extremes for longer periods. Dormancy requires a number of preparations, including building up stores of food, water, and other nutrients, and growing protective structures (such as axillary-bud scales) to shield delicate tissues. Plants must be able to detect the approach of unfavorable growing conditions to prepare for dormancy, as well as to detect the end of unfavorable growing conditions so that they can resume growth as favorable conditions return. Because temperature is closely tied to other weather factors (such as relative humidity, rainfall, and snowfall), gradual temperature changes act as a predictor of upcoming weather. By responding to temperature trends, plants synchronize growth and development to upcoming weather.

In addition to responding to temperature cues, many plants react to variations in photoperiod. In the Northern Hemisphere, the period of daylight gradually lengthens from December until late June and then gradually shortens from late June until early December. Photoperiods are a very reliable indicator. On any given day of the year, the length of daylight and darkness is nearly constant from one year to the next in a particular region. This constancy explains the precise blooming of the saguaro cactus from one year to the next in the southwest deserts of North America, as well as many other plants.

The ability to respond to daylength is known as **photoperiodism.** Photoperiodism enables plants to induce flowering at times when pollinators are active; to initiate dormancy as conditions gradually become less favorable; to correlate seed germination with periods favorable for seedling growth; and to control stem and root development of an emerging seedling.

Natural selection favors those plants that have adaptations that enable them to monitor photoperiod so that reproduction is optimized. Botanists classify plants into three groups, depending on how changes in daylength affect flowering. **Short-day plants,** such as poinsettias, rice, morning glory, chrysanthemum, and ragweed, flower when the length of daylight grows shorter than some critical photoperiod, usually in late summer or fall (Figure 13-16*a*). **Long-day plants** flower when the length of daylight becomes longer than some critical photoperiod, usually in the spring and early summer (Figure 13-16*b*). Larkspurs, spinach, wheat, lettuce, mustard, and petunias are examples of long-day plants. **Day-neutral plants** flower independently of daylength. Most day-neutral plants flower only when they are mature enough to do so, although other factors, such as temperature and water supply may also affect flowering. Day-neutral plants include carnations, roses, snapdragons, sunflowers, and, as most gardeners know, many common weeds. Many agriculturally important plants, such as corn, beans, and tomatoes, are also day-neutral plants.

Plant growth and development are controlled not only by changes in temperature and light but also by internal biological clocks. For example, the leaves of the silk tree (*Albizzia*), as well as other legumes, are oriented horizontally during daylight but fall into a more compact vertical "sleeping" position at night. Even when kept in continuous light and at constant temperature, the leaves fall into the "sleeping" position at exactly the same time each day. The sleep movements of leaves follow a circadian (*circa* = about, *diem* = day) cycle, recurring at 24-hour intervals.

PLANT TROPISMS

Several environmental factors change the rate of production or distribution of growth-regulating hormones, thereby changing the direction of plant growth. Such responses are called **tropisms.** Tropisms are stimulated by light, gravity, contact with objects, chemicals, temperature gradients, wounding, and the presence of water. The three most common tropisms are:

- **Phototropism:** the bending of leaf surfaces and stems toward a light source, often a nearby window.
- **Gravitropism:** changes in plant growth in response to the pull of gravity.
- **Thigmotropism:** changes in plant growth stimulated by contact with another object. For example, many climbing plants have stems and branches that are too weak to support their own weight so they rely on other objects to keep them upright. Using tendrils, some plants either wrap around a supporting structure (Figure 13-17*a*) or produce pads at the tips of branches that fasten the plant to a wall, fence post, or some other solid object for support (Figure 13-17*b*).

Considered together, the diversity of plant tropisms, the mechanisms by which plants monitor and respond to changes in environmental conditions, and the production of hormones that coordinate growth and development help dispel a common misconception that plants are a static, unresponsive form of life. The diversity of plant growth and development adaptations, the complexity by which they are regulated, and the precision by which they operate reveal just the opposite: Plants are dynamic, continually responding and adjusting to their surroundings in ways that promote survival and reproduction.

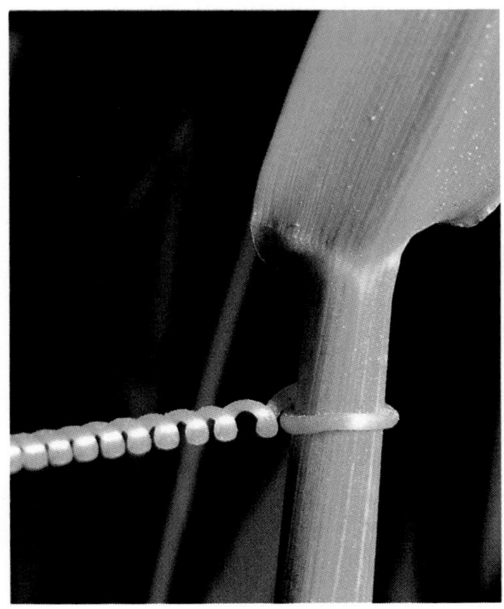

(a)

(b)

FIGURE 13-17

Some climbing vines have very weak stems and rely on a sturdier plant or some other solid object for support. The tendril of a pea plant grows around a corn stem for support ***(a)***, while expanded pads of a Virginia creeper fasten its growing stems to a wall ***(b)***.

EVOLUTION AND ADAPTATION: TYING IT TOGETHER

The interdependence and complementary structures found in plants and their animal pollinators, as well as in plants and the animals that disperse their seeds, illustrate coevolution—the reciprocal evolution of adaptations in different species based on mutual advantages. Part of the great success of flowering plants compared to other plant groups is attributed to the fact that most flowering plants are pollinated by animals—pollen carriers that substantially increase successful crossbreeding over wind or water currents. Increased pollination increases sexual reproduction, enabling flowering plants to proliferate in virtually all the earth's habitats.

Fruits and seeds are adapted to aid in dispersal, thereby reducing competition between newly germinated seedlings or between seedlings and the parent plant. Plant seeds find their way to any number of habitats, each with slightly different environmental circumstances. For example, a seed may be dispersed to the side of a mountain or on a flat plain. To survive, a developing seedling would have to grow differently in each of these habitats. On a mountainside, the shoot cannot grow perpendicular to the surface, as it would on a flat plain, and the root could not grow directly down. A plant disperses its embryos when they are in a very underdeveloped form, enabling them to better adjust their growth to particular environmental circumstances.

SYNOPSIS

The flower is the site of sexual reproduction, seed formation, and eventual fruit development. Variation in flower design are adaptations that promote successful dissemination and collection of pollen for sexual reproduction.

Meiosis leads to the formation of male and female gametes for sexual reproduction. The daughter cells resulting from meiosis in plants form spores. Spores produced in the anthers develop into pollen grains, whereas those formed in the ovary of a pistil develop into embryo sacs contained within an ovule in the ovary of the pistil.

Fertilization follows pollination. Two sperm cells are delivered to each ovule. One sperm fertilizes the egg, while the second fuses with the endosperm mother cell.

Following fertilization, the zygote develops into a plant embryo; the ovule transforms into the seed; and the ovary (and sometimes other flower parts) matures into the fruit. Fruit and seed adaptations help disperse seeds to new habitats and reduce competition between offspring and the parent plant.

Plant growth and development are the result of three interacting levels of control. The intercellular level contains fixed genetic controls, which are regulated by intercellular hormonal controls, which, in turn, are affected by various aspects of the environment.

Most plant growth and development responses are the result of combinations of plant hormones. Four of the five groups of plant hormones (auxins, gibberellins, cytokinins, and ethylene gas) promote growth, whereas the fifth (abscisic acid) inhibits growth. There may be other plant hormones.

Many plants synchronize dormancy and periods of active growth with changing seasons by responding to variations in temperature and day length and by using an internal clock. Photoperiodism helps plants synchronize flowering, seed germination, and renewed growth to favorable periods of environmental conditions.

Review Questions

1. Arrange the following events in the sexual reproduction of a flowering plant in their correct order:

a. pollen dispersal
b. seed germination
c. pollination
d. fertilization
e. seed dispersal
f. fruit development
g. pollen tube growth
h. attraction of animal pollinator

2. Discuss the basic function(s) of fruits.

3. Match the structure to its function:

——1.	embryo sac	a.	attract animal pollinator
——2.	stigma	b.	site of fertilization
——3.	style	c.	pollen deposition
——4.	nectaries	d.	produce pollen grains
——5.	petals	e.	connects stigma and ovary; provides a channel for pollen tube growth

4. Match the hormone with its function (multiple matches are possible).

——auxins	a.	fruit ripening
——cytokinins	b.	apical dominance
——gibberellins	c.	delays senescence
——ethylene gas	d.	stimulate stem elongation
——abscisic acid	e.	dormancy

5. What is the adaptive advantage to photoperiodism in plants?

Critical Thinking Questions

1. Explaining the results of their experiments on light and stem bending, the Darwins reasoned that a growth signal was produced in the coleoptile tip. The signal, they reasoned, traveled down the shoot and triggered the cells below to enlarge differentially which, in turn, caused the young shoot to bend toward light. Describe two ways in which cells could "enlarge differentially" and cause a stem to bend toward light. How can you determine which way was correct?

2. The flower is considered by evolutionary botanists to be the most advanced structure for sexual reproduction in the plant kingdom. Based on what you have learned about the form and function of flower parts, discuss the characteristics of flowers that you believe support this contention.

3. Interpret the graph below that compares the number of seeds and the distance they were dispersed from the parent plant.

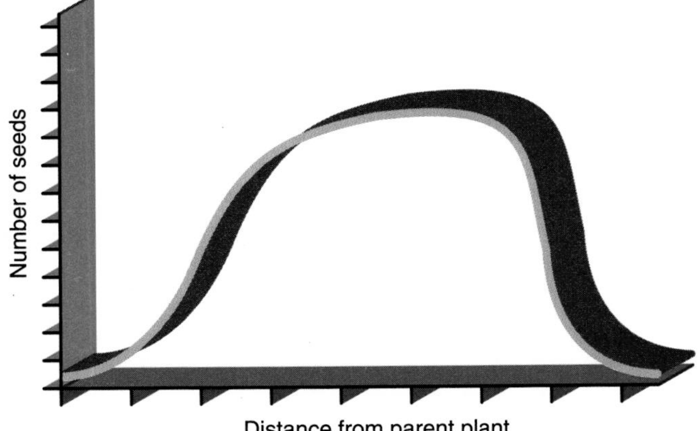

Describe what the graph shows in terms of the relationship between number of seeds and distance travelled. What do you suppose is the dispersal agent—wind, water, bird (carried internally), or animal (carried externally)? Why did you choose this vector? How tall do you think the parent plant is? How large would you suppose the seeds were, and what do you think their relative weight was—extremely light, light, moderate, heavy, very heavy? Explain.

4. A botanist gave wheat seedlings a dose of X-rays sufficient to prevent cell division but which left the seedlings able to function normally otherwise. He gave half of the irradiated seedlings gibberellin, but not the other half. The length of the seedlings is plotted in the graph below. Which of the following conclusions is supported by the experiment, and why? (1) Gibberellins stimulate cell elongation only; (2) Gibberellins stimulate cell division only; (3) gibberellins stimulate both cell elongation and cell division.

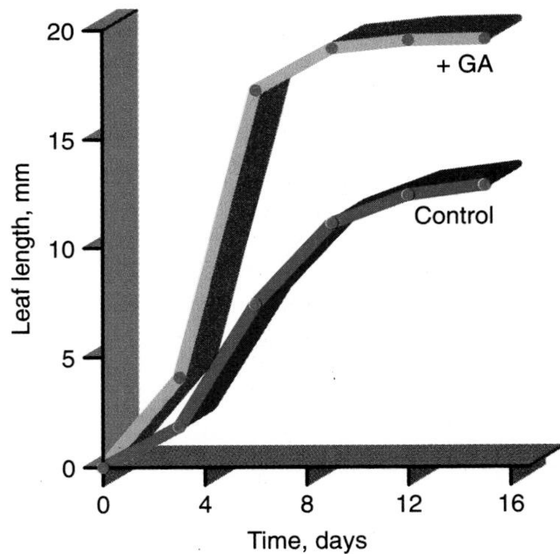

5. Colonizing outer space presents many difficult challenges, not the least of which is how plants can be grown for food and for providing a quality environment for the human inhabitants. Discuss some of the problems you would anticipate in trying to grow plants in a space station. For each problem, try to suggest a solution.

CHAPTER

14

Animal Form and Function

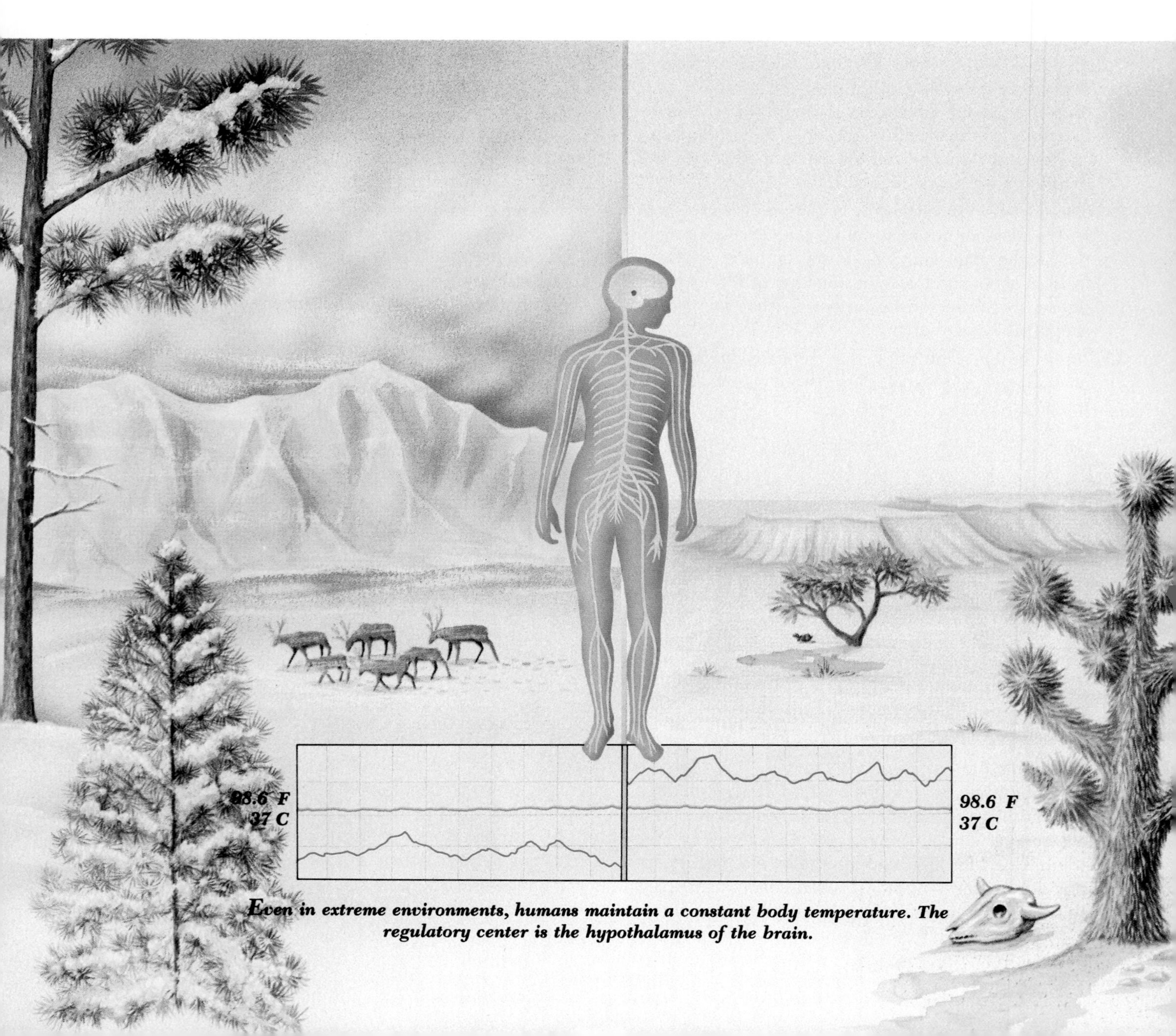

Even in extreme environments, humans maintain a constant body temperature. The regulatory center is the hypothalamus of the brain.

STEPS TO DISCOVERY
The Concept of Homeostasis

In 1843, at the age of 30, Claude Bernard moved from a small French town, where he had been a pharmacist and an aspiring playwright, to Paris, where he planned to pursue his literary career. Bernard's plans changed, however, when he enrolled in medical school, and he went on to become one of the leading physiologists of the nineteenth century. During his prestigious career, Bernard studied body temperature, stomach juices, the role of nerves in controlling the diameter of blood vessels, and the function of the liver, but he focused particular attention on the fluids that bathe the body's internal tissues. Bernard carefully monitored the temperature, acidity, and salt and sugar concentrations within the blood under various conditions and found that the body's fluids resisted change, providing cells with a stable, ordered environment. In his own translated words: "It is the fixity of the internal environment which is the condition of free and independent life. . . . All the vital mechanisms, however varied they may be, have only one object, that of preserving constant the conditions of life in the internal environment."

Bernard's work provided the foundation for a diverse array of studies of the mechanisms by which the body maintains its internal constancy. For example, in the 1880s, Charles Richet in France and Isaac Ott in the United States independently determined that one part of the brain, the hypothalamus, played a key role in maintaining body temperature. If this part of the brain of laboratory animals was damaged, these animals lost their ability to hold their body temperature constant in the face of increasing or decreasing environmental temperatures. In 1912, Henry Barbour of Yale University developed a technique to selectively warm or cool this part of the brain. Barbour accomplished this feat by implanting fine silver tubes into the hypothalamus and then circulating water of particular temperatures through the tubes. Cooling the hypothalamus below normal body temperature caused the animals to increase heat production, just as if they had been outside on a cold winter night. Warming the hypothalamus evoked the opposite response. The animals reacted by losing heat, just as if they had been exposed to a hot desert sun. The results of these studies suggested that one of the functions of the hypothalamus was to determine if the body's temperature was rising or falling and then to elicit an appropriate response that would return the temperature to its normal value.

One of the most influential physiologists of the twentieth century was Walter Cannon of Harvard University. Working in a field hospital in France during the fierce trench warfare of the last few months of World War I, Cannon was struck by the body's ability to withstand the terrible trauma resulting from severe wounds and to restore the orderly conditions necessary for survival. After the war, Cannon returned to Harvard and turned his attention to studying the mechanisms by which the body "fights" to maintain the internal environment. In 1926, he coined the term "homeostasis" (*homeo* = sameness, *stasis* = standing still; balance). Cannon did not mean to suggest that the body was stagnant or incapable of change but, rather, that it had the capacity to respond in dynamic ways to situations that threatened to disrupt its internal stability.

Cannon summarized his views on homeostasis in a book entitled *The Wisdom of the Body*. In this book, Cannon noted that humans could be exposed *for short periods* of time to dry heat above 115°C (239°F) without raising their body temperature, or to high altitudes with greatly reduced oxygen, such as when flying in an airplane or climbing a mountain, without showing serious effects of oxygen deprivation. He also noted that the body could resist disturbances that arise from within. For example, a person running for 20 minutes produces enough heat to "cause some of the albuminous substances of the body to become stiff, like a hard-boiled egg" and enough lactic acid (the acid of sour milk) to "neutralize all the alkali contained in the blood." Yet, the body becomes neither overheated nor poisoned.

The concept of homeostasis identified by Bernard and Cannon pervades virtually every aspect of the physiological sciences. As we consider the function of major organs, such as the lungs, heart, kidney, and liver, we will see how each of these organs helps maintain a stable, ordered internal environment that promotes our health and well-being. We will also examine some of the dangers that lurk should these homeostatic controls fail.

Go without water on a hot summer day and you build up a thirst that has to be quenched. Spend a cold winter night in a bed with inadequate covers and you sleep huddled into a ball, trying to keep warm. Stay up all night studying for an exam and you find yourself so tired you can barely stay awake. All of these common events in our lives illustrate cases where our bodies are "telling" us that we need to take some action to maintain the stability of our internal environment. Most homeostatic activities take place without the need for these types of behavioral responses. For example, without conscious effort, our bodies "automatically" prevent themselves from becoming too salty or dilute, too hot or too cold; our bodies prevent waste products from accumulating to dangerous levels and stop oxygen or ATP concentrations from dropping below required amounts. All organisms have homeostatic mechanisms that enable them to respond to changes that threaten to upset their internal environment. In animals, these responses are carried out by teams of cells, tissues, organs, and organ systems.

▼ ▼ ▼

CONTROLLING BODY TEMPERATURE: AN EXAMPLE OF HOMEOSTASIS

The various activities taking place within your body typically occur at their optimal level when your body temperature is approximately 37°C (98.6°F), which is the temperature at which your body is normally maintained. Maintaining a constant temperature requires following one simple physiological rule: The amount of heat gained by the body must equal the amount of heat lost from the body. Balancing heat gain and heat loss may not seem like a difficult task, but consider the following "fact." If you were to spend 24 hours in bed under conditions where you were unable to either gain any heat from or lose any heat to your environment, the temperature of your body would rise approximately 30°C due to the production of metabolic heat. Of course, you would be dead long before your body temperature increased to 67°C, but this gives you an idea of how much heat your body produces, even when "at rest." If you were engaged in any type of exercise under these conditions, the increased level of metabolism would cause your body temperature to reach the boiling point within the same 24-hour period.

Maintaining a constant body temperature requires a set of components including

1. *receptors* that monitor conditions in both the external and internal environment. The receptors that monitor body temperature are located in the skin, the brain, and various internal organs. Information from the receptors is passed to
2. a *control unit*, or *integrator*, which is usually a part of the brain, where the data is analyzed and signals are sent to the body's
3. *effectors*, which mediate the appropriate response. For temperature regulation, the effectors include blood vessels that carry blood to the skin, muscles that generate heat during shivering, and sweat glands that produce a fluid for evaporative cooling.

HOMEOSTATIC MECHANISMS

The operation of the body's temperature-control system is an example of a **negative-feedback mechanism** in which a change in a property sets in motion a response to reverse that change. For example, a rise in body temperature elicits a response to lower the temperature. In contrast, **positive-feedback mechanisms** operate when a change in the body feeds back to *increase* the magnitude of the change. Blood clotting is an example of positive feedback. Your blood contains proteins that have the potential to form a blood clot that can plug a wound in a blood vessel, should one develop, which can be a life-saving response. The clotting of the blood at the site of an injured vessel begins on a very small scale. First, a small amount of clotted protein is deposited. As the clot forms, it induces a response that triggers the deposition of additional clotted protein, which, in turn, triggers even more deposition. As a result of this positive-feedback mechanism, what began as a small response is rapidly amplified so that further loss of blood is quickly halted and internal stability is maintained.

FOUR TYPES OF TISSUES

The human body contains several hundred recognizably different types of cells, each of which works with other types of cells to accomplish a common function. The simplest of these teams of cells is a **tissue**—an organized group of cells with a similar structure and a common function. Tissues are the "fabrics" from which complex animals are constructed. The use of the word "fabric" to describe tissues is not just a recent invention of biology writers; the term is derived from ***tissu***, an Old English word meaning a "fine cloth." Like intricately woven fabric, tissues are the products of finely organized biological designs and not mere aggregates of cells randomly packed together in a unit. These living fabrics form the foundation of the multicellular organism.

Despite the tremendous diversity that exists in the types of cells present within a single animal (and even greater diversity found among different animals), there is a striking overall unity of function. All of the diverse cells found among animals can be classified as part of one of four fundamental types of tissues of which all organs and organ systems are composed. Each of your organs, for example, is composed entirely of epithelial, connective, muscle, and/or nervous tissue.

EPITHELIAL TISSUE

Epithelial tissue (Figure 14-1) is present as sheets of tightly adhering cells that line the spaces of the body, such as the outer edge of an organ or the inner lining of a blood vessel or duct. The surface of the entire body is covered by an epithelial layer, which constitutes the outer layer *(epidermis)* of the skin. Some epithelia, such as that of the skin or the lining of the mouth, are primarily protective. Other epithelia, such as the lining of the intestine or lungs, regulate the movement of materials from one side of the

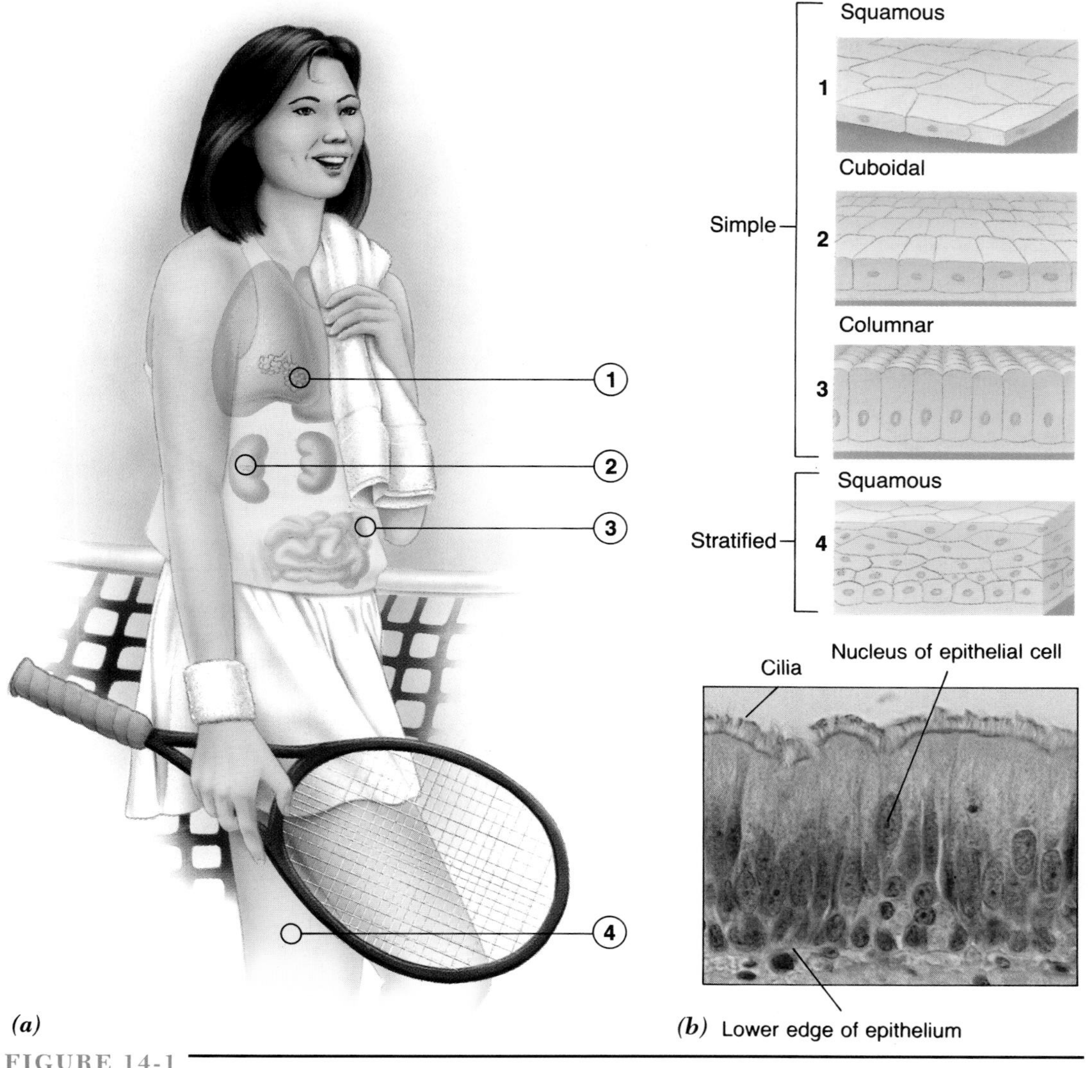

FIGURE 14-1

Epithelial tissue. ***(a)*** Epithelial tissues are categorized as either *simple* or *stratified* depending on whether they consist of one or more than one layer of cells. Epithelia are also categorized as either squamous (flattened), cuboidal, or columnar, depending on the shapes of the cells. Lung tissue is an example of a simple squamous epithelium; mammary glands contain a simple cuboidal epithelium; the inner lining of the small intestine is a simple columnar epithelium; and the outer part of the skin is a stratified epithelium. ***(b)*** A stained section showing the ciliated epithelium that lines the inner surface of the trachea (windpipe).

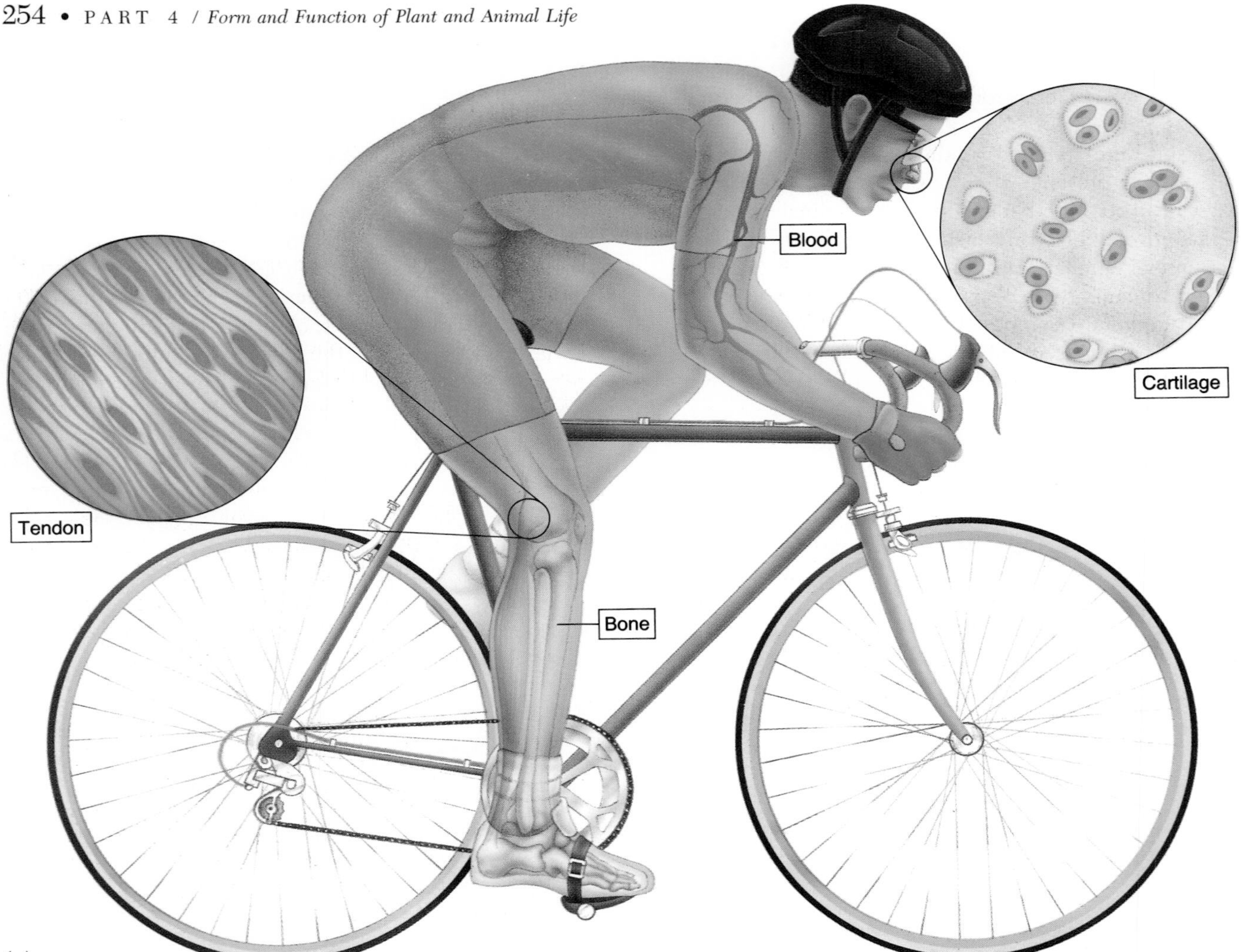

Blood

Cartilage

Tendon

Bone

(a)

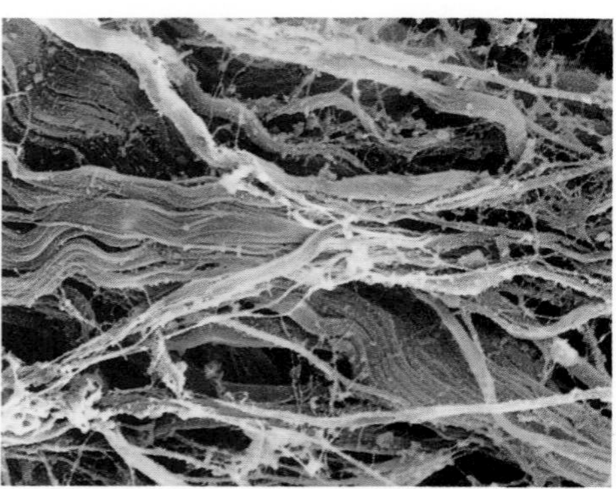

(b)

FIGURE 14-2

Connective tissue. ***(a)*** Connective tissues are quite diverse in structure and function and include bone, cartilage, blood, and tendons. All of these tissues are categorized by extensive extracellular materials. ***(b)*** Bundles of extracellular collagen fibers which give connective tissues their strength.

cell layer to the other. Epithelial cells are often specialized as glandular cells that manufacture and discharge materials outside the surrounding space. The epithelial lining of the trachea (Figure 14-1*b*), for example, contains mucus-secreting cells and ciliated cells that work together to move debris out of the respiratory tract. Other secretory epithelia release hormones, oils, sweat, milk, and digestive enzymes. The structure and functions of epithelia will be discussed in greater detail in later chapters on the skin, respiratory tract, digestive tract, and kidney.

CONNECTIVE TISSUE

The body is held together by **connective tissues** (Figure 14-2), which consist of a loosely organized array of cells surrounded by a nonliving *extracellular matrix*. The properties of connective tissue are due largely to the extracellular matrix, which contains proteins and polysaccharides secreted by the "entrapped" cells. For example, collagen, the most common component of the extracellular matrix (and the most abundant protein in the human body) is an inelastic molecule that provides resistance to pulling forces. Fibers of collagen (Figure 14-2*b*) provide tendons and ligaments with the strength to connect muscles to bones and bones to one another. Looser arrangements of collagen protect delicate structures, such as the eyeball, much like packing material prevents breakage of fragile glassware. Connective tissues may also contain elastic fibers that provide some tissues, such as skin and vocal cords, with the capacity to stretch and recoil.

Other examples of connective tissues include skeletal elements composed of cartilage and bone; sheets *(mesenteries)* that support the visceral (internal) organs; the transparent outer layer *(cornea)* of the eyeball; fat *(adipose)* deposits; and the deeper layer *(dermis)* of the skin. Blood is also a type of connective tissue—blood cells are surrounded by an extracellular matrix with a fluid consistency. The structure and functions of connective tissues will be discussed in greater detail in the chapters on the skin, skeleton and circulatory system.

MUSCLE TISSUE

Muscle tissue (Figure 14-3) consists of muscle cells—cells that contain elongated protein filaments that slide over one another, causing the muscle cells to shorten (contract). When muscle tissue contracts, it generates the forces needed for motion—either to move the animal (or its parts) or to propel substances within the body. Muscle tissue is present in large masses capable of moving the largest bones of the body and is the major element of the heart. Muscle tissue can also be found scattered more subtly throughout the body's internal organs. Among other consequences, when such muscle tissue contracts, blood vessels constrict, food moves through the digestive tract, and urine is emptied from the bladder. The various types of muscle tissue and their particular functions will be described in Chapter 17.

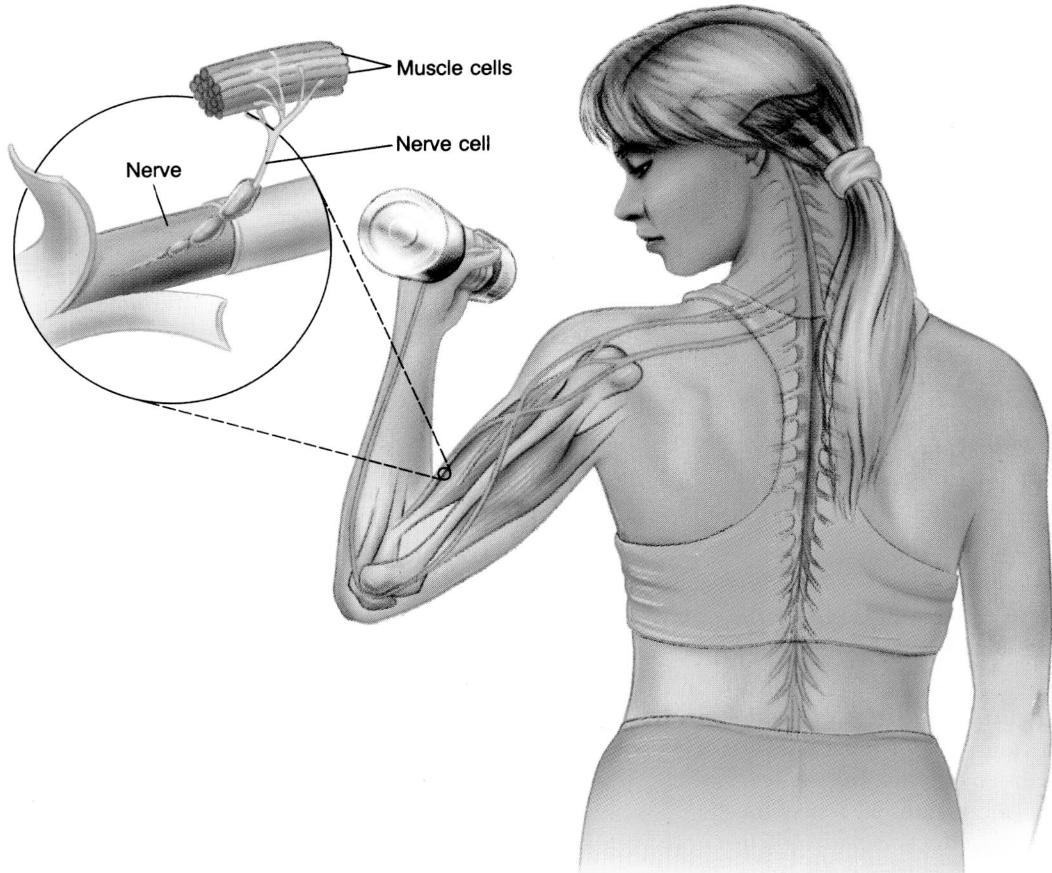

FIGURE 14-3

Muscle and nerve tissues. The structure of one type of muscle tissue is exemplified by a large skeletal muscle of the arm. Other types of muscle tissue are found in the heart or scattered less noticeably throughout the internal organs of the body. Nerve tissue is concentrated within the brain and spinal cord but also extends from these central neural organs into most of the distant regions of the body.

NERVE TISSUE

Nerve tissue (Figure 14-3) consists of highly elongated nerve cells—cells that are specialized for long-distance communication within an animal. Communication is accomplished when a signal is sent along the membrane of a nerve cell to its terminal end and is then relayed to another cell—a nerve cell, muscle cell, or gland cell—thereby evoking a response. Nerve tissue is responsible for coordinating such diverse bodily activities as breathing, sweating, sexual responses, and defecation, as well as providing some animals with memory, emotions, and consciousness. The structure and function of nervous tissues will be discussed in Chapter 15.

ORGANS AND ORGAN SYSTEMS

Animals as diverse as scallops, whales, earthworms, and anteaters have common needs: All animals must obtain oxygen, digest food, eliminate wastes, protect themselves from microbes and predators, and so forth. Each of these basic needs are met by **organs**—structures that are composed of different tissues working together to perform a particular function. A major function of the kidney, for example, is the removal of metabolic waste products from the blood. The elimination of waste products from the body requires additional organs, including the bladder and a series of tubular pathways (the ureter and urethra) that lead to the body surface. All of the various organs that work together to perform a common task, such as the elimination of waste products, make up an **organ system**.

TYPES OF ORGAN SYSTEMS

Figure 14-4 provides a survey of the types of organ systems we will encounter in the following chapters. We will summarize them briefly here.

- *Digestive systems* break down *(digest)* the macromolecules present in food. The breakdown products are absorbed across the lining of the digestive tract, providing the body with nutrients that supply energy and building materials.
- *Excretory/osmoregulatory (urinary) systems* accomplish two interrelated functions: They rid the body of *waste products*, the unusable byproducts of metabolism, and they maintain balanced internal concentrations of salt and water. Waste products include potentially toxic nitrogen-containing compounds, most notably ammonia, which must be either eliminated directly or modified to a less toxic compound, such as urea, and then eliminated. Similarly, many biological processes, ranging from enzyme activity to nerve impulse conduction, are very sensitive to the concentration of dissolved salts, which must be strictly regulated.
- *Respiratory systems* absorb the oxygen animals need to oxidize the organic molecules that fuel their biological activities. The respiratory system may also expel carbon dioxide, a waste product of metabolism. Animals that live in water and absorb the oxygen dissolved in their aquatic medium have different types of respiratory organs than do animals that breathe air.
- *Circulatory systems* transport materials inside an animal's body. Food absorbed by digestive systems, and oxygen absorbed by respiratory systems, must be distributed to those locations in the body where they are needed; wastes generated by cell metabolism must be carried to a site of elimination. Most animals, particularly those of larger size, distribute materials via a system of branched vessels to all parts of the body.
- *Immune systems* protect animals from foreign substances and invading microorganisms. The body's immunological defenses consist of various cells that recognize and ingest foreign materials as well as soluble blood proteins, called *antibodies*, which specifically bind to foreign materials and inactivate them.
- *Integumentary systems* cover the surfaces of animals and provide a barrier between their body and the external environment. These systems protect the animal from dehydration; provide physical support; prevent invasion by foreign microorganisms; and help regulate internal temperature.
- *Skeletal and muscular systems* work together to provide both support and movement. Most animals possess rigid skeletal structures that provide support and maintain body shape. In addition, skeletal structures are often involved in *movement* (a shift in position of a part of the body) and *locomotion* (a shift of the entire animal from one place to another). Movement is often accomplished as rigid skeletal elements are pulled in one direction or another by the contraction of attached muscles. Locomotor structures, such as wings, legs, or flagella, typically project from the body as *appendages* (Figure 14-5) that act on the external environment.
- *Reproductive systems* produce gametes (sperm or eggs) for fertilization and the subsequent development of new individuals of a species. All organisms have a limited lifespan. Individual bacteria live for minutes, a Galapagos tortoise for hundreds of years, and a giant redwood tree for thousands of years. Yet, these *species* have survived for millions of years—the products of continual reproduction. Reproduction is also part of the mechanism by which variability among individuals in an animal population is generated and is thus an important element in the process of biological evolution.

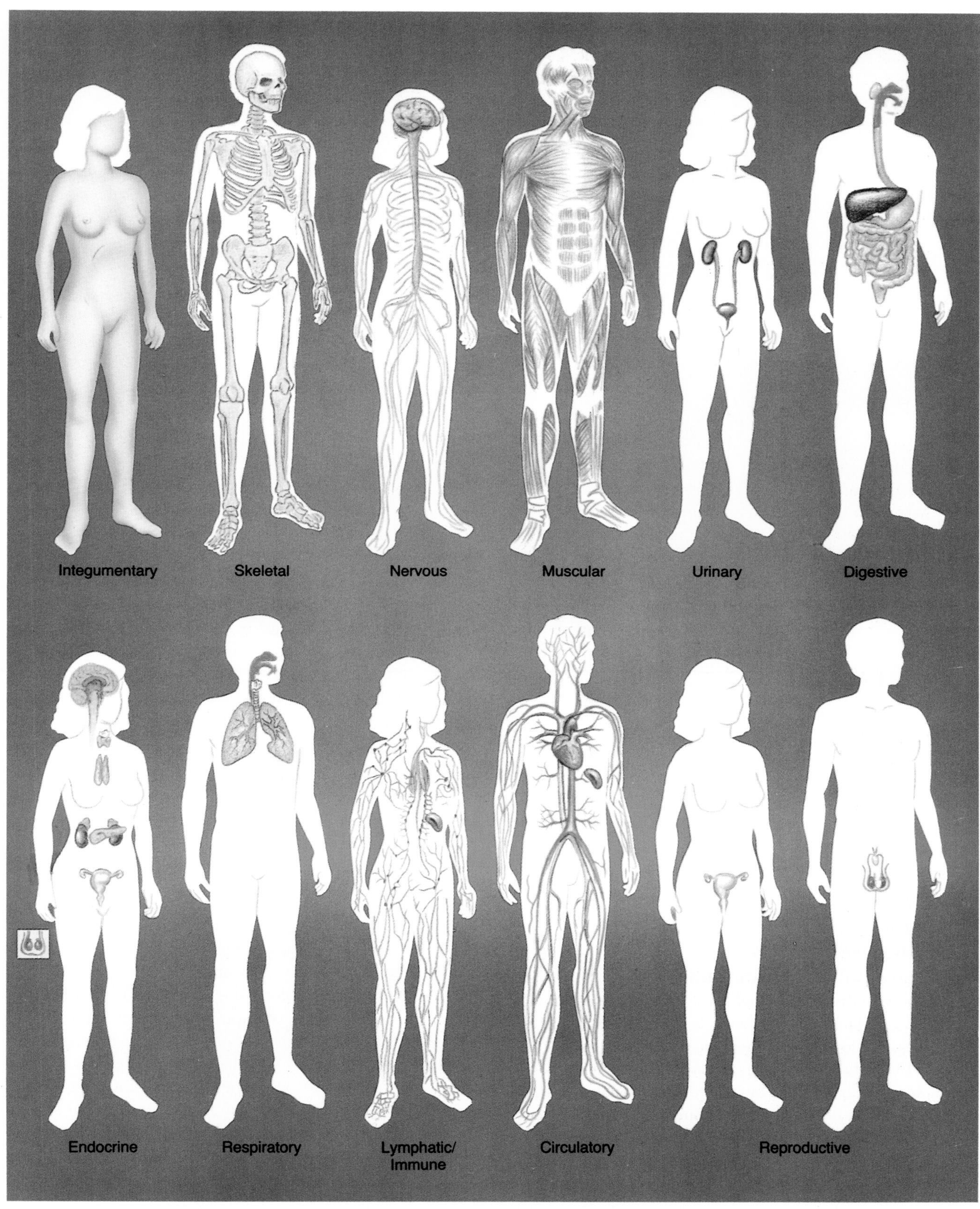

FIGURE 14-4
The types of organs systems found in the human body.

FIGURE 14-5

A gallery of animal appendages. *(a)* A giant Pacific octopus with a shark caught in the suction grip of its tentacles. *(b)* An inhabitant of the South American rain forest, this katydid exhibits a number of appendages, including legs, antenna, and even its mouthparts, which are derived from modified embryonic appendages. *(c)* The wings of this African cape gannet are appendages used for flight. *(d)* The appendages of a chimpanzee are similar in structure and function to our own arms and legs.

The Nervous and Endocrine Systems: Regulating Bodily Functions

The multitude of physiological processes that proceed simultaneously within an animal must be continually monitored and regulated. If, for example, oxygen levels in the body's tissues should drop, mechanisms are triggered that increase the uptake of oxygen from the environment. Or, if an animal is confronted by a potential predator, a protective response is triggered that increases the animal's chance for survival. Two types of systems—the nervous and endocrine systems—regulate and coordinate bodily functions.

- *Nervous systems* are networks of nerve cells that receive information concerning changes in the external and internal environment, integrate the information, and send out directives to the body's muscles and glands to respond in an appropriate manner. Information moves along the pathways of the nervous system in the form of impulses traveling along the plasma membranes of the nerve cells. These impulses provide a mechanism by which all parts of the body can rapidly communicate with one another. Information enters the nervous system via a series of *sensory* structures that detect changes in the external and internal environment.

- *Endocrine systems* also regulate and coordinate many of an animal's internal activities. Endocrine systems consist of a disconnected network of glands that release chemical messengers (*hormones*) into the blood. Hormones circulate through the bloodstream and ultimately interact with their particular target cells, triggering specific responses. Whereas responses triggered by the nervous system tend to occur rapidly, those triggered by the endocrine system occur more slowly and often include metabolic changes. Endocrine responses include changes in the level of glucose in the blood, the metamorphosis of a caterpillar into a butterfly, and sexual maturation (Figure 14-6).

THE EVOLUTION OF ORGAN SYSTEMS

Much of what we know about the evolutionary relationship among animals is derived from studies of fossil remains left behind by individuals living millions of years ago. Fossils almost invariably consist of the hardened skeletal parts of ancient animals; they reveal very little direct information about any of the other systems introduced in this chapter. Consequently, our knowledge about the evolution of most organ systems is based largely on comparative studies of living animals.

A comparison of organ systems in related animals provides some of the clearest insight into how natural selection can lead to the modification of structures to meet different physiological challenges. In some cases, a particular structure has undergone a transformation from one form and function to another. For example, the bones situated just behind your eardrum are derived from bones that formed part of the jaws of your vertebrate ancestors. The evolutionary movement of vertebrates from the water onto the land required certain changes in the way sound vibrations were transmitted from the environment to the sensory receptors in the ear. Fortuitously, one of the bones used to support the jaws of fishes was no longer needed as a jaw brace in the early amphibians. Instead, this bone became "pressed into service" as an ear bone (Figure 14-7*a*). The other two bones in the middle ear of mammals (Figure 14-7*b*) are derived from bones that were previously part of the jaws of our amphibian ancestors. We can see from this

(a)

(b)

FIGURE 14-6
The result of hormones. This emperator fish is transformed from a juvenile ***(a)*** to a sexually mature adult ***(b)*** as the result of hormones secreted by the fish's gonads.

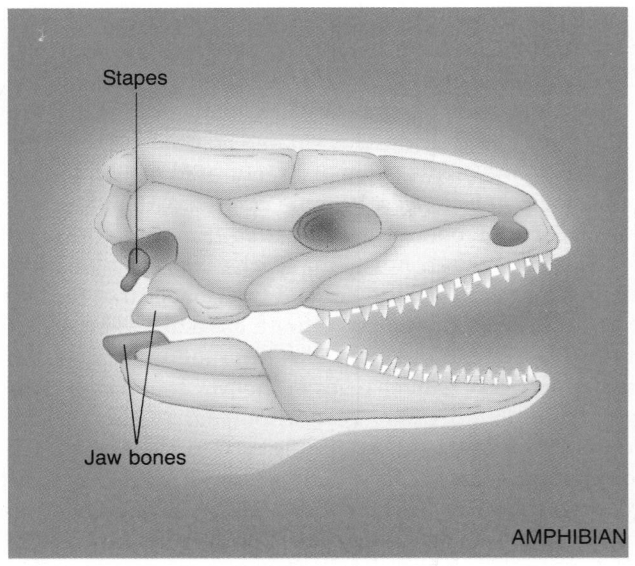

(a)

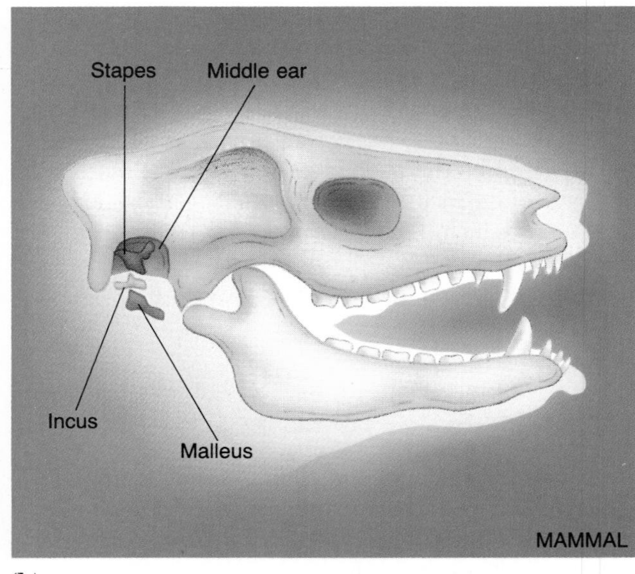

(b)

FIGURE 14-7

The evolution of the bones of the ear. ***(a)*** The middle ear of amphibians contains a single bone, the stapes, which can be traced to a bone in the skull of ancestral fishes. ***(b)*** Mammals have three bones in the middle ear—the incus, malleus, and stapes (often called the anvil, hammer, and stirrup, respectively). The incus and malleus have evolved from bones that are present in the upper and lower jaws of our amphibian ancestors.

description that, during the course of vertebrate evolution, these bones underwent a dramatic change in position, shape, and function. What were once food-gathering structures evolved into transmitters of sound waves.

EVOLUTION AND ADAPTATION: TYING IT TOGETHER

The environment in which an animal lives is a critical factor in determining the types of homeostatic mechanisms that will be adaptive for that animal. For example, sea animals must cope with very salty surroundings and, consequently, possess adaptations that prevent the loss of water by osmosis. In contrast, animals inhabiting a freshwater stream live in a highly dilute environment that has very little available salt. As a result, freshwater animals possess adaptations to eliminate water that floods their body by osmosis. In addition, freshwater animals must cope with marked fluctuations in water temperature, oxygen availability, and the risk of the evaporation of their home. Terrestrial (land) animals face even harsher conditions. They must cope with rapid changes in climate, a shortage of external water, a propensity to lose water to the surrounding air, and the lack of a medium to help support their body weight. Each environment presents unique physiological challenges and selects for unique homeostatic responses.

Common needs often lead to similar, but independent, evolutionary solutions. Even though an earthworm and a whale are so distantly related that their most recent common ancestor *may* have been a single-celled protist, these two animals possess organ systems that have similar overall functions. In the following chapters, we will examine the organ systems found in animals, focusing on human systems. Despite similarities in overall function, such as uptake of oxygen or elimination of waste products, the organ systems of different animals may be constructed very differently.

SYNOPSIS

In order to maintain a homeostatic state, the body must (1) receive information from sensors that detect changes; (2) pass the information to a control unit, typically located in the brain, which is "set" to maintain a certain value; and (3) send out signals to effectors (muscles and glands) to initiate a response that restores conditions to that which are set by the control unit. Maintaining a property at a constant level requires negative feedback. In certain cases, homeostasis requires that a change in the body is amplified, rather than reversed. This can be accomplished by positive feedback, as illustrated by the rapid formation of a blood clot that prevents further loss of blood.

Multicellular animals have similar types of organ systems that meet common needs. These include a digestive system, which disassembles food matter and provides the body with nutrients; an excretory/osmoregulatory system, which eliminates metabolic waste products and maintains a balanced concentration of salt and water; a respiratory system, which absorbs oxygen and often expels carbon dioxide; a circulatory system, which transports materials from place to place in the body; an immune system, which protects against foreign substances and invading microorganisms; an integumentary system, which covers the animal; skeletal and muscular systems, which facilitate support, movement, and locomotion; and a reproductive system, which produces sperm and eggs. Two systems—the nervous and endocrine systems—coordinate the activities of the other organ systems by collecting information about conditions in both the internal and external environment and providing signals to effectors that carry out a particular response.

Although diverse in form and function, cells are organized into four basic types of tissues. Epithelial tissues consist of sheets of cells that act as linings. Their functions include protection, exchange, and secretion. Connective tissues consist of cells surrounded by a nonliving extracellular matrix. As skeletal materials, connective tissues provide support and facilitate movement; as ligaments and tendons, they connect parts of the body and resist stretching; as the cornea, they provide a transparent layer for vision; and as the blood, they distribute materials from place to place. Muscle tissue provides the force for movement of materials within the body and movement of attached skeletal elements. Nervous tissue forms a communication network that transmits information used in coordinating bodily activities.

Review Questions

1. Describe the basic components your body employs to maintain a constant body temperature.
2. Compare and contrast negative- and positive-feedback mechanisms; tissues and organs; connective tissue and epithelial tissue; and the functions of the digestive system and the excretory system.
3. Contrast the basic mechanism of operation of the nervous and endocrine systems in regulating bodily functions.
4. Describe how a body part can undergo dramatic change in shape and function over the course of evolution. Why do such changes occur?

Critical Thinking Questions

1. In the Steps to Discovery vignette, several activities were mentioned that have the potential to disrupt the internal stability of our bodies but are kept from doing so by homeostatic mechanisms. Can you think of any other activities in which you engage that have this potential? Explain your answer.
2. You are cutting an apple with a sharp knife and accidentally cut your finger deeply. Which of the four types of tissues did the knife pass through? Support your answer.
3. Select any five of the organ systems surveyed in this chapter and discuss the effects on your body if one of these systems should suddenly stop operating.
4. Discuss the different types of challenges that would face an animal living on land compared to one living in the ocean or a freshwater pond.

CHAPTER 15

Regulating Bodily Functions: The Nervous and Endocrine Systems

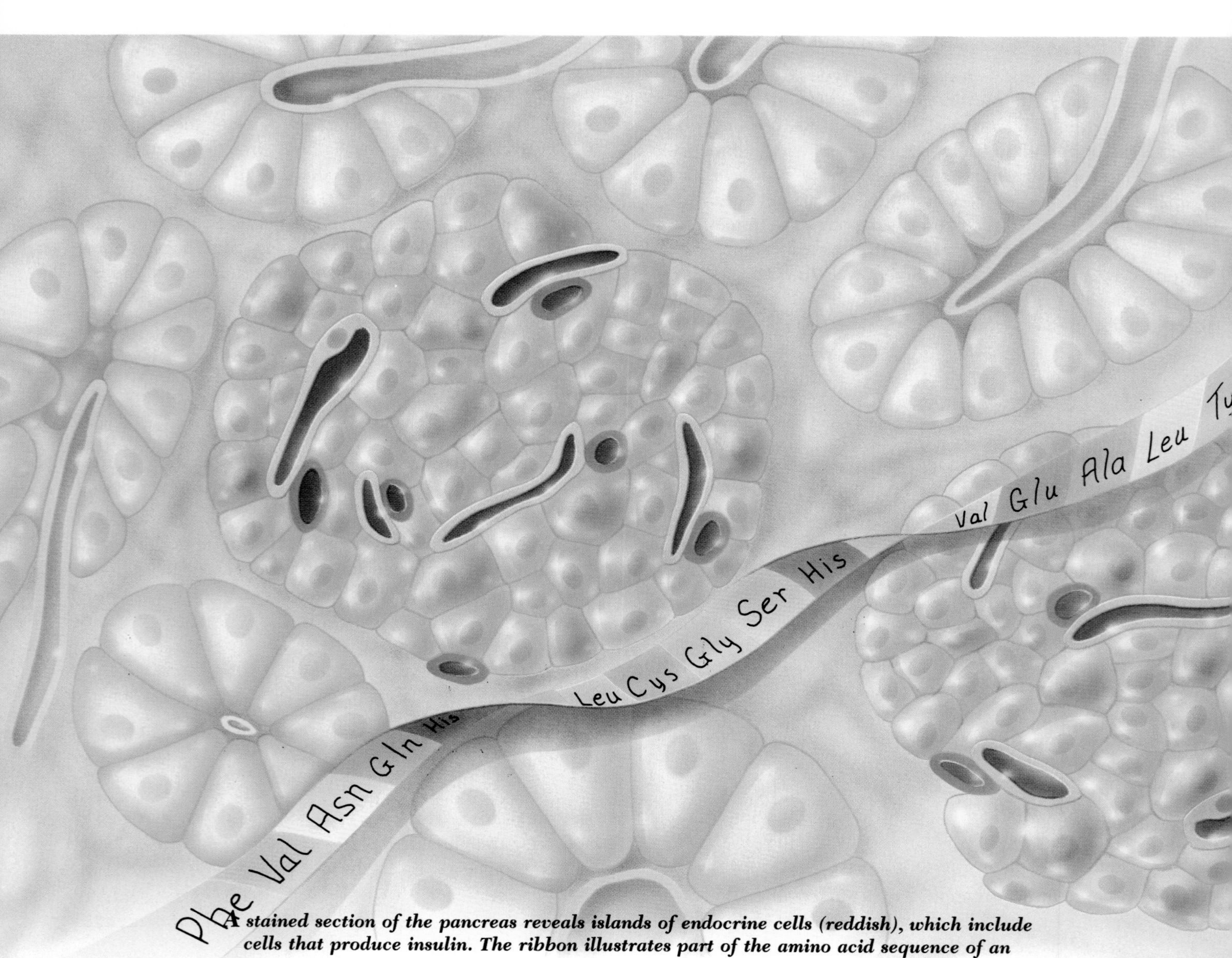

A stained section of the pancreas reveals islands of endocrine cells (reddish), which include cells that produce insulin. The ribbon illustrates part of the amino acid sequence of an insulin molecule.

STEPS TO DISCOVERY
The Discovery of Insulin

The story begins in 1869, when Paul Langerhans, a German doctoral student, discovered that the pancreas had two very different types of secretory cells. One group, the *acinar* cells, produced digestive enzymes that were shipped through ducts from the pancreas to the small intestine. The other group of cells was clustered into islands (later named the *islets of Langerhans*), which possessed neither ducts nor any detectable avenues for the export of the substance these glands secreted; their function was unknown.

Nearly 20 years later, two German physiologists, Oscar Minkowski and Joseph von Mering, attempted to determine the functions of the pancreatic digestive enzymes by removing the pancreas from a dog. Following removal of the glands, the dog, who had previously been house-trained, began urinating frequently on the floor. Minkowski and von Mering were aware that frequent and large-volume urination was one of the key symptoms of diabetes, as was a high level of sugar in the urine. They analyzed the dog's urine and found that it had an elevated sugar content. Removal of the pancreas had made the dog diabetic. The scientists traced the induced effect to the absence of a product from the islets of Langerhans. Yet, for the next 30 years, all attempts to isolate the product of these cells proved unsuccessful. This would soon change when, in 1920, a young doctor lay in bed struggling with insomnia. Unable to sleep, he devised a plan.

The doctor was Frederick Banting, a 29-year-old Canadian surgeon. As part of his preparation for a lecture he was to deliver to a medical class at the University of Western Ontario, Banting had just read a paper describing an autopsy on a person who had died from an obstruction of the major duct leading from the pancreas. The acinar cells of the pancreas had degenerated, leaving the islets of Langerhans unharmed. As Banting tossed around in bed, the contents of the article circulated through his mind. On a pad of paper lying next to his bed he wrote: "Diabetes. Ligate [tie off] pancreatic ducts of dogs. Keep dogs alive till acini degenerate leaving islets. Try to isolate the internal secretion of these to relieve glycosuria [sugar in the urine]." Banting had concluded (as had others) that the inability to isolate the product of the islets of Langerhans was due to the destruction of the substance by the digestive enzymes of the acinar cells. Banting intended to tie off the pancreatic duct, allowing the acinar cells to degenerate. He hoped that this would allow the antidiabetes factor of the Langerhans cells to be extracted in an undigested state.

Many research projects begin with a scientist reading a paper that arouses his or her curiosity or sparks an idea for a new experimental approach to a nagging question. But Banting was not a research scientist, and he had no lab. Instead, he traveled to the nearby University of Toronto to discuss his plan with John Macleod, an authority on carbohydrate metabolism and diabetes. Banting convinced Macleod to give him a bit of lab space and a few dogs to work with. Macleod also asked his physiology class for a volunteer to help out in the summer research project. A 21-year-old medical student named Charles Best came forward. Macleod soon left for a holiday in his homeland of Scotland, leaving Banting and Best to carry out what he thought would be a fruitless project.

Together, the two inexperienced researchers worked out a technique to ligate the duct of the pancreas; they then waited several weeks to allow the dog's gland to degenerate. Meanwhile, the researchers prepared several test subjects—dogs whose pancreas had been removed, causing them to become diabetic. Toward the end of July, Banting and Best prepared an extract from the ligated, shrivelled pancreas and injected the extract into a diabetic dog. The urine and blood of the dog showed a dramatic decrease in sugar content. After repeating the procedure several times over the next few weeks, the researchers noted a clear pattern: Injection of the pancreatic extract temporarily relieved the symptoms of diabetes.

By the time Macleod returned in September, Banting and Best were convinced that they were on the verge of a treatment for diabetes. Macleod remained skeptical, however, insisting that the researchers conduct additional experiments. Slowly, the findings were confirmed, and Macleod turned his entire lab over to the study of the pancreatic secretion. He invited James Collip, a visiting biochemist, onto the project to help develop improved procedures for extracting the antidiabetic substance, which the group had named *insulin*.

In January 1922, a 12-year-old boy lay dying of diabetes in a Toronto hospital bed. The Toronto group extracted insulin from beef pancreas, and the boy's physician injected the insulin into the patient, producing astounding results. After receiving daily injections of the extract, the boy left the hospital and resumed a normal life. In the words of M. Bliss, Banting's biographer, "Those who watched the first starved, sometimes comatose, diabetics receive insulin and return to life saw one of the genuine miracles of modern medicine. They were present at the closest resurrection of the body that our secular society can achieve, ..." Within the year, insulin was being prepared commercially by Eli Lilly, a U.S. pharmaceutical company, and thousands of diabetics were soon receiving the life-saving medication.

In 1923, the Nobel Prize in Medicine and Physiology was awarded to Banting and Macleod, a choice which probably caused more controversy than any other presented. Banting immediately announced that he would share his prize money with Best, whom he felt should have been the corecipient, rather than Macleod. Macleod, not to be outdone, announced that he would share his prize money with Collip. Macleod and Banting remained bitter enemies for the rest of their lives.

*T*he internal environment of the body of a complex animal is remarkably stable. The properties of the blood, for example, are maintained within relatively narrow boundaries; the pH, temperature, salinity, viscosity, sugar level, and so forth, do not fluctuate very widely under normal conditions, despite changes in diet or external conditions.

Two major organ systems have evolved to maintain the constancy of the internal environment as well as to direct the body's daily activities, whether it be playing basketball, eating lunch, or studying for a biology exam. Both of these systems act as communication networks, linking distant parts of the body. One of these networks—the **nervous system**—operates by sending information in the form of coded "electrical" messages across long, cablelike, processes much like the network of telephone lines that links houses, cities, and countries. The other major communication network is the **endocrine system**, which is composed of scattered glandular tissues that release chemical messages, or **hormones**, into the blood. For example, hormones ensure that the glands of the stomach release digestive enzymes following a meal, that blood sugar levels are increased if the body becomes engaged in strenuous activity, and that increased urine is produced if a large volume of fluid is ingested. We will begin with the nervous system.

▼ ▼ ▼

THE NERVOUS SYSTEM

The human nervous system is incredibly complex, consisting of tens of billions of cells that penetrate virtually every nook and cranny of the body. The nervous system gathers information from sensory receptors located around the body, analyzes the information, and sends out command signals to regulate and coordinate the activities of each of the body's parts.

NEURONS AND THEIR TARGETS

Each nerve cell, or **neuron**, is specialized for conducting messages, in the form of moving *impulses*, from one part of the body to another. Messages can be sent along these cellular "transmission lines" at speeds of over 100 meters per second (225 miles per hour). The effect of the impulse depends on two properties: the nature of the neuron, and the type of target cell that responds to the neuron.

- *Nature of the neuron.* Some neurons stimulate their target cells into activity; others inhibit their target cell.
- *Nature of the target cell.* Only three basic types of cells can respond directly to an arriving impulse: (1) muscle cells, which respond by contracting; (2) gland cells, which respond by secreting a substance; and (3) other nerve cells, which may respond by generating an impulse of their own, thereby relaying the message to another target cell.

Describing the form of a "typical" nerve cell is like trying to describe a "typical" human personality. Your body contains more than 100 billion neurons, and no two are exactly alike. Nonetheless, all neurons are composed of the same basic parts (Figure 15-1), which allow them to collect, conduct, and transmit impulses. The nucleus of the neuron is located within an expanded region, called the **cell body**, which is the metabolic center of the cell and the site where most of its material contents are manufactured. Extending from the cell body are a large number of minuscule extensions, called **dendrites**, which receive *incoming* information from external sources, typically other neurons. Also emerging from the cell body is a single, more prominent extension, the **axon**, which conducts *outgoing* impulses away from the cell body and toward the target cell. Axons range in length from a few micrometers to many meters in some of the larger mammals, such as whales and elephants. Most axons are wrapped by a jacket, or *myelin sheath*, consisting of layers of cell membranes. Since cell membranes consist predominantly of lipids, which are poor conductors of electricity, the myelin sheath functions as living "electrical tape," insulating the axon against electrical interference from its neighbors.

Most axons split near their ends into smaller processes, each ending in a **synaptic knob**—a specialized site where impulses are transmitted from neuron to target cell. Some cells in your brain may end in thousands of synaptic knobs, allowing these brain cells to communicate with thousands of potential targets.

Nerve cells can be grouped into three classes—sensory neurons, interneurons, and motor neurons—depending on the direction the impulses are carried and the type of cells to which they are functionally linked (Figure 15-2). **Sensory neurons** carry information about changes in the external and internal environments *toward* the central nervous system (the brain and spinal cord). Once they enter the central nervous system (CNS), impulses from sensory neurons are transmitted to **interneurons**, which transmit impulses from one part of the CNS to another. Interneurons route incoming or outgoing impulses, integrating the millions of messages constantly racing through the CNS. Outgoing impulses are carried by **motor neurons**, which stretch from the CNS to the body's muscles or glands whose contraction or secretion may be stimulated or inhibited.

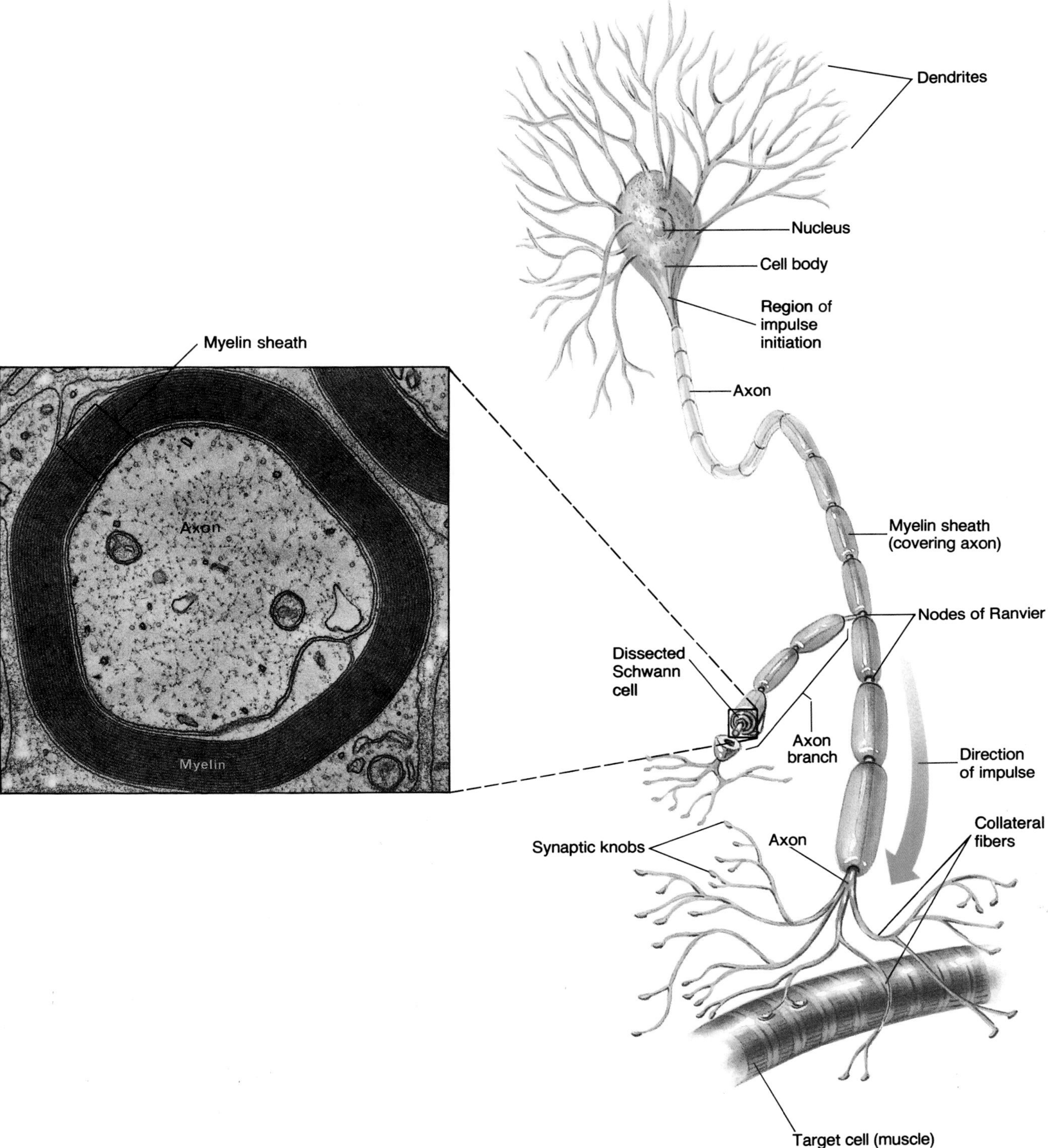

FIGURE 15-1
Anatomy of a neuron. Information enters the nerve cell through a branching network of dendrites. The incoming signals impinge on the cell body, which contains the nucleus and the cell's synthetic machinery. The cell body merges with the elongated axon, which often branches into numerous smaller collateral fibers, each of which ends in a synaptic knob. Most vertebrate axons are wrapped in a myelin sheath composed of Schwann cells. Inset shows an electron micrograph of a cross section through an axon surrounded by a myelin sheath.

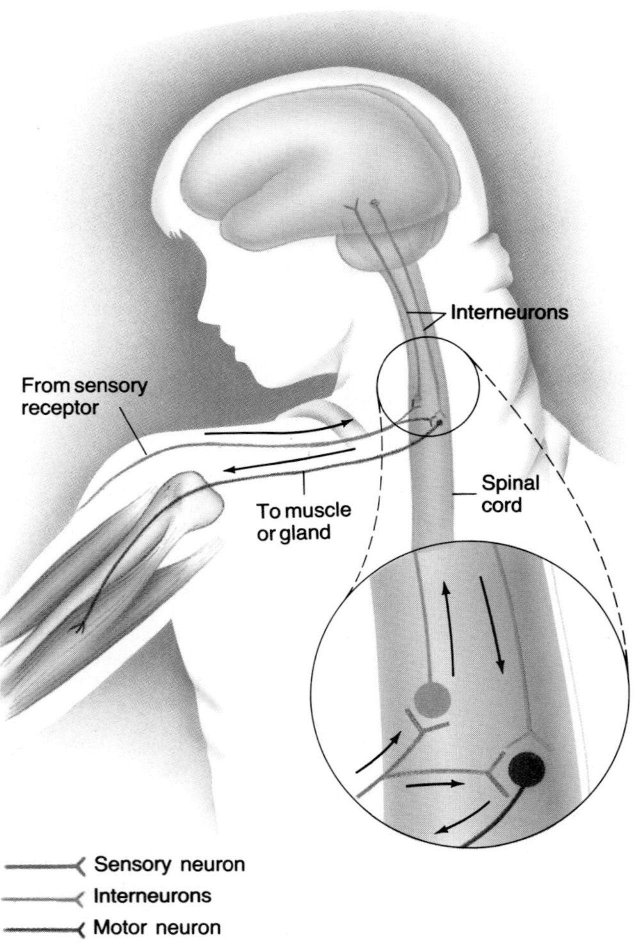

FIGURE 15-2
Three types of neurons. Sensory neurons carry impulses from the periphery (sensory receptors) to the CNS (as indicated by the spinal cord). The various pathways that run up and down the human brain and spinal cord are composed of large numbers of interneurons, which are located entirely within the brain or spinal cord. Motor neurons carry impulses from the CNS to effector cells in the periphery.

GENERATING AND CONDUCTING NEURAL IMPULSES

The concentrations of specific ions on the two sides of the plasma membrane of a "resting" neuron—that is, a neuron that is not conducting an impulse—are very different. The concentration of potassium ions (K^+) is approximately 30 times higher inside the cell than outside, while the concentration of sodium (Na^+) and chloride (Cl^-) ions are approximately 10 to 15 times higher outside the cell than inside. The ability of ions to move across the plasma membrane depends on the permeability of the membrane. Nerve cell membranes have two types of channels: *leak channels*, which are always open, and *gated channels*, which can be either open or closed. In the resting state, potassium ions diffuse out of a cell through potassium leak channels. In contrast, the nerve cell lacks sodium leak channels, and thus the plasma membrane is virtually impermeable to sodium ions.

Potassium ions are positively charged. The movement of positive charges out of the cell leaves the inside of the membrane more negatively charged than the outside. This separation of positive and negative charge is called a **potential difference**. The **resting potential**—the potential difference when the cell is at rest—measures approximately −70 millivolts. The negative value, due largely to the outward diffusion of potassium ions, indicates the negativity of the inside of the cell, relative to the outside (Figure 15-3, left circle).

If a resting neuron is stimulated, the gates of some of the sodium channels in the vicinity of the site of stimulation swing open, allowing a number of sodium ions to move into the cell. This movement of positive charges into the cell reduces the potential difference, making it less negative. Since the reduction in membrane voltage causes a decrease in the potential difference between the two sides of the membrane it is called a *depolarization*.

If the stimulus causes the membrane to depolarize by only a few millivolts, say from −70 to −60 millivolts, the membrane rapidly returns to its resting potential as soon as the stimulus has ceased. If the stimulus is great enough, however, the membrane becomes depolarized beyond a certain point (the *threshold*), triggering a new series of events. The gates on the sodium channels swing open, and sodium ions flood into the cell (Figure 15-3, middle circle). As a result of the inflow of sodium ions, the membrane potential briefly reverses itself (Figure 15-3, lower plot), becoming a positive potential of about +50 millivolts. Then, as the membrane potential reaches its peak positive value, the sodium gates close again, and the gated potassium channels open (Figure 15-3, right circle). As a result, potassium ions flood out of the cell, and the membrane potential swings back to a negative value (Figure 15-3, lower plot). The gated potassium channels then close, leaving only the potassium leak channels open. As a result, the membrane returns to its resting state. These collective changes in membrane potential are called an **action potential**; the entire sequence occurs in a few milliseconds (thousandths of a second).

Certain anesthetics, including Novocaine used by dentists, act by closing the gates of ion channels in the membranes of nerve cells. As long as these ion channels remain closed, the affected neurons are unable to generate action potentials and cannot inform the brain of the painful insults being experienced by your gums or teeth. The next time you are in a dentist's chair listening to the sound of the drill, you might think of the millions of plugged ion channels in the sensory neurons that lead from your teeth.

PROPAGATION OF ACTION POTENTIALS AS AN IMPULSE

Like falling dominoes, the propagation of action potentials along the entire length of the axon is the result of a chain reaction. An action potential at one site of a membrane induces a depolarization at the adjacent site of the membrane further along the neuron, initiating an action potential at that site. This new action potential stimulates the next region of the membrane, and the chain reaction continues, producing a wave of action potentials that travels along the entire length of the excited neuron, creating a neural impulse. Once an impulse is triggered, it passes down the entire length of the neuron without any loss of intensity, arriving at its target cell with the same strength it had at its point of origin.

We noted earlier that axons may be covered with a lipid-containing jacket, or myelin sheath. Notice in Figure 15-1 that the myelin sheath is interrupted along its length by uninsulated gaps, called the *nodes of Ranvier*. Action potentials are able to jump from one unwrapped node to the next, allowing an impulse to speed along a neuron at a velocity approximately 20 times that of a neuron that lacks a myelin sheath. The importance of myelination is dramatically illustrated by multiple sclerosis, a disease that results from the gradual deterioration of the myelin sheath that surrounds axons in various parts of the nervous system. The disease is characterized by progressive muscular dysfunction, often culminating in permanent paralysis.

NEUROTRANSMISSION: JUMPING THE SYNAPTIC CLEFT

Neurons are linked with their target cells at specialized junctions called **synapses**. Careful examination of a synapse reveals that the two cells do not make direct contact but are separated from each other by a narrow gap of about 20 to 40 nanometers. This gap is called the **synaptic cleft** (Figure 15-4). How does an impulse traveling down a presynaptic neuron "jump" across this cleft in order to affect the postsynaptic target cell? It does so by the release of **neurotransmitters**—chemicals that act to stimulate or inhibit target cells.

Neurotransmitter molecules, such as acetylcholine or norepinephrine, are stored in membrane-bound packets called **synaptic vesicles** found inside the synaptic knobs at the termini of the branches of axons (Figure 15-5). When an impulse reaches the end of a presynaptic neuron, calcium channels in the membrane are opened and calcium ions flow into the synaptic knob. The influx of calcium triggers the fusion of synaptic vesicles with the overlying plasma membrane and the consequent discharge of neurotransmitter molecules into the synaptic cleft. The discharged neurotransmitter molecules then diffuse across the narrow gap and bind specifically to receptor proteins

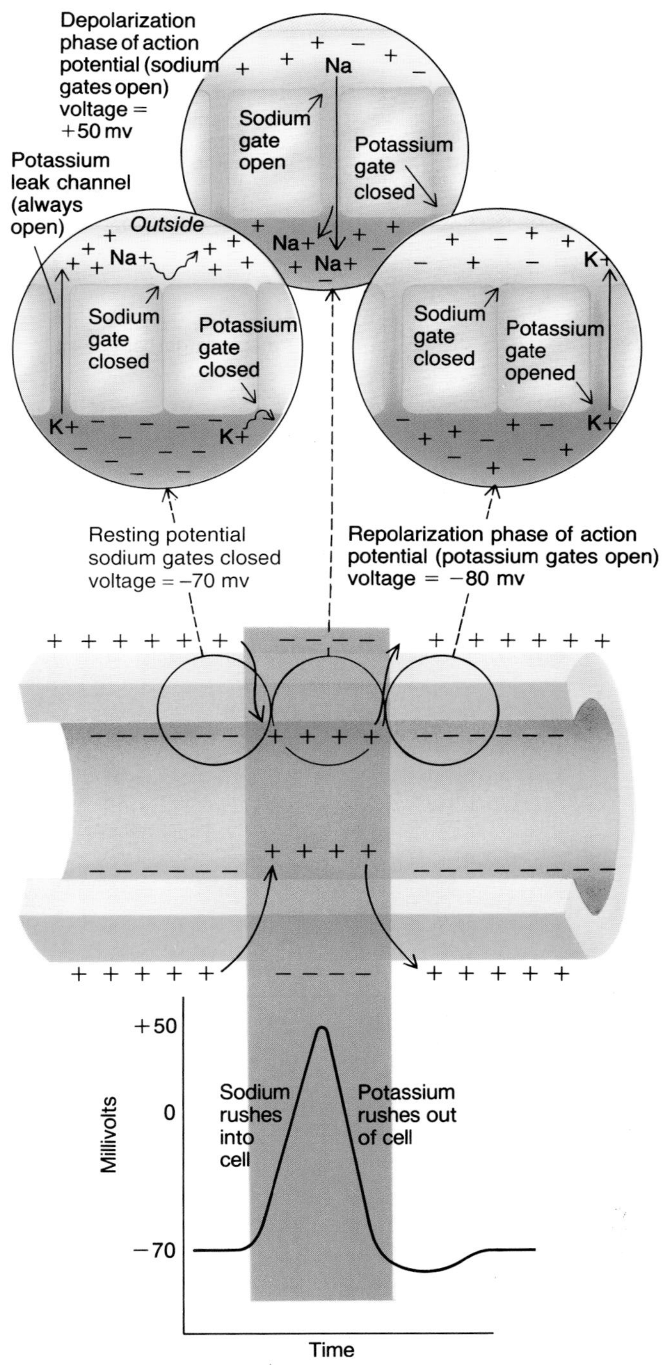

FIGURE 15-3
Formation and propagation of an action potential. When depolarization of the membrane exceeds the threshold value, the membrane's sodium gates open and allow positively charged Na^+ ions to move into the cell. The influx of Na^+ causes a fleeting reversal in the polarity of the membrane potential, from the resting value of −70 millivolts to +50 millivolts. This charge reversal is known as an action potential. Within a brief fraction of a second, the sodium gates close and the potassium gates open, allowing K^+ to diffuse across the membrane and reestablish a negative potential at that location. Almost as soon as they open, the potassium gates close, leaving the potassium leak channels as the primary path of ion movement across the membrane, and reestablishing the resting potential.

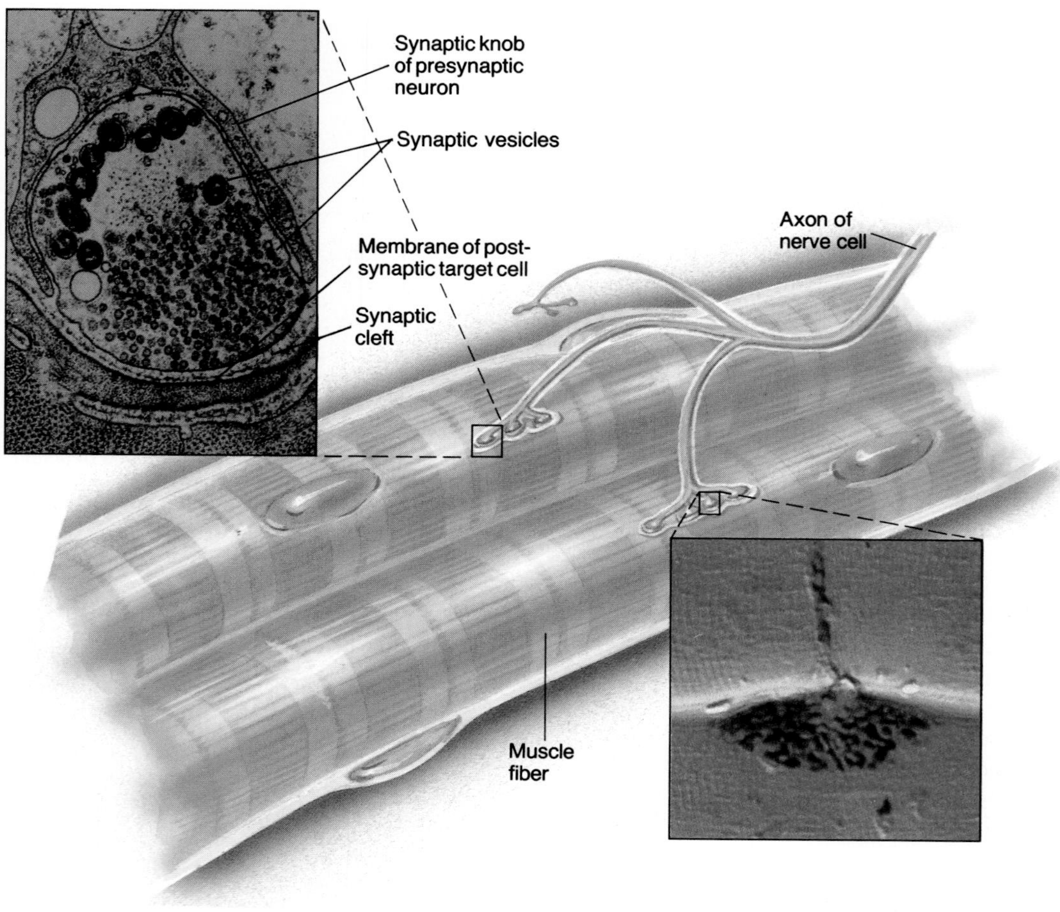

FIGURE 15-4

Synaptic junction between a neuron and a target muscle cell. Each synaptic knob abuts the target cell membrane very closely; the synaptic vesicles within the knob are indicated in the micrograph in the upper left box.

in the membrane of the postsynaptic target cell. The bound neurotransmitter molecules cause a change in the permeability of the postsynaptic membrane, which can lead to the activation (or in some cases, the inhibition) of the target cell. This phenomenon is illustrated by motor nerves that connect the spinal cord with the large skeletal muscles of the arm or leg. When the impulse reaches the tip of the motor neurons, the synaptic knobs release acetylcholine molecules, which bind to the plasma membrane of the muscle cell. Sodium channels in the membrane swing open, initiating an impulse in the target cell, which causes the cell to contract. Neurobiologists have discovered over 30 different chemicals that act as neurotransmitters, many of them within the brain alone. Some of these neurotransmitters may have profound effects on human emotions.

Taken as a group, synapses are more than just connecting sites between adjacent neurons; they are key determinants in the orderly routing of impulses through the nervous system. The billions of synapses that exist in a complex mammalian nervous system act like gates stationed along the various neural pathways; some pieces of information are allowed to pass from one neuron to another, while other pieces are held back or rerouted in some other direction. Such synaptic integration allows us to focus on the book we are reading or the music we are listening to, while simultaneously ignoring all of the distracting background noise with which we are constantly bombarded.

Now that we have described the form and function of neurons and the way they transmit information to other cells, we will examine how nerve cells are organized into a more complex neural structure—the nervous system.

THE HUMAN NERVOUS SYSTEM

The human nervous system, and that of other vertebrates (animals with backbones), is divided into two major divisions: the central nervous system and the peripheral nerv-

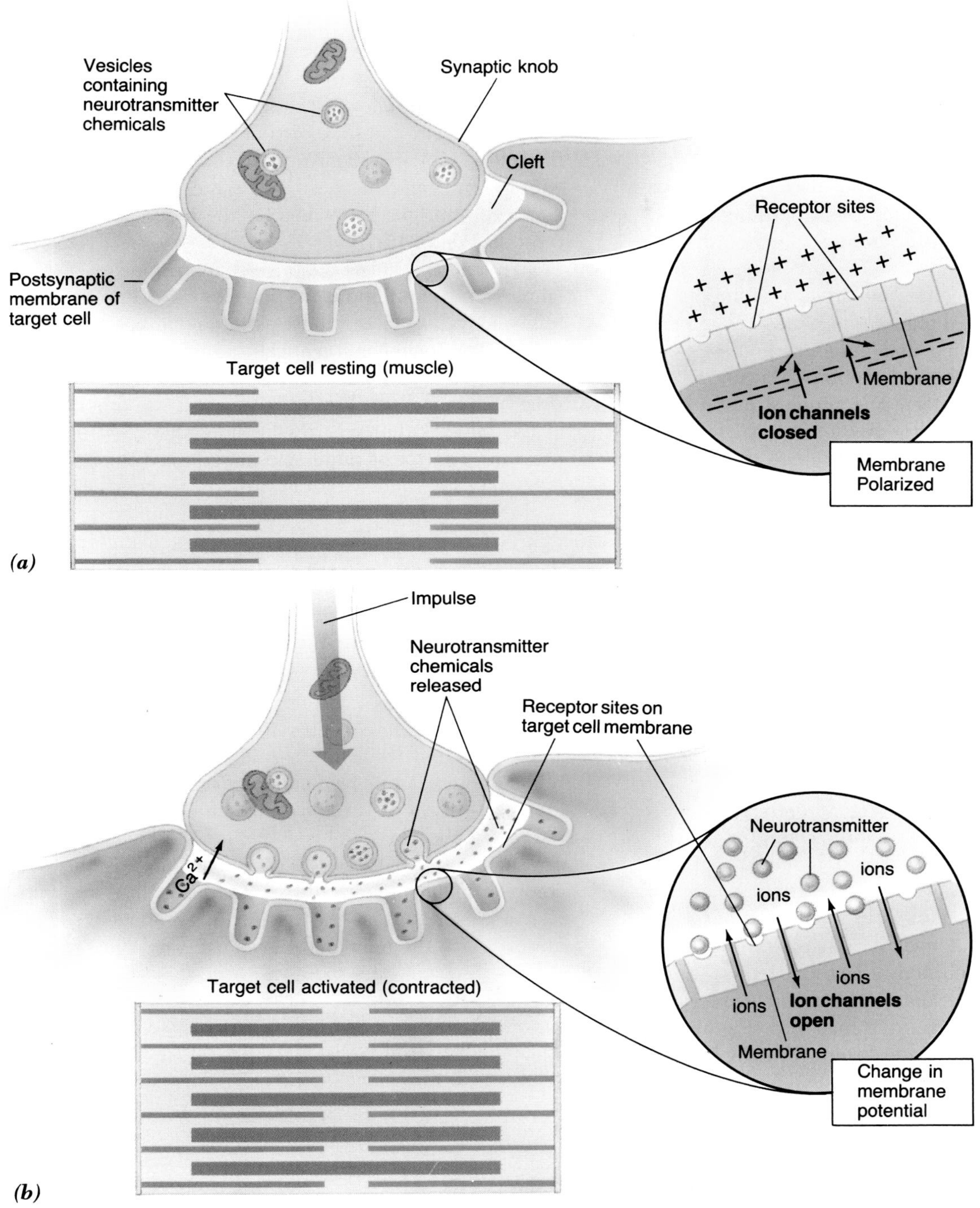

FIGURE 15-5

Synaptic transmission between a neuron and target cell (in this case a muscle fiber). ***(a)*** Neuron at rest. The synaptic cleft contains no neurotransmitter; the receptor sites remain empty; and the target remains unaffected by the neuron. ***(b)*** When the impulse reaches the tip of a neuron, calcium ions enter the synaptic knob; neurotransmitter molecules are released and bind to receptor proteins on the postsynaptic neuron; sodium ion channels swing open; and the muscle fiber contracts.

ous system. The **central nervous system (CNS)** consists of two major parts: the **brain**, which is the center of neural integration, and the **spinal cord**, which contains billions of neurons that run to and from the brain and which mediates many of the body's reflex responses. All neurons, or parts of neurons, situated outside the central nervous system are part of the **peripheral nervous system**. The peripheral nervous system connects the various organs and tissues of the body with the brain and spinal cord. The neurons of the peripheral system are grouped into **nerves**—"living cables" composed of large numbers of individual neurons bundled together in parallel alignment (Figure 15-6). All incoming and outgoing impulses are routed through the CNS, which functions as a centralized "command and control center." The simplest example of neural centralization is the reflex arc.

The Reflex Arc: Shortening Reaction Time

A **reflex** is an involuntary response to a stimulus—a response that occurs "automatically" and requires no conscious deliberation or awareness of the stimulus. Reflex responses occur so rapidly because the impulse travels the shortest route possible: from the site of the stimulus, through the central nervous system, to the responding gland or muscle. The chain of neurons that mediate a reflex make up a **reflex arc**. One of the simplest reflex arcs—that which is responsible for your foot jumping forward when the doctor strikes the area below your knee with a rubber hammer—is illustrated in Figure 15-7.

A reflex arc begins with a **sensory receptor**—a cell that responds to a change in its environment. The *stretch receptor* responsible for the knee-jerk response is actually the end of a sensory neuron embedded within a muscle of the thigh. When the tendon of the knee is tapped, the attached muscle is stretched, activating the receptor, which generates a neural impulse. Some of the terminal processes of the sensory neuron end in the spinal cord (Figure 15-7) in synapses with motor neurons leading directly back to the thigh muscle. Impulses traveling back along these motor neurons cause the muscle to contract, and the leg extends forward.

The stretch reflex just described didn't evolve to help doctors evaluate the state of your nervous system. Rather, the reflex is an adaptive response that helps you maintain your posture and balance. The same stretch reflex that keeps you upright also works for the mountain goats pictured in Figure 15-7b.

Your day is full of adaptive reflex responses. For example, an overambitious sip of steaming coffee triggers a reflex activation of the muscles in your tongue, jaw, and mouth. The tongue automatically jumps to the back of the mouth, which may open and discharge the liquid before your brain can consider the reaction of the other people at the table. Other reflexes include regulation of blood pressure, control of pupil size in response to changes in light intensity, and withdrawal of the hand when it encounters a sharp object. Reflexes are involuntary responses that occur very rapidly in a *stereotyped* manner; that is, the reflex is the same every time the same simple stimulus is encountered.

Architecture of the Human Central Nervous System

The central nervous system performs the most complex neural functions. It collects information about the internal and external environment, "sorts" through the information, relays impulses of its own along different pathways to various parts of the brain and spinal cord, and then acts on the information by sending command messages to peripheral muscles and glands. The human CNS is the most complex, highly organized structure on earth. We will now take a closer look at the two main components of the human CNS: the brain and the spinal cord.

The Brain

The human brain (Figure 15-8) is a mass of nearly 1.5 kilograms (3 pounds) of gelatinlike tissue. It contains a

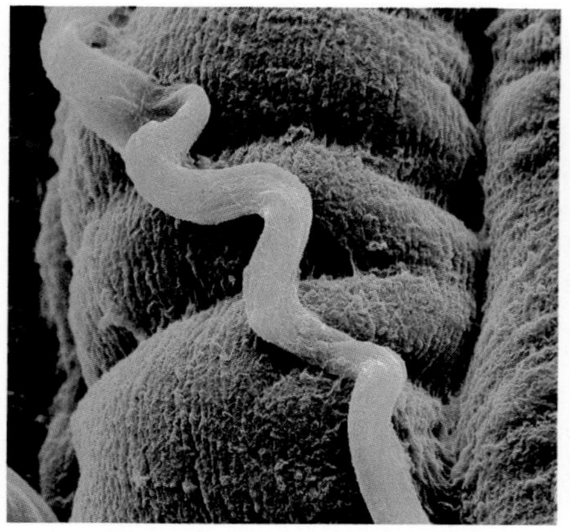
(a)

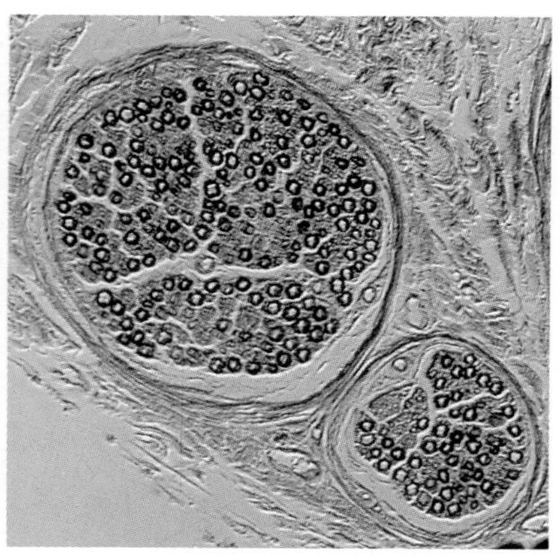
(b)

FIGURE 15-6
Nerves. ***(a)*** A nerve winding across the surface of muscle fibers. ***(b)*** In this cross section, the individual neurons in the nerve are seen bound together in a sheath of connective tissue.

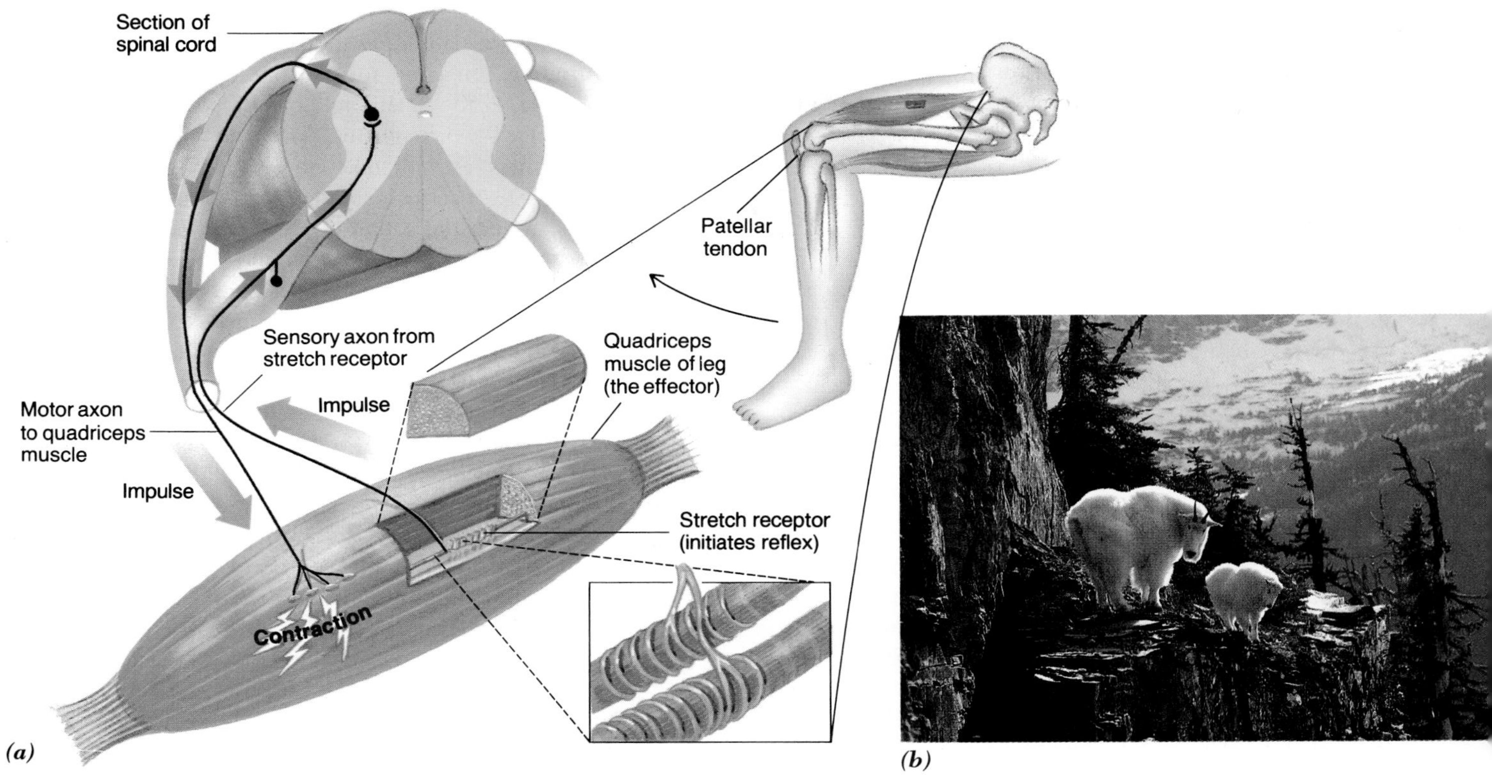

FIGURE 15-7

The reflex arc that mediates the knee-jerk response. ***(a)*** A reflex arc includes a receptor, which generates a nerve impulse after it is activated by some change in the environment; a sensory neuron, which carries the impulse to the CNS; a motor neuron, which sends a command signal back to the periphery; and a muscle or gland, which provides the response. In the case of the knee-jerk response, the receptor is actually the end of a sensory neuron, which is wrapped around a special muscle fiber situated within the quadriceps muscle of the thigh. When the muscle fiber is stretched, it activates the sensory neuron, sending impulses to the spinal cord. In this stretch reflex, impulses are transmitted directly from the sensory neuron to the motor neuron, causing the quadriceps muscle to contract and extend the leg forward. ***(b)*** These mountain goats depend on stretch reflexes to keep their footing on steep mountain trails.

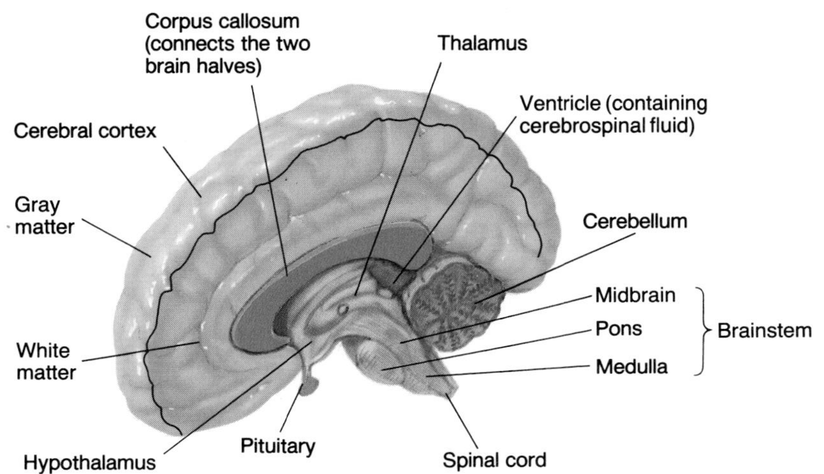

FIGURE 15-8

The human brain. Bisecting the brain reveals some of the brain's structural complexity. The medulla, pons, and midbrain constitute the brainstem, which controls visceral functions, such as breathing and cardiovascular activity. The thalamus is primarily a way station for sensory information that passes to the cerebrum, while the hypothalamus is one of the brain's homeostatic control centers. The right and left halves of the cerebral cortex are connected by a thick mass of nerve fibers that form the corpus callosum. The entire brain is covered by the meninges and bony cranium (not shown).

darker outer region—**gray matter**—in which the cell bodies and dendrites of many of the brain's neurons reside. Gray matter is rich in neuron-to-neuron synapses, places where neural associations are made. The inner region of the brain consists largely of **white matter** composed of myelinated axons; its whitish cast is provided by the light-colored myelin sheaths, which insulate the neurons. The tissue of the brain surrounds a series of distinct but interconnected chambers, called **ventricles**, which are filled with a protein-rich liquid, the **cerebrospinal fluid**, which cushions and nourishes the brain. The fluid also surrounds the delicate brain, cushioning it against injury. The brain and its surrounding fluid is encased in a protective sheath, the *meninges*, and enclosed within a bony case, the *cranium*. In the following discussion, we will describe three major parts of the brain: (1) the cerebrum, (2) the cerebellum, and (3) the brainstem.

The Cerebrum. The cerebrum is the most prominent part of the human brain (Figure 15-8). Its two halves, called cerebral hemispheres, are generally associated with "higher" brain functions, such as speech, memory, and rational thought. These functions are actually attributes of the cerebral cortex, the outer, highly wrinkled layer of the cerebrum (*cortex* = rind). Every cubic inch of this thin layer of gray matter contains 10,000 miles of interconnecting neurons. Each cerebral hemisphere is composed of four lobes—temporal, frontal, occipital, and parietal—each of which has a unique set of functions (Figure 15-9).

Memory Imagine, for a moment that you are standing in front of the main building of your former high school. You walk through the doors, down the halls, and into the room where you took one of your more memorable classes. Your ex-teacher is standing in the front of the class, and you can hear the sound of his or her voice. A minute earlier, none of this was on your mind, and now you have "directed" your brain to recall a series of specific, ordered images about which you can consciously reminisce. How is this possible? Scientists cannot yet answer this question, but they can provide some interesting insights. The types of images you have just brought to "mind" can also be evoked by electrical stimulation of various parts of the brain. When this procedure is performed, the person may see an image of something he or she thought was long forgotten, or hear a voice, or smell an aroma. The memory is somehow "stored" in the cerebral cortex, probably in the pattern of neural circuits formed by the billions of neurons that make up this portion of the brain.

Another way scientists have learned about memory is by studying persons who have suffered specific brain injuries that affect their memory. During the 1950s, a young man (known as H. M.) came to the attention of Brenda Milner of the Montreal Neurological Institute. A portion of the man's brain had been removed during surgery in an

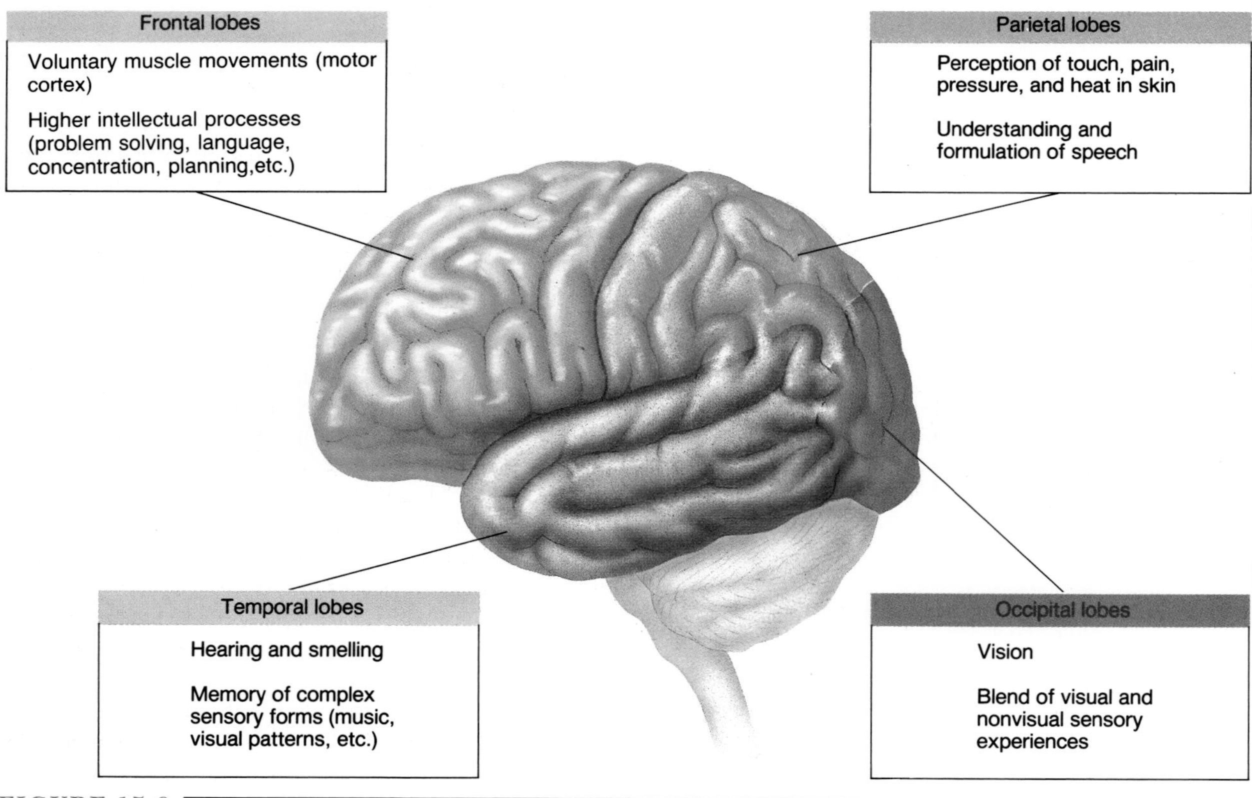

FIGURE 15-9
The four major lobes of the cerebrum and some of their roles in physiology and behavior.

attempt to stop severe seizures. The operation had unexpected and tragic results. H. M. was able to remember events from his childhood and adolescence but was unable to memorize any new information. For example, if he were given a number, he could only remember it as long as he focused all of his attention on it. As soon as he lost his attention, he would forget the number—as well as any recollection of being told about it. Over and over, he would greet as strangers the same doctors and researchers he worked with on a daily basis. Perhaps most tragic, H. M. was not able to grasp the reason for his problem because he could not remember what he was told.

The case of H. M. revealed the dual nature of human memory. H. M. could remember events from the past, similar to your having remembered your high-school classroom, because his *long-term memory* remained basically intact. This type of memory store can be contrasted to *short-term memory,* which allows us to retain a piece of information only for a matter of seconds to minutes. Short-term memory can be considered "working memory," since we use it to carry out our daily activities. Remembering a phone number from a telephone book until we get a chance to write it down is an example of short-term memory. As we go about our activities, certain pieces of information from our short-term memory become processed subconsciously in some unknown way so that they become part of our long-term memory. Short- and long-term memory apparently involve different types of neural processes and different parts of the brain.

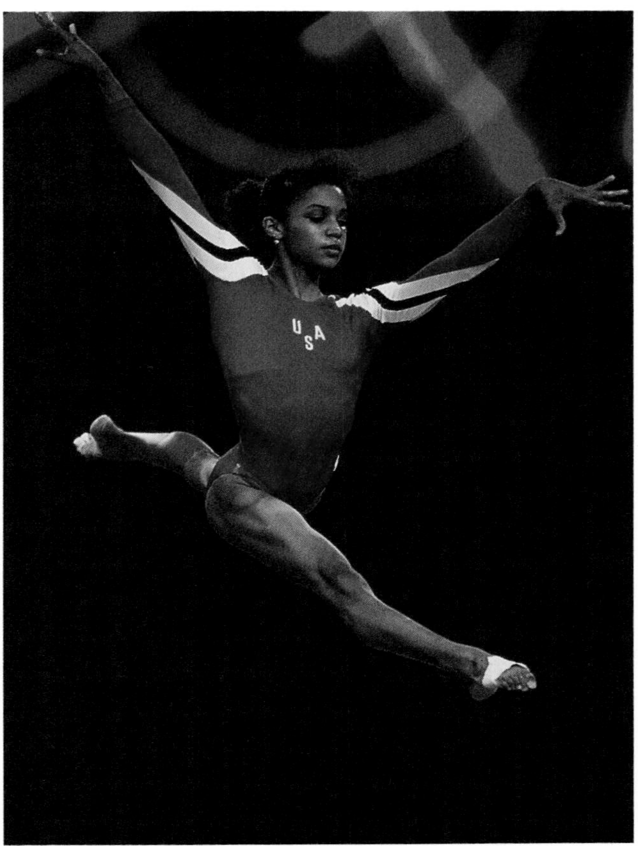

FIGURE 15-10
Betty Okino performing in a floor exercise.

Language Our ability to learn languages is associated with a center located in the frontal lobe of the cerebrum (usually the left frontal lobe). The localization of the language center can be dramatically revealed by anesthesizing the left frontal lobe of the brain, leaving the person awake and alert but unable to speak. The importance of the left frontal lobe in language abilities has been confirmed in studies of deaf children who learn to communicate via sign language at an early age. Sign language is a form of communication performed by motor activities of the hands. Like other motor activities, one might expect that control over signing would emanate from the motor regions of the brain, but researchers have found that it is actually the language center that controls the motions used in sign communication.

The Cerebellum. The cerebellum is a bulbous structure (Figure 15-8) that receives information from receptors located in the muscles, joints, and tendons, as well as from the eyes and ears. The amount of information converging on the cerebellum is revealed by examining the huge number of synaptic contacts made by some of these cells; some individual neurons of the cerebellum receive input from as many as 80,000 other neurons! Vast amounts of sensory information can be processed within a fraction of a second. Messages are then sent to the cerebrum, where they are used in directing such complex motor activities as playing a musical instrument, writing, or competing in gymnastics (Figure 15-10). For this reason, a person who sustains cerebellar damage will have difficulty performing smooth, coordinated movements; motor activities lose their subconscious basis, and the person may have to think about each movement that would otherwise be performed automatically.

The Brainstem, Reticular Formation, and Limbic System. The **brainstem** forms the central stalk of the brain (Figure 15-8), including the *pons* and *medulla*, and is responsible for regulating most of the involuntary, visceral activities of the body, such as breathing and swallowing, as well as maintaining heart rate and blood pressure. Consequently, permanent damage to the brainstem usually leads to coma or death. Whereas the cerebral cortex is the most recently evolved part of the vertebrate brain, the brainstem is probably the oldest part; it makes up the bulk of the brain of "lower" vertebrates, such as fish and amphibians. In humans, the brainstem also contains parts of two interconnected networks of the brain, the reticular formation and the limbic system.

The *reticular formation* is composed of several interconnected sites that selectively arouse conscious activity, producing a state of wakeful alertness. The reticular formation can be activated by any number of internal factors or external stimuli—the sound of an approaching horn; a

FIGURE 15-11
The human spinal cord. The spinal cord extends through hollows in the bony vertebrae and joins with the brain through a hole in the base of the skull. The spinal cord is connected to the periphery by 31 pairs of spinal nerves, each of which contains both sensory and motor neurons. Sensory neurons enter the spinal cord through the dorsal root, while motor neurons enter via the ventral root. The *dorsal root ganglion* is the site of a collection of cell bodies of the sensory neurons of the spinal nerves.

flashlight shining in your eyes; or the middle-of-the-night realization that you forgot to study for tomorrow morning's biology exam. When the reticular system fails to maintain arousal, the brain falls asleep and stays asleep until the reticular formation is activated by sensory input from, say, an alarm clock.

Researchers who study the neural basis of emotion have concentrated on an interconnected group of structures buried in the core of the brain, called the *limbic system*. The most evident component of the limbic system is the **hypothalamus**, a part of the brain responsible for regulating fluid and food intake, as well as body temperature. The hypothalamus is also associated with pleasure and joy, pain and fury, and an ability to balance emotions. Stimulation of the hypothalamus and other parts of the limbic system with electrodes can induce immediate anger or euphoria, sexual arousal, or deep relaxation.

The importance of the limbic system in controlling our emotions is also revealed by mood-altering drugs, such as cocaine. When certain parts of the limbic system are stimulated—either artificially (by electrodes) or naturally (by incoming impulses)—a strong feeling of pleasure is elicited. This feeling results from the release of dopamine, a substance that binds to certain nerve cells in the "pleasure centers" of the limbic system. Normally, the effects of dopamine are short-lived because the chemical is rapidly removed by its quick reuptake into the nerve cell that released it. However, cocaine, a compound extracted from the leaves of the South American coca plant (*Erythroxylum coca*) and inhaled by most users, interferes with the reuptake of dopamine. The sustained presence of dopamine produces a feeling of euphoria—a "high" that lasts for several minutes. The high is typically followed by a "crash," during which the person feels depressed, irritable, and anxious. Since the fastest way to relieve the unpleasant effects of the crash is to inhale more cocaine, the drug can rapidly become addictive.

The Spinal Cord

The brainstem merges with the second major component of the central nervous system—the **spinal cord**—a thick-walled, tubular mass of neural tissue that extends from the top of the neck to the lower back. The diameter of the spinal cord is similar to that of your little finger. Like the brain, the spinal cord is surrounded by cerebrospinal fluid and the meninges, which, in turn, is surrounded by the hollow bony vertebrae that make up the "backbone" (Figure 15-11). A narrow *central canal* filled with cerebrospinal fluid runs the length of the spinal cord and opens into the ventricles of the brain.

Like the brain, the spinal cord is composed of white matter (myelinated axons) and gray matter (dendrites and cell bodies). However, the relative location of white and gray matter is reversed in the spinal cord compared to that in the brain: The spinal cord's white matter surrounds the gray matter (Figure 15-11).

Architecture of the Peripheral Nervous System

The peripheral nervous system provides the neurological bridge between the central nervous system and the various parts of the body. The peripheral nervous system is made up of paired nerves that extend into the periphery from the CNS at various levels along the body. Most of the nerves contain both sensory and motor neurons. The sensory neurons provide the CNS with the sensory input needed for conscious awareness and the ability to respond to changes in the body and in the external environment. The motor neurons carry messages to the body's muscles and glands. These outbound motor neurons constitute two distinct divisions of the peripheral nervous system: the somatic and the autonomic divisions.

The Somatic and Autonomic Divisions of the Peripheral Nervous System. The **somatic division** of the peripheral nervous system carries messages to the skin and to those muscles that move the skeleton. These movements generally follow voluntary orders from the brain, although stereotyped reflex movements, such as the knee-jerk reflex, are also mediated by the somatic division.

The **autonomic division** controls the involuntary, homeostatic activities of the body's internal organs and blood vessels. Autonomic motor neurons regulate heart rate, blood-vessel diameter, respiratory rate, glandular secretions, excretory functions, digestion, sexual responses, and so forth. Although this system's operations are largely involuntary, many autonomic functions can be influenced by conscious control. Biofeedback training, for example, teaches you how to modify certain types of autonomic activity to help reduce blood pressure or lower physiological stress.

The autonomic division can be further divided into two parts: the **sympathetic system** and the **parasympathetic system** (Figure 15-12). Most organs of the body receive neurons from both of these systems, which evoke opposite (*antagonistic*) responses. Your heart rate, for example, is precisely regulated by a balance between continuous stimulatory influences from sympathetic neurons and inhibitory influences from parasympathetic neurons.

A survey of the specific responses evoked by the sympathetic and parasympathetic systems reveal their different adaptive functions (Figure 15-12). When target organs are stimulated by the sympathetic system, an adaptive response is triggered which better enables the body to cope with stressful situations, as might occur if you were suddenly confronted with a person coming toward you with a weapon. In such situations, the heart rate increases, and blood is shunted away from the periphery toward the lungs, for added oxygen uptake, and the skeletal muscles, whose energy demands soar during periods of activity. The liver is stimulated to release additional glucose into the blood, while the activity of the digestive organs is reduced. The level of metabolism is increased, causing a rise in body

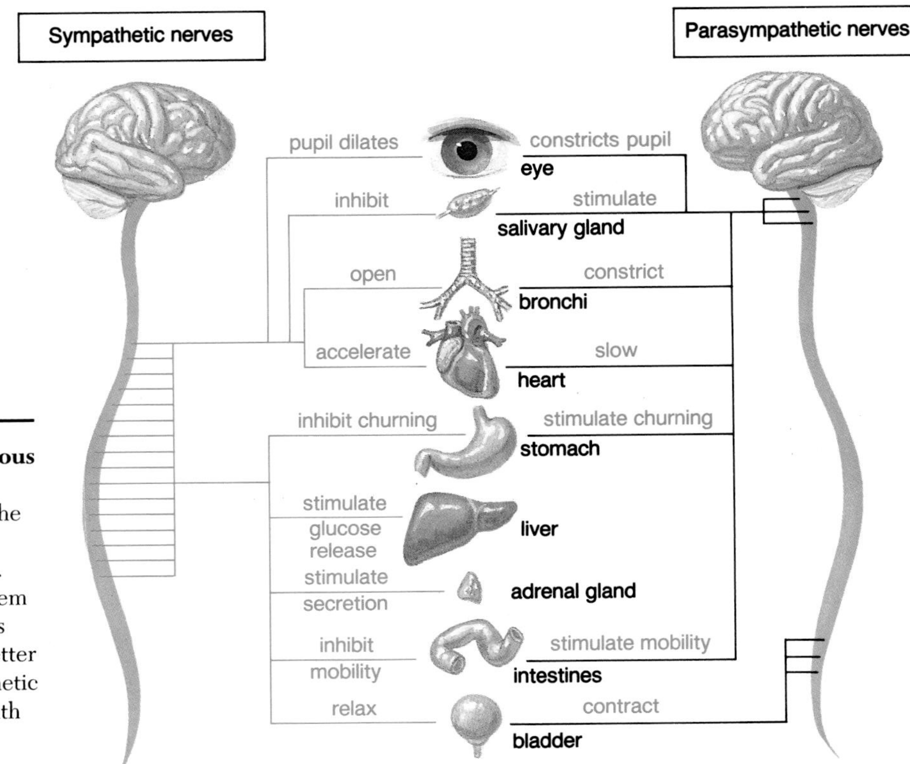

FIGURE 15-12
The two divisions of the autonomic nervous system control most of the same internal organs but under opposite circumstances. The effects of the parasympathetic system predominate during normal, relaxed activity. During times of stress, the sympathetic system predominates, eliciting physiological changes that increase the level of performance to better adapt to the stressful situation. Parasympathetic neurons release acetylcholine at synapses with their target organs, while most sympathetic neurons release norepinephrine.

temperature, which triggers the release of fluids by the sweat glands. Taken together, these changes constitute the "fight-or-flight" response, whereby the body is physiologically prepared either to confront the danger or beat a fast retreat. Sympathetic neurons also stimulate the adrenal gland, which releases hormones that further prepare the body to meet the crisis at hand. This teamwork is one example of the interrelatedness of the nervous and endocrine systems.

In contrast, the parasympathetic system handles the "housekeeping" functions of the body, such as digestion and excretion. For example, emptying the urinary bladder is a reflex that is mediated by the parasympathetic system and can be overridden by voluntary control. A person who has suffered a spinal-cord injury loses this voluntary control and thus cannot prevent voiding urine whenever the stretch receptors in the bladder wall launch the bladder-contraction reflex.

THE ENDOCRINE SYSTEM

Like the nervous system, the endocrine system is a communications network, one that uses chemical messengers, called hormones, to carry information. The hormone molecules are secreted by glandular cells into the surrounding tissue fluid. From there, the hormones diffuse into the bloodstream and are carried to distant parts of the body, where they interact with specific *target cells*—cells that have receptors for that particular hormone. When hormone molecules bind to the receptors of a target cell, the hormone evokes a dramatic change in the target cell's activity. For example, your gonads secrete steroid hormones—testosterone if you are a male, estrogen if you are a female. These hormones circulate in the blood, but only target cells, such as those that comprise your reproductive tract, bind the hormone and are affected by it. If these hormone molecules should stop being produced, as occurs in women during menopause, the reproductive tract loses its ability to carrry out reproductive activities.

Like the nervous system, the endocrine system is also a command-and-control center. Animals face a constant threat of disruption. Fluctuations in hundreds of critical chemicals occur every minute and must be corrected quickly before they disrupt homeostasis and interfere with vital processes. The endocrine system is responsible for detecting many such variations and releasing hormones whose actions help maintain a stable internal environment. Unlike the nervous system, which evokes rapid responses that typically result in muscle contraction, the endocrine organs stimulate slower responses that require changes in metabolic activities of target cells.

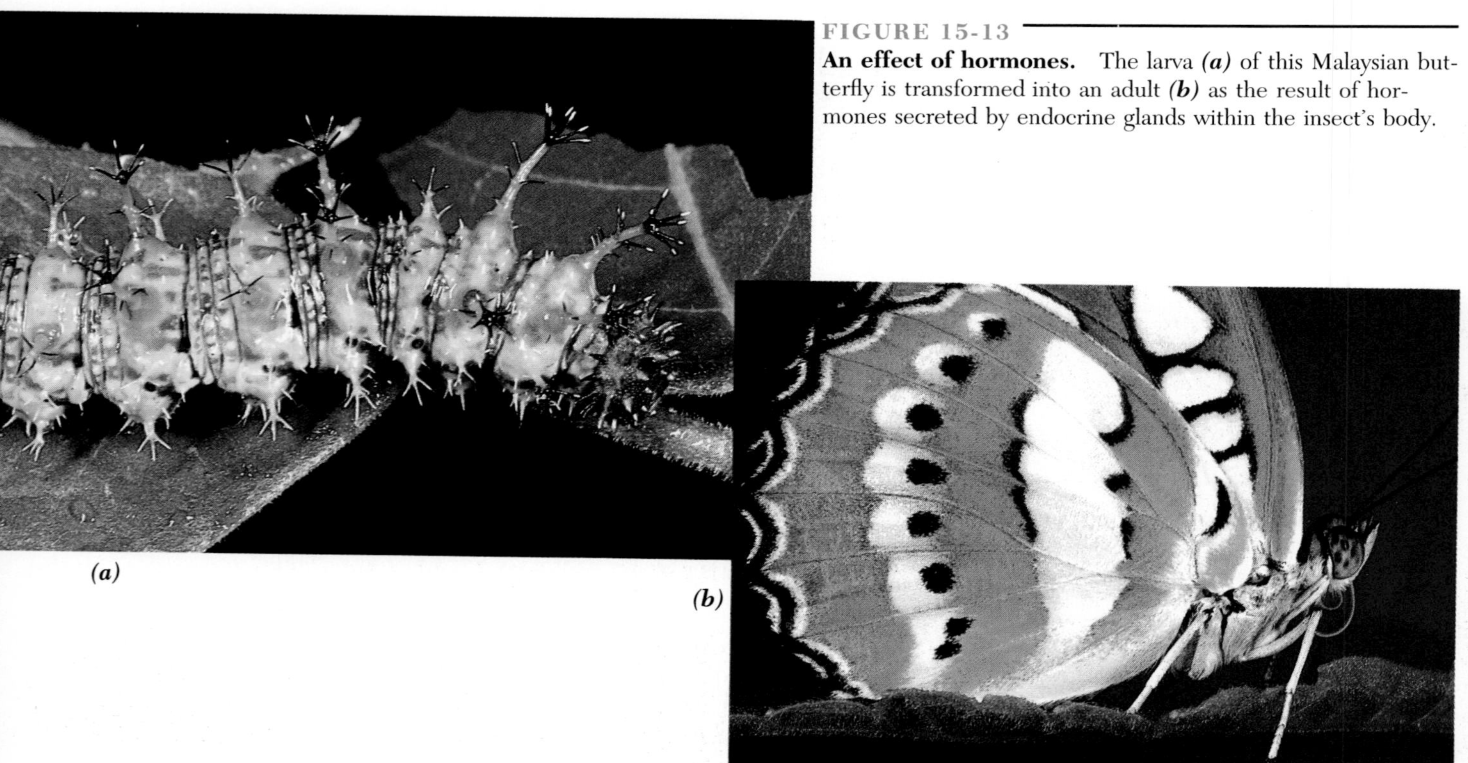

FIGURE 15-13

An effect of hormones. The larva ***(a)*** of this Malaysian butterfly is transformed into an adult ***(b)*** as the result of hormones secreted by endocrine glands within the insect's body.

Not all endocrine functions maintain homeostasis. Some hormones evoke changes that move the body in a new direction, away from a stable, homeostatic condition. For example, hormones "tell" an animal when to become reproductively mature, when to produce sperm, when to release eggs, and, in mammals, when to expel a fully developed fetus and produce milk. In some cases, hormones have the power to transform an animal into a different form that bears little resemblance to its former appearance (Figure 15-13).

The existence of hormones was first glimpsed in 1902 by two English physiologists, W. M. Bayliss and E. H. Starling, in their study of the control of pancreatic secretions. In addition to secreting insulin and other hormones, the pancreas produces digestive juices which pass through ducts into the intestine. These juices only flow when food materials enter the intestine, which suggested to biologists at the time that nerves informed the pancreas of the arrival of food in the intestine. Bayliss and Starling found otherwise. In one of their numerous experiments, they joined together the bloodstreams of two dogs and observed that the pancreas of both animals would secrete their juices when only one animal was being fed. The researchers concluded that a *chemical* message was being sent from the intestine to the pancreas via the circulatory system. They called this blood-borne, chemical messenger *secretin*. Bayliss and Starling further suggested that secretin was probably just one of a number of "hormones" that acted as chemical messengers within the body.

Although endocrine cells have recently been found almost everywhere in the human body, a few organs stand out as major sources of hormones (Figure 15-14). Some of the best-studied human hormones are listed in Table 15-1. We will begin with those parts of the endocrine system that have a close relationship with the nervous system.

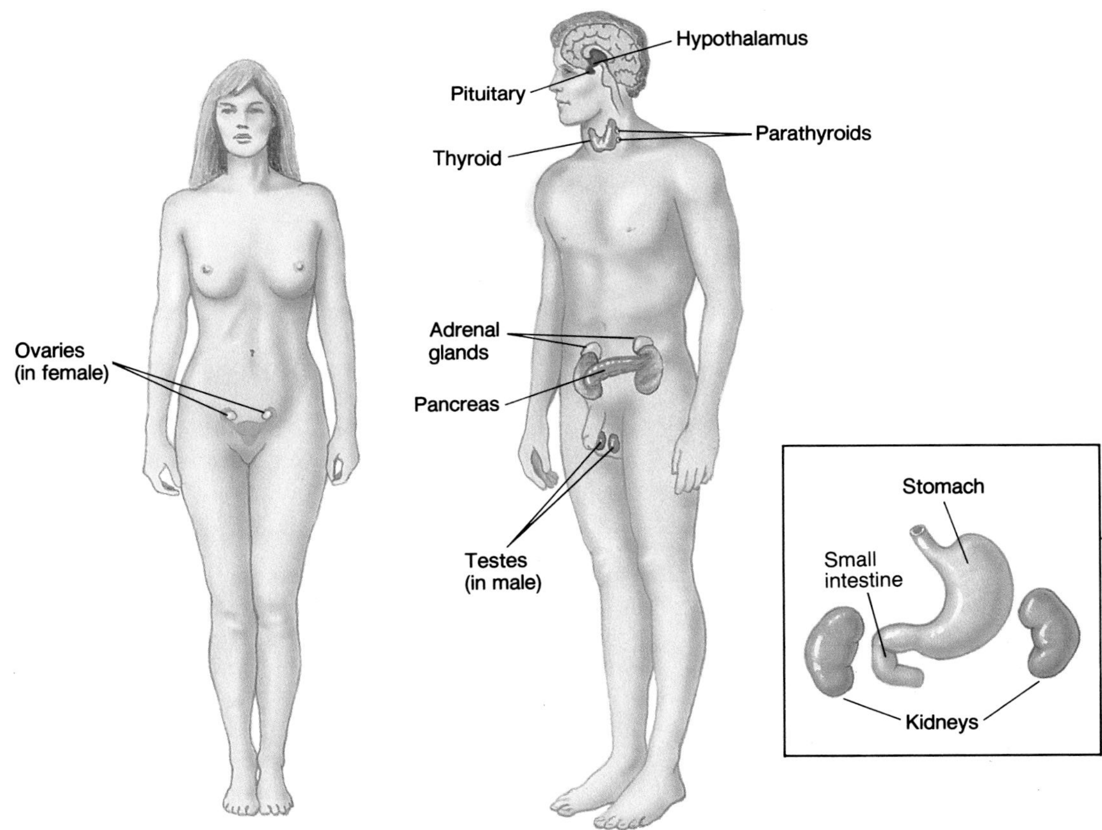

FIGURE 15-14

The major endocrine glands, plus some endocrine cells that are dispersed in nonendocrine organs (inset). Endocrine glands are defined as ductless glands; their secretions diffuse into the bloodstream, where they are carried to a distant target cell. The endocrine activities of the digestive tract, kidneys, and gonads are discussed in Chapters 18, 19, and 21, respectively.

TABLE 15-1
THE MAJOR ENDOCRINE GLANDS AND THEIR HORMONES

Gland	Hormone	Regulates
Anterior pituitary	Growth hormone (GH)	Growth; metabolism
	Thyroid-stimulating hormone (TSH)	Thyroid gland secretions
	Adrenocorticotropic hormone (ACTH)	Adrenal cortex secretions
	Prolactin	Milk production
	Gonadotropic hormones: Follicle-stimulating hormone (FSH); Luteinizing hormone (LH)	Production of gametes and sex hormones by gonads
Posterior pituitary	Oxytocin	Milk secretion; uterine motility
	Antidiuretic hormone (ADH) or vasopressin	Water excretion; blood pressure
Adrenal cortex	Cortisol	Metabolism
	Aldosterone	Sodium and potassium excretion
Adrenal medulla	Epinephrine and norepinephrine	Organic metabolism; cardiovascular function; stress response
Thyroid gland	Thyroxine (T-4)	Energy metabolism; growth
	Triiodothyronine (T-3)	
	Calcitonin	Calcium in blood
Parathyroids	Parathyroid hormone (PTH)	Calcium and phosphate in blood
Ovaries	Estrogen and progesterone	Reproductive system; growth and development; female secondary sex characteristics
Testes	Testosterone	Reproductive system; growth and development; male secondary sex characteristics; sex drive
Pancreas	Insulin and glucagon	Metabolism; blood-glucose concentration
Gastrointestinal tract	Gastrin	Secretory activity of stomach-small intestine; liver; pancreas; gall bladder
	Secretin	
	Cholecystokinin	Release of digestive enzymes
Pineal gland	Melatonin	Biological cycles; sexual maturation

THE RELATIONSHIP BETWEEN THE HYPOTHALAMUS AND THE PITUITARY

The endocrine system has no true central coordinating system comparable to the brain of the nervous system. The endocrine system's pituitary gland has been called the "master gland," however, because it controls the activities of so many other endocrine elements. The master gland has a master of its own—the hypothalamus—which, you will recall, is part of the brain (see Figure 15-8). Information about the state of the body arrives at the hypothalamus from two sources: (1) blood flowing through the hypothalamus provides information on the concentration of various chemicals in the bloodstream; and (2) sensory information arriving by neurons provides information on the conditions of various parts of the body. This information is used to control pituitary secretions through the use of a special type of cell, called a **neurosecretory cell**, that looks like a neuron but acts like an endocrine cell. Like neurons, neurosecretory cells receive information from other nerve cells and have elongated axons that conduct neural impulses. When the impulses reach the tip of the axon, they stimulate the release of stored materials from vesicles. Unlike neurons, however, the substance released from the tip of a neurosecretory cell is a hormone, which diffuses into the bloodstream, rather than a neurotransmitter, which diffuses across a synaptic cleft.

The pituitary is actually two glands in one: the posterior pituitary and the anterior pituitary.

Hormones of the Posterior Pituitary

The posterior pituitary manufactures no hormones of its own but stores and secretes two hormones, both small polypeptides consisting of nine amino acids. Both hormones are produced by neurosecretory cells that originate in the hypothalamus (Figure 15-15*a*). One of the hormones, *antidiuretic hormone (ADH)*, acts on the kidneys to reduce the loss of water during urine formation (Chapter 19). In females, the second hormone, *oxytocin*, triggers uterine contractions during childbirth. Oxytocin also stimulates the release of milk when the breast is stimulated by nursing.

Hormones of the Anterior Pituitary

The anterior pituitary is a true endocrine gland as it manufactures at least six separate hormones. When the hypothalamus senses a need for one of the anterior pituitary hormones, an impulse passes down the axon of the neurosecretory cell, and a stimulatory hormone (called a *releasing factor*) is discharged into tiny blood vessels that flow into the anterior pituitary (Figure 15-15*b*). The releasing factor triggers the secretion of the corresponding pituitary hormone.

Of the six known anterior pituitary hormones, four regulate the activity of other endocrine glands. *Adrenocorticotropic hormone (ACTH)* stimulates the cortex (outer layer) of the adrenal glands to secrete its hormones; *thyroid-stimulating hormone (TSH)* stimulates the thyroid gland to secrete thyroid hormones; and *follicle-stimulating hormone (FSH)* and *luteinizing hormone (LH)* act on the gonads to stimulate production of sex hormones (and promote the formation of sperm and eggs). The other two hormones are *prolactin*, which promotes the production of milk by mammary glands in the breast, and *growth hormone (GH)*, which is produced predominantly during childhood and adolescence and plays a key role in promoting normal body growth. When the anterior pituitary ceases its GH output, overall growth stops.

Extremes in height (giantism or certain kinds of dwarfism) are due to overproduction or underproduction of GH during childhood (Figure 15-16). Children who fail to grow at normal rates due to a deficiency of GH secretion can now be treated with growth hormone produced by recombinant DNA technology (Chapter 11). Excessive secretion

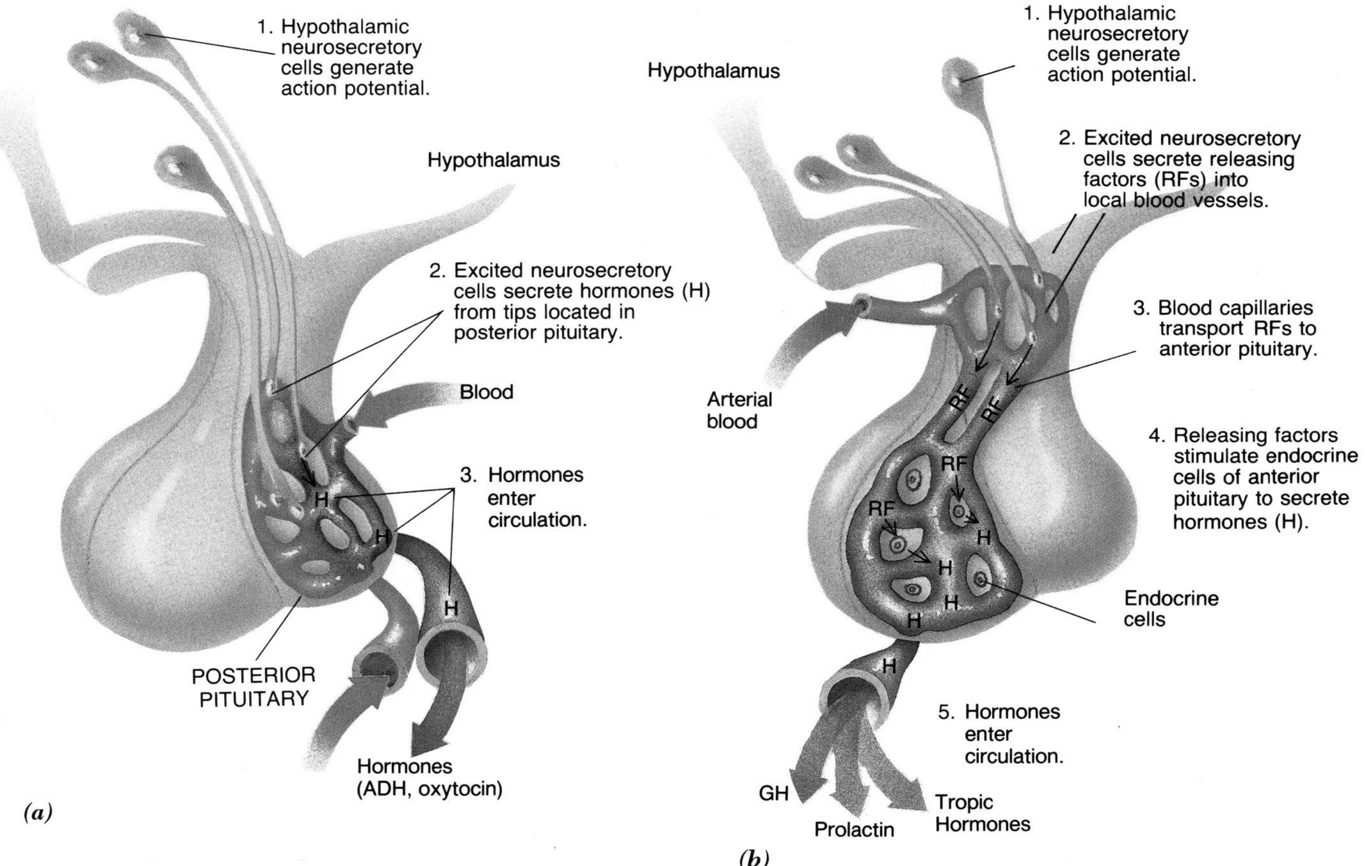

FIGURE 15-15

The pituitary-hypothalmic linkage. ***(a)*** The posterior pituitary is a storage depot for two hormones—ADH and oxytocin—which are produced in the cell bodies of hypothalmic neurons. Hormones are enclosed within vesicles in the cell body and are transported down the length of the axon to the posterior pituitary; from there the hormones are released into the bloodstream. Once in circulation, these hormones are carried to distant sites, where they act on target cells. ***(b)*** Secretion of the various hormones of the anterior pituitary is controlled by releasing and inhibiting factors, which are secreted by neurosecretory cells from the hypothalamus into local blood capillaries that carry the factors to the anterior pituitary. Hormones produced and secreted by the anterior pituitary circulate through the bloodstream and act on target cells around the body.

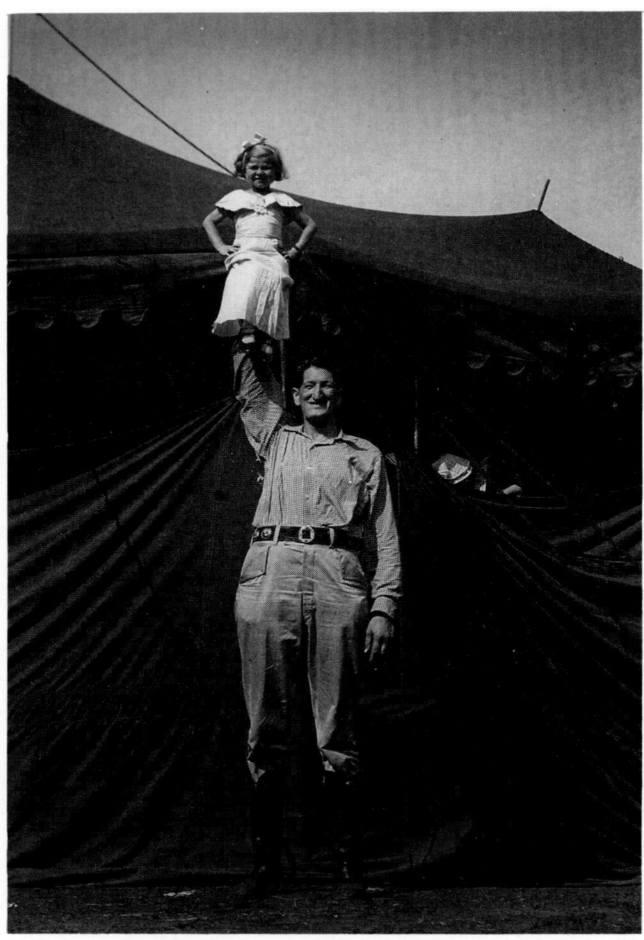

FIGURE 15-16
Circus side shows often included a pituitary dwarf and giant.

of GH *in an adult*, which is typically due to a pituitary tumor, leads to a condition known as *acromegaly*, in which certain parts of the body, including the hands, feet, jaw, and nose become enlarged.

The Adrenal Glands

Perched atop each kidney is an adrenal gland. Like the pituitary gland, each adrenal gland is essentially composed of two independent glands that produce very different hormones.

The gland's outer layer—the **adrenal cortex**—secretes steroid hormones, including:

- *Glucocorticoids*, which promote the conversion of amino acids to glucose and its uptake by the brain. The most important glucocorticoid is *cortisol* (also known as hydrocortisone), whose level is highest during periods of stress, such as that which occurs following a severe physical injury or a period of emotional trauma. Secretion of cortisol helps maintain normal blood-glucose levels and fuels brain activity. Cortisol secretion is stimulated by ACTH released from the pituitary. At very high concentrations, cortisol suppresses the body's normal response to injury, including inflammation. Consequently, cortisol and related synthetic compounds are highly effective in treating persistent inflammation, such as that caused by arthritis or bursitis, and severe allergies or asthma. Glucocorticoids may have serious side effects, however, including suppression of the body's ability to fight infections, hypertension, and ulcers, and must be used with great caution. The reason that extended periods of stress take their toll on the body may be the result of increased secretion of cortisol.
- *Mineralocorticoids*, which regulate the concentrations of sodium and potassium in the blood. The most important mineralocorticoid is *aldosterone*, which stimulates the kidney to reabsorb sodium into the blood, maintaining homeostatic levels of this important ion. If the level of sodium in the blood rises, aldosterone secretion is reduced, increasing excretion of the ion in the urine.

The adrenal cortex is essential for survival. Without aldosterone, for example, the body quickly loses sodium ions needed for neuron activity, muscular contraction, and blood-pressure stability. Without glucocorticoids, glucose concentrations in the blood plummet, crippling cellular energy metabolism. A person whose adrenal cortex fails to produce sufficient levels of these hormones suffers from *Addison's disease*, which is characterized by extreme weakness, weight loss, and impaired heart and kidney function, and is treated by administering synthetic adrenal cortical steroids. John F. Kennedy suffered from Addison's disease prior to the campaign for the presidency in 1960; diagonosis and treatment of the condition transformed a sick, weak-looking candidate into a healthy, vigorous campaigner. Overproduction of adrenal cortical steroids can lead to Cushing's syndrome, which is characterized by fatigue, weakening of bones, deposition of fat (particularly in the back of the neck), and a rounded facial appearance.

The core of the adrenal gland is the **adrenal medulla**, which was mentioned on page 276 in connection with the "fight-or-flight" response. When stimulated by nerve fibers, the adrenal medulla secretes two hormones: **epinephrine** (also called adrenalin) and **norepinephrine** (or noradrenalin). These hormones jolt the body into readiness to escape or confront an emergency. When confronted with an emergency, a person's metabolic rate increases and heart rate accelerates; additional glucose and oxygen are shunted to voluntary muscles, increasing muscular performance; blood vessels to the skin constrict, helping reduce the loss of blood through injured tissues; and red blood cells pour into the bloodstream from their reserves in the spleen, enhancing oxygen transport and replacing blood cells that are lost during bleeding.

THE THYROID GLAND

In the early twentieth century, many people found themselves afflicted with enormous swellings in the front of their necks. This condition, known as *simple goiter*, is the result of an enlargement of the **thyroid gland**, a butterfly-shaped gland that lies just in front of the windpipe (trachea) in humans. The swelling is often accompanied by lethargy, hair loss, slowed heart beat, and mental sluggishness. Goiter initially occurred frequently in geographic regions where natural sources of dietary iodine were scarce. The link between iodine deficiencies and the disease was not established until 1916, at which time it was shown that adding iodine to the diet, such as in iodized table salt, prevented the formation of goiters.

Why should iodine be related to thyroid function? The reason becomes evident when we examine the structure of two hormones produced by the thyroid gland. These hormones—**thyroxine** and **triiodothyronine**—are amino acids that contain iodine. Together, they are called "thyroid hormone." When dietary iodine is inadequate, the thyroid cannot produce enough thyroid hormone, causing the thyroid to enlarge in its perpetual "attempt" to produce enough thyroid hormone.

Normally, thyroid hormone enhances mitochondrial activity in various target cells, which increases energy availability and metabolic rate. In contrast, a person with a lowered output of thyroid hormone—whether due to a decreased availability of iodine or to a sluggish thyroid gland—experiences fatigue and low energy levels. Such thyroid hormone deficits are called *hypothyroidism*, which is corrected by treatment with thyroid hormone supplements. An excess of thyroid hormone (*hyperthyroidism*) may lead to *Grave's disease*, which is characterized by hyperactivity, weight loss, nervousness, and insomnia.

After decades of studying thyroid function, a third thyroid hormone, **calcitonin**, was discovered in the early 1960s. Calcitonin regulates blood-calcium levels in cooperation with another set of endocrine structures, the parathyroid glands.

THE PARATHYROID GLANDS

You probably think of your bones as stable, inert structures. In reality, the inorganic salts that give bones their hardness are continually being dissolved and redeposited by cells embedded in the bone tissue. This is the major reason why calcium is an essential dietary substance; it is needed to replace the calcium that is removed from bone and lost in the urine. Maintaining the calcium concentrations in the blood requires the cooperation of the thyroid gland and four tiny **parathyroid glands** embedded in the back of the thyroid gland.

When calcium levels in the blood are low, the parathyroid glands secrete **parathyroid hormone** (**PTH**), which acts on the bones, kidneys, and intestines to restore normal calcium concentrations. Under PTH influence, calcium is withdrawn from bones, and the released mineral is absorbed into the bloodstream; kidneys retain calcium; vitamin D is activated; and calcium absorption from the intestines is enhanced. In contrast, if blood-calcium levels should rise to abnormally high levels, the thyroid gland is stimulated to secrete calcitonin, which exerts the opposite effects of PTH.

THE PANCREAS

The pancreas is predominantly a producer of digestive enzymes, but this organ also contains tiny endocrine centers, called **islets of Langerhans**, which secrete several protein hormones into the blood. One of these hormones—insulin—is secreted when the concentration of glucose in the blood begins to exceed normal levels, usually as sugar floods the bloodstream following a meal. Insulin acts on numerous organs of the body to stimulate the cellular uptake of glucose, which is necessary in initiating the utilization of the sugar. Insulin also directs the conversion of surplus glucose into glycogen for storage in the liver and muscles. This conversion prevents the loss of surplus sugar, since excess sugar that remains in the blood will be excreted in the urine.

Insulin can do too good a job, however, and deplete the blood of glucose. When the concentration of blood sugar begins to drop below normal, the islets of Langerhans alter their secretory priorities and increase the secretion of **glucagon**, another pancreatic hormone. Glucagon promotes glycogen breakdown in cells in which it is stored and elevates glucose concentration in the blood, especially during times of stress, when increased cellular and physical activity is likely to expend greater amounts of energy.

A deficiency in insulin production can lead to *diabetes mellitus*; complications can include cardiovascular damage, kidney failure, blindness, and susceptibility to life-threatening infections. Because they are unable to take up glucose from the blood, the cells of diabetics must turn to other sources of energy, such as protein and fat reserves. Consequently, some diabetics may become emaciated. Diabetics may also be very thirsty or dehydrated because increased blood-glucose levels promote frequent urination.

In approximately 15 percent of the cases, diabetes begins during childhood as the result of the destruction of the insulin-producing cells of the pancreas, which apparently results from either a viral infection or an attack by the person's own antibodies (Chapter 20). In these cases, which are classified as *juvenile-onset* or *Type 1* diabetes, the person produces little or no insulin and must be treated by daily injection of the hormone. The majority of diabetics are classified as having the less-severe, *adult-onset* or *Type 2* form of the disease, in which insulin levels may be normal but the target cells fail to respond to the hormone because of insulin receptor abnormalities. Type 2 diabetes can often be controlled if the patient adheres to strict dietary recommendations.

The Pineal Gland

Embedded deep within the brain is a tiny, pinecone-shaped organ, the **pineal gland**. In 1898, Otto Huebner, a German physician, reported on a 4-year-old boy who had undergone premature puberty and then died. Autopsy results indicated that the boy had died from a tumor of the pineal gland. Huebner proposed that the pineal gland normally produces a hormone that suppresses sexual development during childhoood. If production of that chemical should cease—if the gland is destroyed by a tumor, for example—the inhibition is removed, and sexual maturation occurs prematurely. The hormone produced by the pineal gland was finally isolated in 1958, a feat requiring 200,000 cattle pineal glands, and was named *melatonin*.

In addition to its role in suppressing the onset of sexual maturation, melatonin is thought to play a key role in regulating our daily rhythms, including sleep, motor activity, and brain waves. Studies of people who fly across several time zones indicate that it takes several days for the normal rhythm of melatonin secretion to reestablish itself, suggesting that the familiar problem of "jet lag" may be due to the time needed to reset the pineal "clock."

GONADS: THE TESTES AND OVARIES

When stimulated by pituitary gonadotropins, the gonads—the male **testes** and the female **ovaries**—secrete the powerful steroid sex hormones **testosterone** and **estrogen**, respectively. These substances promote the development of the reproductive tracts and secondary sex characteristics that distinguish genders, such as deeper voices and facial hair (which are stimulated by testosterone) and breast enlargement (which is triggered by estrogen). The activities and regulation of the sex hormones are discussed in more detail in Chapter 21. The abuse of steroid hormones by some bodybuilders and athletes is discussed in The Human Perspective: Chemically Enhanced Athletes.

ACTION AT THE TARGET

In order for a cell to respond as a target to a particular hormone, the cell must have a protein receptor whose structure allows it to bind that hormone specifically. Depending on the particular hormone, the receptor may be located at the cell surface or within the cytoplasm. The position of the receptor within the cell is a key factor in determining the mechanism of action of the corresponding hormone.

Cell Surface Receptors

Most of the hormones that act via cell surface receptors are water-soluble substances (amino acids and polypeptides) that cannot simply diffuse through the plasma membrane and enter the cytoplasm. All of these hormones act without entering the target cell.

The hormone itself can be considered a "first messenger." It binds to a receptor on the outer surface of the target cell's plasma membrane, promoting a change at the membrane and the release of a "**second messenger**," which enters the cytoplasm and actually triggers the response. The best-studied and most widespread second messenger is a small molecule called *cyclic AMP*. We will illustrate this type of mechanism using the pancreatic hormone glucagon.

When glucagon binds to a glucagon receptor located at the surface of the plasma membrane, the hormone changes the shape of the receptor, which activates an enzyme located on the inner surface of the membrane (Figure 15-17*a*) This enzyme, called *adenylate cyclase*, converts ATP to cyclic AMP (cAMP), one of the most universally important molecules associated with cellular regulation. Cyclic AMP then diffuses into the cytoplasm, where it starts a chain reaction; one enzyme activates another enzyme and so forth, until eventually glycogen is split into its glucose monomers. As a result, glucagon secretion increases the concentration of glucose in the blood. The involvement of two messengers—a hormone and cAMP—amplifies the original signal. The binding of one glucagon molecule at the cell surface promotes the synthesis of thousands of cAMP molecules inside the cell. As a result of this amplification, a very small concentration of hormone in the blood can produce a rapid, massive response within a target cell.

Cytoplasmic Receptors

Steroid hormones have a very different mechanism of action (Figure 15-17*b*). These hydrophobic hormones diffuse through the plasma membrane, where they bind to a specific cytoplasmic receptor molecule. The receptor–hormone complex then enters the nucleus, where it becomes a gene regulatory protein that binds to a particular DNA sequence. Binding of the receptor–hormone complex activates the genes responsible for the hormone-induced changes.

EVOLUTION AND ADAPTATION: TYING IT TOGETHER

The lives of simpler animals, such as flatworms and earthworms, depend just as much on maintaining a relatively constant internal composition as do the lives of more complex animals, such as humans. It is not surprising, therefore, that regulatory systems evolved quite early during animal evolution. In fact, the division of regulatory responsibilities into neural and endocrine mechanisms can be traced back to some of the earliest multicellular animals.

The nervous and endocrine systems have evolved to regulate different types of biological activities. The elec-

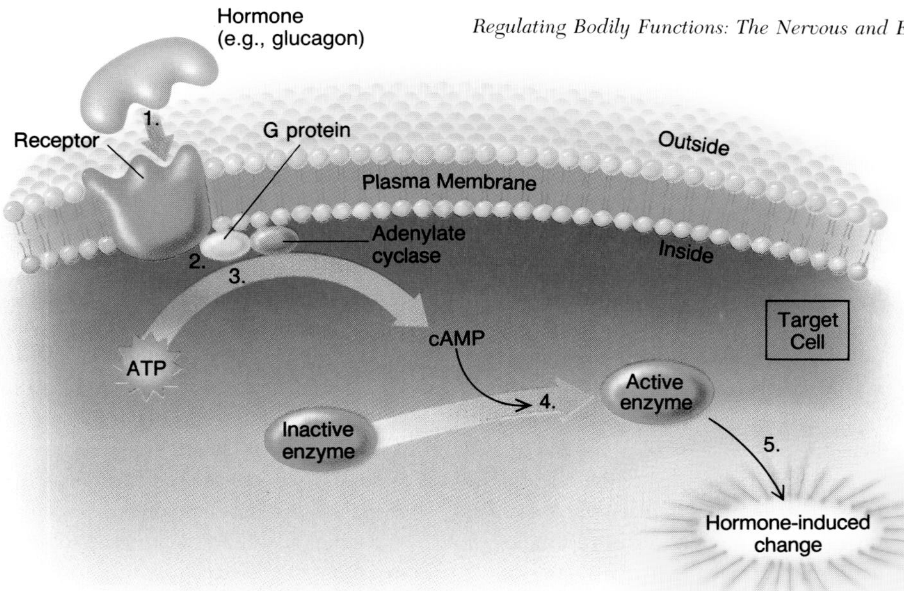

(a)

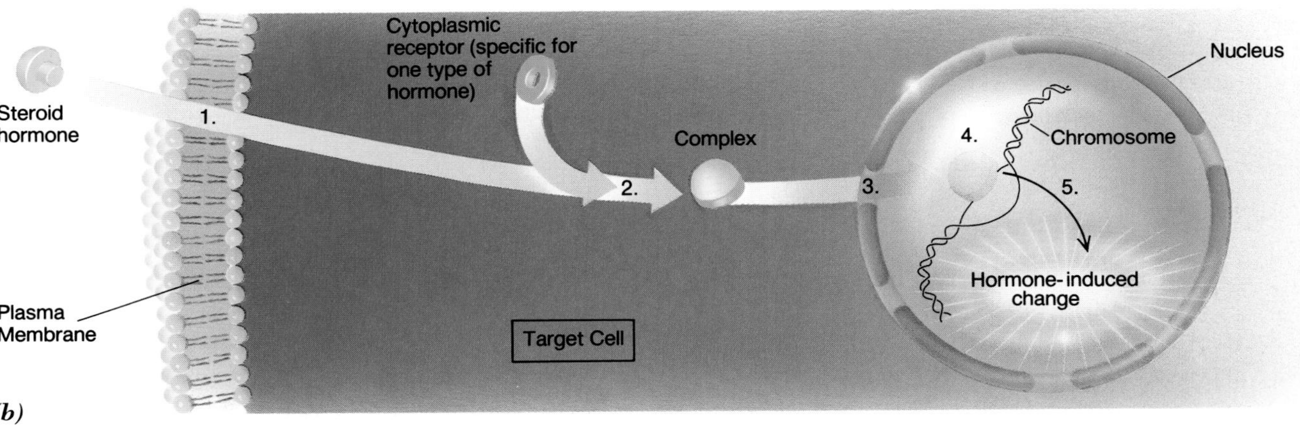

(b)

FIGURE 15-17

Mechanisms of hormone action. ***(a)*** *Second messenger model:* (1) Hormone binds to surface receptor. (2) Binding activates a G protein, which activates adenylate cyclase. (3) Activated adenylate cyclase converts ATP to cAMP (a "second messenger"). (4) cAMP activates (or inhibits) specific enzymes. (5) Activated enzymes produce specific changes in the cell. ***(b)*** *Steroid hormones:* (1) Hormone readily passes through the plasma membrane and (2) reacts with a protein receptor molecule. (3) The hormone–receptor complex enters the nucleus and (4) attaches to a specific site on the chromosome. (5) Attachment activates those genes responsible for the hormone-induced change, such as the genes required for construction of a hair shaft by a previously dormant hair follicle on an adolescent boy's chin.

trical messages of the nervous system are conducted rapidly, and their effect is typically rather short-lived. In contrast, the chemical messages of the endocrine system circulate relatively slowly around the body, and their effects, such as sexual maturation, may last the remainder of the organism's life. The targets of the nervous system are restricted to glands and muscles, while the targets of the endocrine system include virtually every cell in the body. The particular gland or muscle cells that are stimulated by nerve impulses are determined by the particular pathway along which the impulses are conducted; the specific targets stimulated by an endocrine response are determined by the hormones secreted and the presence or absence of receptor molecules in various cells of the body.

While the differences between the nervous and endocrine systems are great, the two cannot be separated from each other, either anatomically or physiologically, and they are best thought of as interrelated systems working together to control bodily functions. For example, most organs of the body receive both neural and hormonal messages, and their overall activity is determined by regulatory influences from both systems.

◁ THE HUMAN PERSPECTIVE ▷

Chemically Enhanced Athletes

In 1896, Ellery Clark leaped nearly 21 feet to win the long-jump event in the Olympic Games; in the 1936 Olympic Games, Jesse Owens won the event with a jump of more than 26 feet; today, the world record is more than 29 feet. While modern athletes are superior to any in history, their achievements have been tainted in recent years by the discovery that too many of them rely on drugs to maximize their physical potential. Topping the list of banned substances taken by some athletes are steroids, commonly referred to as "anabolic" steroids because they tend to increase biosynthesis (anabolism), especially protein synthesis. The most common of these steroids is the male sex hormone testosterone and its synthetic derivatives. These drugs help build muscle, restore energy, and enhance aggressiveness. Their use has grown so common, even among high-school athletes, that suppliers of black-market drugs distribute printed advertising brochures and order forms at local gyms.

The dangers of steroid abuse are considerable. In men, a steroid surplus in the blood, produced by taking external sources of testosterone, causes the body's own testosterone production to drop due to negative feedback and causes the gonadal cells that produce the male hormone to waste away. Other side effects include lowered sperm counts, penis atrophy, reduced resistance against infectious disease, and potentially serious heart irregularities. In women, testosteronelike hormones can stimulate the development of facial hair and other masculine features. Prolonged exposure to steroids suppresses the immune system in both sexes, compromising the body's ability to defend itself against many infectious diseases. Steroid supplements also increase the likelihood of cancer.

The interrelationship between the nervous and endocrine systems is most evident in the hypothalamus, which contains neurosecretory cells that combine neural and hormonal functions in a single structure. Neurosecretory cells are found in some of the most primitive animals, suggesting that the first endocrine cells to evolve were modified neurons. It is a relatively small evolutionary step from a typical neuron secreting a chemical transmitter substance into a synaptic cleft to a neurosecretory cell secreting a hormone into the bloodstream. With the evolution of the neurosecretory cell, new opportunities became available to coordinate events occurring in the outside world with an animal's own internal activities. Information relayed from eyes and ears located on the body surface could be transmitted through the nervous system to neurosecretory cells; chemical messengers could then regulate various internal physiological activities, such as reproduction. As animals became more complex, so did their nervous and endocrine systems.

SYNOPSIS

The nervous system communicates information from one part of the body to another and, in doing so, regulates the body's activities and maintains homeostasis. Nerve cells (neurons) have dendrites that collect information; an axon that conducts information in the form of impulses; and terminal synaptic knobs that transmit information to a target cell (a muscle, gland, or nerve cell).

Neural impulses result from sequential changes in the distribution of ions across the plasma membrane. In the resting neuron, potassium ions leak outward, creating a potential difference. Should the neuron receive a stimulus that causes the potential to drop past the threshold, an action potential is triggered, whereby gated sodium channels open, allowing an inward movement of sodium ions, followed by an opening of gated potassium ion channels, which returns the potential to the resting state. Once an impulse reaches the synaptic knob of a neuron, it is transmitted to the adjacent target cell by the release of a chemical-transmitter substance that diffuses across the narrow synaptic cleft between the two cells. Synapses act as gates that route information in the proper directions through the nervous system.

The nervous system of vertebrates consists of a central nervous system (brain and spinal cord) and a peripheral nervous system (nerves). Sensory neurons carry information about changes in the environment from the periphery toward the CNS, while motor neurons carry command signals to the body's muscles and glands. Interneurons convey impulses within the CNS. The fastest motor reactions to stimuli are mediated by reflex arcs, which consist of at least one sensory and one motor neuron. Motor commands may be sent via the somatic division, which controls voluntary muscle activity, or the autonomic division, which controls involuntary activity. The autonomic division is divided into a parasympathetic branch responsible primarily for "housekeeping" activities, such as digestion and excretion, and a sympathetic branch responsible for readying the body to meet stressful situations.

Hormones are chemical messengers that are secreted into the bloodstream and act on specific target cells that possess receptors for that hormone. Some hormones maintain homeostatic conditions; others cause irreversible bodily changes, including metamorphosis. Hormone action depends on the position of the receptor within the target cell. Most hormones bind to specific receptor molecules on the target cell's surface, causing the release of a second messenger (most often cyclic AMP) into the cytoplasm, which activates an enzyme that initiates the response. Steroid hormones bind to cytoplasmic receptors, forming a complex that activates specific genes whose products trigger the response.

The double-lobed pituitary is the "master gland" of the endocrine system. The hormones of the posterior pituitary—ADH and oxytocin—are products of neurosecretory cells that originate in the hypothalamus. ADH regulates water reabsorption in the kidney, and oxytocin regulates uterine contractions and the release of milk. Of the hormones of the anterior pituitary, ACTH, TSH, FSH, and LH stimulate secretion of hormones by other endocrine glands; growth hormone stimulates body growth; and prolactin induces milk production. The adrenal cortex secretes steroid hormones that regulate sugar metabolism and stabilize sodium and potassium concentrations in the blood. The adrenal medulla secretes epinephrine, which boosts metabolic activity and prepares the body for stressful conditions. The thyroid gland secretes thyroid hormone, which regulates the body's overall rate of metabolism, and calcitonin, which lowers the concentration of calcium in the blood. This latter effect is countered by parathyroid hormone. The pancreas secretes insulin and glucagon which maintain the proper concentration of glucose in the blood.

Review Questions

1. Describe the various parts of a myelinated neuron and the role of each part in neural function.
2. Describe the components required to mediate a simple reflex response, such as the knee-jerk.
3. Compare and contrast the following: somatic and autonomic divisions, sympathetic and parasympathetic systems, resting potential and action potential, the role of leak channels versus gated channels in generating an action potential, the roles of sodium and potassium gated channels, myelinated and nonmyelinated axons.
4. Why is cyclic AMP called a "second messenger"? What is the first messenger? Why is it said that use of a second messenger amplifies the effect of the hormone?
5. How does the anterior pituitary differ from the posterior pituitary?

Critical Thinking Questions

1. If Banting and Best had injected far too strong an extract from a ligated pancreas during their first trials on dogs, what effect do you think the injection of excess insulin would have had on these diabetic animals?
2. Most actions controlled by simple reflex arcs involve protective responses or actions that must be completed frequently. Why is it adaptive that these responses do not depend upon the cerebral cortex?
3. Explain how the existence of two autonomic systems permits fine-tuning of physiological activity.
4. Explain how the endocrine system achieves feedback, amplification, and specificity.
5. What evidence is there for a common origin for nervous and endocrine systems? What reasons can you suggest to explain why the two diverged to form separate systems?

CHAPTER

16

Sensory Perception

By placing invisible barriers in the path of porpoises, biologists discovered that these animals use echolocation to navigate around objects.

STEPS TO DISCOVERY
Echolocation: Seeing in the Dark

For hundreds of years, naturalists have wondered how bats are able to fly through caves in total darkness or through wooded fields at night, capturing fast-flying insects they can't possibly see. Near the end of the eighteenth century, Lazarro Spallanzani, an Italian biologist, caught a number of bats in a cathedral belfry, impaired their vision, and let them go. Returning to the belfry at a later time, Spallanzani found the bats back in their roost at the top of the cathedral. Not only did the bats return home, but their stomachs were full of insects, indicating that they had been able to capture food without their sense of sight. In contrast, bats whose ears had been plugged with wax were reluctant even to take flight and frequently collided into objects when they tried to fly. Hanging objects in front of the bats' mouths also impeded the animals' flying abilities. Spallanzani was baffled: How could an animal's ears and mouth be of more importance than its eyes in guiding flight?

The question remained unanswered until 1938, at which time, Donald Griffin, an undergraduate at Harvard University, brought a cage of bats to one of his physics professors, George Pierce. Pierce had developed an apparatus for converting high-frequency sound waves into a form that could be heard by the human ear. After placing the cage of bats in front of the sound detector, Griffin and Pierce became the first people to "hear" the bursts of sounds emitted by bats. When the bats were set free in the physics laboratory, these same high-frequency sounds could be heard as a bat flew in a straight direction toward the microphone. For Griffin, this was the initial step in a long, illustrious career centered around studying the role of sound in animal navigation.

The mechanism bats use to fly through a pitch-black obstacle course relies on the same principles first employed by the Allied navy to hunt for German submarines during World War II. That mechanism is sonar. The basic principle of sonar is simple: Sound waves are emitted from a transmitter; they bounce off objects in the environment; and the "echoes" are detected by an appropriate receiver. The elapsed time between the emission of the sound pulse and its return provides precise information about the location and shape of the object. Most bats employ a remarkably compact and efficient sonar system to detect insect prey and to avoid colliding with objects during flight. High-frequency sounds—those beyond the range of the human ear—are produced by the bat's larynx (voicebox), and the echoes are received by the animal's large ears. Impulses from the ears are then sent to the brain, where they are perceived as a detailed mental picture of the objects in the environment. Griffin coined the term "echolocation" to describe this phenomenon.

Echolocation by bats is an adaptation appropriate for their nocturnal (nighttime) peak of activity. Using echolocation, a bat flying at night can detect and avoid a wire as thin as 0.3 millimeters in diameter or can hone in on a tiny, moving insect with unerring accuracy. These feats become even more impressive when you realize that the echo returning to the bat from such objects is about 2,000 times less intense than is the sound emitted. Moreover, a bat can pick out this faint echo in a crowd of other bats, each sending its own pulses of sound waves.

During the early 1950s, studies by Winthrop Kellogg, a marine biologist at Florida State University, and others, demonstrated that bats were not the only mammals capable of navigating by echolocation. One day, Kellogg and Robert Kohler, an electronics engineer, were out in a boat in the Gulf of Mexico when they saw a school of porpoises swimming toward them. The scientists lowered a microphone connected to a speaker and tape recorder into the water. As the porpoises swam toward them, all that could be heard above water was the sound of the animals exhaling through their blowholes. "...but the underwater listening gear told a very different story." According to Kellogg, the animals were emitting sequences of underwater clicks and clacks "such as might be produced by a rusty hinge if it were opened slowly. ...By the time the group was about to make its final dive, the crescendo from the speaker in our boat had become a clattering din which almost drowned out the human voice."

To study the echolocating capabilities of these animals, Kellogg persuaded Florida State University to build a special "porpoise laboratory" at its nearby marine station. There, investigators discovered that porpoises possessed a sophisticated sonar system they could use to distinguish between similarly shaped objects, to avoid invisible barriers (such as sheets of glass), to swim through an elaborate obstacle course, to locate food, and to pick up objects off the bottom of their enclosure.

It may surprise you to learn that you probably have some capability to echolocate as well. Over the past hundred years or so, many reports have been written about the remarkable ability of many blind people to detect the presence of obstacles in their path without the use of a cane. The first controlled experiment on this subject was conducted in 1940 by Michael Supa and Milton Kotzin, a pair of graduate students at Cornell University, one of whom was blind himself. Subjects that were either blind or sighted but blindfolded were asked to walk down a long wooden hallway. They were told that a large, fiberboard screen *might* be placed at some random site along their way and were asked to report when they detected the presence of the screen. The subjects were able to detect the screen within 1 to 5 meters of its presence. In those cases when the screen was absent, none of the subjects reported its presence. In contrast, when the subjects' ears were tightly plugged, they collided with the screen in every trial. The authors of the experiment concluded that the echo of the sound emitted from the subjects' footsteps on the floor provided information about the location of the screen.

Consider for a moment what it would be like if you were suddenly deprived of information from your sense organs. Our concept of the world is largely based on information gathered by our eyes, ears, nose, mouth, and the surface of our skin. Without our sense organs, we could not find food, avoid danger, or communicate with others. We would know absolutely nothing about the outside world. Our sense organs also provide us with many of the pleasures in life—the taste of food, the sounds of music, and the touch of another person. Collecting sensory data satisfies a basic human need. In fact, in experiments where people are placed in an environment that is devoid of sensory stimulation—a dark, silent, constant-temperature chamber—the subjects quickly become restless and irrational and often begin to hallucinate.

We also depend on information gathered by the sense organs in our internal environment, which is required to maintain homeostasis. Without information from sensors located in our muscles, joints, tendons, and internal organs, we could not walk, stand, or digest food; we would have no awareness of our body's state of well-being. Without information from sensors that detect chemicals and pressure in the walls of our arteries, we would not be able to maintain proper cardiovascular or respiratory function. In fact, very little in our body would work properly.

▼ ▼ ▼

THE RESPONSE OF A SENSORY RECEPTOR TO A STIMULUS

The first step in the chain of events leading to sensory awareness is the interaction between a *stimulus* and a *sensory receptor*. A stimulus is any change in the internal or external environment, such as a cold breeze or an empty stomach, and a sensory receptor is a cell that can be activated by that change. Each type of stimulus activates only one type of sensory receptor, which include

- *Mechanoreceptors*, which respond to mechanical pressure and detect motion, touch, pressure, and sound.
- *Thermoreceptors*, which detect changes in temperature and react to heat and cold.
- *Chemoreceptors*, which are activated by specific chemicals, such as those that induce a particular taste or smell. Chemoreceptors also monitor concentrations of critical nutrients (glucose, amino acids) or respiratory gases (oxygen and carbon dioxide) in the blood.
- *Photoreceptors*, which respond to light.
- *Pain receptors*, which respond to excess heat, pressure, irritating chemicals, or chemicals that are released from damaged or inflamed tissue.

The structure of each sensory receptor enables the receptor to respond specifically to a particular type of stimulus, such as light of certain wavelengths or particular chemicals. The interaction between a stimulus and a receptor elicits some change in the receptor, which evokes an alteration in the ionic permeability of the plasma membrane, which may trigger an action potential in a sensory neuron. For example, photoreceptors located in the retina of your eye contain membrane pigments that absorb light energy. Absorption of light energy causes a change in the shape of the pigment, which leads to a change in membrane permeability, which can trigger an impulse in a neuron of the optic nerve, which generates electrical activity in parts of the brain concerned with vision. The result is the perception of light.

SOMATIC SENSATION

Somatic sensation (*soma* = body) is a sense of the physiological state of the body. The information is gathered by somatic sensory receptors located in the skin, skeletal muscles, tendons, and joints. Somatic sensory receptors inform the CNS about pressure, stretch, temperature, and whether or not a stimulus is intense or threatening—judgments that often lead to the perception of pain. Information from these receptors is required for bodily movements and for maintaining homeostasis. The nature of somatic sensation is revealed further in The Human Perspective: Perceiving the World: It's All in Your Head, page 292.

VISION

Vision is a sense that is based on light rays reflected into our eyes from objects in the external environment. To most people, vision is the richest form of sensory input. The complexity of our eyes, and the large amount of brain tissue devoted to sight, attests to the evolutionary importance of this sense. We inherited a keen sense of sight from our primate ancestors who lived in trees and got around by jumping from branch to branch. One miscalculation could result in a long fall to the ground. In this environment, natural selection favored those animals with better eyesight.

The main component of the human eye—the eyeball (Figure 16-1)—is roughly spherical in shape and is situated within a protective socket of the skull. Light enters the eyeball by passing through the **cornea**, a delicate, transparent window, whose remarkable structure lets light rays pass into the eye unobstructed.

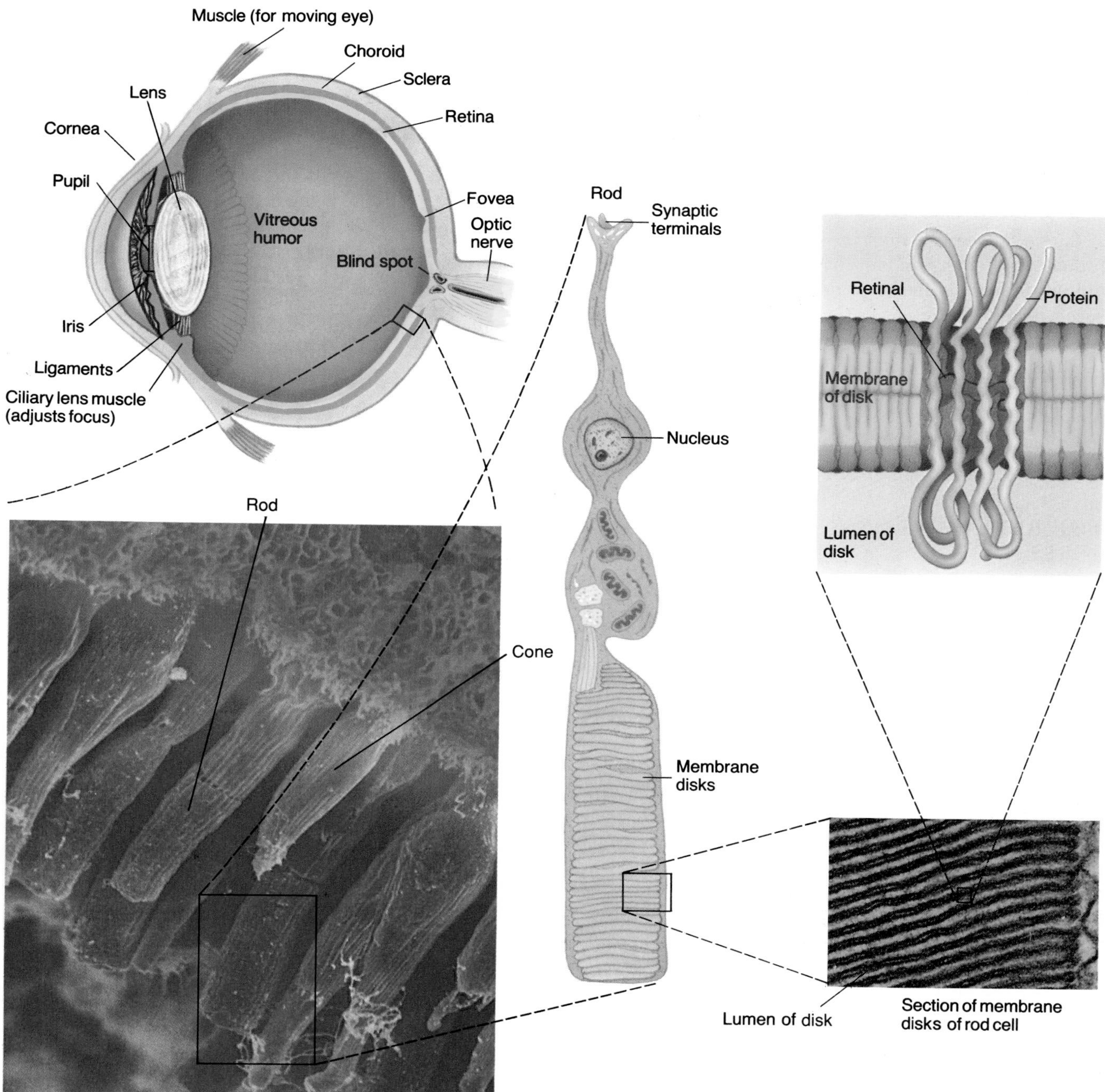

FIGURE 16-1

The human eye. Light enters through the cornea and then travels through the pupil, the hole in the middle of the colored iris. Pupil size changes to regulate the amount of light that enters the eye. Light rays are focused by the lens, whose shape is changed by ciliary lens muscles acting on ligaments. The retina has two kinds of photoreceptors: rods and cones. Activation of the photoreceptors initiates impulses, which are conducted along neurons that comprise the optic nerve. The light-absorbing pigments of the photoreceptors are embedded in membranous disks of the photoreceptor cells. The acute light sensitivity of rods allows us to see in very dim light. Cones allow us to see color. The *fovea* is the site in the retina of the greatest concentration of cones. The hole through which the optic nerve exits the inner surface of the eye lacks photoreceptors and creates a small "blind spot" in the visual field. The rear chamber of the eye is filled with a viscous fluid, the *vitreous humor*.

THE LENS: FOCUSING IMAGES ON THE RETINA

The interior of the eyeball contains the **lens**, which focuses incoming light rays onto the **retina**, a multilayered "screen" containing the light receptors utilized for vision. The focusing abilities of the lens are derived from the glass-like proteins of which it is composed. Contraction of the *ciliary muscles* changes the shape of the lens so that objects at different distances from the eye can be focused on the retina. If this change in lens shape (called *accommodation*) did not occur, the image of nearby objects would be focused behind the retina and would appear blurry. As we age, changes in the structure of the lens proteins makes the lens less elastic. Consequently, our eyes become less able to focus on close objects, which explains why so many people need reading glasses after the age of about 45. In some older individuals, the lens becomes opaque, a condition known as a *cataract*, which can be treated surgically.

THE RETINA: A LIVING PROJECTION SCREEN

The retina contains two different types of photoreceptor cells, the *rods* and *cones* (Figure 16-1, inset), names that reflect their microscopic shape. Both types of cells contain pigment molecules that absorb incoming light. Like the light-absorbing pigments of plant chloroplasts (Chapter 5), photopigments of the rods and cones are embedded in membrane disks. *Rhodopsin*, the best-studied visual pigment, makes its way back and forth across the membranes of rods (most magnified inset of Figure 16-1). The absorption of a single photon of light (the smallest unit of light energy) can cause an alteration in the shape of a single pigment molecule, triggering a sequence of changes that culminate in the closure of several hundred ion channels and a change in membrane potential of about 1 millivolt. The additive changes in potential resulting from the absorption of a number of photons lead to the initiation of an action potential in a sensory neuron of the *optic nerve* leading to the brain.

Rods are much more sensitive to light than are cones and function primarily under low-light conditions. The human retina contains over 100 million rods, which are concentrated toward the peripheral regions of the retina. Rods provide us with night vision, which is characterized by a lack of color and sharpness but a heightened sensitivity. The high sensitivity of the rods can be appreciated when you walk out of bright sunshine into a darkened theater. At first you are virtually blind because only your cones—which respond to bright light—are functioning; the pigment molecules of the rods have been temporarily inactivated by the bright light. As you spend more time in the dark theater, the rod pigments are reactivated, and your eyes become *dark adapted*, enabling you to see the outlines of people in the theater.

Unlike rods, cones are relatively insensitive to light and are essentially useless under conditions of low light intensity. Cones are concentrated in the center of the human retina, where they provide a highly detailed image of the visual field. Unlike rods, cones also provide information on the wavelength of the light, which we perceive as color. Three different types of cones are distinguished by the structure of their pigments and the wavelengths of light they absorb. It is thought that the color we perceive in a particular part of the visual field depends on the ratio of impulses generated by the blue-, green-, and red-absorbing cones in the corresponding area of the retina. The various types of color-blindness can be explained as a loss of one of the three types of cones.

Impulses triggered by the rods and cones of the retina travel to the visual cortex of the occipital lobe of the brain (Figure 15-9). The visual cortex was mapped with considerable precision around the turn of the century by a Japanese physician, Tatsuji Inouye. Inouye interviewed soldiers from the Russo-Japanese War who had recovered from bullet wounds to the back of the head. He was able to correlate blind spots in the soldier's visual field with the location of brain damage determined by the entrance and exit wound of the high-velocity Russian bullets.

HEARING AND BALANCE

The human ear (Figure 16-2) contains two separate sensory systems, one that governs hearing, and the other that allows us to maintain our balance. The sensory receptors for both of these senses are **hair cells**—mechanoreceptors that lie in the interior portion of the ear and contain very fine "hairs." When these hairs are bent or displaced by movements or pressure, a change in the membrane potential of the hair cell is generated, which may launch an

FIGURE 16-2 ▶

The human ear. ***(a)*** The *outer ear* consists of the ear flap and the channel leading to the tympanic membrane, which collects sound waves. The *middle ear* is a small chamber composed of three interconnected ear bones; the stapes (or stirrup), malleus (or hammer), and incus (or anvil) which transmit and amplify sound waves. The *inner ear* contains the cochlea (and its receptors for hearing) and the vestibular apparatus (and its receptors for balance). Motion of the fluid in the cochlea is initiated when the stirrup pushes in at the *oval window*. Waves are transmitted through the cochlear fluid until they reach the *round window*, where they are dissipated. The first electron micrograph shows several rows of hair cells along a portion of the cochlea, and the second micrograph shows a single hair cell. The hair cells translate vibrations into neural impulses that travel to the hearing centers of the brain. ***(b)*** Pete Townshend of "The Who" has experienced severe hearing loss as a result of exposure to amplified music.

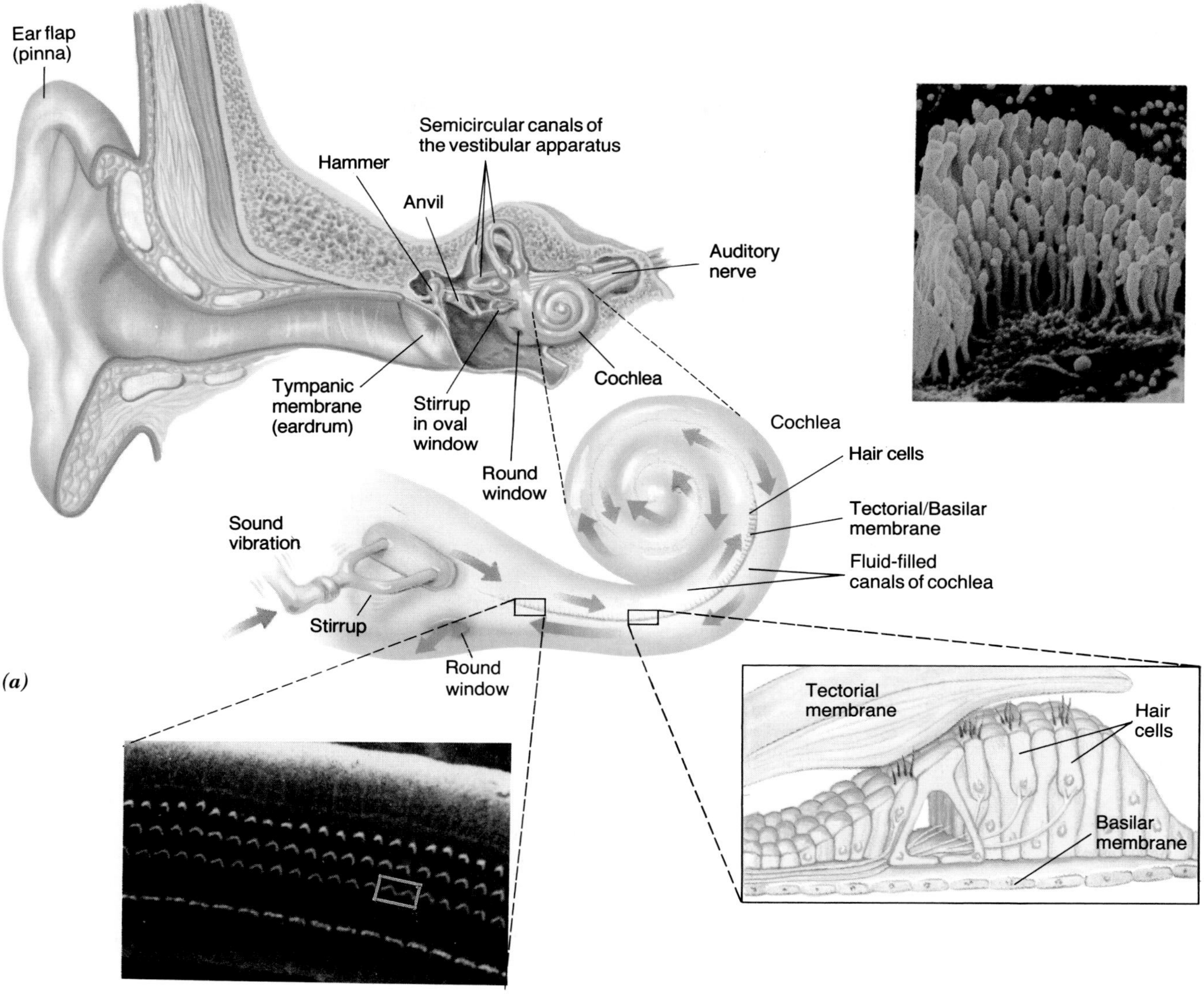

Ear flap (pinna)
Semicircular canals of the vestibular apparatus
Hammer
Anvil
Auditory nerve
Tympanic membrane (eardrum)
Stirrup in oval window
Round window
Cochlea
Cochlea
Hair cells
Tectorial/Basilar membrane
Sound vibration
Fluid-filled canals of cochlea
Stirrup
Round window
Tectorial membrane
Hair cells
Basilar membrane

(a)

(b)

◁ THE HUMAN PERSPECTIVE ▷

Perceiving the World: It's All in Your Head

Imagine for a moment that, while you are sitting reading your text, you have an irrepressible itch on the sole of your right foot. You might simply slip off your shoe, remove your sock, and rub the bottom of your foot against the nearest rough object. Now, imagine that you were to have the same irrepressible itch, but you had lost your right leg in an automobile accident several years earlier. While it may be difficult to imagine, this phenomenon in which amputees perceive "phantom sensations" from limbs that have long been lost, is actually very common. In some cases, a person can actually "feel" a missing hand grasping an object, or a missing foot rubbing along the ground as he or she walks. We have raised these images to illustrate that, while our sense organs may be present in our skin, eyes, or ears, our perception of the world derives from activities occurring within our brain. We can illustrate this concept using tactile (touch) sense as an example.

Information from touch receptors in the skin are transmitted through neural pathways in the spinal cord and the base of the brain to a region of the cerebral cortex known as the *somatic sensory cortex*. A map of the somatic sensory cortex has been obtained by studying individuals undergoing neurosurgery. For example, touching a person on the fingers, stomach, and toes produces electrical activity in three distinct and predictable areas of the cortex. These studies reveal that the somatic sensory cortex is laid out much like a miniature map of the body. For example, impulses from touch receptors in the shoulder are transmitted to a part of the somatic sensory cortex that is adjacent to a region that receives impulses from the upper arm, which, in turn, is adjacent to a region that serves as the destination for impulses from the lower arm.

When we feel a touch against our skin, it is not the pressure on our skin that we perceive, even though that pressure initiates the sensation, it is the accompanying electrical activity in our cerebral cortex that directs our attention to the body surface. In fact, the touch receptors in our skin do not even have to be stimulated for us to perceive the sensation. This is illustrated by the perceptions of the amputees described above. Since these individuals have lost a limb, they obviously lack input from sensory receptors that were originally located in that limb. So why do they experience these phantom sensations? You might think it's because the nerves, which were cut at the time of amputation, are still sending impulses to the somatic sensory cortex. This is not the case, however; the nerves are silent. Given this information, you might conclude that the corresponding area of the sensory cortex is also silent, but this is not the case either. Rather, the part of the somatic sensory cortex that is normally devoted to the limb in question is "taken over" by impulses relayed from another part of the body.

This has been demonstrated by neurobiologist Vilayanur Ramachandran of the University of California in studies on amputees whose skin is stimulated by touch with a cotton swab. One subject, for example, reported feeling a tingling in an index finger of a missing hand as the investigator stroked his upper lip. As the cotton swab moved over the subject's chin, he reported the sensation in his little finger. It was as if the missing hand was laid out in an orderly pattern across the person's lower face. Remarkably, a separate and orderly representation of the missing hand was also present on the surface of the body just below the shoulder. Neurologically speaking, it was as if the person had regrown two new left hands to replace the one that was previously lost. Studies of other amputees revealed similar maps of missing limbs located on the face, torso, or other parts of the body. These results are best explained by concluding that neurons within the brain are capable of making new functional connections that allow impulses from a particular part of the body, such as the face, to be relayed to new areas within the cerebral cortex. Unfortunately, the brain cannot determine that the impulses are coming from some other part of the body, so the person continues to feel a limb that is no longer present. These studies reveal that the neural circuits in the brain are not rigidly fixed like the electrical wires in your house but, rather, are capable of certain adjustments over time. If the brain lacked this flexibility, persons who have suffered brain damage as the result of a severe head injury or a stroke would probably not be able to relearn complex activities that are temporarily lost.

impulse along a sensory neuron carrying the information to the brain. Depending on the part of the ear in which the impulses are generated, we may perceive a particular sound or a feeling of motion.

TRANSLATING VIBRATIONS INTO SOUND

Sound vibrations travel outward in waves from a source, much like ripples in a pond. When a sound wave strikes a solid object, it sets up vibrations in the object. This is evident in the vibrations of a wall in response to the bass sounds from a nearby speaker and is the basis for our sense of hearing. Sound waves are collected by the outer ear flaps, where they pass through the canal of the outer ear and strike the *tympanic membrane* (the eardrum), which oscillates at the same frequency as do the incoming sound waves. These oscillations are transmitted through the middle ear by the movements of three tiny, interconnected bones, which, because of their shape, are commonly known as the hammer, anvil, and stirrup. The bones of the middle ear amplify vibrations by a factor of about 20.

Vibrations enter the inner ear from the stirrup and pass along the **cochlea**, an elongated spiral structure that resembles the shell of a garden snail. The cochlea contains two fluid-filled canals that are separated by a thin **basilar membrane**. Nearly 20,000 hair cells are anchored to the basilar membrane. Sound waves entering the cochlea induce vibrations in the basilar membrane, which activates the hair cells, leading to a change in the voltage across their plasma membrane and to the initiation of impulses in a sensory neuron of the *auditory nerve* leading to the brain. Our ability to discriminate different sound frequencies (from high treble pitches to low bass pitches) is determined by the particular site that is stimulated within the brain's auditory cortex.

Many cases of deafness can be traced to the death of the hair cells of the cochlea. In the past decade, an increasing number of deaf people have had a device implanted into their inner ear which works as an "artificial cochlea." The implant in the ear receives signals from a sound processor that is usually carried on the belt. The sound processor breaks down incoming sounds into their component frequencies, and this information is transmitted to the cochlear implant, which stimulates the appropriate neurons of the auditory nerve. The brain converts this pattern of stimulation into the perception of sounds, just as it would have if the impulses had been triggered by the activation of hair cells.

TRANSLATING FLUID MOVEMENTS INTO A SENSE OF BALANCE

In addition to the cochlea, the inner ear contains the **vestibular apparatus**, which gathers information about the position and movement of the head. This information enables us to maintain balance (even if we are standing on our head), to detect up from down (even with our eyes closed), and to feel changes in the body's position. The vestibular apparatus contains three fluid-filled **semicircular canals** whose walls are lined with hair cells that function as motion receptors. When you turn your head, the walls of the semicircular canals move faster than does the fluid they contain. This movement bends the sensory hairs, which triggers an impulse, informing the brain of the motion. Because the three canals are perpendicular to one another (Figure 16-2), motion in any direction can be detected. Excessive motion, such as that which occurs to a small boat in choppy seas, overstimulates the hair cells of the vestibular apparatus, causing the nausea and vomiting characteristic of motion sickness.

CHEMORECEPTION: TASTE AND SMELL

Chemoreception in most animals, including humans, constitutes two distinct senses—**taste** and **smell** (or **olfaction**). Taste is stimulated by *direct* contact with compounds present in food. In contrast, chemicals that excite the sense of smell arrive from a *distant* source, carried in the air (or in the water, in the case of aquatic animals). In humans, the senses of taste and smell are largely utilized in the selection of palatable food. In addition to heightening our enjoyment of food, these senses protect us from various dangers. For example, toxic plants often evoke a bitter taste that makes them distasteful; food in an advanced state of microbial decomposition is usually detected as spoiled by its aroma before we even taste it; and the acrid smell of smoke alerts us to the presence of fire.

The mechanism of chemoreception is poorly understood. Chemoreceptor cells contain receptor proteins in their plasma membranes which specifically engage the chemicals being sensed. There are a wide variety of different receptor proteins, each having an affinity for distinct chemical groups. Since not everyone is able to detect the same tastes or scents, there are probably genetic differences in the structure of these proteins within the population. The recent identification of the genes for chemoreceptor proteins has opened the door to a better understanding of the mechanism of taste and smell.

TASTE

The receptors for taste reside in the 10,000 or so *taste buds* (Figure 16-3) located on the surface of the tongue and other sites within the mouth. Dissolved chemicals enter the taste bud through an open pore and interact with sensory hairs projecting from each receptor cell. There are only four *primary tastes*: salty, sweet, sour, and bitter. These primary tastes become combined in various ways to give us our rich discrimination of flavors. The perception

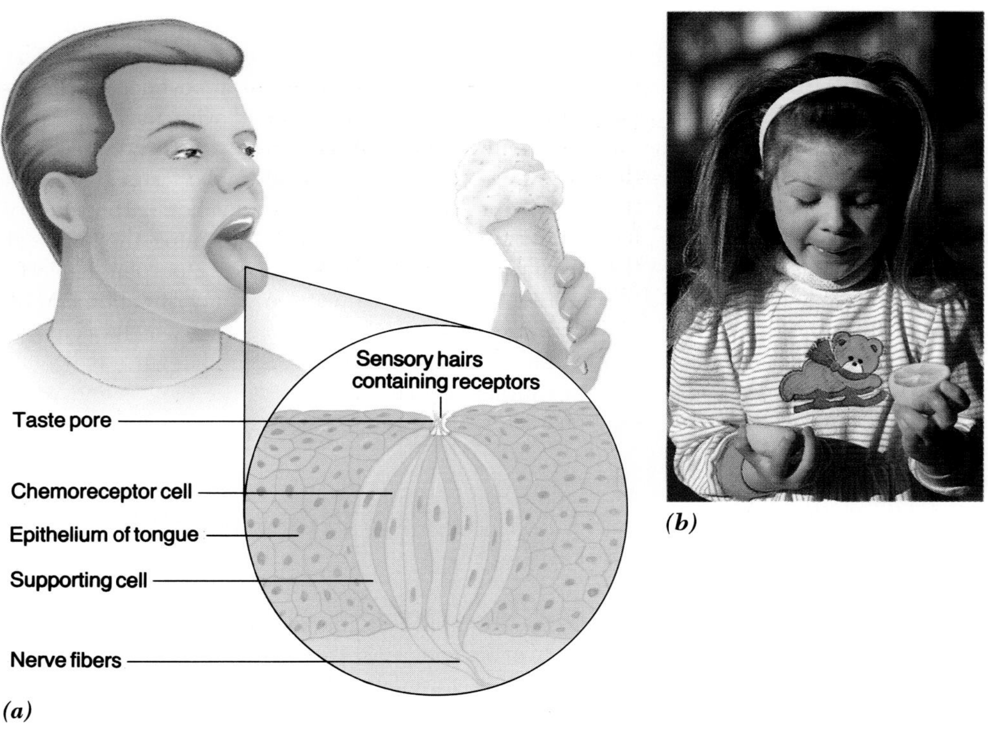

FIGURE 16-3

Taste buds. ***(a)*** Taste buds are present within the epithelium of the tongue. Each taste bud has a pore through which dissolved substances enter and subsequently contact receptor proteins on the "hairs" of chemoreceptor cells. Nonsensory *supporting cells* help maintain the structure of the taste bud. ***(b)*** The citric acid of a lemon evokes one of the primary tastes—sour.

of each primary taste can be elicited by chemicals with very different structures. For example, the sweetest substance yet discovered is thaumatin, a protein isolated from the berry of an African shrub. Thaumatin is several hundred times sweeter than the dipeptide aspartame (Nutrasweet), which is much sweeter than is the disaccharide sucrose (table sugar). Even though taste and smell constitute distinct senses, the taste of our food is greatly affected by its smell. This is readily demonstrated by pinching your nose as you taste different types of foods.

SMELL

The sensory receptors for smell are located within the olfactory epithelium that lines the interior wall of the nasal cavity (Figure 16-4). Even humans, whose sense of smell is relatively poor compared to other mammals (such as mice or dogs), can discern thousands of different types of substances and can detect them at extremely low concentrations. Methyl mercaptan, for example, the substance added to natural gas (which is otherwise odorless) to give it the characteristic aroma that alerts us to gas leaks, can be detected at a concentration of one part methyl mercaptan per billion parts of air. For many substances, a single molecule impinging on the olfactory receptor cell can trigger an impulse to the CNS.

(E)-3-methyl-2-hexenoic acid is a chemical compound that is largely responsible for giving human sweat its characteristic aroma. The substance is not produced by the human body but, rather, is manufactured by bacteria living on the body surface from otherwise odorless molecules produced by the body's sweat glands. George Preti, a biochemist at the Monell Chemical Senses Center in Philadelphia, who originally isolated the offending substance from human underarm sweat, has found that the bacteria can grow without producing an aroma as long as they are provided substitutes for the substances secreted by sweat glands. "Put these substitutes into deodorants, he says, and you could keep the bacteria busy and satiated with their decoys—leaving underarm sweat almost odor-free."

REENTER THE BRAIN: INTERPRETING IMPULSES INTO SENSATIONS

Why does an impulse create pain when delivered from a pain receptor, whereas the exact same type of impulse

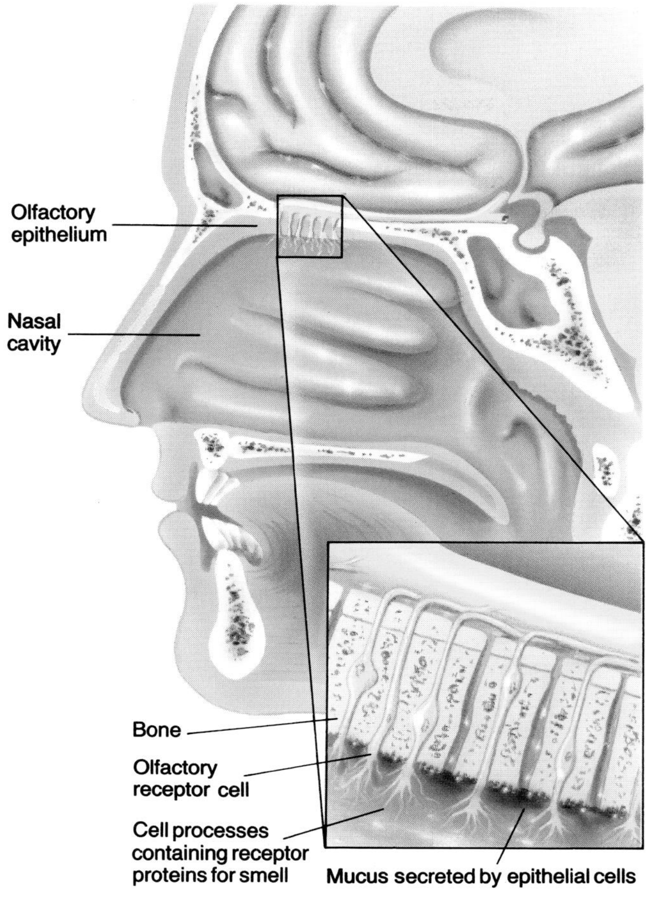

FIGURE 16-4
Olfaction. The olfactory receptor cells are located in patches within the inner lining of the nasal cavity. The receptor proteins themselves are located on processes that extend into a layer of mucus that lines the nasal chamber. Activation of the receptor cells sends impulses along fibers of the olfactory nerve into a nearby portion of the brain, called the olfactory bulb. Information is then transferred from the olfactory bulb to the olfactory centers of the cerebral cortex, where odor is perceived.

creates sight when delivered by a neuron from a photoreceptor? Furthermore, how do we know where the pain is felt, even without looking at the part of the body experiencing it? Just think about the last time a doctor drew a blood sample. Even without looking at the damage, you know what has happened and the exact spot where the needle entered. Your brain has been notified of the disturbance and translates the incoming impulses as "sharp pain in the left arm." The neurons triggered by this stimulus carry no such message, however; all they can transmit is an action potential sweeping along the neuron.

The nature and exact location of the disturbance are determined by the specific portions of the brain that are stimulated. Sensory input from the eyes, ears, nose, and somatic sensory receptors of the skin travel to separate locations within the cerebral cortex (see Figure 15-9). These are the sites in the brain where our perceptions of sight, hearing, smell, and touch are derived. Any time the cells in a particular portion of the brain are activated, the same sensation is perceived in the same location in the body; the brain perceives the sensation, not the arm. This phenomenon is readily demonstrated by electrode stimulation during open-skull surgery. If the electrode stimulates a particular site in the brain, the patient will experience a very real sensation of a sharp object piercing his or her unassaulted finger, for example, just as though the finger were really being pierced.

EVOLUTION AND ADAPTATION: TYING IT TOGETHER

Sense organs provide an animal with information about the external world as well as physiological conditions existing within its own body. The properties of sense organs vary widely among different animals. Our "picture" of the outside world is shaped largely by our sense of sight, but many other animals rely on very different types of sensory organs (Figure 16-5). Consider, for example, the shark, which can find prey buried in the sand by detecting tiny electric potentials generated by the muscles of the buried animal; the bloodhound, which can follow the trail of a specific person hours after he or she has passed; or the rattlesnake, which can strike in total darkness, guided only by the heat emitted from its warm-blooded prey.

Consider for a moment the different types of challenges facing animals that live in different environments. An earthworm or mole that lives underground is exposed to an entirely different set of environmental stimuli than is a fish or shrimp that lives in the sea. Similarly, an aquatic

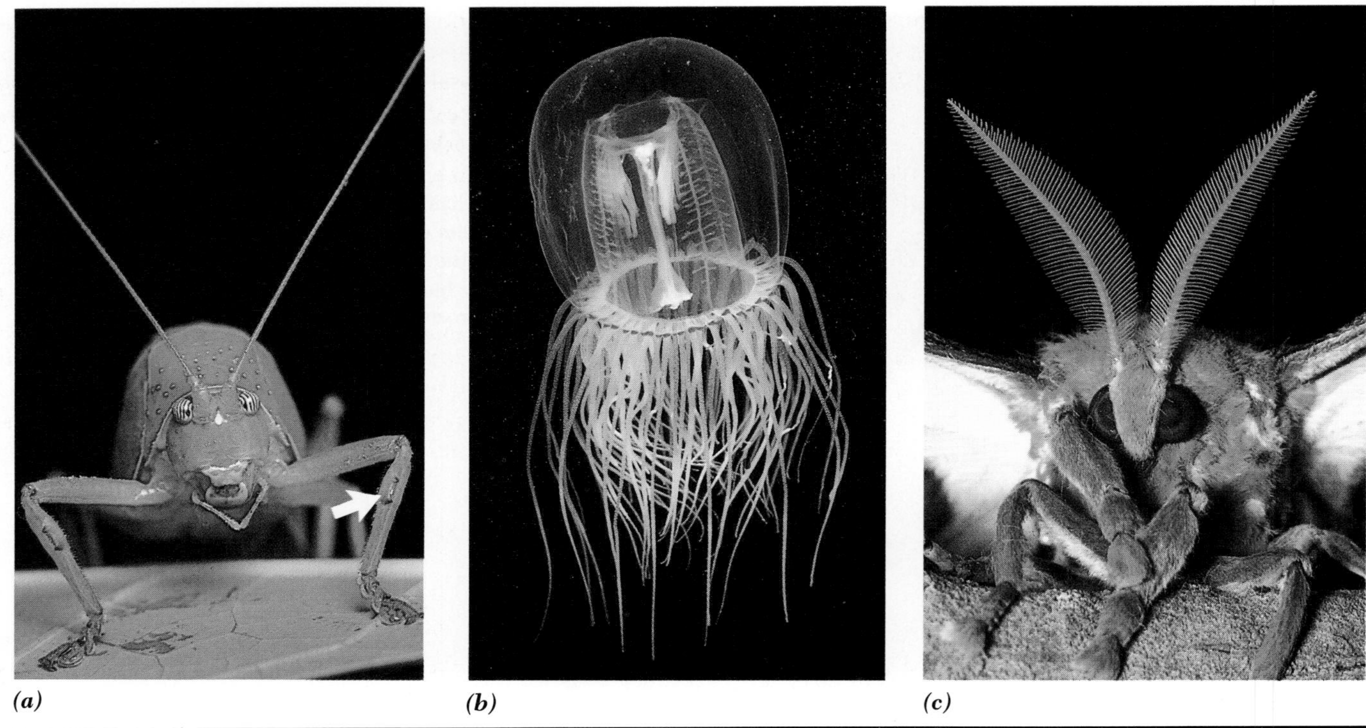

(a) *(b)* *(c)*

FIGURE 16-5

A gallery of sense organs. ***(a)*** The hearing organ of this katydid is located on its legs (arrow). Sound produces vibrations of the exoskeleton. ***(b)*** Motion detectors in the rim of the bell of this jellyfish allow the animal to maintain its orientation as it swims. ***(c)*** Chemoreceptors in the antennae of this male atlas moth allow it to detect the smell of a female moth of the same species from a distance of 11 kilometers.

animal faces entirely different conditions than does a lizard or bird living in a terrestrial habitat. Since sense organs monitor changes in an animal's environment, they must be able to respond to the types of stimuli to which the animal is exposed. In fact, sense organs play a key role in adapting an animal to a particular environment. For example, an owl, which feeds in the dark, has such a keen sense of hearing that it can attack and kill a mouse while blindfolded. In contrast, a hawk, which feeds during the day, has such a keen sense of vision that it can spot a mouse while flying hundreds of feet in the air.

SYNOPSIS

Sense organs collect information about conditions in the external and internal environment and pass the information onto the central nervous system. Each type of sensory receptor is specialized to respond to a particular type of stimulus. The interaction between a stimulus and a receptor induces a change in the receptor, which leads to a change in the permeability of the receptor's plasma membrane, which may initiate an action potential in a sensory neuron leading to the CNS.

Humans rely on several different types of sensory stimuli. Somatic sensation monitors conditions on the surface and within the body, providing information required for movement and homeostasis. Vision is a sense that is based on light rays reflected into our eyes from objects in the external environment. Light rays pass through an outer, transparent cornea and are focused by a glass-like lens onto a living retinal "screen" that contains photoreceptors. The brain interprets impulses from these receptors as a visual image of the environment. The retina contains two types of receptor cells: rods, which help provide an image under conditions of low light levels, and cones, which provide a highly detailed, multicolored image when environmental light levels are high. The inner ear contains a pair of sense organs—the cochlea and the vestibular apparatus—which provide us with the sense of hearing and balance, respectively. Both senses derive from hair cells whose membrane voltage changes when sensory hairs projecting from the receptor cell are displaced. Displacement of hairs in the cochlea occurs when sound vibrations are transmitted from the environment into the basilar membrane of the cochlea, while displacement of hairs in the vestibular apparatus results from fluid movements that occur when the head is moved. Taste and smell are senses that depend on the stimulation of chemoreceptors located in taste buds on the tongue and in the inner lining of the nasal cavity, respectively. In both cases, the chemoreceptor cells contain membrane protein receptors that combine with specific chemicals—either as dissolved food matter (taste) or as air-borne particles (smell).

The location and nature of a stimulus are interpreted by the brain. The cerebral cortex contains different regions devoted to somatic sensation, sight, hearing, and smell. The location and nature of the stimulus are determined by the particular sites within these areas that ultimately receive the impulses. In humans, the two cerebral hemispheres function relatively independently of one another and communicate across the corpus callosum.

Review Questions

1. Trace the events that occur in the eye and the central nervous system that allow you to see the outlines of people seated in a dark theater.
2. Compare and contrast: somatic sensation and visual sensation; the lens and the cornea; the semicircular canals and the cochlea; taste buds and olfactory receptors; rods and cones.
3. How do you know where a stimulus is coming from in the body? Or the strength of the stimulus? Or its nature?
4. Name the sensory structure(s) in humans that contain(s): hair cells, photoreceptors, mechanoreceptors, thermoreceptors, chemoreceptors, olfactory receptors, semicircular canals, tympanic membrane, and somatic receptors.

Critical Thinking Questions

1. What control might the graduate students working on blind or blindfolded human subjects have performed to be certain that sound reception was the basis for avoidance of the screen?
2. The Chinese describe the senses as the "gateway to the mind." Is this an accurate description? Explain your answer.
3. Otosclerosis is a disease that limits the movement of the bones in the middle ear. Why would this condition affect hearing? Can you devise some surgical procedure that might correct the problem?
4. How can you explain the fact that (1) images are projected onto the retina upside-down, but we "see" objects right side up; (2) people who become able to see after being blind from birth do not immediately have a three-dimensional view of the world; (3) we can be fooled by optical illusions?
5. Which sense(s) would you expect would be highly developed in each of the following, and why: a hawk, a monkey, an owl, a bat, a mole, a salmon?

CHAPTER

17

Protection, Support, and Movement

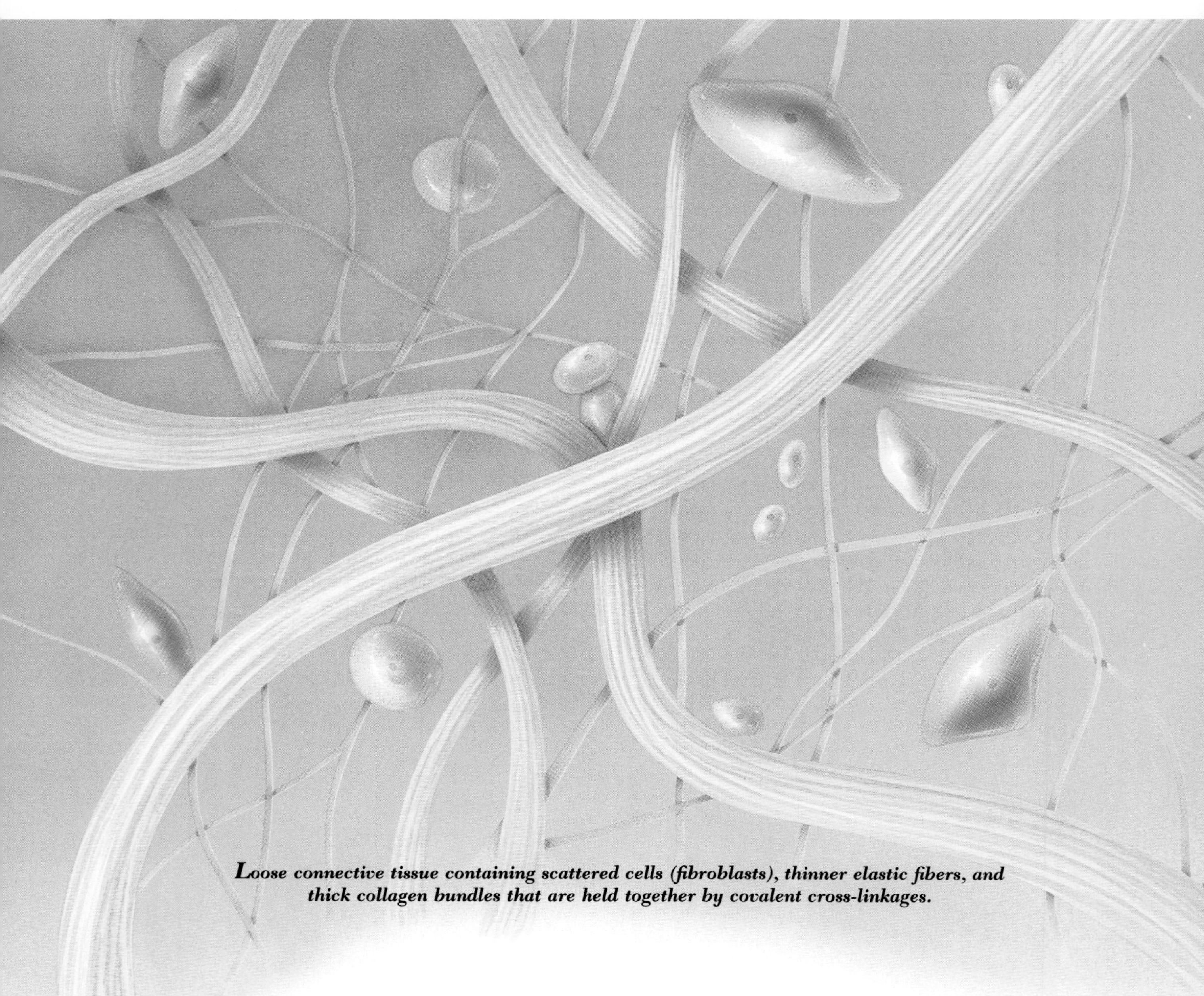

Loose connective tissue containing scattered cells (fibroblasts), thinner elastic fibers, and thick collagen bundles that are held together by covalent cross-linkages.

STEPS TO DISCOVERY
Vitamin C's Role in Holding the Body Together

In 1744, Admiral George Anson of the British Navy set sail with approximately 1,000 men aboard ship. When he returned to England less than a year later, only 144 men remained alive. More than 85 percent of the crew had died of a sailor's disease called *scurvy.* Three years later, James Lind, a surgeon in the British Navy, published a treatise on scurvy, concluding that the dreaded disease was caused by an imbalanced diet. Sailors on these expeditions subsisted almost exclusively on preserved meats and fish. Lind discovered that the addition of citrus fruits, such as lemons and limes, to the sailors' diets would totally prevent the disease (it also earned British sailors the nickname "limeys").

More than 150 years passed before the substance responsible for preventing scurvy was isolated, first from lemon juice, then from cabbage, and, finally, from the adrenal gland of laboratory animals. The structure and properties of the substance, which had been given the name *vitamin C,* were described in the late 1920s and early 1930s, primarily by the Hungarian-born biochemist Albert Szent-Gyorgyi and the British carbohydrate chemist Walter Haworth. Together, Szent-Gyorgyi and Haworth named vitamin C *ascorbic acid,* for its anti-scurvy activity. They both received Nobel prizes for their work in 1937.

Although vitamin C and scurvy are obvious topics in a discussion of nutrition, it is less evident how they relate to the subjects of this chapter, namely skin, bones, and muscles. Victims of scurvy typically suffer from inflamed gums and tooth loss, poor wound healing, brittle bones, and internal bleeding. All of these are consequences of serious defects in the formation and maintenance of connective tissues throughout the body. The properties of connective tissues are determined largely by the properties of the *extracellular matrix,* which contains polysaccharides and proteins secreted by cells into their surroundings. The strength and resiliency of connective tissues often depend on a single protein, collagen. Examination of the connective tissues from laboratory animals suffering from scurvy revealed a marked reduction in the number of collagen fibers that normally fill the spaces between the cells. It appeared that there was some relationship between ascorbic acid and the formation of collagen fibers.

Over the next few decades, a number of laboratories turned their attention to the structure and synthesis of collagen. The great strength of the collagen fibers in the extracellular matrix depends in large part on the formation of cross-linking chemical bonds that bind the collagen polypeptides together into strong fibers. The formation of these cross-links requires that some of the amino acids (specifically lysine and proline) in the collagen polypeptide chains are first modified by the addition of a hydroxyl (—OH) group. During the 1960s, scientists found evidence that ascorbic-acid deficiency, which is the cause of scurvy, decreases the amount of cross-linking that occurs between collagen polypeptides, thereby weakening the entire fabric of the body's connective tissues. Subsequent research revealed that ascorbic acid is, in fact, a coenzyme that is required by the enzymes which add hydroxyl groups to the amino acids. A person suffering from scurvy could not modify his or her amino acids to produce the cross links that strengthen collagen fibers.

Ehlers-Danlos syndrome (EDS) is another condition that has been traced to an inability to add hydroxyl groups to collagen. This inherited disorder is characterized by poor wound healing, tissues that bruise and bleed, extremely flexible joints, and highly extensible skin (which led some EDS sufferers to work as "rubber men" in circus side shows). In 1972, Sheldon Pinnell and his co-workers at Harvard University discovered that a common form of EDS was due to a mutation in an autosomal recessive gene (page 188) that ordinarily directed fibroblasts to produce the enzyme that added a hydroxyl group to the amino acid lysine. A person who is homozygous recessive for this gene fails to form collagen molecules with normal lysine cross-links, a condition that leads to the various symptoms of EDS. Ehlers-Danlos syndrome is less severe than is scurvy, however, because other cross-links (those involving the amino acid proline) are still able to form. In scurvy victims, neither lysine nor proline cross-links are formed, creating a much weaker collagen fiber.

*F*ew people need to be convinced of the importance of their skin, bones, and muscles. Together, they constitute more than 65 percent of your body mass and largely determine your physical appearance, from your body stature to your facial features. Your skin, bones, and muscles protect you from the environment, support you against the effects of gravity, and enable you to move toward food and away from danger. We will begin our discussion of these systems with a look at an animal's first line of defense: its outer body surface.

▼ ▼ ▼

THE INTEGUMENT: COVERING AND PROTECTING THE OUTER BODY SURFACE

The **integument** is the outer-body covering of an animal; in vertebrates, it is the **skin.** Depending on the functions it performs, the integument may be soft, flexible, and permeable (as in an earthworm or a frog) or coarse, stiff, and impermeable (as in a lizard or a fish). Whatever its nature, the integument is strategically located at the boundary between the living animal and its environment. Consequently, the integument must act as a protective barrier; it helps shield the individual's delicate, moist, internal tissues from a changing and often harsh environment that might otherwise infect the body with bacteria, freeze the body's fluids, evaporate the body's water, or mutate the body's genes.

THE HUMAN INTEGUMENT: FORM AND FUNCTION

Your skin is a biological cooperative of the four tissue types—epithelial, connective, muscle, and nerve (Figure 17-1). Thus, skin is an organ. In fact, it is the largest organ of your body. Human skin consists of two distinct layers: the outer **epidermis** and the inner **dermis.** These layers have very different structures that reflect their different functions.

The Epidermis: A Protective Outer Layer

The epidermis is a protective epithelium consisting of many layers of cells. Cells are formed by mitosis in the deepest layer of the epidermis and then move toward the body surface. As they approach the surface, the cells become flattened, and their cytoplasm becomes filled with filaments of the tough, resistant protein *keratin.* By the time they reach the surface, the cells have been transformed into a dead, outer layer of keratin, making your skin airtight, watertight, and resistant to bacteria and most chemicals. Protection from the damaging rays of the sun is afforded by the pigment *melanin,* which gives skin its coloration.

The Dermis: A Complex Inner Layer

Beneath the epidermis lies the dermis (Figure 17-1), which consists of dense connective tissue and a rich supply of blood vessels, nerve fibers, and smooth muscle cells. The border between the dermis and the epidermis is characterized by hills and valleys, which increase the hand's gripping power and form the basis of a person's fingerprints. Bundles of dermal collagen fibers give the skin its strength and cohesion as a continuous thin layer. Elastic fibers provide the skin with the elasticity that allows it to snap back when stretched. The dermal blood vessels provide nutrients and oxygen to the overlying epidermis, which lacks its own blood supply. These vessels also play a key role in maintaining the body's constant temperature by carrying warm blood to the body's surface, where heat can be lost to the environment. Blood flow into the dermis can range from a bare trickle, when heat must be conserved during cold external conditions, to 50 percent of the body's blood supply when heat loss is needed to cool the body. The pinkness of the skin in light-skinned individuals is due to blood flow through the dermis.

Hair: Protecting and Insulating the Skin

Most mammals have considerably more hair (fur) than do humans. Hairs consist of dead, keratinized cells similar to those of the outer layer of the skin. Each hair is formed within a living **follicle** (Figure 17-1). A thick layer of hair provides an excellent cover for protecting the body against abrasion and insulating it against heat loss. Human evolution, which probably occurred in a warm, tropical environment, was accompanied by the loss of body hair, which is thought to have been a result of natural selection favoring those individuals who were better able to lose excess body heat.

The Skin's Various Secretions

Your skin contains large numbers of glands, whose secretions find their way to the surface of your body. The glands of the skin include two broad types—sebaceous glands and sweat glands. **Sebaceous glands** produce a mixture of lipids (*sebum*) that oil the hair and skin, keeping them pliable. **Sweat glands** are distributed over most of the skin's surface, where they secrete a dilute salt solution, whose evaporation cools the body.

As you can see, the skin is a dynamic, living organ. Its importance is dramatically evident when you consider that burns to as little as 20 percent of the body can be fatal if not treated rapidly. The cause of death in such burn cases is dehydration, which results from the loss of water through the damaged, previously waterproof, body cover.

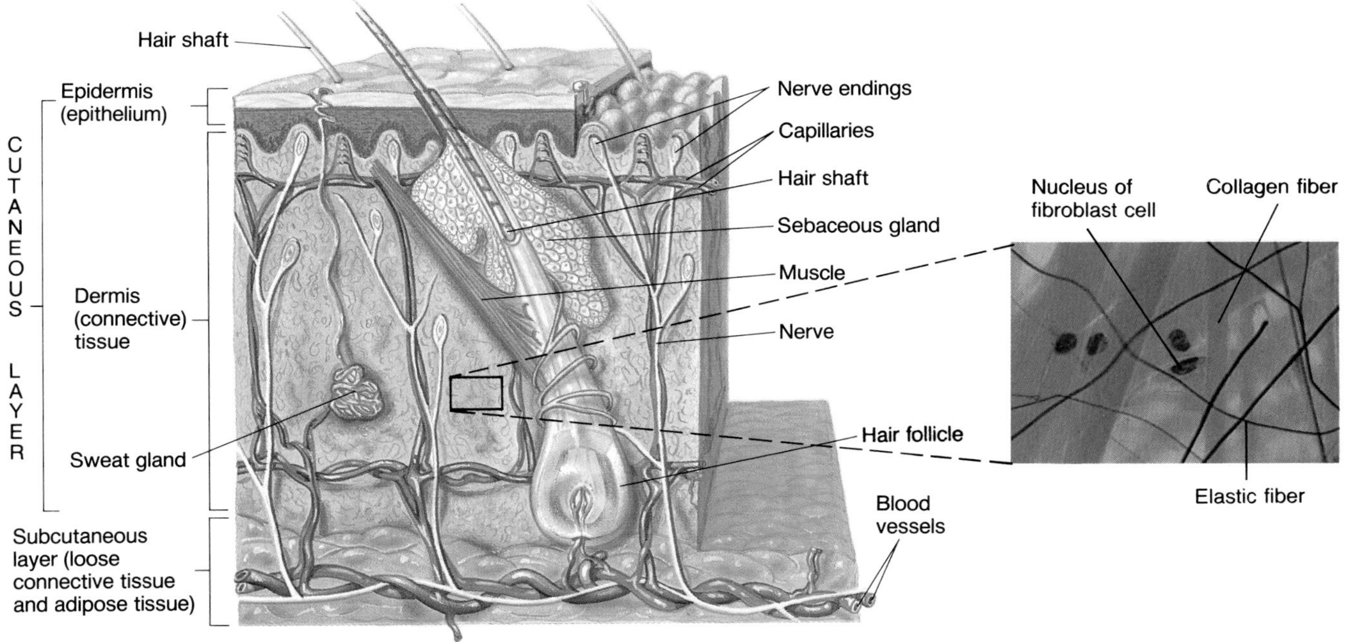

FIGURE 17-1

Human skin. The outer epidermis consists of a superficial layer of dead cells that cover the underlying living epithelial cells. Below the epidermis lies the dermis, where connective tissue predominates. Embedded in the dermis are blood vessels, muscles, nerves, and the basal portions of glands and hair follicles. Hairs are formed from epidermal cells that are generated deep within the follicle. Skin (the *cutaneous layer*) is firmly secured to the underlying layer (the *subcutaneous layer*) by connective tissue. Inset shows the connective tissue of the dermis contains scattered cells (fibroblasts) and extracellular collagen and elastic fibers.

THE SKELETAL SYSTEM: PROVIDING SUPPORT AND MOVEMENT

Most animals possess a rigid form of support which either completely surrounds the body as a protective encasement or forms a system of living girders inside the animal. Even those animals that lack rigid supporting structures still have a way of doing battle with gravity. These animals employ "hydrostatic skeletons."

Hydrostatic Skeletons: Using Fluid to Maintain Rigidity

Although an earthworm doesn't have a single bone, it can push the tip of its body through compact soil, a feat that requires the front end of the worm to remain highly rigid. As it burrows, the worm maintains rigidity by generating pressure inside a closed, fluid-filled chamber. It does so by contracting muscles that encircle the chamber, forming a **hydrostatic skeleton.** You can envision how a hydrostatic skeleton works by filling a balloon with water and exerting pressure by squeezing one end. The balloon becomes rigid and capable of supporting itself. Sea anemones and corals possess a similar fluid-filled internal chamber (Figure 17-2) that acts as a hydrostatic skeleton to maintain the animal's upright stature.

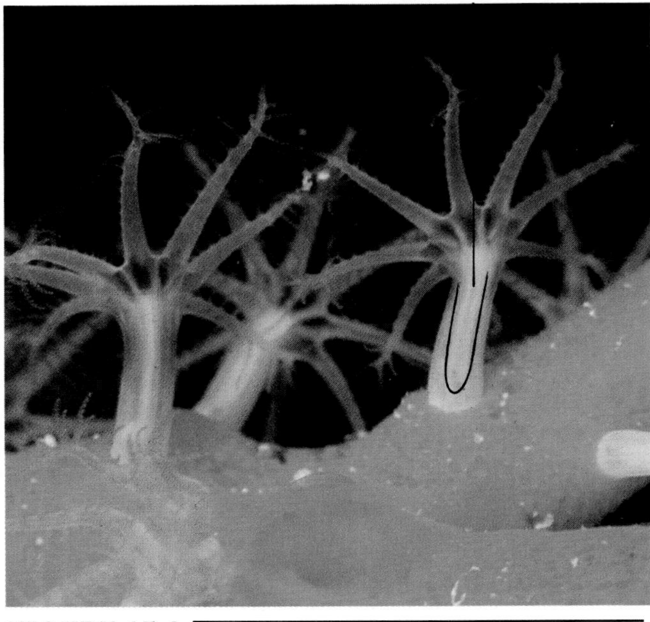

FIGURE 17-2

A hydrostatic skeleton. These coral polyps contain a central, fluid-filled chamber surrounded by rings of muscle tissue. When the mouth is closed, which seals the chamber, and the muscles contract, which pushes against the fluid, the animal generates hydrostatic (water) pressure that allows the animal to stretch tall against the pull of gravity.

EXOSKELETONS: PROVIDING SUPPORT FROM OUTSIDE THE BODY

Anyone who has cracked open a crab or lobster leg at the dinner table, or found the discarded husk of an insect, is familiar with **exoskeletons**—the hard, nonliving external coverings that are secreted by the outer epidermal layer of the animal's body. Exoskeletons provide both protection and support. The hardened exoskeletons of arthropods (Figure 17-3*a*), including those of insects, spiders, and crabs, serve a skeletal function by supporting the weight of the soft tissues of the body. The outer casings of the limbs consist of tubular sections of cuticle which are connected to one another by thin, flexible joints. The limbs are operated by bundles of muscle tissue attached to the inner surface of the cuticle (Figure 17-3*b*). Without a rigid exoskeleton, muscular contraction would merely distort the soft tissue, and the animal wouldn't be able to move.

ENDOSKELETONS: PROVIDING SUPPORT FROM WITHIN THE BODY

Endoskeletons are support structures that reside inside the animal's body. Although also found in a few invertebrates, such as sponges and sea stars, the endoskeletons of vertebrates are the most complex and versatile. Vertebrates are supported by internal skeletons that are composed of two distinct types of connective tissues—bone and cartilage.

Bone

Bone combines two properties—strength and light weight. These properties are rarely found together in a single material. Bone has been called "living concrete," but it is actually four times stronger than concrete and much lighter. For example, a person's skeleton adds only 14 percent to the total weight of the body.

A bone's strength is a product of its architecture. Like other connective tissues, the bulk of a piece of bone consists of an extracellular matrix—materials secreted by cells into the space that surrounds the cells. The extracellular matrix of bone is woven from cables consisting of strands of collagen fibers and is hardened by crystals of calcium phosphate. Together, the protein and mineral salts make bone both hard and resilient. Collagen fibers act as reinforcing rods, similar to the steel rods found in reinforced concrete. Without collagen, your bones would be so brittle that they would shatter under your body's weight. Without calcium, your bones would bend as though they were made of rubber.

The most striking difference between bone and reinforced concrete is that bone is alive, even in its most rock-like parts. Within the solid mass of mineral and collagen, living bone cells thrive. These cells, called **osteocytes** (*osteo* = bone, *cyte* = cell), are engulfed in the calcified matrix that they manufacture. Each bone has a solid, rock-like portion, called **compact bone,** which usually surrounds a honeycombed mass of **spongy bone.** The hollows within spongy bone are filled with **red marrow,** the soft tissue that produces red blood cells. The structure of bone is described in Figure 17-4.

Although the bones of adults no longer increase in size, bone is constantly being disassembled and rebuilt throughout an individual's life (see The Human Perspective: Building Better Bones). The constant interplay be-

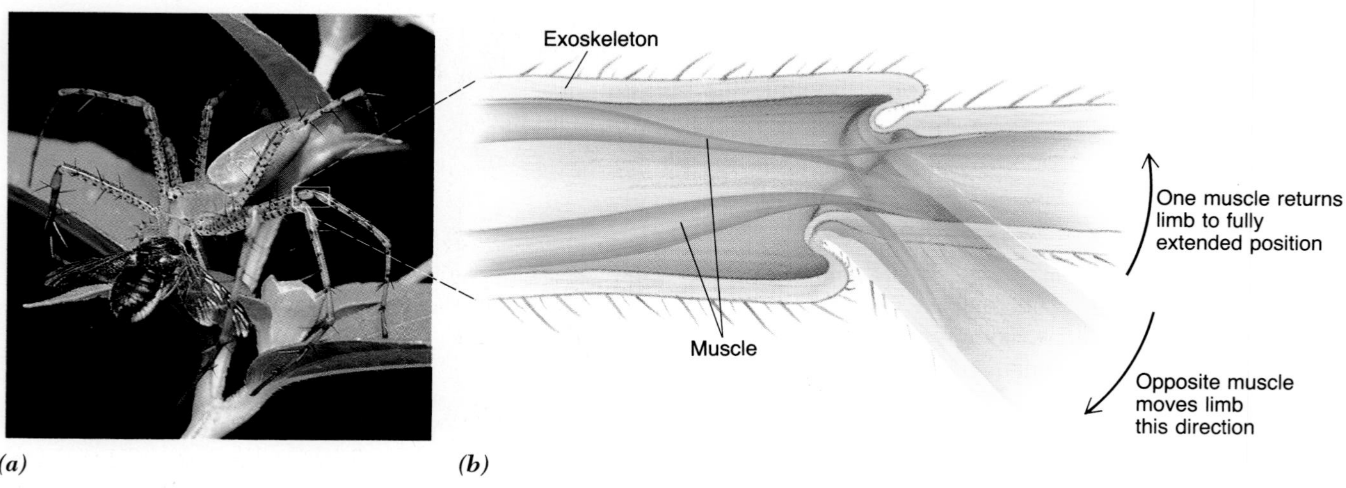

FIGURE 17-3

The arthropod exoskeleton. ***(a)*** The body of an arthropod is covered by an exoskeleton that conforms to its contour. The limbs of arthropods are covered by tubular segments of exoskeleton that are joined by flexible joints. The mobility of the spider and the fly in this photo is provided by dozens of joints in their exoskeletons. ***(b)*** The muscles that move the appendages of an arthropod are attached to the inner surface of the exoskeleton.

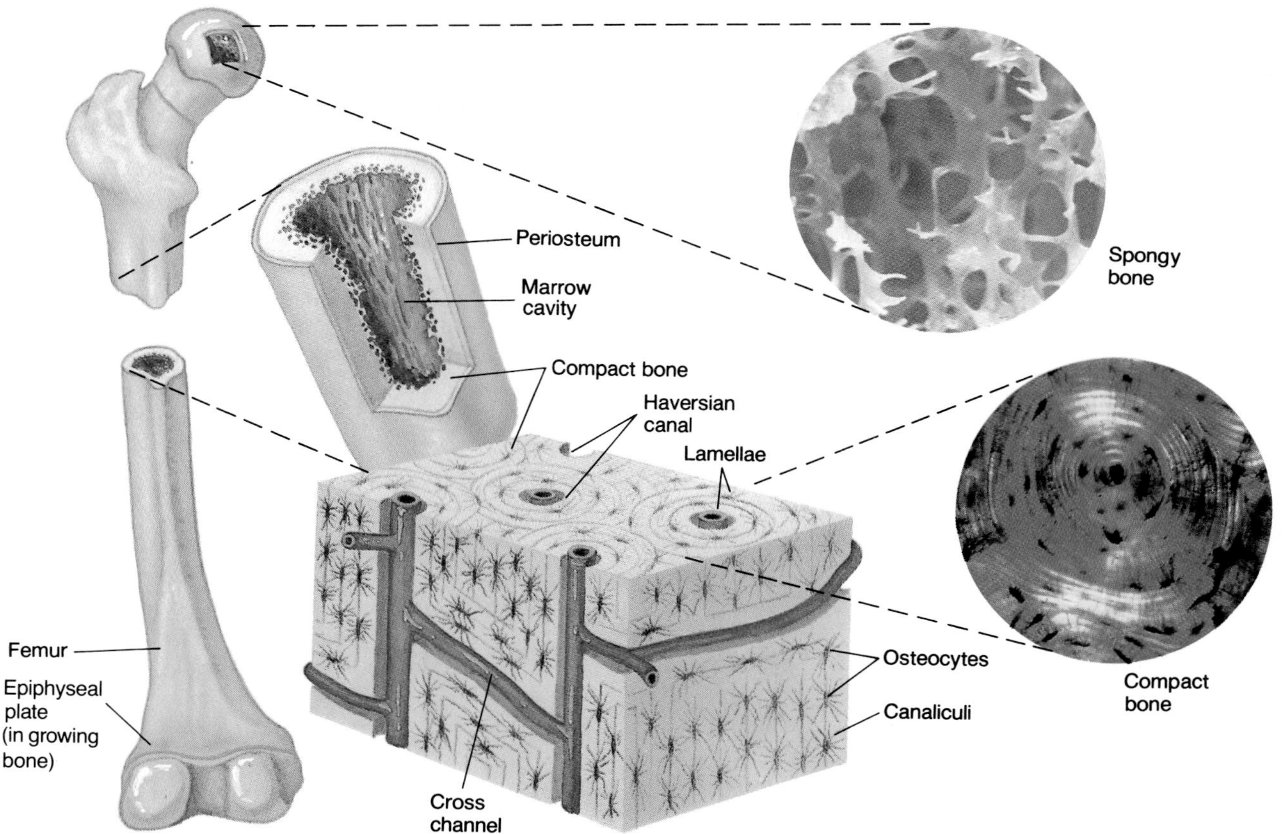

FIGURE 17-4

The architecture of bone. Solid-looking, compact bone consists of living cells, called *osteocytes,* engulfed in an extracellular matrix. This matrix is deposited as concentric cylinders called *lamellae.* The result is a greatly strengthened structure. Osteocytes obtain their nutrients from blood vessels that are threaded through channels in the bone tissue, called *Haversian canals,* which are the lifeline of compact bone. The tiny chambers that house the osteocytes are connected by cross channels and *canaliculi*—microscopic channels that extend outward like a spider web from the central canal. Canaliculi channel nutrients and growth-regulating hormones from the blood to every osteocyte in the Haversian system and evacuate their metabolic wastes. Spongy bone consists of thin, bony elements that surround marrow-filled chambers. The long bone of the leg, shown at the left, grows during childhood and adolescence at *epiphyseal plates* located near each end. The epiphyseal plate is the last part of the bone to become mineralized when elongation of the bone ceases in adulthood. The entire bone is covered by a connective tissue sheath, the *periosteum.*

tween bone assembly and disassembly enables our bodies to strengthen those bones that receive the most use and to diminish the size of those whose services are in less demand. Disuse of bones causes *atrophy* (wasting). Even enclosing a leg in a cast for a few weeks can diminish the size of a bone.

Cartilage

If you were to examine the bones as they initially form in a human embryo, you would find that most of them aren't bones at all but are composed of another structural material—cartilage. Like bone, **cartilage** consists of living cells that secrete an extracellular matrix that envelopes the cells. Like bone, cartilage provides strength and resilience, but its lack of mineral deposits keeps cartilage flexible. To gain a sense of the tough, yet flexible, character of this "building material," wiggle your outer ear or the tip of your nose, both of which are composed largely of cartilage.

STRUCTURE OF THE HUMAN SKELETON

By the time you are born, much of the cartilage present in the early embryo has been transformed into about 350 partially hardened bones. As you grow, many of these bones fuse with one another, producing an adult skeleton consisting of 206 individual bones, linked together by various types of joints. Each bone in the body participates in a specific body movement, supports a particular part of the

◁ THE HUMAN PERSPECTIVE ▷

Building Better Bones

Chances are, when you think of a bone, you picture a dead remnant of a once-living organism. But bones are not dead; they are dynamic, living tissues that require continual maintenance and repair. Maintenance of bone occurs in two stages. First, a portion of existing bone matrix is broken down; second, new bone matrix is deposited, replacing that which was removed. Neither of these complex steps is understood well, and both are regulated by a variety of factors, including hormones (calcitonin, parathyroid hormone, and mineralocorticoids), growth factors, vitamin D, and substances produced by various cells.

Bone matrix is disassembled by *osteoclasts,* specialized cells that emerge from the bone marrow under mysterious circumstances and migrate to sites where bone is being remodeled (Figure 1*a*). Upon its arrival, the osteoclast secretes acid, which dissolves the calcium salts of the bone matrix, and collagenase, an enzyme that digests the collagen fibers. In about 10 days, the osteoclast has finished its work, creating a microscopic crater (Figure 1b) that is ready to be filled in by a newly arriving *osteoblast.* The osteoblast secretes collagen and a number of factors that promote bone mineralization, the precipitation of calcium-phosphate salts.

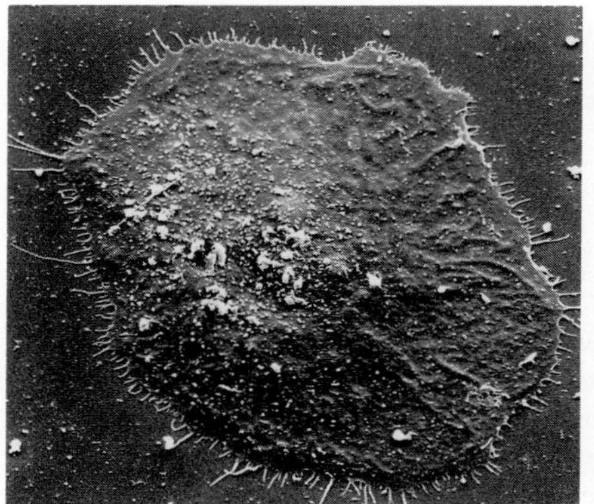

(a)

(b)

FIGURE 1

Repair and remodeling of bone begins with the dissolution of a portion of the existing matrix by large, multinucleated osteoclasts (***a***). A moonlike crater is etched into bone by the action of an osteoclast (***b***).

One of the motivations for researchers to study bone maintenance is to develop better treatments for *osteoporosis,* a bone-weakening condition that is predominant in older women. A person with osteoporosis may have a "hunched-over" appearance and is very susceptible to bone fractures, particularly of the hip or vertebrae. Osteoporosis occurs when the breakdown of bone material by osteoclasts exceeds its reformation by osteoblasts, resulting in a net bone loss. Osteoporosis is particularly common in women who undergo early menopause or who have had their ovaries removed as part of an early-life hysterectomy. The cessation of estrogen production by the ovaries appears to be one of the major contributing factors in the development of the condition. Osteoporosis is best treated by the administration of the sex hormone estrogen, which appears to bind to receptors in the osteoblasts, stimulating them to increase the amount of bone deposition. The mechanism by which this phenomenon occurs remains unknown.

body's weight, or protects an internal organ from damage. The bones of the mammalian skeleton can be divided into two functional groups: the axial skeleton and the appendicular skeleton (Figure 17-5).

The Axial Skeleton

Bones aligned along the long axis of the body—the skull, vertebral column, and rib cage—comprise the **axial skeleton.** This structure assumes the skeleton's protective functions.

The precious 3-pound mass of nervous tissue in your head is enclosed in the *cranium,* an unyielding vault of eight bones. The **skull,** which is composed of all the bones of your head, includes the cranium and 11 additional bones. At birth, the individual bones of the cranium are held together by flexible membranes that allow the head to compress a bit as it passes through the birth canal. By the end of a child's second year, these vulnerable "soft spots" have been replaced by strong, interlocking lines of fusion between adjacent bony plates.

The spinal cord is protected by the **vertebral column,** or backbone. The flexible backbone consists of 33 bones, called *vertebrae,* which are arranged in a gracefully curved line and are cushioned from one another by disks

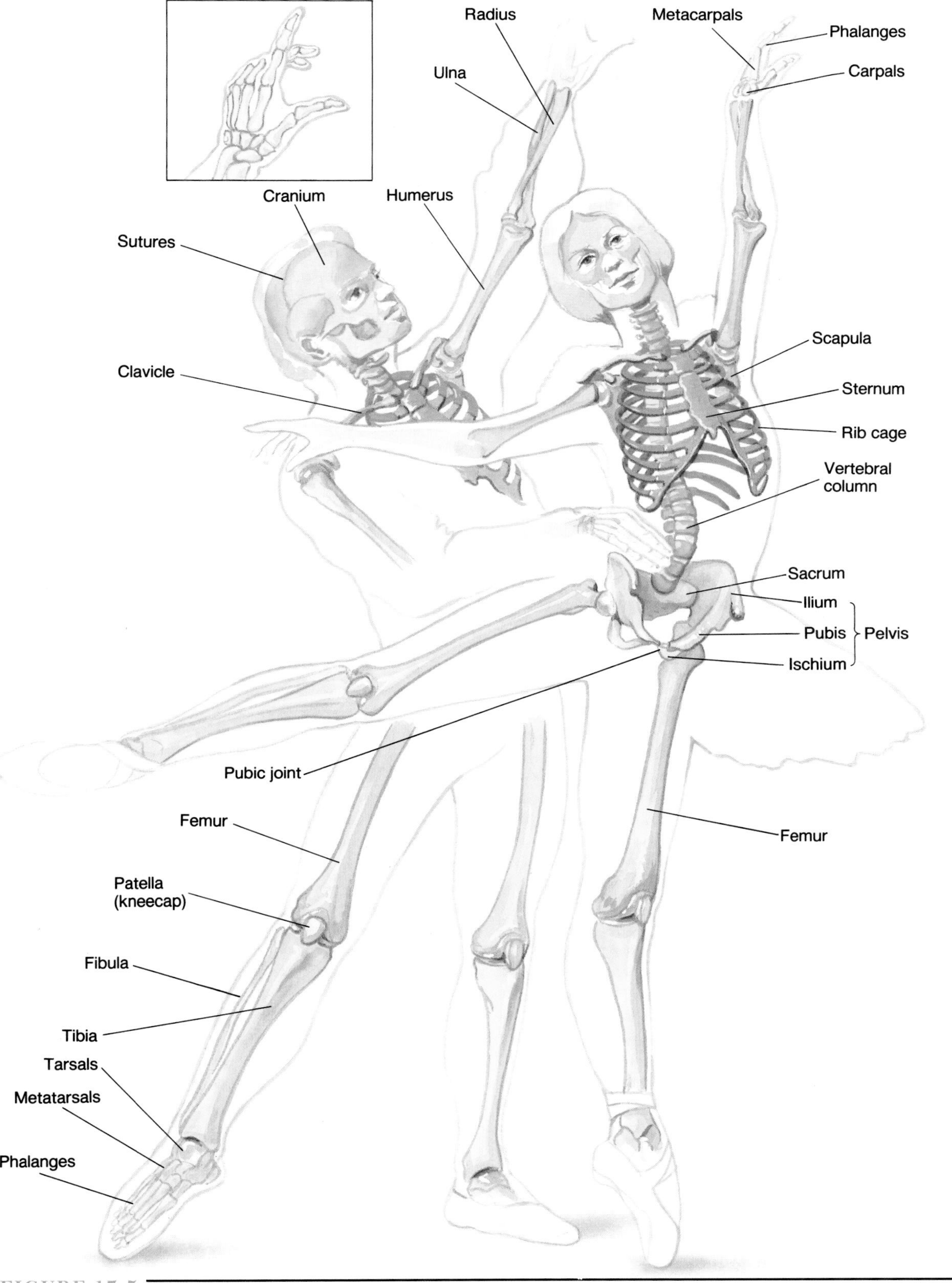

FIGURE 17-5
The human skeleton. Bones of the axial portion are shown in red; those of the appendicular portion are in blue.

of cartilage. The **rib cage** embraces the chest cavity and protects its vital organs. Ribs extend from the vertebrae and form a "cage" by attaching to the *sternum* (the breastbone) in the front.

The Appendicular Skeleton

The movable limbs (*appendages*) attached to the axial skeleton comprise the **appendicular skeleton** (Figure 17-5), which forms a system of levers, providing mobility and dexterity. The **pectoral girdle** holds the arms to the axial skeleton. The two bones (the ***ulna*** and ***radius***) in the forearm allow you to rotate your hand. One of the most extraordinary collection of bones anywhere in the animal kingdom lies at the end of the wrist of humans, apes, and monkeys. The hand combines strength with dexterity, enabling a person to crush objects with the same hand that is delicate enough to insert a contact lens into the eye.

The **pelvic girdle** (or **pelvis**) receives the weight of the upper body from the vertebral column and transmits it either to the bones of the legs or to the surface on which you are now sitting. Although not as dextrous as the hand, the 26 bones that make up each foot are arched to withstand tremendous forces.

Joints

If the skeleton were a single, solid piece of bone, it would be stronger, but movement would be impossible. Strength must be compromised somewhat to provide mobility, producing weaker points that are capable of movement. The site where two bones come together is called a **joint.** In general, the more mobile the joint, the weaker it is. The shoulder joint is the most mobile joint of the body. Any athlete with a shoulder dislocation can painfully testify to both its flexibility and its vulnerability.

The adjoining bones of a joint are held together by strong straps of connective tissue, called **ligaments.** For example, the *cruciate ligaments* hold the bones of the upper and lower legs together at the knee joint. These ligaments are often torn during contact sports, and their repair is usually accomplished by *arthroscopic surgery,* whereby an orthopedic surgeon inserts instruments into the knee joint through an incision only a few centimeters long. The progress of the operation is followed on a television screen using a picture beamed from a miniaturized camera inserted into the knee joint. Recovery usually occurs in days rather than the weeks required after conventional knee surgery.

THE MUSCLES: POWERING THE MOTION OF ANIMALS

Muscle is a highly specialized tissue with one basic function—to generate a pulling force. The larger muscles of

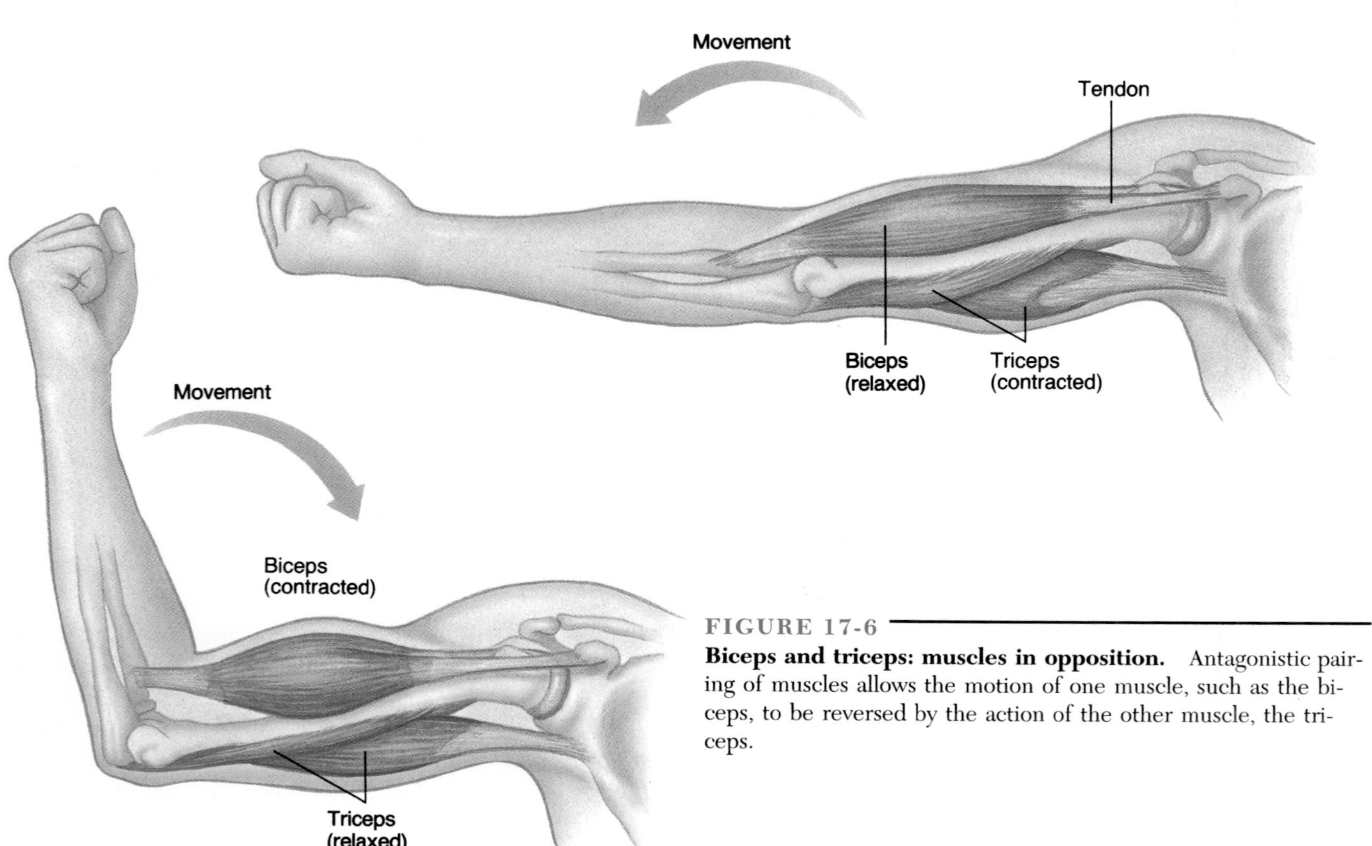

FIGURE 17-6
Biceps and triceps: muscles in opposition. Antagonistic pairing of muscles allows the motion of one muscle, such as the biceps, to be reversed by the action of the other muscle, the triceps.

the body pull on individual bones. But muscles do much more than move parts of the skeleton. In humans, muscles move eyelids and tongues; pump internal fluids through circulatory pipelines; propel nutrients through the digestive thoroughfare; discharge wastes; squeeze secretory products out of glands; and suck oxygen into the lungs.

Muscles require tremendous amounts of ATP to fuel their activities; a person running at full speed burns about 1,000 Calories per hour, which is equivalent to the amount of energy contained in a 6-ounce chocolate bar. Muscles are superbly efficient in the use of this ATP. Muscle cells convert 35 to 50 percent of the energy released into mechanical energy, making them about five times more efficient than most automobile engines. The remainder of the energy is released as heat, which can warm a body to uncomfortable levels.

Vertebrates are equipped with large amounts of muscle. In fact, about half your body weight consists of three types of muscle—*skeletal, smooth,* and *cardiac muscle.* These three types of muscle tissue differ in physical appearance, the types of jobs they perform, the tissue to which they are attached, their speed of contraction, and the manner in which they are excited into action. The most familiar type are the skeletal muscles that bulge beneath the skin and give bodybuilders their characteristic contours.

SKELETAL MUSCLE: RESPONDING TO VOLUNTARY COMMANDS

Skeletal muscles are under voluntary control; they can be consciously commanded to contract. Skeletal muscle makes up nearly 40 percent of a man's body and nearly 23 percent of a woman's. Skeletal muscles derive their name from the fact that most of them are anchored to the bones they move. The muscle tapers at its end, forming a dense connective-tissue cord, or **tendon** (Figure 17-6), that attaches the muscle to the bone.

Muscles only shorten and *pull;* they cannot push. An opposing muscle must be used to provide the opposite movement, so that the limb is not stuck in one position. For example, your *biceps*—the large skeletal muscle of the front of the upper arm—is a muscle whose contraction forces a bone in your forearm to bend at the elbow (Figure 17-6, *lower left*). The biceps, however, cannot move the forearm *away* from the upper arm. The biceps is paired with the *triceps,* the muscle along the back of the upper arm that has just the opposite effect of the biceps, causing the arm to straighten at the elbow (Figure 17-6, *upper right*). Most skeletal muscles are arranged in such *antagonistic pairs,* allowing one muscle to reverse the effects of the other.

The Structure of Skeletal Muscle Cells

If we define a cell as the contents enclosed within a continuous plasma membrane, then the vertebrate skeletal muscle cell is highly unorthodox (Figure 17-7). A single, cylindrically shaped muscle cell may be 100 micrometers thick and run the entire length of a bulky muscle, such as the biceps in your arm. Furthermore, each cell may contain thousands of nuclei; therefore, a skeletal muscle cell is more appropriately called a **muscle fiber.**

Skeletal muscle cells may have the most highly ordered structure of any cell in the body. A cross section of a muscle fiber (Figure 17-7) reveals it to be a cable made

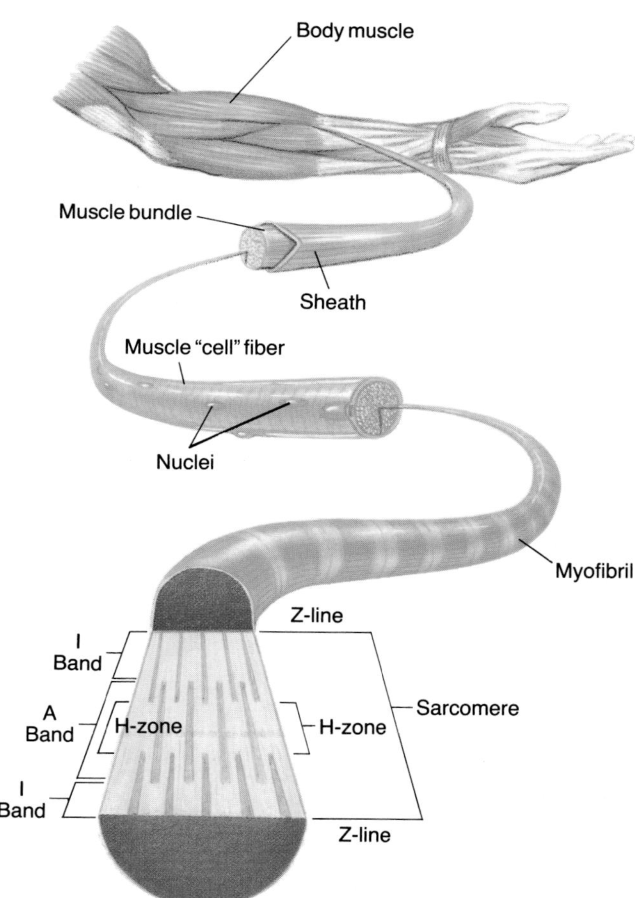

FIGURE 17-7

Skeletal muscle is composed of bundles of parallel, multinucleated cells (muscle fibers). Each muscle fiber is covered by a continuous plasma membrane and is packed with contractile myofibrils. Each myofibril consists of a linear array of sarcomeres—the contractile units of the muscle fiber. Adjacent sarcomeres are separated from one another by dark Z *lines.* Between the Z lines are several dark bands and light zones which make up the sarcomere. The banding pattern results from the overlapping array of contractile protein filaments. Each sarcomere has a pair of lightly staining *I bands* located at its outer edges; a more densely staining *A band* located between the outer I bands; and a lightly staining *H zone* located in the center of the A band. The I band contains only thin filaments, the H zone only thick filaments, and that part of the A band on either side of the H zone represents the region of overlap and contains both types of filaments.

up of hundreds of thinner, cylindrical strands, called **myofibrils,** which are separated from one another by cytoplasm. Each of the fiber's myofibrils consists of a linear array of contractile units called **sarcomeres,** each of which is endowed with a characteristic pattern of bands and lines (Figure 17-7). Examination of muscle sarcomeres with the electron microscope shows the banding pattern to be the result of the partial overlap of two distinct types of filaments, referred to as **thin filaments** and **thick filaments** (Figure 17-8). Thick filaments are composed primarily of the protein myosin, and thin filaments of the protein actin. In addition to forming the thick fibers, myosin is an enzyme that hydrolyzes ATP, releasing the stored energy required for muscle contraction.

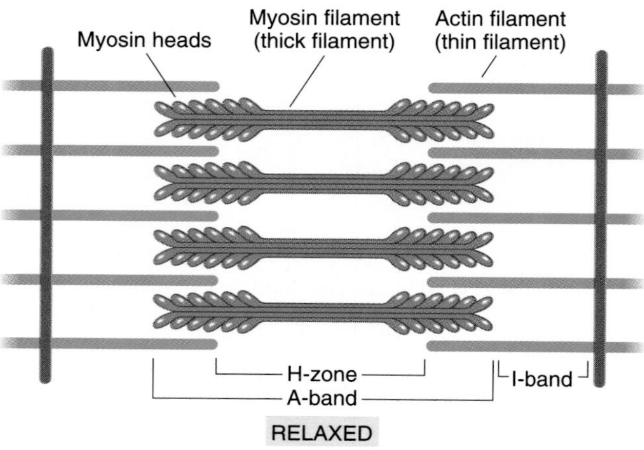

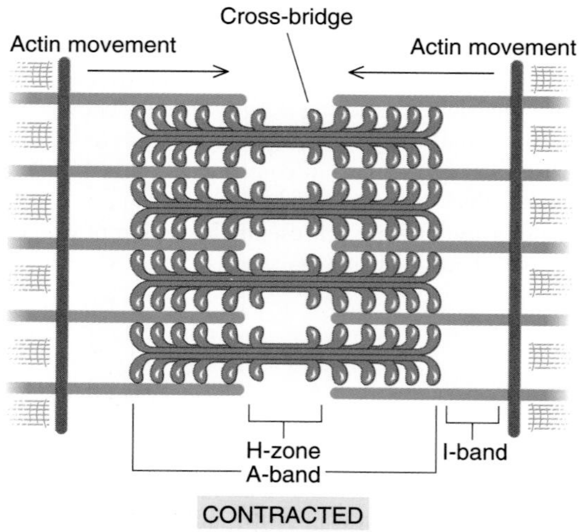

FIGURE 17-8
The mechanism of muscle contraction. The change in banding pattern within a sarcomere during contraction results from the sliding of thin, actin-containing filaments over central, myosin-containing thick filaments. As a result, the sarcomere shortens in length, as evidenced by the decrease in width of the I band and H zone.

Sliding Filaments and Molecular Ratchets

In the 1950s, Hugh Huxley and Jean Hanson of University College in London demonstrated that muscle contraction resulted from the sliding of the thin actin filaments toward the center of the sarcomere. The nature of the force that drove the thin filaments across the sarcomere remained a mystery until the discovery of rounded "bulbs" projecting from the ends of the thick myosin filaments (Figure 17-8). In the relaxed muscle fiber, the bulbs fail to make contact with the nearby, thin filaments. When a muscle fiber is activated following the arrival of a nerve impulse, the myosin bulbs attach to the actin molecules, forming a "cross-bridge" between the thick and thin filaments. These cross-bridges, which form simultaneously along the entire muscle fiber, bend toward the center of the sarcomere. The movements of the myosin bulbs, which require the hydrolysis of ATP, serve as a power stroke that slides the thin actin filaments a perceptible distance over the myosin. The bulbs immediately release the actin and snap back to their original positions, then reattach themselves to the actin filament at a new site further along the thin filament's length and generate another power stroke, moving the filament a bit closer toward the center of the sarcomere. This ratchet mechanism of muscle contraction is analogous to a team of rowers propelling a boat forward with each power stroke, then lifting the oars out of the water so that they can return to their starting position after each cycle. Next time you are engaged in strenuous exercise, you might remember that virtually all of the energy you are expending is being used to bend billions of tiny ratchets within the sarcomeres of your muscles.

Regulating the Strength of Contraction

The same muscles that generate enough power to lift a 400-pound barbell can also gently pick up a newborn baby without hurling it against the ceiling. The strength by which a muscle contracts depends primarily on the number of muscle fibers that are stimulated, which, in turn, depends on the number of neurons that carry impulses into the muscle tissue. In general, the more neurons that are activated, the more muscle fibers that shorten, and the stronger the contraction.

SMOOTH MUSCLE: RESPONDING TO INVOLUNTARY COMMANDS

Smooth muscle consists of nonstriated, spindle-shaped cells (cells with tapering ends) that may be present in small clusters or as part of muscle sheets that surround the body's hollow organs (Figure 17-9). Smooth muscle is *involuntary* because its contraction is regulated by the autonomic nervous system (Fig. 15-12) and is thus independent of conscious control. Smooth muscle in your urinary bladder, for example, automatically contracts in response to the internal pressure exerted on the walls of a full bladder. Fortunately, we have a backup skeletal muscle that is under

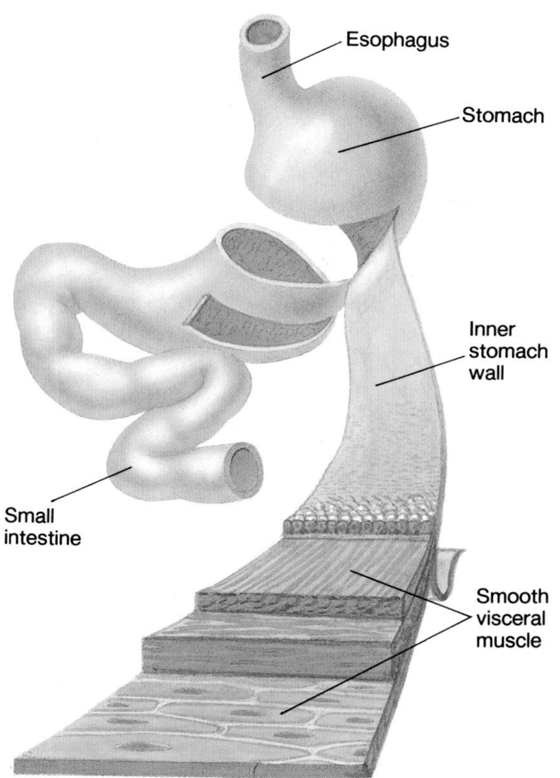

FIGURE 17-9
Smooth, involuntary muscle is often present as sheets of nonstriated muscle cells, such as those found in the digestive tract (pictured here) or along the ducts of the urinary tract. Waves of contraction (*peristalsis*) pass along these muscle sheets, generating a moving constriction that pushes the contents of the channel along the tract.

voluntary control and closes off the exit to prevent voiding urine at inopportune moments. Other functions of smooth muscle include control over the diameter of blood vessels, the diameter of the pupil of the eye, and the movement of food through the digestive tract.

CARDIAC MUSCLE: FORMING THE BODY'S PUMP

Cardiac muscle tissue consists of a type of muscle cell that has a unique combination of properties. Like smooth muscle, cardiac muscle is under involuntary control. Like skeletal muscle, cardiac muscle cells are striated. Unlike most skeletal muscles, however, cardiac muscle does not function anaerobically (page 108). If the oxygen supply to the heart is blocked, as may occur during a heart attack, muscle contraction rapidly ceases, resulting in heart damage.

BODY MECHANICS AND LOCOMOTION

Now that we have seen how bones, joints, and muscles function individually, we can better understand how these components of the skeletomuscular system work together to perform mechanical work. Humans use levers, such as the wheelbarrow depicted in Figure 17-10*a*, to lessen the amount of force required to lift and move an object (in this case, a load of dirt). In this example, the wheel is the *fulcrum*, or pivot site; the dirt is the load to be lifted; and energy is applied as an upward force against the handle.

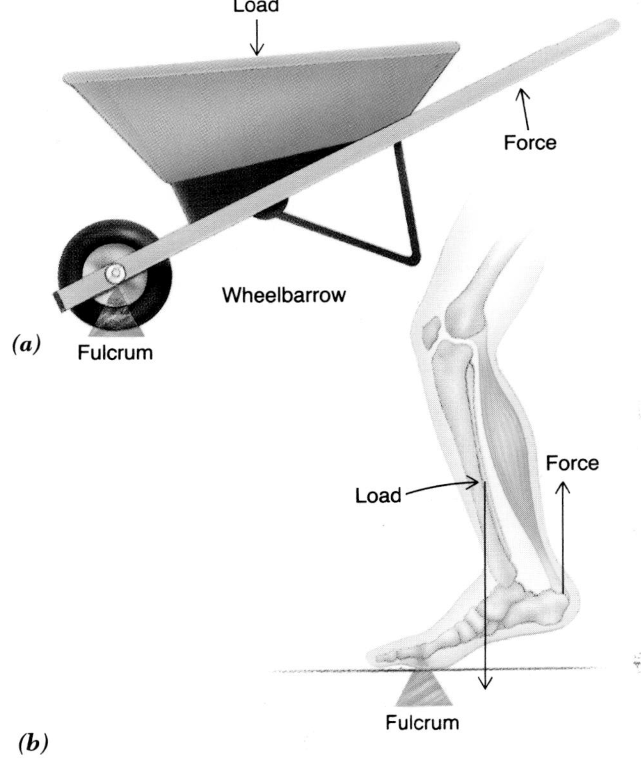

FIGURE 17-10
Bones serve as the body's levers. Pushing a wheelbarrow ***(a)*** and standing on your tiptoes ***(b)*** utilize similar types of levers. In both cases, the fulcrum (pivot point) is at one end of the system, the load to be lifted is in the middle, and the force needed to move the load is exerted on the opposite end. Both the wheelbarrow and the bones of the foot act as levers that allow this type of mechanical work to be performed.

The body uses bones as levers, and joints as fulcrums, to accomplish similar functions. When you stand on your tiptoes, for example, an upward force is generated by the muscle in your calf using the ball joint of the toes as the fulcrum (Figure 17-10*b*) and the bones of the foot as a lever. The load, in this case, is the entire weight of your body.

EVOLUTION AND ADAPTATION: TYING IT TOGETHER

An animal's integument is the protective barrier that shields the organism from its external environment. Accordingly, many of the properties of the integument can be understood by considering the environmental challenges the animal must face. This feature can be illustrated by briefly surveying the types of integuments found in vertebrates, all of which are constructed on a similar epidermal–dermal plan.

The earliest vertebrates were jawless, bottom-dwelling fishes that were clothed in heavy, bony armor that protected them from predators. During subsequent evolution, fishes moved away from the ocean bottom, becoming more buoyant and mobile. The thick plates of bone along the sides of the body were no longer adaptive and became reduced to the thin, familiar bony *scales* that are scraped away when a fish is "cleaned." As vertebrates moved out of the water and onto the land, the integument became adapted to terrestrial habitats. The bony scales of the ancestral fishes were lost, and the dermis became a more fibrous, flexible layer. In amphibians—animals that live both in water and on land—the skin is usually moist and permeable, facilitating oxygen absorption across the body surface (Figure 17-11*a*). Among reptiles and other land vertebrates, the epidermis has become a tough, impervious layer that prevents water loss in harsh, dry, terrestrial environments (Figure 17-11*b*). The role of oxygen uptake in land vertebrates with impermeable integuments was taken over by the lungs.

(a)

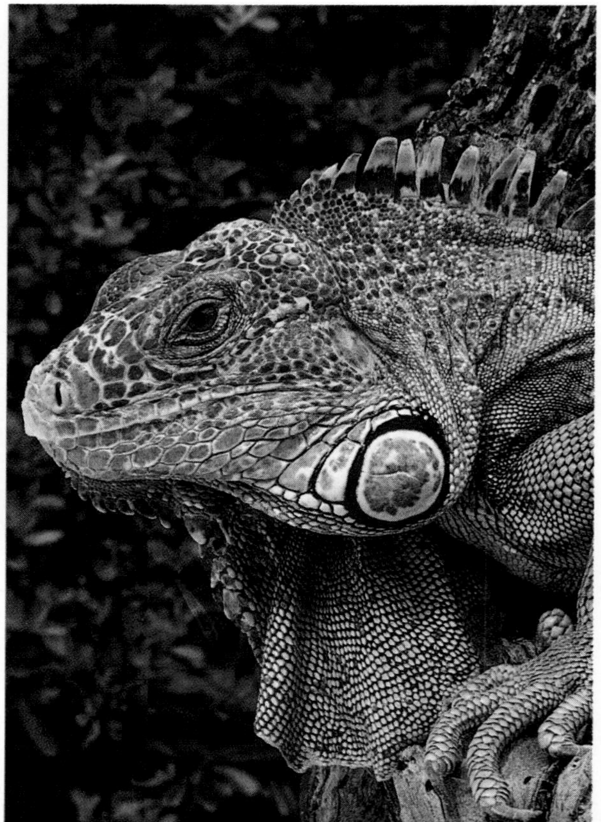

(b)

FIGURE 17-11

Contrasting integuments. The yellow spotted salamander (an amphibian) has thin, moist, permeable skin ***(a)***, while the iguana (a reptile) has thick, dry, impermeable skin ***(b)***.

SYNOPSIS

An animal's body surface is covered by a protective integument. In humans and other vertebrates, the skin is composed of an outer epidermis and an inner dermis. The outer, protective layer of the human epidermis consists of dead, keratinized cells that are continually sloughed and replaced. The dermis contains connective tissue, which provides support and skin cohesion, and blood vessels, which nourish the skin and play a role in heat conservation or heat loss.

Skeletal systems provide the rigidity that is required for support and movement. Some animals use only internal water pressure to support the body; others have a rigid exoskeleton that is external to the animal's living tissues. Still other animals have an internal endoskeleton: The human endoskeleton consists of 206 bones, each with a unique shape that allows it to perform a particular function. Most bones arise in the embryo as cartilaginous structures. They are then converted to bone by the deposition of calcium phosphate to the protein–polysaccharide matrix which surrounds the living cells. The bones of the skeleton are connected by joints that possess varying degrees of flexibilty.

Muscles are composed of specialized cells that contract (shorten) and generate a pulling force. Vertebrate muscle is divided into skeletal, smooth, and cardiac types. Skeletal muscle responds to voluntary commands and is primarily responsible for moving portions of the skeleton. Skeletal muscles are composed of large, multinucleate cells (fibers) containing myofibrils that have a markedly striated appearance. The structural unit of the myofibril is the sarcomere, which contains overlapping sets of thick (myosin-containing) and thin (actin-containing) filaments. When a muscle fiber is activated to contract, the thin filaments of the sarcomeres slide over the thicker filaments. The force required for this motion is fueled by ATP hydrolysis and is generated by the bending movements of the bulbous heads of the myosin molecules when they are attached to the adjacent, thin filaments. The strength by which a muscle contracts depends on the number of fibers that contract, which depends on the number of neurons that carry impulses into the muscle. Smooth muscles, which lack striations, mediate involuntary movements, such as the constriction of blood vessels and the closure of the pupil of the eye. Cardiac muscle, which is striated, makes up the wall of the heart.

Review Questions

1. Compare and contrast skeletal muscle, smooth muscle, and cardiac muscle; osteoclasts and osteoblasts; outer and inner layers of the human epidermis; sebaceous and sweat glands.
2. Compare and contrast compact and spongy bone; hydrostatic skeleton and exoskeleton; cartilage and bone.
3. Describe the functions of the major parts of the human skeleton, including the skull, vertebral column, rib cage, pectoral and pelvic girdles, and limb bones.
4. Describe the structure of a skeletal muscle fiber and the mechanism by which the movement of thick and thin filaments generates the force of muscle contraction.

Critical Thinking Questions

1. Explain, in terms of enzyme activity, why Ehlers-Danlos syndrome would produce less severe effects than would extreme ascorbic-acid deficiency.
2. In addition to protecting the body against the environment, the skin has to receive information from and exchange materials with the environment. How is the skin organized and structured to perform these somewhat contradictory functions?
3. Imagine that your right elbow is resting on the table and your right hand is lifting a heavy weight. Considering the role of the biceps in bending your arm (see Figure 17-6), how do the relative positions of the fulcrum, load, and force differ in this case, compared to the example of standing on your tiptoes (Figure 17-10*b*)?
4. Given the fact that bone is continually being remodeled by the action of osteoblasts and osteoclasts, would you expect to see a change in the architecture of the head of the femur if you were suddenly bedridden for a long period of time? Why?

CHAPTER

18

Nutrition and Digestion

Experiments on chickens revealed that rice kernels contain a vitamin required for normal metabolism. The red bar and bent arrow pinpoint the reaction that is blocked when the substance is absent from the diet.

STEPS TO DISCOVERY
The Battle Against Beri-beri

Unlike so many other human diseases, beri-beri was probably not an ancient scourge, but primarily a product of the Industrial Revolution. The first clearly documented cases of this nervous disorder, which is characterized by fatigue, muscle deterioration, and possible paralysis, appeared in Asia in the nineteenth century. The disease became prevalent among prisoners and soldiers stationed in the Dutch East Indies in the 1880s. The Dutch government dispatched a team of scientists to look into the problem. Among the members of the team was a medical officer named Christiaan Eijkman.

During the 1880s, etiology (the study of disease) was dominated by the findings of Louis Pasteur and Robert Koch, who had been instrumental in proving that diseases are often caused by "germs" that grow in the body. Unfortunately, Pasteur's and Koch's contemporaries believed that this "germ theory" applied to all diseases; that is, they believed that *all* diseases were attributed to either bacterial infections or to the toxins produced by bacteria. Consequently, Eijkman spent 4 fruitless years trying to isolate the bacterium responsible for beri-beri.

One day in 1896, a sudden development provided an unexpected breakthrough in Eijkman's research. For no apparent reason, the chickens that Eijkman was using as experimental animals developed a nerve disease whose symptoms resembled that of human beri-beri. Many of the animals died, but after 4 months, the chickens that had survived the disorder had recovered completely. Upon investigating the matter, Eijkman discovered that the chickens began to recover after a new animal keeper had stopped feeding them leftovers from the military hospital, which consisted largely of polished rice—rice that had been processed by a steam mill until the outer hulls had been removed.

Eijkman used this information in a dietary experiment. He fed some of the chickens a diet of polished rice; these chickens soon developed symptoms of the disease. In contrast, control animals that were fed whole rice remained healthy. Furthermore, the afflicted group could be cured if they were fed either whole rice or polished rice to which the outer hulls had been added. Eijkman concluded that the disease was not due to a bacterial infection but to a dietary deficiency, providing the first evidence that a disease could be caused by the absence of some trace component of the diet. The idea that a disease could result from a dietary deficiency did not initially gain widespread acceptance, however.

In 1911, Casimir Funk, a chemist working at the Lister Institute in London, succeeded in purifying a substance from the hulls of rice that he believed was the same substance that could reverse the symptoms of beri-beri. The substance was an amine (one containing an amino [$—NH_2$] group), which led Funk to coin the word "vitamine," meaning an *amine* that was *vital* to life. Later work showed that the substance crystallized by Funk was not, in fact, the same one that was active against beri-beri. The name caught on, however, and remained the common term used to describe organic substances that are required by the body in trace amounts. Most vitamins, in fact, contain no amine groups.

Finally, in 1926, two Dutch chemists, B. C. Jansen and W. Donath, working in Eijkman's old lab in the East Indies, developed a procedure for purifying the anti-beri-beri factor from rice bran. Crystals of the substance were sent back to Holland, where Eijkman confirmed that this single chemical compound was effective against the nervous disorder exhibited in birds. However, when Jansen and Donath determined the chemical formula for the substance, they overlooked an important feature: the presence of a sulfur atom. This oversight set back the effort to determine the structure of the compound, which had been named vitamin B_1. The presence of the sulfur atom wasn't discovered until 1932; the correct structure was published in 1936 by Robert Williams, an American chemist who had been working on the problem for over 20 years. Within a year, Williams had worked out a complex procedure for synthesizing the compound, which he named thiamin. Soon, thiamin was being manufactured and became available as a widespread vitamin supplement.

The last major step in the story of vitamin B_1 was the discovery of its biological action. In 1937, two German biochemists, K. Lohmann and P. Schuster, found that thiamin was a coenzyme in a key reaction in the oxidation of glucose. Later investigations revealed that virtually all vitamins act in conjunction with enzymes in carrying out one or more crucial metabolic reactions. It was the *failure* to catalyze these reactions that led to the symptoms of the deficiency diseases, such as beri-beri. For his work in establishing the existence of dietary deficiency diseases, Eijkman was awarded the 1929 Nobel Prize in Medicine and Physiology.

*F*or some animals, obtaining food is a relatively simple task. Tapeworms, for example, have no mouth or digestive tract; they simply attach to the wall of an animal's intestine and absorb digested nutrients across their outer body surface. For these animals, eating is not necessary. Unfortunately, as humans, we can't enjoy the same advantage. We can't recline in a bathtub of oatmeal for breakfast and chicken soup for lunch and simply soak up the nutrients. Any such endeavor would only end in starvation. Even if the nutrients could penetrate the epithelial layer of the skin and enter the bloodstream, most of the molecules would be too large to cross the plasma membranes and enter the cells. The molecules would only be wasted since the value of food is unleashed only inside living cells. To meet this requirement for life, most animals possess a team of specialized organs that constitute the **digestive system.**

▼ ▼ ▼

THE DIGESTIVE SYSTEM: CONVERTING FOOD MATTER INTO A FORM AVAILABLE TO THE BODY

At what point does food enter the body? Most people would answer, "as soon as you put it in your mouth," or "when you swallow it." Yet, eating does not actually introduce food into your body. When substances are in the stomach or intestines, they are still outside of you, just as your finger poking through the hole of a doughnut remains outside the pastry. The **digestive tract,** or *gut,* is actually a tubelike continuation of the animal's external surface into or completely through its body. The walls of the tract might be likened to an absorbant version of the skin, one that forms a barrier between the external environment and the internal tissues of the body. Because of this barrier property, the digestive tract can safely provide residence for a large number of bacteria that would be dangerous to the interior, living tissues of the body.

To enter the body itself, nutrients must be absorbed across the epithelium that lines the digestive tract. **Digestion** prepares food to do just that. Digestion is the process of disassembling large food particles into molecules that are small enough to be absorbed by the cells that line the digestive tract. Ultimately, these molecules enter the cytoplasm of every cell in the body, where the nutritive value of food is harvested. Eating only initiates the digestive process, launching food on a journey through tunnels and chambers, where digestion and absorption occur.

THE HUMAN DIGESTIVE SYSTEM: A MODEL FOOD-PROCESSING PLANT

The human digestive tract is approximately 9 meters (30 feet) long. It consists of the *mouth, esophagus, stomach, small intestine, large intestine,* and *anus,* plus a variety of accessory organs (Figure 18-1). The human digestive tract is a model food-processing plant for the stepwise disassembly and absorption of food material. During its journey through the digestive tract, ingested food matter is mixed with various fluids; churned and propelled by the musculature of the wall; broken down by enzymes that are secreted by various glands; and absorbed by cells that line the digestive channel. The indigestible residues are then eliminated from the tract through the anus. All of these complex processes are regulated by the coordinated action of both the nervous and endocrine systems.

The wall of the digestive tract (Figure 18-1) is composed of several layers, including an inner glandular epithelium (*mucosa*), layers of circular and longitudinal smooth muscle, and a connective tissue sheath (*serosa*). Glandular cells in the epithelium secrete mucus, enzymes, ions, and other substances. Contraction of the muscle layers help break up congealed food matter, which is mixed with secreted fluids and moved through the tract. We will begin our journey through the digestive system as food enters the first part of the digestive tract—the mouth.

THE MOUTH AND ESOPHAGUS: ENTRY OF FOOD INTO THE DIGESTIVE TRACT

Digestion of food begins in the **mouth,** where food is cut and ground by the teeth. This action makes the food matter easier to swallow and increases its access to digestive enzymes. While in the mouth, the macerated food is mixed with **saliva**—the secretion of the **salivary glands** that initiates chemical digestion. Saliva contains enzymes that initiate the digestion of starch. Saliva also contains *mucin,* the major protein of mucus, which acts as a lubricant. This feature is best appreciated when we try to swallow something that has not been adequately covered by the slippery fluid.

Mucin also binds the macerated food together into a cohesive mass, called a **bolus,** which is pushed to the back wall of the oral cavity (the *pharynx*). During swallowing, the bolus is forced into the **esophagus,** the tubular channel that leads to the stomach. During the swallowing process, the openings to the respiratory and nasal passages are automatically closed to ensure that food is kept out of these

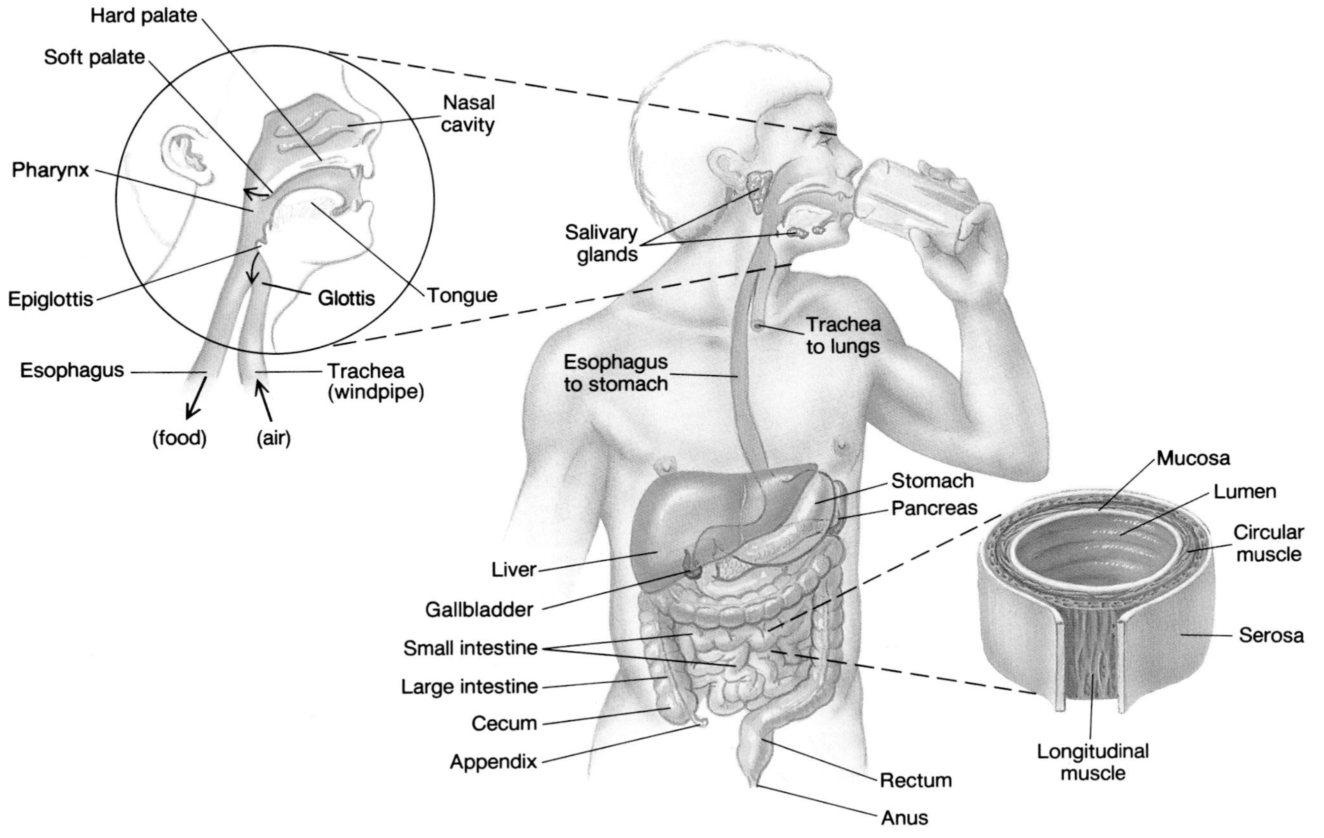

FIGURE 18-1

The human digestive system. In addition to the tubular digestive tract, the digestive system includes accessory organs (the pancreas, liver, and gallbladder) that aid digestion. The mucosa and musculature of the digestive tract are shown in the intestinal cross section. The inset depicts structures involved in swallowing. During swallowing, the *soft palate* elevates and closes off the nasal cavity, while the *glottis* (the opening to the windpipe and lungs) is sealed by the *epiglottis,* leaving the esophagus the only open passageway for the mass of chewed food (bolus).

nondigestive pathways (see inset, Figure 18-1). This explains why you can't breathe while swallowing. The walls of the esophagus contain muscle layers that contract in a rhythmic manner, sending successive waves of contraction, or **peristalsis,** down its length. Peristalsis constricts the channel of the esophagus, pushing the bolus through a thickened muscular valve, or *sphincter,* into the stomach.

THE STOMACH: A SITE FOR STORAGE AND EARLY DIGESTION

Once in the stomach, food matter is churned to a pastelike consistency and mixed with *gastric juice* (*gastro* = stomach), forming a solution called **chyme.** Gastric juice is produced by secretory cells that are located in pits in the wall of the stomach. *Hydrochloric acid* (HCl), one of the compounds that makes up gastric juice, lowers the pH of the stomach contents to around 2.0. This extremely acidic environment kills most microbes in food, including many of those that could cause illness. If the acidic contents of the stomach should leak back into the esophagus, the irritation of the lining is interpreted as "heartburn."

Although most enzymatic digestion occurs in the small intestine, protein digestion begins in the stomach with the action of the enzyme *pepsin.* To prevent the stomach from becoming its own next meal, the protein-digesting enzyme is secreted as an inactive precursor, *pepsinogen,* which is converted to an active pepsin molecule by the hydrochloric acid of the stomach. The living tissue that lines the stomach is protected from both acid and pepsin by a thick layer of alkaline mucus. When protective mechanisms fail, the stomach may begin digesting portions of itself, causing painful and dangerous *peptic ulcers.*

For decades, peptic ulcers have been treated by administration of antacids. Persons suffering from these lesions, however, have found this therapy ineffective; while antacids help to alleviate symptoms, over 95 percent of patients experience a relapse within 2 years. Within the

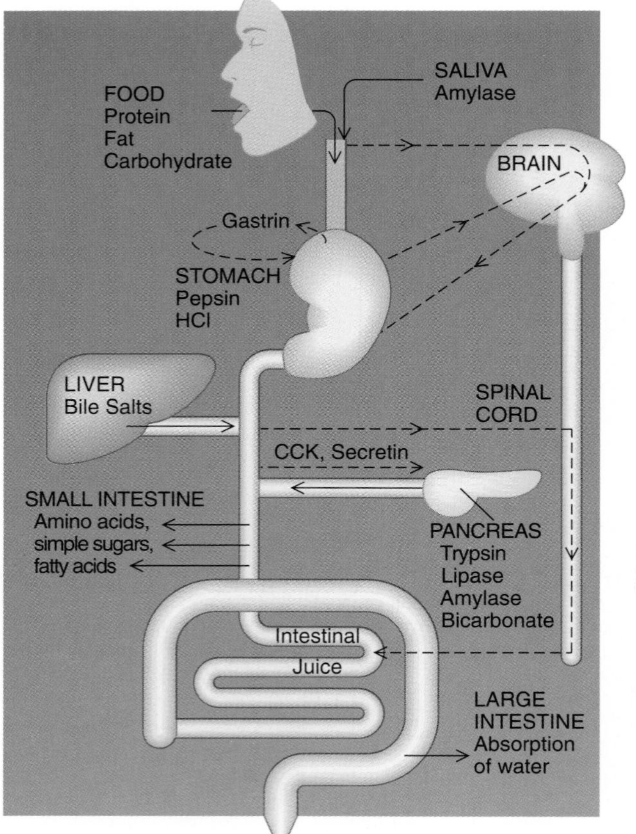

FIGURE 18-2
Control of the processes of digestion by the nervous and endocrine systems.

past few years, a number of studies have shown that, while stomach acidity may be a factor in the development of peptic ulcers, the primary cause is infection by the bacterium *Helicobacter pylori.* Ulcer patients treated with antibiotics that kill the bacteria are much less likely to suffer relapses than are those treated solely with antacids. Don't be surprised if, one day, a vaccine becomes available to prevent ulcers.

The Control of Gastric Secretions

The control of the secretion of gastric juices illustrates the complex communication that occurs during physiological processes (Figure 18-2). The first phase of gastric secretion is stimulated by nerve impulses that reach the stomach from the brain as a result of the smell, taste, or even thought of food. When food actually enters the stomach, two new types of signals are generated which lead to a marked increase in the secretion of gastric juice. One of the signals is carried by sensory neurons from the stomach to the brainstem, which responds by sending impulses down autonomic motor fibers. This stimulates the digestive-gland cells of the stomach wall to release their products. The other signal is a chemical message that is sent by the hormone gastrin, which is released by endocrine cells located in the stomach lining. The message is carried locally in the blood vessels of the stomach wall to the stomach's glandular cells, triggering these cells to release gastric juices. By sending these nervous and endocrine signals, the stomach ensures that gastric juices will be secreted when they are necessary.

Only a few small molecules, such as aspirin and alcohol, enter the bloodstream through the stomach wall. This explains the rapid onset of their effects. Most nutrients are not absorbed until they enter the small intestine.

THE SMALL INTESTINE: A SITE FOR FINAL DIGESTION AND ABSORPTION

Peristaltic waves moving along the wall of the stomach repeatedly push small quantities of chyme into the **small intestine** (Figure 18-1). The small intestine consists of 6 to 7 meters (about 21 feet) of highly coiled muscular tubing, about 2.5 centimeters (1 in.) in diameter. During its stay in the small intestine, macromolecular food substances are digested into small, organic molecules, such as simple sugars, amino acids, and nucleotides, which are absorbed into the bloodstream.

Digestion of materials within the small intestine requires the cooperative activities of several major organs and their secretions, including intestinal juice, pancreatic secretions, and emulsifying lipids from the liver and gallbladder.

Intestinal Secretions

When the inflow of chyme from the stomach stretches the intestinal wall, the action triggers a neural reflex response, whereby the cells of the intestinal lining secrete *intestinal juice* and mucus. Under normal conditions, the small intestine secretes about 2 to 3 liters of intestinal juice per day. This fluid is needed to dissolve the molecules for digestion and to facilitate absorption across the intestinal epithelium. Cholera, one of the most dreaded human diseases, results from a bacterial toxin that greatly increases the fluid released by these intestinal cells. When stimulated by the cholera toxin, the intestine can pour out over 1 liter of fluid per hour, most of which is simply lost through the digestive tract as diarrhea.

Pancreatic Secretions

As we discussed in Chapter 15, the pancreas is a gland that manufactures digestive enzymes and discharges them directly into the small intestine. The pancreas also releases sodium bicarbonate, an alkaline substance that helps neutralize the severe acidity of the chyme entering the small intestine from the stomach. Within the pancreatic secretion are enzymes that digest all of the major types of macromolecules. The secretion of pancreatic enzymes and bicarbonate is stimulated by two hormones—*cholecystokinin* (*CCK*) and *secretin* (Figure 18-2). These hormones are secreted into the blood by endocrine cells in the wall of the small intestine in response to the inflow of chyme from the stomach.

Liver and Gallbladder Secretions

We all know how to remove baby oil or motor grease from our hands, or animal fat from a pan: We wash our hands or the pan with soap and water. Soap contains detergent molecules which, because of their structure, can surround fat molecules and suspend them in water. As a result, the grease or fat comes off your hands or the pan and becomes *emulsified* (suspended) in the surrounding water. A similar approach is taken by your digestive system in dealing with fats in your diet. Suppose you eat a pizza with double cheese. In order for the fat molecules in the cheese to be efficiently hydrolyzed by the pancreatic lipases in the intestine, large globules of fat must be broken apart into much smaller clusters. This process is aided by **bile salts,** which are produced in the liver and stored in the **gallbladder** (see Figure 18-1), a small sac that empties its contents into the intestine through a duct. Bile salts are similar in structure to the detergents present in soap. In the presence of bile salts, fat globules are reduced to stable, microscopic droplets that can be attacked by lipid-digesting enzymes.

Absorption of Products Across the Small-Intestinal Wall

The first step in the absorption of the small, digested food molecules is their movement from the lumen (the space inside a tube) of the small intestine into the epithelial cells that form its lining. The inner surface of the intestinal wall has a "velvety" texture due to the presence of fingerlike projections, called **villi.** Villi increase the absorptive surface of the intestine in the same way that the texture of terrycloth towels enables them to soak up much more water than can smooth, cloth towels. Each villus (singular of villi) is covered with smaller projections, called **microvilli,** which further increase the surface area of the small intestine enormously (Figure 18-3). Together, intestinal villi and microvilli create an interior surface area that is 150 times greater than is the surface of your entire skin.

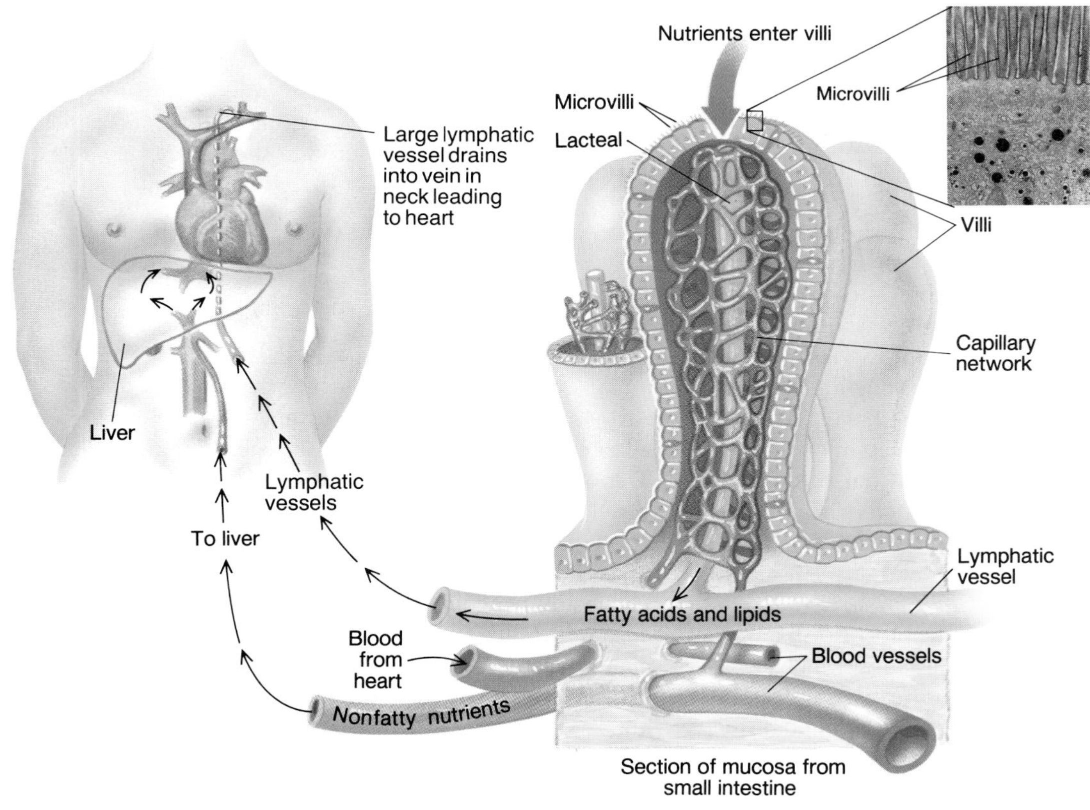

FIGURE 18-3

Structure and function of the small intestine's villi. Nutrients pass from the lumen of the small intestine into capillaries and a central lacteal located within each villus, a projection of the intestinal lining. Tiny projections, called microvilli, extend from the end of each epithelial cell in the villi, further increasing the absorbing surface. Substances absorbed from the small intestine into the bloodstream are carried to the liver, which detoxifies many harmful molecules and stores various nutrients, such as glucose. From the liver, nutrients enter the general circulation.

Packed inside your abdomen is an intestinal surface equivalent to the surface area of a tennis court!

Each villus is laced with a rich network of capillaries surrounding a single, centrally located lymphatic vessel known as a **lacteal** (Figure 18-3). The lacteal absorbs the products of lipid digestion, such as the fatty acids that are produced by lipase digestion of the cheese on that pizza you consumed hours earlier. From the lacteal, microscopic fat droplets are transported through a series of lymphatic vessels that eventually drain into a large vein in the neck.

What about the remainder of the ingredients in that pizza, such as the glucose in the starchy dough, and the amino acids in the protein-rich pepperoni? Most nonfatty nutrients diffuse directly into the blood capillaries of the intestinal villi, where they are carried to the liver and removed from the bloodstream. The liver is the body's primary metabolic regulatory center, producing waste products, controlling blood-glucose levels, and releasing substances into the bloodstream, as needed by the body's tissues.

THE LARGE INTESTINE: PREPARING THE RESIDUE FOR ELIMINATION

By the time digested food material has made its long journey to the end of the small intestine, virtually all of its nutrients have been removed, along with most of the water. The nutrient-depleted chyme is now propelled by peristalsis into the next part of the digestive tract, the **large intestine** (or **colon**) (Figure 18-1).

The most important functions of the large intestine are the reabsorption of water from the digestive tract and the conversion of the remaining contents into a mass, called *feces.* Pressure-sensitive neurons detect when solids accumulate in the terminal (end) portion of the large intestine (the **rectum**) and respond by provoking a *defecation reflex*: Impulses from the large intestine travel to the spinal cord and back to the muscles of the rectum causing the muscles to contract and force the feces out through the anus. Because one of the two anal sphincters is under voluntary control, we can consciously delay expulsion of feces until an appropriate time.

Projecting from the large intestine is a short, blind (dead-ended) tube, the *appendix* (Figure 18-1). Inflammation of the appendix leads to *appendicitis,* which, if untreated, can cause the appendix to rupture, spilling bacteria into the abdominal cavity and creating a potentially fatal infection.

Huge numbers of bacteria reside in the large intestine of the healthy human digestive tract. In fact, bacteria constitute almost half the dry weight of human feces. These intestinal bacteria metabolically attack organic substances in chyme and use them as nutrients, often producing unpleasant-smelling byproducts. These organisms are not freeloaders, however; they manufacture vitamin K, biotin, folic acid, and other nutrients we absorb and utilize. Intestinal bacteria also contribute to our well-being by competing with potentially dangerous microbes for the body's limited space and nutrients. This becomes apparent when we destroy our normal bacterial flora by the extended use of antibiotics, which can cause digestive dysfunction (such as extended bouts of diarrhea) and yeast infections.

NUTRITION

Controversy continues to rage over such topics as the impact of dietary sugar, cholesterol, and saturated fats on our health. Nutritionists generally agree, however, that the healthiest diets are those that balance carbohydrates, triglyceride lipids (fats and oils), and proteins. Foods that provide these three groups of molecules should also contain enough energy, organic building blocks, vitamins, and minerals to satisfy the needs of the average person.

It is estimated that an average person engaged in a relatively sedentary lifestyle requires about 2,500 Calories (2,500 kcals) per day to maintain his or her body at a stable level. A person who engages in frequent strenuous activity, such as a professional athlete, may require over 4,000 Calories per day. Since people differ in their metabolic rates, the amount of calories that will maintain one person's weight may cause another person either to gain or lose pounds. The ingredients that go into making up a "well-balanced" meal are considered in the accompanying Human Perspective: Necessary Ingredients of Human Nutrition.

EVOLUTION AND ADAPTATION: TYING IT TOGETHER

An animal's feeding apparatus and digestive system are adapted to the type of food the animal regularly consumes. The human digestive system is specialized for digesting a variety of foods, ranging from chunks of meat to fibrous vegetables and fruits. Many animals have more specialized diets. For example, a number of mammals, including anteaters and pangolins, feed exclusively on ants (Figure 18-4*a*). Adaptations of ant-eating species include powerful claws that can dig up ant hills; elongated snouts that can extend into an ant nest; and long, sticky tongues that trap ants. Many aquatic animals, including sponges, clams, and blue whales, feed on tiny organisms suspended in the water (Figure 18-4*b*). These *filter feeders,* as they are called, have some type of straining device or sticky, mucus-covered surfaces that screen or trap tiny food particles suspended in the water.

Some animals paradoxically obtain nutrients from fibrous plants that they alone cannot digest. Cattle, antelopes, buffalo, giraffes, and other such *ruminant* animals possess additional stomach chambers, one of which is heav-

◁ THE HUMAN PERSPECTIVE ▷

Necessary Ingredients of Human Nutrition

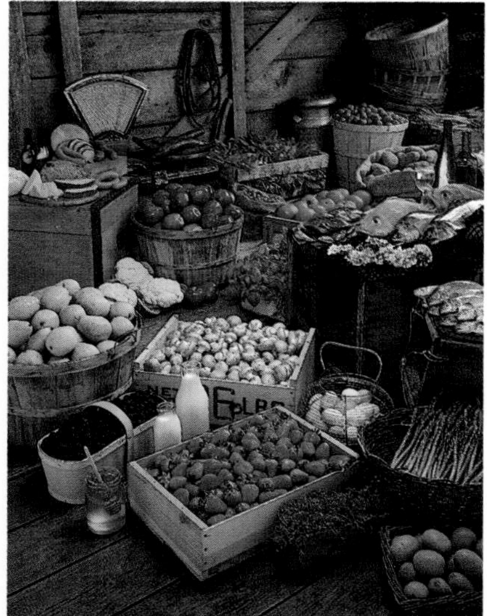

FIGURE 1

CARBOHYDRATES

Carbohydrates provide the most readily available form of glucose and, therefore, the most rapid, readily available form of usable energy. This is why runners, for example, eat large amounts of pasta the night before a marathon; pasta is composed largely of starch, a glucose-containing carbohydrate made by plants. Glucose is the "all-purpose" energy source; it is also the only one that can be used by all brain and nerve cells. When these cells are deprived of energy, the resulting temporary *hypoglycemia* (low blood sugar) may cause unclear thinking, clumsiness, depression, or muscle tremors due to diminished neurological function.

Not all carbohydrates are easily digestible. Cellulose, for example, the polysaccharide of plant cell walls, resists disassembly in the human digestive tract. Yet cellulose (such as that found in celery stalks) promotes health by providing bulk-fiber that assists in the formation and elimination of feces. Low-cellulose diets cause constipation and have been linked to colon cancer. For this reason, foods that are rich in both polysaccharides and undigestible bran, such as fruits, grain products, and legumes, are recommended as part of a balanced diet. In contrast, the simple carbohydrates that are found in sugary products, such as candy and soft drinks, provide calories but lack the fiber necessary for healthy digestion.

LIPIDS

Because of their highly reduced state, fats and oils are rich sources of energy (page 80). Two fatty acids, linolenic and linoleic acid, are *essential fatty acids* since they are needed but cannot be manufactured in the body and are thus required in the diet. These fatty acids are needed for cell-membrane construction and for the synthesis of certain hormones. In addition to being fattening, diets rich in saturated fats and cholesterol (found in butter, animal fats, and eggs) may predispose susceptible

(continues)

(a)

(b)

FIGURE 18-4

Feeding specialists. ***(a)*** Anteaters trap their tiny prey on the sticky surface of their tongue. ***(b)*** Barnacles have modified appendages that strain the surrounding water, collecting microscopic organisms.

individuals to circulatory disease by increasing the likelihood of deposition of lipids in the walls of arteries. In contrast, unsaturated fats, which are found in most vegetable oils, may increase the risk of cancer probably because breakdown of the fatty acids in these molecules produces highly reactive chemicals. Nutritionists generally agree that an ideal diet should be low in fat; that is, fat should provide less than 30 percent of food calories.

PROTEINS

Dietary protein is needed to supply the amino acids from which we assemble our enzymes, antibodies, hormones, and various other types of proteins. Protein can be obtained from virtually any food, including meat, cheese, eggs, and vegetables. We can manufacture all but eight amino acids metabolically. The absence of even one of these eight *essential amino acids* prevents the synthesis of all proteins. All of these required amino acids can be synthesized by plants and microorganisms. The prolonged absence of any of the essential amino acids in the diet can lead to severe protein deficiency and a condition known as *kwashiorkor,* which is one of the world's most serious health problems. We are all too familiar with pictures of listless, malnourished children with swollen bellies (a result of water retention) and arms and legs composed of just skin and bones. These symptoms are a result of the world shortage of dietary protein.

VITAMINS

An organic compound is designated a *vitamin* when it is needed in trace amounts for normal health but cannot be synthesized by the body's own metabolic machinery and thus must be obtained in the diet. As humans, we must acquire 13 vitamins from our diet, or we run the risk of suffering vitamin-deficiency disorders, some of which can be fatal. A well-balanced diet normally provides all the vitamins needed.

Vitamins are usually divided into two groups based on their solubility properties.

TABLE 18-1
DIETARY ESSENTIALS IN HUMAN NUTRITION: VITAMINS

Designation	*Major Mode of Action*	*Major Sources*[a]	*Symptoms of Deficiency*[b]
Retinol (A)	Part of visual pigment, maintenance of epithelial tissues	Egg yolk, butter, fish oils; conversion of carotenes[c]	Nightblindness, corneal and skin lesions, reproductive failure
Calciferol (D)	Ca and P absorption, bone and teeth formation	Fish oils, liver; irradiation of sterols[c]	Rickets, osteomalacia
Tocopherols (E)	Antioxidant	Vegetable oils; green, leafy vegetables	In animals: muscular degeneration, infertility, brain lesions, edema[d]
Vitamin K	Synthesis of blood coagulation factors	Green, leafy vegetables; bacterial synthesis	Slowed blood coagulation
Thiamine (B_1)	Energy metabolism–decarboxylation	Whole grains, organ meats	Beri-beri, polyneuritis
Riboflavin (B_2)	Hydrogen and electron transfer (FAD)	Whole grains, milk, eggs, liver	Cheilosis, glossitis, photophobia
Nicotinic acid (niacin)	Hydrogen and electron transfer (NAD, NADP)	Yeast, meat, liver[e]	Pellagra
Pyridoxine (B_6)	Amino acid metabolism	Whole grains, yeast, liver	Convulsions, hyperirritability
Pantothenic acid	Acetyl group transfer (CoA)	Widely distributed	Neuromotor and gastrointestinal disorders
Biotin	CO_2 transfer	Eggs, liver; bacterial synthesis	Seborrheic dermatitis
Folic acid	One-carbon transfer	Leafy, green vegetables; meat	Anemia
Cobalamine (B_{12})	One-carbon synthesis; molecular rearrangement	Animal products, esp. liver; bacterial synthesis	Pernicious anemia
Ascorbic acid (C)	Hydroxylations, collagen synthesis	Citrus, potatoes, peppers	Scurvy

[a] Most vitamins, especially of the B group, occur in a multitude of foodstuffs and in all body cells.
[b] A variety of symptoms occur with certain vitamin deficiencies; vitamin deficiencies are frequently of a multiple nature, and symptoms similar to those described may have their origin in condtions not related to nutrition.
[c] Certain carotenes, found in green and yellow vegetables, are precursors of vitamin A. Certain sterols, including 7-dehydrocholesterol, which is synthesized in the body, are precursors of vitamin D.
[d] No well-defined syndrome is described for humans. [e] Niacin is one of the end products of normal tryptophan metabolism.
Source: From P. D. Sturkie, *Basic Physiology,* New York, Springer-Verlag, 1981, p. 345.

Vitamins in the first group are water soluble; these include eight different vitamins of the B complex and vitamin C. Vitamins in the second group are insoluble in water but soluble in oil; these include vitamins A, D, E, and K. Most of these vitamins function as coenzymes (page 78) that assist essential enzymatic reactions (Table 18-1). The required daily allowance of each vitamin is relatively low because these molecules are not consumed in the reactions they assist so each molecule is used again and again. Furthermore, the small amount that is required replaces that which is normally excreted.

MINERALS

A dietary supply of certain inorganic elements, or *minerals* (Table 18-2), is just as important for proper nutrition as is that of vitamins. Without calcium and magnesium, for example, large numbers of enzyme-mediated reactions would simply not occur. Calcium is also needed for bone growth and muscle function. Iron forms the functional core of cytochromes, which are needed by all cells for aerobic respiration, and of hemoglobin, the oxygen-transporting protein in blood. Phosphorus is needed for ATP and nucleic-acid synthesis. Sodium, potassium, and chloride are required for maintaining osmotic balance and for propagating nerve impulses. A number of other minerals, including iodine, copper, zinc, cobalt, molybdenum, manganese, and chromium, are required in such tiny amounts that they are referred to as *trace elements*.

TABLE 18-2
DIETARY ESSENTIALS IN HUMAN NUTRITION: MINERALS[a]

Designation	*Major Functions*	*Major Sources*	*Symptoms of Deficiency*[b]
Calcium (Ca)	Bone and teeth, nervous reactions, enzyme cofactor	Dairy products, leafy, green vegetables	Calcium tetany, demineralized bones
Phosphorus (P)	Bone and teeth, intermediary metabolism	Dairy products, grains, meat	Demineralized bones
Magnesium (Mg)	Bone, nervous reactions, enzyme cofactor	Whole grains, meat, milk	Anorexia, nausea, neurological symptoms
Sodium (Na)	Maintenance of osmotic equilibrium and fluid volume	Table salt[c]	Weakness, mental apathy, muscle twitching
Potassium (K)	Cellular enzyme function	Vegetables, meats, dried fruits, nuts	Weakness, lethargy, hyporeflexia
Chlorine (Cl)	Maintenance of fluid and electrolyte balance	Table salt[c]	*d*
Iron (Fe)	Hemoglobin, myoglobin; respiratory enzymes	Meat, liver, beans, nuts, dried fruit	Anemia
Copper (Cu)	Enzyme cofactor (cytochrome-c-oxidase)[e]	Nuts, liver, kidney, dried legumes, raisins	Anemia, neutropenia, skeletal defects
Manganese (Mn)	Enzyme cofactor, bone structure, reproduction	Nuts, whole gains	*d*
Zinc (Zn)	Enzyme cofactor (carbonic anhydrase)[e]	Shellfish, meat, beans, egg yolks	Growth failure, delayed sexual maturation
Iodine (I)	Thyroid hormone synthesis	Iodized table salt, marine foods	Goiter
Molybdenum (Mo)	Enzyme cofactor (xanthine oxidase)[e]	Beef kidney, some cereals and legumes	*d*
Chromium (Cr)	Regulation of carbohydrate metabolism (glucose tolerance factor)	Limited information available	*d*

[a] A human need for the following trace elements is possible but has not been unequivocally established: selenium (Se), fluorine (F), silicon (Si), nickel (Ni), vanadium (V), and tin (Sn). The need for sulfur (S) is satisfied by ingestion of methionine and cystine, and for cobalt (Co) by vitamin B_{12}.
[b] Except for Ca, Fe, and I, dietary deficiency in humans is either unlikely or rare.
[c] Many processed foods contain considerable amounts of sodium chloride. [d] No specific deficiency syndrome described in humans.
[e] Examples of activity as enzyme cofactors. *Source:* From P. D. Sturkie, *Basic Physiology*, New York, Springer-Verlag, 1981, p. 344.

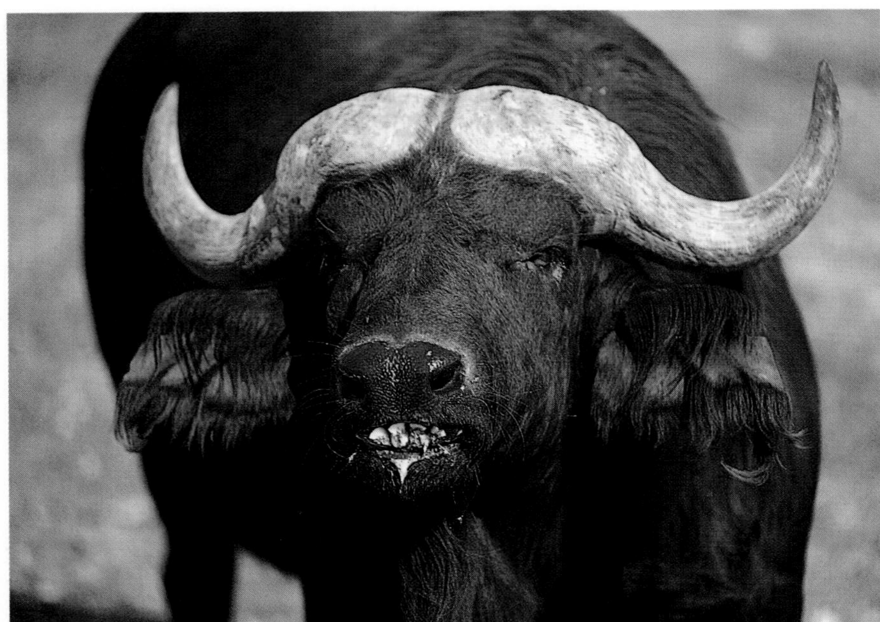

FIGURE 18-5
The cape buffalo; an example of a ruminant.

ily fortified with cellulose-digesting microorganisms (Figure 18-5). In this chamber, called the *rumen,* bacteria and protozoa break down the otherwise indigestible cellulose fibers and use them for growth of more microorganisms. This growing crop of microorganisms is then digested by the animal as the contents of the rumen travel through the rest of the digestive tract. Horses and other grazing animals that lack rumens are able to utilize plant cellulose because they possess an elongated *cecum,* a sac that extends from the large intestine. The cecum serves as a fermenting vat in which microbes disassemble cellulose into products that are digestible by the animal.

SYNOPSIS

An overview of digestion. Large, complex food substances are dismantled into molecules small enough to pass through plasma membranes and enter cells, where their nutritional value is unleashed. The human digestive tract consists of a continuous tube and various accessory organs. Ingested material is forced through the entire tract by peristalsis—waves of contraction of the layers of muscle tissue in the wall of the tract. The muscular and secretory activities that occur at different sites along the digestive tract are coordinated by the actions of both the nervous and endocrine systems.

Each part of the digestive tract is specialized for particular activities that occur in a stepwise fashion as food matter is pushed along the tract. The mouth macerates food and covers it with lubricating mucus. Swallowing pushes the food bolus into the esophagus, where it is moved by peristalsis into the stomach and churned into chyme. Glands in the stomach wall secrete a fluid that contains acid and protein-digesting enzymes. Chyme passes into the small intestine, where its acidity is neutralized by bicarbonate ions from the pancreas; its macromolecules are digested by enzymes also produced in the pancreas. Most of the fluid in which the food matter is suspended is derived from secretions of the wall of the small intestine itself. The fat that is present in the food is emulsified in the form of microscopic droplets by bile salts that are produced in the liver and stored in the gallbladder. As the various macromolecules are enzymatically digested, their component subunits are absorbed across the wall of the intestinal villi into either blood capillaries or lymphatic lacteals. Most of these nutrients are carried to the liver, where they are removed from the bloodstream. Those materials that cannot be digested and absorbed pass into the large intestine, where the remaining water is reabsorbed and the insoluble residues are compacted (together with bacteria) into feces, which are eliminated.

A healthy human diet must contain a variety of components. Chemical energy is most readily supplied by carbohydrates and fats, but it can also be provided by proteins and nucleic acids, which also can be degraded through a cell's oxidative pathways. Of the 20 amino acids incorporated into proteins, eight cannot be synthesized from other compounds and must be obtained from the diet. Protein deficiency is the world's major cause of malnutrition. Two fatty acids are also required as dietary elements. Vitamins are organic compounds that function primarily as coenzymes but cannot be synthesized by the body; thus, they are required in the diet. The diet must also supply a number of minerals. Some of these, such as calcium, magnesium, iron, sodium, and potassium, play major roles in the body and must be present at high levels in the diet. In contrast, the roles of trace elements, such as iodine, copper, and zinc, are more limited.

Review Questions

1. Why isn't food considered to be inside the body immediately after it is swallowed? At what point do nutrients actually enter the body?
2. Trace the fate of a mouthful of food through the entire digestive tract. Describe the changes that occur in each portion of the digestive tract and discuss the activities of saliva, gastric secretions, intestinal secretions, pancreatic secretions, bile, and intestinal bacteria.
3. How are the various activities described in your answer to the previous question regulated by the endocrine and/or nervous system?
4. Compare and contrast the sites of the preliminary digestion of starch, protein, and fats; the role of CCK and secretin; the role of most vitamins compared to minerals, such as calcium and iron.

Critical Thinking Questions

1. Suppose you were studying nutrition and found that laboratory rats that were fed on a diet of raw carrots remained healthy, but those that were fed on a diet consisting only of cooked carrots developed a condition that caused them to lose their hair. What conclusion might you draw about this substance? What kind of experiment would you run to confirm your conclusion? What controls would you use? If you attempted to extract the substance from carrots, how could you determine which of your extracts contained the necessary ingredient?
2. Both mechanical and chemical processes are involved in the breakdown of most foods into small molecules that can be absorbed by cells lining the small intestine. What is the role of each type of activity, and why are both necessary?
3. Why do you think elevating your head when you sleep may help prevent heartburn in the middle of the night? How do you suppose people are able to swallow food even when they are standing on their head? Do you think these two observations contradict each other? If so, how can you resolve the apparent contradiction?
4. What effects might each of the following have on an individual: eating slowly; removal of a cancerous stomach; secretion of excess stomach acid; reversal of peristalsis in the esophagus; gallstones; removing a portion of the small intestine; appendectomy; colostomy (removal of the colon); long treatment with antibiotics?

CHAPTER

19

Circulation, Excretion, and Respiration

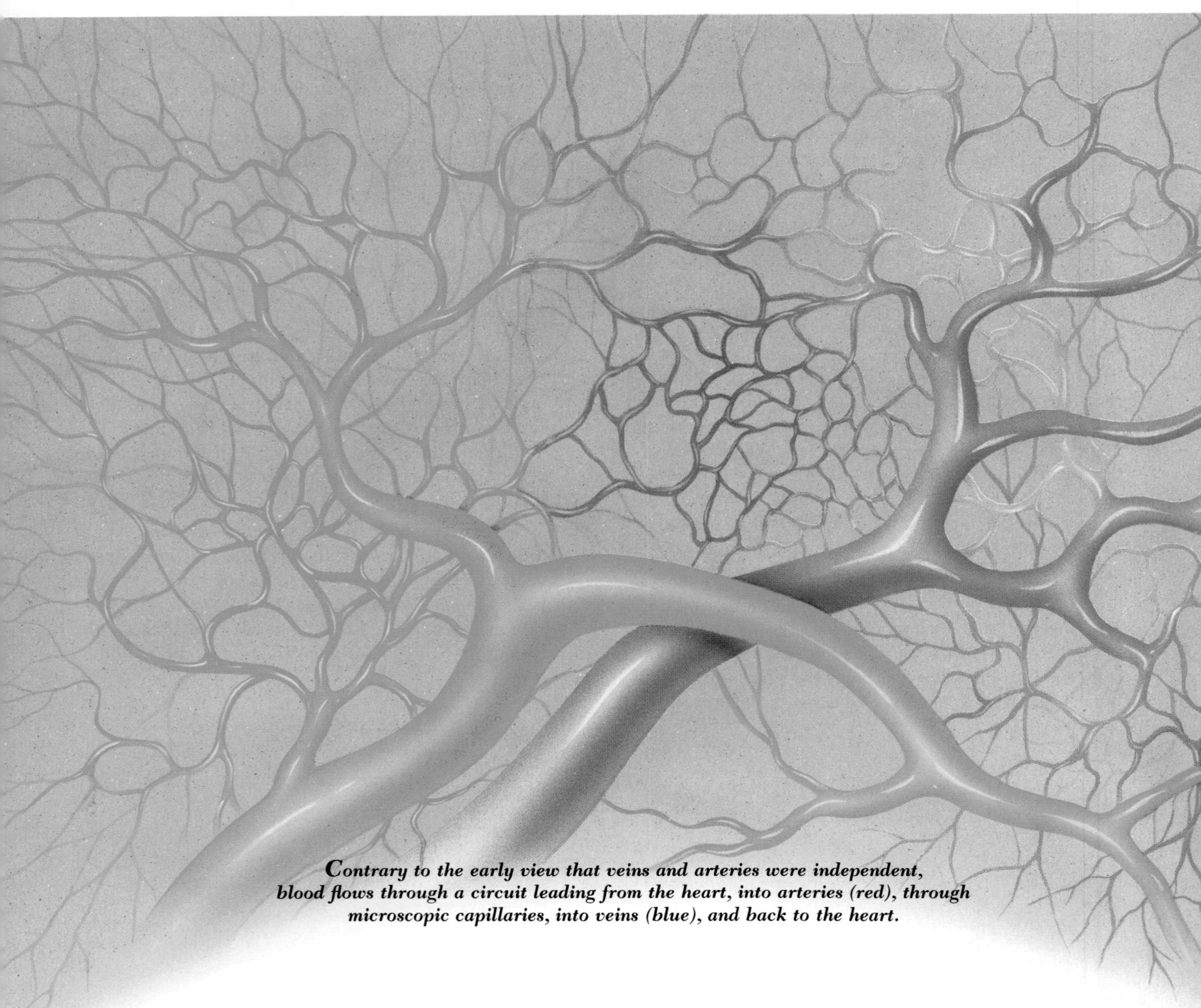

Contrary to the early view that veins and arteries were independent, blood flows through a circuit leading from the heart, into arteries (red), through microscopic capillaries, into veins (blue), and back to the heart.

STEPS TO DISCOVERY
Tracing the Flow of Blood

In 1628, a book entitled *An Anatomical Treatise on the Movement of the Heart and Blood in Animals* was written in Latin by an English physician and published in Germany. It was a small book, published on cheap, crumbling paper and filled with typographical errors. Yet, it has been hailed as the most important work ever published in the field of physiology. In this book, William Harvey, the son of a prosperous English merchant and a graduate of the University of Padua in Italy, described his experiments that led to a new concept of the organization of the human circulatory system.

Prior to Harvey's publication, the established views of blood circulation in humans had been formulated by a Greek named Galen, who had lived 1,400 years earlier and had served as the personal physician for the Roman Emperor Marcus Aurelius. Galen died in A.D., 201, but his views were still accepted in the sixteenth century, and his authority remained unchallenged. A brilliant, though dogmatic scientist, Galen had made a number of important discoveries. But he incorrectly concluded that blood was formed in the liver then flowed to the heart and through the arteries en route to the tissues, where it was entirely absorbed. The blood utilized by the tissues was replaced by new blood from the liver. Galen envisioned the veins as a system of vessels independent of the arteries; blood in the veins simply ebbed back and forth within the same vessels, much the way the tide moves in and out along the shore. The fact that this view survived for 1,400 years is an indication of the stagnation of science throughout the Middle Ages, which were also aptly known as the Dark Ages.

To test Galen's hypothesis of blood flow, Harvey attempted to measure just how much blood is actually pumped out of the heart in a given amount of time. He measured the internal volume of the heart of a cadaver and multiplied this figure by the number of times the heart beats per minute. Using this method, Harvey concluded that it took the heart nearly 30 minutes (which is actually a considerable overestimate) to pump out an amount of blood equivalent to the body's entire blood supply. It was inconceivable that the tissues could actually absorb this amount of blood so rapidly, or that the liver could resupply the blood so quickly. Harvey proposed that, rather than being absorbed by the tissues, blood circulated through the body along some type of "circular" pathway. According to his hypothesis, blood left the heart through the arteries, passed into the various tissues, and then returned to the heart through the veins.

To support his hypothesis, Harvey demonstrated that blood could flow in only one direction through a given vessel. In one simple, but convincing, demonstration, Harvey pressed his finger against one of the major veins of the forearm then moved his depressed finger along the vein toward the individual's hand (away from the heart), pushing the blood out of the vein. If blood flowed in both directions, as was the prevailing view of the time, then the vein should rapidly refill with blood. If the vein carried blood in only one direction—back to the heart—the vein should remain empty. The results were clear: The vein remained empty of blood until Harvey removed his finger. Galen's 1,400-year-old hypothesis of blood flow had been disproved with this simple, but elegant, demonstration.

One of Harvey's great frustrations was his inability to demonstrate just *how* blood flowed from the arteries into the veins. Harvey hypothesized that the tissues contain tiny vessels that complete the circuit from the arteries to the veins, but he had no way of demonstrating the existence of such vessels. The vessels linking the arterial and venous circulation were finally discovered in 1661, 4 years after Harvey's death, by the Italian anatomist Marcello Malpighi, whom we met in Chapter 12 in reference to his experiments on circulation and transport in plants. Using a newly developed instrument, the microscope, Malpighi prepared a thin piece of tissue from the lungs of a frog. He allowed the tissue to hang in the air to dry and then observed the blood vessels with a microscope; the red color of the blood vessels contrasted strikingly against the light background. Malpighi saw the larger arterial vessels branching into smaller vessels and finally giving rise to short, minuscule vessels that merged at the other end into the venous circulation. In this image, Malpighi had discovered the link between the arteries and veins; he named these tiny linking vessels *capillaries,* the Latin word for "hairlike." Malpighi confirmed his observations by examining living tissues, such as the wall of the urinary bladder. He traced the blood as it flowed through the arteries and into capillaries; the blood never spilled out into the spaces of the tissue. Malpighi had demonstrated that blood flowed in a unidirectional, continuous, and uninterrupted cycle.

Just as the towns and cities of a country are connected by roads and rails, the various parts of your body are connected by an extensive system of "living tubes," or vessels, that provide a continuous route for the movement of blood. Like a road or rail system carrying trucks or trains, the blood picks up and delivers materials as it courses through the body. For example, nutrients are picked up as the blood passes along the wall of the small intestine and, ultimately, are delivered to all the tissues of the body; oxygen is picked up as the blood passes through the lungs and is removed by the body's cells; hormones are picked up as the blood passes through the various endocrine glands and are carried to their respective target organs.

Blood vessels, the blood that flows through them, and the heart whose contractions propel the blood, comprise an animal's **circulatory system.** The circulatory system is more than just a conduit for the movement of materials from one organ to another; it is the means by which complex, multicellular animals maintain homeostasis. If ions or other solutes become too concentrated in the body, for example, the blood carries them to the kidney for elimination. If the tissues become too acidic, the blood provides the buffering agents that lower the hydrogen-ion concentration. If the tissues become depleted of oxygen or steeped in carbon dioxide, the blood carries the message to the nervous system to stimulate deeper or more rapid breathing. In birds and mammals, the blood also plays a crucial role in maintaining a constant body temperature; the blood carries heat either to the body surface, where it can be dissipated into the environment, or deep into the body, where it can be conserved. The bloodstream also plays a vital role in an animal's defense against disease-causing organisms (Chapter 20).

We will begin by describing the structure and function of the circulatory system, which provides the transportation routes necessary to maintain homeostasis. We can then examine the way the circulatory system is utilized in regulating the body's salt and water content, excreting bodily wastes, and ensuring that every cell in the body is continually supplied with oxygen.

▼ ▼ ▼

CIRCULATION: THE TRANSPORT OF BODILY FLUIDS

The human circulatory system, or **cardiovascular system** (*cardiac* = heart, *vascular* = vessels), consists of blood vessels, the heart, and blood.

BLOOD VESSELS

The human cardiovascular system contains tens of thousands of kilometers of tubing, which assures that every cell of the body is within diffusion distance of a capillary. Blood vessels are divided into five basic types: arteries, arterioles, capillaries, venules, and veins (Figure 19-1), differing in form and function.

Arteries: Delivering Blood Rapidly Throughout the Body

Blood is pumped out of the heart into **arteries**—large vessels that function as conduits rapidly carrying blood to all parts of the body. Arteries typically are large in diameter and have complex walls that contain concentric rings of elastic fibers (Figure 19-1). When blood is pumped out of the heart, the walls of the arteries are pushed outward, increasing their fluid capacity. As the walls stretch, the rings of elastic fibers respond much as a rubber band does—they recoil and exert pressure against the blood in the arteries. This is what we measure as **blood pressure.**

Blood Pressure. When you have your blood pressure taken, an inflatable cuff is strapped around your upper arm, providing a measure of the fluid pressure in the major arteries in that limb. Blood-pressure readings are expressed as two numbers, one over the other, such as 120/80 (the normal values for a young adult). The first number is the **systolic pressure,** as measured in millimeters of mercury. This is the highest pressure attained in the arteries as blood is propelled out of the heart. The pressure drops rapidly as blood is pushed out of the arteries and the diameter of the arteries decreases. Since the arteries don't have time to return to their resting (unstretched) diameter before the next contraction, the blood in the arteries remains under pressure from the stretched arterial walls. The second number is the **diastolic pressure,** or the lowest pressure in the arteries of the arm recorded just prior to the next heart contraction. If the observed systolic and diastolic values are higher than about 140/90, the person is said to have high blood pressure.

While the causes of high blood pressure, or **hypertension** are still poorly understood, the effects of hypertension can become very evident if the condition continues untreated. The excessive pressure leads directly to a weakening of the walls of the arteries, increasing their chance of rupture. Increased arterial pressure also accelerates the buildup of fatty plaques on the walls of arteries (*athero-*

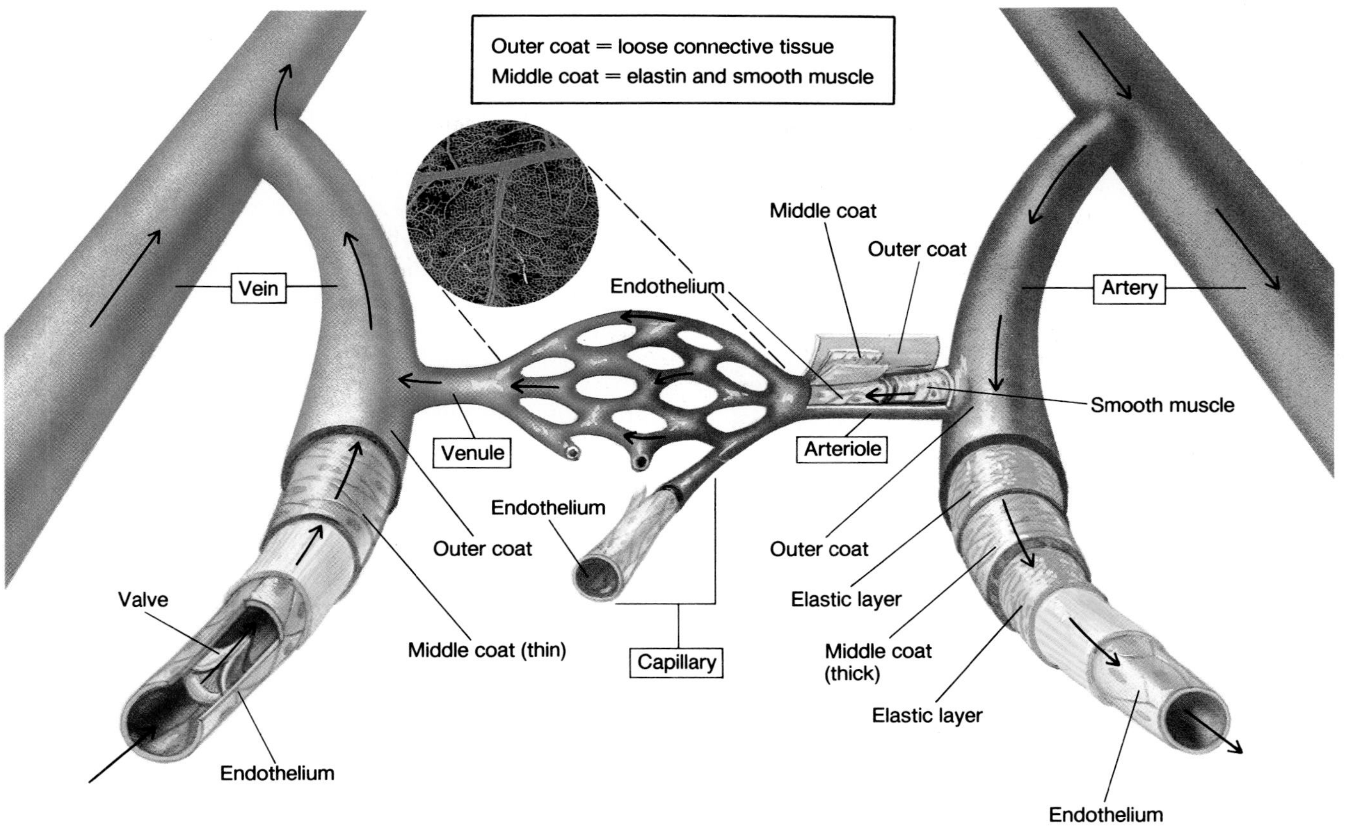

FIGURE 19-1
Form and function of the body's blood vessels. Each of the five categories of vessel has a distinct internal anatomy. The inner lining of all the vessels consists of a layer of flattened, "interlocking" endothelial cells which make up the *endothelium.* The walls of the capillaries—the thinnest of the vessels—consist only of a single, thin endothelial cell layer. The inset shows the delicate latticework of a capillary bed. (Note: this drawing is not to scale.)

sclerosis) and a greatly increased likelihood that blood clots will develop in the vessels. Together, these deleterious effects on the body's arteries promote the rupture or blockage of cerebral vessels, causing a stroke and possible subsequent brain damage. Other risks include the rupture of tiny vessels in the kidney, which can cause kidney failure, and blockage of the coronary arteries, which can cause a heart attack.

Arterioles: Regulating Blood Flow to the Tissues

The major arteries branch into smaller and smaller arteries, eventually giving rise to smaller vessels, called **arterioles.** The amount of blood that flows into a particular tissue depends largely on the diameter of the local arterioles. Arteriolar diameter is determined by the state of contraction of the muscle cells in the walls of these vessels. If an organ needs more oxygen, as does the heart of a person who is engaged in strenuous exercise, the muscle cells in the arterioles of that organ become more relaxed, and the diameter of the vessels increases. Conversely, those organs of the body that operate at a low activity level at a particular time, such as the digestive tract during a time of stress or exercise, receive less blood due to a temporary narrowing of the arterioles.

Capillaries: Exchange Between Tissue and Bloodstream

From the arterioles, blood passes through the **capillaries,** the smallest, shortest, and most porous channels of the vascular network. The lumens of these vessels are just large enough for red blood cells to move along in single file (Figure 19-2). The capillaries are the sites of exchange between the cells and the bloodstream. Your body contains about 40,000 kilometers of capillaries—enough to circumscribe the Earth at its equator—creating an enormous sur-

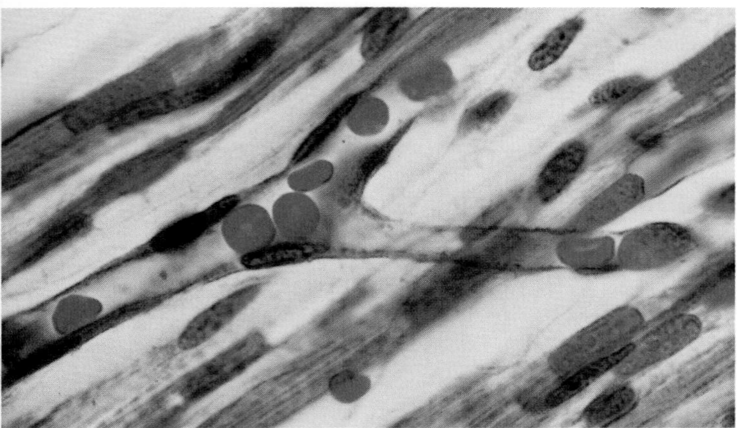

FIGURE 19-2
Red blood cells moving single file through a capillary. The thinness of the capillary walls is evident by the fact that we can see the red blood cells so clearly.

face area for the exchange of materials between the blood and the tissues. Water and its dissolved solutes, including ions, sugars, amino acids, oxygen, and carbon dioxide, readily move back and forth between the bloodstream and the tissues. However, red blood cells and most proteins are too large to penetrate the pores between the cells that form the walls of the capillaries and thus remain in the blood vessel.

Venules and Veins: The Return Trip to the Heart

Blood from capillaries collects in larger vessels, called **venules,** which empty their contents into large, thin-walled **veins;** from there, the contents return to the heart. Blood pressure in veins is very low, having been dissipated by travel through the narrow capillary passageways. Since most human veins direct blood flow upwards to the heart, against the force of gravity, it may seem that blood would simply collect in these large, low-pressure vessels and go nowhere. However, skeletal muscle activity, such as that which occurs during walking, combined with the pressure that is exerted during breathing, squeeze veins and force their contents to travel toward the heart. This direction of blood flow is maintained by one-way flaps, or *valves* (Figure 19-3), which project from the walls of these vessels. Without the valves, blood would simply squirt back and forth in the veins as they were squeezed by skeletal muscles.

THE HUMAN HEART: A MARATHON PERFORMER

The major organ of the circulatory system—the **heart**—is a fist-sized, muscular pump that delivers an astonishing performance. The average person's heart works continually for more than 70 years. Every minute, the heart recirculates the entire volume of blood in the body—approximately 5 liters. The human heart is a two sided, four-chambered model. Each side has a thin-walled **atrium** and a larger, thicker-walled **ventricle** (Figure 19-4). The left and right atria (plural for atrium) function as receiving stations for the blood that flows into the heart from the veins (Figure 19-4*a*). Blood from the left atrium flows into the left ventricle, while blood from the right atrium flows into the right ventricle (Figure 19-4*b*). Contraction of the walls of the ventricles forces the blood out of the heart and into the major arteries (Figure 19-4*c*).

Pumps and Vascular Circuits

The left and right sides of the heart are separated from each other by partitions that effectively divide the organ into two separate pumps, each of which powers blood through a different vascular *circuit* (Figure 19-5). In the **pulmonary circuit,** deoxygenated blood that has returned from its travels through the body's tissues is pumped from the right ventricle into the pulmonary arteries and on to the lungs, where it is oxygenated. Oxygenated blood returns to the left atrium of the heart via the pulmonary veins. (The designation of a vessel as an artery or a vein is based on the direction of blood movement relative to the heart, not whether or not the blood is oxygenated or de-

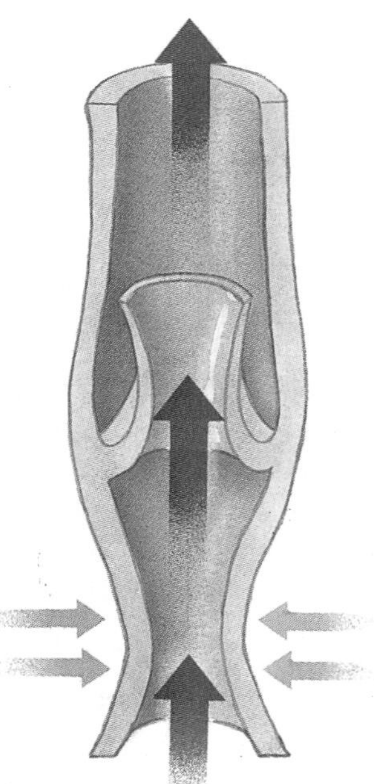

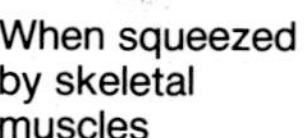

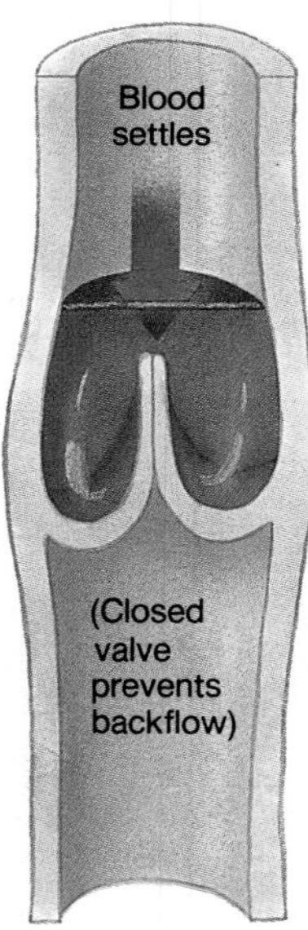

FIGURE 19-3
No turning back. Blood flow is maintained through the veins by one-way valves that open when pressure forces blood toward the heart (indicated by the red arrows) and are forced shut when the flow is reversed.

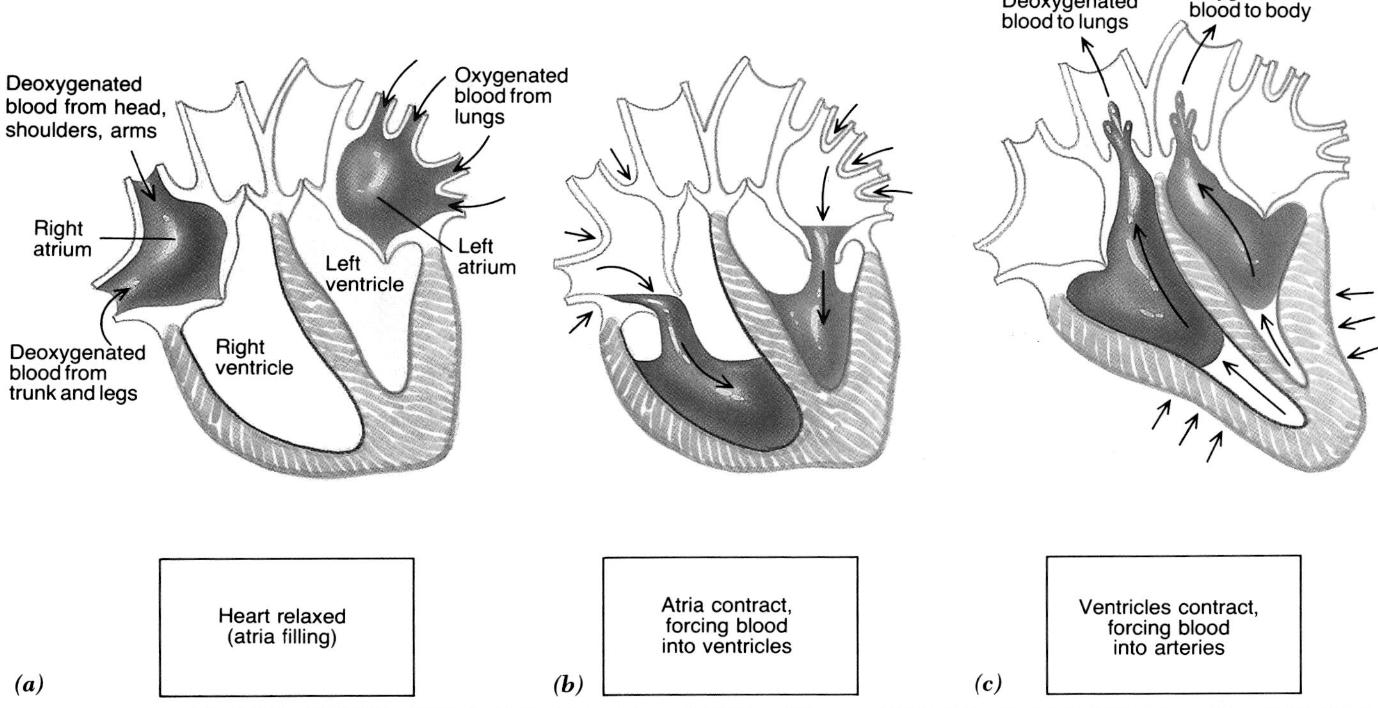

FIGURE 19-4
The movement of blood through the human heart. ***(a)*** When the heart is relaxed, blood flows from the veins into the two thin-walled atria. Deoxygenated blood from the body flows into the right atrium, and oxygenated blood from the lungs flows into the left atrium. ***(b)*** In the next stage of the cycle, the atria contract, forcing blood into the relaxed, thicker-walled ventricles. ***(c)*** In the last part of the cycle, the ventricles contract, forcing blood into the arteries. Oxygenated blood is forced out of the left ventricle into arteries that lead to the tissues of the body, while deoxygenated blood is forced out of the right ventricle into arteries that lead to the lungs. The average duration for the entire cycle in a person at rest is 0.83 seconds (72 beats per minute).

oxygenated. Arteries carry blood from the heart; veins return it to the heart.) Blood returning to the left atrium from the lungs passes into the left ventricle, whose powerful contraction sends it out into the **systemic circuit,** where it nourishes the tissues of the body. In the systemic circuit, blood is initially pumped into the **aorta** (Figure 19-6), the largest artery of the body (approximately 2.5 centimeters [1 inch] in diameter), which feeds into many of the major arterial thoroughfares of the body.

Blood does not nourish the heart as it travels through its chambers; rather, large **coronary arteries** (Figure 19-6) branch from the aorta immediately after the aorta emerges from the heart. Obstructions in the coronary arteries (Figure 19-7*a*) can deprive the heart of the oxygen needed to keep heart tissue alive. Such obstructions are the primary cause of heart attacks. If one or more of the coronary arteries is found to be largely obstructed, the patient is usually treated either by balloon angioplasty or coronary bypass surgery. In *balloon angioplasty,* a tube is threaded into the blocked coronary artery. The tip of the tube is then inflated like a balloon, pushing the walls of the artery outward, opening the vessel for greater flow (Figure 19-7*b*). In *coronary bypass surgery,* a segment of vein taken from the patient's leg is inserted between the aorta and the coronary artery at a point beyond the site of occlusion. In this way, blood is able to flow from the aorta into the heart muscle, bypassing the blocked arterial vessel. Several blocked coronary arteries are usually bypassed in this way during a single operation.

The Heartbeat

The one-way movement of blood through the chambers of the heart and into the appropriate arteries is ensured by the opening and closing of **heart valves.** When the atria contract, the pressure of the atrial blood rises, causing the fluid to push against the valves. The valves are forced open, allowing blood to flow into the ventricles. When the atrial contractions cease, the pressure of the blood in the ventricles becomes greater than that in the atria. This pressure differential forces the valves to close, preventing the flow

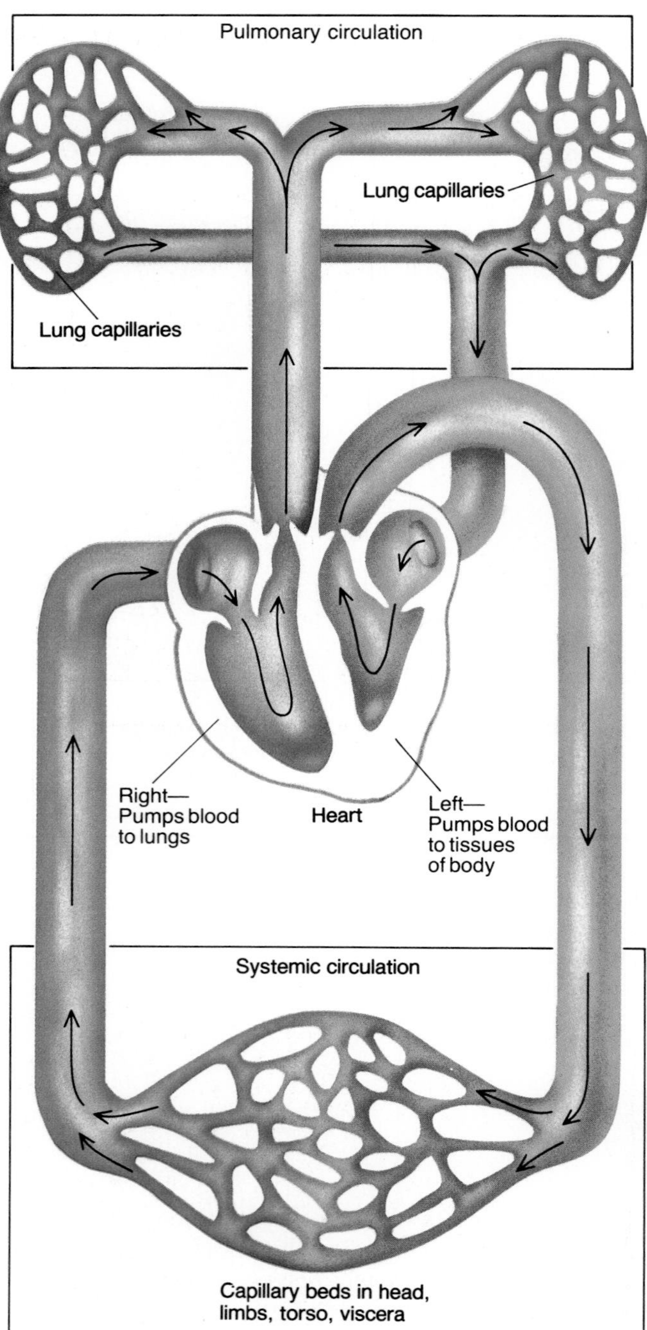

FIGURE 19-5

Our dual circulatory system. Deoxygenated blood is depicted in blue, oxygenated blood in red. In this schematic diagram, the right side of the heart drives the pulmonary circulation, pumping deoxygenated blood into the capillary beds of the lungs, where the blood picks up oxygen and returns to the left side of the heart. The oxygenated blood is then propelled from the left side of the heart through the systemic circulation. Oxygen is given up in the capillary beds of the tissues, and deoxygenated blood returns to the right side of the heart.

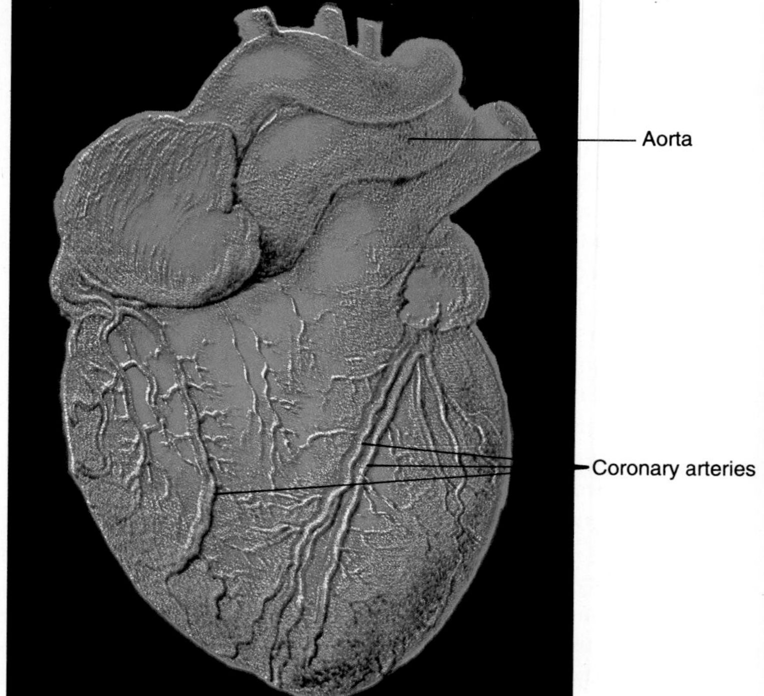

FIGURE 19-6

Blood flow to the heart. A computer-enhanced view of the heart showing the coronary arteries, which are the first major arteries to emerge from the aorta, giving the heart muscle first claim to freshly oxygenated blood.

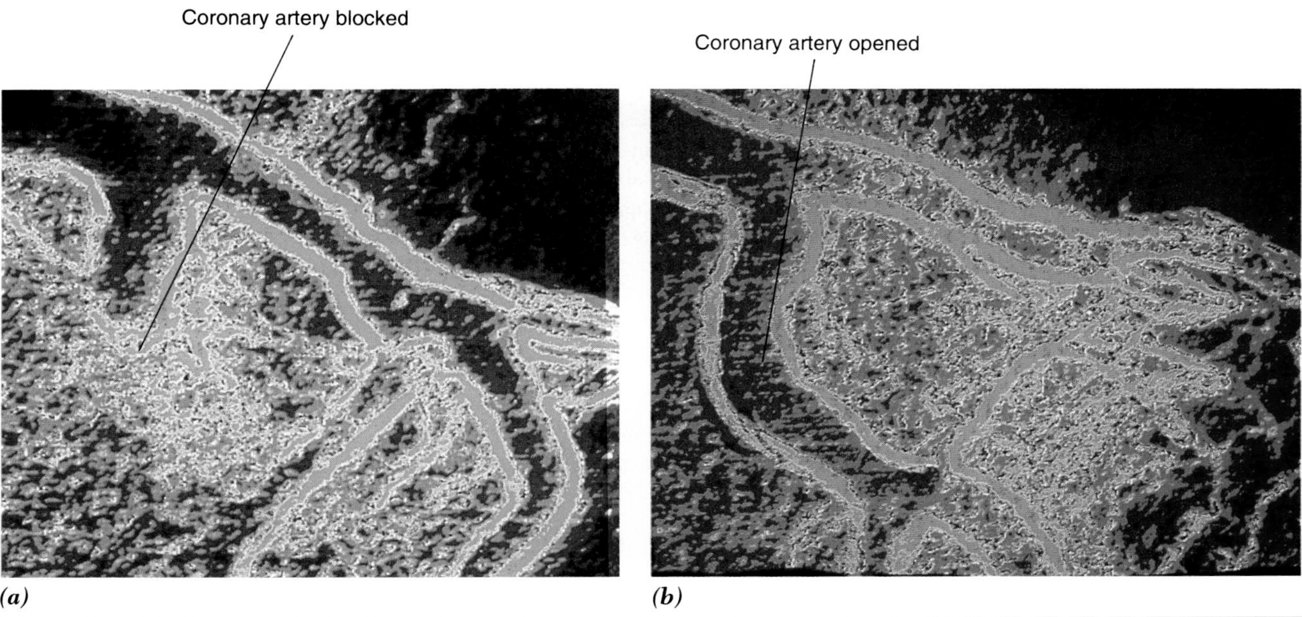

FIGURE 19-7
Coronary artery disease. ***(a)*** An angiogram showing the inability of injected dye to flow normally through an obstruction in a coronary artery. ***(b)*** Angiogram showing the same coronary artery depicted in part ***a*** after being opened by balloon angioplasty.

of blood back into the atria. If you listen to the heartbeat through a stethoscope, you will hear a "lub-dub" sound that repeats itself with each beat of the heart. These sounds are caused by the closure of the heart valves. A number of conditions can cause a disruption in the smooth flow of blood through the heart creating turbulence, which is heard as a *murmur.*

Excitation and Contraction of the Heart

Embedded in the wall of the right atrium is a small piece of specialized cardiac muscle tissue, called the **sinoatrial (SA) node** (Figure 19-8), whose function it is to excite contraction. If a healthy human heart is isolated, as is done during a heart transplant, the isolated heart can beat rhythmically on its own, without any outside stimulation. Each beat of an isolated heart is initiated by a spontaneous electrical discharge that originates every 0.6 seconds or so in the SA node. Because of its role in determining the rate the heart beats, the SA node is called the heart's *pacemaker.* Once an action potential is generated in the SA node, a wave of electrical activity spreads across the walls of the atria, causing the entire atrial wall to contract in a synchronous, coordinated manner. As the wave of electrical excitation reaches the boundary of the atria, it is funneled into the **atrioventricular (AV) node** (Figure 19-8), then spreads over the ventricles, causing the ventricles to contract, forcing blood into the major arteries.

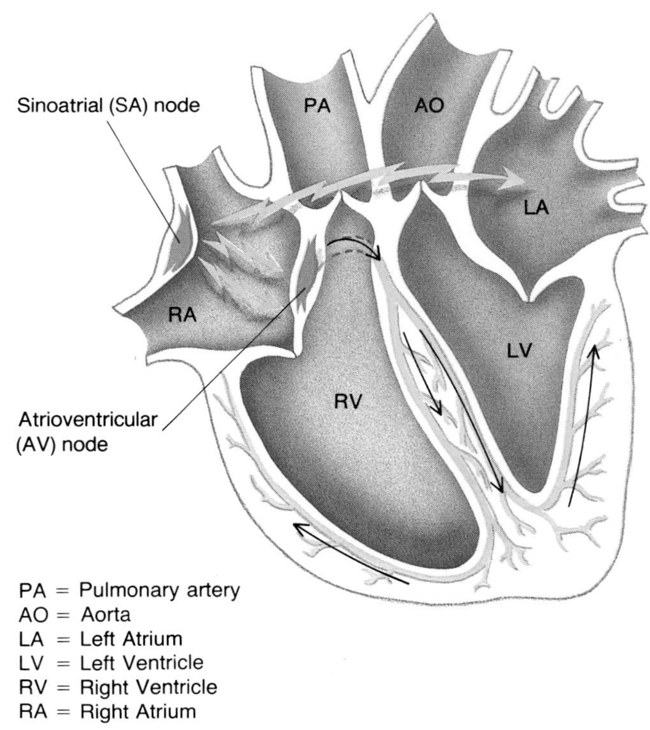

FIGURE 19-8
Synchronizing the beat. Impulses generated in the SA node (the pacemaker) sweep through the muscle cells of the two atria, causing them to contract in unison. When this impulse reaches the AV node, it launches a wave of electrical activity that is channeled through the ventricular walls, causing the two ventricles to contract just as the atria relax.

TABLE 19-1
COMPONENTS OF BLOOD

Component	Percent	Function
Plasma	55	Suspends blood cells so they flow. Contains substances that stabilize pH and osmotic pressure, promote clotting, and resist foreign invasion. Transports nutrients, wastes, gases, and other substances.
White blood cells	<0.1	Allow phagocytosis of foreign cells and debris. Act as mediators of immune response.
Platelets	<0.01	Seal leaks in blood vessels.
Red blood cells	45	Transport oxygen and carbon dioxide.

Blood settles into three distinct layers when treated with substances that prevent clotting.

The rate at which the heart's pacemaker becomes excited is regulated by a number of factors, primarily by signals that arrive from the autonomic nervous system (Chapter 15). In general, impulses that reach the SA node over *parasympathetic nerves* decrease the rate of heart contraction, while impulses that arrive over *sympathetic nerves* increase it. Thus, at any given moment, the heart rate is determined by the balance between these antagonistic influences. If, for example, we become active, angry, or frightened, the sympathetic output to the SA node increases, thereby increasing the heart rate.

COMPOSITION OF THE BLOOD

Human blood is a complex tissue (Table 19-1) that is composed of blood cells and cell-like components suspended in a clear, straw-colored liquid called **plasma.** The average person contains about 4.7 liters (10 pints) of blood, about 55 percent of which is plasma. Most (about nine-tenths) of plasma is water, the blood's solvent. The rest is composed of various dissolved substances, predominantly proteins.

Red Blood Cells

Erythrocytes, which are often called red blood cells due to their high concentration of red-colored hemoglobin molecules, transport about 99 percent of the oxygen carried by the blood. Erythrocytes are flattened, disk-shaped cells with a central depression that gives them the appearance of a doughnut with a depression instead of a hole (Figure 19-9). Not only are red blood cells the most abundant cells in the body (there are approximately 5 billion in each milliliter of blood), but they are also the simplest. Human erythrocytes lack a nucleus, ribosomes, and mito-

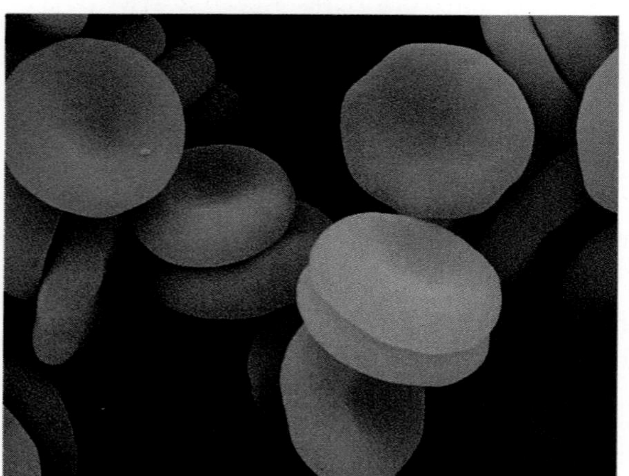

FIGURE 19-9
Red blood cells: biconcave sacks of hemoglobin. The flattened shape of erythrocytes keeps all of the hemoglobin molecules close to the cell's surface for rapid gas exchange.

chondria and constitute little more than nonreproducing sacks of the oxygen-binding protein hemoglobin. The number of erythrocytes produced by the body fluctuates according to availability of oxygen. For example, a person becomes adjusted (*acclimated*) to higher elevations, where the air is thinner (has a lower concentration of oxygen and other gas molecules), by producing more red blood cells, thereby compensating for the reduced amount of oxygen in each breath.

White Blood Cells

If you examine a stained blood smear (Figure 19-10), you will notice a small number of white blood cells, or **leukocytes** (*leuko* = white, *cyte* = cell), scattered among the erythrocytes. Leukocytes defend us against foreign microorganisms that could otherwise invade, overwhelm, and destroy our bodies. These blood cells also function as sanitary engineers, cleaning up dead cells and tissue debris that would otherwise accumulate to obstructive levels.

Five classes of leukocytes are clearly recognized: neutrophils, eosinophils, basophils, monocytes, and lymphocytes. The first three types—*neutrophils, eosinophils,* and *basophils* are characterized by their cytoplasmic granules. These cells protect the body by engulfing foreign intruders by phagocytosis (page 69). In contrast, *monocytes* are nongranular leukocytes that are attracted to sites of inflammation, where they too act as phagocytes, engulfing debris. Monocytes also give rise to phagocytic *macrophages*—huge cells that wander through the body's tissues, ingesting foreign agents. Macrophages also clean up "battlefields" that are littered with the cellular debris of combat between other phagocytes and invading microorganisms. Phagocytosis is not the only mechanism of protection afforded by leukocytes. The *lymphocytes,* a group of nonphagocytic white cells, are the "masterminds" of the immune system, the subject of Chapter 20.

The number and type of leukocytes in the blood can provide an indication of a person's health, which is why your physician may take a "blood cell count" when you are ill. Most infections stimulate the body to release into the bloodstream large numbers of protective leukocytes that are normally held in reserve, causing the "white cell number" to rise. In contrast, certain viruses, including HIV responsible for AIDS, cause infections that *deplete* certain leukocytes to abnormally low levels.

Factors Necessary for Blood Clotting

A hole in an injured vessel can lead to fatal blood loss if the leak is not quickly patched. In other words, a person could bleed to death. The blood is equipped with its own lines of defense that minimize blood loss and maintain homeostasis. Immediately following injury to a vessel, circulating **platelets**—spiny fragments that are released from special blood cells—become trapped at the injury site. As platelets rapidly accumulate, they plug the leak, providing the first step in damage control. The aggregation of platelets then triggers the formation of a **clot,** which forms a more permanent seal at the site of vessel damage.

The resources for clotting circulate in the blood in an inactive form. **Fibrinogen** is a rod-shaped plasma protein that, when converted to the insoluble protein **fibrin,** generates a tangled net of fibers that binds the wound and stops blood loss until new cells replace the damaged vessel (Figure 19-11). The conversion of fibrinogen to fibrin requires a cascade of separate reactions, some of which are indicated in Figure 19-11. The most common clotting disorder is *hemophilia.* Most hemophiliacs have a defective gene that codes for an inactive version of *Factor VIII,* one of the proteins required in the cascade. Episodes of bleeding in hemophiliacs can now be treated with synthetic preparations of Factor VIII, which are produced by recombinant DNA technology. Such preparations eliminate the need for transfusions of blood products, which may be contaminated with viruses that cause AIDS or hepatitis.

While blood clots are one of the body's primary defense mechanisms, they are also responsible for the majority of deaths in industrialized countries. Most heart attacks are due to the formation of a blood clot (*thrombus*) within a coronary artery at a site where the vessel has already become narrowed by the buildup of plaque. While the plaque reduces the flow of blood to the heart, the blood clot totally obstructs the vessel, causing the rapid death of heart tissue which accompanies a heart attack. For this reason, the preferred treatment for heart-attack victims is the injection of a massive dose of a clot-dissolving substance, such as *tissue plasminogen activator* (*TPA*). It is

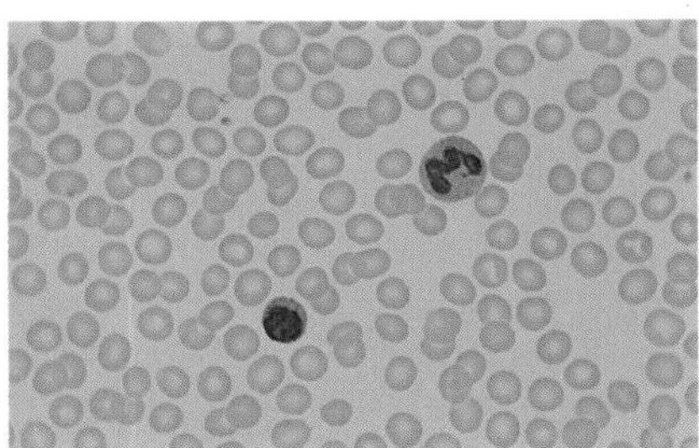

FIGURE 19-10
Human leukocytes and their abundance in the blood relative to erythrocytes. In this blood smear, two leukocytes are seen among a field of erythrocytes.

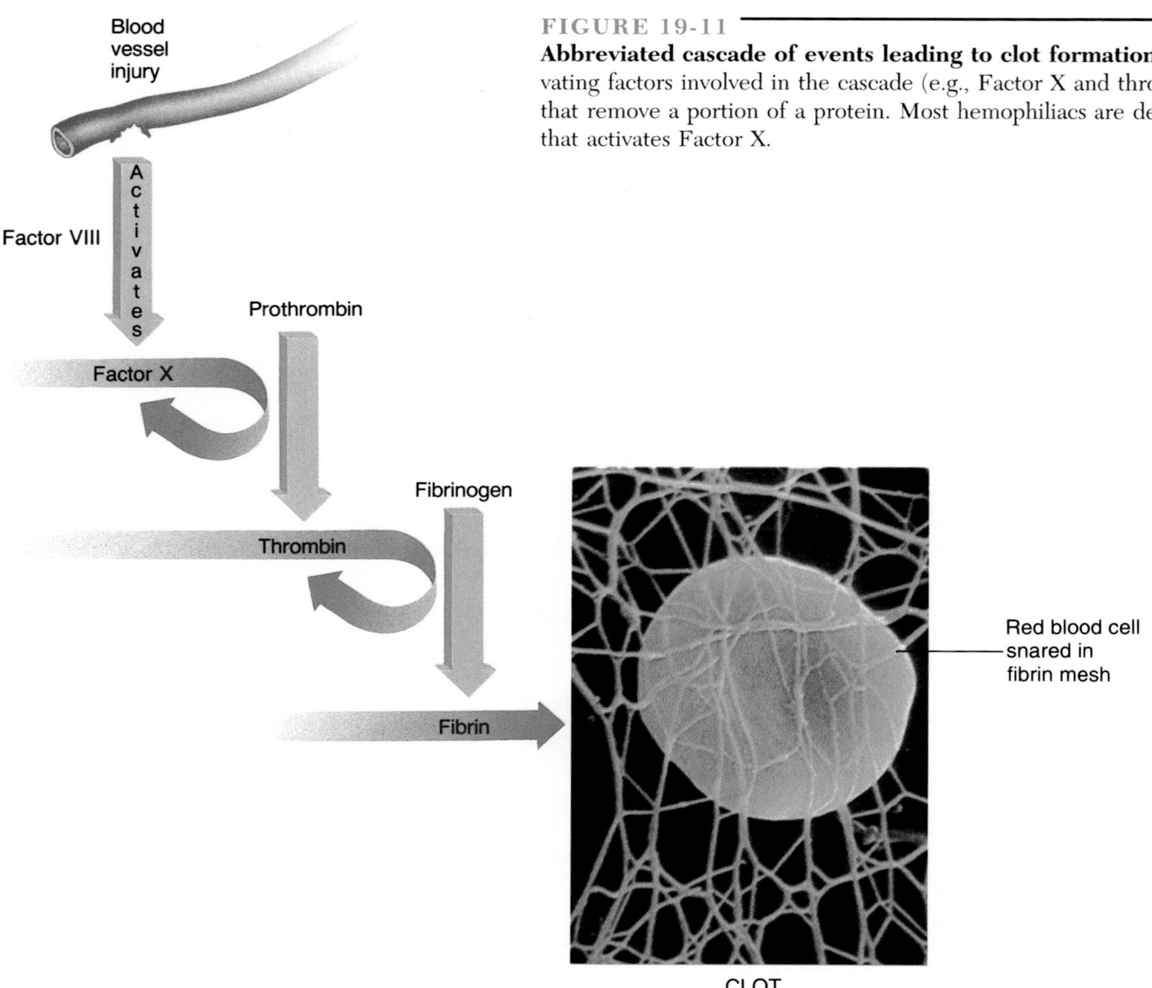

FIGURE 19-11
Abbreviated cascade of events leading to clot formation. Most of the activating factors involved in the cascade (e.g., Factor X and thrombin) are enzymes that remove a portion of a protein. Most hemophiliacs are deficient in the protein that activates Factor X.

important that the "clot buster" be administered very soon after the onset of the attack, otherwise damage to the heart tissue due to oxygen deprivation becomes irreversible.

The Lymphatic System

In addition to the cardiovascular system, humans are equipped with a secondary network of fluid-carrying vessels and associated organs that make up the **lymphatic system.** Lymphatic vessels are a series of one-way channels that originate in the tissues as a bed of *lymphatic capillaries.* Lymphatic capillaries absorb excess tissue fluid that fails to reenter the capillaries. The smaller lymphatic vessels fuse to form larger lymphatic vessels, which ultimately drain into the large veins of the neck. Fats that are absorbed from the intestine following digestion of a fat-containing meal are also collected and delivered to the bloodstream by lymphatic vessels (review Figure 18-3).

Before reentering the bloodstream, lymphatic fluid, or *lymph,* passes through a series of **lymph nodes**—structures that help "purify" the fluid by removing foreign substances and any microorganisms that may be present. During infection, lymph nodes become engorged with white blood cells that are recruited to fight the microbial aggressor.

TYPES OF CIRCULATORY SYSTEMS

Over the course of evolution, two basic types of circulatory systems have appeared: open and closed systems. Some animals, including arthropods (such as insects and spiders) and most molluscs (including snails and clams), have an **open circulatory system** (Figure 19-12a), whereby the blood is pumped through vessels that empty into a large, open space (*hemocoel*), in which most of the body's organs are immersed. The cells receive nutrients directly from the fluid in which they are bathed.

In contrast, the human circulatory system is a **closed circulatory system.** In a closed system, blood surges throughout the body in a *continuous* network of closed vessels. Many simpler animals, including the earthworm (Figure 19-12*b*), also have closed systems. In a closed sys-

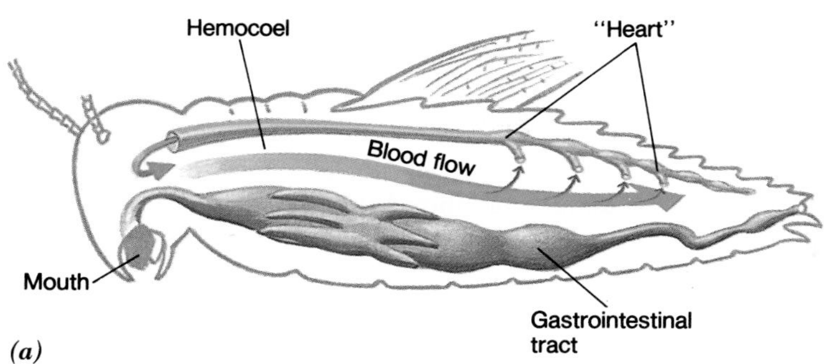

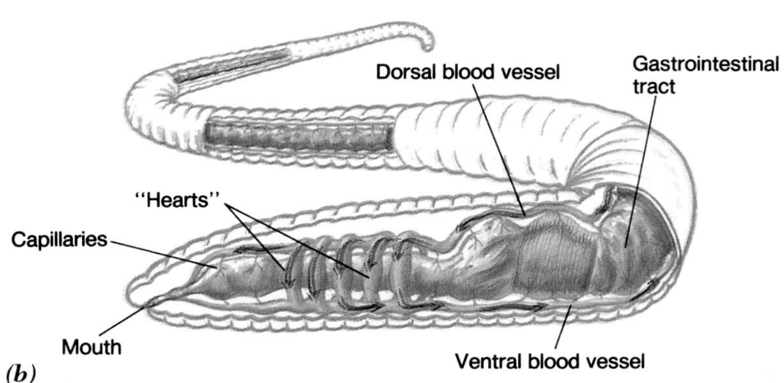

FIGURE 19-12
Circulatory strategies. ***(a)*** The open circulatory system of a grasshopper. Blood is pumped out of blood vessels into spaces (which form the *hemocoel*) that bathe the body's tissues. ***(b)*** The closed circulatory system of an earthworm; blood remains in vessels.

tem, respiratory gases, nutrients, and waste products are exchanged across the porous walls of the finest vessels in the network—the capillaries. The exchange of materials across the capillary walls of a closed circulatory system is illustrated by the activities that take place during excretion and osmoregulation.

EXCRETION AND OSMOREGULATION: REMOVING WASTES AND MAINTAINING THE COMPOSITION OF THE BODY'S FLUIDS

One day you might eat a large bag of salty potato chips, while the next day you may choose relatively salt-free foods. Even though your diet may change drastically from day to day, the composition of your body fluids remains relatively constant. This, once again, is the essence of homeostasis. Maintaining the proper composition of the body's internal fluids requires two related activities: osmoregulation and excretion.

1. **Osmoregulation** is the maintenance of the body's normal salt and water balance. If you were not able to rid yourself of the excess salt you ingested after eating that bag of potato chips, the concentration of sodium, potassium, and chloride ions could increase to dangerous levels. Elevated potassium concentrations in the blood, for example, can lead to a fatal disruption of the rhythmic beating of the heart. Similarly, when you drink more fluid than your body needs to maintain its proper water content, excess water must be eliminated.
2. **Excretion** is the discharge of the body's metabolic waste products. The wastes that tend to accumulate at highest concentration and pose the greatest threat to the delicate physical and chemical balance required for life are products of the metabolic breakdown of nitrogen-containing compounds, notably proteins and nucleic acids. The nitrogen is released during metabolism as poisonous ammonia (NH_3). Many aquatic animals simply excrete ammonia into the surrounding medium. In humans, ammonia is quickly converted by the liver to *urea*, a relatively nontoxic nitrogenous molecule that is tolerated in the tissue fluids until it is eliminated by the kidneys. Many animals, including insects and birds, convert ammonia to *uric acid*, a nontoxic product, which is excreted as a dry paste.

Humans have a single system that carries out both osmoregulation and excretion. This is not the case in all animals.

THE HUMAN EXCRETORY SYSTEM

The human excretory system is illustrated in Figure 19-13. The paired **kidneys** are biological cleaning stations that remove nitrogenous wastes as well as excess salts and water from the blood, forming **urine,** a solution destined for discharge from the body. From the kidneys, urine is moved by peristalsis down a pair of muscular tubes, the **ureters,** and into a holding tank, the **urinary bladder.** Discharge of urine from the bladder through the **urethra** is under control of neural commands.

The Structure of the Kidney

The functional unit of the kidney is the renal tubule, or **nephron** (Figure 19-14), each of which produces a small volume of fluid, or urine. Together, the urine produced by the million or so nephrons of the kidneys makes up the fluid that you void several times a day. The nephron wall consists of a single layer of epithelial cells surrounding a central, hollow lumen. Closely entwined around each nephron is a system of blood vessels, which allows for the exchange of materials back and forth between the urinary fluid in the tubule and the bloodstream. Without such exchange, the formation of urine would be impossible.

Blood enters the kidney through the renal artery, which branches to form smaller vessels, each of which leads into a bundle of capillaries, called a **glomerulus.** Each glomerulus is embedded in a blind, cup-shaped end of a nephron, called **Bowman's capsule** (Figure 19-14). From Bowman's capsule, the nephron begins a winding path, first as the **proximal convoluted tubule,** then into a long, U-shaped portion, called the **loop of Henle,** then into the **distal convoluted tubule,** and finally into a **collecting duct,** which drains the nephron's contents into the ureter.

THREE STAGES OF URINE FORMATION

Urine formation occurs as the result of three distinct processes: glomerular filtration, tubular reabsorption, and tubular secretion (Figure 19-15).

Glomerular Filtration

The capillaries of a glomerulus are particularly porous, and the blood in these vessels is present at unusually high pressure. During **glomerular filtration,** the pressure of the blood in the glomerulus provides the force that pushes fluid out of the capillaries (Figure 19-15, step 1), across the epithelial wall of Bowman's capsule, and into the lumen of the nephron. While proteins and cells are too large to pass through the pores in the walls of the glomerular capillaries, virtually all other blood constituents, including valuable nutrients, water, and wastes, are forced out of the glomerular capillaries and enter the nephron.

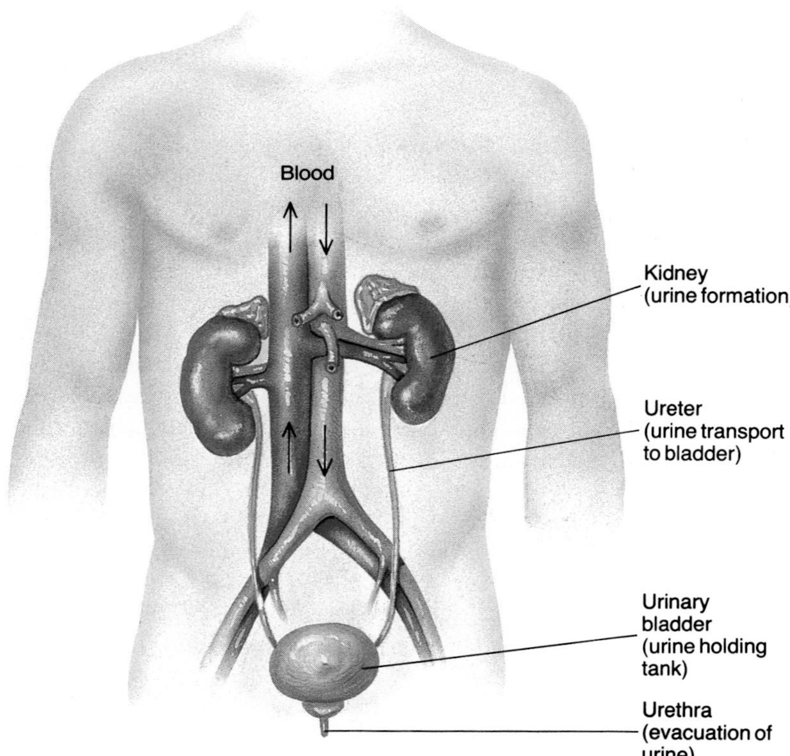

FIGURE 19-13 **The human excretory system.** The kidneys filter the blood, removing waste products and excess salts, forming urine. Ureters provide passageways to the urinary bladder, where urine is held until its release from the body through the urethra.

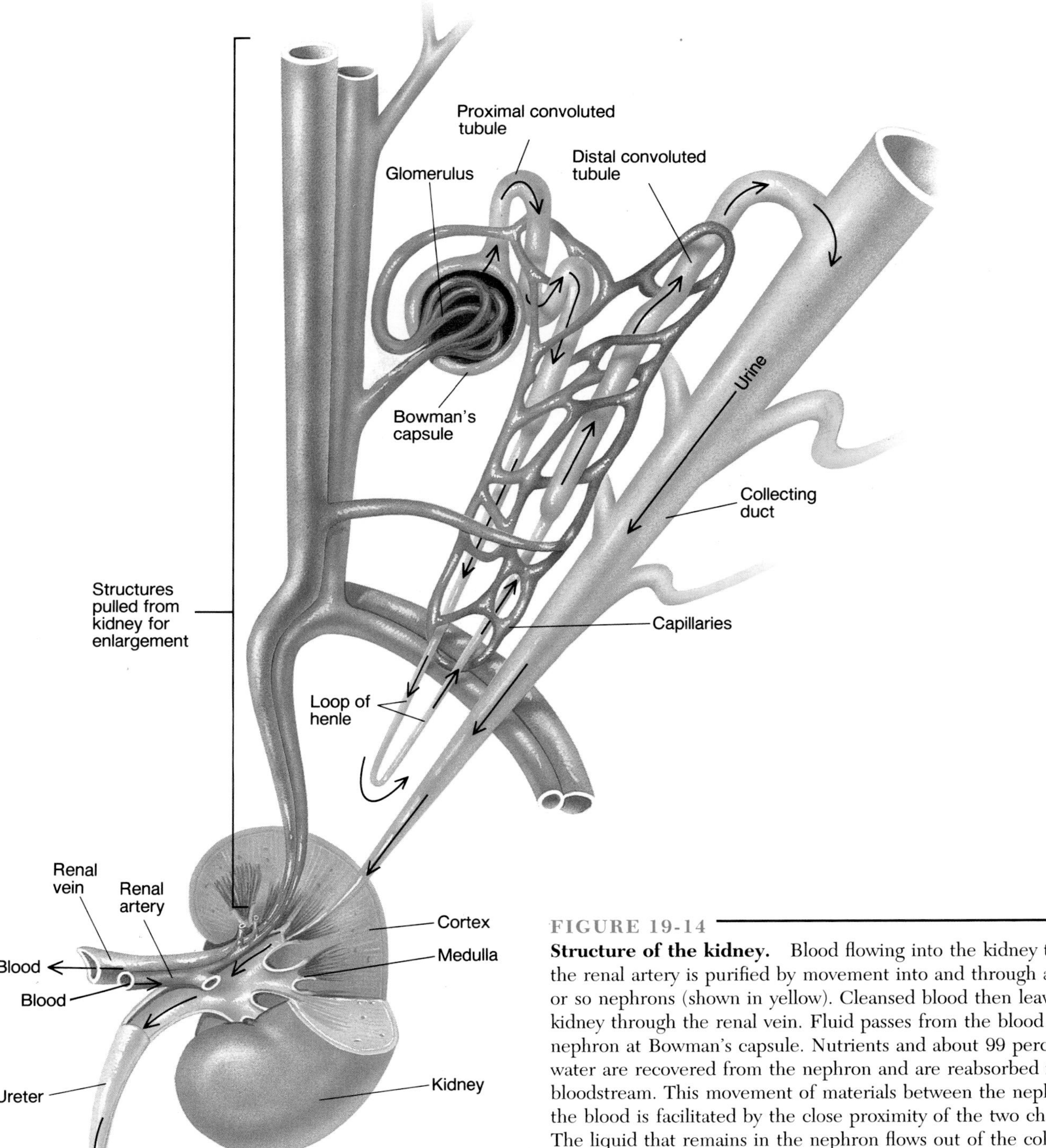

FIGURE 19-14

Structure of the kidney. Blood flowing into the kidney through the renal artery is purified by movement into and through a million or so nephrons (shown in yellow). Cleansed blood then leaves the kidney through the renal vein. Fluid passes from the blood into the nephron at Bowman's capsule. Nutrients and about 99 percent of the water are recovered from the nephron and are reabsorbed into the bloodstream. This movement of materials between the nephron and the blood is facilitated by the close proximity of the two channels. The liquid that remains in the nephron flows out of the collecting ducts as urine, which empties into the ureter.

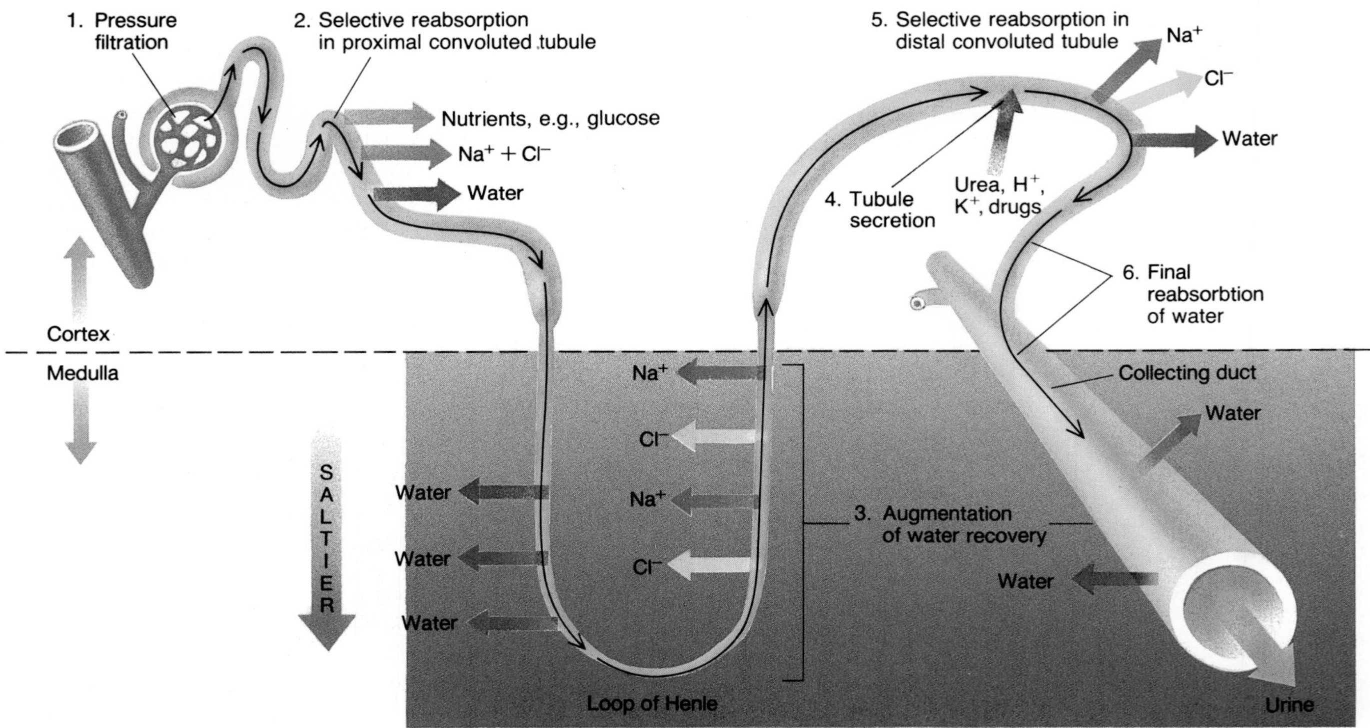

FIGURE 19-15

Urine formation in the nephron. The fluid portion of the blood is forced into the proximal end of the nephron by pressure filtration (step 1). As the fluid passes through the nephron, water and dissolved substances are reabsorbed (steps 2, 3, 5, and 6). Water reabsorption in the loop of Henle requires the presence in the surrounding fluid of a gradient of salt that reaches greatest concentration in the innermost part of the medulla of the kidney. As fluid passes through the loop of Henle, water and salt are reabsorbed. Most of the remaining water is reabsorbed in the distal convoluted tubule and collecting duct. As the fluid moves through the distal tubule, tubular secretion transfers hydrogen and potassium ions, along with waste products from the blood into the nephron (step 4).

Tubular Reabsorption

The kidney is a remarkable purification plant. Approximately 900 liters of blood passes through the kidneys each day. Of this huge quantity, 20 percent (180 liters) is actually forced out of the million or so glomeruli into the adjacent nephrons. Of course, 180 liters is considerably more fluid than you excrete as urine each day. In fact, the average output of urine is 1 to 2 liters per day, or approximately 1 percent of the fluid that enters the nephrons. The remaining 99 percent is *reabsorbed* as the fluid flows along the channel of each nephron (Figure 19-15, steps 2, 3, 5, and 6). Most of the solutes in the fluid that enters the nephrons are also reabsorbed; if they weren't, the body's store of essential substances would soon be discarded in urine. The loss of glucose, for example, as occurs in untreated diabetes (page 281), would deprive a person of his or her readily available fuel supplies. It would be like having a hole in the bottom of the gasoline tank of your car.

Whereas glomerular filtration is a nonselective process, **tubular reabsorption** is highly selective. The kidneys' "strategy" is to push everything into the nephron at its proximal end and then selectively remove those substances, such as salts, sugars, and water, that the body can't afford to lose, leaving behind waste products to be discharged in the urine. Tubular reabsorption is accomplished by transport systems in the membranes of the epithelial cells of the nephron that export glucose, salts, and other valuable nutrients out of the nephron and back into the bloodstream.

Tubular Secretion

The composition of urine is modified in the distal portion of the nephron by **tubular secretion**—the process by which substances are secreted from the blood into the fluid of the distal convoluted tubule (Figure 19-15, step 4). A variety of substances enter the nephron by secretion, including waste molecules, such as urea and urobilin (the yellow pigment derived from hemoglobin breakdown which gives urine its yellow color), certain ions (K^+ and H^+), and a number of medications, such as penicillin and phenobarbitol. Tubular secretion provides one of the final regulatory steps in maintaining homeostasis of the tissue fluids. For example, the kidney regulates blood pH by secreting more hydrogen ions when the blood becomes slightly acidic; this process is reversed when blood becomes too alkaline. The hydrogen-ion concentration of the urine can vary more than 1,000-fold, depending on conditions.

REGULATION OF KIDNEY FUNCTION

Like most physiological activities, urine formation is regulated by both neural and endocrine mechanisms. Two hormones—*aldosterone* and *antidiuretic hormone* (*ADH*)—that act on distal parts of the nephron are particularly important. If you were to ingest a particularly salty meal, the increased level of sodium in your blood would trigger a decrease in the secretion of aldosterone by your adrenal cortex (page 280), leading to a decreased reabsorption of sodium by the cells of the tubule and increased concentration of salt in the urine. Conversely, a drop in the blood's sodium concentration stimulates an increased secretion of aldosterone, leading to increased salt retention. A person suffering from untreated diseases of the adrenal cortex may produce little or no aldosterone and will excrete large quantities of both salt and water.

The volume of the blood is determined by its water content. Blood volume decreases when too little water is reabsorbed from the nephrons back into the bloodstream; this causes the blood to become too concentrated. Receptors in the hypothalamus detect the increase in the blood's osmotic strength and direct the posterior pituitary to release ADH (page 278). ADH increases the permeability of the distal portion of the nephron to water, allowing more water to be reabsorbed from the urine, which dilutes the blood while concentrating the urine. Alcohol interferes with the secretion of ADH, which is the reason why beer induces more trips to the restroom than drinking a comparable volume of a soft drink. In addition to directing the secretion of ADH, the hypothalamus also alerts the cerebral cortex when the water content of blood is low. What you perceive consciously as thirst is your hypothalamus instructing you to consume more water.

RESPIRATION: THE EXCHANGE OF RESPIRATORY GASES

You can survive for weeks without food and for days without water but try surviving without oxygen and you will last only a few minutes. Virtually all organisms depend on oxygen to keep the metabolic "fires" burning. Oxygen unleashes the chemical energy that is stored in food molecules, generating more than 90 percent of the ATP that drives energy-consuming activities. In Chapter 6, we referred to this process as *aerobic respiration.* Here, we will deal with two other aspects of animal respiration—the uptake of oxygen from the environment by a specialized set of organs that make up the **respiratory system,** and the transport of oxygen to the individual cells of the body. We also saw in Chapter 6 how the oxidation of biochemicals generates a waste product—carbon dioxide. The elimination of carbon dioxide from the body accompanies the acquisition of oxygen.

THE HUMAN RESPIRATORY SYSTEM

The human respiratory system, which is typical of mammals, is depicted in Figure 19-16. The system consists of the following components:

1. a branched passageway, or **respiratory tract,** through which air is conducted, and
2. a pair of **lungs** through which oxygen enters the body and is absorbed into the bloodstream. Even though your lungs reside deep within your chest, like your digestive tract, they are actually a part of your body surface; that is, a surface that is exposed to the external environment.

The Path to the Lungs

The entire passageway to the lungs is lined by a mucus-secreting epithelium. The blanket of mucus secreted by these cells keeps the surfaces of the airways moist so that even the driest air is humidified by the time it reaches the gas exchange surface in the lungs. The sticky mucous layer also traps microbes and other dangerous airborne particles in the upper regions of the respiratory tract before they have a chance to enter the lungs and cause serious injury or infection, such as pneumonia.

In Through the Nasal Cavity and Along the Airways

Air enters the respiratory system through either the mouth or *nostrils*—the openings into the **nasal cavity.** Air is warmed and moistened as it passes through the nasal cavity to the **pharynx** (throat), a corridor that is shared by both the respiratory tract and the digestive tract. Food and liquids are routed to the esophagus and are kept out of the airways by a combination of anatomy and reflexes. The passage to the lungs carries the inhaled air through a small opening, called the **glottis,** and into the **larynx,** a short passageway that leads into the lower portion of the respiratory tract. The human larynx is more than just a passageway for air, however. The sides of the larynx contain a pair of muscular folds—the **vocal cords**—from which

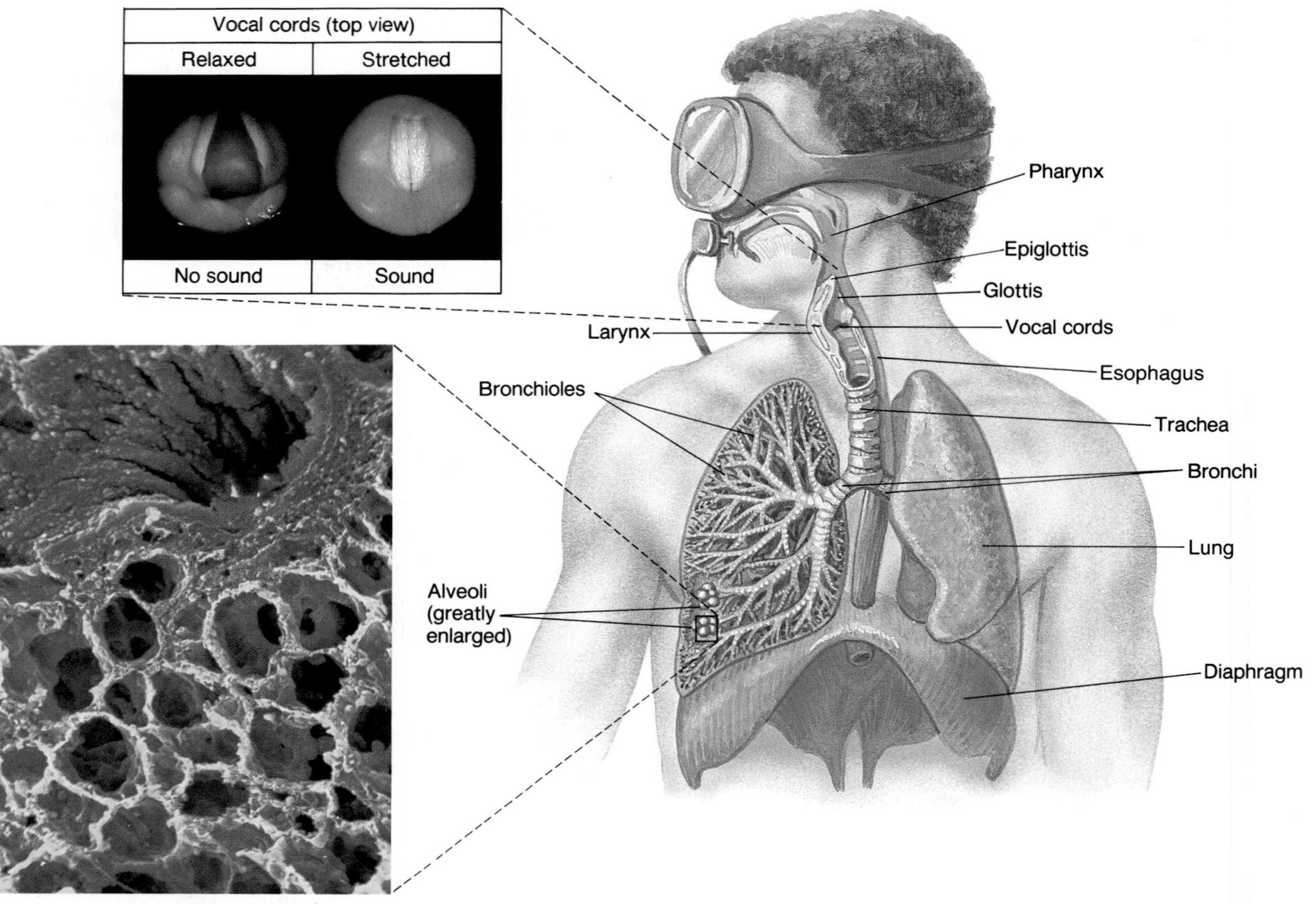

FIGURE 19-16

The human respiratory system. The human respiratory system consists of a system of airways that carry air into a pair of lungs. The airways are lined by a ciliated, mucus-secreting epithelium that moistens the air and traps microscopic debris. The top inset shows the vocal cords, which are located in the larynx. Sounds are produced when exhaled air passes through the vocal cords as they are stretched by muscles. The bottom inset shows a cluster of alveoli that fill the interior of the human lung with air pockets. Each lung is a sac that is filled with smaller sacs, thereby increasing the surface area by a factor of hundreds, with no increase in the space required to house it.

the human voice emanates. The vocal cords are operated by the passage of air through the larynx. When we inhale, air silently rushes through the opening between the relaxed vocal cords (top inset, Figure 19-16). As air escapes from the lungs when we exhale, the stretched vocal cords vibrate, creating the sound of the voice. The loudness of the voice is determined primarily by the force of the exhaled air, while the pitch (highness or lowness of the note) is determined by the level of tension on the vocal cords, which is regulated by the contraction of muscles in the larynx.

Air passes through the larynx and into the tubular **trachea** (or windpipe), which descends into the chest. The trachea divides into a series of smaller and smaller tubes, or *bronchi,* which extend into the various regions of each lung. The bronchi branch further to form a series of even smaller tubules, the **bronchioles.** During an asthma attack or severe allergic reaction, the bronchioles can become clogged with mucus and may constrict due to extended muscular contraction.

Into the Lungs

The ultimate destination of inhaled air is the lungs. The lungs contain millions of microscopic pouches, called **alveoli** (Figure 19-16, bottom inset), which resemble the air pockets in a kitchen sponge. Due to their structure, alveoli are ideally suited for gas exchange. Each alveolus is a hollow sac bounded by an extremely thin wall and surrounded by a network of capillaries (Figure 19-17). Alveoli are the sites of gas exchange in the lung. Gases readily pass through the thin walls of both the alveolus and the capillaries so that blood moving through the lung tissue can quickly pick up a fresh supply of oxygen and unload its cargo of carbon dioxide. Freshly oxygenated blood (depicted in red in Figure 19-17) travels from the alveolar capillaries into larger vessels, until it enters the pulmonary vein, which carries the blood to the heart; from there, the oxygenated blood is pumped to the remainder of the body. The stale air in the lungs that has been depleted of a portion of its oxygen is forced out of the alveoli and back into the bronchioles, where it is expelled from the airways during exhalation.

If your lungs were mere hollow bags, the gas exchange surface would be less than half a square meter (about 5 square feet)—too small a surface to absorb enough oxygen to keep you alive under even the most restful situations. Instead, the spongy interior of your lungs houses more than 300 million alveoli, providing 60 to 70 square meters of an oxygen-collecting, carbon-dioxide-discharging surface. Because of their structure, the lungs are a very delicate tissue and can easily be damaged by inhaled pollutants (see The Human Perspective: Dying for a Cigarette?).

THE EXCHANGE OF RESPIRATORY GASES AND THE ROLE OF HEMOGLOBIN

Two types of gas exchange are constantly occurring in your body—one between the alveoli and the capillaries in the lung, and the other between the capillaries and the various tissues of the body. These events are summarized in Figure 19-18.

Gas Exchange in the Lungs: Uptake of Oxygen into the Blood

Inhaled air is rich in oxygen (about 21 percent) and poor in carbon dioxide (about 0.4 percent). Blood returning from the tissues to the lungs carries the opposite complement of gases—a high concentration of carbon dioxide and

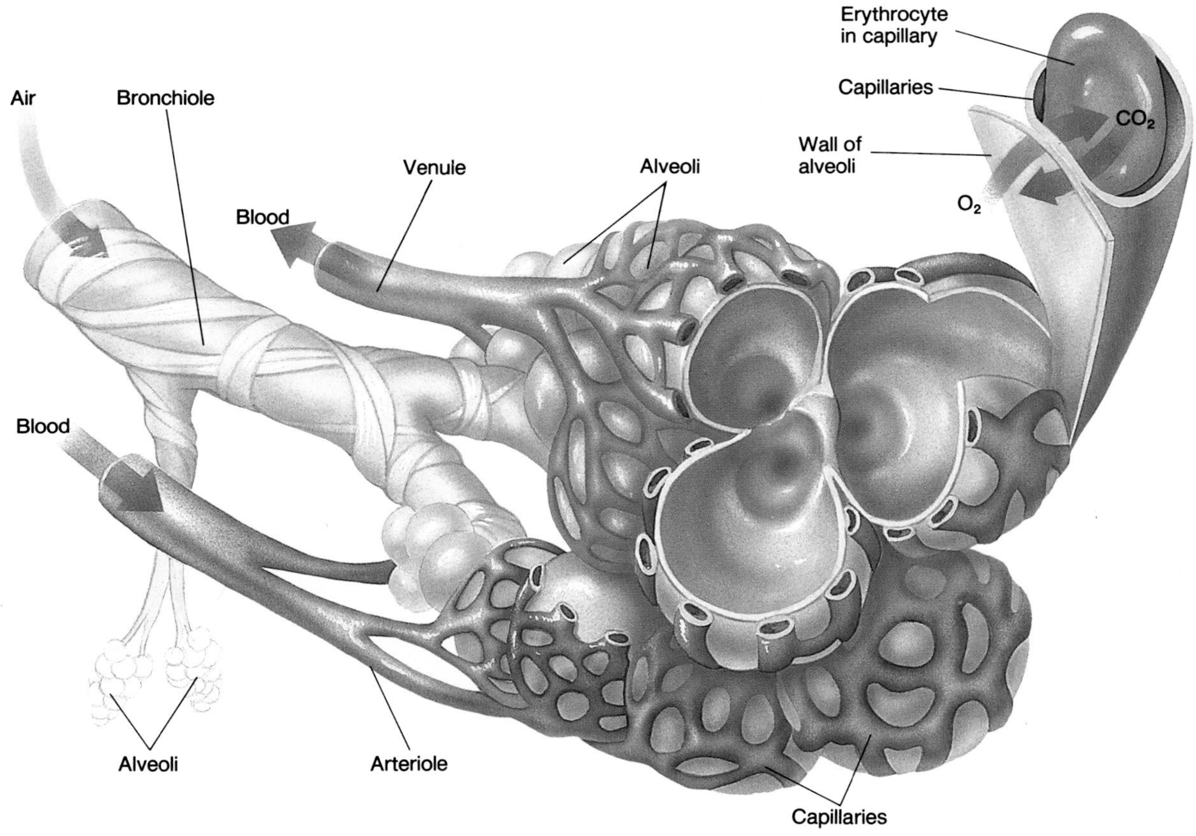

FIGURE 19-17
Alveoli, the lung's trade centers. Each alveolus is a thin-walled, bubblelike chamber (the size of a pin point) that is surrounded by capillaries. Oxygen and carbon dioxide gases quickly move in the directions indicated by the arrows.

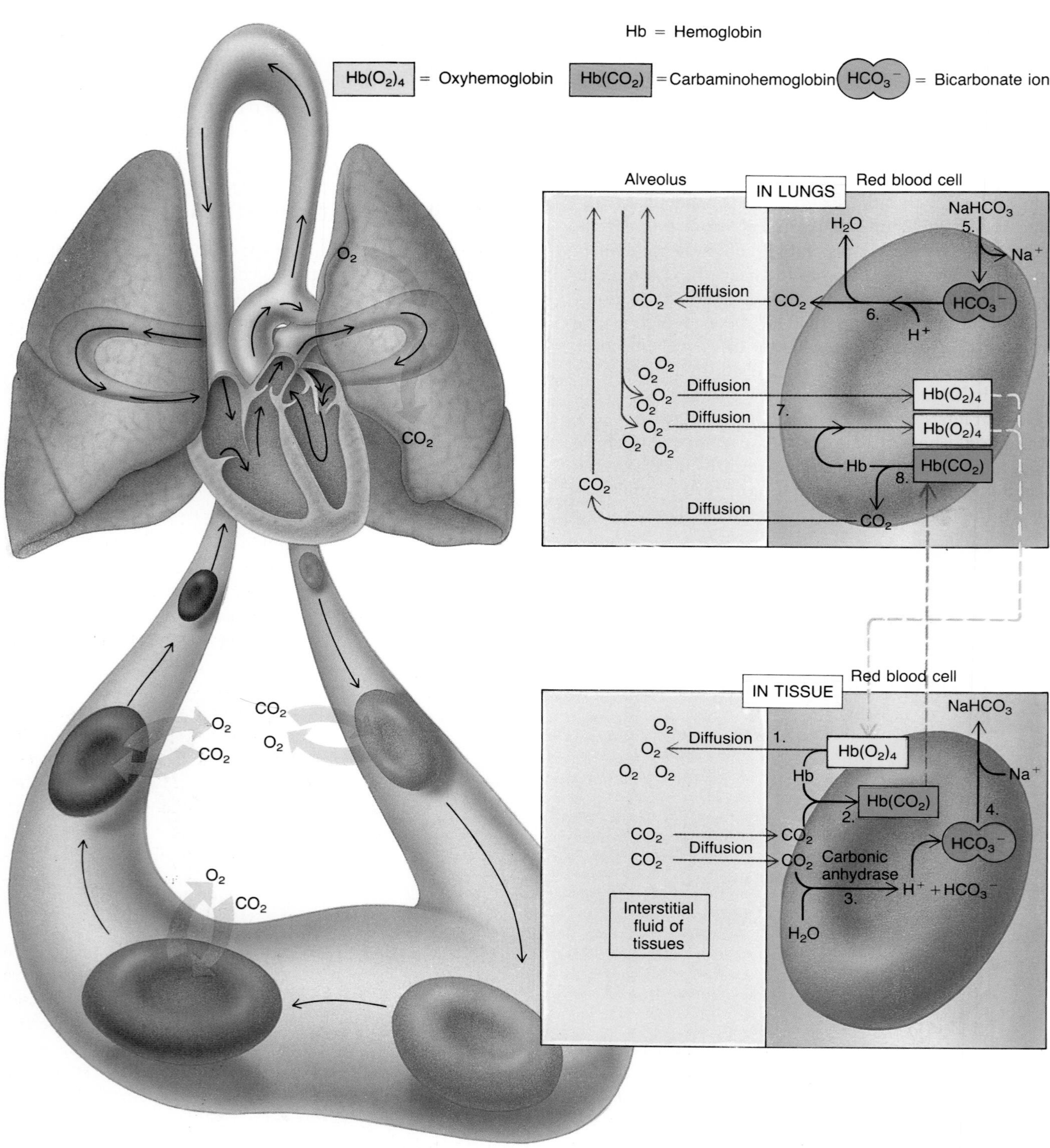

FIGURE 19-18

Transport and exchange of gases. In the tissues: (1) O_2 molecules are released from hemoglobin and diffuse into the cytoplasm of the red blood cell and then into the tissues. (2) As CO_2 diffuses from the tissues into red blood cells, some of it complexes with hemoglobin. (3) Most of the CO_2, however, reacts with water to form bicarbonate ions (HCO_3^-). (4) Some bicarbonate diffuses from the cell and provides the blood with buffering capacity. **In the lungs:** (5) Bicarbonate ions diffuse into the red blood cell (6) and are converted back to CO_2, which diffuses into the alveolus. (7) As O_2 diffuses into the red blood cells, it complexes with hemoglobin, so a steep O_2 concentration gradient is maintained. (8) CO_2 molecules that were bound to hemoglobin molecules are released, freeing the hemoglobin molecules to pick up additional O_2.

◁ THE HUMAN PERSPECTIVE ▷

Dying for a Cigarette?

On the average, smoking cigarettes will cut approximately 6 to 8 years off your life, more than 5 minutes for every cigarette smoked! Cigarette smoking is the greatest cause of preventable death in the United States. According to a recent report by the Center for Disease Control (CDC), nearly 450,000 Americans die each year from smoking-related causes. Smoking accounts for 87 percent of all lung cancer deaths, and smokers are more susceptible to cancer of the esophagus, larynx, mouth, pancreas, and bladder than are nonsmokers. The increased incidence of lung cancer deaths among smokers compared to nonsmokers is shown in Figure 1*a*, and the benefit attained by quitting is shown in Figure 1*b*. Atherosclerosis, heart disease, and peptic ulcers also strike smokers with greater frequency than they do nonsmokers. For example, long-term smokers are 3.5 times more likely to exhibit severe arterial disease than are nonsmokers. Emphysema (a condition caused by destruction of lung tissue producing severe difficulty in breathing) and bronchitis (an inflammation of the airways) are 20 times more prevalent among smokers.

Smokers also endanger other people. Each year, smokers are responsible for the deaths of thousands of "innocent bystanders," nonsmokers who share the same air with smokers. The risks of passive (involuntary) smoking are indisputable; secondhand smoke can make you seriously ill. Children of smokers have double the frequency of respiratory infections as do children who are not exposed to tobacco smoke in the home. Being married to a smoker is especially hazardous; 20 percent of lung cancer deaths among nonsmokers are attributable to inhaling other people's tobacco smoke.

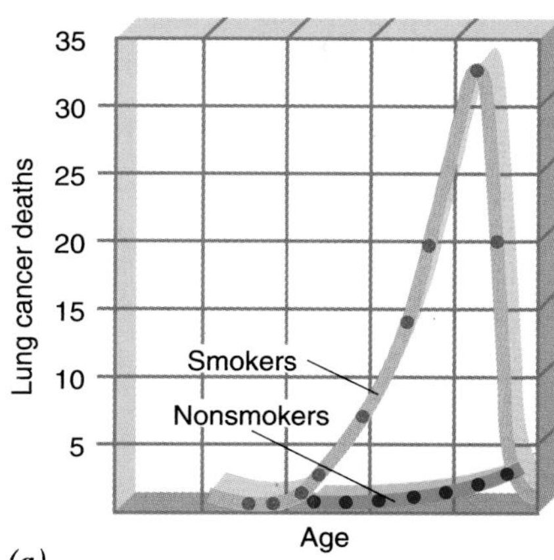

(a)

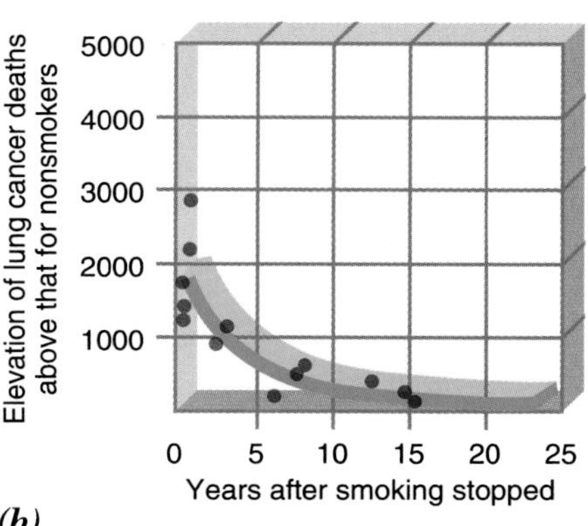

(b)

FIGURE 1

Another "innocent bystander" is a fetus developing in the uterus of a woman who smokes. Smoking increases the incidence of miscarriage and stillbirth and decreases the birth weight of the infant. Once born, these babies suffer twice as many respiratory infections as babies of nonsmoking mothers.

Why is smoking so bad for your health? The smoke emitted from a burning cigarette contains more than 2,000 identifiable substances, many of which are either irritants or carcinogens. These compounds include carbon monoxide, sulfur dioxide, formaldehyde, nitrosamines, toluene, ammonia, and radioactive isotopes. Autopsies of respiratory tissues from smokers (and of nonsmokers who have lived for long periods with smokers) show widespread cellular changes, including the presence of precancerous cells (cells that given time may become malignant) and a marked reduction in the number of cilia that play such a vital role in the removal of bacteria and debris from the airways.

Of all the compounds in tobacco (including smokeless varieties), it is nicotine that makes tobacco so addictive. Nicotine is addictive because it acts like a neurotransmitter by binding to certain acetylcholine receptors (page 268), causing stimulation of postsynaptic neurons. The physiological effects of this stimulation include the release of epinephrine, an increase in blood sugar, an elevated heart rate, and the constriction of blood vessels, causing elevated blood pressure. The nervous system of smokers becomes "accustomed" to the presence of nicotine and decreases the output of the natural neurotransmitter. As a result, when a person tries to stop smoking, the sudden absence of nicotine, together with the decreased level of the natural transmitter, decreases stimulation of postsynaptic neurons, which creates a craving for a cigarette—a "nicotine fit." Even after the physiological addiction disappears, ex-smokers are so conditioned to the act of smoking that the craving for cigarettes can continue for years. If you don't smoke, don't start. If you smoke, quit.

a diminished oxygen cargo. As a result of these differences in concentration, oxygen in the lungs diffuses across the thin, cellular walls of both the alveolus and the capillary and into the bloodstream, while carbon dioxide diffuses in the opposite direction (Figure 19-18). This exchange is completed in about a quarter of a second—the time it takes blood to flow through the site of exchange in a capillary.

If oxygen simply dissolved in the fluid of blood, the blood's oxygen-carrying capacity would be severely limited because oxygen is not very soluble in water. The capacity of the bloodstream to transport oxygen is greatly increased by the presence of hemoglobin. Each hemoglobin molecule contains four polypeptide subunits and binds four oxygen molecules, forming *oxyhemoglobin*. Hemoglobin allows human blood to carry 70 times more oxygen than it could otherwise. Hemoglobin is not simply dissolved in the blood, but rather is encapsulated in erythrocytes (red blood cells). Each milliliter of your blood contains about 5 billion erythrocytes; your entire body contains a total of about 25 trillion.

Gas Exchange in the Tissues: Release of Oxygen from the Blood

Blood leaves the lungs carrying high concentrations of oxygen and low levels of carbon dioxide. When this oxygenated blood reaches a metabolically active tissue, it finds itself in an environment that is low in oxygen and high in carbon dioxide. As a result, oxygen diffuses into the tissues as carbon dioxide enters the blood; both gases move passively from areas of higher concentration to areas of lower concentration (Figure 19-18). The first oxygen molecules to move out of the blood are those that are simply dissolved in the plasma. As these dissolved oxygen molecules move out of the capillary, the concentration of oxygen dissolved in the plasma decreases, promoting the dissociation of oxygen molecules from their binding sites on hemoglobin molecules in the red blood cells. The oxygen molecules released from hemoglobin move out of the erythrocytes and dissolve in the plasma. From there, the dissolved oxygen molecules move out of the capillaries, promoting the release of additional oxygen molecules from the hemoglobin, and so forth. This is the process by which hemoglobin unloads its oxygen cargo in the tissues and readies itself to pick up more oxygen on its upcoming trip through the lungs.

The exchange of gases in the tissues is self-regulating. Those tissues that are metabolizing more actively utilize more oxygen, thereby producing lower oxygen concentrations. The lower the oxygen concentration in the tissue, the steeper the gradient between tissue and blood, which favors the release of more oxygen from the bloodstream. According to this principle, those tissues most "in need" of oxygen receive the most oxygen from the passing blood.

The blood contains buffers—particularly bicarbonate ions—that keep the blood from becoming too acidic or too alkaline. The bicarbonate ions in the blood are formed as the result of the following reaction between carbon dioxide and water which occurs when carbon dioxide is taken into the blood from the tissues.

$$\underset{}{CO_2} + H_2O \xrightarrow{\text{carbonic anhydrase}} \underset{\text{carbonic acid}}{H_2CO_3} \longrightarrow \underset{\text{proton}}{H^+} + \underset{\text{bicarbonate ion}}{HCO_3^-}$$

This first reaction is catalyzed by the enzyme *carbonic anhydrase* that is present in red blood cells. Carbonic acid, the product of the reaction, dissociates into two ions, a hydrogen ion (H^+) and a bicarbonate ion (HCO_3^-).

As the blood passes through the tissues, it gives up its oxygen and takes up carbon dioxide (Figure 19-18). Most of the carbon dioxide simply dissolves in the blood, forming bicarbonate, but a portion of the carbon dioxide molecules becomes bound to hemoglobin for the trip back to the lungs. When the blood reaches the lungs, the process of gas exchange is reversed from that which occurs in the tissues. The more oxygen that is removed during the previous passage of the blood through the body, the greater the number of hemoglobin molecules that will be lacking their full complement of oxygen molecules, and the more oxygen that will be picked up from the alveoli. The lungs have a remarkable capability for gas absorption. Even if you are running at top speed, and your blood is virtually depleted of its oxygen content, the blood will be fully resupplied with oxygen during its short, rapid passage through the lungs.

REGULATING THE RATE AND DEPTH OF BREATHING TO MEET THE BODY'S NEEDS

During the course of a typical day, the respiratory system must make rapid and dramatic changes in its level of activity. One minute, you might be resting quietly on a park bench, and the next minute you might be running after a bus. Three major mechanisms are available to supply the body with an increased supply of oxygen, each of which becomes quite evident during strenuous exercise: You breathe more rapidly; you breathe more deeply; and the rate of your heartbeat increases. All of these changes are controlled by regulatory centers located in the brain.

Although you can hold your breath for a short time, it is impossible to stop breathing voluntarily to the point of severe oxygen deprivation. Eventually, involuntary regulatory mechanisms restart "automatic" breathing. These same regulatory mechanisms keep you breathing while you sleep or when your attention is on other matters. They also cause you to breathe more often and more deeply as your need for oxygen delivery increases. The breathing control center, or **respiratory center,** is located in the medulla, a portion of the brainstem that regulates automatic activities.

ADAPTATIONS FOR EXTRACTING OXYGEN FROM WATER VERSUS AIR

The successful respiratory strategy is one that is adapted to the medium in which the animal lives. Animals that absorb oxygen from water, for example, utilize a different type of respiratory system than do animals that extract oxygen from air.

Extracting Oxygen from Water

The amount of oxygen dissolved in water is at best 21 times less than that which is dissolved in air. As a result, aquatic animals that acquire their oxygen from the surrounding water have less oxygen available than do their air-breathing counterparts. Most animals that extract oxygen from water possess **gills**—outgrowths of the body surface which project into the aqueous environment and are rich with blood vessels. The form of the complex gills of larger, active animals are well-suited for their function. These gills branch into smaller outgrowths that amplify the surface area for gas exchange without increasing the space required to house them. Gill complexity has reached its current pinnacle in fishes (Figure 19-19). Each gill consists of fingerlike projections, called **gill filaments,** from which rows of thin, flattened **lamellae** project into the flowing stream of water. The lamellae are the sites of gas exchange and have a rich supply of capillaries into which oxygen molecules diffuse.

Extracting Oxygen from Air

Terrestrial animals live in an environment that is rich in oxygen but low in water content; thus, oxygen is readily available. However, during the process of gas exchange, the terrestrial animal will invariably lose some of its precious body water. Therefore, the respiratory surfaces of most terrestrial animals—whether a snail, spider, or human—occur as inpocketings of the body surface which are tucked away within the moist, internal environment, thereby minimizing water loss. Two very different types of respiratory systems have evolved in air-breathing animals. Insects, and some of their arthropod relatives, possess *tracheae,* while most other air breathers, including snails, reptiles, birds, and mammals, have *lungs.*

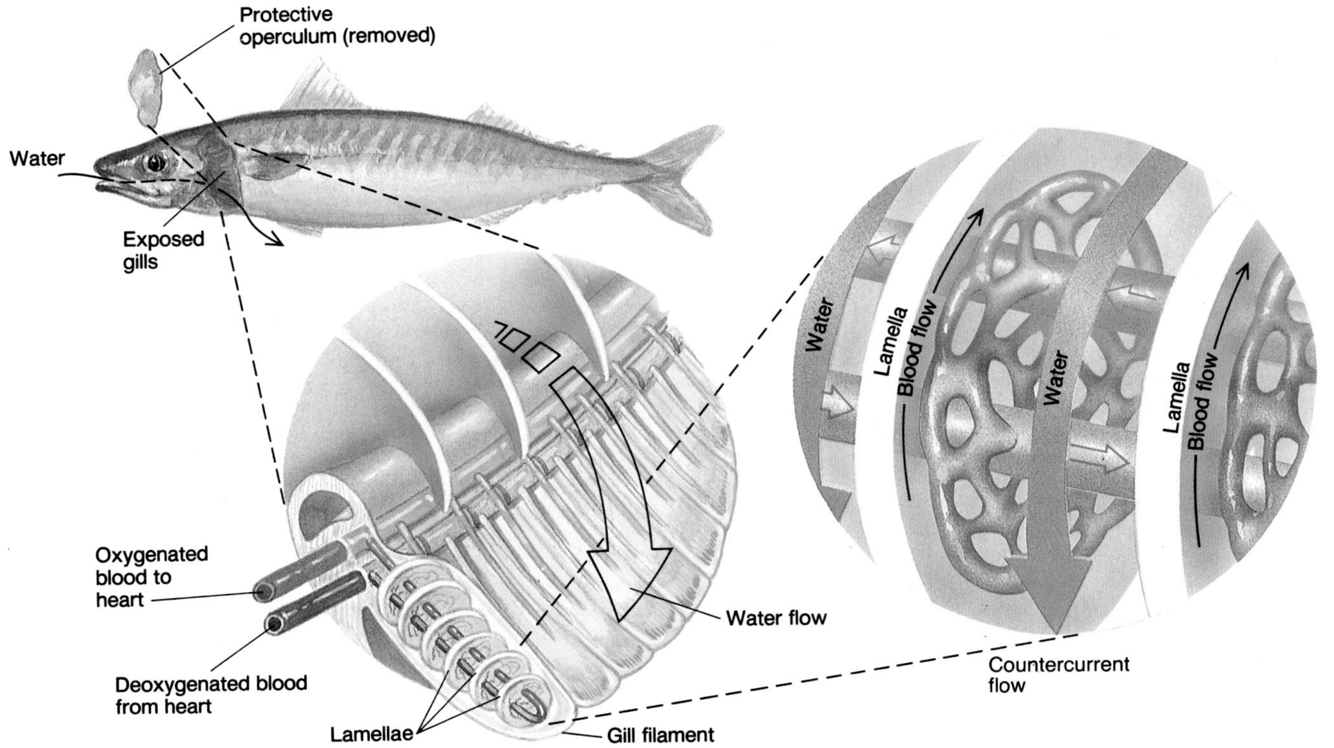

FIGURE 19-19

Complex gills. Fish have complex gills that have an enormous surface area compacted into a small space. In this diagram, blood flowing through each gill filament changes from blue to red as it acquires oxygen from the passing water. The process is enhanced by a countercurrent exchange system, whereby water is forced past the outer surface of the lamellae in a direction counter to that of blood flow within the lamellae.

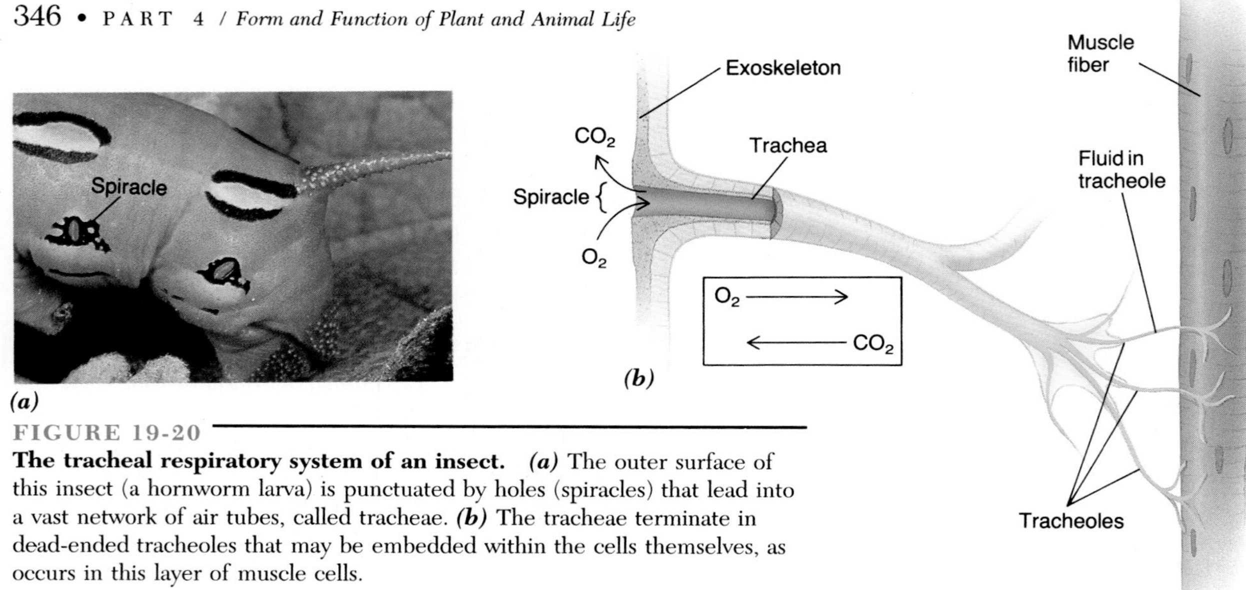

FIGURE 19-20

The tracheal respiratory system of an insect. ***(a)*** The outer surface of this insect (a hornworm larva) is punctuated by holes (spiracles) that lead into a vast network of air tubes, called tracheae. ***(b)*** The tracheae terminate in dead-ended tracheoles that may be embedded within the cells themselves, as occurs in this layer of muscle cells.

Respiration by Tracheae. If you look at the surface of an insect's abdomen under a microscope, you will find a row of tiny openings, called *spiracles* (Figure 19-20*a*). Each spiracle leads into a tube, which, in turn, branches into a network of finer and finer tubes. These tubes, called **tracheae,** carry air to the most remote recesses of the animal's body (Figure 19-20*b*). Ultimately, the finer tracheal tubes give rise to microscopic, dead-end, fluid-filled **tracheoles,** which terminate either very close to or actually within the body's cells. Oxygen travels down this complex network of air tubes directly to the cells, without any help from the animal's circulatory system.

Respiration by Lungs. As we discussed earlier, the lungs are sac-like invaginations of the body surface, which contain an extremely thin respiratory surface and a rich supply of microscopically thin blood capillaries. In less active animals, such as snails, frogs, and snakes, the lungs tend to be relatively simple sacs with little internal surface area. In contrast, the more complex lungs of mammals contain millions of microscopic alveoli, which provide an enormous surface area packed into a relatively small space.

EVOLUTION AND ADAPTATION: TYING IT TOGETHER

Simpler, multicellular animals, including sponges, jellyfish, and flatworms, attain considerable size without a true circulatory system. This is possible because of the way these animals are constructed; that is, because of their *body plan.* The body plan of these simpler animals places virtually all cells close to the source of oxygen and nutrients. This is particularly well illustrated by the flatworm, whose flattened body shape allows oxygen to diffuse directly from the environment to all of the body's cells and allows waste products to diffuse in the opposite direction (Figure 19-21). Similarly, nutrients are able to diffuse to the body's cells directly from the highly branched digestive cavity. As animals became larger and more complex, simple diffusion

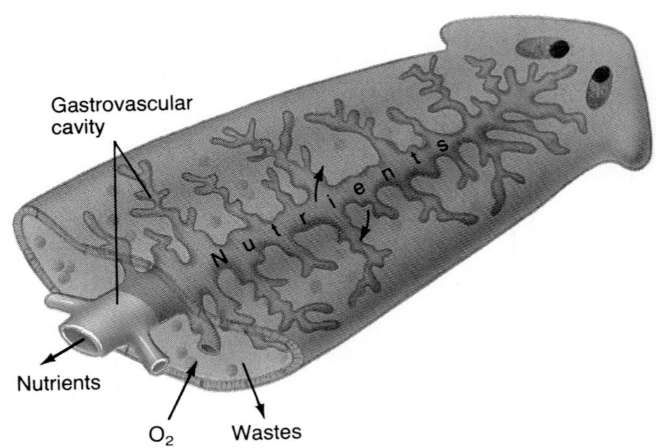

FIGURE 19-21

Flatworm form and function. A flatworm's body is highly flattened, which brings every cell close enough to the outside medium to receive oxygen and expel wastes by diffusion. The digestive chamber of a flatworm branches to form blind (dead-ended) passageways throughout the body. Nutrients are able to diffuse from the digestive channels directly to the body's cells. Because of its dual role as both digestive and circulatory system, the spacious, internal compartment of the flatworm is called a *gastrovascular cavity.*

could no longer deliver oxygen and nutrients to all the cells or carry away their wastes. These challenges were solved by the evolution of a circulatory system—a system containing one or more pumps, long lengths of vascular tubing, and a fluid (blood) in which dissolved gases and solutes could be moved from one part of the animal to another.

One of the primary functions of the circulatory system is to transport respiratory gases to and from the site where they are exchanged with the environment. Many animals, including jellyfish, earthworms, and frogs, respire **cutaneously;** that is, across virtually their entire body surface. Since the respiratory surface must be moist and thin, it follows that animals that respire cutaneously tend to be highly vulnerable to environmental conditions. This is evidenced by the familiar sight of a dehydrated earthworm that managed to make it halfway across a sidewalk before succumbing to the loss of water across its thin, permeable outer surface. Because of the vulnerability associated with cutaneous respiration, selective pressures have favored animals whose delicate respiratory surfaces are restricted to particular sites on or within the body, whether they are the gills of an aquatic animal or the lungs of an animal living on land. Restricting the respiratory surface to a localized organ allows the rest of the body's surface to remain impermeable, as exemplified by the skin that covers your body or the exoskeleton of an insect or spider.

Osmoregulatory and excretory mechanisms are closely correlated with the type of habitat in which an animal lives. Invertebrates that live in marine habitats, such as a sea anemone or an octopus (Figure 19-22*a*), for example, have very salty body fluids that are approximately equal in osmotic concentration to that of the surrounding sea. As a result, these animals neither lose nor gain water by osmosis. Therefore, they are in not in need of an elaborate osmoregulatory system.

In contrast, marine vertebrates are generally much less salty than their environment and thus tend to lose water by osmosis. Most marine fishes (Figure 19-22*b*) regain the water they lose through osmosis by drinking the sea water they live in and excreting concentrated salt solutions from their gills. These animals produce virtually no urine, and many have very rudimentary kidneys. Marine sea birds and sea turtles (Figure 19-22*e,f*) must also obtain their water from the sea; these animals drink sea water and excrete concentrated salt solutions from salt glands located in their heads.

Animals that live in fresh water face just the opposite problem: Their environment is very high in water concen-

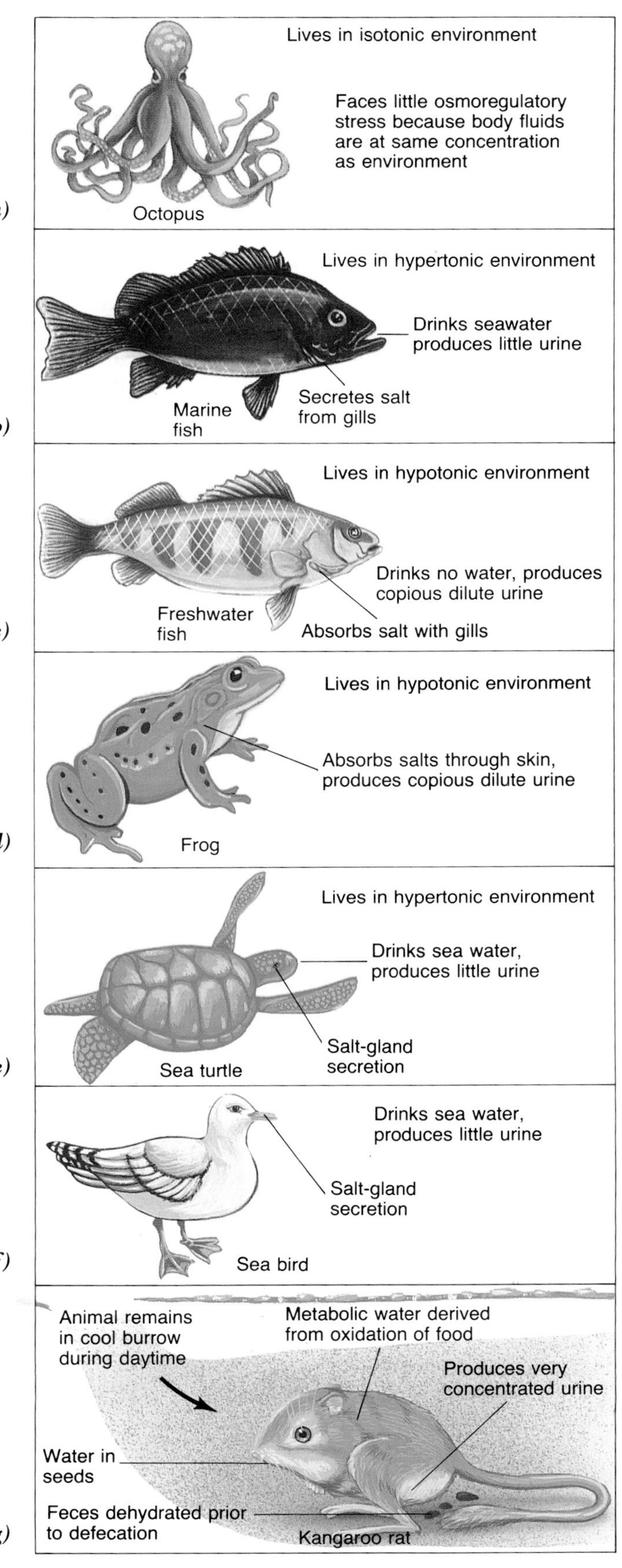
(a)
Lives in isotonic environment
Faces little osmoregulatory stress because body fluids are at same concentration as environment
Octopus
(b)
Lives in hypertonic environment
Drinks seawater produces little urine
Marine fish
Secretes salt from gills
(c)
Lives in hypotonic environment
Drinks no water, produces copious dilute urine
Freshwater fish
Absorbs salt with gills
(d)
Lives in hypotonic environment
Absorbs salts through skin, produces copious dilute urine
Frog
(e)
Lives in hypertonic environment
Drinks sea water, produces little urine
Salt-gland secretion
Sea turtle
(f)
Drinks sea water, produces little urine
Salt-gland secretion
Sea bird
(g)
Animal remains in cool burrow during daytime
Metabolic water derived from oxidation of food
Produces very concentrated urine
Water in seeds
Feces dehydrated prior to defecation
Kangaroo rat

FIGURE 19-22
A diversity of osmoregulatory challenges. After *Animal Physiology* 3/E by Eckert, Randall, and Augustine. Copyright © 1988 by W. H. Freeman and Company. Used with permission.

tration and low in available salts. Consequently, freshwater animals tend to *gain* water by osmosis—water that must be expelled back into the environment. Freshwater fish and frogs (Figure 19-22*c,d*) have well-developed kidneys and produce a very dilute urine. The most serious problem faced by these animals is the loss of valuable salts, which are inevitably washed away in the large volume of urine. Freshwater animals possess highly effective active-transport mechanisms that pump salts back into their bodies, despite the low salt concentration of their environment: Freshwater fish have salt-absorbing gills, while frogs have salt-absorbing skin.

The most serious osmoregulatory problems are faced by terrestrial animals, particularly by those that live in dry, desert habitats. These animals possess physiological and behavioral adaptations that keep water loss to an absolute minimum. One of the best-studied water-savers is the desert kangaroo rat (Figure 19-22*g*). This animal can live its entire life without ever drinking a drop of water; all the water the kangaroo rat needs is either present in the animal's food or formed as a byproduct of metabolic reactions (so-called *metabolic water*). Water conservation is accomplished by the animal's ability to produce an extremely concentrated urine (several times saltier than that which can be excreted by humans) and by its staying out of the desert heat. During the day, these animals remain in their relatively cool burrows; they emerge at night to carry out their feeding and social activities.

SYNOPSIS

Circulatory systems deliver a well-oxygenated nutrient solution to the body's tissues and carry away the wastes. In mammals, the system also contributes to homeostasis by stabilizing pH and osmotic balance, delivering hormones, protecting against foreign intruders, and resisting changes in body temperature.

The anatomy of blood vessels is well suited to their function. The design of the large, elastic arteries allows them to snap back after each surge from the heart, squeezing blood onward. To meet the needs of the tissues they service, muscular arterioles either increase or decrease their diameter. Capillaries are thin and porous and serve as sites of exchange with the surrounding tissue fluid. Venules collect the blood leaving the capillaries and route it to veins. One-way valves assist the blood's return to the heart. Excess fluid returns to the bloodstream via the lymphatic system.

The mammalian heart is equipped with one-way valves to keep blood flowing in a single direction. Acting as a double pump, the right side of the heart pumps blood to lungs for oxygenation; the left side powers blood through the general (systemic) circulation. Although the heart will contract rhythmically without an external stimulus, neural impulses regulate the rate of contraction by stimulating the sinoatrial node, which initiates contraction of the atria and relays the stimulus to the rest of the heart. This intrinsic regulation may be temporarily overridden in times of stress, fear, or exertion.

Nearly half the volume of the blood consists of cells and cell fragments (platelets). Erythrocytes carry oxygen; granular leukocytes and monocytes phagocytize foreign intruders and tissue debris; platelets aid in clotting; and lymphocytes launch the immune response. The fluid phase (plasma) contains a number of proteins that are activated in a specific sequence, leading to the formation of a clot.

Excretion and osmoregulation. The mammalian kidney has two roles: It maintains the body's salt and water balance, and it rids the blood of waste products generated by cellular metabolism. Wastes, excess solutes, and water are voided in the urine—the fluid produced in the nephrons of the kidney as a result of three processes: glomerular filtration, selective reabsorption, and secretion. Blood (minus its cells and proteins) is filtered out of the capillaries of the glomeruli into the proximal ends of the nephrons. From there, the fluid flows down the lengths of the tubule, while salts, sugars, and other nutrients are transported out of the nephron back into the bloodstream. Nitrogenous wastes, hydrogen ions, and excess potassium ions are secreted in the opposite direction. The hormones aldosterone and ADH act on the distal portion of the nephrons to regulate the amount of salt and water voided in the urine.

The need for gas exchange. All animals must obtain oxygen for aerobic respiration and must dispose of carbon dioxide waste. Most complex animals have impermeable outer surfaces with specialized gas exchange structures.

The human respiratory system consists of a branched passageway that leads into a pair of lungs. The lungs consist of millions of microscopic, thin-walled pouches (alveoli) that are underlain by a network of capillaries. The alveoli provide an extensive surface across which gas exchange can occur.

Gas exchange in the lungs and tissues is driven by the diffusion of oxygen and carbon dioxide down their respective concentration gradients. Blood entering the lungs from the tissues is relatively high in carbon dioxide and low in oxygen. In the lungs, oxygen diffuses from the alveoli into the blood, where over 95 percent of the oxygen molecules bind to hemoglobin in the erythrocytes. The blood becomes completely saturated with oxygen, even when we are undergoing strenuous exercise and the blood is moving rapidly through the lung capillaries. When oxygenated blood is pumped through metabolically active tissues, oxygen molecules diffuse out of the plasma, promoting the release of oxygen molecules bound to hemoglobin. The process of gas exchange in the tissues is self-regulating. The more active the tissue, the lower its oxygen concentration and the steeper the oxygen gradient between the blood and that tissue. This relationship favors the release of additional oxygen.

Water breathers and air breathers have different types of respiratory organs. Water breathers typically possess gills, which consist of delicate, fingerlike projections that extend into the surrounding water. Air breathers typically possess either lungs—inpocketings of the external surface into the body where the respiratory surface can be kept moist—or tracheae—blind passageways through which air diffuses into the tissues. Tracheae are found in arthropods and carry air to the tissues without the intervention of the circulatory system.

Review Questions

1. Name and discuss five essential functions of the circulatory system.
2. Compare and contrast: arteries and veins; blood and lymph; atria and ventricles; pulmonary and systemic circuits; sinoatrial and atrioventricular nodes; red blood cells and white blood cells; osmoregulation and excretion; tubular reabsorption and tubular secretion; ADH and aldosterone; gills, lungs, and tracheae; oxygen present in the plasma and in the erythrocytes.
3. Describe the various ways your kidneys help maintain homeostasis.
4. Describe the functional interconnection between the human circulatory system and the respiratory system.
5. Contrast the events that occur during gas exchange in the lungs with that of gas exchange in the tissues.

Critical Thinking Questions

1. Suppose Galen had been right and blood really flowed back and forth in the veins. How would this have affected Harvey's experiment in which he used his finger to stop blood flow? How would the absence of valves in the veins of the arm have affected Harvey's experiment?
2. Why would simple, hollow lungs equipped with blood capillaries along their walls fail to meet the respiratory demands of mammals?
3. Excretion in humans begins with nonselective filtration in the glomeruli. If you were designing an excretory system, would you make this part of the system selective or nonselective? What are the advantages of each approach? Do you think that the existing system is the most efficient design? Why, or why not?
4. After a lengthy visit to a city at a high altitude we adjust to the lowered oxygen content of the air by increasing the number of circulating erythrocytes. Why is this capability referred to as *acclimation* rather than *adaptation?*
5. Trace the path a molecule of oxygen would take from the outside atmosphere to a muscle cell in your leg, naming all of the structures through which it would pass.

CHAPTER 20

Internal Defense: The Immune System

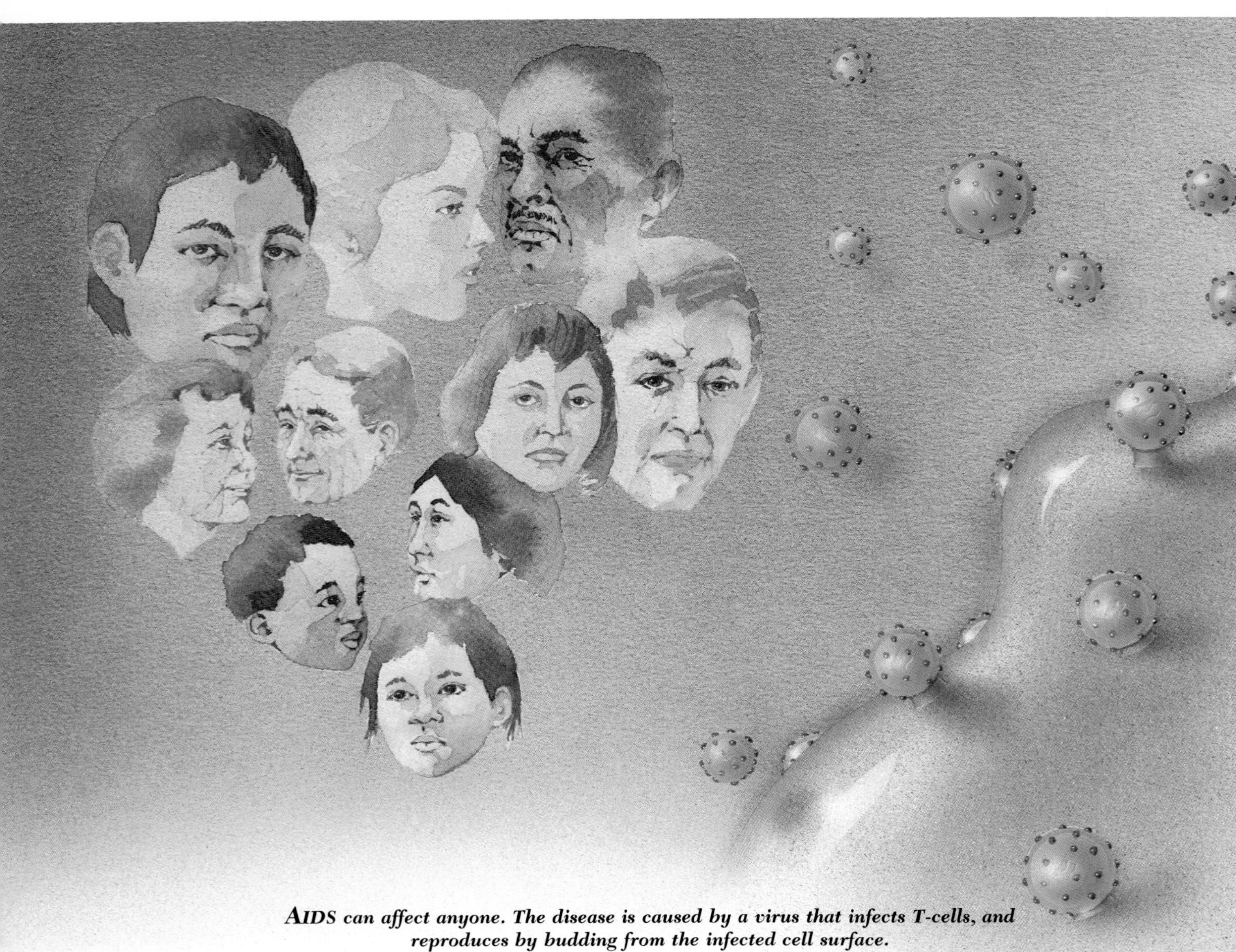

***AIDS** can affect anyone. The disease is caused by a virus that infects T-cells, and reproduces by budding from the infected cell surface.*

STEPS TO DISCOVERY
On the Trail of a Killer: Tracking the AIDS Virus

In the Fall of 1980, the UCLA Medical Center was visited by a 31–year-old-male who was suffering from a persistent fever, weight loss, swollen lymph nodes, and a severe yeast infection in his mouth and throat that usually only strikes people whose immune defenses are crippled. The patient came to the attention of Michael Gottlieb, an immunologist who had recently arrived at UCLA. Gottlieb soon discovered that the patient was also suffering from *Pneumocystis carinii* pneumonia (PCP), a rare type of pneumonia that usually causes lung infections in people suffering from lymphoid cancers. In other words, this patient was suffering from two rare types of infections simultaneously. In most people, these disease-causing microbes would have been attacked and eliminated, killed by a population of white blood cells, called T cells. This particular patient, however, lacked an entire class of T cells (the T helper cells), leaving Gottlieb and his colleagues mystified. By April, four similar patients had come to Gottlieb's attention, all of whom were male homosexuals. Gottlieb submitted a short report to the weekly newsletter of the U.S. Centers for Disease Control (CDC). This letter turned out to be the first report of a new disease, Acquired Immune Deficiency Syndrome (AIDS), that has subsequently grown to epidemic proportions.

Gottlieb's report hit home with Paris physician Willy Rozenbaum, who had treated three similar cases. All three patients had died of similar mixed infections, including PCP. Unlike the patients in Gottlieb's report, however, these victims were not homosexual and had resided in Africa. Rozenbaum, suspecting that all five patients had died from the same disease, concluded that the affliction was the result of an infectious agent rather than an environmental toxin since the disease had appeared in very different populations on two sides of the world. Furthermore, the disease appeared to have a very long *incubation period*—the time between the infection and the appearance of symptoms. The implications of this finding were frightening: The disease could take hold within a population years before any evidence of its presence would be detected.

Because the African cases of the disease occurred before those in the United States, epidemic specialists have concluded that the disease originated in Africa and was transported to the United States by an infected male homosexual, who unwittingly passed it into the homosexual community. Soon, the disease began to appear in other individuals, most notably hemophiliacs who had received blood transfusions and intravenous-drug users who shared hypodermic needles.

As more cases were reported, scientists begin to close in on the agent responsible for AIDS. Don Francis, a scientist at the CDC who had worked with a virus that destroys the immune system of cats, believed that both diseases were caused by similar viruses, called *retroviruses* (a virus whose genetic material is encoded in RNA from which a complementary strand of DNA is synthesized). Just one year earlier, in 1980, Robert Gallo of the National Institutes of Health and a team of researchers showed that retroviruses could infect humans and cause a rare type of leukemia that affected a person's T cells. Evidence pointed squarely at some retrovirus as the causative agent of this new disease. In 1982, Gallo's lab began culturing lymphocytes from patients with AIDS.

Meanwhile, another expert on retroviruses, Luc Montagnier, of the Pasteur Institute in Paris, was conducting similar research, searching for a retrovirus in infected lymphocytes. In 1983, both the U.S. and French teams reported the isolation of the retrovirus responsible for AIDS, setting the stage for one of the major scientific controversies of the decade: Which lab should be credited with the discovery of the virus that causes AIDS (later called Human Immunodeficiency Virus or simply HIV)? For a number of reasons, including the fact that Gallo's virus appears to have been isolated from cells donated by the French scientists, Luc Montagnier is generally credited with the discovery of HIV.

As you read this sentence, you are being attacked. You are, in fact, fighting for your life. The surface of your body is populated by billions of microorganisms—bacteria and other agents that would invade and use your tissues as their next meal if allowed uncontested entry into your body. With every breath you take, you inhale more than a million microorganisms that are suspended in air, many of which would cause fatal infections if you didn't have an arsenal of weapons to protect your respiratory tract against their invasion. Some microorganisms have already entered your bloodstream, perhaps through cuts in your skin so small you were unaware of them. In your internal tissues and fluids, a microorganism has found a bonanza of nutrients and hospitable environmental conditions. The invasion never stops.

The system responsible for repelling this invasion is the **immune system.** It recognizes breaches in security and effectively eliminates invading microorganisms. Before an intruder encounters the immune system, however, it must evade a whole battery of host defenses, the

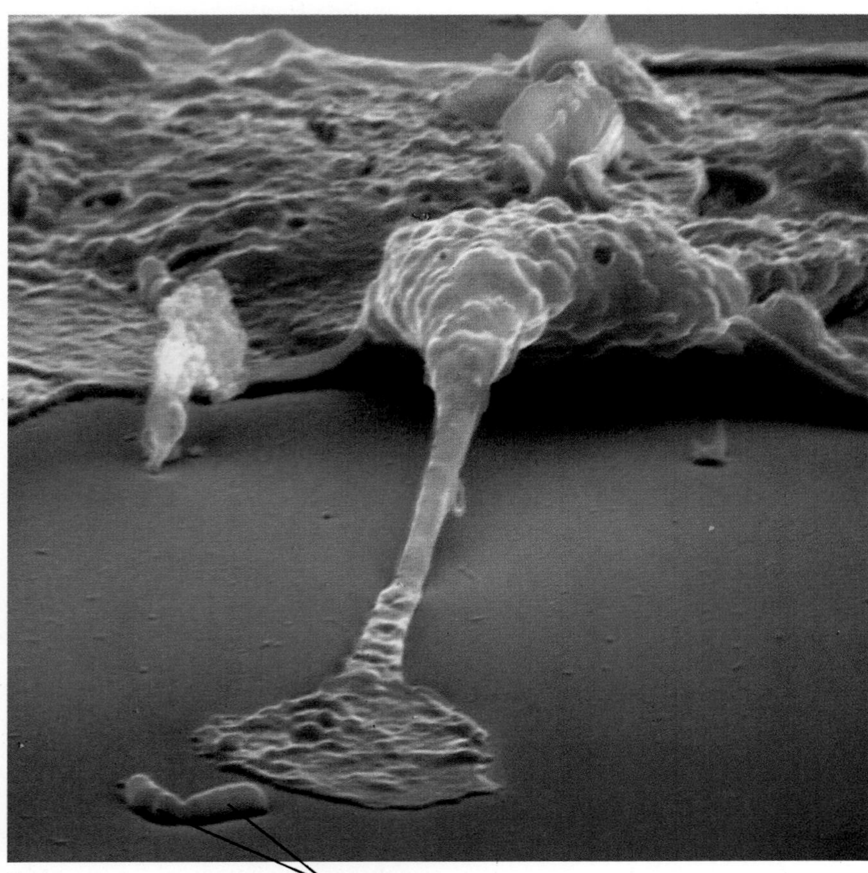

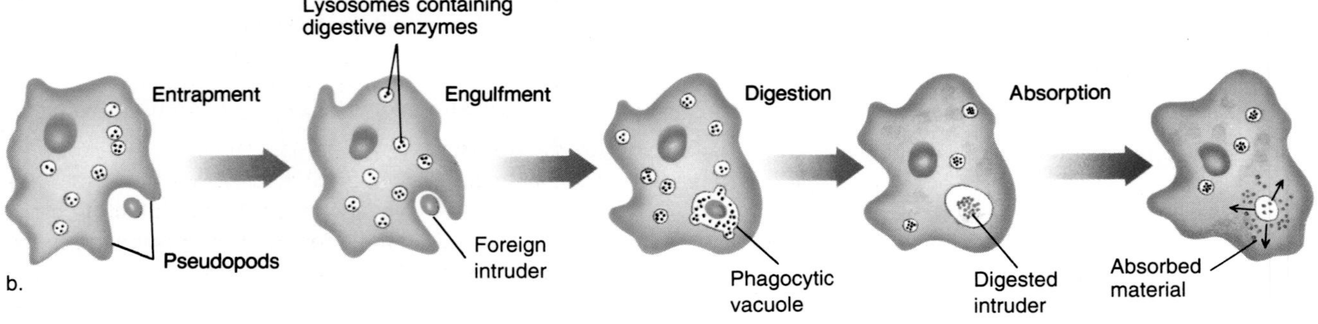

FIGURE 20-1
Death grip. Among your many weapons against disease are your phagocytic cells that engulf and destroy foreign microbes, such as the doomed bacteria about to fall into the clutches of a protective white blood cell ***(a)***. The engulfed intruder is destroyed by the process depicted in part ***(b)***. Trapped in a large membranous vesicle, the intruding cell is digested by enzymes discharged from lysosomes that fuse with the vesicle membrane.

body's first line of defense. These protective mechanisms are not targeted toward any specific organism, but rather form effective mechanical, chemical, and cellular barriers against intruders in general. Because of this "shotgun" approach, these defenses are broadly grouped under the heading "nonspecific protective mechanisms." Immunity, as you will see in a couple of pages, is considerably more focused; each immune attack is specifically directed against a particular intruder.

▼ ▼ ▼

NONSPECIFIC MECHANISMS: A FIRST LINE OF DEFENSE

The body's most evident protective strategy is to keep viruses, bacteria, and other dangerous microorganisms from penetrating into living tissues. The skin forms a relatively impregnable outer body layer, as long as it remains undamaged. The mucous membranes that line the respiratory, digestive, and urinary tracts are protected by a sticky layer of mucus that snares microbes in a trap that is continually shed and replenished. The flushing of your urinary tract each time urine is voided is another nonspecific defense.

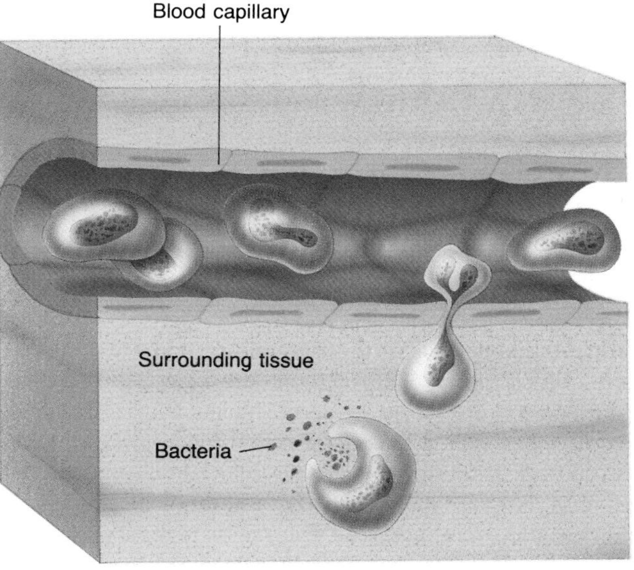

FIGURE 20-2
Arrival of the warriors. Phagocytic leukocytes can squeeze through openings between the cells that line the capillaries and enter an injured or infected area, where they can engulf and destroy invading microbes.

NONSPECIFIC CELLULAR DEFENSES

If a bacterium or other *pathogenic* (disease-causing) microorganism breaches the body surface and reaches the tissues of the body, it will likely encounter a phagocytic cell that can engulf and destroy it (see Figure 20-1). These protective cells are carried to the infected tissues through the blood and lymphatic vessels (Figure 20-2).

Phagocytic cells provide little security unless they can reach the area where protection is needed. The body's overall "strategy" for attracting these and other protective cells to sites of danger, such as an infected wound, is called **inflammation.** Inflammation is initiated when chemicals from cells in the injured tissue are released, causing local blood vessels to dilate, bringing additional blood (and its protective cells) into the affected region. These chemicals attract phagocytic leukocytes, which accumulate in the injured or infected region (Figure 20-2). These cells and the fluid from leaking blood vessels often accumulate to form a yellowish fluid, called *pus.* In fact, inflammation creates many of the uncomfortable symptoms (fever, swelling, and pain) that are signs of protective inflammation. Use of drugs (such as cortisone) to subdue inflammation may provide immediate relief from these symptoms but suppresses this line of defense.

NONSPECIFIC MOLECULAR DEFENSES

Blood contains a group of proteins, collectively called **complement,** that bind to the surface of a bacterium and launch a cascade of reactions that poke holes in the plasma membrane of the foreign cell, which subsequently bursts and dies. Fighting viral infections is spearheaded by another nonspecific molecule, the protein *interferon.* Virus-infected cells secrete interferon, which binds to the surfaces of uninfected, neighboring cells, where it triggers the cell to produce antiviral proteins that block the neighboring cells' ability to support the replication of the virus. This reaction often halts the spread of the infection.

MICROBE-SPECIFIC DEFENSES: THE IMMUNE SYSTEM

Many microorganisms that cause disease stimulate the body's immune system and produce a state of prolonged or permanent protection against further episodes of the same disease. Measles, mumps, and chickenpox, for example, are all diseases that leave the person permanently immune. Any foreign substance that elicits production of an immune response is called an **antigen.** The initial exposure to the antigen of the measles virus, for example, stimulates cells of the immune system to launch an immune response. One component of the immune response is the production of **antibodies**—proteins that bind to the antigen that stimulated their production. Antibodies against the measles virus bind to the virus and prevent it from infecting more cells during the initial infection, and the person recovers. These antibodies also help prevent further episodes of the disease. As long as a person can produce antibodies against this virus, he or she is immune to measles. The protection is highly *specific*; immunity against measles does nothing to protect against chickenpox, polio, or any disease caused by microbes that don't carry the measles antigen.

THE NATURE OF THE IMMUNE SYSTEM

The immune system is composed of cells that are scattered throughout the body and are particularly concentrated in the lymphoid tissues, which include the thymus, spleen, lymph nodes, bone marrow, and tonsils (Figure 20-3). The most prominent cells of the immune system are lymphocytes, which circulate throughout the blood and lymph. Lymphocytes are aided by large phagocytic cells called *macrophages*. When lymphocytes and macrophages come into contact with foreign materials, they launch an *immune response* that mobilizes the body's immunological arsenal. This arsenal is composed of two major types of lymphocytes: *B lymphocytes* (or simply **B cells**) and *T lymphocytes* (**T cells**). Both types of lymphocytes work in cooperation with each other and with macrophages, but they are very different in their mechanisms of protection.

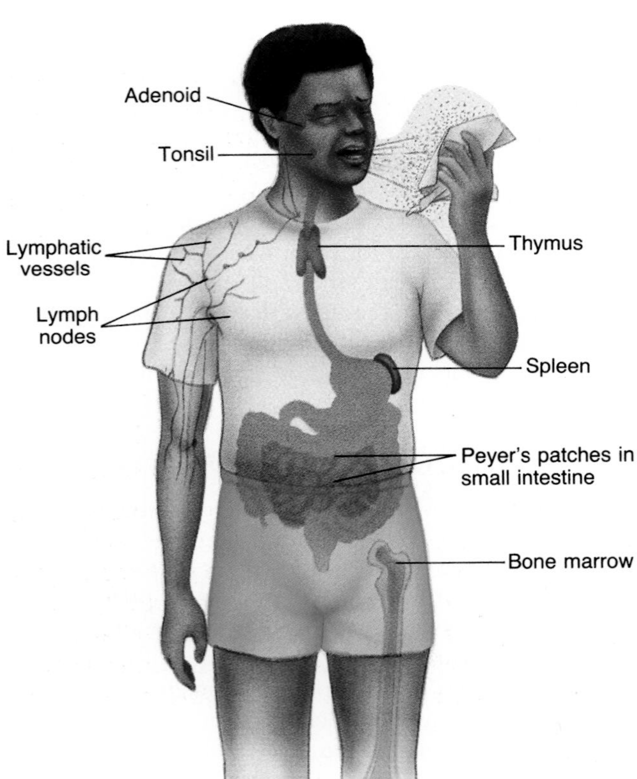

FIGURE 20-3
The human lymphoid system includes various lymphoid organs, such as the thymus, bone marrow, spleen, lymph nodes, and scattered cells located as patches within the small intestine, adenoids, and tonsils. T cells mature (differentiate into immune cells carrying T cell receptors) in the thymus, while B cells mature (differentiate into immune cells carrying membrane-bound antibodies) in the bone marrow.

B Cell Immunity: Protection by Soluble Antibodies

The B cell system is responsible for producing antibodies in response to antigens on a foreign intruder. Antibodies attack the invading microorganism, directly inactivating it or making it more susceptible to ingestion by a patrolling phagocyte. These soluble, blood-borne antibody molecules are proteins that are secreted by **plasma cells,** which are formed from B lymphocytes present within the bone marrow.

When pathogens are *outside* of host cells, they are vulnerable to attack by soluble antibodies in the blood, mucus, and other bodily fluids. Because antibodies cannot penetrate infected cells, however, they have little effect on a pathogen while it is inside of a host cell. Those pathogens that grow *inside* the body's cells are eliminated by the other branch of the immune system, T cell immunity.

T Cell Immunity: Protection by Intact Cells

Because no antibodies are produced by T cells, the type of protection they execute is called **cell-mediated immunity.** Cells, rather than antibodies, are responsible for killing or eliminating the foreign pathogen. T cells recognize infected or abnormal target cells (such as cancer cells) by using membrane proteins, called *T cell receptors,* which are antibody-like proteins embedded in their plasma membranes. The T cell immune response is launched when these T cell receptors lock onto the corresponding foreign antigen. The result is the activation of one of four functionally distinct subclasses of T cells:

1. **Cytotoxic (killer) T cells** (Figure 20-4) recognize aged, malignant, or pathogen-infected cells and release *perforins*—proteins that puncture the membrane of the target cell. By killing infected cells, cytotoxic T cells can eliminate viruses, bacteria, yeast, protozoa, and parasites after they have found their way into a host cell. Cytotoxic T cells may also play a role in destroying cancer cells (see The Human Perspective: Treating Cancer with Immunotherapy).
2. **Effector T cells** release soluble substances that attract other protective cells to the area of antigen stimulation. These substances also activate the cells as they arrive, for example by producing "angry" macrophages that engulf foreign materials with greater killing effectiveness.
3. **Helper T cells** assist in enhancing the effectiveness of other lymphocytes (B cells and cytotoxic T cells), activating them by releasing immune enhancers. The requirement for helper cells to initiate an immune response is likely a fail-safe system that prevents accidental immune reactions against self-tissues. *Inter-*

FIGURE 20-4
Doomed cancer cell. This large cancer cell is being attacked by numerous smaller killer T cells. When specific contact is made between the two cells, the T cell releases a protein that kills the target cell by puncturing its membrane.

leukin II is one such immune enhancer. This activator chemical is currently being produced by recombinant DNA technology and is being studied as a treatment for certain types of cancer. Helper T cells are the cells that are hardest hit by the immune-crippling disease AIDS.

4. **Suppressor T cells** help prevent the potentially fatal consequences of overstimulating the immune system. It does so by *inhibiting* other lymphocytes.

T cells are named for the **thymus gland**—a mass of lymphoid tissue that is situated in the chest cavity, where it programs lymphocytes to become T cells during fetal development and early life. People who lack a functional thymus have no mature T cells, and all the various cell-mediated defensive mechanisms fail to appear. These people quickly die of overwhelming infection.

Organ Transplants and Graft Rejection. The body's T immune system not only repels foreign pathogens, it also rejects foreign tissues, such as those of transplanted organs. Several strategies decrease the likelihood of *graft rejection,* one by minimizing the "foreignness" of the graft, another by suppressing the rejection mechanism:

- By matching the donor to the recipient, the donor's cell-surface proteins (called *histocompatibility antigens*) are as closely matched as possible to the recipient. The closer the match, the less likely the rejection. The risk is always present, however, since no two people other than identical twins have identical histocompatibility antigens.
- The transplant recipient is also treated with drugs (such as the fungal compound cyclosporine) that suppress the person's cell-mediated immunity, thereby reducing the capacity for graft rejection. Even though these drugs are now quite effective, suppression of the immune system leaves the recipient more vulnerable to infection.

ANTIBODY MOLECULES: STRUCTURE AND FORMATION

You possess the immune capacity to respond to nearly every foreign intruder you will encounter during your lifetime. It is estimated that humans can produce millions of different types of antibodies, each type specific against one antigen. You can make antibodies against virtually any antigen of any shape. To understand how antibodies possess such specificity and why there are so many different types of antibodies, it is necessary to examine their structure and formation.

ANTIBODY STRUCTURE

All antibody molecules (also called *immunoglobulins* because they are globulin-type proteins) are composed of two basic types of polypeptide chains, called **heavy chains** and **light chains,** based on their relative sizes. The most common class of immunoglobulins, the IgG (immunoglobulin G) class, is composed of two light chains and two heavy

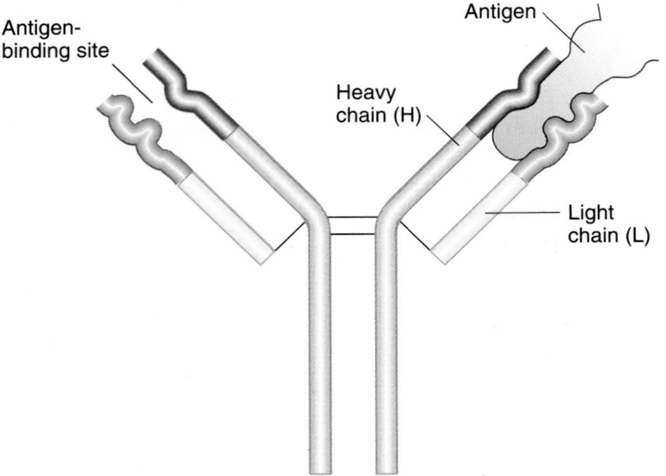

FIGURE 20-5

Structure of an IgG immunoglobulin molecule and its interaction with antigen. Each IgG antibody molecule consists of two light chains and two heavy chains, each of which has a constant and a variable region (shaded green). The chains are held together by covalent bonds (the black lines). Each antibody has two identical antigen-combining sites, one at the end of each branch of the "Y." In this drawing, one site has combined with a complementary-shaped antigen; the other is unbound.

chains, linked together by covalent bonds (Figure 20-5). Each of the chains contains a region that is *constant* (C)—that is, its amino-acid sequence is the same in all that animal's antibodies—and another region that is *variable* (V) from antibody to antibody. Differences in the variable regions account for the specificity an antibody has for its particular antigen.

Because antibody molecules contain two different types of polypeptides (light chains and heavy chains), a tremendous variety of antibodies are possible from a much smaller number of polypeptides. If, for example, there are 1000 different light chains and 1000 different heavy chains, one million (1000 × 1000) different antibody specificities—each with a distinct antigen binding site—can be formed.

ANTIBODY FORMATION

The ability to produce millions of different antibody types generated a difficult question: How could each cell contain the enormous amount of DNA needed for these millions of extra genes? The answer to this question lies in the fact that antibodies are composed of light chains and heavy chains with identical C portions but highly diverse V por-

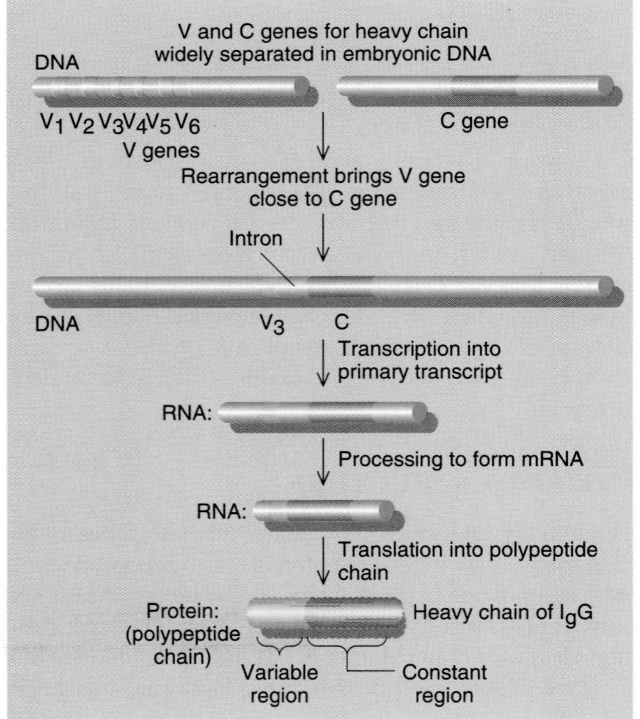

FIGURE 20-6

Antibody genes are formed by DNA rearrangement, which brings the C gene into close proximity with one of the many V genes (in this case, the V_3 DNA segment encoding part of a heavy chain) on the chromosome. The region between the C and V genes after rearrangement represents an intervening sequence (intron). Once DNA rearrangement has occurred, the combined C and V regions are transcribed into one mRNA molecule that directs the formation of a single, combined polypeptide. Such genetic rearrangement during B cell formation generates millions of different antibody-specific B cells.

◁ THE HUMAN PERSPECTIVE ▷

Treatment of Cancer with Immunotherapy

Cancer is traditionally treated in three ways: surgery to remove the malignant cells, radiation of a specific part of the body to kill the tumor, and chemotherapy to poison cancer cells wherever they might have spread throughout the body. Some forms of cancer, such as childhood leukemias and prostate cancer, have responded well to these treatments, while others, such as lung cancer and pancreatic cancer, have not. Furthermore, radiation and chemotherapy kills normal cells and can cause debilitating side effects, such as loss of white blood cells that would otherwise prevent infection. The patient who survives cancer may die of overwhelming infection because the cancer treatment incapacitated the immune system.

For decades, researchers have looked to the immune system for alternative treatments to help fight against cancer. Because cancer cells manufacture new proteins at their cell surfaces, they can be recognized by the immune system as foreign antigens. Although injecting these *tumor-specific antigens* into patients fails to stimulate the immune system to kill the tumor cells, a new more promising approach has been pioneered by Stephen Rosenberg of the National Institutes of Health.

Rosenberg observed that patients with terminal cancer sometimes (although rarely) completely recovered without treatment. Their cancer spontaneously disappeared, convincing him that the immune system has the *potential* to rid the body of its malignant cells. Rosenberg and his colleagues subsequently found that human tumors contained cytotoxic T cells that could specifically attack the tumor cells, although not enough to stem the tide of the growing malignancy. He removed these T cells (called tumor infiltrating lymphocytes) and cultured them to increase their number. Upon reinjecting these T cells into the patient, Rosenberg found that the protective cells accumulated within the tumor. In the first clinical trials on a handful of patients with advanced cancer, more than half of the treated patients experienced a marked reduction in their tumors, some remaining free of the malignancy for years. Moreover, the T cells had no adverse effects on normal tissues, a unique benefit over conventional treatments. To improve the tumor-specific killing power of the tumor infiltrating lymphocytes, Rosenburg has genetically engineered the cells so that they carry a gene for a toxic protein. If the technique proves effective, within just a few years we may be able to use such *immunotherapy* as a fourth weapon in the war against cancer.

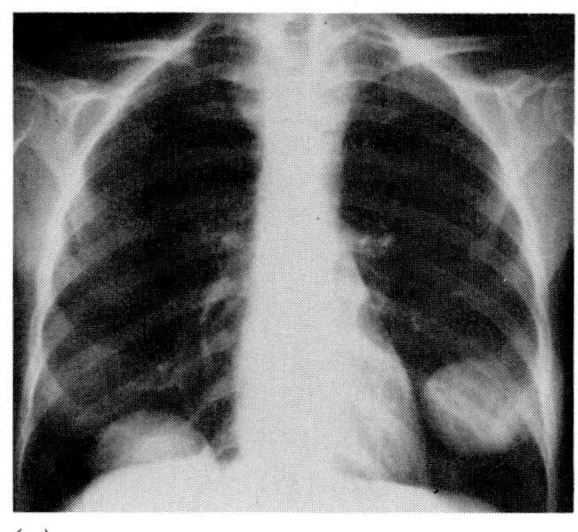

(a)

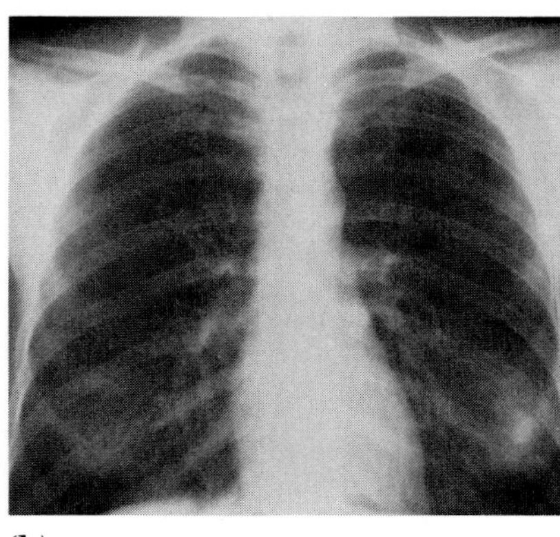

(b)

FIGURE 1

One promise of success against cancer utilizes tumor-infiltrating leukocytes (TILs). ***(a)*** X-ray showing a melanoma that has spread to both lungs. ***(b)*** Two months after treatment with the patient's own TILs, the tumor masses are greatly reduced in size.

tions. Each antibody chain is coded by two separate genes—a C gene and a V gene—that rearrange themselves to form one continuous "gene" encoding a single light or heavy chain. The DNA of a particular chromosome contains a single C gene and a large number of different V genes. During the rearrangement process, which occurrs while the B cell is maturing in the bone marrow, the single C gene is moved very close to one of the V genes, enabling a single messenger RNA (mRNA) to form from these combined genes (Figure 20-6). The mRNA is then translated into a polypeptide that contains both a C and a V amino-acid sequence. With this mechanism, a person can produce a huge number of different antibodies with a minimal amount of genetic information.

Clonal Selection and Antibody Formation

Once your B cells have formed, the cell population contains millions of different specificities. The intrusion of a particular antigen into the body activates only the B cells specific for that antigen. These cells then proliferate into a clone of antibody-producing cells. This mechanism of generating specific antibodies by "clonal selection" following antigen exposure consists of the following fundamental stages (Figure 20-7):

1. **Each B cell becomes committed to producing one type of antibody.** During embryonic development, DNA rearrangements in B cells commit each of these

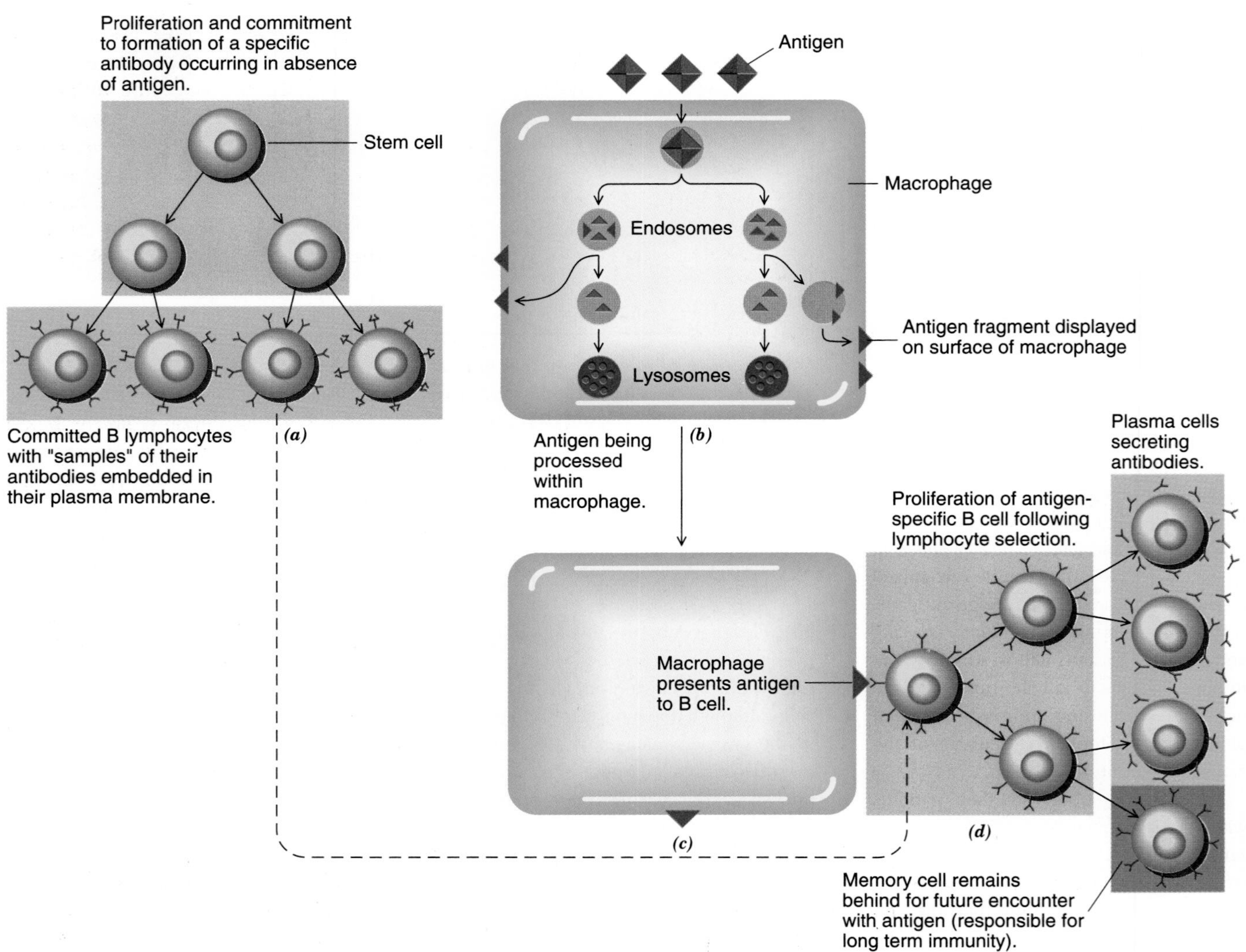

FIGURE 20-7

Generating antibodies by "clonal selection." ***(a)*** *Creating a pool of antigen-specific B cells.* Undifferentiated stem cells undergo proliferation, forming a population of B cells, each committed to the formation of a specific antibody. Antigen commitment occurs as a result of DNA rearrangement. Once committed, the B cell carries "sample" antibodies embedded in its plasma membrane. This phase of B cell development is antigen independent. ***(b)*** *Response to antigen intrusion.* Foreign materials are ingested by wandering macrophages and are packaged into cytoplasmic vesicles (endosomes). Some of the foreign material is completely digested by lysosomal enzymes (the red triangles), while other pieces remain undigested and are moved to the cell surface (the blue triangles), where they can be presented to lymphocytes. ***(c)*** *Activation of an antibody-producing cell.* The plasma cell that can bind to the processed antigen on the macrophage surface is activated. ***(d)*** *Proliferation and production of antibodies.* Activated plasma cells form a clone of antigen-specific cells, most of which become antibody-secreting plasma cells. A few cells, however, remain as antigen-specific memory cells, capable of responding very rapidly should the antigen be reencountered at a later date.

cells to the production of only one type of antibody molecule. Each of these cells embeds some of the antibody molecules in its plasma membrane, providing a recognition site for the corresponding antigen. In this way, the entire diversity of antibody-producing cells an individual will ever possess is already present within the lymphoid tissues *prior to stimulation by an antigen.* While most of these immune cells will never be called on to respond during a person's lifetime, the immune system can respond to virtually any type of antigen to which an individual might be exposed.

2. **Antibody production follows selection of B cells by antigen.** Antigens that enter the body trigger the production of complementary antibodies by *selecting* the appropriate antibody-producing cell. A virus entering the body, for example, is engulfed by a phagocytic macrophage. The virus is disassembled and the fragments moved to the surface of the macrophage, where they are "presented" to the one B cell with the corresponding antibodies displayed at their surface. The two cells bind at their antigen-antibody interface, which activates the B cell to divide and form a clone of cells specific for that antigen. Most of these activated cells differentiate into plasma cells and secrete antibody molecules that combine with the antigen. The time lag between antigen exposure and production of protective levels of antibodies averages between 10 and 14 days. It is during this lag period that disease develops.
3. **Immunologic memory provides long-term immunity.** A few B cells produce no antibodies, but remain in the lymphoid tissues as **memory cells,** which can respond rapidly at a later date if the antigen reappears in the body. Although plasma cells die off following removal of the antigenic stimulus, memory cells persist, often for years, providing *immunological recall.* When stimulated by the same antigen, the memory cells generate a protective immune response in hours rather than the days required for the original response. Consequently, the invader is eliminated before it can cause subsequent disease. The first encounter with the pathogen may produce overt disease but leaves survivors immune to subsequent infection. This type of protection is called **active immunity.**

Distinguishing Between Foreign and "Self"

The immune system must be able to distinguish between those substances that have entered the body from the external world and those that "belong" in the body. That is, the system must avoid launching an immune response against "self" antigens while retaining the ability to attack foreign ("non-self") substances. The immune system becomes *tolerant* of its own tissues by a poorly understood process of *clonal deletion*—killing or inactivating any B and T cells that can react against the body's own tissues. Immune tolerance occasionally fails, however, producing a state of *autoimmunity,* literally immunity to oneself. (See the Human Perspective: Disorders of the Human Immune System.)

IMMUNIZATION

Vaccines have become one of our most potent weapons against infectious disease. Most vaccines are modified forms of disease-causing agents, variants that have lost the ability to cause the disease but retain the same antigens as their dangerous counterparts. An immune system exposed to a vaccine, and often several "booster" doses, builds memory cells against a pathogenic agent without the danger of developing the disease. In this fashion, we can now safely promote active immunity against polio, measles, mumps, diphtheria, tetanus, rabies, and several other dangerous diseases.

Sometimes there is no time to wait for an immune system to produce a protective response. An attack of diphtheria or tetanus or the bite of a poisonous snake requires immediate immunity to neutralize the life-threatening antigens. In these cases, antibodies obtained from the blood of an immune individual (or animal) can be injected into the person who needs immediate protection, producing a temporary state of **passive immunity.** For example, the blood of William Haast, who, as director of the Miami Serpentarium, had been bitten by a variety of poisonous snakes, has been collected to provide passive immunity to individuals bitten by exotic vipers. In mammals, passive immunity is also acquired naturally, when antibodies travel across the placenta from a mother to her developing fetus. Such passive protection is augmented by the presence of antibodies in breast milk, providing breast-fed babies with more disease resistance—particularly to gastrointestinal infections—than bottle-fed babies.

EVOLUTION AND ADAPTATION: TYING IT TOGETHER

An organism's ability to defend itself against the constant onslaught of dangerous microorganisms clearly provides a selective advantage over those that are vulnerable to such attack. Natural selection favors the survival and reproduction of organisms that can protect themselves against these unseen intruders. A sophisticated immune system, however, is a relatively recent evolutionary arrival. Although invertebrates possess phagocytic cells and other mechanisms to resist infectious agents, their defenses lack a high degree of specificity. Immune systems capable of producing specific antibodies and cellular responses are only found in vertebrates. Even among vertebrates, however, there is a marked progression in complexity of the immune system, reaching its present peak in birds and mammals.

◁ THE HUMAN PERSPECTIVE ▷

Disorders of the Human Immune System

A number of disease conditions are associated with the immune system, including allergies, autoimmune diseases, and immunodeficiency disorders.

ALLERGIES

An allergic reaction is triggered when the immune system reacts to a foreign substance in a manner that injures the host. The most common allergic reactions are popularly called "hay fever," which occurs in response to inhaled pollen, dust, mold, or other airborne antigens. The allergic individual produces a special class of antibodies (called IgE antibodies) against one or more of these *allergens* (see Figure 1). When IgE antibodies react with its allergen, it promotes the release of highly active substances, such as *histamine,* that evoke the symptoms of allergy. These chemicals increase the diameter and permeability of blood vessels, leading to watery eyes, nasal congestion, and itching. Antihistamines are therapeutically effective because they block the effects of histamines. Another strategy to prevent the symptoms of allergy is to induce the production of "competing antibodies" by a program of *allergy desensitization.* Weekly injections of the allergen into the patient often stimulates the formation of "blocking antibody," which is simply normal IgG antibody that reacts with the allergen whenever it is naturally encountered, preventing the allergen from launching its IgE-mediated cascade of allergy-producing events.

The most severe type of allergic reaction is **anaphylaxis,** which can occur when an allergen, such as bee venom or penicillin, is introduced directly into the bloodstream of a highly sensitive person. In these cases, the allergen reacts with disseminated IgE molecules, immediately releasing histamine and other active chemicals throughout the body. Airways close and arteries dilate until the blood pressure plunges below that necessary to move the blood through the circulatory system. This is potentially fatal anaphylactic shock. Anaphylaxis can kill a person within 10 minutes of exposure—susceptible people frequently carry a syringe of epinephrine to counter these effects the moment they become evident. Because of anaphylaxis, the common honeybee kills more people in the United States than do sharks, bears, bulls, horses, and poisonous snakes.

Asthma is another form of allergy, one that causes airways to become severely constricted. The quickest relief is obtained by inhaling a mist of epinephrine (adrenaline), which reopens the airways. Antihistamines, however, fail to relieve asthma symptoms.

Some allergies are caused by T cells rather than IgE antibodies. The skin rash following exposure to poison oak or poison ivy is due to absorption of an oil from the plant, which triggers T cells to release substances that cause inflammation at the site of allergen attachment. Cortisone, an immune system suppressant, is often used to treat severe cases.

AUTOIMMUNE DISEASES

Occasionally, the immune system fails to recognize a self-antigen and attacks normal components of the body, resulting in **autoimmune diseases.** In all autoimmune diseases, victims literally reject their own tissues. To understand what happens when the body attacks itself, let's look at two autoimmune diseases, rheumatic fever and systemic lupus erythematosus (often simply called "lupus").

Rheumatic fever may occur following the body's recovery from infections by the streptococcal bacteria that cause "strep" throat. These bacteria possess antigens that are similar in structure to those found in some people's heart valves. Antibodies produced against the bacteria *cross react* with the heart tissue after the microbe has been eliminated, causing the heart valve damage that is responsible for rheumatic fever. Antibodies in people with systemic lupus erythematosus don't merely attack a single tissue, but make antibodies that react with several "self" molecules, including DNA. The reason for the production of these antibodies is unknown. The disease is extremely debilitating and may be fatal, usually as a result of kidney damage.

IMMUNODEFICIENCY DISORDERS

Until the early 1980s, diseases that seriously impaired a person's immune system

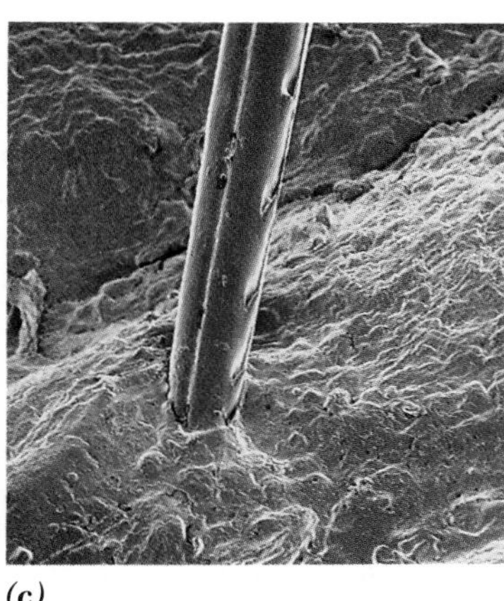

(a) *(b)* *(c)*

FIGURE 1

Common allergens. Scanning electron micrographs of selected materials to which many people are allergic: ragweed pollen ***(a)*** and the house dust mite ***(b)***, a minute arthropod that lives in dust. ***(c)*** The sting of a honeybee is painful but not particularly dangerous for most people. People who are allergic to bee venom, however, suffer allergic responses that can be fatal within 10 minutes of the sting.

were very rare and were usually a result of an inherited disorder. Acquired Immune Deficiency Syndrome (AIDS) has tragically altered that fact, and today immune deficiency is among the top 10 leading causes of death. The AIDS virus attacks and destroys the body's helper T cells, rendering the victim susceptible to cancers and infections that are rare in persons with intact immune systems (details of this disease are discussed in Chapter 25).

Blocking the transmission of the AIDS virus is the primary defense against this disease. The virus is spread by three avenues: sex, blood, and passive transmission from mother to fetus. The spread of AIDS would be greatly reduced if sexually active individuals followed safer sex practices (such as using a condom) and intravenous drug abusers stopped sharing needles or cleaned their needles with bleach, which kills the virus.

Little can be done to save the lives of someone who already has AIDS. Presently, the drug AZT (3'-azido-3'-deoxythymidine) is the primary treatment against AIDS, but other drugs have been approved or are under development. AZT inhibits replication of the AIDS virus (HIV), thereby delaying the appearance of disease symptoms. Controlling secondary infections, the most prominent of which is pneumocystis pneumonia (caused by the protist *Pneumocystis*), allows AIDS patients to live with less suffering.

Developing an AIDS vaccine is one of the major goals in the fight against this disease, but the nature of the virus itself creates some formidable obstacles to overcome before this goal can be achieved. One that most concerns researchers is the fact that some of the virus's antigens might stimulate an autoimmune reaction against T cells. If this is the case, then stimulating immunity to the virus may actually cause an AIDS-like disease (most of the dying T cells in AIDS victims are *not* infected with the virus). Such antigens would have to be identified and meticulously excluded from the vaccine. So far, the results of such efforts have been less than encouraging.

SYNOPSIS

Resistance against invading pathogens are divided into nonspecific and specific defenses. Nonspecific mechanisms include external body surfaces that prevent entry to pathogens, cells that either engulf or poison pathogens that have breached the body's surface, and molecules (such as complement and interferon) that attach to the surface of pathogens or infected cells. With the evolution of the immune system, an animal's ability to eliminate a foreign intruder became much more specific.

The immune system specifically recognizes and destroys foreign substances. The immune system is composed of cells that are scattered throughout the body and concentrated in lymphoid tissues, such as the lymph nodes and bone marrow. Immune responses may be mediated by either (1) soluble antibody molecules secreted by plasma cells that descend from B cells, or (2) cell-mediated immunity produced by T cells. Antibodies attach to specific targets that are present outside of host cells, leading to their destruction, while T cells interact with other cells. Immune responses are stimulated by foreign substances called antigens.

Antibodies contain binding sites that combine specifically with the antigen that stimulated production of that antibody. Antibodies are constructed of two light and two heavy polypeptide chains. Most of the antibody molecule is the same (regardless of the antigen against which it reacts) but has two regions that vary according to antigen specificity. These variable regions are the antigen combining sites. Both heavy and light antibody chains are encoded by genes that form as a result of DNA rearrangement.

Antibodies are formed by clonal selection. Lymphoid tissues contain a population of B cells, each of which is committed to forming a particular antibody. Samples of these antibodies are displayed on the B cell's plasma membrane. When an antigen binds to a B cell's displayed antibody, it stimulates the cell to proliferate into a clone of cells, most of which produce that antibody. The rest become memory cells that can mount a rapid response if the antigen is reencountered at a later time. These memory cells are responsible for long-term immunity. Their production can be induced by natural infection or vaccination, using a safe form of a pathogen's antigens.

T cells are programmed in the thymus. Cytotoxic T cells contact and directly kill their target cells. Other T cells release substances that attract immune cells into the region where protection is needed. Helper and suppressor T lymphocytes have a regulatory function; they specifically activate or inhibit other lymphocytes to enhance protection or to reduce the likelihood of immune-mediated damage to the host.

The immune system distinguishes self from nonself. The body suppresses the formation of T cells and antibodies targeted against the body's own cells and tissues, thereby tolerating "self" antigens. When this process breaks down, the resulting autoantibodies cause serious disease.

Review Questions

1. Distinguish between B cells and T cells, plasma cells and memory cells; complement and antibodies; C genes and V genes; light chains and heavy chains; active and passive immunity.
2. Describe the production of antibody by clonal selection.
3. Describe the steps that occur between the time a cold virus penetrates the nasal epithelium and the time it is eliminated by the immune system.
4. How is it that you have only one C gene per haploid set of chromosomes for a light antibody chain, yet you can synthesize hundreds of thousands of different antibody molecules?
5. The immune system is sometimes called "a double-edged sword." Although we cannot survive without it, what are some hazards of possessing a fully armed immune system?

Critical Thinking Questions

1. When AIDS first appeared, some clinicians speculated that the condition was due to the male homosexual practice of inhaling amyl nitrate, a stimulant of the heart. What evidence might you seek either to support or to refute this suggestion about the cause of AIDS?

2. Why is it important that the immune system operates with such a high degree of specificity? How is this characteristic illustrated by disorders that affect the immune system, such as rheumatic fever?

3. One might argue that it is wasteful for B cells to be produced by a process that is independent of antigen, since most of these B cells will never be needed. Defend or oppose this argument.

4. An unusual characteristic of AIDS is its tendency to change its antigens during the course of infection (mutations alter the amino-acid sequence of the proteins making up its outer coat). Why would this characteristic hamper attempts to produce an AIDS vaccine?

5. If you were to see your doctor after stepping on a rusty nail, you might receive two very different types of shots, one to provide immediate (but temporary) *passive immunity* against tetanus (a wound infection that paralyzes the body with a powerful toxin), the other to provide long-term *active immunity*. What types of materials would be in these different types of shots?

CHAPTER

21

Reproduction and Development

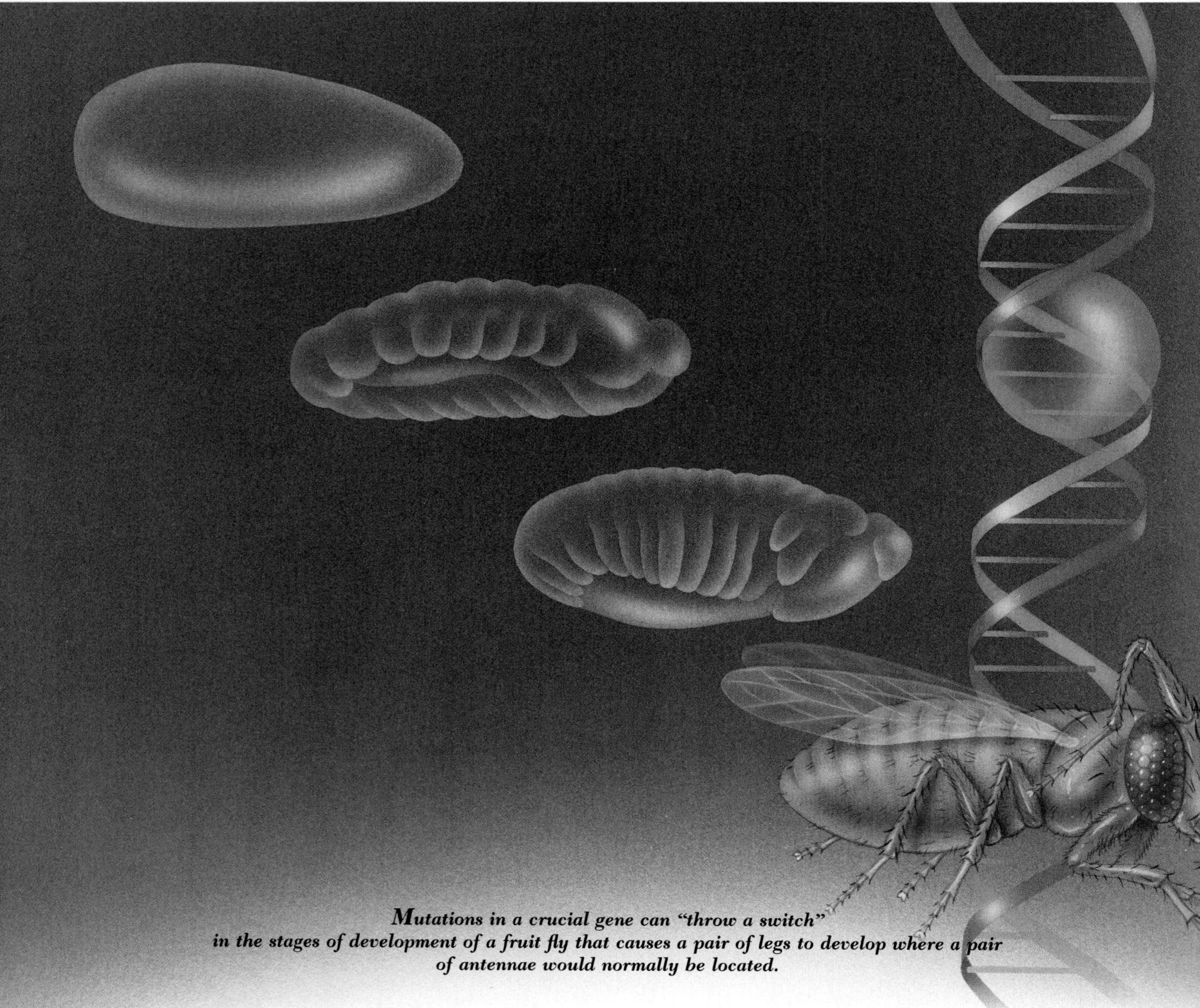

Mutations in a crucial gene can "throw a switch" in the stages of development of a fruit fly that causes a pair of legs to develop where a pair of antennae would normally be located.

STEPS TO DISCOVERY
Genes that Control Development

During the 1940s, Edward B. Lewis, a geneticist at the California Institute of Technology, studied a mutant fruit fly with an abnormal body organization. An insect is composed of a head, thorax, and abdomen, each part containing a defined number of segments. The last segment of the thorax of a fruit fly is usually wingless, but Lewis's mutant fly had a second pair of wings on this segment; thus, the mutant was named *bithorax.*

In following years, other mutant fruit flies were isolated that showed even more profound disturbances in body organization. For example, the mutant *antennapedia*, studied by John Postlethwait and Howard Schneiderman of Case Western Reserve University in the late 1960s, has a pair of legs growing out of its head in the place where antennae are normally found. Genes such as these that affect the spatial arrangement of the body parts are called **homeotic genes.** While humans are not known to suffer such drastic homeotic mutations as *bithorax* and *antennapedia,* there are many examples of serious developmental malformations that could be the result of mutations in homeotic genes. In addition, human embryos with drastic developmental defects tend to abort spontaneously, so the existence of mutant homeotic alleles might easily go undetected.

Homeotic genes are thought to play a role in the basic process by which each part of an embryo becomes committed to developing along a particular pathway—toward forming a leg rather than an antenna, for example. One way homeotic genes might exert such profound influence on the course of development is by acting as a type of "master" gene. As such, homeotic genes would control the transcriptional activity of other genes, whose products (such as collagen or the contractile proteins of muscle) actually form the various tissues. The antennapedia gene, for example, might code for a protein that normally switches on the genes required for antenna formation in the appropriate cells of the developing head of a fruit fly. If the antennapedia gene becomes defective, a different cluster of genes may be switched on, and the cells of that part of the body differentiate into a leg instead of an antenna.

In 1983, Walter Gehring, a Swiss biologist, discovered that a number of homeotic genes in the fruit fly contained a common sequence of about 180 nucleotides; this sequence was named the **homeobox.** Once the homeobox DNA from the fruit fly had been isolated, Gehring and others were able to search the DNAs of other organisms to see if they contained similar DNA segments. The homeobox sequence was soon found to exist within the DNA of many different animals, from worms to humans; in fact, it was even present in plants. This finding suggested that similar types of genetic processes take place during the development of very different types of organisms, but it left an important question unanswered: What was the function of the homeobox?

Deciphering the amino acid sequence encoded by a gene is a relatively simple matter, once the nucleotide sequence of the DNA has been determined. When the amino acid sequence of the homeobox was deciphered and compared to the sequence of other known proteins, it was found to be very similar to a gene regulatory protein found in yeast that was known to bind to DNA. This correlation suggested that homeotic genes encoded DNA-binding proteins. By binding to a specific portion of a particular chromosome, the product of a homeotic gene could activate or repress transcription of nearby genes, much like a repressor protein in bacterial operons. In this way, homeotic genes might control the course of development. This concept was confirmed in 1988, when Patrick O'Farrell and his colleagues at the University of California demonstrated that two of the homeobox-containing genes in the fruit fly actually bound to DNA and altered the rate of transcription of nearby genes.

In the past few years, a number of investigators studying homeotic genes have turned their attention from fruit flies to mice. In doing so, they have invited speculation on the role of these genes in human development. In 1991, for example, Osamu Chisaka and Mario Capecchi of the University of Utah produced mice that carried a genetically engineered version of a homeotic gene in place of the normal gene. The mice that were born with this altered homeotic gene exhibited a variety of severe abnormalities, including deformations of the throat and heart. Interestingly, a similar complex of abnormalities occurs in DiGeorge's syndrome, a rare human disorder that usually causes the death of the affected infant within the first few months of life. It is possible that this human condition is a result of a mutation in a homeotic gene whose expression is required during human development.

Of all animals, humans are the only species apparently aware of their own mortality. We all know that we will die, but we also know that our lineage can continue by having children. Consider for a moment that the genes we possess have been passed down, generation to generation, from distant ancestors. A living descendent of Andrew Jackson, for example, may have the same genetic information for Jackson's distinctive eyebrows. Some of the information encoded in our genes may be billions of years old, such as that which directs the formation of semipermeable cell membranes or oxygen-dependent respiratory chains. Some of these genes have been preserved—with changes—for millions of generations.

The transmission of genetic information from one generation to the next is a crucial part of **reproduction**—the process by which new offspring are generated. All organisms, from bacteria to redwood trees to humans, have a finite lifespan. Reproduction is the process by which the species is perpetuated.

▼ ▼ ▼

REPRODUCTION: ASEXUAL VERSUS SEXUAL

Some animals can reproduce by **asexual reproduction;** that is, they produce more of themselves without the participation of a mate, gametes, or fertilization. Two examples of asexual reproduction are shown in Figure 21-1. During **fission** (Figure 21-1*a*), an animal splits into two or more parts, each of which becomes a complete individual. Fission is common among various groups of worms. Some animals reproduce asexually by **budding,** whereby offspring develop as an outgrowth of some part of the parent. Budding is common in sponges, hydras, and corals (Figure 21-1*b*). Since neither meiosis nor fertilization is part of the process, asexual reproduction produces offspring that are genetically identical to the parent. Most animals, however, produce offspring by **sexual reproduction,** which requires

1. the formation of two different types of haploid gametes—eggs and sperm—by a process that includes meiosis, and
2. the union of a single egg and sperm at fertilization to form a zygote.

(*a*)

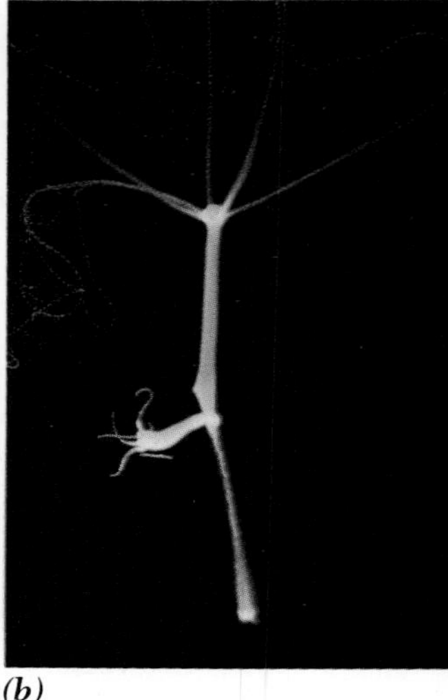
(*b*)

FIGURE 21-1
Types of asexual reproduction in animals. ***(a)*** During fission, this sea anemone splits into two individuals. ***(b)*** In this hydra, a new individual forms as a bud that grows out of the body wall.

Gametes are formed in reproductive organs called **gonads.** Sperm are produced in the male gonads (testes), whereas eggs are produced in the female gonads (ovaries).

THE ADVANTAGE OF SEXUAL REPRODUCTION

Why do organisms engage in sexual reproduction? Why don't all organisms simply reproduce asexually, forming an exact, yet younger, version of themselves? After all, asexual reproduction has some obvious adaptive advantages. First, it generates progeny without the greater investment of energy and resources associated with gamete production, mate seeking, and fertilization. Second, asexual reproduction is very efficient; one individual, living by itself, can generate large numbers of offspring very rapidly.

But sexual reproduction is the predominant mode of reproduction among animals. Even among animal species that reproduce asexually, most do not do so exclusively. Rather, individuals in a population typically reproduce asexually for a period of time; then, in response to some environmental trigger, such as food depletion or overcrowding, these same animals switch to a sexual mode of reproduction. Clearly, there must be some selective disadvantage to a total reliance on asexual reproduction. That disadvantage is presumed to be *genetic monotony;* that is, generation after generation, progeny will be genetically the same. In contrast, sexual reproduction combines traits from two genetically distinct parents in a single individual so that each offspring acquires a unique genetic mix. In addition to gene mixing during fertilization, variation is boosted even further by independent assortment and crossing over during meiosis (Chapter 7).

HUMAN REPRODUCTIVE SYSTEMS

Both the male and female human reproductive systems consist of a pair of gonads in which gametes are formed, and a reproductive tract, possessing various accessory functions that are discussed later in the chapter. The vastly different structures of the human male and female reproductive systems reflect the difference in their roles during reproduction. The male produces huge numbers (trillions during a lifetime) of tiny motile sperm (or *spermatozoa*) and delivers these cells into the female's reproductive tract. In contrast, the female produces a small number of eggs (or *ova*). The female is responsible for more than just gamete production, however. She also provides an environment in which the eggs can be fertilized and the zygote can develop into a fully formed human infant.

The reproductive activities of both sexes are regulated by a battery of hormones. These hormones are responsible for stimulating the initial development of the reproductive system in the embryo, causing its maturation during puberty, and maintaining its day-to-day operation throughout the reproductive years.

THE MALE REPRODUCTIVE SYSTEM

In human males, the paired testes produce sperm throughout adult life at the average rate of 30 million sperm per day. The male reproductive tract is primarily a hollow conduit, equipped with accessory glands, that moves sperm out of the body and into the reproductive tract of the female.

Male External Genitals and Reproductive Tract

The male external genitals (Figure 21-2) consist of the penis and a sac, called the **scrotum,** which houses the testes, or testicles. A passageway, called the **urethra,** runs through the length of the penis, conveying both sperm and urine, although not simultaneously. In order to deposit sperm into the vagina, the penis must have penetrating

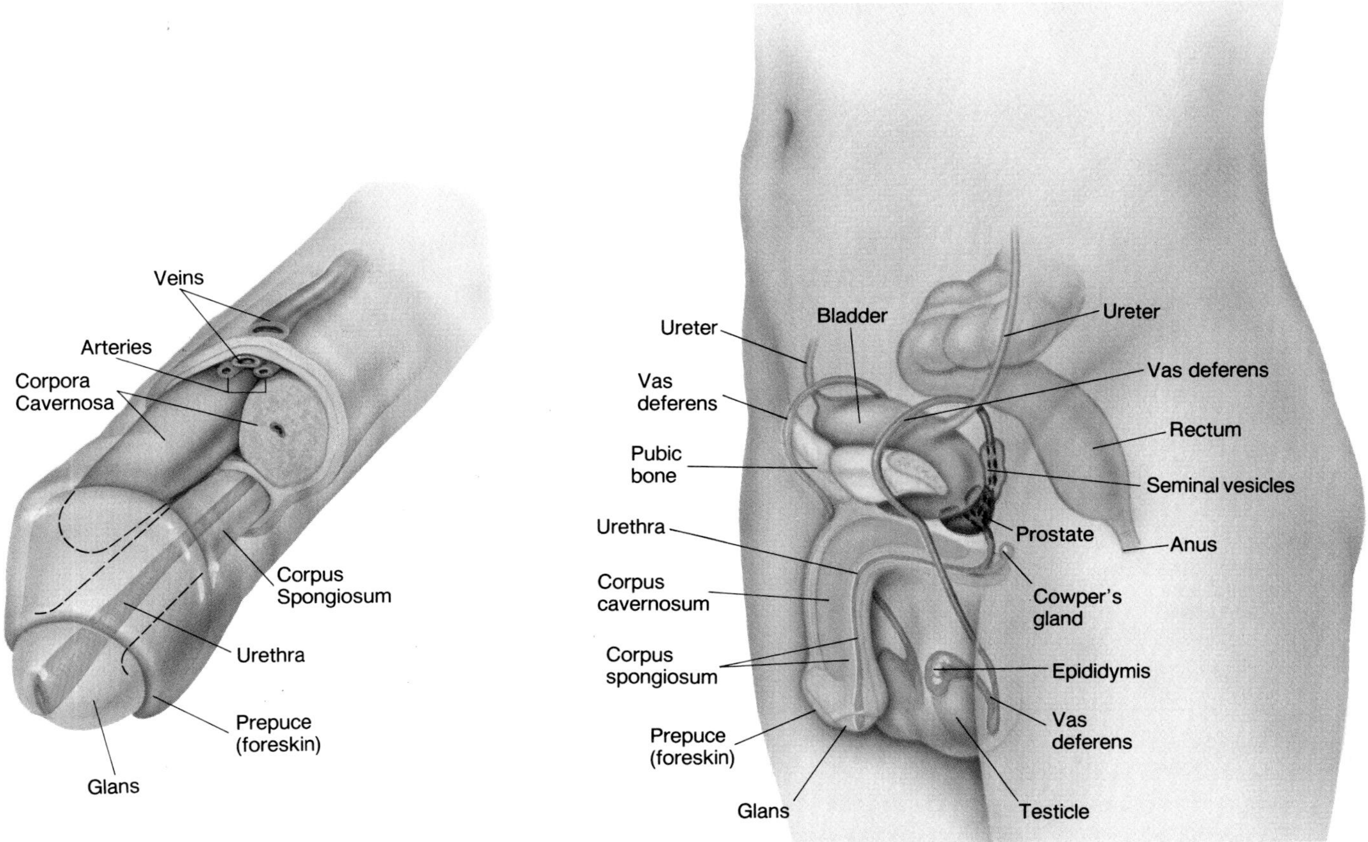

FIGURE 21-2

The male human reproductive system. When the arterioles of the penis become dilated, and the erectile tissues (composed of the *corpus spongiosum* and paired *corpora cavernosa*) become engorged with blood, the penis becomes erect.

capacity, which is accomplished when the organ is engorged with blood. The interior of the penis contains three long cylinders of spongy *erectile tissue.* During periods of sexual excitement, impulses travel along parasympathetic nerves from the brain to the smooth muscle cells that line the arterioles of the penis. Relaxation of the smooth muscles leads to vasodilation of the arterioles, causing blood to engorge the spaces of the erectile tissue, which expands as it fills. As a result, the penis enlarges and becomes rigid enough for vaginal penetration.

The Formation of Sperm

A cross section through a testis reveals that the structure is composed almost entirely of tubular elements, called **seminiferous tubules** (Figure 21-3), whose combined length is equivalent to about four times the length of a football field. Within these tubules, **spermatogenesis,** the formation of the male gametes, takes place. Between the tubules are scattered clusters of *interstitial cells*—the endocrine cells responsible for the production of testosterone, the male sex hormone.

Spermatogenesis takes place in stages, which are revealed in an examination of a cross section of a seminiferous tubule (Figure 21-3). Cells enter spermatogenesis at the outer edge of the tubule and are gradually moved toward its inner lumen as the process continues. The outer edge of the tubule contains a self-perpetuating layer of **spermatogonia**—germ cells that have not yet begun meiosis. In the first step of spermatogenesis, a spermatogonium grows in size and enters meiosis, forming a **primary**

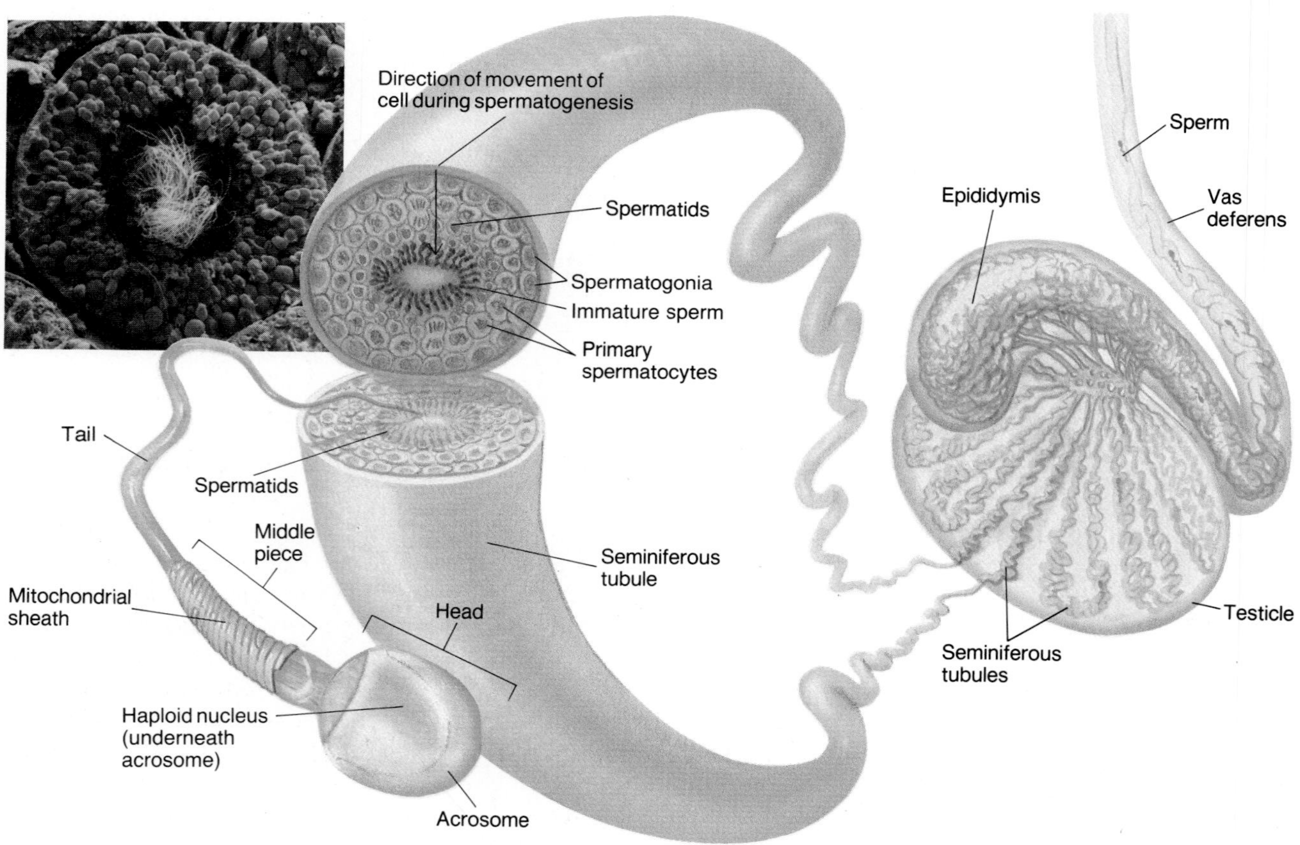

FIGURE 21-3

Spermatogenesis. The cross section through a seminiferous tubule shows cells in different stages of spermatogenesis. The outer layer contains spermatogonia that have not yet begun spermatogenesis. The next layer contains primary spermatocytes that have entered meiosis. Further inward are the spermatids—the products of meiosis—and within the lumen are nearly differentiated sperm. The structure of a single sperm is also shown, illustrating several major features such as the head, the mitochondria-rich middle piece, and the tail. The inset shows a scanning electron micrograph of a cross section through a seminiferous tubule; tails of the sperm are clearly seen extending into the tubule's lumen.

spermatocyte. Each primary spermatocyte undergoes two meiotic divisions to form a total of four haploid **spermatids.** Each spermatid is subsequently transformed into a sperm cell, one of the most specialized cells in the body.

The cells in the central lumen of a seminiferous tubule (inset, Figure 21-3) are not yet able to fertilize an egg. Sperm complete their maturation only after they are pushed out of the seminiferous tubule and into the **epididymis,** a coiled tubule attached to each testis. From the epididymis, sperm then move into the **vas deferens,** a tubule that transports the male gametes to the urethra.

Before their release from the penis, spermatozoa are mixed with secretions from several glands (see Figure 21-2) to form **semen,** the sperm-containing fluid that is expelled from the body as the result of strong muscle contractions that occur during **ejaculation.** The ejaculatory fluid nourishes and protects the sperm; provides a liquid medium needed for sperm motility; and, because of its alkaline pH, temporarily neutralizes vaginal acidity that might otherwise impair sperm-cell activity.

Form and Function of Sperm

A sperm is a compact, streamlined cell (Figure 21-3) whose function it is to move up the female reproductive tract and fuse with an egg. The entire structure of a sperm can be correlated with this function. As it differentiates, the sperm loses nearly all of its cytoplasmic and nuclear fluid, which accounts for more than 90 percent of the volume of the spermatid. The head of a sperm consists of two parts: the nucleus and the acrosome. The nucleus of a sperm is the ultimate in chromosome compactness; the chromosomal material is condensed to a virtual crystalline state. The **acrosome,** which forms a cap over the nucleus, contains digestive enzymes that are released as the sperm "digests" its way through the protective layers that surround an egg. The middle portion of the sperm contains tightly packed mitochondria which, by virtue of the ATP they produce, will power the sperm's movements. The tail of a sperm contains a flagellum that whips against the surrounding fluid, driving the sperm toward and into the egg. Sperm are launched like self-propelled torpedoes with a limited range. If they do not reach the egg within the allotted time (24 to 48 hours in humans), the sperm exhaust their fuel supply and die.

Hormonal Control of Male Reproductive Function

Proper functioning of the testes is maintained by the presence of two hormones that are secreted by the anterior pituitary: **follicle-stimulating hormone (FSH) and luteinizing hormone (LH).** These hormones were originally named because of their effects on the female reproductive system, as we will discuss later in the chapter. The fact that the identical hormones are present in both genders but elicit different responses in males and females illustrates an important principle of endocrine function: The nature of the *target cell* determines the type of response, while the *hormone* itself acts simply as a stimulus or trigger.

FSH and LH are described as *gonadotropins* because they stimulate the activities of the gonads. LH acts primarily on the interstitial cells of the testes, stimulating the production and secretion of testosterone. FSH is required for spermatogenesis. The secretion of both LH and FSH by the pituitary is, in turn, regulated by the level of **gonadotropin-releasing hormone (GnRH)** that is secreted by the hypothalamus.

While the gonadotropins act on the gonads, **testosterone**—the hormone produced by the gonads—acts on the other tissues associated with male sexuality. Testosterone secretion stimulates the differentiation of the male reproductive tract in the embryo, the descent of the testes into the scrotum, the further development of the reproductive tract and penis during puberty, and the development of male *secondary sex characteristics,* including the growth of a beard and chest hair, enlargement of the larynx, and increased muscle mass. Testosterone also plays a role in the development and maintenance of the male *libido,* or sexual desire.

THE FEMALE REPRODUCTIVE SYSTEM

A woman's body performs many activities that are essential to reproductive success: Her ovaries are the sites of **oogenesis,** or formation of ova, and her reproductive tract nourishes, houses, and protects the developing fetus. Even after birth, a woman's body provides breast milk, which nourishes the developing infant. As in the male, each of these aspects of reproductive activity are under the complex control of a number of hormones.

Female External Genitals and Reproductive Tract

The female's external genitals (Figure 21-4) are collectively known as the **vulva.** The most prominent features of the vulva are an outer and inner pair of lips, the **labia majora** and **labia minora,** respectively. The **clitoris** protrudes from the point where the labia minora merge. Rich in sensory neurons and erectile tissue, the clitoris resembles the penis in its sexual sensitivity and erectile capacity. Unlike the penis, however, the clitoris has no urinary or ejaculatory function; its sole function is to receive sexual stimulation, which it transmits to the central nervous system. Thus, the clitoris is the only human structure dedicated exclusively to the enhancement of sexual pleasure.

Also located between the labia minora is the opening to the **vagina,** an elastic channel that receives the sperm that are ejaculated from the penis and forms the birth canal through which an infant leaves its mother's body during childbirth. The vagina leads to the remainder of the female reproductive tract (Figure 21-4), which consists of the uterus and paired oviducts. The vagina is separated from

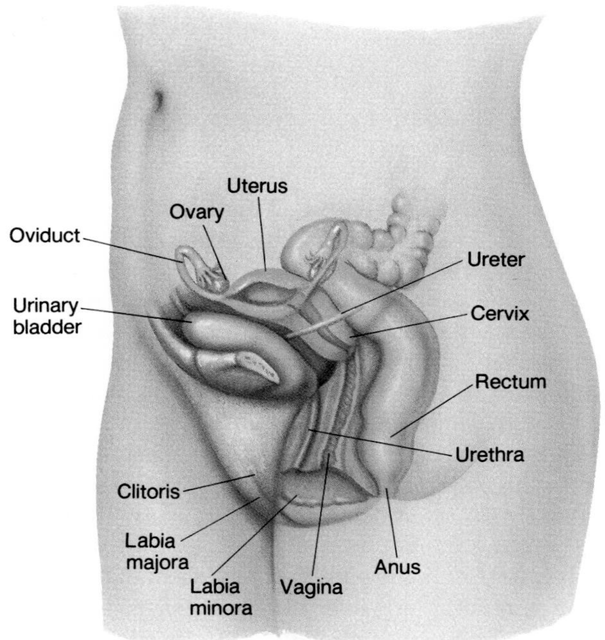

FIGURE 21-4

The female human reproductive system. Eggs are formed in the ovary and swept into the oviducts. Meanwhile, sperm enter the body in the vagina and travel through an opening in the cervix up to the oviducts, where fertilization takes place. The resulting embryo passes down the oviduct and into the uterus, where it is implanted into the uterine wall and develops.

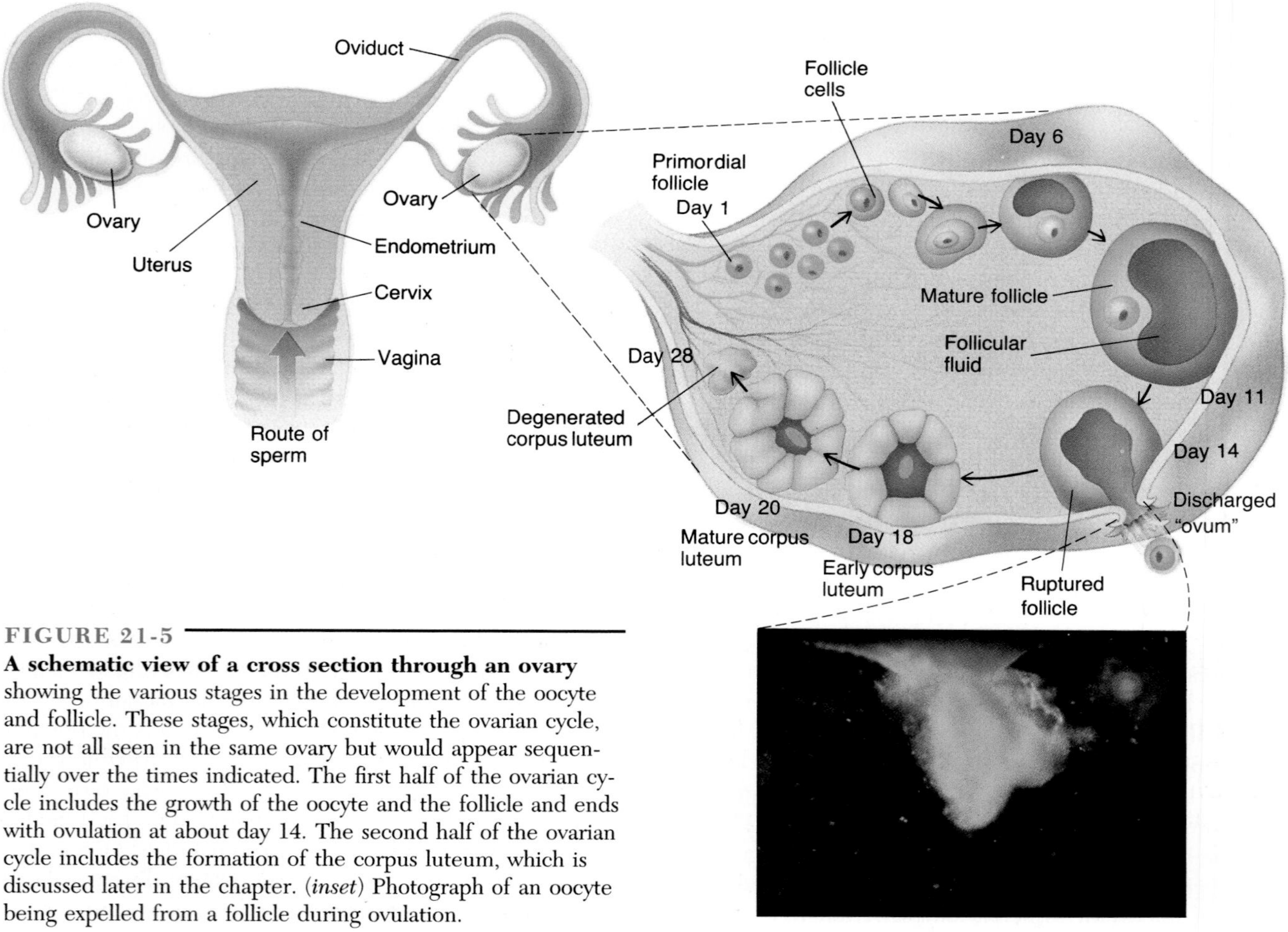

FIGURE 21-5

A schematic view of a cross section through an ovary showing the various stages in the development of the oocyte and follicle. These stages, which constitute the ovarian cycle, are not all seen in the same ovary but would appear sequentially over the times indicated. The first half of the ovarian cycle includes the growth of the oocyte and the follicle and ends with ovulation at about day 14. The second half of the ovarian cycle includes the formation of the corpus luteum, which is discussed later in the chapter. (*inset*) Photograph of an oocyte being expelled from a follicle during ovulation.

the uterus by the **cervix,** which contains the opening through which sperm must pass on their way to fertilize an egg. The **uterus** (or *womb*) is a pear-shaped, thick-walled chamber that houses the embryo and fetus during pregnancy. The **oviducts** (or *fallopian tubes*) emerge from the uterus and serve as the sites where sperm and egg become united during fertilization.

The Formation of Oocytes

A cross section through an ovary (Figure 21-5) shows a very different appearance than that of a testis. There are no tubules in an ovary. Instead, the **oocytes**—the germ cells that have the potential to form ova—are housed within spherical compartments called **follicles** which are scattered throughout the tissue of the ovary. Each follicle contains a single oocyte, surrounded by one or more layers of **follicle cells,** which provide the materials that support the growth and differentiation of the enclosed germ cell. Unlike the male gonad, the ovary of a woman contains no **oogonia,** or premeiotic germ cells. All of the oogonia that are produced during embryonic development have already entered prophase of meiosis I (see Figure 7-8) by the time of birth. Oocytes will remain suspended in meiotic prophase for years, some for several decades. This may be the reason for the increased incidence of chromosomal abnormalities in pregnancies of older women.

The vast majority of the follicles of the adult ovary are small and consist of an undifferentiated oocyte surrounded by a single layer of follicle cells. These *primordial follicles* are storehouses of oocytes that will provide the ova produced during the reproductive life of the woman. A few oocytes will undergo oogenesis during each monthly **ovarian cycle.** During oogenesis, which occurs in the first half of the ovarian cycle, the oocyte increases in size from about 25 to 100 micrometers in diameter as it becomes packed with nutrients that will be utilized by the embryo during the first days following fertilization.

The changes in the oocyte that occur during oogenesis are accompanied by a dramatic transformation of the entire follicle. By the time it completes its growth, a *mature follicle* is large enough to be seen as an obvious bulge at the surface of the ovary. Although several follicles typically undergo maturation within each ovarian cycle, one usually outpaces the others. Ultimately, the wall of this follicle suddenly ruptures, and the enclosed oocyte is released from the ovary (inset, Figure 21-5). This is **ovulation.** The ovulated oocyte is swept into the broad, funnel-shaped opening of the oviduct, where fertilization occurs.

Meiotic Divisions of the Oocyte. Unlike in spermatogenesis, where meiosis occurs before the differentiation of the spermatozoon, meiotic divisions in the female occur after the entire process of growth and differentiation of the ovum has essentially been completed (Figure 21-6). The first meiotic division is completed in the oocyte just before ovulation. Unlike in spermatogenesis, where meiosis I produces two equal-sized cells, the meiotic division in the female produces one large cell and one tiny cell, called a **polar body,** which eventually deteriorates. The second meiotic division begins while the oocyte is in the oviduct, but it does not run to completion. Instead, the meiotic process stops after the chromosomes have lined up for metaphase II. At this stage, the human egg is fertilized. The second meiotic division is completed after fertilization. Meiosis II is also a highly unequal cell division, producing a single, large cell that goes on to develop into a new individual, and another polar body that disintegrates. Thus, unlike meiosis in the male, which produces four equal-sized gametes, meiosis in the female produces only one large ovum, which conserves all the cytoplasmic material in one package.

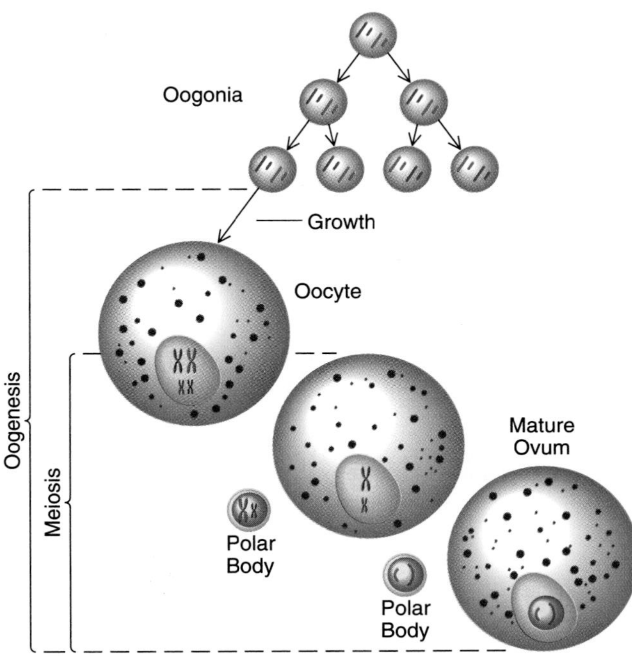

FIGURE 21-6
Gamete formation in the female. Oogonia are premeiotic cells that produce more of their own kind by mitosis. All of these oogonia enter oogenesis (becoming oocytes) during fetal development. In the adults a few oocytes undergo growth and differentiation each month. Meiosis in the female produces only one viable egg containing the nutrient reserves (yolk) that support early embryonic development. The other products of meiosis are polar bodies that disintegrate.

Hormonal Control of Female Reproductive Function

As in the male, the development, maturation, and function of the female reproductive system is under hormonal control. The maturation of the female system during puberty is thought to be initiated within the brain by the production

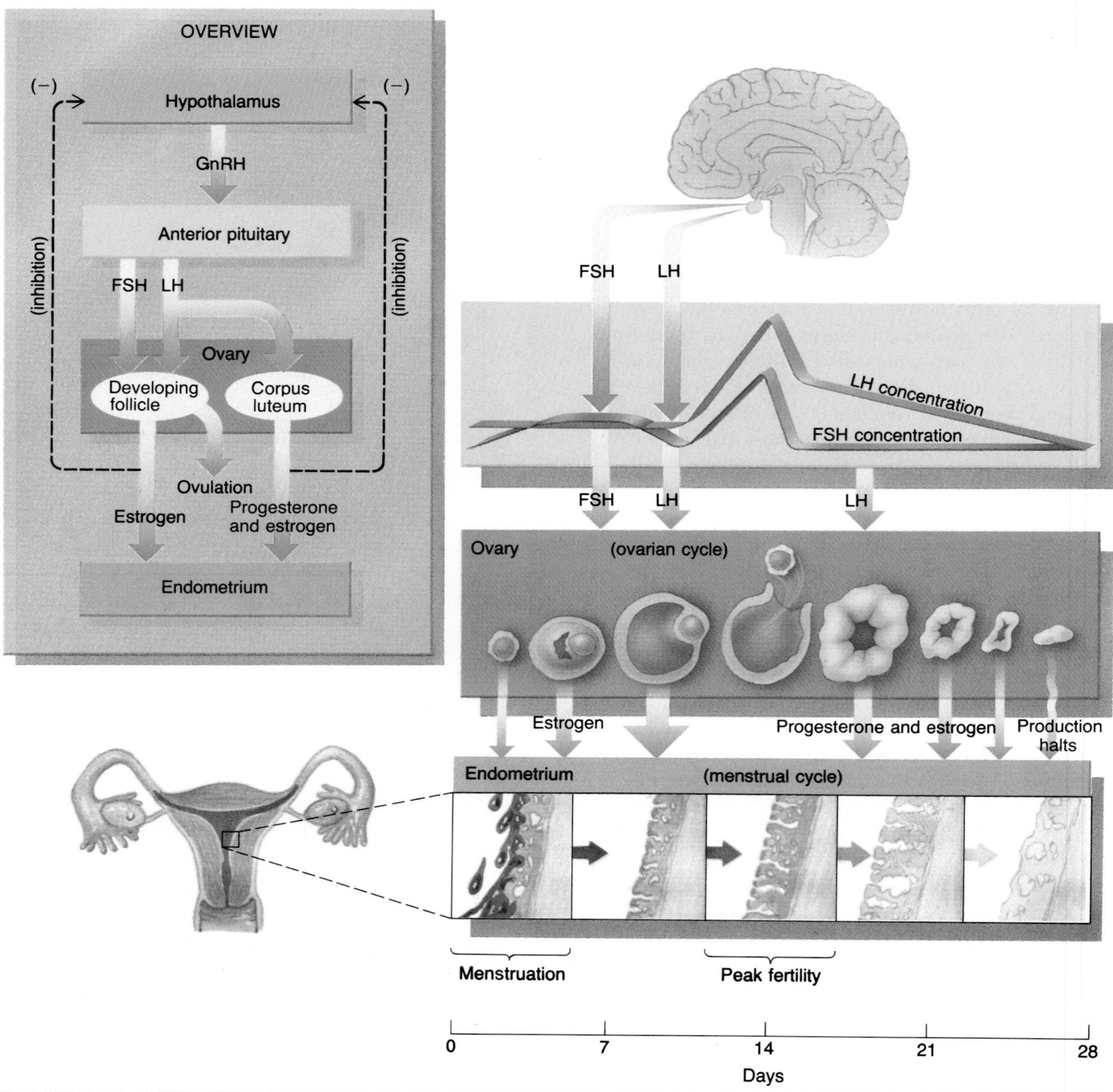

FIGURE 21-7

Synchronizing the ovarian and menstural (uterine) cycles. FSH is released from the anterior pituitary and promotes development of a follicle and the oocyte it contains. The maturing follicle secretes estrogen, which stimulates endometrium development in the uterus. Presumably, when estrogen reaches a critical concentration, it stimulates the hypothalamus and anterior pituitary to flood the body with LH. This surge of LH triggers ovulation. LH also transforms the follicle into a corpus luteum. Progesterone (and some estrogen) produced by the corpus luteum maintain the enriched endometrium, so that it is receptive to implantation. Progesterone from the corpus luteum also inhibits GnRH release, so LH secretion stops, and the gonadotropin declines over the next 14 days. In the absence of implantation, the corpus luteum deteriorates (due to declining LH concentrations) until progesterone and estrogen production halts. Without these two hormones, the extra endometrial tissue cannot be maintained and is sloughed as menstrual fluid. Menstruation signals the beginning of a new cycle. With the disappearance of progesterone, inhibition of GnRH is relaxed and FSH is again released and another follicle begins to mature.

of the hypothalamic hormone GnRH, which stimulates the secretion of LH and FSH by the anterior pituitary. In the female, the gonadotropins LH and FSH act cooperatively on the follicle cells of the ovary, stimulating them to produce the primary female sex hormone, estrogen, which has numerous target tissues. Increased estrogen levels at puberty stimulate maturation of external genitals and the development of secondary sex characteristics, such as the enlargement of breasts and the growth of hair in the armpits and pubic regions. Internally, elevated estrogen levels induce the maturation of the tissues of the uterus, enabling this organ to house and nurture a developing fetus. Puberty is also the time when the ovary begins to produce mature ova on a cyclical basis, a process that will continue until *menopause*, the cessation of a woman's reproductive cycle.

The complex relationship among changing levels of hormones, the ovarian cycle, and the menstrual cycle is shown in Figure 21-7. The first phase of the ovarian cycle is characterized by the growth and differentiation of an oocyte. This maturation process is stimulated by both LH and FSH as well as by estrogen, which is produced within the follicle itself. As the growth of one or a few ovarian follicles nears completion, there is a marked surge in the pituitary's secretion of LH, which triggers ovulation.

The Menstrual (Uterine) Cycle

As the follicles are growing during each ovarian cycle, the uterus undergoes cyclical changes in form which are related to its function as a potential residence for the embryo. These dramatic changes in the uterus constitute the **menstrual cycle,** which takes an average of 28 days to complete (Figure 21-7). The first day of the menstrual cycle (which is defined as the day on which menstruation begins) is characterized by the flow of blood and discarded tissue from the uterus through the vagina. Menstruation takes place when the body "becomes aware" chemically that no fertilization or pregnancy has occurred following the last ovulation. As a result, the interior lining of the uterus, which had been prepared to receive a developing embryo, is broken down and rebuilt. The destruction of the bulk of the uterine wall is initiated by the constriction of arterioles, thereby cutting off the blood supply to the thickened *endometrium*—the inner lining of the uterus. The dead and dying tissue is then expelled from the uterus during menstruation by the contraction of the *myometrium* (uterine muscles).

After the period of menstruation is over, the rebuilding of the uterus begins, preparing the endometrium for the possible reception of a fertilized egg in the upcoming ovarian cycle. The first phase of uterine reconstruction produces the dramatic thickening of the inner glandular endometrium (Figure 21-7), induced by rising levels of estrogen, which are produced by the follicle.

Following ovulation, on about day 14 of the menstrual cycle, the ruptured follicle remaining in the ovary is rapidly converted into a yellowish, glandular structure, called the **corpus luteum** (see Figure 21-5), which secretes estrogen and large quantities of a second ovarian steroid hormone, progesterone. The increasing levels of estrogen and progesterone in the second half of the menstrual cycle act on the uterus to further its growth and development. If the ovulated ovum is fertilized, the resulting embryo will implant itself in the uterus by the eighth day following fertilization. The implanted embryo produces a gonadotropic hormone, called *human chorionic gonadotropin (HCG)*, that acts on the ovary to maintain the activity of the corpus luteum. Biochemical tests for detecting HCG in a woman's blood or urine provide a reliable means of determining pregnancy.

If implantation occurs, the sustained corpus luteum continues to produce estrogen and progesterone, which maintain the endometrium where the embryo is developing. Since progesterone also inhibits the release of FSH, follicles cannot mature during pregnancy, making it highly unlikely that ovulation will occur while a woman is pregnant. Some birth control pills prevent ovulation by artificially increasing the concentrations of these hormones. If the ovulated ovum is not fertilized, there is no embryo available to secrete HCG. In the absence of HCG production, there is a rapid deterioration of the corpus luteum, which quickly loses its ability to produce estrogen and progesterone. When these hormones drop below a critical level, the uterus sloughs its extra endometrial tissue, and menstruation begins.

THE COURSE OF EMBRYONIC DEVELOPMENT

Embryonic development is a programmed course of events that carries an organism along a path that leads from a simple zygote to an increasingly more complex **embryo.** The development of a biologically complex individual is driven by the input of energy and directed by the genetic information in the embryo's cells. In most animals, the energy required to fuel the construction of the embryo is present in the form of **yolk,** a mixture of lipids, proteins, and polysaccharides that are stored within the egg. Yolk provides the embryo with nutrients until the developing animal can obtain food for itself. Some animals, such as sharks, reptiles, and birds, have large yolk supplies that provide nutrients throughout development. For example, a bird hatches from the egg at a relatively advanced anatomical stage, without having had to find its own food prior to that time. Some animals, including sea urchins, produce microscopic-sized eggs that have very little yolk. A sea urchin hatches at an early stage; within a matter of hours, it develops into a **larva**—a self-feeding immature form of an animal. Mammals also develop from very small eggs with very little yolk, but, unlike sea urchins, they receive the required nutrients directly from their mother through an interfacing of the maternal and embryonic bloodstreams.

◁ THE HUMAN PERSPECTIVE ▷

Controlling Pregnancy: Contraceptive Methods

Throughout history, people have devised many ways to prevent reproduction. *Contraceptive* methods have ranged from inserting elephant and crocodile manure into the vagina to block the entrance of sperm into the uterus to "fumigating" the vagina over a charcoal burner after intercourse to kill the sperm. Though primitive, these methods employed the same strategies as do some of the most effective modern contraceptive approaches; that is, they physically blocked the uterine entrance or killed the sperm in the vagina. Other modern methods of avoiding pregnancy prevent ovulation, halt spermatogenesis, trap gametes in the oviducts or vas deferens, or interfere with implantation of the fertilized ovum in the uterus. In the following discussion, *effectiveness* is expressed as the percentage of sexually active women who do not become pregnant in the first year practicing that form of birth control. In the absence of birth control, "effectiveness" would be approximately 10 percent.

CHEMICAL INTERVENTION

Hormonal Contraceptives

Synthetic estrogens and progesterones inhibit the release of FSH and LH, which, in turn, prevent ovulation. Taken orally, either alone or in combination, these two hormones constitute *birth control pills*, the most effective (97 percent) *temporary* contraceptive method known today. Birth control pills must be taken daily to be effective. In the past few years, two non-oral hormonal contraceptives that provide extended protection have become available. Both utilize a progesterone-like hormone and have been shown to be highly effective. The first of these contraceptives is called Norplant and consists of several thin rubber capsules that are implanted beneath the skin of a woman's arm. The capsules release their hormone for a period of 5 years. The second contraceptive is called Depo-Provera and is administered by injection every 3 months.

The most controversial chemical contraceptive is a drug called RU486, which is produced in France. RU486 interferes with the production of progesterone. Because it renders the uterus incapable of receiving an embryo for implantation, the drug has been used in Europe as a "morning after" pill—a pill that blocks pregnancy (which by most definitions begins at implantation) for several days after unprotected sex. RU486 also acts as an "abortion" pill by causing the uterus to reject an implanted embryo. The drug has been prescribed in Europe to terminate pregnancies through about the seventh week. Banned in the United States by the FDA, RU486 became the center of a storm of controversy in 1992 when a woman attempted to bring the drug into the United States aboard a flight from Paris. In 1994, the drug was approved for testing in the United States.

Spermicides

Spermicides are sperm-killing chemicals that are inserted into the vagina before intercourse. When used alone, spermicides are not very reliable (79 percent effective) and thus are often used in combination with a diaphragm or condom.

MECHANICAL INTERVENTION

There are various forms of mechanical intervention to prevent pregnancy. The most popular include the IUD, diaphragm, and condoms.

Intrauterine Device (IUD)

The *intrauterine device,* or *IUD,* is a plastic or metal device that is inserted by a

All eggs undergo similar processes of early development, passing through the stages of fertilization, cleavage, blastulation, and gastrulation. We will begin our discussion as development begins—with an unfertilized egg awaiting a chance encounter with a sperm.

FERTILIZATION: ACTIVATING THE EGG

An unfertilized egg is a cell that is primed and ready to begin development. The egg must first be *activated*, however, a function of the fertilizing sperm. As the tip of the sperm contacts the plasma membrane of the egg (Figure 21-8*a*), the membranes of the two gametes fuse (Figure 21-8*b*) forming a single cell that contains both maternal and paternal sets of chromosomes. Within seconds of sperm contact, a wave of electrical activity sweeps around the surface of the egg, much as an action potential moves along a neuron. This wave makes the egg's plasma membrane instantly unresponsive to the advances of other sperm in the neighborhood besides the one that has already made contact.

physician into the uterus, where it prevents implantation of a developing embryo. In spite of its effectiveness (94 percent) and convenience, once inserted, the IUD may cause serious side effects, the most common of which are painful cramping and irregular bleeding, especially immediately after the device is inserted. Although infrequent, these complications created enough concern to cause IUDs to be taken off the market in the United States in 1987. A few types of IUDs are again being made available, notably those that have few negative side effects.

Diaphragm with Spermicide

Legend has it that the famous lover Casanova used half of a hollowed out lemon to cover the cervix of a lover in an attempt to reduce the number of new little Casanovas he sired. The modern *diaphragm*, a thin rubber dome, works in the same way, although with considerably more success (90 percent effectiveness) because it is precisely fitted to the woman's cervix and is used in conjunction with a spermicide. The diaphragm is inserted just before intercourse and must remain in place for 8 hours following intercourse to be fully effective.

Condoms

Condoms are thin sheaths that are worn over the penis and trap sperm, preventing fertilization. Condoms have the additional advantage of preventing sexually transmitted diseases, which is why they are also called "prophylactics," meaning "disease preventing." Condoms have become an important line of defense in the war against AIDS. Used alone, condoms are 88 percent effective as contraceptives; their effectiveness increases to 95 percent when used with a spermicide. Recently, a female "condom" has become available, which fits into the vagina to form a plastic lining that cannot be penetrated by sperm. The sheath is held in place by a ring that remains outside the vagina. The effectiveness of the female condom remains to be determined.

PERMANENT STERILIZATION

Vasectomy

In a *vasectomy*, a surgical procedure for men, a portion of each vas deferens is removed, and the cut ends of the tubes are tied. Although spermatogenesis continues, sperm are blocked from reaching the penis, so the man's ejaculate contains no sperm cells, making the procedure 99.85 percent effective. The volume of semen and the sensations of orgasm are unaffected by a vasectomy.

Tubal Ligation

In a *tubal ligation*, the woman's oviducts are surgically cut and sealed, preventing an egg from reaching the uterus or even coming in contact with sperm, but allowing ovulation to continue. Tubal ligation is 99.6 percent effective.

NATURAL METHODS

There are two natural methods of birth control: rhythm and coitus interruptus. The *rhythm method* requires abstinence from intercourse during a woman's fertile period, about 12 hours before and 48 hours after ovulation. A major problem with the rhythm method is the difficulty in predicting when ovulation occurs, especially in women with irregular ovarian cycles. (Ovulation occurs 14 days *before* the end of the cycle, not 14 days after the beginning of the cycle, making the prediction difficult.) Even when practiced diligently, failure rates range from 15 to 35 percent and, thus, effectiveness is variable, at best.

During *coitus interruptus*, also known as withdrawal, the man removes his penis from the vagina before ejaculation. Because it requires willpower and reduces gratification, failures are very likely.

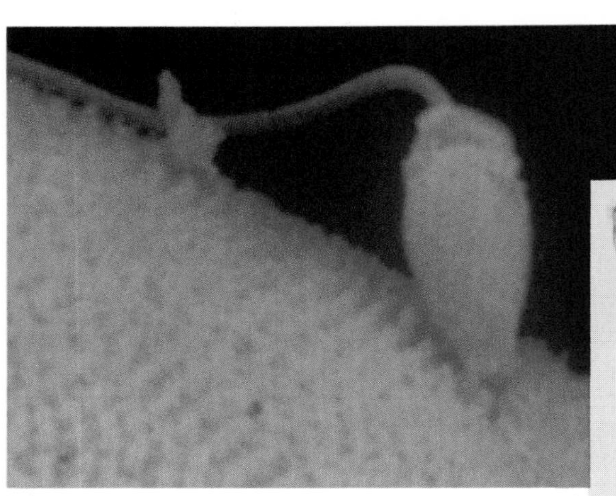

FIGURE 21-8
Fertilization. ***(a)*** The instant of contact between the tip of a sea urchin sperm and the surface of the egg. ***(b)*** A sperm in the process of fusing its plasma membrane with that of the egg. The sperm nucleus will soon be drawn into the egg cytoplasm.

CLEAVAGE AND BLASTULATION: DIVIDING A LARGE ZYGOTE INTO SMALLER CELLS

A fertilized egg is an unbalanced cell; it has a huge amount of cytoplasm but only two sets of homologous chromosomes. This situation changes very rapidly during early development, as the egg undergoes a succession of mitotic divisions, called **cleavage.** Cleavage is not a time of growth but a period in which the oversized egg is divided into a large number of smaller cells, known as **blastomeres** (Figure 21-9). Cleavage ends with the formation of a ball of cells, called a **blastula,** containing an internal, fluid-filled chamber, the **blastocoel,** whose relative size and location depends primarily on the amount of yolk in the egg (Figure 21-10).

GASTRULATION: REORGANIZING THE EMBRYO

Gastrulation is the process whereby the undistinguished blastula is transformed into a much more complex stage of development, called a **gastrula.** Gastrulation is characterized by an extensive series of coordinated cellular movements whereby regions of the blastula are displaced to radically different locations. The dramatic events that take place during gastrulation in a frog (Figure 21-11) provide an overview of the process.

By the time gastrulation has been completed, the embryo (gastrula) can be divided into an inner, middle, and outer layer, which correspond to the three embryonic *germ layers*—the endoderm, mesoderm, and ectoderm. In vertebrates (such as the frog of Figure 21-11), the inner layer

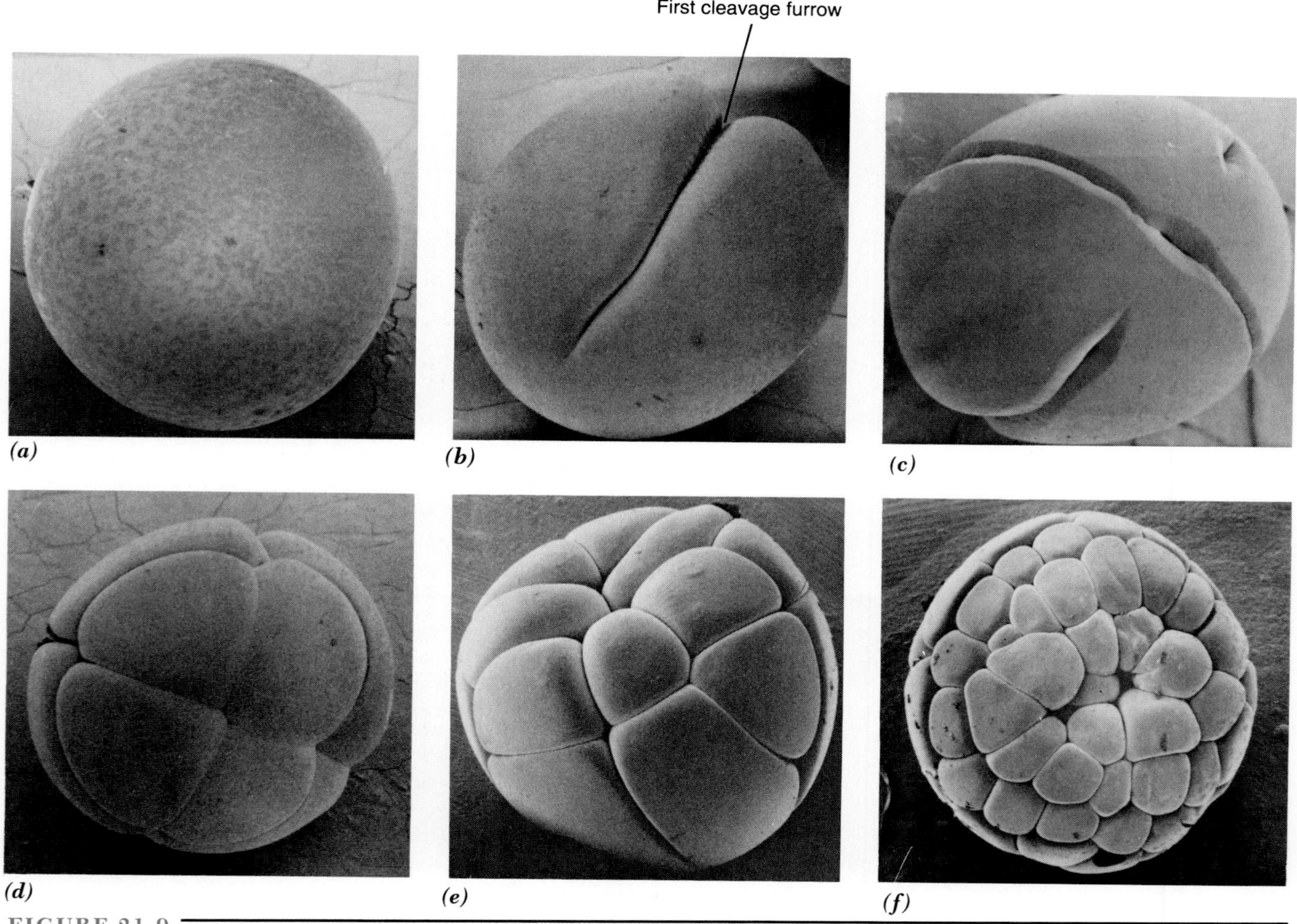

FIGURE 21-9

Cleavage of a frog egg divides the large, single-celled zygote into a number of smaller cells, called blastomeres. Although the egg contains more cells following cleavage, the total volume of cellular material is the same as is that of the original zygote.

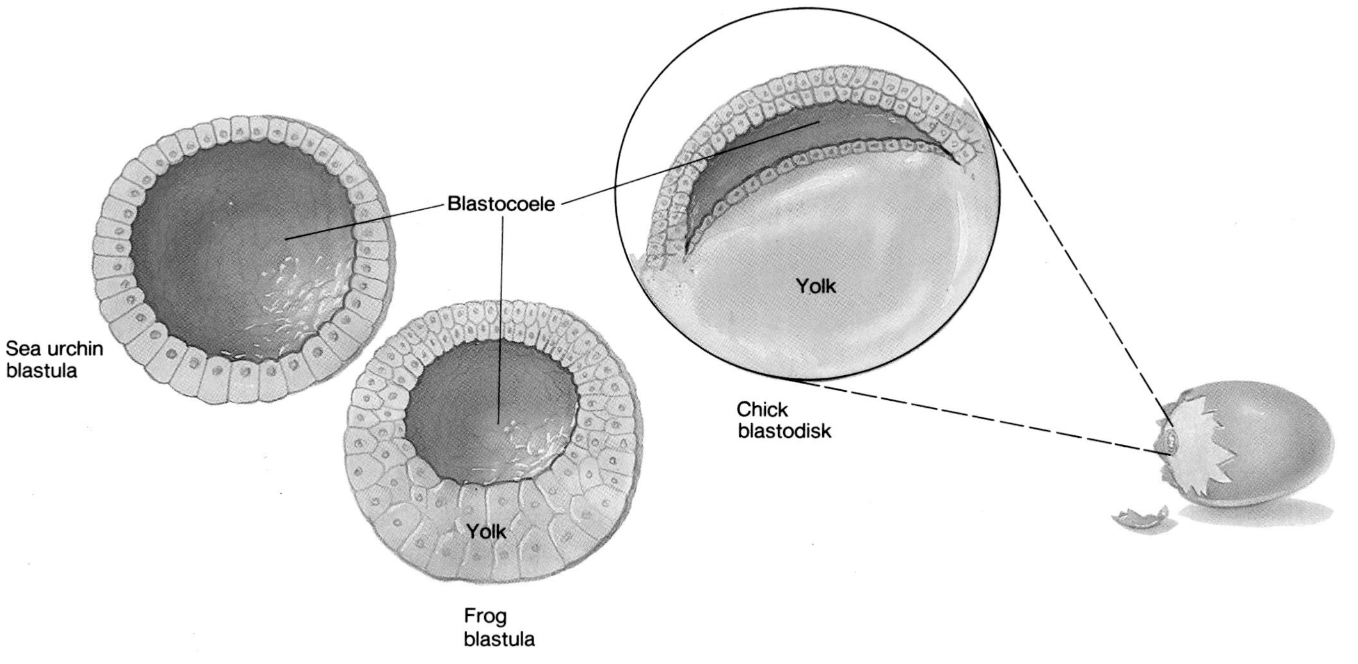

FIGURE 21-10

The blastula stage of a sea urchin, frog, and chick. A sea urchin develops rapidly from a small, relatively yolk-free egg into a feeding larva. The sea urchin's blastula has a large blastocoel, surrounded by a single layer of cells. Frogs and other amphibians have eggs that contain a relatively larger content of yolk, which leaves relatively less room inside the blastocoel. Reptiles and birds—animals whose embryos develop on dry land—must supply their developing offspring with enough food and water to last until they hatch from the shells that enclose them. The enormous yolk and albumen (egg white) provides these resources and occupies most of the egg space. Because of the presence of so much yolk, the growth of early embryonic cells is restricted to a small area on the yolk surface, where an embryonic disk, or *blastodisk,* is formed.

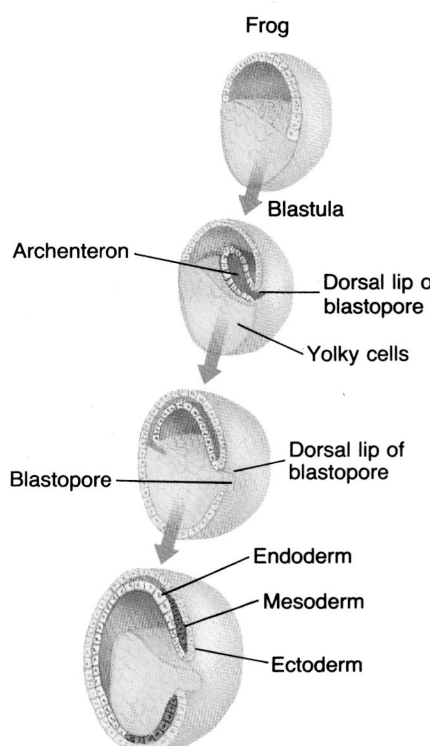

FIGURE 21-11

Gastrulation in the frog. The first indication of the onset of gastrulation in the frog is the appearance of a groove on the dorsal side of the embryo. The opening into the interior of the embryo is the *blastopore,* and the fold above the groove is the *dorsal lip* of the blastopore. During gastrulation, cells at the rim of the blastopore migrate into the interior of the embryo and are replaced by new cells that move over the surface toward the blastoporal lip. Once inside the embryo, the cells move deeper into the interior, away from the blastopore, forming interior walls of an increasingly spacious cavity, called the *archenteron.* The walls of the archenteron consist of endodermal cells (in purple) that will give rise to the digestive tract. The archenteron remains open to the outside through the blastopore, which corresponds in position to the future anus. Between the ectoderm and endoderm, a third group of cells develop into the mesoderm (in green), which will give rise to the skeleton, muscles, and other mesodermal derivatives. Once gastrulation is complete, the entire external layer of the embryo is composed of ectodermal cells. (Note that gastrulation in other animals, such as the sea urchin or chick, occurs by a very different pathway.)

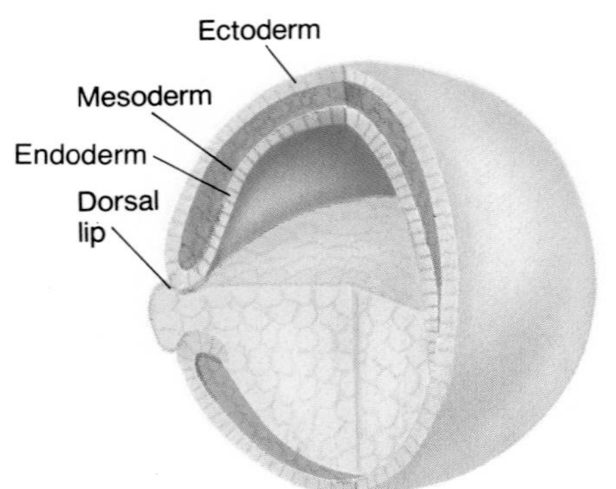

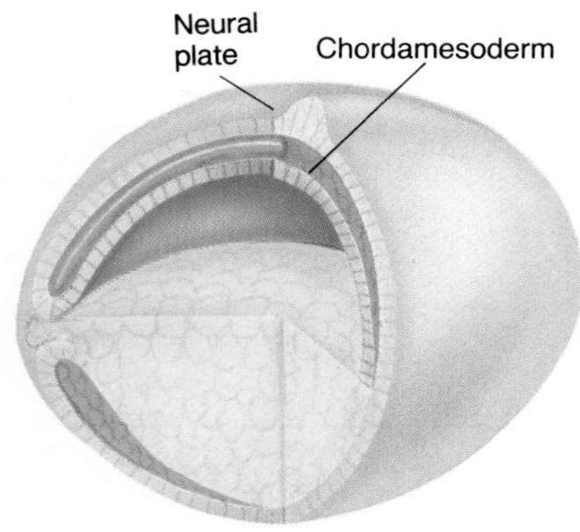

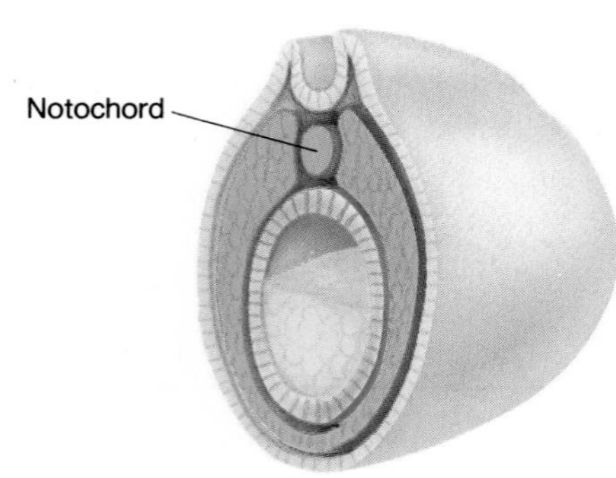

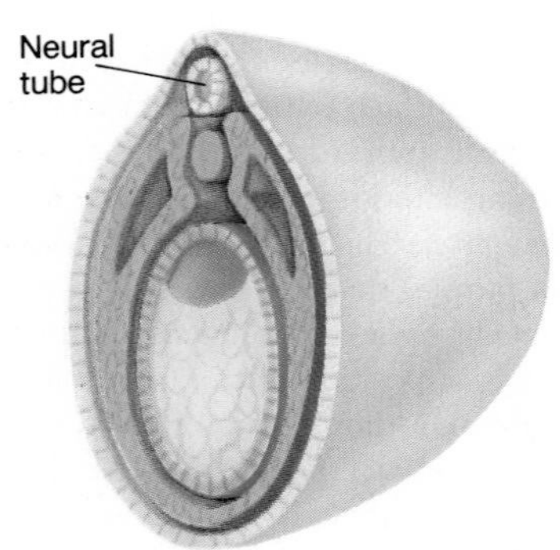

FIGURE 21-12

Successive steps in the formation of the neural tube in a frog. The first sign of the development of the future nervous system is the thickening of the ectoderm along the dorsal midline, which forms the neural plate. The edges of the neural plate roll upward and fuse together, forming a hollow neural tube along the midline. The formation of the neural tube is induced by the underlying mesoderm, which moved into this position during gastrulation.

of the gastrula, or **endoderm,** gives rise to the digestive tract and its derivatives, including the lungs, liver, and pancreas; the outer layer, or **ectoderm,** gives rise to the epidermal layer of the skin and to the entire nervous system; and the middle layer, or **mesoderm,** gives rise to the remaining body components, including the dermal layer of the skin, bones and cartilage, blood vessels, kidneys, gonads, and the inner linings of the body cavities.

NEURULATION: LAYING THE FOUNDATION OF THE NERVOUS SYSTEM

Toward the end of gastrulation in vertebrates, the ectodermal cells situated along the embryo's dorsal surface become elongated, forming a tall epithelial layer called the **neural plate** (Figure 21-12). This single layer of cells will develop into the entire nervous system, in the following manner. First, the neural plate thickens. Then, it rolls upward and fuses with itself to form a cylindrical, hollow **neural tube.** The neural tube is wider in the anterior portion of the embryo, where it will differentiate into the brain, and narrower in the posterior portion, where it gives rise to the spinal cord. On occasion, the neural tube fails to close completely during human development and a baby is born with *spina bifida occulta,* a condition in which a portion of the spinal cord remains open to the outside. Persons born with this type of spina bifida may suffer partial paralysis and are usually confined to a wheelchair.

While the ectoderm on the dorsal surface of the embryo becomes transformed into the neural tube, it cannot accomplish this feat by itself. If, for example, a segment of the underlying mesoderm (the green layer in the top drawing of Figure 21-12) is surgically removed at gastrulation, the nervous system in that portion of the embryo fails to develop. The underlying mesoderm (labeled "chordamesoderm" in Figure 21-12 because it will form the rod-shaped notochord) is said to *induce* the ectoderm to develop into nervous tissue, a developmental path that it would otherwise be unable to follow. Biologists have spent over 50 years trying to determine precisely what it is that the mesoderm passes to the ectoderm to induce it to develop into nervous tissue, but the question remains unanswered.

DEVELOPMENT BEYOND NEURULATION

By the time neurulation has been completed, the ectoderm, mesoderm, and endoderm are in position to begin forming the body's organs. But the cells of the embryo remain undifferentiated at this stage, revealing little of the structural complexity that will soon emerge. The conversion of this organized mass of cells into an organism that contains complex, functional organs is one of the least understood aspects of embyronic development. Organ formation, or **organogenesis,** is a complex process whereby two or more specialized tissues develop in precise relationship with one another. The mechanisms that underlie organogenesis are remarkably similar among different organs in different animals, even though the organs produced are remarkably diverse in structure and function.

Positional Information

What determines the spatial order of the parts that make up an animal's body? Why, for example, does a hand form at the end of a forelimb, and a foot at the end of a hindlimb? Why does a single long bone form in the upper arm, and a pair of long bones in the forearm? These questions concern *positional information*—information that determines the location or spatial organization of parts of the body. If a cell is to "know" what to form, it must receive some type of chemical or physical signal from its surroundings which informs it of its relative position within the whole.

The importance of positional information is readily demonstrated by a simple experiment performed on the developing limb of a chick embryo in 1959 by John Saunders of Marquette University. Although wings and legs have a very different structure and function in the adult bird, both body parts arise from similar-looking *limb buds* in the early embryo. The inner portion of the bud is destined to develop into either the thigh of the leg or the inner portion of the wing, while the outer portion of the bud is destined to develop into either the foot or the wing tip.

Consider what might happen if a small block of undifferentiated tissue is removed from the inner portion of a *leg* bud and transplanted to the *tip* of the wing bud (Figure 21-13). You might expect that this block of transplanted tissue would develop into either (1) thigh tissue, if it ignored the influences from its new surroundings and developed as it would have in its original location, or (2) wing tip tissue, if it developed in harmony with its new environment. In fact, the transplanted tissue does neither; rather, it develops into a toe. The transplanted tissue "remembers" its previous position as part of a leg bud rather than that of a wing. At the same time, the tissue responds to its new location in the outer portion of a limb bud rather than at its base so it forms a toe rather than a thigh. It would appear that the block of transplanted tissue has received two distinct positional signals: An early signal tells the cells that they are part of a leg, while a later signal tells them they are in the outer portion of the bud and thus should develop into a toe.

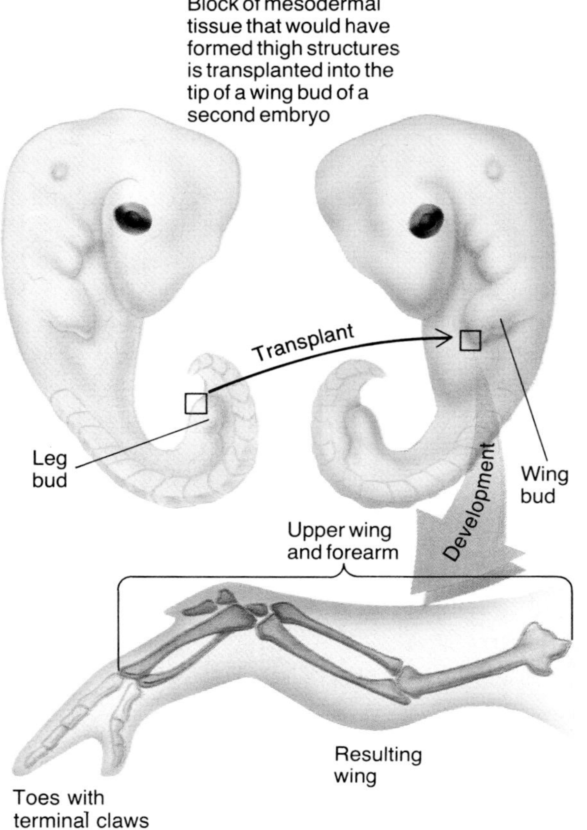

FIGURE 21-13
When a block of prospective thigh mesoderm is grafted into the tip of the wing it differentiates into a toe.

Morphogenesis

Each part of the body has a characteristic shape and internal architecture that can be correlated with its particular function: The spinal cord is basically a hollow tube; the kidney consists of microscopic tubules; the salivary gland consists of groups of secretory cells clustered around a duct; the lungs consist of microscopic air spaces lined by thin cellular sheets; and so forth. The development of form and internal architecture within the embryo is called **morphogenesis** and is the product of several distinct cellular activities. These activities include increased or decreased rates of cell division; changes in the adhesion of cells to their neighbors; changes in cell shape; deposition of extracellular materials; passage of inductive signals from one group of cells to another; the programmed death of cells in certain locations; and movement of cells from one place to another. The ways in which some of these processes can shape embryonic development are illustrated in Figure 21-14.

Cell Differentiation

During **cell differentiation,** the internal contents of a cell become assembled into a structure that allows the cell to carry out a specific set of activities. As a result, the cytoplasm of each type of cell becomes visibly different from that of other cell types. A salivary gland cell, for example, develops an extensive rough endoplasmic reticulum and Golgi apparatus, two specializations necessary for the production and secretion of mucus; a muscle cell develops an extensive array of cytoplasmic filaments, which are required for the generation of tension; and a cell of the adrenal cortex develops an extensive smooth endoplasmic reticulum, in which steroid hormones are synthesized.

While all cells have the same genetic information, different types of cells utilize that inheritance differently. For example, the cells of the salivary gland transcribe those genes that code for the proteins of mucus; the cells of a muscle utilize the genes that code for contractile proteins, such as muscle myosin and actin; and the cells of a differentiating red blood cell utilize the genes that code for hemoglobin. Each of the several hundred different types of cells in your body presumably contains a unique set of gene regulatory proteins that promote the activation of a unique set of genes. As a result, each cell transcribes only those genes whose products are required for the function of that particular cell type.

HUMAN DEVELOPMENT

The development of a human embryo begins with fertilization; 6 to 8 days later, the embryo "burrows" into the prepared endometrium of the uterus (Figure 21-15), implanting itself in the lining. By the time the entry site reseals itself, the embryo is surrounded by a pool of maternal blood, which is continually replenished by fresh

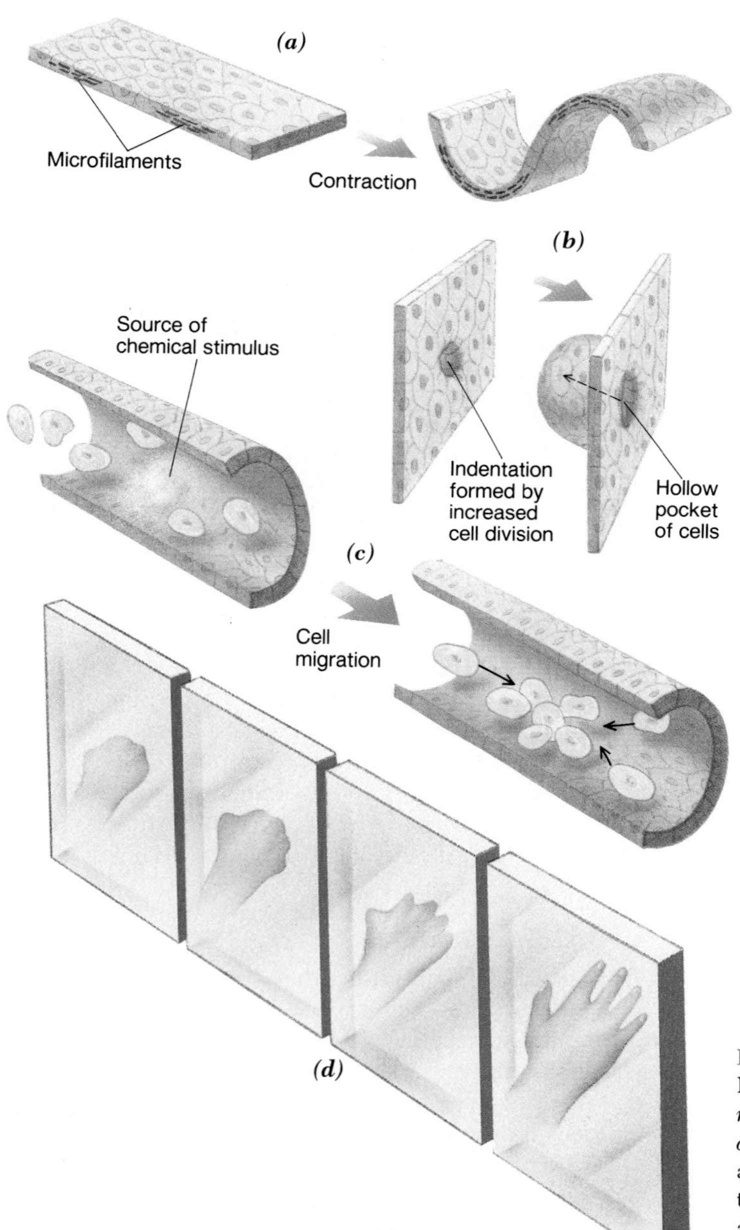

FIGURE 21-14
Morphogenesis: shaping the embryo. ***(a)*** *Contraction of microfilaments* causes the curvature of a sheet of cells. ***(b)*** *Increased cell division* by a group of cells in a sheet can lead to a hollow outgrowth. ***(c)*** *Cell migration*, in this case in response to a chemical, leads to the movement of cells from one site to another. ***(d)*** *Cell death* leads to the elimination of cells from specific parts of a developing limb.

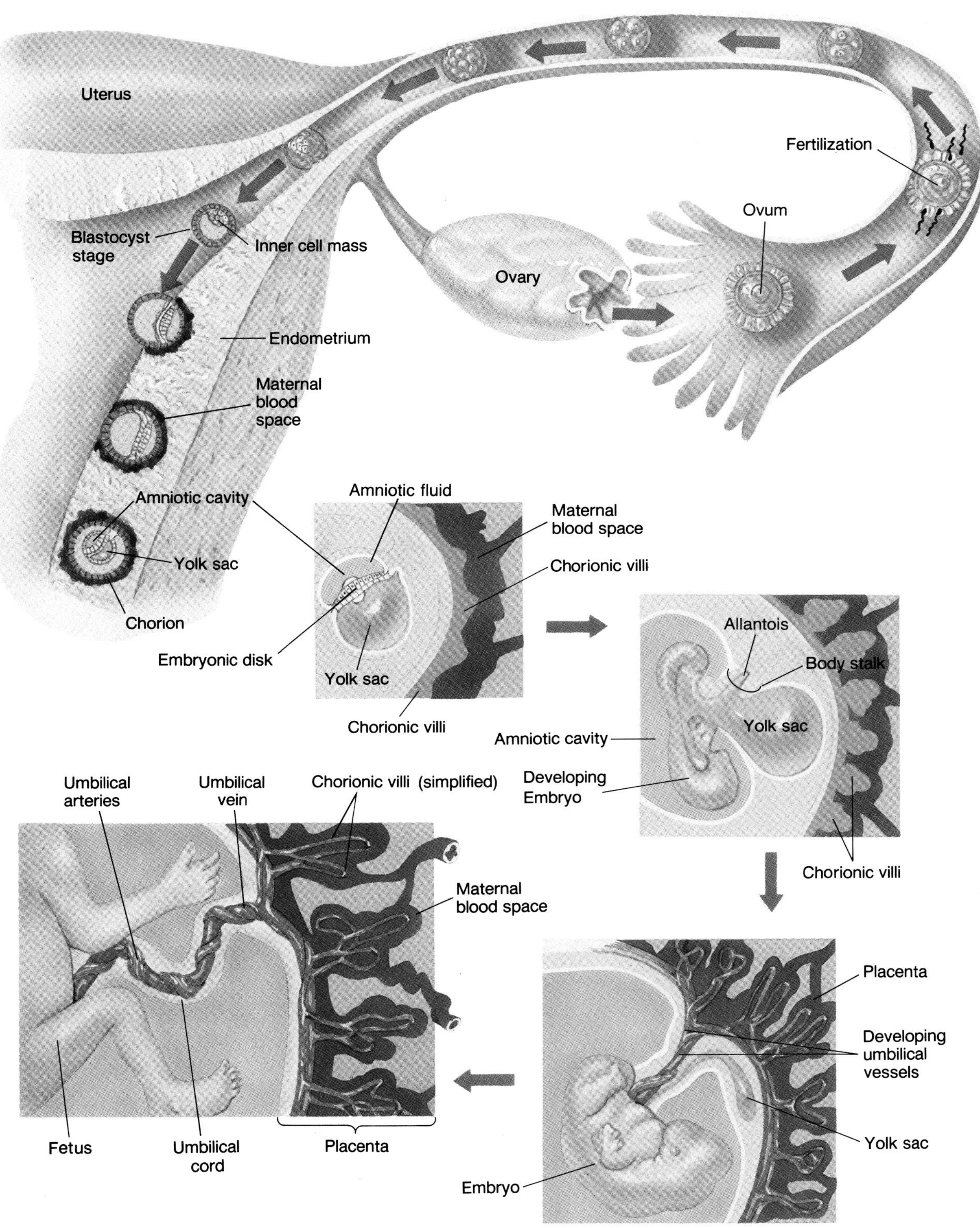

FIGURE 21-15

Implantation and placenta formation. Following fertilization, cleavage, and blastulation, a blastocyst is formed that consists of a single outer layer of cells—the trophoblast—and an inner cell mass. Soon after the embryo has implantated in the uterine wall, the inner cell mass gives rise to the embryonic disk, from which the embryo will develop. The amnion contains the liquid in which the embryo floats as it develops, while the chorion and allantois of the body stalk help form the placenta and umbilical cord. Notice that fetal and maternal blood do not mix but are separated by the selectively permeable membranes of the chorion and the capillaries.

blood from maternal vessels. Branching projections, called *villi,* sprout from the **chorion**—the embryo's outer membrane covering. These chorionic villi provide a large surface for the exchange of respiratory gases, nutrients, and wastes between the embryo and its mother, until a **placenta,** constructed from the embyro's chorion and the mother's uterine lining, takes over the exchange. Veins and arteries develop in the villi, connecting the embryo's early circulatory system with the blood-filled space, which serves as a temporary life-support system. These connecting vessels soon become channeled within an **umbilical cord.** One end of the cord is attached to the embryo's belly, the other end to the placenta. In the placenta, the embryo's vascularized chorionic villi are bathed in maternal blood. The two blood supplies are separated by a thin layer of cells, permeable to nutrients, gases, and wastes.

THE STAGES OF HUMAN DEVELOPMENT

During the first 2 months following fertilization, all the major organs appear, many in a functioning capacity. By the end of the eighth week, the embryo has acquired a distinctively human form. Up to this point, the developing offspring is still called an embryo. During the last 7 months in the uterus, however, it is referred to as a **fetus.** In the fetal stage, organ refinement accompanies overall growth, as the fetus develops the ability to survive outside the uterus.

The 266 or so days between conception and birth are traditionally grouped into three stages, each called a *trimester.* The most dramatic changes occur during the first of these periods (Figure 21-16*a–f*).

The First Trimester

During the first 3 months of development, the embryo is transformed from a single-celled zygote into an ensemble of organ systems that are only slightly less complex than are those found in a full-term baby. Cleavage, blastulation, and implantation take place during the first week or so following fertilization. At the time of implantation, the embryo is called a **blastocyst,** and it contains two different groups of cells (Figure 21-15) with very different fates. Inside the spacious cavity of the blastocyst is a clump of cells called the **inner cell mass,** which will give rise to the tissues of the embryo itself. The outer wall of the blastocyst is called the **trophoblast;** it secretes the enzymes that allow the blastocyst to penetrate the uterine wall and then differentiates into the chorion, the outermost of four *extraembryonic membranes* that do not become part of the embryo and are eventually discarded prior to birth. In addition, the chorion secretes human chorionic gonadotropin (HCG), the hormone that stimulates the corpus luteum to continue to produce progesterone (page 373), which maintains the uterine lining where the embryo is developing.

The innermost extraembryonic membrane, the **amnion,** envelops the young embryo and encloses the *amniotic fluid* that suspends and cushions the developing body throughout its duration in the uterus. A third membrane forms the **yolk sac** which, in humans, simply contains fluid (in birds and reptiles, the yolk sac is filled with nutrient-containing yolk). The fourth membrane, the **allantois,** is rich in blood vessels and eventually helps form the vascular connections between mother and fetus. These same four extraembryonic membranes (chorion, amnion, yolk sac, and allantois) are also present in the eggs of birds and reptiles, revealing the evolutionary relationship among the three classes of "higher" vertebrates—reptiles, birds, and mammals.

A few days after implantation, the embryo begins to gastrulate. The cells of the inner cell mass become rearranged to form a double layer of cells, called an *embryonic disk,* which is separated from the overlying chorion by a newly formed space, called the *amniotic cavity* (Figure 21-15). Your entire body, with the exception of the germ cells that migrate in from the yolk sac, is derived from the cells that make up the embryonic disk. By the end of the first month, the embryo, though still less than half a centimeter (3/16 of an inch) long, has begun forming its nervous system, lungs, liver, and several other internal organs. At this stage, the heart is a four-chambered pump; the first signs of eyes and a nose appear; and four buds protrude from the side of the "C-shaped" embryo. By the end of the following week, these buds will be vaguely recognizable as developing arms and legs.

During the second month, the liver takes over its temporary job as the main blood-producing organ of the embryo, while an intricate cartilage-containing skeleton begins to change into bone. The brain develops cerebral hemispheres, and spinal nerves grow out from the spinal cord. The face acquires a distinctively human look, with its slate-colored eyes and the beginnings of eyelids. Muscles begin to form and assume their permanent relationships. Fingers and toes become evident. The embryo has become a fetus.

The third month is devoted primarily to the growth and development of existing structures, with the exception of the formation of the genitals. The limbs are clearly recognizable, with well-sculpted fingers and toes, complete with nails. The fetus begins to exhibit reflex movements that go undetected by the mother.

The first trimester is unquestionably the most dangerous time for the developing embryo. The magnitude of the developmental changes that occur during this time renders the embryo particularly vulnerable to pathogens and chemicals that would be relatively harmless to an older fetus or a newborn human.

The Second Trimester

As the 7.5-centimeter fetus enters its fourth month, it develops sucking and swallowing reflexes and soon starts kicking. The mother does not usually feel these movements until the fifth month. The developing skeleton is clearly

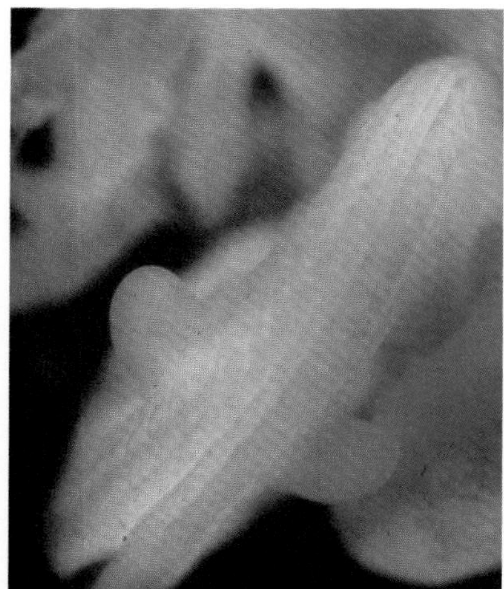

***(a)* 3 weeks**—Organogenesis begins with neural tube formation, seen here as enlarged crests that delineate the neural groove from which the spinal cord and brain develop.
LENGTH—0.25 cm (1/10 in.).

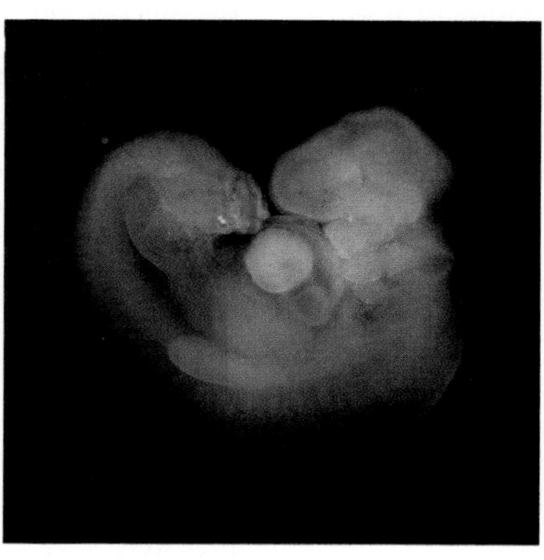

***(b)* 4 weeks**—A pumping heart, developing eye, and arm and leg buds are evident. The embryo has a long tail, gill arches, and a relatively enormous head.
LENGTH—0.7 cm (1/3 in.).

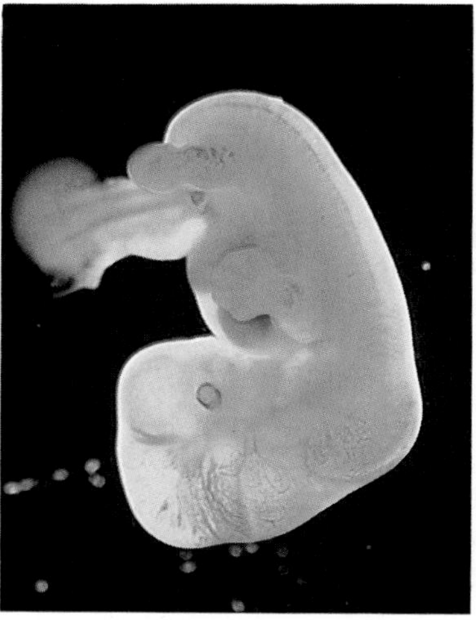

***(c)* 5 weeks**—Internal organ development is well underway. Fingers are faintly suggested. The fetal circulatory system is evident, and the body stalk is now an umbilical cord.
LENGTH—1.2 cm (1/2 in.).

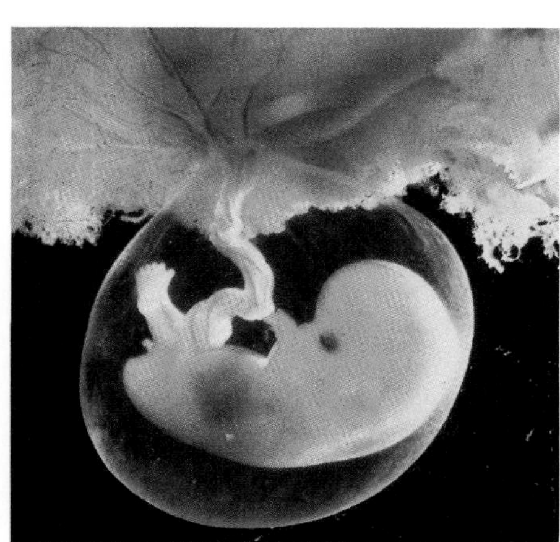

***(d)* 2.5 months**—The fetus, seen here floating in the amniotic cavity, now has all its major organ systems. The umbilical blood vessels and placenta are well defined, and the tail and gill arches have disappeared.
LENGTH—3 cm (1 1/2 in.).

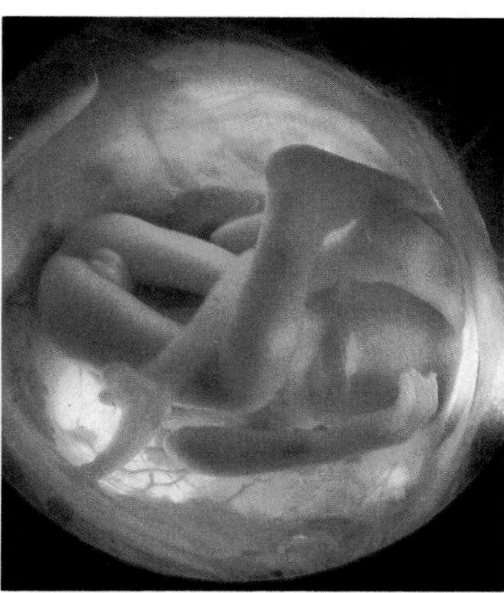

***(e)* 5 months**—The fetus looks very much as she will at birth. Bone marrow is assuming more of the blood-producing duties. The mother begins to feel fetal movements, even hiccupping.
LENGTH—25 cm (10 in.).

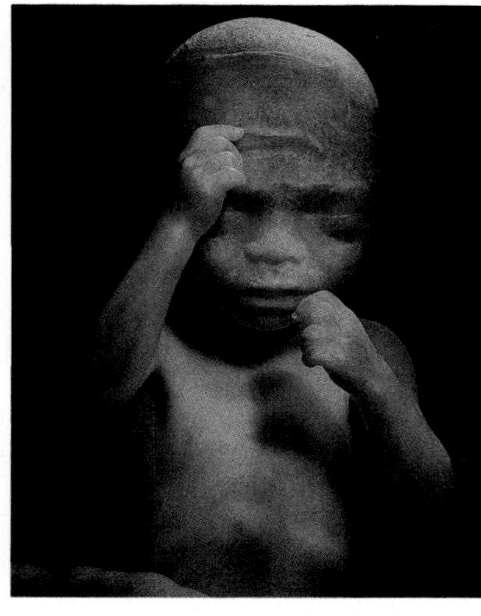

***(f)* 5.5 months**—Internal organs now occupy their permanent positions (except for the testes in males).
LENGTH—30 cm (12 in.).

FIGURE 21-16
Some stages in human embryonic and fetal development.

distinguishable by X-ray examination, and the bone marrow begins blood-cell production. Midway through the pregnancy, two heartbeats become discernible in the mother—her own (about 72 beats per minute) and that of the fetus (up to 150 beats per minute). Convolutions appear in the cerebral cortex of the brain, and sense organs begin supplying the fetus with limited information about its environment (Figure 21-16*e*).

Although most of the organ systems are at least partially functional by the end of the second trimester, the fetus has little chance of survival if it is born before the seventh month. Even with expert medical care, a 23-week-old fetus (Figure 21-16*f*) has only a 10 percent chance of surviving outside the uterus. If born a month later, its chances of survival jump to 50 percent.

The Third Trimester

During the last 3 months of pregnancy, the fetus increases its body weight by 500 to 600 percent. The brain and peripheral nervous system enlarge and mature at an especially rapid rate during the final trimester. Neurological performance later in life, including intelligence, depends on the availability of protein needed for fetal brain development. Women who suffer protein deficiency during this time tend to have babies that are mentally slower throughout life. By the end of the third trimester, the fetus is capable of regulating its temperature and can control its own breathing. This latter ability, together with the degree of lung maturation, generally determines a premature infant's chances of survival.

The formerly lean fetus changes in appearance, as fat deposits form under the skin, imparting the rounded, chubby shape characteristic of many babies. In males, the testes descend into the scrotum. Most fetuses change position in the uterus, becoming aligned for a head-first delivery through the birth canal.

One system that does not mature by the time of birth is the immune system; human babies are born immunologically deficient. During the final fetal month, however, antibodies from the mother's blood cross the placenta and temporarily fortify the baby's defenses against infectious diseases until its own immune arsenal becomes competent. In breast-fed babies, this immunological gift is supplemented by the maternal antibodies and immune cells found in breast milk.

BIRTH

After 9 months of development, a human fetus is 6 billion times larger than was the original zygote. Once labor begins, a cascade of events promote uterine muscle contractions of increasing strength and frequency. The posterior pituitary releases oxytocin, which stimulates muscle contraction in the uterus. This response then triggers a reflex release of even more oxytocin, establishing a positive feedback loop that intensifies contractions. **Birth,** or *parturition,* occurs in three distinct stages. During the first stage, the mucous plug that blocks the cervical canal and prevents microbial invasion of the intrauterine environment is expelled. The amniotic sac ("bag of waters") ruptures during this stage, and its fluid is discharged through the vagina. Contractions during this stage force the cervical canal to dilate until its diameter enlarges to about 10 centimeters (3.9 in.), marking the onset of the second stage of labor—the expulsion of the fetus. Powerful contractions move the fetus out of the uterus and through the vagina, usually head first. Fewer than 4 percent of all deliveries are "breech births," whereby the buttocks or legs come out first. Contractions during the final stage expel the detached placenta, called the *afterbirth,* from the uterus. Additional contractions help stop maternal bleeding by closing vessels that were severed when the placenta detached.

A newborn infant quickly acclimates to its terrestrial environment. With the umbilical cord clamped and severed, the baby rapidly depletes its source of maternal oxygen; still unable to breathe, the baby's respiratory waste (carbon dioxide) accumulates in the blood. The brain's respiratory center responds to this increase by activating the breathing process, and the lungs inflate with air for the first time. But before the lungs can oxygenate the blood, a circulatory modification is required because the unneeded fetal lungs were bypassed by the bloodstream in the uterus. The septum between the right and left atria of the *fetal* heart is perforated by a large hole, called the *foramen ovale,* that allows blood to flow directly from one side of the heart to the other, rather than traveling through the pulmonary circulation. At birth, the oval window is immediately closed by a hinged flap, forcing blood to travel through the lungs to get to the other side of the heart.

EVOLUTION AND ADAPTATION: TYING IT TOGETHER

It would be hard to mistake a fish for a bird, or a turtle for a human, yet the embryos of these vertebrates are remarkably similar (Figure 21-17). Embryos tend to change much more slowly over evolution than do the corresponding adults. For this reason, similarities in embryonic development are often used as evidence for evolutionary relatedness. For example, molluscs (such as snails) and flatworms (such as tapeworms) are so different as adults that there is no reason to think the two groups are more closely related to one another than are molluscs and vertebrates. Yet, the patterns of cleavage of certain molluscs and flatworms are so similar that evolutionary schemes typically show molluscs as descendants of flatworm-like ancestors.

The various parts of the embryos depicted in Figure 21-17 appear so similar because they are *homologous* struc-

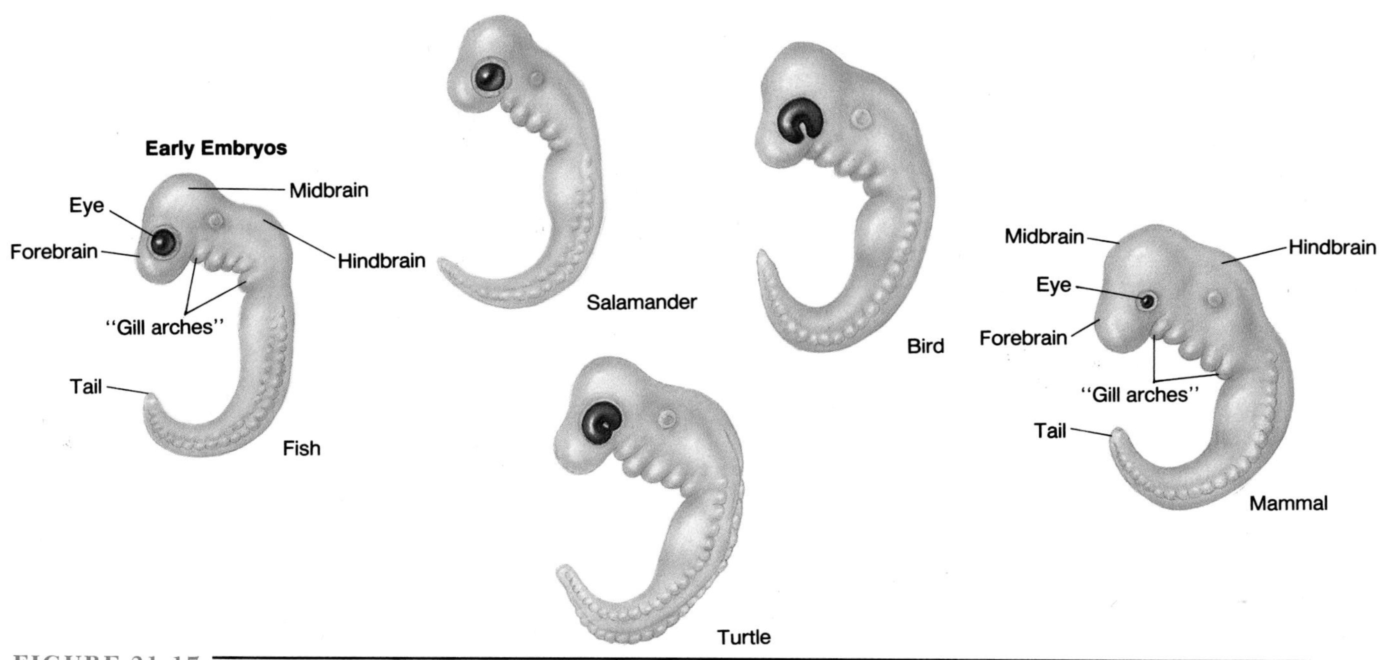

FIGURE 21-17
The striking similarities in the structure of the early embryos of vertebrates reveals their common ancestry.

tures; that is, they are derived from the same structure that was present in a common ancestor. We have seen in this chapter how the resemblance among the four extraembryonic membranes of reptiles, birds, and mammals testifies to the common ancestry of all three groups. Even when the original function of a membrane is no longer needed by a particular species, the membrane remains as a vestigial reminder of the organism's evolutionary origins. For example, humans and other mammals have no yolk in their embryo, yet they still develop a yolk sac—a vestigial remnant left over from an early "yolked" ancestor. Similarly, all vertebrate embryos, including those of humans and other air-breathing mammals, develop "gill slits," even though these structures never become functional, even in the embryo. Instead, the gill slits of the human embryo are only transient structures, appearing rapidly in an early stage and then giving rise to totally different structures, including parts of the jaw, ear, and thyroid gland.

SYNOPSIS

Both asexual and sexual reproduction have advantages. Asexual reproduction is a biologically economical way to generate identical copies of oneself rapidly. In contrast, sexual reproduction is more costly energetically, but it produces offspring with variable characteristics, which makes it much more likely that a species population will be able to survive changing environmental conditions.

Gametes are produced by gonads. In humans, sperm are generated in the seminiferous tubules of the male's testes, and eggs are generated in the follicles of the female's ovaries. Sperm are highly differentiated cells that derive from spermatids that have formed from primary spermatocytes by meiosis. Primary spermatocytes are derived from spermatogonia, whose mitotic divisions provide a continuous source of germ cells throughout life. In the female, all germ cells have entered meiosis by the time of birth. Most of these oocytes are located in small, primordial follicles. During each ovarian cycle, a number of follicles enlarge, and each contained oocyte undergoes enlargement and differentiation to form a cell that contains the nutrients needed to support embryonic development. One of these oocytes is released into an oviduct during each cycle. Meiosis in the female includes highly unequal divisions, producing one cell that retains virtually all of the cytoplasmic material and forms the egg, while the others contain little more than nuclei and eventually disintegrate.

Sexual development and gamete formation in both sexes are regulated by hormones. In both males and females, the hypothalamus produces gonadotropin-releasing hormone (GnRH), which controls the production and release of two gonadotropins, LH and FSH, by the anterior pituitary. In the male, FSH promotes spermatogenesis, while LH stimulates the interstitial cells of the testes to produce testosterone, which maintains male reproductive function. In the female, both FSH and LH control the ovarian cycle and stimulate the production of estrogen. These three hormones, in addition to the progesterone that is produced by the corpus luteum (the follicle from which the oocyte is ovulated) control the menstrual (uterine) cycle. During each cycle, the lining of the uterus is thickened and vascularized in anticipation of an implanted embryo. If the embryo implants itself, HCG is produced and the corpus luteum and uterus are maintained. If the ovulated egg is not fertilized, the failure to produce an implanted embryo leads to the deterioration of the corpus luteum, a drop in progesterone and estrogen production, and the breakdown and sloughing of the uterine wall.

Embryonic development is a programmed course of events that carries an animal from fertilization through cleavage, blastulation, gastrulation, and organ formation. A fertilizing sperm activates an unfertilized egg and donates a set of homologous chromosomes. Cleavage divides the unbalanced egg into a large number of smaller blastomeres. Cleavage leads to the formation of a blastula, a stage that contains an internal chamber, or blastocoel, whose size and location depends primarily on the amount of yolk in the egg.

During gastrulation the various parts of the blastula become rearranged to form an embryo with three defined germ layers. The outer ectoderm gives rise to the epidermis and nervous system; the inner endoderm to the digestive tract and related organs; and the middle mesoderm to the remainder of the embryo. After gastrulation in vertebrates, the dorsal strip of ectoderm becomes thickened into the neural plate, which rolls into a tube that ultimately gives rise to the entire nervous system. This transformation requires induction from the underlying mesoderm.

The formation of organs depends on a number of processes. Cells receive chemical and physical signals from their surroundings, which inform them of their relative postion within the embryo. The shape of the organ formed by developing cells depends on morphogenetic processes, including changes in the rate of cell division, changes in cell shape, changes in cell adhesion, and programmed cell death. During formation of an organ, the internal architecture of the cells assumes a differentiated state, characteristic of that cell type. Each type of cell transcribes a restricted set of genes forming a characteristic set of proteins.

Human development occurs in the uterus. The blastocyst implants itself in the endometrium and gastrulates to form an embryonic disk. Of the four extraembryonic membranes, the chorion and allantois contribute to placenta formation; the amnion protects the fetus; and the yolk sac temporarily manufactures blood cells. By the end of the second month, the formation of virtually all embryonic organs has begun. During the remaining 7 months, the fetus refines these structures and grows in size until uterine contractions expel it through the vagina.

Review Questions

1. Describe the differences in gamete production in men and women.
2. Trace the odyssey of a sperm cell from the spermatogonial stage to fertilization.
3. Describe the effects of LH and FSH in human males and females and how the levels of these gonadotropins are controlled in members of each sex.
4. Arrange these terms in the order of development and briefly describe the role of each process in embryonic development: neural tube formation; birth; blastulation; fertilization; gastrulation; gametogenesis (meiosis).
5. Discuss the role of each of the four extraembryonic membranes in human development and indicate their evolutionary significance.

Critical Thinking Questions

1. Different vertebrae along the backbone can be distinguished by their shape. Occasionally, an infant is born with vertebrae in his or her lower back shaped like those normally found in the neck. In other words, the infant possesses cervical-type vertebrae in the lumbar region of the spine. Do you think this condition could be due to a homeotic mutation? Why or why not?
2. The United States has one of the highest teenage pregnancy rates in the world. What do you think could be done to reduce the number of unwanted pregnancies among teens? (Obviously, there is no right or wrong answer to this question.)
3. What, if any, is the difference in chromosome composition between a spermatogonium and a primary spermatocyte; a spermatid and a sperm; an oocyte in a primodial follicle, an ovulated egg, and an egg awaiting fertilization?
4. Both sea urchins and mammals produce eggs with very little yolk. Sea urchin larvae must find their own food, while mammalian embryos receive nutrients from their mother. What effect do you think this difference has on the number of eggs produced by sea urchins versus those produced by mammals, and on their relative survival rates?
5. Eggs possess mechanisms to prevent their fertilization by more than one sperm. Consider the effects of a human egg that was fertilized by two different sperm. What would happen as the chromosomes line up during metaphase of the first mitotic division? How many sets of chromosomes would the two daughter cells receive? What effect do you think this would have on the survival of the embryo?

PART 5

Evolution and Diversity

Evolution leads to characteristics that improve a species' chances of surviving and producing offspring. This camouflaged coralline sculpin blends in with its environment and with its oversized jaw can seize unwary victims its own size, providing the nutrients needed for growth, development, and reproduction.

CHAPTER 22

Mechanisms of Evolution

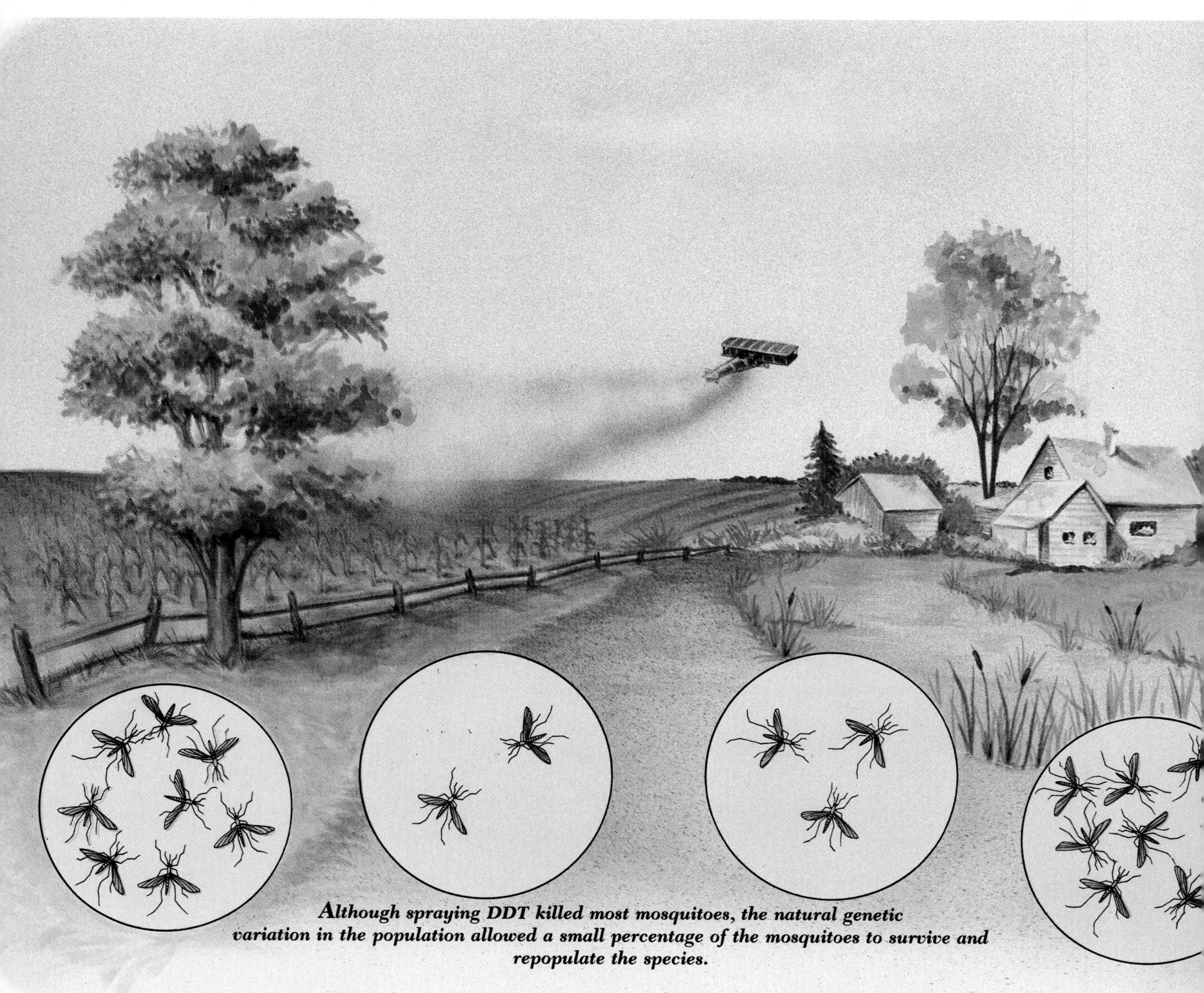

Although spraying DDT killed most mosquitoes, the natural genetic variation in the population allowed a small percentage of the mosquitoes to survive and repopulate the species.

STEPS TO DISCOVERY
Silent Spring Revisited

In 1939, Paul Muller, a researcher at a Swiss pharmaceutical company, discovered that the compound dichloro-diphenyl-trichloro-ethane (DDT) was a very effective insecticide. During World War II, two diseases posed serious threats to U.S. troops: typhus fever, which was common in Europe and was spread by lice; and malaria, which was common in the Pacific and was spread by mosquitos. In Italy in 1943, DDT was sprayed under the clothing of over 3 million troops and civilians to kill the lice and prevent an epidemic of typhus from breaking out. The spread of this deadly disease was arrested. At the same time, in the Pacific, planes sprayed DDT over entire islands, halting the spread of malaria. DDT was hailed as a "miracle" compound.

After the war, DDT use greatly expanded. In 1957, planes sprayed DDT over marshy areas of Massachusetts in order to kill mosquitoes. While spraying, planes happened to spray a 2-acre bird sanctuary. In addition to killing the insects, the insecticide killed the birds. The resident of the sanctuary wrote a letter to a friend, Rachel Carson, a biologist and author of several widely acclaimed books on the sea and its inhabitants. This plea led Carson to read the scientific literature on the effects of DDT and related pesticides. The more Carson read, the more convinced she became that she had to warn the public of the dangers of insecticides.

After 4 years of research, Carson wrote *Silent Spring,* a book that documented the devastating effects of pesticides on the wildlife of the world. The book inspired President Kennedy to establish a commission to regulate the use of pesticides. Congress began holding hearings on the subject, and environmental concern groups were established. These events culminated in the establishment of the Environmental Protection Agency in 1970, which banned the use of DDT in 1972.

A pesticide is a powerful evolutionary agent. In a population of insects, some individuals have a combination of genes that makes them less susceptible to harmful chemicals (such as DDT) than do other individuals. The individuals that are susceptible to the pesticide die, removing the genes that confer susceptibility from the population and leaving only resistant individuals to repopulate the species' ranks. Species that lack the ability to cope with environmental changes will shrink in number or even become extinct, an event chronicled in Carson's book. Carson noted that, while some insect populations were evolving pesticide resistance, many bird populations were being decimated. Studies showed that the birds were eating pesticide-contaminated insects, earthworms, and fish; consequently, the toxins were building to high concentrations in the birds' fatty tissues. The accumulating DDT prevented many birds from producing healthy offspring: Birds that ingested DDT laid fewer eggs; those eggs that were laid had thinner shells and were often broken in the nest; and the chicks that hatched were so loaded with pesticide residues that they often failed to survive. Unlike insects, none of the members of the bird populations possessed genes that conferred resistance to pesticides; therefore, there were no resistant individuals for natural selection to favor.

Concern over the effects of DDT is not limited to birds. Even though DDT has been banned in many countries, the chemical residues of the pesticide remain stored for decades in the fatty tissues of the human body. A report in 1993 indicated that women with high levels of DDT in their body had four times the risk of contracting breast cancer than did women with the lowest levels of the pesticide. The earlier use of DDT may be one of the reasons for the puzzling rise in breast cancer rates in the past few decades.

"A fish out of water" can't survive very long—or so we might think. Yet the mudskipper is a fish that not only survives long treks across mud flats, it even climbs trees (Figure 22-1). In water, the mudskipper's fins and gills work just like those of a typical fish: Its fins propel and steer the mudskipper through the water, and its gills extract dissolved oxygen. So how does the mudskipper remain alive on land with a body and breathing machinery that are so unmistakably adapted for life in the water?

The answer is that the mudskipper's gills and fins have modifications that enable it to survive and move on land. For its respiration, the mudskipper packs a supply of water into its bulging gill pouches, from which it extracts oxygen; the pouches act like a scuba tank in reverse. For motility, the mudskipper's reinforced forefins serve as stubby arms for crawling across the mud or shimmying up tree trunks in search of snails for a meal.

A hobbling mudskipper with water-engorged pouches illustrates how structures originally adapted for one way of life may become refashioned for new functions. The mudskipper's makeshift legs and water bags are modifications of structures that were originally adapted for aquatic life. The fact that the mudskipper possesses these structures is evidence that its ancestors lived strictly underwater and that those ancestors were something other than mudskippers. Every species has a history of ancestors that possessed features and behaviors different from those of the existing species. To trace the history of change in an organism's ancestors is to follow its course of evolution.

FIGURE 22-1
The mudskipper is a fish that can live for hours out of water. Strengthened forefins and modified gill chambers are evidence that the mudskipper evolved from ancestors that were strictly aquatic.

THE BASIS OF EVOLUTIONARY CHANGE

Evolution—the theory that a population of organisms changes as the generations pass—explains how modern organisms evolved from more primitive ancestors, which, in turn, evolved from even more primitive types, and so on, back to the first appearance of life, billions of years ago. The fact that all the diverse organisms present on earth today arose from a common ancestor explains why they have the same basic mechanism for the storage and utilization of genetic information, the same types of cellular organization, and similar enzymes and metabolic pathways: These shared characteristics were present in the earliest organisms and were retained among their descendants.

Individual organisms are born, mature, and eventually die. Along the way, an individual may change, but it does not evolve. Rather, it is the **species**—a group of interbreeding organisms—that evolves. Members of a species form groups, or **populations,** that occupy a particular region. Some species consist of just a single population living in one area, such as the pupfish that lives in a single desert spring in Nevada. Other species, such as manzanitas and ground squirrels, are made up of more than one population, each in a different locality. For evolution to take place, change must occur in the genes that are present in the members of a population. These changes are passed on to the next generation during reproduction. In order to understand this process, biologists have investigated the genetic changes in populations that generate evolutionary change.

Recall from Chapter 9 that each offspring receives one copy of a gene from each of its parents, and that genes can occur in different forms, or alleles. Most alleles are either dominant or recessive, and an individual can be either homozygous (two identical alleles) or heterozygous (two different alleles) at any particular gene locus. Not all individuals that make up a population have the same alleles; consequently, there is genetic variation in the population. In humans, this variety is reflected by differences in pigmentation, body characteristics, and blood types among different individuals and ethnic populations.

The sum of all the various alleles of all the genes in all of the individuals that make up a population is called the population's **gene pool.** If we could count every allele of every gene in every individual of a population, we could measure the genetic variation in a gene pool. The relative occurrence of any allele in the gene pool is expressed as an allele frequency.

FACTORS THAT CAUSE GENE FREQUENCIES TO CHANGE OVER TIME

Evolution occurs when the composition of the gene pool changes. But what causes allele frequencies to change? Sometimes the easiest way to understand a process is first to construct an artificial system or model in which the process does *not* occur. In the case of evolution, by uncovering the conditions that are necessary to keep allele frequencies *constant,* we automatically learn what forces will cause them to change.

In 1908, the British mathematician G. H. Hardy and the German biologist W. Weinberg independently discovered that under certain ideal conditions, allele frequencies will remain constant from generation to generation; this is known as *genetic equilibrium.* Their demonstration is called the Hardy-Weinberg Law and is discussed in detail in Appendix C.

The five conditions that must exist for allele frequency to remain constant are:

1. *No Mutation:* There must be an absence of mutation so that no new alleles appear in the population.
2. *No Immigration or Emigration:* Individuals cannot move into or out of a population so that no alleles enter or leave the population.
3. *Large Population:* The population must be very large so that it is not affected by random changes in allele frequency.
4. *Equal Survival:* All individuals in the population must have an equal chance of survival; that is, there are no genetic traits that give individuals a survival advantage.
5. *Random Mating:* Mating must combine genotypes at random; that is, no preference is shown in the selection of a mate.

Consequently, the factors that disrupt genetic equilibrium and *cause* changes in the frequency of alleles in a population are the opposite—mutation, gene flow, genetic drift, natural selection, and nonrandom mating.

MUTATION

Mutations are random, inheritable changes in DNA that introduce new alleles (or new genes, as the result of chromosome rearrangements) into a gene pool. Mutations occur spontaneously in all the cells of the body, but only those that occur in germ cells contribute to evolutionary change because only they can become gametes and pass the mutation on to the next generation. Thus, mutations add new alleles to the gene pool, supplying the genetic foundation on which the other evolutionary forces operate.

Some mutations are beneficial, some are detrimental, and others are "neutral" and have no apparent effect on the survival or reproductive capacity of an organism. Many harmful mutations are immediately removed from the gene pool because they disrupt the structure and function of a protein whose activity is required for life to continue. Individuals with such lethal mutations typically die during embryonic development. Other harmful mutations are masked by a dominant allele. For example, each of us is believed to carry an average of seven to eight lethal reces-

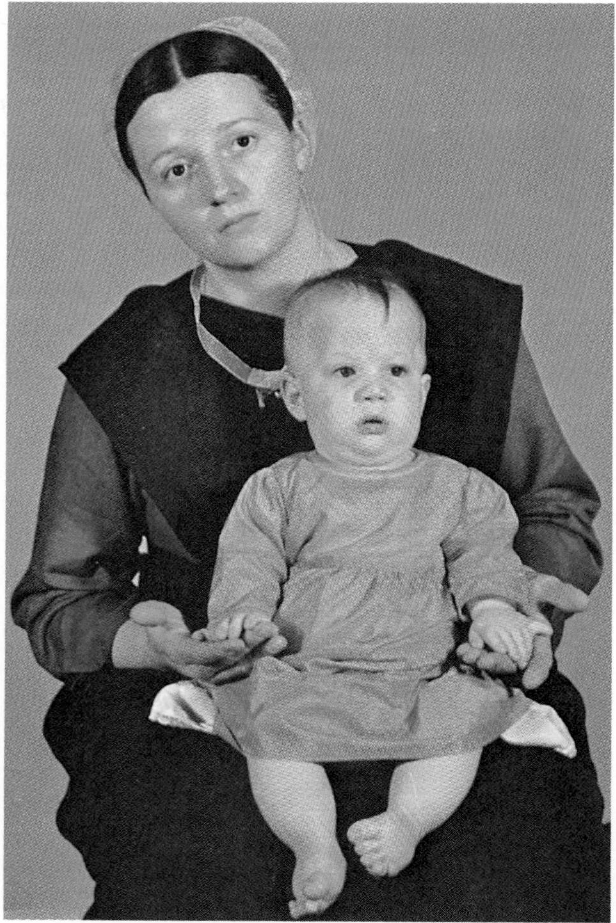

FIGURE 22-2

The founder effect. This Amish woman and her child are descendants of a small group of founding families who immigrated to Pennsylvania in the mid-eighteenth century. As a result of intermarriage, a recessive allele for Ellis-van Creveld syndrome present in one of the founders paired with the same allele, producing homozygotes with the disorder. The child pictured here has the shortened limbs and extra fingers that characterize this syndrome.

sive genes. The fact that we are alive testifies to the role of the dominant allele on the homologous chromosome.

Whether a mutation is beneficial, detrimental, or neutral often depends on the environment in which the organism is living at the time. If the environment changes, the effects of the mutation on survival and reproduction can also change. For example, a mutation that causes an enzyme to function optimally at a higher temperature will be beneficial if the environmental temperature rises and will be detrimental if the temperature falls.

GENE FLOW

Gene flow is the addition or removal of alleles when individuals exit or enter a population from another locality. It is common for animals or their larvae to migrate and for the seeds and pollen of plants to be dispersed by wind or birds to distant locations. Consequently, individuals from one population of a species are moved to another population, creating the opportunity for the transfer of alleles from one population's gene pool to another. Immigrants into a population may add new alleles to the population's gene pool, or they may change the frequencies of alleles that are already present. Emigrants out of a population may completely remove alleles, or they may reduce the frequencies of alleles in the remaining pool.

GENETIC DRIFT

Genetic drift is the result of random changes in allele frequency that occur solely by chance. Chance can affect allele frequency in several ways, but it is especially important during genetic recombination. When gametes are formed by meiosis, the segregation of chromosomes into any particular egg or sperm occurs by chance. When mating takes place, a great many of the gametes are wasted. Only a few combine to form new individuals, representing a random sample of the parents' genes.

Genetic drift occurs in populations of all sizes, but the effects of genetic drift are much more pronounced in small populations. In large populations, chance effects tend to be averaged out. However, even species that normally have large populations sometimes pass through periods when only a small number of individuals survive, such as following a long drought, a disease epidemic, intense predation, or natural catastrophes. During these so-called population *bottlenecks,* allele frequencies can change dramatically due to chance. During the last Ice Age, for example the southward movement of glaciers in North America and Europe squeezed many plant and animal species into small areas, reducing population sizes to very low levels.

When a species expands into another region, a new population is established by a small number of pioneering individuals. The founders are not likely to possess all the alleles found in the original parental population, so the new population that develops is likely to be strongly affected by genetic drift—a phenomenon known as the *founder effect.*

Many examples of the founder effect are seen in isolated locations. For example, when new islands appear, they become colonized by a few members of a species that arrive on the island by chance and are affected by genetic drift. Among animals, a single female arriving on an island carrying fertilized eggs or embryos is all that is required to found a new population. And since many plants can reproduce by either self-fertilization or asexual reproduction, a single seed can colonize a new environment by itself.

The founder effect has also been documented in human populations. For example, in the 1770s, a small number of Germans of the Amish sect emigrated to the United States. For over 200 years, this population remained, for the most part, reproductively isolated, with little outside marrying. One of the members of this founding group apparently carried a recessive allele for a rare form of dwarfism and polydactylism (extra fingers and toes), called the Ellis-van Creveld syndrome (Figure 22-2). A study carried out in the 1960s revealed that of the approximately 8,000 Amish living in the Lancaster area, 43 individuals were homozygous recessive for this allele and exhibited Ellis-van Creveld syndrome, representing more cases of this disorder than in the rest of the world combined! The high occurrence of such a rare disease is an example of the founder effect.

NATURAL SELECTION

Natural selection is the increased reproduction of individuals that have phenotypes that are better suited to survive and reproduce in a particular environment. Not all individuals survive and reproduce equally well in a given environment. As a result, some individuals contribute more offspring to the next generation than do others. As generations pass, those individuals with adaptive characteristics become more common, and those with detrimental characteristics are eliminated. In other words, the environment naturally selects the best adapted individuals. Of all the forces that influence evolution, only natural selection generates populations whose members are better adapted to their environment.

In the Steps to Discovery vignette at the beginning of Chapter 1, we described a well-documented example of natural selection: the change in coloration of peppered moths (*Biston betularia*) that occurred during the Industrial Revolution in England (Figure 22-3). The effects of industrial soot on the frequency of speckled and dark pep-

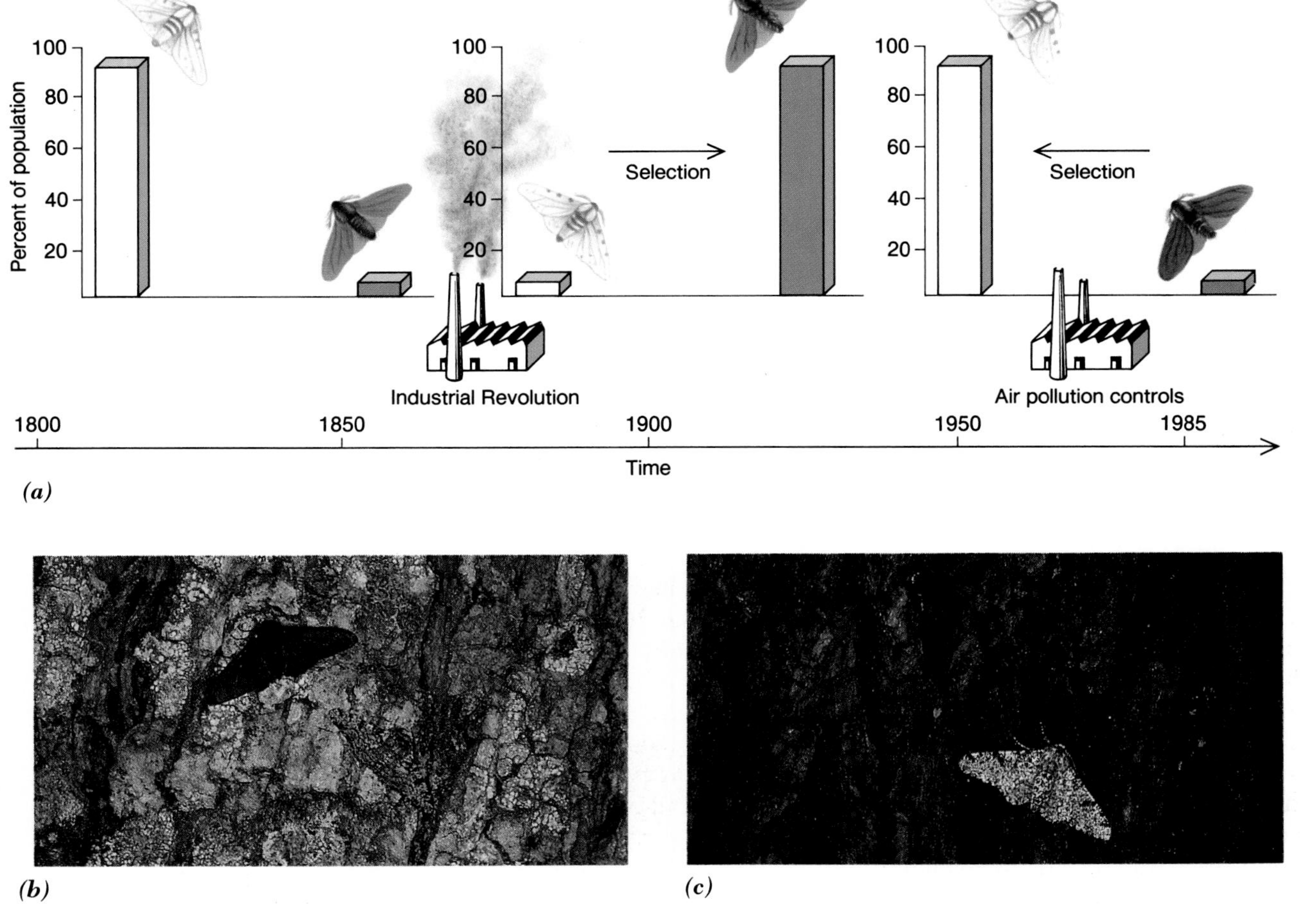

FIGURE 22-3
Natural selection of the peppered moth in England. ***(a)*** The effects of industrial pollution on the prevalence of speckled and dark varieties of peppered moths. ***(b)*** Before the Industrial Revolution, speckled moths were more abundant than dark moths because the darker variety was more easily spotted on light, lichen-covered tree trunks. ***(c)*** During the latter half of the nineteenth century, dark moths became more prevalent because they blended with soot-covered trees. Following recent pollution controls, light speckled individuals are once again more numerous than dark individuals.

TABLE 22-1

COMPARISON OF NUMBER OF RELEASED AND RECAPTURED SPECKLED VERSUS DARK PEPPERED MOTHS IN POLLUTED AND POLLUTION-FREE AREAS

	Number Released		*Number Recaptured*[a]	
Location	*Speckled*	*Dark*	*Speckled*	*Dark*
Birmingham (polluted)	64	154	16(25.0%)	82(53.2%)
Dorset (pollution-free)	393	406	54(13.7%)	19(4.7%)

[a] Percent recaptured is in parentheses.

pered moths were verified experimentally in a study by H.B.D. Kettlewell in the 1950s. Both speckled and pigmented peppered moths were marked with a spot of paint under their wings and were released into both polluted and pollution-free areas. When the survivors were recaptured, the results confirmed that speckled moths were favored in pollution-free areas and that dark moths were favored in polluted areas (Table 22-1). As Darwin had originally suggested, it is the environment that selects which variants in a population will survive to reproduce.

Natural selection leads to three patterns of change—stabilizing selection, directional selection, and disruptive selection (Figure 22-4). All three patterns of natural selection may be acting on any species at any point in time.

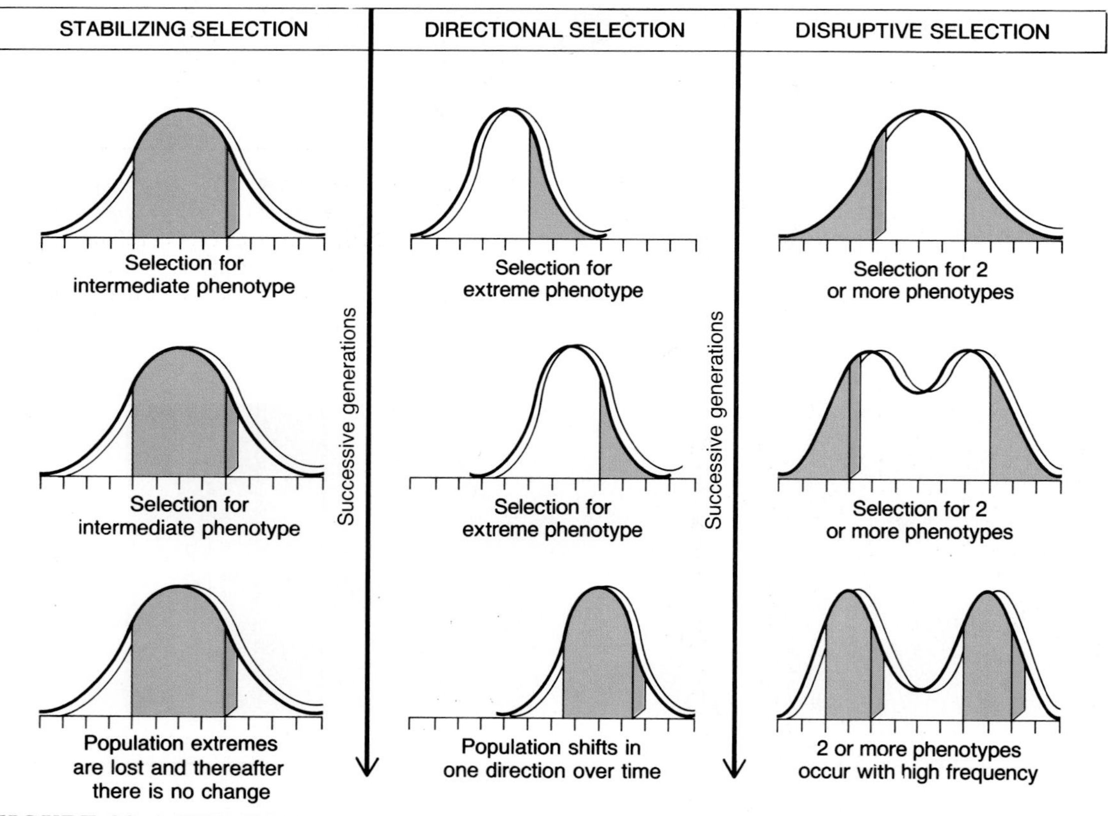

FIGURE 22-4

Three patterns of natural selection. The shaded areas represent the ranges of phenotypes for a particular trait that is favored by natural selection. A selection of the average (intermediate) phenotype ***(a)*** produces no change over three generations during stabilizing selection, whereas selection of extreme phenotypes leads to shifts in the characteristics of a population during directional ***(b)*** and disruptive selection ***(c)***.

Stabilizing Selection

Stabilizing selection occurs when individuals with extreme characteristics die or fail to reproduce, resulting in populations of individuals that possess intermediate characteristics (Figure 22-4*a*). In hummingbirds, for example, birds with very short or long beaks would likely not be able to arvest the nectar of many flowers as well as birds with ermediate-length beaks. A hummingbird with a very ort beak could not reach the nectar at the base of a flower, whereas a hummingbird with a long beak would easily pierce the bottom of a flower or damage the nectary.

Stabilizing selection is most common in unchanging environments, where populations achieve phenotypes that are optimally adapted to their surroundings. This may explain why so-called living fossils, such as the ginkgo (maidenhair tree), chambered nautilus, and horseshoe crab (Figure 22-5), which inhabit stable environments with few competitors, have remained essentially unchanged for tens or hundreds of millions of years.

(a)

(b)

(c)

FIGURE 22-5
"Living fossils." Certain organisms have remained virtually unchanged after many millions of years of stabilizing selection: ***(a)*** maidenhair tree (*Gingko*), essentially unchanged for 100 million years; ***(b)*** chambered nautilus (*Nautilus*), essentially unchanged for 180 million years; ***(c)*** horseshoe crab (*Limulus*), essentially unchanged for 360 million years. (Inset: 145-million-year-old fossil horseshoe crab from Germany.)

Directional Selection

In *directional selection*, phenotypes at only one extreme die or fail to reproduce, while those at the other extreme leave a higher number of offspring, causing the allele frequency to shift gradually in the direction of the favored phenotype (Figure 22-4*b*). Directional selection occurs when there is a change in the environment such that the phenotype at one extreme loses its selective advantage. We have already discussed two examples of directional selection in this chapter: the shift in frequency from the light-colored peppered moth to the dark form during the Industrial Revolution in England, and the increased resistance of mosquitoes to DDT. Two examples from human evolution are increased brain size and loss of body hair, both of which represent directional changes in phenotype.

Disruptive Selection

In *disruptive selection*, extreme phenotypes become more frequent from generation to generation because individuals with intermediate phenotypes die or fail to reproduce (Figure 22-4c). Disruptive selection promotes dimorphism

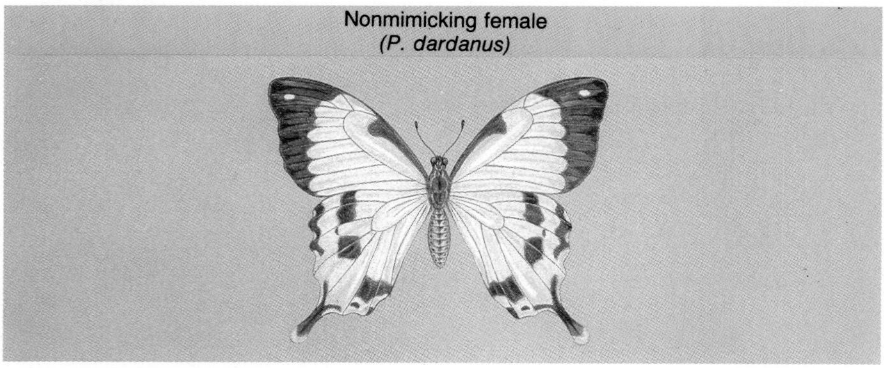

FIGURE 22-6

Altered states. Female African swallowtail butterflies mimic the appearance of local, foul-tasting butterfly species, creating strikingly different female phenotypes, even though they all belong to the same species. These different forms of females provide an example of disruptive selection and polymorphism.

(two forms of a trait) or even polymorphism (several forms of a trait) in a population. This may happen in a diverse or cyclically changing habitat, where different individuals are active at different times or in different parts of the environment. An example of disruptive selection is found in female African swallowtail butterflies (*Papilio dardanus*). Although these butterflies are all members of the same species, the species is widespread, and individuals from one locale are strikingly different in appearance (phenotype) from those in other areas (Figure 22-6). Why would this occur? Some species of butterflies combat predation by concentrating noxious chemicals in their bodies from the plants on which they feed. After one or two nauseating bites, birds learn to recognize these distasteful butterflies and leave them alone. The female African swallowtail butterflies lack these chemicals and would make a tasty meal for a bird, were it not for the fact that they closely resemble (mimic) the distasteful species. Distasteful butterflies tend to live in small populations, however. Female African swallowtails will only be protected by mimicking the species of distasteful butterfly that lives in their own small geographic area. Consequently, natural selection has favored the evolution of several distinct color patterns. Intermediate phenotypes between two local groups would not resemble any distasteful butterflies and would be devoured by the birds.

NONRANDOM MATING

Individuals that have phenotypes that make them more likely to be selected as mates are likely to leave the greatest number of offspring. Because some individuals in a population have a greater chance of mating than others, **nonrandom mating** results. The spectacular tail feathers of a peacock and the spreading antlers of a male deer (Figure 22-7) appear as if they could actually impede the animal's pursuit of food or escape from predators. Since these characteristics improve the chances of attracting females and reproducing, however (and in natural selection, passing on your genes is all that matters), they will be selected for. This form of natural selection is called *sexual selection.*

Sexual selection often leads to sexual dimorphism (Figure 22-8); that is, visible differences between the male and female of the same species. Sexual selection is common among animals because a female's reproductive success is limited by the number of eggs she can produce in her lifetime, and a male's reproductive success is limited by the number of females he can inseminate. Therefore, it is to the female's advantage to choose the most fit male as her mate, and it is to the male's advantage to attract as many females as possible. This leads to natural selection of certain male characteristics, either through male com-

(a)

(b)

FIGURE 22-7
Sexual selection. Although they may actually hinder the individual's mobility, brilliant displays of feathers ***(a)*** or racks of antlers ***(b)*** are products of sexual selection for characteristics that increase an individual's chances of mating.

(a)

FIGURE 22-8

Sexual dimorphism. The larger elephant seal ***(a)*** and the more brightly colored duck ***(b)*** are the males of those species.

(b)

petition with one another or through female choice. Consequently, in these species, males develop characteristics that enable the animal to fight or intimidate other males, like the antlers of a deer or the huge body size of the male elephant seal. At the same time, females choose a mate, so natural selection favors those characteristics that females prefer.

One form of nonrandom mating is inbreeding—mating between close relatives. Inbreeding can have a profound effect on evolution, because most harmful alleles originate as rare mutations and are limited to a small percentage of the population. Usually, these mutations exist as recessive alleles. If inbreeding does not occur, the chances are slight that two organisms will have the same harmful recessive alleles. If the two organisms are closely related and have received their alleles from a common ancestor, however, the chances of their carrying the same harmful recessive alleles are much greater. If these related organisms mate, chances are they will produce offspring with the defective phenotype. When these offspring fail to

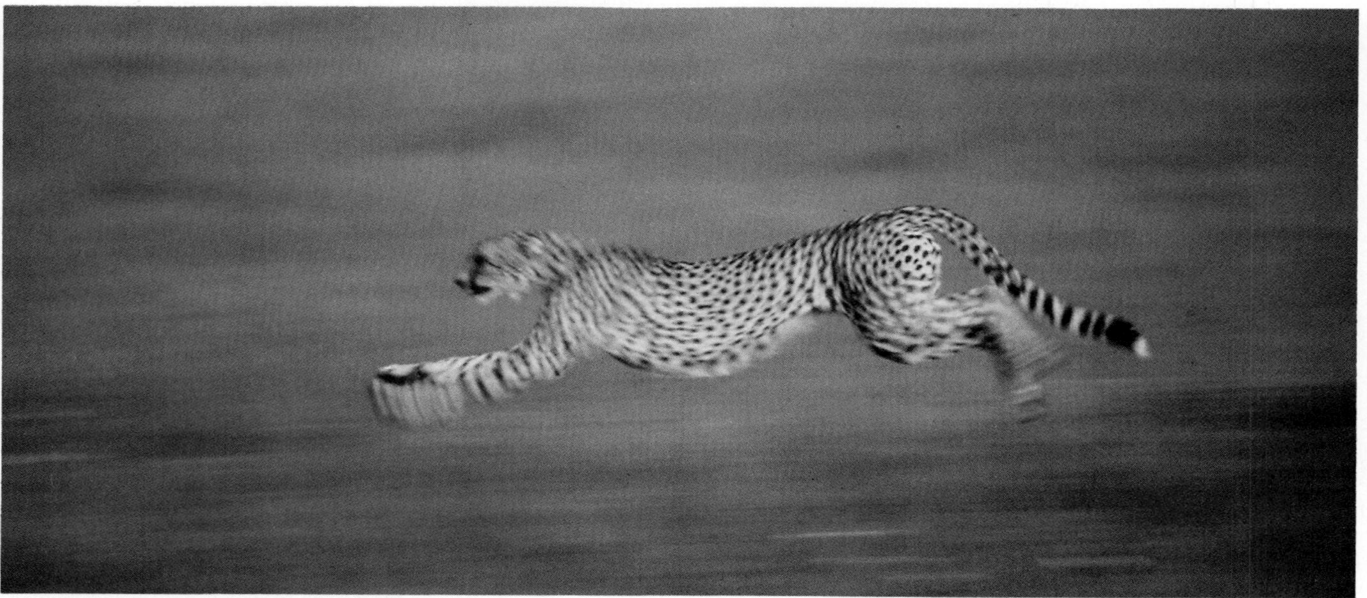

FIGURE 22-9

The cheetah population has lost most of its genetic variability due to inbreeding after the population was drastically reduced in size within the past 20,000 years

reproduce and die, all of their genes (beneficial and deleterious) are removed from the gene pool. Therefore, since inbreeding increases the likelihood of death due to genetic defects, the population will lose more and more of its variability. If the environment changes, the population may lack sufficient variability to adapt to the change and may become extinct.

Smaller populations are affected more dramatically by nonrandom mating (and inbreeding, in particular) than are large ones. The effects of inbreeding are illustrated by the cheetah (Figure 22-9), the world's fastest land animal. At the present time, there are approximately 20,000 cheetahs left in the world, a number that normally would not indicate any danger of extinction. But cheetahs are different. Biologists have discovered that cheetahs are not a healthy species. Cheetah cubs have a much lower survival rate than do the cubs of other species of large cats. Cheetah cubs are more susceptible to diseases, such as distemper, and the adult males typically produce a small number of sperm, most of which have an abnormal shape.

The poor health of cheetahs is attributed to the fact that cheetahs are so genetically similar that they won't even reject tissue grafts from one another, a phenomenon unknown among other mammals. Sometime within the past 20,000 years, conditions arose that apparently decimated cheetah numbers. As the population of cheetahs dwindled to a few individuals, the gene pool was drastically reduced, creating a bottleneck. As the few survivors interbred, the offspring became increasingly homozygous. Consequently, harmful recessive mutations paired more frequently, producing homozygous recessive individuals that died, taking with them some of the desirable alleles at other loci, further reducing genetic variability.

THE EVOLUTIONARY PROCESS: MICROEVOLUTION AND MACROEVOLUTION

The examples we have discussed up to this point—pigmentation in peppered moths, loss of genetic variation in cheetahs, pesticide resistance in mosquitoes—are considered examples of **microevolution** because they result from changes in the allele frequency of a species' gene pool but have not resulted in the appearance of new species, a phenomenon referred to as **macroevolution.** Microevolution reveals the process of evolutionary change over a short enough period of time so that it can be documented and studied. The occurrence of microevolution allows biologists to study the underlying mechanisms—mutation, gene flow, genetic drift, natural selection, and nonrandom mating—which, given sufficient time, lead to macroevolution, the process that results in the origin of new species.

SPECIATION: THE ORIGIN OF NEW SPECIES

Biologists have hypothesized that there are more than 5 million species of organisms alive today, even though only about 1.8 million have been described and named so far. **Speciation**—the process by which new species are formed—occurs when one population splits into separate populations that diverge genetically from one another to the point where they become separate species.

Species is a Latin word meaning "kind." Kinds, or species, of organisms were originally identified by their appearances because members of a species typically look alike. In some cases, identifying an individual as a member of a particular species is not so simple, however, because individuals from distinct species may appear very similar (Figure 22-10). When we consider the more numerous smaller animals, all plants, and all microbes, the problem of identification becomes even more difficult.

Distinguishing between species of similar morphology may require the analysis of biochemical, ecological, and behavioral traits, as well as those visible to the eye. Some species may be difficult to identify because the members of the species population have different appearances due to a high degree of genetic variation. This is illustrated by the polymorphic African butterflies depicted in Figure 22-6.

FIGURE 22-10
Are these elephants members of the same or different species? The elephant on the left is from Africa and is a member of the species *Loxodonta africana,* whereas the elephant on the right is from India and is a member of a different species, *Elephas maximus.* The Indian elephant has smaller ears and more pronounced "bumps" on its head.

The most important criterion for defining a species is reproduction; members of the same species are capable of producing other members by mating. While this definition works well in defining animal species, it does not always hold for plants. Among shrubs and trees, in particular, closely related species may interbreed and form fertile hybrids that then give rise to a population of hybrid individuals.

According to the "reproduction" criterion for a species, members of one species are *reproductively isolated* from members of all other species. That is, reproductive isolation, which prevents the exchange of genes between populations, is the first step leading to the formation of new species. Once the gene pools are isolated, the separated populations inevitably diverge because of differences in mutation, mating patterns, genetic drift, and natural selection. Over time, the isolated populations amass morphological, physiological, and behavioral differences that prevent them from interbreeding. Consequently, even if the original cause of isolation is removed, the populations remain reproductively isolated; they have become different species. Barriers that prevent the exchange of alleles between populations (gene flow) are called isolating mechanisms (Table 22-2).

PATHS OF SPECIATION

The millions of different species that exist today did not emerge by any single sequence of events but have come into existence through five primary paths of speciation:

- **Phyletic speciation** is the gradual accumulation of changes in a lineage through time, until one group is distinct enough to be considered a new species.
- **Allopatric speciation** (*allo* = other, *patri* = habitat) occurs when a physical barrier, such as a mountain range, a river, or even an oil pipeline, geographically separates a population from its parental population,

TABLE 22-2
ISOLATING MECHANISMS

1. Prezygotic Isolating Mechanisms
Ecological Isolation Different habitat requirements separate groups, even though the inhabitants may exist in the same general location. Example: Head and body lice are morphologically very similar, yet they live in different "habitats" on a single human body. Head lice live and lay eggs in the hair on the head of a human, whereas body lice live and lay their eggs in clothing. Both suck blood for nutrition.
Geographical Isolation Emerging mountains, islands, rivers, lakes, oceans, moving glaciers, and other geograhic barriers keep groups isolated. Example: Different tortoises are found on different Galapagos Islands; surrounding oceans keep tortoise populations isolated.
Seasonal Isolation Differences in breeding seasons prevent gene flow, even when populations are found in the same area. Example: Two populations of bigberry manzanitas grow close together in the mountains of southern California, yet the populations do not interbreed because one completes blooming 2 weeks before the other begins to bloom.
Mechanical Isolation Physical incompatibility of genitalia. Example: Genital structures differ in shape for alpine butterfly species, even though these butterflies look nearly identical in all other ways.
Behavioral Isolation Differences in mating behavior prevent reproduction. Example: Many animals have evolved complicated courtship activities before breeding. Some species of fruit flies (*Drosophila*) are indistinguishable to our eyes, yet they do not mate with each other because of differences in courtship behavior.
Gamete Isolation Sperm and egg are incompatible. Gamete isolation is a common isolating mechanism in many plant and animal species.
2. Postzygotic Isolating Mechanisms
Hybrid Inviability Zygotes or embryos fail to reach reproductive maturity. Example: Hybrid embryos formed between two species of fruit flies fail to develop.
Hybrid Sterility Fertilization is successful between two species, but hybrid progeny are sterile. Example: A mule is a sterile hybrid produced from a mating between a horse and a donkey.

thereby cutting off gene flow between the two. While isolated, the separated population develops a number of genetic differences, including a reproductive barrier. At this point, the two populations can be considered separate species.

- **Parapatric speciation** (*para* = beside) occurs when gene pools in populations that lie adjacent to one another diverge because the environment varies sufficiently in the different locales. As a result, different traits are selected in each population.
- **Sympatric speciation** (*sym* = same) occurs in populations where individuals continue to live among one another but some type of biological difference, such as the time of the year when gonads mature, divides the members into different reproductive groups. The best-accepted cases of sympatric speciation occur in plants as a result of polyploidy—an increase in the number of sets of chromosomes per cell. The appearance of tetraploid (4N) offspring from diploid (2N) parents is not an uncommon occurrence among certain types of plants. Once formed, the tetraploid cannot interbreed with diploid members of the population because of chromosome incompatibility. The tetraploid plant must either fertilize another tetraploid individual in the population, or it must produce a population of tetraploid plants on its own (by either self-fertilization or asexual reproduction).
- **Hybridization** occurs when two distinct species come into contact, mate, and produce hybrid offspring that are often reproductively isolated from either parent but not from one another. In just one generation, an entirely new species can be generated by hybridization. Plants are more tolerant of polyploidy than are animals; a single plant with a unique complement of chromosomes can generate a new population by self-fertilization or asexual reproduction. As a result, hybridization is a common means of speciation in plants.

SPECIATION AND PATTERNS OF EVOLUTION

As new species form and adapt to their environments through natural selection, different patterns of evolution emerge. The most common pattern is **divergent evolution;** it occurs when two or more species evolve from a common ancestor and then become increasingly different over time (Figure 22-11*a*). Divergent evolution forms the basis for phylogenetic branches, whereby one ancestral species gives rise to two distinct lines (lineages) of organisms that continue to diverge. Monkeys and apes, for example, diverged from a common ancestor, as did apes and humans (Chapter 23).

Sometimes, when members of a species move into a new area with many diverse environments, new species

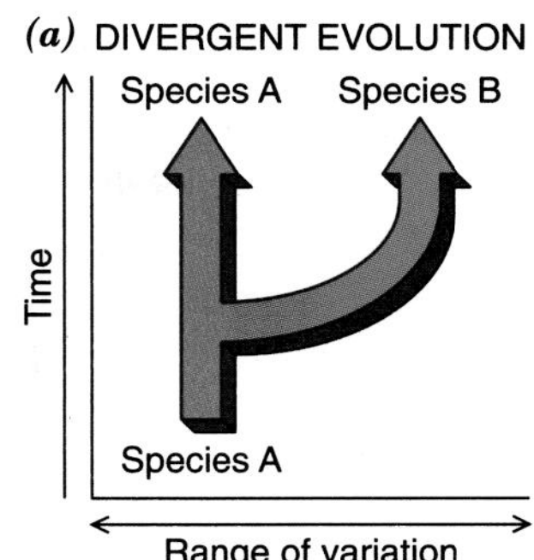

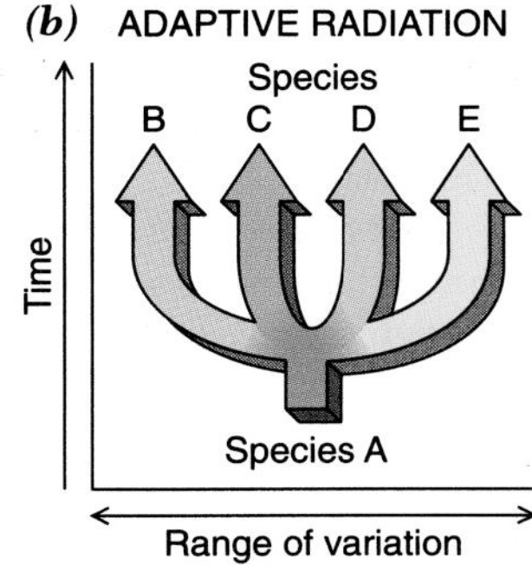

(*c*) CONVERGENT EVOLUTION
Species A Species B
Time
Range of variation

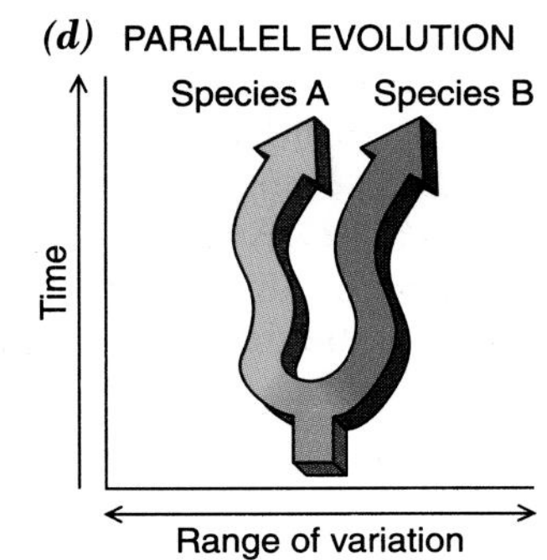

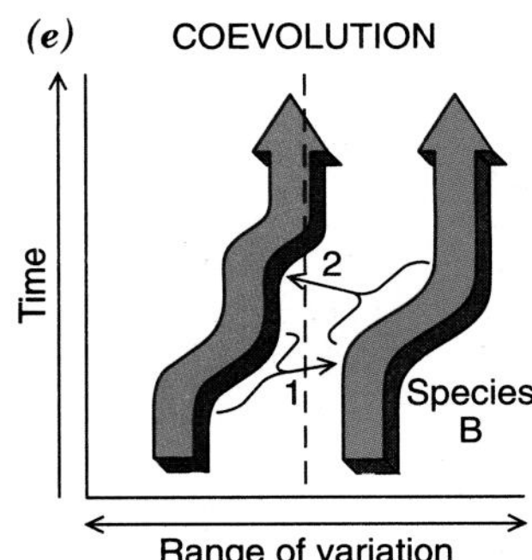

1 Change in species A creates new conditions that cause a change in species B

2 Change in species B creates new conditions that cause a change in species A

FIGURE 22-11

Patterns of evolutionary change. ***(a)*** Divergent evolution: One species splits into two species. ***(b)*** Adaptive radiation: One species gives rise to many new species that are adapted to different types of habitats and/or food sources. ***(c)*** Convergent evolution: Unrelated species evolve similar characteristics as the result of similar selective pressures. ***(d)*** Parallel evolution: Two related species remain similar over long periods of time. ***(e)*** Coevolution: Two species evolve in such a way so that changes in one causes reciprocal changes in the other.

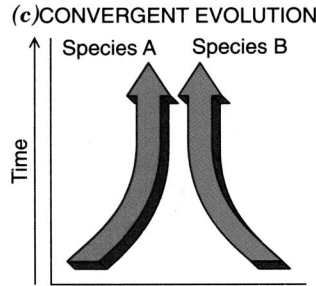

form and rapidly diverge, producing a variety of related species that are adapted to different habitats. This rapid divergent evolution is referred to as **adaptive radiation** (Figure 22-11*b*).

There are a limited number of solutions to any environmental problem. For example, rapid movement through the water requires a streamlined body shape, while movement through the air requires wings. Therefore, when species with different ancestors colonize similar habitats, they may independently acquire similar adaptations and resemble one another superficially. This phenomenon is called **convergent evolution** (Figure 22-11*c*). For example, each of the four marine animals depicted in Figure 22-12 is descended from a different terrestrial ancestor, but they all share certain common adaptations, such as streamlining and paddle-like forelimbs. The similarity among many Australian marsupials and placental mammals on other continents is another example of convergent evolution (Figure 22-13).

Parallel evolution occurs when two species that have descended from the same ancestor remain similar over long periods of time because they independently acquire the same evolutionary adaptations (Figure 22-11*d*). Parallel evolution occurs when genetically related species adapt to similar environmental changes in similar ways. For example, the ancestral arthropod had a segmented body with a pair of legs on each segment. In all three major arthropod lineages that have descended from this ancestor (the crustaceans, insects, and spiders), the number of legs has decreased, and the body segments have become fused, forming larger structures with specialized functions.

Since organisms form part of the natural environment, they can also act as a selective force in the evolution of a species. In nature, species frequently interact so closely that evolutionary changes in one species may cause evolutionary adjustments in others. This evolutionary interaction between organisms is called **coevolution** (Figure 22-11*e*). Flowering plants and their insect pollinators have coevolved for millions of years, leading to many finely tuned structural and behavioral relationships between flowers and pollinators. Other examples of coevolution include predator–prey interactions, where improvements in the hunting ability of a predator favor the survival of a prey with characteristics that increase its ability to escape. Parasites and their hosts also coevolve: Parasites tend to become less destructive of their hosts (a dead host means a dead parasite), and hosts tend to become more resistant to the parasite.

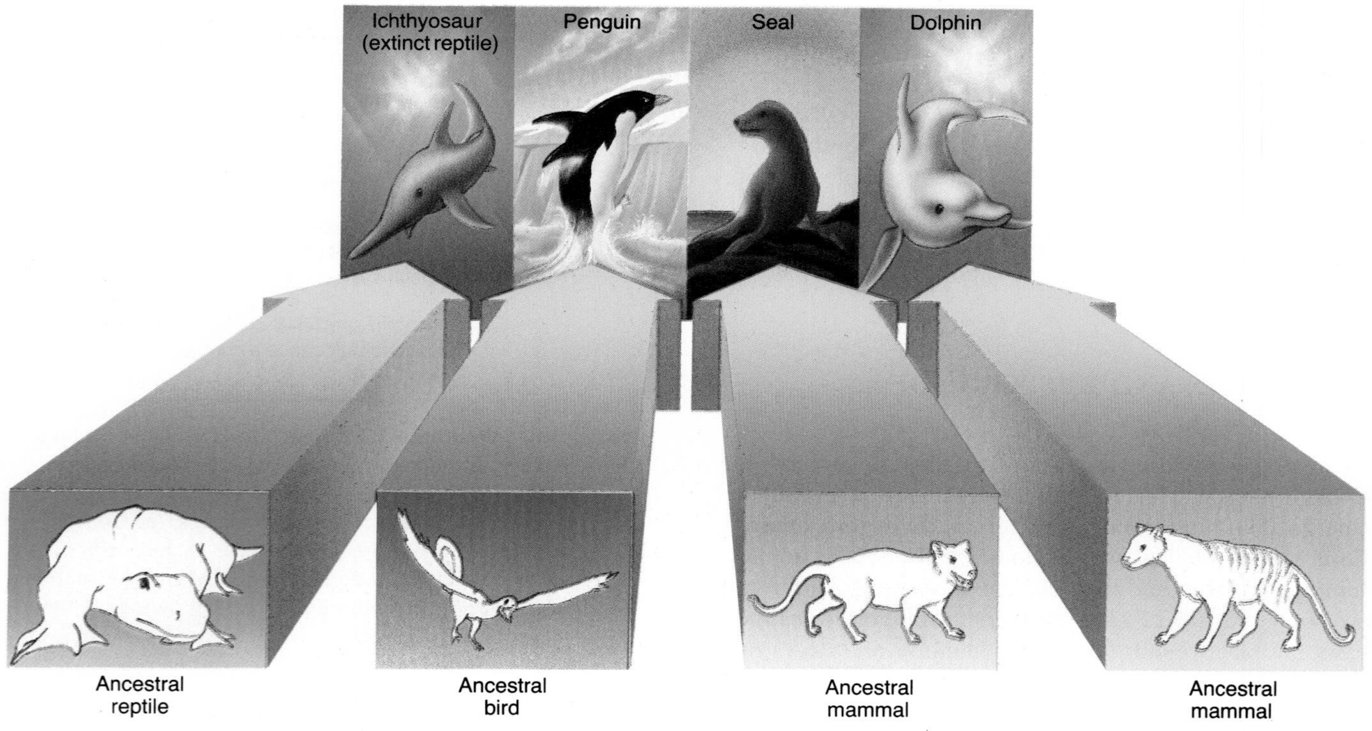

FIGURE 22-12
Convergent evolution. These marine animals are descended from different ancestors. They have all developed similar streamlined bodies and paddle-like front limbs that adapt them to life in the water.

Wolverine

Tasmanian Devil

Southern Flying Squirrel

Sugar Glider

FIGURE 22-13

Living examples of convergent evolution. Australian marsupial mammals and placental mammals on different continents have similar features because they have adapted to similar habitats. Placental mammals are shown on the left, and the Australian marsupial counterpart is on the right.

EXTINCTION: THE LOSS OF SPECIES

When you consider that approximately 99 percent of the species that have existed since the beginning of life on earth are no longer alive, it becomes apparent that **extinction,** the loss of a species, is the ultimate fate of most, if not all, species.

Species may become extinct when they lack genetic variability or when they find themselves in the wrong place at the wrong time. In other words, extinction is due to either bad genes or bad luck. In the first case, a species can become extinct when the environment changes and none of the species' members has the genetic makeup to survive under the new conditions. In the second case, a species may face an unusual catastrophe that essentially eliminates all life in its habitat. Some of the causes that have been attributed to mass extinctions of a multitude of species in the past include asteroid impact, volcanic eruptions, drastic changes in sea level, and radical shifts in the earth's climate. Today, organisms on earth are faced with a new cause of mass extinction: humans. The unbridled destruction of natural habitats by humans has increased the extinction rate from a long-term average of about one species each 1,000 years to hundreds, and perhaps thousands, of species in a single year.

THE PACE OF EVOLUTION: GRADUALISM AND PUNCTUATED EQUILIBRIUM

Darwin viewed evolution by natural selection as a steady, uninterrupted process. He believed that just as natural selection adapted a population to its environment, the process could also turn that population into a new species and eventually found a whole new order, class, or phylum. The discovery of the importance of mutation, gene flow, genetic drift, nonrandom mating, and natural selection seemed to confirm Darwin's view that most evolution occurs in small, adaptive steps. Under such a model, evolution proceeds by **gradualism** (Figure 22-14*a*). This view has been questioned by some paleontologists (biologists who study the fossil remains of animals that lived in the past) who contend that fossil evidence does not often show a gradual succession of forms. Rather, the analysis of fossils of numerous groups indicates that long periods without significant change (periods of "stasis") are interspersed with short periods of very rapid change.

In 1972, paleontologists Niles Eldredge of the American Museum of Natural History in New York and Stephen Jay Gould of Harvard University proposed a hypothesis called **punctuated equilibrium** (Figure 22-14*b*) to explain this pattern of evolution. The punctuated equilibrium model includes two separate proposals. The first states that speciation, when it occurs, is a rapid process. We have already seen how allopatric speciation can lead to new species and how small populations are changed more rapidly than are larger ones. When both phenomena occur together (allopatric separation of small populations), spurts of speciation can occur.

The second proposal is that, once formed, species exist for long periods of time without change, unless the environment is altered in some way. This stasis occurs because, even in semistable environments, species often reach a population size large enough for stabilizing selection and gene flow to operate, preventing the species from changing into a new species.

Although gradual evolution and punctuated equilibrium are alternative explanations for the evolution of new species, one does not necessarily exclude the other. The questions now being debated among many biologists are whether one phenomenon occurs more frequently than the other, and which one most likely occurred in the evolution of a particular group.

EVOLUTION AND ADAPTATION: TYING IT TOGETHER

The most striking result of evolution by natural selection is adaptation. Adaptations can be morphological (such as sharp teeth and claws, horns, or trichomes), behavioral (such as swimming in schools or hunting in packs), or

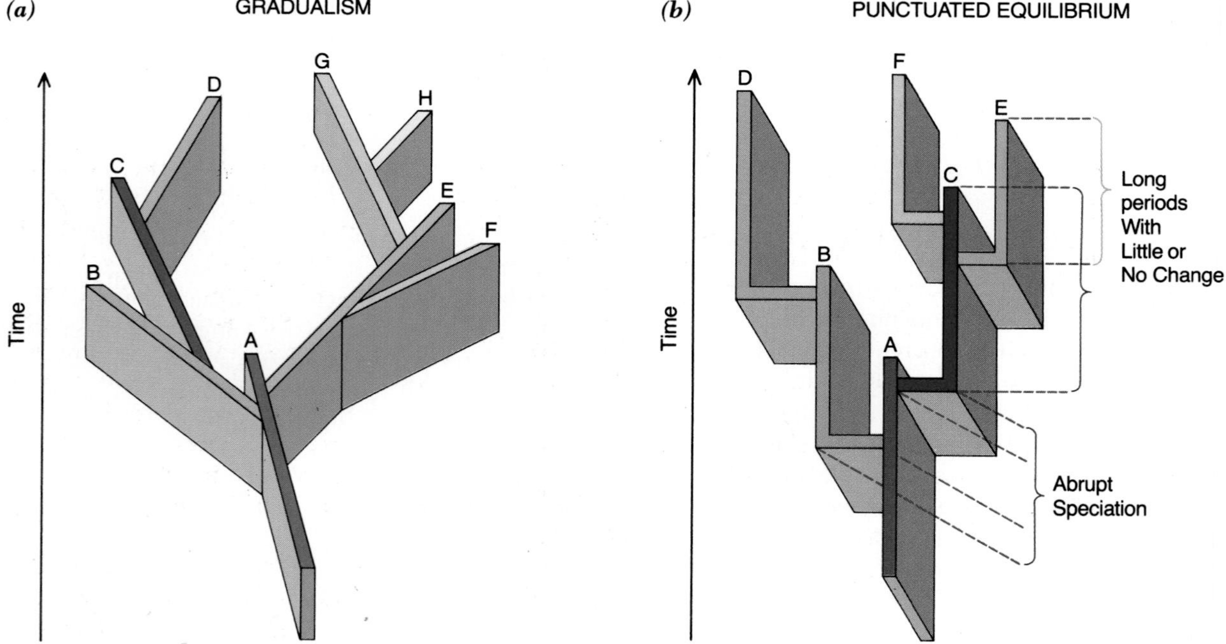

FIGURE 22-14

Gradualism versus punctuated equilibrium. ***(a)*** During gradualism, species arise through gradual, steady accumulation of changes. ***(b)*** During punctuated equilibrium, species arise as a result of the rapid accumulation of changes. Once formed, the species remains relatively unchanged for long periods of time.

FIGURE 22-15

Some remarkable adaptations. ***(a)*** Ringed snake (*Natrix natrix*). ***(b)*** Horned toad (*Certophrys ornata*). ***(c)*** Red mangrove. ***(d)*** Deep-sea angler fish (*Edriolychunus schmidti*).

reproductive (such as nurturing and protecting young or choosing a life-long mate), as the following examples illustrate.

When threatened, the ringed snake (*Natrix natrix*) feigns death by dropping its head, dangling its tongue out of its mouth, and lying completely motionless (Figure 22-15*a*). These behavioral actions help secure the snake's safety because most predators avoid dead organisms.

An effective combination of morphological and behavioral adaptation is revealed in the camouflaged horned toad (*Ceratophrys ornata*). The horned toad buries its body in mud, leaving only its eyes and large jaws protruding (Figure 22-15*b*). Because of its effective camouflage, unwary prey quickly disappear as they move over the concealed head.

The red mangrove demonstrates an effective reproductive adaptation. The seeds of red mangrove trees germinate while they are on the tree (Figure 22-15*c*). When released, the streamlined radicle slices through the water, planting the seedling upright. Seedlings that don't reach the bottom are able to float for months, until they run aground in shallow water and take root. An unusual example of reproductive adaptation in animals involves male and female deep-sea angler fish (*Edriolychunus schmidti*). For the male, life as an independent organism is over almost as soon as it begins. As an adaptation to allow members of the opposite sex to find each other in the blackness of the deep sea, the newly hatched male angler fish permanently attaches itself to the female by sinking its jaws into her body. The female's skin grows over the male's body, and the individuals' circulatory systems connect. The male becomes incorporated into the female body and is reduced to nothing more than a small sperm factory (Figure 22-15*d*).

Natural selection is the only evolutionary force that changes gene frequency in ways that lead to adaptation. In each of these cases, it was natural selection that produced the truly remarkable adaptations that improved the individual's chances of survival and reproduction. Evolution by natural selection often produces adaptations that seem to fit an individual exactly to the world in which it lives, causing us to marvel at the many wonders of nature.

SYNOPSIS

Evolution is the process whereby species become modified over generations. Evolution results when the frequency of alleles in the gene pool of a population changes from one generation to the next. Changes in allele frequency result from: mutation (the introduction of new alleles), gene flow (the addition or removal of alleles when individuals move from one population to another), genetic drift (alterations in allele frequency due to chance), natural selection (increased reproduction of individuals with phenotypes that make them better suited to survive and reproduce in a particular environment), and nonrandom mating (increased reproduction of individuals with phenotypes that make them more likely to be selected as mates).

Mutation and gene flow introduce new genetic material into a population, while genetic drift, natural selection, and nonrandom mating determine which alleles will be passed on to the next generation. Genetic drift is particularly important in small populations, where chance events can have a major impact on a population's gene pool. Genetic drift becomes most important when a population shrinks in size forming a bottleneck, or when a small group of individuals colonize a new habitat.

Natural selection is the only factor that causes a species to adapt to its environment. Natural selection is particularly important when environments change, allowing those individuals that possess favorable phenotypes to survive and reproduce, thereby passing their alleles on to the next generation. Natural selection can have a stabilizing, directional, or disruptive effect on the gene pool of a population.

The diversity of life on earth has arisen through repeated speciation events. For speciation to occur, a population must split into two or more separate populations that can no longer interbreed. Such reproductive isolation usually occurs as a result of the formation of a geographic barrier.

Evolution results in several identifiable patterns. The most common is divergent evolution, whereby one species gives rise to two or more species. Divergent evolution can sometimes result in adaptive radiation, whereby several new species form from a single ancestor. Another evolutionary pattern, convergent evolution, results when unrelated species colonize similar environments and acquire similar adaptations that cause them to resemble one another. During coevolution, changes in one species influence the course of evolution of another species. Extinction is the ultimate fate of most, if not all, species. Extinction occurs when a species lacks the genetic variability needed to adapt to a changing environment or when a sudden catastrophe occurs that essentially eliminates all life in a particular habitat.

The pace of evolution need not be constant. Evolution within a lineage of organisms may occur gradually in small, adaptive steps, or it may occur in spurts, in which species form and remain unchanged for long periods, followed by a period of rapid change.

Review Questions

1. Match the term with its definition.

____1. allopatric speciation
____2. disruptive selection
____3. sympatric speciation
____4. gene pool
____5. gene flow
____6. genetic drift
____7. extinction
____8. convergent evolution

a. formation of a species by geographic isolation
b. formation of a species by ecological isolation
c. result of emigration and immigration
d. change in gene frequencies due to chance
e. organisms resemble each other because of similar adaptive pressures, not common ancestry
f. extreme phenotypes in a species leave more offspring than do average phenotypes
g. the death of every member of a species
h. all of the alleles in all of the members of a species.

2. Of all the factors that cause allele frequency to change over time, why is natural selection the only one that leads to increased adaptation to the environment?
3. Why must gene flow stop before speciation can occur?
4. Why do genetic drift and gene flow have a greater impact in changing the gene frequencies of a small population than of a large one?
5. Did speciation occur in the peppered moth populations of England? Under what conditions might speciation occur in the moth?

Critical Thinking Questions

1. Of the five factors that can affect allele frequencies in a population, which could have been important for insects in developing resistance to pesticides? Which could have been unimportant? Why?
2. Populations that are not changing must meet the five conditions identified by Hardy and Weinberg. Consider each of these conditions for the case of the Amish sect living in the United States. Do any of the five apply? If so, which one(s)?
3. New reproductive technologies, such as improved artificial insemination, have revolutionized the management of domestic animals. For example, many of the dairy cows in the United States have the same father or grandfather. What is the evolutionary disadvantage of having such a small number of fathers for the population?
4. Two very similar squirrels are found on the north and the south rims of the Grand Canyon. The Kaibab squirrel of the north rim is distinctly darker than the Abert squirrel of the south rim, however. Interbreeding occurs rarely, if ever, in nature, but could occur between members of the two squirrel species, producing fertile offspring. Would you consider these squirrels members of the same or different species? Why?
5. Match the following examples with the following patterns of evolution: divergent evolution, adaptive evolution, phyletic evolution, convergent evolution, coevolution, parallel evolution. For each one, explain why the example fits the pattern. (1) Bears and pandas, (2) wolves and foxes, (3) the yucca plant and the yucca moth, (4) *Homo erectus* and *Homo sapiens*, (5) ostrich and emu, (6) 16 species of Hawaiian honeycreepers evolved from a common ancestor, each with a different niche.

CHAPTER 23

Evidence for Evolution

Archaeologists search fossil beds in Africa for remains of early humans illustrated in the time bubble.

STEPS TO DISCOVERY
An Early Portrait of the Human Family

Scientists use the term **hominid** to refer to humans and the various groups of extinct, erect-walking primates that were either our direct ancestors or their relatives. The first hominid fossil was unearthed in 1856 in caves of the Neander Valley in Germany. At first, the remains were dismissed as the bones of a deformed Russian soldier. After the discovery of similar bones in other locations around Europe, however, it became apparent that the earth had been inhabited at one time by "people" that resembled humans. They were called Neanderthals.

Neanderthals lived between 35,000 and 135,000 years ago. Their skulls had a different shape than that of modern humans, with heavy, bony ridges over the eyes. In addition, the bones of Neanderthals were much thicker, with indications of larger attached muscles. Although Neanderthals are often depicted as brutish-looking, grunting, stooped-over cavemen, in reality, if you were to see one of these beings walking down the street in jeans and a T-shirt, you probably wouldn't turn and take notice.

One hominid that might cause you to take notice was discovered in 1891 by Eugene Dubois, a Dutch doctor stationed on the island of Java in the Dutch East Indies. While excavating a site one day, Dubois found a back (molar) tooth that he thought must have belonged to an ape. A meter away, he discovered a skull that possessed characteristics of both human and ape anatomy. The next year, approximately 15 meters from where he had found the skull, Dubois unearthed a thigh bone (femur) that was very similar to that of a modern human. Most importantly, the shape of the femur indicated that the owner had walked erect. Dubois concluded that he had found the "missing link and returned triumphantly to Europe.

Upon seeing Dubois's discoveries, Sir Arthur Keith, one of the most prominent paleontologists of the time, concluded that, even though the size of the Java Man's braincase (the part of the skull that covers the brain) was not much larger than that of an ape, the skull showed definite human features. Keith recommended the Java Man be placed in the same genus as are modern humans. Eventually, Dubois's find became designated *Homo erectus*. However, Dubois never accepted Keith's view that Java Man should be classified as *Homo* and buried the bones of his missing link under the floorboards of his dining room, where they remained for the next 30 years.

Over the next 30 to 40 years, a number of other fossils were found that were similar to that of Java Man and were also assigned to the species *H. erectus*. The most important find was Peking Man, discovered in a cave near Peking, China. Like Java Man, Peking Man had a small, apelike braincase; thick, heavy bones; a prominent, bony ridge above the eyes; and a humanlike lower jaw with humanlike teeth. Most importantly, it was demonstrated that Peking Man had walked with an erect posture, used stone tools, and cooked his dinner over a fire. Both Java Man and Peking Man lived about half a million years ago.

Two fossil finds did not fit the profile of *H. erectus*, however. One was a remarkably complete skull that was discovered by an amateur fossil hunter in 1912 near the town of Piltdown in England. The skull of this so-called Piltdown Man had a large braincase (as large as that of a modern human) and an apelike jaw, characteristics in direct contrast to those of Java Man and Peking Man. Piltdown Man presented a serious problem for interpreting the path of human evolution.

The other perplexing fossil discovery was made in 1924 by Raymond Dart, an Australian on the faculty of a medical school in Johannesburg, South Africa. Dart heard that fossils were being uncovered at a limestone quarry in an area of South Africa called Taung. He asked the owner of the quarry if he might examine some of the fossils; two large boxes were shipped to his house. Dart spotted a dome-shaped piece of stone. As a neuroanatomist, he immediately recognized the stone as the cast of a brain, complete with indications of convolutions and blood vessels. Although the brain was the size and form of an ape's, it revealed distinct humanlike characteristics.

Among the contents of the box, Dart found the remains of the lower jaw and skull, the front of which was covered by an encrusted material. Dart carefully picked away at the crust and slowly revealed an astonishing visage; it was the face of a young "ape," with teeth that showed striking human characteristics. The cranium was slightly larger than that of an ape, and the opening in the skull that allowed entry of the spinal cord was in a position different from that of an ape, suggesting that the individual had walked erect. Based on the other fossils in the box, Dart concluded that the skull was about 1 million years old. He named the creature *Australopithecus africanus* (*Australo* = southern, *pithecus* = ape), but it became known as the Taung Child.

In 1931, Dart traveled to London to attend an anthropological meeting and present his conclusions. Unfortunately, Dart's talk followed a dazzling presentation of the findings that were emerging from China concerning Peking Man. In addition, Dart was a poor speaker, whose evidence was limited to a single skull; he failed to make much of an impact on the skeptical scientists. Discouraged, he went off to dinner with friends while his wife brought the Taung Child back to the hotel. As if the day hadn't gone badly enough, his wife left the infamous skull (wrapped in cloth) in the back seat of the cab, where it traveled around London most of the night. The cab driver finally saw the package and handed it over to the police. Fortunately, Dart was able to recover his package before the police had time to wonder what type of skullduggery they had on their hands.

The "theory of evolution" is no less a fact of life to biologists than the "atomic theory" is to chemists or the "theory of gravitation" is to physicists. For a theory to have gained such widespread acceptance, it must be backed by a tremendous body of evidence. In this chapter we sample only a small portion of this evidence, taken from a variety of different fields. The entire matter can be summed up in a single sentence written by the biologist Theodosius Dobzhansky: "Nothing in biology makes sense except in the light of evolution."

▼ ▼ ▼

DETERMINING EVOLUTIONARY RELATIONSHIPS

Closely related species share a common ancestor and often resemble one another. It might seem, then, that the best way to uncover evolutionary relationships would be to make comparisons of overall similarity between organisms. This approach may be misleading, however, because there are actually two reasons why organisms may have similar characteristics, only one of which is due to evolutionary relatedness (Figure 23-1).

Two species that share a similar characteristic they inherited from a common ancestor are said to share a **homologous feature.** The foot of pigs, camels, deer, and bison is an example of a homologous feature because each species inherited the bone structure for an even-toed foot from a common ancestor. However, when *unrelated* species evolve a similar mode of existence, they may end up resembling one another due to convergent evolution (Chapter 22). This type of shared characteristic is called an **analogous feature.** The paddle-like front limbs and streamlined bodies of many aquatic animals (see Figure 22-12) are examples of analogous features.

Homology is the only evidence that proves that two species are evolutionarily related. But how do biologists tell whether a similarity is homologous or analogous? Years of experimentation and observation have produced the following set of criteria to identify homologies: (1) similarity in detail; (2) similar position in relation to neighboring structures or organs; (3) similarity in embryonic development; and (4) agreement with other characteristics (related animals usually share more than one homology). These criteria are illustrated by mammalian forelimbs (Figure 23-2). At first glance, the wing of a bat, the leg of a cat, the flipper of a whale, the arm of a human, and the leg of a horse may not seem very similar, but they are actually homologous structures. All contain the same type of bones (similar in detail); the forelimb always attaches to the shoulder girdle (similar position in relation to neighboring structures); the forelimb develops from the same tissues in each of the embryos (similar embryonic development); and, in addition to the forelimb, all these animals have hair and mammary glands (other shared homologies).

Evolutionary relationships cannot be reconstructed just by grouping species by their number of shared homologies, however. For example, the hand of the first vertebrates to live on land had five digits (fingers). Many terrestrial vertebrates (such as humans, turtles, lizards, and frogs) also have five digits, which they inherited from this common ancestor. This feature is then homologous in all of these species. In contrast, horses, zebras, and donkeys have only a single digit with a hoof. But humans are more closely related to horses than they are to lizards! The five-digit condition is the *primitive* (original) pattern for the number of digits. This primitive feature has been retained in the line of ancestors leading from early amphibians to humans, but it has been modified and reduced to just one digit in the common ancestor of horses, donkeys, and zebras. While the derived (modified) trait tells us that horses, zebras, and donkeys share a very recent common ancestor, the primitive form (five digits) tells us that species are at least distantly related.

(a)

(b)

FIGURE 23-1
Deceptive similarity. These two "palms" could easily be mistaken for closely related plants based on their similar appearances, but the cycad ***(a)*** is not a palm, or even a flowering plant. It is no more related to the true palm ***(b)*** than is a pine tree to a rose.

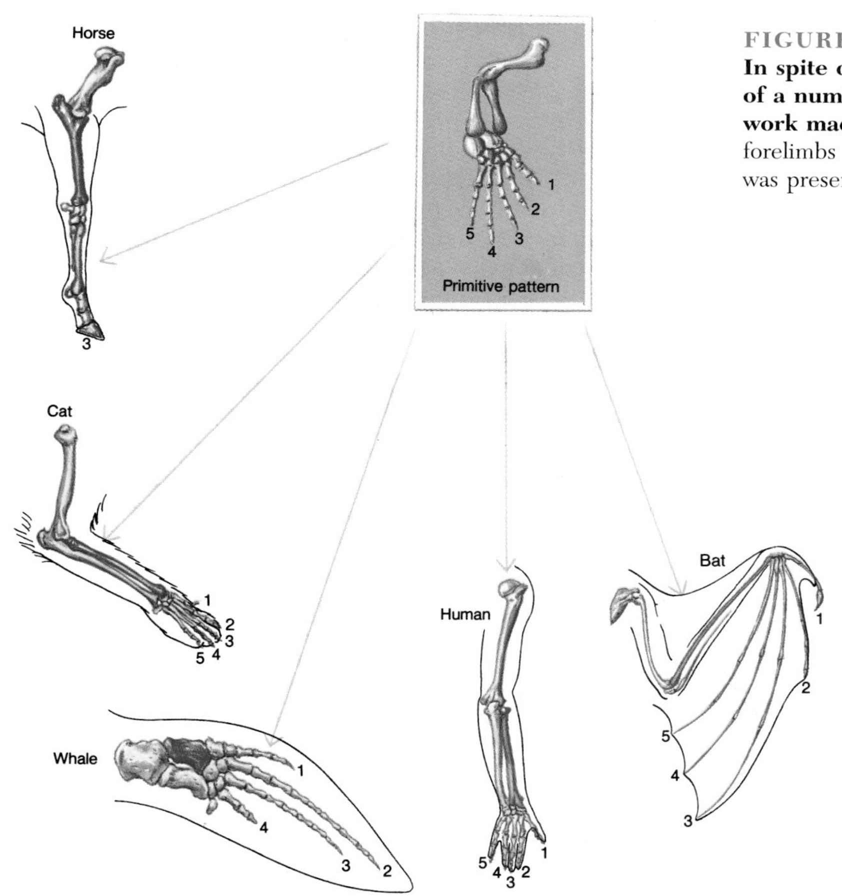

FIGURE 23-2
In spite of different forms and functions, the forelimbs of a number of very different mammals all have a framework made up of the same bones. The bones in these forelimbs are modifications of a primitive skeletal pattern that was present in an early terrestrial vertebrate ancestor.

EVIDENCE FOR EVOLUTION

Evidence supporting the theory of evolution has accumulated from a variety of different biological disciplines, including comparative anatomy, paleontology, comparative embryology, biochemistry and molecular biology, and biogeography.

FOSSILS

A fossil is any remaining evidence of life from the past. Many different types of fossils exist, ranging from preserved "footprints" of animals that walked along a trail (see Figure 23-10) to complete remains, such as frozen mammoths or entombed insects (Figure 23-3*a*), to actual hard parts (teeth and bones), to pieces of petrified trees (Figure 23-3*b*), and even preserved excrement (corpolites). Fossils are formed in several ways: Organisms may be buried in sediments, where they harden and mineralize; trapped in tree sap, which hardens into amber; covered in tar or other natural preservatives, such as the liquid found in peat bogs; or frozen in arctic regions or at high altitudes.

The fossil record consists of a collection of such remains from which paleontologists attempt to reconstruct the biology and history of the organisms. In addition to providing a glimpse of the kinds of organisms that lived in the past, the fossil record provides data that evolutionary biologists can use to describe the lineages by which various groups may have evolved. For example, without fossils, we would not have discovered the close relationship between dinosaurs and birds, and the fact that birds are more closely related to reptiles than they are to mammals.

One of the best known fossils was discovered in 1861 in a limestone quarry in Bavaria, Germany. The fossil suggested that the animal had been a small bipedal dinosaur, but the fine-grained limestone also revealed the imprint of wings with feathers (Figure 23-4*a*). Of all vertebrates, only birds possess feathers; in fact, feathers are a defining characteristic that unites all birds. The fossil was given the name *Archaeopteryx lithographica* (*archaeo* = old, *pteryx* = wing) and was determined to have been a bird that lived 150 million years ago. Yet, unlike modern birds, *Archaeopteryx* had teeth, a long tail containing over 20 vertebrae, free-floating ribs, and wings containing movable fingers with claws, all characteristics of the small, carnivorous reptiles called theropods. Therefore, *Archaeopteryx* provides one of the many pieces of fossil evidence of an evolutionary pathway leading from reptiles to birds.

Knowledge about the half-life of radioisotopes, particularly Carbon 14 (^{14}C) and Potassium 40 (^{40}K), armed scientists with a tool to determine the approximate age of fossils, and thereby identify the geological period in which the organism actually lived. For example, radiodating of a fossilized skull of a saber-toothed tiger determined the skull to be about 1 million years old, placing the saber-toothed tiger in the Pleistocene epoch (the Ice Age), a period of repeated glaciation.

THE ANATOMY OF LIVING ORGANISMS

In order to gather comparative evidence for evolution, biologists study external characteristics, examine bones and teeth, dissect organ systems, study sections of tissue, and examine the finer details of cells and tissues under the electron microscope. Comparing the structures of different organisms is probably the most commonly used evidence of evolution.

VESTIGIAL STRUCTURES

The underdeveloped pelvis and leg bones in snakes, the diminished toe bones in horses, and the appendix in humans are all structures that have little or no function in these organisms (Figure 23-5). Each of these *vestigial structures,* as they are called, is clearly related to a more fully developed, functional structure present in other animals. For example, your appendix is a dwarfed version of the cecum, a part of the digestive tract of many mammals in which food is stored and digested by microorganisms. Similarly, your tailbone is a remnant of the tail present in many of your ancient relatives. If species had been created as they appear today, there would be no explanation for vestigial structures.

COMPARATIVE EMBRYOLOGY

Adult fishes, salamanders, turtles, birds, and humans bear virtually no resemblance to one another. Yet these animals are virtually indistinguishable as embryos (see Figure 21-18). Why would animals that have markedly different adult forms and functions develop from such similar embryos?

(a)

(b)

FIGURE 23-3

Types of fossils. ***(a)*** This ancient pseudoscorpion (a distant relative of spiders) was trapped in a drop of resin that became transformed into hardened amber. ***(b)*** A scene from the Petrified Forest of Arizona.

(a)

(b)

FIGURE 23-4

Archaeopteryx lived 150 million years ago and has many features in common with small, bipedal dinosaurs—features, such as teeth and a long tail, that were eliminated during the evolution of modern birds. ***(a)*** Photograph of the fossil imprint of *Archaeopteryx* in limestone slate. ***(b)*** The skeleton of *Archaeopteryx* lacks the broad, bony breastbone to which the large flight muscles of modern birds are attached. The lack of a breastbone suggests that *Archaeopteryx* was not a strong flyer and may have been primarily a glider. The claws on the toes suggest that the animal could perch on a limb, while the claws on the "fingers" suggest that it may have been able to climb up the trunks of trees. The structure of the pelvis and hind legs suggests that *Archaeopteryx* was capable of running over the ground, its long tail acting as a counterweight to maintain its balance.

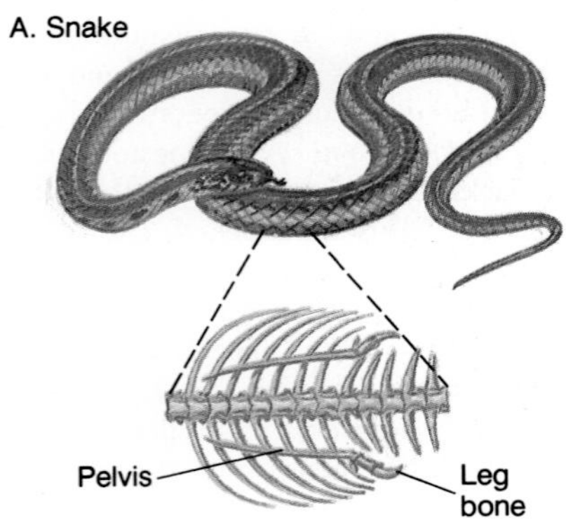

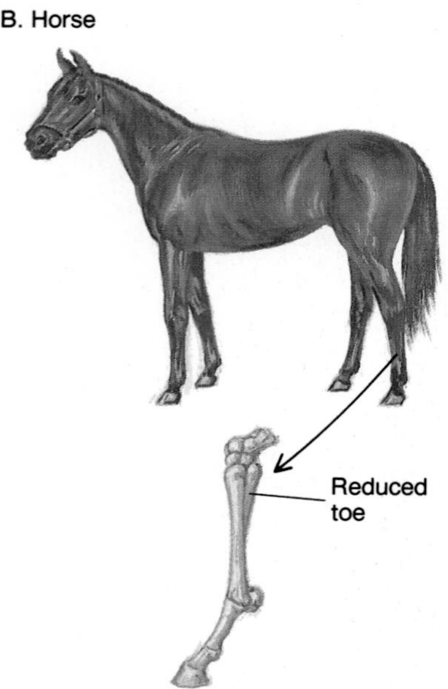

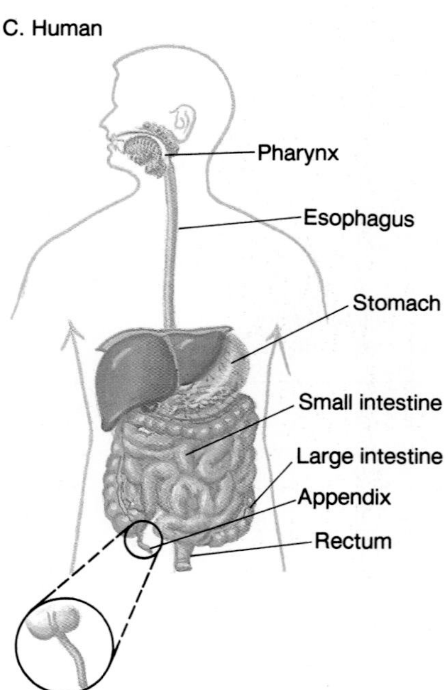

The best scientific explanation is that far back in their history, fishes, salamanders, turtles, birds, and humans all had a common ancestor—probably some type of primitive fish—that developed from a similar type of embryo. As the various types of vertebrates evolved, they each retained this basic vertebrate embryo, although its parts gave rise to different adult organs.

There is another way to illustrate the difficulty in evaluating *comparative embryology,* the study of the similarities and differences of the embryos of different species. Recently, Harvard paleontologist and essayist Stephen J. Gould noted that the question "Why do men have nipples?" heads the list of inquiries from his readers. Gould argues that nipples in men are not adaptive, nor did they evolve from structures that were adaptive in an ancestor. Rather, Gould concludes that nipples in men are the result of a constraint imposed by embryonic development. Male and female mammals of a species pass through identical stages as early embryos; it is not until the secretion of sex hormones that male and female sexual development diverge. That is, nipples are present in human embryos prior to the time of sexual differentiation. Later sexual maturation leads to changes in the female breast and nipple that allow these structures to nourish a newborn infant. In contrast, the male nipple remains as the vestige of an embryonic structure that is simply carried along "for the ride" in the adult. According to Gould, "Male mammals have nipples because females need them . . ."

BIOCHEMISTRY AND MOLECULAR BIOLOGY

Since the characteristics of an organism are determined by its genetic content, changes in organisms over the course of evolution are reflected in changes in nucleotide sequence of DNA (genes) and amino acid sequences of proteins (gene products). In general, the longer the period since two species have diverged from a common ancestor, the greater the number of differences in corresponding genes and proteins. Common ancestry can now be demonstrated just as accurately by homologous molecular information as by homologous anatomical structures. For example, molecular data has been used to determine that humans are more closely related to chimpanzees than are chimpanzees to gorillas (see page 146).

FIGURE 23-5

Vestigial organs: A legacy of ancestors: ***(a)*** Even though they are legless, snakes retain degenerated leg and pelvic bones inherited from four-legged ancestors. ***(b)*** Horses still possess degenerated toe bones left over from their three- and four-toed ancestors. ***(c)*** The human appendix is a degenerated cecum, a chamber used by our vegetarian ancestors for housing cellulose-digesting microorganisms.

BIOGEOGRAPHY: THE GEOGRAPHICAL DISTRIBUTION OF ORGANISMS

Among the evidence that convinced Darwin of the occurrence of evolution were the observations he made on the Galapagos Islands (Chapter 1). Darwin noted that species present on oceanic islands were not found anywhere else in the world. In fact, many were found only on a particular island in the chain and they varied from one island to the next. Recall that Darwin found different species of finches with distinct anatomical differences living on different islands. Although the various finches possessed different-shaped beaks, which were adapted for obtaining different types of foods, the birds were unmistakably similar in overall anatomy, both to one another and to a species found on the mainland. Darwin concluded that individuals from the mainland species had migrated to the islands, where, given the absence of competition from other birds, they had evolved into a variety of different species adapted to different local conditions and food sources. A common origin also explained why the plants and animals of the Galapagos were generally so similar to those species living on the mainland, even though the two regions had totally different climates and terrain.

THE EVIDENCE OF HUMAN EVOLUTION: THE STORY CONTINUES

When we left the discussion of fossil hominids in the chapter-opening Steps to Discovery, the story had become confused by the presence of conflicting data. On the one hand, we were confronted with fossils of *Homo erectus* and *Australopithecus africanus* (Figure 23-6), both characterized by small, apelike brains and humanlike jaws and teeth. On the other hand, we learned of Piltdown Man, with his large, humanlike brain and apelike jaws and teeth.

It was not until the late 1940s and early 1950s that the matter was finally resolved. At that time, a careful analysis was conducted of the jaws of a variety of *Australopithecus* specimens from South Africa, including Dart's Taung Child (Figure 23-7). Sir Wilfrid Le Gros Clark, who had become the foremost British paleontologist of the time, led the analysis. Le Gros Clark established 11 characteristics that clearly distinguished the teeth of humans from those of modern apes. Examination of the *Australopithecus* fossils indicated that, despite the fact that their braincases were small and apelike, their teeth were similar to humans in every one of the 11 criteria. There was no longer any doubt that the australopithecines were hominids. In addition, a new radiodating technique revealed that the australopithecines were very old—up to 2 million years old.

A second important revelation came from a more careful scrutiny of the Piltdown Man. Radiodating provided conclusive evidence that the Piltdown Man was actually a hoax. The Piltdown skull was only a few hundred years old.

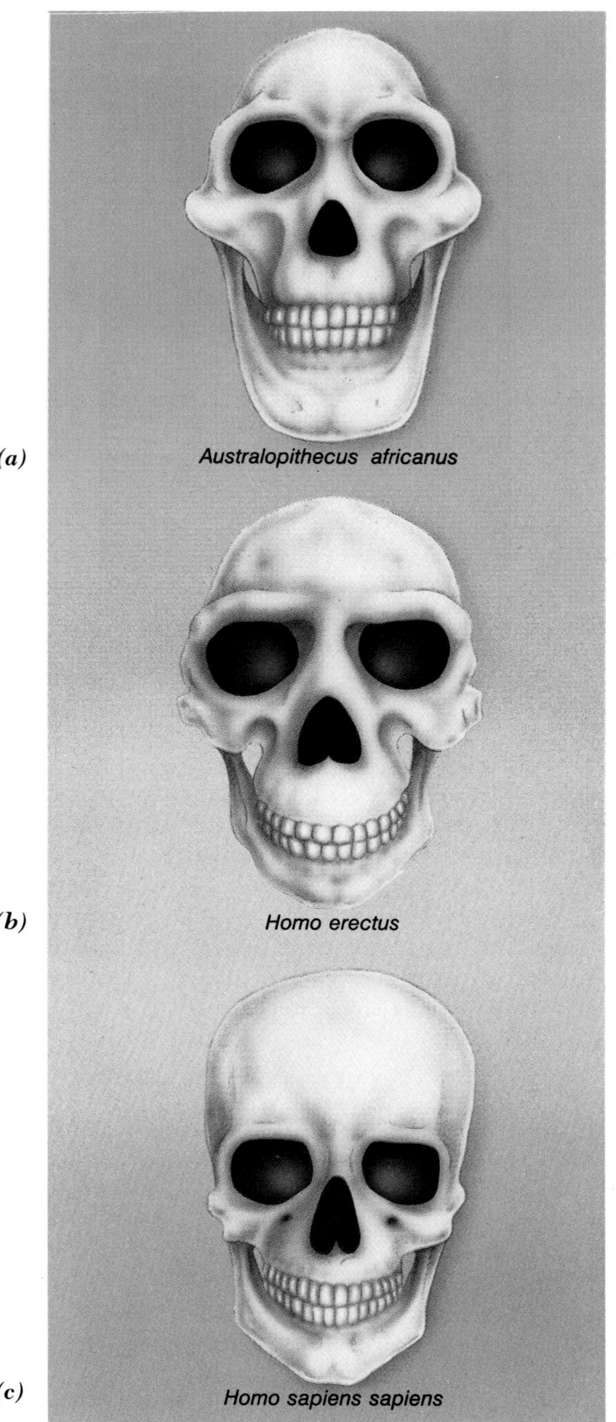

FIGURE 23-6
A comparison of the skulls of modern humans and two hominids: ***(a)*** Skull of the extinct species *Australopithecus africanus* originally represented by the Taung child and later by several other fossils found in South Africa. ***(b)*** Skull of the extinct species *Homo erectus,* which includes Java Man and Peking Man. ***(c)*** Skull of a modern human, *Homo sapiens sapiens.*

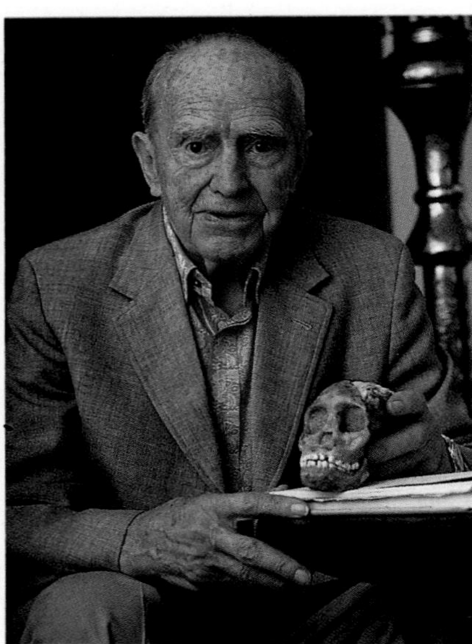

FIGURE 23-7
The skull of the Taung Child in the hands of its discoverer Raymond Dart. Dart died in 1988 at the age of 95.

Someone had taken the skull of a modern human and the lower jaw of a modern orangutan, treated them with chemicals to make them look old, filed down the ape teeth so that they resembled those of a human, broke them into fragments, and buried them alongside one another. With Piltdown Man out of the way, it was evident that the enlargement of the brain occurred during a late stage of human evolution, not an early stage.

HOMO HABILIS AND THE USE OF TOOLS

By the end of the 1950s, hominid excavation had shifted from Europe and Asia to Africa. To understand the more recent evolutionary discoveries, we need to introduce another cast of paleontologists, the most famous of which are Louis and Mary Leakey.

The Leakeys came to East Africa in the 1930s, looking for fossil hominids. They focused their attention on the Olduvai Gorge in Tanzania. The gorge is situated in the Serengeti Plain, on what once was a lakebed. Over time, deposits on the bottom of the lake had created layer upon layer of sediments. The gorge, which is about 100 meters (330 feet) deep, was created by a river that wound through the area, digging deeper and deeper into the layers of sediments. The sides of the gorge revealed the stratifications, while the bottom corresponds to the bottom of the ancient lake as it existed approximately 2 million years ago. The Leakeys were first drawn to Olduvai by the large numbers of primitive tools that were strewn over the bottom of the gorge. It was the maker of these tools for whom the Leakeys were searching.

After 30 years of excavation, the Olduvai Gorge began to reveal bits and pieces of a new hominid. Based on an increased size of the braincase and certain other characteristics, the Leakeys concluded that these hominids were not australopithecines. They named them *Homo habilis* ("the handy man"), implying that these hominids were responsible for making the primitive stone tools found at the bottom of the gorge. According to the Leakeys, the use of tools is just as important (if not more important) than is brain size or tooth structure in defining a fossil hominid as human (*Homo*), as opposed to some other genus. The Leakeys were convinced that the genus *Homo* was older than was generally accepted and that the australopithecines were not our ancestors but our cousins. In other words, they believed that the members of the genus *Homo* went as far back as *Australopithecus* and that the australopithecines were an offshoot that was not on the line leading to modern humans. In addition, the Leakeys argued that members of the two genera lived side by side, a proposal that has since been strengthened by considerable evidence, including radiodating studies that showed that these fossil remains were 1.75 million years old. The age of humans had been pushed back in time from approximately 500,000 years for *H. erectus* to nearly 2 million years for *H. habilis*.

In 1972, a well-preserved skull of the same time period was discovered that was unambiguously *Homo*. The discoverer was, appropriately enough, Richard Leakey, the son of Louis and Mary, who had only recently become immersed in the hominid-hunting family passion. A body of evidence now suggests that *Homo habilis* was indeed walking the earth approximately 2 million years ago, probably in the company of several different species of australopithecines. But what type of ancestor had given rise to *Homo habilis*?

THE DISCOVERY OF LUCY

Our best clue as to the nature of that ancestor came in 1974, when Donald Johanson of the Cleveland Museum of Natural History was searching for hominid fossils at a remote site in northern Africa known as Hadar. Like Olduvai, Hadar was an ancient lake bed. Inhabitants of the area had died and become buried by sediment, only to be unearthed at a later time by torrents of water rushing through newly formed gullies. One morning, Johanson noticed a bone projecting out of the ground. On closer inspection, he identified it as the arm bone of a hominid. Nearby, he saw parts of a skull and a thigh bone. In his words, "An unbelievable, impermissible thought flickered through my mind. Suppose all these fitted together? Could they be parts of a single, extremely primitive skeleton?"

That is exactly what they were. Within a few weeks, all of the bones had been recovered. Together, they constituted approximately 40 percent of the skeleton of an extremely primitive, small-brained female hominid who had stood only about 3.5 feet tall (Figure 23-8). Most importantly, the skeleton indicated that the hominid had walked erect, suggesting that bipedal locomotion was one of the first humanlike traits to have evolved. Radiodating techniques established the age of the hominid to be 3.5 million years. She was the oldest, most complete, and best-preserved hominid fossil that has yet to be recovered. Johanson named her "Lucy," after the Beatles' song "Lucy in the Sky with Diamonds," which was playing in camp on the night of the discovery. The next year, the remains of 13 additional members of the species were found at a nearby site and have become known as "The First Family."

Johanson puzzled over what species name he should assign to Lucy and The First Family. The hominids were too primitive to be considered "human," particularly since there was no evidence that they had used tools. He finally settled on the name *Australopithecus afarensis* (after the Afar region of Ethiopia, where the species had been discovered) and proposed that it was a common ancestor of the other australopithecines and humans. Johanson's proposal for the evolutionary relationships among the various known species of hominids is illustrated in Figure 23-9.

We have just one more "fossil" to describe, not because it tells us what our ancestors looked like but because it provides a mental picture of an event that occurred one rainy day in east Africa nearly 4 million years ago. The fossil in question is a trail of footprints (Figure 23-10) that were made by a pair of fully erect hominids walking together through ash that had been spewed from a nearby volcano and then dampened by a light rain. The wet ash quickly hardened like cement and was covered by additional layers of ash, preserving the footprints (and even a few small craters left by the falling raindrops) until their discovery in 1976 by a team led by Mary Leakey. In the words of Tim White, a member of the team: "They are like modern human footprints. There is a well-shaped modern heel with a strong arch and a good ball of the foot in front of it. . . to all intents and purposes those . . . hominids walked like you and me."

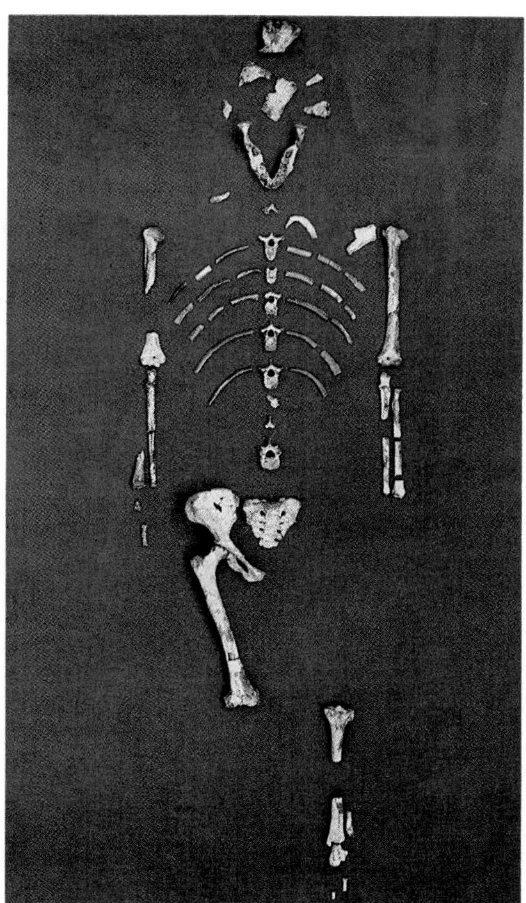

FIGURE 23-8
"Lucy," nearly 40 percent intact, has been assigned to the species *Australopithecus afarensis.* Lucy and other members of "The First Family" may represent a species of hominid that gave rise to the various other species of *Australopithecus* as well as to the genus *Homo.*

EVOLUTION AND ADAPTATION: TYING IT TOGETHER

Evolution is about change. Evidence of change is all around us, and at every level of biological organization. The biosphere changes: An earthquake jolts southern California, causing a fault line to slip and mountains to change elevation; Volcanic eruptions in the Philippines destroy habitats and change the island's topography and size. Global warming and the greenhouse effect permanently alter worldwide weather patterns, changing biomes, ecosystems, and populations. The destruction of tropical rain forests at a rate of 100 acres each minute causes the ex-

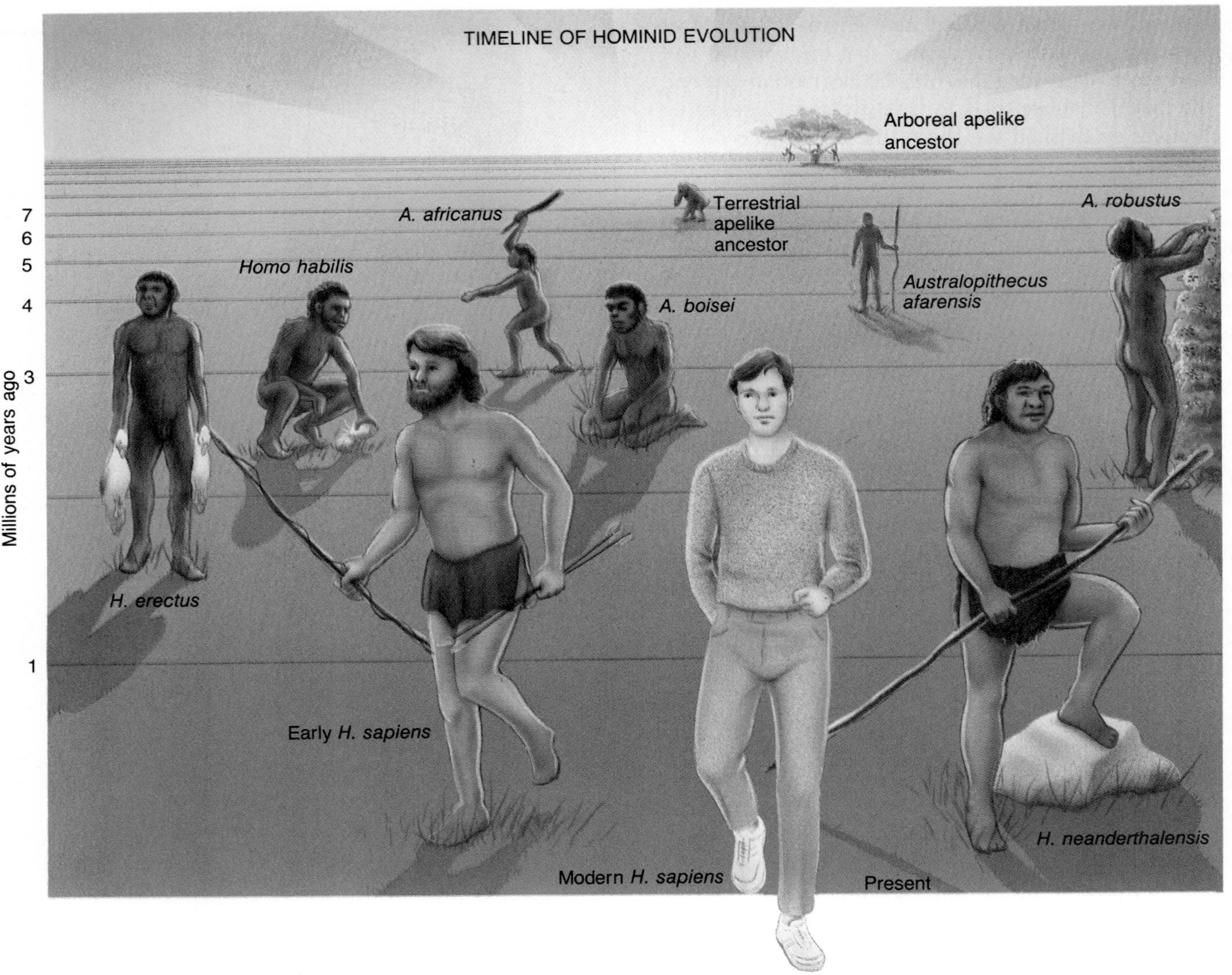

FIGURE 23-9

The human family. According to this scheme, the earliest hominid was *A. afarensis,* represented by Lucy and The First Family. This species gave rise to other australopithecines as well as to the genus of humans (*Homo*). Three species of australopithecines are depicted in the illustration. *A. africanus* had the slightest build and was probably primarily a carnivore. *A. robustus* was the most heavily boned species, with large, thickly enameled molars, suggesting that they lived primarily on a diet of coarse vegetation. *A. boisei* was an australopithecine discovered by Mary Leakey, which helped focus attention on Africa as the cradle of human evolution. The first humans, *H. habilis,* appeared about 2 million years ago. *H. habilis,* like *A. afarensis,* was very small in stature and had unusually long arms, reflecting the arboreal habits of their ancestors. Their brain size was about 700 cubic centimeters, and they were able to make simple tools. *H. erectus,* which was very advanced over its predecessors, may have stood as tall as modern humans, had arms of shorter proportion than its ancestors, had a brain capacity of about 850 to 1,100 cubic centimeters, and was able to make sophisticated tools. There is evidence that these hominids hunted in groups and cooked their prey. While originating in Africa more than 1.6 million years ago, *H. erectus* spread over much of Europe and Asia. By about 500,000 years ago, fossils appear that are intermediate between *H. erectus* and *H. sapiens.* By 300,000 years ago, fossils are found which are unmistakably *H. sapiens.* These early *H. sapiens* had larger brains (1,200 to 1,400 cubic centimeters) and used more sophisticated tools. Modern humans (*H. sapiens sapiens*) arose between 100,000 and 200,000 years ago and are thought to have coexisted with the Neanderthals (*H. sapiens neanderthalensis*), who disappeared about 35,000 years ago.

FIGURE 23-10
Footprints made by a pair of hominids walking together across a plain of wet ash in eastern Africa approximately 4 million years ago.

tinction of at least 1,000 species each year. Desertification in Africa resulting from human activities shrinks available grasslands, reducing precious grazing land in regions surrounding the parched Sahara Desert.

We can even see change over our own lifetime. At a recent family reunion, a great-grandparent surveyed her 142 descendants of great-grandchildren, grandchildren, and children. Not one looked exactly like her (genetic variability, page 120), and the resemblance became less and less obvious with each new generation (genetic recombination resulting from sexual reproduction). On a small scale, this family reunion testifies to the evolutionary forces that trigger inherited variation in a population, the foundation of evolutionary change. This same great-grandmother frequently comments on "how things now-a-days are completely different—nothing is the same as it was when she was a young girl."

Evolution is about how life changes on an ever-changing planet. Evolution describes how populations of different types of organisms become modified over time, sometimes resulting in the formation of new species. Just as new species evolve, others become extinct. And although sometimes incomplete, the combination of the fossil record, changes in the anatomy of organisms, resemblances in embryo development, and similarities in the molecular biology of organisms form a permanent file of the progression of life on earth—a progression that began 3.5 billion years ago with the very simplest kind of life and has led to a grand diversity of millions of forms of life that inhabit the earth today.

SYNOPSIS

Evolution is the greatest unifying concept in biology. While biologists may argue over which mechanisms may have been most important in the evolution of a particular group, there is virtual agreement that all living organisms have arisen by evolution from a common ancestor. The theory of evolution is supported by a mass of evidence gathered from several distinct biological disciplines; no credible scientific evidence has been obtained to suggest that evolution has *not* occurred.

Biologists use similar characteristics to help determine evolutionary relationships. However, homologous features (those that are inherited from a common ancestor) and analogous features (those that result from convergent evolution among unrelated animals) must be distinguished because only homologous features indicate common evolutionary origin.

The evidence for evolution is based on the study of the anatomy of living organisms, comparative embryology, the fossil record, biogeography, and biochemical and molecular data. Comparing the structure of body parts in different organisms provides evidence of evolutionary relationships. Vestigial structures (ones with little or no apparent function, such as your appendix) are remnants of structures that had a function in ancestral species that are in the process of evolutionary disappearance. Fossil analysis provides information on the types of organisms that lived in the past and evidence of the pathways by which various groups might have evolved. Comparisons of the embryos reveals homologies that are not apparent in the adults. Homologies are also revealed by comparing nucleotide and amino acid sequences of different organisms. Geographical information about organisms also helps determine evolutionary relationships. Plants and animals living in nearby areas are more likely to be related than are those living far apart.

Based on fragmentary fossil evidence, humans evolved from an ancestor common to both apes and humans. The first known hominid is *Australopithecus afarensis,* represented by Lucy and The First Family, who lived about 3.5 million years ago. These hominids were small in stature, had small brains, and showed no evidence of tool use, but they walked erect and had jaws with humanlike features. *A. afarensis* may have given rise to a number of other australopithecines as well as to humans (genus *Homo*). The first known humans (*H. habilis*) appeared in Africa about 2 million years ago, and members of the species *H. erectus* appeared about 1.5 million years ago. *H. erectus* survived for over a million years, migrating to diverse regions of the earth. Modern humans (*H. sapiens sapiens*) date back 100,000 to 200,000 years.

Review Questions

1. Name the major anatomical characteristics that underwent change during the course of human evolution over the past few million years. How are these characteristics illustrated in various fossil hominids?
2. What criteria are used to distinguish homologous and analogous features?
3. Why are vestigial structures evidence for the occurrence of evolution?
4. Compare and contrast Java Man, Peking Man, Taung Child, and Piltdown Man; primitive and derived traits; *Archaeopteryx* and fossil reptiles.
5. Why was increased brain size naturally selected during the evolution of modern humans?

Critical Thinking Questions

1. The characteristics used to distinguish *Homo* from other genera, such as *Australopithecus,* have never been universally accepted. List all the fossil characteristics scientists used to identify *Homo* in the Steps to Discovery. Is there one (or a few) characteristics that you feel should be the most important determinant of a "human" hominid?
2. Explain how both analogous and homologous structures provide clues to evolution. Why are homologous structures used to infer common ancestry, but analogous structures are not?
3. List one piece of evidence from each of the categories listed below that links humans with apes and/or other primates: comparative anatomy, fossils, vestigial structures, comparative embryology, comparative biochemistry.
4. Which of the four proposed family trees for hominids in the diagram opposite most closely resembles that described in this chapter? Why are so many different schemes proposed by different scientists working in this field?
5. Why do you think biologists may take offense when someone says "Evolution is *only* a theory"?

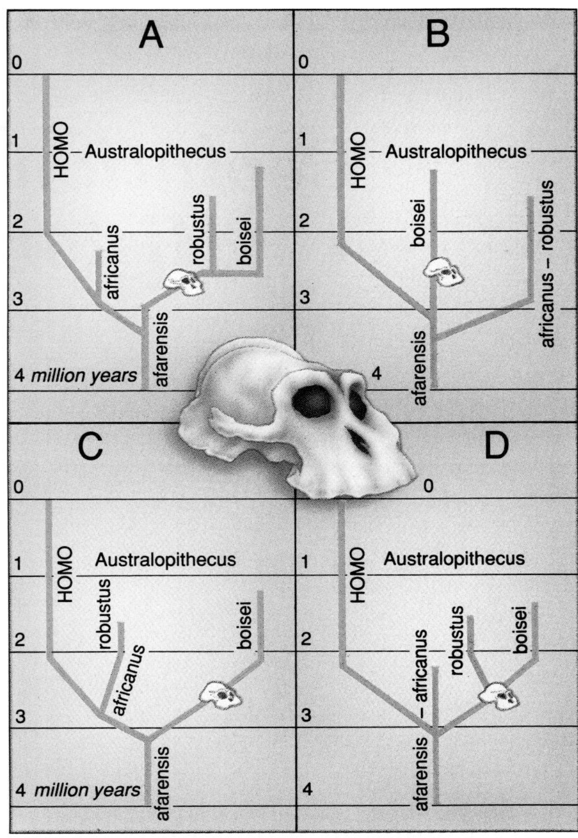

CHAPTER

24

Evolution: Origin and History of Life

Within the primordial seas, the first life forms appeared over 4 billion years ago by the process of chemical evolution.

STEPS TO DISCOVERY
Evolution of the Cell

When Louis Pasteur's research finally laid to rest the idea that living organisms could arise from inanimate materials (Chapter 1), it settled one nagging controversy in biology and began a new one: If life could only arise from other life, how did living organisms initially appear on the planet?

This question was first tackled in the 1920s by the Russian biochemist Aleksandr Oparin, who proposed that life could not have arisen in a single step but only over a long and gradual process of **chemical evolution**—the spontaneous synthesis of increasingly complex organic compounds from simpler organic molecules. In Oparin's view, the formation of life occurred in several distinct stages.

The first step was the formation of simple molecules, such as ammonia (NH_3), methane (CH_4), hydrogen cyanide (HCN), carbon monoxide (CO), carbon dioxide (CO_2), hydrogen gas (H_2), nitrogen gas (N_2), and water (H_2O) during the formation of the earth's crust and atmosphere. The second step involved the spontaneous interaction of these simple molecules to form more complex organic molecules, such as amino acids, sugars, fatty acids, and nitrogenous bases—the building blocks of the macromolecules that characterize life as we know it today. The energy needed to drive the formation of these organic molecules was derived from various sources, including the sun's radiation, electric discharges (lightning), and heat that emanated from beneath the earth's crust. As these organic compounds accumulated to higher concentrations in the shallow lakes and seas that dotted the primitive earth, they formed an "organic soup," in which additional reactions could take place. Some of the simpler organic molecules polymerized to form macromolecules consisting of chains of subunits, similar in basic structure to proteins and nucleic acids. As we discussed in Chapter 10, the first nucleic acids were likely RNA rather than DNA.

Molecules eventually arose that could accomplish functions that were necessary for life. In this step, some of these molecules catalyzed certain chemical reactions, forming the basis for a primitive form of metabolism. Others had the capacity for self-replication and somehow formed copies of themselves. Insoluble lipids that were formed in the organic soup were forced together by hydrophobic interactions, forming membranous sheets made up of lipid bilayers. Some of these self-assembling membranes wrapped around solutions of macromolecules, encapsulating them and forming "precells."

In the final step, precells concentrated organic molecules, allowing the molecules to react more frequently. Within the precell, catalysts directed the synthesis of specific organic polymers; in other words, the precell conducted metabolism. The eventual evolution of a genetic code enabled the precell to pass on naturally selected codes for metabolism, allowing it to reproduce itself. At this point, the precell possessed three of the basic characteristics of life—metabolism, growth, and reproduction—crossing the line between precell and living cell. Earth had its first living organism.

Acceptance of Oparin's theory of chemical evolution received a boost in the early 1950s as a result of a series of experiments conducted by Stanley Miller, a graduate student at the University of Chicago. Miller demonstrated that many of the simpler compounds characteristic of life could be produced in the laboratory under conditions that likely existed soon after the formation of the earth. In a sealed glass vessel, Miller repeatedly jolted a mixture of hydrogen gas, ammonia, methane, and water vapor with electric discharges to simulate lightning strikes. Within a matter of days, a number of organic compounds appeared in the reaction vessel, including several amino acids commonly found in proteins. Soon after, Juan Oro of the University of Houston showed that more complex biochemicals, including the nitrogenous bases that form the building blocks of nucleic acids, could be formed under similar conditions.

In 1969, support for this view of the origin of life crashed in from outer space when a huge meteorite fell close to the town of Murchison, Australia. Analysis of the meteorite fragments revealed the presence of a large variety of organic compounds, including amino acids, pyrimidines, and molecules resembling fatty acids. The fact that these compounds could appear under abiotic (nonbiological), extraterrestrial conditions made it even more likely that similar compounds had formed on the primitive earth. ■

It is one of the greatest detective stories of all time. There were no eyewitnesses and the detectives did not come on the scene until long after the incident had occurred. Although a host of clues had been left behind, they had all become modified over the years; virtually everything had changed.

What was this baffling case? It is nothing less than the origin and history of life on Earth. The events we are trying to understand span the entire duration of the Earth itself, events that continue to unfold today on a modern planet housing an astonishing variety of life. The story begins nearly 4.6 billion years ago with the formation of this remarkable planet.

▼ ▼ ▼

THE EARTH'S BEGINNINGS

Our solar system was once part of a massive cloud of interstellar gases and cosmic dust. As particles of matter were pulled together by gravitational attraction, the vast cloud condensed into a gigantic, spinning disk. Nearly 90 percent of the matter gravitated to the center, causing temperatures to rise high enough to ignite thermonuclear reactions. It was in this scorching center that our sun began to shine.

At the same time the sun was forming, nearly 5 billion years ago, smaller eddies of leftover gases and dust were condensing into the planets of our solar system. Heavier elements, such as iron and nickel, sank inward to form the cores of the planets. In the planets nearest the sun (Mercury, Venus, Earth, and Mars) most of the lighter gases were blasted away when the sun's thermonuclear reactions began, leaving shrunken, dense planets with virtually no gaseous atmospheres.

Immense quantities of thermal energy from gravitational contraction, radioactive decay, meteorite impacts, and solar radiation turned the primitive earth into a red-hot, molten orb. As time passed, heat from radioactive decay lessened, and the earth's contraction slowed. A thin crust of crystalline rock formed on the cooled earth's surface. Below the solid crust, however, the enormous heat of the earth's interior produced massive buildups of hot gases, sparking violent volcanic eruptions that gradually built the earth's rugged land masses and filled the once empty atmosphere with clouds of hot gases and steam. The steam condensed and collected into ponds, eventually accumulating to form the earth's oceans, lakes, rivers, and streams. It was in these bodies of water that life emerged.

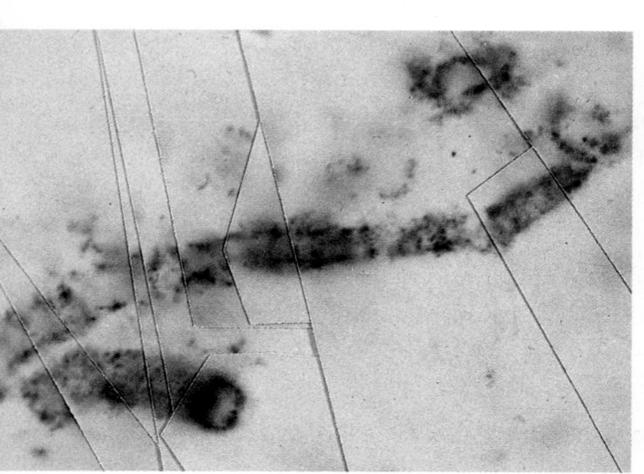

(a)

(b)

FIGURE 24-1

Imprints of the world's most ancient life. ***(a)*** A 3.5-billion-year-old cast of a filamentous cyanobacterium from western Australia. ***(b)*** These stromatolites, which dot the shore in western Australia, consist of dense masses of prokaryotic cells and mineral deposits. Prokaryotic cells have been found in stromatolites that are 3.5 billion years old.

THE HISTORY OF LIFE—A BRIEF OVERVIEW

The *origin* of life is still a hotly debated issue.[1] The *history* of life, however, meets with greater agreement among the scientific community, thanks largely to the existance of fossil evidence. The first fossilized cells were found in rocks from Australia and South Africa that date back 3.5 billion years. These rocks reveal the presence of prokaryotic cells that are not noticably different in appearance from prokaryotes living today. One such fossil consists of chains of cells (Figure 24-1*a*) that resemble modern photosynthetic cyanobacteria. Others consist of rocks formed from stromatolites—dense masses of bacteria and mineral deposits that grow today in warm, shallow seas (Figure 24-1*b*). The fact that such "advanced" prokaryotic cells had already appeared 3.5 billion years ago suggests that the process of chemical evolution that led to the first life forms took place relatively rapidly, probably within the earth's first 600 million years. The first 2 billion years of life was an age of prokaryotes. The evolutionary leap from prokaryotic cells to eukaryotic cells apparently took much longer than did the formation of the first prokaryotic cells themselves. The evolution of single-celled eukaryotes about 2 billion years ago was followed a billion years later by the evolution of the first multicellular eukaryotic organisms. The success of these first multicellular organisms was tremendous, paving the way for a proliferation of complex life forms. Some of these complex organisms moved onto the land and eventually took to the air.

Today, at least 2 million species of organisms populate the earth, and probably millions of others remain to be discovered. However, this is only a fraction of the total number of species that emerged and disappeared during the earth's history.

THE GEOLOGIC TIME SCALE: SPANNING EARTH'S HISTORY

Geologists divide the earth's 4.6-billion-year life span into four great eras: the Proterozoic, Paleozoic, Mesozoic, and Cenozoic Eras (Figure 24-2). With the exception of the Proterozoic Era, the eras are subdivided into periods, which begin and end at times marked by a memorable geologic or biological event, such as an episode of mass extinction. The most recent era, the Cenozoic, is further subdivided into epochs, each of which extends for a few million years (glance ahead at Figure 24-8).

[1] For a discussion of some of these proposals, consult an article entitled "In the beginning," J. Horgan in the February 1991 issue of *Scientific American*.

THE PROTEROZOIC ERA: LIFE BEGINS

The Proterozoic Era begins with the formation of the earth's crust and stretches forward in time for approximately 4 billion years. During the beginning of the Proterozoic Era, the evolution of the first living cells took place. The earliest organisms were probably heterotrophs, totally dependent on the amino acids, sugars, and other organic compounds that had accumulated in the "organic soup" as the result of abiotic chemical reactions. An abundant supply of nutrients probably allowed these early heterotrophs to proliferate at a rapid rate. Such cells may have been similar to modern bacteria that live in the sediments beneath ponds and oceans, where little or no oxygen is available.

Eventually, however, consumption of these organic compounds probably outstripped the rate at which abiotic synthesis could replace them. Organisms that could manufacture their own organic nutrients from inorganic precursors (using energy obtained from their abiotic environment) would not need to compete for this dwindling supply of nutrients. These were the earth's first autotrophs.

The Evolution of Autotrophs

If it were not for the emergence of autotrophs, life might have ended 3.5 billion years ago. The first autotrophs not only supplied themselves with nutrients, they also became food for the heterotrophs. These early autotrophs were probably very similar to present-day anaerobic, photosynthetic bacteria that live in well-lit, oxygen-deficient environments, such as stagnant ponds, marshes, and swamps. The metabolic success of these anaerobic photosynthesizers soon gave way to another group of photosynthetic organisms that split water as a source of electrons (and protons), releasing molecular oxygen as a waste product. These new autotrophs were the first cyanobacteria, a 3-billion-year-old group of photosynthetic bacteria that continue to flourish today. Because water was an enormously plentiful resource, these ancient cyanobacteria flourished, their light-capturing photosynthetic pigments tinting ponds and seas green. These autotrophs released vast amounts of molecular oxygen into the water and air. Molecular oxygen was toxic to most of the anaerobic organisms of the time, fatally oxidizing the life-sustaining molecules in cytoplasm. As the gas accumulated in the atmosphere, it selected for the evolution of mechanisms that protected cells from its potentially deadly oxidizing effects.

From Prokaryotic to Eukaryotic Cells

So far, no fossils of early eukaryotes have been discovered. In the absence of living intermediates, speculations on how eukaryotic cells evolved from simpler prokaryotic ancestors are based on the structures and functions of modern single-celled organisms. Two theories have been proposed to explain the evolution of eukaryotic cells, each of which

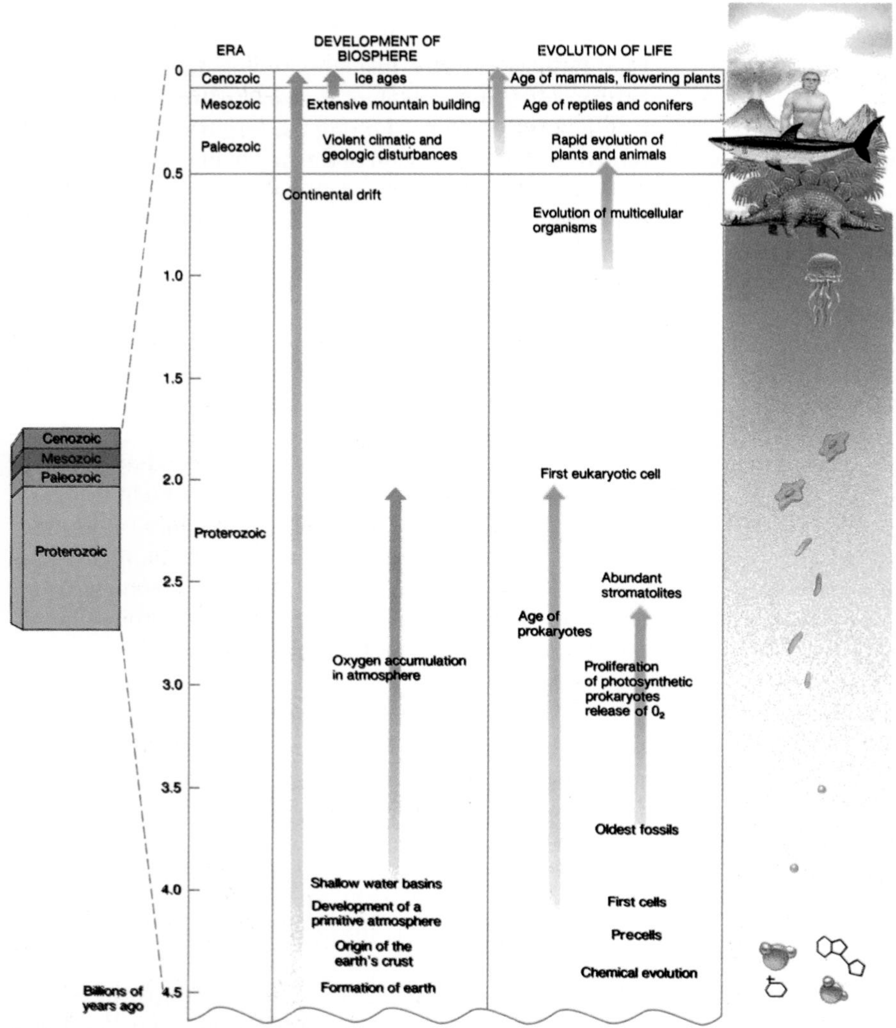

FIGURE 24-2
The geologic time scale. The 4.6-billion-year lifespan of the earth is divided into four eras, based on changes in the fossil record and the earth's terrain. Major changes in the physical makeup of the earth and its atmosphere dramatically affected the evolution of life.

may explain part of the story. Some of the organelles of a eukaryotic cell, such as the mitochondria, chloroplasts, and cilia, may have evolved from small prokaryotic cells that took up residence within larger eukaryotic hosts. Other eukaryotic organelles, such as the endoplasmic reticulum and the nuclear envelope, may have arisen from membranes that invaginated from the cell surface.

From Unicellular Eukaryotes to Multicellular Eukaryotes

Nearly 1 billion years ago, multicellularity evolved independently in a number of different types of eukaryotes. In some cases, multicellularity may have been achieved when a group of independent, single-celled eukaryotes aggregated to form a colony of cells. In other cases, colonies arose when the progeny of a single cell remained together. Eventually, the cells in the colonies lost their capacity for independent life. Gradually, some of the cells became specialized, carrying out specific functions with greater efficiency than other cells in the colony. Some cells constituted locomotor structures, moving the colony from place to place; others acquired proficiency at obtaining and processing nutrients; still others became committed to reproduction. Even though true multicellular plants and animals began this way, there are still a number of organisms that lie somewhere between the single-celled and multicellular state, such as the colonial green alga *Volvox* (glance ahead at Figure 26-4).

By the time the Proterozoic Era ended, members of all the major animal phyla had appeared, leaving fossils in considerable numbers and delineating the transition into the Paleozoic era. Although the animals living prior to this time lacked hardened skeletal parts usually required for fossilization, a few valuable samples of fossilized Proterozoic life reveal the marine organisms, including jellyfish (cnidarians), segmented worms (annelids), and the soft-bodied ancestors of arthropods, that developed during the Proterozoic era about 650 million years ago (Figure 24-3).

THE PALEOZOIC ERA: LIFE DIVERSIFIES

The Paleozoic Era lasted approximately 345 million years. Conditions on earth changed dramatically during this long era, causing episodes of mass extinctions followed by adaptive radiation that generated tremendous diversification of organisms adapted to different habitats (page 403). Many of these changes were the results of shifts in the earth's continents. At times, the drifting continents created broad, shallow seas; at other times, land masses became entirely submerged and later rose to form mountain ranges. These changes created selective pressures that dramatically altered life on earth.

Continental Drift

As you look at a map of the world, you will see that the continents of Africa and South America, which are situated on opposite sides of the Atlantic Ocean, have complementary outlines that fit together spatially. What you don't see from the map is the fact that geologic formations on the opposing coastlines are contiguous, that is, features that end abruptly on one continent continue on the opposite continent an ocean away. Fossil reptiles have been found in Antartica that were virtually identical to fossils found in Africa and India, fossils of "cold-blooded" animals that could never survive in the freezing cold of the polar regions. This is just some of the evidence amassed that revealed a phenomenon known as *continental drift.* The continents were not always in the position they occupy today but drifted there over millions of years. The continents lie atop solid plates that are pulled by currents in the molten rock that makes up the underlying mantle of the earth. As the plates move, they not only relocate the earth's land masses, but they collide with adjacent plates, creating earthquakes, chains of volcanos, and huge "compression wrinkles" that form mountain ranges.

Today's continents were clustered together 200 million years ago, forming a "supercontinent" known as Pangaea (Figure 24-4). Approximately 150 million years ago, Pangaea began to fragment into two large continents, Laurasia and Gondwanaland, causing the separation of many populations of organisms. Eventually, these two large continents split apart; the resulting continents drifted to their present positions, carrying with them fossils of organisms that had once flourished half a world away. This is why the same species of extinct reptiles are spread across land masses ranging from India and Africa to the Antarctic; these continents were once joined together in a common land mass.

Continental drift alone does not explain how reptiles (along with a variety of plant life) were once able to thrive on the Antarctic continent, which today is covered by ice and the scrubby vegetation of the tundra. Antarctica has not moved very far from its original position as part of Pangaea, yet the country has changed from a land of forests to a frozen wilderness. This transformation reflects the dramatic changes in climate that have occurred on earth over the past couple of hundred million years.

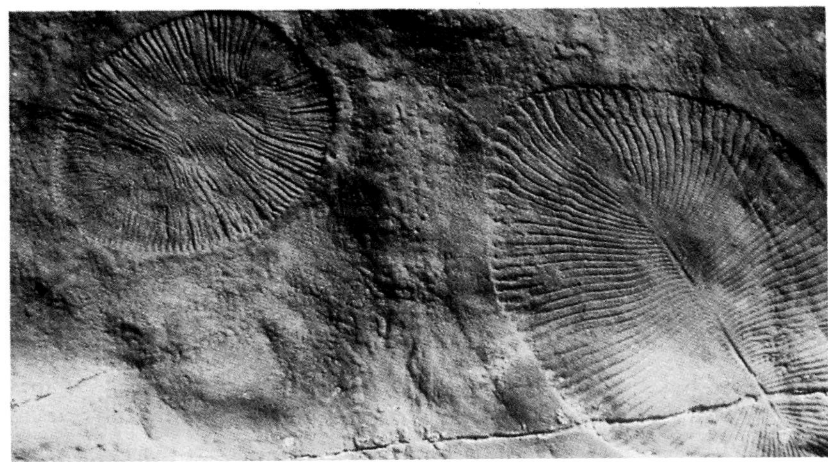

FIGURE 24-3
The imprint of a soft-bodied, multicellular animal that lived approximately 650 million years ago. The animal appears to have been segmented but lacked a head or appendages. It is believed to be an ancestor of the arthropods (insects, spiders, lobsters, centipedes, etc.).

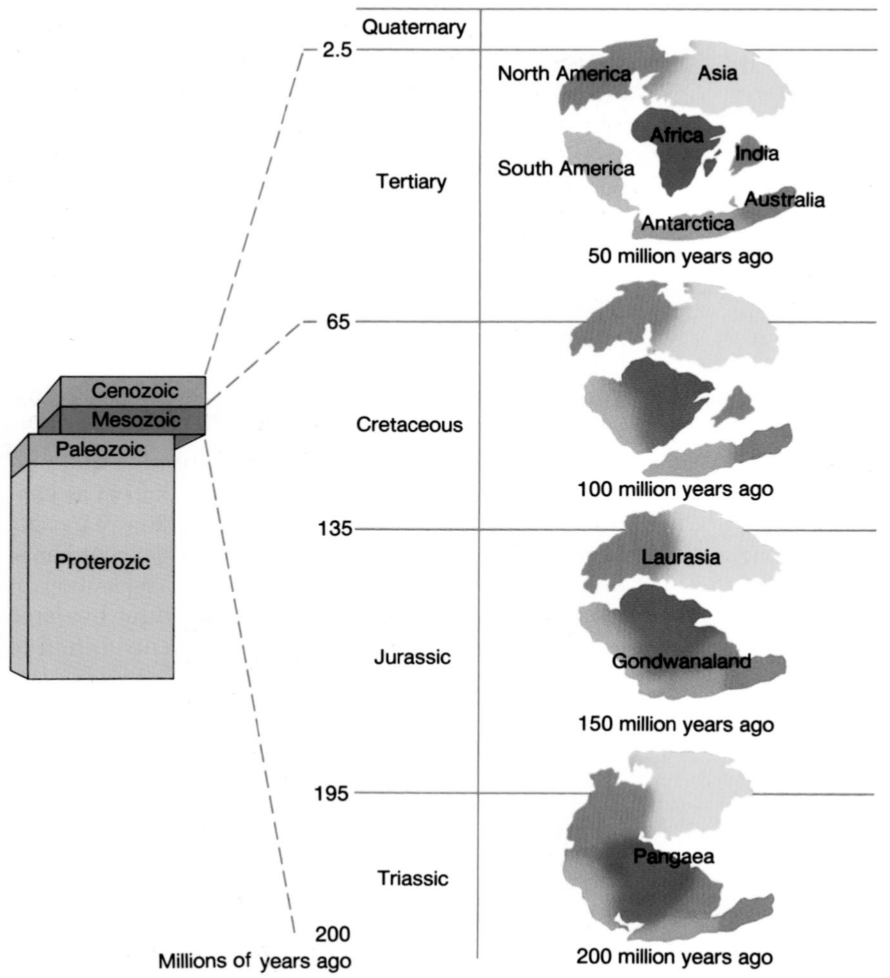

FIGURE 24-4

The consequences of continental drift. Nearly 50 million years after the supercontinent Pangaea formed, it began to fragment into two large continents—Laurasia and Gondwanaland—separating populations of organisms. Over the next 150 million years, these large continents split apart, and the smaller continents drifted to their present positions.

Mass Extinctions

The Paleozoic Era is divided into six periods: the Cambrian, Ordovician, Silurian, Devonian, Carboniferous, and Permian, each of which is characterized by the appearance of major groups of plants and animals (Figure 24-5). For example, jawless fishes, the first vertebrates (animals with backbones), appeared about 475 million years ago. Over the following 200 million years or so, terrestrial habitats were colonized by a succession of multicellular eukaryotes, both plants and animals. The first primitive land plants were followed by mosses and ferns, then seed plants, and eventually gymnosperms. Among the first land animals were scorpion-like arthropods, then wingless insects, amphibians, and, eventually reptiles. The final period of the Paleozoic Era (the Permian Period) was a time of great change, both geologically and biologically. The climate appears to have become much colder and dryer, apparently culminating in a period of glaciation.

About 245 million years ago, as the Paleozoic era neared an end, life on earth experienced the most disastrous mass extinction in its entire history. This was particularly true for sea life; 80 to 90 percent of all marine species disappeared from the earth. The reason for this dramatic decrease in marine organisms is unclear. There is no evidence that it was the result of some sudden catastrophe since careful examination of the fossil record indicates that numbers of species dwindled over a period of several million years. Meanwhile, on land, many amphibian and mammal-like reptile groups also disappeared, creating opportunities for other reptiles to become the dominant terrestrial vertebrates during the Mesozoic Era. Of all life forms, insects and plants appear to have been least affected by the Permian extinctions that ended the Paleozoic Era. As we will see, such episodes of mass extinction occurred during the next 2 eras as well.

THE MESOZOIC ERA: THE AGE OF REPTILES

The Mesozoic Era (65 million to 240 million years ago) was generally a time of stable weather patterns, extensive mountain building that produced much of the earth's present terrain, and rising seas. The Mesozoic Era (Figure 24-6) is divided into three periods, the Triassic, Jurassic, and Cretaceous.

The Mesozoic Era is perhaps most remarkably distinguished by reptiles that dominated virtually all terrestrial habitats and returned to some aquatic habitats. One group, the pterosaurs, became the first vertebrates to evolve the ability to fly. Another group of reptiles evolved into the largest land animals ever to roam the earth — the dinosaurs that dominated the land for 125 million years. But the reign of the dinosaurs collapsed in "sudden" extinction at the end of the Mesozoic Era, creating a fascinating yet unsolved mystery (see The Human Perspective: The Extinction Detectives, page 436). Whatever caused the extinction of

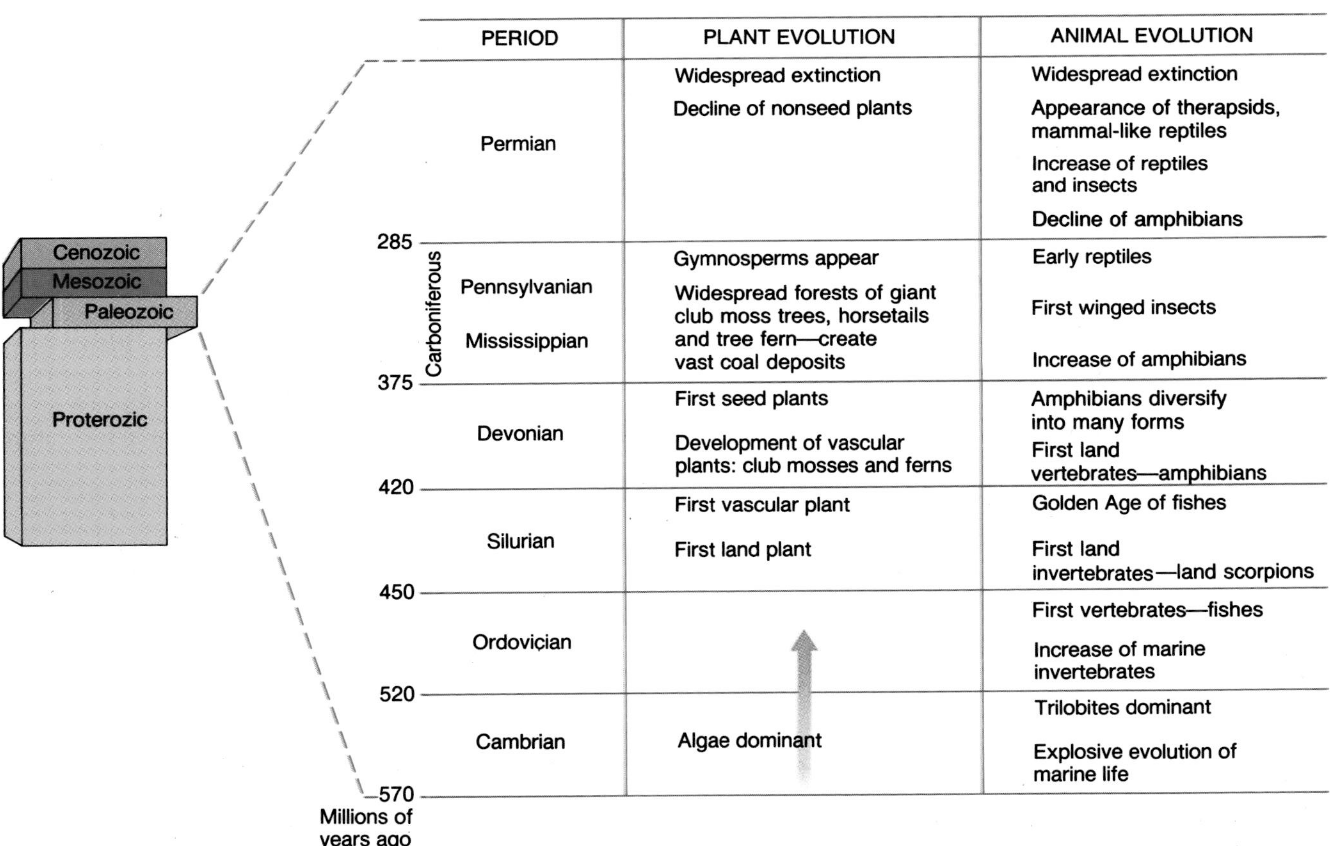

Millions of years ago	PERIOD	PLANT EVOLUTION	ANIMAL EVOLUTION
	Permian	Widespread extinction; Decline of nonseed plants	Widespread extinction; Appearance of therapsids, mammal-like reptiles; Increase of reptiles and insects; Decline of amphibians
285	Pennsylvanian / Mississippian (Carboniferous)	Gymnosperms appear; Widespread forests of giant club moss trees, horsetails and tree fern—create vast coal deposits	Early reptiles; First winged insects; Increase of amphibians
375	Devonian	First seed plants; Development of vascular plants: club mosses and ferns	Amphibians diversify into many forms; First land vertebrates—amphibians
420	Silurian	First vascular plant; First land plant	Golden Age of fishes; First land invertebrates—land scorpions
450	Ordovician		First vertebrates—fishes; Increase of marine invertebrates
520	Cambrian	Algae dominant	Trilobites dominant; Explosive evolution of marine life
570			

FIGURE 24-5

The Paleozoic Era. At the beginning of the Paleozoic Era, the seas were brimming with algae and primitive invertebrates, such as sponges, jellyfish, worms, starfish, and trilobites (an extinct group of arthropods). Later on, plants, and then animals, began to colonize the land. Near the end of the Paleozoic Era, forests of giant tree ferns, club mosses, seed ferns, horsetails, and early conifers (evergreen trees and shrubs) blanketed huge expanses of the earth's surface. Amphibians were the dominant land animal at this time. The end of the Paleozoic Era was marked by the greatest mass extinction ever recorded in the fossil record. Among the survivors were two groups of organisms: the conifers and the reptiles.

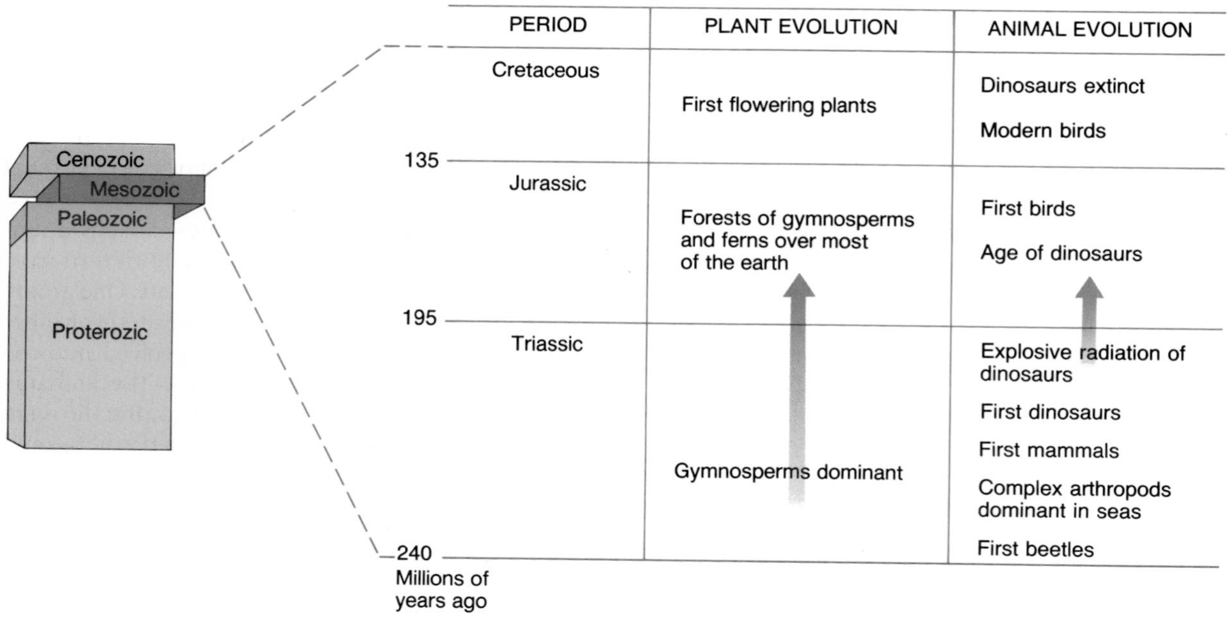

Cenozoic
Mesozoic
Paleozoic
Proterozoic
PERIOD
PLANT EVOLUTION
ANIMAL EVOLUTION
Cretaceous
First flowering plants
Dinosaurs extinct
Modern birds
135
Jurassic
Forests of gymnosperms and ferns over most of the earth
First birds
Age of dinosaurs
195
Triassic
Gymnosperms dominant
Explosive radiation of dinosaurs
First dinosaurs
First mammals
Complex arthropods dominant in seas
First beetles
240
Millions of years ago

FIGURE 24-6

The Mesozoic Era. The Mesozoic Era is often called the Age of Reptiles. Adaptive radiation of reptiles filled the earth's habitats with representatives of this great class of vertebrates. Included among the Mesozoic reptiles were the dinosaurs and the therapsids, the ancestors of mammals. The first flowering plants evolved near the end of the Mesozoic Era, which, like the previous era, culminated in mass extinctions.

these dominant animals also eliminated thousands of other life forms, including many marine organisms.

One group that was not as severely affected by the disaster that ended the Mesozoic era were small, probably nocturnal, animals that had evolved earlier from a group of reptiles known as therapsids (Figure 24-7). These animals had remained in the "shadows" during the success of the reptiles. These animals were the mammals, and they had a number of important biological innovations: They maintained a constant, elevated body temperature that allowed them to remain active at night (although some dinosaurs may also have possessed similar temperature

FIGURE 24-7

A therapsid reptile of the type that gave rise to mammals.These animals' teeth, jaws, and "upright," four-legged posture made them more like mammals than other reptiles.

regulating mechanisms); they were covered with hair that insulated them from colder temperatures; and they gave birth to their young, rather than hatching them from eggs.

The success of birds, insects, and mammals was accompanied by the appearance of another biological innovation, the flower. Flowering plants thrived during the next era and continue to thrive today, well into the Cenozoic Era.

THE CENOZOIC ERA: THE AGE OF MAMMALS

The present era, the Cenozoic Era, began about 65 million years ago and is known as the Age of Mammals (Figure 24-8). Insulating hair and thermoregulatory abilities adapted mammals for the colder climates that became more prevalent during the Cenozoic Era. By the first 10 million to 15 million years of this era, representatives of most of the modern orders of mammals had appeared. During the first half of the Cenozoic Era, primates were represented by the lemurs and tarsiers, small arboreal forms that resembled some of the earliest mammals. Monkeys appear in the fossil record approximately 35 million years ago, and the direct ancestors of the four groups of modern apes (gibbons, orangutans, chimpanzees, and gorillas) appeared about 20 million years ago. The Cenozoic Era also saw the explosive radiation of flowering plants.

The Pleistocene Epoch, which began about 2.5 million years ago, was characterized by intermittent ice ages that caused mass extinctions and migrations, dramatically altering the distribution of life on earth. Tropical vegetation, once widespread over much of the planet, became restricted to warm, wet habitats near the equator. Giant mammoths, ground sloths, and numerous other large mammals disappeared with the last ice age (Figure 24-9). Following this last ice age, which ended about 10,000 years ago, semiarid and arid areas developed, and the surviving plants and animals began the most recent period of adaptive radiation.

The major evolutionary trends not only produced a fascinating history of life on our 4.6-million-year-old planet, it has created at least 2 million species that are alive today. In the following three chapters, we will examine the remarkable diversity of these living organisms that comprise today's five kingdoms of life.

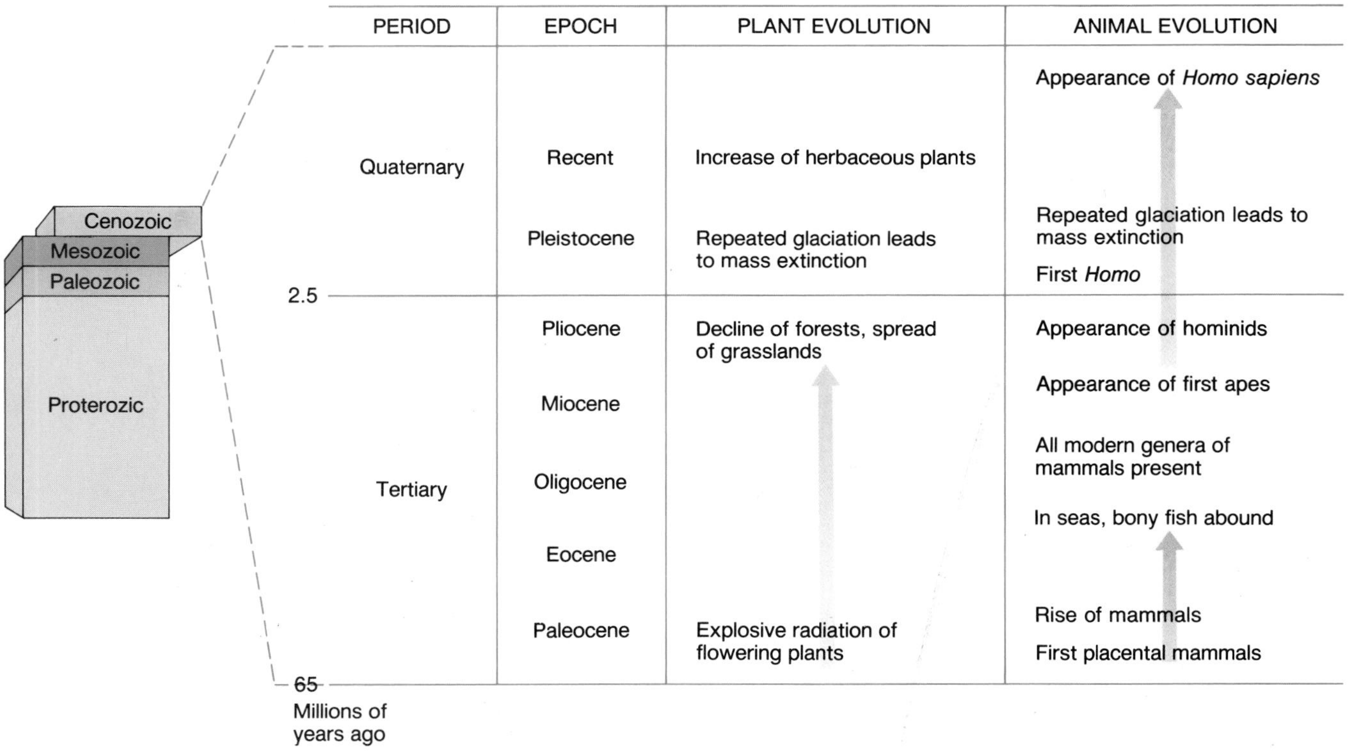

FIGURE 24-8

The Cenozoic Era is divided into two periods, the tertiary and quaternary, which together are subdivided into seven epochs. Mammals and flowering plants evolved rapidly during the Cenozoic Era. Intermittent ice ages dramatically changed the climate, extinguishing many species. The Cenozoic Era is often called the Age of Mammals.

FIGURE 24-9

A gallery of Ice Age animals. From left to right: Jefferson's mammoth, Conkling's pronghorn antelope, giant heron vulture, and a great short-faced bear.

◁ THE HUMAN PERSPECTIVE ▷

The Extinction Detectives

The disappearence of the dinosaurs, the "terrible lizards," poses a mystery like few others in biology. Dinosaurs intrigue amateur and professional biologists alike, leaving most of us wondering what these remarkable creatures were really like and what happened to them. Dinosaurs began their adaptive radiation 250 million years ago, about 100 million years after the appearance of Earth's first reptiles. Many of these animals grew to enormous dimensions, such as *Tyranosaurus rex,* perhaps the largest carnivour ever, and *Diplodocus,* the longest complete dinosaur discovered so far (Figures 1 and 2). After 125 million years of dominance, they fell to extinction, along with a variety of other unrelated organisms, from many species of plants to much of the marine life. All these organisms simply disappeared from the fossil record 65 million years ago. The search for clues to the dinosaurs' departure has lasted for decades.

The proposed explanations for the extinction that ended the Mesozoic Era fall into two broad catagories: those that suggest a gradual process of extinction over a period of a couple of thousand to 3 million years and those who propose a catastrophic extinction that could have occurred "instantly," that is, over a period of months to a few years. Since 1980, opinion has gradually shifted toward the rapid catastrophe hypothesis. At that time, Nobel Laureat Luis Alvarez published a paper describing high concentrations of *iridium* in the rock formations from 65 million years ago, rocks that formed about the time the dinosaurs became extinct. Iridium is a scarce element on Earth but is found in high levels in extraterrestrial bodies, such as asteroids and meteriorites. Alvarez interpreted the 65-million-year-old iridium layer as evidence that a huge asteroid or meteorite struck the earth at that

FIGURE 1
Paleontologists carefully separate the skull of *Tyranosaurus rex* from South Dakota sandstone

EVOLUTION AND ADAPTATION: TYING IT TOGETHER

The history of life on earth is the story of evolution itself. Yet even before the planet's first life forms appeared, evolution was already underway. The chemicals that would eventually combine to form living material were evolving. Some of these were self-replicating molecules that could not only produce copies of themselves, but could also direct the assembly of the inanimate environment's amino acids into proteins. Other new substances appeared, some of which formed sheets that encased the other molecules in a membrane, forming precells. From precells, the first living organisms emerged, using the abiotically produced

time, sending a massive cloud of iridium-containing dust into the atmosphere. The dust was dispersed around the world before settling back to the surface, where it formed a thin layer that has been preserved in the Earth's rocks.

The dust cloud could have dramatically altered the earth's climate by blocking the sun for years, slowing photosynthesis, on which virtually all organisms ultimately depend. Other scientists disagree. They maintain that such an impact would have started global wildfires, releasing so much carbon dioxide as to cause a sudden "greenhouse" warming of the Earth. Still others propose that the atmosphere was poisoned by natural gas when the asteroid's impact released an enormous amount of methane from beneath the earth's crust. It might have even been "acid rain" that killed the dinosaurs, the huge cloud of nitrous oxide thrown skyward by the asteroid collision dissolving in airborne water droplets forming nitric acid.

Whatever ended the reign of these Mesozoic masters, many biologists believe that the collision between Saturn and comet Shoemaker-Levi in 1994 strengthened the catastrophic impact explanation, if for no other reason than to show us how realistic (and catastrophic) such an impact is.

FIGURE 2
A skeleton of the herbivorous dinosaur, *Diplodocus*, along side that of a human, provides some perspective on the size of these huge land animals.

energy-rich molecules in their surroundings for nutrients and energy. Cells acquired more complexity as prokaryotic life gave rise to eukaryotic cells, and later to multicellular organisms. Life diversified as new, more complex species appeared, species that were either better able to compete for nutrients or to use unexploited niches with little competition. The first photosynthesizers, for example, had no competition for available light or carbon dioxide. Their activities, however, changed the earth in two profound ways. One consequence was the mass extinction of oxygen-sensitive organisms, as photosynthesizers saturated the atmosphere with the deadly gas. A second consequence was opposite in effect—photosynthesizers created a vast new food source that threw open the evolutionary "floodgates."

This new source of usable energy and nutrients led to an explosive diversification of organisms that produced millions of species. Today virtually all life on earth depends on photosynthesis.

Over the past 4 billion years, changes in the earth's geography and climate have influenced the history of life. In many ways, however, life directs its own history, creating new selective pressures (such as an oxygen-saturated atmosphere and a predator-stocked environment). These changes force new adaptations to appear. As life becomes more diverse, new organisms emerge with novel adaptations that enable them to better cope with the new biotic conditions. In this way, diversity leads to an even greater variety of life. The "mouse and the mouse-catcher" continue to grow more complex as each evolves mechanisms of coping with a better version of the other.

SYNOPSIS

The earth condensed out of a massive cloud of dust and gases, approximately 4.6 billion years ago. Less than a billion years later, life appeared, the result of spontaneous chemical evolution. The first life forms were likely heterotrophs that used organic compounds in the "primordial soup" for food. The evolution of photosynthetic autotrophs about 3 billion years ago was essential for the continuation of life on earth. Unicellular eukaryotes evolved about 1 billion years ago, paving the way for the evolution of multicellular organisms relatively recently (700 million years ago).

The geologic time scale divides the earth's history into four great eras, beginning with the Proterozoic Era, during which life evolved. Repeated changes in the shape, size, and location of the earth's continents drastically changed the earth's climate and sea level, affecting the evolution of organisms. By the end of the Proterozoic Era, members of all of the major animal phyla had appeared.

Vertebrates evolved in the sea during the Paleozoic Era, then invaded land, as did plants. In addition, adaptive radiation diversified land arthropods, beginning with scorpion-like forms, followed by a variety of insects. Mosses, ferns, and seed plants appeared successively on land. The Paleozoic Era ended with life's greatest episode of mass extinction, extinguishing the vast majority of marine organisms and many terrestrial forms.

During the Mesozoic Era, reptiles and gymnosperms dominated the land. Adaptive radiation among reptiles produced a diversity of organisms, including dinosaurs. Flowering plants and the first mammals and birds evolved during this era. The Mesozoic Era ended with the most famous episode of mass extinction, eliminating the dinosaurs and setting the stage for the adaptive radiation of mammals.

The present Cenozoic Era has produced all modern mammals and flowering plants. In the past 2.5 million years, a series of intermittent ice ages has caused mass extinctions and reshaped the distribution of plants and animals across the earth.

Review Questions

1. Retrace the steps of chemical evolution believed to have led to the first living cell.
2. Discuss the changes in energy sources necessary for biological evolution from the time the earth formed to the present day.
3. Which of the four major geologic eras saw the appearance of the first living cells? The first eukaryotes? The first protists? The first gymnosperms? The first flowering plants? The first vertebrates? The first reptiles? The first mammals? The first humans?
4. How has the distribution of land masses changed over the past 200 million years?
5. Describe how organisms themselves have changed the conditions on earth in a way that has influenced the direction of evolution.

Critical Thinking Questions

1. Why are the characteristics that qualify a structure as a precell so important in the formation of the Earth's first living cell?
2. No signs of life have ever been found on Venus or Mars. Research the environmental conditions on these planets and develop an explanation for why life is unlikely to exist on these planets and why the environment on earth is more hospitable to life.
3. According to the second law of thermodynamics, the universe continues to increase toward a state of increasing disorder. Explain why the evolution of increasingly more complex structures—from simple molecules to highly complex living organisms—does not defy the second law of thermodynamics.
4. Try to envision the future, 10, 50, and 100 years from now. Predict how humans will have affected the evolution and extinction of other organisms on earth. How will that influence the evolution of the human species? What priorities should biologists establish today to circumvent any negative changes in the future?

CHAPTER 25

The Microbial Kingdoms: Viruses, Monerans, Protists, and Fungi

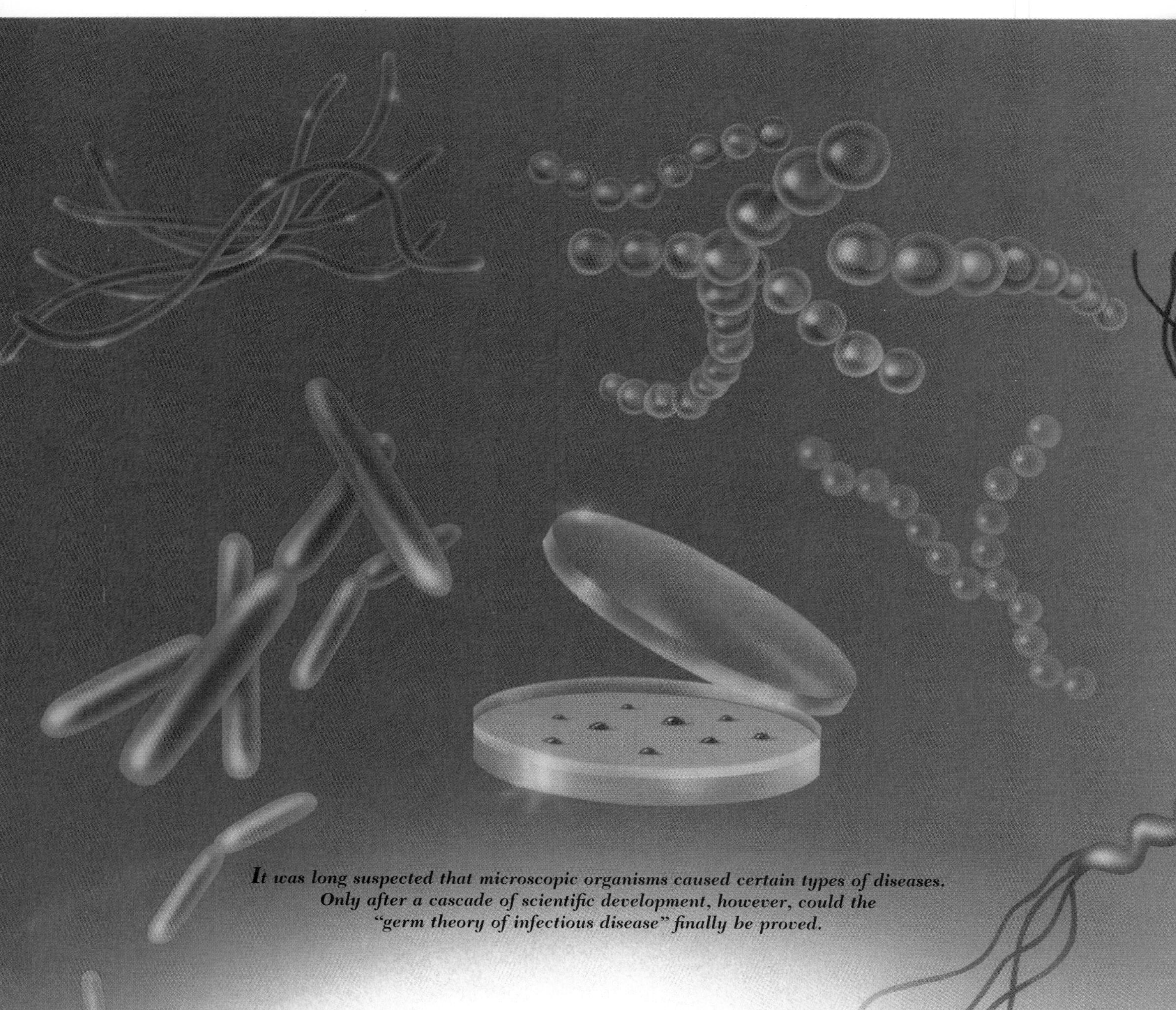

It was long suspected that microscopic organisms caused certain types of diseases. Only after a cascade of scientific development, however, could the "germ theory of infectious disease" finally be proved.

STEPS TO DISCOVERY
The Burden of Proof: Germs and Infectious Disease

The chief of surgery felt helpless. In 1850 nearly half of the women whose babies were delivered by doctors in his hospital contracted "childbed fever," a massive fatal infection of the uterus and bloodstream caused by bacteria introduced into the birth canal during delivery. The cause of the disease was unknown at the time; in fact, no microorganism had ever been proven to cause human disease.

The alarmed chief of surgery, a Hungarian physician named Ignaz Semmelweis, aggressively attacked the situation. He began by observing doctors and patients, trying to identify some factor that might be responsible for the disastrous epidemic in the hospital. He soon found a pattern: The doctors moved from one mother to another with no intervening sanitary precautions such as hand washing (sanitation was not yet recognized as having any practical value). In fact, the doctors often went directly from postmortem dissections to the maternity ward. Semmelweis suspected that the doctors were unwittingly transferring the agent of the fatal disease from one person to another.

Semmelweis conducted a simple but powerful test of his hypothesis. He ordered all physicians in his hospital to wash their hands in a strong chlorine solution before performing each delivery (chlorine kills bacteria). The incidence of childbed fever quickly dropped from 50 percent to 1 percent.

For 17 additional years, however, doctors continued their unsanitary "healing" practices. Then, in 1867, an English surgeon, aware of Semmelweis's success, attacked a similar dilemma. Patients in his hospital were also dying, this time of "postsurgical disease." The doctor, Joseph Lister, believed that microorganisms introduced into the surgical wound were responsible. He recommended using an antimicrobial chemical (phenol) to sanitize surgical instruments, surgeons' hands, and the wound itself. He even sprayed a mist of the chemical in the air of the surgical room. The incidence of postsurgical disease soon plummeted to less than 5 percent. Lister toured the lecture circuit and convinced a reluctant medical community of the value of sanitation. The incidence of hospital-acquired infections decreased dramatically in all hospitals that adopted his sanitary precautions. Lister's work provided evidence that microbes were responsible for certain diseases, but proof that germs actually *cause* disease was still lacking.

Although many scientists and physicians had observed that persons suffering from a particular disease always seemed to have microorganisms in their bodies, the mere presence of a microorganism in a person with the disease did not prove that it causes the disease. In fact, the microorganism's presence may be the *result* of the disease. The challenge was to establish a cause-and-effect relationship between a specific microorganism and a particular disease. Another 15 years went by without proof that microorganisms cause human disease.

Then, in 1882, a German physician named Robert Koch stepped into the picture. Koch's knowledge of the scientific method ultimately provided incontrovertible proof of what we today call *the germ theory of infectious disease.* Before he could execute his plan, however, Koch had to purify the suspected microorganism so he would know that it alone (and no unwanted contaminating microbes) would be responsible for the results when inoculated into experimental animals. The problem was finding a suitable solid medium for growing isolated colonies—groups of cells that are all identical descendants of the one cell deposited on the surface at the spot where the colony would grow. Potato slices and nutrient gelatin proved inadequate because many microbes digested the solid surfaces, turning them to liquid and allowing cells from different colonies to wash together, so different species mixed with one another. Koch needed a solidifying agent that microbes could not digest. The solution was discovered in a household kitchen and delivered to Koch by an alert American woman, Fanny Hesse.

Hesse, the wife of one of Koch's associates, suggested using *agar,* a polysaccharide extracted from red seaweed, which she used for hardening jelly. Bacteria could absorb the nutrients added with the agent but could not digest the agar itself, which remained hard. Agar provided Koch with both a solidifying agent and an easy method for obtaining cultures of bacteria that contain only one kind of microorganism. Working with anthrax, a rapidly fatal disease transmitted from cattle to humans, Koch used solid media to isolate colonies of the suspected bacterium (*Bacillus anthracis*) from a person suffering from the disease. Koch then injected a small amount of these bacteria into a susceptible healthy animal free of the anthrax bacteria. The inoculated animal soon developed anthrax, while the noninoculated animals did not. From the diseased animal, Koch then reisolated huge numbers of *Bacillus anthracis* in pure culture. Robert Koch had proven the germ theory of infectious disease. In doing so, he provided an approach that is still used today to establish the cause of many infectious diseases.

*T*his chapter is devoted to *microorganisms*—living agents too small to be seen without the aid of a microscope. This classification includes several groups of organisms that people tend to brand as pernicious villains. One group, members of the Monera Kingdom, contains *bacteria* and their relatives, organisms that are often automatically associated with human suffering and disease. Another group, the viruses, receives even worse press than do bacteria, especially in the ominous shadow of the virus responsible for today's "modern plague," AIDS (Acquired Immune Deficiency Syndrome). A third group, the Protists, is composed of a "hodgepodge" of microscopic organisms that range from animal-like cells called protozoa to plantlike unicellular algae. A fourth group consists of fungi, a kingdom that includes yeasts, molds, and mushrooms.

Some of these organisms clearly deserve their nasty reputations, but most do not. In fact, some microorganisms are essential for the continued survival of life on earth, as you will see in our examination of the remarkable abilities and surprising influences of the members of the microscopic kingdoms.

▼ ▼ ▼

VIRUSES

Although the cell is the fundamental unit of life, there are a group of "organisms," the **viruses,** that are not cells at all. They have no cytoplasm, no organelles, and no plasma membrane. A fully assembled virus is as inanimate as a crystal of organic chemicals. In fact, some are so simple that they are nothing more than nucleic acid and protein. But once inside a host cell, the virus's biological apparatus "awakens." There, it adopts its "living mode," characterized by two key activities of life—reproduction and heredity. How viruses alternate between these two seemingly paradoxical states is a function of their structural simplicity and the means by which they use their limited resources and their more lavish host cells to complete their life cycle.

VIRAL STRUCTURE

The structural simplicity of viruses is revealed not only by their non-cellular nature but by the unique fact that they possess only *one type of nucleic acid* (either DNA or RNA, but never both). In fact, some viruses consist of nothing more than a molecule of nucleic acid (the viral genes) surrounded by a protein coat (Figure 25-1). This coat protects the genes and allows the virus to attach itself to and infect its specific host cell. Molecules on this outer surface specifically fit with and attach to surface molecules on their particular host cell in a "lock-and-key" type configuration. The most complex viruses are bacteriophages—those that infect bacteria, cells with one of the strongest protective cell walls that the virus must somehow penetrate. Bacteriophage possess a head (containing the viral nucleic acid) and a tail with fibers that recognize and attach to the cell wall of its particular host cell. The tail then contracts and forcefully injects a needle-like hollow tube into the cell, through which the viral nucleic acid enters the cell. This initiates the process of viral infection.

VIRAL REPLICATION

The general strategy of a virus is simply to get its nucleic acid into a susceptible host cell. In fact, all a virus needs for a successful "life" cycle is a set of genes that contain the information needed for producing more viruses, and a way of getting those genes into a host cell (a function of the virus's outer coat). Once the virus abandons its coat, it exists as a single strand of nucleic acid inside the cell—essentially another "chromosome" directing the cell's activities, as shown in Figure 25-2. Although this figure illustrates bacteriophage infection, it typifies the intracellular dynamics of most viral infections, including those of humans. Because the cell's protein synthesis machinery doesn't distinguish between the viral and the cell's nucleic acid, it simply follows the dictates of the newly acquired genetic instructions. In this way, the virus commandeers the cell's metabolic machinery, redirecting it so that instead of producing new cell material, the cell becomes a virus-producing factory.

The host cell replicates the viral nucleic acid, sometimes generating several hundred copies of the viral genes. These genes also instruct the cell to produce huge quantities of coat proteins that automatically aggregate around the new strands of viral nucleic acid, assembling hundreds of progeny viruses, which are released from the cell to infect another susceptible cell.

VIRAL DISEASES

Most viruses that infect humans produce infections (such as chickenpox and mumps) that rapidly run their course and stimulate lifelong immunity in survivors. The resultant immunity is due to the production of immune cells and antibodies that specifically recognize and inhibit subsequent infection by the same types of viruses (see Chapter 20). Yet a number of viral diseases pose special medical problems either because immunity is inadequate or because a patient may die before protective levels of immunity develop. Today, one of the most dreaded viruses is *human immunodeficiency virus* (HIV), the virus that

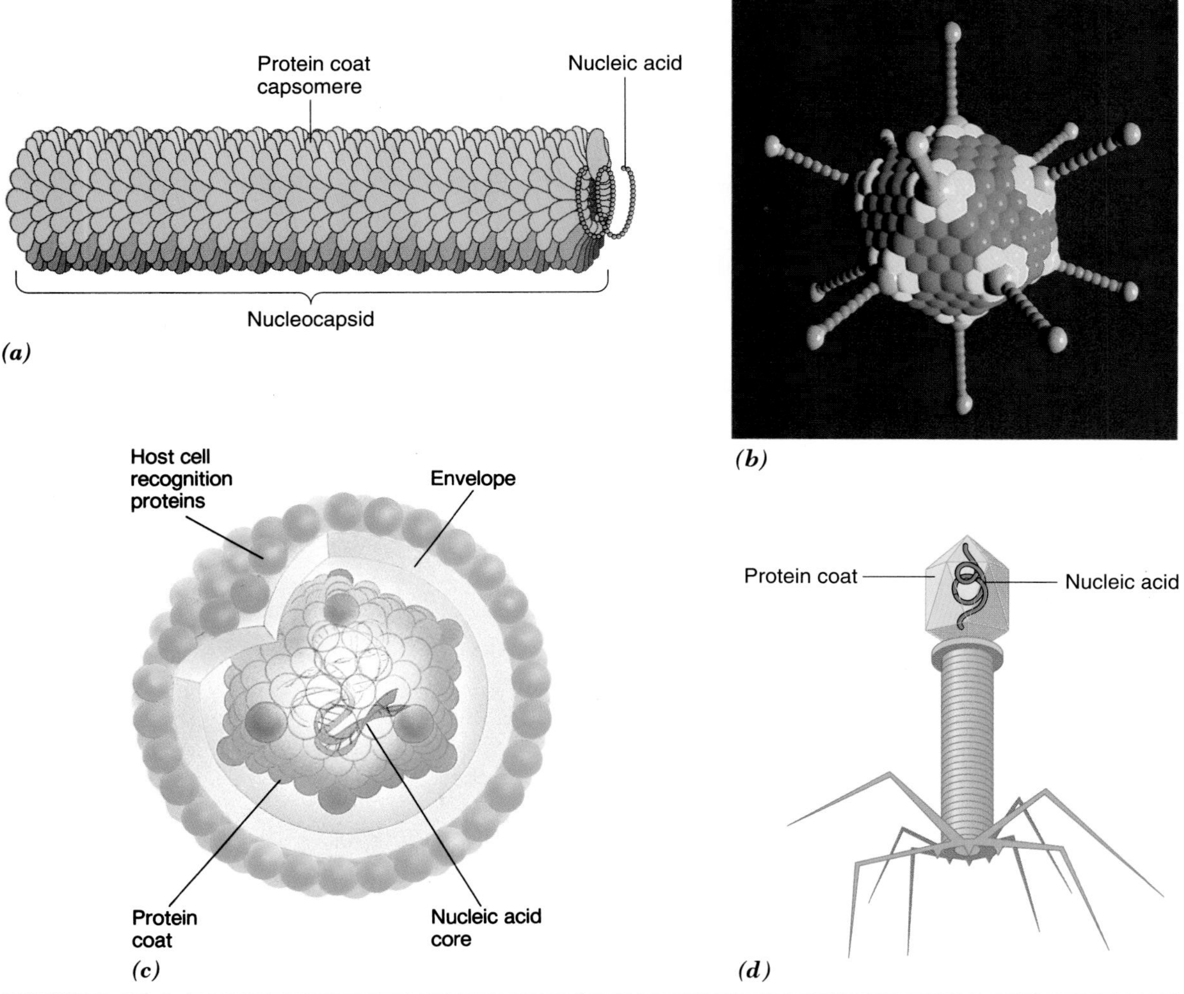

FIGURE 25-1

Four basic forms of viruses. All viruses contain a nucleic acid core housed in a protein coat. Some of these are rod-shaped (proteins attached to a tight spiral of nucleic acid) as in ***(a)***; others are shaped like faceted spheres ***(b)***. Some are surrounded by an outer membrane ***(c)*** that is acquired as the virus escapes the host cell, enclosing itself in a modified portion of the cell's plasma membrane. Bacteria-infecting viruses, called bacteriophages, are the most complex viruses ***(d)***.

causes AIDS. HIV infects and destroys cells of the immune system itself, thereby rendering the infected host defenseless against the virus as well as many other types of potential pathogens.

Another way the AIDS virus can escape the immune system is by establishing **latent** (hidden) **infection,** as do a number of other viruses (such as the herpes viruses). These viruses remain in the host even after disease symptoms disappear and are generally undetectable during these latent periods. Evidence suggests that the DNA of these viruses actually inserts itself into the chromosomal DNA of the host cells, where it resides in a dormant state, replicating only with each division of the host cell. When the viral DNA eventually leaves the host chromosome and begins its viral-productive cycle (similar to that in Figure 25-2), symptoms of disease reappear. Such "latent" diseases may be reactivated periodically by various stimuli. For example, recurrent episodes of Type 1 (oral) herpes or Type 2 (genital) herpes may be triggered by emotional stress, sunburn, menstruation, pregnancy, common colds, or fever (hence the common terms "fever blisters" and "cold sores"). Between episodes, the host appears "cured," as the virus silently resides within the chromosomes of regional nerve cells, poised for its next attack.

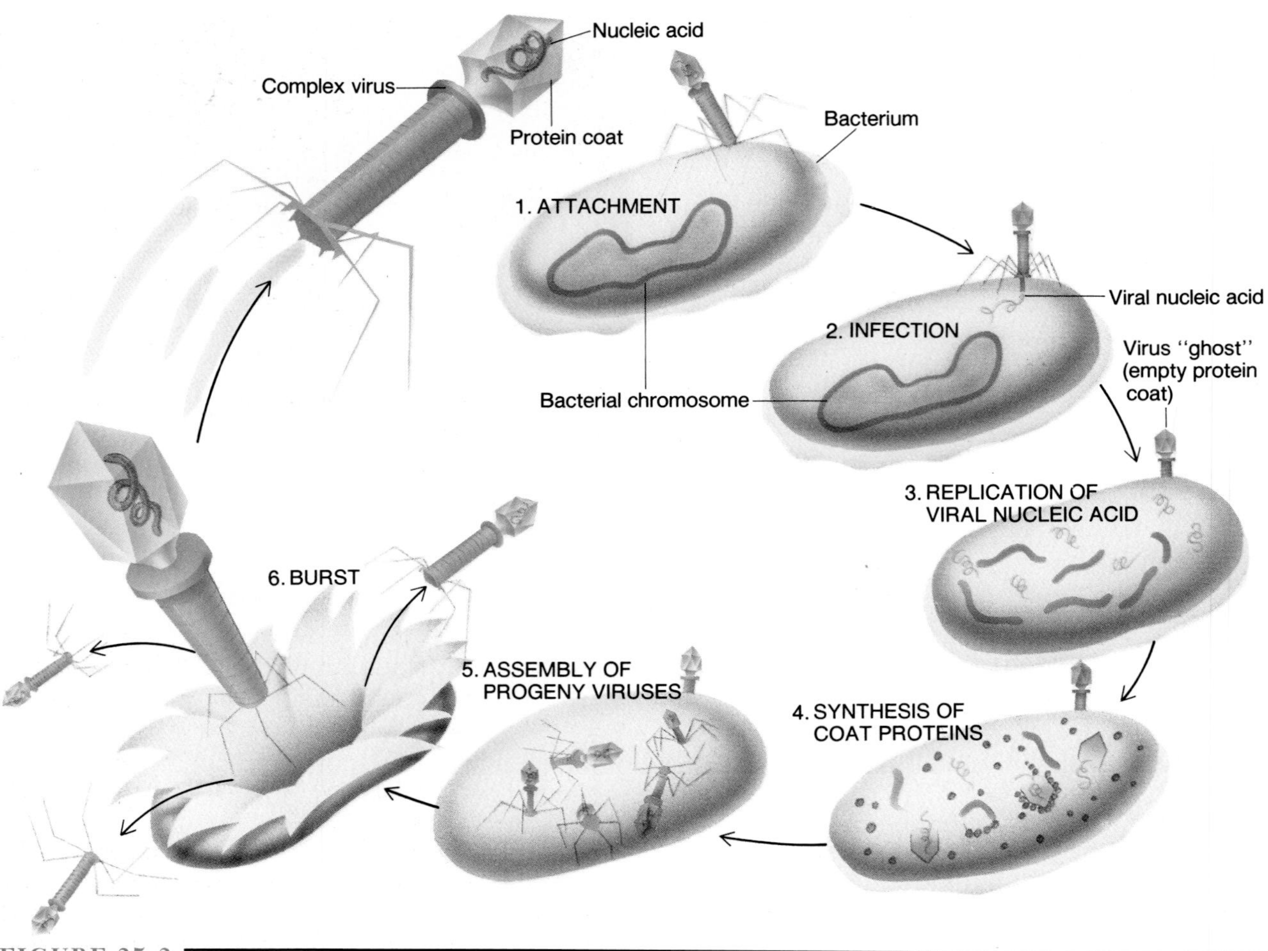

FIGURE 25-2
The life cycle of a typical virus is illustrated here by viral infection of a bacterial cell. Although some variation exists from virus to virus, the basic strategy is the same as shown here. The host cell is not always killed, however. Some infected animal cells "leak" mature viruses through intact plasma membrane.

SUCCESSES AND FAILURES IN THE WAR AGAINST VIRUSES

None of the earth's known species is free of viral infection. Although not always harmful, some cause serious diseases in plants and animals. Viral diseases in humans range in severity from common colds to fatal cases of smallpox, rabies, and AIDS. Even the seemingly innocuous virus that causes diarrhea in children poses a deadly threat in populations ill-equipped to provide adequate care (the diarrhea-causing rotoviruses are the leading killers of children worldwide, causing fatal dehydration). Because no antibiotics are effective against any viral disease, cures still elude medical science. The body's natural immunity, however, eliminates most viral infections if the infected person survives long enough to mobilize defenses and control proliferation of the virus.

Many viral diseases have been controlled, however, with *vaccines*—inactivated or weakened forms of disease-causing viruses that specifically stimulate the immune system (Chapter 20). Vaccination has reduced or eliminated the threat of polio, smallpox, rabies, yellow fever, measles, rubella, and mumps. Some viral diseases still elude attempts to control them, however. The most exasperating of these is the common cold, which is caused by almost 200 different types of viruses. The most frightening of these uncontrolled viral diseases is AIDS.

THE MONERA KINGDOM

All members of the Monera kingdom are single-celled prokaryotes. In other words, they are all bacteria. Typical

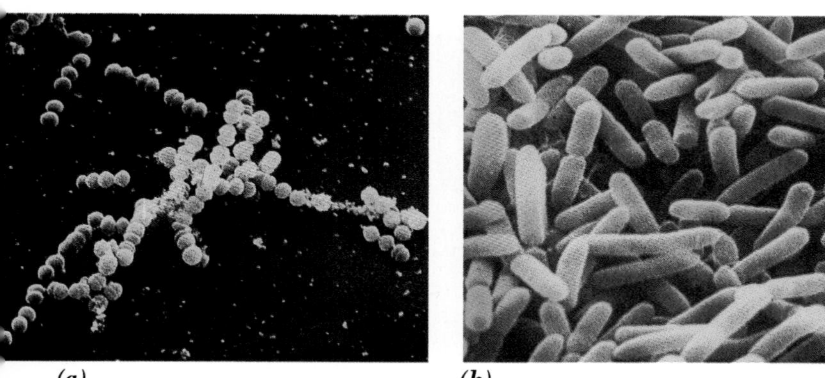

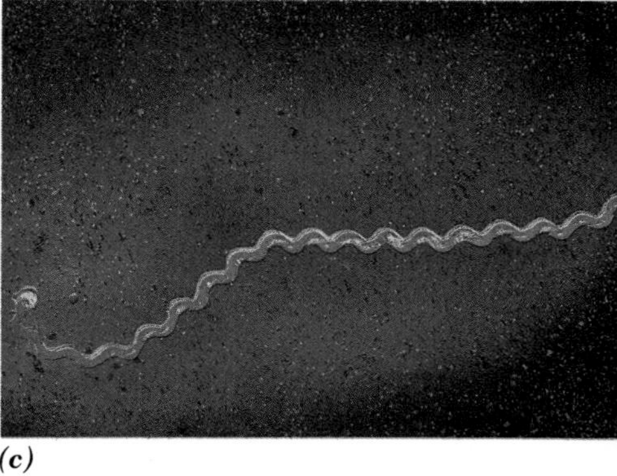

(a) *(b)* *(c)*

FIGURE 25-3

Typical bacterial shapes—*(a)* spheres, ***(b)*** rods, and ***(c)*** spirals. More atypical shapes are generally found among a more recently described group of monerans called "archeobacteria," which will be discussed in the upcoming overview of the kingdom.

bacteria are distinguished from the eukaryotes of the other four kingdoms by some basic characteristics:

- the presence of a **nucleoid** instead of a true nucleus (Chapter 3),
- their unique ribosomes,
- unique cell wall material, and
- the absence of membrane-enclosed organelles (such as mitochondria, chloroplasts, endoplasmic reticulum, and the Golgi apparatus).

Although composed of the same groups of chemical as eukaryotes (proteins, nucleic acids, lipids, and carbohydrates), true bacteria possess several biochemical differences. The most universal distinction is the presence of **peptidoglycan,** a compound found nowhere else in nature other than in the prokaryotic cell wall of bacteria. This polymer extends around the cell, surrounding it with a cross-linked matrix that reinforces the cell wall. Because bacteria have no means of evacuating excess water, they need the strength the peptidoglycan wall provides to protect the fragile cell from swelling and bursting due to the osmotic influx of water. The importance of this protection is illustrated by the effects of the antibiotic penicillin, which prevents bacteria from forming peptidoglycan. Without their cell wall, the bacteria quickly burst, explaining why penicillin can be used to treat bacterial infections without harming the infected host (human cells have neither cell walls nor peptidoglycan).

The rigidity and shape of the cell wall determine the shape of the bacterial cell. Three shapes are most prominent among bacteria: spheres (*cocci*), rods (*bacilli*), and spirals (*spirilla*) (Figure 25-3). Many variations on these three shapes are relatively common.

EVOLUTONARY TRENDS

Of all the organisms alive today, bacteria most resemble our concept of the earth's original life forms, especially anaerobic (oxygen-independent) bacteria such as *Clostridium* (Figure 25-4). A comparison of modern bacteria with ancient fossils reveals that they have not changed much in more than 3.5 billion years. Modern prokaryotes expanded on the metabolic strategies of their primitive ancestors—

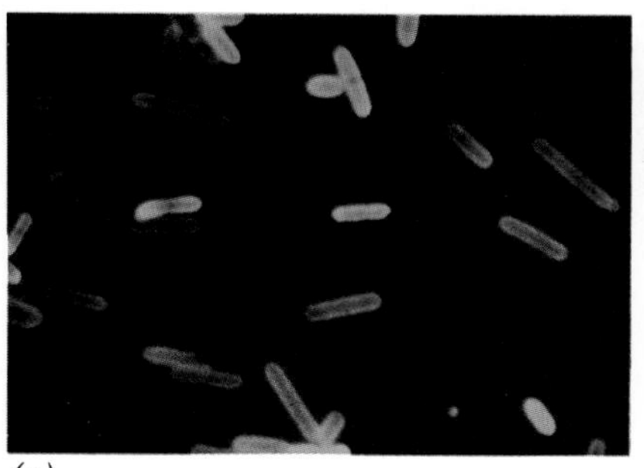

(a)

(b)

FIGURE 25-4

The earth's first life forms may have been very similar to these modern anaerobic bacteria. Species of *Clostridium* ***(a)*** are common in soil and can grow in oxygen-devoid environments, such as injured flesh (causing tetanus or gangrene) and canned foods (the major source of botulism, a fatal form of food poisoning). Even the anaerobic *Bacteroides* ***(b)*** that inhabit your intestine closely resemble early life forms.

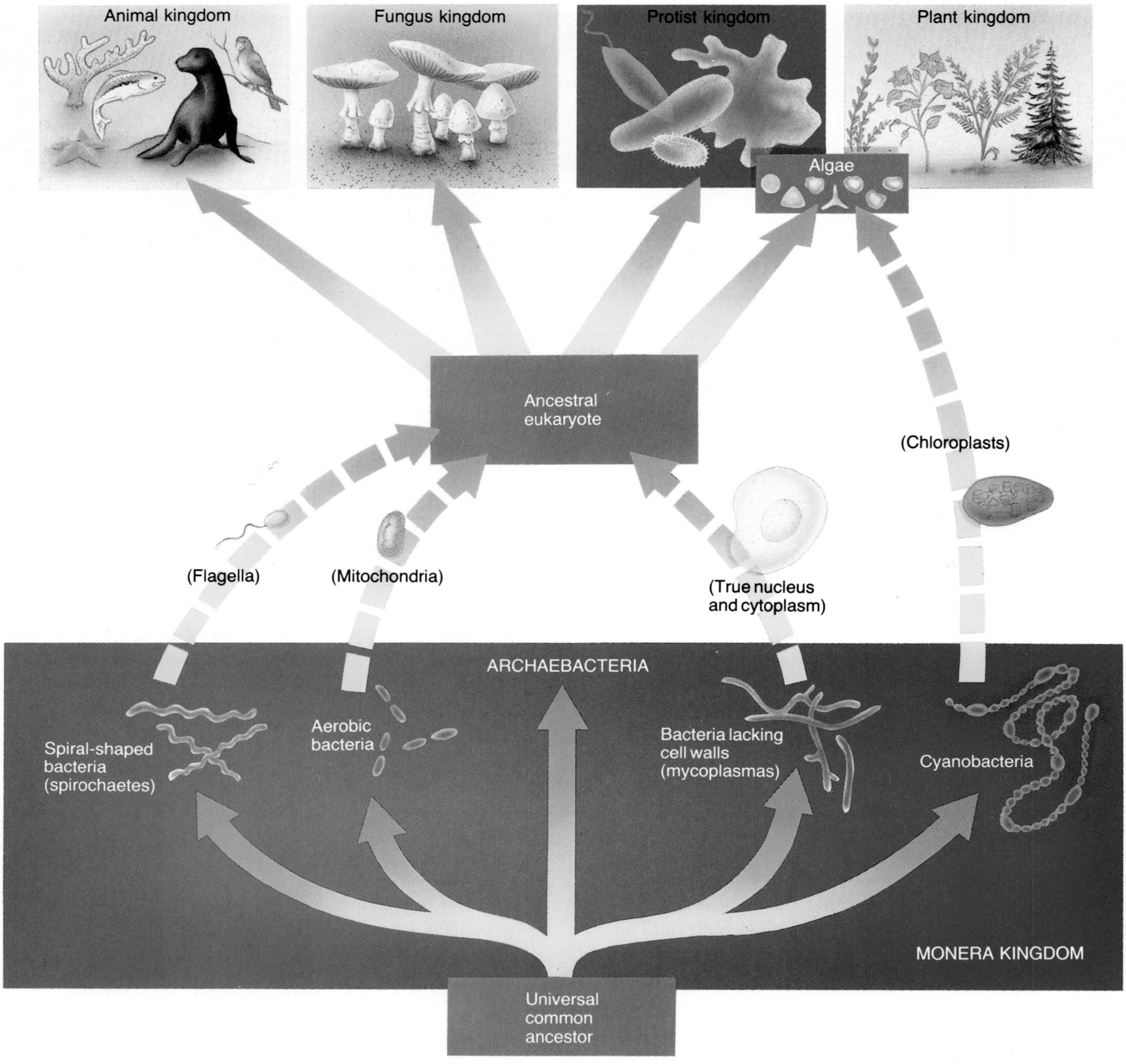

FIGURE 25-5

One proposal of evolutionary descent of all organisms from early prokaryotic cells. The dotted lines represent evolution of eukaryotic organelles from bacteria that formed endosymbiotic relationships with an early prokaryote. Archaebacteria (the most ancient form of life on modern earth) probably evolved as a separate lineage rather than descending directly from prokaryotes.

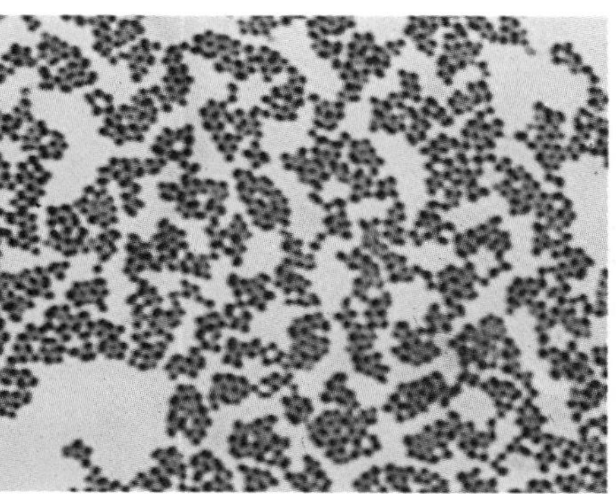
(a)

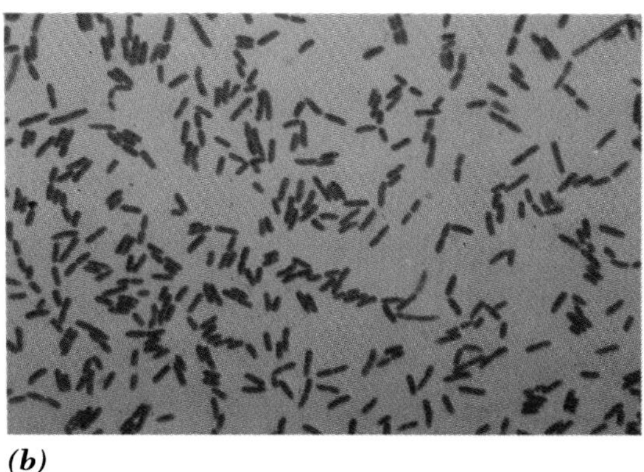
(b)

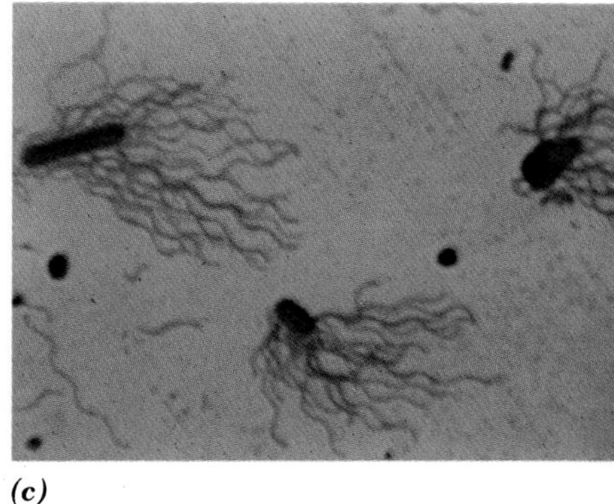
(c)

FIGURE 25-6
Microscopic views of bacteria. Although bacteria are colorless, taxonomists use stains to reveal characteristics that help classify and identify these cells. The most important stain, the Gram stain, identifies a bacterial species as either gram-positive *(a)* or gram negative *(b)* according to the color of the cells after staining. Flagella *(c)* and other cell structures can be detected using special stains that specifically enhance their appearance.

the central metabolism of today's living organisms (including humans) is a variation on some of the original prokaryotic pathways suggesting that eukaryotes evolved from one or more prokaryote ancestors (Figure 25-5).

In spite of their relative simplicity, prokaryotes are extraordinarily successful organisms—bacteria populate virtually all the earth's habitats; in some especially inhospitable places, such as concentrated sulfuric acid pools, they are the *only* inhabitants. One of the cornerstones of their success as relatively simple organisms is their extremely rapid growth rates, their ability to produce enormous numbers of individuals by simple binary fission (one cell dividing into two genetically identical daughter cells).[1] In some cells, a new generation is produced every 20 minutes. The bacterial population's growth is eventually halted when the nutrient supply in the local environment becomes exhausted. Such rapid growth rates often allow bacteria to outcompete the more slow-growing competitors. Many bacteria also transfer genetic material to one another, creating a gene mixing similar to that of sexual reproduction, but at a much faster pace. This mixing results in tremendous diversity among the enormous number of bacteria in virtually every habitat, so that virtually any adverse change will leave a few resistant survivors to repopulate in the new conditions. This is the cornerstone of the tremendous adaptability of bacteria and the reason that there are virtually no regions of the populated earth that are devoid of bacteria.

OVERVIEW OF THE KINGDOM

Because of their structural simplicity, bacteria lack the visual detail needed to categorize them solely on the basis of morphology (form), the principal approach used for classifying protists, fungi, plants, and animals. Thus, bacteriologists must supplement morphological observations with information about the organism's biochemistry, metabolic pathways, growth requirements, and genetic composition.

The starting point in characterizing bacteria is an observation of their appearance on both a microscopic (Figure 25-6) and a macroscopic (Figure 25-7) level. This information is then combined with more subtle criteria, such as that described in Table 25-1. Based on these criteria, taxonomists have described more than 4,000 species of monerans.[2] Although the Monera Kingdom contains the fewest species of any kingdom, the organisms in this kingdom are the most numerous on earth.

The members of the Monera Kingdom are divided into two fundamental types of cells: (1) **eubacteria** (*eu* = true), the group classically called "true bacteria"; and (2) **archaebacteria** (*archeo* = ancient), a prokaryotic group

[1] This form of binary fission is *not* mitosis since no chromosomal condensation occurs and the phases characteristic of mitosis do not occur. The cell simply duplicates its circular chromosome and divides, each cell receiving one of the genomes.

[2] As described in *Bergey's Manual of Determinative Bacteriology* (9th ed.), the authoritative manual on bacterial classification (William & Wilkins, 1993).

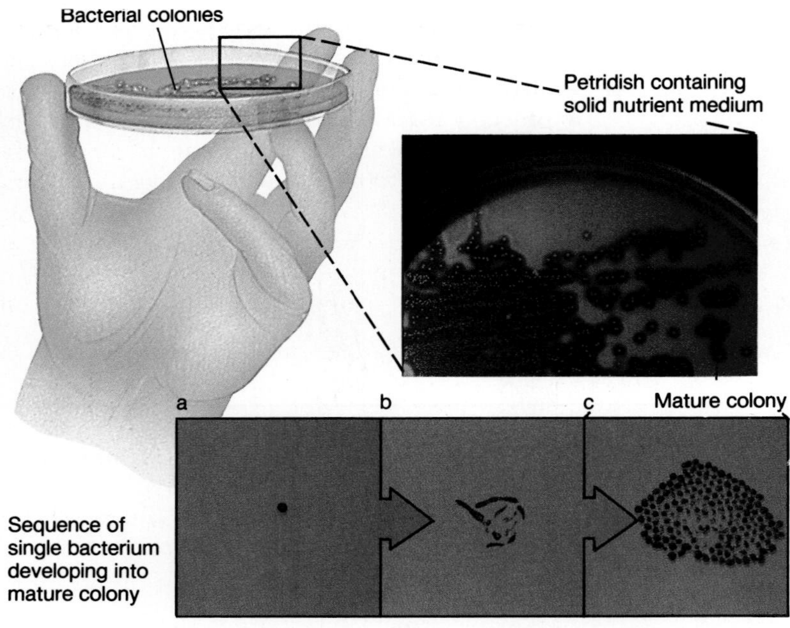

FIGURE 25-7
Colony characteristics are also used by taxonomists to distinguish different species of bacteria from one another. Colonies begin invisibly, as a single cell or a small group of cells, then proliferate into billions of bacteria that form a macroscopic (visible) mass.

TABLE 25-1
CLASSIFYING AND IDENTIFYING BACTERIA

Criteria for Identification and Classification	*Examples*
Microscopic appearance	
Cell shape and arrangement	Spheres in clusters; rods in chains
Differential staining	Gram-positive or gram-negative
Distinguishing structures	Endospores, flagella, capsule
Macroscopic characteristics	Colony texture, size, shape, and color
Biochemical properties	Unique amino acids in cell wall; composition of capsule
Physiological activities	Motility; oxygen requirements; carbon and energy requirements; ability to use specific sugars; metabolic by-products; obligate intracellular growth
Immunological specificity	Unique antigens of the cell, determined by reacting with antibody preparations of known specificity
Genetic analysis	Extraction of DNA, followed by determination of guanine-cytosine (G-C) content and nucleotide sequence
Ribosomal analysis	Comparison of similarities in one of the RNAs in the small subunit of ribosomes; closely related organisms have similar ribonucleotide sequences

that resembles eubacteria but differs from them in their biochemical constitution. One group of eubacteria was incorrectly called "blue-green algae" for years; these photosynthetic eubacteria have been renamed **cyanobacteria.**

EUBACTERIA

Eubacteria comprise well over 90 percent of the moneran species, most of which are "typical bacteria," much like the one shown in Figure 25-8. These cells have the prokaryotic anatomy that characterizes the monerans, the typical features that distinguishes these cells from eukaryotic cells. These features were first observed with the advent of the powerful electron microscope. Although these "typical" bacteria vary little in their external form, they display tremendous diversity in their more subtle features, as illustrated in the following list of features.

- *Diversity of metabolic strategies.* Each bacterium's metabolic pathways are suited for the organism's particular habitat. For example, some bacteria can grow in concentrated sulfuric acid; others thrive in oxygen-devoid environments (such as in your intestines or the sediment at the bottom of a stagnant pond). Still others can survive boiling temperatures too hot for any eukaryote. Bacteria were also the earth's first photosynthetic organisms, using **bacterial chlorophyll** to convert the sun's energy into chemical energy for use as food . The oxygen-generating photosynthetic activities of other photosynthetic bacteria (cyanobacteria) created the oxygen-saturated atmosphere that changed the earth unlike any organism before or since, including humans. These photosynthetic prokaryotes have membrane structures that resemble the grana of chloroplasts, more evidence supporting the evolution

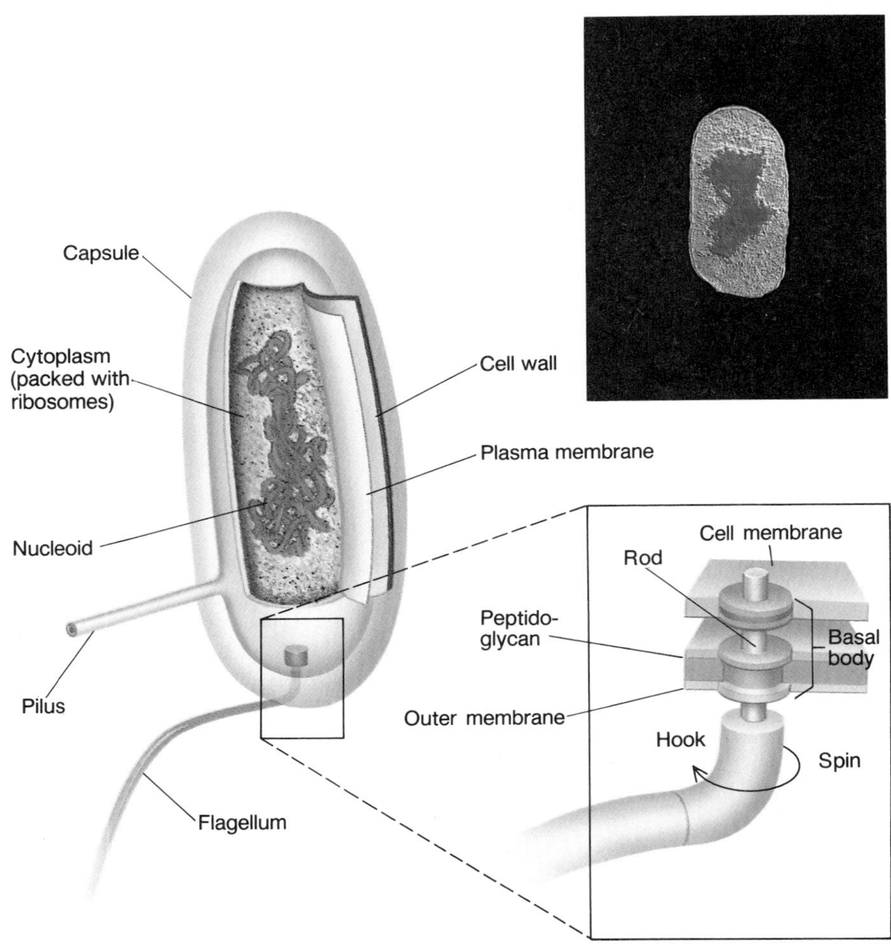

FIGURE 25-8

Anatomy of a typical eubacterium. Compare the simplicity of this prokaryotic cell with the complexity of eukaryotes in Chapter 6. The blow up details the hook and basal body connection to a flagellum. The shaft of the basal body rotates, causing the flagellum to spin.

of modern eukaryotic chloroplasts from endosymbiotic cyanobacteria (Figure 25-9).

- *Unique bacterial structures.* A few types of eubacteria produce a structure that enables the species to survive dehydration, exposure to ultraviolet light and caustic chemicals, and extreme heat. This structure, called an *endospore,* is the reason why boiling medical instruments and foods before canning is an unreliable means of sterilization—any endospores present survive boiling and germinate into actively growing bacteria once conditions have returned to normal. Some of these bacteria can cause serious infections or produce dangerous toxins (such as that which causes botulism, the most deadly of all food-borne illnesses) in canned or vacuum-wrapped foods. Some pathogenic bacteria form a capsule around themselves, that helps them avoid the protective defenses that would otherwise eliminate the bacterium from an infected host. Many types of bacteria that cause human pneumonia, for example, can do so only if they are encapsulated. Without capsules these bacteria are quickly engulfed and destroyed by the body's protective white blood cells. Even the bacterial flagellum (an agent of motility) is different from that of eukaryotic cells. Instead of whipping back and forth as do flexible eukaryotic flagella, the stiff, corkscrew-like bacterial flagella spin like a propeller, rotated by the only circular "motor" known to exist in nature (review Figure 25-8).

- *Pathogenic (disease-causing) bacteria.* Only a relatively few bacterial species can cause human disease. These bacteria live in or on host organisms, causing diseases that range from localized "strep throat" (*Streptococcus* pharyngeal infections) to invasive diseases that can enter the blood and cause "blood poisoning," cross the blood–brain barrier and infect the central nervous system, or attack the lungs (causing pneumonia). Although most pathogenic bacteria live in the tissue fluids *between* host cells, some are **intracellular parasites**—organisms that reproduce only inside host cells. One such group, the **rickettsia,** causes several serious diseases, one of which redirected human history. This disease, epidemic typhus, has incapacitated whole armies. Napoleon, for example, lost more soldiers to typhus than to combat-related injuries, ultimately leading to his defeat. Another group of intracellular bacteria, the **chlamydia,** cause a number of diseases, including trachoma, the world's leading cause of human blindness. Chlamydia infections of the genitals are the most common form of sexually transmitted diseases in the Western world (see The Human Perspective: Sexually Transmitted Diseases). In addition to sexual contact, disease-causing bacteria can be transmitted by respiratory droplets, dust, contaminated objects (such as eating utensils), or the bite of an infected animal. One of the newest bacterial diseases is Lyme disease, a tick-borne infection that is caused by the bacterium *Borrelia burgdorferi.* Victims of Lyme disease (named for Lyme, Connecticut, where the first case occurred in 1975) suffer extreme exhaustion, joint pain, facial paralysis, and heart ailments that often necessitate installation of a pacemaker. A large-diameter rash that resembles a bullseye usually develops at the site where the organism entered. Almost all bacterial diseases are treatable with antibiotics.

- *Cyanobacteria* The possible evolutionary forerunners of chloroplasts in eukaryotes, today's living cyanobacteria are still modifying the biosphere in a profound way. In some species, an occasional cell is enlarged, forming a **heterocyst** (Figure 25-10). These specialized structures "fix" molecular nitrogen, converting it from its atmospheric form (N_2) to a reduced state, such as ammonia (NH_3) or amino groups ($—NH_2$). In this way, cyanobacteria and nitrogen-fixing eubacteria make nitrogen available to other organisms, a process that is critical to the continuation of life on earth. (The nitrogen cycle is discussed in Chapter 29.)

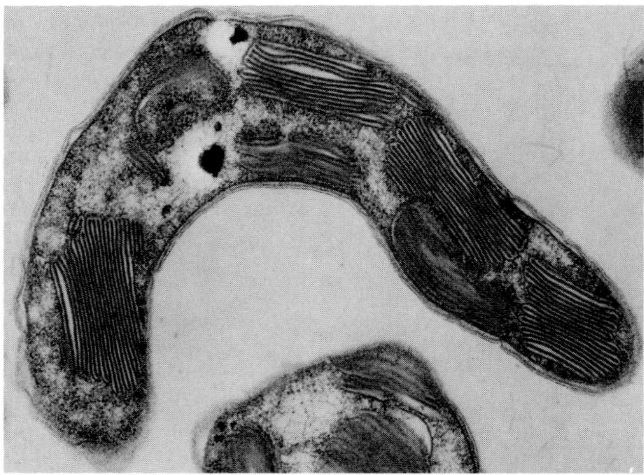

FIGURE 25-9

Forerunner of chloroplasts? The internal membranes of this cyanobacterium contain photosynthetic pigments. They resemble the membrane stacks (grana) of modern chloroplasts.

ARCHAEBACTERIA

Archaebacteria differ from eubacteria (and from typical prokaryotic cells in general) in several ways: They are shaped differently; they possess unusual lipids in their cell membranes; and their cell walls lack peptidoglycan. In fact, archaebacteria resemble eukaryotes in a few biochemical processes, pigments, and proteins. In addition, some archaebacteria live in areas where conditions are too extreme

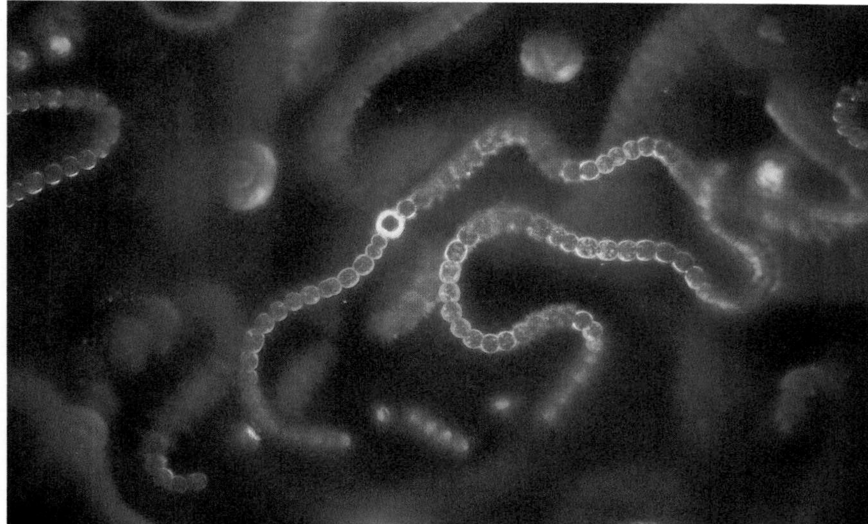

FIGURE 25-10 **Swollen heterocysts** enable chains of *Anabaena,* a cyanobacterium, to "fix" molecular nitrogen (N_2), that is, to convert it to a biologically usable form.

even for the most adaptable eubacteria. For example, *halophilic* archaebacteria flourish in the Dead Sea's supersaturated salt water. *Thermophilic* archaebacteria thrive in hot springs, in superheated marine waters a mile deep on the ocean floor, and in hot volcanoes (although not in molten lava, too hot for any organism to resist incineration). Sulfate-dependent archaebacteria are found in strong sulfuric acid solutions, surviving pHs as low as 1.0, which is acidic enough to dissolve metal.

Other archaebacteria live side by side with eubacteria in habitats devoid of oxygen, such as mammalian intestinal tracts, raw or partially processed sewage, and the smelly ooze at the bottom of stagnant bodies of water. This last group of archaebacteria, called *methanogens,* often generate methane, a flammable organic gas known by many common names, including natural gas or swamp gas. Such methane-producing archaebacteria live in the gastric tracts of cattle, contributing to the greenhouse effect, as cattle belch 100 tons of methane into the atmosphere each year. Like carbon dioxide, methane is a "greenhouse gas" that traps the sun's radiant energy and warms the planet. Excessive global warming may produce severe climate changes, such as drought in important agricultural areas and melting of the polar icecaps, which would raise sea levels and flood coastal communities (see Chapter 29).

Because of the unique properties of archaebacteria, some biologists have proposed creating a new kingdom for these organisms. These biologists maintain that a common cellular ancestor (called a *urkaryote*) gave rise to *three* lines of cellular types: the prokaryotes, the archaebacteria, and the eukaryotes. Evidence for such a proposal comes from the many "unprokaryotic" properties of archaebacteria: Their plasma membranes are unlike any prokaryote; their ribosomal RNA resembles that of eukaryotic cells more than that of prokaryotic cells; and their chromosomal structure contains histones and introns (which true bacteria cells lack). Based on these differences, some biologists believe there should be only three kingdoms—Archae, Prokaryota, and Eukaryota—rather than the current five. Such a scheme would group humans, plants, fungi, and protists together in the same Eukaryote Kingdom. Today, the three groups are considered "domains" rather than kingdoms. Each of the currently accepted five kingdoms fits into one of these domains.

ACTIVITIES OF PROKARYOTES

Monerans perform many activities (such as nitrogen fixation) that have a profound impact on all organisms in the biosphere. Many of these activities are the exclusive domain of prokaryotes, allowing them to occupy habitats that are uninhabitable by eukaryotes. For example, **chemosynthetic bacteria** use carbon dioxide as a source of carbon (as do plants, algae, and other photosynthetic autotrophs), yet they do so in areas devoid of light, the usual energy source of autotrophs. As described in Chapter 5, chemosynthetic bacteria extract their energy from inorganic minerals (such "chemoautotrophs" are found only in the Moneran Kingdom). These organisms form the food source that supports all organisms that live near deep, continually dark ocean trenches (see "Living on The Fringe," page 94).

The vast majority of bacteria are heterotrophic, consuming organic molecules for energy and carbon. Most of these are *saprophytic decomposers*; they extract their energy and carbon from dead organic matter by disassembling organisms and their wastes, molecule by molecule, soon after they die (a single gram of garden soil harbors more than 2 billion bacteria, many of them responsible for the soil's fertility). In fact, living with bacteria is inescapable, as the following section describes.

◁ THE HUMAN PERSPECTIVE ▷

Sexually Transmitted Diseases

As we prepare to enter a new century, we are confronting a group of diseases that have reached uncontrolled epidemic proportions. These are the **sexually transmitted diseases (STDs)** that can infect a person during sexual contact. AIDS represents the most frightening of these diseases, one that is inevitably fatal. Gonorrhea is another example. Although curable with antibiotics, gonorrhea still ranks as America's most common reportable disease, even more prevalent than the common cold among sexually active persons.[3] In combating STDs, the best strategy is to prevent infection. Prevention, however, is complicated by several biological and social factors that continue to encourage the spread of STDs:

- Symptom-free carriers of gonorrhea, syphilis, genital herpes, chlamydia, and human immunodeficiency virus (HIV) serve as hidden sources of infection. Such persons continue to spread the disease since they are usually unaware they are infected.
- Reluctance to seek medical treatment prolongs the amount of time that the pathogen is shed.
- Failure to notify the infected individual's sexual contacts increases the probability that an unsuspecting individual will transmit the disease to others.
- Failure to treat the patient's sex partner often leads to reinfection of the cured person by the untreated partner.
- More convenient methods of birth control have decreased the use of condoms, which is the only effective physical barrier against sexual transmission.
- Sexual activity with multiple partners enhances the probability of exposure to a variety of sexually acquired pathogens.
- The emergence of drug-resistant pathogens represents a potential problem in treating gonorrhea with penicillin. Another disease, recurrent genital herpes, is currently incurable.

Effective precautions include avoiding sexual contact as long as genital lesions are present; seeking prompt medical attention whenever one suspects possible venereal disease; refraining from sexual contact if diagnosed as having an STD; and waiting until medical confirmation of cure before resuming sexual activity. Furthermore, frequent checkups can identify many asymptomatic carriers.

GONORRHEA

Gonorrhea is caused by the bacterium *Neisseria gonorrhoeae,* which is introduced through sexual contact, usually through the genitals but sometimes through the anus or the mouth. Infected persons may either develop symptoms characteristic of classical gonorrhea or become carriers that show no symptoms. Carrier states are especially common among infected women in the early stages of the disease. The large number of gonorrhea carriers in both sexes presents a major obstacle to the control of this disease.

The bacterium typically enters through the urethra of men and the vagina of women. Three to five days following exposure, most infected males experience burning urination and discharge of thick white pus from the penis (painful urination is *not* a symptom in infected women). If untreated, the pathogen may produce permanent scarring in the vas deferens in men and the fallopian tubes in women (such scarring is one of the leading causes of infertility). In addition, an infected mother can transmit the bacteria to her newborn baby as it travels through the birth canal. To prevent infected newborns from developing blinding eye infections from the bacteria, antibacterial solutions (often silver nitrate or antibiotics) are dropped into the eyes of every newborn delivered in U.S. hospitals.

CHLAMYDIA INFECTIONS

Chlamydia trachomatis is the most common sexually transmitted organism in the United States, striking up to 10 million people annually. Because its symptoms mimic gonorrhea, this disease is termed **nongonococcal urethritis (NGU).**[4] Chlamydia can also cause all the complications that gonorrhea can—sterility in untreated men and women, and blindness in newborns, both of which can be avoided with rapid antibiotic treatment.

Syphilis

Untreated syphilis, caused by an especially fragile bacterium called *Treponema pallidum,* can leave its victims mentally degenerated, neurologically incompetent, consumed by destructive lesions, or dead. In fact, people still die of syphilis, in spite of effective antibiotics, and babies continue to be born with congenital syphilis, the tragic result of fetal gestation in an infected mother. Of the 500,000 syphilis cases that occur in the United States each year, only a small percentage progress to the final fatal stage. Most cases (and their contacts) are identified and treated with penicillin while in the early or middle stages. Campaigns to detect the disease in pregnant women help reduce the incidence of congenital syphilis. For example, the mandatory blood test prior to getting a marriage license is strictly for the detection of syphilis. Antibiotic treatment very early in a woman's pregnancy can avert this tragedy.

---

[3] A reportable (or notifiable) disease is one that must be reported to health authorities any time a physician diagnoses it. These data are tallied by the Centers for Disease Control (CDC) in Atlanta Georga and published weekly.

[4] This distinction is very important because NGU and gonorrhea do not respond to the same antibiotic treatment, so a physician cannot determine appropriate treatment based on symptoms alone. The infecting organism must be identified.

Genital Herpes

Genital herpes is characterized by painful blisters on the genitals and surrounding area. It may be transmitted by oral, vaginal, or anal sexual contact. The incidence of genital herpes has dramatically escalated since the late 1960s. More than 20 million Americans currently have the disease, and as many as 500,000 join their ranks each year. Genital herpes is caused by *herpes simplex virus* (HSV), the majority of which are caused by the Type 2 genital virus, although at least 20 percent are caused by HSV Type 1 (the virus that usually causes *oral* fever blisters). It is not uncommon for someone with oral fever blisters to transmit the disease to the genitals of a sex partner, and vice versa. Both genital and oral herpes tend to recur when the virus that has been "hiding" in local nerve fibers are reactivated.

Characteristic blisters are fluid-filled and soon erupt to become open sores teeming with the infectious virus. They occur anywhere on the penis in men and on the vagina, vulva, and cervix in women. The anus perineum, buttocks, and thighs of either sex may also be ulcerated. Although the disease is most communicable while active lesions are present, the virus may also be shed in urine and genital secretions when no lesions are present.

Infected infants exposed to the virus during birth may develop blindness, brain infections, death of much of the skin, or other potentially fatal disorders. Cesarean delivery is necessary to reduce the risk to the newborn whenever the mother has active genital sores.

As with any viral diseases, antibiotics do not cure or even shorten the duration of the genital herpes infection. Acyclovir, however, is a chemotherapeutic drug that can reduce the severity of symptoms and prevent recurrences if used early in the course of the primary infection.

AIDS

Perhaps the most frightening modern medical development emerged for the first time in 1979 with the sudden appearance of a new disease, one that eventually proves fatal to anyone who contracts it.

FIGURE 1

Although physical contact between athletes on a basketball court poses no risk of AIDS transmission, Magic Johnson, perhaps the most celebrated person to have tested positive for HIV, permanently retired from playing his sport in 1992 (although he returned to coaching in 1994). His retirement as a player was precipitated by his concern that competitors afraid of contracting the virus might not play as aggressively against him. Johnson apparently contracted the virus through heterosexual contact with an HIV-infected woman.

The disease, called **Acquired Immune Deficiency Syndrome (AIDS),** continues to increase at an alarming rate. The victims of AIDS can be found worldwide and include heterosexuals, homosexuals, males, females, adults, and children. Transmission among adults primarily occurs by sexual contact—increasingly, *heterosexual* contact. Developing children can acquire the disease across the placenta from an infected mother. Other nonsexual routes of transmission include transfusion with contaminated blood or sharing needles with an infected person, an all-too-common practice among drug abusers. There is no evidence that the disease can be transmitted by casual contact, such as shaking hands.

The disease is caused by the human immunodeficiency virus (HIV), which infects cells of the immune system, including macrophages, helper T cells, and certain B lymphocytes (see Chapter 20). The virus can also infect brain cells, causing AIDS-related dementia. Following exposure, however, it may take as long as 10 years before symptoms of AIDS develop. Onset of the disease is accompanied by a decline in the number of helper T cells, seriously crippling the cell-mediated immune system. The patient is left vulnerable to cancer or other pathogens that are normally eliminated by healthy immune systems. When the helper T cell population falls to 200 cells per cubic millimeter, the person is diagnosed with AIDS (concentrations of T cells in healthy persons exceed 800 cells per cubic millimeter). Persons with AIDS are especially prone to develop otherwise rare opportunistic diseases. *Pneumocystis* pneumonia is by far the most common of these, although oral and respiratory yeast infections, toxoplasmosis, cytomegalovirus (CMV) infections, and tuberculosis are seen with high frequency. About one-third of male AIDS patients contract a formerly rare cancer called Kaposi's sarcoma. All these opportunistic diseases are severe and are often the first indication that HIV infection has progressed to AIDS. They are also usually the cause of death.

HIV infection can usually be diagnosed by testing the blood for the presence of antibody against HIV (the antibodies are not protective, however). There is a period of time, usually 3 months but often more than a year, between infection and the development of enough antibody molecules to be detected by current diagnostic assays. Individuals who test positive for HIV are infected with the virus and can transmit it to others, even if they do not have clinically defined AIDS.

The most common treatment for AIDS is AZT, although other drugs are being speeded through the normally slow approval process. None of these treatments, however, eradicates the virus—they only prolong the life of the AIDS patient. Because there is currently no vaccine and no cure, the threat of AIDS can only be controlled by reducing the transmission of the virus. Nonsexual transmission has been effectively slowed in developed countries by screening blood supplies for the virus and by reducing intravenous drug use (or by the introduction of "sterile needle" programs). Latex condoms with a spermicide containing nonoxynol-9, which kills HIV as well as sperm, reduces sexual transmission of AIDS. Such safer sex practices, are not absolute safeguards, however, because HIV may pass through defects in the condoms.

LIVING WITH BACTERIA

Bacteria used to take a much costlier toll in human lives than they do now. Applications and understanding provided by modern biology have doubled our lifespan and informed us of how to protect ourselves from the "bad guys" in a kingdom that largely consists of beneficial microbes. Here are just a few of the invaluable activities of bacteria:

- Decomposers release organic nutrients from dead organism and their wastes, recycling the earth's limited nutrients (Chapter 30).
- Nitrogen-fixing bacteria and cyanobacteria provide usable nitrogen to the biosphere. Some of these bacteria live in root nodules of legumes (Figure 25-11) such as alfalfa and soybeans. Planting a crop of legumes improves the fertility of agricultural fields that have been exhausted by other crops.

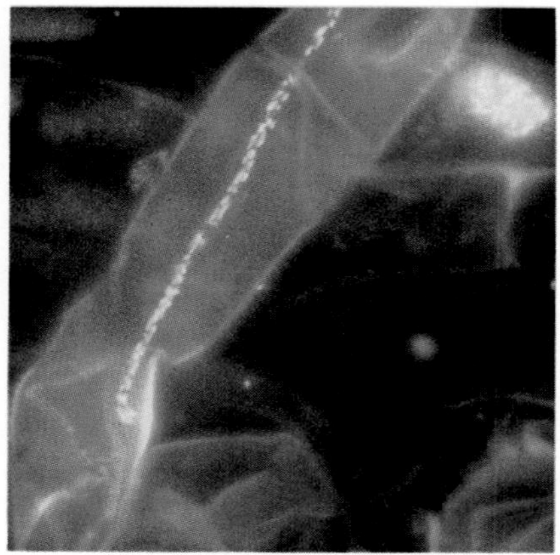

FIGURE 25-11
Orderly invasion of alfalfa root cells by beneficial nitrogen-fixing bacteria. Without the bacteria's ability to convert atmospheric nitrogen to a biologically accessible form, plants (and virtually all animals) would die.

- The billions of bacteria that normally reside on your body's surface constitute your "normal flora." You benefit from their presence every time they successfully compete with bacteria that would otherwise infect your body.
- Cattle and sheep cannot digest cellulose, the chief constituent of their grassy diet. Bacteria (and protozoa) living in the rumen, a special stomach chamber, do the job for them. Without these microbes, there would be no cattle.
- Bacteria are used to produce yogurt, some cheeses, vinegar, and several other food products (page 101).
- Modern waste-water treatment plants employ bacteria to process raw sewage to a safe substance free of disease-causing microorganisms.
- The newest applications use genetically engineered bacteria to produce insulin and other compounds of medial and industrial importance. Chapter 11 presents many of the modern strategies for employing the microbial "work force."

In spite of these benefits, a small minority of bacteria that causes disease, spoils our food, fouls our water, and deteriorates useful products still represents a formidable foe, one that requires sustained efforts to minimize its detrimental impact. Despite these challenges, life on earth could not continue without bacteria.

THE PROTIST KINGDOM

The Protist Kingdom may well consist of the most dissimilar types of species of any of the other four Kingdoms (Figure 25-12). Many protists are unicellular, but some are colonial. Most are microscopic, but some are as large as 5 millimeters (about a quarter of an inch). Some are photo-

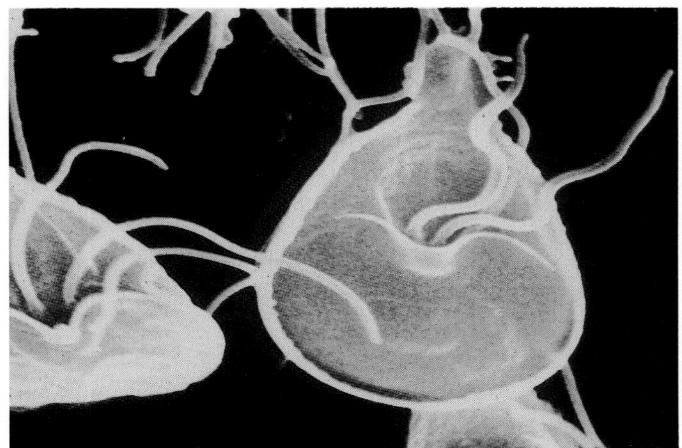

(a)

(b)

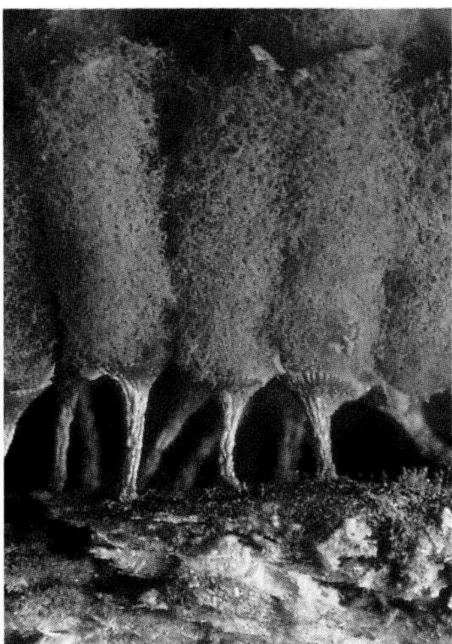

(c)

FIGURE 25-12

A gallery of protists. ***(a)***. Giardia—an animal-like protist that causes a common digestive tract infection of humans. ***(b)***. Diatoms—plant-like protists that support a whole population of marine animals. ***(c)***. Slime mold, a fungus-like protist.

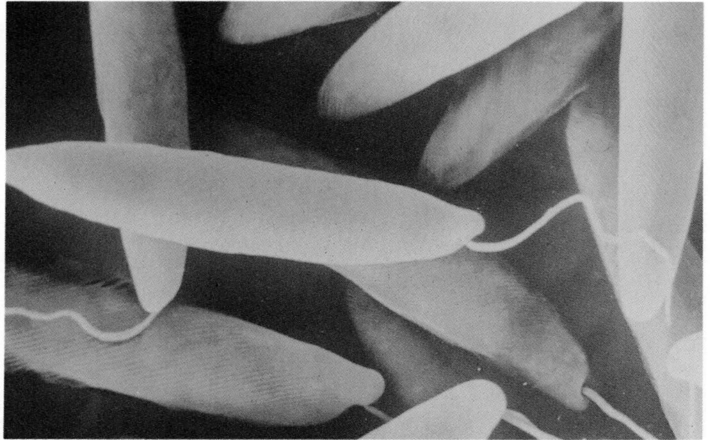

FIGURE 25-13
***Euglena:* animal-like and plant-like.** This actively swimming protozoon possesses the animal-like characteristics of motility (flagella) and lack of cell walls. As long as they are confined to darkness, *Euglena* remain in their animal-like state—heterotrophs that absorb dissolved nutrients from their medium. When exposed to light, however, these versatile protists become autotrophs. Their chloroplasts proliferate, and the cell turns into a green photosynthetic, plantlike cell. This taxonomic "fence straddler" is neither plant nor animal; it is a protist.

synthetic, others are heterotrophic. Some are fungus-like, others are plantlike, still others are animal-like, and many combine characteristics of two or more of these. For example, how would you classify a photosynthetic microorganism that actively swims? It's not a plant (it swims); it's not an animal (it is photosynthetic). It is simply a **protist** (Figure 25-13).

EVOLUTIONARY TRENDS

Protists are credited with several evolutionary advances, not the least of which is sexual reproduction (Figure 25-14). For some protists, sex is not a reproductive strategy. These organisms just conjugate and exchange part of their genetic library with each other. This gene mixing increases genetic diversity of the species, but reproduction continues to be asexual. In contrast, other protists couple sex with reproduction, a process similar to that which occurs in plants and animals, at least from a nuclear standpoint. For example, a diploid protist undergoes meiosis, dividing into gametes of opposite mating types, which then fertilize each other. The resulting diploid zygote develops into a progeny individual.

A major evolutionary "breakthrough" of the protists is the development of eukaryotic cellular features, such as a

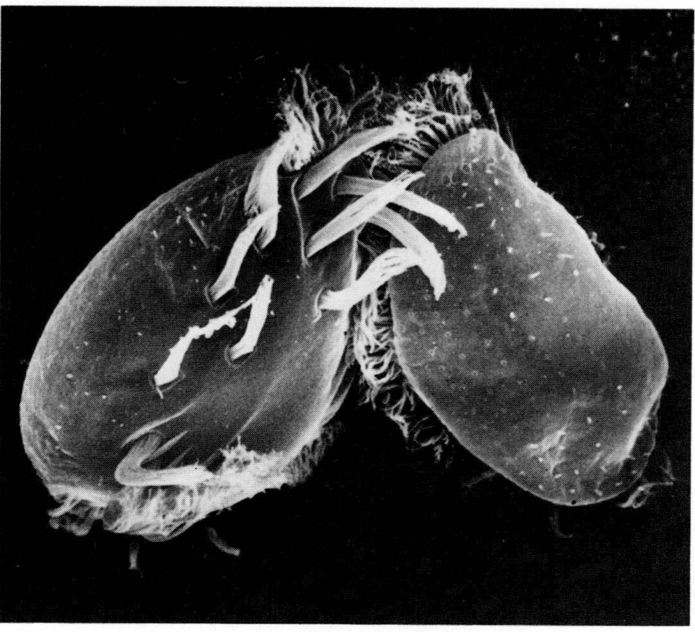

FIGURE 25-14
Sexual interactions are common among some protists. Here, two cells *(Euplotes)* are joined in a preconjugal embrace, holding each other with tufts of fused cilia, called cirri. This "recognition behavior" precedes the exchange of genetic material.

nucleus, mitochondria, chloroplasts, and other double-membraned organelles (Chapter 3). Three varieties of early eukaryotes evolved and are the ancestors of what we today call the "higher" kingdoms—Fungi, Plants, and Animals (Figure 25-15). Modern protists still retain most of the features of these three ancestral types.

Protists also create selective pressures that influence the evolution of other organisms. As either competitors or predators, protists have a significant impact on the species that share their particular microscopic niche. As members of the food chain, protists are essential to many larger organisms that feed on them either directly or indirectly. Protists have also played a major role in human evolution, such as in the development of human sickle cell anemia. People living in areas endemic for malaria, a disease caused by a protist, have an advantage over those who have no sickle cell traits because they are more resistant to the fatal ravages of malaria. Even in malaria-free societies, such as the United States, many people have sickle cell anemia, but here the disease confers only disadvantages.

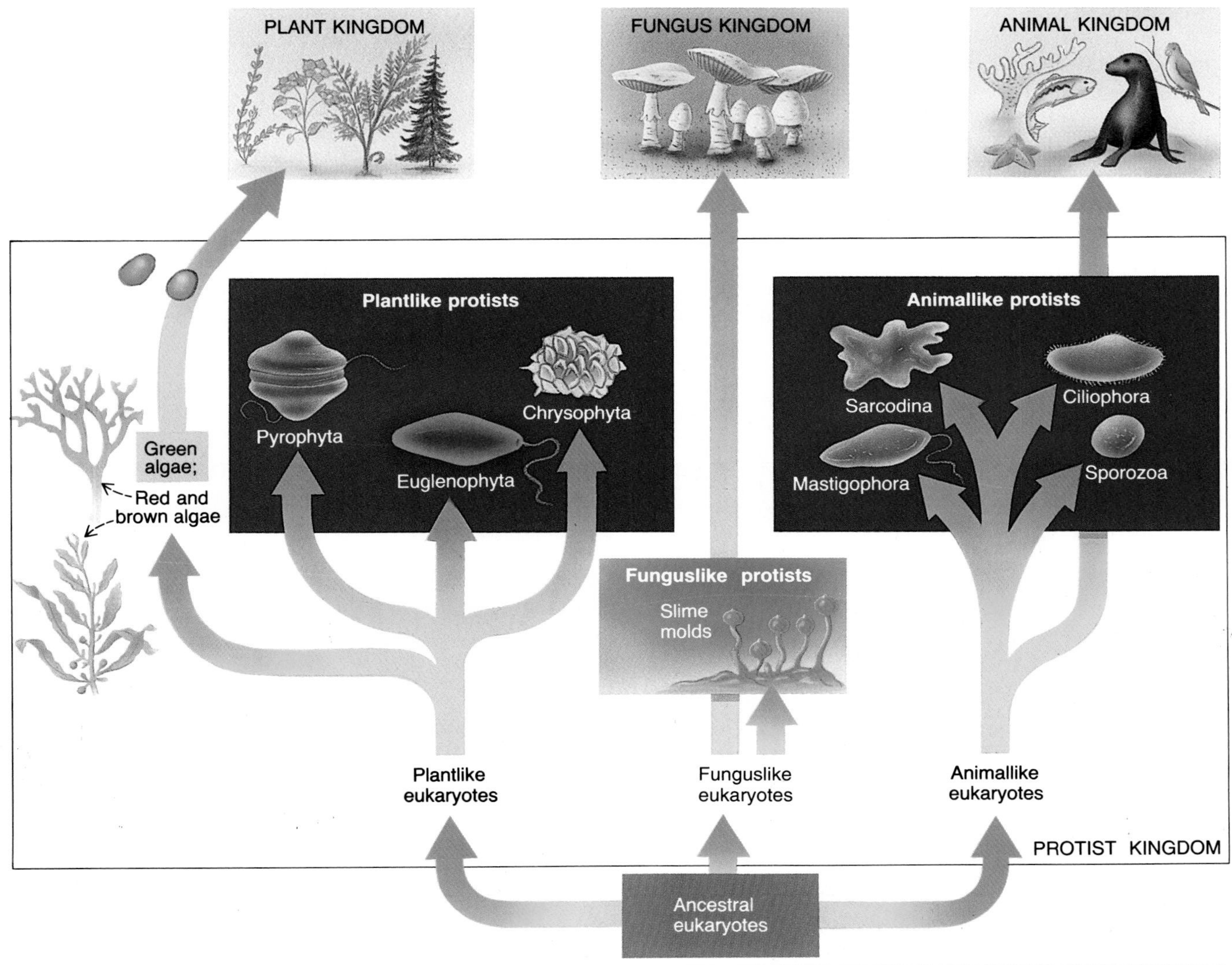

FIGURE 25-15

Evolution of the three major groups of protists. In this proposed lineage, the first eukaryotic cells gave rise to all three major groups of protists.

PROTOZOA: PROTISTS RESEMBLING ANIMAL CELLS

Protozoa (*proto* = first, *zoa* = animals) are unicellular protists, most of which lack cell walls, ingest food particles (or absorb organic molecules), move about freely, and produce no multicellular spore-bearing structures. These animal-like traits distinguish protozoa from the other two groups of protists (the algae and slime molds).

Although most protozoa are harmless to humans, they are often feared because of a few pathogenic species (Figure 25-16). More than half the human population will contract some type of protozoan infection during their lifetimes. Among the most notorious protozoa are four species of *Plasmodium* that cause malaria. These mosquito-born protozoa kill more people than does any other microorganism (Figure 25-17). Another protozoan (*Trypanosoma*) causes deadly encephalitis ("sleeping sickness"), an inflammation of the brain that develops after the protozoa enter the body through the bite of an infected insect. Other dangerous protozoa include those that cause food and water-borne amoebic dysentary, which infects more than a million people in the United States alone. Still another gastrointestinal infection is often contracted by hikers and campers who drink from a "pure" mountain stream. Streams are often contaminated with the *Giardia lamblia* cysts shed in the feces of infected muskrats and beavers (boiling water before drinking it kills the cysts of this protozoon and protects against the disease). This organism may cause more infections than any other protozoan in the United States. Just one infected person can spread the disease throughout a day-care center, a school, or a hospital, especially if hygienic practices (such as hand-washing after diaper changes) are lax. *Giardia*'s microscopic appearance was depicted in Figure 25-13*c*.

Still, protozoa do far more good than harm. They help recycle nutrients of dead animals and wastes and, as heterotrophic **zooplankton,** form an important link in aquatic food chains. Feeding on microscopic algae, zooplankton convert the nutrients of algae (some of which, such as cellulose in cell walls, are nutritionally inaccessible to animals) into protozoan tissue that is digestible by virtually all consumers. Most of the 40,000 species of protozoa live in water, moist soil, or inside other organisms (such as the protozoan that causes the malaria cycle depicted in Figure 25-17).

Classifying protozoa is traditionally based on the organism's mode of motility (Table 25-2). Protozoa move by *cilia, flagella,* or **pseudopodia** (*pseudo* = false, *pod* = foot). Pseudopodia are finger-like extensions of cytoplasm that flow forward from the "body" of an amoeba; the rest of the cell then follows. Even amoebas that are encased in

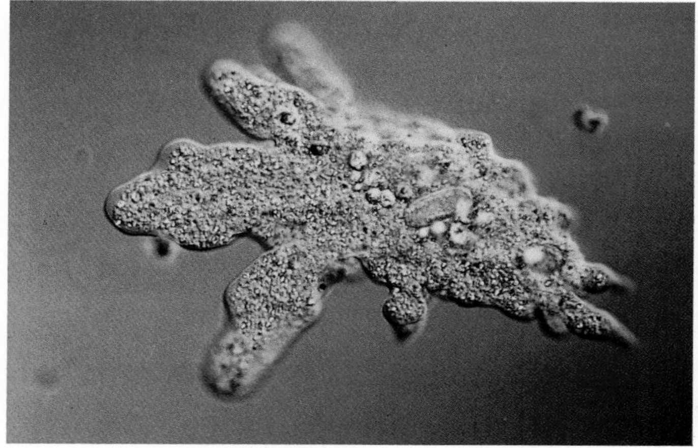

(a)

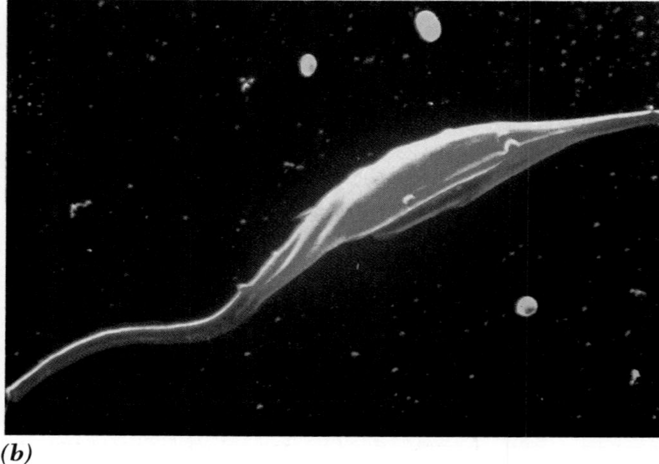

(b)

FIGURE 25-16

The delicate and the deadly. This delicate-looking *Amoeba proteus* ***(a)*** is a harmless resident of stagnant water, a predator that preys on other microbes, which it engulfs by phagocytosis. In contrast, *Trypanosoma gambesia* ***(b)*** is one of the most deadly protozoa. It is an insect-borne flagellate that causes African sleeping sickness (a fatal blood infection that eventually attacks the human brain).

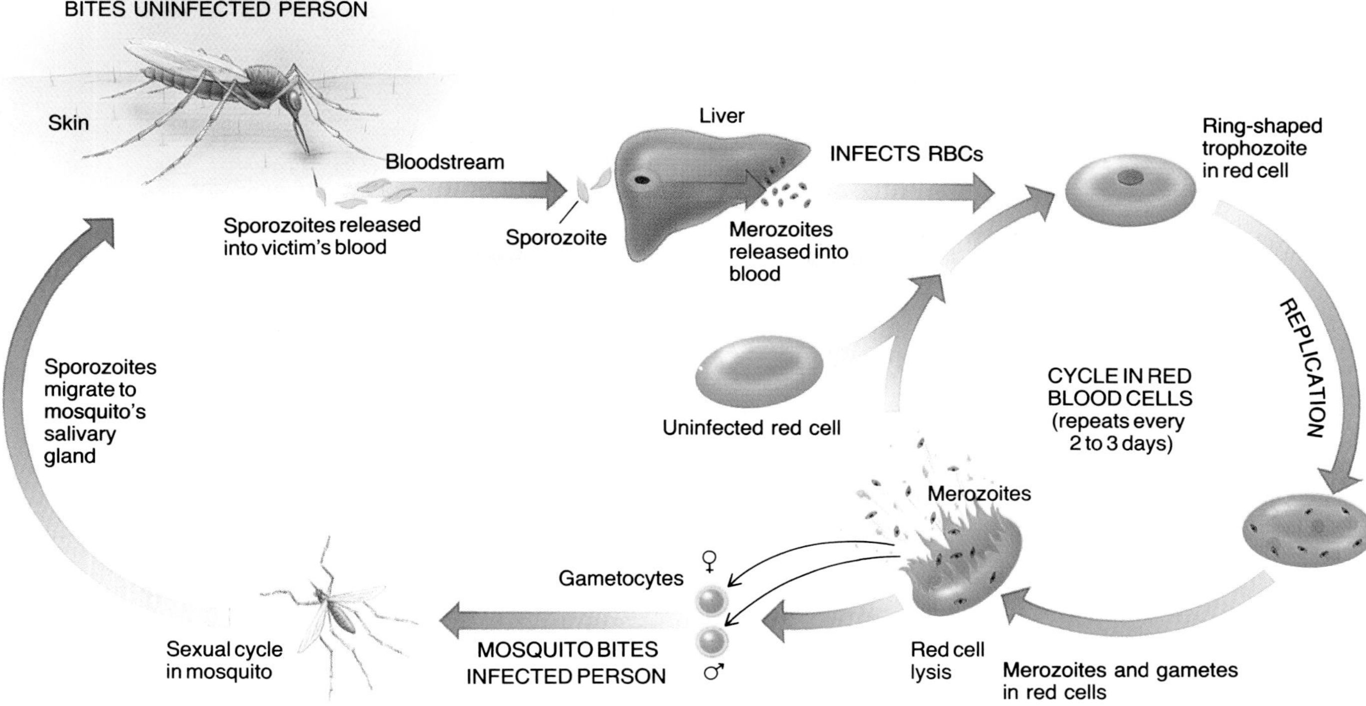

FIGURE 25-17

The world's most deadly microorganism is *Plasmodium,* the agent of malaria. *Plasmodium* kills more people each year than any other genus of microorganisms. During its life cycle, this mosquito-borne parasite alternates in form, depending on its location. Inside red blood cells, the organism proliferates as trophozoites and merozoites, some of which undergo meiosis and become sex cells (gametocytes). The red blood cells burst, releasing both merozoites (which infect other red blood cells) and gametocytes. When sucked into a female *Anopheles* mosquito who bites a malaria victim, the gametocytes fertilize one another, and the resulting zygotes change into sporozoites, which migrate to the mosquito's salivary glands. The sporozoites are then injected into the mosquito's next victim.

TABLE 25-2

CLASSIFICATION OF PROTOZOA ACCORDING TO MOTILITY

Motility group	*Organelles of Locomotion*	*Characteristics*	*Example*
Mastigophora	Flagella	Heterotrophic or, in some cases, photosynthesis; excess water expelled through contractile vacuole; binary fission along long axis; no sexual reproduction.	*Giardia* (agent of intestinal infection)
Ciliata	Cilia	Cilia sweep food particles into mouth; complex cell morphology; many types expel excess water through contractile vacuole; genetic transfer by conjugation; asexual reproduction by fission across long axis	*Paramecium* (predator of smaller organisms)
Sarcodina	Pseudopodia	Amoeba move by extending "fingers" of cytoplasm; feed by phagocytosis; reproduce asexually by binary fission; some species encased in shells of silica or calcium	*Amoeba* (predator of smaller organisms)
Sporozoa	None	Trophozoite stage nonmotile; intracellular parasites; complex life cycles that alternate between sexual and asexual reproductive modes; one species may require two or more different hosts to complete life cycle.	*Plasmodium* (agent of malaria)

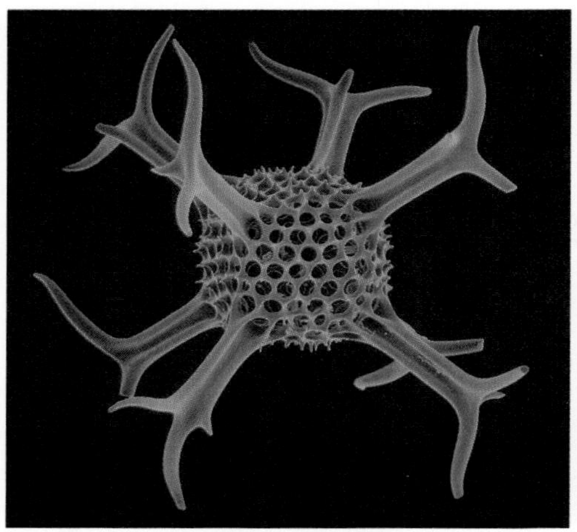

(a) *(b)*

FIGURE 25-18

Shelled amoebas. *(a)* The scanning electron microscope captures the beautiful intricacy of some radiolaria. Pseudopodia protrude through the holes in the shells of these amoeba to gather food particles. The intricate shells are composed of silica; they are literally glass. *(b)* Foramnifera enclose themselves in "snail-like" shells of calcium carbonate. Geologic upheaval raised sediments of enormous numbers of shells deposited millions of years ago when these amoebas abounded in oceans. One striking result of this deposition is the White Cliffs of Dover in England, composed entirely of foramnifera shells.

protective shells (Figure 25-18) can use their pseudopodia by extending them through holes in the encasement.

Pseudopodia, cilia, and flagella often double as food-gathering instruments. *Paramecium* and some other types of ciliated protozoa use their cilia to create currents that sweep food particles into the cytostome (analogous to a "mouth"). Some flagellated protozoa use their flagella as harpoons to spear prey. The role of pseudopodia in phagocytic engulfment was described in Chapter 20.

ALGAE: PROTISTS THAT RESEMBLE PLANTS

Although the classification of algae is controversial, for our purposes **algae** will be considered any unicellular or simple colonial photosynthetic eukaryote (this places the multicellular algae in the Plant Kingdom, which will be discussed in Chapter 26).

The most important role of algae is as **phytoplankton,** microscopic photosynthesizers that live near the surface of seas and bodies of fresh water. Virtually every animal in the sea depends either directly or indirectly on phytoplankton for food, making algae the most important members of the aquatic food chain. About half the world's organic material is produced by algae. In addition, every breath you take depends on these algae; their photosynthetic activities generate 75 percent of the molecular oxygen available on earth.

Algae fall into one of six divisions, three of which contain the photosynthetic unicellular protists considered in this chapter (the others are discussed in the following chapter on plant diversity). Prokaryotic "algae" are not algae at all—they are bacteria and were discussed earlier in this chapter (see cyanobacteria, page 450).

Divisions of Algae

Chrysophytes are golden brown and yellow-green algae. They are the major producers of food for all animals living in marine waters. **Diatoms** are golden-brown algae that are distinguished most dramatically by their intricate silica shells (review Figure 25-13*b*). Diatoms secrete their delicate "glass" coats of armor, each of which is perforated with thousands of tiny holes to allow contact between the enclosed photosynthetic cell and its environment. Diatom shells are of great economic significance, ideal for use in polishing, filtering, and insulating materials. Commercially, they are referred to as "diatomaceous earth," harvested from enormous geologic deposits that resemble fine, white powder. These deposits accumulated for millions of years, the diatom shells forming thick sediment layers on the ocean floor. Many of these deposits have been thrust to the surface by geological activity.

The golden color of chrysophytes is provided by accessory pigments that assist chlorophyll in photosynthesis. These pigments are precursors of the fat-soluble vitamins

A and D, which are concentrated in the liver oils of fishes that dine on chrysophytes. For years, people have harvested fish liver extract as rich sources of these vitamins (such as cod liver oil given as a vitamin supplement).

Pyrrophyta (fire algae) are **dinoflagellates,** single-celled photosynthesizers that have two flagella. One flagellum moves the alga through its medium, the other spins the cell on its axis. Periodic dinoflagellate "blooms" are often hazardous to many members of the food chain. **Blooms** are massive growths of algae that occur when conditions are optimal for algae proliferation. During the bloom called **red tide,** for example, growth of one reddish brown dinoflagellate species is so extensive that it tints the coastal waters and inland lakes a distinctive red color (Figure 25-19). Some red tide dinoflagellates produce a powerful nerve toxin that kills any fish that eats them. During red tide, the water's surface may be blanketed with dead fish. Although shellfish, such as oysters, muscles, and clams, are not killed by the toxin, they concentrate the poisonous substance in their tissues. People who eat these shellfish may consume enough toxin to cause paralysis, which typically sets in within an hour of consuming the toxin. Within 8 hours, the victim may die of *paralytic shellfish poisoning,* even with medical treatment.

Euglenophyta are a group of unicellular algae that includes *Euglena* and similar protists (called euglenoids) that possess characteristics of plant cells and animal cells. Exposure to light induces their plantlike property, the production of chlorophyll *a* and *b* (supplemented with carotenoids). The typical euglenoid (review Figure 25-14) retains its flagella in its alga form; its movement is directed in response to light.

SLIME MOLDS: PROTISTS THAT RESEMBLE FUNGI

Although you may never guess it from their name, **slime molds** are attractive protists. In their sexually mature stage, these organisms display delicate reproductive structures similar to those of molds (a true mold is a fungus, as discussed in the following section).

Slime molds are either *cellular or plasmodial.* **Cellular slime molds** exist as microscopic amoeba-like cells when food and water are plentiful. They patrol their habitat for bacteria, spores, and organic debris, which they engulf and digest, rapidly multiplying as a result. When their food is depleted, the "amoebas" release one of life's most developmentally important chemicals—*cyclic AMP.* This

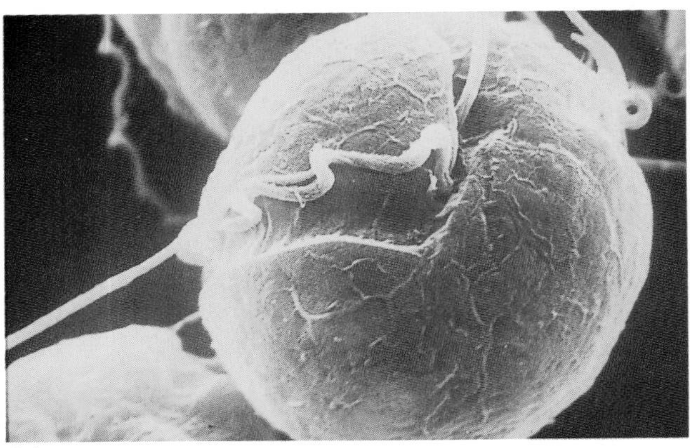

(a)

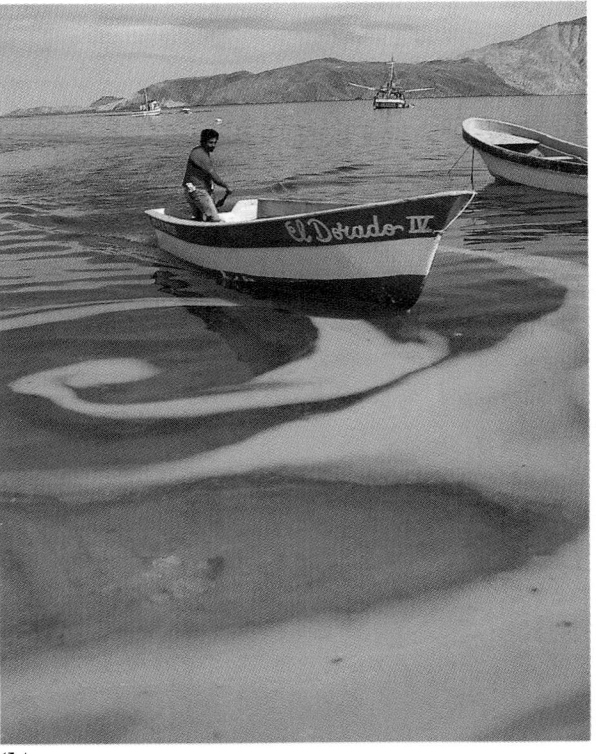

(b)

FIGURE 25-19
Intriguing but deadly. Some dinoflagellates, such as *Gonyaulax* ***(a)***, have killed countless fish and hundreds of people. Dinoflagellate blooms tint the ocean red ***(b)***. Some dinoflagellates are bioluminescent at night, flashing brightly with each breaking wave, inspiring their division name, the "fire algae."

substance is used by virtually all organisms to direct development or to govern gene expression. Hundreds of amoebas migrate up the cAMP concentration gradient to the point of highest concentration. Here, they aggregate into a single slime-covered "slug," that crawls for a while (helping to disperse the species) and then transforms itself into a fungus-like spore-producing structure. When the spores are released, they are carried by the wind or on the surface of animals to new locations. If conditions are suitable, the spores germinate, each forming another amoeba-like cell, and the cycle repeats.

Plasmodial (*acellular*) **slime molds** also have a fungus-like spore-producing stage in their life cycle. These organisms have no migrating "amoebas," however. Instead, they produce a **plasmodium**—a huge multinucleated "cell" that feeds on dead organic matter. A single plasmodium can grow large enough to cover an entire log (Figure 25-20). This giant growing mass then forms reproductive structures that release spores, which germinate into gametes if they can reach a location with suitable conditions. Fertilization between gametes yields a zygote that develops into another multinucleated plasmodium.

THE FUNGUS KINGDOM

In the mid-1800s, thousands of farmers in Ireland began to notice a terrifying transformation. Potatoes, their major source of nutrition, were turning black. The Irish helplessly watched their potato plants rot, and the food crop of an entire nation failed. The disaster of the "Irish potato blight" continued for 14 years, fatally starving more than a million Irish and causing millions to flea to the United States. Today we know the history-shaping killer to be a member of the Fungus Kingdom.

The Irish potato blight is one of several large-scale disasters that have been inflicted on the human population by fungi. The vast majority of fungi, however, are beneficial—not just to people but to the biosphere as a whole. Here are a few of their contributions.

- decomposition of dead organisms and recycling of the nutrients;
- soil formation from solid rock;
- formation of *mycorrhizae* ("fungal roots"), filaments that grow in intimate association with plant roots, increasing the plant's absorption of nutrients and water;
- production of foods, such as cheese, mushrooms, and soy sauce;
- fermentation of sugar to ethyl alcohol, generating beer, wine, and spirits;
- synthesis of valuable chemicals, such as those used in the formation of plastics, industrial solvents, and compounds used for soap production;
- manufacture of medically important compounds, most notably penicillin and many other antibiotics that continue to save lives that would otherwise be lost to bacterial infections (Figure 25-21).

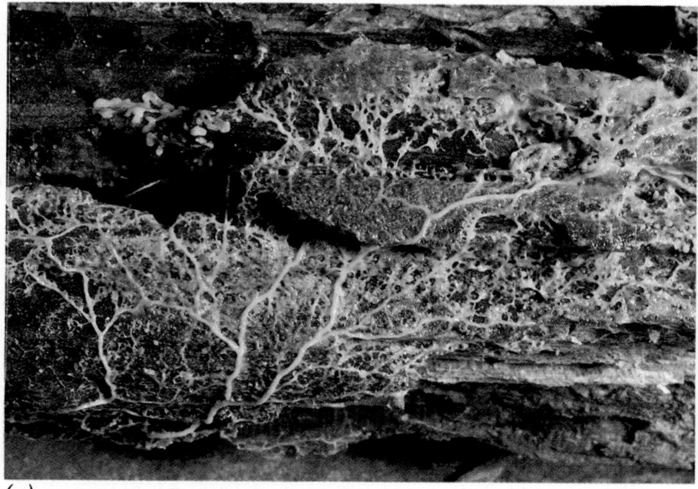
(a)

(b)

FIGURE 25-20

Plasmodial slime molds. ***(a)*** The protozoa-like plasmodium of *Physarium* is an enormous multinucleated "cell" that consumes dead organic matter. ***(b)*** Spore-producing structures of slime molds resemble those of fungi.

FIGURE 25-21

Destroyer and savior. *Penicillium,* the blue-green mold you find on old oranges, bread, and cheddar cheese, spoils millions of tons of food each year. Yet the organism also produces the life-saving antibiotic penicillin. In this photograph, the mold's asexual spores are clearly visible at the tips of special "brushlike" spore-forming structures.

OVERVIEW OF THE KINGDOM

In 1969, a new and separate kingdom was recognized exclusively for classifying these heterotrophic eukaryotes. Some members of the Fungus Kingdom, notably the **yeast,** are unicellular, forming colonies similar to those of bacteria. **Filamentous fungi** constitute the multicellular members of the kingdom, organisms composed mostly of living threads that grow by division of cells at their tips. These filaments, called **hyphae,** are characteristic of **molds,** which typically grow as fluffy masses. Molds are highly efficient exploiters of available nutrients. During peak growth, a mold colony can grow more than half a mile of new hyphae in a single day.

One group of filamentous fungi are called **macrofungi** to signify the large size of their fleshy sexual structures. A mushroom, for example, is so large and elaborate that it is often mistaken for the whole organism. Yet most of the fungus grows as an unseen filamentous mass under the ground or in the tissues of a tree, with only its large reproductive structures showing.

Some filamentous fungi collaborate with algae to form associations called **lichens.** Each member of the lichen contributes to the ability of these "composite organisms" to thrive in a wide range of habitats. In dry environments, or those that are poor in organic nutrients, for example, the lichen's photosynthetic alga uses light energy to generate organic nutrients from inorganic compounds, while the fungal filaments gather and conserve what little water is available. Together, the lichen readily grows in conditions in which neither the fungus nor the alga could survive alone. In some harsh environments, lichens support entire food chains, sustaining such large consumers as the reindeer caribou of frozen northern Alaska.

FUNGAL NUTRITION

Fungi are heterotrophs, most of the kingdom's members being **saprobes,** organisms that obtain their nutrients by decomposing dead organisms. Yet a few types of fungi don't wait until an organism is dead before they start consuming it. Most of these parasitic fungi are pathogenic, causing diseases in plants or animals. In fact, fungal diseases constitute some of the most common and persistent human diseases that continue to challenge modern medicine. More than 20 percent of the world's population suffers from fungal diseases, ranging from irritating maladies of the skin and mucous membranes (for example, athlete's foot, ringworm, and vaginal yeast infections) to such life-threatening systemic infections as histoplasmosis and valley fever, which can affect virtually any organ of the body. Living plants also fall victim to fungi.

FUNGAL REPRODUCTION

Except for mushrooms and other macrofungi, most members of the Fungus Kingdom infrequently resort to sex, relying mainly on asexual reproduction to increase their numbers. Yeast, for example, form small "buds" that enlarge and finally break away from the genetically identical parent. In filamentous fungi, fragments of hyphae may break away from the parent mold and grow mitotically, forming new individuals. Other molds produce asexual spores, which are then released from sporulating structures and germinate into actively growing hyphae.

When fungi do reproduce sexually, the diploid stage is short-lived. The zygote produced by fertilization immediately undergoes meiosis to form haploid spores. All the cells in many types of fungi are therefore haploid, having descended from one haploid spore.

CLASSES OF TERRESTRIAL FUNGI

The type of sexual spore produced by a fungus provides the criterion for placing it into one of four classes (Table 25-3).[6]

Zygomycetes

The black mold growing on an old loaf of bread typifies the **zygomycetes.** These organisms begin as a white, cottony mass of *nonseptate* hyphae, continuous tubes of cytoplasm with no crosswalls separating nuclei of adjacent cells. A day later, the filaments produce dark sporulating structures that resemble balloons on the end of sticks, held upright by a cluster of rootlike *rhizoids.* Each "balloon" is really a sac, or *sporangium,* that is filled with black asexual **sporangiospores** (Figure 25-22). When sporangia rupture, millions of haploid spores are released. Some travel to new locations on the slightest breeze. Spores that settle in favorable conditions germinate into actively growing hyphae, establishing new mold colonies. Zygomycetes infrequently reproduce sexually, when hyphae of opposite mating types fertilize each other. The resulting diploid cell forms a dark warty coat around itself as it develops into a **zygospore.** The zygospore then undergoes meiosis (often after months of dormancy), germinates, and develops into a typical sporangium containing haploid sporangiospores.

At least 50 species of fungi, most of them zygomycetes, trap or snare animals that humans consider pests. Victims include small worms (nematodes) that are trapped when they crawl through special loop-shaped hyphae of the fungus *Arthrobotrys.* The loop clamps shut when stimulated, snaring the worm, which is then digested by fungal enzymes and absorbed into the hyphae. Houseflies, grasshoppers, and other insects also fall prey to "carnivorous" zygomycetes. Scientists are evaluating the feasibility of using such predatory fungi as biological insecticides against many types of insect pests, from aphids that destroy millions of dollars in citrus fruit each year to household cockroaches.

Ascomycetes

Ascomycetes are often referred to as "sac fungi" because they house their sexual **ascospores** in a sac, called an *ascus* (plural = *asci*). In unicellular ascomycetes, such as the common yeast *Saccharomyces cerevisiae* (brewer's and baker's yeast), the cell itself becomes an ascus filled with

[6] Another group of fungi, commonly called "water molds," is not included here because their atypical cell walls have lead many taxonomists to reclassify them in the Protist Kingdom. They do, however, contain important animal and plant pathogens, including the fungus responsible for the Irish potato blight.

TABLE 25-3
THE FOUR CLASSES OF TERRESTRIAL FUNGI

Class	*Sexual spores*	*Asexual spores*	*Hyphae*	*Representative genera*
Zygomycetes	Zygospores	Sporangiospores	Nonseptate*	*Rhizopus* (bread mold)
Ascomycetes	Ascospores	Many different types (conidia, arthrospores, etc.)	Septate	*Saccharomyces* (brewer's yeast) *Penicillium* (antibiotic producer) Several genera of morels and truffles
Basidiomycetes	Basidiospores	Virtually none	Septate	*Agaricus* (supermarket mushrooms), plus hundreds of other mushroom genera; rusts and smuts (plant pathogens)
Deuteromycetes	None	Conidia; arthrospores, etc.	Septate	*Candida* (causes vaginitis and other human infections); many genera of pathogenic fungi that cause serious human diseases

**Nonseptate hyphae are continuous tubes of cytoplasm, whereas septate hyphae have crosswalls that divide the adjacent cells from one another.*

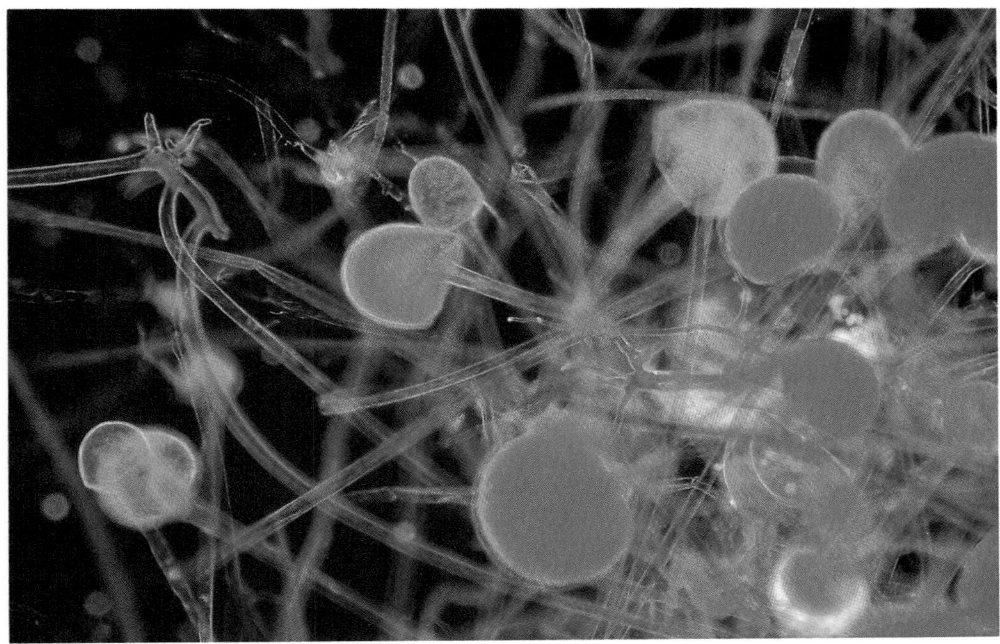

FIGURE 25-22
Packed with sporangiospores, the sporangia of this zygomycete *(Rhizopus)* are ready to burst open and discharge their cargo of reproductive spores. The nonseptate hyphae lack partitions (septa) between adjacent nuclei.

four ascospores. Most ascomycetes are multicellular and filamentous; some form complex fruiting bodies, called ascocarps, that house the spore-filled asci (Figure 25-23). Although some ascomycetes are prized for their flavor, others are more detrimental. Ascomycetes have downed many stately creatures, including millions of chestnut and elm trees. Humans can also fall victim to ascomycetes, which include the agents of such dangerous diseases as histoplasmosis (a systemic infection that initially infects the lungs), and some forms of common "ringworm" infection.

FIGURE 25-23
An ascomycete. The inner surface of these cup fungi are lined with thousands of sacs (asci) that contain sexual ascospores that are released when the sacs rupture.

Basidiomycetes

Also known as "club fungi," members of the **basidiomycete** class are not only the most familiar fungi (Figure 25-24), they are also the most sexually active. Sexual spores are produced by mushrooms and puffballs—"fleshy" structures of compactly intertwined hyphae. The hyphae that constitute the partitions ("gills") in the cap of a mushroom form millions of club-shaped sporulating structures called *basidia*. Each basidium bears four **basidiospores** (Figure 25-25) which are carried by wind or animals to new locations, thereby dispersing the species.

The fleshiness of basidiocarps is often accompanied by a delicious taste, but a few are also endowed with deadly toxins for which there are no antidotes. Distinguishing between safe and poisonous varieties sometimes requires microscopic examination. In addition to poisonous mushrooms, detrimental basidiomycetes include the food-robbing plant pathogens wheat rust and smut, as well as tree-killing bracket fungi.

Deuteromycetes

True **deuteromycetes** are sentenced to a life without sex. By definition, these are fungi in which sexual reproduction has not (yet) been discovered. Each year, more fungi are removed from the ranks of the deuteromycetes when their sexual mechanisms are discovered. The genus *Penicillium*, for example, is traditionally presented as an important example of a deuteromycete. Yet *P. chrysogenum* (the species that produces the antibiotic penicillin) and *P. roqueforti* (the blue mold that imparts the colorful ribbons that flavor Roquefort and blue cheese) were both recently discovered to be ascospore-producing, sexually active fungi. Both are therefore ascomycetes.

The most common deuteromycete is *Aspergillus*, perhaps the most ubiquitous organism in the world. Its spores have even been found in the stratosphere, at an altitude of 100 miles. A spoiler of food, this fungus also makes its unavoidable presence known to hay-fever sufferers, many of whom are allergic to the spores. Another troublesome deuteromycete is the yeast, *Candida albicans*, which commonly causes vaginal yeast infections, diaper rash, and an oral infection called "thrush." Such yeast infections are common following treatment with antibiotics that kill the bacteria that normally live on the body, allowing the yeast to grow to enormous numbers. *Candida* infections also pose a fatal threat to alcoholics and to patients with AIDS or leukemia.

EVOLUTION AND ADAPTATION: TYING IT TOGETHER

Because they are considered neither living nor inanimate, it may be tempting to view viruses as some kind of "missing link"—the ancient transitional form that evolved from inanimate chemicals into the first cells. However, their absolute requirement for a host cell disqualifies them as

(a)

(b)

FIGURE 25-24

Basidiomycetes: fleshy fungi. ***(a)*** These leaf fungi are basidiomycetes that decompose dead organic matter. ***(b)*** Sexual basidiospores are often so abundant that they form a smoky discharge, as from this puffball.

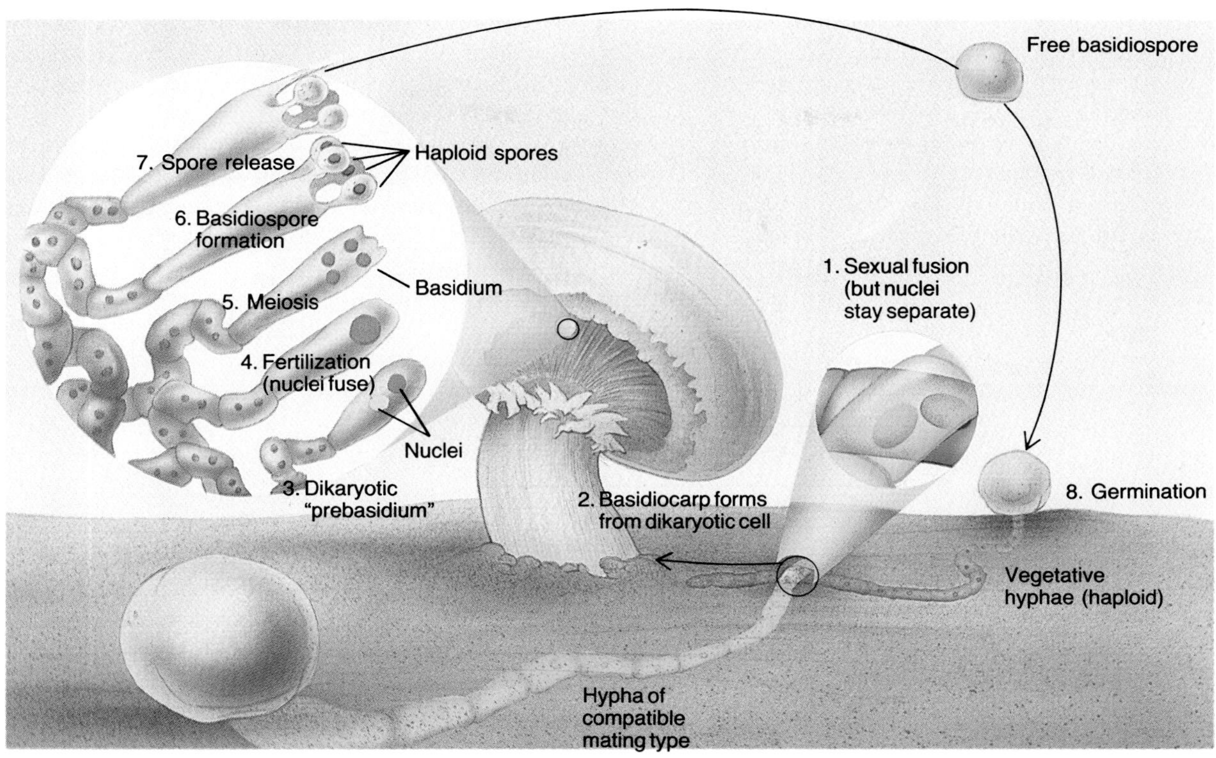

FIGURE 25-25

The mushroom's sexual cycle. A mushroom's fruiting bodies (basidiocarps) develop following fusion of sexually compatible hyphae, creating a *dikaryotic* cell (two nuclei per cell). Proliferation of this cell gives rise to a basidiocarp that is composed of dikaryotic cells. The two nuclei fuse during fertilization in the gills of the basidiocarp, forming a single diploid nucleus. The resulting diploid cell (the basidium) immediately undergoes meiosis, forming four haploid nuclei, each of which ends up in a basidiospore. The cycle begins again when hyphae from a germinated basidiospore come in contact with hyphae from another, sexually compatible basidiospore.

candidates for the earth's initial life forms. Most contemporary biologists believe that viruses are "renegade genes"— pieces of DNA (or RNA) that were once part of a cell's chromosome (or messenger RNA). These pieces of nucleic acid acquired the information for using the host cell machinery to replicate and may have been transferred directly from one cell to its neighbors through adjacent membranes. Eventually, the genetic fragments evolved new genes for making their coat proteins, allowing them to escape their host cells into the inanimate environment and assisting them in infecting new host cells.

The evolutionary impact of viruses on other forms of life is significant in determining which organisms survive, for viruses are as much a selective pressure as is a predator. As disease-causing "menaces," viruses eliminate individuals and species that have little resistance to their lethal attacks. But the evolutionary significance of viruses may be more profound than their role in selective pressure. Viruses transfer genes between bacteria, often carrying advantageous genes that confer traits on recipient cells that allow them to survive or better compete in a hostile habitat. For example, viruses may transfer antibiotic-resistance genes to recipient bacteria, which are no longer susceptible to the drug and continue to proliferate in an infected person in spite of antimicrobial therapy. Most biologists believe that viruses may have played similar gene-transfer roles in higher organisms, supplementing sex as a means of increasing diversity within a species.

The Monera Kingdom contains living bacteria that may be very similar to the original cells from which all life descended. Alternatively, thère may have been a primitive "urkaryotic" cell from which prokaryotes and eukaryotes evolved along separate lines, a cell that also gave rise to archaebacteria. Natural selection favored "team adaptations" that created typical eukaryotic organelles, which were originally independent bacterial cells that formed mutually dependent intracellular relationships inside ancient "pre-eukaryotic" cells.

The similarity of the protists to individual cells of the higher kingdoms suggests how animals, plants, and fungi acquired the adaptations that characterize them. Direct ancestors of each kingdom probably resembled some of the protists alive today. Animals, for example, evolved from an ancient eukaryotic cell that may have been similar to today's flagellated protozoa. Although one such line may have generated the members of the Fungus Kingdom from fungus-like protists, some fossil evidence suggests that the earth's earliest fungi may have directly evolved from non-eukaryotic organisms. In fact, fungi may have been among the original eukaryotes on earth as suggested by their fossil presence in precambrian rock formations. Whichever was their evolutionary history, the fungi were not the ancestors of plants, nor did they descend from plants that lost their chloroplasts.

SYNOPSIS

Viruses are noncellular entities composed of DNA or RNA, protein, and sometimes a surrounding envelope. In a host cell, the viral nucleic acid commandeers the cell and directs it to make a new crop of viruses.

All monerans are unicellular prokaryotes, the simplest living cells. Prokaryotic cells are smaller than eukaryotic cells and lack a nucleus (and other membrane-bound organelles). The Monera Kingdom contains two fundamental types of prokaryotes: eubacteria (true bacteria) and archaebacteria ("ancient" bacteria). Eubacteria contain all disease-causing bacteria, typical "normal flora" bacteria, and the photosynthetic cyanobacteria. Archaebacteria contain organisms that have adapted to harsher environments, such as extremely hot, salty, or acidic habitats. Monerans are found in virtually all habitats and are essential to the continuance of life on earth. As decomposers, they recycle essential nutrients. As nitrogen fixers, they transform molecular nitrogen into a form that can be used by all organisms. As chemosynthetic autotrophs, they form the basis for some marine food chains. However some bacteria cause severe diseases in people, animals, and plants, extracting an enormous cost in lost revenues and lost lives.

Members of the Protist Kingdom have few obvious properties in common; all are eukaryotic, but some are unicellular, and others are relatively simple multicellular complexes.

Protozoa are protists that resemble animal cells. Most members of this group contribute immeasurably to the biosphere. As zooplankton, they are critical links in food chains, forming a nutritional bridge between the producers (algae) and animals that cannot digest some of the components of algae. Protozoa also help decompose and recycle nutrients. Many terrestrial livestock animals require cellulose-digesting protozoa in their rumens before they can harvest the energy and nutrients in their grassy diets. Some protozoa, however, cause serious diseases in humans, such as amebic dysentery, giardiasis, malaria, and encephalitis.

Algae are protists that resemble plants. Algae are indispensable to the biosphere. They produce most of the world's food and oxygen and three-fourths of the living material in aquatic systems. This benefit greatly outweighs the few detrimental activities associated with algae blooms.

Slime molds are protists that resemble fungi. Cellular slime molds have amoeba-like vegetative cells that aggregate and produce a sporulating structure similar to those of fungi. Plasmodial slime molds form a plasmodium—a single giant "cell" with hundreds of nuclei. The plasmodium transforms into spore-filled reproductive fruiting bodies. Slime molds are important decomposers.

The Fungus Kingdom consists of yeasts and filamentous fungi. As decomposers, fungi help sustain the earth's many nutrient cycles. In addition to fertilizing soil by decomposing dead organisms, other fungi, those of lichens, actually help make soil out of rock. Fungi are sources of valuable foods, medicines, and other resources for humans. They also cause mild-to-serious diseases in about a fifth of the world's human population and are responsible for food spoilage and crop losses.

Review Questions

1. List three characteristics that distinguish viruses from all other "organisms." Which of these properties would be considered unique to viruses; that is, not found in cellular organisms?
2. Describe two features that characterize all the members of the Monera Kingdom.
3. Other than wiping out infectious disease, what would be three consequences of eliminating all bacteria from the biosphere?
4. Determine which of the three Protist groups each of these organisms belongs to: *Euglena, Gonyaulax* (the agent of red tide); *Trichomonas vaginalis* (a flagellated heterotroph that causes vaginal infections); *Trichonympha* (a flagellated, cellulose-digesting heterotroph that lives in the gut of termites).
5. Fungi were once classified in the Plant Kingdom. What characteristics distinguish fungi from plants?
6. Discuss the mutual benefits for each organism in the following biological "teams": a "photosynthetic" hydra; mycorrhizae and a pine tree; a lichen; a cow and the cellulose-digesting protozoa in its digestive tract.

Critical Thinking Questions

1. In 1919, a worldwide epidemic of influenza killed 20 million people in less than a year. In nearly every fatal case, the lungs were infected with a bacterium called *Haemophilus influenza,* which was mistakenly identified as the cause of influenza (a viral disease). Use Koch's technique to design a procedure that would prove that this bacterium does *not* cause influenza.
2. Having read this chapter, do you consider viruses living organisms or inanimate collections of self-replicating crystals? Justify your answer.
3. One serious problem in modern medicine is the overuse of antibiotics. Since flooding the environment with antibiotics would constitute an evolutionary selective pressure, why would this practice threaten to decrease the value of antibiotics for use against infectious disease?
4. Human pollution of coastal sea water and oceans is threatening the growth of algae and other protists that thrive there. Explain how the loss of these microscopic organisms would affect: the population of whales; the global economic community; the atmosphere; and the ability of the biosphere to support animal life, including humans.
5. Explorers returning to England from South America fascinated their British sponsors with a novel type of flower they had brought home. Unfortunately, none of these flowers, the orchids, would grow in English conservatories. Eventually it was discovered that if some of the soil in the orchid's original pots was mixed with the European soil, the orchids grew as well as they did in South America. Considering what you've learned about fungal associations with plants, what was lacking in the English soil that was supplied by the soil in which orchids had already grown?

CHAPTER

◂ 26 ▸

The Plant Kingdom

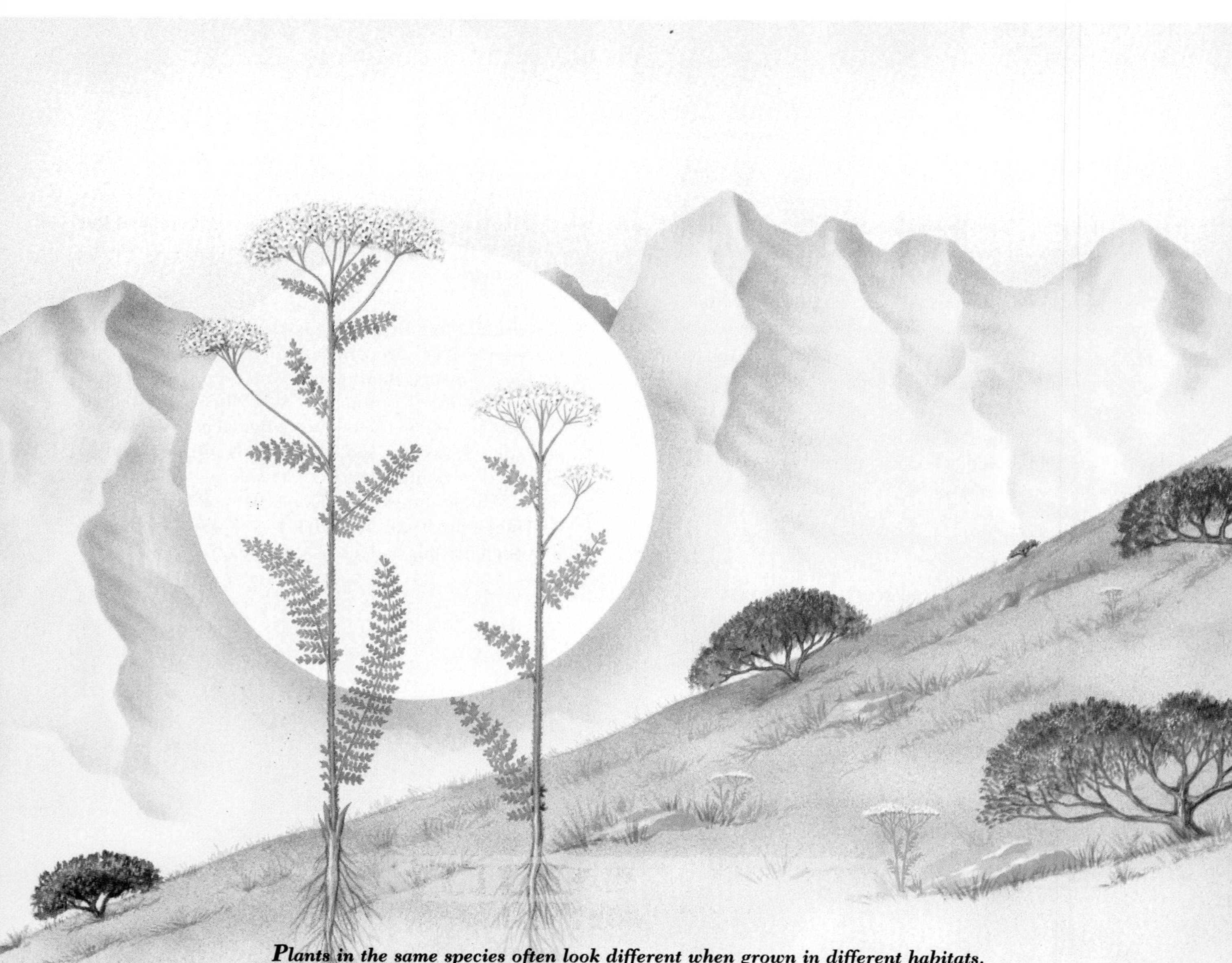

Plants in the same species often look different when grown in different habitats. Are these distinctions due to genetic changes or to environmental effects? Reciprocal growth experiments proved it to be a combination of both.

STEPS TO DISCOVERY
Distinguishing Plant Species: Where Should the Dividing Lines Be Drawn?

Ask several biologists "What is a species?" and you are likely to get different answers, depending on the person's specialty. A zoologist distinguishes animal species solely on the basis of whether individuals can successfully interbreed and produce fertile offspring. Unfortunately, this approach, called the *biological species concept,* works poorly for distinguishing plant species since members of some species can hybridize with individuals of other closely related, but morphologically distinct, species. Botanists must rely on different criteria for determining if two plants belong to the same species. Most botanists use genetically determined morphological traits to define individual plant species. There are two problems with this approach, however. First, are differences in morphology among plants living in different habitats genetically determined or are they induced by differences in environmental conditions? Second, are the genetic differences great enough to justify classifying the two types as separate species?

In the 1920s, Gote Turesson, a Swedish botanist, experimentally attacked the first of these questions by collecting different individuals of the same species from various habitats in Europe and growing them in uniform conditions in a test garden near Akarp, Sweden. Individuals from environments with more extreme conditions were found to be shorter than were members of the same species from environments with more hospitable conditions. If these differences in height were solely the effect of environmental conditions, all plants should be similar when grown in ideal conditions.

When grown side by side, the two plants did not grow equally tall. Turesson concluded that this difference was due at least in part to genetic adaptation to different environments. He coined the term *ecotype* to distinguish genetic varieties within a single species.

Turesson's conclusions were supported by three Stanford scientists, Jens Clausen, David Keck, and William Heisey, who conducted reciprocal transplants of specimens of the perennial herb *Achilla lanulosa* (yarrow). The scientists collected seeds from yarrow specimens growing in dramatically different locations in North America, from the California coasts to the frigid mountain timberlines. They planted these seeds in three experimental plots, one representing low coastal habitats (at Stanford), another simulating the moderately high coniferous forest zone (at Mather), and the third simulating the harsh alpine timberline areas (timberline).

The largest of these plants, an 84-centimeter-tall ecotype from lower altitudes, fared poorly when grown at the timberline plot. Fewer than half of the plants survived, and those that did grew to an average height of only 15 centimeters (compared to the ecotype native to the timberline habitat, which grew to 24 centimeters). In the Stanford coastal plot, however, the timberline ecotype reached only 21 centimeters, in spite of the more hospitable conditions. This plant actually grew better in its harsh, frigid native alpine area than it did in the "ideal" conditions of coastal California. In fact, the plants from all the different regions sampled by Clausen and his co-workers had adapted to their native habitat. These differences were genetically stable and could not be "undone" by moving plants from hostile habitats to less extreme environments.

These studies illustrated that as a plant species becomes widespread, it develops genetic adaptations to the local conditions. In other words, within a particular species, significant genetic variation exists, some so great that different varieties are no longer capable of interbreeding with other varieties. Although we call these variants ecotypes, where do we draw the line and say that the differences are great enough to constitute two separate *species*? The system currently used for distinguishing plant species has a clear advantage: It groups together as the same species those plants that are most similar in their evolutionary history.

A fire rages through a forest. Many animals escape the flames, but the plants cannot flee. Anchored to the ground, they must stand against many life-threatening hazards: violent winds, floods, sleet, herbivores, and even fire. Despite what seems to be an enormous handicap, plants not only manage to survive in virtually all habitats, they have become the most prominent form of life on earth. With more than 400,000 species, the Plant kingdom is second only to the Animal kingdom in diversity.

▼ ▼ ▼

MAJOR EVOLUTIONARY TRENDS

Because ancient plants left so few fossils, it is difficult to document the history of plant evolution. Taxonomists and evolutionary botanists must therefore rely on similarities in biochemistry (mainly photosynthetic pigments), cell structure, growth patterns, and gametes to interpret the course of plant evolution. This evidence suggests that higher plants evolved from a line of filamentous green algae that invaded land over 400 million years ago, during the Silurian period of the Paleozoic Era. Today, only the algae and a few higher plant species live in water; most are terrestrial (Figure 26-1).

CONQUERING DRY LAND

The demands of living on dry land led to a change in the growth pattern of early plants, from single rows of cells to sheets of cells that form ground-hugging mats. Lying flat against the soggy ground, water could diffuse into the cells of these early land plants, quickly replacing water used in metabolism or lost to the atmosphere as vapor. Unobstructed light, abundant carbon dioxide, and a virtual lack of competition allowed these early land plants to flourish as long as they remained close to a continuous supply of water along the edges of beaches, ponds, or streams.

Four types of adaptations enabled plants to make the transition to dry land:

1. *Control of water loss.* The above-ground parts of most plants are coated with a waxy cuticle that retards evaporation. In addition, a plant's outer surfaces are perforated with stomates (page 214), tiny pores that open when the water supply is adequate and close when dehydration threatens the plant.
2. *Vascular tissues.* Plants that lack vascular tissues must grow close to the ground. The evolution of xylem, for transporting water and mineral ions, and phloem, for transporting sugars and other organic molecules, enabled plants to grow taller by providing a means of moving these materials through all parts of the plant (page 214). In addition, thick-walled xylem cells help support numerous leaves and heavy stems.
3. *Resistant spores.* Some plants manufacture **spores,** lightweight cells that are specialized for dispersal and protection in adverse conditions. With thick walls to impede water loss and a virtual lack of metabolism to consume water, dormant spores can survive for long periods without additional moisture.
4. *Protective packaging for gametes and embryos.* Gametes and embryos are protected against dehydration and damage by the following structures: (a) *Multicellular gametangia* (gamete-producing structures) surround reproductive cells and embryos with water-trapping layers of cells. (b) *Pollen grains* encapsulate male gametes in watertight packages that free plants from the need to use water for transferring sperm to an egg for fertilization. (c) *Seeds* serve as drought-resistant enclosures for plant embryos, enabling offspring to be dispersed to new localities. (d) *Fruits* further clothe seeds of flowering plants in additional protective layers, enhancing embryo survival and dispersal.

FROM PROMINENT GAMETOPHYTE TO PROMINENT SPOROPHYTE

During sexual reproduction, all members of the Plant kingdom alternate between a diploid, spore-producing generation—the *sporophyte*—and a haploid, gamete-producing generation—the *gametophyte.* Biologists refer to this sequential change from one generation to the next as *alternation of generations.*

A major trend in the evolution of plants has been toward a prolonged sporophyte (diploid) and a reduced gametophyte (haploid) generation (Figure 26-2). Complex, advanced plants, such as pines and flowering plants, have a prominent sporophyte generation, whereas more primitive terrestrial plants, like mosses, retain a prominent gametophyte, as did their aquatic ancestors. Diploidy offers an important advantage over the haploid state for complex organisms: Since haploid cells contain only one allele for each gene, the effects of a mutant allele cannot be masked by a dominant allele as it can in a diploid organism. Diploidy not only allows an organism to mask lethal mutations without suffering harmful effects, it also enables an organism to harbor neutral mutations that may later prove to be advantageous.

OVERVIEW OF THE PLANT KINGDOM

All multicellular, tissue-forming eukaryotic photosynthesizers are classified within the Plant kingdom. Many botanists also include three divisions of aquatic algae (red,

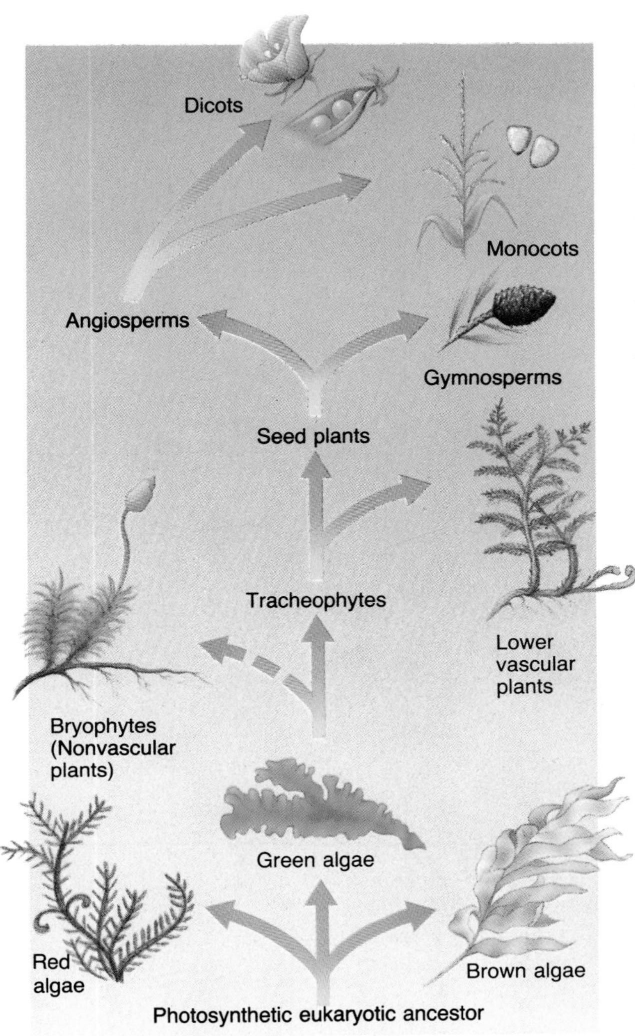

FIGURE 26-1

An overview of the evolution of the major plant groups. From their aquatic photosynthetic ancestors, plants evolved into three divisions of aquatic "plants" (the brown, red, and green algae) and ten divisions of mostly terrestrial (land) plants. (The Plant kingdom is subdivided into divisions, not phyla as in the Animal kingdom.) The green algae were the ancestors of all terrestrial plants. Bryophytes, which includes mosses, lack vascular tissues and are likely to have evolved from green algae or from an ancestral vascular plant that lost its vascular tissue. The remaining plant divisions, which include flowering angiosperms and cone-bearing gymnosperms, have xylem and phloem vascular tissues: Plants with vascular tissues are called *tracheophytes.*

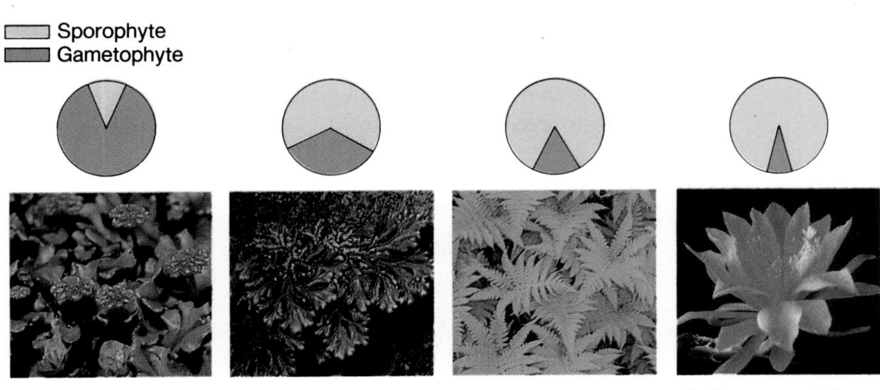

FIGURE 26-2

Terrestrial plants: from prominent gametophyte to prominent sporophyte. Comparing a range of life cycles from primitive to more advanced plants reveals a trend toward a prominent diploid sporophyte.

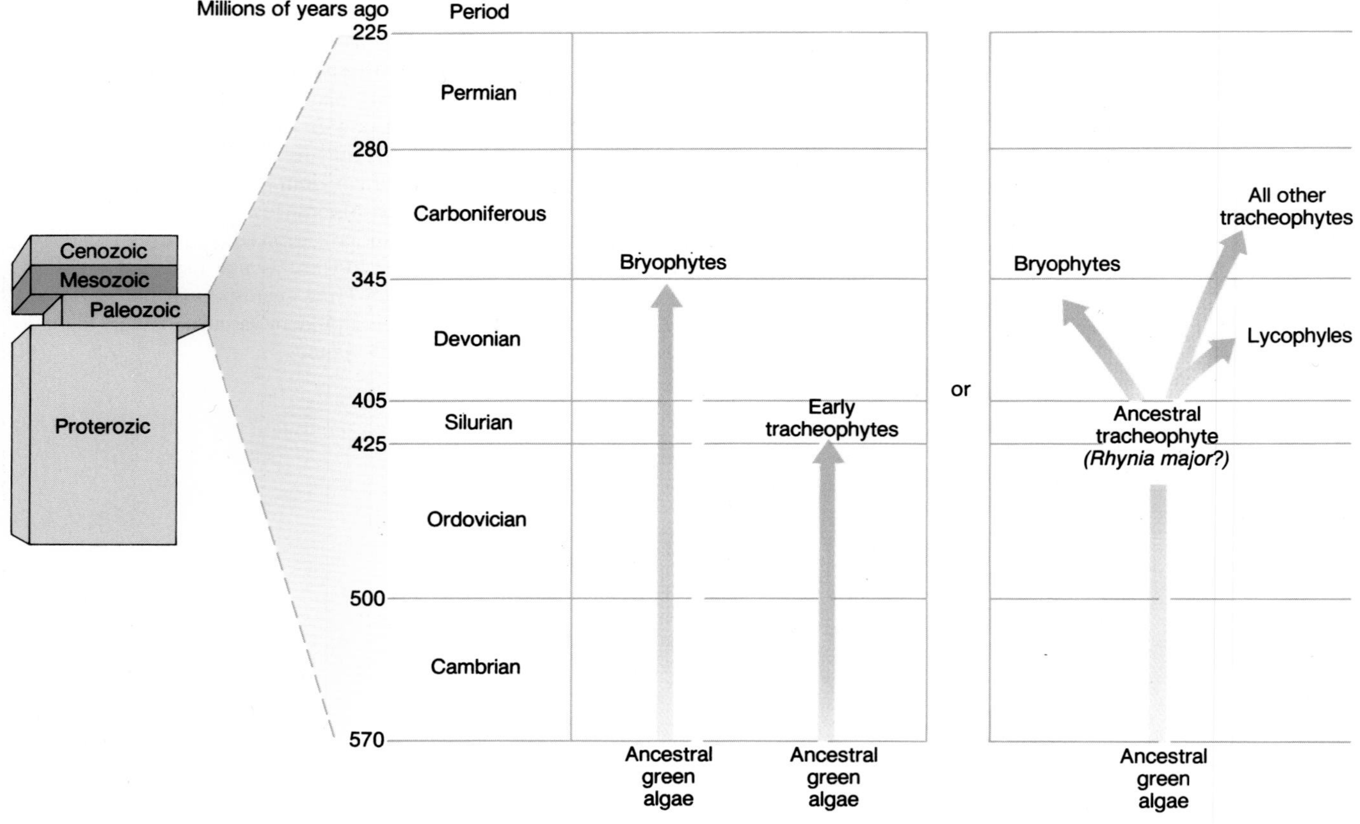

FIGURE 26-3

Plant origins. Two hypotheses have been proposed to explain the origin of plants. Although both agree that plants evolved from an ancestral filamentous green alga, one hypothesis contends that the nonvascular plants (bryophytes) and the vascular plants (tracheophytes) arose from different ancestors, whereas the other asserts that both bryophytes and tracheophytes arose from the same ancient tracheophyte.

brown, and green) in the Plant kingdom, as we have here. Botanists recognize two major groups of land plants, the **bryophytes,** or nonvascular plants, and vascular **tracheophytes** (*tracheo* =, tube, *phyte* = plant). Although the evolutionary origin of these two groups is still being debated (Figure 26-3), botanists agree that today there are ten divisions of terrestrial plants, nine of which are vascular plants (see Table 26-1).

ALGAE: THE FIRST PLANTS

The green, brown, and red algae have been assigned to the Plant kingdom because they contain organisms that are multicellular or colonial (Figure 26-4). Algae are distinguished from one another by the pigments they use to capture different light rays that penetrate water. These photosynthetic pigments tint the organisms the color that characterizes each group.

GREEN ALGAE: ANCESTORS OF HIGHER PLANTS

With 7,000 species, green algae (**Chlorophyta**) is the largest group of algae. Green algae are the only type found mostly in freshwater habitats, although there are a few marine species. Some green algae are even adapted to life on land and blanket trees, soil, porous rocks (or bricks), and snow.

Because green algae share the following characteristics with higher plants, most botanists argue that they are the ancestors to higher plants:

- chlorophyll *a* and *b*,
- carotene accessory pigments,
- the ability to store surplus carbohydrates as starch,
- cell walls of cellulose.

Many green algae are unicellular; others are multicellular or colonial. Colonial forms are intermediate between

TABLE 26-1
CHARACTERISTICS OF DIVISIONS THAT CONTAIN PREDOMINANTLY TERRESTRIAL PLANTS

	Non-vascular	*Tracheophytes*								
		Lower Vascular Plants				*Higher Vascular Plants*				
Characteristics	*Bryophyta* (24,000)[a]	*Psilophyta* (13)	*Lycophyta* (1000)	*Sphenophyta* (15)	*Pterophyta* (12,000)	*Cycadophyta* (100)	*Ginkgophyta* (1)	*Gnetophyta* (71)	*Coniferophyta* (700)	*Anthophyta* (375,000)
Prominent gametophyte										
Prominent sporophyte										
External water needed for fertilization										
Dispersal by spores										
Dispersal by seeds										
True roots, stems, leaves										

[a] Number of described species.

unicellular and multicellular organisms (such as the *Volvox* shown in Figure 26-4*a*). Individual cells in the colony function independently. In true multicellular green algae, cell activities are coordinated, as they are in higher plants. In fact, multicellularity probably originated in the green algae.

The sporophyte and gametophyte stages of some green algae are indistinguishable. For example, when the flagellated gametes of *Ulva* (the sea lettuce shown in Figure 26-4*b*) fuse, the zygote proliferates into a green, leafy seaweed, two cells thick. This diploid sporophyte then forms haploid spores, each of which develops into a haploid seaweed that looks identical to the diploid seaweed. The gametophyte then produces identical-looking gametes, which mate, producing a zygote that develops into a new leafy sporophyte.

In other green algae, the female gametes are large and nonmotile, whereas the male gametes are small and motile (analogous to the situation in higher plants and animals). In most of these algae, the sporophyte is extremely reduced.

BROWN ALGAE: MULTICELLULAR GIANTS

Brown algae (**Phaeophyta**), mostly seaweeds, are distinguished by the presence of an accessory pigment (*fucoxanthin*) that is found in no other group of algae. The brown pigment helps these algae gather blue-green light, the wavelengths that penetrate deep into water. Some brown alga are giants. Underwater "forests" of giant kelp contain individuals that are long enough to attach firmly to the ocean bottom, with their tops reaching the light-rich ocean surface 300 feet above (Figure 26-4*c*). Individual plants grow at an extraordinary rate, about 20 centimeters (1 foot) per day. This makes them particularly productive "crop plants" that can be harvested regularly without depleting their numbers (see The Human Perspective: Brown Algae: Natural Underwater Factories).

Brown algae have developed specialized structures and tissues, such as holdfasts, blades, floats, and conducting vessels. Holdfasts anchor the algae to rocks on the ocean bottom; leaflike blades provide broad surfaces for light absorption, boosting photosynthetic efficiency; hollow, gas-filled floats help keep the seaweed upright in the water; and newly synthesized nutrients are shipped from the blades down to the holdfasts and other parts of the organism through conducting tubes that are reminiscent of the phloem in plants.

The brown algae alternate generations between sporophyte and gametophyte states. The large kelp form is the organism's sporophyte stage, the diploid, spore-producing form. Compared to the sporophyte, the haploid gamete-producing stage is usually very tiny.

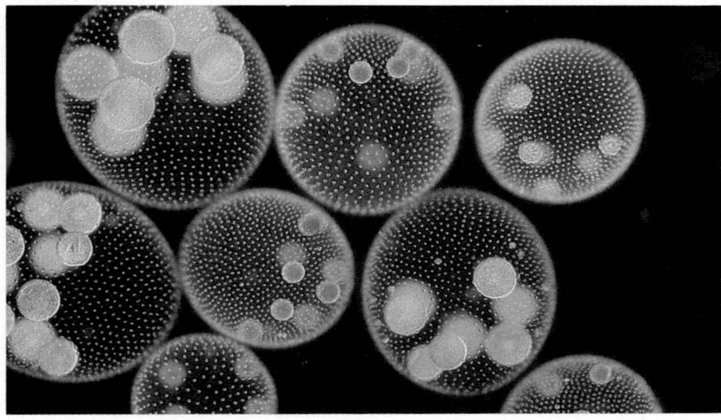

(a)

(b)

FIGURE 26-4

Multicellular algae: the simplest members of the Plant kingdom. *(a)* A typical colonial green alga, these *Volvox* colonies consist of hundreds of attached cells, each spherical colony functioning as a single organism. Daughter colonies inside the mature "adults" are released when the parental colony disintegrates. ***(b)*** Multicellular green algae are exemplified by sea lettuce (*Ulva*). ***(c)*** Brown algae are typified by underwater forests of giant kelp (*Macrocystis*) that stay in place by virtue of holdfasts that anchor each organism to the solid ocean floor. ***(d)*** Red algae live in greater depths than can members of the other divisions.

RED ALGAE: REACHING THE GREATEST DEPTHS

Like brown algae, red algae (**Rhodophyta**) possess supplemental pigments that absorb deep-penetrating light rays, enabling these algae to inhabit an expansive range of the ocean floor. These accessory pigments provide color to the 4,000 species of these seaweeds, which range from red, blue, or purple to a very dark green.

Except for a few freshwater red algae and free-floating marine forms, rhodophytes attach to the bottom of oceans or estuaries, as deep as 300 meters (about 900 feet) if waters are clear enough for sufficient light to penetrate. They are harvested for their many commercial uses, ranging from food (the dark seaweed wrapped around the raw fish of sushi) to the production of agar, the agent used to gel certain types of foods and to solidify nutrient media for growing bacteria, fungi, and other microbes in laboratories.

Red algae reproduce asexually by fragmentation of the body (the fragments develop into new individuals) and production of asexual spores. Sexual reproduction depends on fertilization between two haploid "sex spores"; the resulting zygote develops into a diploid sporophyte. The multicellular diploid sporophyte produces haploid spores that develop into large, free-living gametophytes. The gametophyte produces (by meiosis) the haploid "sex" spores that, when fertilized, develop into another round of diploid sporophytes.

BRYOPHYTES: PLANTS WITHOUT VESSELS

The 24,000 species of nonvascular bryophytes include nearly 15,000 mosses, 9,000 liverworts, and 100 hornworts (Figure 26-5). Since their sperm must swim through water to reach and fertilize an egg, bryophytes are not completely adapted to living on dry land. In addition, bryophytes lack "true" roots for absorbing and conducting water. Instead,

(c)

(d)

(a) (b) (c)

FIGURE 26-5

The bryophytes. The division Bryophyta is divided into three classes: ***(a)*** the Musci (mosses), ***(b)*** the Hepaticae (liverworts), and ***(c)*** the Anthocerotae (hornworts).

water is absorbed directly into outer cells and then slowly diffuses throughout the organism. As a result, bryophytes can grow no taller than a few centimeters.

The gametophyte is the prominent generation in bryophytes. It is the leafy, low-growing form that most people recognize as a moss. During reproduction, the multicellular gametophyte produces either male or female reproductive structures or, in some species, both. The roundish male gametangium (gamete-producing structure) is called an *antheridium*; the flask-shaped female gametangium is an *archegonium* (Figure 26-6). Sperm have flagella, enabling them to swim to an archegonium and fertilize the egg. In unisexual mosses, the sperm must swim to another plant to encounter an archegonium through a film of water.

The formation of a zygote following fertilization begins the sporophyte generation. The zygote divides by mitosis many times to form a multicellular sporophyte, which remains attached to the gametophyte for access to food and water. The *sporangium* (spore-producing structure) of the sporophyte contains cells that undergo meiosis to produce haploid spores, beginning a new gametophyte phase. The spores are dispersed by the wind and grow into a new multicellular gametophyte with archegonia and antheridia, beginning another reproductive cycle.

TRACHEOPHYTES: PLANTS WITH FLUID-CONDUCTING VESSELS

Modern tracheophytes with xylem and phloem conducting tissues are often divided into two general subgroups: the *lower* vascular plants that produce no seeds and have more primitive characteristics, and the *higher* vascular plants that produce seeds and possess more advanced characteristics (see Table 26-1). All lower vascular plants produce separate sporophyte and gametophyte generations. In con-

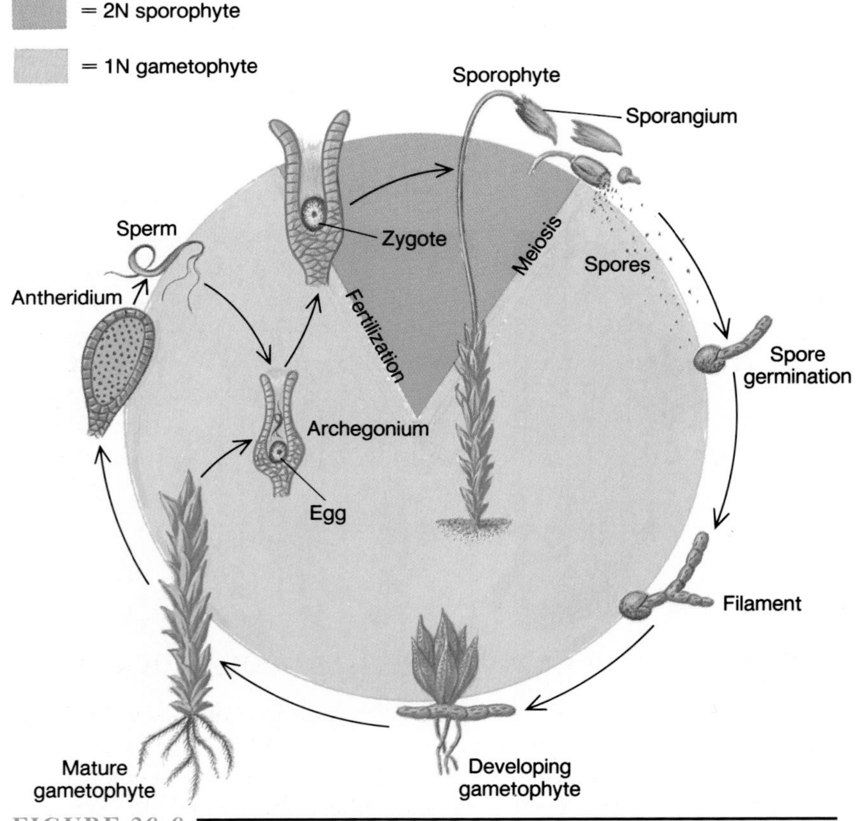

FIGURE 26-6

Moss life cycle. Although some mosses reproduce either male or female gametangia, this moss gametophyte forms both antheridia and archegonia. Following fertilization, the zygote grows into an erect sporophyte that remains permanently attached to the gametophyte body. The single sporangium at the tip of the sporophyte releases haploid spores that grow into new gametophytes.

◁ THE HUMAN PERSPECTIVE ▷

Brown Algae: Natural Underwater Factories

The cell walls and intracellular spaces of brown algae, such as giant kelp (*Macrocystis*) and floating stalks of *Nereocystis*, contain algin, a substance valued by dozens of industries. Algin is a compound that helps the seaweed maintain its structure and oppose the dehydrating effects of saltwater. It is prized by manufacturers because of its ability to regulate the behavior of water, thereby determining the physical properties, such as consistency and texture, of many types of products. Here are just some of the commercial uses of algin employed by dozens of enterprises.[1]

FOOD

1. Thickening agent in toppings, pastry fillings, meringues, potato salad, canned foods, gravies, dry mixes, bakery jellies, icings, dietetic foods, fountain syrups, candies, puddings.
2. Emulsifier and suspension agent in soft drinks and concentrates, salad dressing, barbecue sauces, frozen food batters.
3. Stabilizer in chocolate drinks, eggnog, ice cream, sherbets, sour cream, coffee creamers, party dips, buttermilk, dairy toppings, milk shakes, marshmallows.

PAPER

1. Provides better ink and varnish application on paper surfaces; provides uniformity of ink acceptance, reduction in coating weight, and improved surface characteristics of oil, wax, and solvents in paperboard products. Makes improved coatings for frozen food cartons.
2. Used for coating grease-proof papers.

TEXTILES

1. Thickens print paste and improves dye dispersal. Reduces weaving time and eliminates damage to printing rolls and screens.

PHARMACEUTICALS AND COSMETICS

1. Thickening agent in weight-control products, cough syrups, suppositories, ointments, toothpastes, shampoos, eye makeup.
2. Smoothing agent in lotions, creams, lubricating jellies. Binder in manufacture of pills. Blood anticoagulant.
3. Suspension agent for liquid vitamins, mineral oil emulsions, antibiotics. Gelling agent for facial beauty masks, dental impression compounds.

(ADDITIONAL) INDUSTRIAL USES

1. Used in manufacture of acidic cleaners, films, seed coverings, welding rod flux, ceramic glazes, boiler compounds (prevents minerals from precipitating on tubes), leather finishes, rubber compounds (e.g., automobile tires, electric insulation, form cushions, baby pants), cleaners, polishes, latex paints, adhesives, tapes, patching plaster, crack fillers, wall joint cement, fiberglass battery plates, insecticides, resins, tungsten filaments for light bulbs, digestible surgical gut, oil-well-drilling mud. Used in clarification of beet sugar. Mixed with alfalfa and grain meals in dairy and poultry feeds.

BREWING

1. Helps create creamier beer foam with smaller, longer-lasting bubbles.

In addition to their algin, brown algae are valued for their high potassium and nitrogen content; the seaweed is ground, dried, and used as soil fertilizers. Kelp is also used as livestock feed in many parts of the world and as human food (in sushi, soups, beverages, and other dishes) in the Orient.

[1] From Kingsley R. Stern, *Introductory Plant Biology*, 4th Edition, Copyright 1988, Wm. C. Brown Publishers, Dubuque, Iowa. Reprinted by Permission.

trast, higher vascular plants have a shortened gametophyte phase that depends on the sporophyte for food and water.

The sporophytes of all plants, even bryophytes, produce spores. Lower (seedless) vascular plants mostly manufacture only one type of spore. These spores grow into a gametophyte that produces both antheridia and archegonia. In contrast, higher vascular plants produce two types of spores, one that grows into a female gametophyte, and one that grows into a male gametophyte. By separating gametangia, the chances of cross fertilization are increased, generating greater genetic variability.

LOWER (SEEDLESS) VASCULAR PLANTS

Seedless vascular plants include four divisions: the Psilophyta, Lycophyta, Sphenophyta, and Pterophyta. Between 300 million and 430 million years ago, members of Psilophyta, Lycophyta, and Sphenophyta were the dominant type of vegetation on earth. Today, only about 1,000 species belong to these three divisions. Even after adding the 12,000 species of ferns (division Pterophyta), the four divisions of lower vascular plants account for less than 3 percent of all known plant species.

Psilophyta: Whisk Ferns and Relatives

Some 300 million years ago, lush green forests were filled with species of whisk ferns, or psilophytes. Today, only one family (containing two genera and 13 species) remains of this once diverse division of plants.

Psilotum is the most widespread genus extending from tropical and subtropical areas into a few temperate zones. The most common species of *Psilotum,* and the one frequently displayed in introductory biology laboratories, is the naked-looking sporophyte of *Psilotum nudum* (Figure 26-7). *Psilotum nudum* looks "naked" because its stems lack leaves. Instead, the stem bears small, widely spaced scales. Some scales contain three-chambered sporangia that manufacture and disperse spores. The spores grow into tiny, underground gametophytes that possess antheridia and archegonia. The flagellated sperm produced in the antheridia swim through a film of water to fertilize the egg. The resulting zygote grows and develops into another multicellular sporophyte.

Lychophyta: Club Mosses and Relatives

Like the Psilophyta, the division Lycophyta has seen more glorious days. About 350 million years ago, members of the genus *Lepidodendron* grew as tall as 50 meters (165 feet) and dominated the forests of the Carboniferous period (Figure 26-8). Today, all lycopods are small, herbaceous plants.

The lycophyta includes three families, six genera, and about 1,000 species. Most species are tropical, and the majority are *epiphytes,* plants that grow on other plants. The largest genus, *Selaginella,* contains 700 species. The most common genus in the United States is *Lycopodium,* commonly called club moss. These plants are also called ground pines because the sporophytes look like miniature pine trees (Figure 26-9). The spores of lycopods have been used in firecrackers and flash bars (before flashbulbs were developed) because they ignite explosively, emitting light. Ironically, lycopod spores have also been used as baby powder.

FIGURE 26-7

A "living fossil." The sporophyte of *Psilotum nudum* (division Psilophyta) closely resembles 400-million-year-old fossils of psilophytes. Round, yellow sporangia are produced on the sides of leafless photosynthetic stems.

FIGURE 26-8
The lush Carboniferous forests were dominated by giant psilophytes, sphenophytes, and lycophytes. Lepidodendron was a giant lycopod that produced leaves and sporangia at the ends of its branches. All treelike, seedless vascular plants, including Lepidodendron, are now extinct, the victims of changing environmental conditions.

FIGURE 26-9
Lycopodium complanatum sporophytes with strobili. Each strobilus is a tight cluster of spore-producing leaves.

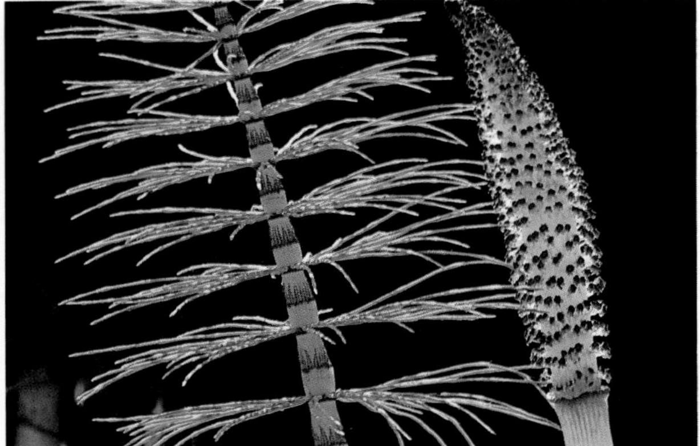

FIGURE 26-10
Horsetails. *Equisetum arvensis* has a horizontal rhizome that supports two types of hollow, upright stems. One is topped with a strobilus (on the right in the photo). The other produces orderly whorls of radiating branches (left). Although the stems appear leafless, small leaves on the stem fuse at the base to form sheaths around each node. *Equisetum* is the only genus in the Sphenophyta, a once widely distributed, diverse division.

Sphenophyta: The Horsetails

Growing alongside the giant lycopods 350 million years ago were woody sphenophytes that grew 15 meters (50 feet) tall. Like the lycopods, all remaining sphenophytes are herbaceous. Today, the Sphenophyta includes 15 species, all of which are classified in one genus, *Equisetum*, the horsetails (Figure 26-10).

The upright sporophytes of horsetails have jointed, hollow stems that arise from a horizontal stem, or rhizome. The vertical stems have distinctive longitudinal grooves and whorls of leaves or branches at the nodes. The leaves of horsetails fuse at the base to form a sheath around each node. Strobili are produced at the ends of stems or from side branches.

Pterophyta: Ferns

Like other seedless vascular plants, ferns are an ancient plant group that arose during the Devonian period between 375 million and 420 million years ago. Two-thirds of the 12,000 fern species living today are tropical. Because

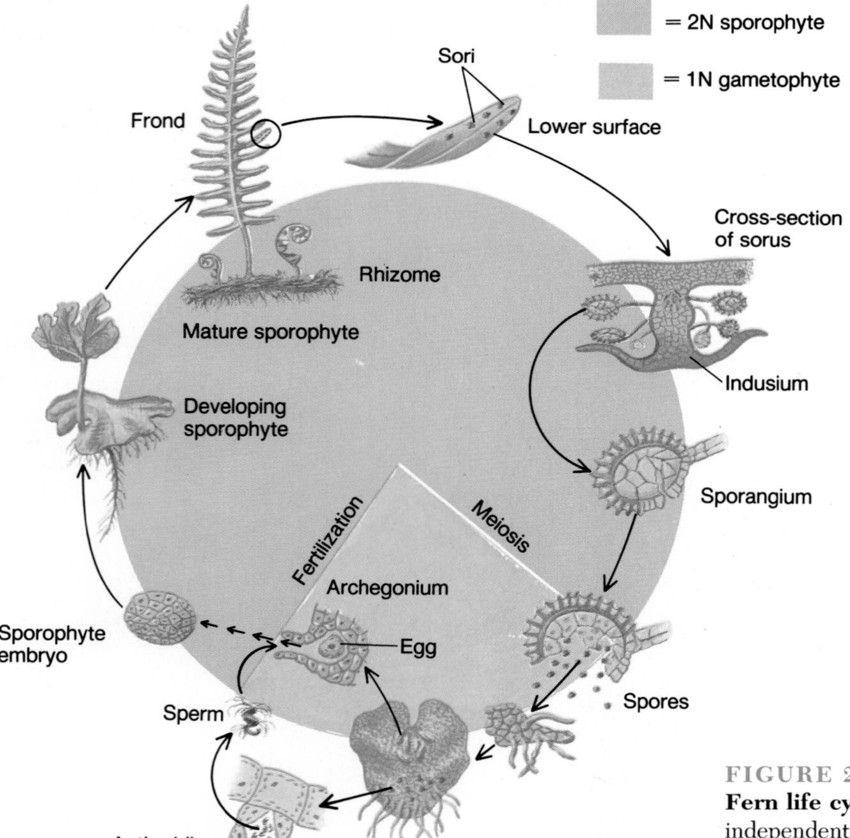

FIGURE 26-11
Fern life cycle. The fern sporophyte and gametophyte live independently. The heart-shaped fern gametophyte produces both antheridia and archegonia.

they produce flagellated sperm that must swim to the egg, all ferns are restricted to habitats that are at least occasionally wet.

The familiar fern sporophyte is comprised of fronds that uncoil from tight spirals that arise from rhizomes or upright stems. The fronds are leaflike structures that, unlike "true" leaves, have an apical meristem and clusters of sporangia called *sori* (Figure 26-11). Spores are released from the sporangia and grow into a small, heart-shaped gametophyte called a *prothallus*. The photosynthetic, independently growing prothallus lacks vascular tissues. Like many gametophytes of lower vascular plants, the prothallus is reminiscent of primitive ground-hugging plants.

SEED PLANTS

The simple conductive tissues of the seedless vascular plants were more than adequate for life in lowland swamps that were so widespread during the Devonian and Permian periods of the Paleozoic Era. Environmental conditions changed dramatically and abruptly at the end of the Permian period, however, shifting the course of plant (and animal) evolution in a new direction.

Gymnosperms (gymno = naked, sperm = seed) were able to colonize the new, drier habitats quickly at the beginning of the Mesozoic Era because they were armed with the following adaptations:

- extensive root systems that reach deeper underground water supplies;
- efficient conducting tissues that deliver water and nutrients to all parts of the plant;
- secondary growth that replaces and adds new vascular tissues and increases support (Chapter 12);
- enclosure of gametes in structures that protect the gametes and prevent drying (sperm in pollen grains, and eggs embedded in sporophyte tissues);
- "dry" fertilization (sperm do not require water for dispersal to an egg);
- encasement of the plant embryo in a protective seed that withstands harsh, dry conditions during dispersal and delays germination until conditions become favorable for growth.

THE GYMNOSPERMS

All modern seed-producing plants are divided into two groups: (1) the gymnosperms (pines, firs, cycads, etc.), or "naked seed" plants, and (2) the angiosperms (*angio* = covered), or flowering plants. The seeds of gymnosperms are referred to as "naked" because they are not surrounded by additional fruit tissues as they are in angiosperms.

The more than 700 species of living gymnosperms are grouped into four divisions: the Cycadophyta, Ginkgophyta, Gnetophyta, and Coniferophyta.

Cycadophyta

Cycads were most abundant and widespread during Mesozoic times, the Age of Dinosaurs. The 100 or so species of cycads that still survive today are native only to tropical and subtropical regions, although they are used as ornamental plants in virtually all regions of the world (Figure 26-12).

Cycads are either male or female, but never both. Pollen is produced in male cones and dispersed by the wind. When a pollen grain lands on a female cone, it grows a pollen tube that reaches to within a few millimeters of the egg. Flagellated sperm then swim the remaining dis-

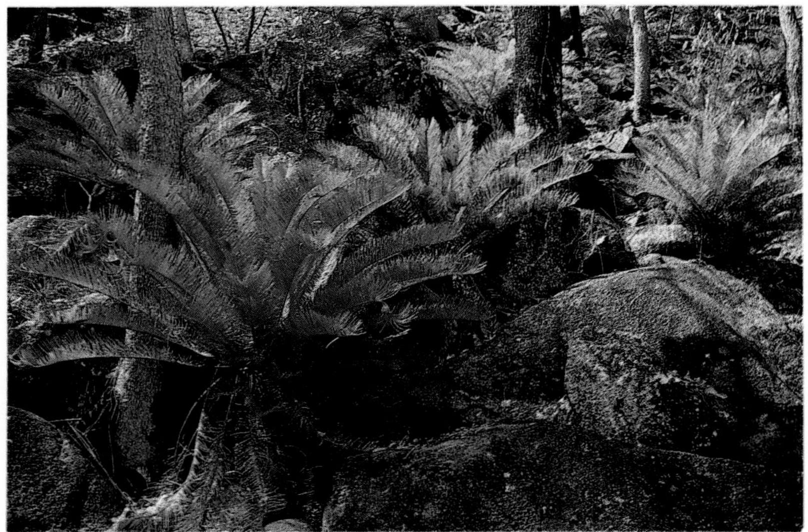

FIGURE 26-12
Mistaken identity. Because of their appearance, cycads are sometimes mistaken for palms. Unlike flowering palms, however, cycads are gymnosperms that produce either male or female cones. Cycads were abundant during the time of the dinosaurs, between 100 million and 200 million years ago.

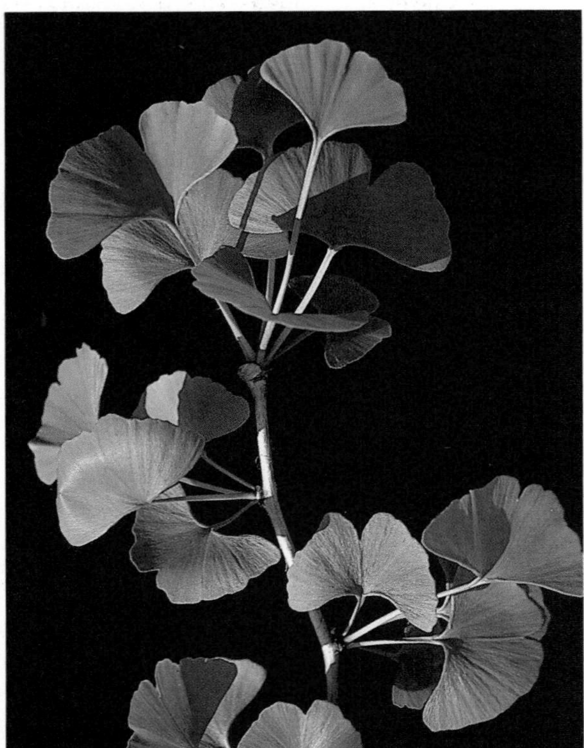

FIGURE 26-13

Saved from extinction, the maidenhair tree (*Ginkgo biloba*) has been cultivated for thousands of years in gardens, parks, and other tended sanctuaries. The maidenhair tree is the only living species in the division Ginkgophyta.

FIGURE 26-14

Less than meets the eye. Although *Welwitschia mirabilis* appears to have many leaves, each plant produces only two broad leaves that shred and curl as they grow. *Welwitschia* grows flat against the hot, sandy soil in the Kalahari Desert of Africa. Seeds are produced in cones along leaf edges.

tance through a miniature water chamber that is provided by the female gametophyte, a remnant of water fertilization.

Ginkgophyta

Like the cycads, a number of ginkgo species grew during the Mesozoic Era, but today only one species of Ginkgophyta remains, *Ginkgo biloba* (the maidenhair tree). *Ginkgo biloba* escaped extinction only because it has been cultivated for thousands of years as an ornamental plant in Asian temples and gardens (Figure 26-13). Today, no natural populations of ginkgos occur anywhere in the world; they survive only where tended by humans.

Gnetophyta

There are 70 species of gnetophyta, grouped into three genera: *Welwitschia, Ephedra,* and *Gnetum. Welwitschia* is the oldest genus; its fossils date back more than 280 million years. The only living species, *Welwitschia mirabilis,* is found in southwestern African deserts and resembles a pile of tattered leaves growing from a wide, blunt stem (Figure 26-14). Although it doesn't look like it, *Welwitschia* actually produces only two leaves during its 100-year lifetime. Each leaf shreds as it grows, giving the appearance of multiple leaves. *Welwitschia's* stubby, succulent stem and deep tap root store water, helping the plant survive long periods of drought.

All 40 species of *Ephedra* are short shrubs, with jointed stems and scalelike leaves. Native Americans made flour and tea from *Ephedra* plants. Today, drugs for treating asthma, emphysema, and hay fever come from extracts of *E. sinica* and *E. equisetina.*

The third genus, *Gnetum,* includes 30 species of vines, woody shrubs, and trees, all of which are found in Asia, Africa, and Central and South America.

Coniferophyta

The most familiar gymnosperms are the **conifers** that dominate many Northern Hemisphere forests. The more than 700 species of conifers are classified into 50 genera, including *Pinus* (pines), *Juniperus* (junipers), *Abies* (firs), and *Picea* (spruces). The largest plants, the giant sequoias (*Sequoiadendron gigantea*), and the tallest plants, the redwoods (*Sequoia sempervirens*), are conifers. The leaves of conifers are either long needles or short scales, both of which are covered with a thick cuticle to retard water vapor loss. Since many conifers retain green leaves throughout the year, they are frequently called evergreens.

Most conifers produce both male and female cones. During the pine life cycle (Figure 26-15), cells in the male

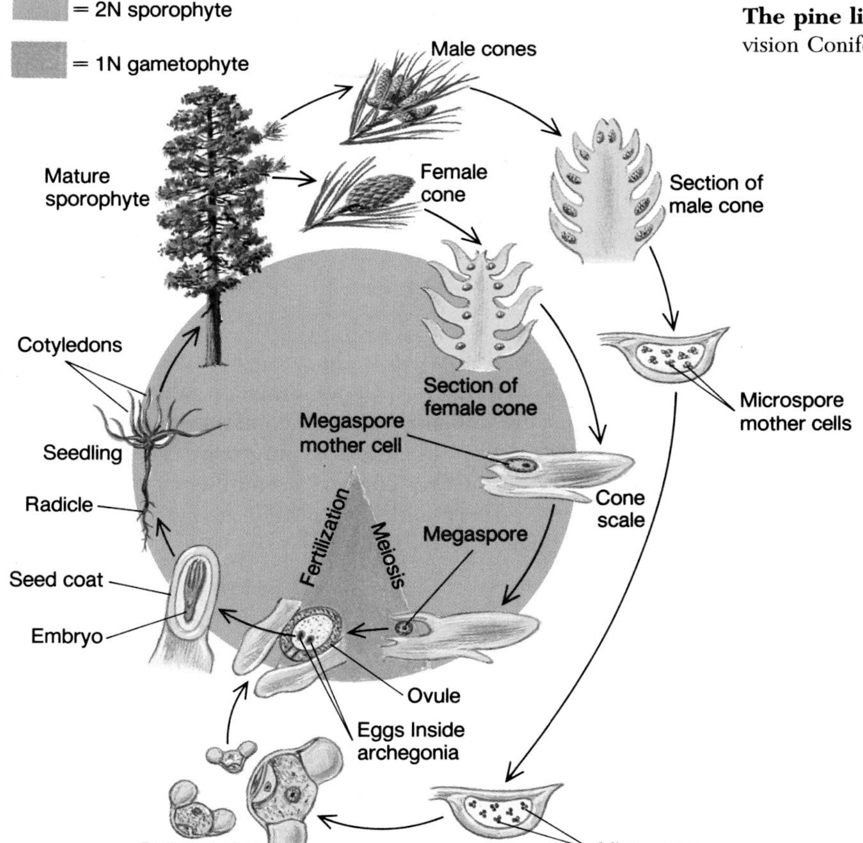

FIGURE 26-15
The pine life cycle represents the typical life cycle of the division Coniferophyta.

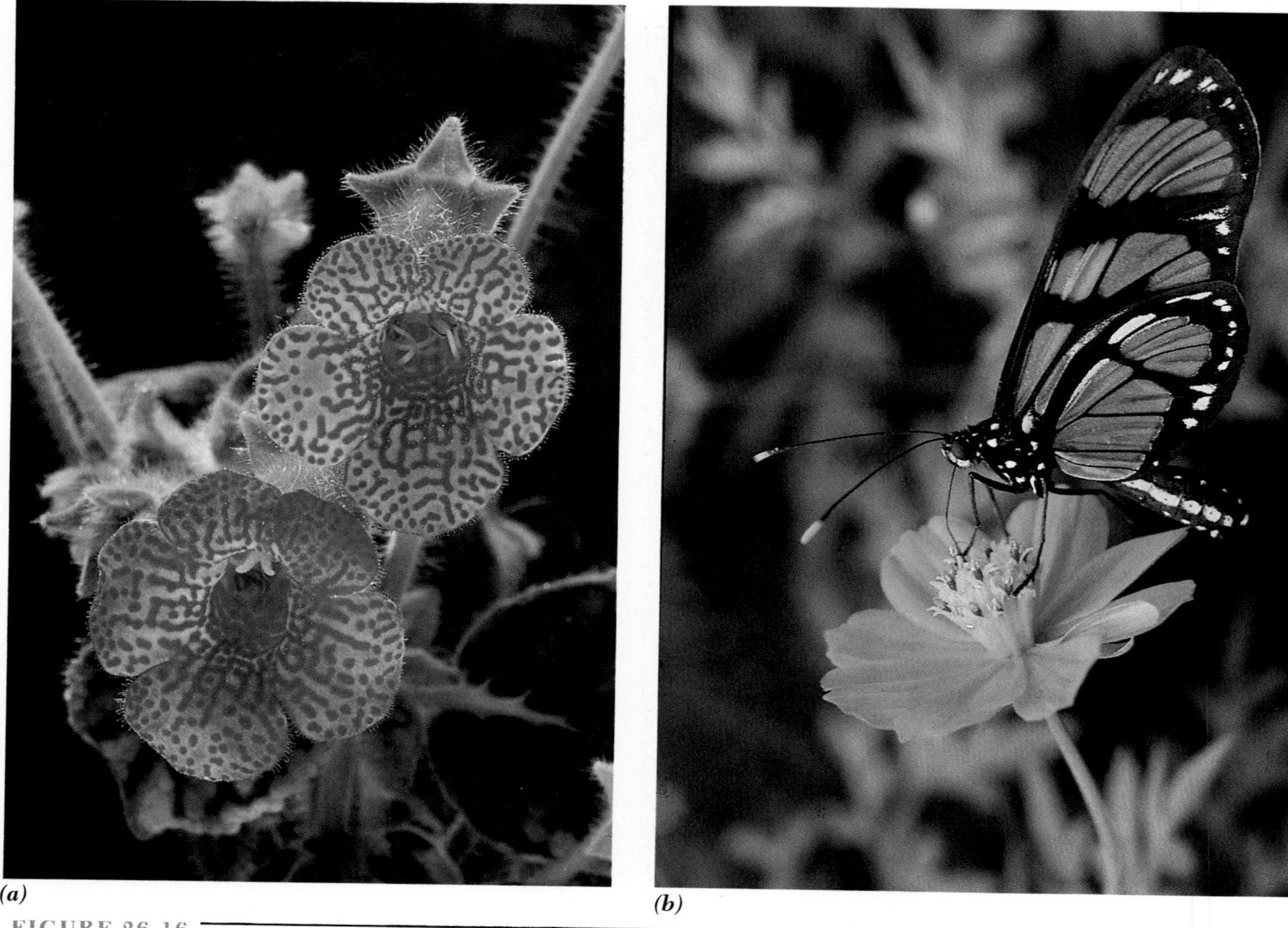

(a) (b)

FIGURE 26-16

A sampler of success. Flowering plants (division Anthophyta) are now the most diverse and abundant types of plants on earth. The Anthophyta are subdivided into dicots (***a*** and ***b***) and monocots (***c*** and ***d***).

cones divide by meiosis to produce spores. The spores divide mitotically to produce the male gametophyte, a pollen grain that contains sperm. The pollen grains of pines develop balloon-like "wings" that help them remain airborne during dispersal. To improve the chances of fertilization, a single pine tree releases billions of pollen grains, sometimes producing clouds of yellow pollen.

Within the female cones, cells divide meiotically, producing four spores, three of which degenerate. The remaining spore divides repeatedly by mitosis to produce a multicellular female gametophyte that contains two archegonia, each with a single egg.

The scales of female cones secrete a sticky substance that captures windborne pollen. Pollen grains then grow a pollen tube, delivering a sperm to the egg. Pine seeds are "winged" to aid dispersal by the wind.

The Coniferophyta is the most advanced division of gymnosperms. Even with an array of adaptations well suited for life on land, however, conifers are not the most diverse group of plants. The **angiosperms** (flowering plants) far outnumber conifers in number of species and, except in northern and high-altitude habitats, in number of individuals.

THE ANGIOSPERMS

All plants that produce flowers belong to one division, the **Anthophyta,** or angiosperms. The Anthophyta is the most recently evolved plant division, appearing about 100 million years ago. The success of flowering plants was swift. Equipped with a more efficient vascular system and two unique characteristics—flowers and fruits—angiosperms

(c)

(d)

rapidly diversified; thousands of angiosperms appear in the fossil record by the end of the Mesozoic Era. Today, the Anthophyta is the most diverse of all plant divisions; nearly 75 percent of all plant species are flowering plants.

Members of the Anthophyta are called angiosperms (meaning "enclosed seed") because their seeds are surrounded by fruit tissues that form from the mature ovary of the flower (Chapter 13). The more than 350,000 species of angiosperms that have been described to date range in size from tiny, stemless duckweeds to towering eucalyptus trees.

The Anthophyta is divided into the monocotyledons, or monocots, and the dicotyledons, or dicots (page 209). There are more than twice as many dicot as monocot species—250,000 versus 100,000 (Figure 26-16). Dicots include flowering trees and shrubs with wood secondary growth as well as many kinds of herbs. Most fruits and vegetables that are available in supermarkets are dicots; these include tomatoes, potatoes, various melons, stone fruits (peaches, cherries, apricots), grapes, zucchini, and broccoli. Many important agricultural staples are monocot grasses, however, such as corn, wheat, rye, oats, barley, and rice. Monocots also include orchids, palms, lilies, and ornamental grasses. (The sexual life cycle of flowering plants was discussed fully in Chapter 13.)

EVOLUTION AND ADAPTATION: TYING IT TOGETHER

The first land plants grew as ground-hugging mats, remaining close to continuous water sources. The evolution of vascular tissues and adaptations to control water loss enabled early plants to disperse into drier habitats and to throw off their height restrictions. Land plants quickly diversified, becoming the predominant form of terrestrial life. The colonization of the land by animals became possible only after plants became established on land, thereby creating the terrestrial habitats and food sources needed by animals.

In addition to adaptations that prevent water loss, seeds that survive fire, and means of transferring gametes and dispersing embryos in nonaquatic habitats, flowering plants coevolved with animals that serve as pollinators, increasing the likelihood of cross-pollination, promoting diversity and accelerating the evolution of advantageous traits and new species.

SYNOPSIS

Terrestrial plants evolved from an aquatic filamentous green algae. Green algae colonized dry land about 400 million years ago. The evolution of adaptations to control water loss and to improve internal transport of water and nutrients, and the evolution of protective structures for gametes and embryos enabled plants eventually to colonize virtually every terrestrial habitat on earth.

The roughly 400,000 species of living plants include three divisions of algae, one division of nonvascular bryophytes, and nine divisions of vascular tracheophytes. The latter two groups constitute the 10 divisions of terrestrial plants.

During sexual reproduction, all plants alternate between two multicellular generations: a diploid sporophyte, and a haploid gametophyte. In sporophytes, meiosis produces spores, which grow into sperm and egg-producing gametophytes. Over the course of evolution, plants have changed from having a prominent gametophyte to a prominent sporophyte, the diploid condition being genetically more advantageous for terrestrial life.

Bryophytes and lower vascular plants still retain characteristics of semiaquatic ancestors. Specifically, bryophytes and lower vascular plants still rely on water for dispersing sperm to female sex organs that contain an egg.

Between 150 million and 350 million years ago, lower (seedless) vascular plants were the predominant plants on earth. Extensive mountain building and shifting continents during the past 150 million years created drier and more variable environmental conditions, causing most seedless vascular plants to become extinct. With more efficient vascular tissues, protected gametes (eggs in ovules, sperm in pollen grains), and enclosed embryos, the early seed plants (the gymnosperms) thrived under these new conditions.

The most recently evolved group of plants—the angiosperms—is now the most diverse. All angiosperms produce flowers and develop a fruit from a ripened ovary. Flowers and fruits provide protection for gametes and embryos and enable pollen and seeds to be dispersed economically by insects and other animals.

Review Questions

1. Complete the classification scheme for the Plant kingdom using the following terms: liverworts, hornworts, Sphenophyta, Cycadophyta, higher vascular plants, bryophytes, tracheophytes, Lycophyta, Ginkgophyta, gymnosperms, angiosperms, Psilophyta, Gnetophyta, mosses, Plant kingdom, lower vascular plants, Pterophyta, Coniferophyta.

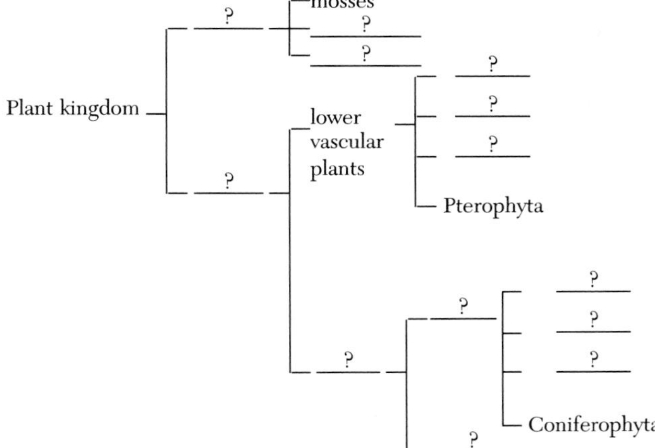

2. Explain why there are fewer plant species than animal species.
3. Match the structures with the appropriate plant group.

   ——1. gymnosperms a. prothallus
   ——2. angiosperms b. strobilus
   ——3. ferns c. rhizoids
   ——4. mosses d. fruit and flowers
   ——5. club mosses e. naked seeds

4. List the divisions of plants that produce the following: pollen, ovules, flowers, antheridia, archegonia.
5. How has the need for liquid water to promote fertilization affected the evolution of plants?

Critical Thinking Questions

1. Since botanists rely on differences in morphology to distinguish plant species, how different do you think two plants should be in order to be considered separate species? For each of the following plant structures, describe the magnitude of differences you believe would justify separate species: leaf characteristics, flower shape, flower size, fruit size, and stem and root characteristics.
2. Discuss the role of water in the evolution of modern plants. How are bryophytes analogous to amphibians in the animal kingdom? How are gymnosperms analogous to reptiles?
3. In some forests of the northern hemisphere, mosses tend to grow in greater profusion on the north-facing surface of the bark. Why do you suppose these mosses do not grow as well on the southernmost faces of trees in these forests?
4. In conifer forests, some gymnosperms rely on periodic episodes of fire to give them a competitive edge over flowering trees. Fire on the forest floor burns the flowering trees, whose lowest branches are readily ignited. The bottom branches of maturing conifers are very high, however—above the level that most fires can reach—and the trunks of these conifers are fairly fire-resistant. Without fire, many conifer forests would soon become forests of flowering trees. What features of these angiosperms would make them so much more successful than these gymnosperms, if not for fire?
5. The perfect plant for a desert environment would possess many of the adaptations discussed in this chapter. List those adaptations plus any additional ones you feel would be beneficial to survival and reproduction in periodically dry, very hot environments. Which adaptations are better suited to very wet habitats, such as a rain forest?

CHAPTER

27

Animal Diversity

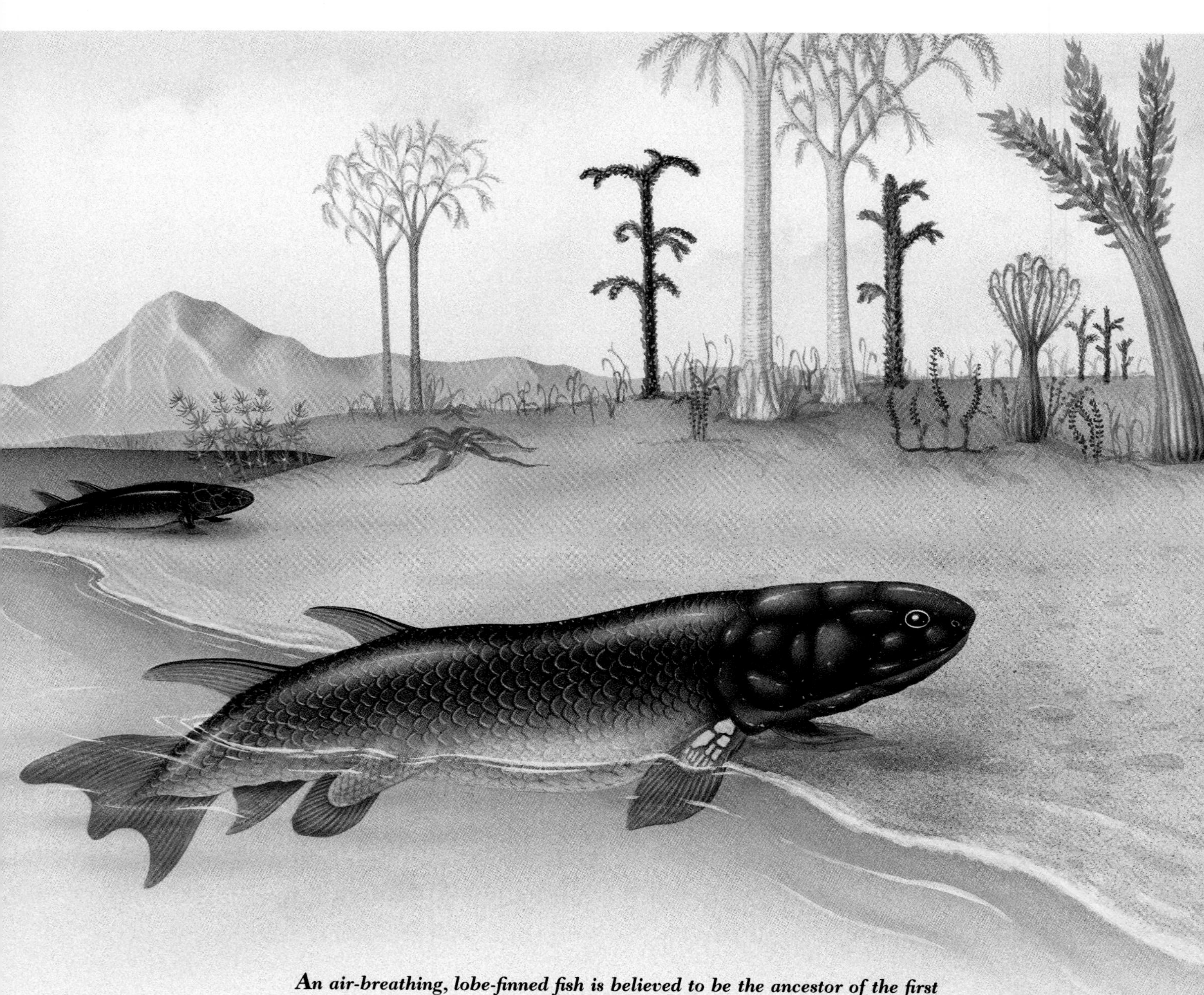

An air-breathing, lobe-finned fish is believed to be the ancestor of the first four-legged amphibian.

STEPS TO DISCOVERY
The World's Most Famous Fish

Ever since the acceptance of Darwin's theory of evolution, biologists have speculated on the origin of four-legged, terrestrial vertebrates, from the earliest amphibians to the more recently evolved reptiles, birds, and mammals. In 1892, Edward Cope of the Academy of Natural Sciences in Philadelphia proposed that the first amphibians evolved during the Devonian Period (approximately 370 million years ago) from a particular type of ancient fish called a rhipidistian. Rhipidistians lived at a time when the land was covered by shallow, swampy bodies of water. Warm, swampy waters often become stagnant and oxygen-deficient. The rhipidistians were adapted to these conditions; they had lungs that could inhale oxygen from the air under conditions of stress. They also had fins situated at the ends of fleshy lobes that contained an internal bony skeleton. The front fins were probably used to support the fish's body as it pushed its head out of the water for a breath of air or dragged its body from one shrinking pond to another.

A fish with lungs and bony, lobed fins is *preadapted* to a terrestrial life. In other words, even though these characteristics were adaptations for life in shallow, freshwater ponds, they could also be put to use by animals that were becoming more and more terrestrial. The lungs are preadapted to the respiratory needs of a terrestrial animal, and the bony lobes are preadapted as terrestrial locomotor structures.

The first amphibian known from the fossil record was primarily aquatic and retained numerous fishlike characteristics, including a long, fishlike tail, complete with a well-developed tail fin. Its body was partially covered with bony scales typical of fishes living at that time. This animal possessed one obvious feature that is not present in any fish. It had four legs and four feet! Cope noticed that the limb bones of this early amphibian were remarkably similar to the bones present within the fleshy lobe of the fins of a rhipidistian. Cope proposed that rhipidistians were the ancestors of amphibians and, consequently, of all four-legged vertebrates.

The rhipidistians were classified in the subclass Crossopterygii, along with another group of fossil fishes, the coelacanths. These two groups of ancient fishes shared a number of features, including the lobed, bony fins. But while the rhipidistians disappeared hundreds of millions of years ago, their cousins, the coelacanths, straggled through the Mesozoic Era until finally disappearing from the fossil record about 70 million years ago, about the same time as the dinosaurs vanished. It would be nearly 50 years before events occurred that would change this viewpoint.

In December 1938, Marjorie Courtenay-Latimer received a telephone message that a trawler had come into dock with a pile of fish for her to examine. Courtenay-Latimer was the curator of a small museum in East London, South Africa, and had asked the local fisherman for their help in preparing exhibits of the fauna of the area. She took a taxi to the wharf and began rummaging through the fish, most of which were sharks. In the pile was a large, bright blue fish with a symmetric tail and fins situated at the ends of fleshy lobes. Despite her years of experience, Courtenay-Latimer had never seen anything like this creature. She wrapped the unusually smelly fish in a bag and, after much discussion, convinced the taxi driver to allow it into his car. When she returned to the museum, she transported the fish in a hand car to the museum's taxidermist. She then drew a rough sketch of the fish, which she mailed with a note to J. L. B. Smith, a friend and zoologist at a South African university, asking for help in identifying the specimen. Smith received the note 11 days later. He was greatly perplexed by the sketch. With its symmetric tail, thick, armored scales, and limblike fins, the fish looked exactly like a coelacanth. This was impossible, however, since coelacanths had disappeared 70 million years ago.

It wasn't until the next month that Smith was able to travel to East London to inspect the remains of the fish; there was no doubt that the animal was indeed a coelacanth. Even the curious intracranial joint characteristic of the skull of rhipidistians and the first amphibians was present in this "living fossil." Smith named the specimen *Latimeria chalumnae* after its discoverer.

As the years passed, and no other coelacanths were found, Smith distributed thousands of leaflets showing a picture of the fish and offering a large reward for its capture. One day, in 1952, Smith and his wife were talking to the captain of a schooner about the fish; 10 days later, the captain sent the Smiths a cable that he had one on his boat. The fish had been caught by a native fisherman, who was attempting to sell it at the local market when it was spotted by a schoolmaster who had seen a copy of the leaflet. In a frantic state, Smith called the prime minister of South Africa, who dispatched a special plane to pick Smith up and fly him to the remote site on a French Comoran Island where the fish was being kept. The subsequent removal of the fish from French territory caused an international uproar.

In many ways, *Latimeria* is highly specialized for life in the ocean depths and lacks certain features, including lungs, that had adapted ancient coelacanths to shallow ponds. In other ways, *Latimeria* retains many of the aspects of its early ancestors. For example, the structure and movement of the fins is similar to that which would be predicted of a tetrapod (four-legged) ancestor. More recent insights on coelacanths have been made by Hans Fricke of the Max Planck Institute in Germany. From a small submersible, Fricke has filmed coelacanths as they slowly swim and feed in their natural habitat. These studies have also revealed the rarity of the species and have raised some concern that the activities of the scientists studying them might lead to the extinction of these invaluable organisms. ■

As individuals, animals are greatly outnumbered by plants, bacteria, and even fungi. Yet there are more *kinds* of animals than any other type of organism. Nearly 75 percent of all known species (a total of about 1.3 million) are included in the Animal kingdom, and the tally is far from complete. Many biologists believe that millions of animal species remain unnamed and unclassified, most of them insects.

Animals range in size and complexity (Figure 27-1), from the microscopic *Trichoplax adhaerens,* thought to be the most primitive animal yet discovered, to the giant blue whale (*Balaenoptera musculus*) that reaches lengths of nearly 40 meters (130 feet) and weighs more than 160 tons. Between these extremes is an immense assortment of animals that range in body form from worms to giraffes and jellyfish to spiders.

Despite their astounding diversity, members of the Animal kingdom typically share several basic characteristics:

- *Animals are multicellular.* All animals are composed of a large number of cells, which are specialized for different functions.
- *Animals are heterotrophic.* Animals are not capable of photosynthesis and must rely on other organisms to provide raw, organic materials.
- *Animals are motile.* Animals are capable of locomotion at some stage of their life. While many aquatic animals remain in one place as adults (they are said to be *sessile*), such animals invariably develop through a motile larval stage.
- *Animals engage in sexual reproduction.* With rare exception, animals consist of diploid cells that contain two sets of homologous chromosomes. During meiosis, which occurs in specialized reproductive tissues, haploid gametes are formed. An egg and a sperm (al-

(a)

(b)

FIGURE 27-1

From one extreme to the other. ***(a)*** *Trichoplax adhaerens* is the least complex animal known. This organism was originally discovered in 1883 and was classified as the larval stage of an unknown animal. In the early 1960s, the creature was rediscovered as a unique species and is now placed in a separate phylum, Placozoa. *T. adhaerens* consists of several types of cells that form a flattened, amorphous mass. ***(b)*** A blue whale, the largest animal that has ever lived, is made up of complex tissues, organs, and organ systems. Blue whales are approximately six times larger than the biggest dinosaur that ever lived. Blue whales can achieve such a massive size because they are supported by the buoyant medium in which they live.

most always from two different individuals) become united during fertilization to form a zygote that undergoes embryonic development, forming a diploid offspring.

▼ ▼ ▼

THE ELEMENTS OF A BODY PLAN

When comparing various types of motor vehicles, such as motorcycles, sports cars, trucks, vans, motor homes, and so forth, one of the first characteristics you may consider is body plan—the vehicle's basic organization, including the nature and arrangement of its major parts. Animals, too, have *body plans* that describe the layout of their body parts. An animal's structural features can be explained in terms of

1. its evolutionary ancestry (organisms invariably retain characteristics of their ancestors whose genes they have inherited); and
2. the environment in which it lives (organisms survive because they possess adaptations to their environment; as a result, related animals living in different environments will possess different adaptations).

Three of the most important properties that distinguish the body plans of animals are symmetry, body cavities, and segmentation.

BODY SYMMETRY: ARRANGEMENT OF THE BODY PARTS

An object has *symmetry* if it can be bisected into two, mirrored halves. An animal's symmetry often reveals information about its mode of existence. Animals exhibiting **radial symmetry** usually have a cylindrical body composed of similar parts that are arranged in a circular fashion around a single central axis (Figure 27-2*a*). Radial symmetry is characteristic of groups of animals that are either sessile (sedentary), such as the sea anemone, or slow moving, such as a jellyfish or sea urchin. Radial symmetry is adaptive for these animals since sensory information and food matter usually comes toward them from all directions.

More motile animals typically exhibit **bilateral symmetry,** whereby only one plane can bisect the animal into mirrored, right and left halves (Figure 27-2*b*). Bilateral symmetry is found among animals that actively search for food and shelter, such as insects and vertebrates. The evolution of bilateral symmetry was accompanied by the concentration of sense organs and nervous tissue at the leading end of the body to form a "head," an evolutionary phenomenon known as **cephalization** (*cephalic* = head). Cephalization allows a motile animal to gather information about the environment into which it is heading, protecting it from proceeding "blindly" into danger.

INTERNAL CAVITIES: THE SPACES WITHIN

The animal depicted in Figure 27-1*a* is essentially a flattened mass of cells; it possesses no internal cavities, other than the simple spaces between the cells. Nearly every other type of animal on earth has at least one internal chamber in which food matter is collected. This chamber, or **digestive tract,** may be a blind sac, as in a sea anemone (see Figure 27-6), or a complete internal tube, the *gut,* with openings at both ends, as in a snail (see Figure 27-10) or a human.

In addition to the digestive tract, most animals have at least one other major body cavity that develops between the digestive tract and the outer body wall. Like symmetry, differences in the nature and location of the internal cavities distinguish major groups of animals from one another.

(a)

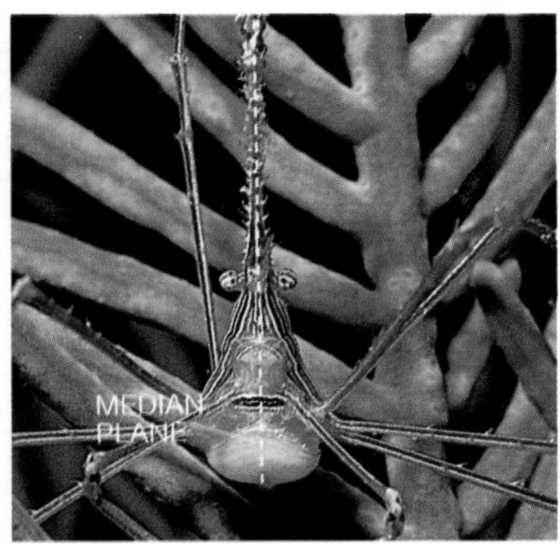

(b)

FIGURE 27-2
Symmetry. ***(a)*** The sea anemone exhibits radial symmetry. Its body can be bisected by a number of planes (dotted lines), as long as the planes pass through a central axis (the round, central dot). The body parts of the sea anemone are arranged in a circle around this central axis. ***(b)*** A crab exhibits bilateral symmetry. Its body can be bisected by only a single plane, shown by the dotted line.

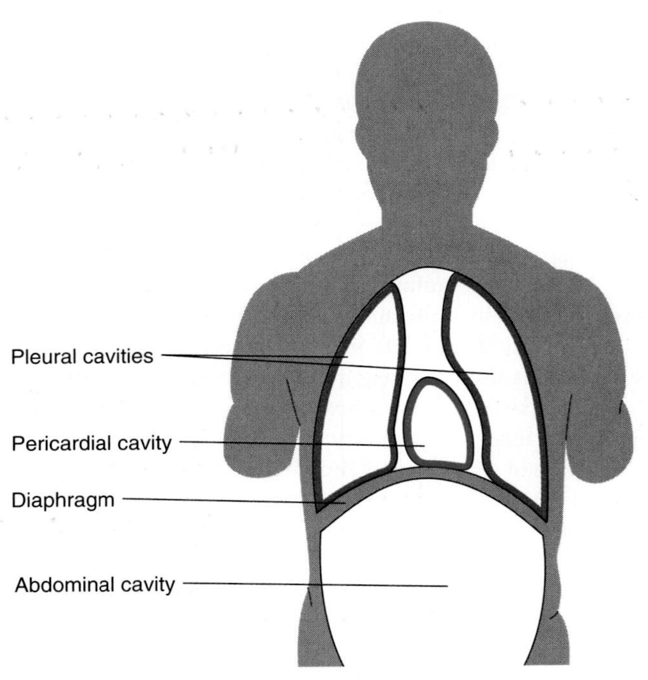

FIGURE 27-3
The coelom is a space that is lined by cells derived from the embryonic mesoderm and situated between the body wall and the digestive tract. In humans, the coelom is partitioned into several distinct cavities: the *abdominal cavity* (which surrounds the visceral organs), the *pleural cavities* (which surround the lungs), and the *pericardial cavity* (which surrounds the heart).

In humans, the major body cavity (or **coelom**) is partitioned into several smaller cavities that house the lungs, heart, and digestive organs (Figure 27-3).

SEGMENTATION: DIVIDING THE BODY INTO REPEATING SEGMENTS

A quick examination of the internal and external anatomy of an earthworm or insect reveals that the body is constructed of repeating subunits, or **segments,** that contain similar organs. The presence or absence of segmentation is one of the most profound characteristics of a body plan.

Three of the largest phyla, the annelids, arthropods, and chordates (which includes vertebrates), consist of segmented animals. As we will discuss later in the chapter, segmentation is thought to have arisen as an adaptation for particular types of locomotion because it allows different parts of the body to engage in different types of activities. Segmentation in the human body is most apparent in your backbone, which consists of a series of bony vertebrae that form from a linear array of segments that develop in the embryo.

DIVIDING ANIMALS INTO PHYLA

The members of the Animal kingdom are organized by phylum. Even though the members of the same phylum may seem highly diverse—such as a snail and an octopus, or a fish and a human—they are all descended from a common ancestor and constructed as a variation on a common body plan. Our exploration of the Animal kingdom includes nine of the approximately 35 animal phyla. Their major characteristics are summarized in Table 27-1. We chose these phyla for two reasons: They contain the greatest numbers of species, and they illustrate the principal evolutionary innovations that occurred during animal evolution.

INVERTEBRATES

For many purposes, particularly college biology courses, animals are divided into two great blocks: **invertebrates,** or animals that lack a backbone, and **vertebrates,** animals that possess a backbone. Considering that there are only about 40,000 vertebrate species and more than a million invertebrate species, this hardly seems like an equitable distribution, yet it is one that reflects our interests. Humans are vertebrates, as are our pets and domestic animals. All of the 35 or so animal phyla contain invertebrates; only one phylum contains vertebrates. We will begin our discussion with the invertebrates.

PHYLUM PORIFERA: SPONGES

There are approximately 10,000 species of sponges. Most sponges are marine animals that live in shallow, unpolluted coastal waters, but there are also numerous deep-sea forms and two families that live in freshwater streams and ponds. Adult sponges are sessile and remain attached to rocks,

TABLE 27-1
SUMMARY OF CHARACTERISTICS FOR ANIMAL PHYLA

Phylum	*Cellular Organization*	*Coelom*	*Circulatory System*	*Nervous System*	*Reproductive System*	*Distinguishing Characteristics*
Porifera (sponges) 10,000 spp.	No tissues	None	None (diffusion)	None	Male, female, or hermaphroditic	Aquatic filter feeders, porocytes, choanocytes
Cnidaria (hydras, corals, jellyfish) 9,000 spp.	Tissues	None	None (diffusion)	Nerve net, sensors over body surface	Male and female medusae, asexual budding	Stinging cnidocytes on tentacles
Platyhelminthes (flatworms) 20,000 spp.	Organs and organ systems	None	None (diffusion)	Ladder-type paired nerve cords, few ganglia	Hermaphroditic	Flatworms with definite head and tail ends
Nematoda (roundworms) 10,000 spp.	Organ systems	Pseudocoelom	None (diffusion)	Simple "brain" dorsal and ventral nerve cords	Male or female, internal fertilization	Tapered body, cuticle covering, tube-within-a-tube digestion
Mollusca (squids, snails, clams) 100,000 spp.	Organ systems	Coelom	Open, 1 heart	Cerebral ganglia, nerve cords	Male, female, or hermaphroditic	Shells in most, free-swimming larvae in some
Annelida (segmented worms) 9,000 spp.	Organ systems	Coelom	Closed, pumping vessels	Simple brain, paired ventral nerve cords	Male, female, or hermaphroditic internal or external fertilization	Aquatic and terrestrial Repeating segments, but continuous digestive, nervous, and circulatory systems
Arthropoda (spiders, crabs, insects) 1,000,000 spp.	Organ systems	Coelom	Open, 1 heart	Simple brain, ventral nerve cord	Male or female, mostly internal fertilization	Jointed appendages, chitinous exoskeleton, specialized segments
Echinodermata (sea urchins, sea stars, crinoids) 5,300 spp.	Organ systems	Coelom	Open, no heart	Nerve ring, radial nerves to each arm	Male or female external fertilization	Radially symmetrical adult; bilaterally symmetrical, free-swimming larvae; spines, water vascular system
Chordata (tunicates, lancelets, vertebrates) 44,300 spp.	Organ systems	Coelom	Closed, 1 or no heart	Anterior brain, dorsal hollow nerve cord	Male, female, or hermaphroditic, internal or external fertilization	Notochord, gill slits, tail, dorsal nerve cord

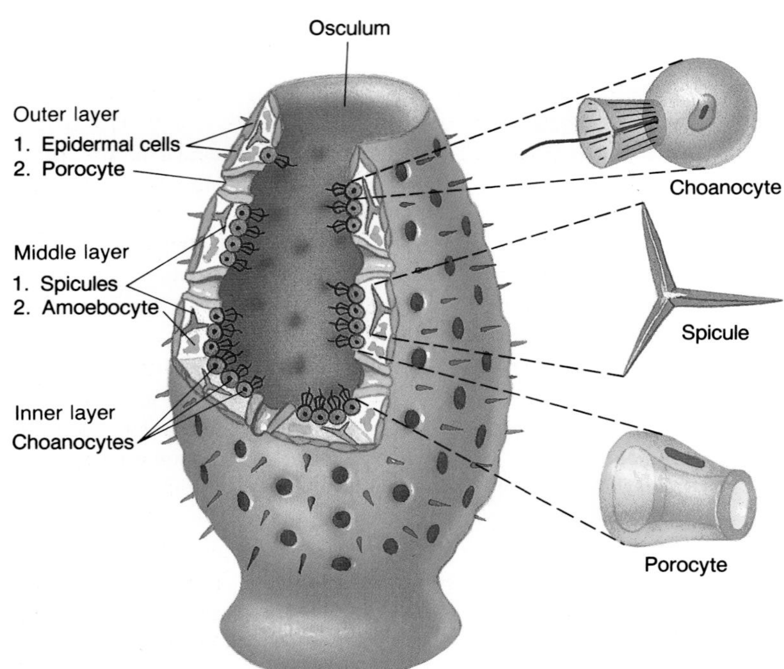

FIGURE 27-4
Sponge anatomy. The simpler, vaselike sponges have a single, central spongocoel chamber surrounded by a body wall that contains three layers.

shells, and other submerged surfaces. All sponges are **filter feeders**; they feed on microscopic particles that are swept into their bodies from the external medium.

The simpler sponges have a vase-shaped body (Figure 27-4) containing a central cavity that opens to the outside through a large aperture located on the upper surface. A thin **body wall,** composed of three distinct layers, surrounds the central cavity (*spongocoel*). The outer layer of the body wall is made up of a sheet of flattened epidermal cells, containing scattered *porocytes*. Porocytes are doughnut-shaped cells, each pierced by a microscopic pore. Water enters the sponge through these pores and exits via the single, large osculum.

The inner layer of the body wall consists of a loosely organized bed of unique cells, called **choanocytes,** each of which has a rounded basal end and an extended collar that surrounds a single undulating flagellum. The combined beating of the flagella of all the choanocytes draws water into the sponge. The fine-mesh collars of the choanocytes then trap the suspended food particles, which are engulfed by phagocytosis and digested in food vacuoles located inside the sponge's cells.

The basal ends of the choanocytes are embedded in a gelatinous middle layer, through which motile cells (*amoebocytes*) wander. The body wall also contains large numbers of pointed **spicules** that function as both a skeleton that supports the animal's soft mass and a protective device that makes the animal less than appetizing to a potential predator.

Most sponges have a more complex internal organization than that depicted in Figure 27-4. The thick body wall of the more complex sponges is riddled with fine channels and hundreds of thousands of microscopic chambers that house the choanocytes. This type of internal architecture increases the surface area available for choanocytes and improves the sponge's efficiency in capturing food, exchanging respiratory gases, and eliminating wastes.

Of all the major groups of animals, sponges have the least complex structure: Their bodies often lack symmetry; they lack sensory cells and nerve cells; the layers of the body wall do not constitute true tissues since the component cells do not function in an integrated manner (each cell carries out its own activities more-or-less independently); and they lack a mouth—the largest opening into the animal serves as an exit for fluid. In addition, sponges have no digestive tract; the internal chambers are simply part of a pathway allowing the environment to flow through the animal, bringing in food and oxygen and carrying out waste products and gametes.

As a group, sponges are diverse and abundant, but they represent an evolutionary dead end; that is, sponges have not given rise to any other form of animal life. Instead, they have remained esentially the same for hundreds of millions of years.

PHYLUM CNIDARIA: HYDRAS, JELLYFISH, SEA ANEMONES, AND CORALS

If you have ever walked along a rocky coastline at low tide, you may have noticed populations of bright, flower-like sea anemones wedged into the crevices in the rocks. Sea anemones are members of the phylum **Cnidaria,** as are jellyfish and the microscopic animals that are responsible for building the great coral reefs, such as the Great Barrier Reef off of Australia's eastern coast. The phylum contains about 9,000 species, including those of Figure 27-5.

Cnidarians are built on a radially symmetric body plan composed of a three-layered body wall surrounding a blind, saclike chamber, called the **gastrovascular cavity.** The simplest cnidarians are the hydras (Figure 27-5*a*). The body wall of a cnidarian is more complex than that of a sponge; it consists of layers of cells whose activities are coordinated to form a tissue. Thus, cnidarians can be considered as having evolved to the *tissue level of organization,* but they lack true organs, such as a heart or a kidney. All of their cells are close enough to the outer medium to receive oxygen and dispel waste products by simple diffusion.

Cnidarians are carnivorous. They stun or kill their prey by the use of poisonous "projectiles" that are fired from cells, called **cnidocytes,** which are concentrated in the tentacles that surround the cnidarian's mouth. Cnidarian toxins are very potent. The Portuguese man o' war produces a venom as potent as that of a cobra; it is capable of raising large welts simply by contact with the skin. The sting of the sea wasp, which lives in the waters of Australia, is powerful enough to kill a human.

Once trapped, the prey is deposited by the tentacles into the gastrovascular cavity, where it is broken down into microscopic particles. The appearance of an internal chamber for storing and disassembling food was an important evolutionary innovation that allowed cnidarians to feed on much larger prey than that which could be ingested by animals that relied strictly on phagocytosis, such as sponges.

As a group, cnidarians have two distinct body forms—the polyp and medusa. **Polyps** are cylindrical cnidarians whose mouth and tentacles are situated at the upward end, as in a hydra, sea anemone, or coral (Figure 27-6*a*). **Medusas** are umbrella-shaped cnidarians whose mouth and tentacles are situated on the lower surface, facing down-

(b)

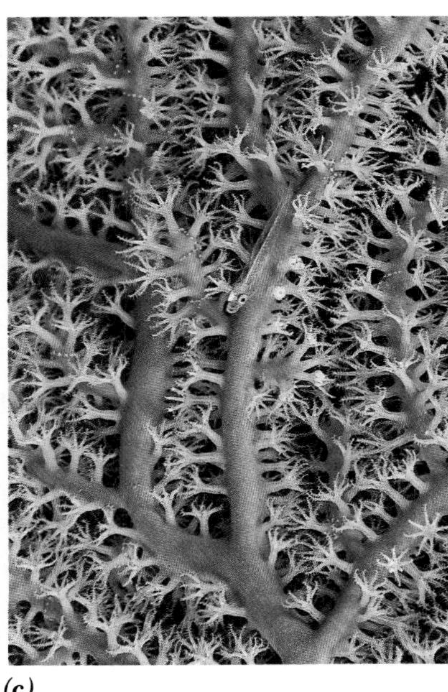

(c)

FIGURE 27-5

Three types of cnidarians. ***(a)*** Hydras are sessile, freshwater members of the class Hydrozoa. ***(b)*** Most jellyfish are members of the class Scyphozoa. The animal moves as a result of pulsating contractions of the muscular bell situated at the top of the animal. The mouth is located at the lower end of the dangling, "frilly" tissue. ***(c)*** These coral polyps, members of the class Anthozoa, construct an outer skeleton composed of calcium carbonate, into which they can retract when threatened. The mass of a coral reef is made up of successive layers of coral skeletons; the living polyps are restricted to the upper surface.

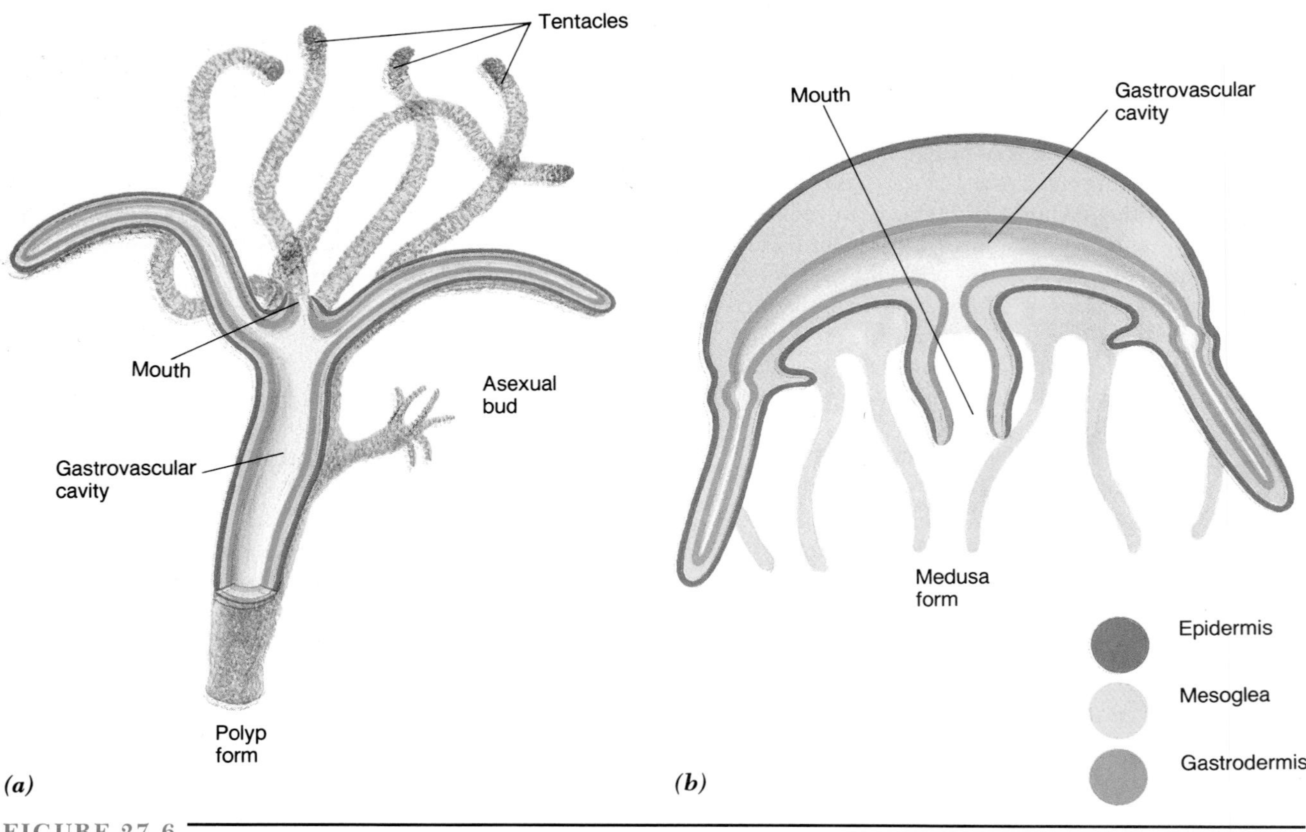

FIGURE 27-6
Two cnidarian body forms: a polyp *(a)* and a medusa *(b)*.

ward, as in a jellyfish (Figure 27-6*b*). These two body forms are adapted for different modes of existence; the polyp is adapted for a sedentary lifestyle, while the medusa is adapted for a free-swimming lifestyle.

PHYLUM PLATYHELMINTHES: FLATWORMS

Among the simplest animals that have a bilaterally symmetric body plan are the flatworms, members of the phylum **Platyhelminthes** (*platy* = flat, *helminthes* = worm). The name of these organisms reflects the flattened body shape (Figure 27-7). The phylum includes about 20,000 species, including *turbellarians,* which are nonparasitic (*free-living*) worms that live primarily under rocks on the bottom of lakes or ocean bed; *trematodes,* or flukes, which are typically leaf-shaped parasites that live as adults inside the bodies of vertebrates; and *cestodes,* or tapeworms, which are highly elongated parasites that live as adults in the intestines of their vertebrate hosts.

Flatworms exhibit numerous "evolutionary advances" compared to the sponges and cnidarians discussed above. These animals possess a simple brain and associated nerve cords which represent the evolutionary beginnings of a central nervous system. The middle layer of the flatworm is well developed and contains layers of muscle tissue and a variety of differentiated organs linked together into organ systems. A turbellarian, for example, has complex osmoregulatory, reproductive, and digestive systems.

Flatworms are described as **acoelomates,** which means that they lack an internal cavity between their digestive tract and their outer body wall (Figure 27-8). Instead, the spaces between the various organs of the flatworm are filled by a mass of cells. How do the cells of this bulky middle layer receive nutrients and oxygen? While more complex animals have a circulatory system to meet these metabolic demands, the flatworm manages as a result of its shape, illustrating the relationship between form and function. Because of the animal's flattened shape, all the cells of the body are close enough to the outside environment to allow the exchange of respiratory gases and wastes by direct diffusion through the body wall.

Trematodes (flukes) and cestodes (tapeworms) are entirely parasitic. The life cycle of trematodes (Figure 27-7*b*)

(a)

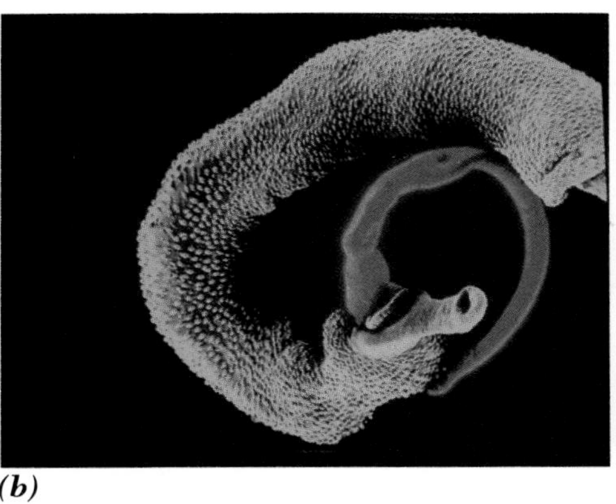

(b)

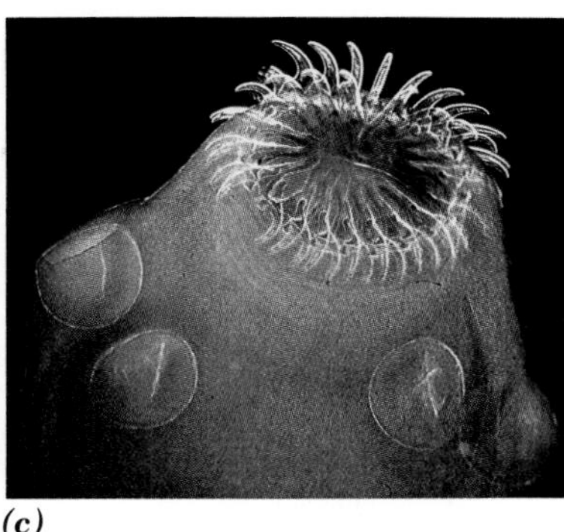

(c)

FIGURE 27-7

Three types of flatworms. ***(a)*** A free-living planarian, member of the class Turbellaria. Note the flattened body shape. ***(b)*** A parasitic blood fluke, *Schistosoma.* As adults, the thinner female lives within a groove in the body wall of the stouter male. The male uses its two anterior suckers to hold onto the wall of a human host's blood vessel. ***(c)*** The anterior end of a tapeworm (class Cestoda) contains a row of hooks and suckers that help the parasite attach itself to the surface of the host's intestinal wall.

involves at least two different host species, one of which is a vertebrate and the other a snail. Among the various flukes that can infect humans, the *schistosomes,* or blood flukes, pose the most serious health hazard, currently infecting as many as 200 million people and causing more fatalities than any other parasites, except for those responsible for malaria. Humans usually become infected with blood flukes by wading in water that contains schistosome larvae.

Humans may become infected with a tapeworm (Figure 27-7c) by eating undercooked fish, pork, or beef that contains a tapeworm larva. Once ingested, the larva develops into an adult within the human intestine. An adult tapeworm may consist of thousands of segments (*proglottids*) and extend several meters in length. Ripe proglottids—those packed with fertilized eggs and developing embryos—typically break off from the posterior end of the worm and pass out of the host with the feces.

PHYLUM NEMATODA: ROUNDWORMS

Members of the phylum **Nematoda,** which includes at least 10,000 species, are the most widespread and abundant animals on earth. We are not aware of the existence of most roundworms because of their microscopic size, but these animals are present in incredible numbers, particularly in the soil. In one count, 90,000 roundworms of several different species were found within a single, rotting apple!

All nematodes have cylindrical, bilaterally symmetric bodies that are basically constructed as a tube within a

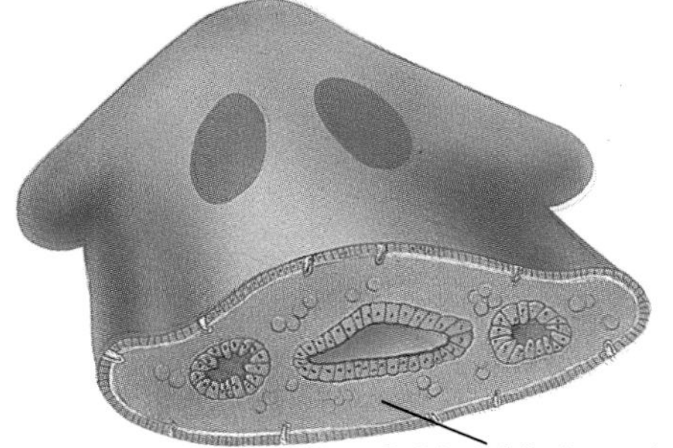

FIGURE 27-8

Flatworms are acoelomate animals. As seen in this cross section, flatworms have a "solid" middle layer (derived from the mesodermal cells of the embryo) that lacks a body cavity.

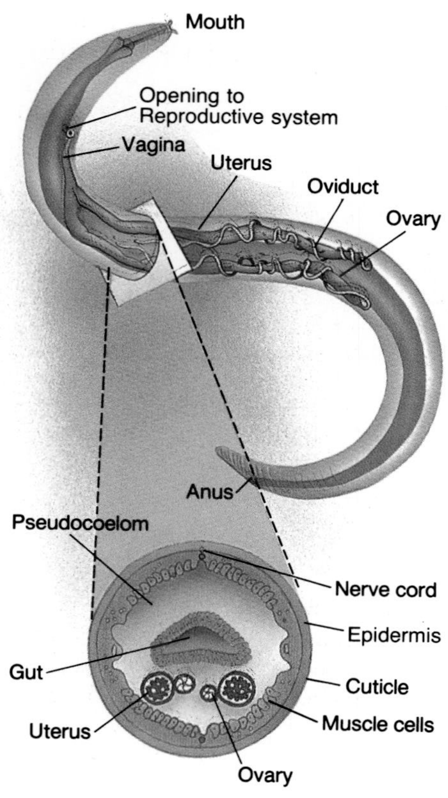

FIGURE 27-9
The anatomy of a female nematode. The transverse section shows the pseudocoelom (an unlined body cavity derived from the blastocoel of the embryo) and the internal organs.

tube (Figure 27-9). The inner tube consists of a relatively simple digestive tract, and the outer tube consists of a relatively complex body wall that is covered by a protective, nonliving (noncellular), outer layer, or *cuticle.* Between the tubes is a fluid-filled body cavity, called a *pseudocoelom* (or "false coelom"), which represents an important evolutionary advance over the body plan of flatworms. This body cavity serves as a hydrostatic skeleton that maintains the rigidity of the body during locomotion (page 301) and as a means of separating the activities of the digestive tract from that of the body wall. Depending on the species, nematodes may live as (1) decomposers, scavenging through moist soil for bits of organic material; (2) predators, feeding on protozoa, earthworms, and one another; or (3) parasites, attacking other animals and many plants. Nematode parasites can quickly devastate an entire agricultural crop or debilitate a large vertebrate.

At least 50 different species of roundworms can inhabit the human body. Hookworms enter the body as larvae, which usually burrow through the skin of the feet. The adult hookworms anchor themselves to a person's intestinal wall, rasp an open sore, and then feed on the host's blood. *Trichinosis* is contracted by eating uncooked pork containing the larvae of the roundworm *Trichinella spiralis.* Once ingested, the larvae mature into adults which reside for a month or so in the human small intestine. During their short life, the adults produce a new generation of tiny young worms that bore through the intestinal wall, enter the circulatory system, and are carried to muscle tissue, where they take up permanent residence. The burrowing of worms into the muscles may be accompanied by excruciating pain.

PHYLUM MOLLUSCA: MOLLUSCS

It is hard to imagine that animals as diverse as a snail, a scallop, and a squid would have much in common, yet they are all classified in the phylum **Mollusca.** With over 100,000 representatives, Mollusca is the second largest phylum in the Animal kingdom. While the basic body plan of molluscs has been "molded" by evolution into a great variety of body forms, most species share certain common features (Figure 27-10):

1. a **head-foot** portion that is concerned primarily with sensory reception, feeding, and locomotion;
2. a dorsal **visceral mass** that includes the major organs of circulation, excretion, respiration, and digestion; and
3. a **mantle,** a fold of tissue that surrounds the visceral mass and secretes the shell. Because the mantle and its covering shell hang over the edge of the visceral mass, they form a space, or **mantle cavity,** that serves as a site of respiratory gas exchange.

Most molluscs have a unique feeding structure known as a **radula,** which is a ribbon-like organ that contains rows of curved teeth made of chitin (Figure 27-10, inset). Its presence in diverse molluscs illustrates how a unique structure may undergo evolutionary modifications as members of a group become adapted to different habitats and modes of existence. For example, many marine snails use the radula to scrape algae from the rocks on which they feed. One of the biggest pests of the oyster industry is *Urosalpinx,* a mollusc that uses its radula to drill through the oyster's shell before feeding on the internal contents. In contrast, an octopus uses its radula to tear the flesh from its prey. The radula of tropical snails of the genus *Conus* is composed of a number of hollow, barbed teeth that are filled with a highly potent neurotoxin capable of killing a human. The snail can detach each tooth from the radula and eject it like a harpoon into the body of its prey.

The most complex molluscs are the cephalopods, particularly octopuses. Octopuses are rapid swimmers and voracious predators, capable of learning complex behaviors. The complexity of a cephalopod's brain and eyes is unsurpassed anywhere in the invertebrate world. Remarkably, the eyes of an octopus are quite similar in organization to those of vertebrates, providing one of the best examples of convergent evolution—the formation of similar structures to meet similar needs by totally independent paths of evolution.

PHYLUM ANNELIDA: SEGMENTED WORMS

The phylum **Annelida** includes about 9,000 species of worms, including earthworms, polychaetes, and leeches (Figure 27-11). The two most important features (Figure 27-12) to appear during the evolution of annelids were

1. the division of the embryonic body into a linear succession of units, or *segments,* having very similar internal structures, and
2. the formation of a true coelom, an internal body cavity lined by a *peritoneum* (a layer of cells derived from the mesoderm).

Segmentation increases flexibility, allowing different parts of the body to bend independently of other parts. Increased flexibility improves locomotion. The coelom aids swimming or burrowing activities by serving as a hydrostatic skeleton.

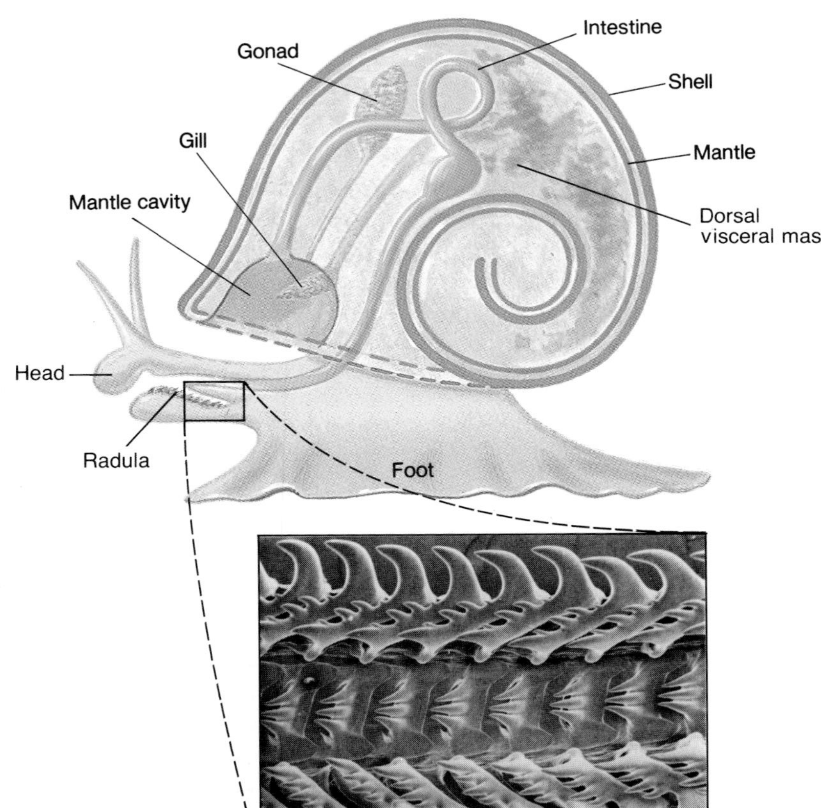

FIGURE 27-10
The major molluscan characteristics are seen in this aquatic snail. In this mollusc, the foot is adapted for creeping over the substrate, and the radula (inset photo) is adapted for scraping algae off the surface of rocks. The visceral mass is coiled and packaged under the coiled shell. The mantle cavity is located near the front of a snail and serves as a chamber into which the digestive, reproductive, and excretory systems empty their products. The incurrent portion of the mantle cavity contains a *gill* for removing oxygen as well as sense organs that monitor the composition of the medium.

(a)

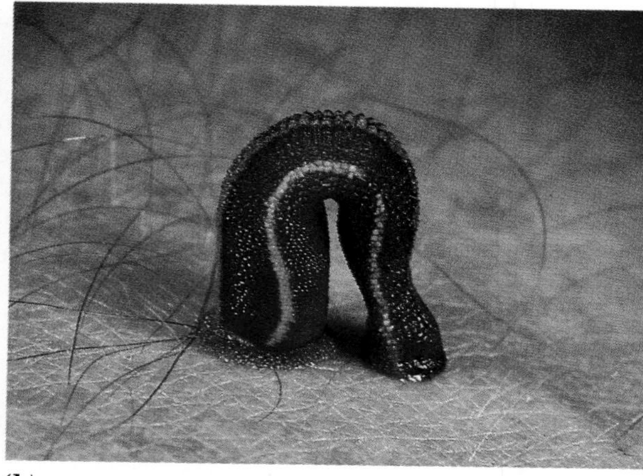

(b)

FIGURE 27-11

Annelids. ***(a)*** A bristle worm, member of the annelid class Polychaeta. ***(b)*** This leech, shown as it feeds on human skin, is a member of the class Hirudinea. The other major annelid class (Oligochaeta) contains the earthworms, illustrated in the next figure.

Another innovation that appeared during annelid evolution was the development of paired appendages. Many of the polychaetes, such as the bristleworm shown in Figure 27-11*a*, have a pair of fleshy lobes, called **parapodia,** that project from each segment. These appendages function as swimming paddles and sites of gas exchange. Many polychaetes are sessile and live in tubes. In these species, the movements of the appendages generate currents that sweep food particles into the animal's mouth. While earthworms lack paired appendages (which would interfere with their burrowing activities), they possess short bristles called **setae,** which act as anchors to prevent backward slippage as the worm pushes its advancing head through the soil.

PHYLUM ARTHROPODA: ARTHROPODS

Biologists estimate that there are a billion billion (1×10^{18}) arthropods living at any given moment, making them close rivals to nematodes as the most abundant type of animal. It is not only their enormous abundance that makes the arthropods an important group, however. Arthropods are also the most diverse organisms on earth. More than 900,000 species are included in the phylum **Arthropoda,** a term that refers to the **jointed appendages** (*arthros* = jointed, *poda* = feet) of these animals. The phylum is divided into a number of major groups (Figure 27-13) including (1) the *crustaceans* (such as shrimp, barnacles, and crabs); (2) the *insects* (such as beetles, flies, and mosquitoes); (3) the *chelicerates* (such as spiders, scorpions, ticks, and mites); and (4) the *myriapods* (such as millipedes and centipedes).

Arthropods exhibit a number of unique features (Figure 27-14). Some of the most important changes during the early stages of arthropod evolution took place at the surface of the animal, where the body comes into contact with the external environment. Arthropods are covered by

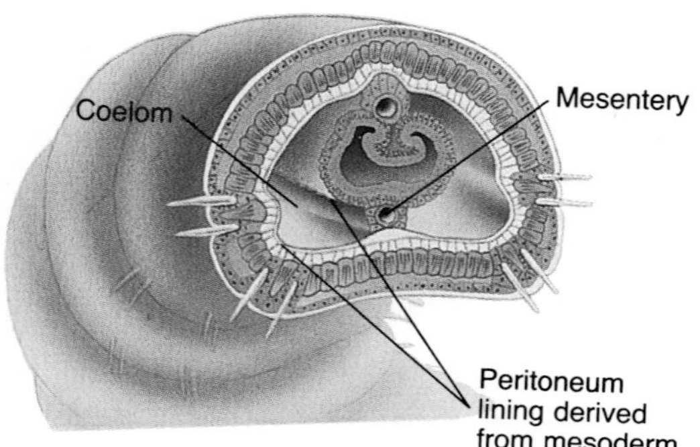

FIGURE 27-12

Like other annelids, earthworms are constructed of repeating segments, each with its own set of internal organs. The cross section shows that the coelom is lined by mesodermally derived cells, forming a *peritoneum,* and sheets of tissue, called *mesenteries,* which support the internal organs suspended in the coelom.

(a) (b) (c) (d)

FIGURE 27-13

Arthropods. ***(a)*** A barnacle (a crustacean) straining the water for food. ***(b)*** A female praying mantis (an insect) standing atop her egg case. The animal is displaying a threatening posture triggered by the intrusion of the camera. ***(c)*** A jumping spider (a chelicerate). ***(d)*** A centipede (a myriapod) crawling over a rock.

a thick, impermeable, hardened, nonliving cuticle (or **exoskeleton**). It is likely that protection from predators was a primary selective pressure that influenced the development of the arthropod exoskeleton.

Being covered by a hardened, nonexpansible, nonliving cuticle is somewhat like living in a suit of armor. As an arthropod grows, it periodically sheds its cuticle and replaces it with a larger one. This process, called **molting,** is under hormonal control. Even though it is made of relatively lightweight materials (consider the weight of an empty crab leg), the weight of an outer body covering greatly increases with increasing body size, which is a major limiting factor in the size of arthropods, particularly the terrestrial species. Terrestrial arthropods are also limited in size by the nature of their respiratory system, which requires that air diffuse down long tubules, called **tracheae,** into the depths of the body (Figure 19-20).

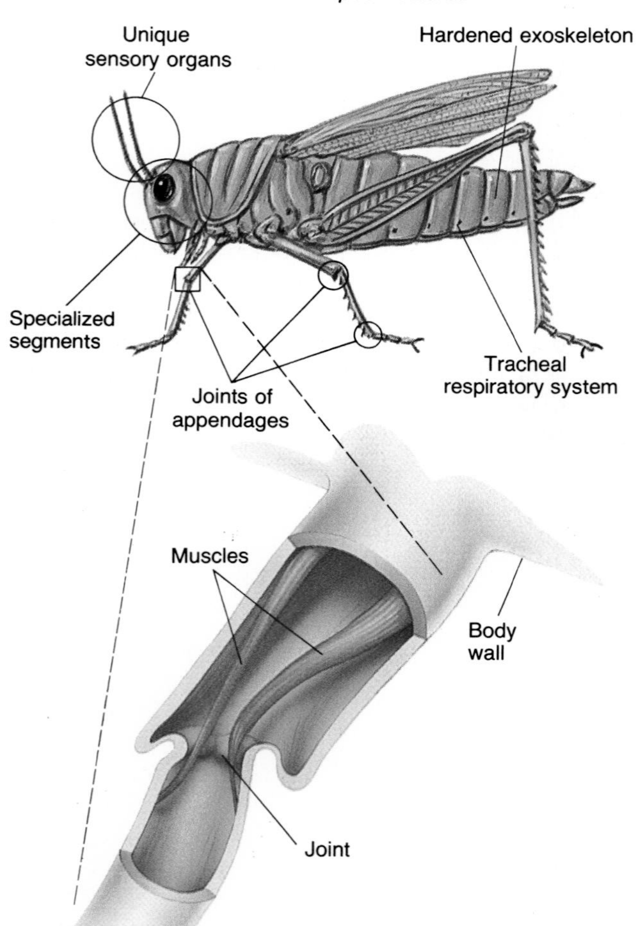

FIGURE 27-14

The arthropod body plan, as illustrated by the grasshopper. A number of unique arthropod features can be seen by examining the animal's external anatomy. The inset shows the attachment of bundles of muscle fibers to the inside surface of the exoskeleton on either side of a flexible joint.

Arthropods are bilaterally symmetrical animals with segmented bodies and jointed appendages. The efficiency and speed of the movement of arthropod limbs is strikingly evident when you watch a cockroach scurry for cover, a spider run across a web, or a tiny flea perform a standing broad jump of more than 25 centimeters. But the appendages of arthropods are much more than simply locomotor structures. Over the course of evolution, the shapes of these structures have been modified to perform many different activities, including food handling, gas exchange, gamete transfer, and sensory reception. For example, you might think that being covered with a nonliving exoskeleton would prevent arthropods from gathering information about their environment. In fact, just the opposite is true: The cuticle has become an intimate part of a variety of sense organs, transmitting stimuli from the environment to the underlying receptor cells.

Insects

The largest group of arthropods are the **insects,** which have three thoracic segments, each bearing a pair of legs (Figure 27-14). Insects can be found at the tops of mountains, in the hottest deserts, the binding of a book, a sack of the driest flour, and nearly every other conceivable terrestrial habitat on earth. The first insects appeared approximately 400 million years ago. They are thought to have been inconspicuous, wingless animals that lived under the leaves, probably resembling modern, wingless bristletails (Figure 27-15). These ancestors underwent a remarkable adaptive radiation, spurred by the evolution of flight. The ability to fly is of immeasurable importance; it allows insects to avoid predators, to cross barriers, and to exploit widely scattered resources. In most insects, both pairs of wings are used in flight, but there are many variations. Some insects are wingless, either because they evolved from primitive, wingless insects (as in springtails and bristletails), or because the wings were secondarily lost during evolution (as in lice and fleas).

In some insects, including grasshoppers, the egg develops into a larva that is essentially a small version of the adult. However, most insects progress through a course of development that involves *metamorphosis*—dramatic changes in body form. In these species, the egg develops into a larva that bears no resemblance to the adult. The larva is typically a wormlike form (such as a maggot or a caterpillar) that feeds voraciously and grows rapidly. The larva then transforms into a *pupa,* in which the winged adult form takes shape. Once it emerges from the pupal casing, the adult is specialized for mating, dispersal, and egg laying. In general, each developmental stage has its

FIGURE 27-15
A wingless bristletail, one of the most primitive insects.

own habitat, environmental requirements, and adaptations that differ from those of other stages.

Crustaceans, Chelicerates, and Myriapods

If you watch a barnacle in a saltwater aquarium, you will see it repeatedly fan the water with a "net" of hairy-looking appendages (Figure 27-13*a*). In carrying out this movement, the barnacle is straining its environment for small organisms and other food particles. This is how most crustaceans make a living—as filter feeders. Larger crustaceans, such as crabs and lobsters, are typically bottom-dwelling scavengers or predators. The most numerous crustaceans are tiny copepods that live in the upper layers of lakes and oceans. Copepods and other small crustaceans are important links in aquatic food chains, serving as the primary food source for small fish as well as for giant filter-feeding blue whales and basking sharks.

The most familiar chelicerates (named after the first pair of appendages, the *chelicerae*) are the spiders (Figure 27-13*c*), arthropods whose trademark is the use of silk. Silk is produced as a liquid within silk glands located in the abdomen. The liquid silk is converted into a solid thread as the result of a change in intermolecular arrangement of the silk molecules as they are forced through the fine tubular channels located at the tip of the spider's abdomen. The chelicerae in spiders take the form of a sharp curved fang used to pierce the body of its prey and inject venom. Spiders have no jaws to chew their food, and their mouths do not take in solid material. Instead, spiders are primarily fluid-feeders. In order to fully extract the nutritive contents of their prey, spiders exude digestive enzymes into the prey. The enzymes liquify the tissues, which are then pumped into the digestive tract of the spider.

Myriapods (*myria* = many; *poda* = feet) are worm-like arthropods with large numbers of legs. Among the group, the centipedes are fast-moving, carnivorous animals with long legs and a long stride (Figure 29-13*d*). Millipedes have more numerous, shorter legs than centipedes and are almost exclusively vegetarians. Many of the millipedes are burrowers whose legs provide the animal with the power to push through the topsoil.

PHYLUM ECHINODERMATA: ECHINODERMS

The phylum **Echinodermata** includes about 5,000 species of sea urchins, sand dollars, sea stars, feather stars, and sea cucumbers. Echinoderms have no head or brain, and there is no concentration of sensory, respiratory, or excretory structures; all of these functions are carried out by groups of cells that are scattered throughout the animal's body.

Echinoderms possess a number of traits that set them apart from all other invertebrates. Adult echinoderms are basically radially symmetric but in a different way from the cnidarians, the other major group of radial animals. The radial symmetry of echinoderms is fivefold, or *pentamerous*; that is, parts of the body (such as arms or gonads) tend to be repeated five times around the circle (Figure 27-16).

Another feature unique to echinoderms is their **water vascular system,** which consists of a network of channels and tiny "tube feet" (Figure 27-16) that are used in locomotion, respiration, and sensory perception as well as for clinging to surfaces and attacking prey. The tube feet, which protrude through the skeleton, are present in paired rows and are operated by hydraulic pressure that is generated in the water vascular system. The coordinated extension and withdrawl of thousands of these tiny feet allow the sea urchin or sea star to glide slowly over the substrate or to attach itself by its suckers to one particular spot.

PHYLUM CHORDATA: CHORDATES

Even though vertebrates form a distinct and phylogenetically related group, they are not placed in a phylum of their own. Rather, they are grouped together with a number of invertebrates (called **protochordates**) in the phylum **Chordata.**

Chordates are distinguished from all other animals by a number of characteristics (Figure 27-17), including the following:

1. **Pharyngeal gill slits.** At some stage in their lives, even if only temporarily during embryonic stages, chordates possess perforations in the wall of the pharynx (the anterior portion of the digestive tract). In protochordates and fishes, these openings in the embryo give rise to the gills. In terrestrial vertebrates, they close over and disappear during development.
2. **A dorsal hollow tubular nerve cord.** The central nervous system of a chordate develops from a hollow neural tube (page 378). The wall of the tube develops into the brain and spinal cord, while the central channel becomes filled with fluid.

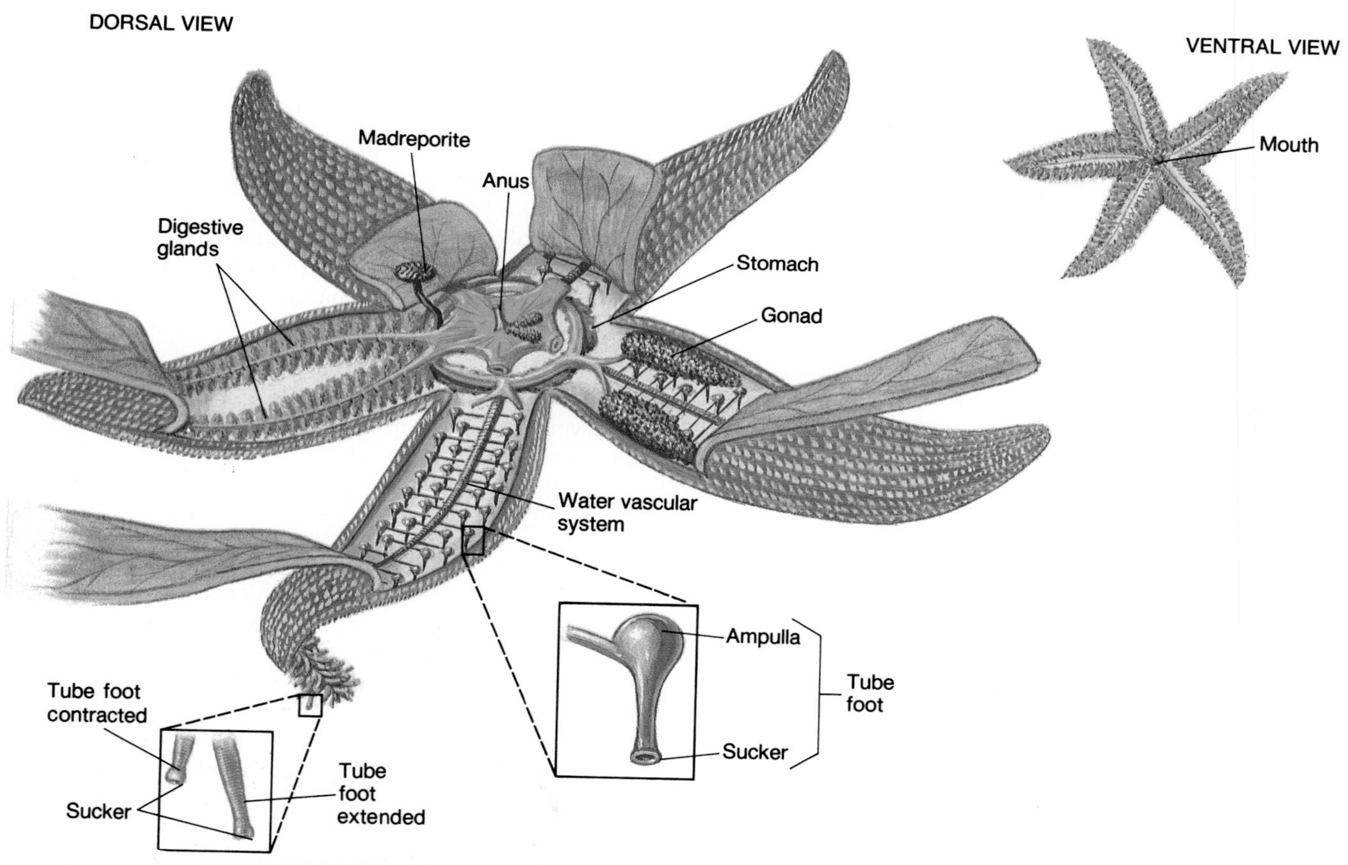

FIGURE 27-16

The anatomy of a sea star, detailing the water vascular, reproductive, and digestive systems. The inset shows a single tube foot. Water enters the water vascular system through the sievelike *madreporite,* passes along a network of channels, which connect with the five paired rows of *ampullae.* When muscles in the wall of the ampulla contract, fluid is forced into the hollow, cylindrical tube foot, extending the foot in the direction of movement. When the muscles of the ampulla relax, the foot is withdrawn. When the tip of the foot is pressed against the substrate, it acts like a sucker, holding the animal to the surface.

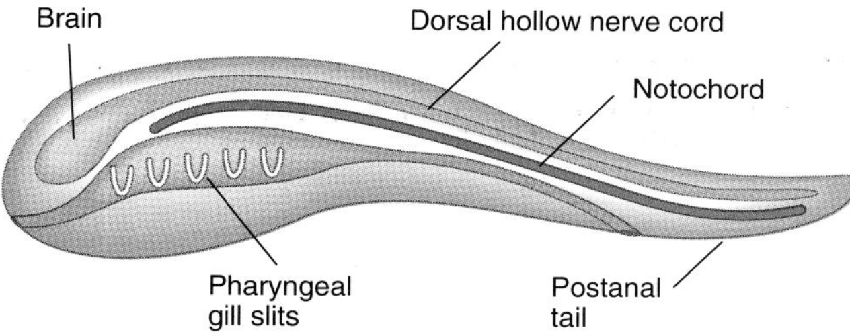

FIGURE 27-17
A generalized chordate showing the four major chordate characteristics; pharyngeal gill slits, a dorsal hollow nerve cord, a notochord, and a postanal tail.

3. **A notochord.** The notochord is a flexible rod that forms in the embryo just beneath the neural tube. The notochord persists throughout life in the protochordates, where it functions as a flexible skeletal rod, stiffening the body during swimming or burrowing. In vertebrates, the notochord is replaced by a bony vertebral column.
4. **A postanal tail.** The anus is not situated at the posterior tip of chordates as it is in most animals; rather, the anus is followed by a tail. In many vertebrates, the tail disappears during embryonic development.

Protochordates

There are two major groups of protochordates: tunicates and lancelets, both of which provide interesting insights into vertebrate evolution.

Tunicates are saclike marine animals enclosed in a gelatinous outer coat, or "tunic," which they secrete. If biologists were able to observe only adult tunicates (Fig. 27-18*a*), it is unlikely that these animals would ever have been classified in the same group as vertebrates. However, when the development of a tunicate is observed, the fertilized egg is seen to transform itself into a larva of the type shown in Figure 27-18*b*, which possesses all of the basic chordate characteristics. When these "tadpole larvae" settle onto a suitable substrate, they metamorphose into an adult; in the process, they lose their notochord, dorsal nerve cord, and postanal tail. Of the major chordate characteristics, only the pharyngeal gill slits remain in the adult.

Since it is the tunicate larva that resembles vertebrates, rather than the adult, biologists speculate that the *larva* of an ancient tunicate-like organism acquired the ability to produce gametes for sexual reproduction and acted as the ancestor of all other chordate groups. The phenomenon whereby a larval form becomes a sexually mature adult occurs quite often among animals. For example, many salamanders fail to undergo metamorphosis but develop into sexually mature adults in a larval body form.

(a)

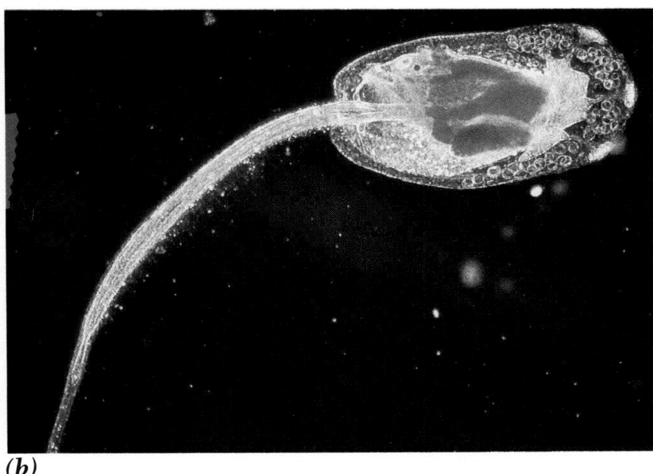
(b)

FIGURE 27-18
Tunicates. ***(a)*** Adult tunicates are filter feeders that draw seawater in through an incurrent siphon and filter out suspended food particles as the seawater passes through its gill slits. (***b***) The tunicate larva possesses all four of the major chordate characteristics: a dorsal hollow nerve cord, notochord, gill slits, and a postanal tail.

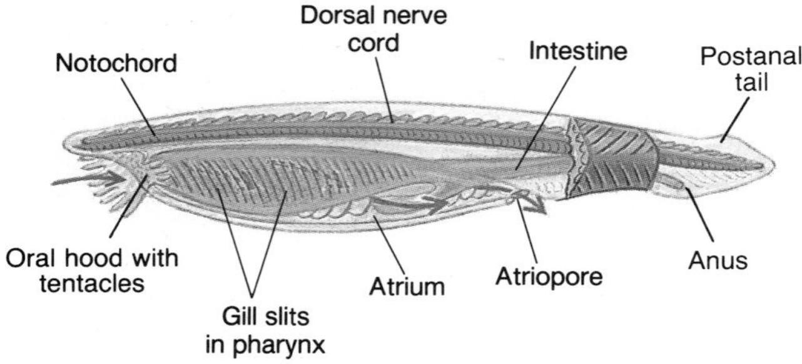

FIGURE 27-19
Knifelike adult lancelets wriggle into the seabed, their head protruding for feeding. Sea water enters through the oral hood, passes through the gill slits and into the atrium, exiting via the atriopore. Food particles become trapped in the mucus that lines the pharynx.

Lancelets are small, fishlike invertebrates that live in shallow oceans throughout the world. Unlike adult tunicates, adult lancelets (Figure 27-19) show all the major chordate characteristics. Like tunicates, lancelets are filter feeders that pump water through their pharyngeal gill slits, trapping suspended food particles. On each side of the notochord are blocks of muscle tissue that are separated from one another by thin sheets of connective tissue. This organization of muscle and connective tissue clearly foreshadows the similar arrangement found in fishes.

VERTEBRATES

The vertebrates are an immensely successful group of animals, occupying habitats in virtually every terrestrial, marine, and freshwater environment. All vertebrates have a vertebral column—a backbone composed of separate vertebrae that are constructed of bone or cartilage, which surround and protect the delicate spinal cord. The vertebral column replaces the embryonic notochord and forms the main axis of the internal skeleton. Living vertebrates are grouped into seven classes, three of which are composed of fishes.

FISHES

The first traces of vertebrates occur as 550-million-year-old flakes of bone that were found in sediments from an ancient sea that once spread over much of the western United States. The most primitive living fishes are members of the class Agnatha, which include hagfish and lampreys (Figure 27-20). These animals bear a superficial resemblance to eels, but they lack jaws, fins, and scales. The mouth of a lamprey is surrounded by a round sucker, armed with teeth. The lamprey attaches itself to the outside of another fish, then rasps a hole in the prey's flesh with its tooth-bearing tongue. Along with pollution, lampreys have contributed to the devastation of the fishing industry of the Great Lakes of the United States and Canada; the eradication of this animal has been one of the primary goals of the industry for several decades.

Fishes with jaws appeared about 415 million years ago. Jaws allowed fishes to become predators, pursuing prey rather than feeding on sediments on the bottom of their aquatic environment. Jaws also provided a means of defense, reducing the need for heavy protective armor and increasing mobility. Along with the evolution of jaws came the appearance of paired **fins** on the side of the body, which must have greatly increased a fish's maneuverability. From these early jawed fishes, two great lines of modern fishes arose: the cartilaginous fish (class Chondrichthyes) and the bony fish (class Osteichthyes).

Cartilaginous Fishes

The **cartilaginous fishes** include the sharks, skates, and rays. While these fish arose from ancestral species with heavy bony skeletons, bone formation was lost during evolution, and their skeleton consists entirely of cartilage. Most sharks are aggressive predators that possess extremely keen sense organs to help them locate their prey. The whale shark (*Rhincodon typus*) is the largest of all fishes, weighing up to 20 tons and growing to lengths of 18 meters (60 feet). Ironically, this giant shark doesn't consume large prey; a sieve in its pharynx filters minute crustaceans and drifting plankton. Skates and rays have a flattened shape adapted for living close to the ocean bottom.

Bony Fishes

The **bony fishes** include species with skeletons that are at least partly composed of bone. These vertebrates have

FIGURE 27-20
Living, jawless fishes of the class Agnatha include this lamprey, which feeds on the flesh of other living fishes.

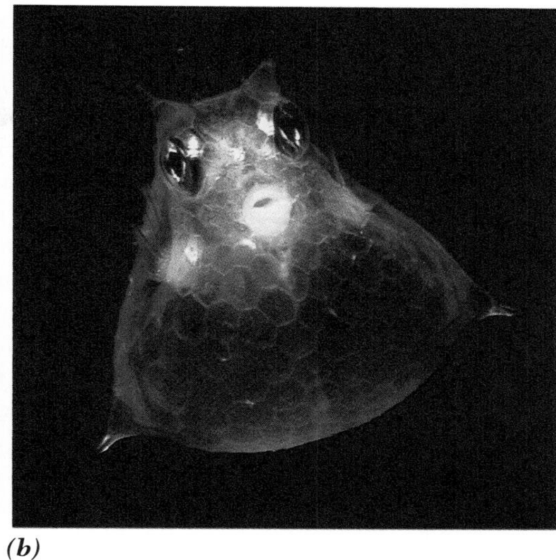

(a) *(b)*

FIGURE 27-21

Two representatives of the diverse class Osteichthyes (the bony fishes). ***(a)*** The blenny has eyes on top of its head which enable it to scan for danger. ***(b)*** A juvenile cowfish.

been extremely successful and number over 18,000 diverse species today (Figure 27-21). Because water is much denser than air, it provides an organism with buoyancy (upward support), eliminating the need for a heavy, supportive skeleton. (This is evident when you try to remove the delicate bones from a trout or salmon steak before eating it.) Although the density of water is high, the density of bone and muscle is even greater; therefore, fish must have some adaptation for keeping them from sinking to the bottom. Most fishes have a gas-filled *swim bladder* that gives the fish an overall density equal to its surroundings, allowing the animal to remain at any desired depth. Over the course of their evolutionary history, two major groups of bony fishes—the *ray-finned fishes* and the *lobe-finned fishes*—have appeared. The latter group, represented today by only a few species, such as the coelacanth discussed at the beginning of the chapter, were once the predominant fishes and are the apparent ancestors of terrestrial vertebrates. The ray-finned fishes, whose fins contain sharp bony supports, comprise over 99 percent of the species of bony fishes living today.

AMPHIBIANS

The class Amphibia includes about 3,000 species of frogs and toads, salamanders, and the less familiar wormlike caecilians (Figure 27-22). Amphibians were the first terrestrial vertebrates, but they never became fully adapted

(a) *(b)*

FIGURE 27-22

Representative amphibians. Amphibians include salamanders and newts, frogs and toads ***(a)***, and the less familiar caecilians ***(b)***.

to life in dry, terrestrial habitats. For example, unlike reptiles, birds, and mammals, modern amphibians have a moist, thin, permeable skin, so they rapidly lose body water. As a result, adults must remain in damp habitats or stay close to water to prevent dehydration. Similarly, amphibian eggs lack a protective impermeable shell and lose water rapidly in dry environments. No matter where they live as adults, most amphibians return to the water to mate and lay their eggs. Within the past few years, the numbers of individuals in many amphibian populations have decreased drastically. The reason for the decimation of these species' populations, which is thought to be due to human alteration of the environment, is the subject of current research.

Amphibians have played a central role in the history of vertebrates. They were the first vertebrates to leave the water and walk with four legs on the land. Nearly 250 million years ago, a new type of vertebrate appeared that had well-developed lungs and scaly, water-resistant skin. This new animal laid "land eggs" that resisted drying and were equipped with an internal supply of food and water for the developing embryo. Freed of the need to be near water or to return to water for reproduction, these animals were better adapted for life on land than were their amphibian ancestors. These new vertebrates were the reptiles.

REPTILES

The earliest reptiles appeared approximately 300 million years ago. They remained as small, inconspicuous, lizard-like animals for over 100 million years before beginning an adaptive radiation of unprecedented splendor. Many of them grew to enormous size; if they were alive today, some would be tall enough to peer over the top of a three-story building. The reptiles of the Mesozoic Era are classified into more than 15 distinct orders. Today, the class **Reptilia** (with approximately 6,000 species) contains only four orders: lizards and snakes; turtles (Figure 27-23*a*); crocodiles and alligators (Figure 27-23*b*); and one lizard-like animal, the tuatara, which is the sole survivor of an ancient order.

The most important terrestrial adaptations in reptiles are those that restrict water loss. A reptile's skin is much thicker than that of amphibians, and it is very dry. In addition, the epidermis contains horny *scales,* which protect the skin's surface and help prevent water loss. Reptiles also possess mechanisms that prevent water loss during excretion. For example, reptiles convert nitrogenous wastes to uric acid which, because it is virtually insoluble in water, can be excreted as a nearly dry paste. Finally, unlike amphibians, reptiles reproduce by internal fertilization, and a reptile's eggs are enclosed in a waterproof outer casing—either a leathery coat or a thin, brittle shell. The embryo that develops within this enclosed environment forms in association with several extraembryonic membranes, whose functions we discussed in Chapter 21. One of these membranes, the *amnion,* surrounds a fluid-filled chamber that provides the embryo with a protective, aquatic environment, even though the egg is situated on dry land.

BIRDS

The class **Aves** contains about 9,000 species of *birds,* ranking this group of vertebrates second only to bony fishes in number of species. The most distinctive characteristic of birds is their possession of feathers (Figure 27-24), a feature that is directly involved in two of the most important

(a)

(b)

FIGURE 27-23
Representative reptiles. ***(a)*** In murky waters, this flattened turtle attracts little attention from an unwary fish, which quickly disappears into the reptile's wide mouth. ***(b)*** A flick of its head, and this crocodile will soon be dining on a young frog.

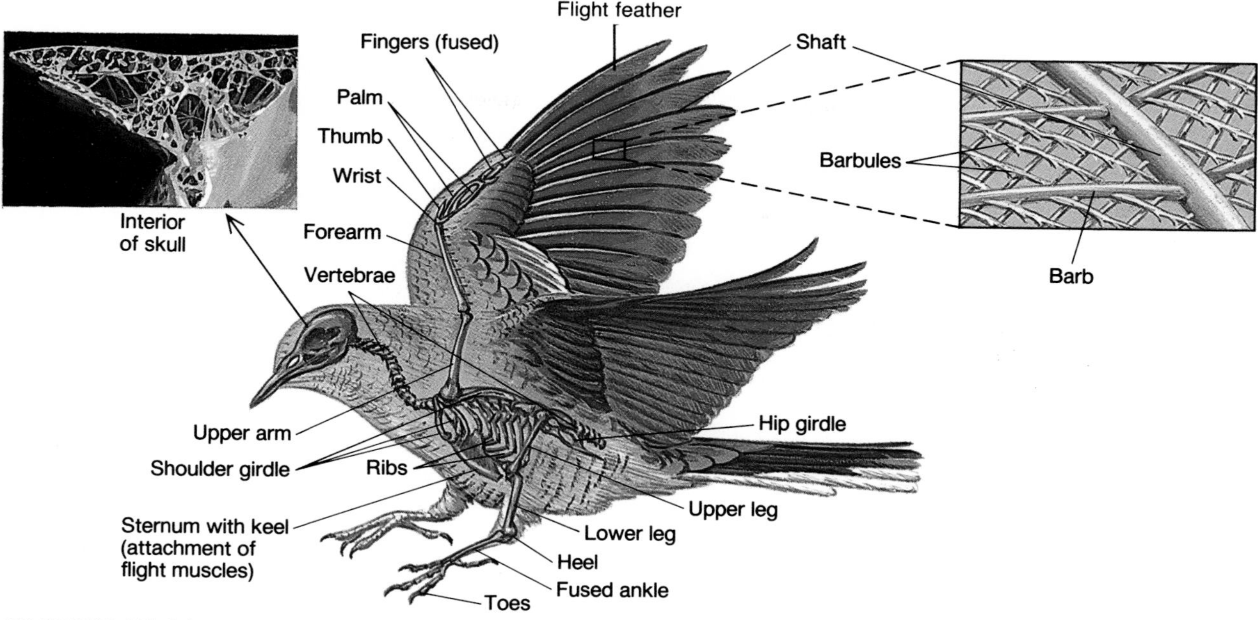

FIGURE 27-24
Class Aves—birds. The anatomy of a bird is geared toward making it an efficient flying "machine." The bird's bones are extremely light and are filled with air spaces; its feathers contain lightweight protein and consist of a shaft with projecting barbs and interlocking barbules. The large *flight feathers* play a key role in providing the thrust and lift necessary for flight.

aspects of avian (bird) biology: maintenance of a constant, elevated body temperature, and the ability to fly.

The average body temperature of a bird is 41°C (106°F), several degrees higher than that of mammals. Feathers provide insulation, helping the bird maintain a body temperature higher than that of the environment. Because of their high level of metabolic activity, birds have been able to colonize some of the world's harshest environments, as is exemplified by the emperor penguins that lay their eggs on the bare Antarctic ice in the middle of winter.

Flight gives birds a freedom that is unmatched among other vertebrates. Being airborne, birds have far greater visibility of the terrain and its offerings than do animals that remain on the ground. Flight also allows small, defenseless animals to escape from land-based predators, and it enables all birds to take advantage of distantly separated food sources.

The ancestors of birds were probably small, bipedal, insect-eating dinosaurs that walked on their hindlimbs; the forelimbs were free to evolve into wings. The evolution of flight required sweeping anatomic and physiologic changes, including a reduction in the weight of the skeleton. For example, the skeleton of a frigate bird weighing 2 kilograms and possessing a wingspan of 2 meters weighs only about 110 grams, or 5 percent of the bird's body weight—less than the weight of its feathers. The reduction in a bird's skeletal weight is largely a result of the hollow construction of the skeleton (Figure 27-24). In order for the flight muscles to deliver the power required to fly, a bird's metabolism is very high; its body contains numerous large air sacs; its circulatory and respiratory systems are extremely efficient; and its digestive system is capable of processing large amounts of food in a very short period of time.

The skull of a bird is also very lightweight (Figure 27-24, inset). It bears a light, horny, toothless *beak* that is composed primarily of the protein keratin. Since the bird's jaws have no teeth to support, the jaw can also be reduced in mass. The beak is a bird's primary "tool." A knowledgeable biologist can look at a bird's beak and tell if it is used for digging, probing, piercing, chiseling, straining, cracking, tearing, pecking, or some other activity. The varying functions of the beak are reflected in the striking differences in its shape among different types of birds.

MAMMALS

The class **Mammalia** includes about 4,500 species whose body form and lifestyle are highly diversified. Two characteristics distinguish mammals from other vertebrates:

(a)

(b)

FIGURE 27-25

The class Mammalia contains three different groups. ***(a)*** Monotremes include strange-looking spiny anteaters and this duckbilled platypus, the only mammals that lay eggs. ***(b)*** Young marsupials are born at an extremely immature stage and complete development in their mother's pouch. A baby koala is born after only 30 days of development. Blind and with no back legs, the underdeveloped koala uses miniature front legs to crawl into its mother's pouch, where it stays for about 6 months. When a young koala becomes too large for the pouch, it clings to its mother's back, returning to the pouch for an occasional drink of milk. ***(c)*** Most mammals, such as this mountain gorilla, nourish their embryos and fetuses through a placenta, then continue to nourish their newborn with maternal milk from mammary glands.

1. Mammalian skin is endowed with hair—thin, elongated filaments made largely of the protein keratin. Most mammals are covered with a relatively dense layer of hair, though it is reduced to patches in humans. Like feathers, hair traps air, insulating the body surface and reducing the loss of body heat.
2. Mammals nourish their young with milk produced by the female's mammary glands. Milk typically contains large amounts of casein, a phosphate-containing protein, as well as calcium and other salts, the sugar lactose, and suspended globules of fat.

"True" mammals, which are distinguished from their reptile ancestors by having a lower jaw composed of a single bone, appeared approximately 210 million years ago. These early mammals remained relatively scarce in number until the end of the Mesozoic Era (65 million years ago), at which time the ruling reptiles disappeared from the earth. The earliest mammals are thought to have been small (rat-sized), nocturnal forms that ate insects and depended on a keen sense of smell and an acute sense of hearing.

Mammals are classified into three groups (Figure 27-25).

1. **Monotremes.** Unlike other mammals, monotremes lay eggs which, like birds, are incubated outside the female's body. Unlike birds, newly hatched monotremes are nourished with milk from their mother. Today, monotremes are represented by only two groups—the platypus and the spiny anteaters.
2. **Marsupials.** Marsupials give birth to their young at a very early stage of development. The remainder of the developmental process occurs outside the uterus in a special pouch that contains the mammary glands. The most striking feature of the marsupials is the incredible diversity of forms among such a small number of species. Included within the approximately 250 species are carnivores, insectivores, herbivores, and omnivores; terrestrial, arboreal (tree-dwelling), and fossorial (underground) species; and species that run, hop, burrow, swim, climb, and glide. Examples of marsupials include kangaroos, koala bears, and opposums.
3. **Placentals.** Placental mammals nourish developing offspring internally through a placenta, the structure through which nutrients and wastes are exchanged between the mother and fetus. Following birth, most

(c)

mammalian mothers keep their young with them for a period of months up to a year or two, during which time much of their behavior may be learned. The extended period of dependence in the human species is unique within the Animal kingdom, reflecting the long lifespan of our species and the complexity of the behavior we must learn.

Placental mammals are typically divided into about 16 different orders, several of which contain only a single genus. We will confine our discussion to the order in which humans are classified—the order Primates.

Primates

The **primates** are a diverse taxonomic group, ranging from primitive, insect-eating tree shrews, to monkeys, apes, and humans. Primates are one of the oldest mammalian orders, having evolved from small, arboreal (tree-dwelling) ancestors. The early primates remained in the trees; this fact, more than any other, seems to have shaped the course of primate evolution. While smaller arboreal animals, such as squirrels, depend on their claws to hold them to the bark as they scamper through the trees, primates have evolved a different approach; they grasp the branches with their hands and feet. The development of a grasping hand, with an opposable thumb that could close to meet the fingertips, has allowed primates to become much larger than other arboreal mammals since they can hold onto the limbs rather than just balance themselves on them. As an added evolutionary "bonus," grasping hands and feet have allowed primates to manipulate objects to a degree unachieved elsewhere among animals.

As their own unique style of arboreal locomotion became perfected, primates became more agile, gaining the ability to hang from branches, jump from tree to tree, and perform all the other amazing gymnastic feats performed by monkeys. These feats require a great deal of *hand–eye coordination.* The animal must be able to judge distances with great precision, learn just how far to jump, and make instantaneous decisions while in mid-air. Such attributes require a keen sense of vision, a sensitive touch, and a great deal of intelligence and motor control, all characteristics that we recognize in ourselves. If it weren't for the fact that our ancestors lived in trees, it is very unlikely that we would be living in houses.

EVOLUTION AND ADAPTATION: TYING IT TOGETHER

It is convenient in a chapter on animal diversity to organize groups of animals in an order of increasing structural complexity. We have organized the chapter in this way not to convince you that humans are the highest forms of life on earth but because it allows us to construct a body of knowledge about animals which gradually builds upon itself from one group to the next. For this reason, we began with the simplest multicellular animals, the sponges, which are essentially constructed as a collection of cells, and worked our way through to the birds and mammals, which are arguably the most complex animals on earth. Keep in mind that the simpler animals, including the sponges, have survived on earth for hundreds of millions of years. The continued existence of less complex animals attests to their ability to compete successfully for food and shelter with so-called "higher" forms.

One of the foremost goals of zoologists is to describe the evolutionary pathways that led to the appearance of the various animal groups. Some of these phylogenetic conclusions are based on the study of fossil remains; others are based on comparisons among living animals and their embryos. Interpretations of such data are inevitably speculative since they attempt to describe evolutionary events that occurred long ago. It is assumed that all members of a group—whether a phylum, such as chordates, a class, such as mammals, or an order, such as primates—arose from a single ancestral species by a series of adaptive radiations. Even though all living species have been separated from the common ancestral form for the same amount of time, not all will have diverged to the same degree. Some lines of evolution show very little change, while others produce animals that are hardly recognizable as related to the ancestral stock. Even within lines that have undergone rapid evolutionary change, many features of the group may remain unchanged. In fact, some characteristics of animals, such as the structure of cilia and flagella (see Figure 3-13), remain unchanged through the entire Animal kingdom.

SYNOPSIS

Members of the Animal kingdom are multicellular, heterotrophic, sexually reproductive, and capable of locomotion at some stage of their life. An animal's body plan can be explained on the basis of its evolutionary ancestry and the environment in which it lives. Three of the most important characteristics that distinguish body plans are symmetry, body cavities, and the presence or absence of segmentation. Symmetry concerns the arrangement of the body parts relative to an axis or plane through the animal. All animals but sponges have a digestive cavity. In addition, most animals have a secondary cavity between their digestive tract and body wall. The bodies of annelids, arthropods, and chordates are segmented; they develop from repeating blocks of tissues.

The members of a phylum are all descended from a common ancestor and are construed as a variation on a common body plan. Sponges are aquatic animals that circulate water through their bodies, trapping and phagocytizing microscopic food particles. Water enters the body through myriad microscopic pores; food matter is trapped by choanocytes that line the water channels; and the nutrient- and oxygen-depleted water exits through one or several large openings. Cnidarians are aquatic, radially symmetric animals that are composed of a three-layered body wall surrounding a blind gastrovascular chamber. There are two basic cnidarian body forms: polyps, which tend to be sessile, and medusas, which tend to be motile. Cnidarians feed on other living animals that are stunned by projectiles fired from cells located on the tentacles surrounding the cnidarian's mouth. Flatworms are bilaterally symmetric animals that exhibit the beginnings of cephalization; they have a head with a concentration of nervous tissues and sense organs. Unlike cnidaria, flatworms have well-developed organ systems. Their flattened shape allows them to nourish and oxygenate their cells without the need for a circulatory system. Roundworms include the most abundant microscopic organisms of the soil as well as some of the worst human parasites. Roundworms are bilaterally symmetric, cylindrical animals, whose body contains a fluid-filled pseudocoelom, which acts as a hydrostatic skeleton. Molluscs are bilaterally symmetric, nonsegmented animals whose bodies can be divided roughly into a head-foot, a dorsal visceral mass, and a mantle. The ribbon-shaped, tooth-bearing radula is a unique molluscan structure. The cephalopod molluscs contain the most advanced nervous system of any invertebrates. Annelids are bilaterally symmetric, segmented animals with a well-developed head and organ systems. Their segmented body provides flexibility during swimming or burrowing, while their internal, fluid-filled coelom acts as a hydrostatic skeleton. Polychaetes have paired appendages that function as swimming paddles and sites of gas exchange. Arthropods are the most abundant and diverse phylum. Arthropods are bilaterally symmetric, segmented animals that are covered by a complex, jointed cuticle (exoskeleton), and jointed appendages. The exoskeleton provides support, protection, locomotion, food handling, gas exchange, gamete transfer, and sensory reception. It is shed periodically and replaced during periods of growth. Echinoderms are radially symmetric animals that have a unique water vascular system complete with tiny tube feet used for attachment and locomotion.

Chordates include a small number of invertebrate protochordates and the vertebrates (fishes, amphibians, reptiles, birds, and mammals). Chordates are bilaterally symmetric animals that are characterized by pharyngeal gill slits, a dorsal hollow tubular nerve cord, a notochord, and a postanal tail in at least one stage of their development.

Review Questions

1. Match the following structures with the appropriate phylum:

——notochord	a. Cnidaria
——cnidocytes	b. Chordata
——radula	c. Porifera
——tube feet	d. Mollusca
——choanocytes	e. Annelida
——parapodia	f. Arthropoda
——tracheae	g. Echinodermata

2. What characteristics distinguish mammals from other chordates? Birds from other chordates? Amphibians from reptiles?
3. What characteristics are shared by the arthropods and the vertebrates, which enable them to be fully terrestrial?
4. Which animals lack tissues? Which lack organs? Which lack a complete digestive system with a mouth and an anus?
5. Which animal phylum probably has the greatest number of individuals? The greatest number of species? The largest-sized members?

Critical Thinking Questions

1. In what ways does *Latimeria* illustrate the concept that an animal's characteristics reflect both its ancestry and its environment?
2. Annelids are segmented animals, while molluscs are nonsegmented animals, yet embryological evidence indicates that the two groups are evolutionarily related. How can you explain these differences in segmentation? Draw a phylogenetic scheme that explains the relationship among arthropods, molluscs, and annelids, using segmentation as the key characteristic.
3. Which of the human organ systems illustrated in Figure 14-4 are the most bilaterally symmetrical and which are the least?
4. Why do you suppose the largest fishes (and the largest mammals) are filter feeders? Why do you suppose the largest arthropods (a species of crab) are aquatic? Why do you suppose the largest sponges contain large numbers of internal chambers rather than simple channels?
5. Which characteristics of protochordates allowed these animals to give rise to the vertebrates?

PART 6

Ecology and Animal Behavior

The two major arenas of life are the earth's lands and waters. Lichens slowly transform solid rock into soil that supports a multitude of organisms, including the trees from which these vibrant leaves have fallen. The water teams with microorganisms that break down the leaves. By decaying dead organisms, even simple microorganisms form a critical link in the recycling of chemical nutrients in all ecosystems.

CHAPTER

28

Networks of Life: Ecology and the Biosphere

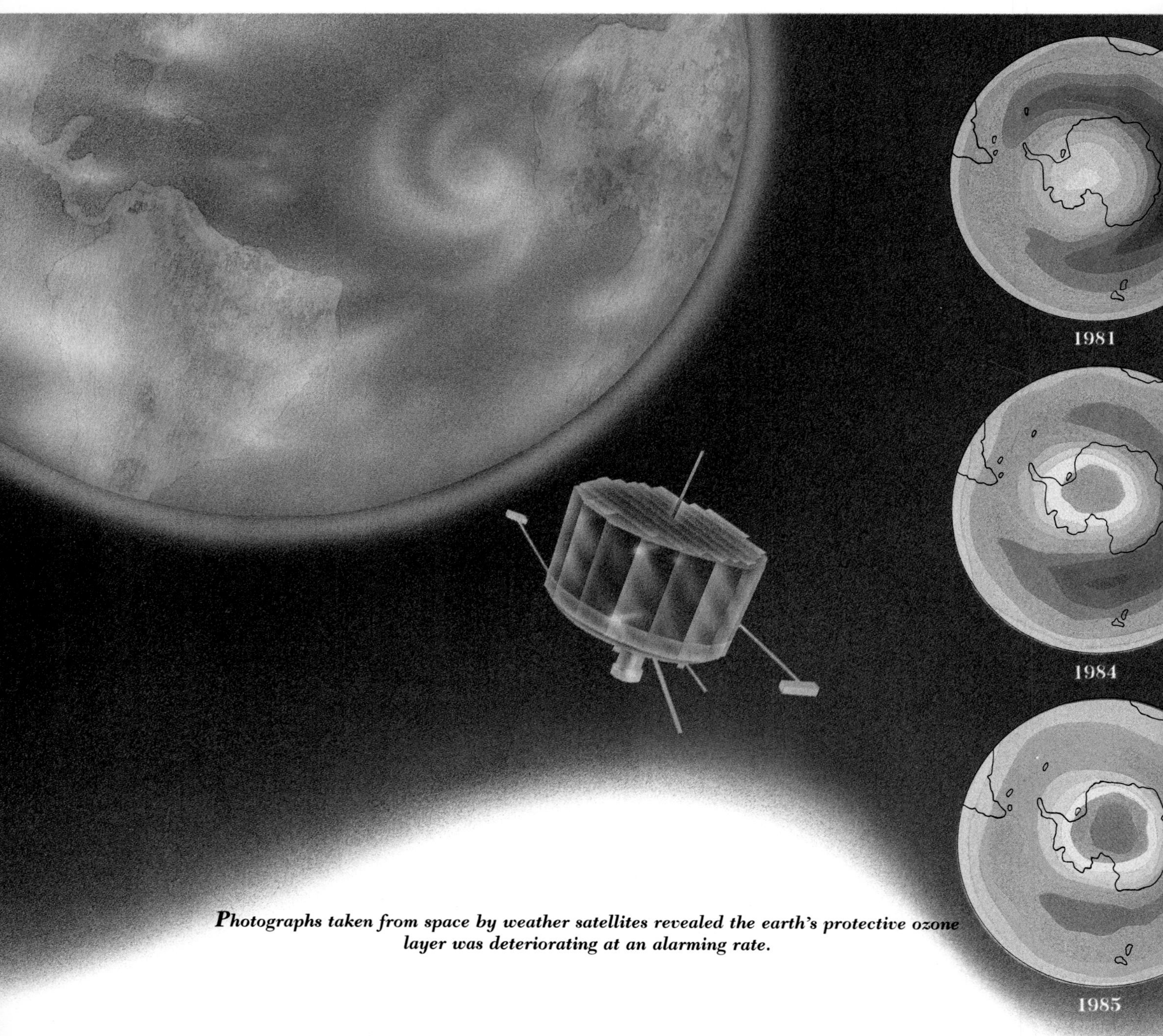

Photographs taken from space by weather satellites revealed the earth's protective ozone layer was deteriorating at an alarming rate.

STEPS TO DISCOVERY
The Antarctic Ozone Hole

Virtually all the earth's organisms owe their lives to the filtering effects of the ozone layer 20 to 50 kilometers (12 to 30 miles) above. About 99 percent of the sun's lethal ultraviolet rays are absorbed by this invisible layer; as a result, organisms are spared overexposure to these killer rays that destroy many biological molecules, including DNA. Studies reveal that the ozone layer is rapidly being destroyed, however. As this protective layer becomes depleted, we are likely to see more cases of skin cancer, cataracts, and immune deficiencies, as well as reduced crop yields and other serious consequences.

In 1974, F. Sherwood Rowland and Mario Molina, two atmospheric chemists at the University of California, Irvine, created laboratory conditions resembling those found in the earth's mid- to outer stratosphere, where the protective ozone layer is located. The chemists examined the effects of a group of humanmade compounds, called chlorofluorocarbons (CFCs), on ozone molecules (O_3). CFCs had been widely used as propellants in aerosol products, such as deodorants and air fresheners, as refrigerants, and to inflate the bubbles in Styrofoam. Rowland and Molina discovered that CFCs destroyed ozone molecules with alarming efficiency. They projected that CFCs could destroy between 20 and 30 percent of the ozone layer, threatening all life on earth.

At first many scientists were skeptical of Rowland and Molina's projections. But, in the early 1980s, satellite studies confirmed that the CFC concentration in the stratosphere had doubled in only 10 years. This led the Environmental Protection Agency (EPA) to project a 60 percent decline in ozone levels by 2050 if CFC use and production continued to grow at the current annual rate of 4.5 percent. Because of these findings, 24 nations signed an agreement in 1987, known as the Montreal Protocol, to cut CFC production in half by 1999.

At the same time, scientists began to measure changes in the thickness of the ozone layer. The British Antarctic Survey collected monthly samples of ozone and found that each spring, ozone concentrations fell sharply and that ozone depletion was growing worse each year. Rapid ozone depletion was correlated with periods of increased usage of CFCs. In the late 1980s, satellite photos confirmed that ozone had been depleted by as much as 60 percent at certain altitudes over Antarctica. By 1990, ozone levels had dropped by as much as 95 percent. In 1992, the World Meteorological Organization reported that there were regions over Antarctica where no ozone could be detected at all—the infamous *ozone hole.* The ozone hole over Antarctica had enlarged to a record size of over 9 million square miles, about three times the size of the continental United States. This was about 25 percent larger than reported in previous years.

These findings are very important because they reveal that the rate of ozone depletion is even more rapid than originally forecasted by Rowland and Molina. In a recent interview, Rowland commented: "What we are looking at is ozone depletion caused by CFCs that were released back in 1987 and 1988. The expectation is that it will probably continue to get worse in the stratosphere for another decade or so." Ultimately, life hangs in the balance.

Life as we know it does not exist beyond the earth's atmosphere, nor is life found deep beneath the earth's solid surface. All life is restricted to a relatively narrow zone of air, water, and land, called the **biosphere,** the thin envelope in which all living organisms are found (see Chapter 1). The biosphere is only 22.4 kilometers (14 miles) thick, from the upper limits of life in the atmosphere to the depths of the dark ocean trenches. In relation to the size of the earth, the biosphere is only about as thick as the skin on an apple. All life as we know it exists within this thin layer that envelops the earth (Figure 28-1). The area of biology that investigates how all organisms in the biosphere interact with one another and with their surroundings is called **ecology.**

▼ ▼ ▼

ECOLOGY DEFINED

Although the term "ecology" is a familiar one, the word itself is somewhat new. It was coined a little more than 125 years ago by the German zoologist Ernst Haeckel to refer to the total relations between an animal and its organic and inorganic environment. Many biologists simply define ecology as the "study of ecosystems" because, as you may recall from Chapter 1, an ecosystem includes all living organisms (the biotic community) and the inanimate physical (abiotic) environment. Put simply, ecology studies:

- where organisms are found,
- how many organisms occur there, and
- why organisms occur where they do.

The distribution and abundance of organisms are affected by the ways in which organisms interact with one another in the biotic community (e.g., competition) and with the surrounding physical abiotic environment (e.g., availability of light or nutrients). These are the topics we will explore in the final chapters of this text.

THE EARTH'S CLIMATES

Climate—the prevailing weather in an area—is the chief environmental factor determining where organisms are distributed within the biosphere. Not only does climate differ greatly over the surface of the earth today, but global climates have changed dramatically over the almost 4 billion years since life originated on earth. These climatic changes have resulted in the formation of a great diversity of large ecosystems (known as biomes), as well as the evolution of millions of kinds of organisms.

The earth's climates are shaped by major circulation patterns that develop in the earth's atmosphere and oceans. Air and ocean currents result from three primary factors: (1) differing amounts of incoming solar radiation to various parts of the earth; (2) the daily rotation of the earth and its annual orbit around the sun; and (3) the distribution and elevation of the earth's land masses.

Because the earth is round, different parts of the earth receive different amounts of solar radiation (Figure 28-2*a*). Differences in incoming radiation heat the earth unevenly, producing warm tropics near the equator, where sunlight hits the earth more directly, and progressively cooler regions in higher latitudes, as the intensity of sunlight is reduced. The north and south poles are the coldest regions in the biosphere because they receive the least amount of sunlight, about five times less than the amount that reaches the equator. This variation produces different climates and biomes at different latitudes. In addition, the earth's annual orbit around the sun and its constant 23.5-degree tilt cause annual changes in incoming solar radiation to those parts of the earth that lie farther away from the equator (Figure 28-2*b*). These annual differences trigger a progression of seasons away from the equator.

The variation in the amount of sunlight that strikes different parts of the earth at different times of the day and year heats the earth's air and oceans unevenly. Warm air near the equator rises and flows toward the poles, where cooler air sinks. Because of the earth's rotation, however, the moving air mass breaks into six circulating coils—three in the Northern Hemisphere, and three in the Southern Hemisphere (Figure 28-2*c*).

As air rises at the equator, it cools, releasing its moisture and producing abundant rainfall. The cool poleward-moving air masses sink and become reheated at about 30 degrees north and south latitude, creating dry desert regions. As air coils move over the surface of the rotating earth, prevailing winds are formed. Trade winds result as air moves back toward the equator, and the westerlies form as air moves northward from 30 degrees toward 60 degrees latitude. At 60 degrees north and south latitude, the air rises and cools, releasing its moisture as rainfall. This region of cool temperatures and relatively high rainfall produces the expansive temperate forests of North and South America, Europe, and Asia.

Prevailing winds blowing over the ocean surface create currents in the sea's upper layers. Continents deflect these water movements, creating slow, circular ocean currents. Circulation mixes the oceans' waters, bringing warmer waters that originated in tropical areas to higher latitudes, and cooler waters from higher latitudes to tropical areas. These exchanges redistribute heat around the earth, contributing to the formation of the earth's varied climates and biomes, each of which contains thousands of ecosystems (Figure 28-2*d*).

FIGURE 28-1
The earth from space. Portions of Africa, Madagascar, and Antarctica can be seen beneath the swirling cloud layer. Seemingly calm at this distance, the earth teems with millions of species of organisms.

Normally, ocean currents in the South Pacific circulate in a counterclockwise pattern; that is, warmer waters from Tahiti flow southeastward and cool and then flow north along the coast of South America. But for reasons that are not yet completely understood, this circulation pattern reverses itself every 5 years or so, bringing warm waters to the coast of South America. This reverse circulation produces what is referred to as an El Niño. An El Niño not only changes the temperature of the ocean and climate near South America, it also triggers climate changes even as far away as Africa and Australia. The widespread global impact of an El Niño is further evidence that all components of the biosphere are ultimately interconnected and a part of one, integral system.

AQUATIC ECOSYSTEMS

Nearly 75 percent of the earth's surface is covered with water. Most of this water (71 percent) is salty, forming the earth's seas and oceans. Other aquatic habitats include freshwater lakes, ponds, rivers, streams, and estuaries of mixed fresh and salt water.

OPEN OCEANS

Marine ecosystems are found throughout the earth's oceans, in shallow coastal waters around continents, islands, and reefs; in intertidal zones (the area between high and low tides); and in the **pelagic zone,** or open oceans (Figure 28-3). Biologists subdivide the vast pelagic zone into three vertical layers: (1) the upper, sunlit **epipelagic zone**; (2) the dimly lit, intermediate **mesopelagic zone**; and (3) the continually dark, bottom **bathypelagic zone.** The sea floor itself is called the *benthic zone.*

Sunlight does not usually penetrate the ocean below 100 meters (320 feet). Considering that the average ocean depth is 3.9 kilometers (12,500 feet), the sunlit epipelagic zone is indeed very thin, permitting photosynthesis in only the upper 2 percent of the ocean's volume. In regions of the epipelagic zone where nutrients are abundant, large populations of *phytoplankton* flourish. Phytoplankton are microscopic photosynthetic bacteria and algae that drift with the ocean currents. Most phytoplankton are eaten by *zooplankton,* tiny crustaceans (mostly copepods and shrimplike krill), larvae of invertebrates, and fish that are small enough to be swept along by ocean currents. Larger fishes and other animals feed on the tiny zooplankton or

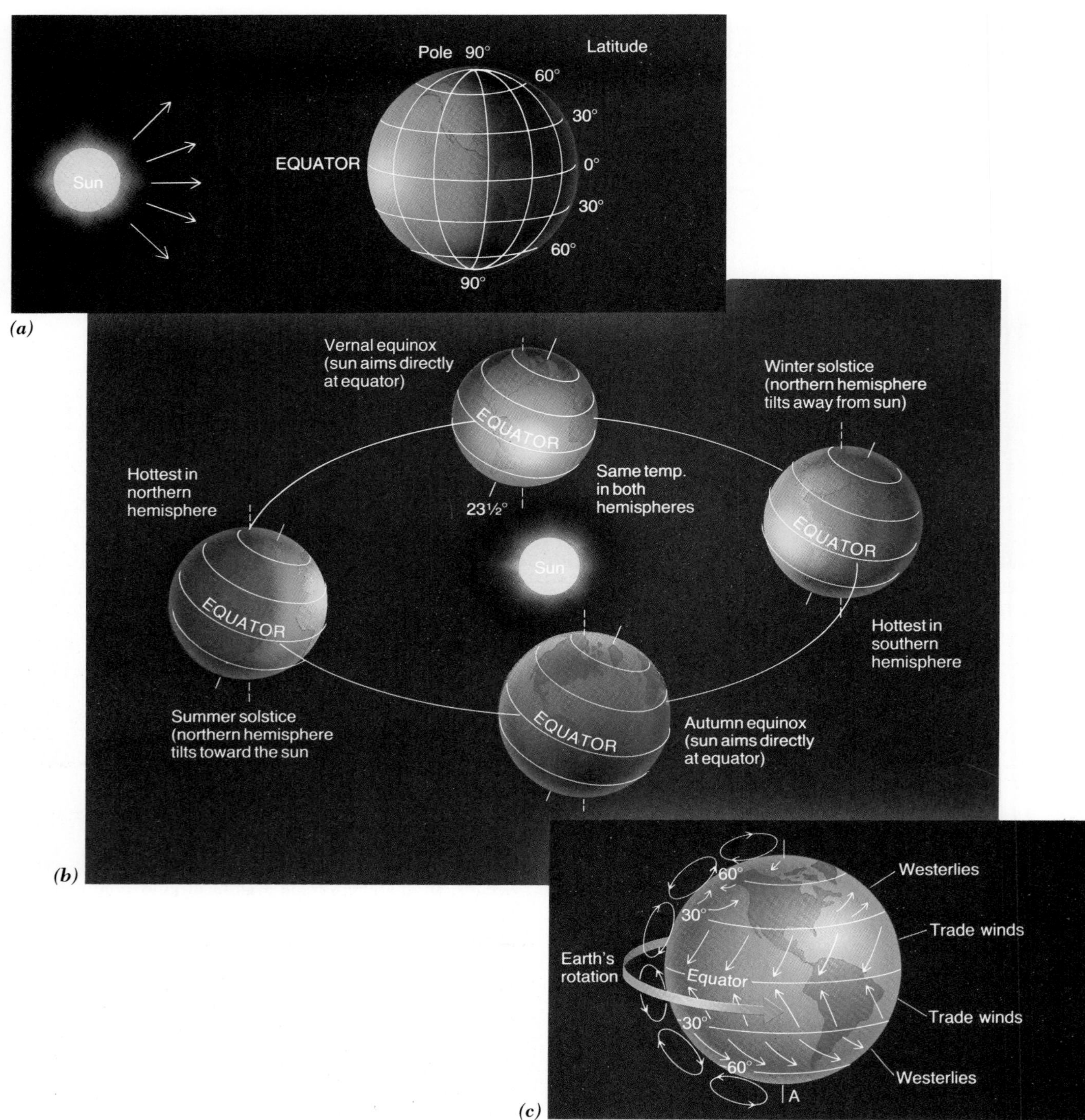

Pole 90°
Latitude
60°
30°
0°
30°
60°
90°
EQUATOR
Sun
(a)
Vernal equinox
(sun aims directly
at equator)
Winter solstice
(northern hemisphere
tilts away from sun)
EQUATOR
Hottest in
northern
hemisphere
Same temp.
in both
hemispheres
23½°
Sun
Hottest in
southern
hemisphere
Summer solstice
(northern hemisphere
tilts toward the sun
Autumn equinox
(sun aims directly
at equator)
(b)
Westerlies
Trade winds
Earth's
rotation
Equator
Trade winds
Westerlies
60°
30°
30°
60°
A
(c)

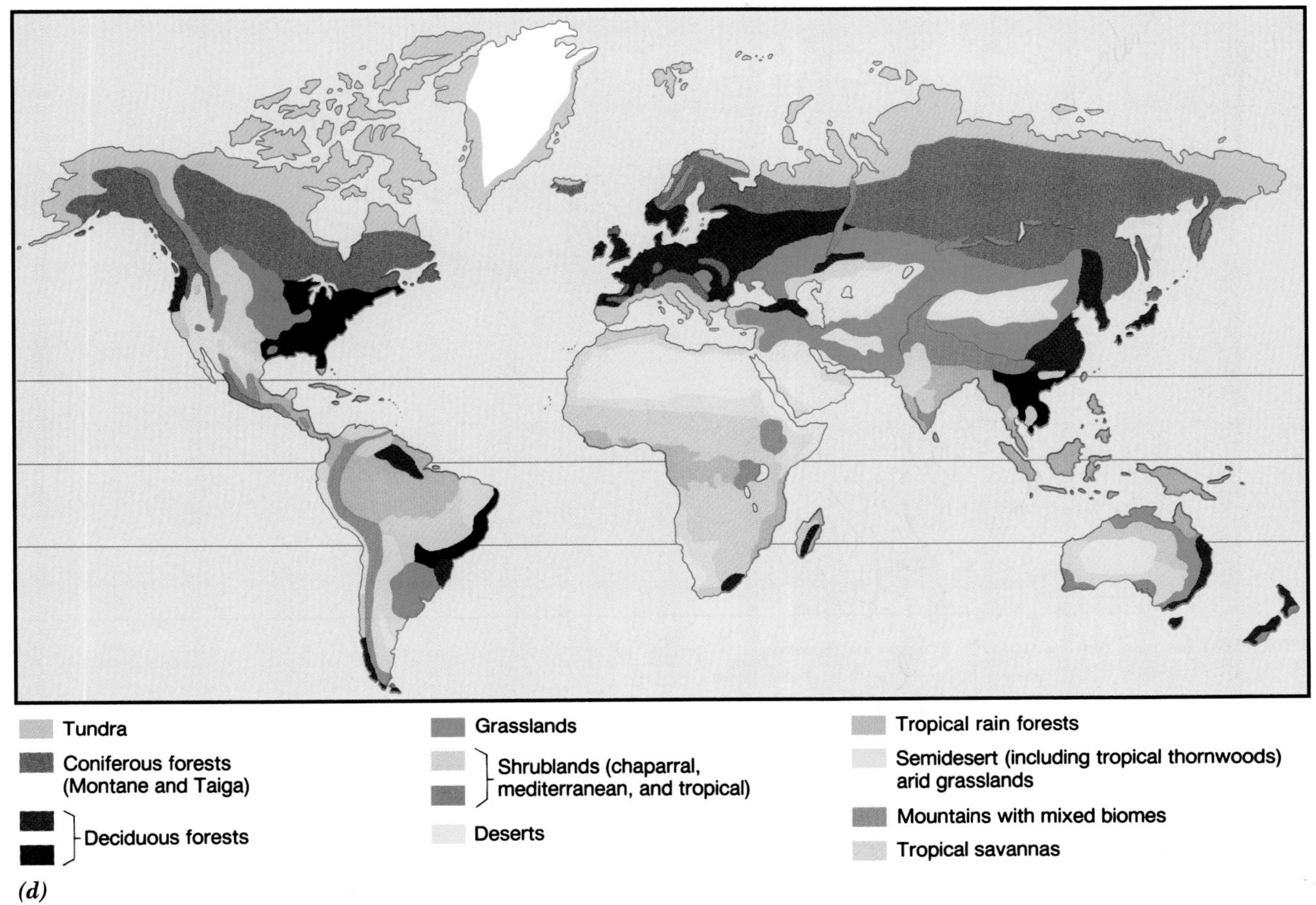

FIGURE 28-2

Climate and the earth's biomes. ***(a)*** The round shape of the earth results in differences in the amount of incoming sunlight in different regions. ***(b)*** The earth's yearly orbit around the sun and its constant tilt result in differences in the amount of incoming sunlight at different times of the year, generating the earth's seasons. ***(c)*** Six coils of air circulation create the earth's prevailing winds, as well as zones of high and low rainfall. ***(d)*** Climate is the principal factor determining where plants grow, which, in turn, is the principal factor determining where animals live. Not surprisingly, the locations of the earth's biomes closely follow its 11 main climate types.

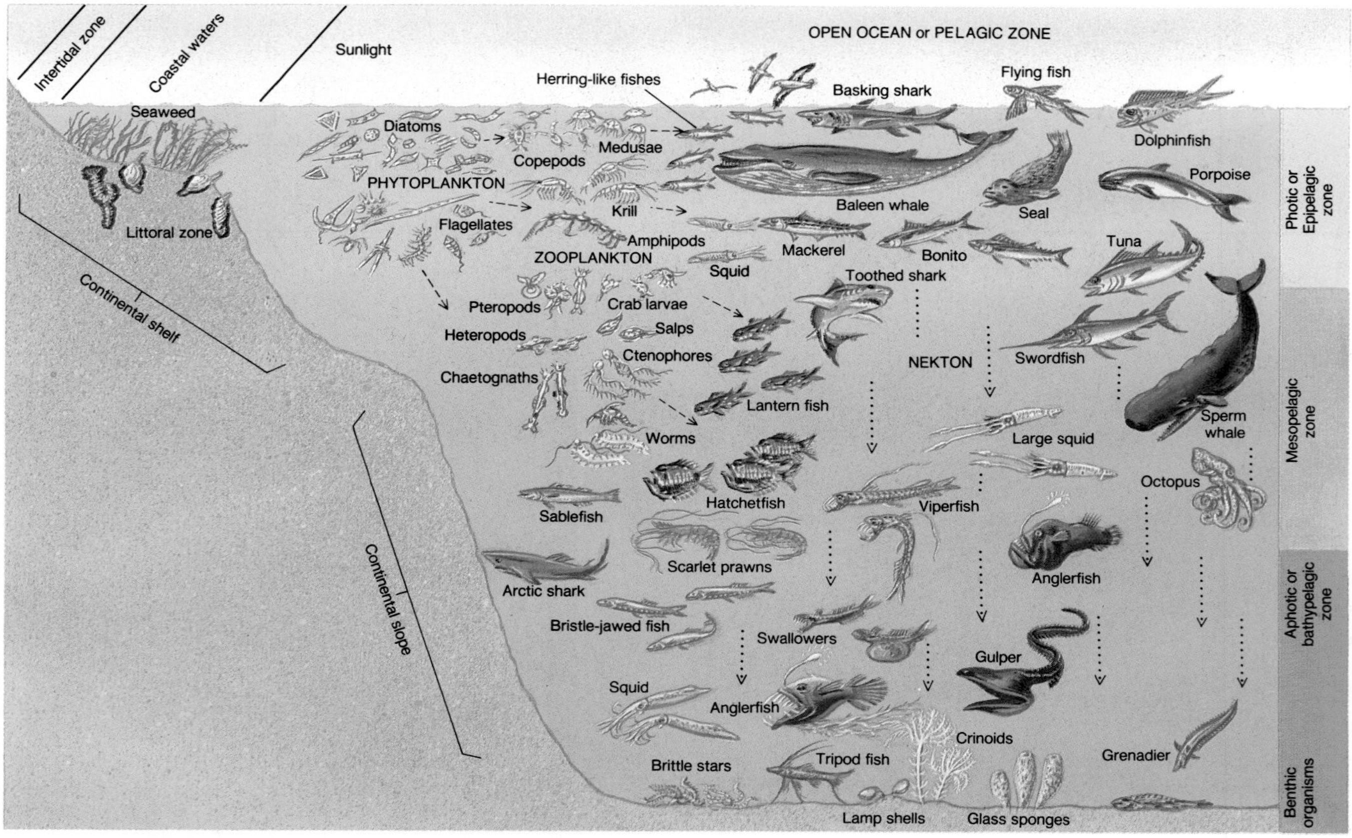

FIGURE 28-3

Marine habitats and representative marine organisms. Note the different zones and associated organisms. (Sizes of organisms are not drawn to same scale.)

FIGURE 28-4

Generating its own light, the suspended lantern on this anglerfish attracts this larval fangtooth in the black of the ocean's bathypelagic zone. Backward-pointing teeth prevent even large fishes from escaping this predator's grasp.

on both phytoplankton and zooplankton (or simply, plankton). An adult blue whale, for example, guzzles an average of 3 tons of plankton in a single day.

Since the dim midwaters of the mesopelagic zone do not receive adequate light to power photosynthesis, phytoplankton do not reside there. Inhabitants of this zone must therefore make daily migrations either up to the epipelagic zone or down to the bathypelagic zone to feed. Large fishes, whales, and squid are the principal animals of the mesopelagic zone. They eat smaller fishes that feed on the wastes or carcasses of organisms from the sunlit zone.

The pitch-black benthic zone is populated primarily by heterotrophic bacteria and scavengers that feed on a constant rain of organic debris, wastes, and corpses that settle to the bottom, as well as by predators that eat the scavengers and one another. These bottom dwellers include sponges, sea anemones, sea cucumbers, worms, sea stars, and crustaceans, as well as a collection of odd-looking fish, some with dangling lanterns that light up to attract a meal or a potential mate (Figure 28-4).

COASTAL WATERS

The greatest concentration of marine life inhabits the shallow **coastal waters** along the edges of continents and reefs (Figure 28-5). In addition to abundant light, coastal waters are generally rich in nutrients that are available as a continuous drain from the surrounding land. Waves, winds, and tides constantly stir coastal waters, distributing the nutrients. Bathed in light and nutrients, photosynthetic organisms grow at fantastic rates and in great profusion, providing ample food and habitats for a multitude of fishes, arthropods, molluscs, worms, and mammals.

The richest coastal waters occur in regions of *upwelling*, where nutrient-laden water from below circulates to the surface. With abundant light and nutrients, upwellings along the coasts of Peru, Portugal, Africa, and California form the earth's most fertile fishing waters.

Coastal waters present some unique problems for organisms, however. Rough, surging waters can shred or bash organisms against rocks. In these turbulent conditions, animals that remain in burrows or possess tough, protective shells are favored and naturally selected. And since surging waters can quickly wash organisms onto the shore or pull them out to sea, a number of coastal organisms have evolved adaptations for clinging to stationary objects. For example, algae have holdfasts that enable them to cling to rocks; mussels anchor themselves with powerful cords; and abalone hold tight with their large, muscular foot.

Shallow waters are also found along coral reefs (Figure 28-5*b*). In these shallow, warm, tropical waters, exceptionally diverse communities flourish. For example, F. H. Talbot and his colleagues identified nearly 800 species of fish around one small island on the southern edge of the Great

(a)

(b)

FIGURE 28-5
The ocean's bounty. Abundant light and the constant circulation of nutrients make coastal waters ***(a)*** and coral reefs ***(b)*** the ocean's richest habitats.

Barrier Reef in Australia. At the northern edge of the Great Barrier Reef, over 1,500 species of fish were recorded. Coral reefs develop only in tropical areas (between 30 degrees north and 30 degrees south latitudes), where water temperatures never fall below 16°C (60°F).

THE INTERTIDAL ZONE

Unlike life in coastal waters, where organisms are continually submerged in shallow waters, the organisms that inhabit intertidal zones must be able to survive in both water and air, as tides rhythmically submerge and expose their habitats. As its name suggests, the **intertidal zone** lies between high and low tides at the interface between the ocean and the land.

The distribution of organisms in the intertidal zone is a striking example of the dynamic interplay between physical factors (temperature and dehydration) and interactions between organisms (competition and predation). This interplay produces four distinct strata of organisms (Figure 28-6). The uppermost *splash zone* is where organisms receive only sprays of water at high tides. Below this is the

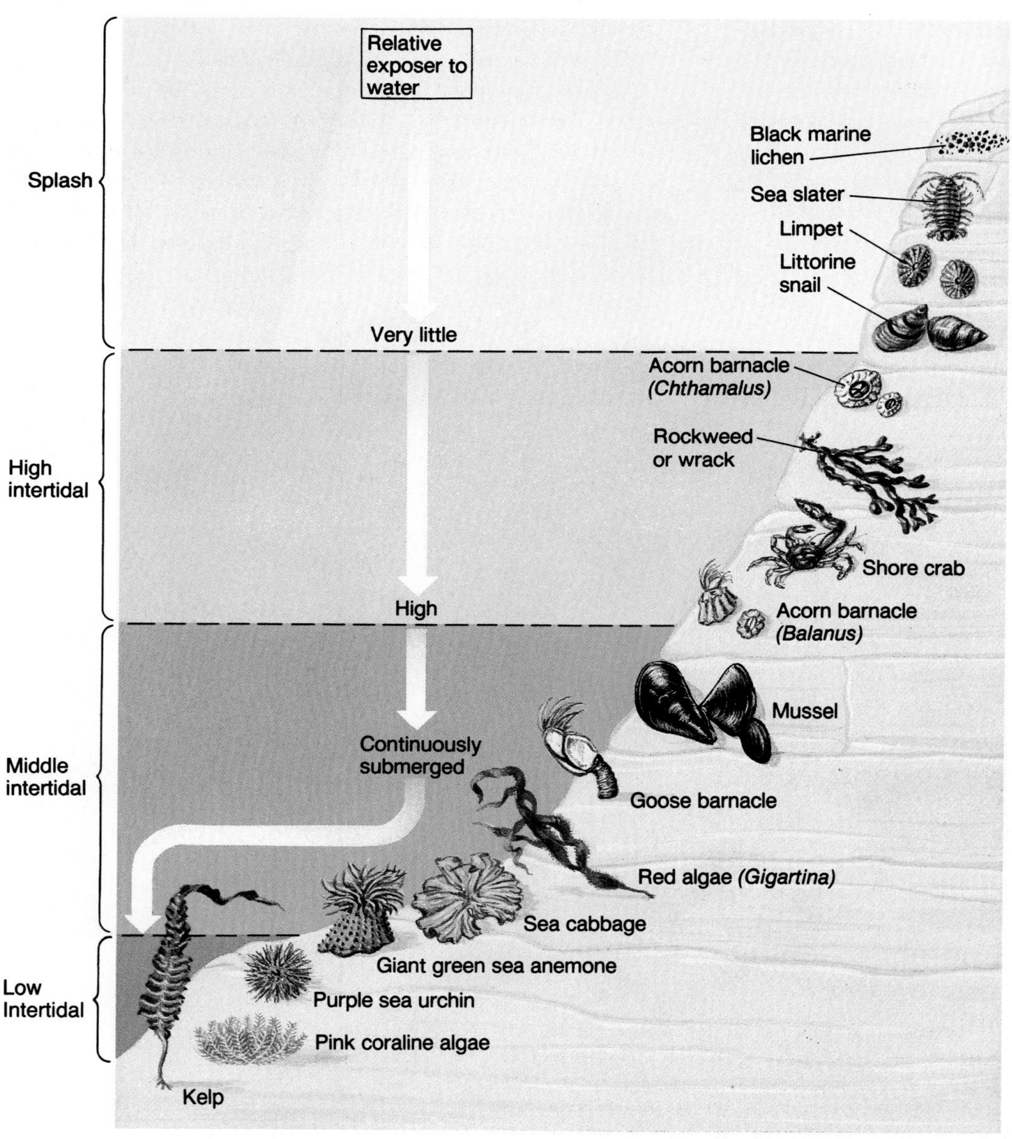

FIGURE 28-6

Intertidal zones. Organisms that live in the splash zone must survive with only periodic sprays of moisture. Organisms are submerged in water about 10 percent of the time in the high intertidal zone, 50 percent of the time in the middle intertidal zone, and 90 percent of the time in the low intertidal zone.

high intertidal zone, followed by the *middle intertidal zone*, and finally the *low intertidal zone.* Organisms in the high intertidal zone must be able to withstand exposure to air for longer periods than do those that inhabit the middle or low intertidal zones.

ESTUARIES

Estuaries form where rivers and streams empty into oceans, mixing fresh water with salt water (Figure 28-7). An organism's location in an estuary depends on its ability to tolerate different concentrations of salt. Salinity changes as tides bring in new influxes of salty sea water or as storms increase the flow of freshwater. At any one spot in an estuary, salinity may change within minutes, from low concentrations (equal to that of freshwater) to very high concentrations (the salinity of sea water), creating an ever-changing environment for the organisms that live there.

Despite such fluctuations, estuaries are tremendously fertile habitats; they are continually enriched with nutrients from rivers and from the debris that is washed in by tides. Because of this richness, phytoplankton and rooted plants flourish along the estuary's edges and provide abundant food for crustaceans, fishes, shellfishes, and the young of many open ocean animals that use rich estuaries for spawning.

FRESHWATER HABITATS

Oceans contain 97.2 percent of the earth's total water. The remaining 2.8 percent is freshwater: 2 percent is permanently frozen in ice caps and glaciers; 0.6 percent is groundwater; 0.017 percent is concentrated in lakes and rivers; and 0.001 percent is in the atmosphere as ice and water vapor. Since the quantity of water on earth is so enormous, even a small percentage like 0.017 percent produces more than 52,000 cubic miles of freshwater habitats on earth. The principal freshwater habitats are flowing rivers and streams and standing lakes and ponds.

Rivers and Streams

Size determines whether flowing freshwater forms a **stream** or a **river**; smaller streams converge into larger rivers. The constant flow of water erodes the land, sculpting landscapes and creating unique habitats that support particular groups of organisms. The velocity of a current

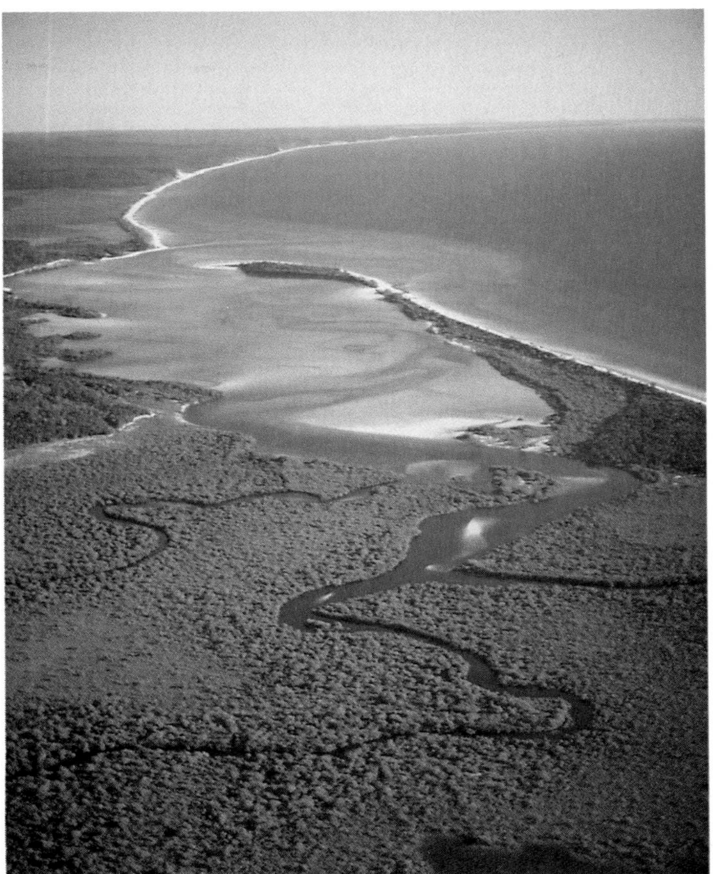

FIGURE 28-7

Estuaries are shallow inlets, where freshwater mixes with salty seawater. Nearly one-half of the ocean's living matter is found in such nutrient-rich estuaries. This estuary is on Fraser Island in Australia.

affects not only erosion but also the deposition of sediments and the supply of oxygen, carbon dioxide, and nutrients—factors that determine which organisms inhabit a stream or river.

In a river's open waters, the never-ending tug of the current presents unique challenges for animals. Some fishes use their fins to force themselves down to the less turbulent bottom waters. Conversely, powerful trout and salmon can swim fast enough to oppose the current or even to swim upstream.

Lakes and Ponds

Lakes and **ponds** are standing bodies of water that form in depressions of the earth's crust; lakes are larger and usually deeper than ponds. A typical lake has three zones: a shallow **littoral zone** along the water's edge; a **limnetic zone** that encompasses the lake's open, lighted waters; and a dark **profundal zone** at depths below which light is unable to penetrate. Rooted plants and floating algae are the characteristic photosynthesizers of the littoral zone. In the limnetic zone, phytoplankton and photosynthetic plants provide food for zooplankton, fishes, and other animals. And, as in the deep, dark waters of oceans, heterotrophic predators and scavengers inhabit the lake's profundal zone. Since oxygen often becomes scarce near the bottom of a lake, anaerobic bacteria and other species that can survive anaerobically are found in the profundal zone.

In many lakes, bursts of biological activity occur in the spring and autumn, when the lake's waters naturally circulate, bringing a supply of rich nutrients to the surface (Figure 28-8). Circulation is triggered in the spring as the sun warms the upper waters and winds stir the lake, driving heat and oxygen down into the deep waters and bringing nutrients up to the surface. This *spring overturn* stimulates biological activity. As spring passes into summer, heated surface waters become more buoyant and resist mixing. Water circulation stops altogether. Without mixing, oxygen is quickly depleted on the lake's bottom, and nutrients begin to accumulate, slowing down biological activity in the lake throughout the summer.

Circulation is triggered again in the autumn as heat is dissipated from warm surface waters to the cold air. Cooler surface waters sink, and the autumn winds increase, restoring circulation in the lake. Rising warmer water from the bottom brings nutrients upward and warms the surface waters, producing another burst of biological activity during this *autumn overturn*.

Unlike flowing rivers and streams, lakes and ponds may eventually fill in with sediment and organic matter and become dry land (Figure 28-9). A recently formed young lake is termed *oligotrophic* (little nourished) because it contains relatively few nutrients. Oligotrophic lakes support very little life; as a result, they are usually crystal clear. Middle-aged lakes are *mesotrophic* (moderately nourished) and support large populations of organisms. Finally, the nutrient-rich waters of older, *eutrophic* (fully nourished) lakes support the largest populations and the greatest diversity of species. As organisms continually add increasing supplies of organic matter to the water, the filling rate is accelerated. Outside sources of nutrients, such as human sewage or runoff from nutrient-soaked agricultural lands, also promote lake filling.

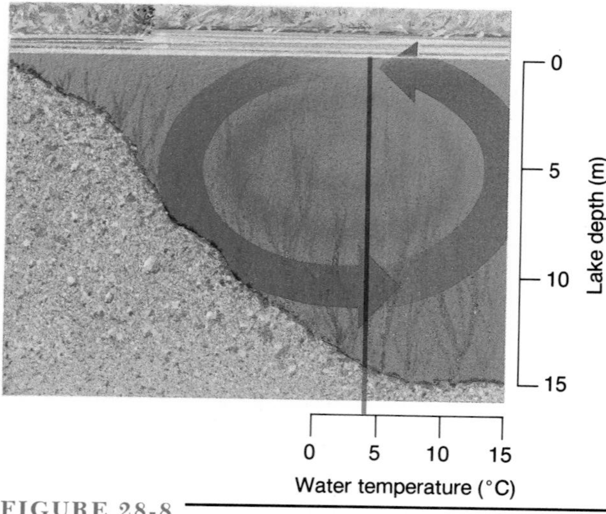

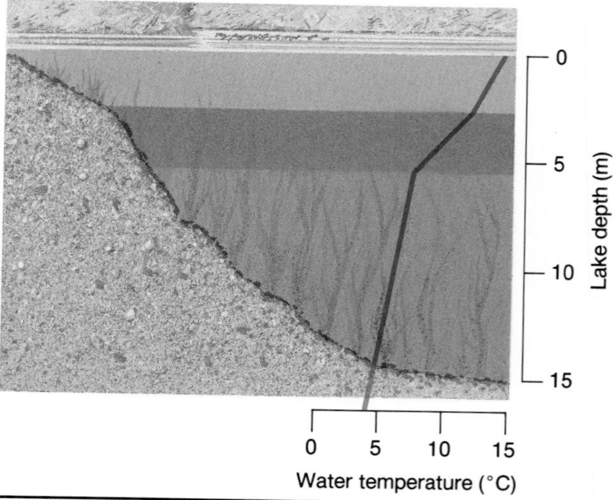

FIGURE 28-8

Seasonal changes in a temperate lake. Water temperature is nearly uniform during spring and autumn overturns. In the summer, warmer surface waters resist mixing, forming distinct temperature zones.

(a)

(b)

(c)

FIGURE 28-9
A lake's life cycle. ***(a)*** Newly formed *oligotrophic lakes* are clear. Oligotrophic lakes eventually begin to fill with sediment, providing nutrients for greater numbers of organisms. ***(b)*** As sediment and life increases, oligotrophic lakes become *mesotrophic lakes,* which then become *eutrophic lakes* **(c)**.

BIOMES: PATTERNS OF LIFE ON LAND

Large terrestrial (land) ecosystems are called **biomes.** Plants form the bulk of the living mass in a biome; as a result, each biome is characterized by the predominant type of plant that grows there. Dense, tall trees form forest biomes; short, woody plants form shrublands; and grasses and herbs form grasslands. Since the prevailing climate (especially temperature and moisture) is the primary factor in determining the types of plants that grow in an area, the earth's terrestrial biomes tend to follow global climate patterns (Figure 28-2). And since climate changes with elevation, layers of biomes are found on mountainsides (Figure 28-10). Climate affects terrestrial biomes in another, unexpected way: Prevailing climate patterns spread pollutants from human activities over very wide distances, causing disruptions several hundred and thousands of miles away (see The Human Perspective: Acid Rain and Acid Snow: Global Consequences of Industrial Pollution).

FORESTS

Over 30 percent of the earth's land is covered by forests. Biologists recognize three main forest biomes: (1) lush *tropical rain forests* that grow in a broad belt around the equator; (2) *deciduous forests,* whose trees drop their leaves during unfavorable seasons; and (3) *coniferous forests* that are dominated by evergreen conifers.

Tropical Rain Forests

Warm temperature, abundant rainfall (over 250 millimeters, or 100 inches, a year), and roughly constant daylength throughout the year produce rich, **tropical rain forests** in Central and South America, Africa, India, Asia, and Australia (Figure 28-2). Over half of the earth's forests are tropical rain forests. They contain more species of plants and animals than do all other biomes combined.

Towering trees over 50 meters high (160 feet) form the upper story (overstory) of the tropical rain forest (Figure 28-11). Below these giants is a layer of tall trees (the

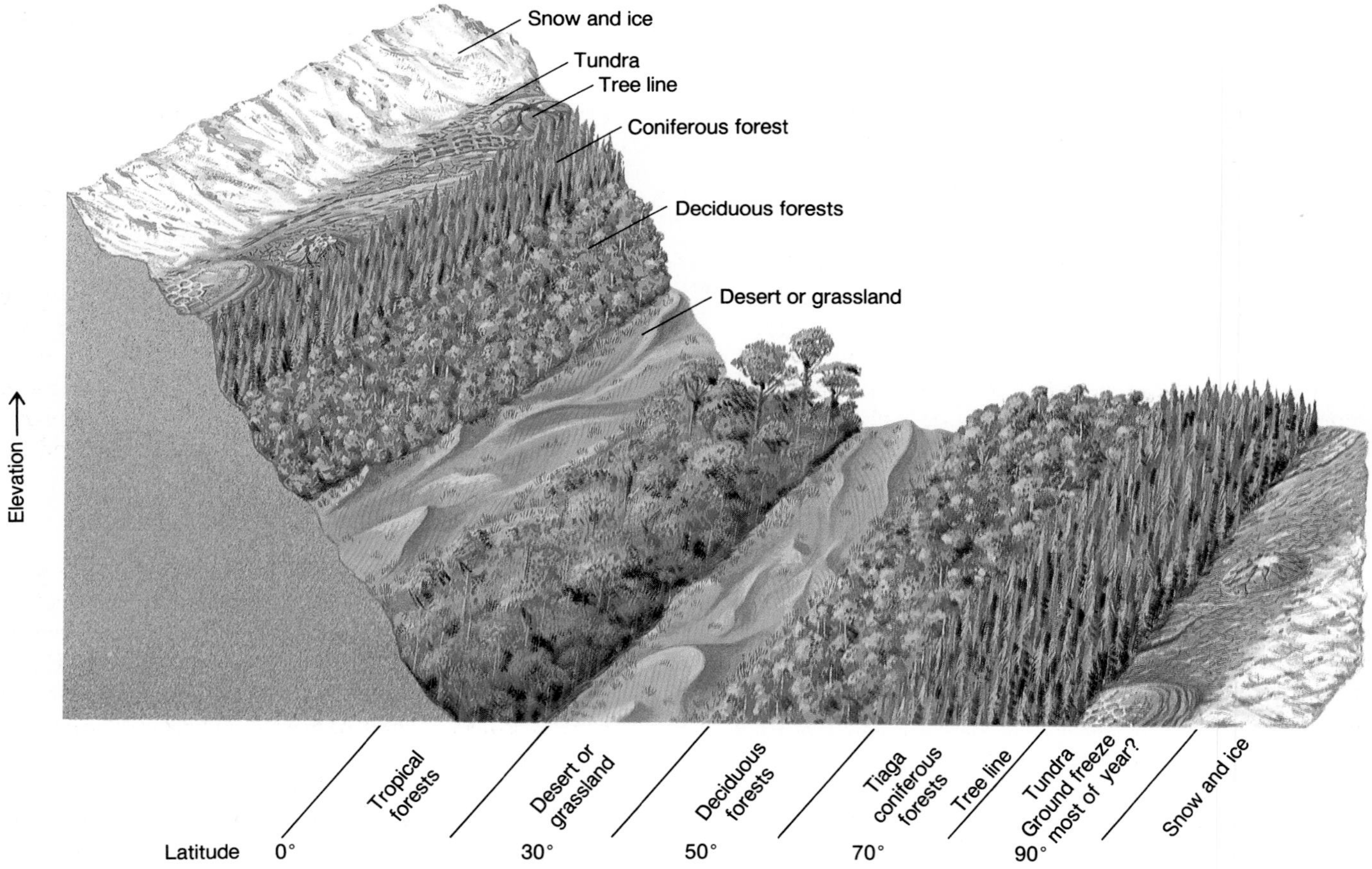

Snow and ice
Tundra
Tree line
Coniferous forest
Deciduous forests
Desert or grassland
Elevation ⟶

Tropical forests
Desert or grassland
Deciduous forests
Tiaga coniferous forests
Tree line
Tundra Ground freeze most of year?
Snow and ice

Latitude 0° 30° 50° 70° 90°

FIGURE 28-10

Elevation and latitude affect the distribution of biomes. Terrestrial biomes change according to elevation, as well as with distance from the equator.

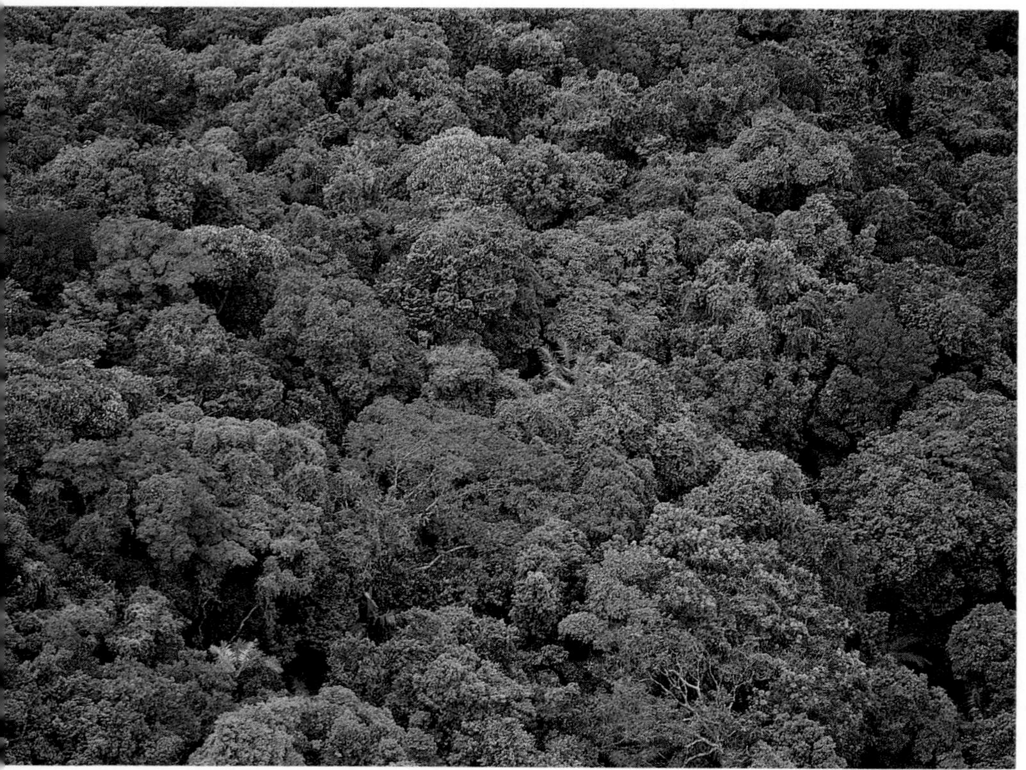

FIGURE 28-11

A picture of unrivaled diversity. Life in the tropical rain forest is more varied and more abundant than is that found in any other biome on earth. Immense, broad-leafed evergreen trees drip moisture onto masses of clinging epiphytes (plants such as bromiliads and orchids that grow on other plants) and enormously long vines, some over 200 meters long. These giant trees have widespreading buttresses that keep them upright, even in shallow tropical soil.

understory) so densely packed that very little sunlight penetrates through their canopy. Since much of the light is blocked, plant growth is greatly suppressed below these two tree layers. Where a glimmer of sunlight does penetrate, ferns, shrubs, and mosses crowd the forest floor.

Of all the biomes, the rate of decomposition is fastest in the tropical rain forest. A dead animal or fallen tree can swiftly be cleared from the forest floor by hordes of fungi and bacteria that promptly carry out decomposition. Despite such rapid decomposition, virtually no nutrients accumulate in the soil; the nutrients are either absorbed immediately by plants or washed away by steady rains. Surprisingly, the earth's lushest forests have poor soils, a condition that has contributed to the rapid destruction of the earth's tropical rain forests by humans (see Chapter 13, The Human Perspective: Saving Tropical Rain Forests, page 243).

Deciduous Forests

In those areas of the earth that have distinct seasons, biological activity follows the seasonal pattern (Figure 28-12). Unfavorable growing conditions during one or more seasons produce **deciduous forests** with trees, such as maple, beech, hickory, and oak, that produce leaves during warm, wet periods and lose their leaves at the onset of either the dry season (in tropical deciduous forests) or the cold season (in temperate deciduous forests).

The trees found in deciduous forests are less dense than are those found in tropical rain forests, enabling sunlight to reach the forest floor. In the presence of adequate light, many layers of plants develop beneath the top layer of deciduous trees. Understory plants receive varying amounts of sunlight throughout the year, as overstory deciduous trees bear fully expanded leaves and then drop their leaves at the beginning of the unfavorable season. The growth and reproductive cycles of understory plants coincide with brief periods of maximum sunlight and favorable conditions.

Coniferous Forests

Some of the most extensive forests in the world are populated by evergreen conifers (cone-bearing trees), such as pines, firs, and spruces (Figure 28-13). Belts of **coniferous forests** girdle the huge continents in the Northern Hemisphere (Figure 28-2) and blanket higher elevations of mountain ranges in North, South, and Central America, and in Europe.

The transition from deciduous forests to coniferous forests is a result of colder winters. During the spring and summer, melting snow fills the lakes and forms watery bogs

(a)

(b)

FIGURE 28-12

The yearly rhythm of a deciduous forest. Deciduous trees in temperate areas produce leaves in the spring, when temperatures are warm. During the summer, fully expanded leaves of the overstory trees shade the plants below ***(a)***. Protected terminal buds enable dormant deciduous trees to withstand the bite of freezing winds and snows during the winter ***(b)***.

◁ THE HUMAN PERSPECTIVE ▷

Acid Rain and Acid Snow: Global Consequences of Industrial Pollution

Two alarming trends were recently documented in the biomes of North America and Europe. The first was a change in the color of several lakes, from murky green to crystal clear. Sounds good, right? Not really, for a green lake is a biologically active lake, teaming with microscopic algae, which are eaten by small aquatic animals, which, in turn, are eaten by fish. A clear lake is biologically sterile, devoid of aquatic life. How widespread is lake sterility? In eastern Canada, nearly 100 lakes have become sterile, as have more than 1,000 lakes in the northeastern United States and approximately 20,000 lakes in Sweden.

The second trend was the premature death of an excessive number of trees, especially those on high slopes that face prevailing winds. More than 17 million acres (7 million hectares) of trees in North America and Europe look as if they've been burned, but there have been no fires. Over 50 percent of the forests in Germany alone are affected in this way, impacting more than 1.2 million acres (500,000 hectares).

Just how are dead trees and dead lakes related? The destruction of both trees and lakes is caused by acid rain or runoff from acid snow. Acid deposition not only destroys forests and kills lakes, it also damages crops, alters soil fertility, and erodes statues and buildings.

The chemicals that create acid rain and snow (sulfur oxides and nitrogen oxides) come primarily from human activities. Although sulfur oxides are released during volcanic eruptions, forest fires, and from bacterial decay, quantities of sulfur oxides from human activities far exceed those that come from natural sources. Nearly 70 percent of sulfur oxides comes from electrical generating plants, most of which burn coal. Most nitrogen oxides come from motor vehicles and industries, including electrical generation. When sulfur and nitrogen oxides mix with the water in the air, they form acids:

$$SO_2 + H_2O \rightarrow H_2SO_4 \text{ (sulfuric acid)}$$
$$NO_2 + H_2O \rightarrow HNO^{-3} \text{ (nitric acid)}$$

Acid rain or acid snow has a pH below 5.7, the pH of unpolluted rain. Over the past 25 years, rains in the northeastern United States dropped to an average pH 4.0. The lowest recorded pH for rainfall was 2.0, reported in Wheeling, West Virginia. The rain in Wheeling was more acidic than lemon juice!

Acids that create acid rain and snow remain airborne for up to 5 days, during which time they can travel over great distances. For example, the acid rainfall that killed many of the lakes in Sweden was caused by pollutants that were released in England. The acid rain that is damaging trees and lakes in the Adirondack Mountains of New York originated in the upper Mississippi and Ohio River Valleys. Since these acids circulate in large air masses, acid deposition is widespread, spreading from Japan to Alaska, from New Jersey to Canada, to name just a few places.

The rate of destruction caused by acid rain and snow is increasing. A 1988 survey of U.S. lakes lists 1,700 lakes as having high acidity. Another 14,000 lakes were identified as becoming acidified. Scientists estimate that by the turn of the century, over half of the 48,000 lakes in Quebec, Canada, will have been destroyed.

In 1979, the U.S. Congress passed the "Acid Precipitation Act" to identify sources of acid deposition. Congress is also considering taking steps to cut sulfur oxide emissions by nearly 50 percent, and nitrogen oxides by 10 percent by the year 2000. Such steps might include: (1) installing scrubbers on power plant smoke stacks; (2) using coal that is low in sulfur; (3) using coal that has been pretreated to remove sulfur; or (4) reducing auto and truck use.

FIGURE 28-13
A coniferous forest in the icy grip of winter. Many animals hibernate or migrate to less severe habitats. Only a few, like this moose, remain active during harsh winter months, forced to travel over large areas in search of scarce food.

and marshes. This is why the northern coniferous forests are called **taigas,** which is Russian for "swamp forests." The growing season is relatively short; overall biological activity is restricted to a period of only 3 to 4 months. As a result, decomposition by fungi and bacteria is limited to these relatively brief warm periods. Consequently, the forest floor accumulates a thick layer of needles, producing acidic and relatively infertile soils.

Coniferous forests typically have two stories of plants: a dense overstory of trees, and an understory of shrubs, ferns, and mosses. Unlike the tropical rain forests, coniferous forests are usually not made up of a mixture of many tree species but of expanses of many individuals of a few species.

TUNDRA

The timberline is a zone in which trees thin and eventually disappear; it marks the boundary between the coniferous forest and the **tundra** biome (Figures 28-2 and 28-10). In Russian, "tundra" means a marshy, unforested area, describing the frigid, treeless landscape. There are two types of tundra: *alpine tundra,* which are found at high elevations of mountain ranges; and *arctic tundra,* which are found to the north at high latitudes in Alaska, Canada, and in northern Europe and Asia (Figure 28-14). Both types have low-growing plants (often only 10 centimeters, or about 4 inches tall) that often belong to the same species. Tundra plants include mosses, lichens. perennial forbs, grasses, sedges, and dwarf shrubs.

The climate that produces a tundra is brutal. The growing season lasts only 50 to 90 days per year, with an average temperature in the warmest month of less than 10°C (50°F). Thick snow covers the ground, and icy winds blow in the winter months. During the brief summer period, melting snows create frigid marshes and ponds. In the arctic tundra, only the top 0.5 meters (1.5 feet) thaw, leaving permanently frozen soil, or *permafrost,* that halts root growth, restricts drainage, and impairs decomposition. Under such harsh conditions, it is not surprising that relatively few plants and animals have evolved adaptations that enable them to survive the rigorous tundra climate. In fact, there are only about 600 plant species in the entire arctic tundra region of North America, which covers thousands of square miles, fewer species than you would find in a single square mile of tropical rain forest.

GRASSLANDS

At one time, over 30 percent of the earth's lands were covered by **grasslands** of densely packed grasses and herbaceous plants (Figure 28-2 and 28-15). In North America, grasslands were once more widespread than any other biome. But the combination of rich soil and favorable growing (and living) conditions made grassland habitats prime targets for agriculture, livestock grazing, and urbanization. Today, most of the earth's grasslands have been cleared for farming and human development.

Grasslands naturally develop in regions that have cold winters, hot summers, and seasonal rainfall. Natural fires periodically clear grasslands, opening up space and releasing nutrients for new growth. Rainfall pattern greatly affects the nature of the grassland. For example, as precipitation decreases from east to west, *tallgrass prairies,* with plants reaching 2 meters (6.5 feet) in height, gradually give way to *mixed grass prairies,* with grasses growing no more than 1 meter (3.28 feet) tall, which, in turn, give way to *shortgrass prairies,* with bunch grasses less than 0.3 meter (1 foot) in height.

FIGURE 28-14

Life is fleeting in the tundra. During the brief summer growing period, ground-hugging tundra plants erupt with new growth, produce flowers, and set seeds, before severe weather returns. In the arctic tundra shown here, permanently frozen soil (*permafrost*) prevents drainage of water from melting snow, producing multitudes of shallow ponds.

FIGURE 28-15

Grassland in New South Wales, Australia.

SAVANNAS

A combination of grassland and scattered or clumped trees forms a **savanna** biome. *Tropical savannas* are found in South America, Africa, Southeast Asia, and Australia and cover nearly 8 percent of the earth's land (Figure 28-16). In many temperate areas, pockets of savannas are found sandwiched between grasslands and forests.

Like grasslands, savannas are characterized by seasonal rainfall, punctuated with a dry season. During the dry season, the above-ground stems of the grasses and herbs die, providing fuel for fast-moving surface fires. The grasses and herbs have evolved adaptations that enable them to recover quickly from fires by resprouting from underground roots and stems, or if they are killed, by fast-growing seedlings. Since ground fires move rapidly through the savanna, the trees are not usually damaged.

The tropical savannas of Africa are able to support large populations of herbivores, including wildebeest, gazelles, impalas, zebras, and giraffes. However like many grasslands, tropical savannas are now being used for grazing. Overgrazing reduces grass cover and allows trees to invade the area, reducing the number of animals that can inhabit overgrazed savannas. In addition, the loss of ground cover allows nutrient-rich soil to be blown or washed away, leaving only coarse sand and gravel behind that cannot hold sufficient water for plant seeds to be able to germinate and grow. As a result, neighboring deserts expand into areas that were once tropical savannas, a process called **desertification.** Desertification is especially pronounced in Africa, where the Sahara desert has expanded southward over 200 miles from its original position, claiming fertile tropical savannas.

SHRUBLANDS

Woody, drought-tolerant shrubs predominate in **shrublands.** In regions of the earth with a Mediterranean-type climate (hot, dry summers and cool, wet winters), shrubs grow very close together and have small, leathery leaves with few stomates and thick cuticles to retard water vapor loss (Chapter 12). Remarkably, similar shrublands develop in areas with the same climate, from areas around the Mediterranean Sea, to coastal mountains in California and Chile, to the tip of South Africa, to southwestern Australia. In California, this type of shrubland is called a *chaparral* (Figure 28-17).

Organisms that live in Mediterranean-type shrublands must survive many stressful periods. For example, over the long, hot summers, when water is needed most by organisms, there is little or no rainfall. Since the supply and

FIGURE 28-16
The African savanna has flat-topped acacias and dry grassland. Some tropical savannas in central Africa are studded with palm trees.

FIGURE 28-17

The chaparral is a shrubland that forms in those regions of California that have a Mediterranean-type climate. In addition to being tolerant to drought, most chaparral plants are adapted to fire and are able to regrow rapidly after a fire. Layers of charcoal testify that fire is a natural component of the chaparral, and of forests, grasslands, savannas, and other shrublands.

demand for water are directly out of sync, most plant and animal activity is restricted to the spring, when temperatures are warm and the soil is still moist from winter rains.

Fires are common in Mediterranean-type shrublands; the accumulation of dry, woody stems and highly flammable litter greatly increases the chances of fire during the hot summer. Most shrubland plants have evolved adaptations—underground stems or seeds that germinate only after being burned—that enable them to recover from recurring fires. In the chaparral of California, a burned area often recovers within just a few years.

In tropical regions that have a short wet season, another type of shrubland develops: the **tropical thornwood.** Most thornwood plants lose their small leaves during the dry season, reducing transpiration and exposing sharp thorns that discourage even the hungriest browser. One type of thornwood, the acacia plant, has coevolved with certain ants, forming a close partnership that helps both species survive. The plant provides food and shelter for the ants, and the ants patrol the ground and stems of the plant, aggressively warding off hungry herbivores and removing flammable debris that may accumulate under the plant (Figure 28-18).

DESERTS

Intense solar radiation, lashing winds, and little moisture (less than 25 centimeters, or 10 inches, per year) create some of the harshest living conditions in the biosphere. These conditions characterize the **deserts,** which cover about 30 percent of the earth's land and occur mainly near 30 degree north and south latitude, where global air currents create belts of descending dry air. Deserts may also be produced in the rainshadows of high mountain ranges (a rainshadow is a reduction in rainfall on the leeward slopes of mountains, slopes that face away from incoming storms). The skies over the deserts are generally cloudless,

FIGURE 28-18
Mutual aid. Swollen thorns of this acacia tree provide a home and brooding place for ants, while nectar and nutrient-rich plant structures produced at the tips of leaves provide the ants with a balanced diet. Ants protect the acacia by attacking, stinging, or biting herbivores (or scientists). Ants also remove flammable debris from around the base of the tree, forming a natural fire break. Any plant seedling that grows within this "bare zone" is quickly clipped and discarded, protecting the acacia from competing plants, including its own offspring.

so the sun quickly heats the desert by day, producing the highest air temperatures in the biosphere. High daytime temperatures and persistent winds accelerate water evaporation and transpiration of water vapor from plants. High evapotranspiration and low annual rainfall characterize all deserts (Figure 28-19*a*).

Plants have evolved many adaptations that allow them to survive the rigors of the desert. For example, following brief spring and summer rains, the desert floor often becomes carpeted with masses of small, colorful annual plants (Figure 28-19*b*). These annuals germinate, grow, flower, and release seeds, all within the brief period when water

(a)

(b)

FIGURE 28-19
The desert. ***(a)*** During dry periods, a few hardy perennial bushes, cacti, and Joshua trees are scattered over the scorching desert floor. Many animals remain underground, in moist, cool burrows. ***(b)*** Following a spring rain, the desert floor blooms with colorful annual growth.

is available and temperatures are warm. By remaining dormant as seeds the remainder of the year, annuals avoid the most severe stresses of the desert.

Many desert animals rely on learned and instinctive behavior to avoid the desert heat and dryness. Some, such as kangaroo rats and ground squirrels, remain in cool, humid underground burrows during the day and search for food at night or in the early morning or late afternoon. If they venture out during the day, their underground burrows act as heat sinks, quickly removing the heat they acquired while scurrying across the desert in search of food. Many birds avoid stressful desert conditions by simply flying to less hostile areas, while other animals *aestivate*; that is, they sleep through the driest part of the year. With watertight skins, desert snakes and lizards are able to conserve water even during the heat of the day. The desert toad (*Bufo punctatus*) uses a survival strategy similar to that employed by succulent plants that have internal water-storing tissues: It stores water in its urinary bladder, carrying its own version of an internal well.

EVOLUTION AND ADAPTATION: TYING IT TOGETHER

No organism can ever be separated from its environment. As a result, organisms are continually affecting and, in turn, are continually being affected by their environment. This constant interplay is the foundation for evolution, and it is also the study of ecology. In other words, evolution takes place in the ecological arena.

Organisms occur where they do because, through evolution, they have acquired adaptations that enable them to survive and reproduce in particular habitats. These adaptations are the result of natural selection, the process that drives evolution. Recall that natural selection is the result of a number of factors, including limited resources, competition for space or mates, adverse changes in the physical environment, the introduction of new organisms into an ecosystem, and so on. These "natural selection factors" are also ecological factors. That is, natural selection is ecology in action.

SYNOPSIS

The biosphere is composed of the earth's lands, waters, and air, which support all life as we know it. Although organisms are scarce in extreme environments, the biosphere contains all of the resources and conditions necessary for life. Since resources for life are limited, recycling of matter within the biosphere is necessary for the continuation of life throughout time.

The earth's oceans support the greatest quantity of life. Most aquatic life is concentrated in shallow coastal waters along the fringes of land and coral reefs.

Biomes cover vast expanses of the earth's land and are characterized by particular plants. Since prevailing climate is the primary factor that controls plant distributions, terrestrial biomes follow global climate patterns and elevational gradients.

Forests cover more than 30 percent of the earth's land. The major forest biomes include tropical rain forests, deciduous forests, and coniferous forests. Tropical rain forests contain more species of organisms than do all other biomes combined.

Originally, grasslands covered more than 30 percent of the earth's land. Today, most grasslands are used for agriculture, grazing, and other types of development.

Nearly 8 percent of the earth's lands are covered by savannas of grass and scattered trees. Overgrazing is changing the savannas, reducing their ability to support animal life.

Dense shrublands develop in land areas that have a relatively long dry season. Because fires are frequent, many shrubland organisms have evolved adaptations that enable them to recover from recurring fire.

Hot, dry deserts create harsh living conditions that have resulted in many unique adaptations in desert organisms. Some desert organisms avoid harsh conditions by remaining in moist, cool burrows or by restricting their activities to brief periods in the day or year when conditions are more mild. For example, many desert plants germinate, grow, produce flowers, and set seed only during brief periods in the spring when water is available and temperatures are warm.

Review Questions

1. Rank the earth's terrestrial biomes, starting with that which covers the greatest land surface and ending with that which covers the smallest area. Why do some biomes cover more land than others?
2. Explain how lakes have alternating periods of low biological activity and high biological activity.
3. List the strata of organisms found in the intertidal zone. What kinds of adaptations are necessary to help organisms withstand increasing lengths of exposure to dry air?
4. In what ways are the following similar?
   a. lakes and rivers.
   b. profundal zone and desert.
   c. forest understory and bathypelagic zone.
   d. coastal waters and littoral zone.
5. For each of the following, list the major adaptations needed for survival. Are there distinct differences for plants and animals? Make a list of different plant and animal adaptations for each habitat and briefly explain why such differences exist.
   a. prolonged hot and dry season.
   b. surging waters against a rocky coastline.
   c. calm, warm fresh water with abundant nutrients and many species of organisms.
   d. dramatic fluctuations in water salinity.

Critical Thinking Questions

1. Recently, F. Sherwood Rowland projected that the ozone hole depletion will continue for at least another decade despite attempts to cut CFC use by 50 percent before 1999. What was the basis for Rowland's ominous projection? In addition to international accords such as the Montreal Protocal, what other measures could be taken to slow the rate of ozone depletion?

2. Prepare a chart, similar to the one below, showing the ways in which each organism affects its environment and how it is affected by the environment.

Organism	How It Affects Its Enviroment	How It Is Affected by Its Environment
earthworm		
moss growing on a rock		
maple tree		
elephant		
reef building coral		
bacterium of decay		

3. The earth's diverse climates create the myriad habitats that support a tremendous variety of organisms. In what ways would the diversity of climate (and therefore the diversity of life) change if conditions were different? For each of the following conditions describe the impact on climate change and then the impact on life:
   a. the earth does not rotate on its axis once every 24 hours.
   b. the earth is not tilted 23.5°
   c. all the earth's land mass is contained in a single, large continent. (Is the location of this land mass significant? If so, be sure to state its location and effects.)
4. What environmental and evolutionary factors account for the fact that species diversity is greater in terrestrial biomes, yet the abundance of organisms is greater in aquatic habitats?
5. For each of the human activities listed, describe its effects on the biosphere: deforestation, automobile travel, agricultural use of fertilizers, use of pesticides, heavy industry, destruction of wetlands, building large cities.

CHAPTER 29

Natural Order of Life: Ecosystems and Communities

Hundreds of random samples (small circles) of insects and plants in the Great Smoky Mountains showed that both plants and insects distribute themselves independently. Data from different samples are shown by histograms.

STEPS TO DISCOVERY
The Nature of Communities

Conflicting views arise even among brilliant scientists in the same field. Each unresolved conflict heralds a research opportunity for an alert and eager student. When Robert H. Whittaker began graduate school in 1946 at the University of Illinois at Urbana to study insect ecology, he soon found himself embroiled in a 30-year-old conflict among plant ecologists regarding the nature of communities of organisms.

Before Whittaker's time, many scientists had already observed that groups of plants and animals tended to form repeatable, discrete communities of organisms. Indeed, distinct communities have been described at least as far back as 300 B.C. by Theophrastus, the Greek philosopher who studied with Aristotle. Much later, during the 1800s, Carl L. Willdenow, one of the first botanists to study the distributions of plants, noted that similar climates tended to produce similar plant communities, even in regions that are widely separated.

But not everyone shared Willdenow's view. By the turn of this century, botanists had still not agreed on the principal cause of plant communities. For example, two eminent botanists, Frederick E. Clements of the University of Nebraska and Carnegie Institution of Washington, D.C., and Henry A. Gleason of the New York Botanical Gardens, formulated opposing views on the nature of plant communities. Clements considered a plant community a "super organism;" that is, a well-defined association of plant species that are always found together, much as the cells, tissues, and organs of an organism are always found together. According to Clements, the groups of species in a community all have identical distribution limits, so they recur together in distinct communities. In contrast, Gleason argued that each plant species was distributed individualistically; that is, instead of all species in a community having identical distribution limits, the limits of each species were determined by the species' own genetics. As a result, each species' distribution changed gradually, making it difficult to divide vegetation into discrete associations or communities. Although Clements proposed his view in 1916, and Gleason in 1926, this conflict remained unresolved until the mid-1940s, the time Robert Whittaker entered graduate school.

Whittaker referred to the conflict between Clements and Gleason as "exciting confusion" for a new graduate student. Whittaker then decided to test these conflicting views on insects. He sampled the distribution of insects at different elevations in the Great Smoky Mountains National Park in Tennessee and North Carolina. The hypothesis Whittaker formulated was in line with Clements' view. That is, Whittaker proposed that he would find distinct groups of insect species along the continuous elevational gradient, each group more or less separated by abrupt transitions. Instead, he discovered that insect populations changed irregularly with elevation, disproving his (and Clements') hypothesis.

Whittaker decided to retest the same hypothesis, this time on plant species in the same forest. He took 300 random samples of plants over the mountain range. Again, he failed to find definite groups of plant species. Whittaker finally rejected his hypothesis and reanalyzed his results to see whether Gleason's individualistic view applied to the vegetation in the Great Smoky Mountains. When he plotted plant distributions on a chart with elevation and moisture gradients as axes (an analysis technique he'd invented), Whittaker verified that plants distributed themselves individualistically. In his words, Whittaker found "a subtly wrought tapestry of differently distributed species populations variously combining to form the mantle of plant communities covering the mountains."

- A long drought destroys vast expanses of grass and trees, causing thousands of animals to starve and leaving others frail and malnourished.
- The cold runoff from an exceptionally heavy winter snow keeps streams cool throughout most of the summer, slowing hatching and larval development of blackflies, mayflies and caddisflies, which, in turn, reduce the number and size of fishes that normally eat these larvae.
- Acrid Los Angeles smog blows into mountain forests, where it reduces photosynthesis in pines by as much as 80 percent. Unable to manufacture enough resin (sap) to protect themselves against burrowing insects, thousands of pines are dying from bark beetle infestations.

These examples illustrate a fundamental ecological principle: Organisms are continually affected by other organisms as well as by the surrounding physical environment. Because of this constant interplay, the organisms and the physical environment in a particular area function as an organized unit known as an **ecosystem.** This level of biological organization includes both the community of living organisms and the physical, nonliving environment.

▼ ▼ ▼

ECOSYSTEM BOUNDARIES

The variety of ecosystems in the biosphere is enormous. Each expansive biome contains countless ecosystems (Figure 29-1). The boundaries that separate ecosystems are not always sharp delineations, however, since ecosystems are linked with other ecosystems, to some degree. For example, consider a lake and a cave, two ecosystems that at first seem clearly independent. However, the rate at which the lake fills with sediment may affect the number of beetles that live in a cave, even if the lake and the cave are separated by several miles.

This connection is possible because bats link the two "separate" ecosystems. Bats forage for insects around the lake, where insects are plentiful, and then return to the cave during the day to rest. As the lake fills with sediment and organic material, more plants grow along the margins of the lake, providing more food for greater numbers of plant-eating insects, the dietary mainstay for the bats. When well-fed bats return to the cave, they defecate large amounts of feces onto the cave floor. The bat feces, in turn, nourish the growth of mold, a staple of the cave beetles' diet. Thus, the amount of sediment in the lake eventually affects the number of beetles in the cave. Furthermore, since beetles, crickets, flies, springtails, moths, spiders, centipedes—indeed, nearly every organism that makes up the cave community—depend on the supply of bat droppings, the lake ecosystem affects many aspects of the cave ecosystem, and the cave affects the lake, as bats regulate the number of insects.

Furthermore, a lake may not be linked only to cave ecosystems but to other ecosystems as well. For example, runoff from a nearby forest ecosystem provides nutrients and debris to the lake, while the river ecosystem that drains the lake removes sediments and nutrients. In fact, because of these interconnections, many biologists view the entire biosphere as one giant global "ecosystem" of tremendous complexity and order.

THE STRUCTURE OF ECOSYSTEMS

Each ecosystem is a functional unit in which energy and nutrients flow between the physical, abiotic environment and a community of living organisms that make up the biotic environment. The abiotic and biotic environments continually affect each other, producing interdependent connections. Often even a slight modification in one factor can disrupt an entire ecosystem.

It is generally easier to envision how the abiotic environment changes the biotic community than vice versa. For instance, a long period of freezing temperatures or a devastating fire may kill many organisms outright. Extremely strong winds can uproot plants and blow flying insects and birds far away from their natural habitat. The relationship between the abiotic and the biotic environment is two-way, however; organisms change the abiotic environment as well. For instance, recall from Chapter 24 that the earth's original atmosphere was devoid of oxygen. Ancient organisms released oxygen during photosynthesis, gradually adding this important gas to the atmosphere. In other words, organisms dramatically altered the abiotic environment, building up the oxygen of the earth's atmosphere to its current level. The biotic environment affects the abiotic environment in various other ways as well: Plants contribute to the formation and fertility of soils; coral reefs change the flow and temperature of oceans; dense forests modify humidity, temperature, and the amount of light and rain that reaches the forest floor; and, today, human activities produce pollutants that are changing the earth's climate.

COMPONENTS OF THE BIOTIC AND ABIOTIC ENVIRONMENTS

Biologists divide ecosystems into five subcomponents, two of which comprise the abiotic environment, and three that

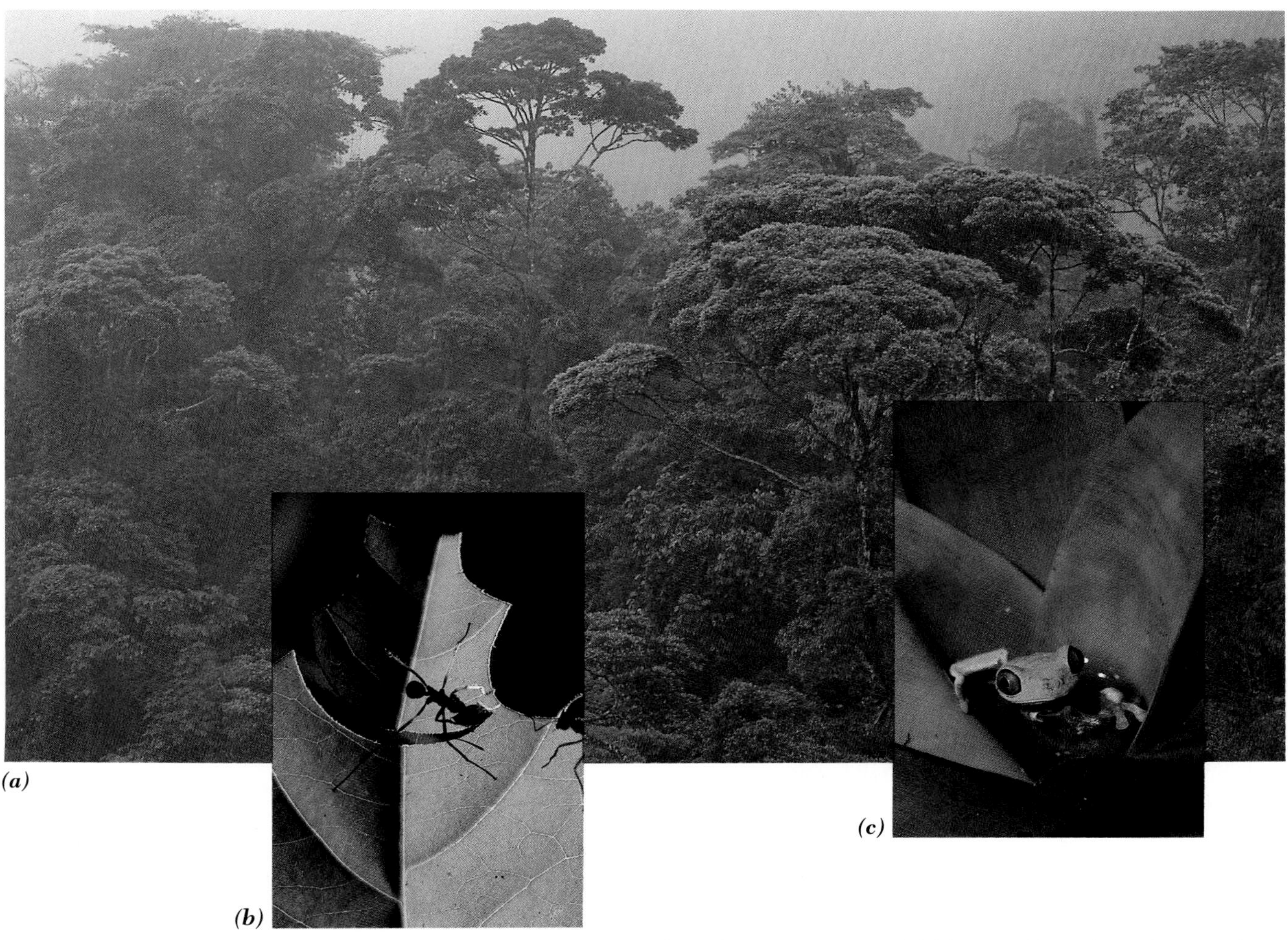

(a)

(b)

(c)

FIGURE 29-1

Ecosystems within ecosystems. ***(a)*** A South American tropical rain forest contains many ecosystems, each with a unique community of organisms and a unique set of physical factors. ***(b)*** The ecosystem supports colonies of leaf cutter ants that harvest leaves to cultivate a fungus garden ecosystem for food. ***(c)*** Perched high on a tree branch, the top of a bromeliad collects water, forming a tiny pond ecosystem, the home of this red-eyed tree frog.

comprise the biotic environment. The two abiotic subcomponents are:

1. *abiotic resources:* the energy and inorganic substances (nitrogen, carbon dioxide, water, phosphorus, potassium, and so on) needed by organisms for the construction of organic compounds; and
2. *abiotic conditions:* the substrate and/or medium (air, water, and soil) in which organisms live, and the surrounding conditions, such as temperature and water currents.

The three subcomponents of the biotic environment are:

1. **primary producers:** autotrophs (algae, bacteria, and plants) that use sunlight or chemical energy to manufacture food from inorganic substances;
2. **consumers:** heterotrophs that feed on other organisms and
3. **decomposers and detritovores:** heterotrophs that get their nutrition by breaking down the organic compounds found in waste organic matter and dead organisms (**detritus**). Decomposers are primarily microscopic bacteria and fungi, whereas detritovores are typically larger animals, such as some worms, nematodes, insects, lobsters, shrimp, and birds that feed on detritus.

TOLERANCE RANGE

For each abiotic resource or condition, an organism is able to survive and reproduce only within a certain maximum and minimum limit. For example, the lethal temperature limits for many land animals, such as humans, are a minimum of 0°C (32°F) and a maximum of about 42°C (107°F). This range is known as an organism's **tolerance range** (Figure 29-2). Tolerance ranges can also be determined for an entire species by determining the tolerance ranges of all members of the species and then setting the tolerance range limits at the maximum and minimum levels found. The *Theory of Tolerance,* as this concept came to be called, was proposed in 1913 by Victor Shelford, an animal ecologist at the University of Chicago.

LIMITING FACTORS

When the limits of tolerance for one or a few factors are approached, those factors usually take on greater importance than do all the others in determining where or how well an organism will survive. These more critical factors become **limiting factors** because they alone impose restraints on the distribution, health, or activities of the organism. Since each species has a multitude of tolerance ranges (one for each abiotic factor), the ability to identify a few key limiting factors helps ecologists understand otherwise extremely complex ecosystems.

This concept of limiting factors was first proposed in 1840 by Justus von Liebig of the University of Heidelberg and has become known as Liebig's **Law of the Minimum.** An agriculturist and physiologist, Liebig discovered that the yield of a crop was restricted by the soil nutrient most limited in amount. His Law of the Minimum states that the growth and/or distribution of a species is dependent on the one environmental factor that is available in the shortest supply.

Examples of Liebig's law are common: Low amounts of phosphate in a lake drastically reduce algae growth, which, in turn, limits the growth of all consumers; the amount of zinc in soil is usually so scarce that it alone limits the yield of many agricultural crops; the limited rainfall in deserts severely restricts plant growth; low light intensity on the floor of a tropical rain forest limits photosynthesis in understory plants.

Exceptions to the Law of the Minimum occur when one factor changes the tolerance for another. For example, changes in the oxygen content of water can change the temperature tolerance of American lobsters. When the oxygen content of water is low, lobsters tolerate temperatures only as high as 29°C (84°F), whereas when the oxygen content is high, the temperature tolerance rises to 32°C (89°F). Another exception to the Law of the Minimum is found when one factor substitutes for another, such as when molluscs use strontium to build their shells when calcium is limiting. But no matter what their limiting factors, all organisms have specific requirements for space and nutrients that must be provided by its ecosystem.

ECOLOGICAL NICHES

The organisms that comprise a community possess structural and behavioral adaptations that enable them to grow, reproduce, and survive in a particular ecosystem. Each

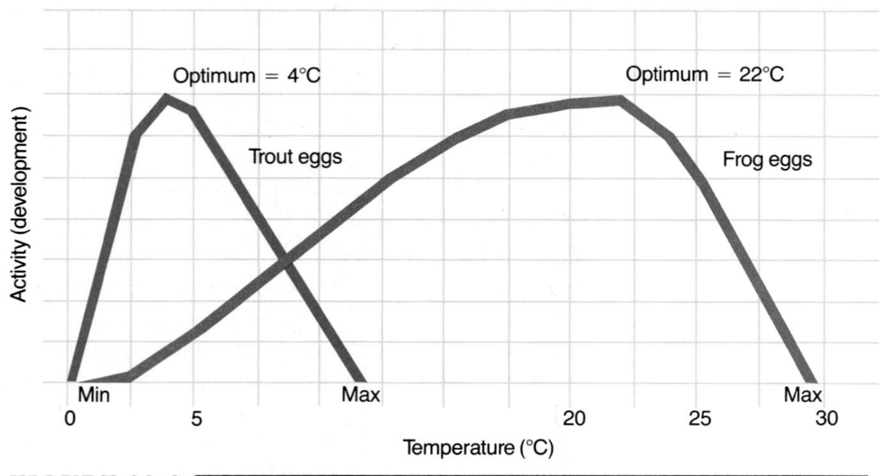

FIGURE 29-2

The limits of life. For each physical factor, organisms have minimum and maximum tolerance limits. Within most tolerance ranges, such as the temperature ranges for the development of trout and frogs, lies an optimum, at which growth is greatest.

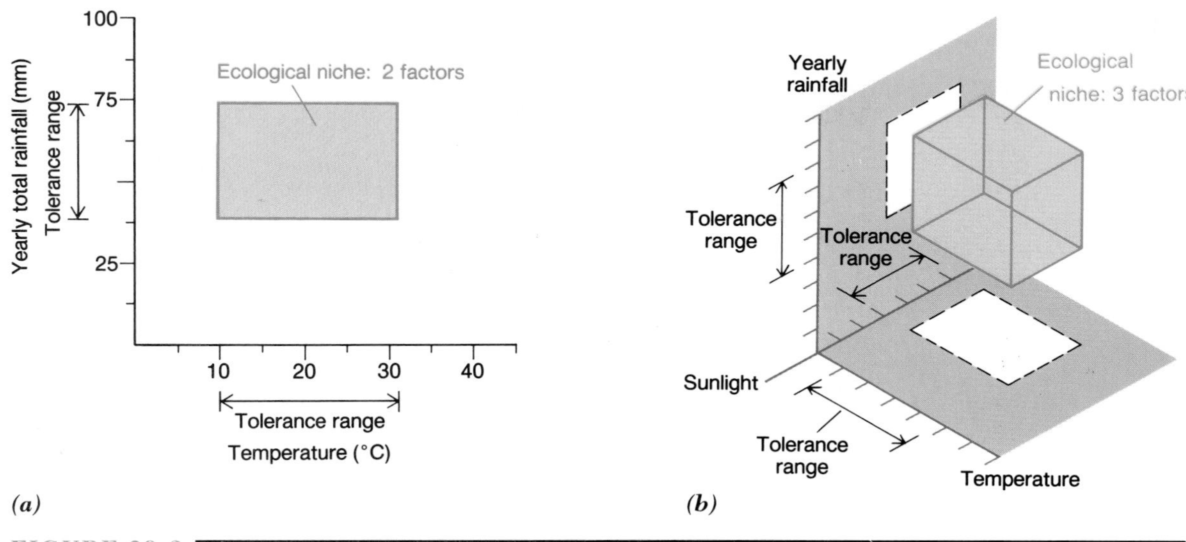

FIGURE 29-3
An ecological niche based on the tolerance range of two ***(a)*** and three ***(b)*** factors.

organism in an ecosystem occupies a specific *habitat,* the physical location in which the organism lives and reproduces (the bottom of a lake, under a rock, inside another organism, and so on). In addition to needing space, organisms also require energy and nutrients. The processes by which organisms acquire these resources partly define their "role(s)" in an ecosystem.

Together, an organism's habitat, role, requirements for environmental resources, and tolerance ranges for each abiotic factor comprise its **ecological niche.** Since the ecological niche includes all aspects of an organism's existence, it too is the outcome of evolution through natural selection. That is, the full scope of adaptations an organism acquires through natural selection establishes the range and boundaries of an organism's—or a species'—ecological niche.

If only two or three components are considered, it is possible to plot an ecological niche on a graph, each axis representing a different factor (Figure 29-3). It is impossible to draw a graph that includes *all* the components of the ecological niche (all tolerance ranges, the total range of habitats, and all functional roles), however, because such a representation would circumscribe a multidimensional area, or what ecologists refer to as a *hypervolume.* The hypervolume represents the potential niche, or the **fundamental niche.** Most organisms never realize their fundamental niche because interactions with other organisms (such as competition for limited resources or mates) often reduce the range of available habitats or possible functional roles. (We will discuss the range of organism interactions in the next chapter.) The remaining portion of the hypervolume, in which an organism actually exists and functions, is its **realized niche.** In other words, the fundamental niche represents all possible ranges under which an organism or species can exist, while the realized niche is the range of conditions in which the organism actually lives and reproduces.

NICHE BREADTH

Ecologists speak of *niche breadth* as a measure of the relative dimensions of an organism's ecological niche. Some species have very broad niches. Hawks, for example, have a relatively broad niche because they visit many habitats over large areas and have evolved adaptations that enable them to feed on several kinds of organisms (Figure 29-4*a*). Conversely, the cotton boll weevil has a narrow niche because it lives, feeds, and reproduces solely on cotton plants (Figure 29-4*b*). Species with narrow niches possess adaptations for very specific habitat and environmental requirements. For example, the cotton boll weevil has a long, tubular mouth specifically adapted for feeding on a cotton plant; female boll weevils have adaptations for laying eggs in cotton buds (bolls); and boll weevil larvae can only survive by eating the internal tissues of flower buds and developing cotton bolls.

Although they may not always be apparent, practical benefits sometimes arise from studying an organism's niche breadth. For example, consider the constant battle between farmers and weeds. All farmers want to stop weeds from invading and overgrowing their agricultural fields, since weeds contaminate harvests and cut down yields by competing with the crops for space and limited nutrients. One way to stop weed invasion is to apply a chemical weed killer that destroys only the weeds, not the crops. Unfortunately, such "selective chemical killers" are

(a)

(b)

FIGURE 29-4

Opposite ends of the spectrum. ***(a)*** The boll weevil's specific adaptations for living and reproducing only on cotton plants results in a narrow niche breadth. ***(b)*** Adaptations that enable a hawk to harvest a variety of animals for food from a number of habitats result in a broad niche breadth.

not available for every kind of weed. Alternatively, weeds could be stopped by using a **biological control,** whereby another organism destroys the weeds. Of course, it is important that the organism kill only the weed and not the crop plants; in other words, the weed-controlling organism must have a narrow niche.

Many examples of biological weed control exist. One in particular involves the unlikely pairing of the klamath weed (*Hypericum perforatum*) and a leaf-feeding beetle (*Chrysolina quadrigemina*). After studying the "natural history" of *C. quadrigemina,* scientists discovered that the feeding habits of the beetle were very specific: These beetles fed only on plants that had leaves of the same surface texture and margin as that of the klamath weed. Fortunately, the beetle was active during the same period that the klamath weed germinates and enters its period of rapid growth. The beetle's life history pattern was ideal for controlling the klamath weed. Unfortunately, the two organisms did not naturally occur together; they didn't even exist on the same continent. Once the leaf-feeding beetle was introduced into the United States, however, it became a very effective biological control agent against the klamath weed. This example reaffirms the importance of "basic research" in science. If separate scientists had not been studying the "natural history" of both of these organisms, and if they had not shared this information with each other, effective control of the klamath weed would still be a fantasy.

NICHE OVERLAP

Niche overlap occurs when organisms share the same habitat, have the same functional roles, or have identical environment requirements in some other way. Niche overlap leads to competition for the same needed resource; the greater the overlap, the more intense the competition. When two species have identical niches, competition can become so intense that both species are unable to coexist in the same ecosystem. We discuss this phenomenon, called the competitive exclusion principle, in more detail in Chapter 30.

ENERGY FLOW THROUGH ECOSYSTEMS

All organisms must secure a supply of energy and nutrients from their environment in order to remain alive and reproduce. Biologists categorize organisms according to their energy-acquiring strategy: Primary producers harvest energy through photosynthesis or chemosynthesis, forming a direct link between the abiotic and biotic environments, whereas consumers and decomposers obtain energy and nutrients from other organisms.

TROPHIC LEVELS

Tracking the transfer of food (energy and nutrients) among the organisms in an ecosystem is relatively easy if you follow a single feeding path. You simply count how many leaves a caterpillar eats, how much phytoplankton goes into a copepod, or how many prairie dogs a hawk consumes. Energy and nutrients are transferred step by step between organisms. For example, in a streamside community, a hawk eats a snake that ate a frog that ate a moth that sipped nectar from the flower of a periwinkle plant that converted the sun's energy into chemical energy during photosynthesis.

This linear feeding pathway has the same organization in all ecosystems: It begins with a *primary producer* (the periwinkles in the streamside community), which provides food for a *primary consumer* (the moth), which is eaten by a *secondary consumer* (the frog), which is eaten by a *tertiary consumer* (the snake), and so on, until the final consumer dies and is disassembled by decomposers. Each step along a feeding pathway is referred to as a **trophic level** (*trophic* = feeding).

The sequence of trophic levels maps out the course of energy and nutrient (food) flow among functional groups of organisms (primary producers, primary consumers, secondary consumers, and so on). As Figure 29-5 illustrates, trophic levels are numbered consecutively to indicate the order of energy flow. Trophic level 1 is always populated by primary producers, and Trophic level 2 is always populated by primary consumers. Levels of consumers beyond the primary consumer are then numbered sequentially. The final carnivore, called the **ultimate carnivore,** and those organisms that escape being eaten (such as humans), eventually die and are consumed by decomposers.

FOOD CHAINS, MULTICHANNEL FOOD CHAINS, AND FOOD WEBS

The transfer of food from organism to organism forms a food chain (Figure 29-5). Like a trophic level diagram, a food chain is a flowchart that follows the course of energy and nutrients through an ecosystem. Unlike trophic level diagrams, food chains name the organisms (rather than the group) that occupy each link. The example we gave earlier of periwinkle plants, moths, frogs, snakes, and hawks is a food chain because the organism that occupies each step is identified.

Most ecosystems contain many food chains. When more than one food chain originates from the same primary producer, a **multichannel food chain** is formed. In Figure 29-6, for example, different parts of a single manzanita shrub provide energy and nutrients for at least five independent food chains.

Linkages often form among food chains, creating networks of connections. In the streamside ecosystem, for example, frogs may sometimes eat flies instead of moths, and snakes may eat small rodents instead of frogs (the flies and rodents originating from other food chains). When all the interconnections between food chains are mapped out for an ecosystem, they form a **food web** (Figure 29-7). A food web illustrates all possible transfers of energy and nutrients among the organisms in an ecosystem, whereas a food chain traces only one pathway in the food web.

Food webs may intersect within ecosystems. For example, in a grassland, energy and nutrients may flow between a *grazing food web* and a *detritus food web.* In the

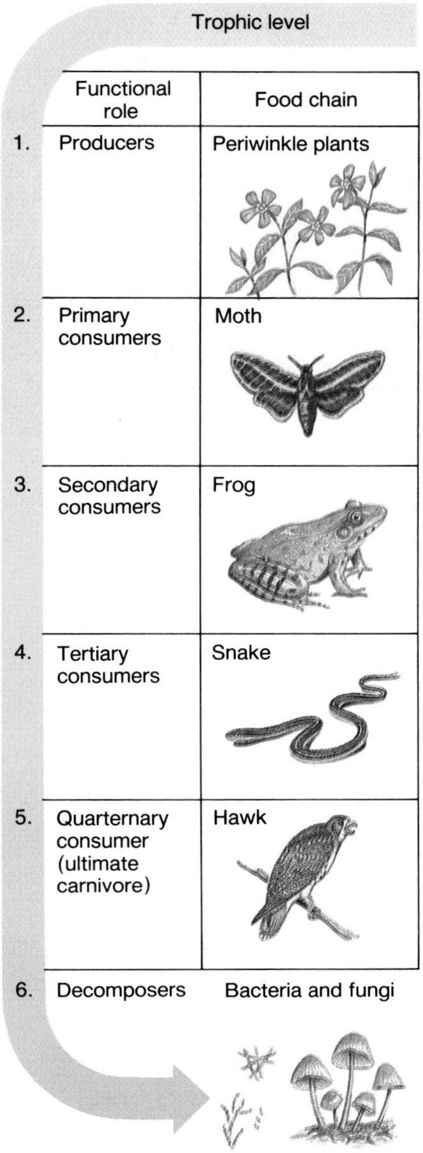

Trophic level	Functional role	Food chain
1.	Producers	Periwinkle plants
2.	Primary consumers	Moth
3.	Secondary consumers	Frog
4.	Tertiary consumers	Snake
5.	Quarternary consumer (ultimate carnivore)	Hawk
6.	Decomposers	Bacteria and fungi

FIGURE 29-5
A feeding path in a streamside ecosystem. Energy and nutrients flow through the biotic community, as food passes from trophic level to trophic level. Trophic level numbers indicate the order of flow.

FIGURE 29-6

Multichannel food chains. This manzanita (*Arctostaphylos pringlei* var. *drupacea*) is the primary producer for many food chains, including the five drawn here.

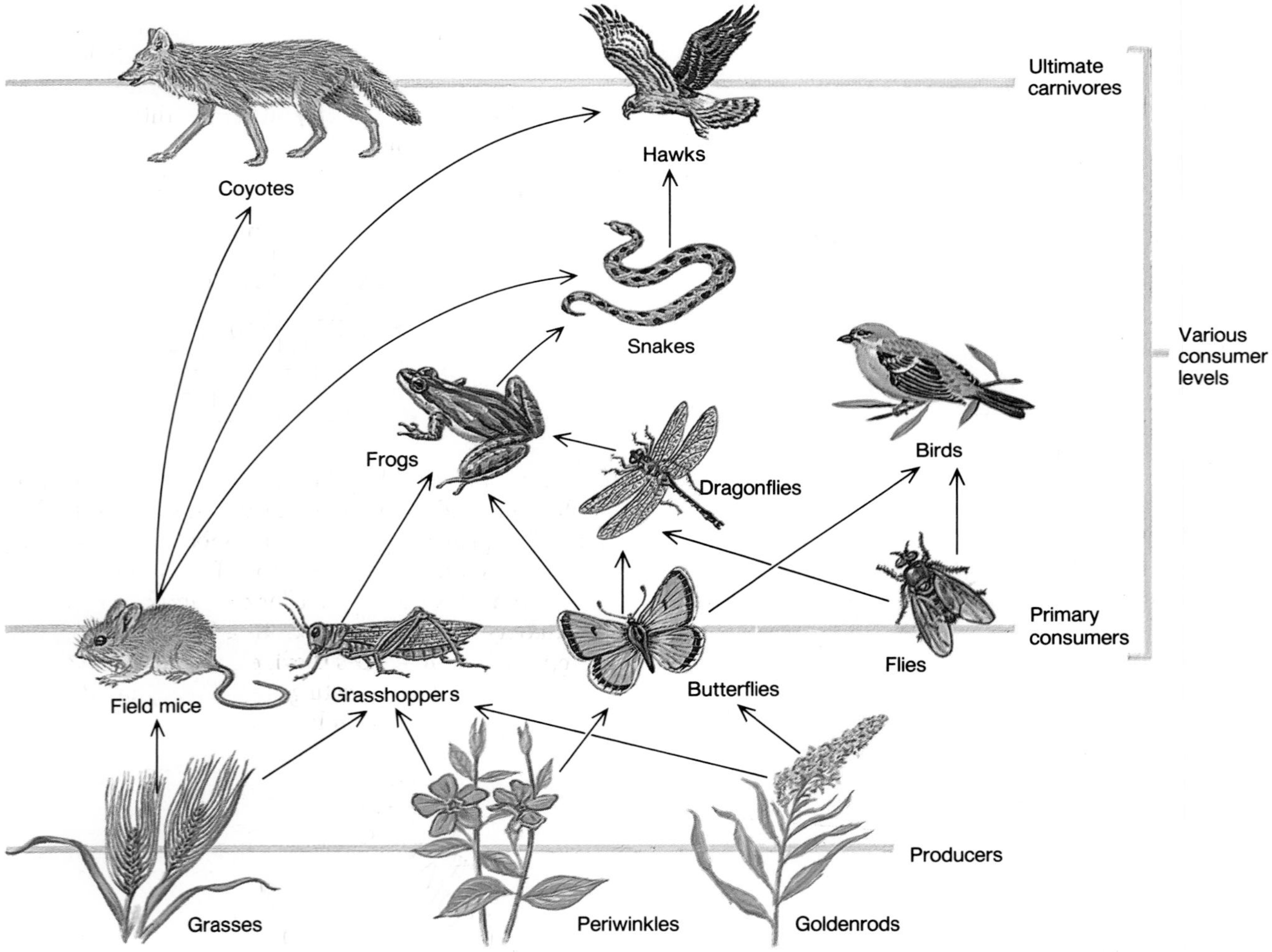

FIGURE 29-7

A simplified food web. A complete food web charts all possible feeding transfers within an ecosystem.

grazing food web, living tissues of photosynthetic grasses are consumed by a variety of herbivores, which may then be consumed by a variety of carnivores. But not all plants or all herbivores and carnivores are eaten, so when these organisms die, their bodies enter the detritus web. Even before they die, animals excrete organic wastes that also enter the detritus web, where earthworms, insects, millipeds, fungi, and bacteria eventually break down organic matter to simple inorganic molecules.

ECOLOGICAL PYRAMIDS

Feeding relationships are graphically represented by plotting the energy content, number of organisms, or biomass—the total weight of organic material—at each trophic level. Such graphs are called *ecological pyramids* because of their triangular shape. Each trophic level of a food chain forms a tier on the pyramid; that is, each successive trophic level is stacked on top of the level that represents its food source.

Pyramid of Energy

Energy always flows one way through a food chain; it never recycles. All the energy that enters a food chain is eventually dissipated as unusable heat energy. Since energy does not recycle, there must be an outside resource of energy to refuel life continually. For more than 99 percent of the earth's ecosystems, this extraterrestrial energy source is radiant energy from the sun.

A *pyramid of energy* illustrates the rate at which the energy in food moves through each trophic level of an ecosystem (measured in kilocalories of energy transferred per square meter of area per year—kcal/m^2/year). A typical pyramid of energy for a food chain in a grassland ecosystem during the spring (excluding decomposers) looks like this:

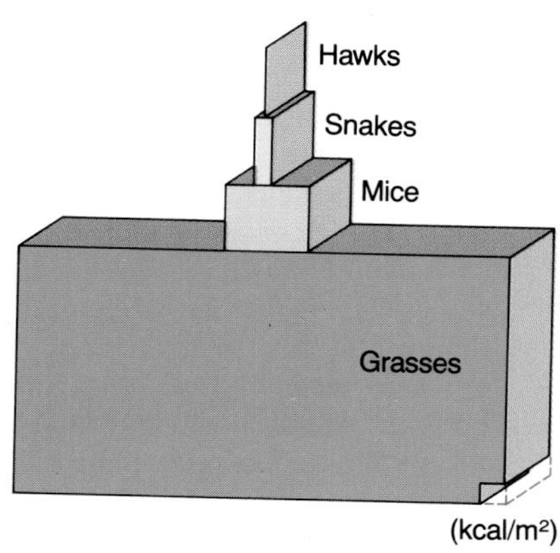

As you can see, there is a dramatic reduction in usable energy as food is transferred from one trophic level to the next. On average, only about 10 percent of the energy in any trophic level is converted into organic matter (*biomass*) in the next level. The least efficient transfer is between trophic levels 1 and 2; as much as 99 percent of the energy available in the producers is not transferred to the primary consumer level.

What happens to this huge amount of lost energy? Some energy is lost as heat during chemical conversions (the second law of thermodynamics); some energy is used by the organisms in each trophic level for their own metabolism and biological processes; not all food in one trophic level is eaten by organisms in the next trophic level; not all of the food that is eaten is usable (some energy is excreted or defecated as waste); and some powers nonbiological phenomena, such as fire. Since such a small fraction of energy gets passed on to successive trophic levels, the number of links in a food chain is limited and rarely exceeds more than four or five. Supporting a food chain with more than five trophic levels would require an enormous energy base at trophic level 1, a situation that does not occur very frequently.

Pyramid of Numbers

The rate of energy flow through a trophic level determines how many individuals can be supported in an ecosystem as well as the overall biomass of the organisms. Because energy decreases with each successive trophic level, the number of organisms and their biomass also decrease. The following *pyramid of numbers* illustrates this principle for a grassland ecosystem:

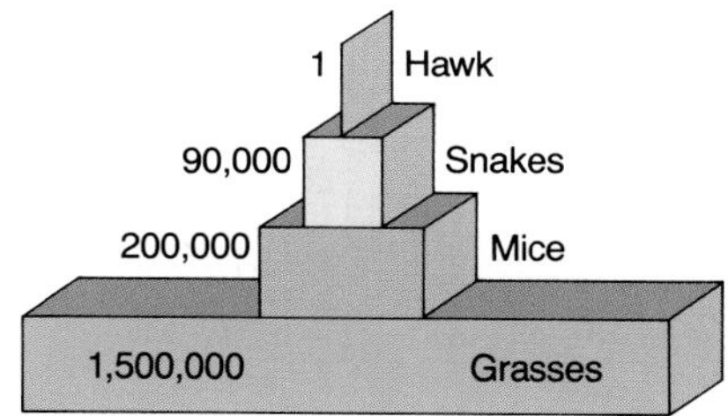

In grasslands, all primary producers are relatively small, compared to primary consumers, so it takes a large number of plants to provide enough food for big consumers, such as wildebeests or bison. In ecosystems where larger producers support smaller consumers, the pyramid of numbers becomes inverted. In a forest ecosystem, for example, a few large trees provide enough food to support many insects and insect-eating birds. The pyramid of numbers for such a food chain looks like this:

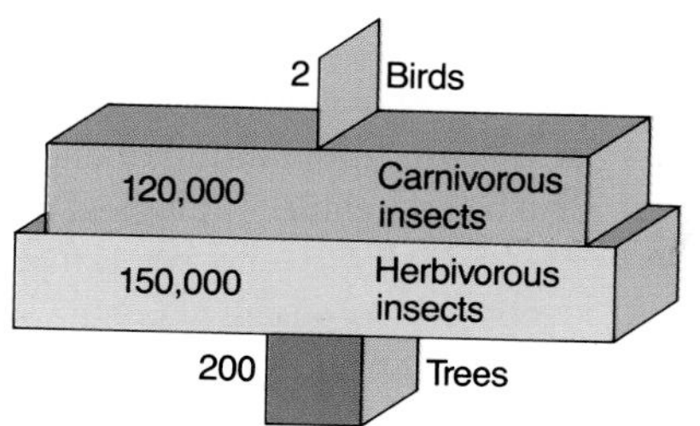

Pyramid of Biomass

A *pyramid of biomass* represents the total dry weight (in grams of dry weight per square meter of area—g/m^2) of the organisms in each trophic level at a particular time. Although most pyramids of biomass are "upright," they may become inverted, depending on when the samples are taken. For example, in the open oceans, where producers are microscopic phytoplankton, and consumers range all the way up to massive blue whales, the biomass of consumers may temporarily exceed that of the primary producers if data are taken when the number of phytoplankton is low. During such sampling periods, the pyramid of biomass could look like this:

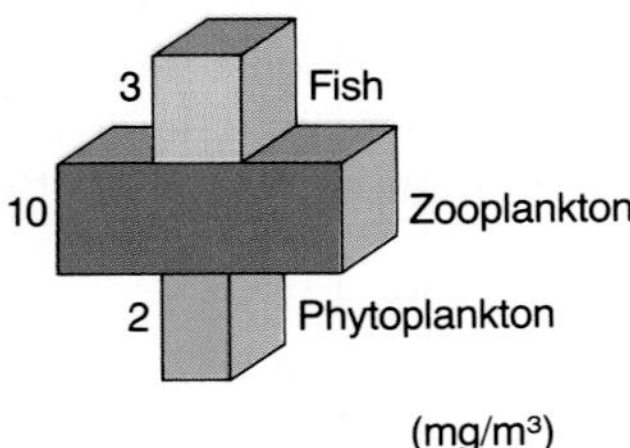

If samples are taken during the spring, however, when phytoplankton populations are immensely large, or if multiple generations of phytoplankton are included, the pyramid of biomass assumes the upright pyramid shape:

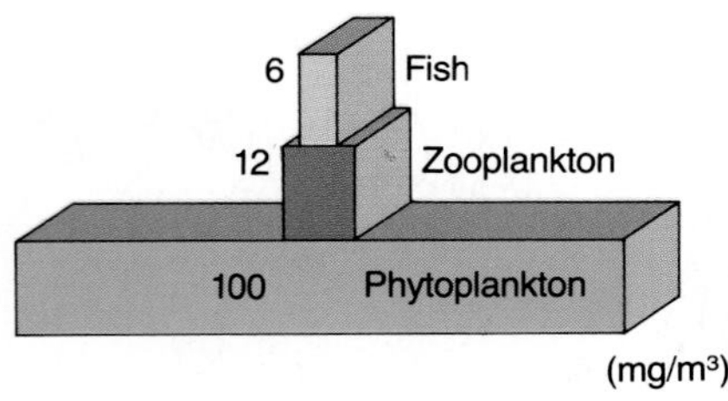

Although it is possible for pyramids of biomass and numbers to be inverted, an energy pyramid can never be inverted; there must always be more energy at lower levels to maintain life at higher trophic levels.

Bioconcentration and Biological Magnification

Ecologists often use ecological pyramids to illustrate how toxic chemicals in the environment can gradually accumulate in the bodies of the organisms in an ecosystem, a phenomenon known as **bioconcentration.** Bioconcentration of a toxic chemical may build to a high enough level to kill the organism. Bioconcentration may also lead to **biological magnification,** the buildup of chemicals in the organisms that form a food chain. Biological magnification exposes organisms toward the end of a food chain (at higher trophic levels) to potentially dangerous levels of chemicals. The use of DDT as a pesticide provides a clear example of both bioconcentration and biological magnification (Figure 29-8).

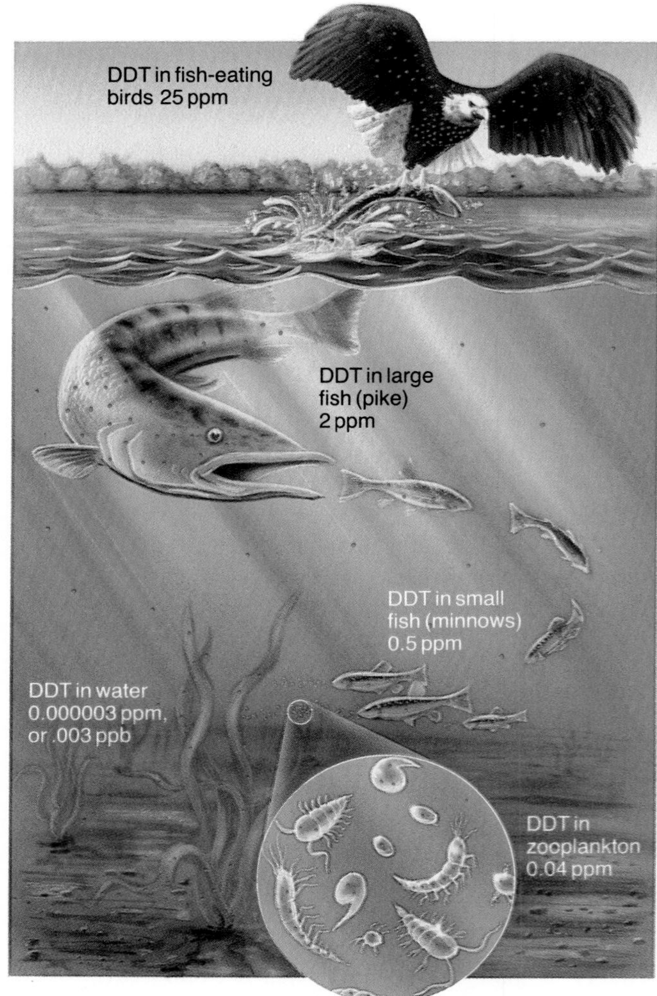

FIGURE 29-8
DDT, bioconcentration, and biological magnification. DDT was so effective in killing pests in agricultural fields it became widely used. Unfortunately, runoff from these fields contaminated streams, rivers, and oceans with DDT. The algae growing in these aquatic ecosystems absorbed the DDT. As the algae were eaten by small aquatic animals, the DDT was passed on. As these animals were eaten by progressively larger fish, the DDT continued to pass along the entire food chain. DDT accumulates at each feeding level because organisms cannot metabolize DDT (such bioconcentration is represented by the red dots). When birds eat larger fish, the high doses of DDT interfere with calcium deposition in the birds' eggs, making the egg shells so thin and weak that they can no longer protect developing birds. Because of its harmful effects on wildlife, DDT was banned in the United States in 1968. The chemical is still widely used in other nations, however, and continues to contaminate much of the earth's waters.

BIOGEOCHEMICAL CYCLES: RECYCLING NUTRIENTS IN ECOSYSTEMS

All organisms are composed of chemical elements (Chapter 2). Of the more than 90 naturally occurring elements found in the biosphere, only about 30 are used to make organisms. Some elements, such as carbon, hydrogen, oxygen, phosphorus, sulfur, and nitrogen, are needed in large supplies, whereas others, including sodium, manganese, iron, zinc, copper, and boron, are required in small, or even minute, amounts.

Unlike radiant energy, which showers on the biosphere daily, there is no outside source to supply the elements essential for life. Since there is a finite amount of each, essential elements must be recycled (reused) over and over again for life to continue. Organisms today are using the same atoms that were present in the primitive earth. Some of the atoms that comprise your body may have been part of an ancient bacterium, a long-extinct tree fern, or a dinosaur.

During recycling, elements pass back and forth between the biotic and abiotic environments, forming **biogeochemical cycles.** Primary producers typically introduce elements into the biotic environment by incorporating them into organic compounds, and consumers and decomposers release the elements back into the abiotic environment by breaking down complex organic molecules into simple inorganic forms. The rate of nutrient recycling depends primarily on where the element is found in the abiotic environment. For example, elements that cycle through the atmosphere or hydrosphere, forming gaseous nutrient cycles, recycle much faster than do those of sedimentary nutrient cycles, which cycle through the earth's soil and rocks.

GASEOUS NUTRIENT CYCLES

Gaseous nutrient cycles are those in which the element occurs as a gas at some phase in its cycle, and a large proportion of the element resides in the earth's atmosphere. Although several elements have gaseous cycles, we will discuss only the water (hydrologic), carbon, and nitrogen cycles here because of their central role in living organisms.

The Hydrologic Cycle

Nutrient cycles may involve more than one element. For example, in the **hydrologic cycle,** both hydrogen and oxygen cycle together in the form of water molecules, which often change from one phase to another (liquid water to water vapor, or vice versa) as they move between the earth's oceans, land, organisms, and atmosphere (Figure 29-9).

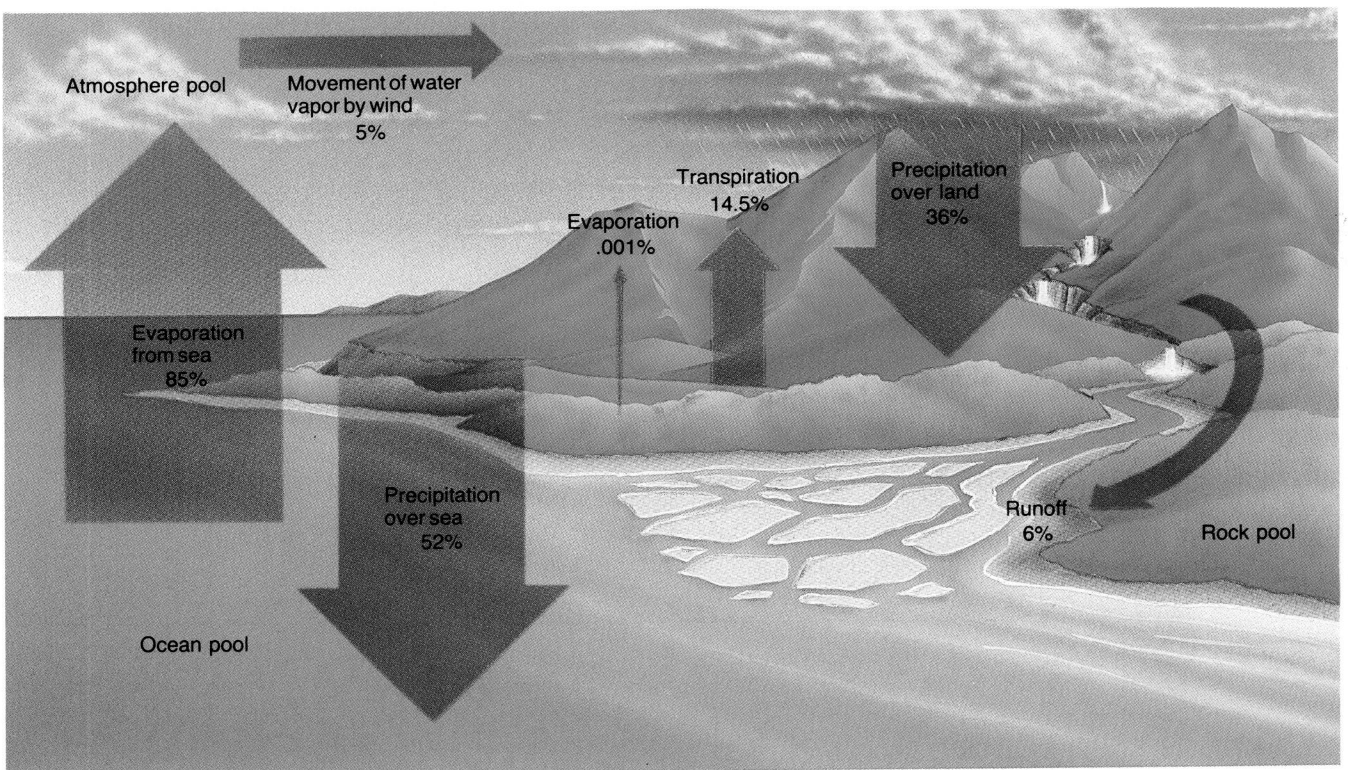

FIGURE 29-9
The hydrologic (water) cycle.

The largest reservoir of water, about 97 percent of the earth's available water, is in the oceans; only about 0.001 percent is present as water vapor in the atmosphere. Yet, despite this huge difference, the greatest quantities of water are exchanged between the oceans and the atmosphere via evaporation and precipitation. Over 80 percent of the liquid water that is evaporated during the hydrologic cycle comes from the oceans and moves into the atmosphere. About 52 percent of water in the atmosphere falls back into the oceans as precipitation, or rain. The remainder either stays in the atmosphere in the form of clouds, ice crystals, and water vapor, or falls back to earth as rain over the land.

Although some of the rain that falls on the land is intercepted by vegetation, generally rain reaches ground level and either percolates into the soil or runs off the soil surface into streams and rivers, where it may eventually flow back to the ocean. Some water is pulled downward into the ground by gravity, contributing to ground water supplies, most of which eventually flows back to the ocean. Of course, some of the water in the ground is absorbed by plants. For many plants, more than 90 percent of the water taken in by roots is released back into the atmosphere as water vapor through transpiration. The remainder hydrates plant cells and tissues or is used in biochemical reactions, particularly photosynthesis. In comparison with other biogeochemical cycles, organisms seem to play a relatively minor role in the hydrologic cycle.

The Carbon Cycle

Large amounts of carbon are continually exchanged between the atmosphere and the community of organisms. As a result, the recycling rate of carbon is especially rapid (Figure 29-10). In the **carbon cycle,** primary producers constantly extract carbon dioxide from the atmosphere and

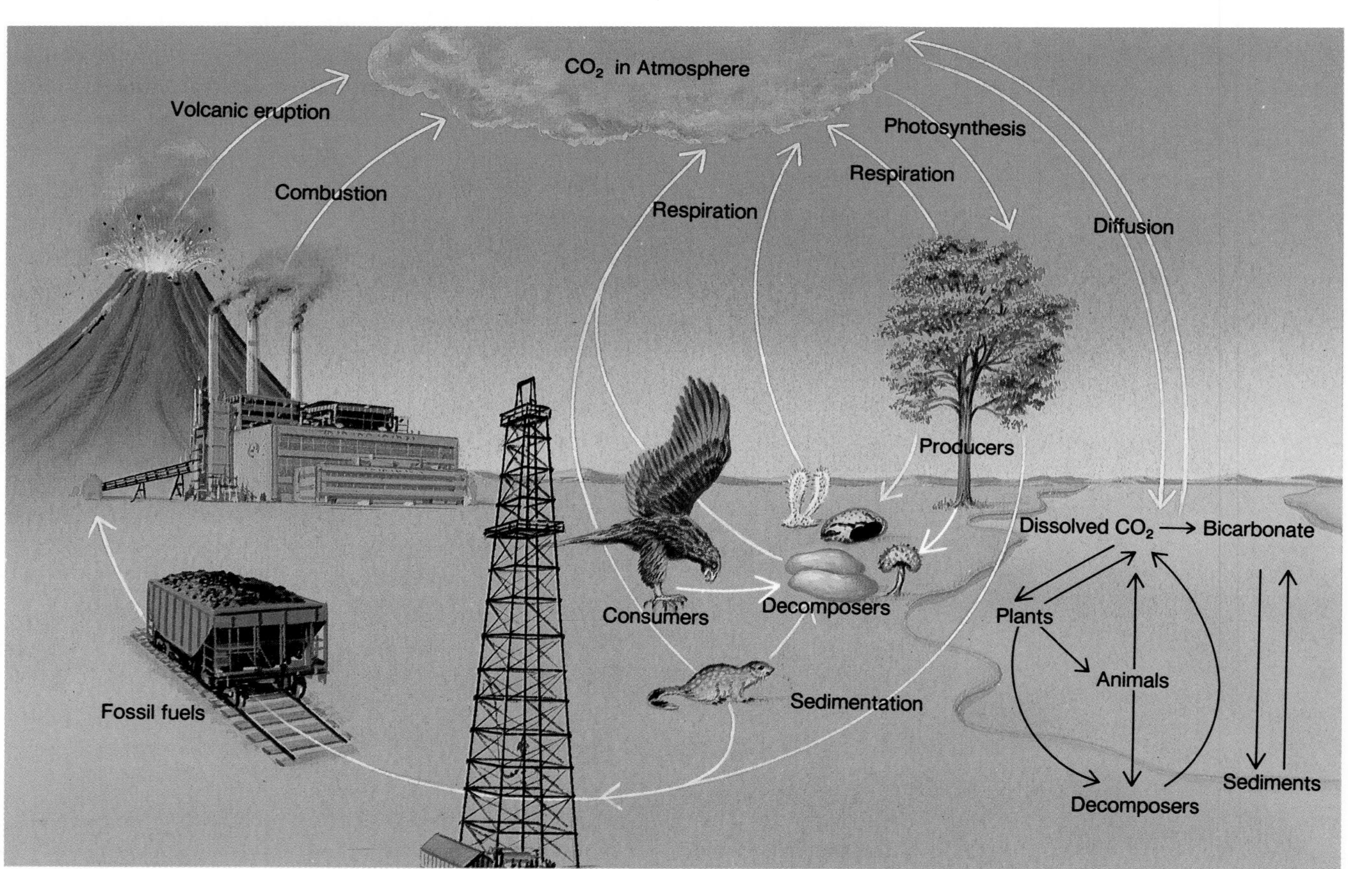

FIGURE 29-10

The carbon cycle. Atmospheric and dissolved carbon dioxide are used by primary producers to make energy-rich organic compounds during photosynthesis. When producers are eaten, carbon-containing organic compounds are passed to consumers and, eventually, to the decomposers. All organisms (producers, consumers, and decomposers) release carbon as carbon dioxide during respiration, most of which is returned to the atmosphere. Organisms that are not decomposed may eventually form fossil fuels. Combustion of fossil fuels also releases carbon as carbon dioxide back into the atmosphere.

use the carbon to form the chemical backbone for building organic molecules. Organic molecules are then disassembled by all organisms during cellular respiration, releasing carbon as carbon dioxide.

On land, producers extract carbon dioxide gas directly from the atmosphere. In water, gaseous carbon dioxide must dissolve before it can be incorporated into organic compounds by aquatic autotrophs. Not all of the carbon dioxide dissolved in water is used in photosynthesis. Resulting carbon-containing bicarbonates and carbonates, compounds with low solubility, eventually settle to the bottom of oceans, streams, and lakes. Additional carbon sediments form as the skeletons and shells of organisms accumulate. The settling of carbonate, shells, and skeletons can tie up carbon for long periods of time in sediment, limestone, and several other forms of rock.

Even on land, dead organisms and organic matter may not be decayed by decomposition, a process that would ordinarily release carbon dioxide back into the atmosphere. Nondecayed organic material may build up deposits that turn into fossil fuels, producing a carbon reservoir within the earth. Most of the world's fossil fuels were formed during the Carboniferous period, between 285 million and 375 million years ago, when shallow seas repeatedly covered vast forests, preventing decomposition. As we burn fossil fuels, we return this carbon to the atmosphere (see The Human Perspective: The Greenhouse Effect: Global Warming).

The Nitrogen Cycle

Although the earth's atmosphere is 79 percent nitrogen gas (N_2), only a few microorganisms are able to tap this huge reservoir, initiating a series of conversions and transfers that produces the **nitrogen cycle** (Figure 29-11). These "nitrogen-fixing" microorganisms include bacteria and cyanobacteria.

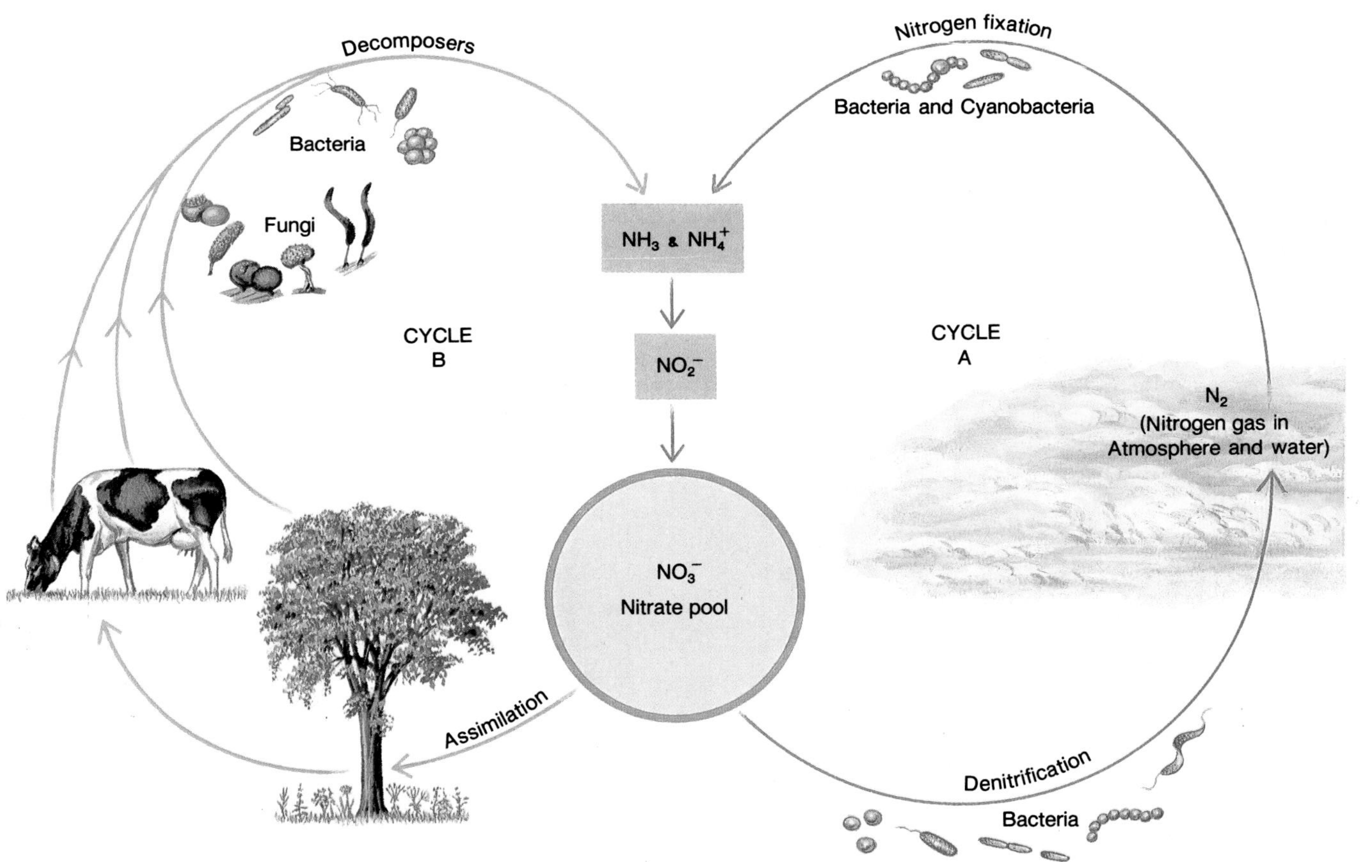

FIGURE 29-11

The nitrogen cycle. A pool of nitrates joins together two cycles that exchange nitrogen between the biotic community and the abiotic environment. In Cycle A, nitrogen gas from the atmosphere is converted to ammonia or organic compounds by bacteria during nitrogen fixation. The ammonia is converted to nitrites and then nitrates by nitrifying bacteria. The nitrates are either converted by other bacteria into nitrogen gas during denitrification (completing Cycle A), or they are absorbed by primary producers and the nitrogen incorporated into nitrogen-containing organic compounds (beginning Cycle B). The ammonia that is excreted by consumers or released during decay is converted into nitrites and then into nitrates, returning nitrogen to the nitrate pool (completing Cycle B).

◁ THE HUMAN PERSPECTIVE ▷

The Greenhouse Effect: Global Warming

The greenhouse effect is a term that has been coined to describe the rise in global temperature that is triggered by increased amounts of gaseous pollutants, mainly carbon dioxide, that trap heat within the atmosphere. Gaseous pollutants absorb heat-generating infrared radiation from the earth, preventing it from escaping into space at night. The greenhouse effect in the atmosphere is similar to what happens when a car is left parked in the sun with its windows closed. The car's windows are like the gases in the atmosphere: They allow the sun's radiant energy to enter the car, heating the interior but preventing the inside heat from escaping; the car gets hotter and hotter.

As levels of carbon dioxide and pollutants increase in the atmosphere, they trap more and more heat, causing global temperatures to rise gradually, a phenomenon referred to as "global warming." By analyzing air bubbles that formed in the ice of Antarctica and Greenland over the past 1,000 years, scientists have documented an increase in atmospheric carbon dioxide of over 25 percent; 85 percent of this increase occurred between 1870 and 1989, mostly from the burning of fossil fuels. At the current rate, the global carbon dioxide level is expected to double by 2050, raising the average global temperature between 2°C and 5°C (3.5°F and 9°F). Some of the anticipated consequences of global warming include the following:

- **The sea level is expected to rise during the next century.** Using computer models, scientists predict that rising sea levels may cover the homes of more than 20 million Americans who live on the East Coast and ruin rice production in Asia. Most of the rice grows in low-lying regions that would be flooded with salt water. Increases in sea temperature will also begin melting the polar ice caps, contributing to rising sea levels. Satellite photographs show that the polar ice caps have shrunk by 6 percent over the past 15 years.
- **Reduced rainfall and rising temperatures will amplify the world's hunger crisis.** Most of the world's food supply is currently produced in the band of agricultural land found in North America, Europe, and Russia. Climatic changes in these regions will cause crop production to fall, shifting farming northward to mountainous areas that are more difficult to farm.
- **Estuaries, marshes, and swamps will be flooded.** As seas rise, coastal estuaries, marshes, and swamps will be flooded, and many plant and animal species will be lost, as will much of our seafood supplies.

Since carbon dioxide has the biggest impact on global warming, reducing activities that produce carbon dioxide will help curb the problem. Much of the carbon dioxide that is released into the atmosphere originates from burning fossil fuels to produce electricity and burning of huge expanses of tropical rain forests. Thus, reducing energy consumption and saving tropical rain forests (see the Human Perspective, Chapter 13) could make a significant difference.

Nitrogen fixers convert atmospheric nitrogen gas to ammonia (NH_3) in a process called *nitrogen fixation.* Once nitrogen is converted, other groups of soil bacteria, collectively known as the nitrifying bacteria, convert ammonia into nitrites (NO_2) and then nitrates (NO_3). Plants absorb nitrates and incorporate the inorganic nitrogen into organic molecules: nucleotides and amino acids, the fundamental building blocks of DNA, RNA, and proteins. When the producers are eaten or die, nitrogen is passed on to the consumers and decomposers in a food chain. The decomposers then convert the nitrogen-containing organic molecules into inorganic ammonia, which is also released directly from the consumers (in urea) as a means of eliminating the excess nitrogen that would otherwise accumulate and poison the organism.

Once again, ammonia is converted into nitrites and then into nitrates by nitrifying bacteria, making nitrogen available for absorption by plants. Some nitrogen is converted to nitrogen gas during *denitrification* by denitrifying bacteria; nitrogen fixation is therefore needed to renew usable nitrogen resources.

The nitrogen cycle illustrates the critical role microorganisms play in the biosphere. Without bacteria, there wouldn't be sufficient nitrogen recycled to support the diversity of life on earth. Without bacteria-dependent nitrogen fixation, there wouldn't be any source of usable nitrogen for producers or the consumers they support. In other words, without bacteria, life on earth would cease.

THE PHOSPHORUS CYCLE: A SEDIMENTARY NUTRIENT CYCLE

Some elements never, or only rarely, exist as a gas. Such elements accumulate in the soil or rocks and, as a result, have a sedimentary nutrient cycle. Elements with sedimentary cycles include calcium, iron, magnesium, sodium, and phosphorus. We use the **phosphorus cycle** (Figure 29-12) to illustrate the general features of a sedimentary nutrient cycle.

All organisms require large amounts of phosphorus to construct ATP, DNA, RNA, and cellular membranes. Most phosphorus is contained in rock deposits. Erosion and runoff from rain dissolve the phosphorus in rocks and form phosphates (PO_4^{-2}). Plants and other primary producers absorb phosphates and use the phosphorus to build organic molecules. When these primary producers are eaten, phosphorus is passed to the primary consumer and then to the other organisms in a food chain. Decomposers break down phosphorus-containing compounds and release it back into the environment as phosphates, which are either reabsorbed by plants or leached out of the soil, where they accumulate in sediments. Since phosphorus is easily leached from soil, it is one of the least available essential elements in the biosphere. The large quantities of phosphorus needed by organisms often makes this element a major limiting factor in many ecosystems.

SUCCESSION: ECOSYSTEM CHANGE AND STABILITY

Ecosystems are constantly changing as energy and elements flow from the abiotic environment to the biotic community, as organisms interact within the biotic community, and as elements flow from the biotic to the abiotic environment. Ecosystems also change as the seasons trig-

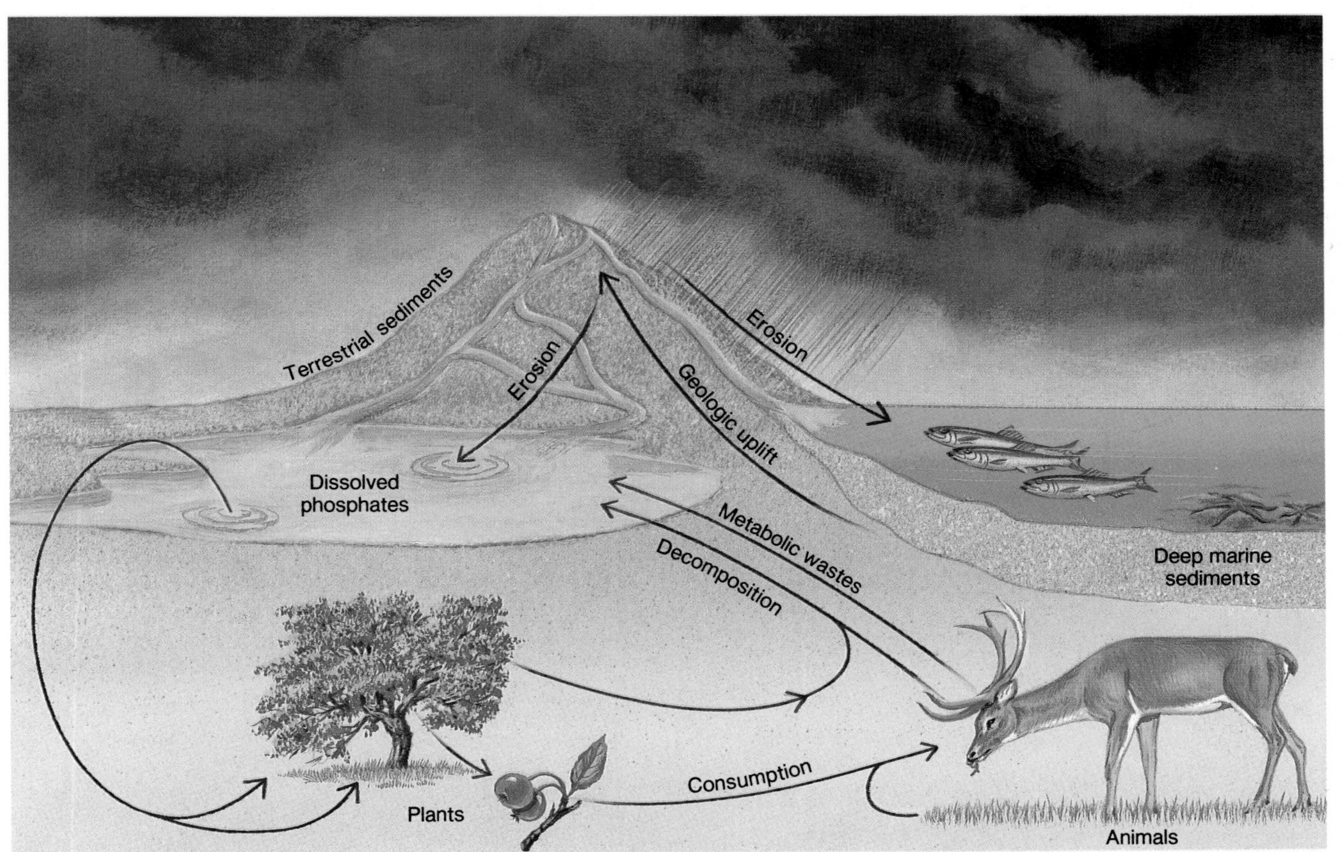

FIGURE 29-12

The phosphorus cycle. Phosphorus has a sedimentary nutrient cycle because most of the element is found in sedimentary rocks. Geological uplift raises phosphate sediment, and erosion caused by waves and rain dissolves the phosphorus. Some dissolved phosphates are absorbed by primary producers, incorporated into organic compounds, and then passed to the other organisms in a food chain. However, the majority of dissolved phosphates runs off and accumulates as precipitated solids at the bottom of streams, lakes, and oceans, where it becomes part of new sediments. Decomposition and wastes from some animals, such as birds, also add to the dissolved phosphate pool.

ger fluctuations in both the abiotic and biotic environments. Changes triggered by regular, seasonal fluctuations make an ecosystem dynamic, but they do not cause permanent changes in the composition and organization of organisms in the biotic community. However, permanent changes in the biotic community do occur, such as in newly formed habitats and in areas disturbed by fire, floods, hurricanes, drought, or the activities of humans. Such large-scale changes trigger the process of **succession,** a progression of distinct communities that eventually leads to a **climax community** that remains stable and perpetuates itself over time. The populations of organisms that make up a climax community are in equilibrium with their abiotic environment. Thus, the kinds of organisms and their abundance remain relatively constant over long periods.

Climax communities tend to contain many species in a highly organized trophic structure. Generally, large amounts of organic compounds are manufactured by producers of climax communities, but consumers and decomposers use nearly all of the excess, so the total biomass does not increase.

Severe or long-term changes in either the abiotic or biotic environments can permanently change the organization and composition of the biotic community, however. For example, cycles of declining temperatures during the Pleistocene Epoch produced a series of ice ages that destroyed and permanently altered most communities in the biosphere. Another large-scale factor that often causes permanent changes in a community is fire. Fires not only kill many organisms outright, they also modify conditions so much that different organisms predominate in the affected area.

When a community has been permanently changed, or when an entirely new habitat is formed (such as following a volcanic eruption or when a glacier retreats), a variety of species invade the area, forming a *pioneer community.* As pioneer species take hold, they modify the environment by changing the soil, the temperature of the ground, the amount of light that penetrates to ground level, and many other environmental characteristics. Eventually, the pioneer community changes conditions so much that new species invade the community. The new group gradually displaces the pioneer community and forms its own community. The process continues—one community replacing another—until a stable climax community develops. This orderly, directional sequence of communities that leads to

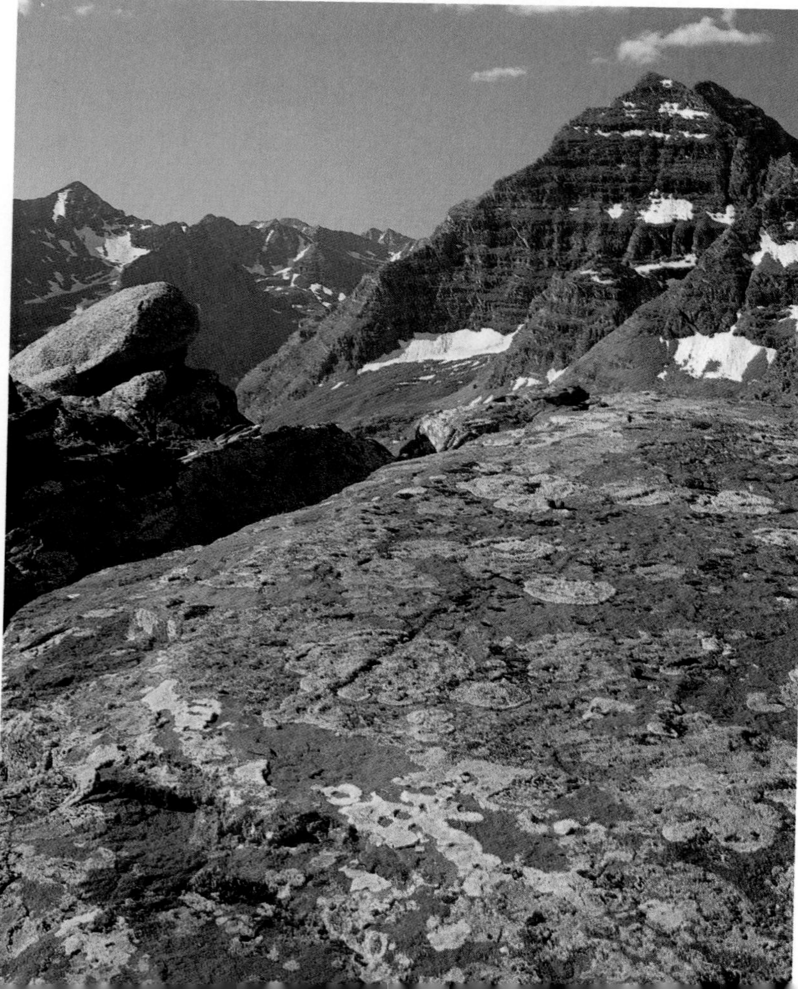

FIGURE 29-13

Primary succession. New habitats do not remain bare for very long. Even a dry rock is soon colonized by lichens, a "compound organism" composed of a fungus and resident algae. The hyphae of the fungus are able to grow into even the tiniest rock fissures, prying the rock open. At the same time, hyphae secrete chemicals that help erode the rock. The combination of intrusive growth and chemical erosion, together with abrasion from the wind and water and repeated heating and cooling, gradually crumbles the rock into small fragments. The process continues over many years, until what was once bare rock may eventually support a desert, woodland, or even a forest, depending on the climate.

FIGURE 29-14
The eruption of Mount St. Helens in Washington on May 18, 1980, initiated both primary and secondary succession. Near the center of eruption, species of bacteria quickly populated warm, standing pools that formed from rains and melting snows, initiating primary succession in each pool. Away from the eruption center, the force of the eruption severed trees at their base, seared off branches, and blew over burnt trees like toothpicks. Soil remained in both these outer areas, where secondary succession is now occurring. This series of photographs were taken from the same spot ***(a)*** 3 months, ***(b)*** 2 years, ***(c)*** 4 years, and ***(d)*** 9 years after the May 1980 eruption.

a climax community is succession. The entire series of successional communities, from pioneer to climax, forms a successional *sere.* Each community in a sere is called a *stage.*

Ecologists recognize two types of succession: **Primary succession** occurs in areas where no community existed before (new volcanic islands, deltas, dunes, bare rocks, or lakes) (Figure 29-13); **secondary succession** occurs in disturbed habitats where some soil, and perhaps some organisms, still remain after the disturbance (Figure 29-14). Fires, floods, drought, and many human practices (such as clearing forests for agriculture and construction projects) would prompt secondary succession. Secondary succession also occurs on abandoned farmlands, in overgrazed areas, and in forests cleared for lumber.

EVOLUTION AND ADAPTATION: TYING IT TOGETHER

Since all organisms are part of some ecosystem, and since the ecosystem contains both the abiotic and biotic environments, it is the ecosystem that is the basic unit of evolution. Traits that enable an organism to acquire more nutrients, to outrun a predator, to find a mate, or any other feature that increases survivorship and reproduction are naturally selected by elements of the ecosystem. In this way, natural selection leads to adaptations for a particular habitat, with defined role(s) and tolerance ranges for each abiotic factor in a particular ecosystem. In other words, an organism's ecological niche is the result of natural selection.

SYNOPSIS

Ecosystems are dynamic, self-sustaining units. They are composed of a community of organisms and the surrounding physical environment. Ecosystems are connected to other ecosystems to varying degrees.

Organisms (or species) have a tolerance range for each physical factor. When the maximum or minimum tolerance is approached or exceeded for any given factor, that factor limits the distribution, health, or activities of the organism. Each organism has a suite of adaptations that defines its ecological niche—the combination of an organism's habitat, functional role(s), and total environmental requirements and tolerances.

Energy flows through an ecosystem. Trophic levels, food chains, and food webs track the flow of energy and nutrients between the members of the biotic community.

Essential nutrients cycle between the abiotic and biotic environment, forming biogeochemical cycles. Producers and decomposers are required for exchanging nutrients between the biotic and abiotic environments.

Carbon, water (hydrogen and oxygen), and nitrogen form gaseous nutrient cycles because the largest proportion of the element resides in the earth's atmosphere. Hydrogen and oxygen cycle together as water, forming the hydrologic cycle where the vast quantities of water are exchanged via evaporation and rainfall between the oceans and the atmosphere. In the carbon cycle, photosynthesis removes carbon from the atmosphere, whereas the respiration of organisms releases it back to the atmosphere. The nitrogen cycle requires microscopic bacteria both to remove nitrogen from the atmosphere and to release it back again.

Phosphorus cycling is an example of a sedimentary cycle because phosphorus accumulates in sediments (soil and rocks). Plants absorb phosphates and use the phosphorus to build organic molecules. Decomposers break down organic molecules and release phosphorus back into the environment as phosphate.

Ecosystems inevitably change. Changes in an ecosystem are triggered by permanent changes that occur in the abiotic or biotic environments.

Review Questions

1. Give an example of each of the following:
   a. realized niche
   b. Law of the Minimum
   c. secondary succession
   d. food web
2. List as many components as you can think of for the fundamental niche and realized niche of a whale (a relatively broad niche) and a tapeworm (a narrow niche). Now try it for humans.
3. Using your experience with house and garden plants, list some effects of limiting factors that you have observed for yourself.
4. Refer to the nitrogen cycle on page 553. Describe at least three separate pathways that would enable nitrogen to be recycled through the biotic community.
5. Compare the carbon cycle (a gaseous nutrient cycle) and the phosphorus cycle (a sedimentary nutrient cycle). In what ways are they similar? How do these similarities affect the relative rates of recycling for each?

Critical Thinking Questions

1. Describe the kind of plant and animal data that Whittaker would have to have found in order to support Clement's original hypothesis. Why do you think it took biologists so long to unravel the nature of communities?
2. If organisms are so dependent on their physical environment, how do you explain the following? (1) Many plants and animals thrive when introduced into new areas; for example, many European wildflowers thrive as weeds in North America. (2) When environments undergo change, some plants and animals survive, while others are wiped out; for example, removing maple and beech trees from eastern forests promotes the growth of birch and aspen.
3. Like all biomes, a desert encompasses several ecosystems. One type of desert ecosystem is a "wash." Although dry most of the year or often over several years, a desert wash forms as water from rain is channeled, producing a "river," sometimes only the size of only a small trickle, and other times the size of a large flood. Describe as many components of the ecological niches of five organisms (two plants, two animals, and one protist or fungus) that you would expect to find in a desert wash. Are there similarities in niche breadth between the different types of organisms? In which ways do niches overlap between the plants? Between the animals? Does niche breadth and overlap change over time as water alternates between abundance and scarcity?
4. All ecosystems depend on a flow of energy through the living system and cycling of material elements. Prepare a diagram showing these characteristics of a generalized ecosystem and the role of producers, consumers, and decomposers in the system.
5. Explain why nutrients must be recycled in an ecosystem. What are the natural recyclers? How are human activities affecting natural cycles? Give two specific examples.

CHAPTER

30

Life's Interactions: Communities

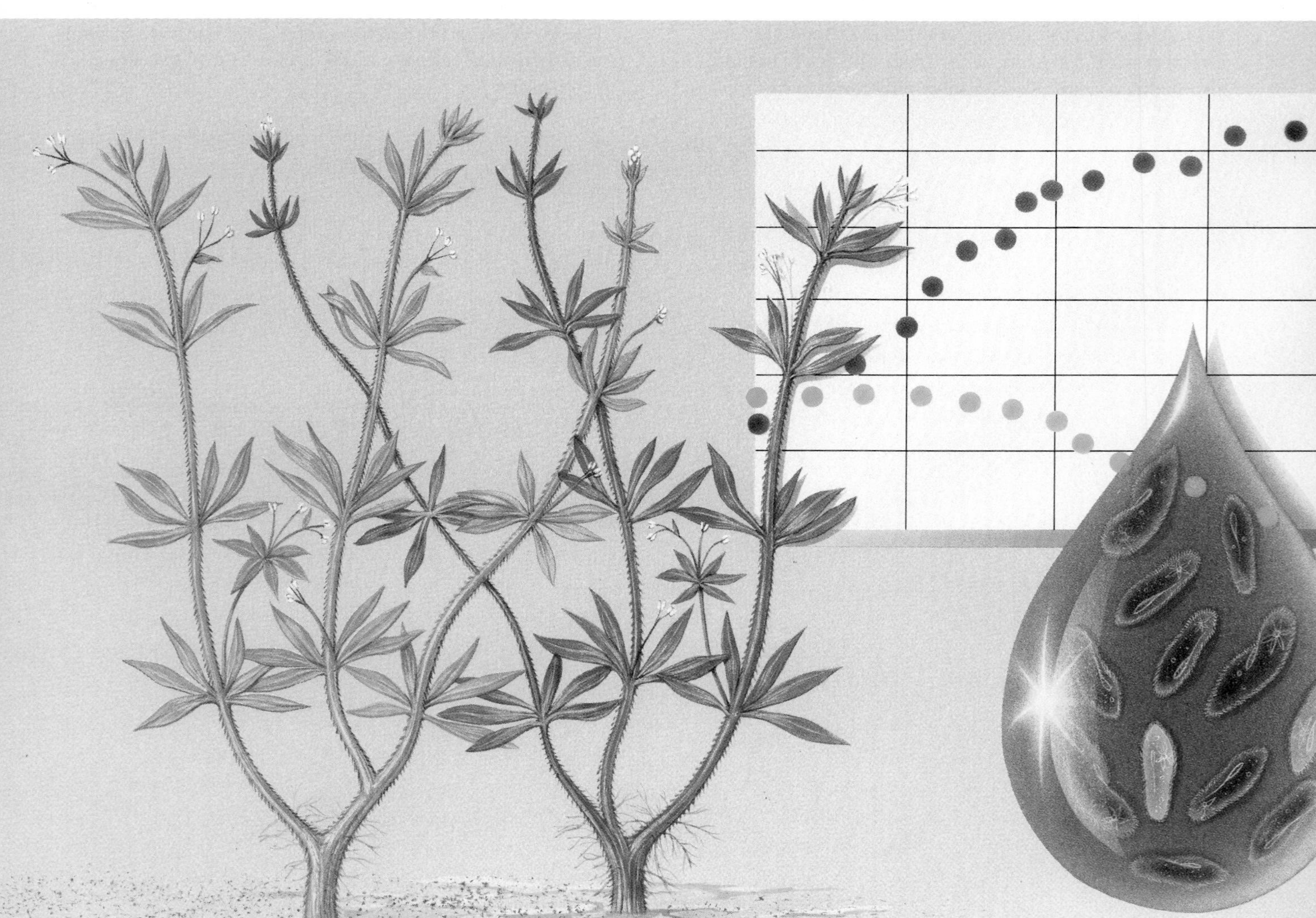

As the graphs illustrate, competition between similar species having identical requirements, whether between bedstraw plants or paramecia (in droplet), results in one species completely outcompeting the other.

STEPS TO DISCOVERY
Species Coexistence: The Unpeaceable Kingdom

Charles Darwin was an inexhaustible thinker and worker. Not only did he explore the process of evolution, he pursued research in several other areas of biology, including orchid pollination, selective breeding, animal taxonomy, plant movements, and competition. Darwin observed that when two individuals lived in the same area and required the same limited resources—space, food, whatever—those individuals would compete with one another for that resource. He also noted:

> *"As species of the same genus have usually, though by no means invariably, some similarity in habits and constitution, and always in structure, the struggle will generally be more severe between species of the same genus, when they come into competition with one another, than between species of distinct genera."*

In other words, the more closely related the competing individuals, the more similar their needs and the more intense the competition for limited resources.

Since Darwin's time, many scientists have studied competition. These studies eventually led to the formulation of a mathematical principle regarding competition. At the center of this discovery was G. Gause, a Russian microbiologist at the University of Moscow. As in many scientific investigations, Gause acquired insight not only from his own research efforts but from the research of other biologists, mathematicians, and physicists. In particular, Gause acknowledged his debt to botanists (plant ecologists, in particular). For example, in 1917, Sir Arthur G. Tansley, who founded the British Ecological Society, reported the results of his studies on the competition between two species of bedstraw plants, *Galium saxatile* and *G. sylvestre.* Each of these species of bedstraw is more abundant in different soil types: *G. saxatile* grows best in silica-rich soils, while *G. sylvestre* thrives in limestone soils.

Tansley grew both plants in various soils, including silica-rich soil and limestone, and then monitored germination, seedling survival, and competition. He found that *G. sylvestre* had a higher germination rate, grew more vigorously, and outcompeted *G. saxatile* in lime-rich soils. In contrast, in lime-poor soils, *G. saxatile* eventually outcompeted *G. sylvestre.* Tansley was one of the first to show that some plants were better able than others to compete in certain soils, suggesting that competition in natural communities influences the distribution and abundance of plants (organisms) in an ecosystem.

From Tansley's work, and that of other plant ecologists, Gause concluded that light, nutrients, water, and pollinators were common limiting resources for which plants compete. Sources of animal competition include water, food, mates, nesting sites, wintering sites, and sites that are safe from predators.

During the 1920s, researchers began formulating mathematical models to account for what happens when two species require the same limited resource or when one species preys on or parasitizes another. One such model, the Lotka-Volterra equation, was derived independently by Alfred J. Lotka at Johns Hopkins University in 1925 and V. Volterra in Italy in 1926. According to the Lotka-Volterra equation, one possible outcome of competition is the complete displacement of one competitor by another, resulting in the extinction of the weaker species.

To test whether this "winner-takes-all" outcome really occurs in nature, Gause initiated a number of studies in 1932 to test competition between microorganisms, first between competing species of yeast and then between competing species of protozoa. In Gause's best-known experiment, he monitored two species of *Paramecium* (*P. caudata* and *P. aurelia*). Each species was first grown in a separate culture and then in a mixed culture, where the species competed for a limited food supply. When grown separately, the number of individuals of both paramecia increased rapidly and then leveled off and remained constant. When cultured together, however, competition for limited food supplies resulted in the elimination of *P. caudatum,* which were outcompeted by the more rapidly reproducing "winner," *P. aurelia.*

Gause concluded from this and other similar experiments that "the process of competition under our conditions has always resulted in one species being entirely displaced by another." Gause's experiments supported the Lotka-Volterra equation and the "winner-take-all" outcome of competition and eventually became known as Gause's *Principle of Competitive Exclusion.*

Although seemingly calm, ecosystems actually bustle with activity. During warmer months in a forest, for example, the soil teems with bacteria, fungi, nematodes, springtails, amoebas, mites, slugs, worms, beetles, spiders, and scores of other organisms that churn the ground as they move about, grow, and reproduce. As they erupt through the soil surface, delicate plant seedlings absorb the nutrients recycled by microorganisms and fungi. These seedlings eventually grow into the herbs, shrubs, and trees that create the habitats and manufacture food for countless herbivores, which are, in turn, devoured by an assortment of carnivores.

Although the participants vary from one ecosystem to the next, all ecosystems are similar to the forest described above in that the organisms that live together often interact with one another. Some of these interactions benefit one or both participants; some have neutral consequences; and some harm either or both participants (Table 30-1).

Some organisms interact because they physically live together or because they live in very close association with one another. A close, long-term relationship between two individuals of different species is called **symbiosis,** which literally means "to live together." There are many examples of symbiotic relationships, such as between fungi and plants (mycorrhizae, page 224), plants and bacteria (root nodules, page 226), fungi and algae (lichens, page 463), jellyfish and algae, and aphids and ants (Figure 30-1). Symbiotic interactions are often naturally selected, particularly when both participants benefit from the close association.

▼ ▼ ▼

COMPETITION: INTERACTIONS THAT HARM BOTH ORGANISMS

Recall from Chapter 29 that species can coexist in a community as long as they have slightly different ecological niches, even though their niches may overlap. When a shared resource is abundant, such as oxygen in the air of terrestrial habitats, or water in aquatic habitats, there is more than enough for all. But in general, most resources are limited so organisms with overlapping niches enter into **competition,** a form of interaction in which two or more individuals or species utilize the same limited resource. Competition always harms both participants because each competitor reduces the other's supply of a needed, limited resource. The more similar the requirements of the organisms (such as between members of similar species or between members of the same species), the greater their niches overlap; the greater the niches overlap, the more intense the competition (Figure 30-2).

Organisms can be equally powerful agents of natural selection as environmental factors, if not more so. Competition between organisms is a powerful natural selection force. As we will see, competition can lead to the extinction of one competitor, to the exclusion of one competitor from an ecosystem, or to rapid evolutionary changes in the characteristics of the competitors.

EXPLOITATIVE AND INTERFERENCE COMPETITION

Organisms compete either *directly* or *indirectly* for a limited resource. Indirect competition occurs when competitors have equal access to a limited resource but one species manages to get more of the resource, reducing the competitor's supply. This form of indirect competition is called **exploitative competition.** An example of exploitative competition is currently taking place in the California deserts between deep-rooted native plants and newly introduced tamarisk trees. Tamarisk trees were brought to

TABLE 30-1
INTERACTIONS BETWEEN ORGANISMS IN A COMMUNITY

Kind of Interaction	*Organism 1*	*Organism 2*
Competition	Harmed	Harmed
Predation (including herbivory)	Benefited	Harmed
Parasitism[a]	Benefited	Harmed
Allelopathy	Benefited	Harmed
Commensalism[a]	Benefited	Unaffected
Protocooperation	Benefited	Benefited
Mutualism[a]	Benefited	Benefited

[a] May include symbiotic interactions.

FIGURE 30-1

Symbiosis: living together. Some ants live with groups of aphids. The aphids feed on the sugary juices of plants, often taking in more than their bodies can hold. The excess juice passes through the aphid's digestive system and out its anus, forming a honeydew drop that is lapped up by the ants. The ants, in turn, protect the aphids by aggressively keeping predators away (such as ladybird beetles and syrphid fly larvae).

the southwest from the Middle East to act as windbreaks along freeways and railroad tracks. The rapidly growing and reproducing tamarisk trees are better able to tap deep groundwater supplies, interfering with water availability to the deserts' native trees, such as mesquite and desert willows, and reducing their populations.

In contrast to exploitative competition, **interference competition** is direct: One species directly interferes with the ability of a competing species to gain access to a resource. Aggressive behavior in animals, as when hyenas drive away vultures from the remains of a zebra, or by **territoriality,** as when male bighorn sheep establish and defend an area against other males of their species, are examples of interference competition.

COMPETITIVE EXCLUSION: "WINNER-TAKES-ALL"

Although species with small niche differences are often able to live in the same community, those with identical

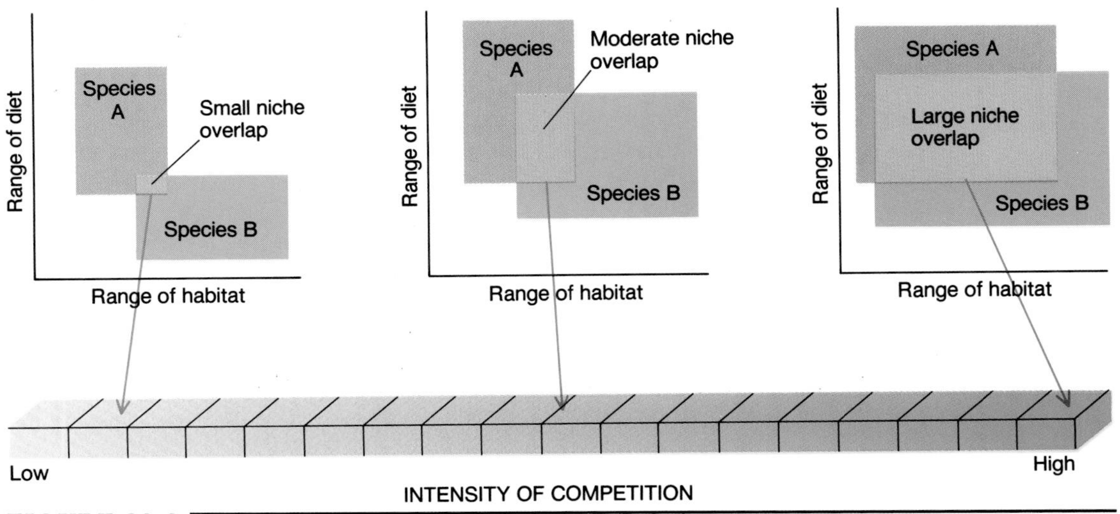

FIGURE 30-2

Niche overlap and competition. In each of the three graphs, habitat and diet requirements are plotted for two species. When niches overlap, species compete for limited resources. The greater the similarity in the requirements of the species, the greater the niche overlap, and the more fierce the competition.

ecological niches cannot do so, even if they share only one scarce resource and many abundant resources. Competition becomes so intense in this case that one species eventually eliminates the other from the community, either by taking over its habitat and displacing the species from the community or by causing the species' extinction. As we mentioned in the Steps To Discovery vignette at the beginning of this chapter, this "winner-takes-all" outcome is referred to as Gause's **Principle of Competitive Exclusion.**

Competitive exclusion is not always evident in natural communities. For instance, six species of leafhoppers (*Erythoneura*) are able to live on a single sycamore tree, feeding side by side on the same leaves. Not only are the habitats and food source the same for all these insects, but the species' life cycles are virtually identical as well, a seemingly perfect set-up for competitive exclusion. Yet, investigators found no evidence whatsoever that these species harm one another, much less exhibit competition that results in exclusion. Perhaps competitive exclusion is avoided in this case because shared resources are so abundant.

RESOURCE PARTITIONING AND CHARACTER DISPLACEMENT—ALTERNATIVES TO COMPETITIVE EXCLUSION

Competitive exclusion is not the only outcome of competition. Sometimes, a shared resource becomes partitioned in a way that allows competitors to use different portions of the same resource. For example, five species of North American warblers feed in slightly different zones on the same spruce tree, enabling these very similar birds to coexist with minimum competition (Figure 30-3). In addition to such spatial partitioning, a shared resource may be exploited at different times, producing temporal partitioning. An example of temporal partitioning occurs in a grassland ecosystem, where a species of buttercup (*Ranunculus*) grows only in early spring, before competing perennial grasses begin to grow. Dividing a resource in either time or space is known as **resource partitioning.**

Another alternative to competitive exclusion is **character displacement,** whereby intense competition dra-

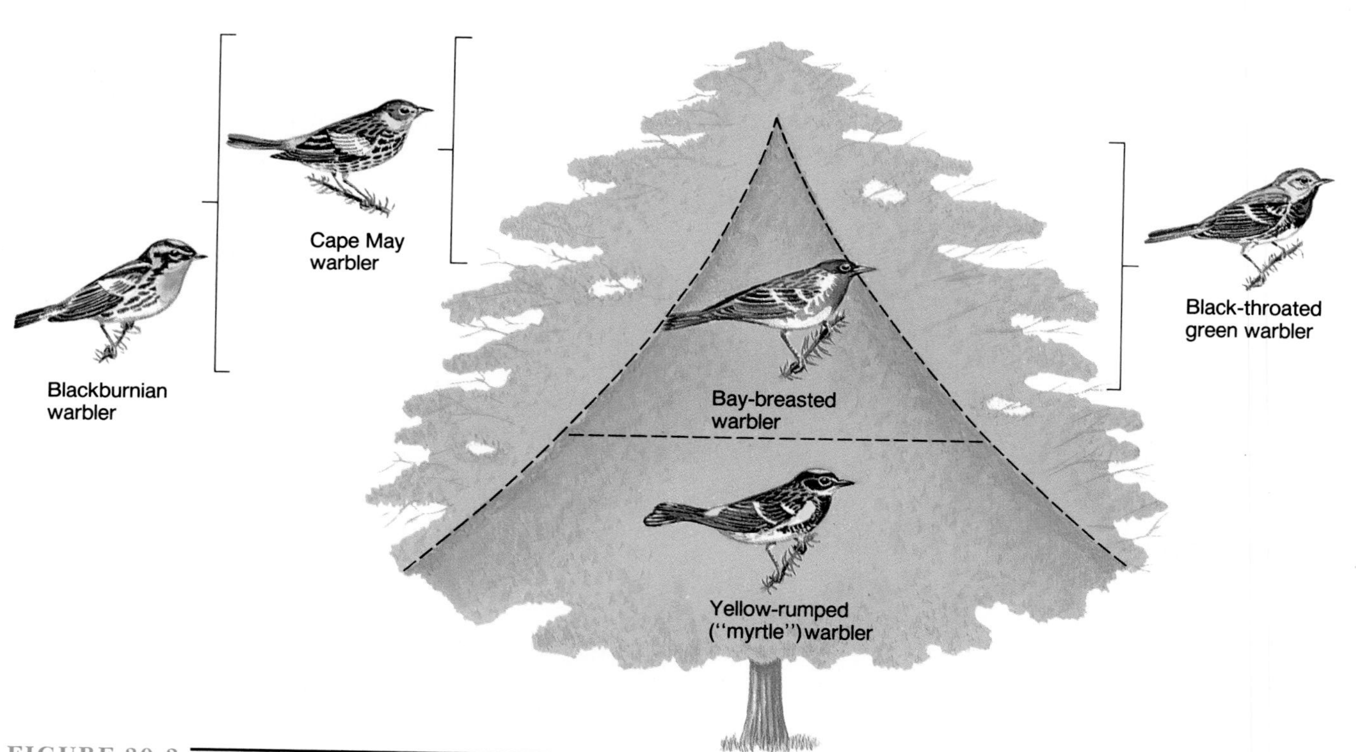

FIGURE 30-3

Resource partitioning: dividing the ecological pie. Five species of North American warblers feed in the same spruce tree, but each feeds in a slightly different zone. Foraging in different areas of a common resource at the same time is one form of resource partitioning. Resource partitioning reduces competition, enabling species with similar ecological niches to coexist in a community.

matically affects evolution, leading to changes in one or more characteristics of a species. Imagine two bird species that, because of the size and shapes of their beaks, both harvest the same kind and size of fruit. If, as a result of natural selection, one bird species evolves a different-sized bill—a bill that can harvest larger fruits—competition between the two species would decrease.

INTERACTIONS THAT HARM ONE ORGANISM AND BENEFIT THE OTHER

All organisms require a source of energy and nutrients to survive and reproduce. When one organism supplies a resource to the other, or is itself the resource, the two organisms must interact. Natural selection has produced a battery of adaptations that help organisms secure needed resources from others and that help organisms defend themselves from becoming a resource.

PREDATION AND HERBIVORY

During **predation,** one organism (the **predator**) acquires its needed resources by eating another organism (the **prey**). If the prey is a primary producer, such as an herb, shrub, or tree, the interaction is called **herbivory**; plant-eating animals are called **herbivores.** Organisms that eat other animals for energy and nutrients are **carnivores,** and those, like humans, that eat a mixed diet of plants and animals are **omnivores.**

Predator and Prey Dynamics

Although some predators limit their diets to one type of prey, most rely on more than one species for nourishment. The choice often depends on the abundance and accessibility of prey. During the summer, for example, a red fox mainly eats meadow mice. As the availability of meadow mice dwindles in the cooler seasons, however, the fox shifts to eating the more abundant white-footed mice.

As the availability of prey increases in an area, so does the number of predators; more prey feed more predators. More predators consume greater numbers of prey, however, reducing the availability of prey. As a result, the number of predators drops. This reciprocal interaction generates recurring cycles of increases and decreases in predator and prey abundance, striking a balance between the number of prey and the number of predators (Figure 30-4).

Predator and Prey Adaptations and Defenses

Coevolution between predators and prey over long periods of time has produced an array of effective adaptations;

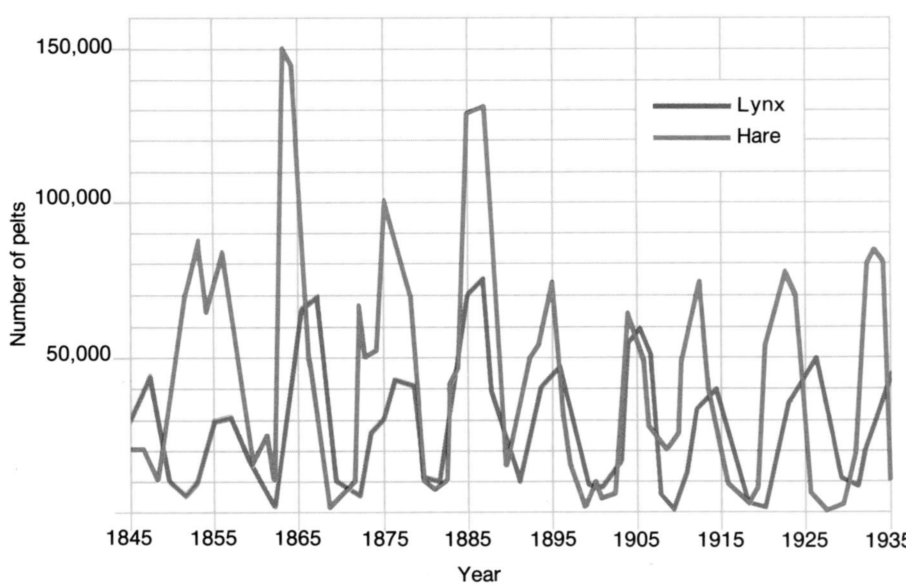

FIGURE 30-4

The ups and downs in numbers of predators and prey. This graph plots the number of lynx and snowshoe hare furs sold by trappers to the Hudson's Bay Company in Canada between 1845 and 1930. As you can see, increases and decreases in the number of prey (the hare) triggered increases and decreases in the number of predators (the lynxes), and vice versa. Other factors may also have affected the lynx and hare populations. Overbrowsing and fluctuations in plant growth may have altered the availability of food for the hares, while outbreaks of disease and climate changes could have affected the size of the lynx population.

TABLE 30-2
HOW SOME ORGANISMS AVOID BECOMING PREY

Escape Adaptations	Effect
Camouflage	
Cryptic coloration	Hides from predator
Disruptive coloration	Distorts shape and confuses predator
Individual responses	
Startle behavior	Confuses predator
Playing dead	Confuses predator
Shedding body parts	Escapes capture
Outdistancing predators	Escapes capture
Group responses	Warn, protect, and confuse
Herds, packs	Potects young and weak
Schools	Confuses predator
Defense Adaptations	
Physical defenses	
Armor	Deters an attack
Aposematic	Advertises noxious trait
Mimicry	
Müllerian mimicry	Noxious species avoided
Batesian mimicry	Harmless or palatable species avoided
Chemical Defenses	
Poisons	Kills predators
Hormones	Disrupts predator development
Allelochemicals	Repels predator

some improve the skills of predators in capturing prey, while others improve the prey's chances of escaping predators (Table 30-2).

Camouflage. Some prey go unnoticed by predators because they blend in with their surroundings or because they appear inanimate (like a dried twig) or inedible (like bird droppings). Such adaptations are called **camouflage** (Figure 30-5) because the color, shape, and behavior of an organism make it difficult to detect, even when in plain sight. Camouflage is not reserved exclusively for prey, however; predators also rely on camouflage to help conceal themselves while waiting to ambush prey.

The camouflage of some organisms helps these individuals resemble their background. This type of adaptation is called **cryptic coloration** because the camouflaged organism is hidden from view (Figure 30-6). **Disruptive coloration** disguises the *shape* of an organism, as in the coloration of the moth shown in Figure 30-7. The color pattern breaks up the outline of the moth when the individual is resting on a dark tree trunk, concealing its shape. Cryptic and disruptive color adaptations are not used solely by animals. Cryptic coloration in some plants helps them resemble less palatable plants, and cryptic and disruptive coloration helps camouflage some plants from herbivores (Figure 30-8). In addition to coloration, the shape and behavior of an organism can also provide a successful disguise. The crab spider and potoo bird in Figure 30-9 remain motionless, reducing the chance that they will be detected by prey.

Individual Responses. When confronted with a predator, some prey rely on sudden escape responses. Generally,

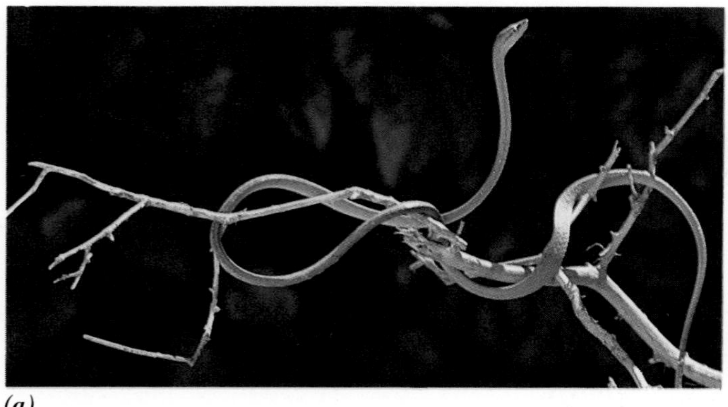
(a)

(b)

FIGURE 30-5
Camouflage: nature's masqueraders. ***(a)*** The "vine" "growing" on this branch is really a grass green whip snake (*Dryophis*). ***(b)*** The Malaysian horned frog has adaptations that help the animal blend in with dead leaves lying on the forest floor. Such adaptations include shading (which conceals the frog's eyes) and curly horns that resemble drying leaf tips. These characteristics help make the horned frog virtually invisible to its prey.

(a)

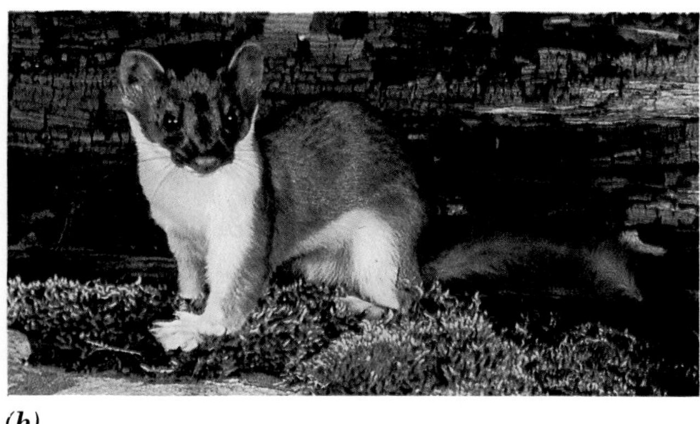

(b)

FIGURE 30-6

Keeping a "low profile." The fur of the long-tailed weasel (*Mustela frenata*) changes from white in winter ***(a)*** to brown in summer ***(b)***. Both cryptic colors help the weasel integrate into its surroundings and escape detection by predators.

FIGURE 30-7

Disruptive coloration disguises the shape of this moth as it rests on a tree trunk, creating an image that goes unrecognized by its sharp-eyed predators, the birds.

FIGURE 30-8

Appearing more like stones than plants, pebble plants (*Dinteranthus*) usually go unnoticed by passing herbivores.

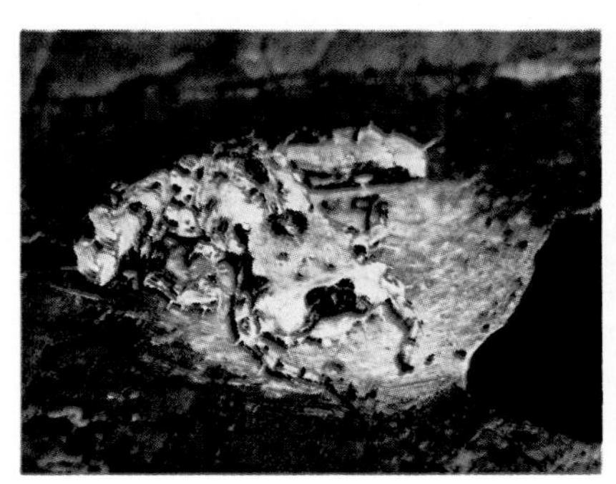

(a)

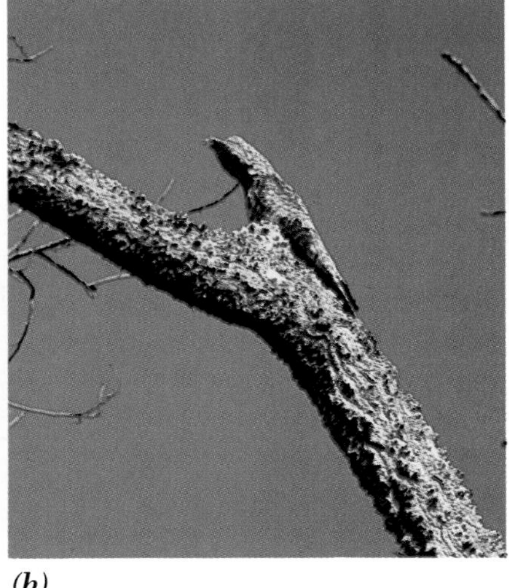

(b)

FIGURE 30-9

Nature's impostors. ***(a)*** The Borneo crab spider resembles bird droppings, a disguise that lures butterflies and other insect prey that eat genuine droppings. At the same time, birds, the predators of the spider, stay clear of what appears to be their own wastes. ***(b)*** Unwary animals fall easy prey to this dead tree trunk (really a motionless potoo bird), with its eyes half open and neck outstretched. The potoo's single-spotted egg also blends with the broken tree stump, camouflaging it from predators.

the predator is momentarily stopped by the unexpected response, especially if the escape response seems dangerous. This moment's hesitation may give the prey a chance to escape. Examples of such "last-ditch" responses include the following.

- An owl fluffs its feathers and spreads out its wings, a last-minute bluff that usually startles an attacking hawk.
- A mosquito fish frantically splashes on the surface of a pond when approached by a voracious pickerel (a small fish), making it difficult for the pickerel to launch a pinpoint attack.
- A tiny bombardier beetle sprays hot chemical irritants at a rodent, thwarting the attack.

A few organisms may escape a predator's clutches by releasing the seized part of their body. For example, a lizard quickly detaches its tail, which continues to move for several minutes, keeping the predator occupied while the lizard scurries to safety to begin regenerating a new tail.

Finally, some animals escape becoming prey simply by outrunning their predator. A healthy antelope or impala, for instance, can usually outdistance a lion. Like many predators, lions generally capture the young or the weak. By removing the young, injured, and weak from the breeding population, predators act as a powerful natural selection agent for the prey population.

Group Responses. Schools, packs, colonies, and herds typically defend themselves more effectively than can a single organism. For example, the first smelt fish to notice an approaching predator releases chemicals into the water, which immediately send the school of smelt fleeing in various directions. The confused predator does not know which way to turn. In grazing herds, stronger individuals generally surround the younger and weaker, protecting them from an advancing predator. There is indeed safety in numbers; a predator is less able to pick out a single target among a swarming group.

Physical Defenses. Organisms have evolved an arsenal of anatomical features to help protect themselves against direct attacks. Many have protective shells (molluscs and turtles), barbed quills (porcupines), needlelike spines (sea urchins), or piercing thorns, spines, or stinging hairs (plants) that can discourage even the hungriest predator. In fact, your skin is a protective armor against the daily invasions of millions of microbes.

Many foul-tasting, poisonous, stinging, smelly, biting, or in other ways obnoxious animals ironically have striking colors, or bold stripes and spots. This type of defense, called **aposematic coloring,** or warning coloration, is the opposite of camouflage; it makes an organism stand out from its surroundings. (*Aposematic* refers to anything that serves to warn off potential attackers.) The distinctive aposematic stripes of a skunk, for example, advertise to potential predators that this animal can yield an obnoxiously smelly counterattack.

Mimicry. If species that appear similar to one another are equally obnoxious, the resemblance is called **Müllerian mimicry.** In the tropics, for example, many species of beetles have bright orange wingcases with bold, black tips. When attacked, the beetles release drops of their own foul-tasting blood, just a taste of which deters a predator, sparing the beetle. Consequently, predators learn to stay away from other similarly colored beetle species. Examples of Müllerian mimicry abound in many tropical butterfly species, such as between monarch and *Acraea* butterflies (Figure 30-10).

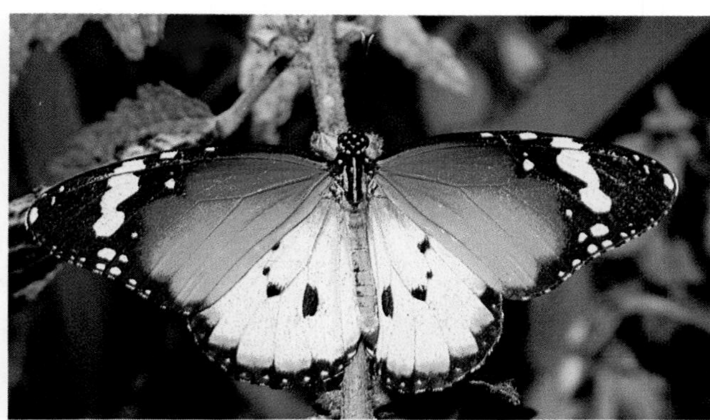

FIGURE 30-10

Equally distasteful, the *Acrea* butterfly (left) and the African monarch (right) are Müllerian mimics. These butterflies are not even closely related, yet they resemble each other almost exactly in color, pattern, behavior, and flavor.

FIGURE 30-11
The deadly moray eel (left) and the harmless plesiops fish (right) are an example of Batesian mimicry. When pursued by a predator, the plesiops fish swims head first into a rock crevice. The shape, color, pattern, and false eye of its exposed tail strongly resemble the head of the dangerous spotted moray eel, frightening predators away.

In some cases of mimicry, a harmless or palatable species gets a "free ride" by resembling a vile-tasting or stinging species. When a good-tasting or harmless species (the *mimic*) resembles a species with unpleasant, predator-deterring traits (the *model*), the similarity is called **Batesian mimicry** (Figure 30-11).

Chemical Defenses. Some plants and animals release **allelochemicals**—chemicals that deter, kill, or in some other way discourage predators. Such defenses must be swift and effective; otherwise, the predator may inflict fatal damage before the allelochemical has a chance to take effect. Some tropical toads and frogs secrete extremely poisonous chemicals. South American tribesmen simply need to touch the tip of their arrows to the skin of poisonous toads to produce a lethal missile that can kill an animal (including a human) within minutes. Other swift-killing poisons are manufactured by the Japanese puffer fish, the Asian goby fish, and the American newt. Poisonous animals frequently exhibit aposematic coloration, which serves as a blatant signal of the consequences of an attack to experienced predators.

PARASITISM

Parasitism is another type of interaction that benefits one organism and harms the other. A **parasite** secures its nourishment by living on or inside another organism, called the **host.** Most parasites are *host-specific*; that is, their anatomy, morphology, metabolism, and life history are adapted specifically to those of their host. For example, the human tapeworm lacks eyes, a digestive tract, and muscular systems. The combination of adaptations it evolved are suited for living inside human intestines, however. They include

- an outer cover that protects the tapeworm from powerful digestive enzymes yet allows the absorption of nutrients;
- a long, flat shape that creates a maximum absorptive surface area yet prevents obstruction of the host's intestine;
- hooks on its "head," which anchor it to the host's intestinal lining;
- a reproductive system with both male and female parts, allowing for self-fertilization. Self-fertilization is an important reproductive strategy in a location where contact with another tapeworm is highly unlikely. (Internal parasites are often little more than reproduction "machines," producing millions of offspring, increasing the chances of infecting a new host.)

Animals are hosts to a huge battery of parasites, including viruses, bacteria, fungi, protozoa, and other animals (flatworms, flukes, tapeworms, nematodes, mites, fleas, and lice, for example). A single bird may be host to 20 different parasites, of which there may be hundreds of individuals. The range of plant parasites is equally broad and includes viruses, bacteria, fungi, nematodes, and other plants, such as mistletoe and dodder.

Although most parasites rely on their host only as a source of nutrients, some parasites also use their host as a haven for protection from predators. For example, the pearl fish (*Carapus*) develops and lives in a safe, but very unusual place: the anus of a sea cucumber (*Actinopygia*). As a sea cucumber draws in water through its anus for gas exchange, a newly hatched pearl fish swims in (Figure 30-12). The pearl fish feeds on its host's tissues, taking periodic excursions outside its host to supplement its diet and

FIGURE 30-12
One very unusual habitat for a fish is the anus of a sea cucumber. A young pearl fish has poor eyesight and is barely able to swim, making this animal quite vulnerable to predators. Soon after hatching, the pearl fish locates a sea cucumber. When the sea cucumber opens its anus to draw in water, the parasitic pearl fish enters this unusual, but effective, shelter.

to reproduce. It returns to the sea cucumber for protection from predators.

Some parasites exploit the behavior of their host, an interaction called **social parasitism.** Examples of social parasitism are provided by European cuckoos, American cowbirds, and African honey guides. After a host bird builds a nest and lays eggs, the parasite bird destroys one of the host's eggs and replaces it with her own. The egg is usually so similar in size and coloration that the host bird fails to recognize it as an alien egg and incubates the parasite's egg as its own. After hatching, the parasitic baby bird instinctively shoves all solid objects out of the nest, including the host's babies and any unhatched eggs. After clearing the nest of its rivals, the parasite snatches up all of the food brought to the nest by its duped "foster parents."

ALLELOPATHY

Some organisms wage chemical warfare on other members of the community. **Allelopathy** is a type of interaction whereby one organism releases allelochemicals that harm another organism. Some of the animal chemical defenses we described earlier, such as the spray of the bombardier beetle or skunk, are examples of allelopathy.

Some plants manufacture allelochemicals that kill herbivores or competing plants. For instance, the chemicals released by some chaparral plants accumulate in the soil beneath the plants, blocking the germination and growth of other plants and reducing competition for scarce water and nutrients. Members of the crucifer family (cabbages, broccoli, brussels sprouts, mustards, radishes, and so on) produce mustard oils, chemicals that are lethal to many herbivores and disease-causing fungi and bacteria.

During the 1960s, investigators accidentally discovered that plants produce allelochemicals that disrupt the normal growth and development of insect herbivores, preventing the development of adult insects with reproductive organs. Researchers have discovered similar allelochemicals in some ferns, conifers, and flowering plants. Once again, these chemicals were virtually identical to those insect hormones that coordinate development during metamorphosis. As larvae consume these plants, the allelochemicals cause premature metamorphosis or produce sterile adults. Either way, herbivore reproduction is disrupted, illustrating a very effective plant adaptation for protection against increasing numbers of herbivores. Some of these hormone-mimicking allelochemicals are being considered for use as natural pesticides because they would cause considerably less environmental damage than do synthetic insecticides.

COMMENSALISM: INTERACTIONS THAT BENEFIT ONE ORGANISM AND HAVE NO EFFECT ON THE OTHER

The benefits of commensalism are one-sided: Only one of the participants (the commensal) profits, while the other is virtually unaffected. Nature exhibits many examples of commensalism. For instance, remoras are fish that attach themselves by suckers to the undersides of sharks and gather food scraps as the sharks feed. Remoras benefit from this interaction, but their presence apparently has little or no impact on the shark.

Some commensals simply live in a habitat that is created by another organism. The burrows of large "innkeeper" sea worms, for instance, house an array of "guests" that use the burrow for shelter but do not hinder or benefit the innkeeper worm in any way (Figure 30-13). Epiphytes are commensal plants that grow on the branches of taller plants. Being higher up in the forest canopy, the epiphyte captures more light than it could if it occupied a position lower in the canopy. Barnacles encrusted on a humpback whale are also commensals who gain a habitat as well as a means of transportation to new sources of food.

INTERACTIONS THAT BENEFIT BOTH ORGANISMS

Throughout this text, we have seen how natural selection favors characteristics that improve an organism's survival and reproductive success. Interactions between organisms in an ecosystem which provide benefits to both participants are favored by natural selection because the positive interactions contribute to each organism's survival or reproduction. There are many examples of beneficial interactions, some of which are compulsory to both participants, and others that are optional.

PROTOCOOPERATION

Protocooperation interactions benefit both participants, but they are noncompulsory. For example, protocooperation between a fungus and algae forms a lichen (page 463). The fungus uses some of the food produced by the algae, while the algae gains a watery habitat as well as some minerals absorbed by the fungus. Both organisms benefit

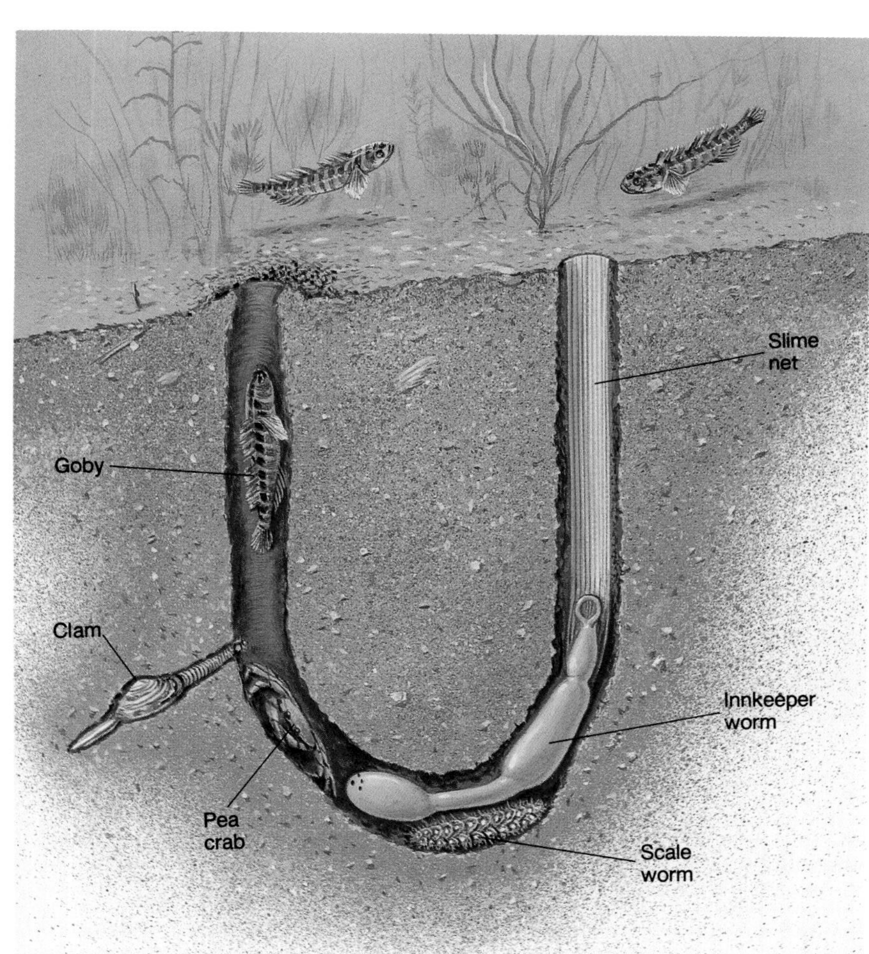

FIGURE 30-13
Commensalism: One benefits, the other is unaffected. The innkeeper worm bores a tunnel in the mud of shallow coastal waters. The worm then spins a slime net that traps minute organisms as the worm pumps water into the tunnel and through the net. When the net is full, the innkeeper gulps down the whole thing—net, trapped food, and all. But the innkeeper worm is not the only occupant of its tunnel. The goby uses the burrow for protection, while the pea crab, clam, and scale worm feast on the innkeeper's leftovers. Although the guests benefit from the association, the innkeeper worm apparently neither profits nor suffers from their presence.

FIGURE 30-14
Protocooperation between oxtail birds and rhinos. Although neither animal requires the other for its survival, both benefit from their interaction. The bird removes bloodsucking ticks from the rhino, while the rhino supplies the bird with an abundant food supply and warmth.

from this form of symbiosis, yet they could live successfully on their own as well.

Another example of protocooperation is the relationship between an oxtail bird and a rhinoceros (Figure 30-14). The bird perches on the back of the rhinoceros and removes pests that may land there (bloodsucking ticks and flies). The oxtail benefits by receiving food, warmth, and protection from predators; the rhino benefits by receiving protection against parasites. The sharp-eyed oxtail also alerts the dim-visioned rhino of approaching intruders. Again, the relationship is optional because both animals are capable of surviving on their own.

MUTUALISM

Mutualism is another form of interaction in which both participants benefit. Unlike protocooperation, however, the mutualistic interaction is essential to the survival or reproduction of both participants. Many mutualistic interactions are symbiotic, involving a close association between the participants. The pollination of some flowers by specific insects, birds, or bats (Chapter 13), and the interaction between ants and the *Acacia* plant found in the tropics (Chapter 28) are examples of mutualism that have been discussed earlier in this book.

EVOLUTION AND ADAPTATION: TYING IT TOGETHER

Through evolution, the adaptations of many competitors, and symbiotic, protocooperative and mutualistic partners have become functionally interlocked. The partnership between many species of termites and their intestinal protozoa, for instance, goes beyond the termite's simply providing food and housing for the protozoa, and the protozoa's digesting the cellulose in wood for the termite (Figure 30-15). Coevolution has led to synchronized life cycles between these organisms. In fact, the synchrony is so precise that the internal protozoa are transmitted from one developmental stage of the termite to the next during molting. The same hormones that trigger the termite to molt also trigger the protozoa to form a cyst. When the termite reingests its gut lining after molting, it "reinfects" itself with its mutually beneficial partner.

All interactions between organisms in an ecosystem which enhance both the participant's survival and reproduction are strongly favored by natural selection. Camouflage, escape responses, chemical and physical defenses against predators, allelopathy, and mimicry are all examples of adaptations that help organisms survive. Each adaptation evolved as a result of repeated selection of individuals with traits that increase survivability. For example, cryptic coloration allows individuals with coloration and patterns that harmonize best with the background to escape hungry predators more easily than can individuals that stand out. These more cryptically colored individuals will survive and produce more offspring than those less cryptically colored individuals, passing on the adaptive traits. Behaviors can also be adaptive. For example, individual and group escape behaviors are adaptations that result from natural selection. Again, ecology is evolution in action; since all organisms are a part of some community and ecosystem, evolution therefore takes place within communities and ecosystems.

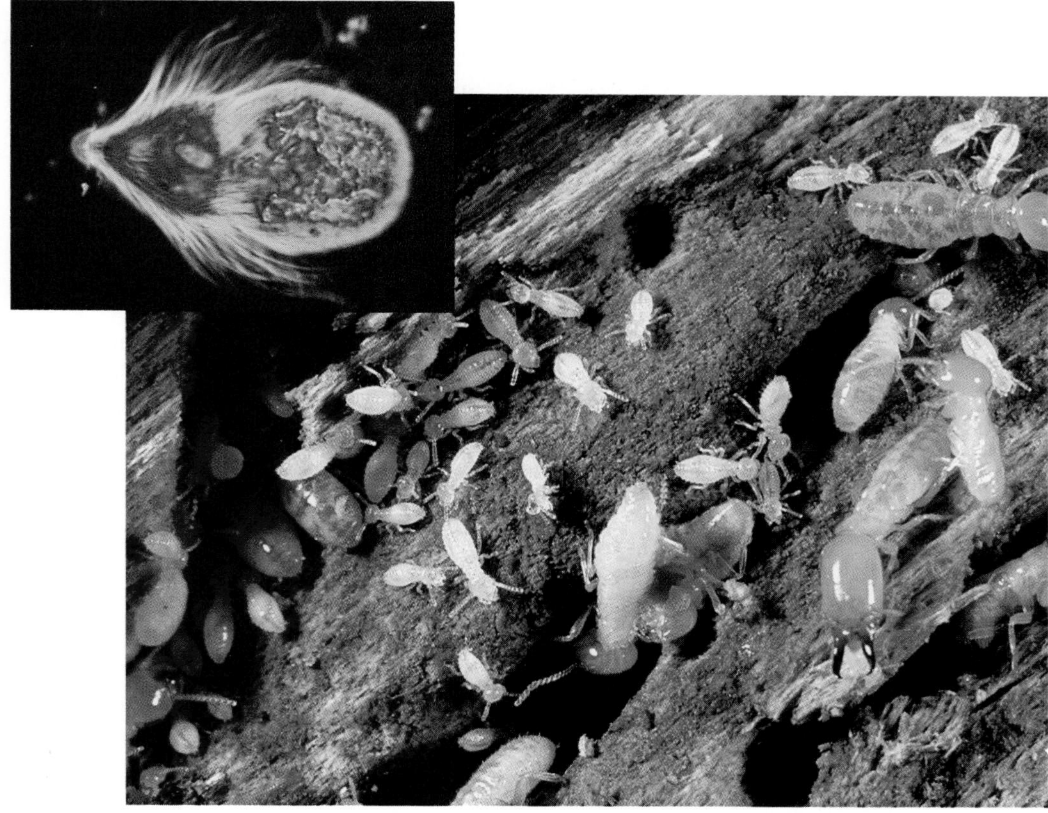

FIGURE 30-15

Evolution of protozoa and termites. Without internal protozoa (inset), termites would starve to death because they are unable to digest the wood they consume. Linked through coevolution, protozoa inhabit the gut of termites and obtain a habitat and food supply, while the termites receive a supply of usable nutrients from the digestion of wood by the protozoa.

SYNOPSIS

The organisms that make up the biotic community interact with one another in a variety of ways:

- Competition between organisms harms both participants.
- Predators gain energy and nutrients by consuming prey.
- Parasites live in or on a host organism, damaging or killing the host in the process.
- Commensalism interactions benefit one organism but do not harm the other.
- Both organisms benefit from protocooperation, yet each is able to survive independently.
- Mutualism benefits both interacting organisms, but neither can survive without the other.

When the niches of two species overlap, members of both species compete for the limited resources they require. Intense competition may lead to the sharing of different parts of a resource or of the entire resource at different times or to evolutionary changes in characteristics that reduce competition.

When the ecological niches of two species in a community are identical, competition between the two species results in one rival eliminating or excluding the other from the community. Species can coexist in the same community when they have slightly different ecological niches.

Evolution has resulted in a number of physical and behavioral adaptations that enhance organisms' predatory skills or help organisms escape predators. These adaptations include camouflage, which helps organisms blend in with their surroundings, conceal their shape, or protect vital parts; individual or group behaviors that confuse or distract attackers; anatomical features, such as shells, spines, or armor, that discourage attackers; a foul taste; and chemicals that kill or discourage predators.

Review Questions

1. Consider two seedlings of different plant species growing right next to each other in a community. Both grow at about the same rate and develop roots to the same depth. List the resources for which the seedlings will compete as they grow. What is the probable outcome of this situation? What will happen if one plant suddenly outgrows the other?

2. Is there greater opportunity for resource partitioning in a tropical rain forest, a deciduous forest, or a desert? Why? How does each of these terrestrial biomes compare to potential resource partitioning in the pelagic zone of oceans?

3. Use examples to distinguish between cryptic coloration and aposematic coloring. How do these adaptations help prey escape predators?

4. With the exception of parasitoids, the vast majority of parasites do not kill their host. What advantage is there to killing a host? Must there be an advantage, from a natural selection/evolutionary point of view, in order for there to be any host-killing parasitoids at all?

5. For each of the following pairs of terms, state how they are similar and how they are different:
   a. predation and allelopathy
   b. mutualism and protocooperation
   c. Müllerian and Batesian mimicry

Critical Thinking Questions

1. In what sense is competition, which is always harmful to the organisms involved and may result in competitive exclusion, good for the species? How does this concept connect ecology with evolution?

2. Many symbioses are very specific and permanent. The pollination of the Spanish dagger (*Yucca whipplei*) by only female pronuba moths is an example of such a relationship. Neither the plant nor the moth can reproduce without the other. From an evolutionary point of view, there are advantages and disadvantages to such compulsory and exclusive interactions. List and explain as many advantages and disadvantages as you can. Since there are disadvantages, why would narrow and binding relationships be favored by natural selection at all?

3. More than 70 percent of species obtain their energy and nutrients by consuming all or part of another organism, while only about 30 percent of all species on earth harvest energy and nutrients from the physical environment. These proportions are not always the same for all ecosystems, however. In fact, in some ecosystems (or biomes) the percentages may even be reversed. In which ecosystems would you expect the percentages to be the same, and in which would you expect the percentages to be reversed? Are there any ecosystems in which the biotic community is completely one or the other? With energy being so abundant in the abiotic environment of most ecosystems, explain why 70 percent of species consume other organisms for energy.

4. A biologist who has heard the phrase "nature abhors a vacuum" on many occasions wants to test whether this idea is true or not. As stated, is this a testable hypothesis? If so, design an experiment or series of experiments to test the hypothesis. If not, how could the phrase be reworded so that it could be tested? Design an experiment to test your new hypothesis.

5. Study the following graphs on competition between two grain beetles living in wheat at 29.1°C (Graph A) and at 32.3°C (Graph B). Is the principle of competitive exclusion supported by these data? Altering only one factor (temperature) changed the outcome of competition. Can you offer an explanation for this change?

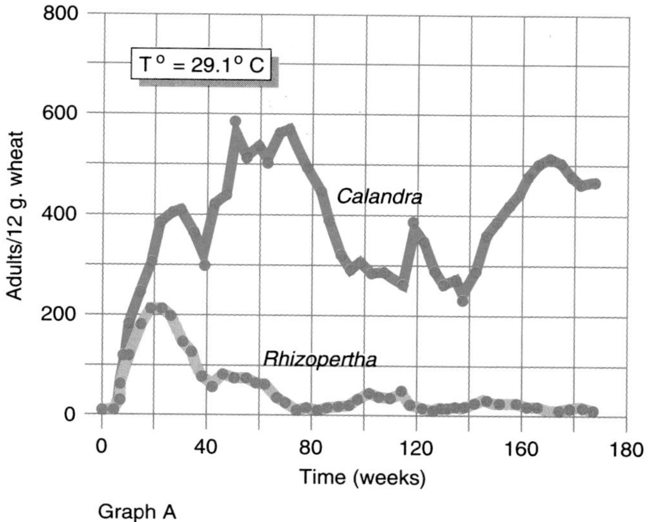

Graph A

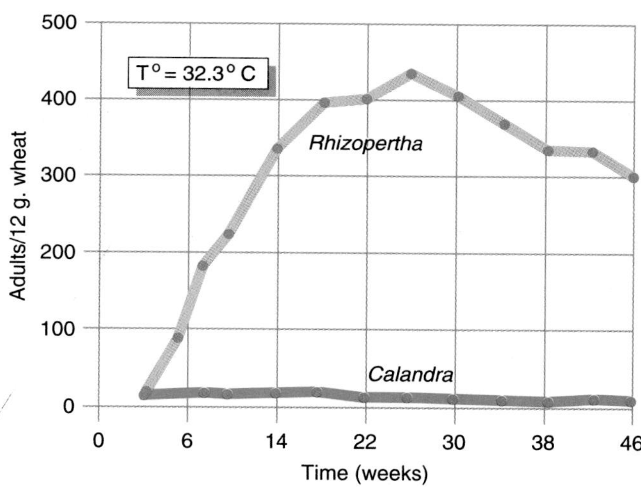

Graph B

CHAPTER

31

Animal Behavior

With limited resources, competition is greater among funnel spiders living in the desert as opposed to funnel spiders living in rich riparian habitats.

STEPS TO DISCOVERY
Mechanisms and Functions of Territorial Behavior

To many people, the behavior of spiders evokes feelings of terror and retreat. A few people, however, are transfixed by spiders, fascinated by what these arachnids do and why they do it. Susan Reichert falls into the "fascinated" category, having spent her professional life exploring the territorial behavior of the funnel-web building spider (*Agelenopsis aperta*). These spiders compete for web sites and fight for the territory around the web (a territory is an area defended against intruders, usually to protect a resource). A spider's territory must be large enough to provide sufficient food for survival and reproduction. The intensity of the disputes differs among spiders that live in different territories. For example, desert grassland spiders defend their areas more aggressively than do spiders that live in relatively lush *riparian* territories (near lakes and streams). Threat displays of grassland spiders are more likely to escalate into battles, which tend to result in physical injury.

In examining animal behavior, including that of humans, the question of "nature versus nurture" always arises. That is, how much of a behavior is genetically determined, and how much is determined by environmental influences? The territorial behavior in funnel-web building spiders seems to have a strong genetic component. In one study, Reichert collected spiders from a desert grassland environment in New Mexico. The spiders were allowed to mate only with each other, establishing pure-bred lines. Similarly, pure-bred lines for riparian spiders were produced. After the spiderlings emerged from the eggs, each was raised separately on a mixed diet of all it could eat, so that even the grassland spiders from naturally austere conditions were lavishly supplied with food. Did their ample food source teach the spiders to be less competitive? No, they still defended the same broad territory when placed into experimental enclosures where they could build webs. The study suggests that territoriality is an inherited characteristic and is not determined by hunger or learned from previous territorial disputes.

Working with John Maynard Smith, Reichert continued to explore the genetic mechanisms underlying territoriality and aggression in these spiders. Pure-bred lines of grassland spiders were mated with pure-bred lines of riparian spiders. Smith and Reichart found that fights between the hybrid offspring were even more likely to end in injury or death than were those between pure grassland spiders. Mating homozygous "aggressive" individuals with homozygous "low aggression" spiders, the researchers discovered that aggressive territorial behavior is determined by two conflicting tendencies: "aggression," and what Reichert called "fear" (actually the tendency was *retreat.* Even though the retreat *seemed* to be motivated by fear, emotions cannot be ascribed to a spider's behavior). They concluded that each tendency is controlled by a gene or, more likely, a gene complex. The allele(s) for high aggression (A) is dominant to that for low aggression (a), and the allele(s) for low fear (B) is dominant to that for high fear (b). Researchers proposed that aggression and fear are low in riparian spiders and high in grassland spiders. Thus, grassland spiders would be homozygous for high aggression (AA) and high fear (bb). In contrast, riparian spiders would be homozygous for low aggression (aa) and low fear (BB). As a result, the hybrid offspring (AaBb) would have high aggression and low fear, a genetic combination that, according to Reichert and Smith's findings, led to very costly fights. Additional mating experiments revealed that aggression is inherited on a sex chromosome whereas the gene for "fear" was located on an autosome.

The differences in territorial behavior are well suited to the spiders' local environments, a product of natural selection. Prey are more abundant in relatively lush riparian areas found along rivers and lakes, so spiders can capture adequate prey in a smaller area. In severe desert grassland environments, in contrast, prey are scarce, and larger territories are needed to provide enough food, so the value of each web is increased. The most aggressive individuals would be expected to have the distinct advantage as a species, as long as the web site fought for increased reproductive success. And it apparently does. An investigation of web-building spiders in New Mexico (along side a lava bed rather than grassland) supports the importance of web site quality to the animal's reproductive success. Those spiders with high-quality web sites had 13 times the reproductive potential of similar spiders in poor-quality sites.

"Why is that animal doing that?" This is the fundamental question of **ethology,** the study of animal behavior. This seemingly simple question has been interpreted in several ways. For example, the Dutch biologist Niko Tinbergen, a corecipient of the Nobel Prize in medicine and psychology in 1973, identified four related questions embedded in this larger query: (1) What are the mechanisms that cause the behavior? (2) How does the behavior develop? (3) What is its survival value? (4) How did it evolve? These four questions form the organizational basis for this chapter's approach to animal behavior.

▼ ▼ ▼

FIGURE 31-1
Reproductive success depends on the proper courtship ritual. Among sticklebacks, the behavior is built from a sequence of fixed action patterns.

MECHANISMS OF BEHAVIOR

Mechanisms that dictate behavior include not only genetically determined traits but also the inherited *potential* for learning and even the learned behaviors themselves. Genes generally encode a range of potential phenotypes. Sometimes genes specify a precise behavior, leaving little room for modification by learning. Behaviors that are precisely specified by genes are often those that must be expressed in nearly perfect form, even on the very first trial. For example, if an animal fails to respond appropriately the first time it encounters a predator, it may not get a second chance to refine its escape response. Genes also play an important role in determining the actions of animals that have little opportunity to learn, either due to a short lifespan or because there are no "teachers" (e.g., parents) around. In other cases, the behavioral blueprint is more general so that the behavior is almost entirely shaped by experience. These two extremes represent primary innate behavior versus learning.

PRIMARILY INNATE BEHAVIOR

Innate behaviors are those determined by fairly precise genetic control. Innate behaviors are often species-specific and highly *stereotyped*—the behavior is the same regardless of the animal's previous experience.

FIXED ACTION PATTERNS

Among the primarily innate behaviors are **fixed action patterns (FAPs).** These are motor responses that are triggered by some environmental stimulus. Once started, FAPs continue to completion without external input. For example, a brooding female greylag goose will retrieve an egg that has rolled just outside her nest by reaching beyond it with her bill and rolling it toward her with the underside of the bill. Once the rolling behavior has begun, if the egg is experimentally removed, the goose will continue the retrieval response until the now imaginary egg is safely returned to the nest. The egg retrieval response of the female greylag goose illustrates other characteristics that are generally true of most FAPs. An FAP is performed by all appropriate members of a species. Furthermore, each time the behavior is exhibited, the sequence of actions is virtually identical, modified very little by experience. As evidence, an FAP will be exhibited even in inappropriate circumstances. For example, a brooding female greylag goose will retrieve a beverage bottle or any small object outside the nest as if it were her egg.

STIMULI AND TRIGGERS

A fixed action pattern is produced in response to an environmental stimulus. Ethologists called such a stimulus a **sign stimulus.** If the sign stimulus is given by a member of the same species, it is termed a **releaser.** For example, a male European robin will attack another male robin that enters its territory. Experiments have shown, however, that a tuft of red feathers is attacked as vigorously as is an intruding male. Of course, in the world of male robins, red feathers usually appear on the breast of a competitor.

One way ethologists can identify a sign stimulus from the barrage of other information reaching an animal is by using an object in which only one trait is presented at a time. The object (called a *model*) is presented to an individual in the appropriate physiological state to see whether it will respond as it would to the normal stimulus. For example, a male stickleback fish in reproductive condition defends his territory from any intruding males. By constructing models of sticklebacks of varying degrees of likeness to the real male and painting them red, pale silver, or green, Niko Tinbergen and his co-workers demonstrated that a red tint on the undersurface of the trespasser releases an aggressive territorial response in male sticklebacks. A very realistic replica lacking the red color was not attacked, but a model barely resembling a fish, whose underside was painted red, provoked an assault.

CHAIN OF REACTIONS

More complex behaviors can be built from sequences of FAPs. The final product is an intricate pattern called a *chain of reactions,* whereby each component FAP brings the animal into the situation that triggers the next FAP. An early analysis of a chain of reactions was conducted on the courtship ritual of the three-spined stickleback. This sequence of behaviors culminates in synchronized gamete release, an event of obvious adaptive value in an aquatic environment. Each female behavior is triggered by the preceding male behavior, which, in turn, was triggered by the preceding feminine behavior (Figure 31-1).

If a female stickleback enters a male's territory and exposes her egg-swollen abdomen, the male will begin his courtship with a zig-zag dance. This dance encourages the approach behavior of the female. Her movement induces the male to turn and swim rapidly toward the nest, an action that entices the female to follow. Once at the nest, the male stickleback lies on his side and makes a series of rapid thrusts with his snout into the entrance to the nest while raising his dorsal spines toward his mate. This action is the releaser for the female to enter the nest. The presence of the female in the nest, in turn, is the releaser for the male to begin to prod the base of the female's tail with his snout, causing the female to release her eggs. The female then swims out of the nest, making room for the male to enter and fertilize the eggs. We can see that this complex sequence is largely a chain of FAPs, each triggered by its own sign stimulus or releaser.

GENES AND BEHAVIOR

Such behaviors have a strong genetic basis. Genes do, in fact, influence all behavior to some extent in that an animal with a certain gene can perform the behavior whereas an animal that lacks the gene does not. Genes direct the synthesis of proteins, that may affect some of the connections within the nervous system or may act as regulators, such as certain enzymes that have a regulatory function. In a few cases, we know the link between the protein product of a specific gene and a behavior. For example, the gene behind egg-laying behavior of the sea hare *Aplysia* directs the synthesis of a long chain of amino acids that is later cleaved into many proteins, three of which are known to be important in egg-laying behavior (Figure 31-2). One of

FIGURE 31-2
The sea hare Aplysia lays eggs in a stereotyped sequence of actions that is controlled by a single gene that codes for a long chain of amino acids. This chain is cleaved into three shorter proteins, three of which are important in orchestrating the behavior.

FIGURE 31-3

Territorial fish. A male blue gourami that is classically conditioned to cues signaling the approach of a rival has a competitive edge in territorial disputes over an intruder that has not yet been conditioned.

the three proteins is egg-laying hormone (ELH), which stimulates the reproductive system to contract and expel a string of eggs. The two other proteins function as neurotransmitters that increase or decrease the activity of neurons involved in egg-laying behavior.

LEARNING

Unlike perfectly stereotyped behavior, learning is a process in which the animal benefits from experience so that its behavior is better suited to environmental conditions. It is often useful to group learning types into categories, including habituation, classical conditioning, operant conditioning, insight learning, social learning, and play.

Habituation

The simplest form of learning is **habituation,** whereby the animal learns *not* to show a characteristic response to a particular stimulus because that stimulus was shown to be unimportant during repeated encounters. For example, the marine clamworm that lives in underwater burrows partially emerges from the burrow while feeding. The presence of a shadow that could herald the approach of a predator causes the clamworm to withdraw quickly for protection. But if no adverse consequences follow the shadow's presence (for example, if the shadow is cast by kelp drifting into the sun's path), the withdrawal response gradually wanes. Habituation is beneficial in that it eliminates responses to frequently occurring stimuli that have no bearing on the animal's welfare, without diminishing reactions to significant stimuli.

Classical Conditioning

In **classical conditioning,** an animal learns a new association between a novel stimulus and a response. The new stimulus repeatedly occurs before a usual stimulus; gradually, the new stimulus begins to serve as a "substitute" for the normal stimulus and eventually elicits the response that ordinarily follows the natural stimulus. The most familiar example of classical conditioning is that of Pavlov's dogs, who learned to associate the sound of a bell with the presence of food. During training, Pavlov rang a bell immediately before feeding a hungry dog. When the dog saw the food, it began to salivate. The procedure was then repeated many times; eventually, the dog began to salivate at the mere sound of the bell, even in the absence of food. This response is an example of a **conditioned reflex.**

The *unconditioned stimulus* (in this case, food) elicits an automatic unlearned response from the animal. A second stimulus (the bell) is called the **conditioned stimulus** since the animal was conditioned by learning to respond to it. The adaptive function of classical conditioning in nature is to prepare animals for important events, such as mating (Figure 31-3). In the laboratory, fish that have been classically conditioned to a signal that indicates the approach of a rival are more successful at defending their territory and preparing for battle even before the unconditioned stimulus (intrusion into the territory) triggers the defensive response. The approaching male inadvertently sends signals of impending territorial invasion even before it enters the other male's domain. The conditioned response may increase the release of androgen, which enhances aggressiveness. A more aggressive male has a better chance of winning the battle and defending his territory, thereby increasing his chances of mating (females only mate with males that have an established territory).

Operant Conditioning

Operant conditioning differs from classical conditioning in that the conditioning stimulus *follows* the affected behavior. If a behavior has favorable consequences, the probability that the act will be repeated is increased. Teaching a dog to roll over by rewarding it after it performs the desired behavior is an example of operant conditioning. In nature, the behavior is spontaneously performed (unlike classical conditioning), and the favorable result, or *reinforcement*, must follow soon after the behavior for operant conditioning to be successful. The timing of events is critical, establishing a cause-and-effect relationship between the performance of the act and the delivery of the reinforcement.

B. F. Skinner devised an apparatus used to study operant conditioning in the laboratory. Typically, a hungry animal is placed into a "Skinner box" and must learn to manipulate a mechanism that yields food. For example, a hungry rat placed in a Skinner box will move about randomly, investigating each nook and cranny. Eventually, the rat will put its weight on a lever provided in the box. When the lever is pressed, a bit of food drops into a tray. The rat will usually press the lever again within a few minutes. In other words, the rat first presses the lever as a random act; then, when the action is rewarded, the probability of its being repeated increases.

Insight Learning

Insight learning is a sudden solution to a problem without obvious trial-and-error procedures. For example, captive chimpanzees have been known to stack boxes in order to climb up and reach a banana hanging from the ceiling of their cage. One interpretation of the chimps' problem-solving abilities was that they saw new relationships among events—relationships that were not specifically learned in the past. It may be that the chimps formed a mental image of the solution or created new associations between previously learned behaviors. It has been argued, for instance, that chimps that moved boxes and then climbed on them to reach a banana had already acquired two separate behaviors: moving climbing structures beneath targets, and climbing on an object to reach another object.

FIGURE 31-4
Snow monkeys washing food. The tradition of washing sweet potatoes was begun by a young Japanese snow monkey and spread rapidly throughout the troop.

Social Learning

Some organisms can learn from others, especially members of social species who spend more time close to others. It is often more efficient, and perhaps less dangerous, to learn about the world from others of the same species. For example, rhesus monkeys can learn to fear and avoid snakes by watching other monkeys show fear of snakes. Many types of birds perfect their territorial defense songs by singing with a neighboring adult. There are literally hundreds of such socially learned behaviors, from food washing in raccoons and monkeys (Figure 31-4) to the hundreds of skills and activities taught to human children by their parents.

Play

Play is a learned behavior that borrows pieces of other behavior patterns, usually incomplete sequences and often in an exaggerated form. Although the advantage of play is speculative, one hypothesis suggests that it builds physical training for strength, endurance, and muscular coordination. In a safe environment, the sensory and motor stimulation of play promotes the formation of a network of synapses in the cerebellum, a part of the brain responsible for sensory-motor coordination. Another hypothesis maintains that play allows individuals to practice social skills, such as grooming and sexual behavior, that are important in establishing and maintaining social bonds. Finally, play may be a mechanism for learning specific skills or improving overall perceptual abilities.

FILIAL IMPRINTING

Young chicks, ducklings, and goslings generally follow their mother wherever she goes. Konrad Lorenz, an Austrian biologist and corecipient of the 1973 Nobel prize, was the first to systematically study this behavior, working with newly hatched goslings. In one experiment, Lorenz divided a clutch of eggs laid by a greylag goose into two groups. One group was hatched by its mother; as expected, these goslings trailed behind her. The second group was hatched in an incubator. The first moving object these goslings encountered was Lorenz, and the goslings "imprinted" on him, responded to him as they normally would to their mother. Lorenz then marked the goslings so that he could determine to which group they belonged and placed both groups together under the same box. When the box was removed, the goslings streamed toward their respective "parents"—normally reared goslings sought out their mother goose, and incubator- reared youngsters went directly to Lorenz. The attachment was unfailing, and from that point on Lorenz had geese following in his footsteps. Today, the process by which young birds develop a preference for following their mother is called *filial imprinting*. Imprinting is relatively quick (virtually instantaneous) and occurs only during a limited time (called the *critical period*). Imprinting also occurs without any obvious reward. The behavior prevents young birds from associating with adults that may harm them.

DEVELOPMENT OF BEHAVIOR

The development of a behavior begins during embryonic development with the expression of genes needed for behaviors that are performed as soon as an animal hatches or is born (for example, suckling in mammals, or imprinting in birds). Some behavior-directing genes, however, are delayed in their expression, for example, genes that determine mating behavior. Learning is another aspect of behavior development. Genes and experience work together throughout the organism's development to produce most behaviors. As the organism changes over time, however, the nature of the stimulus-response will vary. Consider song development in birds (Figure 31-5). Young males must learn to sing by hearing adult males sing. However, if a young male white-crowned sparrow is isolated from other members of its species and is allowed to hear recordings of bird songs, including one of its own species, it will learn to sing correctly. How does the sparrow know which song is correct? It must have inherited a template, or genetic ability to recognize its own species' song. But the experience of hearing the song is also needed for the male to sing properly. Most complex behaviors cannot be divided neatly into either the innate or learned categories; rather they are influenced by both genetics and prior experience.

SURVIVAL VALUE OF SOCIAL BEHAVIORS

Most behaviors have obvious survival value, for example, the web-building behavior of a spider that enables it to catch its dinner. The value of some behaviors, however, is less immediately obvious, especially those exhibited by animals that live in social groups. Solitary animals, such as spiders and tigers, have little need for group interactions, but baboons, ants, and hundreds of other species depend on intraspecies cooperation and "teamwork" for their survival. Why do these species opt for "group living"?

COSTS AND BENEFITS OF GROUP LIVING

Living in groups has both costs and benefits. The costs include increased competition from other members of the group, increased exposure to parasites or disease, increased visibility to predators or prey, and an increased risk of wasted energy in raising offspring that are not one's own. But the benefits are numerous. Two of the most important advantages are cooperative hunting and additional protection. Predators that engage in cooperative hunting generally have a higher capture success rate and average energy intake per individual than do solitary hunters (consider a single wolf trying to bring down a moose). In addition, group behavior helps detect, confuse, and repel predators.

FIGURE 31-5
Song development in this species depends on both genes and experience. A young male must learn to sing its song by hearing the song of adult males. However, it inherits a genetic "template" of its own species' song, which is perfected by learning.

FIGURE 31-6
Male red deer roaring. Male red deer challenge one another in a roaring contest. Because roaring is strenuous, it provides an honest cue for assessing a rival's fighting ability.

With all the members of the group on alert, a predator is likely to be spotted by at least one individual, who alerts the whole group. Furthermore, many prey species, such as flocks of starlings or schools of fish, form even tighter clusters when a predator is detected, preventing even one member from being singled out. Alternatively, the group may engage in mobbing behavior, successfully attacking an otherwise superior predator in mass. A group of small birds may attack an owl. Baboons and chimpanzees repel leopards by screaming, charging, and even throwing sticks at the predator.

Another advantage of group living is greater success in defending their young, food, or space against competing groups of the same species. Mating efficiency may also be increased (especially in vertebrates), since within groups mating displays among males trigger the release of hormones in one another as well as in the females, improving fertility and mating behavior.

ANIMAL COMMUNICATION

Social animals communicate with one another for many reasons. Species recognition is important to ensure that reproductive efforts are not wasted on members of the wrong species and that aggression is directed toward those individuals who are competing for the same resources. Sexual reproduction often depends on communication, such as mate attraction and courting. During courtship, the male often advertises himself, and the female assesses his qualities. Aggression (and cannibalistic behavior) is reduced, so the mates conceive rather than fighting or eating each other. Resulting offspring may communicate their desire to be fed, or the parent may have to indicate its willingness to feed the young.

Communication is also important in aggressive interactions. Most of these interactions are stereotyped threat displays, the submissive individual accepting its loss and leaving without risking injury. Rivals are usually kept "honest" in that their displays exhibit some intimidating physical characteristic that cannot be faked, such as size or strength. Male red deer stags, for example, challenge each other in a vocal duel, in which each male takes turn roaring at the other (Figure 31-6). As the pace of the bellowing increases, one contestant usually gives up, the intensity of bellowing a reliable indicator of the rival's fighting ability.

Other signals promote recruitment, bringing individuals together to perform a specific duty. For example, fire ants leave an odor trail that will guide recruits to a food source. Honeybees have one of the most studied forms of recruitment, a dance performed by a scout that has located

(a)

Fly 20° from sun's direction

Fly toward the sun

1.

2.

3.

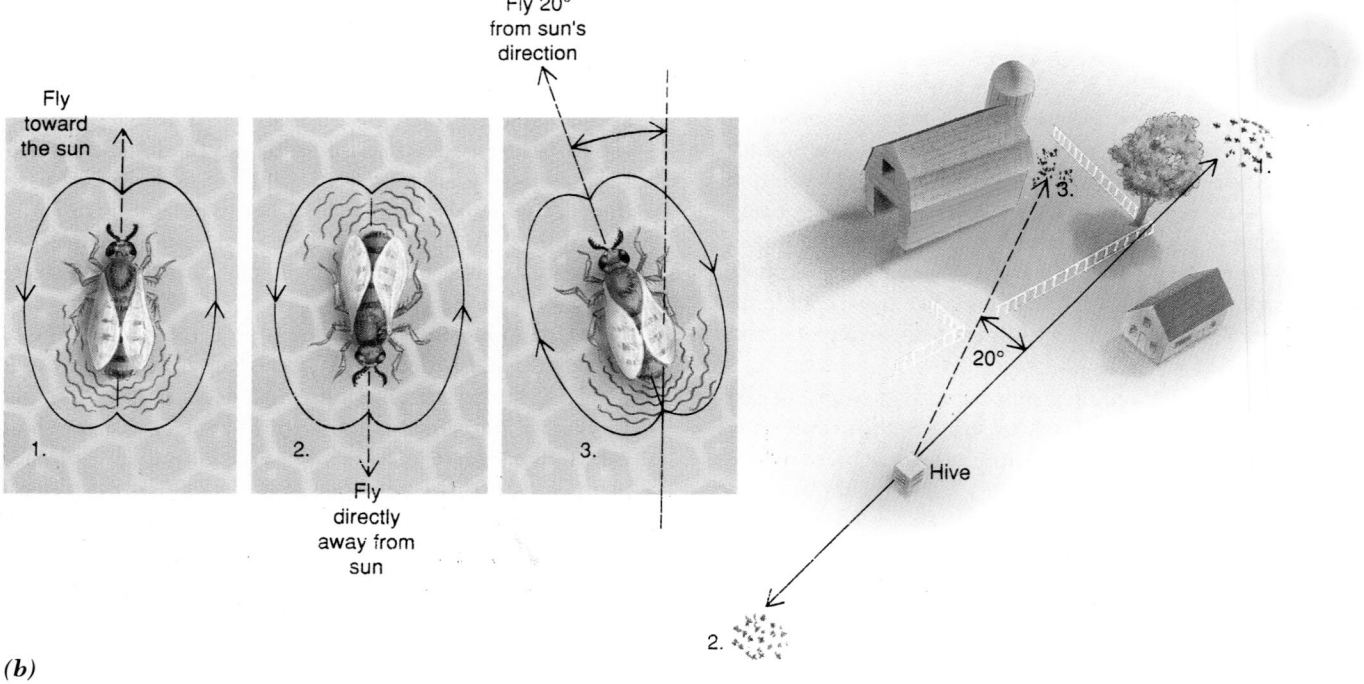

Fly directly away from sun

20°

Hive

2.

3.

1.

(b)

◀ FIGURE 31-7

Dance of a honeybee scout. ***(a)*** After finding food close to the hive, a scout returns and does a round dance on the vertical surface of the comb. The dance consists of circling alternately to the left and right. The dance informs recruits that food can be found within a certain distance of the hive. ***(b)*** Waggle dance of honeybee is performed when a scout finds food at some distance from the hive. The dancer traces the pattern of a figure eight and waggles her abdomen during the central straight part of the dance. Aspects of the dance correlate with the distance and direction to the food source. The dancer indicates direction by the orientation of her waggle run. The direction in which recruits should fly parallels the angle of the waggle run *relative to vertical.* When the food is in the direction of the sun, the dancer waggles straight up; when the recruits should fly directly way from the sun, the waggle run is oriented straight down. A food source located 20 degrees to the left of the sun would be indicated by a dance oriented so that the waggle run were 20 degrees to the left of vertical.

a food source (Figure 31-7). By feeling the dancer with their antennae, other bees get the message and fly to the indicated location.

TYPES OF COMMUNICATION SIGNALS

Any sensory avenue may be used for communication. Visual signals that are enriched with color and brightness are easiest to localize. Therefore, visual displays are frequently used by animals that are active during the day for short-range communication in open environments. At night, sound or scent is generally employed (with the notable exception of fireflies and a few other luminescent organisms). The croaks of frogs and the "fiddling" of crickets are perhaps the most familiar examples of these mating serenades. Water is a much better transmitter of audio signals than is air. The songs of whales, for example, can be heard hundreds of miles away.

Some species communicate with smells rather than (or in addition to) sounds. Pheromones—chemical signals released by an animal that influence the behavior of other members of the same species—may be the best signal when long-distance communication is desired. Perhaps the best known pheromones are sex attractants, such as those produced by moths. The atlas moth sex attractant is carried by the wind and may attract mates more than 21 kilometers (13 miles) away from the female. Chemical signals are also the most durable, which is why chemicals are often used to mark territories. Not all chemicals are long-lived, however; some, like the alarm signals of some ants, deteriorate within 30 seconds.

Light, an occasionally used social signal, is most effective in terrestrial animals. Light is rapidly attenuated in water, so visual displays are generally not used by aquatic animals if a signal must be sent over a long distance. Likewise, in a dense forest or other environment where long-distance vision is limited, animals relying solely on visual signals meet with little success.

In some social species bodily contact forms an important mode of communication. Social bonds may be cemented by physical contact. Some species of animals reduce tensions by touching. Greeting ceremonies are common in social animals and include touching and embracing. It is not uncommon to observe nonhuman primates sitting with their arms around one another or kissing. (The tendency to attribute human-like intelligence to animal species is discussed in **The Human Perspective:** Just How Human-like are Other Animals?) Grooming (Figure 31-8) is a form of social behavior that is found in a variety of animals, but it is especially prominent among higher

FIGURE 31-8

Grooming in primates. Although the original function of grooming may have been only skin care, its social functions have now become more important. Primates spend a major portion of their day grooming; it helps form and maintain social bonds.

primates. One effect of such behavior is to rid the animal being groomed of parasites, hardened skin secretions, and debris. Equally important are the social bonds cemented by grooming.

ALTRUISM

Altruism is the performance of a service that benefits another member of the species at some cost to the altruist (the one who does the deed). On the surface, altruism seems to contradict evolutionary theory in that the altruist would seemingly be less successful in generating offspring that bear its genetic traits. In other words, the altruistic trait would "eliminate itself" from the population. But the altruistic individual is the beneficiary of the altruistic behavior of others, thereby increasing its survival and reproductive efficiency. A look at several classes of altruistic behavior will help us understand its advantage and "survivability." In some cases, without altruistic behavior, a species may well become extinct.

1. *Individual selection.* In this form of altruism, the altruistic individual benefits from the altruistic act. Although the benefit may not be immediate, the altruist often gains in future reproductive potential, so natural selection favors the survival of the altruistic trait. A gift-bearing male spider who sacrifices its own food in order to present its potential mate with a "present" benefits by increasing the likelihood of mating and leaving offspring.
2. *Kin selection.* If family members are assisted in a way that increases their reproductive success, the alleles that the altruist has in common with its kin survive to the next generation, just as they would if the altruist reproduced personally. (Of course the altruist doesn't know this, but natural selection promotes the survival of these altruistic genes as long as the beneficiary of altruism also bears the altruistic allele. The closer the relationship, the greater the likelihood the two will share such a common gene.) Aiding a cousin, who shares an average of only one-eighth of the same alleles, is less productive than is assisting a brother or sister, who is likely to share half of its alleles with the altruist. Several cues help individuals determine the degree of relatedness. Individuals who share one's home are likely to be closer kin. Alternatively, individuals might be identified as kin because they are recognized from prior social contact or as a result of their mutual association with a known relative. Certain traits that characterize family members can also be used for comparison; that is, an image of a family member may be developed and matched to the appearance of a stranger. Genetically determined "labels" may also trigger altruistic behavior, automatically stimulating the altruist to assist others who bear the label.
3. *Reciprocal Altruism.* Altruism may evolve, despite the initial cost to the altruist, if the service is repaid with interest by other members of the population. In other words, altruism is favored if the final gain to the altruist exceeds the initial cost. But reciprocal altruism only works in species that can recognize and discriminate against individuals who fail to make restitution.

Having examined some of the advantages of altruism, let's look at several examples of this "selfless" behavior.

Alarm Calls

Belding's ground squirrels warn the other members of their colony by issuing an alarm calls (whistles) at the sight of a predator; the alarm is relayed by the other individuals, and all scurry to shelter. In some cases, the caller is assuming the risk of attracting the predator (coyotes and other terrestrial predators tend to attack callers over noncallers). However, those saved by the warning are likely to be the caller's relatives. Furthermore, females are more likely to sound an alarm than are males. This is consistent with kinship theory since females are more likely to have nearby relatives who would benefit from the warning, and reproductive females *with living relatives* call more frequently than do reproductive females with no living family. This behavior promotes the continuation of the altruistic allele, even if the caller is eaten and cannot directly pass its genes to offspring, promoting passage of the altruistic trait.

Helping

A helper is an individual who provides food and protection for offspring that are not its own, usually those of siblings (kin selection). The helpers are not always relatives, however. Helping may be a means of obtaining permission to remain in a high-quality territory, of finding a mate, or of obtaining help in protecting one's young from predators.

Cooperation in Mate Acquisition

The mate-acquiring activities of male lions typify this type of altruistic behavior. The group of males, called a coalition, generally consists of brothers, half-brothers, and cousins who left their natal pride as a group. These lions challenge the males of other prides (Figure 31-9); the larger coalition usually wins and is rewarded with a harem of lionesses and reproductive success. But the coalition may also accept an unrelated male, increasing the size of the coalition. Larger coalitions have a better chance of ousting competing groups of males.

Food Sharing

Vampire bats share food with familiar roostmates even if they are not related. This generosity may mean the difference between life and death for the recipient. If a vampire bat fails to find food on two successive nights, it will starve

FIGURE 31-9
Male lions from different prides. Coalitions of male lions, most of them relatives, fight other coalitions for control of a harem and reproductive success of all members of the coalition.

to death, unless another bat feeds it part of its recently obtained blood meal. Although the benefit to the recipient is great, the cost to the donor is small. The recipient gains 12 hours of life and, therefore, another chance to find food, but the donor loses about only 12 hours of time until starvation, leaving it 2 full nights of hunting. Usually only individuals who have had a prior association share food, so the benefit might be considered a reciprocal exchange, since even unrelated bats helped each other. Helpers might then be beneficiaries on the next night's hunt.

Eusociality ("true" sociality)

Eusocial species are those that have sterile workers, engage in cooperative care of the young, and have an overlap of generations so that colony labor is a family affair. Eusocial insects (ants, bees, and wasps, for example) behave altruistically in several ways: food is shared, some colony members sacrifice their lives defending the colony, and some sterile members of the colony perform food-gathering chores or care for the young of the colony's royalty. Even some mammals exhibit eusociality, such as African naked mole-rats. As in honeybees, breeding of these mammals is restricted to a single female (the queen), and there is differentiation of labor among individuals within the colony. Both honeybees and naked mole-rats may have been predisposed to a eusocial lifestyle by a close genetic relationship among colony members and ecological factors that maximize the benefits of group life. In honeybees, the system of sex determination, called haplodiploidy (fertilized eggs develop into females and nonfertilized eggs develop into males), produces a closer relationship between sister workers and their siblings than between a female and her own offspring. The female workers are likely to share 75 percent of their alleles with the reproductively capable siblings they helped to raise, but they share only 50 percent of their alleles with the offspring they produced. As a result, a female worker makes greater evolutionary gains by raising siblings than she would by producing offspring.

EVOLUTION AND ADAPTATION: TYING IT TOGETHER

The fourth—and overriding—consideration in discussing animal behavior lies in understanding how a behavior evolved and in determining its adaptiveness. Why do those animals that behave in a certain way survive and reproduce better than those that behave in some other way? For example, why do nesting sea gulls expend energy removing broken eggshells from their nests? Since broken eggshell bits attract predators, the risk of predation decreases with increasing distance of eggshell pieces from the nest. Thus, fastidious parents leave more offspring to perpetuate their genes. Eggshell removal is clearly adaptive.

The costs of an action must be less than its benefits in terms of increased fitness (greater reproductive success, which, in turn, improves the chances of retaining the gene for the behavior in subsequent generations). For example, an individual may be territorial if the benefits are greater than the cost of defending the resource. If the nest's resources are too scarce, an individual may not gain enough to pay the defense bill; it may be economically "wiser" to look for greener pastures. On the other hand, it would be energetically wasteful and economically unsound to defend territories in environments that are so rich in resources that every animal gets what it needs. For example, female marine iguanas don't defend their nesting territories on most of the biologically "generous" Galapagos Islands (Figure 31-10), but they ferociously defend their territories on Hood Island, the only Galapagos island where nest sites are in short supply.

Some species depend on the behavior of other members of the population, an optimal course of action called an *evolutionary stable strategy* (ESS). Once most members of a population have adopted it, an ESS cannot be replaced by any another strategy. It results in maximum reproductive success for the individuals employing it, so it is both unbeatable and uncheatable. Insights into ESS have improved our understanding of the advantage of animal combat, especially in species that possess anatomical weapons, such as horns or sharp teeth. In such potentially lethal instances, contests rarely escalate from display to full battle (Figure 31-11). Conversely when the risk of injury is low

FIGURE 31-10
Marine iguanas on the Galapagos Islands. Female iguanas need not defend their nesting sites on most Galapagos islands (where nest sites are abundant), except on parsimonious Hood Island, which provides limited nest sites.

◁ THE HUMAN PERSPECTIVE ▷

Animal Cognition: Just How Human-like are Other Animals?

Listen to a fisherman talk about a particularly old and legendary fish. "It's smart—too crafty to be caught. It can figure out any new trick you throw at it and avoid the hook." Such statements attribute cognition to the fish, assuming it has intelligence, thoughts, and feelings, not unlike our own. We also tend to ascribe similar attributes to our pets. But how do we go about determining if non-human animals are really cognitive beings?

If nonhuman animals have thoughts and feelings, they may use their communication signals to inform others of their cognitive disposition. If we could learn to speak their language, we could "eavesdrop" and thereby glimpse into their minds. One sign of cognition could be the ability to form mental representations of objects or events that are out of sight. We might ask, then, whether animals can refer to things that are not present. Some apes can learn a language that uses symbols. Kanzi, the pygmy chimpanzee, for instance, can communicate by using a computer keyboard that has over 250 symbols, called lexigrams. In addition, Alex, an African gray parrot, can request more than 80 different items vocally, even if they are out of sight. Alex can also quantify and categorize those objects. He has shown an understanding of the concepts of color, shape, and "same versus different" with both familiar and novel objects. But these animals have been taught to use language. Are the communication signals that animals use in nature symbolic?

The apparent simplicity of this question is deceptive because observations are often open to alternative interpretations. For example, the waggle dance of the honeybee that was described earlier communicates information regarding the direction and how far it is to a distant food source. However, most scientists agree that this behavior is not evidence of thought since the dances (and the responses) are genetically preprogrammed; bees perform and react perfectly to these dances without previous experience. Neither the dance nor the response is learned. A sign of cognition might be whether or not animals adjust signals according to existing conditions. For example, when an individual sees a predator, does it always sound an alarm, or does this action depend on the individual's present company? In domestic chickens, a cock is more likely to issue its "predator present" alarm when at least one companion is around. This may be interpreted as an indication that the cock chooses whether or not to call, which in turn is interpreted by some ethologists as evidence of cognition. Learning studies may also shed some light on the issue of animal cognition. Some scientists believe that insight learning shows that the animal is thinking; an animal that thinks about objects or events can be said to experience a simple level of consciousness. An animal that thinks must also form mental representations of objects or events. Therefore, insight has been used as evidence of animal cognition.

But not every ethologist agrees that animals might show cognition. Others might be willing to accept the idea in a chimp but not in a pigeon that shows similar behavior.

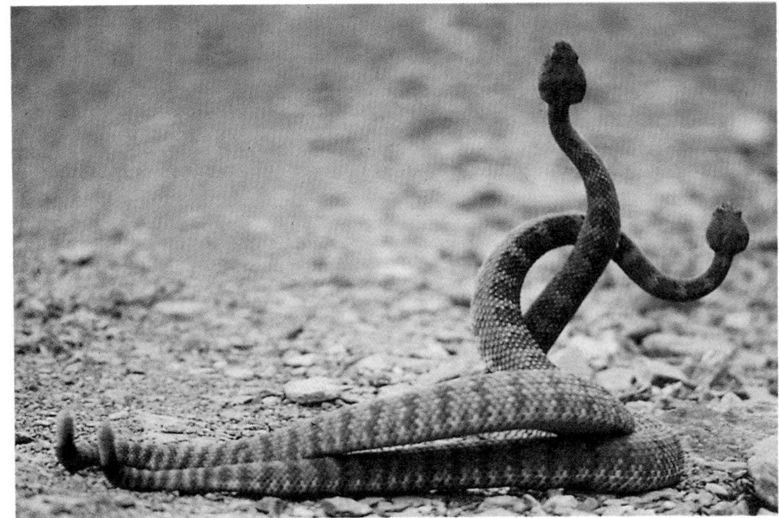

FIGURE 31-11

Limited warfare. Fights are not likely to escalate when the cost of injury is great. These male rattlesnakes could kill one another, but their aggressive encounters rarely escalate to such lethal actions.

(as in toads and other animals without weapons), the combatants fight fiercely. One exception to this involves the defense of offspring or potential offspring, a direct measure of fitness and therefore of great evolutionary value. A mother will fight fiercely to defend her young. Male elephant seals fight brutal and often bloody battles for the right to mate. Such battles can become titanic, reflecting the magnitude of their importance (a male's entire reproductive success is at stake). All matings are performed by a few winners of these battles—dominant males who successfully defend harems of females. Such behaviors are adaptive because they increase the chances of survival and reproduction of the offspring, which inherit the dominant male's traits.

SYNOPSIS

Four considerations underlie our study of animal behavior. These address the behavior's mechanisms, its development, its survival value, and its evolution.

Innate behavior. Innate behaviors (stereotyped; controlled primarily by genes), once triggered, will continue to completion without further stimulation. One such response may trigger another, creating a chain of reactions.

Learned behaviors. Learning allows an animal to benefit from its experience. Habituation, a simple form of learning whereby an animal learns to ignore incidental stimuli, allows an animal to focus its attention on important aspects of the environment. Classical conditioning enables an animal to learn a new association between a stimulus and a response, creating a conditioned reflex in response to a conditioned stimulus. In operant conditioning, however, an animal learns to repeat a behavior because the action is followed by a reward. Insight learning provides a solution to a problem without obvious trial-and-error procedures. When an animal learns from others of the same species, it is called social learning.

Genes and experience interact during development to produce each behavior. Some birds learn the correct song by hearing adult males of its species sing. The young male inherits a template of its species' song, but the experience of hearing the song provides the guidance for him to perfect it.

Optimality theory considers the relative advantages versus disadvantages of each available behavioral alternative as affected by natural selection. For example, optimality theory predicts that an animal would defend a territory when the benefits (increased reproductive success) from enhanced access to the resource within the territory are greater than the costs of defending the territory.

Social behavior. Communication is essential to social life. The functions of communication include species and mate recognition, courtship, aggressive interactions, and recruitment. Some displays allow individuals to select the mate with the most adaptive qualities. The evolutionary selection of a sensory channel for a particular communication signal is influenced by both the nature of the message and the ecological situation of the organism.

Altruism. The advantages of altruism include (1) individual selection, in which there is a net gain directly to the altruist; (2) kin selection, which indirectly increases the altruist's genetic success by improving the likelihood of survival of its relatives; and (3) later repayment for services by other members of the population.

Adaptive advantage and evolution of behavior. Behavior is a trait as susceptible to natural selection as is any physical trait; that is, it is subject to the forces of natural selection. Gulls that remove eggshell pieces from the nest experience lower predation rates. Species that engage in play learn specific skills or improve overall perceptual abilities that better enable them to capture prey, escape predators, or mate. Imprinting improves chick survivability by helping the hatchlings avoid hostile adults. Food-hiding behavior provides a source of nutrition during times when competitors may starve. The appropriate behavior improves success at mate-seeking, fertilization, food gathering, protecting resources and young, and countless other advantages.

Review Questions

1. Define and give an example of each of the following: fixed action pattern, releaser, and chain of reactions. Explain how the three are related.
2. What is the difference between an unconditioned stimulus (US) and a conditioned stimulus (CS)? Describe a situation in which a conditioned reflex might be beneficial to an animal.
3. Explain how genes and experience interact during the development of song in male white-crowned sparrows.
4. What are some costs and benefits of defending a territory? What conditions would favor territorial defense? What conditions would make territorial defense economically unsound?
5. Explain the three hypotheses for the evolution of altruism. Explain why each of the following may be considered examples of altruism: alarm calling by ground squirrels; helping; male lions cooperating in acquiring a mate; food sharing among vampire bats; and eusociality.
6. Explain how the close genetic relationship among honeybees within a colony develops. How is this different from the way that a close genetic relationship among naked mole-rats within a colony develops?

Critical Thinking Questions

1. What would you predict would happen to the territorial behavior of riparian funnel-web building spiders during a drought in which their normally ample resources become temporarily scarce? Explain why.
2. Herring gull chicks peck at the adult's beak, which is yellow with a red spot near the tip, until the adult regurgitates food into the chick's beak. Describe how you would determine which characteristic(s) of an adult herring gull's head serve(s) to release pecking behavior in the chicks. Would you expect this behavior to be innate or learned?
3. In *The Life and Times of Archie and Mehitabel* by Don Marquis, Archie the cockroach says, "as a representative of the insect world I have often wondered on what man bases his claims to superiority. Everything he knows he has had to learn whereas insects are born knowing everything we need to know." Based on what you have learned in this chapter, write a reaction to this statement.
4. Chimpanzees in the wild can be observed stripping leaves from a stem and poking the stem into termite or ant hills. They then withdraw the stem and lick off any insects clinging to it. Describe how you would determine whether this behavior is genetically determined or learned.
5. Discuss the probable survival value of each of the following behaviors: (a) Butterfly courtship involves a series of signals in a particular order between male and female. Failure to produce the right signal at the right time interrupts the courtship. (b) Toads exhibit a striking and swallowing reflex to any elongated shape moving lengthwise. (c) Male bower birds, which lack colorful feathers, decorate nests with brightly colored objects. (d) Tawny owls, which are long-lived, territorial predators, can lay up to four eggs a year. Typically, however, not all pairs in an area breed every year, and some that do breed fail to incubate the eggs.

CHAPTER

32

Population Dynamics

The primary factors that have led to the decreased reproduction of the Saguaro cactus include ants harvesting seeds, rodents consuming seedlings, cattle trampling small individuals, and decreased protective shade.

STEPS TO DISCOVERY
Threatening a Giant

Driving across Arizona, you will see forests of giant saguaro cactuses (*Carnegiea gigantea*) that extend for miles in all directions. It seems inconceivable that scientists argue that saguaro populations are declining and that this majestic giant may someday disappear from earth. But after conducting a study for the federal government in 1910, Forrest Shreve, a research associate with the Carnegie Institute in Washington, D.C., made just such a conclusion. During his studies, Shreve found no saguaros younger than 15 years of age and no young seedlings anywhere in the population. He concluded that the saguaro was not reproducing enough offspring to replace the older, dying individuals.

Shreve's conclusion has been verified by other Southwest desert scientists. In an effort to understand why saguaros may be dying out, and in order to provide a safe haven for saguaros while researchers determined the causes, The Saguaro National Monument was established in 1933 near Tucson, Arizona. Research within the monument also documented population decline. Since then, a great deal of research has focused on analyzing every aspect of the saguaro's life cycle in an effort to pinpoint the reason(s) why saguaros are failing to reproduce in adequate numbers.

A saguaro may live for 175 years or more and grow more than 15 meters (45 feet) tall. Although a saguaro may not bloom until it is at least 30 to 50 years old, its reproductive lifetime still stretches well over 100 years, which helps explain why saguaros have such an enormous reproductive potential. Each year, a single saguaro produces an average of 200 fruits, containing a total of 400,000 seeds. Over its reproductive lifetime, an individual saguaro produces some 50 million seeds. It takes only one of these seeds to grow, become established, and reach reproductive maturity in order to maintain stable saguaro populations. But despite prodigious seed production, this is not happening.

The problem clearly does not lie with flowering, pollination, or fruit and seed development since each individual generally produces several million seeds in its lifetime. The explanation must therefore lie with seed survival, germination, and/or seedling and adult survival.

In 1969, Warren Steenbergh and Charles Lowe of the Saguaro National Monument and the University of Arizona, Tucson, documented that saguaro seeds disappear at a high rate. During a 5-week period, mammals, birds, and insects (particularly harvester ants) consumed nearly all of the seeds produced in the researchers' study site. Only 4 in 1,000 seeds survived to germination. To make matters even worse, saguaro seeds do not survive from one year to the next so reproduction is always limited to the current year's seed supply.

Of the seeds that did germinate, all but a few died within the first year. To study the causes of seedling death, R. Turner, S. Alcorn, and G. Olin transplanted 1,600 young saguaro seedlings in the Saguaro National Monument in 1957. The researchers enclosed some of the young saguaros with cages to protect the seedlings from grazing ground squirrels, rodents, and rabbits; other saguaros were left uncaged. All the uncaged plants were killed within just 1 year by grazing animals; only 1.9 percent of the caged seedlings remained alive after 10 years.

Turner, Alcorn, and Olin also found that drought caused high losses of saguaro seedlings during the first few years of life. Small saguaros are usually found in the shade of desert trees or shrubs, places where water loss is reduced. Small saguaros have a small water-storage capacity and become dehydrated easily in direct sun. In another study, Turner, Alcorn, and Olin protected seedlings from rodents and studied the effects of shading on these protected seedlings. All 1,200 unshaded seedlings died within 1 year, while 35 percent of 1,200 shaded seedlings survived. The researchers concluded that the survival of young saguaro seedlings is closely tied to that of other perennials that provide the seedlings with shade. This relationship led to speculation that the rapidly growing cattle industry may also be contributing to saguaro population declines since trampling by cattle reduces tree and shrub cover and crushes young saguaro seedlings.

In 1976, Steenbergh and Lowe identified another important factor associated with saguaro seedling survival: freezing weather. Young saguaros freeze at −3°C (26°F) to −12°C (10°F) or when exposed to more than 19 hours of freezing temperatures. This means that during certain years, all saguaro seedlings may be killed by freezing.

Many saguaro populations may be declining primarily because virtually all the seeds and seedlings are quickly eaten by mammals, birds, and ants. If weather conditions are just right, only a few of the remaining seeds germinate and grow, provided that they are in the shade of a tree or shrub and are not eaten by rodents and rabbits or trampled by cattle. Periodic freezing can also kill saguaro seedlings. Like all organisms, saguaro reproduction is affected by a number of environmental factors.

"When we consider . . . how soon some species of trees would equal in mass the earth itself, if all their seeds became full-grown trees, how soon some fishes would fill the ocean if all their ova became full-grown fishes, we are tempted to say that every organism, whether animal or vegetable, is contending for the possession of the planet. Nature opposes to this many obstacles, as climate, myriads of brute and also human foes, and of competitors . . . Each suggests an immense and wonderful greediness and tenacity of life. . ."

—Henry David Thoreau, journal entry, March 22, 1861.

Elephants are among the slowest reproducers on earth. Over its lifetime, a female elephant can give birth to a maximum of only six babies. Even so, the number of possible descendants from just one pair of mating elephants could total 5 billion (5×10^9) after just 1,000 years. After 100,000 years, the number of potential descendants from one mating pair would theoretically pack the visible universe with elephants.

If such outlandish growth is possible for a slow reproducer like the elephant, imagine what could happen with organisms that have faster reproduction rates, such as house flies. In less than 1 year, the number of possible descendants from a single pair of house flies would exceed 5.5 trillion (5.5×10^{12})!

Clearly, animals have a tremendous capacity to reproduce, and the reproductive potential of plants is often even greater. Yet, the world is not tightly packed with elephants, flies, or any other kind of organism. Disease, parasitism, predation, and food and space limitations curb the potential number of individuals, often leading to a balance between the number of individuals living in an area and the availability of resources to support them.

POPULATION STRUCTURE

Most communities contain many **populations,** each of which consists of the individuals of the same species that live in the same area at the same time. For example, a mountain forest not only contains a population of yellow pine trees but also populations of sugar pine trees, white fir trees, brown bears, Anna's hummingbirds, and more.

To understand the structure and dynamics of each of the populations that make up a community, ecologists examine three fundamental properties of populations:

- *population density* (the size of the population, expressed as the number of individuals that live in a given area at a particular time);
- *distribution of individuals throughout the habitat.* The distribution of individuals is typically categorized into one of three patterns: *clumped, uniform,* or *random* (Figure 32-1); and
- *growth rate* (increases or decreases in population density per unit of time).

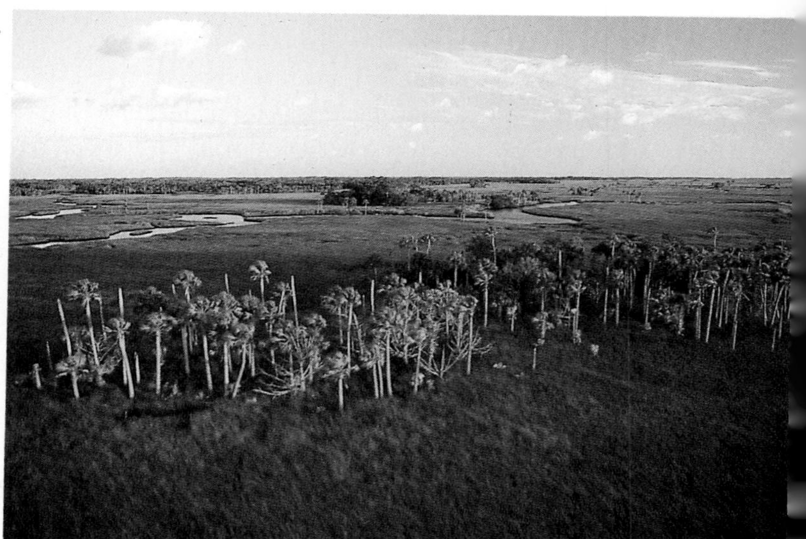

(a)

FIGURE 32-1
Distribution patterns. ***(a)*** **Clumped:** A grove of clumped palms. A school of fish or a herd of elephants are other examples of clumped distributions. Clumped distributions result when members of a population have some effect on one another or when environmental conditions favor growth in patches. ***(b)*** **Uniform**: Uniform spaces result when members of a species repel one another. For example, the oaks in this photograph secrete chemicals that prevent growth of nearby oaks, creating more-or-less equal distances between trees. Similarly, when animals defend their territories, the individuals remain separated, producing a uniform distribution. ***(c)*** **Random:** Joshua trees may have random distributions in some locations. Random distributions result when the location of one individual has no effect on another individual of the same species.

FACTORS AFFECTING POPULATION GROWTH

Like ecosystems, populations inevitably change. Four events trigger increases or decreases in the density of a group of organisms:

1. **Natality** increases density, as new individuals are born into a population.
2. **Immigration** increases density, as new individuals permanently move into the area.
3. **Mortality** decreases density, as individuals in the population die.
4. **Emigration** decreases density, as individuals permanently move out of the area.

If the combination of natality and immigration exceeds that of mortality and emigration, the population grows and density increases. Conversely, when the combination of mortality and emigration exceeds that of natality and immigration, population density decreases. The growth rate of a population equals the rate at which the population size changes. **Zero population growth** occurs when the combined additions and losses to a population are equal; when this happens, population density remains the same.

AGE–SEX RATIOS AND SURVIVORSHIP CURVES

The age and sex of individuals in a population affect population growth. The age of the members of a population is a predictor of both natality and mortality. For example, red alder trees live to be about 100 years old. If the majority of red alders are older than 95 years, the mortality rate will likely be high and the population of trees will likely decline over the next 5 years. In addition, since each individual reproduces only during part of its lifetime, the ages of the members of a population can also be used to predict natality. In general, the saguaro reproduces between the ages of 40 and 150 years; female humans reproduce between ages 15 and 44.

Population biologists plot the number of individuals of a certain age and sex to determine the **age–sex structure** of a population, which helps them predict future population changes. When a large proportion of individuals in a population are at reproductive age (or younger), the age–sex structure tends to be shaped like a pyramid:

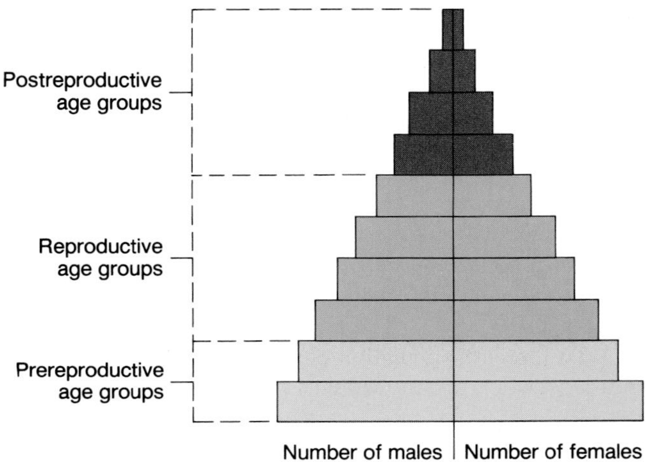

A broad-based population like this one will increase in size; in general, the broader the base, the more rapidly the population will increase. In contrast, a population with an inverted pyramid will decline because most individuals are past reproductive age.

(b)

(c)

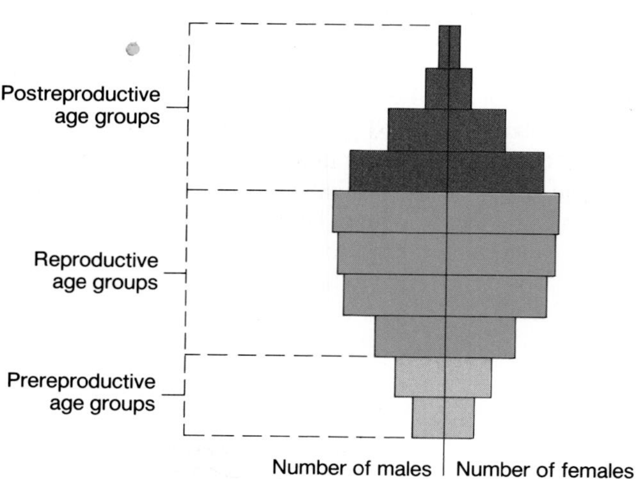

In this stable population, the number of prereproductive individuals balances the number of older individuals, so natality equals mortality. In this stable population, there is either no migration at all, or emigration and immigration are equal.

To forecast population changes accurately, ecologists must consider the individual's life expectancy as well as the age–sex structure of the population. When life expectancy is plotted on a graph, a *survivorship curve* is produced. Ecologists identify three general types of survivorship curves: Type I, Type II, and Type III; these are illustrated in Figure 32-2.

BIOTIC POTENTIAL

All species have the capacity to produce tremendously large numbers of descendants eventually, as long as there are no restrictions to curb population growth. Organisms find themselves in such conditions only on rare occasion, however, such as when organisms colonize a new favorable habitat, when environmental conditions suddenly change for the better, or when organisms are introduced into new habitats in which there are no natural competitors or predators (Figure 32-3). The innate capacity to increase in number under ideal conditions is called the **biotic potential** of a population.

EXPONENTIAL GROWTH AND THE j-SHAPED GROWTH CURVE

When populations grow at their maximum rate of increase, the number of individuals increases exponentially. That is, the number increases by a fixed proportion, such as when

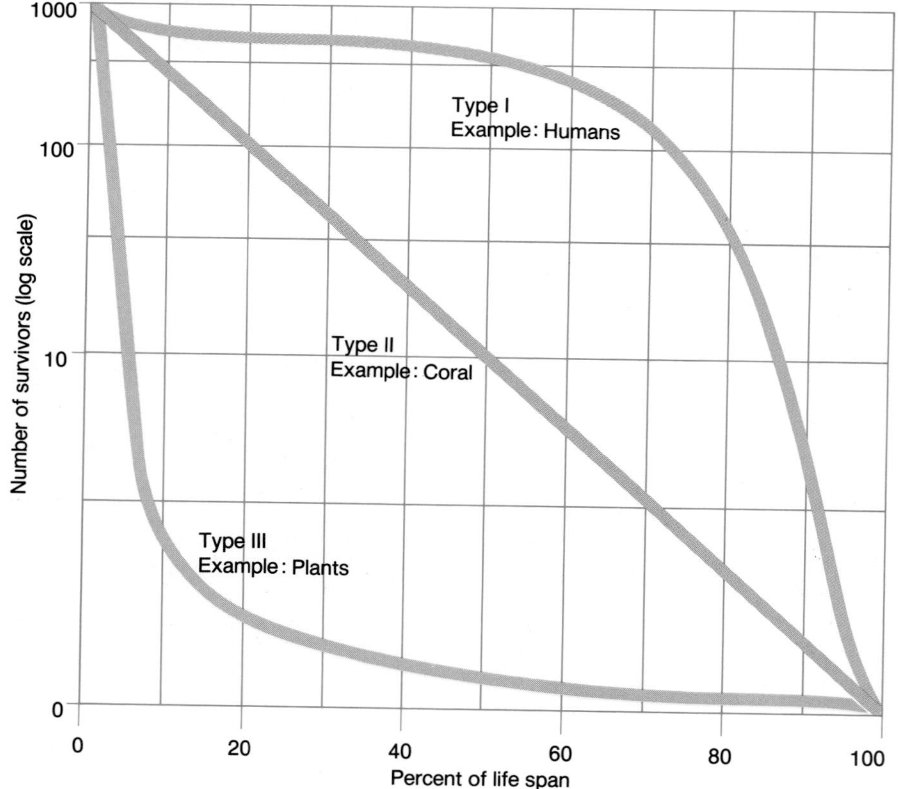

FIGURE 32-2

Three types of survivorship curves are produced by plotting the number of survivors (on a log scale) against the percentage of the lifespan of a species. In the Type I curve, the mortality rate is low in the first years of life and then becomes higher at old age. Humans exhibit a Type I curve. In the Type II curve, the chances of surviving or dying are virtually the same throughout an organism's entire life. Organisms with a Type II curve include lizards, songbirds, and pines and firs. In a Type III curve, nearly all of the young die quickly, but the mortality rate is quite low for the few survivors, until they reach old age. Oysters, some insects, and weedy plants exhibit a Type III curve.

a population doubles in size with each new generation (2, 4, 8, 16, 32, 64, 128, and so on). Such **exponential growth** can be demonstrated by placing a single *E. coli* bacterium into a nutrient culture (Figure 32-4). The bacterium divides into two cells after 20 minutes; the two divide into four cells in another 20 minutes; and the four divide into eight cells 20 minutes later. The population continues to double every 20 minutes. After only 5 hours, 32,768 bacteria have been produced from the original bacterium. After 7 hours, there are more than 2 million bacteria, and after 36 hours, there would be enough bacteria to blanket the entire earth's surface with 28.8 centimeters (1 foot) of bacteria.

The pattern of exponential growth is always the same: The size of a population increases gradually at first and then grows larger and larger in progressively shorter periods of time, as more and more individuals are added to the population. Plotting exponential growth on a graph

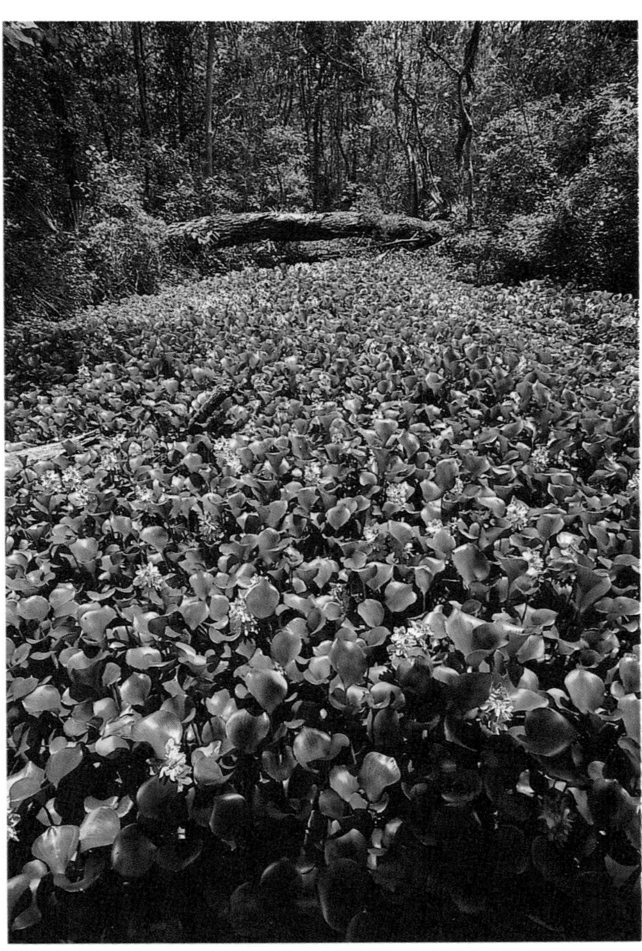

FIGURE 32-3
Although it may have seemed like a great idea at the time, importing plants into new habitats without natural population controls has produced some unexpected overpopulation disasters. For example, water hyacinth plants, with their orchid-like flowers and attractive stems and leaves, were brought to the United States from Venezuela to be displayed at the 1884 New Orleans Cotton Exposition. Many of the exposition visitors were given clippings of the charming plant to place in ponds and streams near their homes. With no competitors, few predators, and plenty of available space and nutrients, the growth of the water hyacinth spread rapidly. As this photo dramatically shows, today, water hyacinths are choking streams, rivers, irrigation systems, and hydroelectric installations.

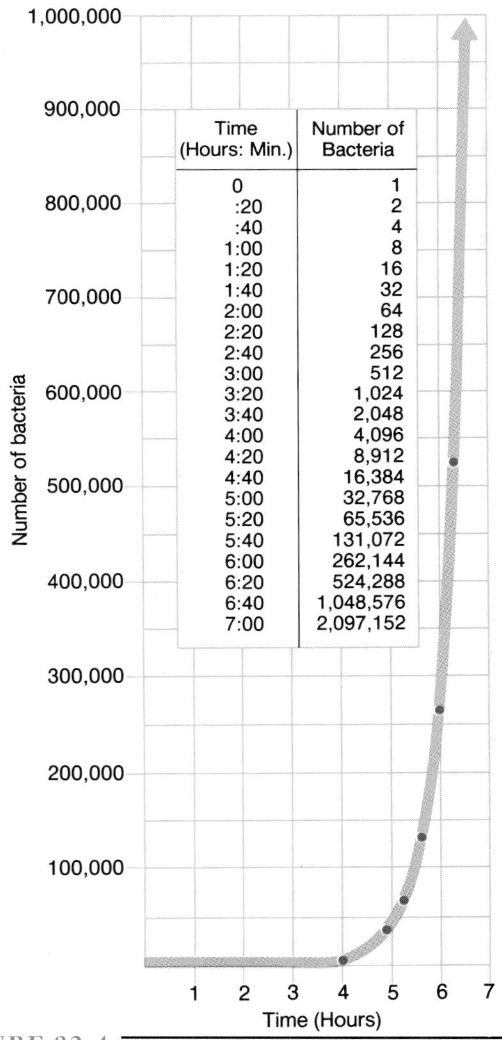

Time (Hours: Min.)	Number of Bacteria
0	1
:20	2
:40	4
1:00	8
1:20	16
1:40	32
2:00	64
2:20	128
2:40	256
3:00	512
3:20	1,024
3:40	2,048
4:00	4,096
4:20	8,912
4:40	16,384
5:00	32,768
5:20	65,536
5:40	131,072
6:00	262,144
6:20	524,288
6:40	1,048,576
7:00	2,097,152

FIGURE 32-4
Exponential growth of an *E. coli* bacterial culture produces a J-shaped growth curve. After rounding the "bend" of the J, the curve gets steeper and steeper, until some limitation, such as depletion of food or the buildup of contaminating waste products, curbs further growth.

produces a curve that resembles the letter "J"; not surprisingly, it is called a **J-shaped curve.** As Figure 32-4 illustrates, once a population "rounds the bend" of a J-shaped curve, the number of individuals added to a population begins to skyrocket.

ENVIRONMENTAL RESISTANCE AND CARRYING CAPACITY

No natural ecosystem can support continuous exponential growth for any species; eventually, some environmental limitation imposes a restriction. The factor(s) that eventually limit the size of a population create an **environmental resistance** to population growth. Environmental resistances include competition, predation, hostile weather, limited food or water, restricted space, depleted soil nutrients, and the buildup of toxic byproducts.

The combined limitations imposed by the environment establish a ceiling for the number of individuals that can be supported in an area. The size of a population that can be supported indefinitely is called the **carrying capacity** of the environment. Populations undergoing exponential growth sometimes overshoot their carrying capacity before environmental resistances are able to curb their growth. The greater the reproductive momentum, the more likely it is that the population will exceed the environment's carrying capacity.

Laboratory experiments and field observations identify three possible fates for populations that overshoot the carrying capacity (Figure 32-5). All the repercussions begin with a *dieback,* the death of a portion of the population. The least dramatic of the three fates is also the least common: The population simply dies back to the level of the original carrying capacity (curve 1 on the graph). In most cases, however, the excess population damages the environment in some way, which, in turn, reduces the carrying capacity. For example, an excessively large population of caterpillars could consume all of the leaves on a tree, weakening or possibly killing the plant. When the carrying ca-

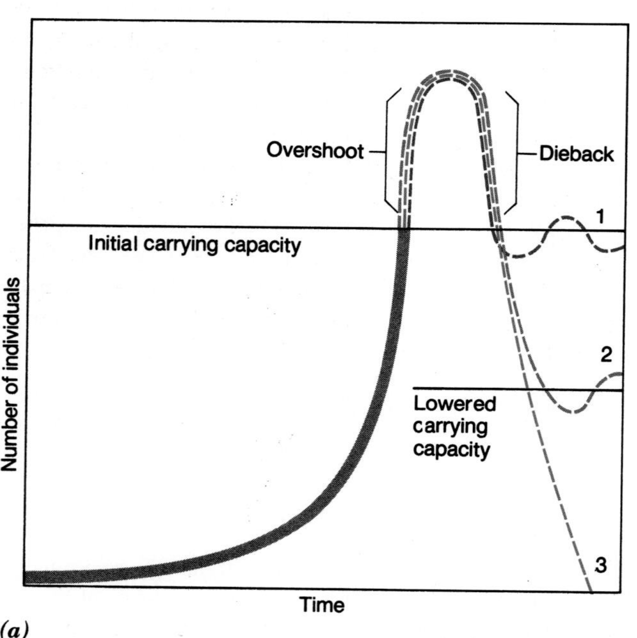

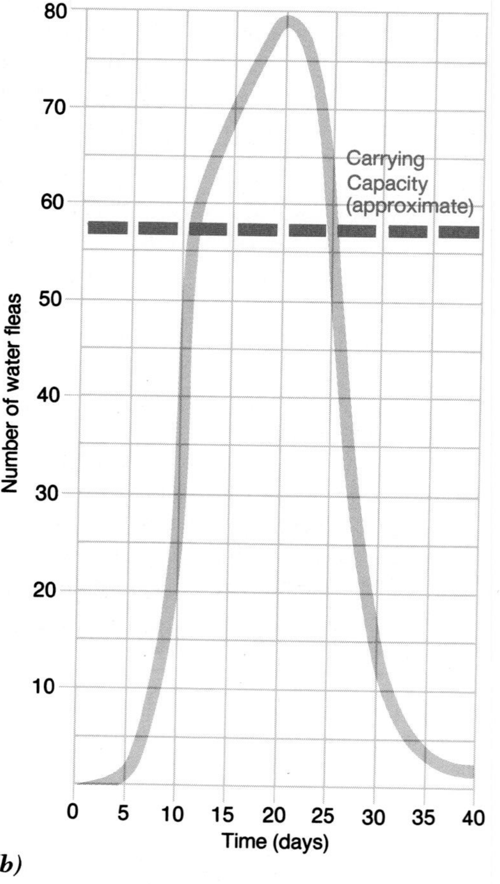

FIGURE 32-5

The risks of overpopulation. ***(a)*** A population undergoing exponential growth may overshoot the carrying capacity of the environment. When this happens, the population experiences a dieback to one of three levels: (1) to the original carrying capacity (the least common outcome), (2) to a population density at a lower carrying capacity, or (3) to extinction or near extinction levels (the most common outcome). ***(b)*** A natural population of water fleas (*Daphnia*) illustrates the third fate: extinction.

◁ THE HUMAN PERSPECTIVE ▷

Impacts of Poisoned Air, Land, and Water

On December 3, 1984, in Bhopal, India, a huge cloud of methyl isocyanate gas leaked out from a storage tank at the Union Carbide chemical plant, killing 3,000 people. It is estimated that an additional 2,000 people will die from side effects by 1995, and 17,000 of the 200,000 people injured from the gas leak have been permanently disabled as a result of lung ailments.

This event dramatically underscores the hazards of toxic substances that surround us in modern society. We are being exposed to more and more toxic substances in the air, in our water, and on the land—chemicals that adversely affect living organisms. In the United States alone, 60,000 chemicals are added to our food or are used to make cosmetics or to combat pests. Hundreds of these chemicals are known to be hazardous. Each year, 700 to 1,000 new chemicals enter the marketplace; fewer than 10 percent are ever tested to assess their health effects. Over 170 million metric tons (378 billion pounds) of potentially hazardous chemicals are manufactured each year in the United States alone, exposing people to hazardous chemicals in their homes, at schools, at work, and even while playing outdoors.

Toxic chemicals can affect virtually every cell in an organism and can cause cancer, mutations, birth defects, or reproductive impairment. These chemicals can affect cells in several ways: (1) by disturbing enzyme activities that regulate critical chemical reactions (e.g., mercury and arsenic inactivate enzymes); (2) by binding directly to cells or to essential molecules in the cell (e.g., carbon monoxide binds with hemoglobin in the blood, preventing it from carrying oxygen to cells); or (3) by releasing chemicals that have an adverse effect (e.g., in addition to its immediate destructive effects, carbon tetrachloride triggers nerve cells to release large amounts of epinephrine, which is believed to cause long-term liver damage).

Some chemicals gradually accumulate in the bodies of organisms, eventually building to toxic levels. The buildup of chemicals in the organisms in a food chain exposes organisms toward the top of a food chain to potentially dangerous levels of chemicals. The use of DDT as a pesticide is a good example of this phenomenon (refer to Figure 29-8).

Like air pollutants, water pollutants can cause a physical or chemical change that adversely affects life. In the United States, the water in 40 states is already hazardously polluted, and more than half of Poland's water supply is so polluted that it cannot be used even by industry. Water pollutants include poisonous toxic chemicals, such as mercury, nitrates, and chlorine; microscopic bacterial pathogens that cause disease; various physical agents, such as soil sediments; and excess nutrients and organic matter, such as the remains of plants and animals, feces, debris from food processing plants, and runoff from feedlots, sewage treatment plants, and fertilized agricultural land.

Pollutants are either dumped into or seep into all the earth's water sources, including surface waters (lakes, ponds, streams, and rivers), groundwater aquifers (which, in the United States, supply more than one-quarter of the annual water demands), and oceans, especially biologically rich coastlines, coastal wetlands (bays, swamps, marshes, and lagoons), and estuaries. Not only are coastal zones rich with myriad organisms, they are also the most vulnerable of the ocean's regions to numerous sources of pollution, including wastes from sewage plants and factories, sediment from erosion, and oil spills. Combine these sources with the fact that many cities draw huge quantities of fresh water from streams during droughts, diminishing water flow into important regions and increasing pollutant concentrations. As a result, coastal zones are being destroyed by pollution, water loss, sedimentation, dredging, and filling, at alarming rates. In the United States alone, more than 40 percent of estuaries have been destroyed to date, despite the implementation of state and federal laws to protect them.

pacity is lowered, the population plunges. If the damage caused by overpopulation is not too severe, the population eventually comes into balance with a lower carrying capacity (curve 2), or suffers the third fate: It disappears altogether (curve 3).

Occasionally a population exceeds the carrying capacity because a limiting factor does not take effect until a threshold level is reached, at which point the factor causes a sharp decrease in population size. For example, the gradual buildup of toxic waste may have no effect until these chemicals reach a critical level, at which point a large number of individuals will die. Some biologists warn that this might be the case for the human population, as our air, soil, and water become more and more polluted (see The Human Perspective: Impacts of Poisoned Air, Land, and Water).

SIGMOID GROWTH CURVE

An alternative to unbridled exponential growth occurs when environmental resistance increases as a population approaches the carrying capacity of the environment, slowing growth. When this happens, a **sigmoid growth curve** is produced (Figure 32-6). The "Lazy S"-shaped sigmoid curve begins the same as does the J-shaped curve; that is, the number of individuals increases slowly at first, then, as the reproductive base builds, the population goes into a period of exponential growth. During sigmoid growth, however, the rate of growth slows down as the population approaches the carrying capacity, forming the top of the S-shaped curve.

FACTORS CONTROLLING POPULATION GROWTH

The impact of some population controls increases or decreases in intensity as the size, or density, of the population

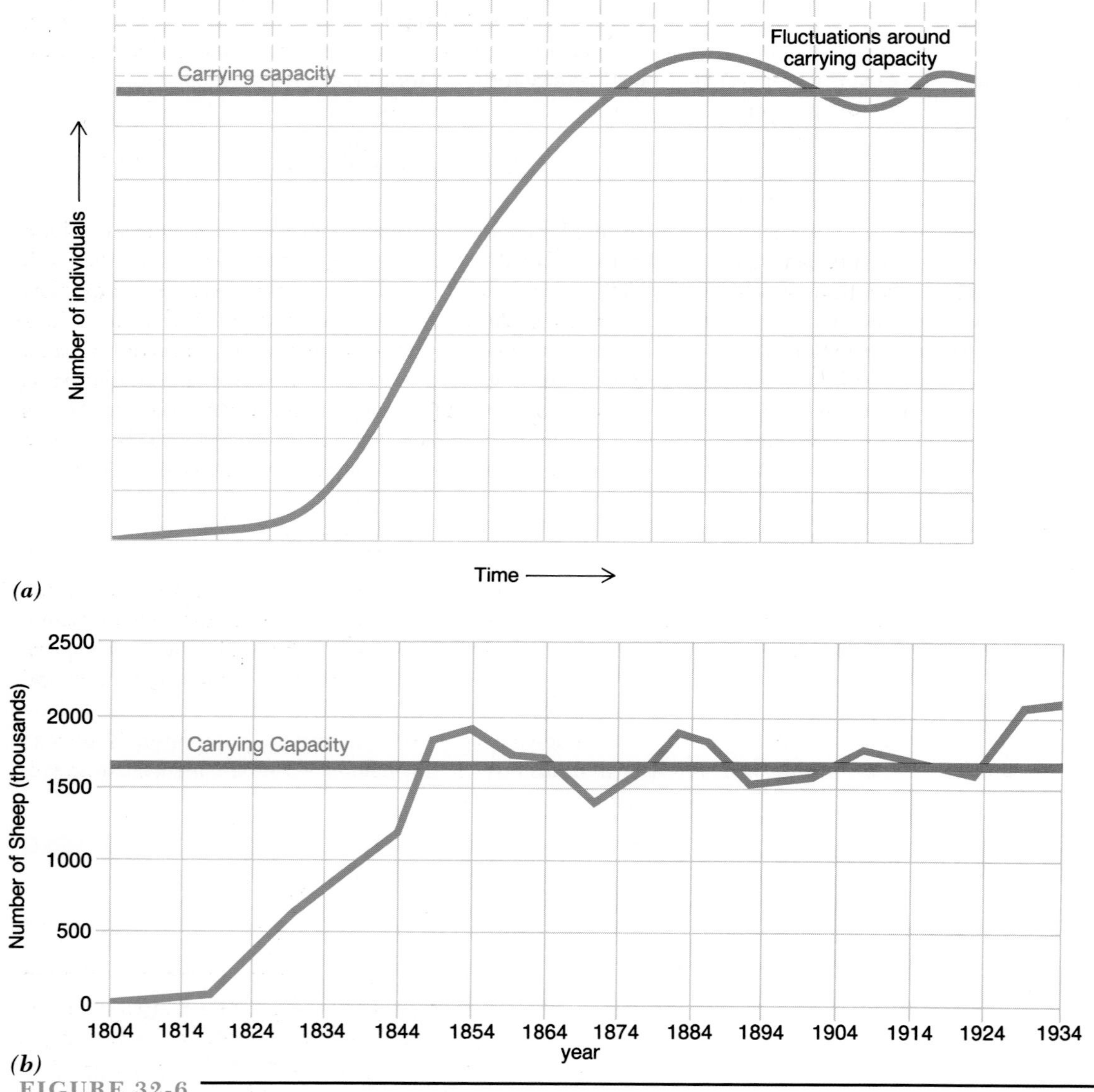

FIGURE 32-6

The sigmoid growth curve. ***(a)*** In a new habitat, a population increases slowly at first and then undergoes exponential growth, until environmental resistances, such as limited space and nutrients, begin to reduce growth rates and stabilize population density around the carrying capacity. This common growth pattern produces a sigmoid growth curve. ***(b)*** Sheep were introduced on Tasmania in the early 1800s. The population growth pattern produced a sigmoid growth curve, with the population oscillating between 1.5 million and 2.0 million individuals, the result of food and space limitations.

changes. The intensity of other population controls remain the same, regardless of the population size.

DENSITY-DEPENDENT FACTORS

As the size of a population increases, the increasing density of the number of individuals within the population sometimes limits the population growth rate. For example, when population density is low, young locusts develop normal-length wings. When locust density is high, however, hormonal changes trigger the development of longer wings in offspring. Long wings increase emigration, which, in turn, reduces the population density in that area. Such factors that are influenced by the number of individuals in a population and ultimately affect population density are called **density-dependent factors.**

There is a direct relationship between the intensity of density-dependent factors and population size: As population density rises, the intensity of density-dependent regulatory mechanisms increases, dampening population growth by reducing natality or by boosting mortality or emigration (as in locust populations). Conversely, when population density falls, density-dependent mechanisms decrease in intensity, allowing population growth to accelerate. Density-dependent mechanisms explain why there are small fluctuations around the carrying capacity of the environment during sigmoid growth (Figure 32-6*b*).

Crowding is another example of a density-dependent factor. Laboratory studies on overcrowding in rats show that crowding increases aggressive behavior, delays sexual maturation, reduces sperm production in male rats, and causes irregular menstrual cycles in females. Overcrowding also reduces sexual contacts between males and females and increases homosexual contact between males. Such density-dependent factors quickly curb population growth by reducing natality. Whether such dramatic effects occur in populations outside the laboratory or in other species is not yet known. However, many demographers (scientists who study human population dynamics) believe that increased crime, drug abuse, and suicide may in part be the result of overcrowding, a situation that worsens as the world's human population swells by more than 250,000 people each day (an increase of three persons per second).

DENSITY-INDEPENDENT FACTORS

Not all regulatory factors are affected by the density of a population. Factors that are not influenced by population size are called **density-independent factors.** A deadly earthquake, such as that which jolted Los Angeles in January 1994, measuring 6.3 and killing over 50 people, is an example of a density-independent factor: The population density neither caused the earthquake nor affected the magnitude of the quake or the percentage of individuals killed. Many catastrophic events, including fires, floods, hurricanes, tornadoes, volcanic eruptions, and avalanches, are density-independent factors.

HUMAN POPULATION GROWTH

In 1987, the world human population reached 5.0 billion people. By the end of 1993, there were more than 5.5 billion people, and the number continues to climb quickly. More than 380,000 people are born each day (over four babies every second), and nearly 130,000 people die each day. This means that the world's human population is growing by some 250,000 people every single day. At this rate, nearly 100 million people are added to the human population every year. How did we get to this point?

HISTORICAL OVERVIEW

Modern humans appeared on earth less than 1 million years ago. The size of early human populations remained more or less in balance with the available food supply, increasing slowly as our early ancestors spread out and discovered new lands.

When the size of the human population is plotted against time on a log scale (Figure 32-7), three surges of

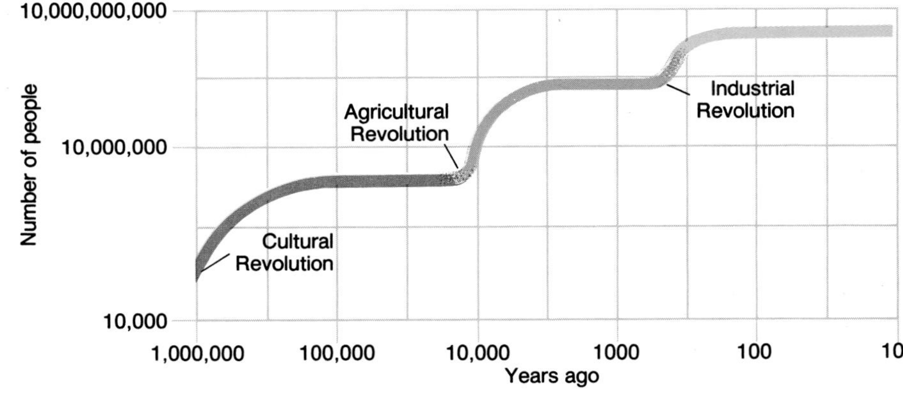

FIGURE 32-7 **Surges of human population growth** become clear when the size of the human population is plotted against time on a log–log scale. (In log–log graphs, both population size and time are plotted on log scales, concentrating huge numbers of individuals and enormous time spans on a single sheet of graph paper.) Bursts of exponential growth occurred on three occasions, corresponding to the cultural, agricultural, and industrial revolutions.

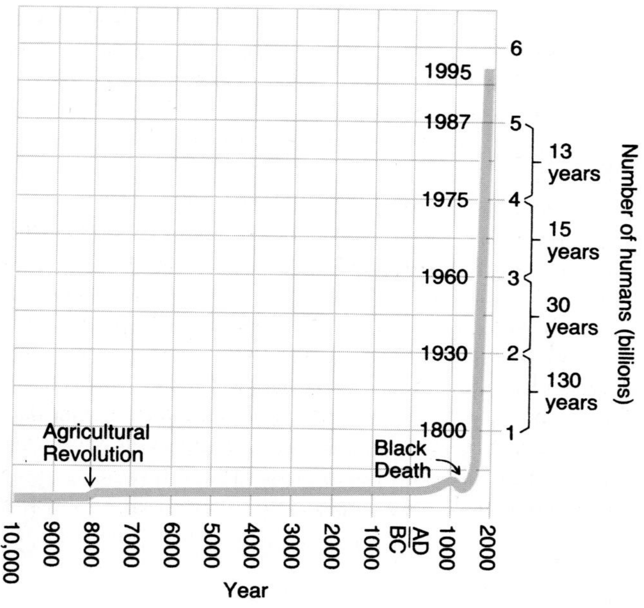

FIGURE 32-8
The human population growth curve. The world's human population grew slowly for 2 million years, until the 1800s, at which time population growth began to soar. By the year 2000, there will be an estimated 7 billion to 14 billion people on earth, if our population continues to grow at its present rate.

population growth become evident. Each surge is the result of a major technological invention that improved food supply and/or human health: Surge 1—the use of tools and a stationary society; Surge 2—the Agricultural Revolution; and, Surge 3—the Industrial Revolution. Throughout this time, humans continued to make advances in agriculture, medicine, and hygiene, decreasing the death rate and increasing the average lifespan. These advances lessened some of the former controls on human population growth, mainly food shortages and disease.

The overall growth curve for the world's human population is clearly a J-shaped curve (Figure 32-8). It took hundreds of thousands of years for the world human population to reach 1 billion people. Once a reproductive base of 1 billion people was established in 1800, however, it took progressively less time to add another 1 billion people to the human population: 130 years to reach 2 billion people; 30 years to reach 3 billion; 15 years to reach 4 billion; and finally only 13 years to reach 5 billion people in 1987. The world human population is projected to reach between 7 billion and 14 billion by the year 2000.

GROWTH RATES AND DOUBLING TIMES

The current human birth rate is 27.7 babies per 1,000 people per year. The death rate is 9.5 people per 1,000 per year, making the current human population growth rate equal to 18.2 humans per 1,000 people per year: 27.7 (birth rate) − 9.5 (death rate) = 18.2 (growth rate).

Population increases are often expressed as the number of people added to the population per 100 individuals, yielding the **percent annual increase** in population. The average annual increase for all nations is currently 1.8 percent (see Table 32-1). This means that it will take less than 56 years for the world's human population to double. That is, if you add 1.8 people to a population of 100 each year, it will take 55.6 years for the population to reach 200 (55.6 × 1.8 = 100). As we discussed earlier, however, population size affects the rate of increase; thus, a growth rate of 1.8 actually has a doubling time of only 39 years instead of 55.6 years because of the large reproductive base.

TABLE 32-1

PERCENT GROWTH RATE FOR VARIOUS COUNTRIES (1990-95)

Country	*Percent Annual Increase*
Developing countries	
India	1.9
Brazil	1.6
Uganda	3.0
Mexico	2.2
Kenya	3.3
Developed countries	
United Kingdom	0.2
United States	0.7
Japan	0.3

Source: World Resources 1994–95, World Resources Institute *in collaboration with* The United Nations Environment Programme *and* The United Nations Development Programme. Data for Percent Annual Increase calculations taken from Tables 16.1 and 16.2. Oxford University Press, 1994.

CURRENT AGE–SEX STRUCTURE AND FERTILITY RATES

There is a big difference between the age-sex structure of developing nations (many countries in Asia, Africa, and South America) and that of developed countries (Figure 32-9). The populations of developing nations are growing rapidly because large numbers of individuals in these populations are moving into their reproductive years. Since developing countries are heavily populated, they have a large effect on the world's human population growth. As a result, the world's human population will continue to increase at a rapid rate.

Although age–sex structure diagrams help predict population growth, they do not take into consideration another factor that affects human population growth: fertility rates. A **fertility rate** is the average number of children born to each woman between 15 and 44 years of age (the reproductive years). The average fertility rate for all nations in the world is now slightly below 2.1 births per woman. In developed countries, however, fertility rates average 1.9 births, compared to 4.5 births in developing nations.

Given the current mortality rates, fertility rates of between 2.1 and 2.5 are required to maintain zero population growth. With an average fertility rate of 1.9 in developed countries, populations in these countries will decline. In contrast, with an average fertility rate of 4.5, populations in developing countries will rapidly increase. As a result, the world's human population will continue to increase by greater and greater numbers each year. Some scientists believe that the world's human population will begin leveling off to between 8 billion and 14 billion people, while others believe that the human population will not reach

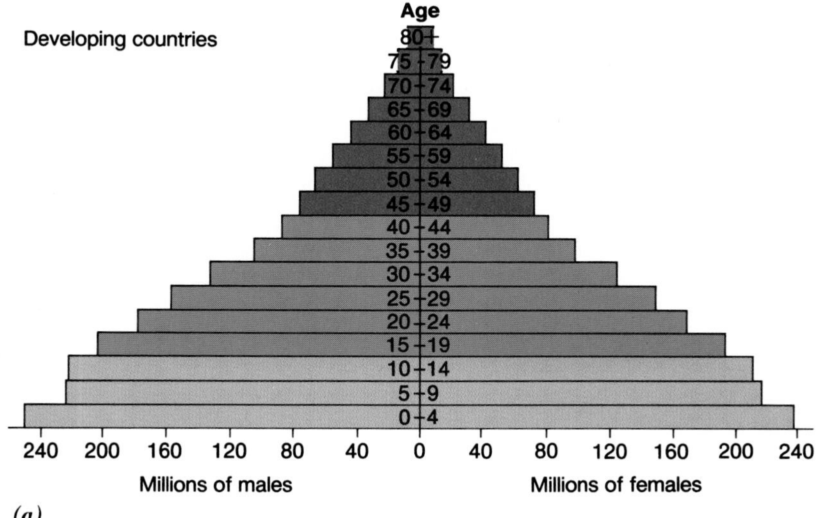

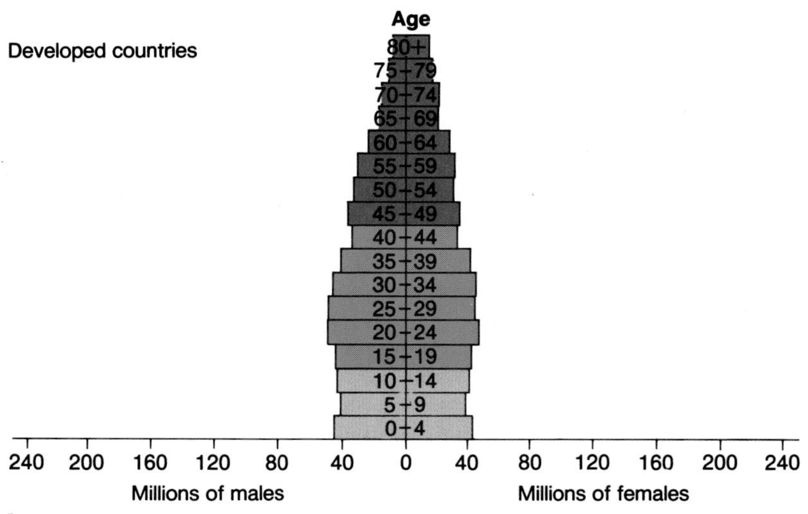

FIGURE 32-9

Age–sex structure diagrams for ***(a)*** developing and ***(b)*** developed countries differ radically. With a tremendous reproductive base, developing countries will continue to grow rapidly, while developed countries will decline in population. Since 75 percent of humans live in developing countries, the world's human population is expected to continue its perilous increase.

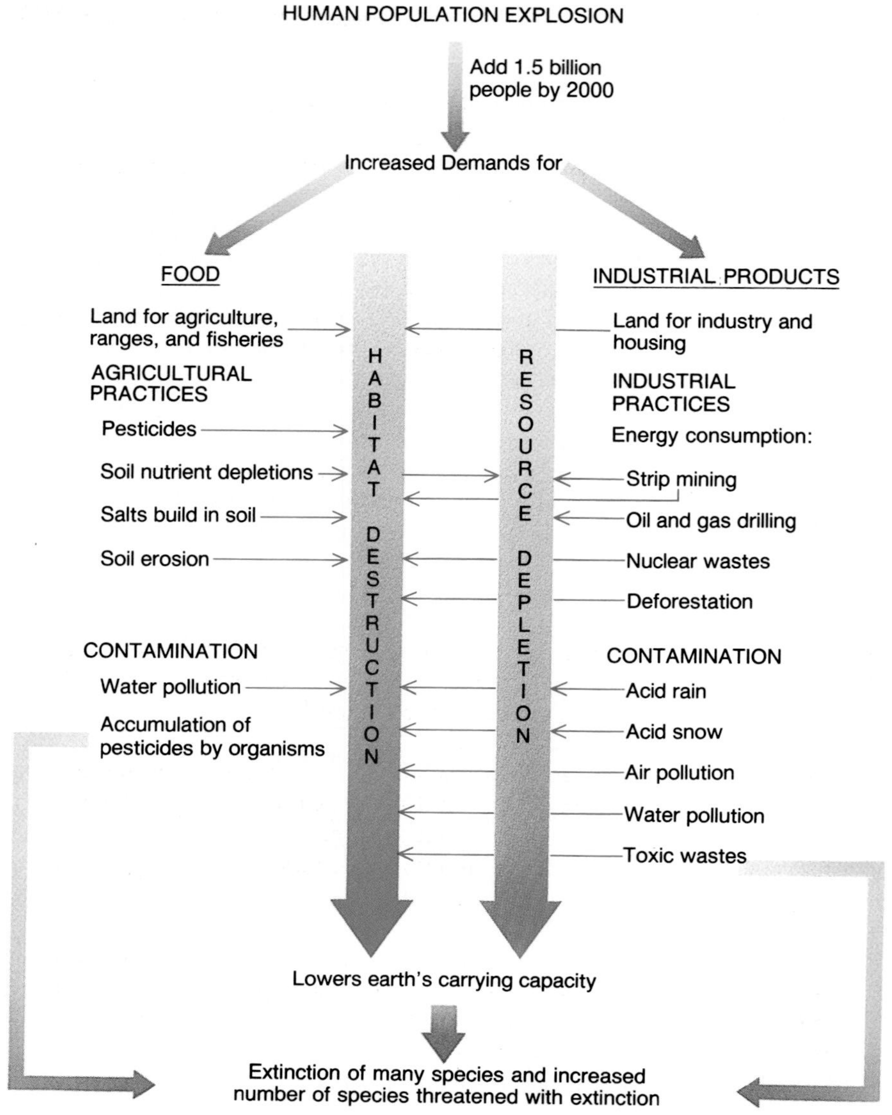

HUMAN POPULATION EXPLOSION

Add 1.5 billion people by 2000

Increased Demands for

FOOD

Land for agriculture, ranges, and fisheries

AGRICULTURAL PRACTICES

Pesticides

Soil nutrient depletions

Salts build in soil

Soil erosion

CONTAMINATION

Water pollution

Accumulation of pesticides by organisms

HABITAT DESTRUCTION

RESOURCE DEPLETION

INDUSTRIAL PRODUCTS

Land for industry and housing

INDUSTRIAL PRACTICES

Energy consumption:

Strip mining

Oil and gas drilling

Nuclear wastes

Deforestation

CONTAMINATION

Acid rain

Acid snow

Air pollution

Water pollution

Toxic wastes

Lowers earth's carrying capacity

Extinction of many species and increased number of species threatened with extinction

FIGURE 32-10 **Impacts of the human population explosion.** Increased numbers of people mean increased demands for food and products. These demands deplete resources and destroy habitats, which, in turn, lower the earth's carrying capacity and increase the number of species threatened with extinction.

these levels because we are very close to, or have already exceeded, the earth's carrying capacity for humans.

THE EARTH'S CARRYING CAPACITY AND FUTURE POPULATION TRENDS

Density-dependent controls are already beginning to curb human population growth, as the ever-burgeoning human population is reducing the ability of the earth to support life (Figure 32-10). Between 10 million and 20 million people—mostly children—die each year from starvation or malnutrition-related disease. Even in affluent countries, overpopulation has accelerated the rate of environmental deterioration, lowering the quality of life and most likely reducing the environment's carrying capacity. Is it possible that the world's human population is already in overshoot? If so, what lies ahead? A dieback to the original carrying capacity, a dieback to a lower carrying capacity, or a dieback to extinction?

Many biologists believe that these undesirable consequences can be avoided through immediate efforts to curb population growth and to protect the environment; these are the challenges facing us today. Biotechnological advances for increasing food production, methods for disposing of wastes that otherwise contaminate the biosphere, and other advances that would help control growth rates may someday help us avert the disasters of overpopulation.

EVOLUTION AND ADAPTATION: TYING IT TOGETHER

All organisms, from the slowest reproducer to the fastest, have the potential eventually to produce huge numbers of descendants. This capacity for reproduction enables species to colonize new habitats, to recover from disasters that kill a large proportion of the population, and to provide the variability upon which natural selection operates.

Reproduction is a basic characteristic of life. Organisms have evolved a range of reproductive strategies, from species that grow fast, reach reproductive maturity early, and invest the greatest amount of energy and nutrients they acquire into a single, large reproductive event to species that grow slowly, delay reproduction, invest the largest proportion of energy and nutrients into growth, increasing competitive abilities, and have small, multiple reproductive events. Between these extremes lies a full range of species with intermediate reproductive strategies, including that of humans.

The world human population is currently growing exponentially, and, as with all species, ultimately a population ceiling for humans will be reached. Humans are unique in their ability to analyze and anticipate future trends as well as to modify their environment to their advantage. All of these skills will be necessary in order to solve the problem of the human population explosion.

SYNOPSIS

Communities are made up of populations. Each population contains all of the individuals of the same species that occupy an area at the same time.

Every population has the potential to reproduce large numbers of offspring. A population growing at its maximum rate increases in size exponentially and produces a J-shaped growth curve. Eventually, some factor or combination of factors halts continued exponential growth, causing a population dieback, sometimes to extinction.

For most populations, growth starts off slowly, increases rapidly, and then slows down again and levels off, producing a Sigmoid Growth Curve. The population levels off at the carrying capacity of the environment, where growth remains in dynamic balance with available nutrients and appropriate conditions.

Populations grow when natality and immigration exceed mortality and emigration. In populations where there is no immigration or emigration, the rate of growth equals the difference between the birth rate and the death rate.

The world's human population exceeded 5.5 million people in 1993. The human population is continuing to grow exponentially, mainly as a result of growth in developing countries.

Review Questions

1. Match the process (numbered column) with the resulting outcome (lettered column).

1. sigmoid growth	a. J-shaped growth curve
2. maximum rate of increase	b. S-shaped growth curve
3. exponential growth	c. a population dieback
4. immigration	d. increased density

2. Check those conditions that would cause a population to decrease in density.

______1. natality that greatly exceeds mortality, immigration, and emigration

______2. pyramid-shaped age–sex structure

______3. a Type III survivorship curve, high emigration, and high mortality

______4. a population that has greatly exceeded the carrying capacity of the environment

3. List two density-dependent and two density-independent factors that would affect human population growth. For example, a collision between two planes as a result of crowded airways would be a density-dependent factor, whereas a plane crash into a mountain as a result of adverse weather would be a density-independent factor.

4. Carrying capacity of the environment, biotic potential, and environmental resistance all affect the rate of population growth. Consider two distinctly different types of organisms in two distinctly different habitats: elephants in an African savanna, and phytoplankton in the open ocean. Describe the similarities and differences between population growth factors for these two organisms. Be sure to consider all aspects.

5. Name four factors or events that have blocked or postponed traditional limits on human population growth. Given the current rate of growth, name two ways humans can continue to block traditional limits.

Critical Thinking Questions

1. As in all plants, the critical phases in the life cycle of the saguaro cactus include seed germination, seedling growth and development, growth to reproductive maturity, flowering, pollination, fertilization, fruit and seed development, and seed dispersal. Which critical phases were tested by the research presented in the chapter opening Steps To Discovery vignette, and which phases were not? Of those phases that were tested, are there any other elements that should be looked at to be sure that nothing was missed? Of those phases that were not tested, choose one that you believe may explain why saguaros are not reproducing adequate numbers of offspring to maintain stable populations, and devise an experiment to test whether you are correct.
2. Each ecosystem supports many populations of species, each growing at different rates. Since all populations share the same abiotic environment, explain how the growth rates of each population can be different. Could any populations have the same growth rate? How?
3. The growth rates of populations can be calculated from birth rates and death rates, assuming there is no emigration or immigration. By convention, birth and death rates are given as the number per 1,000 people, whereas growth rate is given as a percent (i.e., per 100). Thus, the calculation for growth rate is

   growth rate (%) = birth rate − death rate/10.

   You can estimate the time in years that it will take a population to double by dividing 70 by the annual growth rate.

   Calculate the growth rate and doubling time of the countries in the table below:

Country	*Birth Rate*	*Death Rate*	*Growth Rate*	*Doubling Time*
United States	17	9		
India	31	10		
China	21	7		
Somalia	49	19		
Poland	15	10		
France	14	9		

4. Wildebeests and birds migrate from one location to another in search of food and other resources. Barnacles and mussels are sessile and cannot travel to locate richer sources of needed materials. Yet, the populations of both migratory and sessile species are subject to limitations to growth. List some of the differences in the ways limiting factors would affect the size of migratory populations, compared to sessile ones. Are there common features to the limiting factors in each category that may lead you to draw a general conclusion about how environmental resistances differ between migratory and sessile organisms?
5. Describe your opinion concerning the future trends in human population growth. Listen to the news and read a major newspaper over the next week and document which stories support your opinion and which stories do not. Based on these stories, should your projection of future population trends be modified? In what way? Choose one of the following statements that you feel best supports your revised opinion, and devise an experiment for testing whether the statement is correct or not:
   a. Humans are fundamentally different from all other organisms (behaviorally, intellectually, and physiologically) and are not subject to the same population controls as are other species.
   b. Humans are governed by the same population controls as are all other species. Although humans cannot change the types of natural population controls, their intellect and ingenuity empower them to modify levels of natural controls.

APPENDIX A

Metric and Temperature Conversion Charts

Metric Unit (symbol)		*Metric to English*	*English to Metric*
Length			
kilometer (km)	= 1,000 (10^3) meters	1 km = 0.62 mile	1 mile = 1.609 km
meter (m)	= 100 centimeters	1 m = 1.09 yards = 3.28 feet	1 yard = 0.914 m 1 foot = 0.305 m
centimeter (cm)	= 0.01 (10^{-2}) meter	1 cm = 0.394 inch	1 inch = 2.54 cm
millimeter (mm)	= 0.001 (10^{-3}) meter	1 mm = 0.039 inch	1 inch = 25.4 mm
micrometer (μm)	= 0.000001 (10^{-6}) meter		
nanometer (nm)	= 0.000000001 (10^{-9}) meter		
angstrom (Å)	= 0.0000000001 (10^{-10}) meter		
Area			
square kilometer (km^2)	= 100 hectares	1 km^2 = 0.386 square mile	1 square mile = 2.590 km^2
hectare (ha)	= 10,000 square meters	1 ha = 2.471 acres	1 acre = 0.405 ha
square meter (m^2)	= 10,000 square centimeters	1 m^2 = 1.196 square yards = 10.764 square feet	1 square yard = 0.836 m^2 1 square foot = 0.093 m^2
square centimeter (cm^2)	= 100 square millimeters	1 cm^2 = 0.155 square inch	1 square inch = 6.452 cm^2
Mass			
metric ton (t)	= 1,000 kilograms = 1,000,000 grams	1 t = 1.103 tons	1 ton = 0.907 t
kilogram (kg)	= 1,000 grams	1 kg = 2.205 pounds	1 pound = 0.454 kg
gram (g)	= 1,000 milligrams	1 g = 0.035 ounce	1 ounce = 28.35 g
milligram (mg)	= 0.001 gram		
microgram (μg)	= 0.000001 gram		
Volume Solids			
1 cubic meter (m^3)	= 1,000,000 cubic centimeters	1 m^3 = 1.308 cubic yards = 35.315 cubic feet	1 cubic yard = 0.765 m^3 1 cubic foot = 0.028 m^3
1 cubic centimeter (cm^3)	= 1,000 cubic millimeters	1 cm^3 = 0.061 cubic inch	1 cubic inch = 16.387 cm^3
Volume Liquids			
kiloliter (kl)	= 1,000 liters	1 kl = 264.17 gallons	
liter (l)	= 1,000 milliliters	1 l = 1.06 quarts	1 gal = 3.785 l 1 qt = 0.94 l 1 pt = 0.47 l
milliliter (ml)	= 0.001 liter	1 ml = 0.034 fluid ounce	1 fluid ounce = 29.57 ml
microliter (μl)	= 0.000001 liter		

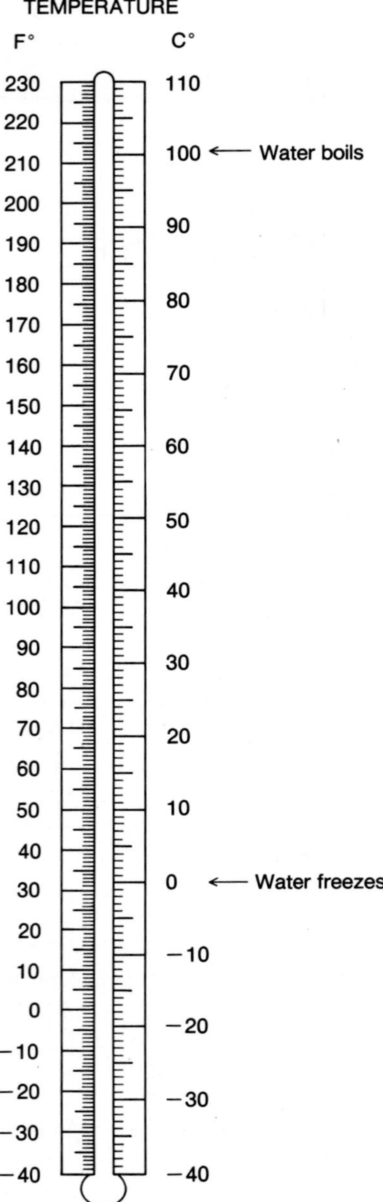

Fahrenheit to Centigrade: $°C = \frac{5}{9}(°F - 32)$

Centigrade to Fahrenheit: $°F = \frac{9}{5}(°C + 32)$

APPENDIX B

Microscopes: Exploring the Details of Life

Microscopes are the instruments that have allowed biologists to visualize objects that are vastly smaller than anything visible with the naked eye. There are broadly two types of specimens viewed in a microscope: whole mounts which consist of an intact subject, such as a hair, a living cell, or even a DNA molecule, and thin sections of a specimen, such as a cell or piece of tissue.

THE LIGHT MICROSCOPE

A light microscope consists of a series of glass lenses that bend (refract) the light coming from an illuminated specimen so as to form a visual image of the specimen that is larger than the specimen itself (a). The specimen is often stained with a colored dye to increase its visibility. A special phase contrast light microscope is best suited for observing unstained, living cells because it converts differences in the density of cell organelles, which are normally invisible to the eye, into differences in light intensity which can be seen.

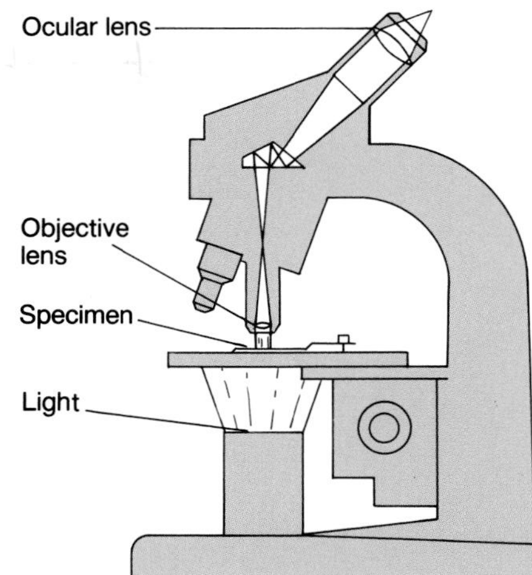

(a)

All light microscopes have limited *resolving power*—the ability to distinguish two very close objects as being separate from each other. The resolving power of the light microscope is about 0.2 μm (about 1,000 times that of the naked eye), a property determined by the wave length of visible light. Consequently, objects closer to each other than 0.2 μm, which includes many of the smaller cell organelles, will be seen as a single, blurred object through a light microscope.

THE TRANSMISSION ELECTRON MICROSCOPE

Appreciation of the wondrous complexity of cellular organization awaited the development of the transmission electron microscope (or TEM), which can deliver resolving powers 1000 times greater than the light microscope. Suddenly, biologists could see strange new structures, whose function was totally unknown—a breakthrough that has kept cell biologists busy for the past 50 years. The TEM (b) works by shooting a beam of electrons through very thinly sliced specimens that have been stained with heavy metals,

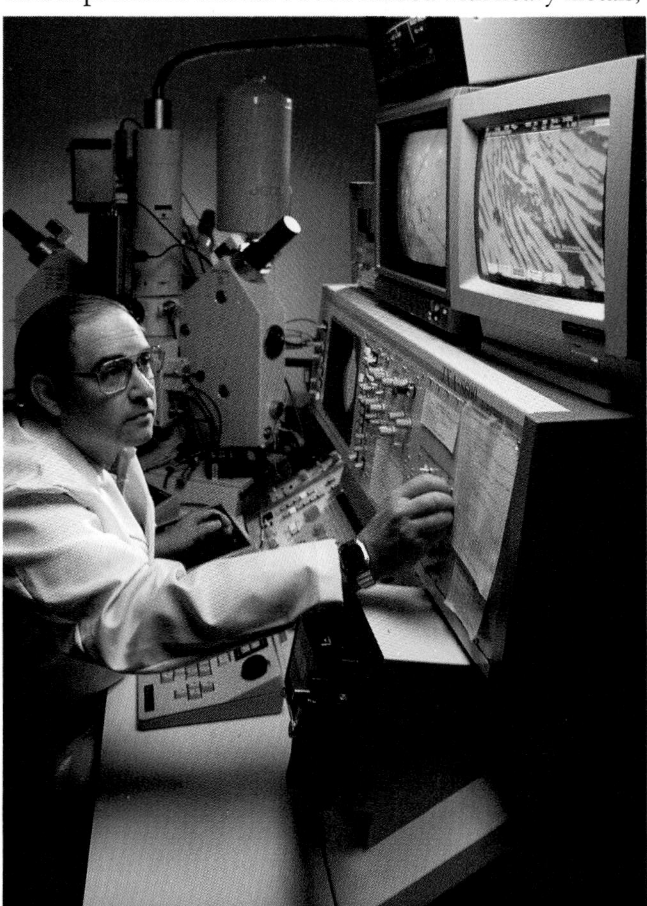

(b)

such as uranium, capable of deflecting electrons in the beam. The electrons that pass through the specimen undeflected are focused by powerful electromagnets (the lenses of a TEM) onto either a phosphorescent screen or high-contrast photographic film. The resolution of the TEM is so great—sufficient to allow us to see individual DNA molecules—because the wavelength of an electron beam is so small (about 0.0005 μm).

THE SCANNING ELECTRON MICROSCOPE

Specimens examined in the scanning electron microscope (SEM) are whole mounts whose surfaces have been coated with a thin layer of heavy metals. In the SEM, a fine beam of electrons scans back and forth across the specimen and the image is formed from electrons bouncing off the hills and valleys of its surface. The SEM produces a three-dimensional image of the surface of the specimen—which can range in size from a virus to an insect head (c,d)—with remarkable depth and clarity. The SEM produces black and white images; the colors seen in many of the micrographs in the text have been added to enhance their visual quality. Note that the insect head (d) is that of an antennapedia mutant as described on p. 365.

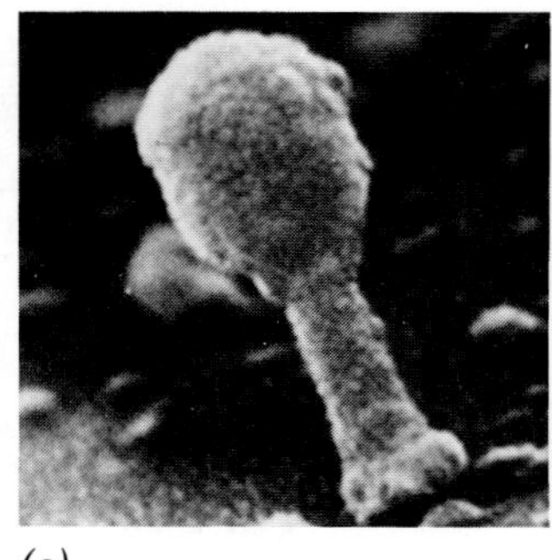

(c)

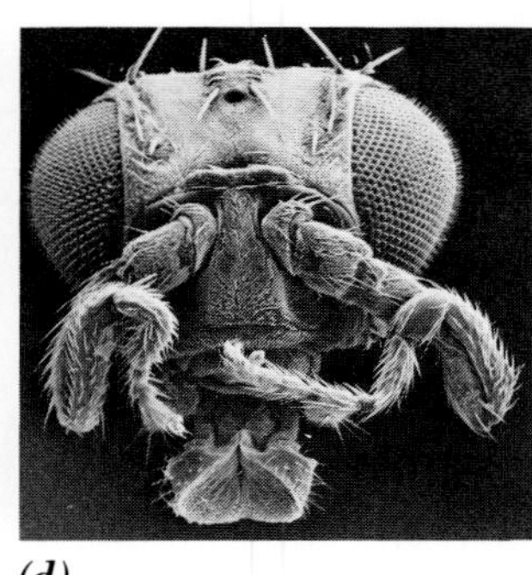

(d)

APPENDIX C

The Hardy-Weinberg Principle

If the allele for brown hair is dominant over that for blond hair, and curly hair is dominant over straight hair, then why don't all people by now have brown, curly hair? The **Hardy-Weinberg Principle** (developed independently by English mathematician G. H. Hardy and German physician W. Weinberg) demonstrates that the frequency of alleles remains the same from generation to generation unless influenced by outside factors. The outside factors that would cause allele frequencies to change are mutation, immigration and emigration (movement of individuals into and out of a breeding population, respectively), natural selection of particular traits, and breeding between members of a small population. In other words, unless one or more of these forces influence hair color and hair curl, the relative number of people with brown and curly hair will not increase over those with blond and straight hair.

To illustrate the Hardy-Weinberg Principle, consider a single gene locus with two alleles, *A* and *a*, in a breeding population. (If you wish, consider *A* to be the allele for brown hair and *a* to be the allele for blond hair.) Because there are only two alleles for the gene, the sum of the frequencies of A and *a* will equal 1.0. (By convention, allele frequencies are given in decimals instead of percentages.) Translating this into mathematical terms, if

p = the frequency of allele A, and
q = the frequency of allele a,

then $p + q = 1$.

If A represented 80 percent of the alleles in the breeding population ($p = 0.8$), then according to this formula the frequency of a must be 0.2 ($p + q = 0.8 + 0.2 = 1.0$).

After determining the allele frequency in a starting population, the predicted frequencies of alleles and genotypes in the next generation can be calculated. Setting up a Punnett square with starting allele frequencies of $p = 0.8$ and $q = 0.2$:

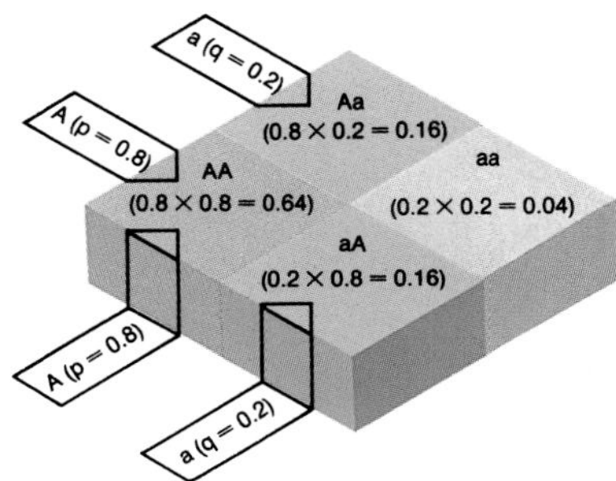

The chances of each offspring receiving any combination of the two alleles is the product of the probability of receiving one of the two alleles alone. In this example, the chances of an offspring receiving two A alleles is $p \times p = p^2$, or $0.8 \times 0.8 = 0.64$. A frequency of 0.64 means that 64 percent of the next generation will be homozygous dominant (*AA*). The chances of an offspring receiving two a alleles is $q^2 = 0.2 \times 0.2 = 0.04$, meaning 4 percent of the next generation is predicted to be *aa*. The predicted frequency of heterozygotes (*Aa* or *aA*) is 0.32 or $2pq$, the sum of the probability of an individual being $Aa(p \times q = 0.8 \times 0.2 = 0.16)$ plus the probability of an individual being $aA(q \times p = 0.2 \times 0.8 = 0.16)$. Just as all of the allele frequencies for a particular gene must add up to 1, so must all of the possible genotypes for a particular gene locus add up to 1. Thus, the Hardy-Weinberg Principle is

$$p^2 + 2pq + q^2 = 1$$
$$(0.64 + 0.32 + 0.04 = 1)$$

So after one generation, the frequency of possible genotypes is

$$AA = p^2 = 0.64$$
$$Aa = 2pq = 0.32$$
$$aa = q^2 = 0.04$$

Now let's determine the actual allele frequencies for *A* and *a* in the new generation. (Remember the original allele frequencies were 0.8 for allele A and 0.2 for allele *a*. If the Hardy-Weinberg Principle is right, there will be no change in the frequency of either allele.) To do this we sum the frequencies for each genotype containing the allele. Since heterozygotes carry both alleles, the genotype frequency must be divided in half to determine the frequency of each allele. (In our example, heterozygote *Aa* has a frequency of 0.32, 0.16 for allele A, plus 0.16 for allele *a*.) Summarizing then:

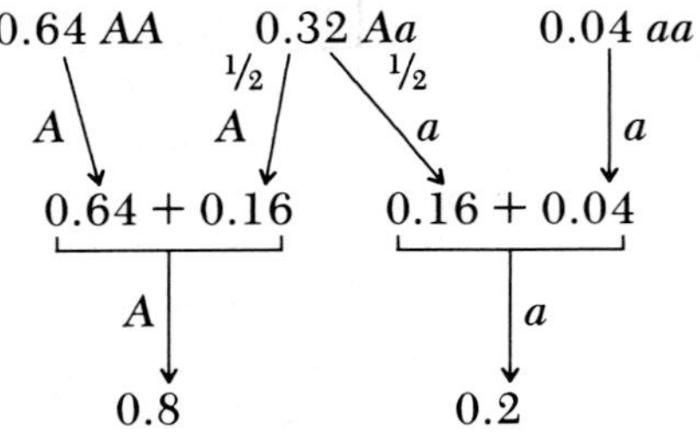

As predicted by the Hardy-Weinberg Principle, the frequency of allele *A* remained 0.8 and the frequency of allele *a* remained 0.2 in the new generation. Future generations can be calculated in exactly the same way, over and over again. As long as there are no mutations, no gene flow between populations, completely random mating, no natural selection, and no genetic drift, there will be no change in allele frequency, and therefore no evolution.

Population geneticists use the Hardy-Weinberg Principle to calculate a starting point allele frequency, a reference that can be compared to frequencies measured at some future time. The amount of deviation between observed allele frequencies and those predicted by the Hardy-Weinberg Principle indicates the degree of evolutionary change. Thus, this principle enables population geneticists to measure the rate of evolutionary change and identify the forces that cause changes in allele frequency.

APPENDIX D

Careers in Biology

Although many of you are enrolled in biology as a requirement for another major, some of you will become interested enough to investigate the career opportunities in life sciences. This interest in biology can grow into a satisfying livelihood. Here are some facts to consider:

- Biology is a field that offers a very wide range of possible science careers
- Biology offers high job security since many aspects of it deal with the most vital human needs: health and food
- Each year in the United States, nearly 40,000 people obtain bachelor's degrees in biology. But the number of newly created and vacated positions for biologists is increasing at a rate that exceeds the number of new graduates. Many of these jobs will be in the newer areas of biotechnology and bioservices.

Biologists not only enjoy job satisfaction, their work often changes the future for the better. Careers in medical biology help combat diseases and promote health. Biologists have been instrumental in preserving the earth's life-supporting capacity. Biotechnologists are engineering organisms that promise dramatic breakthroughs in medicine, food production, pest management, and environmental protection. Even the economic vitality of modern society will be increasingly linked to biology.

Biology also combines well with other fields of expertise. There is an increasing demand for people with backgrounds or majors in biology complexed with such areas as business, art, law, or engineering. Such a distinct blend of expertise gives a person a special advantage.

The average starting salary for all biologists with a Bachelor's degree is $22,000. A recent survey of California State University graduates in biology revealed that most were earning salaries between $20,000 and $50,000. But as important as salary is, most biologists stress job satisfaction, job security, work with sophisticated tools and scientific equipment, travel opportunities (either to the field or to scientific conferences), and opportunities to be creative in their job as the reasons they are happy in their career.

Here is a list of just a few of the careers for people with degrees in biology. For more resources, such as lists of current openings, career guides, and job banks, write to Biology Career Information, John Wiley and Sons, 605 Third Avenue, New York, NY 10158.

A SAMPLER OF JOBS THAT GRADUATES HAVE SECURED IN THE FIELD OF BIOLOGY*

Agricultural Biologist
Agricultural Economist
Agricultural Extension Officer
Agronomist
Amino-acid Analyst
Analytical Biochemist
Anatomist
Animal Behavior Specialist
Anticancer Drug Research Technician
Antiviral Therapist
Arid Soils Technician
Audio-neurobiologist
Author, Magazines & Books
Behavioral Biologist
Bioanalyst
Bioanalytical Chemist
Biochemical/Endocrine Toxicologist
Biochemical Engineer
Pharmacology Distributor
Pharmacology Technician
Biochemist
Biogeochemist
Biogeographer
Biological Engineer
Biologist
Biomedical Communication Biologist
Biometerologist
Biophysicist
Biotechnologist
Blood Analyst
Botanist
Brain Function Researcher
Cancer Biologist
Cardiovascular Biologist
Cardiovascular/Computer Specialist
Chemical Ecologist
Chromatographer
Clinical Pharmacologist
Coagulation Biochemist
Cognitive Neuroscientist
Computer Scientist
Dental Assistant
Ecological Biochemist
Electrophysiology/ Cardiovascular Technician
Energy Regulation Officer
Environmental Biochemist
Environmental Center Director
Environmental Engineer
Environmental Geographer
Environmental Law Specialist
Farmer
Fetal Physiologist
Flavorist
Food Processing Technologist
Food Production Manager
Food Quality Control Inspector
Flower Grower
Forest Ecologist
Forest Economist
Forest Engineer
Forest Geneticist
Forest Manager

Forest Pathologist
Forest Plantation Manager
Forest Products Technologist
Forest Protection Expert
Forest Soils Analyst
Forester
Forestry Information Specialist
Freeze-Dry Engineer
Fresh Water Biologist
Grant Proposal Writer
Health Administrator
Health Inspector
Health Scientist
Hospital Administrator
Hydrologist
Illustrator
Immunochemist
Immunodiagnostic Assay Developer
Inflammation Technologist
Landscape Architect
Landscape Designer
Legislative Aid
Lepidopterist
Liaison Scientist, Library of Medicine
Computer Biologist
Life Science Computer Technologist
Lipid Biochemist
Livestock Inspector
Lumber Inspector
Medical Assistant
Medical Imaging Technician
Medical Officer
Medical Products Developer
Medical Writer
Microbial Physiologist
Microbiologist
Mine Reclamation Scientist
Molecular Endocrinologist
Molecular Neurobiologist
Molecular Parasitologist
Molecular Toxicologist
Molecular Virologist
Morphologist
Natural Products Chemist
Natural Resources Manager
Nature Writer
Nematode Control Biologist
Nematode Specialist
Nematologist
Neuroanatomist
Neurobiologist
Neurophysiologist
Neuroscientist
Nucleic Acids Chemist
Nursing Aid
Nutritionist
Occupational Health Officer
Ornamental Horticulturist
Paleontologist
Paper Chemist
Parasitologist
Pathologist
Peptide Biochemist
Pharmaceutical Writer
Pharmaceutical Sales
Pharmacologist
Physiologist
Planning Consultant
Plant Pathologist
Plant Physiologist
Production Agronomist
Protein Biochemist
Protein Structure & Design Technician
Purification Biochemist
Quantitative Geneticist
Radiation Biologist
Radiological Scientist
Regional Planner
Regulatory Biologist
Renal Physiologist
Renal Toxicologist
Reproductive Toxicologist
Research and Development Director
Research Technician
Research Liaison Scientist
Research Products Designer
Research Proposal Writer
Safety Assessment Sanitarian
Scientific Illustrator
Scientific Photographer
Scientific Reference Librarian
Scientific Writer
Soil Microbiologist
Space Station Life Support Technician
Spectroscopist
Sports Product Designer
Steroid Health Assessor
Taxonomic Biologist
Teacher
Technical Analyst
Technical Science Project Writer
Textbook Editor
Theoretical Ecologist
Timber Harvester
Toxicologist
Toxic Waste Treatment Specialist
Urban Planner
Water Chemist
Water Resources Biologist
Wood Chemist
Wood Fuel Technician
Zoning and Planning Manager
Zoologist
Zoo Animal Breeder
Zoo Animal Behaviorist
Zoo Designer
Zoo Inspector

*Results of one survey of California State University graduates. Some careers may require advanced degrees

Glossary

◀ A ▶

Abiotic Environment Components of ecosystems that include all nonliving factors.

Abscisic Acid (ABA) A plant hormone that inhibits growth and causes stomata to close. ABA may not be commonly involved in leaf drop.

Abscission Separation of leaves, fruit, and flowers from the stem.

Acclimation A physiological adjustment to environmental stress.

Acetyl CoA Acetyl coenzyme A. A complex formed when acetic acid binds to a carrier coenzyme forming a bridge between the end products of glycolysis and the Krebs cycle in respiration.

Acetylcholine Neurotransmitter released by motor neurons at neuromuscular junctions and by some interneurons.

Acid Rain Occurring in polluted air, rain that has a lower pH than rain from areas with unpolluted air.

Acids Substances that release hydrogen ions (H^+) when dissolved in water.

Acid Snow Occurring in polluted air, snow that has a lower pH than snow from areas with unpolluted air.

Acoelomates Animals that lack a body cavity between the digestive cavity and body wall.

Acquired Immune Deficiency Syndrome (AIDS) Disease caused by infection with HIV (Human Immunodeficiency Virus) that destroys the body's ability to mount an immune response due to destruction of its helper T cells.

Actin A contractile protein that makes up the major component of the thin filaments of a muscle cell and the microfilaments of nonmuscle cells.

Action Potential A sudden, dramatic reversal of the voltage (potential difference) across the plasma membrane of a nerve or muscle cell due to the opening of the sodium channels. The basis of a nerve impulse.

Activation Energy Energy required to initiate chemical reaction.

Active Site Region on an enzyme that binds its specific substrates, making them more reactive.

Active Transport Movement of substances into or out of cells against a concentration gradient, i.e., from a region of lower concentration to a region of higher concentration. The process requires an expenditure of energy by the cell.

Adaptation A hereditary trait that improves an organism's chances of survival and/or reproduction.

Adaptive Radiation The divergence of many species from a single ancestral line.

Adenosine Triphosphate (ATP) The molecule present in all living organisms that provides energy for cellular reactions in its phosphate bonds. ATP is the universal energy currency of cells.

Adenylate Cyclase An enzyme activated by hormones that converts ATP to cyclic AMP, a molecule that activates resting enzymes.

Adrenal Cortex Outer layer of the adrenal glands. It secretes steroid hormones in response to ACTH.

Adrenal Medulla An endocrine gland that controls metabolism, cardiovascular function, and stress responses.

Adrenocorticotropic Hormone (ACTH) An anterior pituitary hormone that stimulates the cortex of the adrenal glands to secrete cortisol and other steroid hormones.

Adventitious Root System Secondary roots that develop from stem or leaf tissues.

Aerobe An organism that requires oxygen to release energy from food molecules.

Aerobic Respiration Pathway by which glucose is completely oxidized to CO_2 and H_2O, requiring oxygen and an electron transport system.

Afferent (Sensory) Neurons Neurons that conduct impulses from the sense organs to the central nervous system.

Age-Sex Structure The number of individuals of a certain age and sex within a population.

Aggregate Fruits Fruits that develop from many pistils in a single flower.

AIDS **See Acquired Immune Deficiency Syndrome.**

Albinism A genetic condition characterized by an absence of epidermal pigmentation that can result from a deficiency of any of a variety of enzymes involved in pigment formation.

Alcoholic Fermentation The process in which electrons removed during glycolysis are transferred from NADH to form alcohol as an end product. Used by yeast during the commercial process of ethyl alcohol production.

Aldosterone A hormone secreted by the adrenal cortex that stimulates reabsorption of sodium from the distal tubules and collecting ducts of the kidneys.

Algae Any unicellular or simple colonial photosynthetic eukaryote.

Algin A substance produced by brown algae harvested for human application because of its ability to regulate texture and consistency of products. Found in ice cream, cosmetics, marshmellows, paints, and dozens of other products.

Allantois Extraembryonic membrane that serves as a repository for nitrogenous wastes. In placental mammals, it helps form the vascular connections between mother and fetus.

Allele Alternative form of a gene at a particular site, or locus, on the chromosome.

Allele Frequency The relative occurrence of a certain allele in individuals of a population.

Allelochemicals Chemicals released by some plants and animals that deter or kill a predator or competitor.

Allelopathy A type of interaction in which one organism releases allelochemicals that harm another organism.

Allergy An inappropriate response by the immune system to a harmless foreign substance leading to symptoms such as itchy eyes, runny nose, and congested airways. If the reaction occurs throughout the body (anaphylaxis) it can be life threatening.

Allopatric Speciation Formation of new species when gene flow between parts of a population is stopped by geographic isolation.

Alpha Helix Portion of a polypeptide chain organized into a defined spiral conformation.

Alternation of Generations Sequential change during the life cycle of a plant in which a haploid (1N) multicellular stage (gametophyte) alternates with a diploid (2N) multicellular stage (sporophyte).

Alternative Processing When a primary RNA transcript can be processed to form more than one mRNA depending on conditions.

Altruism The performance of a behavior that benefits another member of the species at some cost to the one who does the deed.

Alveolus A tiny pouch in the lung where gas is exchanged between the blood and the air; the functional unit of the lung where CO_2 and O_2 are exchanged.

Alzheimer's Disease A degenerative disease of the human brain, particularly affecting acetylcholine-releasing neurons and the hippocampus, characterized by the presence of tangled fibrils within the cytoplasm of neurons and amyloid plaques outside the cells.

Amino Acids Molecules containing an amino group ($-NH_2$) and a carboxyl group (-COOH) attached to a central carbon atom. Amino acids are the subunits from which proteins are constructed.

Amniocentesis A procedure for obtaining fetal cells by withdrawing a sample of the fluid that surrounds a developing fetus (amniotic fluid) using a hypodermic needle and syringe.

Amnion Extraembryonic membrane that envelops the young embryo and encloses the amniotic fluid that suspends and cushions it.

Amoeba A protozoan that employs pseudopods for motility.

Amphibia A vertebrate class grouped into three orders: Caudata (tailed amphibians); Anura (tail-less amphibians); Apoda (rare worm-like, burrowing amphibians).

Anabolic Steroids Steroid hormones, such as testosterone, which promote biosynthesis (anabolism), especially protein synthesis.

Anabolism Biosynthesis of complex molecules from simpler compounds. Anabolic pathways are endergonic, i.e., require energy.

Anaerobe Organism that does not require oxygen to release energy from food molecules.

Anaerobic Respiration Pathway by which glucose is completely oxidized, using an electron transport system but requiring a terminal electron acceptor other than oxygen. (Compare with fermentation.)

Analogous Structures (Homoplasies) Structures that perform a similar function, such as wings in birds and insects, but did not originate from the same structure in a common ancestor.

Anaphase Stage of mitosis when the kinetochores split and the sister chromatids (now termed chromosomes) move to opposite poles of the spindle.

Anatomy Study of the structural characteristics of an organism.

Angiosperm (Anthophyta) Any plant having its seeds surrounded by fruit tissue formed from the mature ovary of the flowers.

Animal A mobile, heterotrophic, multicellular organism, classified in the Animal kingdom.

Anion A negatively charged ion.

Annelida The phylum which contains segmented worms (earthworms, leeches, and bristleworms).

Annuals Plants that live for one year or less.

Annulus A row of specialized cells encircling each sporangium on the fern frond; facilitates rupture of the sporangium and dispersal of spors.

Antagonistic Muscles Pairs of muscles whose contraction bring about opposite actions as illustrated by the biceps and triceps, which bends or straightens the arm at the elbow, respectively.

Antenna Pigments Components of photosystems that gather light energy of different wavelengths and then channel the absorbed energy to a reaction center.

Anterior In anatomy, at or near the front of an animal; the opposite of posterior.

Anterior Pituitary A true endocrine gland manufacturing and releasing six hormones when stimulated by releasing factors from the hypothalamus.

Anther The swollen end of the stamen (male reproductive organ) of a flowering plant. Pollen grains are produced inside the anther lobes in pollen sacs.

Antibiotic A substance produced by a fungus or bacterium that is capable of preventing the growth of bacteria.

Antibodies Proteins produced by plasma cells. They react specifically with the antigen that stimulated their formation.

Anticodon Triplet of nucleotides in tRNA that recognizes and base pairs with a particular codon in mRNA.

Antidiuretic Hormone (ADH) One of the two hormones released by the posterior pituitary. ADH increases water reabsorption in the kidney, which then produces a more concentrated urine.

Antigen Specific foreign agent that triggers an immune response.

Aorta Largest blood vessel in the body through which blood leaves the heart and enters the systemic circulation.

Apical Dominance The growth pattern in plants in which axillary bud growth is inhibited by the hormone auxin, present in high concentrations in terminal buds.

Apical Meristems Centers of growth located at the tips of shoots, axillary buds, and roots. Their cells divide by mitosis to produce new cells for primary growth in plants.

Aposematic Coloring Warning coloration which makes an organism stand out from its surroundings.

Appendicular Skeleton The bones of the appendages and of the pectoral and pelvic girdles.

Aquatic Living in water.

Archaebacteria Members of the kingdom Monera that differ from typical bacteria in the structure of their membrane lipids, their cell walls, and some characteristics that resemble those of eukaryotes. Their lack of a true nucleus, however, accounts for their assignment to the Moneran kingdom.

Archenteron In gastrulation, the hollow core of the gastrula that becomes an animal's digestive tract.

Arteries Large, thick-walled vessels that carry blood away from the heart.

Arterioles The smallest arteries, which carry blood toward capillary beds.

Arthropoda The most diverse phylum on earth, so called from the presence of jointed limbs. Includes insects, crabs, spiders, centipedes.

Ascospores Sexual fungal spore borne in a sac. Produced by the sac fungi, Ascomycota.

Asexual Reproduction Reproduction without the union of male and female gametes.

Association In ecological communities, a major organization characterized by uniformity and two or more dominant species.

Asymmetric Referring to a body form that cannot be divided to produce mirror images.

Atherosclerosis Condition in which the inner walls of arteries contain a buildup of cholesterol-containing plaque that tends to occlude the channel and act as a site for the formation of a blood clot (thrombus).

Atmosphere The layer of air surrounding the Earth.

Atom The fundamental unit of matter that can enter into chemical reactions; the smallest unit of matter that possesses the qualities of an element.

Atomic Mass Combined number of protons and neutrons in the nucleus of an atom.

Atomic Number The number of protons in the nucleus of an atom.

ATP (see **Adenosine Triphosphate**)

ATPase An enzyme that catalyzes a reaction in which ATP is hydrolyzed. These enzymes are typically involved in reactions where energy stored in ATP is used to drive an energy-requiring reaction, such as active transport or muscle contractility.

ATP Synthase A large protein complex present in the plasma membrane of bacteria, the inner membrane of mitochondria, and the thylakoid membrane of chloroplasts. This complex consists of a baseplate in the membrane, a channel across the membrane through which protons can pass, and a spherical head (F_1 particle) which contains the site where ATP is synthesized from ADP and P_i

Atrioventricular (AV) Node A neurological center of the heart, located at the top of the ventricles.

Atrium A contracting chamber of the heart which forces blood into the ventricle. There are two atria in the hearts of all vertebrates, except fish which have one atrium.

Atrophy The shrinkage in size of structure, such as a bone or muscle, usually as a result of disuse.

Autoantibodies Antibodies produced against the body's own tissue.

Autoimmune Disease Damage to a body tissue due to an attack by autoantibodies. Examples include thyroiditis, multiple sclerosis, and rheumatic fever.

Autonomic Nervous System The nerves that control the involuntary activities of the internal organs. It is composed of the parasympathetic system, which functions during normal activity, and the sympathetic system, which operates in times of emergency or prolonged exertion.

Autosome Any chromosome that is not a sex chromosome.

Autotrophs Organisms that satisfy their own nutritional needs by building organic molecules photosynthetically or chemosynthetically from inorganic substances.

Auxins Plant growth hormones that promote cell elongation by softening cell walls.

Axial Skeleton The bones aligned along the long axis of the body, including the skull, vertebral column, and ribcage.

Axillary Bud A bud that is directly above each leaf on the stem. It can develop into a new stem or a flower.

Axon The long, sometimes branched extension of a neuron which conducts impulses from the cell body to the synaptic knobs.

◄ B ►

Bacteriophage A virus attacking specific bacteria that multiplies in the bacterial host cell and usually destroys the bacterium as it reproduces.

Balanced Polymorphism The maintenance of two or more alleles for a single trait at fairly high frequencies.

Bark Common term for the periderm. A collective term for all plant tissues outside the secondary xylem.

Base Substance that removes hydrogen ions (H^+) from solutions.

Basidiospores Sexual spores produced by basidiomycete fungi. Often found by the millions on gills in mushrooms.

Basophil A phagocytic leukocyte which also releases substances, such as histamine, that trigger an inflammatory response.

Batesian Mimicry The resemblance of a good-tasting or harmless species to a species with unpleasant traits.

Bathypelagic Zone The ocean zone beneath the mesopelagic zone, characterized by no light; inhabited by heterotrophic bacteria and benthic scavengers.

B Cell A lymphocyte that becomes a plasma cell and produces antibodies when stimulated by an antigen.

Benthic Zone The deepest ocean zone; the ocean floor, inhabited by bottom dwelling organisms.

Bicarbonate Ion $HCO\text{-}_3$.

Biennials Plants that live for two years.

Bilateral Symmetry The quality possessed by organisms whose body can be divided into mirror images by only one median plane.

Bile Salts Detergentlike molecules produced by the liver and stored by the gallbladder that function in lipid emulsification in the small intestine.

Binomial A term meaning "two names" or "two words". Applied to the system of nomenclature for categorizing living things with a genus and species name that is unique for each type of organism.

Biochemicals Organic molecules produced by living cells.

Bioconcentration The ability of an organism to accumulate substances within its' body or specific cells.

Biodiversity Biological diversity of species, including species diversity, genetic diversity, and ecological diversity.

Biogeochemical Cycles The exchanging of chemical elements between organisms and the abiotic environment.

Biological Control Pest control through the use of naturally occurring organisms such as predators, parasites, bacteria, and viruses.

Biological Magnification An increase in concentration of slowly degradable chemicals in organisms at successively higher trophic levels; for example, DDT or PCB's.

Bioluminescence The capability of certain organisms to utilize chemical energy to produce light in a reaction catalyzed by the enzyme luciferase.

Biomass The weight of organic material present in an ecosystem at any one time.

Biome Broad geographic region with a characteristic array of organisms.

Biosphere Zone of the earth's soil, water, and air in which living organisms are found.

Biosynthesis Construction of molecular components in the growing cell and the replacement of these compounds as they deteriorate.

Biotechnology A new field of genetic engineering; more generally, any practical application of biological knowledge.

Biotic Environment Living components of the environment.

Biotic Potential The innate capacity of a population to increase tremendously in size were it not for curbs on growth; maximum population growth rate.

Blade Large, flattened area of a leaf; effective in collecting sunlight for photosynthesis.

Blastocoel The hollow fluid-filled space in a blastula.

Blastocyst Early stage of a mammalian embryo, consisting of a mass of cells enclosed in a hollow ball of cells called the trophoblast.

Blastodisk In bird and reptile development, the stage equivalent to a blastula. Because of the large amount of yolk, cleavage produces two flattened layers of cells with a blastocoel between them.

Blastomeres The cells produced during embryonic cleavage.

Blastopore The opening of the archenteron that is the embryonic predecessor of the anus in vertebrates and some other animals.

Blastula An early developmental stage in many animals. It is a ball of cells that encloses a cavity, the blastocoel.

Blood A type of connective tissue consisting of red blood cells, white blood cells, platelets, and plasma.

Blood Pressure Positive pressure within the cardiovascular system that propels blood through the vessels.

Blooms are massive growths of algae that occur when conditions are optimal for algae proliferation.

Body Plan The general layout of a plant's or animal's major body parts.

Bohr Effect Increased release of O_2 from hemoglobin molecules at lower pH.

Bone A tissue composed of collagen fibers, calcium, and phosphate that serves as a means of support, a reserve of calcium and phosphate, and an attachment site for muscles.

Botany Branch of biology that studies the life cycles, structure, growth, and classification of plants.

Bottleneck A situation in which the size of a species' population drops to a very small number of individuals, which has a major impact on the likelihood of the population recovering its earlier genetic diversity. As occurred in the cheetah population.

Bowman's Capsule A double-layered container that is an invagination of the proximal end of the renal tubule that collects molecules and wastes from the blood.

Brain Mass of nerve tissue composing the main part of the central nervous system.

Brainstem The central core of the brain, which coordinates the automatic, involuntary body processes.

Bronchi The two divisions of the trachea through which air enters each of the two lungs.

Bronchioles The smallest tubules of the respiratory tract that lead into the alveoli of the lungs where gas exchange occurs.

Bryophyta Division of non-vascular terrestrial plants that include liverworts, mosses, and hornworts.

Budding Asexual process by which offspring develop as an outgrowth of a parent.

Buffers Chemicals that couple with free hydrogen and hydroxide ions thereby resisting changes in pH.

Bundle Sheath Parenchyma cells that surround a leaf vein which regulate the uptake and release of materials between the vascular tissue and the mesophyll cells.

◄ C ►

C_3 Synthesis The most common pathway for fixing CO_2 in the synthesis reactions of photosynthesis. It is so named because the first detectable organic molecule into which CO_2 is incorporated is a 3-carbon molecule, phosphoglycerate (PGA).

C_4 Synthesis Pathway for fixing CO_2 during the light-independent reactions of photosynthesis. It is so named because the first detectable organic molecule into which CO_2 is incorporated is a 4-carbon molecule.

Calcitonin A thyroid hormone which regulates blood calcium levels by inhibiting its release from bone.

Calorie Energy (heat) necessary to elevate the temperature of one gram of water by one degree Centigrade (1° C).

Calvin Cycle The cyclical pathway in which CO_2 is incorporated into carbohydrate. See C_3 synthesis.

Calyx The outermost whorl of a flower, formed by the sepals.

CAM Crassulacean acid metabolism. A variation of the photosynthetic reactions in plants, biochemically identical to C_4 synthesis except that all reactions occur in the same cell and are separated by time. Because CAM plants open their stomates at night, they have a competitive advantage in hot, dry climates.

Cambium A ring or cluster of meristematic cells that increase the width of stems and roots when they divide to produce secondary tissues.

Camouflage Adaptations of color, shape and behavior that make an organism more difficult to detect.

Cancer A disease resulting from uncontrolled cell divisions.

Capillaries The tiniest blood vessels consisting of a single layer of flattened cells.

Capillary Action Tendency of water to be pulled into a small-diameter tube.

Carbohydrates A group of compounds that includes simple sugars and all larger molecules constructed of sugar subunits, e.g. polysaccharides.

Carbon Cycle The cycling of carbon in different chemical forms, from the environment to organisms and back to the environment.

Carbon Dioxide Fixation In photosynthesis, the combination of CO_2 with carbon-accepting molecules to form organic compounds.

Carcinogen A cancer-causing agent.

Cardiac Muscle One of the three types of muscle tissue; it forms the muscle of the heart.

Cardiovascular System The organ system consisting of the heart and the vessels through which blood flows.

Carnivore An animal that feeds exclusively on other animals.

Carotenoid A red, yellow, or orange plant pigment that absorbs light in 400-500 nm wavelengths.

Carpels Central whorl of a flower containing the female reproductive organs. Each separate carpel, or each unit of fused carpels, is called a pistil.

Carrier Proteins Proteins within the plasma membrane that bind specific substances and facilitate their movement across the membrane.

Carrying Capacity The size of a population that can be supported indefinitely in a given environment.

Cartilage A firm but flexible connective tissue. In the human, most cartilage originally present in the embryo is transformed into bones.

Casparian Strip The band of waxy suberin that surrounds each endodermal cell of a plant's root tissue.

Catabolism Metabolic pathways that degrade complex compounds into simpler molecules, usually with the release of the chemical energy that held the atoms of the larger molecule together.

Catalyst A chemical substance that accelerates a reaction or causes a reaction to occur but remains unchanged by the reaction. Enzymes are biological catalysts.

Cation A positively charged ion.

Cecum A closed-ended sac extending from the intestine in grazing animals lacking a rumen (e.g., horses) that enables them to digest cellulose.

Cell The basic structural unit of all organisms.

Cell Body Region of a neuron that contains most of the cytoplasm, the nucleus, and other organelles. It relays impulses from the dendrites to the axon.

Cell Cycle Complete sequence of stages from one cell division to the next. The stages are denoted G_1, S, G_2, and M phase.

Cell Differentiation The process by which the internal contents of a cell become assembled into a structure that allows the cell to carry out a specific set of activities, such as secretion of enzymes or contraction.

Cell Division The process by which one cell divides into two.

Cell Fusion Technique whereby cells are caused to fuse with one another producing a large cell with a common cytoplasm and plasma membrane.

Cell Plate In plants, the cell wall material deposited midway between the daughter cells during cytokinesis. Plate material is deposited by small Golgi vesicles.

Cell Sap Solution that fills a plant vacuole. In addition to water, it may contain pigments, salts, and even toxic chemicals.

Cell Theory The fundamental theory of biology that states: 1) all organisms are composed of one or more cells, 2) the cell is the basic organizational unit of life, 3) all cells arise from pre-existing cells.

Cellular Respiration (See **Aerobic respiration**)

Cellulose The structural polysaccharide comprising the bulk of the plant cell wall. It is the most abundant polysaccharide in nature.

Cell Wall Rigid outer-casing of cells in plants and other organisms which gives support, slows dehydration, and prevents a cell from bursting when internal pressure builds due to an influx of water.

Central Nervous System In vertebrates, the brain and spinal cord.

Centriole A pinwheel-shaped structure at each pole of a dividing animal cell.

Centromere Indented region of a mitotic chromosome containing the kinetochore.

Cephalization The clustering of neural tissues and sense organs at the anterior (leading) end of the animal.

Cerebellum A bulbous portion of the vertebrate brain involved in motor coordination. Its prominence varies greatly among different vertebrates .

Cerebral Cortex The outer, highly convoluted layer of the cerebrum. In the human, this is the center of higher brain functions, such as speech and reasoning.

Cerebrospinal Fluid Fluid present within the ventricles of the brain, central canal of the spinal cord, and which surrounds and cushions the central nervous system.

Cerebrum The most dominant part of the human forebrain, composed of two cerebral hemispheres, generally associated with higher brain functions.

Cervix The lower tip of the uterus.

Chapparal A type of shrubland in California, characterized by drought-tolerant and fire-adapted plants.

Character Displacement Divergence of a physical trait in closely related species in response to competition.

Chemical Bonds Linkage between atoms as a result of electrons being shared or donated.

Chemical Evolution Spontaneous synthesis of increasingly complex organic compounds from simpler molecules.

Chemical Reaction Interaction between chemical reactants.

Chemiosmosis The process by which a pH gradient drives the formation of ATP.

Chemoreceptors Sensory receptors that respond to the presence of specific chemicals.

Chemosynthesis An energy conversion process in which inorganic substances (H, N, Fe, or S) provide energized electrons and hydrogen for carbohydrate formation

Chiasmata Cross-shaped regions within a tetrad, occurring at points of crossing over or genetic exchange.

Chitin Structural polysaccharide that forms the hard, strong external skeleton of many arthropods and the cell walls of fungi.

Chlamydia Obligate intracellular parasitic bacteria that lack a functional ATP-generating system.

Chlorophyll Pigments Major light-absorbing pigments of photosynthesis.

Chlorophyta Green algae, the largest group of algae; members of this group were very likely the ancestors of the modern plant kingdom.

Chloroplasts An organelle containing chlorophyll found in plant cells in which photosynthesis occurs.

Cholecystokinin (CCK) Hormone secreted by endocrine cells in the wall of the small intestine that stimulates the release of digestive products by the pancreas.

Chondrocytes Living cartilage cells embedded within the protein-polysaccharide matrix they manufacture.

Chordamesoderm In vertebrates, the block of mesoderm that underlies the dorsal ectoderm of the gastrula, induces the formation of the nervous system, and gives rise to the notochord.

Chordate A member of the phylum Chordata possessing a skeletal rod of tissue called a notochord, a dorsal hollow nerve cord, gill slits, and a post-anal tail at some stage of its development.

Chorion The outermost of the four extraembryonic membranes. In placental mammals, it forms the embryonic portion of the placenta.

Chorionic Villus Sampling (CVS) A procedure for obtaining fetal cells by removing a small sample of tissue from the developing placenta of a pregnant woman.

Chromatid Each of the two identical subunits of a replicated chromosome.

Chromatin DNA-protein fibers which, during prophase, condense to form the visible chromosomes.

Chromatography A technique for separating different molecules on the basis of their solubility in a particular solvent. The mixture of substances is spotted on a piece of paper or other material, one end of which is then placed in the solvent. As the solvent moves up the paper by capillary action, each substance in the mixture is carried a particular distance depending on its solubility in the moving solvent.

Chromosomes Dark-staining structures in which the organism's genetic material (DNA) is organized. Each species has a characteristic number of chromosomes.

Chromosome Aberrations Alteration in the structure of a chromosome from the normal state. Includes chromosome deletions, duplications, inversions, and translocations.

Chromosome Puff A site on an insect polytene chromosome where the DNA has unraveled and is being transcribed.

Cilia Short, hairlike structures projecting from the surfaces of some cells. They beat in coordinated ways, are usually found in large numbers, and are densely packed.

Ciliated Mucosa Layer of ciliated epithelial cells lining the respiratory tract. The beating of cilia propels an associated mucous layer and trapped foreign particles.

Circadian Rhythm Behavioral patterns that cycle during approximately 24 hour intervals.

Circulatory System The system that circulates internal fluids throughout an organism to deliver oxygen and nutrients to cells and to remove metabolic wastes.

Class (Taxonomic) A level of the taxonomic hierarchy that groups together members of related orders.

Classical Conditioning A form of learning in which an animal develops a response to a new stimulus by repeatedly associating the new stimulus with a stimulus that normally elicits the response.

Cleavage Successive mitotic divisions in the early embryo. There is no cell growth between divisions.

Cleavage Furow Constriction around the middle of a dividing cell caused by constriction of microfilaments.

Climate The general pattern of average weather conditions over a long period of time in a specific region, including precipitation, temperature, solar radiation, and humidity.

Climax Final or stable community of successional stages, that is more or less in equilibrium with existing environmental conditions for a long period of time.

Climax Community Community that remains essentially the same over long periods of time; final stage of ecological succession.

Clitoris A protrusion at the point where the labia minora merge; rich in sensory neurons and erectile tissue.

Clonal Selection Mechanism The mechanism by which the body can synthesize antibodies specific for the foreign substance (antigen) that stimulated their production.

Clones Offspring identical to the parent, produced by asexual processes.

Closed Circulatory System Circulatory system in which blood travels throughout the body in a continuous network of closed tubes. (Compare with open circulatory system).

Clumped Pattern Distribution of individuals of a population into groups, such as flocks or herds.

Cnidaria A phylum that consists of radial symmetrical animals that have two cell layers. There are three classes: 1) Hydrozoa (hydra), 2) Scyphozoa (jellyfish), 3) Anthozoa (sea anemones, corals). Most are marine forms that live in warm, shallow water.

Cnidocytes Specialized stinging cells found in the members of the phylum Cnidaria.

Coastal Waters Relatively warm, nutrient-rich shallow water extending from the high-tide mark on land to the sloping continental shelf. The greatest concentration of marine life are found in coastal waters.

Coated Pits Indentations at the surfaces of cells that contain a layer of bristly protein (called clathrin) on the inner surface of the plasma membrane. Coated pits are sites where cell receptors become clustered.

Cochlea Organ within the inner ear of mammals involved in sound reception.

Codominance The simultaneous expression of both alleles at a genetic locus in a heterozygous individual.

Codon Linear array of three nucleotides in mRNA. Each triplet specifies a particular amino acid during the process of translation.

Coelomates Animals in which the body cavity is completely lined by mesodermally-derived tissues.

Coenzyme An organic cofactor, typically a vitamin or a substance derived from a vitamin.

Coevolution Evolutionary changes that result from reciprocal interactions between two species, e.g., flowering plants and their insect pollinators.

Cofactor A non-protein component that is linked covalently or noncovalently to an enzyme and is required by the enzyme to catalyze the reaction. Cofactors may be organic molecules (coenzymes) or metals.

Cohesion The tendency of different parts of a substance to hold together because of forces acting between its molecules.

Coitus Sexual union in mammals.

Coleoptile Sheath surrounding the tip of the monocot seedling, protecting the young stem and leaves as they emerge from the soil.

Collagen The most abundant protein in the human body. It is present primarily in the extracellular space of connective tissues such as bone, cartilage, and tendons.

Collenchyma Living plant cells with irregularly thickened primary cell walls. A supportive cell type often found inside the epidermis of stems with primary growth. Angular, lacunar and laminar are different types of collenchyma cells.

Commensalism A form of symbiosis in which one organism benefits from the union while the other member neither gains nor loses.

Community The populations of all species living in a given area.

Compact Bone The solid, hard outer regions of a bone surrounding the honey-combed mass of spongy bone.

Companion Cell Specialized parenchyma cell associated with a sieve-tube member in phloem.

Competition Interaction among organisms that require the same resource. It is of two types: 1) intraspecific (between members of the same species); 2) interspecific (between members of different species).

Competitive Exclusion Principle (Gause's Principle) Competition in which a winner species captures a greater share of resources, increasing its survival and reproductive capacity. The other species is gradually displaced.

Competitive Inhibition Prevention of normal binding of a substrate to its enzyme by the presence of an inhibitory compound that competes with the substrate for the active site on the enzyme.

Complement Blood proteins with which some antibodies combine following attachment to antigen (the surface of microorganisms). The bound complement punches the tiny holes in the plasma membrane of the foreign cell, causing it to burst.

Complementarity The relationship between the two strands of a DNA molecule determined by the base pairing of nucleotides on the two strands of the helix. A nucleotide with guanine on one strand always pairs with a nucleotide having cytosine on the other strand; similarly with adenine and thymine.

Complete Digestive Systems Systems that have a digestive tract with openings at both ends—a mouth for entry and an anus for exit.

Complete Flower A flower containing all four whorls of modified leaves—sepels, petals, stamen, and carpels.

Compound Chemical substances composed of atoms of more than one element.

Compound Leaf A leaf that is divided into leaflets, with two or more leaflets attached to the petiole.

Concentration Gradient Regions in a system of differing concentration representing potential energy, such as exist in a cell and its environment, that cause molecules to move from areas of higher concentration to lower concentration.

Conditioned Reflex A reflex ("automatic") response to a stimulus that would not normally have elicited the response. Conditioned reflexes develop by repeated association of a new stimulus with an old stimulus that normally elicits the response.

Conformation The three-dimensional shape of a molecule as determined by the spatial arrangement of its atoms.

Conformational Change Change in molecular shape (as occurs, for example, in an enzyme as it catalyzes a reaction, or a myosin molecule during contraction).

Conjugation A method of reproduction in single-celled organisms in which two cells link and exchange nuclear material.

Connective Tissues Tissues that protect, support, and hold together the internal organs and other structures of animals. Includes bone, cartilage, tendons, and other tissues, all of which have large amounts of extracellular material.

Consumers Heterotrophs in a biotic environment that feed on other organisms or organic waste.

Continental Drift The continuous shifting of the earth's land masses explained by the theory of plate tectonics.

Continuous Variation An inheritance pattern in which there is graded change between the two extremes in a phenotype (compare with discontinuous variation).

Contraception The prevention of pregnancy.

Contractile Proteins Actin and myosin, the protein filaments that comprise the bulk of the muscle mass. During contraction of skeletal muscle, these filaments form a temporary association and slide past each other, generating the contractile force.

Control (Experimental) A duplicate of the experiment identical in every way except for the one variable being tested. Use of a control is necessary to demonstrate cause and effect.

Convergent Evolution The evolution of similar structures in distantly related organisms in response to similar environments.

Cork Cambium In stems and roots of perennials, a secondary meristem that produces the outer protective layer of the bark.

Coronary Arteries Large arteries that branch immediately from the aorta, providing oxygen-rich blood to the cardiac muscle.

Corpus Callosum A thick cable composed of hundreds of millions of neurons that connect the right and left cerebral hemispheres of the mammalian brain.

Corpus Luteum In the mammalian ovary, the structure that develops from the follicle after release of the egg. It secretes hormones that prepare the uterine endometrium to receive the developing embryo.

Cortex In the stem or root of plants, the region between the epidermis and the vascular tissues. Composed of ground tissue. In animals, the outermost portion of some organs.

Cotyledon The seed leaf of a dicot embryo containing stored nutrients required for the germinated seed to grow and develop, or a food digesting seed leaf in a monocot embryo.

Countercurrent Flow Mechanism for increasing the exchange of substances or heat from one stream of fluid to another by having the two fluids flow in opposite directions.

Covalent Bonds Linkage between two atoms which share the same electrons in their outermost shells.

Cranial Nerves Paired nerves which emerge from the central stalk of the vertebrate brain and innervate the body. Humans have 12 pairs of cranial nerves.

Cranium The bony casing which surrounds and protects the vertebrate brain.

Cristae The convolutions of the inner membrane of the mitochondrion. Embedded within them are the components of the electron transport system and proton channels for chemiosmosis.

Crossing Over During synapsis, the process by which homologues exchange segments with each other.

Cryptic Coloration A form of camouflage wherein an organism's color or patterning helps it resemble its background.

Cutaneous Respiration The uptake of oxygen across virtually the entire outer body surface.

Cuticle 1) Waxy layer covering the outer cell walls of plant epidermal cells. It retards water vapor loss and helps prevent dehydration. (18) 2) Outer protective, nonliving covering of some animals, such as the exoskeleton of anthropods.

Cyanobacteria A type of prokaryote capable of photosynthesis using water as a source of electrons. Cyanobacteria were responsible for initially creating an O_2-containing atmosphere on earth.

Cyclic AMP (Cyclic adenosine monophosphate) A ring-shaped molecular version of an ATP minus two phosphates. A regulatory molecule formed by the enzyme adenylate cyclase which converts ATP to cAMP. A second messenger.

Cyclic Pathways Metabolic pathways in which the intermediates of the reaction are regenerated while assisting the conversion of the substrate to product.

Cyclic Photophosphorylation A pathway that produces ATP, but not NADPH, in the light reactions of photosynthesis. Energized electrons are shuttled from a reaction center, along a molecular pathway, back to the original reaction center, generating ATP en route.

Cysts Protective, dormant structure formed by some protozoa.

Cytochrome Oxidase A complex of proteins that serves as the final electron carrier in the mitochondrial electron transport system, transferring its electrons to O_2 to form water.

Cytokinesis Final event in eukaryotic cell division in which the cell's cytoplasm and the new nuclei are partitioned into separate daughter cells.

Cytokinins Growth-producing plant hormones which stimulate rapid cell division.

Cytoplasm General term that includes all parts of the cell, except the plasma membrane and the nucleus.

Cytoskeleton Interconnecting network of microfilaments, microtubules, and intermediate filaments that serves as a cell scaffold and provides the machinery for intracellular movements and cell motility.

Cytotoxic (Killer) T Cells A class of T cells capable of recognizing and destroying foreign or infected cells.

Day Neutral Plants Plants that flower at any time of the year, independent of the relative lengths of daylight and darkness.

Deciduous Trees or shrubs that shed their leaves in a particular season, usually autumn, before entering a period of dormancy.

Deciduous Forest Forests characterized by trees that drop their leaves during unfavorable conditions, and leaf out during warm, wet seasons. Less dense than tropical rain forests.

Decomposers (Saprophytes) Organisms that obtain nutrients by breaking down organic compounds in wastes and dead organisms. Includes fungi, bacteria, and some insects.

Deletion Loss of a portion of a chromosome, following breakage of DNA.

Denaturation Change in the normal folding of a protein as a result of heat, acidity, or alkalinity. Such changes result in a loss of enzyme functioning.

Dendrites Cytoplasmic extensions of the cell body of a neuron. They carry impulses from the area of stimulation to the cell body.

Denitrification The conversion by denitrifying bacteria of nitrites and nitrates into nitrogen gas.

Denitrifying Bacteria Bacteria which take soil nitrogen, usable to plants, and convert it to unusable nitrogen gas.

Density-Dependent Factors Factors that control population growth which are influenced by population size.

Density-Independent Factors Factors that control population growth which are not affected by population size.

Deoxyribonucleic Acid (DNA) Double-stranded polynucleotide comprised of deoxyribose (a sugar), phosphate, and four bases (adenine, guanine, cytosine, and thymine). Encoded in the sequence of nucleotides are the instructions for making proteins. DNA is the genetic material in all organisms except certain viruses.

Depolarization A decrease in the potential difference (voltage) across the plasma membrane of a cell typically due to an increase in the movement of sodium ions into the cell. Acts to excite a target cell.

Dermal Bone Bones of vertebrates that form within the dermal layer of the skin, such as the scales of fishes and certain bones of the skull.

Dermal Tissue System In plants, the epidermis in primary growth, or the periderm in secondary growth.

Dermis In animals, layer of cells below the epidermis in which connective tissue predominates. Embedded within it are vessels, various glands, smooth muscle, nerves, and follicles.

Desert Biome characterized by intense solar radiation, very little rainfall, and high winds.

Detrivore Organism that feeds on detritus, dead organisms or their parts, and living organisms' waste.

Deuterostome One path of development exhibited by coelomate animals (e.g., echinoderms and chordates).

Diabetes Mellitus A disease caused by a deficiency of insulin or its receptor, preventing glucose from being absorbed by the cells.

Diaphragm A sheet of muscle that separates the thoracic cavity from the abdominal wall.

Diastolic Pressure The second number of a blood pressure reading; the lowest pressure in the arteries just prior to the next heart contraction.

Diatoms are golden-brown algae that are distinguished most dramatically by their intricate silica shells.

Dicotyledonae (Dicots) One of the two classes of flowering plants, characterized by having seeds with two cotyledons, flower parts in 4s or 5s, net-veined leaves, one main root, and vascular bundles in a circular array within the stem. (Compare with Monocotylenodonae).

Diffusion Tendency of molecules to move from a region of higher concentration to a region of lower concentration, until they are uniformly dispersed.

Digestion The process by which food particles are disassembled into molecules small enough to be absorbed into the organism's cells and tissues.

Digestive System System of specialized organs that ingests food, converts nutrients to a form that can be distributed throughout the animal's body, and eliminates undigested residues.

Digestive Tract Tubelike channel through which food matter passes from its point of ingestion at the mouth to the elimination of indigestible residues from the anus.

Dihybrid Cross A mating between two individuals that differ in two genetically-determined traits.

Dimorphism Presence of two forms of a trait within a population, resulting from diversifying selection.

Dinoflagellates Single-celled photosynthesizers that have two flagella. They are members of the pyrophyta, phosphorescent algae that sometimes cause red tide, often synthesizing a neurotoxin that accumulates in plankton eaters, causing paralytic shellfish poisoning in people who eat the shellfish.

Dioecious Plants that produce either male or female reproductive structures but never both.

Diploid Having two sets of homologous chromosomes. Often written 2N.

Directional Selection The steady shift of phenotypes toward one extreme.

Discontinuous Variation An inheritance pattern in which the phenomenon of all possible phenotypes fall into distinct categories. (Compare with continuous variation).

Displays The signals that form the language by which animals communicate. These signals are species specific and stereotyped and may be visual, auditory, chemical, or tactile.

Disruptive Coloration Coloration that disguises the shape of an organism by breaking up its outline.

Disruptive Selection The steady shift toward more than one extreme phenotype due to the elimination of intermediate phenotypes as has occurred among African swallowtail butterflies whose members resemble more than one species of distasteful butterfly.

Divergent Evolution The emergence of new species as branches from a single ancestral lineage.

Diversifying Selection The increasing frequency of extreme phenotypes because individuals with average phenotypes die off.

Diving Reflex Physiological response that alters the flow of blood in the body of diving mammals that allows the animal to maintain high levels of activity without having to breathe.

Division (or Phylum) A level of the taxonomic hierarchy that groups together members or related classes.

DNA (see **Deoxyribonucleic Acid**)

DNA Cloning The amplification of a particular DNA by use of a growing population of bacteria. The DNA is initially taken up by a bacterial cell—usually as a plasmid—and then replicated along with the bacteria's own DNA.

DNA Fingerprint The pattern of DNA fragments produced after treating a sample of DNA with a particular restriction enzyme and separating the fragments by gel electrophoresis. Since different members of a population have DNA with a different nucleotide sequence, the pattern of DNA fragments produced by this method can be used to identify a particular individual.

DNA Ligase The enzyme that covalently joins DNA fragments into a continuous DNA strand. The enzyme is used in a cell during replication to seal newly-synthesized fragments and by biotechnologists to form recombinant DNA molecules from separate fragments.

DNA Polymerase Enzyme responsible for replication of DNA. It assembles free nucleotides, aligning them with the complementary ones in the unpaired region of a single strand of DNA template.

Dominant The form of an allele that masks the presence of other alleles for the same trait.

Dormancy A resting period, such as seed dormancy in plants or hibernation in animals, in which organisms maintain reduced metabolic rates.

Dorsal In anatomy, the back of an animal.

Double Blind Test A clinical trial of a drug in which neither the human subjects or the researchers know who is receiving the drug or placebo.

Down Syndrome Genetic disorder in humans characterized by distinct facial appearance and mental retardation, resulting from an extra copy of chromosome number 21 (trisomy 21) in each cell.

Duodenum First part of the human small intestine in which most digestion of food occurs.

Duplication The repetition of a segment of a chromosome.

◄ E ►

Ecdysis Molting process by which an arthropod periodically discards its exoskeleton and replaces it with a larger version. The process is controlled by the hormone ecydysone.

Ecdysone An insect steroid hormone that triggers molting and metamorphosis.

Echinodermata A phylum composed of animals having an internal skeleton made of many small calcium carbonate plates which have jutting spines. Includes sea stars, sea urchins, etc.

Echolocation The use of reflected sound waves to help guide an animal through its environment and/or locate objects.

Ecological Equivalent Organisms that occupy similar ecological niches in different regions or ecosystems of the world.

Ecological Niche The habitat, functional role(s), requirements for environmental resources and tolerance ranges for each abiotic condition in relation to an organism.

Ecological Pyramid Illustration showing the energy content, numbers of organisms, or biomass at each trophic level.

Ecology The branch of biology that studies interactions among organisms as well as the interactions of organisms and their physical environment.

Ecosystem Unit comprised of organisms interacting among themselves and with their physical environment.

Ecotypes Populations of a single species with different, genetically fixed tolerance ranges.

Ectoderm In animals, the outer germ cell layer of the gastrula. It gives rise to the nervous system and integument.

Ectotherms Animals that lack an internal mechanism for regulating body temperature. "Cold-blooded" animals.

Edema Swelling of a tissue as the result of an accumulation of fluid that has moved out of the blood vessels.

Effectors Muscle fibers and glands that are activated by neural stimulation.

Efferent (Motor) Nerves The nerves that carry messages from the central nervous system to the effectors, the muscles, and glands. They are divided into two systems: somatic and autonomic.

Egg Female gamete, also called an ovum. A fertilized egg is the product of the union of female and male gametes (egg and sperm cells).

Electrocardiogram (EKG) Recording of the electrical activity of the heart, which is used to diagnose various types of heart problems.

Electron Acceptor Substances that are capable of accepting electrons transferred from an electron donor. For example, molecular oxygen (O_2) is the terminal electron acceptor during respiration. Electron acceptors also receive electrons from chlorophyll during photosynthesis. Electron acceptors may act as part of an electron transport system by transferring the electrons they receive to another substance.

Electron Carrier Substances (such as NAD^+ and FAD) that transport electrons from one step of a metabolic pathway to the next or from metabolic reactions to biosynthetic reactions.

Electrons Negatively charged particles that orbit the atomic nucleus.

Electron Transport System Highly organized assembly of cytochromes and other proteins which transfer electrons. During transport, which occurs within the inner membranes of mitochondria and chloroplasts, the energy extracted from the electrons is used to make ATP.

Electrophoresis A technique for separating different molecules on the basis of their size and/or electric charge. There are various ways the technique is used. In gel electrophoresis, proteins or DNA fragments are driven through a porous gel by their charge, but become separated according to size; the larger the molecule, the slower it can work its way through the pores in the gel, and the less distance it travels along the gel.

Element Substance composed of only one type of atom.

Embryo An organism in the early stages of development, beginning with the first division of the zygote.

Embryo Sac The fully developed female gametophyte within the ovule of the flower.

Emigration Individuals permanently leaving an area or population.

Endergonic Reactions Chemical reactions that require energy input from another source in order to occur.

Endocrine Glands Ductless glands, which secrete hormones directly into surrounding tissue fluids and blood vessels for distribution to the rest of the body by the circulatory system.

Endocytosis A type of active transport that imports particles or small cells into a cell. There are two types of endocytic processes: phagocytosis, where large particles are ingested by the cell, and pinocytosis, where small droplets are taken in.

Endoderm In animals, the inner germ cell layer of the gastrula. It gives rise to the digestive tract and associated organs and to the lungs.

Endodermis The innermost cylindrical layer of cortex surrounding the vascular tissues of the root. The closely pressed cells of the endodermis have a waxy band, forming a waterproof layer, the Casparian strip.

Endogenous Plant responses that are controlled internally, such as biological clocks controlling flower opening.

Endometrium The inner epithelial layer of the uterus that changes markedly with the uterine (menstrual) cycle in preparation for implantation of an embryo.

Endoplasmic Reticulum (ER) An elaborate system of folded, stacked and tubular membranes contained in the cytoplasm of eukaryotic cells.

Endorphins (Endogenous Morphinelike Substances) A class of peptides released from nerve cells of the limbic system of the brain that can block perceptions of pain and produce a feeling of euphoria.

Endoskeleton The internal support structure found in all vertebrates and a few invertebrates (sponges and sea stars).

Endosperm Nutritive tissue in plant embryos and seeds.

Endosperm Mother Cell A binucleate cell in the embryo sac of the female gametophyte, occurring in the ovule of the ovary in angiosperms. Each nucleus is haploid; after fertilization, nutritive endosperm develops.

Endosymbiosis Theory A theory to explain the development of complex eukaryotic cells by proposing that some organelles once were free-living prokaryotic cells that then moved into another larger such cell, forming a beneficial union with it.

Endotherms Animals that utilize metabolically produced heat to maintain a constant, elevated body temperature. "Warm-blooded" animals.

End Product The last product in a metabolic pathway. Typically a substance, such as an amino acid or a nucleotide, that will be used as a monomer in the formation of macromolecules.

Energy The ability to do work.

Entropy Energy that is not available for doing work; measure of disorganization or randomness.

Environmental Resistance The factors that eventually limit the size of a population.

Enzyme Biological catalyst; a protein molecule that accelerates the rate of a chemical reaction.

Eosiniphil A type of phagocytic white blood cell.

Epicotyl The portion of the embryo of a dicot plant above the cotyledons. The epicotyl gives rise to the shoot.

Epidermis In vertebrates, the outer layer of the skin, containing superficial layers of dead cells produced by the underlying living epithelial cells. In plants, the outer layer of cells covering leaves, primary stem, and primary root.

Epididymis Mass of convoluted tubules attached to each testis in mammals. After leaving the testis, sperm enter the tubules where they finish maturing and acquire motility.

Epiglottis A flap of tissue that covers the glottis during swallowing to prevent food and liquids from entering the lower respiratory tract.

Epinephrine (Adrenalin) Substance that serves both as an excitatory neurotransmitter released by certain neurons of the CNS and as a hormone released by the adrenal medulla that increases the body's ability to combat a stressful situation.

Epipelagic Zone The lighted upper ocean zone, where photosynthesis occurs; large populations of phytoplankton occur in this zone.

Epiphyseal Plates The action centers for ossification (bone formation).

Epistasis A type of gene interaction in which a particular gene blocks the expression of another gene at another locus.

Epithelial Tissue Continuous sheets of tightly packed cells that cover the body and line its tracts and chambers. Epithelium is a fundamental tissue type in animals.

Erythrocytes Red blood cells.

Erythropoietin A hormone secreted by the kidney which stimulates the formation of erythrocytes by the bone marrow.

Essential Amino Acids Eight amino acids that must be acquired from dietary protein. If even one is missing from the human diet, the synthesis of proteins is prevented.

Essential Fatty Acids Linolenic and linoleic acids, which are required for phospholipid construction and must be acquired from a dietary source.

Essential Nutrients The 16 minerals essential for plant growth, divided into two groups: macronutrients, which are required in large quantities, and micronutrients, which are needed in small amounts.

Estrogen A female sex hormone secreted by the ovaries when stimulated by pituitary gonadotrophins.

Estuaries Areas found where rivers and streams empty into oceans, mixing fresh water with salt water.

Ethology The study of animal behavior.

Ethylene Gas A plant hormone that stimulates fruit ripening.

Etiolation The condition of rapid shoot elongation, small underdeveloped leaves, bent shoot-hook, and lack of chlorophyll, all due to lack of light.

Eubacteria Typical procaryotic bacteria with peptidoglycan in their cell walls. The majority of monerans are eubacteria.

Eukaryotic Referring to organisms whose cellular anatomy includes a true nucleus with a nuclear envelope, as well as other membrane-bound organelles.

Eusocial Species Social species that have sterile workers, cooperative care of the young, and an overlap of generations so that the colony labor is a family affair.

Eutrophication The natural aging process of lakes and ponds, whereby they become marshes and, eventually, terrestrial environments.

Evolution A process whereby the characteristics of a species change over time, eventually leading to the formation of new species that go about life in new ways.

Evolutionarily Stable Strategy (ESS) A behavioral strategy or course of action that depends on what other members of the population are doing. By definition, an ESS cannot be replaced by any other strategy when most of the members of the population have adopted it.

Excitatory Neurons Neurons that stimulate their target cells into activity.

Excretion Removal of metabolic wastes from an organism.

Excretory System The organ system that eliminates metabolic wastes from the body.

Exergonic Reactions Chemical reactions that occur spontaneously with the release of energy.

Exocrine Glands Glands which secrete their products through ducts directly to their sites of action, e.g., tear glands.

Exocytosis A form of active transport used by cells to move molecules, particles, or other cells contained in vesicles across the plasma membrane to the cell's environment.

Exogenous Plant responses that are controlled externally, or by environmental conditions.

Exons Structural gene segments that are transcribed and whose genetic information is subsequently translated into protein.

Exoskeletons Hard external coverings found in some animals (e.g., lobsters, insects) for protection, support, or both. Such organisms grow by the process of molting.

Exploitative Competition A competition in which one species manages to get more of a resource, thereby reducing supplies for a competitor.

Exponential Growth An increase by a fixed percentage in a given time period; such as population growth per year.

Extensor Muscle A muscle which, when contracted, causes a part of the body to straighten at a joint.

External Fertilization Fertilization of an egg outside the body of the female parent.

Extinction The loss of a species.

Extracellular Digestion Digestion occurring outside the cell; occurs in bacteria, fungi, and multicellular animals.

Extracellular Matrix Layer of extracellular material residing just outside a cell.

◀ F ▶

F_1 First filial generation. The first generation of offspring in a genetic cross.

F_2 Second filial generation. The offspring of an F_1 cross.

Facilitated Diffusion The transport of molecules into cells with the aid of "carrier" proteins embedded in the plasma membrane. This carrier-assisted transport does not require the expenditure of energy by the cell.

FAD Flavin adenine dinucleotide. A coenzyme that functions as an electron carrier in metabolic reactions. When it is reduced to $FADH_2$, this molecule becomes a cellular energy source.

Family A level of the taxonomic hierarchy that groups together members of related genera.

Fast-Twitch Fibers Skeletal muscle fibers that depend on anaerobic metabolism to produce ATP rapidly, but only for short periods of time before the onset of fatigue. Fast-twitch fibers generate greater forces for shorter periods than slow-twitch fibers.

Fat A triglyceride consisting of three fatty acids joined to a glycerol.

Fatty Acid A long unbranched hydrocarbon chain with a carboxyl group at one end. Fatty acids lacking a double bond are said to be saturated.

Fauna The animals in a particular region.

Feedback Inhibition (Negative Feedback) A mechanism for regulating enzyme activity by temporarily inactivating a key enzyme in a biosynthetic pathway when the concentration of the end product is elevated.

Fermentation The direct donation of the electrons of NADH to an organic compound without their passing through an electron transport system.

Fertility Rate In humans, the average number of children born to each woman between 15 and 44 years of age.

Fertilization The process in which two haploid nuclei fuse to form a zygote.

Fetus The term used for the human embryo during the last seven months in the uterus. During the fetal stage, organ refinement accompanies overall growth.

Fibrinogen A rod-shaped plasma protein that, converted to fibrin, generates a tangled net of fibers that binds a wound and stops blood loss until new cells replace the damaged tissue.

Fibroblasts Cells found in connective tissues that secrete the extracellular materials of the connective tissue matrix. These cells are easily isolated from connective tissues and are widely used in cell culture.

Fibrous Root System Many approximately equal-sized roots; monocots are characterized by a fibrous root system. Also called diffuse root system.

Filament The stalk of a stamen of angiosperms, with the anther at its tip. Also, the threadlike chain of cells in some algae and fungi.

Filamentous Fungus Multicellular members of the fungus kingdom comprised mostly of living threads (hyphae) that grow by division of cells at their tips (see molds).

Filter Feeders Aquatic animals that feed by straining small food particles from the surrounding water.

Fitness The relative degree to which an individual in a population is likely to survive to reproductive age and to reproduce.

Fixed Action Patterns Motor responses that may be triggered by some environmental stimulus, but once started can continue to completion without external stimuli.

Flagella Cellular extensions that are longer than cilia but fewer in number. Their undulations propel cells like sperm and many protozoans, through their aqueous environment.

Flexor Muscle A muscle which, when contracted, causes a part of the body to bend at a joint.

Flora The plants in a particular region.

Florigen Proposed A chemical hormone that is produced in the leaves and stimulates flowering.

Fluid Mosaic Model The model proposes that the phospholipid bilayer has a viscosity similar to that of light household oil and that globular proteins float like icebergs within this bilayer. The now favored explanation for the architecture of the plasma membrane.

Follicle (Ovarian) A chamber of cells housing the developing oocytes.

Food Chain Transfers of food energy from organism to organism, in a linear fashion.

Food Web The map of all interconnections between food chains for an ecosystem.

Forest Biomes Broad geographic regions, each with characteristic tree vegetation: 1) tropical rain forests (lush forests in a broad band around the equator), 2) deciduous forests (trees and shrubs drop their leaves during unfavorable seasons), 3) coniferous forest (evergreen conifers).

Fossil Record An entire collection of remains from which paleontologists attempt to reconstruct the phylogeny, anatomy, and ecology of the preserved organisms.

Fossils The preserved remains of organisms from a former geologic age.

Fossorial Living underground.

Founder Effect The potentially dramatic difference in allele frequency of a small founding population as compared to the original population.

Founder Population The individuals, usually few, that colonize a new habitat.

Frameshift Mutation The insertion or deletion of nucleotides in a gene that throws off the reading frame.

Free Radical Atom or molecule containing an unpaired electron, which makes it highly reactive.

Freeze-Fracture Technique in which cells are frozen into a block which is then struck with a knife blade that fractures the block in two. Fracture planes tend to expose the center of membranes for EM examination.

Fronds The large leaf-like structures of ferns. Unlike true leaves, fronds have an apical meristem and clusters of sporangia called sori.

Fruit A mature plant ovary (flower) containing seeds with plant embryos. Fruits protect seeds and aid in their dispersal.

Fruiting Body A spore-producing structure that extends upward in an elevated position from the main mass of a mold or slime mold.

FSH Follicle stimulating hormone. A hormone secreted by the anterior pituitary that prepares a female for ovulation by stimulating the primary follicle to ripen or stimulates spermatogenesis in males.

Functional Groups Accessory chemical entities (e.g., —OH, —NH_2, —CH_3), which help determine the identity and chemical properties of a compound.

Fundamental Niche The potential ecological niche of a species, including all factors affecting that species. The fundamental niche is usually never fully utilized.

Fungus Yeast, mold, or large filamentous mass forming macroscopic fruiting bodies, such as mushrooms. All fungi are eukaryotic nonphotosynthetic heterotrophics with cell walls.

◄ G ►

G_1 Stage The first of three consecutive stages of interphase. During G_1, cell growth and normal functions occur. The duration of this stage is most variable.

G_2 Stage The final stage of interphase in which the final preparations for mitosis occur.

Gallbladder A small saclike structure that stores bile salts produced by the liver.

Gamete A haploid reproductive cell—either a sperm or an egg.

Gas Exchange Surface Surface through which gases must pass in order to enter or leave the body of an animal. It may be the plasma membrane of a protistan or the complex tissues of the gills or the lungs in multicellular animals.

Gastrovascular Cavity In cnidarians and flatworms, the branched cavity with only one opening. It functions in both digestion and transport of nutrients.

Gastrula The embryonic stage formed by the inward migration of cells in the blastula.

Gastrulation The process by which the blastula is converted into a gastrula having three germ layers (ectoderm, mesoderm, and endoderm).

Gated Ion Channels Most passageways through a plasma membrane that allow ions to pass contain "gates" that can occur in either an open or a closed conformation.

Gel Electrophoresis (See **Electrophoresis**)

Gene Pool All the genes in all the individuals of a population.

Gene Regulatory Proteins Proteins that bind to specific sites in the DNA and control the transcription of nearby genes.

Genes Discrete units of inheritance which determine hereditary traits.

Gene Therapy Treatment of a disease by alteration of the person's genotype, or the genotype of particular affected cells.

Genetic Carrier A heterozygous individual who shows no evidence of a genetic disorder but, because they possess a recessive allele for a disorder, can pass the mutant gene on to their offspring.

Genetic Code The correspondence between the various mRNA triplets (codons, e.g., UGC) and the amino acid that the triplet specifies (e.g., cysteine). The genetic code includes 64 possible three-letter words that constitute the genetic language for protein synthesis.

Genetic Drift Random changes in allele frequency that occur by chance alone. Occurs primarily in small populations.

Genetic Engineering The modification of a cell or organism's genetic composition according to human design.

Genetic Equilibrium A state in which allele frequencies in a population remain constant from generation to generation.

Genetic Mapping Determining the locations of specific genes or genetic markers along particular chromosomes. This is typically accomplished using crossover frequencies; the more often alleles of two genes are separated during crossing over, the greater the distance separating the genes.

Genetic Recombination The reshuffling of genes on a chromosome caused by breakage of DNA and its reunion with the DNA of a homologoue.

Genome The information stored in all the DNA of a single set of chromosomes.

Genotype An individual's genetic makeup.

Genus Taxonomic group containing related species.

Geologic Time Scale The division of the earth's 4.5 billion-year history into eras, periods, and epochs based on memorable geologic and biological events.

Germ Cells Cells that are in the process of or have the potential to undergo meiosis and form gametes.

Germination The sprouting of a seed, beginning with the radicle of the embryo breaking through the seed coat.

Germ Layers Collective name for the endoderm, ectoderm, and mesoderm, from which all the structures of the mature animal develop.

Gibberellins More than 50 compounds that promote growth by stimulating both cell elongation and cell division.

Gills Respiratory organs of aquatic animals.

Globin The type of polypeptide chains that make up a hemoglobin molecule.

Glomerular Filtration The process by which fluid is filtered out of the capillaries of the glomerulus into the proximal end of the nephron. Proteins and blood cells remain behind in the bloodstream.

Glomerulus A capillary bundle embedded in the double-membraned Bowman's capsule, through which blood for the kidney first passes.

Glottis Opening leading to the larynx and lower respiratory tract.

Glucagon A hormone secreted by the Islets of Langerhans that promotes glycogen breakdown to glucose.

Glucocorticoids Steroid hormones which regulate sugar and protein metabolism. They are secreted by the adrenal cortex.

Glycogen A highly branched polysaccharide consisting of glucose monomers that serves as a storage of chemical energy in animals.

Glycolysis Cleavage, releasing energy, of the six-carbon glucose molecule into two molecules of pyruvic acid, each containing three carbons.

Glycoproteins Proteins with covalently-attached chains of sugars.

Glycosidic Bond The covalent bond between individual molecules in carbohydrates.

Golgi Complex A system of flattened membranous sacs, which package substances for secretion from the cell.

Gonadotropin-Releasing Hormone (GnRH) Hypothalmic hormone that controls the secretion of the gonadotropins FSH and LH.

Gonadotropins Two anterior pituitary hormones which act on the gonads. Both FSH (follicle-stimulating hormone) and LH (luteinizing hormone) promote gamete development and stimulate the gonads to produce sex hormones.

Gonads Gamete-producing structures in animals: ovaries in females, testes in males.

Grasslands Areas of densely packed grasses and herbaceous plants.

Gravitropisms (Geotropisms) Changes in plant growth caused by gravity. Growth away from gravitational force is called negative gravitropism; growth toward it is positive.

Gray Matter Gray-colored neural tissue in the cerebral cortex of the brain and in the butterfly-shaped interior of the spinal cord. Composed of nonmyelinated cell bodies and dendrites of neurons.

Greenhouse Effect The trapping of heat in the Earth's troposphere, caused by increased levels of carbon dioxide near the Earth's surface; the carbon dioxide is believed to act like glass in a greenhouse, allowing light to reach the Earth, but not allowing heat to escape.

Ground Tissue System All plant tissues except those in the dermal and vascular tissues.

Growth An increase in size, resulting from cell division and/or an increase in the volume of individual cells.

Growth Hormone (GH) Hormone produced by the anterior pituitary; stimulates protein synthesis and bone elongation.

Growth Ring In plants with secondary growth, a ring formed by tracheids and/or vessels with small lumens (late wood) during periods of unfavorable conditions; apparent in cross section.

Guard Cells Specialized epidermal plant cells that flank each stomated pore of a leaf. They regulate the rate of gas diffusion and transpiration.

Guild Group of species with similar ecological niches.

Guttation The forcing of water and mineral completely out to the tips of leaves as a result of positive root pressure.

Gymnosperms The earliest seed plants, bearing naked seeds. Includes the pines, hemlocks, and firs.

◄ H ►

Habitat The place or region where an organism lives.

Habituation The phenomenon in which an animal ceases to respond to a repetitive stimulus.

Hair Cells Sensory receptors of the inner ear that respond to sound vibration and bodily movement.

Half-Life The time required for half the mass of a radioactive element to decay into its stable, non-radioactive form.

Haplodiploidy A genetic pattern of sex determination in which fertilized eggs develop into females and non-fertilized eggs develop into males (as occurs among bees and wasps).

Haploid Having one set of chromosomes per cell. Often written as 1N.

Hardy-Weinberg Law The maintenance of constant allele frequencies in a population from one generation to the next when certain conditions are met. These conditions are the absence of mutation and migration, random mating, a large population, and an equal chance of survival for all individuals.

Haversian Canals A system of microscopic canals in compact bone that transport nutrients to and remove wastes from osteocytes.

Heart An organ that pumps blood (or hemolymph in arthropods) through the vessels of the circulatory system.

Helper T Cells A class of T cells that regulate immune responses by recognizing and activating B cells and other T cells.

Hemocoel In arthropods, the unlined spaces into which fluid (hemolymph) flows when it leaves the blood vessels and bathes the internal organs.

Hemoglobin The iron-containing blood protein that temporarily binds O_2 and releases it into the tissues.

Hemophilia A genetic disorder determined by a gene on the X chromosome (an X-linked trait) that results from the failure of the blood to form clots.

Herbaceous Plants having only primary growth and thus composed entirely of primary tissue.

Herbivore An organism, usually an animal, that eats primary producers (plants).

Herbivory The term for the relationship of a secondary consumer, usually an animal, eating primary producers (plants).

Heredity The passage of genetic traits to offspring which consequently are similar or identical to the parent(s).

Hermaphrodites Animals that possess gonads of both the male and the female.

Heterosporous Higher vascular plants producing two types of spores, a megaspore which grows into a female gametophyte and a microspore which grows into a male gametophyte.

Heterozygous A term applied to organisms that possess two different alleles for a trait. Often, one allele (A) is dominant, masking the presence of the other (a), the recessive.

High Intertidal Zone In the intertidal zone, the region from mean high tide to around just below sea level. Organisms are submerged about 10% of the time.

Histones Small basic proteins that are complexed with DNA to form nucleosomes, the basic structural components of the chromatin fiber.

Homeobox That part of the DNA sequence of homeotic genes that is similar (homologous) among diverse animal species.

Homeostasis Maintenance of fairly constant internal conditions (e.g., blood glucose level, pH, body temperature, etc.)

Homeotic Genes Genes whose products act during embryonic development to affect the spatial arrangement of the body parts.

Hominids Humans and the various groups of extinct, erect-walking primates that were either our direct ancestors or their relatives. Includes the various species of *Homo* and *Australopithecus*.

Homo the genus that contains modern and extinct species of humans.

Homologous Structures Anatomical structures that may have different functions but develop from the same embryonic tissues, suggesting a common evolutionary origin.

Homologues Members of a chromosome pair, which have a similar shape and the same sequence of genes along their length.

Homoplasy (see **Analogous Structures**)

Homosporous Plants that manufacture only one type of spore, which develops into a gametophyte containing both male and female reproductive structures.

Homozygous A term applied to an organism that has two identical alles for a particular trait.

Hormones Chemical messengers secreted by ductless glands into the blood that direct tissues to change their activities and correct imbalances in body chemistry.

Host The organism that a parasite lives on and uses for food.

Human Chorionic Gonadotropin (HCG) A hormone that prevents the corpus luteum from degenerating, thereby maintaining an adequate level of progesterone during pregnancy. It is produced by cells of the early embryo.

Human Immunodeficiency Virus (HIV) The infectious agent that causes AIDS, a disease in which the immune system is seriously disabled.

Hybrid An individual whose parents possess different genetic traits in a breeding experiment or are members of different species.

Hybridization Occurs when two distinct species mate and produce hybrid offspring.

Hybridoma A cell formed by the fusion of a malignant cell (a myeloma) and an antibody-producing lymphocyte. These cells proliferate indefinitely and produce monoclonal antibodies.

Hydrogen Bonds Relatively weak chemical bonds formed when two molecules share an atom of hydrogen.

Hydrologic Cycle The cycling of water, in various forms, through the environment, from Earth to atmosphere and back to Earth again.

Hydrolysis Splitting of a covalent bond by donating the H^+ or OH^- of a water molecule to the two components.

Hydrophilic Molecules Polar molecules that are attracted to water molecules and readily dissolve in water.

Hydrophobic Interaction When nonpolar molecules are "forced" together in the presence of a polar solvent, such as water.

Hydrophobic Molecules Nonpolar substances, insoluble in water, which form aggregates to minimize exposure to their polar surroundings.

Hydroponics The science of growing plants in liquid nutrient solutions, without a solid medium such as soil.

Hydrosphere That portion of the Earth composed of water.

Hydrostatic Skeletons Body support systems found usually in underwater animals (e.g., marine worms). Body shape is protected against gravity and other physical forces by internal hydrostatic pressure produced by contracting muscles encircling their closed, fluid-filled chambers.

Hydrothermal Vents Fissures in the ocean floor where sea water becomes superheated. Chemosynthetic bacteria that live in these vents serve as the autotrophs that support a diverse community of ocean-dwelling organisms.

Hyperpolarization An increase in the potential difference (voltage) across the plasma membrane of a cell typically due to an increase in the movement of potassium ions out of the cell. Acts to inhibit a target cell.

Hypertension High blood pressure (above about 130/90).

Hypertonic Solutions Solutions with higher solute concentrations than found inside the cell. These cause a cell to lose water and shrink.

Hypervolume In ecology, a multidimensional area which includes all factors in an organism's ecological niche, or its' potential niche.

Hypocotyl Portion of the plant embryo below the cotyledons. The hypocotyl gives rise to the root and, very often, to the lower part of the stem.

Hypothalamus The area of the brain below the thalamus that regulates body temperature, blood pressure, etc.

Hypothesis A tentative explanation for an observation or a phenomenon, phrased so that it can be tested by experimentation.

Hypotonic Solutions Solutions with lower solute concentrations than found inside the cell. These cause a cell to accumulate water and swell.

◄ I ►

Immigration Individuals permanently moving into a new area or population.

Immune System A system in vertebrates for the surveillance and destruction of disease-causing microorganisms and cancer cells. Composed of lymphocytes, particularly B cells and T cells, and triggered by the introduction of antigens into the body which makes the body, upon their destruction, resistant to a recurrence of the same disease.

Immunoglobulins (IGs) Antibody molecules.

Imperfect Flowers Flowers that contain either stamens or carpels, making them male or female flowers, respectively.

Imprinting A type of learning in which an animal develops an association with an object after exposure to the object during a critical period early in its life.

Inbreeding When individuals mate with close relatives, such as brothers and sisters. May occur when population sizes drastically shrink and results in a decrease in genetic diversity.

Incomplete Digestive Tract A digestive tract with only one opening through which food is taken in and residues are expelled.

Incomplete (Partial) Dominance A phenomenon in which heterozygous individuals are phenotypically distinguishable from either homozygous type.

Incomplete Flower Flowers lacking one or more whorls of sepals, petals, stamen, or pistils.

Independent Assortment The shuffling of members of homologous chromosome pairs in meiosis I. As a result, there are new chromosome combinations in the daughter cells, which later produce offspring with random mixtures of traits from both parents.

Indoleatic Acid (IAA) An auxin responsible for many plant growth responses including apical dominance, a growth pattern in which shoot tips prevent axillary buds from sprouting.

Induction The process in which one embryonic tissue induces another tissue to differentiate along a pathway that it would not otherwise have taken. (32) Stimulation of transcription of a gene in an operon. Occurs when the repressor protein is unable to bind to the operator.

Inflammation A body strategy initiated by the release of chemicals following injury or infection which brings additional blood with its protective cells to the injured area.

Inhibitory Neurons Neurons that oppose a response in the target cells.

Inhibitory Neurotransmitters Substances released from inhibitory neurons where they synapse with the target cell.

Innate Behavior Actions that are under fairly precise genetic control, typically species-specific, highly stereotyped, and that occur in a complete form the first time the stimulus is encountered.

Insight Learning The sudden solution to a problem without obvious trial-and-error procedures.

Insulin One of the two hormones secreted by endocrine centers called Islets of Langerhans; promotes glucose absorption, utilization, and storage. Insulin is secreted by them when the concentration of glucose in the blood begins to exceed the normal level.

Integumentary System The body's protective external covering, consisting of skin and subcutaneous tissue.

Integuments Protective covering of the ovule.

Intercellular Junctions Specialized regions of cell-cell contact between animal cells.

Intercostal Muscles Muscles that lie between the ribs in humans whose contraction expands the thoracic cavity during breathing.

Interference Competition One species' direct interference by another species for the same limited resource; such as aggressive animal behavior.

Internal Fertilization Fertilization of an egg within the body of the female.

Interneurons Neurons situated entirely within the central nervous system.

Internodes The portion of a stem between two nodes.

Interphase Usually the longest stage of the cell cycle during which the cell grows, carries out normal metabolic functions, and replicates its DNA in preparation for cell division.

Interstitial Cells Cells in the testes that produce testosterone, the major male sex hormone.

Interstitial Fluid The fluid between and surrounding the cells of an animal; the extracellular fluid.

Intertidal Zone The region of beach exposed to air between low and high tides.

Intracellular Digestion Digestion occurring inside cells within food vacuoles. The mode of digestion found in protists and some filter-feeding animals (such as sponges and clams).

Intraspecific Competition Individual organisms of one species competing for the same limited resources in the same habitat, or with overlapping niches.

Intrinsic Rate of Increase (r_m) the maximum growth rate of a population under conditions of maximum birth rate and minimum death rate.

Introns Intervening sequences of DNA in the middle of structural genes, separating exons.

Invertebrates Animals that lack a vertebral column, or backbone.

Ion An electrically charged atom created by the gain or loss of electrons.

Ionic Bond The noncovalent linkage formed by the attraction of oppositely charged groups.

Islets of Langerhans Clusters of endocrine cells in the pancreas that produce insulin and glucagon.

Isolating Mechanisms Barriers that prevent gene flow between populations or among segments of a single population.

Isotopes Atoms of the same element having a different number of neutrons in their nucleus.

Isotonic Solutions Solutions in which the solute concentration outside the cell is the same as that inside the cell.

◄ J ►

Joints Structures where two pieces of a skeleton are joined. Joints may be flexible, such as the knee joint of the human leg or the joints between segments of the exoskeleton of the leg of an insect, or inflexible, such as the joints (sutures) between the bones of the skull.

J-Shaped Curve A curve resulting from exponential growth of a population.

◄ K ►

Karyotype A visual display of an individual's chromosomes.

Kidneys Paired excretory organs which, in humans, are fist-sized and attached to the lower spine. In vertebrates, the kidneys remove nitrogenous wastes from the blood and regulate ion and water levels in the body.

Killer T Cells A type of lymphocyte that functions in the destruction of virus-infected cells and cancer cells.

Kinases Enzymes that catalyze reactions in which phosphate groups are transferred from ATP to another molecule.

Kinetic Energy Energy in motion.

Kinetochore Part of a mitotic (or meiotic) chromosome that is situated within the centromere and to which the spindle fibers attach.

Kingdom A level of the taxonomic hierarchy that groups together members of related phyla or divisions. Modern taxonomy divides all organisms into five Kingdoms: Monera, Protista, Fungi, Plantae, and Animalia.

Klinefelter Syndrome A male whose cells have an extra X chromosome (XXY). The syndrome is characterized by underdeveloped male genitalia and feminine secondary sex characteristics.

Krebs Cycle A circular pathway in aerobic respiration that completely oxidizes the two pyruvic acids from glycolysis.

K-Selected Species Species that produce one or a few well-cared for individuals at a time.

◄ L ►

Lacteal Blind lymphatic vessel in the intestinal villi that receives the absorbed products of lipid digestion.

Lactic Acid Fermentation The process in which electrons removed during glycolysis are transferred from NADH to pyruvic acid to form lactic acid. Used by various prokaryotic cells under oxygen-deficient conditions and by muscle cells during strenuous activity.

Lake Large body of standing fresh water, formed in natural depressions in the Earth. Lakes are larger than ponds.

Lamella In bone, concentric cylinders of calcified collagen deposited by the osteocytes. The laminated layers produce a greatly strengthened structure.

Large Intestine Portion of the intestine in which water and salts are reabsorbed. It is so named because of its large diameter. The large instestine, except for the rectum, is called the colon.

Larva A self-feeding, sexually, and developmentally immature form of an animal.

Larynx The short passageway connecting the pharynx with the lower airways.

Latent (hidden) Infection Infection by a microorganism that causes no symptoms but the microbe is well-established in the body.

Lateral Roots Roots that arise from the pericycle of older roots; also called branch roots or secondary roots.

Law of Independent Assortment Alleles on nonhomologous chromosomes segregate independently of one another.

Law of Segregation During gamete formation, pairs of alleles separate so that each sperm or egg cell has only one gene for a trait.

Law of the Minimum The ecological principle that a species' distribution will be limited by whichever abiotic factor is most deficient in the environment.

Laws of Thermodynamics Physical laws that describe the relationship of heat and mechanical energy. The first law states that energy cannot be created or destroyed, but one form can change into another. The second law states that the total energy the universe decreases as energy conversions occur and some energy is lost as heat.

Leak Channels Passageways through a plasma membrane that do not contain gates and, therefore, are always open for the limited diffusion of a specific substance (ion) through the membrane.

Learning A process in which an animal benefits from experience so that its behavior is better suited to environmental conditions.

Lenticels Loosely packed cells in the periderm of the stem that create air channels for transferring CO_2, H_2O, and O_2.

Leukocytes White blood cells.

LH Luteinizing hormone. A hormone secreted by the anterior pituitary that stimulates testosterone production in males and triggers ovulation and the transformation of the follicle into the corpus luteum in females.

Lichen Symbiotic associations between certain fungi and algae.

Life Cycle The sequence of events during the lifetime of an organism from zygote to reproduction.

Ligaments Strong straps of connective tissue that hold together the bones in articulating joints or support an organ in place.

Light-Dependent Reactions First stage of photosynthesis in which light energy is converted to chemical energy in the form of energy-rich ATP and NADPH.

Light-Independent Reactions Second stage of photosynthesis in which the energy stored in ATP and NADPH formed in the light reactions is used to drive the reactions in which carbon dioxide is converted to carbohydrate.

Limb Bud A portion of an embryo that will develop into either a forelimb or hindlimb.

Limbic System A series of an interconnected group of brain structures, including the thalamus and hypothalamus, controlling memory and emotions.

Limiting Factors The critical factors which impose restraints of the distribution, health, or activities of an organism.

Limnetic Zone Open water of lakes, through which sunlight penetrates and photosynthesis occurs.

Linkage The tendency of genes of the same chromosome to stay together rather than to assort independently.

Linkage Groups Groups of genes located on the same chromosome. The genes of each linkage group assort independently of the genes of other linkage groups. In all eukaryotic organisms, the number of linkage groups is equal to the haploid number of chromosomes.

Lipids A diverse group of biomolecules that are insoluble in water.

Lithosphere The solid outer zone of the Earth; composed of the crust and outermost portion of the mantle.

Littoral Zone Shallow, nutrient-rich waters of a lake, where sunlight reaches the bottom; also the lakeshore. Rooted vegetation occurs in this zone.

Locomotion The movement of an organism from one place to another.

Locus The chromosomal location of a gene.

Logistic Growth Population growth producing a sigmoid, or S-shaped, growth curve.

Long-Day Plants Plants that flower when the length of daylight exceeds some critical period.

Longitudinal Fission The division pattern in flagellated protozoans, where division is along the length of the cell.

Loop of Henle An elongated section of the renal tubule that dips down into the kidney's medulla and then ascends back out to the cortex. It separates the proximal and distal convoluted tubules and is responsible for forming the salt gradient on which water reabsorption in the kidney depends.

Low Density Lipoprotein (LDL) Particles that transport cholesterol in the blood. Each particle consists of about 1,500 cholesterol molecules surrounded by a film of phospholipids and protein. LDLs are taken into cells following their binding to cell surface LDL receptors.

Low Intertidal Zone In the intertidal zone, the region which is uncovered by "minus" tides only. Organisms are submerged about 90% of the time.

Lumen A space within an hollow organ or tube.

Luminescence (see **Bioluminescence**)

Lungs The organs of terrestrial animals where gas exchange occurs.

Lymph The colorless fluid in lymphatic vessels.

Lymphatic System Network of fluid-carrying vessels and associated organs that participate in immunity and in the return of tissue fluid to the main circulation.

Lymphocytes A group of non-phagocytic white blood cells which combat microbial invasion, fight cancer, and neutralize toxic chemicals. The two classes of lymphocytes, B cells and T cells, are the heart of the immune system.

Lymphoid Organs Organs associated with production of blood cells and the lymphatic system, including the thymus, spleen, appendix, bone marrow, and lymph nodes.

Lysis (1) To split or dissolve. (2) Cell bursting.

Lysomes A type of storage vesicle produced by the Golgi complex, containing hydrolytic (digestive) enzymes capable of digesting many kinds of macromolecules in the cell. The membrane around them keeps them sequestered.

◄ M ►

M Phase That portion of the cell cycle during which mitosis (nuclear division) and cytokinesis (cytoplasmic division) takes place.

Macroevolution Evolutionary changes that lead to the appearance of new species.

Macrofungus Filamentous fungus so named for the large size of its fleshy sexual structures; a mushroom, for example.

Macromolecules Large polymers, such as proteins, nucleic acids, and polysaccharides.

Macronutrients Nutrients required by plants in large amounts: carbon, oxygen, hydrogen, nitrogen, potassium, calcium, phosphorus, magnesium, and sulfur.

Macrophages Phagocytic cells that develop from monocytes and present antigen to lymphocytes.

Macroscopic Referring to biological observations made with the naked eye or a hand lens.

Mammals A class of vertebrates that possesses skin covered with hair and that nourishes their young with milk from mammary glands.

Mammary Glands Glands contained in the breasts of mammalian mothers that produce breast milk.

Marsupials Mammals with a cloaca whose young are born immature and complete their development in an external pouch in the mother's skin.

Mass Extinction The simultaneous extinction of a multitude of species as the result of a drastic change in the environment.

Maternal Chromosomes The set of chromosomes in an individual that were inherited from the mother.

Mechanoreceptors Sensory receptors that respond to mechanical pressure and detect motion, touch, pressure, and sound.

Medulla The center-most portion of some organs.

Medusa The motile, umbrella-shaped body form of some members of the phylum Cnidaria, with mouth and tentacles on the lower, concave service. (Compare with polyp.)

Megaspores Spores that divide by mitosis to produce female gametophytes that produce the egg gamete.

Meiosis The division process that produces cells with one-half the number of chromosomes in each somatic cell. Each resulting daughter cell is haploid (1N)

Meiosis I A process of reductional division in which homologous chromosomes pair and then segregate. Homologues are partitioned into separate daughter cells.

Meiosis II Second meiotic division. A division process resembling mitosis, except that the haploid number of chromosomes is present. After the chromosomes line up at the meta-phase plate, the two sister chromatids separate.

Melanin A brown pigment that gives skin and hair its color

Melanoma A deadly form of skin cancer that develops from pigment cells in the skin and is promoted by exposure to the sun.

Memory Cells Lymphocytes responsible for active immunity. They recall a previous exposure to an antigen and, on subsequent exposure to the same antigen, proliferate rapidly into plasma cells and produce large quantities of antibodies in a short time. This protection typically lasts for many years.

Mendelian Inheritance Transmission of genetic traits in a manner consistent with the principles discovered by Gregor Mendel. Includes traits controlled by simple dominant or recessive alleles; more complex patterns of transmission are referred to as Nonmendelian inheritance.

Meninges The thick connective tissue sheath which surrounds and protects the vertebrate brain and spinal cord.

Menstrual Cycle The repetitive monthly changes in the uterus that prepare the endometrium for receiving and supporting an embryo.

Meristematic Region New cells arise from this undifferentiated plant tissue; found at root or shoot apical meristems, or lateral meristems.

Meristems In plants, clusters of cells that retain their ability to divide, thereby producing new cells. One of the four basic tissues in plants.

Mesoderm In animals, the middle germ cell layer of the gastrula. It gives rise to muscle, bone, connective tissue, gonads, and kidney.

Mesopelagic Zone The dimly lit ocean zone beneath the epipelagic zone; large fishes, whales and squid occupy this zone; no phytoplankton occur in this zone.

Mesophyll Layers of cells in a leaf between the upper and lower epidermis; produced by the ground meristem.

Messenger RNA (mRNA) The RNA that carries genetic information from the DNA in the nucleus to the ribosomes in the cytoplasm, where the sequence of bases in the mRNA is translated into a sequence of amino acids.

Metabolic Intermediates Compounds produced as a substrate are converted to end product in a series of enzymatic reactions.

Metabolic Pathways Set of enzymatic reactions involved in either building or dismantling complex molecules.

Metabolic Rate A measure of the level of activity of an organism usually determined by measuring the amount of oxygen consumed by an individual per gram body weight per hour.

Metabolic Water Water produced as a product of metabolic reactions.

Metabolism The sum of all the chemical reactions in an organism; includes all anabolic and catabolic reactions.

Metamorphosis Transformation from one form into another form during development.

Metaphase The stage of mitosis when the chromosomes line-up along the metaphase plate, a plate that usually lies midway between the spindle poles.

Metaphase Plate Imaginary plane within a dividing cell in which the duplicated chromosomes become aligned during metaphase.

Microbes Microscopic organisms.

Microbiology The branch of biology that studies microorganisms.

Microevolution Changes in allele frequency of a species' gene pool which has not generated new species. Exemplified by changes in the pigmentation of the peppered moth and by the acquisition of pesticide resistance in insects.

Microfibrils Bundles formed from the intertwining of cellulose molecules, i.e., long chains of glucose molecules in the cell walls of plants.

Microfilaments Thin actin-containing protein fibers that are responsible for maintenance of cell shape, muscle contraction and cyclosis.

Micrometer One millionth (1/1,000,000) of a meter.

Micronutrients Nutrients required by plants in small amounts: iron, chlorine, copper, manganese, zinc, molybdenum, and boron.

Micropyle A small opening in the integuments of the ovule through which the pollen tube grows to deliver sperm.

Microspores Spores within anthers of flowers. They divide by mitosis to form pollen grains, the male gametophytes that produce the plant's sperm.

Microtubules Thin, hollow tubes in cells; built from repeating protein units of tubulin. Microtubules are components of cilia, flagella, and the cytoskeleton.

Microvilli The small projections on the cells that comprise each villus of the intestinal wall, further increasing the absorption surface area of the small intestine.

Middle Intertidal Zone In the intertidal zone, the region which is covered and uncovered twice a day, the zero of tide tables. Organisms are submerged about 50% of the time.

Migration Movements of a population into or out of an area.

Mimicry A defense mechanism where one species resembles another in color, shape, behavior, or sound.

Mineralocorticoids Steroid hormones which regulate the level of sodium and potassium in the blood.

Mitochondria Organelles that contain the biochemical machinery for the Krebs cycle and the electron transport system of aerobic respiration. They are composed of two membranes, the inner one forming folds, or cristae.

Mitosis The process of nuclear division producing daughter cells with exactly the same number of chromosomes as in the mother cell.

Mitosis Promoting Factor (MPF) A protein that appears to be a universal trigger of cell division in eukaryotic cells.

Mitotic Chromosomes Chromosomes whose DNA-protein threads have become coiled into microscopically visible chromosomes, each containing duplicated chromatids ready to be separated during mitosis.

Molds Filamentous fungi that exist as colonies of threadlike cells but produce no macroscopic fruiting bodies.

Molecule Chemical substance formed when two or more atoms bond together; the smallest unit of matter that possesses the qualities of a compound.

Mollusca A phylum, second only to Arthropoda in diversity. Composed of three main classes: 1) Gastropoda (spiral-shelled), 2) Bivalvia (hinged shells), 3) Cephalopoda (with tentacles or arms and no, or very reduced shells).

Molting (Ecdysis) Shedding process by which certain arthropods lose their exoskeletons as their bodies grow larger.

Monera The taxonomic kingdom comprised of single-celled prokaryotes such as bacteria, cyanobacteria, and archebacteria.

Monoclonal Antibodies Antibodies produced by a clone of hybridoma cells, all of which descended from one cell.

Monocotyledae (Monocots) One of the two divisions of flowering plants, characterized by seeds with a single cotyledon, flower parts in 3s, parallel veins in leaves, many roots of approximately equal size, scattered vascular bundles in its stem anatomy, pith in its root anatomy, and no secondary growth capacity.

Monocytes A type of leukocyte that gives rise to macrophages.

Monoecious Both male and female reproductive structures are produced on the same sporophyte individual.

Monohybrid Cross A mating between two individuals that differ only in one genetically-determined trait.

Monomers Small molecular subunits which are the building blocks of macromolecules. The macromolecules in living systems are constructed of some 40 different monomers.

Monotremes A group of mammals that lay eggs from which the young are hatched.

Morphogenesis The formation of form and internal architecture within the embryo brought about by such processes as programmed cell death, cell adhesion, and cell movement.

Morphology The branch of biology that studies form and structure of organisms.

Mortality Deathrate in a population or area.

Motile Capable of independent movement.

Motor Neurons Nerve cells which carry outgoing impulses to their effectors, either glands or muscles.

Mucosa The cell layer that lines the digestive tract and secretes a lubricating layer of mucus.

Mullerian Mimicry Resemblance of different species, each of which is equally obnoxious to predators.

Multicellular Consisting of many cells.

Multichannel Food Chain Where the same primary producer supplies the energy for more than one food chain.

Multiple Allele System Three or more possible alleles for a given trait, such as ABO blood groups in humans.

Multiple Fission Division of the cell's nucleus without a corresponding division of cytoplasm.

Multiple Fruits Fruits that develop from pistils of separate flowers.

Muscle Fiber A muiltinucleated skeletal muscle cell that results from the fusion of several pre-muscle cells during embryonic development.

Muscle Tissue Bundles and sheets of contractile cells that shorten when stimulated, providing force for controlled movement.

Mutagens Chemical or physical agents that induce genetic change.

Mutation Random heritable changes in DNA that introduce new alleles into the gene pool.

Mutualism The symbiotic interaction in which both participants benefit.

Mycology The branch of biology that studies fungi.

Mycorrhizae An association between soil fungi and the roots of vascular plants, increasing the plant's ability to extract water and minerals from the soil.

Myelin Sheath In vertebrates, a jacket which covers the axons of high-velocity neurons, thereby increasing the speed of a neurological impulse.

Myofibrils In striated muscle, the banded fibrils that lie parallel to each other, constituting the bulk of the muscle fiber's interior and powering contraction.

Myosin A contractile protein that makes up the major component of the thick filaments of a muscle cell and is also present in nonmuscle cells.

◄ N ►

NADPH Nicotinamide adenine dinucleotide phosphate. NADPH is formed by reduction of $NADP^+$, and serves as a store of electrons for use in metabolism (see Reducing Power).

NAD^+ Nicotinamide adenine dinucleotide. A coenzyme that functions as an electron carrier in metabolic reactions. When reduced to NADH, the molecule becomes a cellular energy source.

Natality Birthrate in a population or area.

Natural Killer (NK) Cells Nonspecific, lymphocytelike cells which destroy foreign cells and cancer cells.

Natural Selection Differential survival and reproduction of organisms with a resultant increase in the frequency of those best adapted to the environment.

Neanderthals A subspecies of Homo sapiens different from that of modern humans that were characterized by heavy bony skeletons and thick bony ridges over the eyes. They disappeared about 35,000 years ago.

Nectary Secretory gland in flowering plants containing sugary fluid that attracts pollinators as a food source. Usually located at the base of the flower.

Negative Feedback Any regulatory mechanism in which the increased level of a substance inhibits further production of that substance, thereby preventing harmful accumulation. A type of homeostatic mechanism.

Negative Gravitropism In plants, growth against gravitational forces, or shoot growth upward.

Nematocyst Within the stinging cell (cnidocyte) of cnidarians, a capsule that contains a coiled thread which, when triggered, harpoons prey and injects powerful toxins.

Nematoda The widespread and abundant animal phylum containing the roundworms.

Nephridium A tube surrounded by capillaries found in an organism's excretory organs that removes nitrogenous wastes and regulates the water and chemical balance of body fluids.

Nephron The functional unit of the vertebrate kidney, consisting of the glomerulus, Bowman's capsule, proximal and distal convoluted tubules, and loop of Henle.

Nerve Parallel bundles of neurons and their supporting cells.

Nerve Impulse A propagated action potential.

Nervous Tissue Excitable cells that receive stimuli and, in response, transmit an impulse to another part of the animal.

Neural Plate In vertebrates, the flattened plate of dorsal ectoderm of the late gastrula that gives rise to the nervous system.

Neuroglial Cells Those cells of a vertebrate nervous system that are not neurons. Includes a variety of cell types including Schwann cells.

Neuron A nerve cell.

Neurosecretory Cells Nervelike cells that secrete hormones rather than neurotransmitter substances when a nerve impulse reaches the distal end of the cell. In vertebrates, these cells arise from the hypothalamus.

Neurotoxins Substances, such as curare and tetanus toxin, that interfere with the transmission of neural impulses.

Neurotransmitters Chemicals released by neurons into the synaptic cleft, stimulating or inhibiting the post-synaptic target cell.

Neurulation Formation by embryonic induction of the neural tube in a developing vertebrate embryo.

Neutrons Electrically neutral (uncharged) particles contained within the nucleus of the atom.

Neutrophil Phagocytic leukocyte, most numerous in the human body.

Niche An organism's habitat, role, resource requirements, and tolerance ranges for each abiotic condition.

Niche Breadth Relative size and dimension of ecological niches; for example, broad or narrow niches.

Niche Overlap Organisms that have the same habitat, role, environmental requirements, or needs.

Nitrogen Fixation The conversion of atmospheric nitrogen gas N_2 into ammonia (NH_3) by certain bacteria and cyanobacteria.

Nitrogenous Wastes Nitrogen-containing metabolic waste products, such as ammonia or urea, that are produced by the breakdown of proteins and nucleic acids.

Nodes The attachment points of leaves to a stem.

Nodes of Ranvier Uninsulated (nonmyelinated) gaps along the axon of a neuron.

Noncovalent Bonds Linkages between two atoms that depend on an attraction between positive and negative charges between molecules or ions. Includes ionic and hydrogen bonds.

Non-Cyclic Photophosphorylation The pathway in the light reactions of photosynthesis in which electrons pass from water, through two photosystems, and then ultimately to $NADP^+$. During the process, both ATP and NADPH are produced. It is so named because the electrons do not return to their reaction center.

Nondisjunction Failure of chromosomes to separate properly at meiosis I or II. The result is that one daughter will receive an extra chromosome and the other gets one less.

Nonpolar Molecules Molecules which have an equal charge distribution throughout their structure and thus lack regions with a localized positive or negative charge.

Notochord A flexible rod that is below the dorsal surface of the chordate embryo, beneath the nerve cord. In most chordates, it is replaced by the vertebral column.

Nuclear Envelope A double membrane pierced by pores that separates the contents of the nucleus from the rest of the eukaryotic cell.

Nucleic Acids DNA and RNA; linear polymers of nucleotides, responsible for the storage and expression of genetic information.

Nucleoid A region in the prokaryotic cell that contains the genetic material (DNA). It is unbounded by a nuclear membrane.

Nucleoplasm The semifluid substance of the nucleus in which the particulate structures are suspended.

Nucleosomes Nuclear protein complex consisting of a length of DNA wrapped around a central cluster of 8 histones.

Nucleotides Monomers of which DNA and RNA are built. Each consists of a 5-carbon sugar, phosphate, and a nitrogenous base.

Nucleous (pl. nucleoli) One or more darker regions of a nucleus where each ribosomal subunit is assembled from RNA and protein.

Nucleus The large membrane-enclosed organelle that contains the DNA of eukaryotic cells.

Nucleus, Atomic The center of an atom containing protons and neutrons.

◄ O ►

Obligate Symbiosis A symbiotic relationship between two organisms that is necessary for the survival or both organisms.

Olfaction The sense of smell.

Oligotrophic Little nourished, as a young lake that has few nutrients and supports little life.

Omnivore An animal that obtains its nutritional needs by consuming plants and other animals.

Oncogene A gene that causes cancer, perhaps activated by mutation or a change in its chromosomal location.

Oocyte A female germ cell during any of the stages of meiosis.

Oogenesis The process of egg production.

Oogonia Female germ cells that have not yet begun meiosis.

Open Circulatory System Circulatory system in which blood travels from vessels to tissue spaces, through which it percolates prior to returning to the main vessel (compare with closed circulatory system).

Operator A regulatory gene in the operon of bacteria. It is the short DNA segment to which the repressor binds, thus preventing RNA polymerase from attaching to the promoter.

Operon A regulatory unit in prokaryotic cells that controls the expression of structural genes. The operon consists of structural genes that produce enzymes for a particular metabolic pathway, a regulator region composed of a promoter and an operator, and R (regulator) gene that produces a repressor.

Order A level of the taxonomic hierarchy that groups together members of related families.

Organ Body part composed of several tissues that performs specialized functions.

Organelle A specialized part of a cell having some particular function.

Organic Compounds Chemical compounds that contain carbon.

Organism A living entity able to maintain its organization, obtain and use energy, reproduce, grow, respond to stimuli, and display homeostatis.

Organogenesis Organ formation in which two or more specialized tissue types develop in a precise temporal and spatial relationship to each other.

Organ System Group of functionally related organs.

Osmoregulation The maintenance of the proper salt and water balance in the body's fluids.

Osmosis The diffusion of water through a differentially permeable membrane into a hypertonic compartment.

Ossification Synthesis of a new bone.

Osteoclast A type of bone cell which breaks down the bone, thereby releasing calcium into the bloodstream for use by the body. Osteoclasts are activated by hormones released by the parathyroid glands.

Osteocytes Living bone cells embedded within the calcified matrix they manufacture.

Osteoporosis A condition present predominantly in postmenopausal women where the bones are weakened due to an increased rate of bone resorption compared to bone formation.

Ovarian Cycle The cycle of egg production within the mammalian ovary.

Ovarian Follicle In a mammalian ovary, a chamber of cells in which the oocyte develops.

Ovary In animals, the egg-producing gonad of the female. In flowering plants, the enlarged base of the pistil, in which seeds develop.

Oviduct (Fallopian Tube) The tube in the female reproductive organ that connects the ovaries and uterus and where fertilization takes place.

Ovulation The release of an egg (ovum) from the ovarian follicle.

Ovule In seed plants, the structure containing the female gametophyte, nucellus, and integuments. After fertilization, the ovule develops into a seed.

Ovum An unfertilized egg cell; a female gamete.

Oxidation The removal of electrons from a compound during a chemical reaction. For a carbon atom, the fewer hydrogens bonded to a carbon, the greater the oxidation state of the atom.

Oxidative Phosphorylation The formation of ATP from ADP and inorganic phosphate that occurs in the electron-transport chain of cellular respiration.

Oxyhemoglobin A complex of oxygen and hemoglobin, formed when blood passes through the lungs and is dissociated in body tissues, where oxygen is released.

Oxytocin A female hormone released by the posterior pituitary which triggers uterine contractions during childbirth and the release of milk during nursing.

◄ P ►

P680 Reaction Center (P = Pigment) Special chlorophyll molecule in Photosystem II that traps the energy absorbed by the other pigment molecules. It absorbs light energy maximally at 680 nm.

Palisade Parenchyma In dicot leaves, densely packed, columnar shaped cells functioning in photosynthesis. Found just beneath the upper epidermis.

Pancreas In vertebrates, a large gland that produces digestive enzymes and hormones.

Parallel Evolution When two species that have descended from the same ancestor independently acquire the same evolutionary adaptations.

Parapatric Speciation The splitting of a population into two species' populations under conditions where the members of each population reside in adjacent areas.

Parasite An organism that lives on or inside another called a host, on which it feeds.

Parasitism A relationship between two organisms where one organism benefits, and the other is harmed.

Parasitoid Parasitic organisms, such as some insect larvae, which kill their host.

Parasympathetic Nervous System Part of the autonomic nervous system active during relaxed activity.

Parathyroid Glands Four glands attached to the thyroid gland which secrete parathyroid hormone (PTH). When blood calcium levels are low, PTH is secreted, causing calcium to be released from bone.

Parenchyma The most prevalent cell type in herbaceous plants. These thin-walled, polygonal-shaped cells function in photosynthesis and storage.

Parthenogenesis Process by which offspring are produced without egg fertilization.

Passive Immunity Immunity achieved by receiving antibodies from another source, as occurs with a newborn infant during nursing.

Paternal Chromosomes The set of chromosomes in an individual that were inherited from the father.

Pathogen A disease-causing microorganism.

Pectoral Girdle In humans, the two scapulae (shoulder blades) and two clavicles (collarbones) which support and articulate with the bones of the upper arm.

Pedicel A shortened stem carrying a flower.

Pedigree A diagram showing the inheritance of a particular trait among the members of the family.

Pelagic Zone The open oceans, divided into three layers: 1) photo- or epipelagic (sunlit), 2) mesopelagic (dim light), 3) aphotic or bathypelagic (always dark).

Pelvic Girdle The complex of bones that connect a vertebrate's legs with its backbone.

Penis An intrusive structure in the male animal which releases male gametes into the female's sex receptacle.

Peptide Bond The covalent bond between the amino group of one amino acid and the carboxyl group of another.

Peptidoglycan A chemical component of the prokaryotic cell wall.

Percent Annual Increase A measure of population increase; the number of individuals (people) added to the population per 100 individuals.

Perennials Plants that live longer than two years.

Perfect Flower Flowers that contain both stamens and pistils.

Perforation Plate In plants, that portion of the wall of vessel members that is perforated, and contains an area with neither primary nor secondary cell wall; a "hole" in the cell wall.

Pericycle One or more layers of cells found in roots, with phloem or xylem to its' inside, and the endodermis to its' outside. Functions in producing lateral roots and formation of the vascular cambium in roots with secondary growth.

Periderm Secondary tissue that replaces the epidermis of stems and roots. Consists of cork, cork cambium, and an internal layer of parenchyma cells.

Peripheral Nervous System Neurons, excluding those of the brain and spinal cord, that permeate the rest of the body.

Peristalsis Sequential waves of muscle contractions that propel a substance through a tube.

Peritoneum The connective tissue that lines the coelomic cavities.

Permeability The ability to be penetrable, such as a membrane allowing molecules to pass freely across it.

Petal The second whorl of a flower, often brightly colored to attract pollinators; collectively called the corolla.

Petiole The stalk leading to the blade of a leaf.

pH A scale that measures the concentration of hydrogen ions in a solution. The pH scale extends from 0 to 14. Acidic solutions have a pH of less than 7; alkaline solutions have a pH above 7; neutral solutions have a pH equal to 7.

Phagocytosis Engulfing of food particles and foreign cells by amoebae and white blood cells. A type of endocytosis.

Pharyngeal Pouches In the vertebrate embryo, outgrowths from the walls of the pharynx that break through the body surface to form gill slits.

Pharynx The throat; a portion of both the digestive and respiratory system just behind the oral cavity.

Phenotype An individual's observable characteristics that are the expression of its genotype.

Pheromones Chemicals that, when released by an animal, elicit a particular behavior in other animals of the same species.

Phloem The vascular tissue that transports sugars and other organic molecules from sites of photosynthesis and storage to the rest of the plant.

Phloem Loading The transfer of assimilates to phloem conducting cells, from photosynthesizing source cells.

Phloem Unloading The transfer of assimilates to storage (sink) cells, from phloem conducting cells.

Phospholipids Lipids that contain a phosphate and a variable organic group that form polar, hydrophilic regions on an otherwise nonpolar, hydrophobic molecule. They are the major structural components of membranes.

Phosphorylation A chemical reaction in which a phosphate group is added to a molecule or atom.

Photoexcitation Absorption of light energy by pigments, causing their electrons to be raised to a higher energy level.

Photolysis The splitting of water during photosynthesis. The electrons from water pass to Photosystem II, the protons enter the lumen of the thylakoid and contribute to the proton gradient across the thylakoid membrane, and the oxygen is released into the atmosphere.

Photon A particle of light energy.

Photoperiod Specific lengths of day and night which control certain plant growth responses to light, such as flowering or germination.

Photoperiodism Changes in the behavior and physiology of an organism in response to the relative lengths of daylight and darkness, i.e., the photoperiod.

Photoreceptors Sensory receptors that respond to light.

Photorespiration The phenomenon in which oxygen binds to the active site of a CO_2-fixing enzyme, thereby competing with CO_2 fixation, and lowering the rate of photosynthesis.

Photosynthesis The conversion by plants of light energy into chemical energy stored in carbohydrate.

Photosystems Highly organized clusters of photosynthetic pigments and electron/hydrogen carriers embedded in the thylakoid membranes of chloroplasts. There are two photosystems, which together carry out the light reactions of photosynthesis.

Photosystem I Photosystem with a P700 reaction center; participates in cyclic photophosphorylation as well as in noncyclic photophosphorylation.

Photosystem II Photosystem activated by a P680 reaction center; participates only in noncyclic photophosphorylation and is associated with photolysis of water.

Phototropism The growth responses of a plant to light.

Phyletic Evolution The gradual evolution of one species into another.

Phylogeny Evolutionary history of a species.

Phylum The major taxonomic divisions in the Animal kingdom. Members of a phylum share common, basic features. The Animal kingdom is divided into approximately 35 phyla.

Physiology The branch of biology that studies how living things function.

Phytochrome A light-absorbing pigment in plants which controls many plant responses, including photoperiodism.

Phytoplankton Microscopic photosynthesizers that live near the surface of seas and bodies of fresh water.

Pineal Gland An endocrine gland embedded within the brain that secretes the hormone melatonin. Hormone secretion is dependent on levels of environmental light. In amphibians and reptiles, melatonin controls skin coloration. In humans, pineal secretions control sexual maturation and daily rhythms.

Pinocytosis Uptake of small droplets and dissolved solutes by cells. A type of endocytosis.

Pistil The female reproductive part and central portion of a flower, consisting of the ovary, style and stigma. May contain one carpel, or one or more fused carpels.

Pith A plant tissue composed of parenchyma cells, found in the central portion of primary growth stems of dicots, and monocot roots.

Pith Ray Region between vascular bundles in vascular plants.

Pituitary Gland (see **Posterior and Anterior Pituitary).**

Placenta In mammals (exclusive of marsupials and monotremes), the structure through which nutrients and wastes are exchanged between the mother and embryo/fetus. Develops from both embryonic and uterine tissues.

Plant Multicellular, autotrophic organism able to manufacture food through photosynthesis.

Plasma In vertebrates, the liquid portion of the blood, containing water, proteins (including fibrinogen), salts, and nutrients.

Plasma Cells Differentiated antibody-secreting cells derived from B lymphocytes.

Plasma Membrane The selectively permeable, molecular boundary that separates the cytoplasm of a cell from the external environment.

Plasmid A small circle of DNA in bacteria in addition to its own chromosome.

Plasmodesmata Openings between plant cell walls, through which adjacent cells are connected via cytoplasmic threads.

Plasmodium Genus of protozoa that causes malaria.

Plasmodium A huge multinucleated "cell" stage of a plasmodial slime mold that feeds on dead organic matter.

Plasmolysis The shrinking of a plant cell away from its cell wall when the cell is placed in a hypertonic solution.

Platelets Small, cell-like fragments derived from special white blood cells. They function in clotting.

Plate Tectonics The theory that the earth's crust consists of a number of rigid plates that rest on an underlying layer of semimolten rock. The movement of the earth's plates results from the upward movement of molten rock into the solidified crust along ridges within the ocean floor.

Platyhelminthes The phylum containing simple, bilaterally symmetrical animals, the flatworms.

Pleiotropy Where a single mutant gene produces two or more phenotypic effects.

Pleura The double-layered sac which surrounds the lungs of a mammal.

Pneumocytis Pneumonia (PCP) A disease of the respiratory tract caused by a protozoan that strikes persons with immunodeficiency diseases, such as AIDS.

Point Mutations Changes that occur at one point within a gene, often involving one nucleotide in the DNA.

Polar Body A haploid product of meiosis of a female germ cell that has very little cytoplasm and disintegrates without further function.

Polar Molecule A molecule with an unequal charge distribution that creates distinct positive and negative regions or poles.

Pollen The male gametophyte of seed plants, comprised of a generative nucleus and a tube nucleus surrounded by a tough wall.

Pollen Grain The male gametophyte of conifers and angiosperms, containing male gametes. In angiosperms, pollen grains are contained in the pollen sacs of the anther of a flower.

Pollination The transfer of pollen grains from the anther of one flower to the stigma of another. The transfer is mediated by wind, water, insects, and other animals.

Polygenic Inheritance An inheritance pattern in which a phenotype is determined by two or more genes at different loci. In humans, examples include height and pigmentation.

Polymer A macromolecule formed of monomers joined by covalent bonds.. Includes proteins, polysaccharides, and nucleic acids.

Polymerase Chain Reaction (PCR) Technique to amplify a specific DNA molecule using a temperature-sensitive DNA polymerase obtained from a heat-resistant bacterium. Large numbers of copies of the initial DNA molecule can be obtained in a short period of time, even when the starting material is present in vanishingly small amounts, as for example from a blood stain left at the scene of a crime.

Polymorphic Property of some protozoa to produce more than one stage of organism as they complete their life cycles.

Polymorphic Genes Genes for which several different alleles are known, such as those that code for human blood type.

Polyp Stationary body form of some members of the phylum Cnidaria, with mouth and tentacles facing upward. (Compare with medusa.)

Polypeptide An unbranched chain of amino acids covalently linked together and assembled on a ribosome during translation.

Polyploidy An organism or cell containing three or more complete sets of chromosomes. Polyploidy is rare in animals but common in plants.

Polysaccharide A carbohydrate molecule consisting of monosaccharide units.

Polysome A complex of ribosomes found in chains, linked by mRNA. Polysomes contain the ribosomes that are actively assembling proteins.

Polytene Chromosomes Giant banded chromosomes found in certain insects that form by the repeated duplication of DNA. Because of the multiple copies of each gene in a cell, polytene chromosomes can generate large amounts of a gene product in a short time. Transcription occurs at sites of chromosome puffs.

Pond Body of standing fresh water, formed in natural depressions in the Earth. Ponds are smaller than lakes.

Population Individuals of the same species inhabiting the same area.

Population Density The number of individual species living in a given area.

Positive Gravitropism In plants, growth with gravitational forces, or root growth downward.

Posterior Pituitary A gland which manufactures no hormones but receives and later releases hormones produced by the cell bodies of neurons in the hyopthalamus.

Potential Energy Stored energy, such as occurs in chemical bonds.

Preadaptation A characteristic (adaptation) that evolved to meet the needs of an organism in one type of habitat, but fortuitously allows the organism to exploit a new habitat. For example, lobed fins and lungs evolved in ancient fishes to help them live in shallow, stagnant ponds, but also facilitated the evolution of terrestrial amphibians.

Precells Simple forerunners of cells that, presumably, were able to concentrate organic molecules, allowing for more frequent molecular reactions.

Predation Ingestion of prey by a predator for energy and nutrients.

Predator An organism that captures and feeds on another organism (prey).

Pressure Flow In the process of phloem loading and unloading, pressure differences resulting from solute increases in phloem conducting cells and neighboring xylem cells cause the flow of water to phloem. A concentration gradient is created between xylem and phloem cells.

Prey An organism that is captured and eaten by another organism (predator).

Primary Consumer Organism that feeds exclusively on producers (plants). Herbivores are primary consumers.

Primary Follicle In the mammalian ovary, a structure composed of an oocyte and its surrounding layer of follicle cells.

Primary Growth Growth from apical meristems, resulting in an increase in the lengths of shoots and roots in plants.

Primary Immune Response Process of antibody production following the first exposure to an antigen. There is a lag time from exposure until the appearance in the blood of protective levels of antibodies.

Primary Oocyte Female germ cell that is either in the process of or has completed the first meiotic division. In humans, germ cells may remain in this stage in the ovary for decades.

Primary Producers All autotrophs in a biotic environment that use sunlight or chemical energy to manufacture food from inorganic substances.

Primary Sexual Characteristics Gonads, reproductive tracts, and external genitals.

Primary Spermatocyte Male germ cell that is either in the process of or has completed the first meiotic division.

Primary Succession The development of a community in an area previously unoccupied by any community; for example, a "bare" area such as rock, volcanic material, or dunes.

Primary Tissues Tissues produced by primary meristems of a plant, which arise from the shoot and root apical meristems. In general, primary tissues are a result of an increase in plant length.

Primary Transcript An RNA molecule that has been transcribed but not yet subjected to any type of processing. The primary transcript corresponds to the entire stretch of DNA that was transcribed.

Primates Order of mammals that includes humans, apes, monkeys, and lemurs.

Primitive An evolutionary early condition. Primitive features are those that were also present in an early ancestor, such as five digits on the feet of terrestrial vertebrates.

Prions An infectious particle that contains protein but no nucleic acid. It causes slow diseases of animals, including neurological disease of humans.

Processing-Level Control Control of gene expression by regulating the pathway by which a primary RNA transcript is processed into an mRNA.

Products In a chemical reaction, the compounds into which the reactants are transformed.

Profundal Zone Deep, open water of lakes, where it is too dark for photosynthesis to occur.

Progesterone A hormone produced by the corpus luteum within the ovary. It prepares and maintains the uterus for pregnancy, participates in milk production, and prevents the ovary from releasing additional eggs late in the cycle or during pregnancy.

Prokaryotic Referring to single-celled organisms that have no membrane separating the DNA from the cytoplasm and lack membrane-enclosed organelles. Prokaryotes are confined to the kingdom Monera; they are all bacteria.

Prokaryotic Fission The most common type of cell division in bacteria (prokaryotes). Duplicated DNA strands are attached to the plasma membrane and become separated into two cells following membrane growth and cell wall formation.

Prolactin A hormone produced by the anterior pituitary, stimulating milk production by mammary glands.

Promoter A short segment of DNA to which RNA polymerase attaches at the start of transcription.

Prophase Longest phase of mitosis, involving the formation of a spindle, coiling of chromatin fibers into condensed chromosomes, and movement of the chromosomes to the center of the cell.

Prostaglandins Hormones secreted by endocrine cells scattered throughout the body responsible for such diverse functions as contraction of uterine muscles, triggering the inflammatory response, and blood clotting.

Prostate Gland A muscular gland which produces and releases fluids that make up a substantial portion of the semen.

Proteins Long chains of amino acids, linked together by peptide bonds. They are folded into specific shapes essential to their functions.

Prothallus The small, heart-shaped gametophyte of a fern.

Protists A member of the kingdom Protista; simple eukaryotic organisms that share broad taxonomic similarities.

Protocooperation Non-compulsory interactions that benefit two organisms, e.g., lichens.

Proton Gradient A difference in hydrogen ion (proton) concentration on opposite sides of a membrane. Proton gradients are formed during photosynthesis and respiration and serve as a store of energy used to drive ATP formation.

Protons Positively charged particles within the nucleus of an atom.

Protostomes One path of development exhibited by coelomate animals (e.g., mollusks, annelids, and arthropods).

Protozoa Member of protist kingdom that is unicellular and eukaryotic; vary greatly in size, motility, nutrition and life cycle.

Provirus DNA copy of a virus' nucleic acid that becomes integrated into the host cell's chromosome.

Pseudocoelamates Animals in which the body cavity is not lined by cells derived from mesoderm.

Pseudopodia (psuedo = false, pod = foot). Pseudopodia are fingerlike extensions of cytoplasm that flow forward from the "body" of an amoeba; the rest of the cell then follows.

Puberty Development of reproductive capacity, often accompanied by the appearance of secondary sexual characteristics.

Pulmonary Circulation The loop of the circulatory system that channels blood to the lungs for oxygenation.

Punctuated Equilibrium Theory A theory to explain the phenomenon of the relatively sudden appearance of new species, followed by long periods of little or no change.

Punnett Square Method A visual method for predicting the possible genotypes and their expected ratios from a cross.

Pupa In insects, the stage in metamorphosis between the larva and the adult. Within the pupal case, there is dramatic transformation in body form as some larval tissues die and others differentiate into those of the adult.

Purine A nitrogenous base found in DNA and RNA having a double ring structure. Adenine and guanine are purines.

Pyloric Sphincter Muscular valve between the human stomach and small intestine.

Pyrimidine A nitrogenous base found in DNA and RNA having a single ring structure. Cytosine, thymine, and uracil are pyrimidines.

Pyramid of Biomass Diagrammatic representation of the total dry weight of organisms at each trophic level in a food chain or food web.

Pyramid of Energy Diagrammatic representation of the flow of energy through trophic levels in a food chain or food web.

Pyramid of Numbers Similar to a pyramid of energy, but with numbers of producers and consumers given at each trophic level in a food chain or food web.

◄ Q ►

Quiescent Center The region in the apical meristem of a root containing relatively inactive cells.

◄ R ►

R-Group The variable portion of a molecule.

r-Selected Species Species that possess adaptive strategies to produce numerous offspring at once.

Radial Symmetry The quality possessed by animals whose bodies can be divided into mirror images by more than one median plane.

Radicle In the plant embryo, the tip of the hypocotyl that eventually develops into the root system.

Radioactivity A property of atoms whose nucleus contains an unstable combination of particles. Breakdown of the nucleus causes the emission of particles and a resulting change in structure of the atom. Biologists use this property to track labeled molecules and to determine the age of fossils.

Radiodating The use of known rates of radioactive decay to date a fossil or other ancient object.

Radioisotope An isotope of an element that is radioactive.

Radiolarian A prozoan member of the protistan group Sarcodina that secretes silicon shells through which it captures food.

Rainshadow The arid, leeward (downwind) side of a mountain range.

Random Distribution Distribution of individuals of a population in a random manner; environmental conditions must be similar and individuals do not affect each other's location in the population.

Reactants Molecules or atoms that are changed to products during a chemical reaction.

Reaction A chemical change in which starting molecules (reactants) are transformed into new molecules (products).

Reaction Center A special chlorophyll molecule in a photosystem (P_{700} in Photosystem I, P_{680} in Photosystem II).

Realized Niche Part of the fundamental niche of an organism that is actually utilized.

Receptacle The base of a flower where the flower parts are attached; usually a widened area of the pedicel.

Receptor-Mediated Endocytosis The uptake of materials within a cytoplasmic vesicle (endocytosis) following their binding to a cell surface receptor.

Receptor Site A site on a cell's plasma membrane to which a chemical such as a hormone binds. Each surface site permits the attachment of only one kind of hormone.

Recessive An allele whose expression is masked by the dominant allele for the same trait.

Recombinant DNA A DNA molecule that contains DNA sequences derived from different biological sources that have been joined together in the laboratory.

Recombination The rejoining of DNA pieces with those of a different strand or with the same strand at a point different from where the break occurred.

Red Marrow The soft tissue in the interior of bones that produces red blood cells.

Red Tide Growth of one of several species of reddish brown dinoflagellate algae so extensive that it tints the coastal waters and inland lakes a distinctive red color. Often associated with paralytic shellfish poisoning (see dinoflagellates).

Reducing Power A measure of the cell's ability to transfer electrons to substrates to create molecules of higher energy content. Usually determined by the available store of NADPH, the molecule from which electrons are transferred in anabolic (synthetic) pathways.

Reduction The addition of electrons to a compound during a chemical reaction. For a carbon atom, the more hydrogens that are bonded to the carbon, the more reduced the atom.

Reduction Division The first meiotic division during which a cell's chromosome number is reduced in half.

Reflex An involuntary response to a stimulus.

Reflex Arc The simplest example of central nervous system control, involving a sensory neuron, motor neuron, and usually an interneuron.

Regeneration Ability of certain animals to replace injured or lost limbs parts by growth and differentiation of undifferentiated stem cells.

Region of Elongation In root tips, the region just above the region of cell division, where cells elongate and the root length increases.

Region of Maturation In root tips, the region above the region of elongation; cells differentiate and root hairs occur in this region.

Regulatory Genes Genes whose sole function is to control the expression of structural genes.

Releaser A sign stimulus that is given by an individual to another member of the same species, eliciting a specific innate behavior.

Releasing Factors Hormones secreted by the tips of hypothalmic neurosecretory cells that stimulate the anterior pituitary to release its hormones. GnRH, for example, stimulates the release of gonadotropins.

Renal Referring to the kidney.

Replication Duplication of DNA, usually prior to cell division.

Replication Fork The site where the two strands of a DNA helix are unwinding during replication.

Repression Inhibition of transcription of a gene which, in an operon, occurs when repressor protein binds to the operator.

Repressor Protein encoded by a bacterial regulatory gene that binds to an operator site of an operon and inhibits transcription.

Reproduction The process by which an organism produces offspring.

Reproductive Isolation Phenomenon in which members of a single population become split into two populations that no longer interbreed.

Reproductive System System of specialized organs that are utilized for the production of gametes and, in some cases, the fertilization and/or development of an egg.

Reptiles Members of class Reptilia, scaly, air-breathing, egg-laying vertebrates such as lizards, snakes, turtles, and crocodiles.

Resolving Power The ability of an optical instrument (eye, microscopes) to discern whether two very close objects are separate from each other.

Resource Partitioning Temporal or spatial sharing of a resource by different species.

Respiration Process used by organisms to exchange gases with the environment; the source of oxygen required for metabolism. The process organisms use to oxidize glucose to CO_2 and H_2O using an electron transport system to extract energy from electrons and store it in the high-energy bonds of ATP.

Respiratory System The specialized set of organs that function in the uptake of oxygen from the environment.

Resting Potential The electrical potential (voltage) across the plasma membrane of a neuron when the cell is not carrying an impulse. Results from a difference in charge across the membrane.

Restriction Enzyme A DNA-cutting enzyme found in bacteria.

Restriction Fragment Length Polymorphism (RFLP) Certain sites in the DNA tend to have a highly variable sequence from one individual to another. Because of these differences, restriction enzymes cut the DNA from different individuals into fragments of different length. Variations in the length of particular fragments (RFLPs) can be used as genetic signposts for the identification of nearby genes of interest.

Restriction Fragments The DNA fragments generated when purified DNA is treated with a particular restriction enzyme.

Reticular Formation A series of interconnected sites in the core of the brain (brainstem) that selectively arouse conscious activity.

Retroviruses RNA viruses that reverse the typical flow of genetic information; within the infected cell, the viral DNA serves as a template for synthesis of a DNA copy. Examples include HIV, which causes AIDS, and certain cancer viruses.

Reverse Genetics Determining the amino acid sequence and function of a polypeptide from the nucleotide sequence of the gene that codes for that polypeptide.

Reverse Transcriptase An enzyme present in retroviruses that transcriibes a strand of DNA, using viral RNA as the template.

Rhizoids Slender cells that resemble roots but do not absorb water or minerals.

Rhodophyta Red algae; seaweeds that can absorb deeper penetrating light rays than most aquatic photosynthesizers.

Rhyniophytes Ancient plants having vascular tissue which thrived in marshy areas during the Silurian period.

Ribonucleic Acid (RNA) Single-stranded chain of nucleotides each comprised of ribose (a sugar), phosphate, and one of four bases (adenine, guanine, cytosine, and uracil). The sequence of nucleotides in RNA is dictated by DNA, from which it is transcribed. There are three classes of RNA: mRNA, tRNA, and rRNA, all required for protein synthesis.

Ribosomal RNA (rRNA) RNA molecules that form part of the ribosome. Included among the rRNAs is one that is thought to catalyze peptide bond formation.

Ribosomes Organelles involved in protein synthesis in the cytoplasm of the cell.

Ribozymes RNAs capable of catalyzing a chemical reaction, such as peptide bond formation or RNA cutting and splicing.

Rickettsias A group of obligate intracellular parasites, smaller than the typical prokaryote. They cause serious diseases such as typhus.

River Flowing body of surface fresh water; rivers are formed from the convergence of streams.

RNA Polymerase The enzyme that directs transcription and assembling RNA nucleotides in the growing chain.

RNA Processing The process by which the intervening (noncoding) portions of a primary RNA transcript are removed and the remaining (coding) portions are spliced together to form an mRNA.

Root Cap A protective cellular helmet at the tip of a root that surrounds delicate meristematic cells and shields them from abrasion and directs the growth downward.

Root Hairs Elongated surface cells near the tip of each root for the absorption of water and minerals.

Root Nodules Knobby structures on the roots of certain plants. They house nitrogen-fixing bacteria which supply nitrogen in a form that can be used by the plant.

Root Pressure A positive pressure as a result of continuous water supply to plant roots that assists (along with transpirational pull) the pushing of water and nutrients up through the xylem.

Root System The below-ground portion of a plant, consisting of main roots, lateral roots, root hairs, and associated structures and systems such as root nodules or mycorrhizae.

Rough ER (RER) Endoplasmatic reticulum with many ribosomes attached. As a result, they appear rough in electron micrographs.

Ruminant Grazing mammals that possess an additional stomach chamber called rumen which is heavily fortified with cellulose-digesting microorganisms.

◄ S ►

S Phase The second stage of interphase in which the materials needed for cell division are synthesized and an exact copy of cell's DNA is made by DNA replication.

Sac Body The body plan of simple animals, like cnidarians, where there is a single opening leading to and from a digestion chamber.

Saltatory Conduction The "hopping" movement of an impulse along a myelinated neuron from one Node of Ranvier to the next one.

Sap Fluid found in xylem or sieve of phloem.

Saprophyte Organisms, mainly fungi and bacteria, that get their nutrition by breaking down organic wastes and dead organisms, also called decomposers.

Saprobe Organism that obtains its nutrients by decomposing dead organisms.

Sarcolemma The plasma membrane of a muscle fiber.

Sarcomere The contractile unit of a myofibril in skeletal muscle.

Sarcoplasmic Reticulum (SR) In skeletal muscle, modified version of the endoplasmic reticulum that stores calcium ions.

Savanna A grassland biome with alternating dry and rainy seasons. The grasses and scattered trees support large numbers of grazing animals.

Scaling Effect A property that changes disproportionately as the size of organisms increase.

Scanning Electron Microscope (SEM) A microscope which operates by showering electrons back and forth across the surface of a specimen prepared with a thin metal coating. The resultant image shows three-dimensional features of the specimen's surface.

Schwann Cells Cells which wrap themselves around the axons of neurons forming an insulating myelin sheath composed of many layers of plasma membrane.

Sclereids Irregularly-shaped sclerenchyma cells, all having thick cell walls; a component of seed coats and nuts.

Sclerenchyma Component of the ground tissue system of plants. They are thick walled cells of various shapes and sizes, providing support or protection. They continue to function after the cell dies.

Sclerenchyma Fibers Non-living elongated plant cells with tapering ends and thick secondary walls. A supportive cell type found in various plant tissues.

Sebaceous Glands Exocrine glands of the skin that produce a mixture of lipids (sebum) that oil the hair and skin.

Secondary Cell Wall An additional cell wall that improves the strength and resiliency of specialized plant cells, particularly those cells found in stems that support leaves, flowers, and fruit.

Secondary Consumer Organism that feeds exclusively on primary consumers; mostly animals, but some plants.

Secondary Growth Growth from cambia in perennials; results in an increase in the diameter of stems and roots.

Secondary Meristems (vascular cambium, cork cambrium) Rings or clusters of meristematic cells that increase the width of stems and roots when the divide.

Secondary Sex Characteristics Those characteristics other than the gonads and reproductive tract that develop in response to sex hormones. For example, breasts and pubic hair in women and a deep voice and pubic hair in men.

Secondary Succession The development of a community in an area previously occupied by a community, but which was disturbed in some manner; for example, fire, development, or clear-cutting forests.

Secondary Tissues Tissues produced to accommodate new cell production in plants with woody growth. Secondary tissues are produced from cambia, which produce vascular and cork tissues, leading to an increase in plant girth.

Second Messenger Many hormones, such as glucagon and thyroid hormone, evoke a response by binding to the outer surface of a target cell and causing the release of another substance (which is the second messenger). The best-studied second messenger is cyclic AMP which is formed by an enzyme on the inner surface of the plasma membrane following the binding of a hormone to the outer surface of the membrane. The cyclic AMP diffuses into the cell and activates a protein kinase.

Secretion The process of exporting materials produced by the cell.

Seed A mature ovule consisting of the embryo, endosperm, and seed coat.

Seed Dormancy Metabolic inactivity of seeds until favorable conditions promote seed germination.

Secretin Hormone secreted by endocrine cells in the wall of the intestine that stimulates the release of digestive products from the pancreas.

Segmentation A condition in which the body is constructed, at least in part, from a series of repeating parts. Segmentation occurs in annelids, arthropods, and vertebrates (as revealed during embryonic development).

Selectively Permeable A term applied to the plasma membrane because membrane proteins control which molecules are transported. Enables a cell to import and accumulate the molecules essential for normal metabolism.

Semen The fluid discharged during a male orgasm.

Semiconsevative Replication The manner in which DNA replicates; half of the original DNA strand is conserved in each new double helix.

Seminal Vesicles The organs which produce most of the ejaculatory fluid.

Seminiferous Tubules Within the testes, highly coiled and compacted tubules, lined with a self-perpetuating layer of spermatogonia, which develop into sperm.

Senescence Aging and eventual death of an organism, organ or tissue.

Sense Strand The one strand of a DNA double helix that contains the information that encodes the amino sequence of a polypeptide. This is the strand that is selectively transcribed by RNA polymerase forming an mRNA that can be properly translated.

Sensory Neurons Neurons which relay impulses to the central nervous system.

Sensory Receptors Structures that detect changes in the external and internal environment and transmit the information to the nervous system.

Sepal The outermost whorl of a flower, enclosing the other flower parts as a flower bud; collectively called the calyx.

Sessile Sedentary, incapable of independent movement.

Sex Chromosomes The one chromosomal pair that is not identical in the karyotypes of males and females of the same animal species.

Sex Hormones Steroid hormones which influence the production of gametes and the development of male or female sex characteristics.

Sexual Dimorphism Differences in the appearance of males and females in the same species.

Sexual Reproduction The process by which haploid gametes are formed and fuse during fertilization to form a zygote.

Sexual Selection The natural selection of adaptations that improve the chances for mating and reproducing.

Shivering Involuntary muscular contraction for generating metabolic heat that raises body temperature.

Shoot In angiosperms, the system consisting of stems, leaves, flowers and fruits.

Shoot System The above-ground portion of an angiosperm plant consisting of stems with nodes, including branches, leaves, flowers and fruits.

Short-Day Plants Plants that flower in late summer or fall when the length of daylight becomes shorter than some critical period.

Shrubland A biome characterized by densely growing woody shrubs in mediterranean type climate; growth is so dense that understory plants are not usually present.

Sickle Cell Anemia A genetic (recessive autosomal) disorder in which the beta globin genes of adult hemoglobin molecules contain an amino acid substitution which alters the ability of hemoglobin to transport oxygen. During times of oxygen stress, the red blood cells of these individuals may become sickle shaped, which interferes with the flow of the cells through small blood vessels.

Sieve Plate Found in phloem tissue in plants, the wall between sieve-tube members, containing perforated areas for passage of materials.

Sieve-Tube Member A living, food-conducting cell found in phloem tissue of plants; associated with a companion cell.

Sigmoid Growth Curve An S-shaped curve illustrating the lag phase, exponential growth, and eventual approach of a population to its carrying capacity.

Sign Stimulus An object or action in the environment that triggers an innate behavior.

Simple Fruits Fruits that develop from the ovary of one pistil.

Simple Leaf A leaf that is undivided; only one blade attached to the petiole.

Sinoatrial (SA) Node A collection of cells that generates an action potential regulating heart beat; the heart's pacemaker.

Skeletal Muscles Separate bundles of parallel, striated muscle fibers anchored to the bone, which they can move in a coordinated fashion. They are under voluntary control.

Skeleton A rigid form of support found in most animals either surrounding the body with a protective encasement or providing a living girder system within the animal.

Skull The bones of the head, including the cranium.

Slow-Twitch Fibers Skeletal muscle fibers that depend on aerobic metabolism for ATP production. These fibers are capable of undergoing contraction for extended periods of time without fatigue, but generate lesser forces than fast-twitch fibers.

Small Intestine Portion of the intestine in which most of the digestion and absorption of nutrients takes place. It is so named because of its narrow diameter. There are three sections: duodenum, jejunum, and ilium.

Smell Sense of the chemical composition of the environment.

Smooth ER (SER) Membranes of the endoplasmic reticulum that have no ribosomes on their surface. SER is generally more tubular than the RER. Often acts to store calcium or synthesize steroids.

Smooth Muscle The muscles of the internal organs (digestive tract, glands, etc.). Composed of spindle-shaped cells that interlace to form sheets of visceral muscle.

Social Behavior Behavior among animals that live in groups composed of individuals that are dependent on one another and with whom they have evolved mechanisms of communication.

Social Learning Learning of a behavior from other members of the species.

Social Parasitism Parasites that use behavioral mechanisms of the host organism to the parasite's advantage, thereby harming the host.

Solute A substance dissolved in a solvent.

Solution The resulting mixture of a solvent and a solute.

Solvent A substance in which another material dissolves by separating into individual molecules or ions.

Somatic Cells Cells that do not have the potential to form reproductive cells (gametes). Includes all cells of the body except germ cells.

Somatic Nervous System The nerves that carry messages to the muscles that move the skeleton either voluntarily or by reflex.

Somatic Sensory Receptors Receptors that respond to chemicals, pressure, and temperature that are present in the skin, muscles, tendons, and joints. Provides a sense of the physiological state of the body.

Somites In the vertebrate embryo, blocks of mesoderm on either side of the notochord that give rise to muscles, bones, and dermis.

Speciation The formation of new species. Occurs when one population splits into separate populations that diverge genetically to the point where they become separate species.

Species Taxonomic subdivisions of a genus. Each species has recognizable features that distinguish it from every other species. Members of one species generally will not interbreed with members of other species.

Specific Epithet In taxonomy, the second term in an organism's scientific name identifying its species within a particular genus.

Spermatid Male germ cell that has completed meiosis but has not yet differentiated into a sperm.

Spermatogenesis The production of sperm.

Spermatogonia Male germ cells that have not yet begun meiosis.

Spermatozoa (Sperm) Male gametes.

Sphinctors Circularly arranged muscles that close off the various tubes in the body.

Spinal Cord A centralized mass of neurons for processing neurological messages and linking the brain with that part of peripheral nervous system not reached by the cranial nerves.

Spinal Nerves Paired nerves which emerge from the spinal cord and innervate the body. Humans have 31 pairs of spinal nerves.

Spindle Apparatus In dividing eukaryotic cells, the complex rigging, made of microtubules, that aligns and separates duplicated chromosomes.

Splash Zone In the intertidal zone, the uppermost region receiving splashes and sprays of water to the mean of high tides.

Spleen One of the organs of the lymphatic system that produces lymphocytes and filters blood; also produces red blood cells in the human fetus.

Splicing The step during RNA processing in which the coding segments of the primary transcript are covalently linked together to form the mRNA.

Spongy Parenchyma In monocot and dicot leaves, loosely arranged cells functioning in photosynthesis. Found above the lower epidermis and beneath the palisade parenchyma in dicots, and between the upper and lower epidermis in monocots.

Spontaneous Generation Disproven concept that living organisms can arise directly from inanimate materials.

Sporangiospores Black, asexual spores of the zygomycete fungi.

Sporangium A hollow structure in which spores are formed.

Spores In plants, haploid cells that develop into the gametophyte generation. In fungi, an asexual or sexual reproductive cell that gives rise to a new mycelium. Spores are often lightweight for their dispersal and adapted for survival in adverse conditions.

Sporophyte The diploid spore producing generation in plants.

Stabilizing Selection Natural selection favoring an intermediate phenotype over the extremes.

Starch Polysaccharides used by plants to store energy.

Stamen The flower's male reproductive organ, consisting of the pollen-producing anther supported by a slender stalk, the filament.

Stem In plants, the organ that supports the leaves, flowers, and fruits.

Stem Cells Cells which are undifferentiated and capable of giving rise to a variety of different types of differentiated cells. For example, hematopoietic stem cells are capable of giving rise to both red and white blood cells.

Steroids Compounds classified as lipids which have the basic four-ringed molecular skeleton as represented by cholesterol. Two examples of steroid hormones are the sex hormones; testosterone in males and estrogen in females.

Stigma The sticky area at the top of each pistil to which pollen adheres.

Stimulus Any change in the internal or external environment to which an organism can respond.

Stomach A muscular sac that is part of the digestive system where food received from the esophagus is stored and mixed, some breakdown of food occurs, and the chemical degradation of nutrients begins.

Stomates (Pl. Stomata) Microscopic pores in the epidermis of the leaves and stems which allow gases to be exchanged between the plant and the external environment.

Stratified Epithelia Multicellular layered epithelium.

Stream Flowing body of surface fresh water; streams merge together into larger streams and rivers.

Stretch Receptors Sensory receptors embedded in muscle tissue enabling muscles to respond reflexively when stretched.

Striated Referring to the striped appearance of skeletal and cardiac muscle fibers.

Strobilus In lycopids, terminal, cone-like clusters of specialized leaves that produce sporangia.

Stroma The fluid interior of chloroplasts.

Stromatolites Rocks formed from masses of dense bacteria and mineral deposits. Some of these rocky masses contain cells that date back over three billion years revealing the nature of early prokaryotic life forms.

Structural Genes DNA segments in bacteria that direct the formation of enzymes or structural proteins.

Style The portion of a pistil which joins the stigma to the ovary.

Substrate-Level Phosphorylation The formation of ATP by direct transfer of a phosphate group from a substrate, such as a sugar phosphate, to ADP. ATP is formed without the involvement of an electron transport system.

Substrates The reactants which bind to enzymes and are subsequently converted to products.

Succession The orderly progression of communities leading to a climax community. It is one of two types: primary, which occurs in areas where no community existed before; and secondary, which occurs in disturbed habitats where some soil and perhaps some organisms remain after the disturbance.

Succulents Plants having fleshy, water-storing stems or leaves.

Suppressor T Cells A class of T cells that regulate immune responses by inhibiting the activation of other lymphocytes.

Surface Area-to-Volume Ratio The ratio of the surface area of an organism to its volume, which determines the rate of exchange of materials between the organism and its environment.

Surface Tension The resistance of a liquid's surface to being disrupted. In aqueous solutions, it is caused by the attraction between water molecules.

Survivorship Curve Graph of life expectancy, plotted as the number of survivors versus age.

Sweat Glands Exocrine glands of the skin that produce a dilute salt solution, whose evaporation cools in the body.

Symbiosis A close, long-term relationship between two individuals of different species.

Symmetry Referring to a body form that can be divided into mirror image halves by at least one plane through its body.

Sympathetic Nervous System Part of the autonomic nervous system that tends to stimulate bodily activities, particularly those involved with coping with stressful situations.

Sympatric Speciation Speciation that occurs in populations with overlapping distributions. It is common in plants when polyploidy arises within a population.

Synapse Juncture of a neuron and its target cell (another neuron, muscle fiber, gland cell).

Synapsis The pairing of homologous chromosomes during prophase of meiosis I.

Synaptic Cleft Small space between the synaptic knobs of a neuron and its target cell.

Synaptic Knobs The swellings that branch from the end of the axon. They deliver the neurological impulse to the target cell.

Synaptonemal Complex Ladderlike structure that holds homologous chromosomes together as a tetrad during crossing over in prophase I of meiosis.

Synovial Cavities Fluid-filled sacs around joints, the function of which is to lubricate and separate articulating bone surfaces.

Systemic Circulation Part of the circulatory system that delivers oxygenated blood to the tissues and routes deoxygenated blood back to the heart.

Systolic Pressure The first number of a blood pressure reading; the highest pressure attained in the arteries as blood is propelled out of the heart.

◀ T ▶

Taiga A biome found south of tundra biomes; characterized by coniferous forests, abundant precipitation, and soils that thaw only in the summer.

Tap Root System Root system of plants having one main root and many smaller lateral roots. Typical of conifers and dicots.

Taste Sense of the chemical composition of food.

Taxonomy The science of classifying and grouping organisms based on their morphology and evolution.

T Cell Lymphocytes that carry out cell-mediated immunity. They respond to antigen stimulation by becoming helper cells, killer cells, and memory cells.

Telophase The final stage of mitosis which begins when the chromosomes reach their spindle poles and ends when cytokinesis is completed and two daughter cells are produced.

Tendon A dense connective tissue cord that connects a skeletal muscle to a bone.

Teratogenic Embryo deforming. Chemicals such as thalidomide or alcohol are teratogenic because they disturb embryonic development and lead to the formation of an abnormal embryo and fetus.

Terminal Electron Acceptor In aerobic respiration, the molecule of O_2 which removes the electron pair from the final cytochrome of the respiratory chain.

Terrestrial Living on land.

Territory (Territoriality) An area that an animal defends against intruders, generally in the protection of some resource.

Tertiary Consumer Animals that feed on secondary consumers (plant or animal) or animals only.

Test Cross An experimental procedure in which an individual exhibiting a dominant trait is crossed to a homozygous recessive to determine whether the first individual is homozygous or heterozygous.

Testis In animals, the sperm-producing gonad of the male.

Testosterone The male sex hormone secreted by the testes when stimulated by pituitary gonadotropins.

Tetrad A unit of four chromatids formed by a synapsed pair of homologous chromosomes, each of which has two chromatids.

Thallus In liverworts, the flat, ground-hugging plant body that lacks roots, stems, leaves, and vascular tissues.

Theory of Tolerance Distribution, abundance and existence of species in an ecosystem are determined by the species' range of tolerance of chemical and physical factors.

Thermoreceptors Sensory receptors that respond to changes in temperature.

Thermoregulation The process of maintaining a constant internal body temperature in spite of fluctuations in external temperatures.

Thigmotropism Changes in plant growth stimulated by contact with another object, e.g., vines climbing on cement walls.

Thoracic Cavity The anterior portion of the body cavity in which the lungs are suspended.

Thylakoids Flattened membrane sacs within the chloroplast. Embedded in these membranes are the light-capturing pigments and other components that carry out the light-dependent reactions of photosynthesis.

Thymus Endocrine gland in the chest where T cells mature.

Thyroid Gland A butterfly-shaped gland that lies just in front of the human windpipe, producing two metabolism-regulating hormones, thyroxin and triodothyronine.

Thyroid Hormone A mixture of two iodinated amino acid hormones (thyroxin and triiodothyronine) secreted by the thyroid gland.

Thyroid Stimulating Hormone (TSH) An anterior pituitary hormone which stimulates secretion by the thyroid gland.

Tissue An organized group of cells with a similar structure and a common function.

Tissue System Continuous tissues organized to perform a specific function in plants. The three plant tissue systems are: dermal, vascular, and ground (fundamental).

Tolerance Range The range between the maximum and minimum limits for an environmental factor that is necessary for an organism's survival.

Totipotent The genetic potential for one type of cell from a multicellular organism to give rise to any of the organism's cell types, even to generate a whole new organism.

Trachea The windpipe; a portion of the respiratory tract between the larynx and bronchii.

Tracheal Respiratory System A network of tubes (tracheae) and tubules (tracheoles) that carry air from the outside environment directly to the cells of the body without involving the circulatory system.

Tracheid A type of conducting cell found in xylem functioning when a cell is dead to transport water and dissolved minerals through its hollow interior.

Tracheophytes Vascular plants that contain fluid-conducting vessels.

Transcription The process by which a strand of RNA assembles along one of the DNA strands.

Transcriptional-Level Control Control of gene expression by regulating whether or not a specific gene is transcribed and how often.

Transduction A type of genetic recombination resulting from transfer of genes from one organism to another by a virus.

Transfer RNA (tRNA) A type of RNA that decodes mRNA's codon message and translates it into amino acids.

Transgenic Organism An organism that possesses genes derived from a different species. For example, a sheep that carries a human gene and secretes the human protein in its milk is a transgenic animal.

Translation The cell process that converts a sequence of nucleotides in mRNA into a sequence of amino acids in a polypeptide.

Translational-Level Control Control of gene expression by regulating whether or not a specific mRNA is translated into a polypeptide.

Translocation The joining of segments of two nonhomologous chromosomes

Transmission Electron Microscope (TEM) A microscope that works by shooting electrons through very thinly sliced specimens. The result is an enormously magnified image, two-dimensional, of remarkable detail.

Transpiration Water vapor loss from plant surfaces.

Transpiration Pull The principle means of water and mineral transport in plants, initiated by transpiration.

Transposition The phenomenon in which certain DNA segments (mobile genetic elements, or jumping genes) tend to move from one part of the genome to another part.

Transverse Fission The division pattern in ciliated protozoans where the plane of division is perpendicular to the cell's length.

Trimester Each of the three stages comprising the 266-day period between conception and birth in humans.

Triploid Having three sets of chromosomes, abbreviated 3N.

Trisomy Three copies of a particular chromosome per cell.

Trophic Level Each step along a feeding pathway.

Trophozoite The actively growing stage of polymorphic protozoa.

Tropical Rain Forest Lush forests that occur near the equator; characterized by high annual rainfall and high average temperature.

Tropical Thornwood A type of shrubland occurring in tropical regions with a short rainy season. Plants lose their small leaves during dry seasons, leaving sharp thorns.

Tropic Hormones Hormones that act on endocrine glands to stimulate the production and release of other hormones.

Tropisms Changes in the direction of plant growth in response to environmental stimuli, e.g., light, gravity, touch.

True-Breeder Organisms that, when bred with themselves, always produce offspring identical to the parent for a given trait.

Tubular Reabsorption The process by which substances are selectively returned from the fluid in the nephron to the bloodstream.

Tubular Secretion The process by which substances are actively and selectively transported from the blood into the fluid of the nephron.

Tumor-Infiltrating Lymphocytes (TILs) Cytotoxic T cells found within a tumor mass that have the capability to specifically destroy the tumor cells.

Tumor-Suppressor Genes Genes whose products act to block the formation of cancers. Cancers form only when both copies of these genes (one on each homologue) are mutated.

Tundra The marshy, unforested biome in the arctic and at high elevations. Frigid temperatures for most of the year prevent the subsoil from thawing, which produces marshes and ponds. Dominant vegetation includes low growing plants, lichens, and mosses.

Turgor Pressure The internal pressure in a plant cell caused by the diffusion of water into the cell. Because of the rigid cell wall, pressure can increase to where it eventually stops the influx of more water.

Turner Syndrome A person whose cells have only one X chromosome and no second sex chromosome (XO). These individuals develop as immature females.

◀ U ▶

Ultimate (Top) Consumer The final carnivore trophic level organism, or organisms that escaped predation; these consumers die and are eventually consumed by decomposers.

Ultracentrifuge An instrument capable of spinning tubes at very high speeds, delivering centrifugal forces over 100,000 times the force of gravity.

Unicellular The description of an organism where the cell is the organism.

Uniform Pattern Distribution of individuals of a population in a uniform arrangement, such as individual plants of one species uniformly spaced across a region.

Urethra In mammals, a tube that extends from the urinary bladder to the outside.

Urinary Tract The structures that form and export urine: kidneys, ureters, urinary bladder, and urethra.

Urine The excretory fluid consisting of urea, other nitrogenous substances, and salts dissolved in water. It is formed by the kidneys.

Uterine (Menstrual) Cycle The repetitive monthly changes in the uterus that prepare the endometrium for receiving and sustaining an embryo.

Uterus An organ in the female reproductive system in which an embryo implants and is maintained during development.

◀ V ▶

Vaccines Modified forms of disease-causing microbes which cannot cause disease but retain the same antigens of it. They permit the immune system to build memory cells without diseases developing during the primary immune response.

Vacoconstriction Reduction in the diameter of blood vessels, particularly arterioles.

Vacuole A large organelle found in mature plant cells, occupying most of the cell's volume, sometimes more than 90% of it.

Vagina The female mammal's copulatory organ and birth canal.

Variable (Experimental) A factor in an experiment that is subject to change, i.e., can occur in more than one state.

Vascular Bundles Groups of vascular tissues (xylem and phloem) in the shoot of a plant.

Vascular Cambrium In perennials, a secondary meristem that produces new vascular tissues.

Vascular Cylinder Groups of vascular tissues in the central region of the root.

Vascular Plants Plants having a specialized conducting system of vessels and tubes for transporting water, minerals, food, etc., from one region to another.

Vascular Tissue System All the vascular tissues in a plant, including xylem, phloem, and the vascular cambium or procambium.

Vasodilation Increase in the diameter of blood vessels, particularly arterioles.

Veins In plants, vascular bundles in leaves. In animals, blood vessels that return blood to the heart.

Venation The pattern of vein arrangement in leaf blades.

Ventricle Lower chamber of the heart which pumps blood through arteries. There is one ventricle in the heart of lower vertebrates and two ventricles in the four-chambered heart of birds and mammals.

Venules Small veins that collect blood from the capillaries. They empty into larger veins for return to the heart.

Vertebrae The bones that form the backbone. In the human there are 33 bones arranged in a gracefully curved line along the bone, cushioned from one another by disks of cartilage.

Vertebral Column The backbone, which encases and protects the spinal cord.

Vertebrates Animals with a backbone.

Vesicles Small membrane-enclosed sacs which form from the ER and Golgi complex. Some store chemicals in the cells; others move to the surface and fuse with the plasma membrane to secrete their contents to the outside.

Vessel A tube or connecting duct containing or circulating body fluids.

Vessel Member A type of conducting cell in xylem functioning when the cell is dead to transport water and dissolved minerals through its hollow interior; also called a vessel element.

Vestibular Apparatus A portion of the inner ear of vertebrates that gathers information about the position and movement of the head for use in maintaining balance and equilibrium.

Vestigial Structure Remains of ancestral structures or organs which were, at one time, useful.

Villi Finger-like projections of the intestinal wall that increase the absorption surface of the intestine.

Viroids are associated with certain diseases of plants. Each viroid consists solely of a small single-stranded circle of RNA unprotected by a protein coat.

Virus Minute structures composed of only heredity information (DNA or RNA), surrounded by a protein or protein/lipid coat. After infection, the viral mucleic acid subverts the metabolism of the host cell, which then manufactures new virus particles.

Visible Light The portion of the electromagnetic spectrum producing radiation from 380 nm to 750 nm detectable by the human eye.

Vitamins Any of a group of organic compounds essential in small quantities for normal metabolism.

Vocal Cords Muscular folds located in the larynx that are responsible for sound production in mammals.

Vulva The collective name for the external features of the human female's genitals.

◀ W ▶

Water Vascular System A system for locomotion, respiration, etc., unique to echinoderms.

Wavelength The distance separating successive crests of a wave.

Waxes A waterproof material composed of a number of fatty acids linked to a long chain alcohol.

White Matter Regions of the brain and spinal cord containing myelinated axons, which confer the white color.

Wild Type The phenotype of the typical member of a species in the wild. The standard to which mutant phenotypes are compared.

Wilting Drooping of stems or leaves of a plant caused by water loss.

Wood Secondary xylem.

◀ X ▶

X Chromosome The sex chromosome present in two doses in cells of a female, and in one dose in the cells of a male.

X-Linked Traits Traits controlled by genes located on the X chromosome. These traits are much more common in males than females.

Xylem The vascular tissue that transports water and minerals from the roots to the rest of the plant. Composed of tracheids and vessel members.

◀ Y ▶

Y Chromosome The sex chromosome found in the cells of a male. When Y-carrying sperm unite with an egg, all of which carry a single X chromosome, a male is produced.

Y-Linked Inheritance Genes carried only on Y chromosomes. There are relatively few Y-linked traits; maleness being the most important such trait in mammals.

Yeast Unicellular fungus that forms colonies similar to those of bacteria.

Yolk A deposit of lipids, proteins, and other nutrients that nourishes a developing embryo.

Yolk Sac A sac formed by an extraembryonic membrane. In humans, it manufactures blood cells for the early embryo and later helps to form the umbilical cord.

◀ Z ▶

Zero Population Growth In a population, the result when the combined positive growth factors (births and immigration).

Zooplankton Protozoa, small crustaceans and other tiny animals that drift with ocean currents and feed on phytoplankton.

Zygospore The diploid spores of the zygomycete fungi, which include Rhizopus, a common bread mold. After a period of dormancy, the zygospore undergoes meiosis and germinates.

Zygote A fertilized egg. The diploid cell that results from the union of a sperm and egg.

Photo Credits

Table of Contents Page xii: Courtesy R.S. Wilcox, Biology Department, SUNY Binghamton. Page xiv (top): Nancy Kedersha. Page xiv (bottom): Stephen Frink/AllStock, Inc. Page xv: G.I. Bernard/Earth Scenes/Animals Animals. Page xvi (top): Nancy Sefton/Photo Researchers. Page xvi (bottom): Seaphot Limited/Planet Earth Pictures. Page xvii: R. Konig/Jacana/The Image Bank. Page xviii (bottom): Carr Cliffton. Page xviii (top): Roy P. Fontaine/Photo Researchers. **Part 1 Opener:** Norbert Wu. **Chapter 1** Figure 1.1: Jeff Gnass. Figure 1.2a: Frans Lanting/Minden Pictures, Inc. Figure 1.2b: David Muench Photography. Figure 1.3a: Courtesy Dr. Alan Cheetham, National Museum of Natural History, Smithsonian Institution. Figure 1.3b: Manfred Kage/Peter Arnold, Inc. Figure 1.3c: Stephen Dalton/Natural History Photographic Agency. Figure 1.3d: Charles Summers, Jr. Figure 1.3e: Larry West/ Photo Researchers. Figure 1.3f: Bianca Lavies. Figure 1.3g: Steve Allen/Peter Arnold, Inc. Figure 1.3h: George Grall. Figure 1.3i: Michio Hoshino/Minden Pictures, Inc. Figure 1.6a: CNRI/Science Photo Library/Photo Researchers. Figure 1.6b: M. Abbey/Visuals Unlimited. Figure 1.6c: Steve Kaufman/Peter Arnold, Inc. Figure 1.6d: Willard Clay. Figure 1.6e: Jim Bradenburg/Minden Pictures, Inc. Figure 1.7: Paul Chesley/Photographers Aspen. Figure 1.8: Rainbird/Robert Harding Picture Library. Figure 1.9: Dick Luria/FPG International. Figure 1.11: Courtesy Institut Pasteur. Figure 1.13a: Ted Horowitz/The Stock Market. Figure 1.13b: Laurence Gould/Earth Scenes. **Part 2 Opener:** Nancy Kedersha. **Chapter 2** Figure 2.1: Franklin Viola. Figure 2.2: Courtesy Pachyderm Scientific Industries. Figure 2.7b: Courtesy R.S. Wilcox, Biology Department, SUNY Binghamton. Figure 2.8 (inset): Courtesy Stephen Harrison, Harvard Biochemistry Department. Figure 2.9: Gary Milburn/Tom Stack & Associates. Figure 2.12: Alastair Black/Tony Stone World Wide. Figure 2.15a: Don Fawcett/Visuals Unlimited. Figure 2.15b: Jeremy Burgess/Photo Researchers. Figure 2.15c: Cabisco/Visuals Unlimited. Figure 2.16: Tony Stone World Wide. Figure 2.17: Zig Leszczynski/Animals Animals. Figure 2.21a: Frans Lanting/AllStock, Inc. Figure 2.21b: Mantis Wildlife Films/Oxford Scientific Films/ Animals Animals. Figure 2.25: Stanley Flegler/Visuals Unlimited. Figure 2.26a: Courtesy Nelson Max, Lawrence Livermore Laboratory. Figure 2.26b: Tsuned Hayashida/The Stock Market. **Human Perspective** Figure 1a: Photofest. Figure 1b: Bill Davila/Retna. **Chapter 3** Figure 3.1a: Granger Collection. Figure 3.1b : Bettmann Archive. Figure 3.2a: CNRI/Photo Researchers. Figure 3.2b: Dr. Jeremy Burgess/Photo Researchers. Figure 3.3: Omikron/Photo Researchers. Figure 3.5a: Courtesy Richard Chao, California State University at Northridge. Figure 3.6: Courtesy G.F. Bohr. Figure 3.7: Courtesy D.W. Fawcett, Harvard Medical School. Figure 3.8a: Courtesy U.S. Department of Agriculture. Figure 3.9: Courtesy Dr. Birgit H. Satir, Albert Einstein College of Medicine. Figure 3.10a: K.R. Porter/Photo Researchers. Figure 3.10d: Courtesy Lennart Nilsson, from *A Child Is Born.* Figure 3.11: Courtesy T. Elliot Weier. Figure 3.12: Courtesy J.V. Small. Figure 3.13a: Courtesy E. Vivier, from *Paramecium* by W.J. Wagtendonk, Elsevier North Holland Biomedical Press, 1974. Figure 3.13b: David Phillips/Photo Researchers. Figure 3.13c: Courtesy L.G. Tilney and K. Fujiwara. Figure 3.14a: Courtesy W. Gordon Whaley, University of Texas, Austin. Figure 3.14c: Courtesy R.D. Preston, University of Leeds, London. Figure 3.18a: Photo by Lennart Nilsson, from *The Incredible Machine.* Figure 3.18b: Courtesy Dr. Ravi Pathak, Southwestern Medical Center, University of Texas. **Human Perspective** Figure 1a: Martin Rotker/Phototake. Figure 1b: Cabisco/Visuals Unlimited. **Chapter 4** Figure 4.1: Marty Stouffer/Animals Animals. Figure 4.2: Kay Chernush/The Image Bank. Figure 4.6a: Courtesy Stan Koszelek, PhD., University of California, Riverside. **Chapter 5** Figure 5.1a: Carr Clifton. Figure 5.1b: Shizuo Lijima/Tony Stone World Wide. Figure 5.4: Joe Englander/Viesti Associates, Inc. Figure 5.5 (inset): Courtesy T. Elliot Weier. Figure 5.10: Courtesy Woods Hole Oceanographic Institution. **Chapter 6** Figure 6.1: Stephen Frink/AllStock, Inc. Figure 6.8a: Grafton M. Smith/The Image Bank. **Human Perspective** Top left: Jerry Cooke. Top and bottom inset: Courtesy MacDougal. Bottom left: Richard Kane/Sports Chrome Inc. **Chapter 7** Figure 7.2: Courtesy Dr. Andrew Bajer, University of Oregon. Figure 7.3: CNRI/Science Photo Library/Photo Researchers. Figure 7.4: Dr. G. Shatten/Science Photo Library/ Photo Researchers. Figure 7.5: Courtesy Professor R. G. Kessel. Figure 7.6: David Phillips/ Photo Researchers. Figure 7.7a: Dr. R. Vernon/Phototake. Figure 7.7b: CNRI/Science Photo Library/Photo Researchers. Figure 7.9: Courtesy Science Software Systems. **Part 3 Opener:** David Scharf. **Chapter 8** Figure 8.6a: Doc Pele/Retna. Figure 8.6b: FPG International. Figure 8.8: Sydney Freelance/Gamma Liaison. Figure 8.10: Hans Reinhard/Bruce Coleman, Inc. **Chapter 9** Figure 9.3: Courtesy Computer Graphics Group, Lawrence Livermore Laboratory. Figure 9.13: Fawcett/ Olins/Photo Researchers.**Human Perspective** Courtesy Skin Cancer Foundation. **Chapter 10** Figure 10.1a (left): Oxford Scientific Films/ Animals Animals. Figure 10.1a (right): F. Stuart Westmorland/Tom Stack & Associates. Figure 10.1b: Courtesy Dr. Cecilio Barrera, Department of Biology, New Mexico State University. Figure 10.7a: Courtesy Stephen Case, University of Mississippi Medical Center. Figure 10.8: Roy Morsch/The Stock Market. **Chapter 11** Figure 11.3: Tom Raymond/Medichrome/The Stock Shop. Figure 11.6a: Will & Deni McIntyre/Photo Researchers. Figure 11.7: Art Wolfe/AllStock, Inc. Figure 11.9b: Stanley N. Cohen/Photo Researchers. Figure 11.9b: Courtesy R.L. Brinster, Laboratory for Reproductive Physiology, University of Pennsylvania. **Human Perspective 1** Gamma Liaison. **Human Perspective 2** Photograph by Anita Corbin and John O'Grady, courtesy The British Council. **Part 4 Opener:** J. H. Carmichael, Jr./The Image Bank. **Chapter 12** Figure 12.1a: Ralph Perry/Black Star. Figure 12.1b: Jeffrey Hutcherson/DRK Photo. Figure 12.4 (inset): Courtesy Ray Evert, University of Wisconsin, Madison. Figure 12.5: Walter H. Aodge/Peter Arnold, Inc. Figure 12.5 (inset): Courtesy Carolina Biological Supply Company. Figure 12.7: Dr. Jeremy Burgess/Science Photo Library/ Photo Researchers. Figure 12.8a: Carr Clifton. Figure 12.8b: Ed Reschke. Figure 12.9: Courtesy Thomas A. Kuster, Forest Products Laboratory/ USDA. Figure 12.10: Courtesy DSIR Library Centre. Figure 12.12a: Biophoto Associates/Photo Researchers. Figure 12.12b: Alfred Pasieka/Peter Arnold, Inc. Figure 12.12c: Courtesy DSIR Library Centre. Figure 12.14: Biophoto Associates/Photo Researchers. Figure 12.14 (inset): Courtesy C.Y. Shih and R.G. Kessel, reproduced from *Living Images* with permission, Science Books International, 1982. Figure 12.15: Martin Zimmerman. Figure 12.19: Robert & Linda Mitchell. Figure 12.19 (inset): From *Botany Principles* by Roy H. Saigo and Barbara Woodworth Saigo, c. 1983 Prentice-Hall, Inc. Used by permission. Figure 12.22: Courtesy C.P. Reid, School of Natural Resources, University of Arizona. Figure 12.24: D. Cavagnaro/DRK Photo. Figure 12.25: Dr. J. Burgess/Photo Researchers. **Chapter 13** Figure 13.1a: G.I. Bernard/ Earth Scenes/Animals Animals. Figure 13.1b: Phil Degginger. Figure 13.1c: G. I. Bernard/

Animals Animals. Figure 13.2a: Dwight R. Kuhn/DRK Photo. Figure 13.3: E.R. Degginger. Figure 13.4a: Runk/Schoenberger/Grant Heilman Photography. Figure 13.4b: William E. Ferguson. Figure 13.13: Gary Milburn/Tom Stack & Associates. Figure 13.15: Scott Camazine/Photo Researchers. Figure 13.16a: E.R. Degginger. Figure 13.16b: Fritze Prenze/Earth Scenes/Animals Animals. Figure 13.17a: Pam Hickman/Valan Photos. Figure 13.17b: R.F. Head/Earth Scenes/Animals Animals. **Chapter 14** Figure 14.1: Dr. Mary Notter/ Phototake. Figure 14.2b: Department of Anatomy, University of "La Sapienza", Rome/ Photo Researchers. Figure 14.5a: Fred Bavendam/Peter Arnold, Inc. Figure 14.5b: Michael Fogden/DRK Photo. Figure 14.5c: Peter Lamberti/Tony Stone World Wide. Figure 14.5d: Gerry Ellis Nature Photography. Figure 14.6a: Mike Severns/Tom Stack & Associates. Figure 14.6b: Norbert Wu. **Chapter 15** Figure 15.1(inset): Fawcett/Coggeshall/Photo Researchers. Figure 15.4 (top left inset): Vu/T. Reese and D.W. Fawcett/Visuals Unlimited. Figure 15.6: Courtesy Lennart Nilsson, *Behold Man,* Little Brown and Co., Boston. Figure 15.7b: Alan & Sandy Carey. Figure 15.10: Sports Chrome Inc. Figure 15.13: Robert & Linda Mitchell. Figure 15.16: Photograph reproduced with permission of Ringling Brothers-Barnum & Bailey Circus, courtesy Circus World Museum. **Chapter 16** Figure 16.1(left): Omikron/Photo Researchers. Figure 16.1(right): Don Fawcett/ K.Saito/ K. Hama/ Photo Researchers. Figure 16.2a (top): Courtesy Lennart Nilsson, from *Behold Man*, p. 178. Figure 16.2a (center): Don Fawcett/ K. Saito/ K. Hama/Photo Researchers. Figure 16.2b: Star File. Figure 16.3: Jerome Shaw. Figure 16.5a: Anthony Bannister/© Natural History Photographic Agency. Figure 16.5b: Giddings/ The Image Bank. Figure 16.5c: Michael Fogden/Bruce Coleman, Inc. **Chapter 17** Figure 17.1: Ed Reschke/Peter Arnold, Inc. Figure 17.2: Wallin/Taurus Photos . Figure 17.3a: E.S. Ross/Phototake. Figure 17.4 (top inset): Lennart Nilsson, *Behold Man,* Little Brown & Company, Boston. Figure 17.4 (bottom inset): Michael Abbey/Science Source/Photo Researchers. Figure 17.11a: Joe Devenney/The Image Bank. Figure 17.11b: John Cancalosi/ DRK Photo. **Human Perspective** Courtesy Professor Philip Osdoby, Department of Biology, Washington University, St. Louis. **Chapter 18** Figure 18.3 (inset): Micrograph by S.L. Palay, courtesy D.W. Fawcett, from *The Cell,* c. W.B. Saunders Co. Figure 18.4a: Warren Garst/Tom Stack & Associates. Figure 18.4b: Herve Chaumeton/Jacana. Figure 18.5: Sari Levin. **Human Perspective** Derik Muray Photography. **Chapter 19** Figure 19.1 (inset): Biophoto Associates/Photo Researchers. Figure 19.2: Courtesy Lennart Nilsson, *Behold Man,* Little Brown & Company, Boston. Figure 19.6: CNRI/ Science Photo Library/Photo Researchers. Figure 19.7: Howard Sochurek, Inc. Figure 19.9: Dr. Tony Brain/Science Photo Library/ Photo Researchers. Figure 19.10: Ed Reschke/Peter Arnold, Inc. Figure 19.11: Manfred Kage/Peter Arnold, Inc. Figure 19.16 (left and right): Lennart Nilsson, from *Behold Man,* Little Brown & Company, Boston. Figure 19.16 (bottom): CNRI/ Science Photo Library/ Photo Researchers. Figure 19.20: Robert & Linda Mitchell. **Chapter 20** Figures 20.1a and 2.4: Courtesy Lennart Nilsson, Boehringer Ingelheim Internations GmbH. **Human Perspective 1** Figure 1a: David Scharf/Peter Arnold, Inc. Figure 1b: Courtesy Acarology Laboratory, Museum of Biological Diversity, Ohio State University, Columbus. Figure 1c: Scott Camazine/Photo Researchers. **Human Perspective 2** Courtesy Steven Rosenberg, National Cancer Institute. **Chapter 21** Figure 21.1a: R. La Salle/Valan Photos. Figure 21.1b: Biological Photo Service. Figure 21.3 (inset): Courtesy Richard Kessel and Randy Kandon, from *Tissues and Organs,* W.H. Freeman. Figure 21.5: C. Edelmann/Photo Researchers. Figure 21.8a: G. Shih and R. Kessel/Visuals Unlimited. Figure 21.8b (inset): Courtesy A.L. Colwin and L.H. Colwin. Figure 21.9: Courtesy Richard Kessel and Gene Sinh. Figure 21.16: Courtesy Lennart Nilsson, from *A Child Is Born.* **Part 5 Opener:** Joe Devenney/The Image Bank. **Chapter 22** Figure 22.1: Ivan Polunin/ Natural History Photographic Agency. Figure 22.2: Courtesy Victor A. Mc Kusick, Medical Genetics Department, Johns Hopkins University. Figure 22.3: Courtesy Professor Lawrence Cook, University of Manchester. Figure 22.5a: COMSTOCK, Inc. Figure 22.5b: Tom McHugh/AllStock, Inc. Figure 22.5c: Kevin Schafer & Martha Hill/Tom Stack & Associates. Figure 22.5c (inset): Courtesy K.W. Barthel, Museum beim Solenhofer Aktienverein, Germany. Figure 22.7a: John Garrett/Tony Stone World Wide. Figure 22.7b: Heather Angel. Figure 22.8a: Frans Lanting/Minden Pictures, Inc. Figure 22.8b: S. Nielsen/DRK Photo. Figure 22.9: Art Wolfe/AllStock, Inc. Figure 22.10: Zig Leszczynski/Animals Animals. Figure 22.13 (top left): Gary Milburn/ Tom Stack & Associates. Figure 22.13 (top right): Tom McHugh/Photo Researchers. Figure 22.13 (bottom left): Nick Bergkessel/Photo Researchers. Figure 22.13 (bottom right): C.S. Pollitt/Australasian Nature Transparencies. Figure 22.15a: Jane Burton/Bruce Coleman, Inc. Figure 22.15b: R. Konig/Jacana/The Image Bank. Figure 22.15c: Nancy Sefton/Photo Researchers. Figure 22.15d: Seaphot Limited/ Planet Earth Pictures. **Chapter 23** Figure 23.1a: Leonard Lee Rue III/ Animals Animals. Figure 23.1b: Bradley Smith/Earth Scenes/Animals Animals. Figure 23.3a: Courtesy George Poinar, U.C. Berkeley. Figure 23.3b: David Muench Photography. Figure 23.4a: William E. Ferguson. Figure 23.7: David Brill. Figure 23.8: Courtesy Institute of Human Origins. Figure 23.10: John Reader/Science Photo Library/Photo Researchers. **Chapter 24** Figure 24.1a: Courtesy Dr. S.M. Awramik, University of California, Santa Barbara. Figure 24.1b: William E. Ferguson. Figure 24.3: Courtesy Professor Seilacher. Figures 24.7 and 24.9: Carl Buell. **Human Perspective** Louie Psihoyos/ Matrix. **Chapter 25** Figure 25.1b: Richard Feldman/Phototake. Figure 25.3 a,b: David M. Phillips/Visuals Unlimited. Figure 25.3c: Omikron/Science Source/Photo Researchers. Figure 25.4a: Courtesy Wellcom Institute for the History of Medicine. Figure 25.4b: Courtesy Searle Corporation. Figure 25.6a: A.M. Siegelman/Visuals Unlimited. Figure 25.6b: Science Vu/Visuals Unlimited. Figure 25.6c: John D. Cunningham/Visuals Unlimited. Figure 25.7 (top): Courtesy Dr. Edward J. Bottone, Mount Sinai Hospital. Figure 25.7 (bottom): Courtesy R.S. Wolfe and J.C. Ensign. Figure 25.8: CNRI/ Science Source/Photo Researchers. Figure 25.9: Courtesy C.C. Remsen, S.W. Watson, J.N. Waterbury and H.S. Tuper, from J. Bacteriology, vol. 95, p.2374, 1968. Figure 25.10: Sinclair Stammers/Science Source/Photo Researchers. Figure 25.11: Courtesy B. Ben Bohlool, NiftAL. Figure 25.12a: Jerome Paulin/Visuals Unlimited. Figure 25.12b: Stanley Flegler/ Visuals Unlimited. Figure 25.12c: Victor Duran/Sharnoff Photos. Figure 25.13: R. Kessel and G. Shih/Visuals Unlimited. Figure 25.14: Courtesy Romano Dallai, Department of Biology, Universita' di Siena. Figure 25.16a: M. Abbey/Visuals Unlimited. Figure 25.16b: James Dennis/CNRI/Phototake. Figure 25.18a: Manfred Kage/Peter Arnold, Inc. Figure 25.18b: Eric Grave/ Science Source/Photo Researchers. Figure 25.19a: David M. Phillips/ Visuals Unlimited. Figure 25.19b: Kevin Schafer/Tom Stack & Associates. Figure 25.20a: E.R. Degginger. Figure 25.20b: Waaland/Biological Photo Service. Figure 25.21: Dr. Jeremy Burgess/ Science Source/Photo Researchers. Figure 25.22: Herb Charles Ohlmeyer/Fran Heyl Associates. Figure 25.23: Michael Fogden/Earth Scenes/Animals Animals. Figure 25.24a: Michael Fogden/Earth Scenes/ Animals Animals. Figure 25.24b: Stephen Dalton/© Natural History Photographic Agency. **Human Perspective** Rick Rickman/Duomo Photography, Inc. **Chapter 26** Figure 26.2a: Michael P. Gadomski/Earth Scenes/Animals Animals. Figure 26.2b: Courtesy Edward S. Ross, California Academy of Sciences. Figure 26.2c: Rod Planck/Photo Researchers. Figure 26.2d: Kurt Coste/Tony Stone World Wide. Figure 26.4a: Kim Taylor/Bruce Coleman, Inc. Figure 26.4b: Breck Kent. Figure 26.4c: Flip Nicklin/Minden Pictures, Inc. Figure 26.4d: D.P. Wilson/Eric & David Hosking/Photo Researchers. Figure Figure 26.5a: Runk/Schoenberger/Grant Heilman Photography. Figure 26.5b: John Gerlach/Tom Stack & Associates. Figure 26.5c: Robert A. Ross/ R.A.R.E. Photography. Figure 26.7: John H. Trager/Visuals Unlimited. Figure 26.9: John D. Cunningham/ Visuals Unlimited. Figure 26.10: Milton Rand/ Tom Stack & Associates. Figure 26.12: Cliff B. Frith/Bruce Coleman, Inc. Figure 26.13: Leonard Lee Rue/Photo Researchers. Figure 26.14: Anthony Bannister/Earth Scenes. Figure

26.16a: J. Carmichael/The Image Bank. Figure 26.16b: Sebastio Barbosa/The Image Bank. Figure 26.16c: Holt Studios/Earth Scenes. Figure 26.16d: Gwen Fidler/COMSTOCK, Inc. **Chapter 27** Figure 27.1a: Courtesy Karl G. Grell, Universitat Tubingen Institut fr Biologie. Figure 27.1b: Franois Gohier/Photo Researchers. Figure 27.2a: Larry Ulrich/DRK Photo. Figure 27.2b: Christopher Newbert/Four by Five/SUPERSTOCK. Figure 27.5a: Robert & Linda Mitchell. Figure 27.5b: Runk/Schoenberger/Grant Heilman Photography. Figure 27.5c: Fred Bavendam/Peter Arnold, Inc. Figure 27.7a: Carl Roessler/Tom Stack & Associates. Figure 27.7b: Goivaux Communication/Phototake. Figure 27.7c: Biomedia Associates. Figure 27.10(inset): Courtesy D. Phillips. Figure 27.11a: Marty Snyderman/Visuals Unlimited. Figure 27.11b: Robert & Linda Mitchell. Figure 27.13a: Biological Photo Service. Figure 27.13b: Edward S. Ross, National Geographic,*The Praying of Predators,* February 1984,p.277. Figure 27.13c: John Shaw/Tom Stack & Associates. Figure 27.13d: Patrick Landman/Gamma Liaison. Figure 27.15: Robert & Linda Mitchell. Figure 27.18a: Dave Woodward/Tom Stack & Associates. Figure 27.18b: Terry Ashley/Tom Stack & Associates. Figure 27.20: Russ Kinne/COMSTOCK, Inc. Figure 27.21a: W. Gregory Brown/Animals Animals. Figure 27.21b: Chris Newbert/Four by Five/SUPERSTOCK. Figure 27.22a: Zig Leszczynski/Animals Animals. Figure 27.22b: Chris Mattison/Natural Science Photos. Figure 27.23a: Michael Fogden/Animals Animals. Figure 27.23b: Jonathan Blair/Woodfin Camp & Associates. Figure 27.25a,b: Dave Watts/Tom Stack & Associates. Figure 27.25c: Boyd Norton. **Part 6** Opener: David Doubilet, National Geographic Society. **Chapter 28** Figure 28.1: Courtesy Earth Satellite Corporation. Figure 28.4: Norbert Wu/Tony Stone World Wide. Figure 28.5a: Stephen Frink/AllStock, Inc. Figure 28.5b: Nicholas De Vore III/Bruce Coleman, Inc. Figure 28.7: Joel Arrington/Visuals Unlimited. Figure 28.9a: K. Gunnar/Bruce Coleman, Inc. Figure 28.9b: Claude Rives/Agence C.E.D.R.I. Figure 28.9c: Maurice Mauritius/Serrailer-Rapho/Black Star. Figure 28.11: Jim Zuckerman/West Light. Figure 28.12: James P. Jackson/Photo Researchers. Figure 28.13: Steve McCutcheon/Alaska Pictorial Service. Figure 28.14: Linda Mellman/Bill Ruth/Bruce Coleman, Inc. Figure 28.15: Bill Bachman/Photo Researchers. Figure 28.16: Nigel Dennis/© Natural History Photographic Agency. Figure 28.17: Bob Daemmrich/The Image Works. Figure 28.18: B.G. Murray, Jr./Animals Animals. Figure 28.19a: Breck Kent. Figure 28.19b: Carr Clifton. **Human Perspective** Will McIntyre/AllStock, Inc. **Chapter 29** Figure 29.1a: Gary Braasch/AllStock, Inc. Figure 29.1b: George Bernard/© Natural History Photographic Agency. Figure 29.1c: Stephen Krasemann/© Natural History Photographic Agency. Figure 29.4a: Anthony Mercieca/Natural Selection. Figure 29.4b: E.R. Degginger/Photo Researchers. Figure 29.6: Walter H. Hodge/Peter Arnold, Inc. Figure 29.13: Larry Ulrich/DRK Photo. Figure 29.14: © Gary Braasch. **Chapter 30** Figure 30.1: Andy Callow/© Natural History Photographic Agency. Figure 30.5a: Cosmos Blank/Photo Researchers. Figure 30.5b: Michael Fogden/Bruce Coleman, Inc. Figure 30.6: B&C Calhoun/Bruce Coleman, Ltd. Figure 30.7: Kjell B. Sandved/Photo Researchers. Figure 30.8: Anthony Bannister/© Natural History Photographic Agency. Figure 30.9a: Adrian Warren/Ardea London. Figure 30.9b: Michael Fogden/Bruce Coleman, Inc. Figure 30.10: P.H. & S.L. Ward/Natural Science Photos. Figure 30.11: Tom McHugh/Photo Researchers. Figure 30.12: Dr. J.A.L. Cooke/Oxford Scientific Films/Animals Animals. Figure 30.14: Stephen G. Maka. Figure 30.15a: Eric Grave/Phototake. Figure 30.15b: Raymond A. Mendez/Animals Animals. **Chapter 31** Figure 31.1: Dwight R. Kuhn. Figure 31.2: Mike Severns/Tom Stack & Associates. Figure 31.3: Robert Maier/Animals Animals. Figure 31.4: Courtesy Masao Kawai, Primate Research Institute, Kyoto University. Figure 31.5: Glenn Oliver/Visuals Unlimited. Figure 31.6: Hans Reinhard/Bruce Coleman, Ltd. Figure 31.7: Oxford Scientific Films/Animals Animals. Figure 31.8: Roy P. Fontaine/Photo Researchers. Figure 31.9: K. and K. Ammann/Bruce Coleman, Inc. Figure 31.10: Konrad Wothe/Bruce Coleman, Ltd. Figure 31.11: Gordon Wiltsie/Bruce Coleman, Inc. **Chapter 32** Figure 32.1a: Steve Krasemann/DRK Photo. Figure 32.1b: Carr Clifton. Figure 32.1c: Willard Clay. Figure 32.3: Robert & Linda Mitchell. **Appendix B** Figure b: Bob Thomason/Tony Stone World Wide. Figure c: Dr. Gennavo/ Science Photo Library/Photo Researchers. Figure d: F.R. Turner, Indiana Univeristy/ Biological Photo Service.

Index

Note: A t following a page number denotes a table, f denotes a figure, and n denotes a footnote.

◄ A ►

◄ D ►

◄ F ►

◄ H ►

◄ T ►

◄ U ►